THE BIOLOGY OF SEAWEEDS

BOTANICAL MONOGRAPHS

BOTANICAL MONOGRAPHS · VOLUME 17

THE BIOLOGY OF SEAWEEDS

EDITED BY

CHRISTOPHER S. LOBBAN
Division of Sciences,
University of New Brunswick,
Saint John, New Brunswick, Canada

AND

MICHAEL J. WYNNE
Division of Biological Sciences,
University of Michigan,
Ann Arbor, Michigan, U.S.A.

UNIVERSITY OF CALIFORNIA PRESS
BERKELEY AND LOS ANGELES 1981

UNIVERSITY OF
CALIFORNIA PRESS
Berkeley and Los Angeles

Library of Congress
Catalog Card Number
81–69858

ISBN 0–520–04585–8

Printed in Great Britain

CONTENTS

CONTRIBUTORS

SUSAN BRAWLEY, Division of Paleobotany, U.S. National Museum, Washington, D.C. 20560, U.S.A.

RICHARD G. BUGGELN, Marine Sciences Research Laboratory, Memorial University of Newfoundland, St. John's, Newfoundland, Canada A1C 5S7.

JAMES A. DEBOER, Division of Allied Health and Life Sciences, The University of Texas at San Antonio, San Antonio, Texas 78285, U.S.A.

LOUIS D. DRUEHL, Department of Biological Sciences, Simon Fraser University, Burnaby, British Columbia, Canada V5A 1S6.

ALAN J. FIELDING, Department of Biological Sciences, John Dalton Faculty of Technology, Manchester Polytechnic, Manchester, Lancashire, England.

MAX H. HOMMERSAND, Department of Botany, University of North Carolina, Chapel Hill, North Carolina 27514, U.S.A.

BRUNO P. KREMER, Seminar für Didaktik der Biologie, Universität zu köln, D-5000 köln 41, Federal Republic of Germany.

GERALD T. KRAFT, School of Botany, University of Melbourne, Parkville, Victoria 3052, Australia.

KLAUS LÜNING, Biologische Anstalt Helgoland, Meeresstation Helgoland, Palmaille 9, 2000 Hamburg 50, Federal Republic of Germany.

ARTHUR C. MATHIESON, Department of Botany and Plant Pathology, University of New Hampshire, Durham, New Hampshire 03824, U.S.A.

ESTHER L. McCANDLESS, Department of Biology, McMaster University, Hamilton, Ontario, Canada L8S 4K1.

DIETER G. MÜLLER, Fachbereich Biologie, Universität Konstanz, Postfach 7733, D-7750 Konstanz, Federal Republic of Germany.

MICHAEL NEUSHUL, Department of Biological Sciences, University of California, Santa Barbara, California 93106, U.S.A.

TREVOR A. NORTON, Department of Botany, The University, Glasgow G12 8QQ, Scotland.

POUL MØLLER PEDERSEN, Københavens Universitet, Institut for Sporeplanter, Ø. Farimagsgade 2D, 1353 København, Denmark.

MARK A. RAGAN, Atlantic Research Laboratory, National Research Council of Canada, 1411 Oxford St., Halifax, Nova Scotia, Canada B3H 3Z1.

JOSEPH RAMUS, Duke University, Marine Laboratory, Pivers Island, Beaufort, North Carolina 28516, U.S.A.

GEORGE RUSSELL, Department of Botany, The University of Liverpool, P.O. Box 147, Liverpool L69 3BX, Lancashire, England.

KLAUS SCHMITZ, Botanisches Institut der Universität Köln, Gyrhofstrasse 15, 5 Köln 41, Federal Republic of Germany.

CHRISTOPHER E. TANNER, Division of Natural Sciences and Mathematics, St. Mary's College of Maryland, St. Mary's City, Maryland 20680, U.S.A.

C. K. TSENG, Director, Institute of Oceanology, Academia Sinica, 7 Nan-Hai Road, Tsingtao, People's Republic of China.

C. VAN DEN HOEK, Biological Centre, Department of Systematic Botany, Kerklaan 30, Post Box 14, Haren (GR), The Netherlands.

J. ROBERT WAALAND, Department of Botany, University of Washington, Seattle, Washington 98195, U.S.A.

JOHN A. WEST, Department of Botany, University of California, Berkeley, California 94720, U.S.A.

RICHARD WETHERBEE, School of Botany, University of Melbourne, Parkville, Victoria 3052, Australia.

MICHAEL J. WYNNE, Division of Biological Sciences, University of Michigan, Ann Arbor, Michigan 48109, U.S.A.

PREFACE

During the past decade several important multi-author treatises of interest to marine phycologists appeared on the market, among them Otto Kinne's five volume *Marine Ecology* (starting in 1969), W. D. P. Stewart's Botanical Monograph, *Algal Physiology and Biochemistry* (1974) and I. Morris' *Physiological Ecology of Phytoplankton* (1980). In addition, there were several books covering selected aspects of marine phycology, such as Dixon's book on the morphology and life histories of red algae (1973) and Santelices' summary (in Spanish) of the ecology of benthic marine algae. There seemed to be a need for a book which under one cover would deal with all three major divisions of marine macrophytes, including aspects of experimental work and industrial application as well as the more classical systematic and life history studies.

For the present book, authors were asked to exclude material on freshwater algae and phytoplankton, as well as benthic marine diatoms and blue green algae, except where such material might illuminate discussion of seaweeds. (Diatoms and blue green algae have been subjects of Botanical Monographs edited by Werner (1977) and Carr & Whitton (1973) respectively.) Still, in order to compress into one volume the range of topics we wanted to cover, it was inevitable that certain subjects would be discussed very briefly or not at all. Among the subjects which received short shrift were ionic relations, calcification, photosynthetic pigments, and phototaxis. In spite of some major editing, the book is considerably longer than had been planned; yet many authors—especially those who were asked to trim as much as 50% off their detailed manuscripts—may justifiably feel that their chapters are too brief.

The editors wish to thank first and most especially the authors for their hard work and cooperation. We also want to thank the following colleagues for critical reading of some of the chapters: R. G. S. Bidwell, A. R.O. Chapman, M. J. Duncan, R. L. Fletcher, P. J. Harrison, M. H. Hommersand, J. W. LeClaire II, D. Mauzerall, T. F. Mumford, Jr., G. F. Papenfuss, J. D. Pringle, J. Scott, G. R. South, S. D. Waaland, W. Yaphe, and C. F. Yarish. Finally, our thanks to Robert Campbell at Blackwell Scientific Publications for his advice and support throughout the preparation of this book.

January 1981

C. S. Lobban
M. J. Wynne

SECTION I
STRUCTURE AND REPRODUCTION

INTRODUCTION

MICHAEL J. WYNNE

The chapters of this initial section summarize recent advances in our understanding of the systematics of red, brown, and green algae. This forward progress has been based upon studies involving morphology, development, cytology, and life histories. Although biochemical data and flagellar ultrastructure serve as the primary criteria for delineating algal divisions, it is these other characteristics that bear directly on the categorization of algae into classes, orders, families, and genera. It is readily apparent that the 'pioneer (or exploratory) phase' proposed by Davis & Heywood (1963) is still alive and well, as evidenced by the many newly described red algal genera (referred to in Chapter 1). The 'consolidation phase' of Davis and Heywood is exemplified by the publications of Woelkerling (1971, 1973), Dixon (1976), Garbary (1979), and others, who argue for a return to a two-genus or even one-genus concept for the very confusing and ill-defined members of the family Acrochaetiaceae.

Exciting discoveries are still being made, sporadically and unpredictably, in our times. For example, the clever use of a green-coloured mutant by van der Meer & Todd (1980) to demonstrate the existence of a sexual cycle in *Palmaria palmata* is such a break-through. It is noteworthy that the life-history pattern of this species is so distinctive that recognition of the Palmariales seems assured, despite the fact that the grounds for Guiry and Irvine's recent establishment of this order in Guiry (1978) did not rest upon the character of its sexual cycle. Similarly, the unequivocal occurrence of a sexual cycle in a species of *Porphyra*, as shown by Hawkes (1978), should end the long-nagging doubt about sexuality in this genus. A subsequent study of 24 species of Bangiaceae by Cole & Conway (1980) confirmed sexual reproduction for some species. But they also revealed that such a cycle is not obligatory and that both gametophytic and sporophytic phases can duplicate themselves directly by accessory asexual mechanisms. It has become apparent that the life history is very adaptable in these algae and others.

The occurrence of sexual reproduction in *Scytosiphon*, another frequently studied alga, has been recently reinforced, even though it is a phenomenon restricted to a relatively short span during the year, namely, late winter–early spring (Clayton 1980). Morphological gametophytes do not function sexually during the balance of the growing season. The lesson here is that field work has to be carried out even when conditions are least amenable to such work. Boney (1978) made reference to a 'culture-based taxonomy' and cautioned in regard to the well acknowledged variation obtainable under culturing conditions. Yet meticulous studies, supplemented by chromosomal counts, such as that by Ajisaka (1979) on *Acrothrix*, certainly help unravel the story of algae with heteromorphic phases.

New tools of investigation are being applied to the study of seaweeds. Among these innovative approaches is numerical taxonomy, an investigation of the tropical green alga *Chlorodesmis* (Ducker *et al.* 1965) being an early example. Subsequent treatments of this type include investigations on *Alaria* (Widdowson 1972), British Phaeophyceae (Russell & Fletcher 1975; Russell & Garbary 1978; Pankhurst & Tittley 1978), European Sphacelariales (Prud'homme van Reine 1978), and the Acrochaetiaceae (Garbary 1978). Such efforts, which take into account the fullest scope of characteristics possible, help push algal systematics closer toward the biosystematic and encyclopaedic phases envisioned by Davis & Heywood (1963).

Before turning over the floor to the contributors, I need to say a word about the need to respect differences of opinion. Indeed we are still far from the day when even a handful of workers will concur on issues and/or terminology. I found this to be true during my editing chores. Readers will note, for example, that in Chapter 6 two distinct green algal orders, the Cladophorales and the Siphonocladales, are treated, whereas in Chapter 3 the single order Cladophorales (including the family Siphonocladaceae) is recognized. These differences in taxonomic viewpoints are respected. Likewise among the red algae equally good cases can be made for the recognition of the orders Bonnemaisoniales and Gelidiales as distinct from the Nemalionales or to contain these as families within the Nemalionales. At these higher taxonomic levels judgments seem to become more arbitrary and perhaps subjective. A perusal of some recent accounts on the origin of the higher taxa of organisms and their interrelationships (Fox *et al.* 1980; Maruyama 1980; Stewart & Mattox 1980) shows that this field of synthesis is rife with new ideas, speculation, and honest differences of opinion on matters of interpretation.

REFERENCES

AJISAKA T. (1979) The life history of *Acrothrix pacifica* Okamura et Yamada (Phaeophyta, Chordariales) in culture. *Jap. J. Phycol.*, **27**, 75–81.

BONEY A. D. (1978) Taxonomy of red and brown algae. In *Modern Approaches to the Taxonomy of Red and Brown Algae*, ed. Irvine D. E. G. & Price J. H., pp. 1–19. Systematics Assoc. Spec. vol. 10, Academic Press, London.

CLAYTON M. N. (1980) Sexual reproduction—a rare occurrence in the life history of the complanate form of *Scytosiphon* (Scytosiphonaceae, Phaeophyta) from southern Australia. *Br. Phycol. J.*, **15**, 105–18.

COLE K. & CONWAY E. (1980) Studies in the Bangiaceae : reproductive modes. *Bot. Mar.*, **23**, 545–53.

DAVIS P. H. & HEYWOOD V. H. (1963) *Principles of Angiosperm Taxonomy*. Van Nostrand Co. Inc., Princeton, N.J.

DIXON P. S. (1976) Appendix. In Check-list of British marine algae—third revision, ed. Parke M. & Dixon P. S. *J. Mar. Biol. Ass. U.K.*, **56**, 527–94.

DUCKER S. C., WILLIAMS W. T. & LANCE G. N. (1965) Numerical taxonomy of the Pacific forms of *Chlorodesmis* (Chlorophyta). *Austr. J. Bot.*, **13**, 489–99.

FOX G. E. *et al.* (1980) The phylogeny of prokaryotes. *Science*, **209**, 457–63.

GARBARY D. (1979) Numerical taxonomy and generic circumscription in the Acrochaetiaceae (Rhodophyta). *Bot. Mar.*, **22**, 477–92.

GUIRY M. D. (1978) The importance of sporangia in the classification of the Florideophyceae. In *Modern Approaches to the Taxonomy of Red and Brown Algae*, ed. Irvine D. E. G. & Price J. H., pp. 111–44. Systematics Assoc. Spec. vol. 10, Academic Press, London.

HAWKES M. W. (1978) Sexual reproduction in *Porphyra gardneri* (Smith et Hollenberg) Hawkes (Bangiales, Rhodophyta). *Phycologia*, **17**, 329–53.

MARUYAMA K. (1980) A cellular approach to phyletic relations in eucaryotic algae. *Jap. J. Phycol.*, **28**, 1–18.

PANKHURST R. J. & TITTLEY I. (1978) The application of computers to the identification of the Phaeophyta. In *Modern Approaches to the Taxonomy of Red and Brown Algae*, ed. Irvine D. E. G. & Price J. H., pp. 325–37. Systematics Assoc. Spec. vol. 10, Academic Press, London.

PRUD'HOMME VAN REINE W. F. (1978) Criteria used in systematic studies in the Sphacelariales. In *Modern Approaches to the Taxonomy of Red and Brown Algae*, ed. Irvine D. E. G. & Price J. H., pp. 301–23. Systematics Assoc. Spec. vol. 10, Academic Press, London.

RUSSELL G. & FLETCHER R. L. (1975) A numerical taxonomic study of the British Phaeophyta. *J. Mar. Biol. Ass. U.K.*, **55**, 763–84.

RUSSELL G. & GARBARY D. (1978) Generic circumscription in the family Ectocarpaceae (Phaeophyceae). *J. Mar. Biol. Ass. U.K.*, **58**, 517–25.

STEWART K. D. & MATTOX K. (1980) Phylogeny of phytoflagellates. In *Phytoflagellates*, ed. Cox E. R., pp. 433–62. Elsevier/North Holland, New York.

VAN DER MEER J. P. & TODD E. R. (1980) The life history of *Palmaria palmata* in culture. A new type for the Rhodophyta. *Can. J. Bot.*, **58**, 1250–6.

WIDDOWSON T. B. (1972) A statistical analysis of variation in the brown alga *Alaria*. *Syesis*, **4**, 125–43.

WOELKERLING W. J. (1971) Morphology and taxonomy of the *Audouinella* complex (Rhodophyta) in southern Australia. *Austral. J. Bot.* Suppl., **1**.

WOELKERLING W. J. (1973) The morphology and systematics of the *Audouinella* complex (Acrochaetiaceae, Rhodophyta) in northeastern United States. *Rhodora*, **75**, 529–621.

CHAPTER 1
RHODOPHYTA: MORPHOLOGY AND CLASSIFICATION

GERALD T. KRAFT
School of Botany, University of Melbourne,
Parkville, Victoria, Australia

1 INTRODUCTION

Certain partisans have been known to maintain that 'all literature is a footnote to Shakespeare', and more than a few people could probably be found who would say much the same about red algal taxonomy *vis-à-vis* Harald Kylin. For however true it may be that Kylin owed heavy debts to predecessors like Schmitz and Oltmanns, his compendium of red algal genera, families and orders (Kylin 1956) remains the single most authoritative work on the group for most of us and the point of departure for any revisionist attempts to alter the classification of red algae at and above the genus level.

Compared to other algal divisions, the current grouping of Rhodophytan genera into orders and families has been relatively untouched by data from recent life history, biochemical or ultrastructural studies. The mechanics of red algal zygote initiation and development that are so unique an attribute of most of the Division have more or less insured that the group's classification remains largely based on features which (even at their most obscure) are at least potentially viewable in fortuitous hand-sections under medium-power light microscopy. Nevertheless, the more esoteric disciplines are asserting their influences ever more strongly, particularly in such groups as the Bangiales, Nemalionales, Palmariales, Gigartinaceae and Phyllophoraceae.

The range of forms, life histories and reproductive intricacies is greater in the Rhodophyta than in any other algal division. Searles (1980) has even written an *apologia* on behalf of the group, proposing that its lack of flagellated gametes may be responsible for the complexities of life history and reproductive anatomy found in it. Some recent texts which have reviewed and illustrated red algal morphology are those by Bold & Wynne (1978), Dixon & Irvine (1977) and Kraft & Woelkerling (1981), and it is assumed here that readers are familiar with basic red algal anatomy and terminology. Although mentions of taxonomic traits are made throughout this review, its primary purpose is to touch on some of the issues and high spots of the systematic and morphological work done since Kylin's monograph of genera and Dawson's (1962) census of species.

2 THE SUBCLASSES AND ORDERS OF RED ALGAE

At (my) last count, the Rhodophyta appears to have something like 4100 species in 675 genera. At the highest taxonomic level, there is mild dispute over whether the group consists of two classes, the Florideophyceae and Bangiophyceae, or the single class Rhodophyceae and the two subclasses Florideophycidae and Bangiophycidae.* The former is argued by Dixon (1973) and followed by Abbott & Hollenberg (1976) and Parke & Dixon (1976), among others, while the latter is adopted by Kylin (1956, as the Florideae and Bangioideae) and others such as Cole & Conway (1975), Bold & Wynne (1978) and Kraft & Woelkerling (1981). Arguments in favour of considering the Division as a single class with the first taxonomic split at the subclass level are based on recent blurrings of what were supposed to be major separating factors between bangiophycean and florideophycean algae. The close links between the 'conchocelis' phase of *Bangia* and *Porphyra* and the 'lower' Nemalionales in regard to chloroplast structure (Cole & Conway 1975), pit connections (Lee 1971), and cell wall composition (Gretz *et al.* 1980) are striking, and many recent authors follow Cole & Conway (1975) in opting for the fundamental unity of the Division at the class level.

Although the ordinal makeup of the Bangiophycidae either still follows Kylin (Abbott & Hollenberg 1976; Bold & Wynne 1978) or differs only by the deletion of the Goniotrichales (Parke & Dixon 1976), a number of changes have been urged and some adopted within the Florideophycidae, mostly having to do with the order Nemalionales. Dixon (1973) has argued strongly for returning the two families of the Gelidiales to the Nemalionales, rejecting Kylin's (1923) elevation of the Gelidiaceae to ordinal status. This opinion is partly based on recent culture/life history work which has shown an alternation of gametophytic and tetrasporophytic generations in most Nemalionales wholly comparable to that of the Gelidiaceae. Recent workers (e.g. Abbott & Hollenberg 1976; Parke & Dixon 1976; Bold & Wynne 1978; Guiry 1978a) have tended to accept Dixon's recommendation. Because it is so heterogeneous an assemblage, many families of the Nemalionales have been elevated to orders in themselves at one time or other. Thus the Acrochaetiales, Bonnemaisoniales and Chaetangiales

* Christensen (1979) promotes a change of nomenclature that leaves the reds with the single class Bangiophyceae and substitutes the Nemaliophycidae for the Florideophycidae. This debatable application of naming rules manages to gratuitously replace clarity with confusion. Note, for example, how nonsensical Christensen's maneuver renders the title of Gretz *et al.*'s (1980) paper.

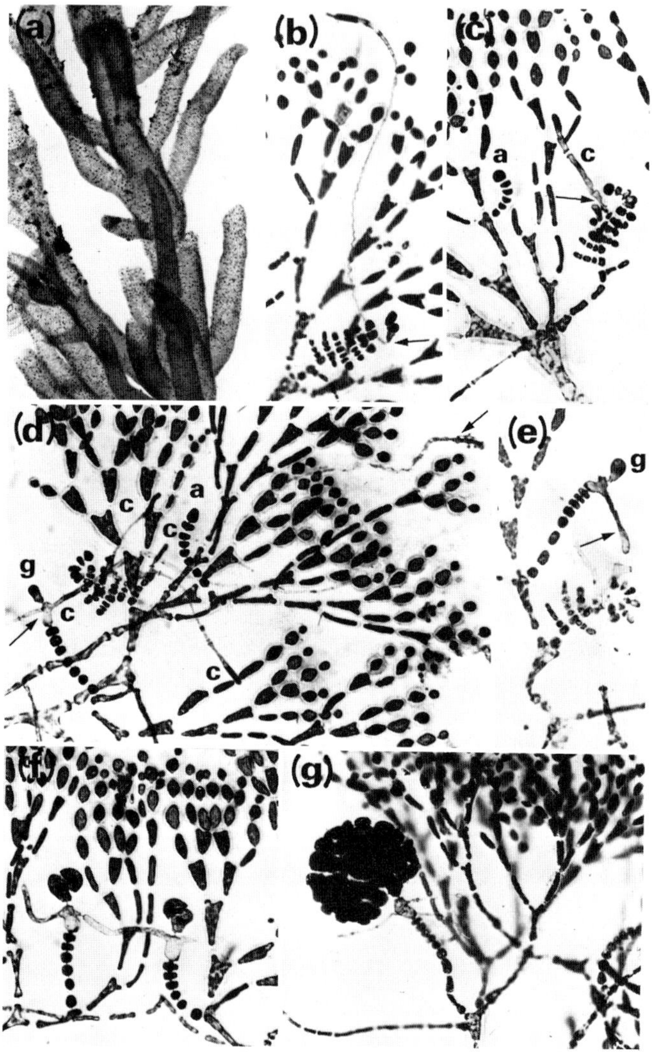

Fig. 1.1 The reproductive sequence in an *Acrosymphyton* species (Dumontiaceae) from Australia results in the 'simple' type of cystocarp in which all of the cells of the carposporophyte appear to mature into carposporangia. Algae with similar cystocarps occur in other families of the Cryptonemiales, as well as some Gigartinales and Ceramiales.

(a) The habit of some cystocarpic branches of a plant (MELU, K9603) collected at Lord Howe Island, off New South Wales (× 3).

(b) A carpogonial branch terminated by the female gamete, or carpogonium (arrow), with its recurved trichogyne extending beyond the cortex. A unique feature of the genus is the opposite side branchlets on cells of the carpogonial branch (× 270).

(c) Following fertilization, a protruberance from the carpogonium fuses with a lateral cell of a subtending carpogonial branch cell (arrow) and initiates the first connecting filament (c) which conveys a zygote-nucleus derivative. Also present on the cortical filament is an auxiliary cell branch bearing the terminal auxiliary cell (a) (× 280).

have all been recently argued for (Chihara & Yoshizaki 1972b; Garbury 1978; see Dixon 1973 for a review of other proposals), but have tended to be regarded as premature or unnecessary by many authors. A recently introduced order (segregated from the Rhodymeniales) which is gaining acceptance is the Palmariales, which Guiry & Irvine (in Guiry 1978a) established for three genera whose major characteristics were an apparent lack of female gametophytes and a radically different manner of tetrasporogenesis from other Rhodophyta.

There is continuing speculation as to which are the 'primitive' and which the 'advanced' groups within the red algae. It is generally agreed that the Ceramiales, with their extremely uniform carposporophyte initiation and development, represent a sort of culmination or end point of specialization, but differences of opinion occur regarding the 'primitive' red alga. Garbary (1978) presents evidence for the pivotal importance of the Acrochaetiaceae, from which the Bangiaceae and the rest of the Bangiophycidae may have diverged at an early stage. That the Bangiophycidae, with their relatively simple morphology and numerous freshwater taxa, represent a reduction from, rather than ancestors to, the more complex Florideophycidean algae is an opinion expressed by early workers such as Fritsch (1945) and held by many today. The discovery of conchocelis-like fossils from the Upper Silurian (Campbell 1980), however, points to the great antiquity of the Bangiophycean group.

Umezaki (1977) bases speculation about red algal evolution on life history and cytological considerations, and opts for a hypothetical ancestor of three phases in which the carposporophyte is free-living rather than attached to the parent gametophyte as in all known cases today. On this scheme the *Polysiphonia*-type life cycle of isomorphic gametophytes and tetrasporophytes is the surviving primitive form from which the heteromorphic, carpotetrasporangial or somatic alternations of generations characteristic of some red algae have evolved. As admitted by Umezaki, a classification based on life history patterns does not conform to the present ordinal system, and it seems that very little can really be made from this approach by way of identifying those precise families that might

---

(d) An overview of a later stage of postfertilization development. Remnants of the trichogyne have several attached spermatia (arrow), showing the relative size of these diminutive gametes. The carpogonial branch has initiated over 4 connecting filaments (c), one of which has fused with an auxiliary cell (arrow) and grown beyond. The primary gonimoblast (g) arises from the diploidized auxiliary cell, and an undiploidized auxiliary cell (a) occurs in the same section ($\times$ 200).

(e) Detail of a carpogonial branch in which the carpogonium has fused with several lower cells of the branch and formed two connecting filaments, one of which (arrow) has reached a neighbouring auxiliary cell and resulted in a primary gonimoblast (g) ($\times$ 267).

(f) Not only can a single fertilization result in a number of connecting filaments, each bearing zygote derivatives, but each connecting filament can effect more than one diploidization, as shown by these auxiliary cells with their two-celled gonimoblasts which are linked by a common connecting filament ($\times$ 260).

(g) The mature cystocarp is a spherical mass of carposporangia to which incoming and outgoing connecting filament remnants are still attached ($\times$ 210).

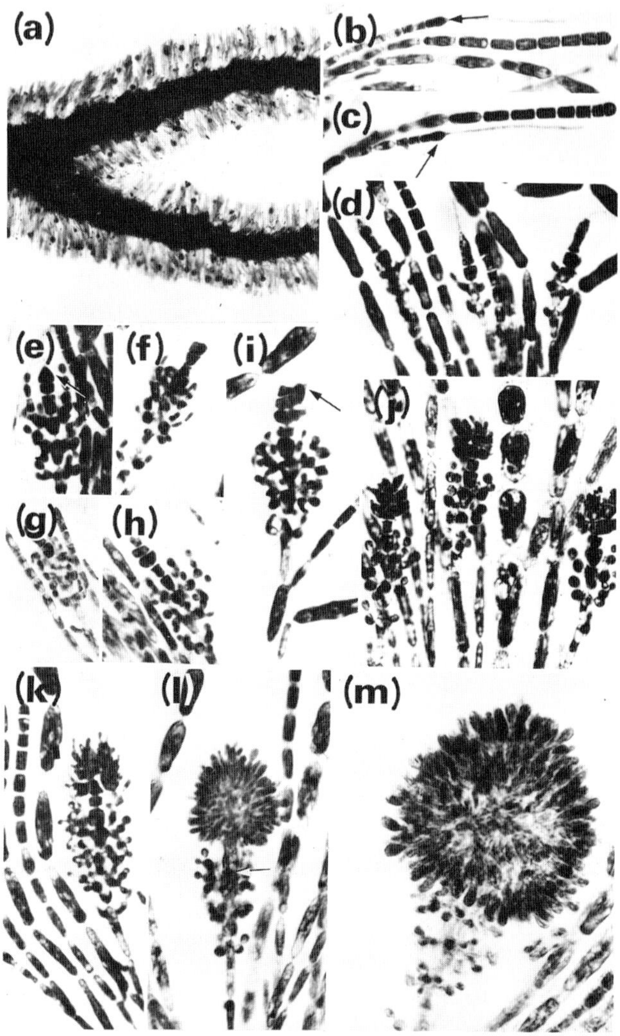

Fig. 1.2 A *Trichogloea* species (Nemalionaceae) from the Great Barrier Reef represents the type of cystocarp in which the inner cells of the carposporophyte remain sterile while only the peripheral layer produces carposporangia. This is the most common of the red algal cystocarp architectures and occurs as a result of very disparate developmental processes in all the Florideophycidean orders (save the Palmariales).

(a) Habit of a cystocarpic branch on a plant (MELU, K15375) from One Tree Island, Queensland. A central core of calcium carbonate crystals parallels the medullary filaments and defines their boundary with the cortex (× 7.2).

(b, c) Carpogonial branches with their terminal carpogonia (arrows) and trichogynes prior to fertilization (both × 264).

(d) A series of 3 carpogonial branches showing early growth of the characteristic sterile lateral cells borne on the sub-hypogynous cells (× 264).

(e–h) Stages in the first (e, arrow) and subsequent divisions (f–h) of the zygote, which are contain 'primitive' members or represent 'primitive' levels of reproductive morphology. My own tendency (at the moment) is to look for such algae among the 'lower' families of the Gigartinales and Cryptonemiales (Gymnophloeaceae, Calosiphoniaceae, Dumontiaceae) in which plants produce 'simple' carposporophytes (e.g. Fig. 1.1), maximize each fertilization by producing a myriad of connecting filaments which can each lead to production of more than one cystocarp, lack the capacity (generally) to form secondary pit connections, can possess structures such as rhizoids and auxiliary cell filaments which seem convertible at times into cortical cell systems, and which generally hang around in deep warm waters which of all habitats on earth were probably the most constant and least shifted around geographically over the ages. Although many of these algae do not possess isomorphic gametophytic and tetrasporangial generations a number do, and I would follow Feldmann (1952) in supposing that there was an isomorphic alternation of generations in whatever reds actually were the earliest to employ a sexual cycle.

The survey which follows is very piecemeal but attempts to review the orders and most marine families of Rhodophyta, saying what has happened to them and some of their constituent genera since their treatment by Kylin (1956) or their later introduction. Although space restrictions decree that this will not be it, the time is getting ripe for a major updating of Kylin's monograph, which, brilliant as it is, in effect represents the state of the art as it was some 30 years ago.

3 CRITERIA USED IN CLASSIFICATION OF HIGHER TAXA

The characteristic oogamy, the retention and development of the zygote on the mother gametophyte, and the intercalation of a distinctive spore-bearing generation (the carposporophyte) between the gametophytic and meiosporangial phases of the life history in most reds form the bases for defining the orders and many families of the Florideophycidae. The location of zygote formation and the site of its ultimate growth are usually separate (Figs 1.1 and 1.3) and often far apart (Fig. 1.1), so that the presence, nature and position of a generative auxiliary cell are important taxonomic features in any given species of alga. The

---

initiated by the first horizontal division of the fertilized carpogonium and remain attached to the carpogonial branch throughout development (all × 272).

(i–k) Early proliferations of the gonimoblast tend to be unilateral (i, arrow), but become uniformly radial in later stages (j, k) (i × 336; j, k × 268).

(l) As the carposporophyte matures, walls between cells of the carpogonial branch break down to form an elongate fusion cell (arrow) (× 256).

(m) The ultimate result is a spherical cystocarp with a sterile-celled interior and a surface layer of terminal carposporangia. Progressive atrophy of the cell contents of the subtending sterile laterals on the carpogonial branch suggests their possible nutritive function for the growing carposporophyte (× 264).

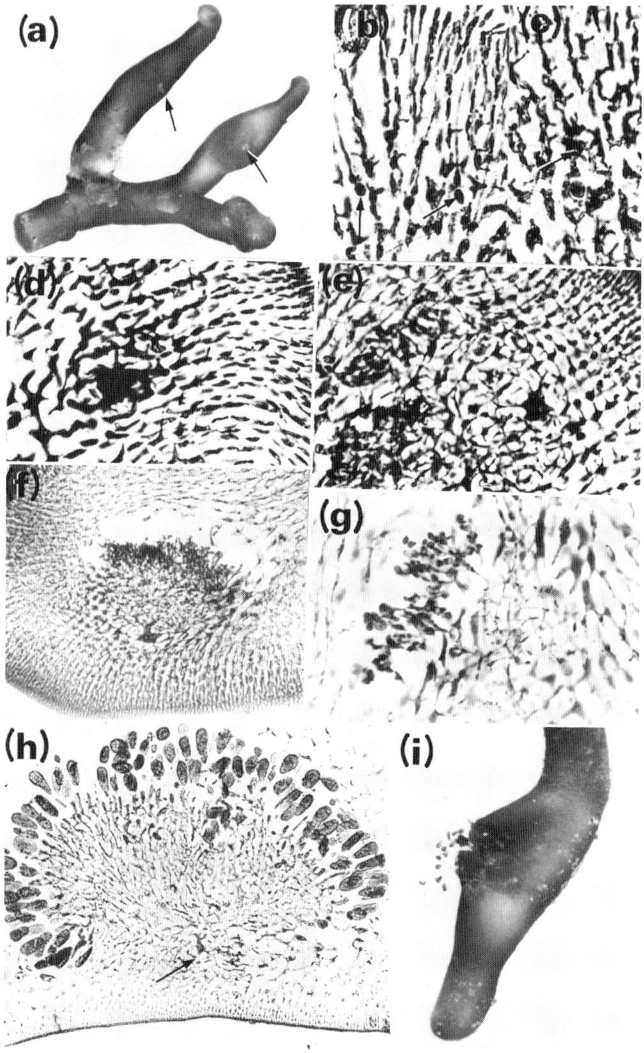

Fig. 1.3 The genus *Dicranema* (Dicranemaceae) represents the most morphologically complex cystocarp type in the Rhodophyta. Here diploid cells of the developing carposporophyte pass through a region of intimate contact and interconnection with the haploid tissue of the gametophyte before initiating gonimoblasts which consolidate and produce a surface layer of carposporangia. Cystocarps approaching this complexity are found in only a few families of the Cryptonemiales and Gigartinales.

(a) Habit of a cystocarpic branch tip from a plant (MELU, K4569) collected in South Australia. Arrows indicate the ostioles ($\times$ 9.2).

(b) The two-celled carpogonial branches (arrows) lie within the compact cortex near the branch tips ($\times$ 320).

(c) Fertilization of the carpogonium appears to be followed by direct fusion of that cell with an undifferentiated cortical cell towards the plant surface (arrow) which acts as the auxiliary cell ($\times$ 312).

orientation and number of the first cell derivatives of zygote germination, the 'primary gonimoblasts' (Figs 1.1–1.3), are significant, as is the ensuing form of the mature carposporophyte. At the simplest extreme (represented by Fig. 1.1), carposporophytes in a number of families and orders consist almost completely of cells which become carposporangia. In many other groups (represented by Fig. 1.2), the mature gonimoblast is a consolidated or diffuse arrangement of sporophytic cells of which the inner layers remain sterile while only the surface produces a single layer or chains of carposporangia. In some of the most specialized types (represented by Fig. 1.3), carposporophytes can achieve great complexity in morphology and development and may grow in intimate contact with large amounts of host gametophyte tissue before issuing carposporangia. Of lesser but still some relevance at the family level are division patterns of tetrasporangia (zonate, cruciate, tetrahedral), and vegetative structure (uniaxial *v.* multiaxial). Of the four to five orders of the Bangiophycidae, sexual reproduction is known for certain in only two to three genera, so that elements of vegetative structure, localization of cell division and type of monospore production assume more importance than gamete and zygote features in classification.

4 RECENT TAXONOMIC DEVELOPMENTS IN THE RED ALGAL ORDERS

4.1 The Florideophycidae

THE NEMALIONALES*

This is undoubtedly the most heterogeneous of all the red algal orders and the one in which the most controversy reigns over the status of the various families.

* Often called the 'Nemaliales'; this review follows Bold & Wynne (1978) in adopting the orthography used by Kylin and justified by Feldmann (1976).

(d) Early growth of the gonimoblast follows formation of a small fusion cell and the emitting of several primary gonimoblast filaments toward the center of the fertile branch (× 216).

(e) For some distance from the fusion cell the gonimoblasts pass through parent tissue and form interconnections with it. At a certain point the gonimoblasts abut gametophyte cells that begin to break down, forming a cavity over the developing carposporophyte (× 190).

(f) As the cavity develops, gonimoblasts at the surface of the mixed haploid and diploid tissue emit dense chains of gonimoblast cells into the chamber (× 180).

(g) At a mid stage of gonimoblast development the cystocarp consists of the inner cortical fusion cell with its inwardly radiating gonimoblast filaments growing centripetally through the host tissue and erupting in a dense mass into the chamber formed by lysis of the gametophyte medullary cells (× 88).

(h) The mature cystocarp shows the fusion cell remnant (arrow) dwarfed by the generation to which it has given rise. Terminal carposporangia cover the surface of the carposporophyte (× 62.4).

(i) Release of the mature carpospores is effected through an ostiole that develops on the side opposite the auxiliary cell, and is proceded by breakdown of the wall surrounding the cystocarp which begins at the ostiole (× 21) (photos (a) and (i) courtesy of Robert W. Ricker).

In the broad sense adopted by Dixon (1973) and many others the group contains 12 families, of which three (with some eight genera and 90 species) are exclusively freshwater. Of the other nine families with their 40–48 genera (depending on how one views the Acrochaetiaceae) and roughly 510 species, eight are completely marine while one, the Acrochaetiaceae, has both marine and freshwater members. The only feature to tenuously unify this diverse assemblage and set it apart from the other orders is the growth of the carposporophyte from some point on the carpogonial branch, usually the carpogonium itself.

The Acrochaetiaceae enjoys a privileged spot in many speculations on red algal phylogeny. As species in a number of families and orders display 'acrochaetioid' stages, the temptation is strong to consider it an ancestral family or organizational level to these higher groups. It is worth noting, however, that acrochaetioid stages in other taxonomic groups are invariably tetrasporophytes rather than the gametophytes where the most conservative red algal processes are generally thought to reside.

The major taxonomic activities in the Nemalionales tend to have centred on the Acrochaetiaceae and on a group of mostly tropical, gelatinous and often calcified genera considered by Kylin to belong to the Helminthocladiaceae. Through the efforts of Abbott, Doty and others, a number of family and generic distinctions have been drawn among the 'helminthoclads' which bring out the reproductive subtlety and diversity of this group.

In regard to the several families of the Nemalionales which have been nominated by various authors for elevation to orders, signs now seem to point to a time not too far off when a revision of the entire assemblage of the sort called for by Dixon (1973) will divide the Nemalionales into more phylogenetically uniform entities with the separation off of some or all of these groups.

The Acrochaetiaceae

The taxonomy of this group (called the Chantransiaceae by Kylin), which includes both freshwater and marine members of generally simple morphology but often complex life histories, is presently being approached from either of two opposing points of view. One extreme is to regard the seven genera recognized by Kylin (omitting *Conchocelis*, a life phase of some Bangiophycean algae) as belonging to the single large genus *Audouinella*. This idea was first approached rigorously by Woelkerling (1971), who reduced the family to the form-genus *Colaconema* (for species in which only monosporic reproduction is known) and *Audouinella* (for all the rest with gametophytes and/or tetrasporophytes). Dixon & Irvine (1977) and Garbary (1979, 1980) treat the family as consisting of the one genus *Audouinella*, thus following Woelkerling in rejecting such prior generic criteria as chromatophore morphology, presence of spermatangiophores, and orientation of the first zygote division.

On the other hand, a number of authors still maintain some or most of the taxa treated as genera by Kylin, the epitome being Stegenga & Vroman (1977) who recognize seven genera based on postfertilization details, type of life history, mode of spore germination and chloroplast structure. Stegenga & van Wissen (1979) make further generic distinctions based on life history features in culture.

Regardless of its number of genera, the family contains a large roster of species, with over 160 credited to it by Kylin and others since. Critical work by Woelkerling (1971, 1973a, b) and Dixon & Irvine (1977) suggests that the number of valid species will be far fewer than this, although new ones of distinctive morphology continue to be added (Lee & Lindstrom 1979). Still unresolved is the placement of *Halosacciocolax*, a parasite on a member of the Palmariaceae. Variously placed in the Cryptonemiales (Lund 1959) or Rhodymeniales (Parke & Dixon 1976), it has recently been suggested by Cabioch & Guiry (1976) to be anatomically in line with the Acrochaetiaceae, despite the fact that it is said to display cell fusions and secondary pit connections which otherwise seem to be absent from the family. Garbary (1979) strongly disputes placement of the genus in the Acrochaetiaceae.

The relatively simple vegetative and reproductive structures of the group lend it to being interpreted as either highly primitive or greatly reduced. Garbary (1978) reviews evidence and arguments for both points of view and concludes that the family probably consists of the world's most primitive extant Florideophycidean algae. He advocates raising the Acrochaetiaceae to the rank of an order (as several authors had previously suggested), a result of which would be to emphasize its likely importance in red algal phylogeny whichever way it is considered.

The Nemalionaceae

The Nemalionaceae is a group restricted by Abbott (1976) to the three genera *Nemalion, Trichogloea* and *Trichogloeopsis*, in which the carpogonial branches occupy positions homologous to vegetative cortical filaments, as opposed to accessory positions on inner cortical cells as in the Helminthocladiaceae.

Nemalion is an uncalcified, temperate water alga of some five to six species, the most common being the type from Europe and several other world regions. Vegetative and reproductive morphology are detailed by Womersley (1965a) and Yoshizaki (1977). *Trichogloea* (Fig. 1.2), with five to seven species, contains *Nemalion*-like calcified algae from tropical seas. A detailed study of reproductive anatomy is given by Yoshizaki (1979). *Trichogloeopsis* Abbott & Doty (1960), with three tropical species, is also calcified and exhibits a combination of *Trichogloea* and *Nemalion* features, as well as a unique sterile-cell investment of the fertilized carpogonial branch that descends from the developing gonimoblast.

The Helminthocladiaceae

The Helminthocladiaceae contains five genera and around 80 species, of which perhaps 60 belong to the calcified, mostly tropical genus *Liagora*. Fan & Wang (1974) have recently removed one pantropical species (*L. farinosa*) to the new genus *Ganonema* based on the position of the carpogonial branches on special 'fertile filaments' issued from the bases of assimilatory filaments. Post-fertilization and reproductive morphology in *Liagora* have been studied by Chiang (1972), Chihara & Yoshizaki (1972a) and Womersley (1965a).

The non-calcified genera *Helminthocladia* and *Helminthora* occur in both temperate (Abbott 1965; Womersley 1965a) and tropical (Doty & Abbott 1961) regions. Bodard (1971) has described an African species of *Helminthocladia* with carpotetraspores, and Umezaki (1960) and Womersley (1965a) have studied reproductive morphology in other members of the genus. The most recent addition to the family, *Izziella,* was created by Doty (1978a) for a *Liagora*-like Hawaiian alga with a uniquely derived sterile cell investment associated with hypogenous cells of the carpogonial system.

The Dermonemataceae

The Dermonemataceae was elevated to a family from a subfamily of the Helminthocladiaceae by Abbott (1976), based on the issuing of protruberances (rather than simple divisions of the carpogonium) following fertilization, production of diffuse gonimoblasts with terminal carposporangia, and the almost complete lack of sterile filaments in the gonimoblast. The five genera and seven species are almost all tropical, and three of the genera (*Dotyophycus, Yamadaella* and *Liagoropsis*) are calcified.

The genus *Dermonema* has been studied by Yoshizaki (1978), who details the anatomy of the pan-tropical type species and a second Japanese representative. Doty & Abbott (1964) have monographed the genus *Liagoropsis* from the Caribbean and western Pacific and have reduced the three described species to one.

Yamadaella was created by Abbott (1970) for a single tropical Pacific and Indian Ocean species of *Liagora* with filamentous gonimoblasts and 'carpotetrasporangia'. *Dotyophycus* is an Abbott (1976) genus from Hawaii in which the carpogonial branch is situated on a medullary, rather than a cortical, filament. Like *Yamadaella*, it produces a diffuse carposporophyte with terminal carpotetrasporangia.

The monotypic Californian genus *Cumagloia* is the one exception to the warm-water distribution of the group. Desikachary (1962) has detailed its reproductive anatomy.

The Remaining Families

The Bonnemaisoniaceae is a family of six genera and around 22 species which seem to be concentrated in the Australasian region and northern Pacific. Chihara & Yoshizaki (1972b) review the morphology and life history information available for the family and conclude that it should be raised to an order, as had previously been done by J. and G. Feldmann (1942). Besides morphological and gonimoblast features, the authors present cytological and biochemical evidence for separating the family (and possibly the Naccariaceae) from the rest of the Nemalionales and placing it near the Ceramiales.

Since Kylin's treatment, the genus *Pleuroblepharidella* (Wynne 1970b, 1972, 1980) has been added, a distinctive north Pacific alga in which only compound tetrasporangial stichidia are known. Chihara & Yoshizaki (1978) critically examine the supposed differences between *Ptilonia* and *Delisea* and conclude that the single genus *Delisea* should contain the 11 or so species that were formerly in both. Hudson & Wynne (1969) show the difficulty of finding sound bases to separate *Asparagopsis* from *Bonnemaisonia*, finally resorting to a suite of fairly cryptic reproductive features. Chihara & Yoshizaki (1972b), however, suggest that *Bonnemaisonia* itself might be split, based on two differing morphologies of the diminutive tetrasporophytes in the several present species.

The small family Naccariaceae has been studied by Womersley & Abbott (1968), who sink the Australian genus *Neoardissonia* into *Naccaria* and present criteria for distinguishing the two genera *Naccaria* and *Atractophora* which comprise the group.

The Chaetangiaceae is a family of eight genera and nearly 80 species of predominantly tropical to warm-temperate distribution, the most notable exception being the Puget Sound genus *Whidbeyella* (Scagel 1962b). The calcified genus *Galaxaura* has had over 70 species attributed to it. Papenfuss & Chiang (1969) review the history of the genus and maintain that over half the recorded species are synonyms owing to the heteromorphic tetrasporophyte and gametophyte generations that are known in some cases and presumed to exist in the rest. Itono (1977a) has treated 20 *Galaxauras* from southern Japan. *Actinotrichia*, which differs from *Galaxaura* in having isomorphic generations, contains one pan-tropical species and a second recently added by Itono (1979). A close second to *Galaxaura* in numbers is *Scinaia*, a non-calcified genus of some 25 species to which the most recent addition has been made by Eiseman (1979). Studies of gonimoblast development in *Scinaia* have been made by Chiang (1970a) and Kaneko (1975). Reproduction and life history studies of a Californian species of *Pseudogloiophloea* have been made by Ramus (1969), and gonimoblast development of a Japanese representative has been detailed by Chihara & Yoshizaki (1969). Desikachary (1963) has argued that the Chaetangiaceae should be elevated to an order based on auxiliary cell and apical features.

The Gelidiaceae consists of nine genera and around 115 species, the largest genus being *Gelidium*. Santelices (1976) notes the poor criteria in use for setting up taxa in both *Gelidium* and the second largest genus of the family, *Pterocladia*. The two are impossible to tell apart when non-cystocarpic, as type specimens and field collections usually are. Although Kylin credits around 40 species to *Gelidium*, there are at least 85 species names floating about. *Pterocladia*, which differs from *Gelidium* only in having one rather than two ostioles in the cystocarp, has around 14 species. Santelices (1976, 1977) and Silva (1978) discuss the taxonomy of several central and eastern Pacific members of the family.

The major work which treats the bulk of the Gelidiaceae is that of Fan (1961), who divides it into three tribes. Besides *Gelidium* and *Pterocladia*, Fan discusses the remaining genera, many of which are monotypic and previously little-known.

The Gelidiellaceae was established by Fan (1961) for gelidioid algae lacking the internal hyphae of the Gelidiaceae and in which cystocarpic plants are unknown. The family consists of the one mostly tropical genus *Gelidiella* to which some 21 species are now linked. The most widely distributed and commercially valuable species (*G. acerosa*) has been cultured by Chihara & Kamura (1963), who found the germination stages typically gelidiaceous.

The Cryptonemiales

The 14 families, approximately 130 genera and around 900–950 species of the Cryptonemiales form a morphologically diverse group whose major linking feature involves the position of the auxiliary cell, generally one of the most difficult structures to find and interpret in any alga. In female gametophytes of this order, the auxiliary cells from which the gonimoblasts arise and the cystocarps develop are located on or within 'accessory' branches (as in Fig. 1.1) which are supernumerary in location and time of origin to the non-specialized filaments which constitute the vegetative thallus. In this way members of the order are distinguished from those of the Gigartinales, in which the auxiliary cells occupy intercalary positions in otherwise ordinary vegetative filaments. Since Kylin distributed families between the two orders and attributed 12 to the Cryptonemiales, there has been emphasis on determining the accessory versus the non-accessory nature of auxiliary cell filaments in groups where this distinction has been particularly unclear. As a result, the Polyideaceae and Rhizophyllidaceae have been transferred from the Cryptonemiales to the Gigartinales (Papenfuss 1966; Wiseman 1975).

Names of two of the families in Kylin, the Grateloupiaceae and Squamariaceae, have been altered to the Halymeniaceae and Peyssonneliaceae to conform to various rules of the Botanical Code, while four completely new families (the Corynomorphaceae, Pseudoanemoniaceae, Pterocladiophyllaceae and Weeksiaceae) have been added.

Recent research has tended to centre on four families, the Corallinaceae, Halymeniaceae, Kallymeniaceae and Peyssonneliaceae, which are also the largest of the order. The group of greatest life history interest is the Peyssonneliaceae, as so many of its members are known only as tetrasporophytes and have proved or may prove to be parts of life histories of algae in other families and even orders. The Corallinaceae, with an importance extending well beyond academic botany to impinge on economics, paleontology, geology and reef physiognomy, continues to draw researchers from many disciplines and is a paradigm for showing the importance of basic taxonomy to a variety of human endeavors.

Phylogenetically the order's lower forms appear to belong to the Dumontiaceae, Weeksiaceae and Gloiosiphoniaceae, these being closest in structure and reproduction to the lowest families of the Gigartinales. Specialization in both vegetative and reproductive morphology may have progressed through such intermediate levels as the Halymeniaceae and Peyssonneliaceae, and reached its heights in the Corallinaceae, Kallymeniaceae and the parasitic group Choreocolacaceae. The five genera of the latter family are of considerable biological interest and have led some authors (Lee 1971; Demoulin 1975) to consider them possible evidence for the evolution of Ascomycetes and Basidiomycetes from red algal ancestors.

Recent broad discussions of the Cryptonemialean families from a phylogenetic standpoint have been made by Norris (1957), Balakrishnan (1960), Chiang (1970b) and Lebednik (1977).

The Corallinaceae

Forty per cent of the genera and around half of the species of the Cryptonemiales belong to the Corallinaceae—the largest, most studied and taxonomically hardest family of the order. Several reviews have covered most aspects of the group's taxonomy and biology, including those by Littler (1972) for the encrusting forms and Johansen (1974) for the articulated genera. Johansen (1976) lists the genera of the family and surveys taxonomic criteria and concepts. A forthcoming book on the family by Johansen (in press) will comprehensively survey its biology, until which time the review by Adey & MacIntyre (1973) remains the indispensable illustrated primer on the terminology, classification and habitat ranges of the crustose members of the group.

Considerable change has taken place since Kylin put 36 genera in the family. Some 42–50 living and about 30 exclusively fossil genera are now recognized (Woelkerling personal communication), although the number of species properly belonging to the group is still largely guesswork, being somewhere between 400 and 500. An attempt to reinterpret coralline algal carposporophyte development and to place the family in an evolutionary framework within the Cryptonemiales has been made by Lebednik (1977).

Some of his concepts and particularly his terminology have been criticized by Woelkerling (1980).

The Halymeniaceae

Running a distant second to the Corallinaceae in numbers, the Halymeniaceae contains 18 genera and around 160 reasonably delimited species, with perhaps another 60 or so loose ends tagged mainly onto the genera *Carpopeltis, Grateloupia* and *Halymenia*. This group was termed the Grateloupiaceae by Kylin, and has gone under the name Cryptonemiaceae until recently changed again by Guiry (1978b).

Differences in the make-up of the family since Kylin include the removal of three genera (*Corynomorpha, Estebania* and *Lobocolax*) and the addition of *Pachymeniopsis* (Kawabata 1954, 1962) and *Dactylymenia*, the latter having been removed from the Kallymeniaceae and provisionally transferred to the Halymeniaceae by Womersley & Norris (1971). The major work to treat a large number of genera and attempt a revision of phylogenetic criteria within the family is that of Chiang (1970b), who compares ampullar types and proposes an evolutionary sequence based on their relative complexity. Kraft (1977a) found a seeming exception to Chiang's scheme in a New Zealand species of *Grateloupia* and has questioned the uniform importance of ampullar types in the evolution of the group.

Regional surveys illustrating large numbers of the family include those for California by Abbott & Hollenberg (1976), the Caribbean by Taylor (1960), New Zealand by Chapman & Parkinson (1974) and Peru by Acleto (1973).

The Peyssonneliaceae

A family of about eight genera and 65–70 species, the Peyssonneliaceae of Denizot (1968) replaces the Squamariaceae to which Kylin credited nine genera. The group contains many problematical taxa in which gametophytes are unknown, and has been the object of much generic shifting, partly as a result of recent culture work. Genera now dropped from Kylin's roster are *Ethelia, Contarinia* and *Herpophyllon* (Denizot 1968), *Erythrodermis* and *Porphyrodiscus*, the latter two of which are considered to be tetrasporic stages of algae in the Phyllophoraceae (Gigartinales) by Dixon & Irvine (1977). Added to the family have been *Cruoriella* and *Polystrata* (Denizot 1968), *Metapeyssonnelia* (Boudouresque *et al.* 1976), and *Pulvinia* (Hollenberg 1970), while *Pseudorhododiscus* and *Rhodophysemopsis* (Masuda 1976a, b) and *Ethelia, Chevaliericrusta, Riquetophycus* and *Sonderopelta* (Denizot 1968) are all genera of peyssonnelioid anatomy which have not formally been placed in the family for lack of female gametophytes. Although Abbott & Hollenberg (1976) retain *Cruoriopsis* in the family, Dixon & Irvine (1977) consider it a probable part of the *Gloiosiphonia* life history. *Rhodophysema* (as *Rhododermis* in Kylin) was considered of uncertain family position by Denizot (1968), but is retained in the family by Abbott & Hollenberg (1976) and Parke & Dixon (1976). Like many other members of the family, it has been the subject of a number of culture studies. *Rhododiscus* is maintained in the family by Parke & Dixon (1976), who suspect its generic status and anticipate that it will prove related to *Thuretellopsis* in the Dumontiaceae.

The largest and about the only unquestioned member of the family is *Peyssonnelia* itself, for which Denizot (1968) lists 49 species in his monograph of the world's encrusting non-corallinaceous algae. The morphology and reproduction of various species are illustrated by Abbott & Hollenberg (1976), Hollenberg & Abbott (1968), Nozawa (1972a,b) and Yoshida (1975). Boudouresque & Denizot (1973) discuss and evaluate species criteria in the genus.

The Kallymeniaceae

A family of 18 genera and some 110 species, the Kallymeniaceae has been one of the more intensively studied groups of the Cryptonemiales. Basic understanding of essential features and reproductive variation has been advanced by Norris (1957), who detailed both procarpic and non-procarpic, as well as polycarpogonial and monocarpogonial members of the family. Hommersand & Ott (1970) have revealed great complexity in the relationship of gametophyte to carposporophyte tissue in the gonimoblast initiation and development of the type species of *Kallymenia*, while Hansen (1977) has detailed a bizarre apomixis in a species of *Cirrulicarpus*.

Several regional monographs have appeared, including those for Australia by Womersley & Norris (1971), New Zealand by Chapman & Parkinson (1974) and South Africa by Norris (1964).

The genus *Euthora* has been merged with *Callophyllis* by Hooper & South (1974), but several new genera have been added to the family, including *Cirrulicarpus* (Tokida & Masaki 1956; Norris *et al.* 1960), *Thamnophyllis* (Norris 1964; Womersley & Norris 1971) and *Austrophyllis* (Womersley & Norris 1971). The need for so large a suite of genera to accommodate variances from the two major genera of the group (*Kallymenia* and *Callophyllis*) has been questioned by Codomier (1978).

Kallymeniopsis was created by Perestenko (1975) for three Siberian species and was placed, along with *Erythrophyllum*, *Cirrulicarpus* and the resurrected Ruprecht genus *Crossocarpus* in the separate family Crossocarpaceae, which Perestenko distinguishes from the Kallymeniaceae on seemingly minor details of the fusion-cell formation. The validity of the Crossocarpaceae has not been generally accepted. Kylin considered *Crossocarpus* itself to be a synonym of *Callophyllis*. *Beringia* Perestenko (1975), although apparently similar to *Crossocarpus*, was not specifically included in the Crossocarpaceae.

Callophyllis, the largest genus in the family, is credited with around 50 species by Kylin and probably contains something like 40. Major studies have been made by Abbott & Norris (1965), Chapman & Parkinson (1974), Norris (1957) and Womersley & Norris (1971). The life history of *C. firma* was followed by Murray & Dixon (1972), who claim it to be the first documented instance of the *Polysiphonia*-type in the Cryptonemiales.

The Remaining Families

The Dumontiaceae is a distinctive family of 13 genera and around 37 species and is unusual in having both uniaxial and multiaxial members and both cruciately and zonately tetrasporangiate species. The strangely shaped *Gibsmithia* was described by Doty (1963), who attached it to the family with some reservation as certain crucial reproductive stages were lacking. The genus has been confirmed as dumontiaceous by Karam-Keriman (1976), although it is still anomalous in having internal rhizoids which branch and form secondary pit connections. *Abbottia*, from the northwest Pacific, is a multiaxial blade described by Perestenko (1975), who places it in the Dilseaceae. Bert (1965) erected the Dilseaceae for the genera *Dilsea* and *Neodilsea* solely because of their multiaxial rather than uniaxial structure, but the family has not been accepted by other authors, who continue to treat the genera as part of the Dumontiaceae.

Studies on some of the individual genera have been made by Chihara & Yoshizaki (1971), Mitchell (1966) and Abbott (1962a, 1968). The largest genus of the family is *Dudresnaya*, with some 10–11 mostly warm water species. Reproductive anatomy is illustrated by Kawashima (1957), and new species have most recently been added by Lee (1963) and Littler (1974). The closely related genus *Acrosymphyton* has been studied by Abbott (1962b), Cortel-Breeman and his colleagues (e.g. Cortel-Breeman & ten Hoopen 1978), and is illustrated in this review (Fig. 1.1).

The Weeksiaceae is a family of three genera and nine species established by Abbott (1968) for plants with basically dumontiaceous gametophytes but in which the auxiliary cell branches are non-functional and gonimoblasts develop from a 'nutritive' cell of the carpogonial branch. Previously Norris (1957) had disputed the importance of these features as family criteria. The Weeksiaceae is maintained by Abbott & Hollenberg (1976), where many of the species are illustrated.

Formerly included in the Cryptonemiaceae, the genus *Corynomorpha* was placed in its own family by Balakrishnan (1960, 1962) because of the polycarpogonial ampullae of the Indian type species. The move has been supported by Chiang (1970a). A presumed second species occurs in the Caribbean and Brazil, but is not known cystocarpically (Joly *et al.* 1965).

Composed of six genera and around 15 species, the Gloiosiphoniaceae is a predominently cool-temperate Northern Hemisphere family. Its largest genus is *Schimmelmannia*, with six somewhat poorly differentiated eastern and northwestern Pacific species. Gonimoblast development and taxonomy are studied by Acleto (1972). Abbott (1961) has reduced the monotypic genus *Baylesia* of the Dumontiaceae to synonymy with *Schimmelmannia*. Lee & Yoo (1979) contrast several features of the family's genera in establishing the monotypic genus *Gloeophycus* from Korea. Similar in morphology to *Thuretella*, it differs in the position of the auxiliary cell, formation of the auxiliary cell branch, and lack of an involucre around the carposporophyte. Kaneko *et al.* (1980) demonstrate an unusual form of connecting-cell formation in Japanese specimens of *Gloeophycus* in which the carpogonium divides into three cells following fertilization, the middle of which transfers the zygote nucleus to the auxiliary cell.

In the Hildenbrandiaceae, a family of two genera and 10–11 species, gametophytes are unknown and regularly to irregularly zonate tetrasporangia are formed in conceptacular pits. DeCew & West (1977) record nine species of *Hildenbrandia* and give evidence that life histories in two of them are apomeiotic, resulting in the continuous recycling of tetrasporangial crusts. *Apophloea*, with two New Zealand species, is included in the family by Denizot (1968) and Chapman & Parkinson (1974).

The Pterocladiophyllaceae was founded by Fan & Papenfuss (1959) for a single New Zealand species alloparasitic on *Pterocladia*. *Pterocladiophylla* is distinguished from the Choreocolacaceae by its zonate tetrasporangia and conceptacle-produced male and female gametes, and from the Corallinaceae by its non-calcified thalli and catenate spermatangia.

The Choreocolacaceae contains five genera and nine species of problematic alloparasites now tentatively placed in the Cryptonemiales. The most intensely studied genus of the family has been *Harveyella*. One in a series of papers by Goff (Goff & Cole 1975) concerns reproductive anatomy and taxonomy but leaves the ordinal placement in doubt owing to difficulty in interpreting the position of the auxiliary cell. *Gelidiocolax* was transferred to the family from the Gelidiaceae by Fan & Papenfuss (1959), although Stewart (in Abbott & Hollenberg 1976) retains the Californian type species in the Gelidiaceae without explicit justification. Ganesan (1970) keys the six species and summarizes their comparative features.

The Pseudoanemoniaceae contains the congeneric (Hawkes & Hommersand personal communication) *Hummbrella* (Earle 1969) and *Pseudoanemonia* (Chapman 1969). The group has yet to be defined reproductively (Chapman 1969), and its inclusion in the Cryptonemiales is still tentative.

The Gigartinales

Almost half of the Florideophycidean families belong to the Gigartinales (currently 28), a diverse group of around 100 genera and 700 species. This great family profusion is sometimes taken as an indication of the order's disarray, although Kraft (1977b) has implied that this is largely a proper recognition of the evolution of the Gigartinales into specialized offshoots from a core developmental line leading from the Gymnophloeaceae through the Solieriaceae.

Since Kylin credited it with 21 families, the group has lost the Stictosporaceae, seen nomenclatural alterations of the Nemastomaceae to the Gymnophloeaceae, the Rhabdoniaceae to the Caulacanthaceae, and the Rhodophyllidaceae to the Cystocloniaceae, had the Polyideaceae and Rhizophyllidaceae transferred to it from the Cryptonemiales, and the five families Blinksiaceae, Cubiculosporaceae, Mychodeophyllaceae, Nizymeniaceae and Wurdemanniaceae described as new.

Major research emphasis has centred on the Phyllophoraceae and Gigartinaceae, where the discovery of unusual and complex life histories in some species has resulted in major taxonomic changes. Searles (1968), Kraft (1973, 1975a, b, 1977b, c, 1978) and Min-Thein & Womersley (1976) have studied a large number of other families and have done most to further phylogenetic speculation within the order. These authors have called a number of Kylin's original family criteria into question and have substituted others based primarily on carposporophyte ontogeny. Members of the Gigartinales exhibit a range of carposporophyte types from the most morphologically simple (Umezaki 1967; Kraft & John 1976) to the most complex (Kraft 1975b, 1977b, c, 1978) in the red algae, and it is in this order that the precise and often cryptic details of gonimoblast development become most taxonomically crucial.

Although it is complex, the Gigartinales seems a generally uniform and defensible group, with the exception of a few families that now sit rather uneasily in it. The Polyideaceae, which Papenfuss (1966) transferred from the Cryptonemiales, does not have the three-celled carpogonial branch typical of the bulk of the present Gigartinales, nor do its connecting filaments behave like the rest of the order. As claimed by Nozawa (1970), connecting filaments in *Rhodopeltis* contact cells of the carpogonial branch before proceeding on to the auxiliary cells, suggesting affinities (as Kylin maintained) to the Dumontiaceae rather than the Gigartinales. Much the same seems to be true of the Rhizophyllidaceae, which Wiseman (1975) also transfers from the Cryptonemiales. In both these groups the auxiliary cell filaments occur in nemathecia, and although Wiseman (1977) shows convincingly that the auxiliary cell filaments in *Octodes* arise as ordinary nemathecial filaments do, it may be that there is an inherent blurring of the accessory versus non-accessory origin of filaments in situations where carpogonial and auxiliary cell branches develop in nemathecial palisades. Given our present definition of the Gigartinales, however, the transfers seem to be justified and call for closer study of the links between the Gigartinales and Cryptonemiales.

If the Polyideaceae and Rhizophyllidaceae merit transfer to the Gigartinales, several other cryptonemialean families seem equally vulnerable to that move.

This is particularly true of the Peyssonneliaceae, in which auxiliary cell and carpogonial branch filaments also originate in nemathecia. It has been suggested by some that the Corallinaceae should also go to the Gigartinales, although Lebednik (1977) argues against this, based on his particular definition of what constitutes an accessory auxiliary cell system. Still, if the supporting cell of the carpogonial branch in the Corallinaceae is really the first recipient of the zygote nucleus and the starting point for the fusion cell, as stated by Lebednik (1977), this could be interpreted as analogous to the condition in several other gigartinalean families (e.g. the Gigartinaceae, Mychodeaceae and Plocamiaceae). Two other groups in which the supporting cell of the carpogonial branch also functions as the auxiliary cell are the Pseudoanemoniaceae (Hommersand & Earle 1969) and Choreocolacaceae (Goff & Cole 1975). The parasitic genus *Kintokiocolax* (Tanaka & Nozawa 1960) was ascribed to the Gigartinales but not assigned to a family. It too apparently has supporting cells which act as auxiliary cells, but its unique combination of vegetative and reproductive features seems to indicate that it will ultimately require placement in a family of its own.

Although the Gigartinales (unlike the presently constituted Cryptonemiales) has few calcified members and no reef-building types (*Titanophora* being the only calcified genus, plus *Rhodopeltis* if the Polyideaceae is admitted to the order), it is an important group if for no other reason than its many phycocolloid producers, found mainly in the Furcellariaceae, Gigartinaceae, Gracilariaceae and Solieriaceae. For this reason some members are in much commercial demand, these often belonging to the most taxonomically difficult genera (e.g. *Eucheuma, Gigartina, Gracilaria, Hypnea*). Such algae provide some of the strongest incentives and opportunities for the application of newer chemotaxonomic and culture techniques, as well as the classical morphological ones, and again point to the need for basic taxonomy in the real world.

The Gigartinaceae

The Gigartinaceae was credited with five genera and around 130 species by Kylin. Kim (1976) has monographed the group and reduced it to two genera (*Chondrus* and *Gigartina*) with about 100 species. Kim merges *Rhodoglossum* with *Iridaea* and makes the latter a subgenus of *Gigartina*. The existing subgenus of *Gigartina*, *Mastocarpus*, is raised to generic status and removed to its own family (the Mastocarpaceae) which is, like the genus itself however, never properly validated. Kim's criteria for inclusion in *Mastocarpus* rather than *Gigartina* are lack of isomorphic gametophyte and tetrasporophyte generations, presence of sterile laterals in the procarp, and production of few rather than a dozen or so gonimoblast initials. Kim's treatments of both the *Iridaea* and mastocarpoid complexes of species seem yet to have gained support (Abbott & Hollenberg 1976; Dixon & Irvine 1977).

Other proposed changes to the family are the removal of *Besa* to the Phyllophoraceae (Kim 1976) and the addition of *Petrocelis* from the Cruoriaceae. Denizot (1968) considered *Petrocelis*, with its three to four species, to belong to a distinct family. Cystocarps had been reported to be of the *Cruoria* type, but Abbott & Hollenberg (1976) and Dixon & Irvine (1977) suspect that these records are based on unrelated taxa. The Petrocelidaceae is recognized by Abbott & Hollenberg (1976), but rejected by Parke & Dixon (1976) and Dixon & Irvine (1977) on the bases of several culture studies by other workers which have shown that some mastocarpoid *Gigartinas* produce crustose tetrasporophytes identifiable with species of *Petrocelis*.

Other monographic work on elements of the Gigartinaceae includes that of Mikami (1965) for Japan, and Abbott (1971), who reduces 18 *Iridaea* species from the eastern and northeastern Pacific to seven. Because of the economic importance of this group as carrageenan sources, much work on its biology has taken place, such as that reported for *Chondrus crispus* (Harvey & McLachlan 1973) and eastern Pacific Gigartinaceae by Abbott (1980).

The Gracilariaceae

The Gracilariaceae is a family of six genera and around 130 species, of which 100 or so belong to the worldwide agarophyte *Gracilaria*. Kylin records nine genera, of which *Tylotus* has been transferred to the Dicranemaceae (Kraft 1977c) and *Corallopsis* synonymized with *Gracilaria* (Dawson 1954). *Gracilariopsis*, which was created by Dawson (1949) for *Gracilaria*-like species lacking nutritive filaments in the cystocarp, is rejected by Papenfuss (1967) and Yamamoto (1975) because of the feature's inconsistency.

Major regional studies of *Gracilaria* include those of Yamamoto (1978) for Japan, Chang & Xia (1976) for mainland China, Umamaheswara Rao (1972) for India, and Kim (1970) for Chile. Studies of reproductive morphology have been made by Kim (1970), Yamamoto (1975, 1978), Oza (1976) and Edelstein *et al.* (1978). Kraft (1977d) illustrates a New Zealand species and suggests a change in terminology for the so-called 'nutritive' filaments of the cystocarp, calling them 'traversing' filaments instead.

Chang & Xia (1963) place *Gracilaria*-type algae with compound spermatangial cavities in the separate genus *Polycavernosa*, to which they ascribe two new species and transfer two others. Their generic criterion seems a bit weak in light of the wide and rather continuous range of variation known in *Gracilaria* male sorus types (Yamamoto 1975), and *Polycavernosa* has not been adopted by other writers.

The remaining genera of the family have not been recently studied. *Curdiea* and *Melanthalia* have a number of mostly Australasian species (Chapman 1979), while *Gelidiopsis* and *Ceratodictyon* are tropical. Placement of the latter two needs to be confirmed. The parasitic genus *Gracilariophila* has two Californian species which are illustrated by Abbott & Hollenberg (1976). A second parasite of *Gracilaria, Gracilariocolax*, was described from Java and is poorly known. Kylin put it among the unplaced genera.

The Phyllophoraceae

This is the gigartinalean family in which the most taxonomic change has taken place since Kylin, who credited it with five genera and 64 species. With the addition of *Erythrodermis, Porphyrodiscus, Ozophora, Petroglossum, Besa* and *Schottera*, the group now contains 11 genera and some 70 species. Guiry (1978b) advocates substitution of the obscure but earlier name Tylocarpaceae for the well-known family name in current use.

Classical taxonomic studies of the Phyllophoraceae have been made in Japan by Mikami (1965) and in Europe by Schotter (1968). The largest genus of the group is *Gymnogongrus*, with about 40 species, to which recent additions have been made by Chapman (1979) and Taylor (1971). Newroth (1971) has studied five North Atlantic species of *Phyllophora*, and Dixon & Irvine (1977) summarize the nomenclatural and taxonomic changes which have affected some of the eight or so species of this mostly European genus. Dixon & Irvine (1977) have also argued for the transfers of *Erythrodermis* and *Porphyrodiscus* from the Peyssonneliaceae to the Phyllophoraceae based on the anatomy and tetraspore distribution of the former and the seeming appearance of the latter in culture studies of *Ahnfeltia*. Schotter (1968) credits *Ahnfeltia* with nine species, the majority occurring in Japan (Mikami 1965). *Ozophora*, which Kylin considered a synonym of *Phyllophora*, is reinstated by Abbott (1969) for plants with male and female gametes on lateral leaflets. *Petroglossum* was also placed in synonymy with *Phyllophora* by Kylin but is recognized by Abbott & Hollenberg (1976), based partly on nemathecial spermatangia and tetrasporangia. Because tetrasporangia remain unknown, the placement of *Besa* in the Phyllophoraceae is somewhat tentative (Kim 1976).

Schottera, whose single species Schotter (1968) had earlier transferred from *Gymnogongrus* to *Phyllophora*, was described by Guiry & Hollenberg (1975) as a distinct genus based on the cystocarp morphology. It was known only from Europe until recorded in Australia by Lewis & Kraft (1979). *Stenogramme*, its type species widely distributed in cool-temperate regions of both hemispheres, contains two additional species of limited distribution in Australia and Chile, the latter described by Joly & Alveal (1969). *Ceratocolax* is a monotypic parasite on *Phyllophora* and has not been recently studied. Its circumpolar distribution is documented by Newroth & Taylor (1968).

The Solieriaceae

The most generically diverse family of the Gigartinales is the Solieriaceae, with its 15 genera and around 50 species. Changes since Kylin include the loss of *Predaea* to the Gymnophloeaceae, the nomenclatural alteration of *Thysanocladia* to *Callophycus*, and the addition of *Placentophora*.

Kraft (1977b, c, 1978) considers this family a pivotal one in the possible evolution of several specialized groups of the Gigartinales, and links a number of other families to it. The boundaries of the Solieriaceae were expanded with the inclusion of the endemic New Zealand genus *Placentophora* (Kraft 1975b), in which the carposporophyte develops in closer association with gametophytic tissue than apparently happens in other genera of the family. The limits are even more expanded with proposals by Gabrielson (1980) that several uniaxial genera now in the Rhabdoniaceae be transferred to the Solieriaceae.

Recent work includes a monograph of Australian species of the group by Min-Thein & Womersley (1976). Five of seven *Callophycus* species are treated while a sixth, from South Africa, was previously studied by Hewitt (1960). The Indian Ocean genus *Sarconema* has been monographed by Papenfuss & Edelstein (1974), who reduce it to two species. Four out of the five circumboreal species of *Turnerella* are reduced to one by South *et al.* (1972), who discovered that the hitherto unknown tetrasporophytes are crustose phases of the life history. Goff (1977) establishes the validity of the parasitic genus *Gardneriella*, countering suggestions that it may represent nothing more than host gall tissue. The Australian species of *Solieria* has been studied by Min-Thein & Womersley (1976), and a new species has recently been described by Yoshida (1980) from Japan.

The largest and most commercially valuable genus of the family is the carrageenan weed *Eucheuma*, with at least ten species widely dispersed through the world's tropics. The economic utilization and potential of Asian taxa are described by Doty (1973, 1978b). Studies of the taxonomy of Caribbean representatives have been made by Cheney & Babbel (1978), who remove two species from the genus on morphological and electrophoretic grounds. Kraft (1972) has described the habit and distribution of a unique 'coral-mimicking' species from the Philippines, and Mshigeni (1979) has studied the range of forms and habitats of five members of the genus from Tanzania.

Particular attention has been paid to *Agardhiella* in North America, where distinguishing marks between its anatomy and that of *Solieria* are defined by Taylor & Rhyne (1969) and Wynne & Taylor (1973). Abbott (1978) has separated the eastern Pacific species from the western Atlantic one on the basis of careful anatomical comparisons. Nomenclatural considerations led Wynne & Taylor (1973) to rename the genus *Neoagardhiella*, but *Agardhiella* has been restored after reappraisal by Kraft & Wynne (1979). Gabrielson & Hommersand (1980) consider *Agardhiella* and two other genera (*Sarcodiotheca* and *Meristiella*, the latter not yet formally described) to share sufficient apical, cystocarpic and distributional features to warrant inclusion in a distinct tribe, the Agardhielleae.

The remaining families

The Gymnophloeaceae (Nemastomaceae in Kylin) is now a family of six genera and about 50 species of mostly tropical to warm-temperate algae. Kraft (1975a, 1978) considers this the probable most primitive family of the order and a possible point of departure for the evolution of the Cryptonemiales as well as the higher families of the Gigartinales. The Australian genus *Adelophyton* (Kraft 1975a) is suggestive in this regard as it seems to combine features of both orders in its uniquely situated auxiliary cells. Kraft & Abbott (1971) have surveyed the genera of the family as a whole, and Kraft (1975a) has reviewed its critical features in comparison to other gigartinalean families. The largest genus in the family is *Nemastoma*, with some 18 mostly warm-temperate species (Kraft & John 1976). *Platoma*, with seven species, is very similar to *Nemastoma* and often inconsistently distinguished from it. Kraft & John (1976) and Norris & Bucher (1977) have discussed this problem without, however, resolving it. The unusual morphology and distribution of the six *Predaea* species are canvassed by Kraft & Abbott (1971), while a seventh species from Japan has been added recently by Yoshida (1980). The one cold-water genus of the Gymnophloeaceae, *Schizymenia*, has been studied taxonomically by Abbott (1967). Ardre (1977) presents culture results and Sciuto *et al.* (1979) give chemical evidence which suggests that the encrusting genus *Haematocelis*, formerly placed in the Cruoriaceae, is the tetrasporophyte of *Schizymenia*.

The Calosiphoniaceae is a small family of two genera and four species which differs from the Gymnophloeaceae mainly only in being uniaxial rather than multiaxial. It is best known from the morphological studies of Feldmann (1954) and Mayhoub (1974, 1975).

Containing a single Californian species in the genus *Blinksia*, the Blinksiaceae was described by Hollenberg & Abbott (1968) for encrusting plants distinct from the Gymnophloeaceae on habit and from the Cruoriaceae by the production of outwardly developing gonimoblasts from auxiliary cells rather than of both inwardly and outwardly developing gonimoblasts from connecting filaments. The Cruoriaceae is a family in which much change has taken place and now consists of the single encrusting genus *Cruoria* following removal of *Petrocelis, Cruoriopsis* (Dixon & Irvine 1977) and *Haematocelis* (Ardre 1977) to other families. *Cruoria* contains about seven species, some of which are not known sexually and thus questionably associated (Denizot 1968).

The Caulacanthaceae (formerly the Rhabdoniaceae, see Guiry 1978b) now contains 11 genera and about 30 mostly Southern Hemisphere species. The latest revision of a large portion of the family is that of Min-Thein & Womersley (1976), who concentrate on southern Australia where the group is mainly centred. Work by Searles (1968) has extended the definition of the family through the inclusion of some slightly atypical genera. The Cubiculosporaceae (Kraft 1973), with its single genus and species, has affinities with the

Caulacanthaceae but differs in being multiaxial and by lacking any sort of a fusion cell in the cystocarp.

The Phacelocarpaceae was created by Searles (1968) for *Phacelocarpus* and a second genus (*Stenocladia*) which Kylin had included in the Sphaerococcaceae, plus *Nizymenia* which Kylin placed in the Sarcodiaceae. Womersley (1971a) removed *Stenocladia* and *Nizymenia* to the new family Nizymeniaceae based mainly on their spermatangial sori and production of cruciate tetrasporangia on uniseriate filaments.

A family of ten genera and about 35 species, the Cystocloniaceae (formerly the Rhodophyllidaceae, see Guiry 1978b) is largely confined to the Australasian region where it has been monographed by Min-Thein & Womersley (1976). The circumscription of the family was expanded by Searles (1968), who removed the monotypic genus *Stictosporum* from the family created for it by Kylin and transferred it to the Rhodophyllidaceae. In *Stictosporum* the group gains a member with outward, rather than inward, gonimoblast development, thus blurring one of Kylin's major family criteria in the Gigartinales. The recently described genus *Fimbrifolium* (Hansen 1980) also initiates its gonimoblasts outwardly and additionally shows some distinctive cystocarp and tetrasporangial features.

The Acrotylaceae, Dicranemaceae, Mychodeaceae and Mychodeophyllaceae, all relatively obscure families from the Southern Hemisphere, have recently been monographed by Kraft (1977b, c, 1978). *Mychodeophyllum* (Kraft 1978) was described and made the basis of its own family on features having mainly to do with the 'placentation' of the cystocarp. These groups are of interest in that they appear to be highly specialized and display some of the most complex cystocarps of any of the red algae (e.g. Fig. 1.3).

The two genera *Plocamium* and *Plocamiocolax* constitute the small, anomalous family Plocamiaceae. For while the tetrasporangial division pattern and procarp are gigartinalean, the vegetative structure and stichidia are very atypical, so that the family's links to other groups of the order are obscure. Recent work on *Plocamium* has tended to be on a regional basis, with floras produced for New Zealand (South & Adams 1979), southern Australia (Womersley 1971b) and South Africa (Simons 1964). As a result, species numbers have declined, and probably fewer than 40 will ultimately remain. Kylin credited the genus with around 50.

The Wurdemanniaceae was created by Taylor (1960) for a single western Atlantic and Mediterranean genus (*Wurdemannia*) and is a problematical family of multiaxial structure and zonate tetrasporangia in which gametophytes are unknown. Taylor treats it as a member of the Gelidiales, and others (like Guiry 1978) follow by placing it in the Nemalionales. Kylin himself left the genus as a taxon of uncertain affinity, and Joly (1967) refrained from placing the family in an order. The zonate tetrasporangia and multiaxial, densely cellular thallus are not characteristic of the Nemalionales but are similar to some Cryptonemiales and Gigartinales (Papenfuss 1961). The structure is very like *Gelidiopsis* of the Gracilariaceae, which differs in having cruciate tetrasporangia.

The Rhodymeniales

The Rhodymeniales is a small but distinct order of three families, some 38 genera and 265 species of multiaxial algae with rather uniform procarps and a narrow range of cystocarp morphologies. The major event in the order's recent history has been the discovery by Guiry (1974) that its most famous species, *Rhodymenia palmata*, does not belong to *Rhodymenia* at all but is the type of the genus *Palmaria* and the basis of a distinct family and order.

The Rhodymeniaceae is a large family of 31 genera and about 190 species. Recent studies which canvass several members in detail are those of Sparling (1957) and Lee (1978), while summaries of features and commentaries on the group have been made by Moe (1979) and Ricker & Kraft (1979). Changes since Kylin include the removal of five genera (*Cryptarachne* by Abbott & Littler 1969; *Halosaccion* by Guiry 1974; *Dendrymenia* by Sparling 1957; *Myrioglossa* by Papenfuss 1958; and *Weberella* by Huvé & Huvé 1977), and the description of five new genera (*Cenacrum* by Ricker & Kraft 1979; *Gloiodermatopsis* by Lindauer 1949; *Gloiocolax* by Sparling 1957; *Maripelta* by Dawson 1963; and *Minium* by Moe 1979). Two genera, *Cordylecladia* and *Drouetia*, have been transferred to the family from other orders (Feldmann 1967; Denizot 1968).

The two largest genera of the family are *Rhodymenia*, with somewhere between 40–60 often poorly-defined species, and *Botryocladia*, with around 25. Sparling (1957) describes the anatomy of four *Rhodymenia* species, while detailed studies of others have been made by Lee (1978) in Japan and Guiry (1977) in Europe. Feldmann & Bodard (1965) give a key to the 20 *Botryocladia* species known then, although at least five more have been described since (e.g. Taylor & Abbott 1973; Chiang & Xia 1978).

Kylin divided the Rhodymeniales into the two families Rhodymeniaceae and Lomentariaceae, the latter of which (as the Champiaceae) Sparling (1957) characterized by the longitudinally running medullary filaments bordering the hollow thallus interiors which is a combination lacking in the Rhodymeniaceae. Lee (1978) maintains both the Champiaceae and the Lomentariaceae. As defined by Lee, the four genera (*Champia, Chylocladia, Coeloseira* and *Gastroclonium*) and approximately 35 species of the Champiaceae possess fronds in which single-layered partitions regularly divide the thalli, carpogonial branches are four-celled, carposporangia are terminal, and tetra- or polysporangia are scattered. Taxonomy of the large genus *Champia* is studied on a regional basis by Lee (1978), Reedman & Womersley (1976), and Chapman & Dromgoole (1970), while *Chylocladia* is studied and new species added by Ercegovic (1956) and Reedman & Womersley (1976). Guiry (1978a) remarks on the occurrence of polysporangia in both *Gastroclonium* and *Coeloseira*, and suggests that the type species of the latter be transferred to the former.

Lee (1978) characterizes the Lomentariaceae by its multilayered partitioning of hollow parts, three-celled carpogonial branches, carposporophytes composed almost entirely of carposporangia, and tetrasporangia aggregated into sori around cavities in the outer cortex. Besides *Lomentaria*, the family contains *Binghamia* (Lee & Kurogi 1973) and the recently described California genus *Binghamiopsis* (Lee *et al.* 1981).

The Palmariales

The Palmariales was established by Guiry & Irvine (in Guiry 1978a) for the single family Palmariaceae, four genera and around 15 species of mostly circumboreal algae which produce isomorphic male gametophytic and tetrasporophytic plants and whose tetrasporangia are produced successively from a generative stalk cell within the persistent tetrasporangial wall. Female gametophytes were unknown until produced in culture by van der Meer & Todd (1980), who found them to be relatively diminutive free-living thalli bearing apparently sessile carpogonia. Zygotes directly initiate the tetrasporophytes, which remain attached to and eventually overgrow the mother plants. Besides *Palmaria*, with its two boreal species (Guiry 1976; Lee 1978), the group contains the two to three antarctic and subantarctic species of *Leptosarca*, 10–11 species of the circumboreal *Halosaccion*, and the monotypic parasitic genus *Neohalosacciocolax* (Lee & Kurogi 1978) from the Aleutians. A fifth genus, *Halosacciocolax,* has been placed in the family by Garbary (1979) on its supposed resemblances to *Neohalosacciocolax*. Lee & Kurogi (1978), however, specifically contrast the two genera and point out differences which would seem to exclude *Halosacciocolax* from the Palmariales and leave its actual affinities uncertain.

Although the vegetative and male gamete features of the Palmariales strongly suggest affinities to the Rhodymeniales, the events surrounding tetrasporogenesis and development of the zygote are unlike that or any other order, making the taxonomic position of the group highly problematical.

The Ceramiales

Almost half the red algal genera and a third of the species belong to the Ceramiales, a group of extreme reproductive uniformity when compared to the other orders in the Division (Hommersand 1963). As a result, the criteria for separating families, tribes, genera and species tend to rely heavily on such features as vegetative structure, procarp position, and the amounts and sorts of sterile cells with which the procarp and cystocarps are associated.

The approximately 325 genera and 1500 species of the Ceramiales are all uniaxial (at least initially) and have procarps consisting of four-celled carpogonial branches and auxiliary cells which arise from the supporting cells of the carpogonial branches only after fertilization has taken place in any given procarp. The one anomaly in this system is the few known or suspected cases (Brauner 1979; Gordon-Mills & Kraft 1981) where the zygote nucleus is passed directly to the supporting cell, which then gives rise to a gonimoblast initial rather than an auxiliary cell.

The Ceramiales is a highly successful group in terms of numbers and world distribution, and some of its genera are among the largest in the red algae. The order is considered to be an advanced one at both the morphological (Fritsch 1945) and ultrastructural (Wetherbee 1980) levels, judgments based on the uniformity of reproductive processes and parsimony of auxiliary cell production as well as in organelle complexity during carposporogenesis. Although members of the group are seldom among the largest red algae in any region (many being diminutive turf components or epiphytes), and few have any immediate economic importance, the Ceramiales has probably attracted a greater proportion of recent taxonomic efforts than any other order.

The group is composed of four families, three of which are subdivided into a bewildering array of tribes.

The Ceramiaceae

The Ceramiaceae contains 100+ genera and around 500 species which lack pericentral cells (*sensu* Min-Thein & Womersley 1976) and (with the exception of *Lejolisia*) consolidated pericarps. It has undergone extensive change since Kylin, mainly through the efforts of several Australian researchers whose complete works are tabulated by Kraft & Woelkerling (1981). Of the 70 genera listed by Kylin for the group, four have since been reduced to existing genera* and at least 34 new or resurrected genera have been added. With this increase has come a number of new tribes, of which seven have been described since Kylin. The two most recent are based on newly-discovered genera from Antarctica (Moe & Silva 1979) and southern Australia (Gordon-Mills & Kraft 1981). The Ceramiaceae itself was divided into two subfamilies by Hommersand (1963), whose criteria were found by Wollaston (1968) and Gordon (1972) to be inconsistent when applied to the rich hoard of Australasian taxa. Moe & Silva (1980) characterize the family as one in which 'explosive evolution' is probably taking place.

Major regional treatments of the Ceramiaceae are those of Itono (1977b) for southern Japan, Abbott & Hollenberg (1976) for California, and Mendoza (1969c) for Argentina.

Almost 200 species of the family belong to the four genera *Antithamnion, Callithamnion, Ceramium* and *Griffithsia*. The largest genus is *Ceramium*, to

* *Bracebridgia, Callithamniella, Ceramiella, Ceramothamnion.*

which over 70 species have been credited. Critical studies of the genus and several close relatives have been made by Hommersand (1963). The chaos of *Ceramium* taxonomy is discussed by Dixon (1960), who distinguishes a number of acceptable versus unacceptable species criteria. Womersley (1978) makes further contributions to firm species distinctions in a monograph of 15 southern Australian representatives.

Antithamnion, with over 40 species described, has been studied in southern Japan by Itono (1969), in France by L'Hardy-Halos (1968), and in southern Australia and Pacific North America by Wollaston (1968, 1971). Whittick & Hooper (1977) and Abbott (1979) have reported species with the first apparently zonate tetrasporangia recorded in the Ceramiales.

Based on her studies of genera placed in the Crouanieae by Kylin, Wollaston (1968, 1971) has distinguished not only a number of new and segregate genera but also established the additional tribe Heterothamnieae. Not all of this is universally acccepted, however, and Moe & Silva (1980) have questioned some of Wollaston's generic distinctions and the validity of separating the Antithamnieae and Heterothamnieae.

Although little major work has been published recently on the taxonomy of *Callithamnion* (Price 1978), to which over 50 species are credited, *Griffithsia* (with over 30 species described) has been examined with great elegance in southern Australia by Baldock (1976). As a result the Griffithsieae now consists of three genera, including the reborn Naegeli genus *Anotrichium* and the form-genus *Monosporus* in which only propagules (*sensu* Guiry 1978a) are known. Another regional monograph of the tribe occurs in Itono (1977b).

The Delesseriaceae

Whether because of the beauty of its membranous fronds or the weatlh of its morphological diversity, the Delesseriaceae (with some 90 genera and 300 species*) is one of the most keenly investigated families in the red algae. As pointed out by Wynne *et al.* (1973), it seems to be a rapidly evolving group, particularly in the north Pacific. A distinctive component of most floras, particularly subtidally, the family has been the subject of a number of regionally-based studies (e.g. Zinova 1965, 1972, 1976; Mendoza 1969a,b, 1974; Wynne 1970a; Mikami 1971, 1973, 1978; Abbott & Hollenberg 1976; Yoshida 1979). A major impetus to investigation of the Delesseriaceae was provided by Wagner (1954), who reviewed the group's taxonomy and studied 13 species in several little-known genera. Some 20 genera and five tribes have been added since Kylin, a reflection perhaps of the tendency to separate genera on the basis of single-feature differences which has developed among students of the family (as suggested by Schneider & Eiseman 1979).

* Genera listed in Kylin since lost through synonymy are *Implicaria, Opephyllum, Papenfussia* and perhaps *Rhodoseris*. *Sonderella* has been transferred to the Rhodomelaceae by Womersley (1965).

Although there is general agreement on the division of the family into the two subfamilies Delesserioideae and Nitophylloideae (Wagner 1954; Hommersand 1963), based mainly on the location of the procarps and to a lesser extent on cell division features, some dispute centres on the taxonomic position of the tribe referred to as the Sarcomenia Group by Kylin, as this assemblage shows several characters intermediate between the Delesseriaceae and Rhodomelaceae. Womersley & Shepley (1958) present arguments (based on the order of pericentral cell initiation and the branching pattern of the gonimoblasts) for including the group in the Rhodomelaceae and considering it a subfamily coequal there with the remainder of the rhodomelaceous genera. Hommersand (1963) and Wynne (1969) present counter arguments and opt strongly for maintaining the Sarcomenieae in the Delesseriaceae, where Kylin had put it. Although Womersley & Shepley think the Sarcomenieae reflects a possible link between the Delesseriaceae and Rhodomelaceae, Hommersand (1963) considers its similarities to the Rhodomelaceae to be cases of convergent or parallel evolution.

The Dasyaceae

The Dasyaceae is a small but distinctive family of 12 genera and around 100 species of generally robust, at times highly intricate (Aregood & Hackett 1971) morphology. The genus *Colacodasya* (Mendoza 1970; Hollenberg 1970) contains two diminutive and presumably parasitic species.

The taxonomic history of the group has been reviewed by Parsons (1975), who has made a critical and detailed analysis of the family based mostly on Australian representatives. Parsons confirms a characteristic postfertilization development within the species studied in which the auxiliary cell and the central cell of the fertile segment are the initial components of the fusion cell. Sympodial growth of the axes is found to be consistent throughout the family, and Parsons elucidates the essential features of this often obscure type of development. The genera *Haplodasya*, placed in the family by Kylin, and *Heterodasya* (Joly & Oliveira 1966) were found to be monopodial and to lack the dasyaceous fusion cell, so are removed to the Rhodomelaceae by Parsons.

The most recent addition to the family is the genus *Dasysiphonia* (Lee & West 1979) from Korea, an alga with features of both the two major genera of the Dasyaceae, *Heterosiphonia* and *Dasya*.

The Rhodomelaceae

The Rhodomelaceae consists of around 125 genera (almost 130 if the Sarcomenia Group is included) and somewhere between 600–650 species, making it the largest family of the Rhodophyta. Kylin divided it into 16 groups corresponding to tribes on such bases as the nature of the trichoblasts, pericentral cells, degree of dorsiventrality, and presumed parasitism. Scagel (1953) studied five of the dorsiventral genera and arrived at a somewhat different ordering from Kylin's of 14 groups which are also equivalent to tribes. Hommersand (1963) made detailed analyses of species in three genera and a critical evaluation of vegetative features in the family as a whole, and on these bases plus a survey of what was known of the rest of the Rhodomelaceae divided it into three subfamilies and 13 tribes. Parsons (1975) criticises Hommersand's subfamily criteria on the basis of a precise study of the rhodomelacean tribe Lophothalieae, and opts for the temporary abandonment of such artificial divisions until more is known about a greater percentage of the family's genera. At present the Rhodomelaceae consists of some 15–16 generally agreed upon tribes, the most recent of which are the Rhodolachneae (Womersley & Bailey 1970) and Brongniartelleae (Parsons 1975).

The Rhodomelaceae has lost three genera (*Euzonia*—Scagel 1962a; *Orcasia*—Scagel 1957; *Vertebrata*—Silva 1952) out of Kylin's roster of 102 while picking up an additional 25 or so new or transferred genera. *Polysiphonia, Chondria* and *Laurencia* together make up nearly half the species of the family and, like *Ceramium* and *Callithamnion* of the Ceramiaceae, are found in some numbers in almost every geographic region. With over 200 species credited to it, *Polysiphonia* is the largest genus of red algae. Recent regional studies depicting large numbers of species and critically evaluating taxonomic criteria have been compiled by Hollenberg (1968a, b), Hollenberg & Norris (1977), Kapraun (1977) and Lauret (1970). Womersley (1979) has further elucidated diagnostic features of the genus and illustrated 26 species, several of them new, from southern Australia.

Laurencia, with over 80 described species, is best known through a number of regional monographs by Saito (1967, 1969a, b), Saito & Womersley (1974), Durairatnam (1963) and Jaasund (1970). *Chondria*, with over 30 species, is particularly well represented in Australia and the tropics but has not had recent broad taxonomic treatment. Fair numbers of species are illustrated in regional studies by Taylor (1960) and Abbott & Hollenberg (1976).

4.2 The Bangiophycidae

The second subclass of the Rhodophyta, the Bangiophycidae, is a rather small group of four to five orders, around 30 genera and 110 species. Lacking both the vegetative and reproductive complexity of the Florideophycidae, its members have presented long-standing taxonomic difficulties. With the work of Nichols & Lissant (1967), Heerebout (1968) and others, there has been a tendency for some major merging of species distinctions among the simpler filamentous representatives. The genus *Porphyra*, on the other hand, continues to grow despite already being the largest genus of the subclass, with at least 15 new species added to it since Dawson (1962).

Despite its small size, the Bangiophycidae has an extensive literature devoted to life history, cytology and culture work in the economically important family Bangiaceae. And one *Porphyra* species is of particular significance as the subject of one of the major papers on red algae in recent times, Hawkes' (1978) confirmation of the existence of a sexual process in the genus.

The subclass has a large percentage of freshwater or subaerial representatives and one order, the Compsopogonales, is exclusively non-marine. Bourrelley (1970) presents a good illustrated compilation of these taxa.

The Rhodochaetales

This order is composed of the monotypic European and Caribbean (Taylor 1971) genus *Rhodochaete*, in many ways an anomalous and unusual alga. Besides exhibiting the only consistent apical growth of the subclass, *Rhodochaete* possesses a distinctive type of pit connection (Boillot 1978) and produces zygotes that arguably result from the most simple sexual process in the red algae (Magne 1960). The plants lack a trichogyne and differentiated carpogonium, the female gametes being an intercalary vegetative cell which fuses with an adhering male gamete. Chadefaud (1965) considers *Rhodochaete* an important developmental level in the evolution of sexuality in the Rhodophyta, and later (1975) incorporates this reproductive anatomy into a broader theory for the possible evolution of higher fungi from red algal-like forerunners.

The Bangiales

The Bangiales is divided into the three families Erythropeltidaceae, Bangiaceae and Boldiaceae, of which the latter consists solely of the freshwater genus *Boldia*. The order contains nine genera and about 65 species, of which around 45 belong to the cosmopolitan genus *Porphyra*.

Despite some reports of sex, members of the Erythropeltidaceae are known with certainty to reproduce only by monospores. Besides the genera listed by Kylin the family now contains *Smithora* (Hollenberg 1959) and *Membranella* (Hollenberg & Abbott 1968). Taxonomy of the group is in some flux. Nichols & Lissant (1967) surveyed 14 species of *Erythrocladia* and concluded that it consists of a single polymorphic species. In a study of the genus on a worldwide basis, Heerebout (1968) credits it with two species. Although Dangeard (1968) lists 29 species for the largest genus in the family, *Erythrotrichia*, Heerebout (1968) reduces it to three. Despite this tendency to consolidate taxa, a number of new species have been described for *Erythrotrichia* and *Erythrocladia* (e.g. Dangeard 1968; Noda 1970, 1971; South & Adams 1976). Womersley & Conway (1975) have added a third species of *Porphyropsis* to the two that were previously recognized.

Conchocelis-like stages have been discovered in several of the genera, including *Smithora* by Richardson & Dixon (1969), *Porphyropsis* by Murray *et al.* (1972) and *Erythrotrichia* by Heerebout (1968). In all cases these stages lack conchospores, the implication being that they may not represent alternate phases in a sexual life cycle that they presumably do in the Bangiaceae. In *Porphyropsis* (Murray *et al.* 1972), the conchocelis is also known to differ from those of *Bangia* and *Porphyra* in lacking pit connections.

Since Drew's discovery of the link between 'conchocelis' algae and some *Porphyra* species, the life history and cytology of the Bangiaceae has attracted much attention. The anatomy of the conchocelis phase has been a continuing study, with Lee & Fultz (1970) first demonstrating conclusively the presence of 'Floridean'-style pit connections in this stage of the life history. Lee (1971) considers these pits evidence for the derivation of the family from ancestors in the lower Nemalionales, and argues against the counter-proposal that the higher reds evolved from bangiacean precursors. Cole & Conway (1975) consider the conchocelis phase a possible link between the two subclasses of red algae which points to the overall unity of the Division. They see the divergence of the subclasses at the level of the 'primitive' Nemalionales which, like the Bangiaceae have uninucleate cells with similar stellate to band-shaped chloroplasts, often with pyrenoids.

That sexual fusion takes place in *Porphyra* was long accepted on the basis of 19th and early 20th century reports and mostly taken on faith. In recent years a number of workers questioned this assumption due to lack of confirming evidence, and a whole new reproductive terminology evolved that avoided sexual connotations. This longrunning puzzle has been summarized and resolved by Hawkes (1978), who demonstrates morphologically and cytologically the gametic natures of the 'carpogonium' and 'beta-spores' in one *Porphyra* species.

The second common genus of the family, *Bangia*, has been shown by Richardson & Dixon (1968) and Sommerfeld & Nichols (1970) to produce conchocelis phases in its life history, although demonstration of sexual fusions has yet to be made. *Bangia* has one or more marine species and a freshwater representative which Geesink (1973) has shown can adapt to marine conditions in the laboratory over a period of gradual acclimatization and is thus probably conspecific with the marine form.

The substantial increase in *Porphyra* species numbers has largely come about through several regionally-based studies, including those of Coll & Cox (1977), Coll & Oliveira Filho (1976), Krishnamurthy (1973), Miura (1967, 1968), Mumford (1975), Munda & Pedersen (1978), Noda (1964) and Oliveira Filho & Coll (1975). Other important local studies include those by Conway (1964) for Great Britain, Conway *et al.* (1975) for British Columbia and Washington, Cordeiro (1974) for the Philippines and Womersley & Conway (1975) for southern Australia.

The third genus of the Bangiophyceae listed by Kylin, *Porphyrella*, has been merged with *Porphyra* by Hawkes (1977).

The Porphyridiales

An order originally restricted to the unicellular or irregularly colonial genera, this group is now considered by Dixon (1973) and others to include the colonial forms of quasi-filamentous morphology previously placed in the separate order Goniotrichales, based on studies showing that some species of the latter can revert to porphyridialean forms under certain culture conditions. The merging of the orders was first urged by Feldmann (1955), but has been disputed by Krishnamurthy (1957) and is still not wholly accepted (e.g. Abbott & Hollenberg 1976). The three families of the former two orders, the Porphyridiaceae, Goniotrichaceae and Phragmonemataceae remain, however, more or less as Kylin treated them.

The only genera possibly qualifying as 'seaweeds' among these diminutive algae are the common marine epiphytes *Goniotrichum* and *Chroodactylon*, the latter of which appears to have at least a marine and a freshwater strain, if not two distinct species (Lewin & Robertson 1971). A summary of the nomenclature of *Chroodactylon*, referred to as *Asterocytis* by Kylin and others, is given by John *et al.* (1979). The apparently uncommon Californian genus *Goniotrichopsis* differs from *Goniotrichum* only by the presence of multiple, rather than single, chloroplasts per cell.

5 POSTSCRIPT

For us closet Platonists, compulsively seeking to hang labels on the world's phenomena, the red algae must hold a particular attraction, there being so many poorly-known and 'new' ones still cached in both the mundane and exotic spots of that world to whet our joy of discovery and abet our armchair phylogenies. The taxonomy of red algae is obviously not for everybody, but in this benighted age it is truly heartening to know that not only do some of Ralph Lewin's best friends still press such plants (Lewin 1972), but that they also presumably take some modicum of pleasure in putting the results into reasoned pigeonholes both in their minds and at the herbarium.

I thank Professor Maxwell S. Doty, University of Hawaii, for the kind support he gave with his library and laboratory space, and Dr. Richard Wetherbee for keeping me company on this project and showing that it takes more than just razor blades to do justice to the reds. Financial support has come from the University of Melbourne.

6 REFERENCES

Abbott I. A. (1961) On *Schimmelmannia* from California and Japan. *Pacif. Nat.*, **2**, 1–8.

Abbott I. A. (1962a) Structure and reproduction of *Farlowia* (Rhodophyceae). *Phycologia*, **2**, 29–37.

Abbott I. A. (1962b) Morphological studies in a new species of *Acrosymphyton* (Rhodophyceae). *Am. J. Bot.*, **49**, 845–9.

ABBOTT I. A. (1965) *Helminthora* and *Helminthocladia* from California. *Hydrobiologia*, **25**, 88–98.
ABBOTT I. A. (1967) Studies in the foliose red algae of the Pacific coast. II. *Schizymenia*. *Bull. S. Calif. Acad. Sci.*, **66**, 161–74.
ABBOTT I. A. (1968) Studies in some foliose red algae of the Pacific coast. III. Dumontiaceae, Weeksiaceae, Kallymeniaceae. *J. Phycol.*, **4**, 180–98.
ABBOTT I. A. (1969) Some new species, new combinations, and new records of red algae from the Pacific coast. *Madroño*, **20**, 42–53.
ABBOTT I. A. (1970) *Yamadaella*, a new genus in the Nemaliales (Rhodophyta). *Phycologia*, **9**, 115–23.
ABBOTT I. A. (1971) On the species of *Iridaea* (Rhodophyta) from the Pacific coast of North America. *Syesis*, **4**, 51–72.
ABBOTT I. A. (1976) *Dotyophycus pacificum* gen. et sp. nov., with a discussion of some families of Nemaliales (Rhodophyta). *Phycologia*, **15**, 125–32.
ABBOTT I. A. (1978) Morphologic and taxonomic observations on *Neoagardhiella* (Gigartinales, Rhodophyta), with emphasis on Pacific populations. *J. Phycol.*, **14**, 48–53.
ABBOTT I. A. (1979) Some tropical species related to *Antithamnion* (Rhodophyta, Ceramiaceae). *Phycologia*, **18**, 213–27.
ABBOTT I. A. (1980) Some field and laboratory studies on colloid-producing red algae in central California. *Aquat. Bot.*, **8**, 255–66.
ABBOTT I. A., & DOTY M. S. (1960) Studies in the Helminthocladiaceae. II. *Trichogloeopsis*. *Am. J. Bot.*, **47**, 632–40.
ABBOTT I. A., & HOLLENBERG G. J. (1976) *Marine Algae of California*. Stanford University Press, Stanford.
ABBOTT I. A. & LITTLER M. M. (1969) Some Rhodymeniales from Hawaii. *Phycologia*, **8**, 165–9.
ABBOTT I. A., & NORRIS R. W. (1966) Studies on *Callophyllis* (Rhodophyceae) from the Pacific coast of North America. *Nova Hedw.*, **10**, 67–84.
ACLETO O. C. (1972) Structure and reproduction of *Schimmelmannia dawsonii* sp. nov. (Rhodophyceae, Cryptonemiales). *Phycologia*, **11**, 5–9.
ACLETO O. C. (1973) Las algas marinas del Peru. *Bol. Soc. Peru. Bot.*, **6**, 1–164.
ADEY W. H. & MACINTYRE I. G. (1973) Crustose coralline algae: a re-evaluation in the geological sciences. *Geol. Soc. Am. Bull.*, **84**, 883–904.
ARDRE F. (1977) Sur le cycle du *Schizymenia dubyi* (Chauvin ex Duby) J. Agardh (Nemastomacée, Gigartinale). *Rev. Alg.*, N.S. **12**, 73–86.
AREGOOD C. C. & HACKETT H. E. (1971) A new *Dictyurus* (Rhodophyceae—Dasyaceae) from the Maldive Islands, Indian Ocean. *J. Elisha Mitchell Sci. Soc.*, **87**, 91–6.
BALAKRISHNAN M. S. (1960) Reproduction in some Indian red algae and their taxonomy. *Proc. Symp. Algology*, 85–98. New Delhi.
BALAKRISHNAN M. S. (1962) Studies on Indian Cryptonemiales—II. *Corynomorpha* J. Ag. *Phytomorphology*, **12**, 77–86.
BALDOCK R. N. (1976) The Griffithsieae group of the Ceramiaceae (Rhodophyta) and its southern Australian representatives. *Aust. J. Bot.*, **24**, 509–93.
BERT J.-J. (1965) Sur la structure et le développement de l'appareil reproducteur femelle de *Dilsea carnosa* (Schmidel) Kuntze et la position systématique du genre *Dilsea*. *C.R. Acad. Sci. Paris*, **261**, 2702–4.
BODARD M. (1971) Étude morphologique et cytologique d'*Helminthocladia senegalensis* (Rhodophycées), Nemalionale nouvelle à carpotetraspores et à cycle haplodiplophasique. *Phycologia*, **10**, 361–74.
BOILLOT A. (1978) Les ponctations intercellulaires du *Rhodochaete parvula* Thuret (Rhodophycées, Bangiophycées). *Rev. Alg.*, N.S. **13**, 251–8.
BOLD H. C. & WYNNE M. J. (1978) *Introduction to the Algae*. Prentice-Hall, Englewood Cliffs.
BOUDOURESQUE C.-F., COPPEJANS E. & MARCOT J. (1976) Un nouveau genre de Peyssonneliaceae, *Metapeyssonnelia* (Rhodophyta). *Phycologia*, **15**, 283–8.
BOUDOURESQUE C.-F. & DENIZOT M. (1973) Note preliminaire sur les *Peyssonnelia* de Mediterranée et sur la valeur systématique des critères specifiques utilisées. *Bull. Soc. Phycol. Fr.*, **18**, 1–7.

BOURRELLY P. (1970) *Les Algues d'Eau Douce, Vol. 3, Algues Bleues et Rouges*. Boubée, Paris.

BRAUNER J. F. (1979) Developmental morphology and taxonomy of *Medeiothamnion flaccidum* (Hooker f. & Harvey) comb. nov. (Ceramiaceae, Rhodophyta) from South America. *Phycologia*, **18**, 338–46.

CABIOCH J. & GUIRY M. D. (1976) *Halosacciocolax kjellmanii* Lund, Rhodophyceae parasite nouvelle sur les côtes de France. *Trav. Stat. Biol. Roscoff*, N.S. **23**, 27–9.

CAMPBELL S. E. (1980) *Palaeoconchocelis starmachii*, a carbonate boring microfossil from the Upper Silurian of Poland (425 million years old): implications for the evolution of the Bangiaceae (Rhodophyta). *Phycologia*, **19**, 25–36.

CHADEFAUD M. (1965) Sur les organes réproducteurs et l'évolution des Rhodophycées: une hypothèse. *Rev. Alg.*, N.S. **6**, 225–69.

CHADEFAUD M. (1975) L'origine 'parafloridéenne' des Eumycètes et l'archétype ancestral de ces champignons. *Ann. Sci. Nat. Bot. Paris*, **16**, 217–47.

CHANG C. F. & XIA B. M. (1963) *Polycavernosa*, a new genus of the Gracilariaceae. *Stud. Mar. Sin.*, **3**, 119–26, pls. 1, 2.

CHANG C. F. & XIA B. M. (1976) Studies on Chinese species of *Gracilaria*. *Stud. Mar. Sin.*, **11**, 91–163.

CHANG C. F. & XIA B. M. (1978) Studies on some marine red algae of the Xisha Islands, Guandong Province, China. I. *Stud. Mar. Sin.*, **12**, 27–39.

CHAPMAN V. J. (1969) A remarkable new seaweed from New Zealand. *J. Phycol.*, **5**, 259–65.

CHAPMAN V. J. (1979) *The marine algae of New Zealand. Part III: Rhodophyceae. Issue 4: Gigartinales*, pp. 279–509. Cramer, Lehre.

CHAPMAN V. J. & DROMGOOLE F. I. (1970) *The marine algae of New Zealand. Part III: Rhodophyceae. Issue 2: Florideophycidae: Rhodymeniales*, pp. 115–54. Cramer, Lehre.

CHAPMAN V. J. & PARKINSON P. G. (1974) *The marine algae of New Zealand. Part III: Rhodophyceae. Issue 3: Cryptonemiales*, pp. 155–278. Cramer, Lehre.

CHENEY D. P. & BABBEL G. R. (1978) Biosystematic studies of the red algal genus *Eucheuma*. I. Electrophoretic variation among Florida populations. *Mar. Biol.*, **47**, 251–64.

CHIANG Y.-M. (1970a) Observations on the development of the carposporophyte of *Scinaia pseudojaponica* Yamada et Tanaka (Nemaliales, Chaetangiaceae). *J. Phycol.*, **6**, 289–92.

CHIANG Y.-M. (1970b) Morphological studies of red algae of the family Cryptonemiaceae. *Univ. Calif. Publ. Bot.*, **58**, 1–83.

CHIHARA M., & KAMURA S. (1963) On the germination of tetraspores of *Gelidiella acerosa*. *Phycologia*, **3**, 69–74.

CHIHARA M. & YOSHIZAKI M. (1969) Morphological study on the thallus structure and the reproductive organs of *Pseudogloiophloea okamurai* (Florideophycidae, Rhodophyta). *Bot. Mag. Tokyo*, **82**, 397–402.

CHIHARA M. & YOSHIZAKI M. (1971) The thallus structure and the reproductive system of *Hyalosiphonia caespitosa* (Cryptonemiales, Rhodophyta). *Bot. Mag. Tokyo*, **84**, 319–25.

CHIHARA M. & YOSHIZAKI M. (1972a) Reproductive system of *Liagora japonica* (Nemaliales, Rhodophyta). *Bull. Nat. Sci. Mus. (Tokyo)*, **15**, 395–401.

CHIHARA M. & YOSHIZAKI M. (1972b) Bonnemaisoniaceae: their gonimoblast development, life history and systematics. In *Contributions to the Systematics of Benthic Marine Algae of the North Pacific*, eds. Abbott I. A. & Kurogi M., pp. 243–52. Jap. J. Phycol.

CHIHARA M. & YOSHIZAKI M. (1978) Anatomical and reproductive features of *Ptilonia okadai* (Rhodophyta, Bonnemaisoniaceae). *Phycologia*, **17**, 382–7.

CHRISTENSEN T. (1979) Annotations to a textbook of phycology. *Bot. Tidsskr.*, **73**, 65–70.

CODOMIER L. (1978) Recherches sur les *Kallymenia* (Cryptonemiales). III. Problèmes de la delimitation du genre *Kallymenia*. *Rev. Alg.*, N.S. **13**, 241–49.

COLE K. & CONWAY E. (1975) Phenetic implications of structural features of the perennating phase in the life history of *Porphyra* and *Bangia* (Bangiophyceae, Rhodophyta). *Phycologia*, **14**, 239–45.

COLL J. & COX J. (1977) The genus *Porphyra* C. Ag. (Rhodophyta, Bangiales) in the American north Atlantic. I. New species from North Carolina. *Bot. Mar.*, **30**, 155–9.

COLL J. & OLIVEIRA FILHO E. C. DE (1976) The genus *Porphyra* (Rhodophyta-Bangiales) in the American south Atlantic. II. Uruguayan species. *Bot. Mar.*, **19**, 191–6.

CONWAY E. (1964) Autecological studies of the genus *Porphyra*; 1. The species found in Britain. *Br. Phycol. Bull.*, **2**, 342–6.

CONWAY E., MUMFORD T. F. Jr. & SCAGEL R. F. (1975) The genus *Porphyra* in British Columbia and Washington. *Syesis*, **8**, 185–244.

CORDERO P. A. (1974) Genus *Porphyra* of the Philippines, its species and their occurrences. *Bull. Jap. Soc. Phycol.*, **22**, 134–42.

CORTEL-BREEMAN A. M. & TEN HOOPEN A. (1978) The short day response in *Acrosymphyton purpuriferum* (J. Ag.) Sjöst. (Rhodophyceae, Cryptonemiales). *Phycologia*, **17**, 125–32.

DANGEARD P. (1968) Recherches sur quelques Bangiophycées (Protofloridées). *Botaniste*, **51**, 5–57.

DAWSON E. Y. (1949) Studies of Northeast Pacific Gracilariaceae. *Allan Hancock Found. Publ. Occ. Pap.*, **7**, 1–54.

DAWSON E. Y. (1954) Notes on tropical Pacific marine algae. *Bull. S. Calif. Acad. Sci.*, **53**, 1–7.

DAWSON E. Y. (1962) *New Taxa of Benthic Green, Brown and Red Algae Published since DeToni 1889, 1895, 1924 respectively, as Compiled from the Dawson Algal Library*. Beaudette Foundation, Santa Ynez.

DAWSON E. Y. (1963) Marine red algae of Pacific Mexico. IV. Rhodymeniales. *Nova Hedw.*, **5**, 437–76.

DECEW T. C. & WEST J. A. (1977) Culture studies on the marine red algae *Hildenbrandia occidentalis* and *H. prototypus* (Cryptonemiales, Hildenbrandiaceae). *Bull. Jap. Soc. Phycol.*, **25** (suppl.) (Mem. Issue Yamada), 31–41.

DEMOULIN V. (1975) The origin of ascomycetes and basidiomycetes. The case for a red algal ancestry. *Bot. Rev.*, **40**, 315–45.

DENIZOT M. (1968) *Les Algues Floridées Encroutantes*. Lab. de Cryptogamie, Mus. Nat. Hist., Paris.

DESIKACHARY T. V. (1962) *Cumagloia* Setchell et Gardner and *Dermonema* (Grev.) Harv. *Ind. Bot. Soc. J.*, **41**, 132–47.

DESIKACHARY T. V. (1963) Status of the order Chaetangiales (Rhodophyta). *Ind. Bot. Soc. J.*, **42A**, 16–26.

DIXON P. S. (1960) Studies on marine algae of the British Isles: the genus *Ceramium*. *J. Mar. Biol. Assoc. U.K.*, **39**, 331–74.

DIXON P. S. (1973) *Biology of the Rhodophyta*. Oliver & Boyd, Edinburgh.

DIXON P. S. & IRVINE L. M. (1977) *Seaweeds of the British Isles. Vol. 1, Rhodophyta. Part 1, Introduction, Nemaliales, Gigartinales*. British Museum, London.

DOTY M. S. (1963) *Gibsmithia hawaiiensis* gen. n. et sp. n. *Pacif. Sci.*, **17**, 458–65.

DOTY M. S. (1973) Farming the red seaweed, *Eucheuma*, for carrageenans. *Micronesica*, **9**, 59–73.

DOTY M. S. (1978a) *Izziella abbottae*, a new genus and species among the gelatinous Rhodophyta. *Phycologia*, **17**, 33–9.

DOTY M. S. (1978b) *Eucheuma*—current marine agronomy. In *The Marine Plant Biomass of the Pacific Northwest Coast*, ed. Krauss R. W., pp. 203–14. Oregon State Univ. Press, Corvallis.

DOTY M. S. & ABBOTT I. A. (1961) Studies in the Helminthocladiaceae (Rhodophyta): *Helminthocladia*. *Pacif. Sci.*, **15**, 56–63.

DOTY M. S. & ABBOTT I. A. (1964) Studies in the Helminthocladiaceae, III. *Liagoropsis*. *Pacif. Sci.*, **18**, 441–52.

DURAIRATNAM M. (1963) Some marine algae from Ceylon—2. *Laurencia* Lamouroux. *Bull. Fish. Res. Stn. Ceylon*, **16**, 19–28.

EARLE S. A. (1969) *Hummbrella*, a new red alga of uncertain taxonomic position from the Juan Fernandez Islands. *Occ. Pap. Farlow Herb. Crypt. Bot. Harvard*, **June 1969**, 1–5.

EDELSTEIN T., CHEN L. C.-M. & MCLACHLAN J. (1978) Studies on *Gracilaria* (Gigartinales, Rhodophyta): reproductive structures. *J. Phycol.*, **14**, 92–100.

EISEMAN N. J. (1979) Marine algae of the east Florida continental shelf. I. Some new records of Rhodophyta, including *Scinaia incrassata* sp. nov. (Nemaliales; Chaetangianceae). *Phycologia*, **18**, 355–61.

FAN K.-C. (1961) Morphological studies of the Gelidiales. *Univ. Calif. Publ. Bot.*, **32**, 315–68.

FAN K.-C. & PAPENFUSS G. F. (1959) Red algal parasites occurring on members of the Gelidiales. *Madroño*, **15**, 33–8.

FAN K.-C. & WANG Y.-C. (1974) Studies on the marine algae of Hsisha Islands, China. 1. *Ganonema* gen. nov. *Acta Phytotax. Sin.*, **12**, 489–93.

FELDMANN G. (1967) Le genre *Cordylecladia* J. Ag. (Rhodophycées, Rhodymeniales) et sa position systematique. *Rev. Gén. Bot.*, **74**, 357–75.

FELDMANN G. & BODARD M. (1965) Une nouvelle espèce de *Botryocladia* des côtes du Senegal. *Bull. Inst. Oceanogr. Monaco*, **65**, 1–14.

FELDMANN J. (1952) Les cycles de reproduction des algues et leurs rapports avec la phylogenie. *Rev. Cytol. Biol. Vég.*, **13**, 1–49.

FELDMANN J. (1954) Recherches sur la structure et le développement des Calosiphoniacées (Rhodophycées—Gigartinales). *Rev. Gén. Bot.*, **61**, 453–99.

FELDMANN J. (1955) Un nouveau genre de Protofloridée: *Colacodictyon* nov. gen. *Bull. Soc. Bot. Fr.*, **102**, 23–8.

FELDMANN J. (1967) Sur une Bangiophycée endozoique (*Neevea repens* Batters) et ses affinities (Rhodophyta). *Blumea*, **15**, 25–29.

FELDMANN J. (1976) Review of Dixon P. S., *Biology of the Rhodophyta*. *Br. Phycol. J.*, **11**, 202.

FELDMANN J. & FELDMANN G. (1942) Recherches sur les Bonnemaisoniacées et leur alternance de générations. *Ann. Sci. Nat. Bot.*, **11**, 77–169.

FRITSCH F. E. (1945) *The Structure and Reproduction of the Algae*, vol. 2. Cambridge University Press, Cambridge.

GABRIELSON P. W. (1980) Phylogenetic relationships of selected genera in the family Solieriaceae and taxonomic implications. *J. Phycol.*, **16** (suppl.), 11.

GABRIELSON P. W. & HOMMERSAND M. H. (1980) Taxonomy and phytogeography of the tribe Agardhielleae (Solieriaceae). *J. Phycol.*, **16** (suppl.), 12.

GANESAN E. K. (1970) A new species of *Gelidiocolax* Gardner (Choreocolacaceae, Rhodophyta) from the Caribbean Sea. *Bol. Inst. Oceanogr. Univ. Oriente*, **12**, 93–102.

GARBARY D. (1978) On the phylogenetic relationships of the Acrochaetiaceae (Rhodophyta). *Br. Phycol. J.*, **13**, 247–54.

GARBARY D. (1979) Numerical taxonomy and generic circumscription in the Acrochaetiaceae (Rhodophyta). *Bot. Mar.*, **22**, 477–92.

GARBARY D. (1980) On the systematic position of *Liagorophila* (Acrochaetiaceae, Rhodophyta). *Taxon*, **29**, 67–9.

GEESINK R. (1973) Experimental investigations on marine and freshwater *Bangia* (Rhodophyta) from The Netherlands. *J. Exp. Mar. Biol. Ecol.*, **11**, 239–47.

GOFF L. J. (1977) Developmental studies of the red algal adelphoparasite *Gardneriella tubifera* (Gigartinales). *J. Phycol.*, **13** (suppl.), 24.

GOFF L. J. & COLE K. (1975) The biology of *Harveyella mirabilis* (Cryptonemiales, Rhodophyceae). II. Carposporophyte development as related to the taxonomic affiliation of the parasitic red alga, *Harveyella mirabilis*. *Phycologia*, **14**, 227–38.

GORDON E. M. (1972) Comparative morphology and taxonomy of the Wrangelieae, Sphondylothamnieae and Spermothamnieae (Ceramiaceae, Rhodophyta). *Aust. J. Bot.*, suppl. ser. **4**, 1–180.

GORDON-MILLS E. M. & KRAFT G. T. (1981) The morphology of *Radiathamnion speleotis* gen. et sp. nov., representing a new tribe in the Ceramiaceae (Rhodophyta) from southern Australia. *Phycologia*, (in press).

GRETZ M. R., ARONSON J. M. & SOMMERFELD M. R. (1980) Cellulose in the cell walls of the Bangiophyceae (Rhodophyta). *Science*, **207**, 779–81.

GUIRY M. D. (1974) A preliminary consideration of the taxonomic position of *Palmaria palmata* (Linnaeus) Stackhouse = *Rhodymenia palmata* (Linnaeus) Greville. *J. Mar. Biol. Ass. U.K.*, **54**, 509–28.

GUIRY M. D. (1975) An assessment of *Palmaria palmata* forma *mollis* (S. et G.) comb. nov. (= *Rhodymenia palmata* forma *mollis* S. et G.) in the eastern north Pacific. *Syesis*, **8**, 245–61.

GUIRY M. D. (1977) Studies on marine algae of the British Isles. 10. The genus *Rhodymenia*. *Br. Phycol. J.*, **12**, 385–425.

GUIRY M. D. (1978a) The importance of sporangia in the classification of the Florideophyceae. In *Modern Approaches to the Taxonomy of Red and Brown Algae*, eds. Irvine D. E. G. & Price J. H. pp. 111–44. Academic Press, London, New York.

GUIRY M. D. (1978b) Notes on some family names of Florideophyceae (Rhodophyta). *Taxon*, **27**, 191–5.

GUIRY M. D. & HOLLENBERG G. J. (1975) *Schottera* gen. nov. and *Schottera nicaeensis* (Lamour. ex Duby) comb. nov. (= *Petroglossum nicaeense* (Lamour, ex Duby) Schotter) in the British Isles. *Br. Phycol. J.*, **10**, 149–64.

HANSEN G. I. (1977) *Cirrulicarpus carolinensis*, a new species in the Kallymeniaceae (Rhodophyta). *Farlow Herb. Harvard Univ. Occ. Pap.*, **12**, 1–22.

HANSEN G. I. (1980) A morphological study of *Fimbrifolium*, a new genus in the Cystocloniaceae (Gigartinales, Rhodophyta). *J. Phycol.*, **16**, 207–17.

HARVEY M. J. & MCLACHLAN J. (eds.) (1973) *Chondrus Crispus*. Nova Scotia Inst. Sci., Halifax.

HAWKES M. W. (1977) A field, culture and cytological study of *Porphyra gardneri* (Smith & Hollenberg) comb. nov. (= *Porphyrella gardneri* Smith & Hollenberg), (Bangiales, Rhodophyta). *Phycologia*, **16**, 457–69.

HAWKES M. W. (1978) Sexual reproduction in *Porphyra gardneri* (Smith & Hollenberg) Hawkes (Bangiales, Rhodophyta). *Phycologia* **17**, 326–50.

HEEREBOUT G. R. (1968) Studies on the Erythropeltidaceae (Rhodophyceae—Bangiophycidae). *Blumea*, **16**, 139–57.

HEWITT F. E. (1960) A morphological study of three South African Gigartinales. *Univ. Calif. Publ. Bot.*, **32**, 195–234.

HOLLENBERG G. J. (1959) *Smithora*, an interesting new algal genus in the Erythropeltidaceae. *Pacif. Nat.*, **1**, 3–11.

HOLLENBERG G. J. (1968a) An account of the species of *Polysiphonia* of the central and western tropical Pacific Ocean. I. *Oligosiphonia*. *Pacif. Sci.*, **22**, 56–98.

HOLLENBERG G. J. (1968b) An account of the species of the red alga *Polysiphonia* of the central and western tropical Pacific Ocean. II. *Polysiphonia*. *Pacif. Sci.*, **22**, 198–207.

HOLLENBERG G. J. (1970) Phycological notes IV, including new marine algae and records for California. *Phycologia*, **9**, 61–72.

HOLLENBERG G. J. & ABBOTT I. A. (1968) New species of marine algae from California. *Can. J. Bot.*, **46**, 1235–51.

HOLLENBERG G. J. & NORRIS J. N. (1977) The red alga *Polysiphonia* (Rhodomelaceae) in the northern Gulf of California. *Smithson. Contrib. Mar. Sci.*, **1**, 1–21.

HOMMERSAND M. H. (1963) The morphology and classification of some Ceramiaceae and Rhodomelaceae. *Univ. Calif. Publ. Bot.*, **35**, 165–366.

HOMMERSAND M. H. & EARLE S. A. (1969) A new red alga from the Juan Fernandez Islands. *VI Int. Bot. Cong.*, **94** (*abs.*).

HOMMERSAND M. H. & OTT D. W. (1970) Development of the carposporophyte of *Kallymenia reniformis* (Turner) J. Agardh. *J. Phycol.*, **6**, 322–31.

HOOPER R. & SOUTH G. R. (1974) A taxonomic appraisal of *Callophyllis* and *Euthora* (Rhodophyta). *Br. Phycol. J.*, **9**, 423–8.

HUDSON P. R. & WYNNE M. J. (1969) Sexual plants of *Bonnemaisonia geniculata* (Nemaliales). *Phycologia*, **8**, 207–13.

HUVÉ P. & HUVÉ H. (1977) Notes de nomenclature algale. I. Le genre *Halichrysis* (J. Agardh 1851 emend J. Agardh 1876) Schousboe mscr. in Bornet 1892 (Rhodymeniales, Rhodophycées). *Soc. Phycol. Fr. Bull.*, **22**, 99–107.

ITONO H. (1969) The genus *Antithamnion* (Ceramiaceae) in southern Japan and adjacent waters—1. *Mem. Fac. Fish. Kagoshima Univ.*, **18**, 29–45.

ITONO H. (1977a) Studies on the southern Japanese species of *Galaxaura* (Rhodophyta). *Micronesica*, **13**, 1–26.

ITONO H. (1977b) Studies on the ceramiaceous algae (Rhodophyta) from southern parts of Japan. *Bibl. Phycol.*, **35**, 1–499.

ITONO H. (1979) *Actinotrichia robusta*, a new species of the Chaetangiaceae (Nemaliales, Rhodophyta). *Jap. J. Phycol.*, **27**, 137–41.

JAASUND E. (1970) Marine algae in Tanzania II. *Bot. Mar.*, **13**, 59–64.

JOHANSEN H. W. (1974) Articulated coralline algae. *Oceanogr. Mar. Biol. Ann. Rev.*, **12**, 77–127.

JOHANSEN H. W. (1976) Current status of generic concepts in coralline algae (Rhodophyta). *Phycologia*, **15**, 221–44.

JOHN D. M., PRICE J. H., MAGGS C. A. & LAWSON G. W. (1979) Seaweeds of the western coast of tropical Africa and adjacent islands: a critical assessment. III. Rhodophyta (Bangiophyceae). *Bull. Br. Mus. Nat. Hist. (Bot.)*, **7**, 69–82.

JOLY A. B. (1967) *Gêneros de Algas Marinhas da Costa Atlântica Latino-Americana*. Univ. São Paulo, São Paulo.

JOLY A. B. & ALVEAL K. (1969) Notes on Chilean algae. II. *Stenogramme rhodymenioides*, a new entity of the American south Pacific flora. *Rickia*, **4**, 47–60.

JOLY A. B., CORDEIRO-MARINO M., YAMAGUISHI-TOMITA N., UGADIM Y., OLIVEIRA FILHO E. C. DE, & FERREIRA M. M. (1965) Additions to the marine flora of Brazil. V. *Arq. Est. Biol. Mar. Univ. Cearà*, **5**, 65–78.

JOLY A. B. & OLIVEIRA FILHO E. C. DE (1966) *Spyridiocolax* and *Heterodasya*, two new genera of the Rhodophyceae. *Sellowia*, **18**, 115–25.

KANEKO T. (1975) On *Scinaia japonica* Setchell from the Okhotsk Sea coast of Hokkaido. *Bull. Jap. Soc. Phycol.*, **23**, 8–13.

KANEKO T. MATSUYAMA K. & YAMADA I. (1980) On *Gloeophycus koreanum* I. K. Lee & Yoo (Rhodophyta, Gloiosiphoniaceae) in Hokkaido. *Jap. J. Phycol.*, **28**, 97–104.

KAPRAUN D. F. (1977) The genus *Polysiphonia* in North Carolina, U.S.A. *Bot. Mar.*, **20**, 313–31.

KARAM-KERIMAN T. B. (1976) Structure, reproduction et discussion sur la position systématique du genre *Gibsmithia* (Rhodophyceae). *Bull. Mus. Nat. Hist. Nat.*, **365**, 21–32.

KAWABATA S. (1954) On the structure of the frond, and the reproductive organ of *Pachymeniopsis lanceolata* Yamada (*Aeodes lanceolata* Okam.). *Bull. Jap. Soc. Phyc.*, **2**, 67–71.

KAWABATA S. (1962) A contribution to the systematic study of Grateloupiaceae from Japan (1). *J. Hokkaido Gakugei Univ.*, **13**, 22–51.

KAWASHIMA S. (1959) On the structure of the female organ and its development of *Dudresnaya minima* Okamura. *Bull. Jap. Soc. Phycol.*, **7**, 11–18.

KIM D. H. (1970) Economically important seaweeds in Chile—1. *Gracilaria. Bot. Mar.*, **13**, 140–62.

KIM D. H. (1976) A study of the development of cystocarps and tetrasporangial sori in Gigartinaceae (Rhodophyta, Gigartinales). *Nova Hedw.*, **27**, 1–146.

KRAFT G. T. (1972) Preliminary studies of Philippine *Eucheuma* species (Rhodophyta). Part 1, taxonomy and ecology of *Eucheuma arnoldii* Weber-van Bosse. *Pacif. Sci.*, **26**, 318–34.

KRAFT G. T. (1973) The morphology of *Cubiculosporum koronicarpis* gen. et sp. nov., representing a new family in the Gigartinales (Rhodophyta). *Am. J. Bot.*, **60**, 872–82.

KRAFT G. T. (1975a) Consideration of the order Cryptonemiales and the families Nemastomataceae and Furcellariaceae (Gigartinales, Rhodophyta) in light of the morphology of *Adelophyton corneum* (J. Agardh) gen. et comb. nov. from southern Australia. *Br. Phycol. J.*, **10**, 279–90.

KRAFT G. T. (1975b) Morphology of *Placentophora* (Solieriaceae, Gigartinales: Rhodophyta), a new genus based on *Sarcodiotheca colensoi* from New Zealand. *J. Phycol.*, **11**, 399–410.

KRAFT G. T. (1977a) The morphology of *Grateloupia intestinalis* from New Zealand, with some thoughts on generic criteria within the family Cryptonemiaceae (Rhodophyta). *Phycologia*, **16**, 43–51.

KRAFT G. T. (1977b) Studies of marine algae in the lesser-known families of the Gigartinales (Rhodophyta). I. The Acrotylaceae. *Aust. J. Bot.*, **25**, 97–140.

KRAFT G. T. (1977c) Studies of marine algae in the lesser-known families of the Gigartinales (Rhodophyta). II. The Dicranemaceae. *Aust. J. Bot.*, **25**, 219–67.

KRAFT G. T. (1977d) Transfer of the New Zealand red alga *Tylotus proliferus* (Graciliariaceae, Gigartinales) to the genus *Gracilaria*. *N.Z. J. Bot.*, **15**, 495–502.

KRAFT G. T. (1978) Studies of marine algae in the lesser-known families of the Gigartinales (Rhodophyta). III. The Mychodeaceae and Mychodeophyllaceae. *Aust. J. Bot.*, **26**, 515–610.

KRAFT G. T. & ABBOTT I. A. (1971) *Predaea weldii*, a new species of Rhodophyta from Hawaii, with an evaluation of the genus. *J. Phycol.*, **7**, 194–202.

KRAFT G. T. & JOHN D. M. (1976) The morphology and ecology of *Nemastoma* and *Predaea* species (Nemastomataceae, Rhodophyta) from Ghana. *Br. Phycol. J.*, **11**, 331–44.

KRAFT G. T. & WOELKERLING W. J. (1981) Rhodophyta: systematics and biology. In *Marine Botany, an Australasian Perspective*, eds. Clayton M. & King R. Longman Cheshire, Melbourne (in press).

KRAFT G. T. & WYNNE M. J. (1979) An earlier name for the Atlantic North American red alga *Neoagardhiella baileyi* (Solieriaceae, Gigartinales). *Phycologia*, **18**, 325–9.

KRISHNAMURTHY V. (1957) The genus *Bangiopsis* Schmitz from south India. *Phytomorphology*, **7**, 102–12.

KRISHNAMURTHY V. (1973) A revision of the species of the algal genus *Porphyra* occurring on the Pacific coast of North America. *Pacif. Sci.*, **26**, 24–49.

KYLIN H. (1923) Studien uber die Entwicklungsgeschichte der Florideen. *K. Svenska Vetensakad. Handl.*, **63**, 1–139.

KYLIN H. (1956) *Die Gattungen der Rhodophyceen*. Gleerups, Lund.

LAURET M. (1970) Morphologie, phenologie, répartition des *Polysiphonia* marins du littoral languedocien. *Nat. Monsp. Bot.*, **21**, 121–63.

LEBEDNIK P. A. (1977) Postfertilization development in *Clathromorphum, Melobesia* and *Mesophyllum* with comments on the evolution of the Corallinaceae and the Cryptonemiales (Rhodophyta). *Phycologia*, **16**, 379–406.

LEE I. K. (1978) Studies on Rhodymeniales from Hokkaido. *J. Fac. Sci. Hokkaido Univ.*, ser. **5** (11), 1–203.

LEE I. K. & KUROGI M. (1973) The development and structure of vegetative and reproductive organs of *Binghamia californica* (Rhodophyta). *Bot. Mag. Tokyo*, **86**, 253–66.

LEE I. K. & KUROGI M. (1978) *Neohalosacciocolax aleutica* gen. et sp. nov. (Rhodophyta), parasitic on *Halosaccion minjaii*, I. K. Lee from the north Pacific. *Br. Phycol. J.*, **13**, 131–9.

LEE I. K. & WEST J. A. (1979) *Dasysiphonia chejuensis* gen. et sp. nov. (Rhodophyta, Dasyaceae) from Korea. *Syst. Bot.*, **4**, 115–29.

LEE I. K., WEST J. A. & HOMMERSAND M. H. (1981) *Binghamiopsis caespitosa* gen. et sp. nov. (Lomentariaceae, Rhodophyta) from the eastern Pacific. *Phycologia*, **20**, (in press).

LEE I. K. & YOO S. A. (1979) *Gloeophycus koreanum*, gen. et sp. nov. (Rhodophyta, Gloiosiphoniaceae) from Korea. *Phycologia*, **18**, 347–54.

LEE R. E. (1971) The pit connections of some lower red algae: ultrastructure and phylogenetic significance. *Br. Phycol. J.*, **6**, 29–38.

LEE R. E. & FULTZ S. A. (1970) Ultrastructure of the Conchocelis stage of the marine red alga *Porphyra leucosticta*. *J. Phycol.*, **6**, 22–8.

LEE R. K. S. (1963) The structure and reproduction of *Dudresnaya hawaiiensis* sp. nov. (Rhodophyta). *Am. J. Bot.*, **50**, 315–19.

LEE Y. P. & LINDSTROM S. C. (1979) *Audouinella kurogii*, a new marine red alga (Acrochaetiaceae) from eastern Hokkaido. *Jap. J. Phycol.*, **27**, 115–22.

LEWIN R. A. (1972) Book review of *Selected papers in phycology*, ed. Rosowski J. & Parker B. *Phycologia*, **11**, 301.

LEWIN R. A. & ROBERTSON J. A. (1971) Influence of salinity on the form of *Asterocytis* in pure culture. *J. Phycol.*, **7**, 236–8.

LEWIS J. A. & KRAFT G. T. (1979) Occurrence of a European red alga (*Schottera nicaeensis*) in southern Australian waters. *J. Phycol.*, **15**, 226–30.

L'HARDY-HALOS M.-T. (1968) Les Ceramiaceae (Rhodophyceae, Florideae) des côtes de Bretagne; 1.—Le genre *Antithamnion* Nägeli. *Rev. Alg.*, N.S. **9**, 152–82.

LINDAUER V. W. (1949) Additions to the marine algae of New Zealand. *Trans. Roy. Soc. N.Z.*, **77**, 390–3.

LITTLER M. M. (1972) The crustose corallinaceae. *Oceanogr. Mar. Biol. Ann. Rev.*, **10**, 311–47.

LITTLER M. M. (1974) The structure and reproduction of *Dudresnaya lubrica* sp. nov. (Rhodophyta, Dumontiaceae). *Br. Phycol. J.*, **9**, 149–56.

LUND S. (1959) The marine algae of east Greenland. I. Taxonomical part. *Medd. Gronl.*, **156**, 1–247.

MAGNE F. (1960) Le *Rhodochaete parvula* Thuret (Bangioidae) et sa reproduction sexuée. *Cah. Biol. Mar.*, **1**, 407–20.

MASUDA M. (1976a) On a new red algal genus *Pseudorhododiscus*. *Acta Phytotax. Geobot.*, **27**, 123–32.

MASUDA M. (1976b) Taxonomic notes on *Rhodophysemopsis* gen. nov. (Rhodophyta). *J. Jap. Bot.*, **51**, 175–87.

MAYHOUB H. (1973) Cycle du développement de *Calosiphonia vermicularis* (J. Agardh) Schmitz (Rhodophycée, Gigartinale). *C. R. Acad. Sci. Paris*, **277**, 1137–40.

MAYHOUB H. (1975) Nouvelles observations sur le cycle de développement du *Calosiphonia vermicularis* (J. Ag.) Sch. (Rhodophycée, Gigartinale). *C.R. Acad. Sci. Paris*, **280**, 2441–3.

MENDOZA M. L. (1969a) Las Delesseriaceae (Rhodophyta) de Puerto Deseado, Provincia de Santa Cruz, Argentina. II. Estudio sistemático del género *Myriogramme* Kylin. *Physis*, **29**, 245–60.

MENDOZA M. L. (1969b) Las Delesseriaceae (Rhodophyta) de Puerto Deseado, Provincia de Santa Cruz, Argentina. I. Estudio sistemático y ecológico de los géneros *Schizoseris* Kylin, *Cladonta* Skottsberg e *Hymenena* Greville. *Physis*, **28**, 419–41.

MENDOZA M. L. (1969c) Estudio sistemático y ecológico de las Ceramiaceae (Algae, Rhodophyta) de Puerto Deseado, Provincia de Santa Cruz (Argentina). *Darwiniana*, **15**, 287–362.

MENDOZA M. L. (1970) Dos especies de algas rojas hemiparasitas de Puerto Deseado, Provincia de Santa Cruz, Argentina. *Physis*, **29**, 363–71.

MENDOZA M. L. (1974) El estado sexual masculino de *Delesseria fuegiensis* Skottsberg y la presencia de los géneros *Delesseria* Lamouroux y *Schizoseris* Kylin (Rhodophyta) en Tierra del Fuego e isla de los Estados. *Physis*, **33**, 483–504.

MIKAMI H. (1965) Systematic study of Phyllophoraceae and Gigartinaceae from Japan and its vicinity. *Fac. Sci. Hokkaido Univ. Inst. Alg. Res. Sci. Pap.*, **5**, 181–285.

MIKAMI H. (1971) *Congregatocarpus*, a new genus of the Delesseriaceae (Rhodophyta). *Bot. Mag. Tokyo*, **84**, 243–6.

MIKAMI H. (1973) Key to the species of Delesseriaceae in Hokkaido, Japan. *Bull. Jap. Soc. Phycol.*, **21**, 65–9.

MIKAMI H. (1978) On *Laingia hookeri* (Rhodophyceae. Delesseriaceae) from New Zealand. *Jap. J. Phycol.*, **26**, 65–8.

MIN-THEIN U. & WOMERSLEY H. B. S. (1976) Studies on southern Australian taxa of Solieriaceae, Rhabdoniaceae and Rhodophyllidaceae (Rhodophyta). *Aust. J. Bot.*, **24**, 1–166.

MITCHELL E. A. (1966) The southern Australian genera of the Dumontiaceae (Rhodophyta). *Nova Hedw.*, **11**, 209–20.

MIURA A. (1967) Two new species and a new record of *Porphyra* from Enoshima, Sagami Bay. *J. Tokyo Univ. Fish.*, **53**, 65–71.

MIURA A. (1968) *Porphyra katadai*, a new species from Japanese coast. *J. Tokyo Univ. Fish.*, **54**, 55–9.

MOE R. L. (1979) *Minium parvum* gen. et sp. nov., a crustose member of the Rhodymeniales (Rhodophyta). *Phycologia*, **18**, 38–46.

MOE R. L. & SILVA P. C. (1979) Morphological and taxonomic studies on Antarctic Ceramiaceae (Rhodophyceae). I. *Antarcticothamnion polysporum* gen. et sp. nov. *Br. Phycol. J.*, **14**, 385–405.

MOE R. L. & SILVA P. C. (1980) Morphological and taxonomic studies on Antarctic Ceramiaceae (Rhodophyceae). II. *Pterothamnion antarcticum* (Kylin) comb. nov. (*Antithamnion antarcticum* (Kylin). *Br. Phycol. J.*, **15**, 1–17.

MSHIGENI K. E. (1979) The economic algal genus *Eucheuma* (Rhodophyta, Gigartinales): observations on the morphology and distribution ecology of Tanzanian species. *Bot. Mar.*, **22**, 437–45.

MUMFORD T. F. Jr. (1975) Observations of the distribution and seasonal occurrence of *Porphyra schizophylla* Hollenberg, *Porphyra torta* Krishnamurthy, and *Porphyra brumalis* sp. nov. (Rhodophyta, Bangiales). *Syesis*, **8**, 321–32.

Munda I. M. & Pedersen P. M. (1978) *Porphyra thulaea* sp. nov. (Rhodophyceae, Bangiales) from East Iceland and West Greenland. *Bot. Mar.*, **21**, 283–8.

Murray S. N. & Dixon P. S. (1972) The life history of *Callophyllis firma* (Kylin) Norris in laboratory culture. *Br. Phycol. J.*, **7**, 165–8.

Murray S. N., Dixon P. S. & Scott J. L. (1972) The life history of *Porphyropsis coccinea* var. *dawsonii* in culture. *Br. Phycol. J.*, **7**, 323–33.

Newroth P. R. (1971) Redescriptions of five species of *Phyllophora* and an artificial key to the North Atlantic Phyllophoraceae. *Br. Phycol. J.*, **6**, 225–30.

Newroth P. R. & Taylor A. R. A. (1968) The distribution of *Ceratocolax hartzii*. *Br. Phycol. Bull.*, **3**, 543–6.

Nichols H. W. & Lissant E. K. (1967) Developmental studies of *Erythrocladia* Rosenvinge in culture. *J. Phycol.*, **3**, 6–18.

Noda M. (1964) On the *Porphyra* from Sado Island in the Japan Sea. *J. Fac. Sci. Niigata Univ.*, ser. 2, **4**, 1–13.

Noda M. (1970) Some marine algae collected on the coast of Iwagasaki Prov., Echigo, facing the Japan Sea. *Sci. Rep. Niigata Univ.*, ser. D, **7**, 27–35.

Noda M. (1971) Some new species of marine algae from the northeastern coast of Japan Sea. *Sci. Rep. Niigata Univ.*, ser. D, **8**, 53–9.

Norris J. N. & Bucher K. (1977) The genus *Platoma* (Gigartinales, Rhodophyta) with a description of *P. abbottiana* sp. nov. *J. Phycol.*, **13**, 155–62.

Norris R. E. (1957) Morphological studies on the Kallymeniaceae. *Univ. Calif. Publ. Bot.*, **28**, 251–334.

Norris R. E. (1964) The morphology and taxonomy of South African Kallymeniaceae. *Bot. Mar.*, **7**, 90–129.

Norris R. E., Tokida J. & Masaki T. (1960) Further studies on *Cirrulicarpus gmelini* (Grunow) Tokida & Masaki. *Bull. Fac. Fish. Hokkaido Univ.*, **11**, 29–36.

Nozawa Y. (1970) Systematic anatomy of the red algal genus *Rhodopeltis*. *Pacific. Sci.*, **24**, 99–133.

Nozawa Y. (1972a) Systematic anatomy of the Squamariaceae in the southern islands of Japan. 3. *Peyssonnelia squamaria* (Gmelin) Decaisne. *Bull. Jap. Soc. Phycol.*, **20**, 1–6.

Nozawa Y. (1972b) Systematic anatomy of the Squamariaceae in the southern islands of Japan. 4. *Peyssonnelia conchicola* Picc. & Grun., and *P. rubra* var. *orientalis* Weber van Bosse. *Bull Jap. Soc. Phycol.*, **20**, 41–7.

Oliveira Filho E. C. de & Coll J. (1975) The genus *Porphyra* (Rhodophyta—Bangiales) in the American South Atlantic. I. Brazilian species. *Bot. Mar.*, **18**, 191–7.

Oza R. M. (1976) Studies on Indian *Gracilaria*. II. The development of reproductive structures of *Gracilaria corticata*. *Bot. Mar.*, **19**, 107–14.

Papenfuss G. F. (1958) Notes on algal nomenclature IV. *Taxon*, **7**, 104–9.

Papenfuss G. F. (1961) Review of W. R. Taylor, 1960. *Bull. Torrey Bot. Club*, **88**, 421–3.

Papenfuss G. F. (1966) A review of the present system of classification of the Florideophycidae. *Phycologia*, **5**, 247–55.

Papenfuss G. F. (1967) Notes on algal nomenclature—V. Various Chlorophyceae and Rhodophyceae. *Phykos*, **5**, 95–105.

Papenfuss G. F. & Chiang Y.-M. (1969) Remarks on the taxonomy of *Galaxaura* (Nemaliales, Chaetangiaceae). *Proc. Int. Seaweed Symp.*, **6**, 303–14.

Papenfuss G. F. & Edelstein T. (1974) The morphology and taxonomy of the red alga *Sarconema* (Gigartinales: Solieriaceae). *Phycologia*, **13**, 31–44.

Parke M. & Dixon P. S. (1976) Check-list of British marine algae—third revision. *J. Mar. Biol. Ass. U.K.*, **56**, 527–94.

Parsons M. J. (1975) Morphology and taxonomy of the Dasyaceae and Lophothalieae (Rhodomelaceae) of the Rhodophyta. *Aust. J. Bot.*, **23**, 549–713.

Perestenko L. P. (1975) The red algae of the far eastern seas of the U.S.S.R. Foliose Cryptonemiacean algae (Cryptonemiales, Rhodophyta). *Bot. Zhurn.*, **60**, 1676–89.

Price J. H. (1978) Ecological determinations of adult form in *Callithamnion*: its taxonomic implications. In *Modern Approaches to the Taxonomy of Red and Brown Algae*, eds. Irvine D. E. G. & Price J. H. pp. 263–300. Academic Press, London, New York.

RAMUS J. (1969) The developmental sequence of the marine red alga *Pseudogloiophloea* in culture. *Univ. Calif. Publ. Bot.*, **52**, 1–28.

REEDMAN D. J. & WOMERSLEY H. B. S. (1976) Southern Australian species of *Champia* and *Chylocladia* (Rhodymeniales: Rhodophyta). *Trans. Roy. Soc. S. Aust.*, **100**, 75–104.

RICHARDSON N. & DIXON P. S. (1968) Life history of *Bangia fuscopurpurea* (Dillw.) Lyngb. in culture. *Nature (Lond.)*, **218**, 496–7.

RICHARDSON N. & DIXON P. S. (1969) The conchocelis phase of *Smithora naiadum* (Anders.) Hollenb. *Br. Phycol. J.*, **4**, 181–3.

RICKER R. W. & KRAFT G. T. (1979) Morphology of the subantarctic red alga *Cenacrum subsutum* gen. et sp. nov. (Rhodymeniales) from Macquarie Island. *J. Phycol.*, **15**, 434–44.

SAITO Y. (1967) Studies on Japanese species of *Laurencia*, with special reference to their comparative morphology. *Mem. Fac. Fish. Hokkaido Univ.*, **15**, 1–81.

SAITO Y. (1969a) The algal genus *Laurencia* from the Hawaiian Islands, the Philippine Islands, and adjacent areas. *Pacif. Sci.*, **23**, 148–60.

SAITO Y. (1969b) On morphological distinctions of some species of Pacific North American *Laurencia*. *Phycologia*, **8**, 85–90.

SAITO Y. & WOMERSLEY H. B. S. (1974) The southern Australian species of *Laurencia* (Ceramiales: Rhodophyta). *Aust. J. Bot.*, **22**, 815–74.

SANTELICES B. (1976) Taxonomic and nomenclatural notes on some Gelidiales (Rhodophyta). *Phycologia*, **15**, 165–73.

SANTELICES B. (1977) A taxonomic review of Hawaiian Gelidiales (Rhodophyta). *Pacif. Sci.*, **31**, 61–84.

SCAGEL R. F. (1953) A morphological study of some dorsiventral Rhodomelaceae. *Univ. Calif. Publ. Bot.*, **27**, 1–108.

SCAGEL R. F. (1957) An annotated list of the marine algae of British Columbia and northern Washington. *Bull. Nat. Mus. Canada*, **150**, 1–289.

SCAGEL R. F. (1962a) The genus *Dasyclonium* J. Agardh. *Can. J. Bot.*, **40**, 1017–40.

SCAGEL R. F. (1962b) A morphological study of the red alga *Whidbeyella cartilaginea* Setchell et Gardner. *Can. J. Bot.*, **40**, 1217–22.

SCHNEIDER C. W. & EISEMAN N. J. (1979) *Searlesia*, a new genus from the western Atlantic based on *Membranoptera subtropica* (Rhodophyta, Delesseriaceae). *Phycologia*, **18**, 319–24.

SCHOTTER G. (1968) Recherches sur les Phyllophoracées. *Bull. Inst. Oceanogr. Monaco*, **67**, 1–99.

SCIUTO S., PIATTELLI M., CHILLEMI R. FURNARI G. & CORMACI M. (1979) The implications of *Haematocelis rubens* J. Agardh in the life history of *Schizymenia dubyi* (Chauvin) J. Agardh (Rhodophyta, Gigartinales): a chemical study. *Phycologia*, **18**, 296–302.

SEARLES R. B. (1968) Morphological studies of red algae of the order Gigartinales. *Univ. Calif. Publ. Bot.*, **43**, 1–86.

SEARLES R. B. (1980) The strategy of the red algal life history. *Am. Nat.*, **115**, 113–20.

SILVA P. C. (1952) A review of nomenclatural conservation in the algae from the point of view of the type method. *Univ. Calif. Publ. Bot.*, **25**, 241–324.

SILVA P. C. (1978) Type specimens of Gelidiaceae (Rhodophyceae) described by Nina H. Loomis. *Phycologia*, **17**, 256–60.

SIMONS R. H. (1964) Species of *Plocamium* on the South African coast. *Bothalia*, **8**, 183–93.

SOMMERFELD M. R. & NICHOLS H. W. (1970) Developmental and cytological studies of *Bangia fuscopurpurea* in culture. *Am. J. Bot.*, **57**, 640–8.

SOUTH G. R. & ADAMS N. M. (1976) *Erythrotrichia foliiformis* sp. nov. (Rhodophyta, Erythropeltidaceae) from New Zealand. *J. Roy. Soc. N.Z.*, **6**, 399–405.

SOUTH G. R. & ADAMS N. M. (1979) A revision of the genus *Plocamium* Lamouroux (Rhodophyta, Gigartinales) in New Zealand. *Phycologia*, **18**, 120–32.

SOUTH G. R., HOOPER R. G., & IRVINE L. M. (1972) The life history of *Turnerella pennyi* (Harv.) Schmitz. *Br. Phycol. J.*, **7**, 221–33.

SPARLING S. R. (1957) The structure and reproduction of some members of the Rhodymeniaceae. *Univ. Calif. Publ. Bot.*, **29**, 319–96.

STEGENGA H., & VAN WISSEN M. J. (1979) Remarks on the life histories of three Acrochaetioid algae (Rhodophyta, Nemaliales). *Acta Bot. Neerl.*, **28**, 97–115.

STEGENGA H. & VROMAN M. (1977) The significance of life history characters for the systematics of the Acrochaetiaceae. *J. Phycol.*, **13** (suppl.), 64.

TANAKA T. & NOZAWA Y. (1960) One red algal parasite from Japan. *Mem. Fac. Fish. Kagoshima Univ.*, **9**, 107–12.

TAYLOR W. R. (1960) *Marine Algae of the Eastern Tropical and Subtropical Coasts of the Americas.* University of Michigan Press, Ann Arbor.

TAYLOR W. R. (1971) Notes on algae from the tropical Atlantic Ocean—VI. *Br. Phycol. J.*, **6**, 145–56.

TAYLOR W. R. & ABBOTT I. A. (1973) A new species of *Botryocladia* from the West Indies. *Br. Phycol. J.*, **8**, 409–12.

TAYLOR W. R. & RHYNE C. F. (1970) Marine algae of Dominica. *Smithson. Contrib. Bot.*, **3**, 1–16.

TOKIDA J. & MASAKI T. (1956) Studies on the reproductive organs of red algae II. On *Erythrophyllum gmelini* (Grun.) Yendo. *Bull. Fac. Fish. Hokkaido Univ.*, **7**, 63–71.

UMAMAHESWARA RAO M. (1972) On the Gracilariaceae of the seas around India. *J. Mar. Biol. Ass. India*, **14**, 671–96.

UMEZAKI I. (1960) On the development of reproductive organs of the genus *Helminthocladia* found in Japan—*H. australis* Harv. and *H. macrocephala* Yamada. *Acta Phytotax. Geobot.*, **18**, 169–77.

UMEZAKI I. (1967) Notes on some marine algae from Japan (1). *J. Jap. Bot.*, **42**, 169–74.

UMEZAKI I. (1977) Life histories in the Florideophyceae and their evolution. *Acta Phytotax. Geobot.*, **28**, 1–18.

VAN DER MEER J. P. & TODD E. R. (1980) The life history of *Palmaria palmata* in culture. A new type for the Rhodophyta. *Can. J. Bot.*, **58**, 1250–6.

WAGNER F. S. (1954) Contributions to the morphology of the Delesseriaceae. *Univ. Calif. Publ. Bot.*, **27**, 279–346.

WETHERBEE R. (1980) Postfertilization development in the red alga *Polysiphonia*. 1. Proliferation of the carposporophyte. *J. Ultrastr. Res.*, **70**, 259–74.

WHITTICK A., & HOOPER R. G. (1977) The reproduction and phenology of *Antithamnion cruciatum* (Rhodophyta: Ceramiaceae) in insular Newfoundland. *Can. J. Bot.*, **55**, 520–4.

WISEMAN D. R. (1975) On the status of the red algal family, the Rhizophyllidaceae (Gigartinales). *Taxon*, **24**, 489–90.

WISEMAN D. R. (1977) Observations on the reproductive morphology of the red algal genus *Ochtodes* J. Agardh (Rhizophyllidaceae, Gigartinales). *Phycologia*, **16**, 1–8.

WOELKERLING W. J. (1971) Morphology and taxonomy of the *Audouinella* complex (Rhodophyta) in southern Australia. *Aust. J. Bot.*, Suppl. ser. **1**, 1–91.

WOELKERLING W. J. (1973a) The *Audouinella* complex (Rhodophyta) in the western Sargasso Sea. *Rhodora*, **75**, 78–101.

WOELKERLING W. J. (1973b) The morphology and systematics of the *Audouinella* complex (Acrochaetiaceae, Rhodophyta) in northeastern United States. *Rhodora*, **75**, 529–621.

WOELKERLING W. J. (1980) Studies on *Metamastophora* (Corallinaceae, Rhodophyta). 1. *M. flabellata*: morphology and anatomy. *Br. Phycol. J.*, **15**, 201–25.

WOLLASTON E. M. (1968) Morphology and taxonomy of southern Australian genera of Crouanieae Schmitz (Ceramiales, Rhodophyta). *Aust. J. Bot.*, **16**, 217–417.

WOLLASTON E. M. (1971) *Antithamnion* and related genera occurring on the Pacific coast of North America. *Syesis*, **4**, 73–92.

WOMERSLEY H. B. S. (1965a) The Helminthocladiaceae (Rhodophyta) of southern Australia. *Aust. J. Bot.*, **13**, 451–87.

WOMERSLEY H. B. S. (1965b) The morphology and relationships of *Sonderella* (Rhodophyta, Rhodomelaceae). *Aust. J. Bot.*, **13**, 435–50.

WOMERSLEY H. B. S. (1971a) The relationships of *Nizymenia* and *Stenocladia* (Gigartinales, Rhodophyta). *Phycologia*, **10**, 199–203.

WOMERSLEY H. B. S. (1971b) The genus *Plocamium* in southern Australia. *Trans. Roy. Soc. S. Aust.*, **95**, 9–27.

Womersley H. B. S. (1978) Southern Australian species of *Ceramium* Roth (Rhodophyta). *Aust. J. Mar. Freshwat. Res.*, **29**, 205–57.

Womersley H. B. S. (1979) Southern Australian species of *Polysiphonia* Greville (Rhodophyta). *Aust. J. Bot.*, **27**, 459–528.

Womersley H. B. S. & Abbott I. A. (1968) Structure and reproduction of *Neoardissonea* Kylin (Rhodophyta—Naccariaceae). *J. Phycol.*, **4**, 173–7.

Womersley H. B. S. & Bailey A. (1970) Marine algae of the Solomon Islands. *Phil. Trans. Roy. Soc. London Biol. Sci.*, **259**, 257–352.

Womersley H. B. S. & Conway E. (1975) *Porphyra* and *Porphyropsis* (Rhodophyta) in southern Australia. *Trans. Roy. Soc. S. Aust.*, **99**, 59–70.

Womersley H. B. S. & Norris R. E. (1971) The morphology and taxonomy of Australian Kallymeniaceae (Rhodophyta). *Aust. J. Bot.*, Suppl. ser. **2**, 1–62.

Womersley H. B. S. & Shepley E. A. (1959) Studies on the *Sarcomenia* group of the Rhodophyta. *Aust. J. Bot.*, **7**, 168–223.

Wynne M. J. (1969) *Platysiphonia decumbens* sp. nov., a new member of the *Sarcomenia* group (Rhodophyta) from Washington. *J. Phycol.*, **5**, 190–202.

Wynne M. J. (1970a) Marine algae of Amchitka Island (Aleutian Islands). 1. Delesseriaceae. *Syesis*, **3**, 95–144.

Wynne M. J. (1970b) Marine algae of Amchitka Island (Aleutian Islands). II. Bonnemaisoniaceae. *Pacif. Sci.*, **24**, 433–8.

Wynne M. J. (1972) Notes on the distribution of *Pleuroblepharis* (Bonnemaisoniaceae) and the status of *Odonthalia japonica* Okamura. *Bull. Jap. Soc. Phycol.*, **20**, 48–53.

Wynne M. J. (1980) *Pleuroblepharidella* nom. nov. (Bonnemaisoniaceae, Rhodophyceae) proposed for *Pleuroblepharis* Wynne. *Taxon*, **29**, 325–6.

Wynne M. J., McBride D. L. & West J. A. (1973) *Polyneuropsis stolonifera* gen. et sp. nov. (Delesseriaceae, Rhodophyta) from the Pacific coast of North America. *Syesis*, **6**, 243–53.

Wynne M. J. & Taylor W. R. (1973) The status of *Agardhiella tenera* and *Agardhiella baileyi* (Rhodophyta, Gigartinales). *Hydrobiologia*, **43**, 93–107.

Yamamoto H. (1975) The relationship between *Gracilariopsis* and *Gracilaria* from Japan. *Bull. Fac. Fish. Hokkaido Univ.*, **26**, 217–22.

Yamamoto H. (1978) Systematic and anatomical study of the genus *Gracilaria* in Japan. *Mem. Fac. Fish. Hokkaido Univ.*, **25**, 97–152.

Yoshida T. (1975) On the structural characteristics of *Peyssonnelia caulifera* Okamura (Rhodophyceae, Squamariaceae) from Japan. *Bull. Jap. Soc. Phycol.*, **23**, 1–7.

Yoshida T. (1979) A new genus *Kurogia* (Delesseriaceae, Rhodophyta) from Hokkaido, northern Japan. *Jap. J. Phycol.*, **27**, 83–9.

Yoshida T. (1980) Two new species of red algae from the west coast of Kyushu, Japan. *Jap. J. Phycol.*, **28**, 69–74.

Yoshizaki M. (1977) Morphology and taxonomy of the Japanese representatives of Nemaliales. 1. Thallus structure and reproductive organs of *Nemalion multifidum*. *J. Jap. Bot.*, **52**, 8–14.

Yoshizaki M. (1978) Morphology and taxonomy of the Japanese representatives of the Nemaliales. 2. Thallus structure and reproductive organs of *Dermonema dichotomum* and *D. pulvinatum*. *J. Jap. Bot.*, **53**, 17–25.

Yoshizaki M. (1979) Morphology and taxonomy of the Japanese representatives of the Nemaliales 3. Thallus structure and reproductive organs of *Trichogloea requienii*. *J. Jap. Bot.*, **54**, 225–33.

Zinova A. D. (1965) Representatives of the family Delesseriaceae (Rhodophyta) in the northern part of the Pacific Ocean. *Nov. Sist. Nizsh. Rast.*, **2**, 78–97.

Zinova A. D. (1972) Representatives of the family Delesseriaceae (Rhodophyta) in the northern part of the Pacific Ocean. 2. *Nov. Sist. Nizsh. Rast.*, **9**, 65–82.

Zinova A. D. (1976) Some species of the family Delesseriaceae (Rhodophyta) in the northern part of the Pacific Ocean. *Nov. Sist. Nizsh. Rast.*, **13**, 7–10.

CHAPTER 2
PHAEOPHYTA: MORPHOLOGY AND CLASSIFICATION

MICHAEL J. WYNNE
Division of Biological Sciences, University of Michigan,
Ann Arbor, Michigan, U.S.A.

1 INTRODUCTION

Since the Linnaean epoch when the known brown algae were comfortably categorized within the broad confines of a mere handful of genera (*Fucus, Conferva, Ulva, Tremella*), a steady process of erosion has dissected this early group into smaller but more meaningful units. The continual discovery of new algae and the establishment of new taxa as segregates of known genera have combined to bring about the present-day approximate total of 265 genera for this algal division. Much progress has been made since the tentative early steps of Lamouroux and Harvey in recognizing the integrity of the brown algae as a distinct assemblage, separable from other seaweeds mainly by their dark brown pigmentation. Equally significant advances were made by Thuret, who determined that sexual reproduction occurred in brown algae, and Reinke, who discovered that an alternation of morphological generations was present in certain brown algae.

Observations and descriptions by numerous early workers—on thallus organization, mode of growth, type of sexual reproduction, and life history pattern—formed the basis of the modern system of classification for this division. By and large, this system of classification has been accepted for the past half century. Yet, even a cursory survey of contemporary literature will reveal the dynamic nature of this subject. Today, through *in situ* and *in vitro* observation and experimentation, workers continue to unravel the complex interplay of genotype and environment with morphology and life history patterns. Thus a better understanding of the phylogenetic relationships within this division has been attained.

2 MORPHOLOGY

2.1 Vegetative organization

Thalli of brown algae collectively exhibit a broad morphological variety, ranging from the simple organization of creeping branched filaments lacking a heterotrichous organization (e.g. *Phaeostroma*) to the very complex organization seen in the giant kelps (Laminariales) (Figs 2.1 and 2.2), provided with sieve tubes conducting products of photosynthesis (see Chapter 15). Diverse intermediate forms are also known (Figs 2.3–2.8), but yet the Phaeophyta do not approach the elaborate morphological diversity expressed within the Rhodophyta. Rather, we witness a theme of basic thallus types appearing in the various orders: tufts of uniseriate branched filaments; adherent discs or crusts; and erect axes, which may be simple or branched, terete or ligulate, saccate or foliose, or a combination of these shapes.

Filamentous brown algae refer to those having a thallus organisation in which a vertical series of cells, or axis, undergoes transverse septation only. A uniseriate chain of cells, that is, a filament, results. Branching of this filament can occur by a cell expanding outward, this protuberance or bud then being cut off from the initiating cell by a septation oriented transverse to this new branch primordium. Cell division may occur in almost any cell; this pattern of growth is referred to as diffuse or generalized. Or cell division may be localized within a discrete region; this pattern of growth is an intercalary meristem.

Filamentous brown algae typically have a heterotrichous organization, meaning that the thallus consists of an erect part arising from a basal part, the latter made up of horizontally growing filaments that serve to attach the algae to the substrate. Larger, more massive forms can be achieved by the consolidation of filaments into pseudoparenchymatous systems. However, they lack the longitudinal septations that characterize the truly parenchymatous forms but resemble them by their size.

The criterion of parenchymatous *versus* filamentous or pseudoparenchymatous construction continues to be used to distinguish brown algal orders (however, see p. 77 for further discussion of this point). But whether a thallus is genuinely parenchymatous is not always an easily decipherable trait, particularly when it is composed of compactly aligned filaments (i.e. pseudoparenchymatous). Several examples of erroneous taxonomic assignments have been uncovered in recent years. *Scytothamnus australis*, which had been placed in the

Fig. 2.1 *Macrocystis integrifolia.*

Dictyosiphonales (*sensu* Papenfuss 1947), was found to be pseudoparenchymatous and thus transferred to the Chordariales (Price 1969). Another example is *Myelophycus*, which had been placed in the Dictyosiphonales, but *M. simplex*, the type species, was shown to be more properly assigned to the Chordariales (Wynne 1969). On the other hand, a second species of this genus, *Myelophycus intestinalis*, has a parenchymatous construction, and the genus *Melanosiphon* was established to accommodate it. *Ishige* had originally been classified as a parenchymatous alga, despite its obviously filamentous medulla; the current opinion is to assign it to the Chordariales (Hori 1972), perhaps in its own family. On the other hand, *Endarachne*, a genus with a foliose form but with a

Fig. 2.2 *Nereocystis luetkeana.*

pronounced filamentous medulla, is placed in the Scytosiphonales, the parenchymatous order to which it clearly belongs.

Another common morphological expression is the disc or crust. Members of the Myrionemataceae are adherent epiphytic discs, in which the basal layer is only loosely coalescent, whereas members of the Ralfsiaceae are lithophytic polystromatic crusts, in which the basal layer is quite compact. But this distinction is blurred by exceptions and intermediates (Fletcher 1978), and thus some authors treat this complex as a single family, the Myrionemataceae (Loiseaux 1968). The construction of these discs or crusts is pseudoparenchymatous, and the filaments arising from the basal layer range in consistency from very firm to easily separable. This criterion of crust texture varies within the genus *Pseudolithoderma* (Sears & Wilce 1973). Crusts may have a smooth contour, be composed of overlapping lobes, or be divided into narrow, cylindrical anastomozing branches (Hollenberg 1969).

A prostrate form also occurs in members of orders with parenchymatous

Fig. 2.3 *Costaria costata.*

organization, e.g. *Lobophora* (Dictyotales), *Battersia* (Sphacelariales), and *Zanardinia* and the sporophyte (or *Aglaozonia*-stage) of *Cutleria* (Cutleriales). The degree of attachment of these prostrate forms to the substrate varies from loose to very firm.

Most brown algae produce an erect axis, in which growth may take place by various methods: a diffuse or generalized pattern, an intercalary meristem, or apical cell growth. As previously pointed out, it is not always an easy matter to distinguish thalli as parenchymatously or pseudoparenchymatously constructed. Species of *Desmarestia* range in form from terete to foliose axes, but all have a similar trichothallic (= intercalary) meristem, in which only transverse

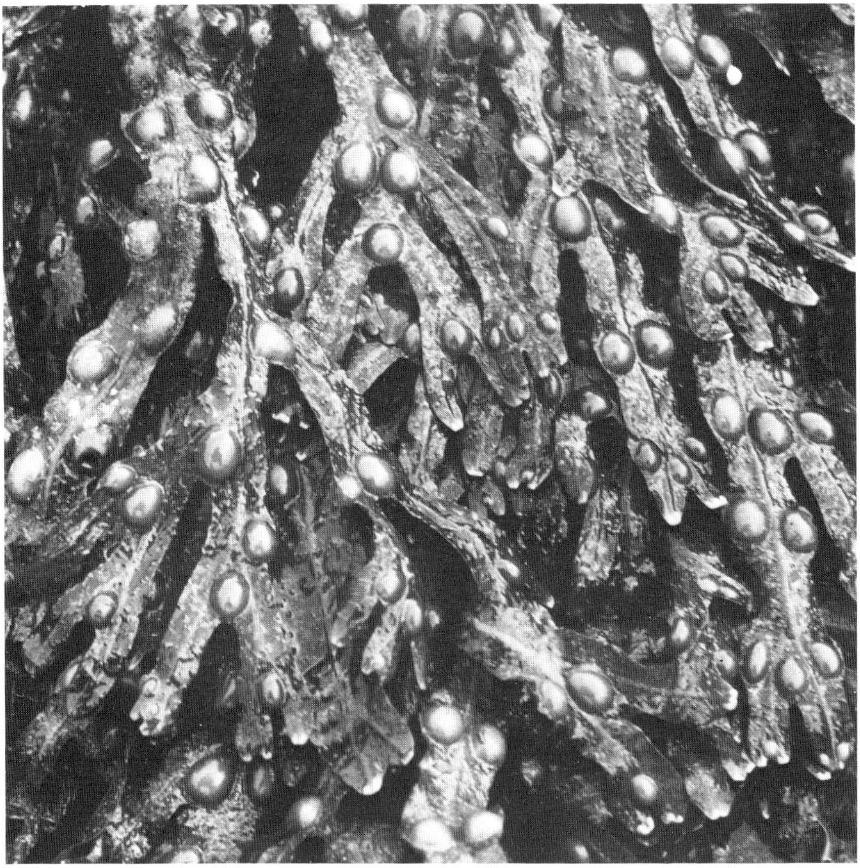

Fig. 2.4 *Fucus vesiculosus.*

segmentation of the axes occurs, and thus the thallus organization is strictly pseudoparenchymatous. On the other hand, the ligulate thalli of *Dictyota* have a single apical cell terminating each branch, and the occurrence of both transverse and longitudinal segmentation within the products from the apical cell bring about a parenchymatously constructed thallus.

Some brown algae have a combination of a well developed prostrate system, which is typically perennial, and erect axes that are produced seasonally. An example is *Analipus*, which has pseudoparenchymatous erect axes that may be simple or branched and that arise from a sometimes very expanded crustose system (Wynne 1971). *Stschapovia* consists of simple, flagelliform parenchymatous axes arising from a compact crustose base (Zinova 1954). Interestingly, in *Caepidium* the pseudoparenchymatous erect system is dimorphic: compact terete axes as well as hollow, vesicular outgrowths.

The blade is frequently expressed among the Phaeophyta, and it may be formed also by various methods of growth. The foliose thallus of *Zonaria* is

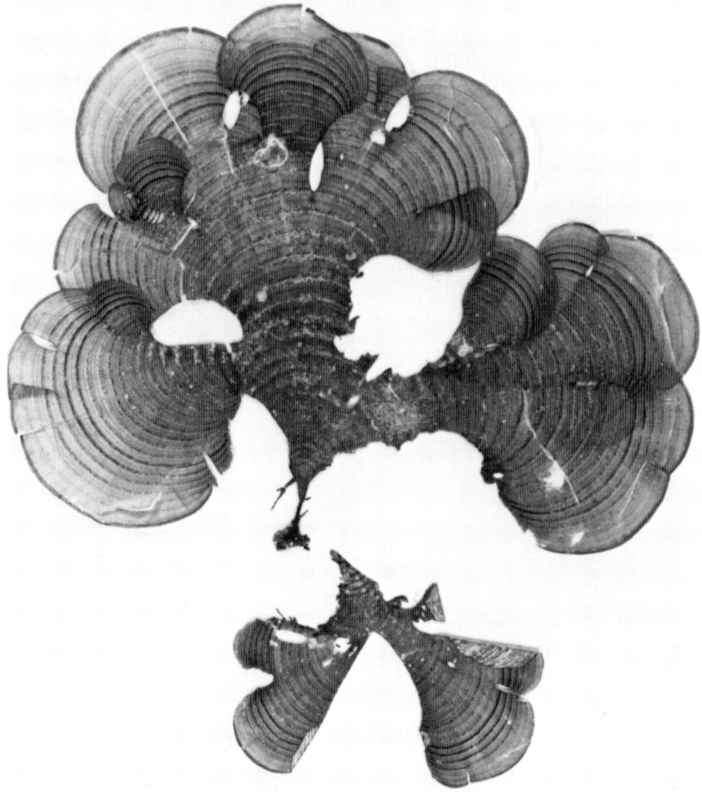

Fig. 2.5 *Padina tenuis.*

organized into columns of cells closely appressed in one plane, each column terminating in an apical cell (Dahl 1971). The anticlinal division of an apical cell effects branching of these columns, or filaments, and also increases thallus width. A uniformly advancing thallus margin is maintained by the co-ordination of apical cell division, which seems to be brought about by their close contact, and by some control from more proximal portions of the thallus (Dahl 1971; Neushul & Dahl 1972).

In contrast, the thallus of *Cutleria*, composed of columns, or filaments, of cells arranged in a plane as in *Zonaria*, grows by means of a trichothallic meristem. Distal to the meristematic cells is a loose fringe of filaments, but proximal to the meristematic cells (where lateral adhesion of the cell files occurs) each column of cells undergoes both periclinal and anticlinal divisions to

Fig. 2.6 Flattened sacs of *Coilodesme plana* epiphytic on *Cystoseira osmundacea.*

produce parenchymatous tissue. Thus, the thallus is a combination of parenchymatously constructed filaments held together pseudoparenchymatously. Species of *Cutleria* embody various shapes: a fan-shaped blade in *C. hancockii*, branched, ribbon-like axes in *C. multifida*, and branched, terete axes in *C. cylindrica* (Hollenberg 1978).

It is evident that parallel evolution must have been involved in the development of these similar thallus types. For example, the shape of a simple, cylindrical axis (solid or hollow) of macroscopic size characterized 29 genera belonging to five different orders (Wynne 1973). This shape again is the result of various growth processes (apical, diffuse or intercalary) and may be parenchymatously or pseudoparenchymatously organized. These basic differences come into play as taxonomic tools.

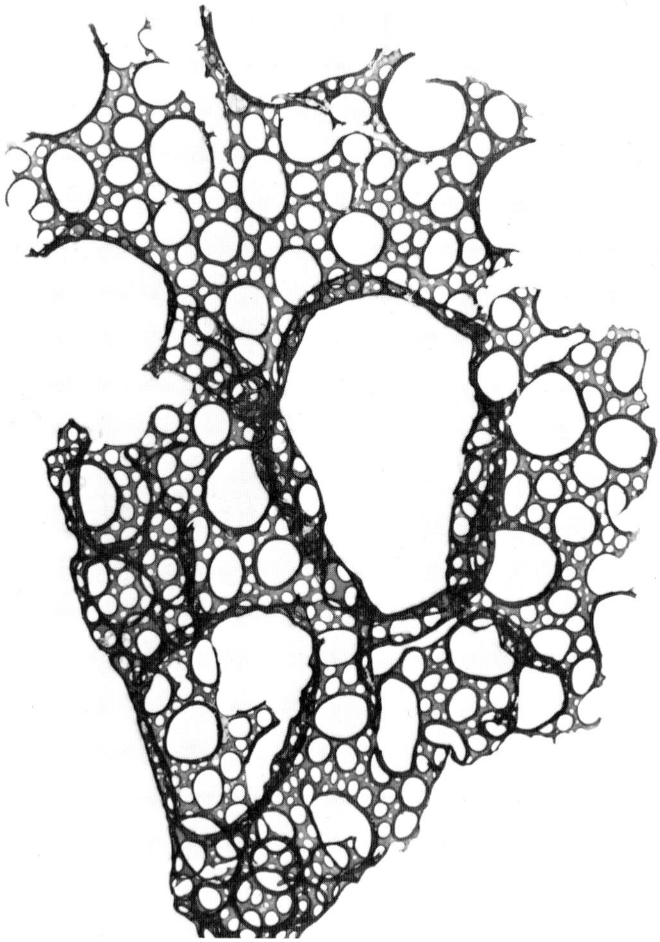

Fig. 2.7 *Hydroclathrus clathratus.*

2.2 Reproductive structures

Although the two major types of reproductive structures in the brown algae are well known, our understanding of their role has been modified by some recent contributions. Unilocular sporangia (Fig. 2.9a, b) are usually the site of meiosis, and synaptonemal complexes have been reported (Toth & Markey 1973). However, many examples of apparent apomeiosis are known (Wynne & Loiseaux 1976). A sexual capability of the zoids released from unilocular sporangia continues to be reported (Caram 1965; Loiseaux 1967b), but such a behaviour in *Ectocarpus siliculosus* has been refuted by Müller (1975).

A variation from the usual picture of unilocular sporangia is present in *Syringoderma abyssicola.* The contents of the apparent unilocular sporangia

Fig. 2.8 *Sargassum filipendula.*

undergo cytokinesis into numerous products, and walls are laid down contiguous to one another but not to the original sporangial wall. Each of these walled cells divides to form a tetrad of flagellated cells that are released (Walker & Henry 1978). The nature of these flagellated cells is unknown.

Various forms of plurilocular structures are recognized and are of taxonomic value. The most frequently encountered type has a single apical region for discharging the zoids or gametes (Fig. 2.9c, d). The type in the Sphacelariales and Cutleriales has locules with peripheral openings (Fig. 2.9e). Lengthy intercalary series of plurilocular sporangia are characteristic of some genera (*Pilayella, Leptonematella, Litosiphon*, and monosiphonous parts of *Stictyosiphon*). (See Kornmann & Sahling 1977, for excellent photographs of such examples.)

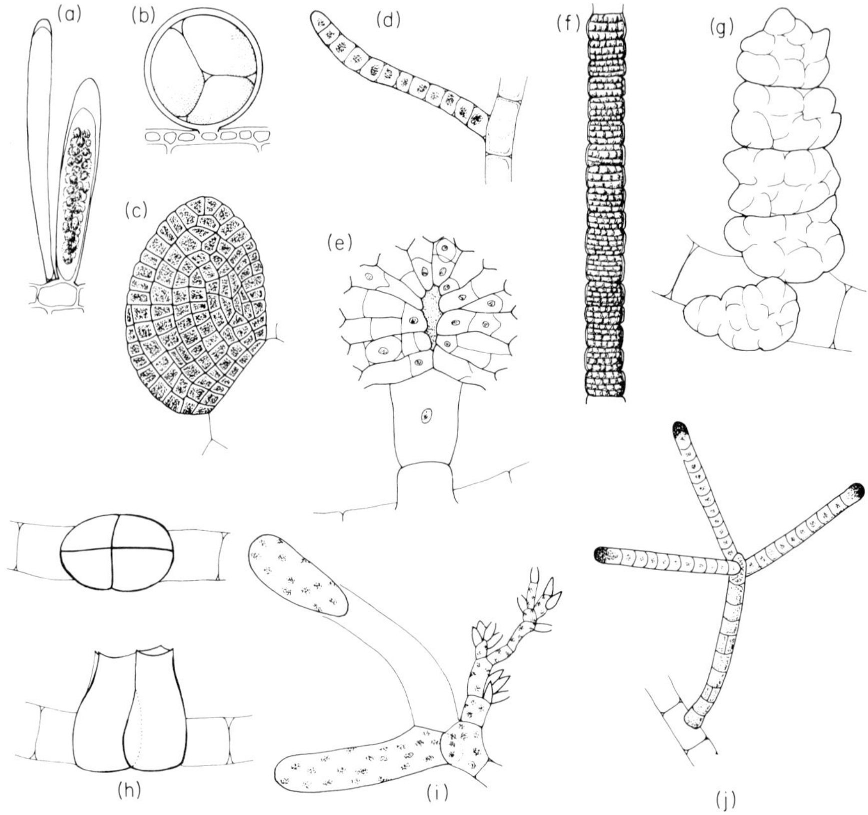

Fig. 2.9 Examples of brown algal reproductive structures. (a) unilocular sporangium (with paraphysis) of *Laminaria*; (b) unilocular sporangium (tetrasporangium) of *Dictyota*; (c) plurilocular sporangium of *Giffordia*; (d) plurilocular sporangium of *Laminariocolax*; (e) plurilocular macrozoidangium of *Sphacelaria*; (f) series of plurilocular sporangia of *Pilayella*; (g) sorus of plurilocular sporangia of *Polytretus*; (h) plurilocular sporangium of *Porterinema* (top and side views); (i) gametophyte of *Desmarestia* bearing oogonia and antheridia; (j) propagule of *Sphacelaria*. ((d), (f) & (j) redrawn from Kornmann & Sahling; (e) from van den Hoek & Flinterman; (g) from Pedersen; (h) from Dop; (i) from Nakahara and Nakamura.)

What appears to be plurilocular structures in *Sorocarpus* and *Polytretus* are interpreted to be a cluster or sorus of plurilocular sporangia (Fig. 2.9g) (Pedersen 1977), and this feature has been proposed as a family criterion for the Sorocarpaceae. Very simple sporangia composed of usually only four or eight locules (Fig. 2.9h) are present in *Porterinema* (Dop 1979). In the case of the oogamous orders Desmarestiales and Laminariales the gametangia are reduced to single cells (Fig. 2.9i) but are equivalent to plurilocular structures in that each cell (or locule) produces a single gamete.

3 CRITERIA USED IN CLASSIFICATION

Before discussing the traits that are employed in distinguishing the higher taxa of brown algae, it is useful to refer to the problems that sometimes arise in distinguishing some of the simpler Phaeophyta from those Chrysophyceae that would be considered among the most vegetatively complex representatives of that class. For example, *Giraudyopsis* was described by Dangeard (1965) as a heterotrichous brown alga that develops a parenchymatous tissue. But closer scrutiny by Loiseaux (1967a) revealed its affinities with the Chrysophyceae on the basis of its walls being entirely pectic in composition, the pyrenoids being included within the chloroplast rather than being exserted*, the ill-defined nature of the sporangia, and the formation of encysted cells. Another instance of an alga that was described as phaeophycean is *Phaeosaccion*. This alga has been shown to be more properly assigned to the Chrysophyceae by the total absence of alginate in the walls as well as additional characteristics (McLachlan *et al.* 1971).

On the other hand, the reverse situation has also occurred. Dop (1979) presented evidence suggesting that a species of the Chrysophycean genus *Apistonema, A. expansum*, a filamentous freshwater alga, is most likely the widely occurring, euryhaline *Porterinema fluviatile*, one of the relatively few brown algae occurring in freshwater habitats. Typical phaeophycean hairs occur in this species. These endogenously developed, multicellular structures, with a basal meristem and often a collar-like sheath, are unique to the Phaeophyta, thus providing a useful means of identification when they do occur. These examples of confusion between the brown algae and the golden algae nonetheless suggest a possible phylogenetic relationship between these two algal groups (Loiseaux & West 1970).

3.1 Morphology

Extensive field collections have been used to demonstrate the inherent morphological variation in some brown algae. Both exposure and salinity have been shown to have pronounced effects on the morphology. *Papenfussiella callitricha* exhibits fairly thick, strongly percurrent axes with little or no branching in habitats with the greatest wave exposure but shows much narrower, laxly organized axes with branching to several orders in the most protected sites (Hooper & South 1977). A similar phenomenon was recorded for *Chordaria flagelliformis* in Icelandic waters (Munda 1979). This boreal species can tolerate a wide range of exposures, as reflected by its polymorphic nature. At least five forms have been recognized. Two of them, f. *flagelliformis* with a deliquescent aspect and f. *densa* with a percurrent main axis densely clothed

* See p. 66 for a subsequent qualification of this characteristic.

with numerous short branches, are restricted to lower eulittoral levels. Forma *firma*, with a rigid and reduced aspect, is present on exposed slopes, whereas the essentially unbranched f. *chordaeformis* occurs in sandy lagoons. Forma *ramusculifera,* which has slender axes with sparsely placed, short laterals and a loose consistency, occurs at high levels and also in tide pools.

A correlation between morphology and salinity has also been observed for some brown algae. In a study of the benthic algae occurring in the fjords of Iceland, dwarf forms of *Pilayella littoralis* and *Dictyosiphon chordaria* were observed where salinities dropped below 2‰; the latter species becomes larger and more heavily branched with increasing salinities (Munda 1978). Thalli of *Fucus vesiculosus* lacking vesicles and usually with a narrow width dominate in habitats different from those of this species in which vesicles are present and in which the frond is usually broad (Fig. 2.4); yet they can also be found to occur side by side. In the hyposaline waters of the Baltic Sea off southwestern Finland this species was observed to be the most conspicuous element of the vegetation (Ravanko 1968). Earlier, Waern (1952) had observed evesiculate individuals to be the most common form of this species in the vicinity of the Öregrund Peninsula of eastern Sweden. But all intergrades between the two morphs were noted.

A fuller discussion of such environmental influences on the morphological expression is presented in Chapter 12, including the observation that one often encounters a disparity between the appearance of a particular alga in culture and in the field. The implications of such investigations on systematics are obvious.

3.2 Life histories

Recent developments regarding this fundamental criterion in the taxonomy of brown algae are treated in Chapter 5. It suffices here to refer to the fact that the traditional concept of a particular life history pattern being applicable to a particular order has come under much modification in recent years. Excepting the Fucales* and Durvillaeales, brown algal orders exhibit an alternation of morphological generations, typically involving isomorphic or heteromorphic phases. In only very general terms can we state that an order has a specific life history. A 'direct type of development', in which the expected alternation of phases is bypassed or 'short-circuited' in some way, is known for the members of several orders (Wynne & Loiseaux 1976). Furthermore, heteromorphic life histories are known to occur that bear no relationship to a sexual cycle (e.g. see Roeleveld *et al.* 1974).

* See Caplin (1968) and Jensen (1974) for arguments that interpret the Fucales as having an alternation of morphological generations.

3.3 Development

In lieu of culture studies, it has often proved helpful to correlate juvenile and developmental stages with the mature forms by careful field studies at different times during the growing season. An example is Lund's (1959) study of marine algae in Greenland. He was able to detect very young stages of *Delamarea* that consisted of monosiphonous filaments, which subsequently underwent longitudinal and transverse divisions to convert the axis into a solid parenchyma. Additionally, in two newly described species of *Litosiphon*, Lund observed intercalary series of plurilocular sporangia in juvenile filaments, which showed some resemblance to *Stictyosiphon soriferus*.

Pedersen (1976a) demonstrated that fertility in the erect plants of several members of the Dictyosiphonales is not dependent on size or age but is simply a matter of environmental conditions. Thus, a precocious production of sporangia is a routine event for such genera as *Stictyosiphon, Myriotrichia, Litosiphon* and *Coelocladia*.

Members of the Fucales in particular the fertilized eggs of *Fucus*, have long served as systems for developmental studies, such as the establishment of zygote polarity (Quatrano 1974; Nakazawa 1975). Developmental differences in regard to the number of primary rhizoids, the time of their initiation, and alternative modes of attachment prior to rhizoid initiation have been noted, and these differences have taxonomic implications at the familial and generic levels. A single primary rhizoid is produced by members of the Fucaceae, but two to four or a tuft of primary rhizoids characterize the families Cystoseiraceae and Sargassaceae, respectively (Nizamuddin 1962).

Although the zygote rapidly produces a primary rhizoid (*c.* 14 h) in *Fucus* and *Ascophyllum*, rhizoid initiation is significantly delayed in *Halidrys, Pelvetia* and *Himanthalia*. The zygote initially becomes a multicellular spheroid germling, and several days or weeks may elapse before the rhizoids appear. In the meantime, the zygote is able to remain attached to the substrate by other means. The mesochiton layer of the original oogonial wall functions to attach the zygote of *Pelvetia canaliculata* (Moss 1974; Hardy & Moss 1979). A rigid, adhesive zygote wall serves this same function in *Halidrys siliquosa*, this wall later being shed as a cap when the four primary rhizoids emerge (Hardy & Moss 1978).

An important aspect of development is the relationship between environmental influences and morphological expression. For *Petalonia fascia* and *Scytosiphon lomentaria* an alternation between a macrothallus (a blade or tube) and a crustose or filamentous microthallus has been recognized, but the relationship between these two dissimilar phases has been variously interpreted (Wynne & Loiseaux 1976). The erect thalli of *S. lomentaria* develop in an $8:\overline{16}$ photoperiod, whereas the crustose phase develops in a $16:\overline{8}$ photoperiod. A sharply defined critical daylength between 12 and 13 h is involved in the stimulation of one or the other expression according to Dring & Lüning (1975).

(See also Chapter 9.) Other workers present strong evidence supporting the view that the two dissimilar expressions in some of these algae are correlated with a sexual–asexual alternation (Nakamura & Tatewaki 1975; Clayton 1979).

Certain species of *Sphacelaria* can reproduce either by asexual propagules (Fig. 2.9j) or by the release of motile cells (anisogametes by the gametophytes or zoids by the sporophytes). The mode of reproductive development has been shown to be controlled by temperature and photoperiod (van den Hoek & Flinterman 1968; Colijn & van den Hoek 1971). Propagules are produced at the higher temperatures and under long-day or day-neutral conditions; however, gametangia are produced at lower temperatures or under short-day conditions.

3.4 Cytology

The number of chloroplasts per cell and the presence or absence of pyrenoids have served as reasonably useful cytological features among members of the Phaeophyta. Two basic conditions of the chloroplasts in brown algal cells have been recognized by certain workers: archeoplastidy (i.e. a single chloroplast) and neoplastidy (i.e. multiple chloroplasts). Within this framework eight types of chloroplast arrangements have been depicted (Asensi *et al.* 1977) on the basis of electron microscope observations of Southern Hemisphere representatives, combining such characteristics as archeoplastidy or neoplastidy, presence or absence of pyrenoids, pyrenoids with or without membranous invaginations, and aggregation or randomness of the chloroplasts.

The shape of the chloroplast has also been taken into account: stellate, rod-like, plate-like, discoid or lenticular, or a branching/dissected ribbon. Although such genera as *Splachnidium, Bachelotia* and *Asteronema* all appear to have a stellate chloroplast, a closer examination reveals that *Splachnidium* indeed has a single, axile chloroplast with a single pyrenoid (Asensi & Delépine 1976), but the stellate configuration in the other two genera is derived from numerous rod-shaped chloroplasts radiating from a central point in the cell (Delépine & Asensi 1975; Magne 1976). The many chloroplasts in these latter two genera each contain a pyrenoid at their converging ends, separated by a thin area of cytoplasm crossed by interpyrenoid ducts (Delépine *et al.* 1976).

The phyletic significance of the pyrenoid has been pointed out by Hori (1971, 1972), its presence being regarded as characteristic of the less advanced groups within this algal division. In the more advanced orders (Sphacelariales, Dictyotales, Laminariales and Fucales) a pyrenoid has been detected rarely, sometimes as a rudimentary structure, or present only in reproductive cells (Evans 1968; Chi 1971; Hori & Ueda 1975). Its vestigial nature is evident, and thus the phyletic interpretation still has some validity.

Although the pyrenoid is typically a stalked or exserted structure in those brown algae having it (Dodge 1973), a few examples, including *Petroderma, Splachnidium* and *Scytothamnus*, of a sessile type, in which the pyrenoid is embedded in the chloroplast stroma, have been cited (Asensi *et al.* 1977). Intrapyrenoidal invaginations of membranes have also been noted in a few genera, but the significance of this is not fully understood (Delépine *et al.* 1976; Asensi *et al.* 1977).

In a monograph of the Ectocarpaceae of the English Channel, Cardinal (1964) employed chloroplast characteristics as the primary criterion for differentiating the various genera in his key. The degree of chloroplast variation expressed among genera of the Ectocarpales, Chordariales and Dictyosiphonales suggests the heterogeneity of these orders (Asensi *et al.* 1977). In orders such as the Sphacelariales, Dictyotales, Laminariales, and Fucales, the chloroplasts are invariably present in the cells as many small discs. Feldmann (1949) separated a group of genera from the Dictyosiphonales *s. l.* which have a single parietal chloroplast with a conspicuous pyrenoid and recognized the group as the Scytosiphonales almost two decades before the distinctive nature of the life history was realized. Other significant observations on the Phaeophyta based upon cytological methods are covered in greater detail in Chapter 7.

4 SURVEY OF SELECTED ORDERS

The brown algal orders treated below were selected to highlight some of the recent trends affecting systematics and morphology.

4.1 Ectocarpales

A persistent problem in the systematics of this order has been appreciating the potential for morphological variation in these algae. The pleiomorphism of some of these genera is a common theme in the literature, e.g. *Pilayella* (Russell 1963) and *Feldmannia* (Knoepffler-Peguy 1970). A balanced perspective can be achieved by a comparison of careful morphological investigations (e.g. Kuckuck's classic series 'Ectocarpaceen-Studien') and equally precise culturing studies under controlled conditions (e.g. Russell's studies on both free-living and attached forms of *Ectocarpus*). On the basis of these studies Russell (1966, 1967) was able to conclude that only two species of *Ectocarpus* (*E. siliculosus* and *E. fasciculatus*) occurred in Britain despite the fact that many other species had been recorded.

Another example of synthesizing field collection studies with observations made on living material is Clayton's (1974) work with members of this order occurring in southern Australia. She discussed a '*Feldmannia globifer–F. simplex* species complex' and included other species in synonymy. A total of 34 collections of this group were analysed for filament diameter and the length:width ratio of the plurilocular sporangia. The results showed a continuous range of variation in these two characteristics, causing her to conclude that *Feldmannia globifer* and *F. simplex* are not distinct species. The culture studies corroborated this conclusion and provided evidence that the variation was genetically based. On the basis of culture experiments, Knoepffler-Peguy (1972, 1974, 1977) offered evidence that *Acinetospora* represented a morphological state of either *Feldmannia* or *Giffordia* that was expressed independently of any alternation of nuclear or morphological phases.

In a recent numerical taxonomic treatment of *Pilayella littoralis*, a cosmopolitan species, Bolton (1979) analysed 163 plants from a wide range of habitats in Britain and examined 74 morphological and ecological traits by computer clustering techniques. The morphological variation caused by salinity (estuarine *versus* marine), substrate (epiphytic *versus* epilithic), exposure, seasonality and aging were all reflected in the clustering patterns. But Bolton concluded that it was not possible to distinguish any reliable taxonomic entities within this highly variable species.

Russell (1978) reviewed British taxa of Ectocarpaceae and concluded on the basis of his numerical taxonomic analysis that there has been an over-liberal recognition of genera. Indeed, along these same lines Russell & Garbary (1978) stated that all members of the British Ectocarpaceae could be reasonably accommodated within a single broadly defined genus, namely, *Ectocarpus*. However, most recent workers continue to uphold the integrity of most genera in this family and would not support the contention of Ravanko (1970) that many Ectocarpacean genera and species merely represent developmental or environmental stages of a limited number of species.

In Russell's (1978) studies of British Ectocarpaceae two major categories emerged based simply on thallus size: microscopic or macroscopic. Criteria that have been traditionally used in separating genera of filamentous brown algae include chloroplast morphology, position of reproductive structures, presence or absence of monosporangia, and presence or absence of crampons, true hairs, or pseudo-hairs. In regard to hairs in brown algae, a very different type of hair is known, distinct from the typical phaeophycean hair with its characteristic basal meristem. This unusual type of hair has a long and proximally constricted basal cell with a meristematic zone of two to four barrel-shaped cells above it. This distinctive hair is known in three different forms belonging to three separate families: *Phaeostroma pustulosum* (Ectocarpaceae), *Symphyocarpus longisetus* (but not for *S. strangulans*!) (Ralfsiaceae), and *Halonema subsimplex* (Chordariaceae). In Lund's opinion (1959) this feature should not be given excessive taxonomic importance, and he cautioned against placing such forms into a single family merely on the basis of this unusual hair type.

Müller attempted to answer the question whether a species such as the widely occurring *Ectocarpus siliculosus* is actually capable of inter-breeding throughout its range. Although his initial study (Müller 1976) revealed a sterility barrier existing between Mediterranean and New England strains of this species, subsequent studies demonstrated interfertility among European strains and between these strains and those from North Carolina, Florida, and southern

Australia (Müller 1979). Only strains from Massachusetts showed complete sexual isolation; strains from Texas showed partial isolation. (See also Chapter 19.)

With reference to the Ralfsiaceae, more evidence has accumulated to argue against the separation of these crustose brown algae into a separate order. The criteria that were employed, namely, an alleged isomorphic life-history pattern, the presence in the cells of a single parietal plate-like chloroplast lacking a pyrenoid, and a 'discal type' of germination have been shown to be untenable on the basis of investigations of the ten members of this family occurring in the British Isles (Fletcher 1978). Various life-history investigations afford no real case for a sexual process, excepting *Nemoderma tingitana*, despite the usual presence of separate plants bearing unilocular sporangia and plurilocular structures. The available evidence suggests that the unilocular sporangia are apomeiotic and that development is of the direct type (Loiseaux 1968; Edelstein *et al.* 1971; Fletcher 1978).

A single parietal chloroplast without a pyrenoid characterizes *Ralfsia*, but many discoid chloroplasts occur in the cells of *Pseudolithoderma*. Although the germlings of the various representatives generally develop into small discs, different ontogenetic patterns have been observed (Fletcher 1978). The initial phase was either a circular disc with an empty spore wall attached, as in *Ralfsia verrucosa*, or an elongate filament lacking any differentiated original spore wall, as in *Pseudolithoderma roscoffensis*. Furthermore, neither pattern of germination is unique to members of this group. One is left with the conclusion that these crustose genera are best retained as a family within the Ectocarpales or that the group includes some genera that would fit better elsewhere.

Finally, the Sorocarpaceae was recently established on the basis of the occurrence of sympodial branching, the termination of the filaments by a phaeophycean hair, and the clustering or fusion of the plurilocular sporangia into a sorus (Fig. 2.9g) (Pedersen 1977). Except for its plurilocular structures not being fused into sori, the gametophyte of *Hummia onusta*, which is placed in the Dictyosiphonales, similarly shows sympodial axes terminating in phaeophycean hairs (Fiore 1977).

4.2 Sphacelariales

The British members of this reasonably discrete order were subjected to numerical taxonomic methods by Russell & Fletcher (1975), and their work was subsequently amplified by Prud'homme van Reine (1978), who expanded the project to include all European taxa of this order. Russell & Fletcher's studies demonstrated the basic integrity of the order except for the 'outlying' genera *Buttersia, Choristocarpus* and *Sphacella*. (*Sphacella* was later excluded from the British flora.) The crustose *Battersia* and the branching, filamentous *Choristocarpus* are admittedly rather atypical by Sphacelarialean standards. Yet the results of the similarity matrix comparing all taxa of British Sphacelariales placed *Battersia* and *Choristocarpus* in closest proximity to species of *Sphacelaria*.

Prud'homme van Reine (1978) proposed an outline of a system of classification for the European taxa that was based primarily on morphological traits. Using more than 100 traits in one of his sets, he then proceeded to analyse these taxa by several numerical methods, in which the intuitive weighting of characters is deliberately absent. He concluded that at least for the large genus *Sphacelaria* numerical taxonomy corroborated the standard taxonomic practice of weighting characters. He also admitted the possible merit in Sauvageau's earlier suggestion that *Halopteris* and *Stypocaulon* should be merged.

The similarity matrix analysis by Prud'homme van Reine showed *Sphacella subtilissima* to be placed most closely with the Choristocarpaceae, whose members also have axes in which the cells do not undergo longitudinal segmentation. An Australian species of this genus, *S. ferruginea*, was transferred (Clayton 1974) to the Ectocarpales, despite its distinctive apical growth, whereas Delépine & Asensi (1975) included this same species in their new genus *Asteronema*, which they related to the Ectocarpales.

4.3 Desmarestiales

The extreme pleiomorphism in the ligulate forms of *Desmarestia* occurring on the Pacific coast of North America was recognized by the statistical methods of Chapman (1972). He followed several morphological traits, such as main axis width, overall length, holdfast diameter, primary branch density and number of branch orders, in natural populations in relation to seasonal changes. A continuous overlapping series of these former diagnostic criteria became evident. Chapman therefore reduced the ten taxa that had been recognized to two species, one with two varieties. A more workable taxonomic treatment was thus achieved.

A different approach to solving taxonomic problems is illustrated by the discovery by Moe & Silva (1977) of juvenile stages of *Himantothallus*, a large seaweed occurring in Antarctic waters. This genus had been thought to be a kelp and was even recognized as a separate family and possible order (Zinova 1959). However, the germlings and juvenile plants had the trichothallic meristem and the pseudoparenchymatous construction characteristic of the Desmarestiales, and the transfer was accordingly effected. Moe & Silva also reduced to synonyms of *Himantothallus* two other monotypic genera (*Phaeoglossum* and *Phyllogigas*) from the Antarctic.

4.4 Dictyosiphonales

Observations of algal clones under *in vitro* conditions have allowed us to better appreciate the full potential for morphological variation in a given species. An example of this approach is Pedersen's (1978a) study of two Danish clones of *Myriotrichia*. He found that the three species that had been previously recognized could be reduced to a single species (*M. clavaeformis*); in addition, fertile, prostrate systems of these clones greatly resembled species of *Streblonema* (Ectocarpales).

The presence or absence of hairs, which has been used as a criterion at the generic level within the Ectocarpales, was employed by Pedersen (1978a) in resurrecting *Pogotrichum* as a separate taxon from *Litosiphon*. Hairs are not produced by *P. filiforme* either in nature or in culture, whereas they are commonly produced in *Litosiphon*. In addition, a vegetative cell is converted into a single sporangium (unilocular or plurilocular) in *Pogotrichum*; however, in *Litosiphon* an individual vegetative cell undergoes vertical and transverse divisions to form the cells that in turn initiate the sporangia. Pedersen attached sufficient significance to these differences to call for the establishment of the family Pogotrichaceae as distinct from the Punctariaceae, in which *Litosiphon* remains. Evidence was offered to suggest that *Omphallophyllum*, also hairless and with a similar direct conversion of vegetative cells into sporangia, is probably best assigned to the Pogotrichaceae.

Pedersen (1976b) has asserted that *Coilodesme* should be transferred from the Dictyosiphonaceae to the reinstated Coilodesmaceae of Setchell and Gardner. The justification for this proposal is that the mode of early development in the formation of the macrothallus of *Coilodesme* is very different from that of *Dictyosiphon*.

Descriptive work continues in this order with the appearance of new genera such as *Platysiphon* (Wilce 1962) and *Melanosiphon* (Wynne 1969) and new species in *Punctaria* (Iwamoto 1960; Umezaki 1961) and *Stictyosiphon* (Taylor 1975), for example.

An additional family that has been set up in this order is the Coelocladiaceae (Pedersen 1976a). *Coelocladia* was separated from the Delamareaceae because of the unusual branching system of the sporangia, giving them a clustered or crown-like appearance, and the occasional sympodial branching of the primary filament.

4.5 Scytosiphonales

Fresh reports verify that the life history for at least some members of this order consists of an alternation of dissimilar phases. Fletcher (1978) demonstrated that in Britain unispores from *Ralfsia clavata* developed into germlings that eventually produced blades identifiable as *Petalonia fascia* and that unispores from *Microspongium gelatinosum* ultimately produced narrow, terete thalli identifiable as *Scytosiphon lomentaria*. These two distinct expressions of an erect thallus (foliose or terete, depending on the genus) and a basal microthallus (crust, disc, branched filaments, knot-filaments) appeared to be facultatively related rather than dependent on a sexual link. Winter-like growth conditions enhanced the production of erect thalli; summer-like conditions enhanced the production of the microthalli, especially the crust. These results supported those of earlier workers, such as Roeleveld *et al.* (1974) and others.

On the other hand, the contention that these two phases are connected by a sexual process, which has also been maintained by certain workers, has received support. Clayton (1979) observed that a single late-winter collection of *Colpomenia peregrina* in Australia yielded sexual thalli (anisogamous and dioecious) and that a sporophytic microthallus bearing unilocular sporangia was developed, its unispores subsequently returning the saccate *C. peregrina*. Most of the populations of *C. peregrina* examined, however, behaved asexually.

From their culture studies of members of this family occurring in Japan, Nakamura & Tatewaki (1975) concluded that the microthalli (i.e. the phase bearing unilocular sporangia) could be haploid or diploid for *Colpomenia bullosa, Endarachne binghamiae, Petalonia zosterifolia*, and *Scytosiphon lomentaria*. Clayton (1979) expressed the same opinion for *Colpomenia peregrina* in Australia. She regarded the macrothallus as a potentially dioecious gametophyte. One difference between the results of these workers is that Nakamura and Tatewaki observed that the parthenogametes for the four Japanese species underwent different developmental pathways (namely, macrothallus formation is favoured by relatively cool conditions and microthallus formation is favoured by relatively warm conditions, but Clayton observed no such environmental control of the proportion of the two expressions developed from parthenogametes.

Clayton (1978) has shown that Australian *Scytosiphon lomentaria* is asexual and that its microthallus is more myrionematoid in appearance than ralfsioid. This microthallus stage can produce both unilocular and plurilocular sporangia. It was not possible for her to distinguish any subspecific taxa for this species.

Although the genera in the Scytosiphonaceae seem to be easily distinguishable on the basis of their type species, some of the problems that arise with generic delimitations are apparent and have been discussed by Wynne & Norris (1976). The morphological and ecological characteristics of four species of *Colpomenia* from the Gulf of California, Mexico, were analysed in their study. The type species, *C. sinuosa*, is roughly spherical, and this shape is generally regarded as the trademark for the genus. However, *C. tuberculata* by its thick, warty structure resembles *Iyengaria*; *C. phaeodactyla* with its finger-like sacs resembles *Scytosiphon*; and *C. ramosa* by its branching habit resembles *Rosenvingea*. But the justification for assigning these species to *Colpomenia* was presented by Wynne and Norris, namely, that *Colpomenia* includes forms ranging from spherical to saccate to prostrate and that the erect cylindrical sacs formed in some species (*C. bullosa* and *C. phaeodactyla*) arise as hollow evaginations that push outward from the prostrate basal system. *Utriculidium*, a Southern Hemisphere genus whose status has been uncertain, has been re-collected and reported on by Acleto (1973).

Two species of *Colpomenia* occurring in Australia were investigated statistically by Clayton (1975). Several traits that had been used to distinguish *C. sinuosa* from *C. peregrina*, such as form of thallus, colour, brittleness, sporangial length, and wall thickness, were found to be insufficient for taxonomic value. But the shape of the sori and the presence or absence of a cuticle over the plurilocular sporangia were noted to adequately separate the two species.

4.6 Laminariales

Among the representatives of this order, the genus *Laminaria* has continued to be the focus of attention of many recent workers, owing to its widespread distribution and the unsettled nature of its species delimitations. On the basis of extensive collections from the North Atlantic, Wilce (1960, 1965) concluded that mucilage duct distribution in *Laminaria* is both genetically and environmentally controlled but not a specific characteristic. Within the Simplices section, for example, he used it to differentiate ecotypes within two species (*L. longicruris* and *L. saccharina*). Each of these species was regarded as having three ecotypes: those with ducts in blade and stipe, those with ducts in blade but not in stipe, and those without ducts in blade or stipe. Wilce attributed the presence or absence of the mucilage ducts, or 'canals', to water temperature, an idea supported by Burrows (1964).

On the other hand, Druehl (1968) separated the species of *Laminaria* occurring in the northeast Pacific by such criteria as, among others, presence or absence of mucilage ducts, their occurrence in blade and/or stipe, and their location near the periphery of the stipe or in a more deep-seated zone. Subsequent efforts have been carried out to re-evaluate these criteria and others by testing them experimentally with interfertility tests and transplant experiments and by growing the plants under uniform conditions. Chapman (1978) has recently reviewed the current status of this research not only in reference to *Laminaria* but also to other genera in this order.

Chapman (1973) offered evidence for the conspecificity of two reported species of *Laminaria* with simple blades occurring in Nova Scotia. He took into account both seasonal changes and spatial relationships (effects of depth). Although all young plants have solid stipes, he observed that decreasing exposure was correlated with an increasing occurrence of hollow stipes and increased stipe length for mature plants. No differentiation between mature plants was observable on the basis of depth. The variation was overlapping, and the two putative species could not be separated.

In other experiments involving interfertility tests, Chapman (1974, 1975) demonstrated that members of populations from the extremes of the morphological range were able to cross and produce fertile hybrids. He observed a clinal differentiation of stipe traits (hollow v. solid) in the populations. The research also indicated a significant influence of the environment on the nature of mucilage duct development. Mucilage duct types could not be applied to the biological species concept and thus was effectively eliminated as a suitable taxonomic characteristic. Chapman confirmed his earlier conclusion of the conspecificity between taxa in the Simplices section of *Laminaria* from eastern Canada, with the exclusion of *L. solidungula*.

To further verify the interfertility of these same taxa, successful crosses were carried out, involving plants of *Laminaria longicruris* from eastern Canada and two forms of European *L. saccharina*. Hybrid sporophytes were obtained. Mucilage ducts occurred in blades of the offspring of the European strains, whereas ducts were absent in the Canadian strains. All the hybrid sporophytes between these two parents lack mucilage ducts (Lüning *et al.* 1978).

Two blade forms of *Laminaria saccharina* are present in Europe: bullate and non-bullate. The former occurs along the Atlantic coast, whereas the latter is restricted to the Baltic Sea, the Swedish west coast, and Helgoland in the North Sea. Lüning (1975) demonstrated the interfertility of these two forms and observed that the hybrid sporophytes produce a bullate blade, which he interpreted to be a dominant trait.

In a study of the Simplices section of *Laminaria* occurring in the Shetland Isles, Kain (1976) observed blades with long, hollow stipes, attributable to *Laminaria faeroensis*. She noted that these populations were restricted to sheltered habitats and shallow water, and that these plants intergraded with solid-stipe specimens identifiable as *L. saccharina*. An increase in stipe length, accompanied by the appearance of a hollow stipe and increased stipe diameter, was correlated with sheltered situations. The most plausible explanation in her viewpoint was that only a single species is involved, which is affected by an inflow of genes causing the development of long stipes under fairly sheltered conditions and hollow stipes in extreme shelter. She called for experiments to test this hypothesis.

Laminaria sinclairii and *L. longipes*, members of the Simplices section and occurring in the North Pacific, both produce multiple stipes from an extensive rhizomatous holdfast. The main distinction is the presence of mucilage ducts in the stipes of *L. sinclairii* and their absence in those of *L. longipes*. In light of the previous discussion regarding the suspicious validity of mucilage ducts as a criterion in distinguishing species, Markham (1972) carried out reciprocal transplant experiments and determined that none of the plants showed any change in mucilage ducts from the original condition. He also detected five additional differences, dealing with blade width, seasonal loss of blades, morphology of gametophytes, temperature tolerances, and wave exposure of the respective habitats. The conclusion of this study was to retain *L. sinclairii* and *L. longipes* as two distinct species.

Less controversy marks the delineation of species of the Digitatae section of *Laminaria*, perhaps because there have been fewer experimental analyses of the taxonomic criteria applied to these species. Druehl (1979) has stressed that the characteristic blade-splitting occurs during young developmental stages. He also noted that for at least the Pacific plants the stipes of the Digitatae section are sufficiently strong to hold the blade away from the substratum but that in the Simplices section the stipes are quite flexible.

Two similar digitate species, *Laminaria digitata* and *L. hyperborea*, both of the North Atlantic, occupy different depth zones in the sublittoral region, older plants of *L. digitata* occurring not more than 2 m deep and *L. hyperborea* inhabiting the deeper zone of 8 m (Lüning 1979). Less frequently, these two species are present in sheltered sites, and under such conditions of calm, deep water they independently produce a blade of similar morphology called f. *cucullata*, which at one time was regarded as a distinct species. Hybridization tests and transplant experiments confirmed that this form is merely an ecad for each of these species (Svendsen & Kain 1971). In a compilation of information concerning *L. hyperborea*, Kain (1971) has provided a detailed account of the genus as a whole. In her key she cautioned against dependence upon the use of mucilage duct characteristics.

In reference to the genus *Agarum*, Nakahara & Yamada (1974) were able to determine that the four morphologically distinct forms of *A. cribrosum* occurring in northern Japan were interfertile. A 73% average interfertility was demonstrated, the range running from 32 to 96%.

Searles (1978) analysed subtidal populations of *Lessonia* from southern South America. He compared such features as the type of blade margin; blade width, length and thickness; depth of collection; angle of blade base; and the solid *versus* lacunate nature of the cortex. From scatter diagrams of these characteristics, he was able to distinguish two species, which were usually separated spatially by a bed of *Macrocystis*. The shallow-water species, *Lessonia vadosa*, has narrow blades with many small marginal dentations and a lacunate cortex; the deep-water species, *L. flavicans*, has wider blades with margins ranging from entire to coarsely serrate and a solid cortex. Both species have a single stipe arising from the holdfast system. Searles also recognized a third species in the study area, *L. nigrescens*, which differs from the other two in having numerous stipes arising from a massive holdfast.

Alaria, a kelp widely distributed in the temperate waters of the Northern Hemisphere, has been subjected to a taxonomic revision by Widdowson (1972a), who was able to distinguish 14 species on the basis of external morphology. By means of statistical techniques, Widdowson (1972b) treated ten of the 14 species. He was able to distinguish five from each of the other nine species; however, intergrades existed between *Alaria marginata* and *A. tenuifolia*. Three other species, all occurring in Japan, also proved to be indistinguishable on the basis of this type of analysis. On Soviet shores Petrov (1973) recognized only five species of *Alaria*, two of which (*A. angusta* and *A. marginata*) were broadly defined to include a number of species regarded as distinct by Widdowson (1972a).

In his treatment of the members of the Laminariales and Fucales occurring in Soviet seas, Petrov (1974) established the Arthrothamnaceae for *Arthrothamnus*, a genus that is usually assigned to the Laminariaceae (Papenfuss 1951a). *Costularia, Feditia* and *Phyllariella*, three western Pacific genera that had been previously described, were included in Petrov's concept of the Laminariaceae.

4.7 Fucales

Members of the Fucales have provided useful tools for attempting to understand the effect of the environment on morphological variation, the traditional source of taxonomic bewilderment. Using three British examples (*Halidrys siliquosa, Pelvetia canaliculata* and *Fucus vesiculosus*), Russell (1978) presented a series of scatter diagrams comparing combinations of morphological traits (e.g. frond length and width, number of dichotomies, and number of vesicles) with exposure or habitat (sublittoral *versus* littoral rock pool). The correlation of morphological variation with change in environmental parameters was readily apparent. Russell also raised the question of the differing goals of classical and experimental systematics and the lack of consistency in the use of infraspecific taxa (ecotype, ecad, forma and varietas).

The highly pleiomorphic nature of *Fucus* has been recognized by several authors over the years. Powell (1963) adopted a fairly conservative treatment of this genus, recognizing only six species but these with infraspecific categories. Part of the phenotypic variability is the result of the influence of salinity and exposure differences as has been shown for the degree of vesiculation and branching in *F. vesiculosus* on the coast of Maine (Jordan & Vadas 1972). The alleged ease of hybridization among these species was also regarded as contributing to this pleiomorphism (Burrows & Lodge 1951, 1953). However, recent studies of two dioecious species of *Fucus* and *Ascophyllum nodosum* demonstrated that inherent barriers exist preventing inter-generic or inter-specific cross-fertilization (Bolwell *et al.* 1977). It was admitted that these barriers became less pronounced for freshly released eggs from plants that had been stored for five to seven days, suggesting a break-down in the membrane receptors of the older egg cell. It has been speculated, as they noted, that the specificity in sexuality is dependent on surface-localized complementary macromolecular recognition phenomena. (See also Chapter 19.)

Several morphological and taxonomic treatments of other genera have been published in recent years: *Turbinaria* (Taylor 1963), *Cystophora* (Womersley 1964), *Myriodesma* (Nizamuddin & Womersley 1966), *Cystoseira* and *Cystophyllum* (Papenfuss & Jensen 1967), and Cystoseiraceae and Sargassaceae (Jensen 1974). Roberts (1978) presented an in depth study of *Cystoseira*, a large genus of more than 50 species and particularly well represented in the Mediterranean and the adjacent Atlantic coastline. She arranged most of the species into five fairly well defined groups and furnished evidence that the genus is undergoing active speciation.

4.8 Durvillaeales

Hay (1979) reported on the taxonomy of *Durvillaea*, the only genus in this Southern Hemisphere order. *Durvillaea*, rather than *Durvillea*, was held to be the proper spelling. Of the four species recognized, *D. antarctica* is the most variable. It is separated from the other species by a distinctive honeycomb-like tissue permeating its blade.

5 FORMULATION OF CLASSIFICATION SCHEMES

From about 1890 into the 1930's most of the fundamental information emerged, upon which present-day classification of the brown algae is based, thanks to the efforts of phycologists like Kjellman, Sauvageau, Kuckuck, Oltmanns, Setchell and Gardner, and Kylin. By the time Kylin introduced the concept of three classes distinguished by the pattern of the life history in 1933, most of the currently recognized phaeophycean orders had already been established. Relatively few major changes have subsequently been made, causing Russell (1973) to refer to brown algal taxonomy as being in a state of 'stagnation'. Within this taxonomic framework, our understanding of both ordinal and supra-ordinal taxa arises from a combination of life history strategies, types of sexuality, modes of growth, and the type of vegetative organization.

But just as the fluctuations of tide and time affect these organisms in their natural setting, so also do the human notions about their presumed systematic relationships exhibit a cyclical nature. Within the recent past, several new orders have been proposed, such as the Scytosiphonales, Delamareales, Durvillaeales, Ralfsiales and Ascoseirales, representing decisions to segregate groupings that came to be regarded as sufficiently distinct. Yet other workers have simultaneously presented data and arguments to define certain ordinal limits more broadly. Thus, in the recent scheme of Russell & Fletcher (1975) the Ectocarpales would encompass the Chordariales, Ralfsiales, Dictyosiphonales, Tilopteridales and Scytosiphonales of other authors, therefore including forms with isomorphic and heteromorphic life histories as well as with nonparenchymatous and parenchymatous constructions. On the basis of numerical taxonomy Russell & Fletcher (1975) argue that some species that had been regarded as Ectocarpales in the original restricted sense (e.g. *Hecatonema maculans, Microspongium globosum, Ralfsia californica, R. clavata* and *Streblonema sphaericum*) have been identified to be microthalli of members of the Dictyosiphonales and Scytosiphonales, whereas other species (or apparently even strains of the same species) seem to lead completely autonomous lives. These same authors incorporated the Sporochnales within the Desmarestiales. But other authors, like Pedersen (1978a), counter-argue that parenchyma formation or lack of parenchyma formation is a valid criterion for distinguishing some of these orders from the Ectocarpales.

A modification of Kylin's tripartite scheme (Isogenerateae, Heterogenerateae and Cyclosporeae) was offered by Papenfuss (1951b), who recognized the Ectocarpales as the possible ancestral stock from which various other brown algal groups evolved. This close agreement between the Ectocarpales and the Chordariales, Dictyosiphonales, and Scytosiphonales was later reiterated by Loiseaux (1972). Perhaps the relationship suggested in Fig. 2.10 from van den Hoek & Jahns (1978) even more aptly depicts the central position of the Ectocarpales and the lack of any clear-cut separation from the Chordariales, Dictyosiphonales and Scytosiphonales. It also does justice to the presumably separate evolutionary developments of oogamy from isogamy, heteromorphic life histories from isomorphic patterns, and parenchymatous construction from non-parenchymatous construction.

Of the various classificatory schemes offered in recent years, the main difference has been how the orders are clustered together into subclasses. In the tradition of Kylin and other earlier students of the brown algae, Chadefaud (1960) based his primary distinction on the presence or absence of an alternation of generations, the class Cyclosporeae (= Cyclosporées) including only the Fucales, the order generally conceded to be the most distinct among the members of the Phaeophyta. His class Phéosporées, or Phaeosporeae, was subdivided into three assemblages on the basis of degree of specialization of the gametophytic and sporophytic phases: the Proto-phéosporées showing the least specialization in the two phases; the Eu-phéosporées showing significant specialization only in the sporophyte, and the Para-phéosporées showing comparable specialization in the two phases. His scheme is presented in Table 2.1.

Nakamura (1972) employed a different emphasis in his subdivision of the class Phaeosporeae, using three sub-classes that were recognized on the mode of construction (non-parenchymatous, parenchymatous or both modes), whereas the life-history pattern was used at the ordinal level within this framework. The arrangement proposed by Nakamura is shown in Table 2.1.

Perestenko (1972) stressed germination patterns and the early stages of development to trace phylogenetic relationships and offered yet another realignment of these orders (Table 2.1). It is evident that a great discrepancy exists between the system proposed by Perestenko and those of the other authors, who maintain a solitary position for the Fucales. Perestenko's system also recognized the orders Ascoseirales and Durvillaeales, groups which are incorporated within the Fucales by both Chadefaud and Nakamura.

These various schemes with their differing emphases and their ever-changing

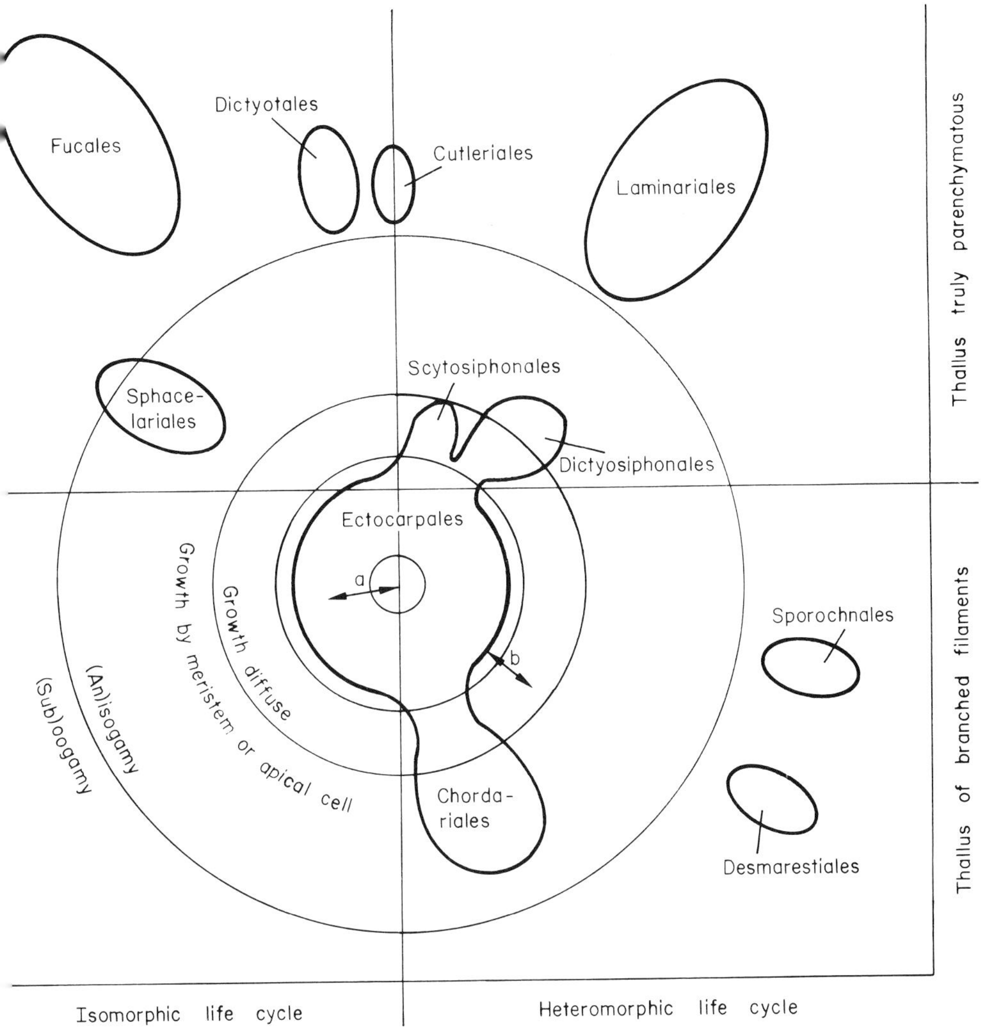

Fig. 2.10 Distribution of characteristics and interrelationships among the brown algae. (a) Within the circle no clear differentiation between creeping and erect filaments, outside the circle this distinction is present; (b) within this circle the thallus is filamentous, outside the circle the thallus is (pseudo-)parenchymatous. (Redrawn and permission obtained from van den Hoek & Jahns 1978.)

perspectives appear to point to a capricious and unstable classification system. Instead, I feel that they reflect a dynamic, malleable attitude of the workers toward the art of classifying. It is a healthy sign that the basic tenets of any taxonomic system are continually scrutinized and critically reviewed. In short, progress is being made.

Table 2.1 Classification Schemes of the Phaeophyta.

Chadefaud (1960)	Nakamura (1972)	Perestenko (1972)
Class Phéosporées	**Class Phaeosporeae**	**Class Phaeosporophycidae**
Sub-class Proto-phéosporées	Sub-class Haplostichidae	Sub-class Ectocarpophycidae
O. Ectocarpales	*O. Ectocarpales*	*O. Ectocarpales*
Sub-class Eu-phéosporées	*O. Chordariales*	*O. Chordariales*
O. Chordariales	*O. Ralfsiales*	*O. Scytosiphonales*
O. Scytosiphonales	*O. Desmarestiales*	*O. Dictyosiphonales*
O. Dictyosiphonales	*O. Sporochnales*	*O. Cutleriales*
O. Desmarestiales	Sub-class Haplopolystichidae	Sub-class Laminariophycidae
O. Sporochnales	*O. Scytosiphonales*	*O. Desmarestiales*
O. Laminariales	*O. Dictyosiphonales*	*O. Sporochnales*
Sub-class Para-phéosporées	*O. Laminariales*	*O. Laminariales*
O. Tilopteridales	Sub-class Polystichidae	
O. Sphacelariales	*O. Tilopteridales*	**Class Cyclosporophycidae**
O. Cutleriales	*O. Sphacelariales*	*O. Tilopteridales*
O. Dictyotales	*O. Cutleriales*	*O. Sphacelariales*
	O. Dictyotales	*O. Dictyotales*
Class Cyclosporées		*O. Ascoseirales*
O. Fucales	**Class Cyclosporeae**	*O. Durvillaeales*
	O. Fucales	*O. Fucales*

6 REFERENCES

ACLETO O. C. (1973) Las algas marinas del Perú. *Bol. Soc. Peruana Bot.*, **6**, 1–164.

ASENSI A. O. & DELÉPINE R. (1976) Ecophysiologie et cycle de développement de la Phéophycée australe *Splachnidium rugosum* (L.) Grev. *Bull. Soc. Phycol. Fr.*, **21**, 35–47.

ASENSI A. O., DELÉPINE R. & GUGLIELMI G. (1977) Nouvelles observations sur l'ultrastructure du plastidome des Phéophycées. *Bull. Soc. Phycol. Fr.*, **22**, 192–205.

BOLTON J. J. (1979) The taxonomy of *Pilayella littoralis* (L.) Kjellm. (Phaeophyta, Ectocarpales) in the British Isles, a numerical approach. *Br. Phycol. J.*, **14**, 317–25.

BOLWELL G. P., CALLOW J. A., CALLOW M. E. & EVANS L. V. (1977) Cross-fertilisation in fucoid seaweeds. *Nature (Lond.)*, **268**, 626–7.

BURROWS E. M. (1964) An experimental assessment of some of the characters used for specific delimitation in the genus *Laminaria*. *J. Mar. Biol. Assn. U.K.*, **44**, 137–43.

BURROWS E. M. & LODGE S. M. (1951) Autecology and the species problem in *Fucus*. *J. Mar. Biol. Assn. U.K.*, **30**, 161–76.

BURROWS E. M. & LODGE S. M. (1953) Culture of *Fucus* hybrids. *Nature*, **172**, 1009–10.

CAPLIN S. M. (1968) *Fucus* life cycle: time to correct the text books. *BioScience*, **18**, 193.

CARAM B. (1965) Recherches sur la reproduction et le cycle sexué del quelques Phéophycées. *Vie et Milieu*, **16** (1A), 21–226.

CARDINAL A. (1964) Étude sur les Ectocarpacées de la Manche. *Beihefte z. Nova Hedw.*, **15**.

CHADEFAUD M. (1960) Les végétaux non vasculaires (Cryptogamie). In *Traité de Botanique Systematique*, vol. 1, eds. Chadefaud M. & Emberger L. Masson et Cie, Paris.

CHAPMAN A. R. O. (1972) Morphological variation and its taxonomic implications in the ligulate members of the genus *Desmarestia* occurring on the west coast of North America. *Syesis*, **5**, 1–20.

CHAPMAN A. R. O. (1973) Phenetic variability of stipe morphology in relation to season, exposure, and depth in the non-digitate complex of *Laminaria* Lamour (Phaeophyta, Laminariales) in Nova Scotia. *Phycologia*, **12**, 53–7.

CHAPMAN A. R. O. (1974) The genetic basis of morphological differentiation in some *Laminaria* populations. *Mar. Biol.*, **24**, 85–91.

CHAPMAN A. R. O. (1975) Inheritance of mucilage ducts in *Laminaria* (section Simplices) in eastern Canada. *Br. Phycol. J.*, **10**, 219–23.

CHAPMAN A. R. O. (1978) Experimental and numerical taxonomy of the Laminariales: a review. In *Modern Approaches to the Taxonomy of Red and Brown Algae*, eds. Irvine D. E. G. & Price J. H., pp. 423–32. Academic Press, London.

CHI E. Y. (1971) Brown algal pyrenoids. *Protoplasma*, **72**, 101–4.

CLAYTON M. N. (1974) Studies in the development, life history and taxonomy of the Ectocarpales (Phaeophyta) in southern Australia. *Aust. J. Bot.*, **22**, 743–813.

CLAYTON M. N. (1975) A study of variation in Australian species of *Colpomenia* (Phaeophyta, Scytosiphonales). *Phycologia*, **14**, 187–95.

CLAYTON M. N. (1978) Morphological variation and life history in cylindrical forms of *Scytosiphon lomentaria* (Scytosiphonaceae: Phaeophyta) from southern Australia. *Mar. Biol.*, **47**, 349–57.

CLAYTON M. N. (1979) The life history and sexual reproduction of *Colpomenia peregrina* (Scytosiphonaceae, Phaeophyta) in Australia. *Br. Phycol. J.*, **14**, 1–10.

COLIJN F. & VAN DEN HOEK C. (1971) The life-history of *Sphacelaria furcigera* Kütz. (Phaeophyceae). II. The influence of daylength and temperature on sexual and vegetative reproduction. *Nova Hedw.*, **21**, 899–922.

DAHL A. L. (1971) Development, form and environment in the brown alga *Zonaria farlowii* (Dictyotales), *Bot. Mar.*, **14**, 76–112.

DANGEARD P. (1965) Sur un nouveau genre de Phéophycées: *Giraudyopsis* nov. gen. (*G. stellifer* nov. sp.). *C. R. Acad. Sci. Paris*, **261**, 2699–701.

DELÉPINE R. & ASENSI A. (1975) *Asteronema* nov. gen. nouveau genre de Phéophycée australe. *Bull. Soc. Bot. Fr.*, **122**, 295–304.

DELÉPINE R., ASENSI A. & GUGLIELMI G. (1976) Nouveaux types d'ultrastructure plastidiale chez les Phéophycées. *Phycologia*, **15**, 425–34.

DODGE J. D. (1973) *The Fine Structure of Algal Cells*. Academic Press, London.

DOP A. J. (1979) *Porterinema fluviatile* (Porter) Waern (Phaeophyceae) in the Netherlands. *Acta Bot. Neerl.*, **28**, 449–58.

DRING M. J. & LÜNING K. (1975) A photoperiodic response mediated by blue light in the brown alga *Scytosiphon lomentaria*. *Planta*, **125**, 25–32.

DRUEHL L. D. (1968) Taxonomy and distribution of northeast Pacific species of *Laminaria*. *Can. J. Bot.*, **46**, 539–48.

DRUEHL L. D. (1979) On the taxonomy of California *Laminaria* (Phaeophyta). *J. Phycol.*, **15**, 337–8.

EDELSTEIN T., CHEN L. & MCLACHLAN J. (1971) On the life history of some brown algae from eastern Canada. *Can. J. Bot.*, **49**, 1247–51.

EVANS L. V. (1968) Chloroplast morphology and fine structure in British Fucoids. *New Phytol.*, **67**, 173–8.

FELDMANN J. (1949) L'ordre des Scytosiphonales. *Mém. Hist. Nat. Afr. Nord,* Hors-sér., **2**, 103–15.

FIORE J. (1977) Life history and taxonomy of *Stictyosiphon subsimplex* Holden (Phaeophyta, Dictyosiphonales) and *Farlowiella onusta* (Kützing) Kornmann in Kuckuck (Phaeophyta, Ectocarpales). *Phycologia*, **16**, 301–11.

FLETCHER R. L. (1978) Studies on the family Ralfsiaceae (Phaeophyta) around the British Isles. In *Modern Approaches to the Taxonomy of Red and Brown Algae*, eds. Irvine D. E. G. & Price J. H., pp. 371–98. Academic Press, London.

HARDY F. G. & MOSS B. L. (1978) The attachment of zygotes and germlings of *Halidrys siliquosa* (L.) Lyngb. (Phaeophyceae, Fucales). *Phycologia*, **17**, 69–78.

HARDY F. G. & MOSS B. L. (1979) Attachment and development of the zygotes of *Pelvetia canaliculata* (L.) Dcne. et Thur. (Phaeophyceae, Fucales). *Phycologia*, **18**, 203–12.

HAY C. H. (1979) Nomenclature and taxonomy within the genus *Durvillaea* Bory (Phaeophyceae: Durvilleales Petrov). *Phycologia*, **18**, 191–202.

HOLLENBERG G. J. (1969) An account of the Ralfsiaceae (Phaeophyta) of California. *J. Phycol.*, **5**, 290–301.

HOLLENBERG G. J. (1978) Phycological notes VIII. Two brown algae (Phaeophyta) new to California. *Bull. Southern Calif. Acad. Sci.*, **77**, 28–35.

HOOPER R. & SOUTH G. R. (1977) Distribution and ecology of *Papenfussiella callitricha* (Rosenv.) Kylin (Phaeophyceae, Chordariaceae). *Phycologia*, **16**, 153–7.

HORI T. (1971) Survey of pyrenoid distribution in brown algae. *Bot. Mag., Tokyo*, **84**, 231–42.

HORI T. (1972) Further survey of the pyrenoid distribution in Japanese brown algae. *Bot. Mag., Tokyo* **85**, 125–34.

HORI T. & UEDA R. (1975) The fine structure of algal chloroplasts and algal physiology. In *Advance of Phycology in Japan*, eds. Tokida J. & Hirose H., pp. 11–42. W. Junk, The Hague.

IWAMOTO K. (1960) Marine algae from Lake Saroma, Hokkaido. *J. Tokyo Univ. Fish.*, **46**, 21–49.

JENSEN J. B. (1974) Morphological studies in Cystoseiraceae and Sargassaceae (Phaeophyceae) with special reference to apical organization. *Univ. Calif. Pub. Bot.*, **68**, 61.

JORDAN A. J. & VADAS R. L. (1972) Influence of environmental parameters on intraspecific variation in *Fucus vesiculosus*. *Mar. Biol.*, **14**, 248–52.

KAIN J. M. (1971) Synopsis of biological data on *Laminaria hyperborea*. *FAO Fish. Synopsis*, No. 87, Food & Agriculture Organization of the United Nations, Rome.

KAIN J. M. (1976) New and interesting marine algae from the Shetland Isles II. Hollow and solid stiped *Laminaria* (Simplices). *Br. Phycol. J.*, **11**, 1–11.

KNOEPFFLER-PÉGUY M. (1970) Quelques *Feldmannia* Hamel, 1939 (Phaeophyceae-Ectocarpales) des côtes d'Europe. *Vie et Milieu*, Sér. A, **21**, 137–88.

KNOEPFFLER-PÉGUY M. (1972) Comportement de deux espèces suédoises de *Feldmannia* cultivées en diverses conditions de température et de salinité. *Soc. Bot. Fr., Mém.*, **1972**, 101–4.

KNOEPFFLER-PÉGUY M. (1974) Le genre *Acinetospora* Bornet 1891 (Phaeophyceae–Ectocarpales). *Vie et Milieu*, Sér. A, **24**, 43–72.

KNOEPFFLER-PÉGUY M. (1977) Polymorphisme et environment chez *Feldmannia* (Ectocarpacées). *Rev. Algol.*, N.S., **12**, 111–28.

KORNMANN P. & SAHLING P.-H. (1977) Meeresalgen von Helgoland. Benthische Grün-, Braun- und Rotalgen. *Helgol. Wiss. Meeresunters.*, **29**, 1–289.

LOISEAUX S. (1967a) Sur la position systématique du genre *Giraudyopsis* P. Dangeard. *Rev. Gén. Bot.*, **74**, 389–96.

LOISEAUX S. (1967b) Recherches sur les cycles de développement des Myrionématacées (Phéophycées). I-II. Hécatonématées et Myrionématées. *Rev. Gén. Bot.*, **74**, 529–76.

LOISEAUX S. (1968) Recherches sur les cycles de développement des Myrionématacées (Phéophycées). III. Tribu des Ralfsiées. IV. Conclusions générales. *Rev. Gén. Bot.*, **75**, 295–318.

LOISEAUX S. (1972) Variations des cycles chez les Myrionématacées et leur signification phylogenetique. *Soc. Bot. Fr. Mém.*, **1972**, 105–16.

LOISEAUX S. & WEST J. A. (1970) Brown algal mastigonemes: comparative ultrastructure. *Trans. Am. Microsc. Soc.*, **89**, 524–32.

LUND S. (1959) The marine algae of East Greenland. I. Taxonomical part. *Medd. Grönl.*, **156**, 1–247.

LÜNING K. (1975) Kreuzungsexperimente an *Laminaria saccharina* von Helgoland und von der Isle of Man. *Helgol. Wiss. Meeresunters.*, **27**, 108–14.

LÜNING K. (1979) Growth strategies of three *Laminaria* species (Phaeophyceae) inhabiting different depth zones in the sublittoral region of Helgoland (North Sea). *Mar. Ecol.*, Progr. ser. **1**, 195–207.

LÜNING K., CHAPMAN A. R. O. & MANN K. H. (1978) Crossing experiments in the non-digitate complex of *Laminaria* from both sides of the Atlantic. *Phycologia*, **17**, 293–8.

MCLACHLAN J., CHEN L. C.-M., EDELSTEIN T. & CRAIGIE J. S. (1971) Observations on *Phaeosaccion collinsii* in culture. *Can. J. Bot.*, **49**, 563–6.

MAGNE F. (1976) Quelques caractères cytologiques particuliers du *Bachelotia antillarum* (Phéophycées, Ectocarpales). *Phycologia*, **15**, 309–19.

MARKHAM J. W. (1972) Distribution and taxonomy of *Laminaria sinclairii* and *L. longipes* (Phaeophyceae, Laminariales). *Phycologia*, **11**, 147–57.

MOE R. L. & SILVA P. C. (1977) Antarctic marine flora: uniquely devoid of kelps. *Science*, **196**, 1206–8.

MOSS B. L. (1974) Attachment and germination of the zygotes of *Pelvetia canaliculata* (L.) Dcne. et Thur. (Phaeophyceae, Fucales). *Phycologia*, **13**, 317–22.

MÜLLER D. G. (1975) Experimental evidence against sexual fusions of spores from unilocular sporangia of *Ectocarpus siliculosus* (Phaeophyta). *Br. Phycol. J.*, **10**, 315–21.

MÜLLER D. G. (1976) Sexual isolation between a European and an American population of *Ectocarpus siliculosus* (Phaeophyta). *J. Phycol.*, **12**, 252–4.

MÜLLER D. G. (1979) Genetic affinity of *Ectocarpus siliculosus* (Dillw.) Lyngb. from the Mediterranean, North Atlantic and Australia. *Phycologia*, **18**, 312–8.

MUNDA I. (1978) Salinity dependent distribution of benthic algae in estuarine areas of Icelandic fjords. *Bot. Mar.*, **21**, 451–68.

MUNDA I. (1979) A note on the ecology and growth forms of *Chordaria flagelliformis* (O. F. Müll.) C. Ag. in Icelandic waters. *Nova Hedw.*, **31**, 567–91.

NAKAHARA H. & YAMADA I. (1974) Crossing experiments between four local forms of *Agarum cribrosum* Bory (Phaeophyta) from Hokkaido, northern Japan. *J. Fac. Sci., Hokkaido Univ.*, Ser. V (Bot.), **10**, 49–54.

NAKAMURA Y. (1972) A proposal on the classification of the Phaeophyta. In *Contributions to the Systematics of Marine Algae of the North Pacific*, eds. Abbott I. A. & Kurogi M., pp. 147–56. Japanese Soc. Phycology, Kobe.

NAKAMURA Y. & TATEWAKI M. (1975) The life history of some species of Scytosiphonales. *Sci. Papers Inst. Algol. Res., Hokkaido Univ.*, **6**, 57–93.

NAKAZAWA S. (1975) Physiology of *Fucus*. In *Advance of Phycology in Japan*, eds. Tokida J. & Hirose H., pp. 160–70. W. Junk, The Hague.

NEUSHUL M. & DAHL A. L. (1972) Zonation in the apical cell of *Zonaria*. *Am. J. Bot.*, **59**, 393–400.

NIZAMUDDIN M. (1962) Classification and the distribution of the Fucales. *Bot. Mar.*, **4**, 191–203.

NIZAMUDDIN M. & WOMERSLEY H. B. S. (1966) The morphology and taxonomy of *Myriodesma* (Fucales). *Nova Hedw.*, **12**, 373–83.

PAPENFUSS G. F. (1947) Extension of the brown algal order Dictyosiphonales to include the Punctariales. *Bull. Torrey Bot. Club*, **74**, 398–402.

PAPENFUSS G. F. (1951a) Phaeophyta. In *Manual of Phycology*, ed. Smith G. M., pp. 119–58. Chronica Botanica, Waltham, Mass.

PAPENFUSS G. F. (1951b) Problems in the classification of the marine algae. *Svensk Bot. Tidskr.*, **45**, 4–11.

PAPENFUSS G. F. & JENSEN J. B. (1967) The morphology, taxonomy and nomenclature of *Cystophyllum trinode* (Forsskål) J. Agardh and *Cystoseira myrica* (S. G. Gmelin) C. Agardh (Fucales: Cystoseiraceae). *Blumea*, **15**, 17–24.

PEDERSEN P. M. (1976a) Culture studies on marine algae from West Greenland II. *Coelocladia arctica* (Dictyosiphonales, Coelocladiaceae fam. nov.). *Norw. J. Bot.*, **23**, 243–9.

PEDERSEN P. M. (1976b) Marine, benthic algae from southernmost Greenland. *Medd. Grönl.*, **199 (3)**.

PEDERSEN P. M. (1977) *Polytretus reinboldii*, a rare brown alga in culture (Ectocarpales, Sorocarpaceae fam. nov.). *Bot. Notiser*, **130**, 35–40.

PEDERSEN P. M. (1978a) Culture studies on marine algae from West Greenland III. The life histories and systematic positions of *Pogotrichum filiforme* and *Leptonematella fasciculata* (Phaeophyceae). *Phycologia*, **17**, 61–8.

PEDERSEN P. M. (1978b) Culture studies in the pleomorphic brown alga *Myriotrichia clavaeformis* (Dictyosiphonales, Myriotrichiaceae). *Norw. J. Bot.*, **25**, 281–91.

PERESTENKO L. P. (1972) Ontogenesis of Phaeophyta and ontogenetic principle of constructing phylogenetic systems. *Botan. Zhurnal* (*Leningrad*), **57**, 750–64. (In Russian.)

PETROV J. E. (1973) Genus *Alaria* Grev. in maribus URSS. *Novit. System. Plantarum non Vascul.*, **10**, 49–59. (In Russian.)

PETROV J. E. (1974) Clavis synoptica Laminarialium et Fucalium e maribus URSS. *Novit. System. Plantarum non Vascul.*, **11**, 153–69. (In Russian.)

POWELL H. T. (1963) Speciation in the genus *Fucus* L. and related genera. In *Speciation in the Sea*, eds. Harding J. P. & Tebble N., pp. 63–77. Systematics Assn. Publ. No. 5, London.

PRICE I. R. (1969) The structure and classification of *Scytothamnus australis* (J. Agardh) J. D. Hooker et Harvey (Phaeophyta). *Phycologia*, **8**, 37–41.

PRUD'HOMME VAN REINE W. F. (1978) Criteria used in systematic studies in the Sphacelariales. In *Modern Approaches to the Taxonomy of Red and Brown Algae*, eds. Irvine D. E. G. & Price J. H., pp. 301–23. Academic Press, London.

QUATRANO R. S. (1974) Developmental biology: development in marine organisms. In *Experimental Marine Biology*, ed. Mariscal R. N., pp. 303–46. Academic Press, London & New York.

RAVANKO O. (1968) Macroscopic green, brown, and red algae in the southwestern archipelago of Finland. *Acta Bot. Fennica*, **79**.

RAVANKO O. (1970) Morphological, developmental, and taxonomic studies in the *Ectocarpus* complex (Phaeophyceae). *Nova Hedw.*, **20**, 179–252.

ROBERTS M. (1978) Active speciation in the taxonomy of the genus *Cystoseira* C. Ag. In *Modern Approaches to the Taxonomy of Red and Brown Algae*, eds. Irvine D. E. G. & Price J. H., pp. 399–422. Academic Press, London.

ROELEVELD J. G., DUISTEROF M. & VROMAN M. (1974) On the year cycle of *Petalonia fascia* in the Netherlands. *Netherlands J. Sea Res.*, **8**, 410–26.

RUSSELL G. (1963) A study in populations of *Pilayella littoralis*. *J. Mar. Biol. Assn. U.K.*, **43**, 469–83.

RUSSELL G. (1966) The genus *Ectocarpus* in Britain. I. The attached forms. *J. Mar. Biol. Assn. U.K.*, **46**, 267–94.

RUSSELL G. (1967) The genus *Ectocarpus* in Britain. II. The free-living forms. *J. Mar. Biol. Assn. U.K.*, **47**, 233–50.

RUSSELL G. (1973) The Phaeophyta: a synopsis of some recent developments. *Oceanogr. Mar. Biol. Ann. Rev.*, **11**, 45–88.

RUSSELL G. (1978) Environment and form in the discrimination of taxa in brown algae. In *Modern Approaches to the Taxonomy of Red and Brown Algae*, eds. Irvine D. E. G. & Price J. H., pp. 339–69. Academic Press, London.

RUSSELL G. & FLETCHER R. L. (1975) A numerical taxonomic study of the British Phaeophyta. *J. Mar. Biol. Assn. U.K.*, **55**, 763–83.

RUSSELL G. & GARBARY D. (1978) Generic circumscription in the family Ectocarpaceae (Phaeophyceae). *J. Mar. Biol. Assn. U.K.*, **58**, 517–23.

SEARLES R. B. (1978) The genus *Lessonia* Bory (Phaeophyta, Laminariales) in southern Chile and Argentina. *Br. Phycol. J.*, **13**, 361–81.

SEARS J. R. & WILCE R. T. (1973) Sublittoral benthic marine algae of southern Cape Cod and adjacent islands: *Pseudolithoderma paradoxum* sp. nov. (Ralfsiaceae, Ectocarpales). *Phycologia*, **12**, 75–82.

SVENDSEN P. & KAIN J. M. (1971) The taxonomic status, distribution, and morphology of *Laminaria cucullata* sensu Jorde and Klavestad. *Sarsia*, **46**, 1–22.

TAYLOR W. R. (1963) The genus *Turbinaria* in eastern seas. *J. Linn. Soc. London, Bot.*, **58**, 475–90.

TAYLOR W. R. (1975) Notes on algae from the tropical Atlantic Ocean—VIII. *Rev. Algol.*, N.S. **11**, 58–71.

TOTH R. & MARKEY D. R. (1973) Synaptonemal complex in brown algae. *Nature (Lond.)*, **243**, 236–7.

UMEZAKI I. (1961) Some new and noteworthy species of genus *Punctaria* (brown algae) from Japan. *J. Jap. Bot.*, **36**, 362–7.

VAN DEN HOEK C. & FLINTERMAN A. (1968) The life-history of *Sphacelaria furcigera* Kütz. (Phaeophyceae). *Blumea*, **16**, 193–242.

VAN DEN HOEK C. & JAHNS H. M. (1978) *Algen. Einführung in die Phykologie*. G. Thieme, Stuttgart.

WAERN M. (1952) Rocky-shore algae in the Öregrund Archipelago. *Acta Phytogeographica Suecia*, **30**.

WALKER D. C. & HENRY E. C. (1978) Unusual reproductive structures in *Syringoderma abyssicola* (S. & G.) Levring. *New Phytol.*, **80**, 193–7.

WIDDOWSON T. B. (1972a) A taxonomic revision of the genus *Alaria* Greville. *Syesis*, **4**, 11–49.

WIDDOWSON T. B. (1972b) A statistical analysis of variation in the brown alga *Alaria*. *Syesis*, **4**, 125–43.

WILCE R. T. (1960) Studies in the genus *Laminaria* II. *Laminaria groenlandica* L. K. Rosenv. *Bot. Notiser*, **113**, 203–9.

WILCE R. T. (1962) A new member of the Punctariaceae: *Platysiphon verticillatus* gen. nov., sp. nov. *Bot. Tidsskr.*, **58**, 35–42.

WILCE R. T. (1965) Studies in the genus *Laminaria* III. A revision of the North Atlantic species of the Simplices section of *Laminaria*. *Proc. Fifth Mar. Biol. Symp. Botan. Gothoburgensia*, **3**, 247–56.

WOMERSLEY H. B. S. (1964) The morphology and taxonomy of *Cystophora* and related genera (Phaeophyta). *Austr. J. Bot.*, **12**, 53–110.

WYNNE M. J. (1969) Life history and systematic studies of some Pacific North American Phaeophyceae (brown algae). *Univ. Calif. Pub. Bot.*, **50**, 1–88.

WYNNE M. J. (1971) Concerning the Phaeophycean genera *Analipus* and *Heterochordaria*. *Phycologia*, **10**, 169–75.

WYNNE M. J. (1973) A key to unbranched, cylindrical brown algae. *Contrib. Mar. Sci., Univ. Texas*, **17**, 133–52.

WYNNE M. J. & LOISEAUX S. (1976) Recent advances in life history studies of the Phaeophyta. *Phycologia*, **15**, 435–52.

WYNNE M. J. & NORRIS J. N. (1976) The genus *Colpomenia* Derbès et Solier (Phaeophyta) in the Gulf of California. *Smithsonian Contrib. Bot.*, **35**.

ZINOVA A. D. (1954) New family, genus and species of brown algae. *Trudy Bot. Inst. Komarov, Akad. Nauk SSSR*, Ser. 2, **9**, 223–44. (In Russian.)

ZINOVA A. D. (1959) About two brown algae from Antarctica—*Phyllogigas* and *Himantothallus*. *Botan. Zhurnal (Leningrad)*, **44**, 372–9. (In Russian.)

CHAPTER 3
CHLOROPHYTA: MORPHOLOGY AND CLASSIFICATION

C. VAN DEN HOEK
Department of Systematic Botany, Biological Centre of the University,
P.O. Box 14, 9750 AA Haren, The Netherlands

1 CIRCUMSCRIPTION OF THE DIVISION CHLOROPHYTA

The division Chlorophyta is here defined by the following characteristics (cf. van den Hoek & Jahns 1978; Moestrup 1978):

1 The zoids are isokontan, which means that the flagella of one zoid are similar, although they may differ in length.

2 The flagella lack the tinsel (mastigonemes) of the heterokontan algae but may be covered by various types of very delicate hairs and scales.

3 The transition region between flagellum and flagellar base contains a star-shaped body (Fig. 3.2e) (Moestrup 1978; van der Veer 1979).

4 The chloroplast is enclosed only by a double chloroplast-envelope and not by an additional envelope of endoplasmic reticulum.

5 The thylakoids form lamellae (stacks) of various numbers (two to six or more), pseudograna (stacks of partly overlapping thylakoids) or real grana; girdle lamellae (peripheral encircling stacks) are lacking.

6 The chloroplasts contain chlorophylls *a* and *b*.

7 Common xanthophylls in the chloroplasts are lutein, violaxanthin, antheraxanthin, and neoxanthin.

8 The pyrenoid (if present) is situated within the chloroplast; thylakoids often penetrate into the pyrenoid, which is surrounded by starch grains.

9 The most important reserve-polysaccharide is starch; starch grains are formed within the chloroplast.

The above circumscription of the Chlorophyta is based upon the concept that cell architecture of zoids (flagellate cells) is conservative because the typical eukaryotan flagellum, with its '9 + 2' structure, occurs in the large majority of plant divisions and animal phyla and is consequently considered as a vestige of a primeval eukaryotic flagellate cell. The zoids therefore provide characteristics having much taxonomic weight for the distinction of groups at a high hierarchical level, such as divisions and classes. After the origin of the primeval eukaryotic cell the first phylogenetic radiation possibly took place on the basis of cell architecture and resulted in a multitude of various cell-types, many of which were flagellate. Extant cell types, such as the Rhodophytan (non-flagellate), Heterokontophytan, Dinophytan, Euglenophytan, Chytridiomycotan and Eumycotan (non-flagellate) cell types, are possibly remnants of such an early radiation (Leedale 1974; Taylor 1976; van den Hoek & Jahns 1978). The ancestral Chlorophytan cell can possibly be envisaged as an isokontan, originally colourless and heterotrophic flagellate which later acquired, as endosymbionts, prokaryotic green cells with chlorophylls *a* and *b*. According to this hypothesis these endosymbionts later developed into Chlorophytan chloroplasts. Extant relatives of these hypothetical ancestral Chlorophytan chloroplasts are perhaps the recently discovered green unicellular prokaryotes associated with tropical ascidians. These prokaryotes resemble Cyanophyta in most respects but differ by the presence of chlorophylls *a* and *b*, the absence of phycobilisomes and the phycobiliproteins contained in them, and by the stacking of thylakoids (which are evenly spaced in Cyanophytes) (Lewin 1975; Lewin & Withers 1975; Schulz-Bades & Lewin 1976; Whatley 1977; Thinh 1979). They share these differentiating characters with Chlorophytan chloroplasts and for this reason have recently been classified in the proposed new prokaryotan division Prochlorophyta (Lewin 1976, 1977; but see Antia 1977).

Although the Chlorophyta, as defined above, also include stoneworts, mosses, liverworts and vascular plants (Fott 1974; van den Hoek & Jahns 1978), only those organisms traditionally considered to be Chlorophyta will be discussed in this paper. However, the discussion will not be restricted to seaweeds, since it is often necessary to discuss freshwater and microscopic taxa in order to fully appreciate the significance and breadth of a particular characteristic.

2 CHARACTERISTICS USED FOR THE SUBDIVISION OF CHLOROPHYTA

Five categories of characteristics are currently important for the subdivision of the Chlorophyta into classes and of the class Chlorophyceae into orders, namely:

1, the type of flagellate cell architecture; 2, the type of mitosis and cytokinesis; 3, the organization level of the thallus and its elaboration; 4, the life-history type; 5, the type of sexual fusion. Categories 1, 2 and 4 have only recently become important, whereas category 3 (and, to a certain extent 5) has been traditionally used and is still important for the systematic subdivision of the Chlorophyta. The following three categories of characteristics have additional value for the distinction of Chlorophycean orders: 6, the structure of chloroplasts and pyrenoids; 7, the chemical composition and structure of the cell walls; 8, the presence or absence of the xanthophylls siphonein and siphonoxanthin.

Categories 1–5 and 7 are listed in Table 3.1 with their supposedly 'primitive' and 'derived' states. The states are graphically depicted in Fig. 3.1 and regions that agree with the character states for each green algal order can be demarcated in the figure. In this figure the regions occupied by the stoneworts (class Charophyceae), the mosses and liverworts (division Bryophyta) and vascular plants (division Tracheophyta) are also given to indicate the phylogenetic relationships of these groups of green plants with the green algae.

Table 3.1 Six categories of characters used for the subdivision of Chlorophyta, and their supposedly 'primitive' and 'derived' states.

Category of characters	Primitive	Derived	Strongly derived
Type of zoid	Cruciate type		1 Unilateral type 2 Intermediate (*Trentepohlia*) type
Type of mitosis–cytokinesis	The telophase spindle is persistent, and keeps the daughter nuclei far apart; cytokinesis by simple invagination of plasmalemma (Type I in Fig. 3.4)		1 The telophase is non-persistent and collapses soon so that the daughter nuclei approach each other; cytokinesis by invagination, or a cell-plate in a phycoplast (Types II and III in Fig. 3.4) 2 The persistent telophase spindle develops into a phragmoplast. Cytokinesis by a cell-plate in the phragmoplast (Type IV in Fig. 3.4).
Organization level	Flagellate	1 Coccoid 2 Sarcinoid 3 Filamentous 4 Thallose	1 Siphonous 2 Stem, root and leaves with conducting tissues
Life history	Haplontic	Diplohaplontic	1 Diplontic 2 Diplohaplontic with strong reduction of haploid gametophytic generation (nearly diplontic)
Sexual fusion	Isogamy	Anisogamy	Oogamy
Cell-wall polysaccharides (supporting layer)	Cell-covering of diamond-shaped, *c.* 0.05 μm large scales	Non-crystalline polysaccharides	1 Highly crystalline cellulose I 2 Xylan, mannan

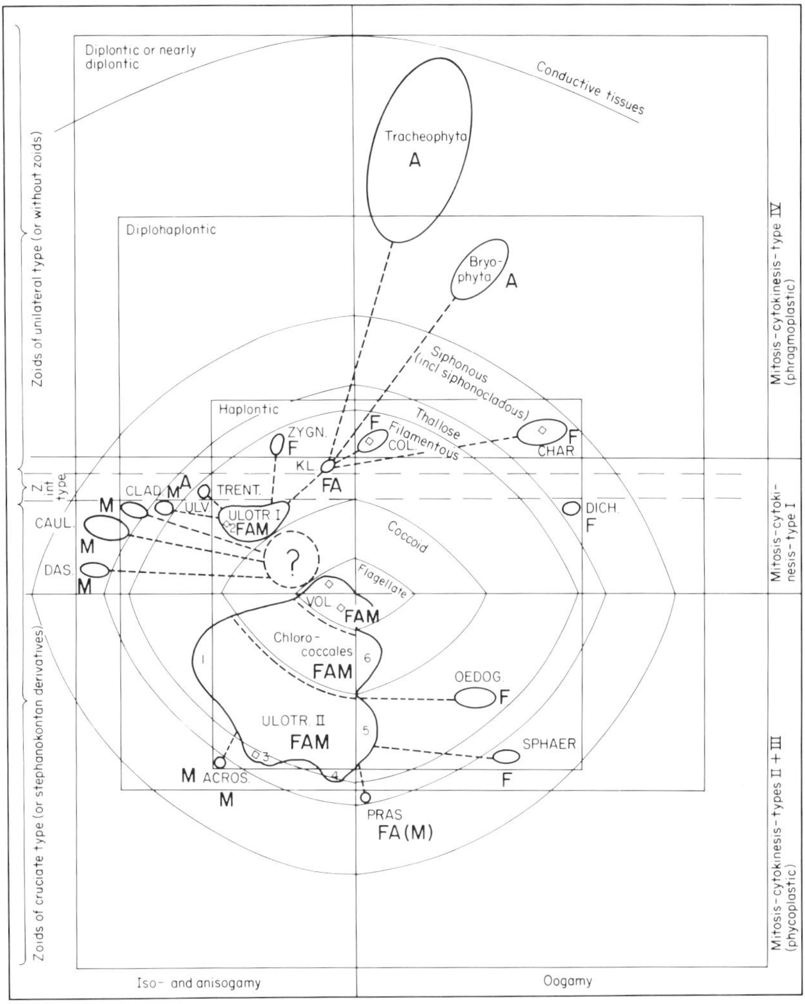

Fig. 3.1 Distribution of 'primitive' and 'derived' character-states (cf. Table 3.1), and phylogenetic relationships within Chlorophyta. ACROS = Acrosiphoniales; CAUL = Caulerpales; CHAR = Charophyceae; CLAD = Cladophorales; COL = Coleochaetales; DAS = Dasycladales; DICH = Dichotomosiphonales; KL = Klebsormidiales; OEDOG = Oedogoniales; PRAS = Prasiolales; SPHAER = Sphaeropleales; ULOTR I = Ulotrichales I (*Ulothrix, Stichococcus, Radiofilum, Raphidonema, Microthamnion*); ULOTR II = Ulotrichales II (*Microspora, Uronema, Chaetophora, Stigeoclonium, Draparnaldia, Fritschiella, Schizomeris*); ULV = Ulvales; VOL = Volvocales; ZYGN = Zygnematales; z. int. type = zoids of intermediate type. 1 = *Draparnaldiopsis indica* (Singh 1942, 1945), *Entocladia viridis* (O'Kelly & Yarish 1981) and probably *Eugomontia* (Kornmann 1960); 2 = *Pseudendoclonium basiliense, Trichosarcina polymorpha, Ulothrix zonata*; 3 = *Monostroma* (*M. grevillei* with diamond scales); 4 = *Fritschiella tuberosa* (Singh 1947); 5 = *Cylindrocapsa, Chaetonema, Chaetonemopsis*; 6 = *Golenkinia, Micractinium, Eremosphaera viridis*. ◇ = zoids with diamond scales (0.05 μm); A = aerophytic; F = freshwater; M = marine.

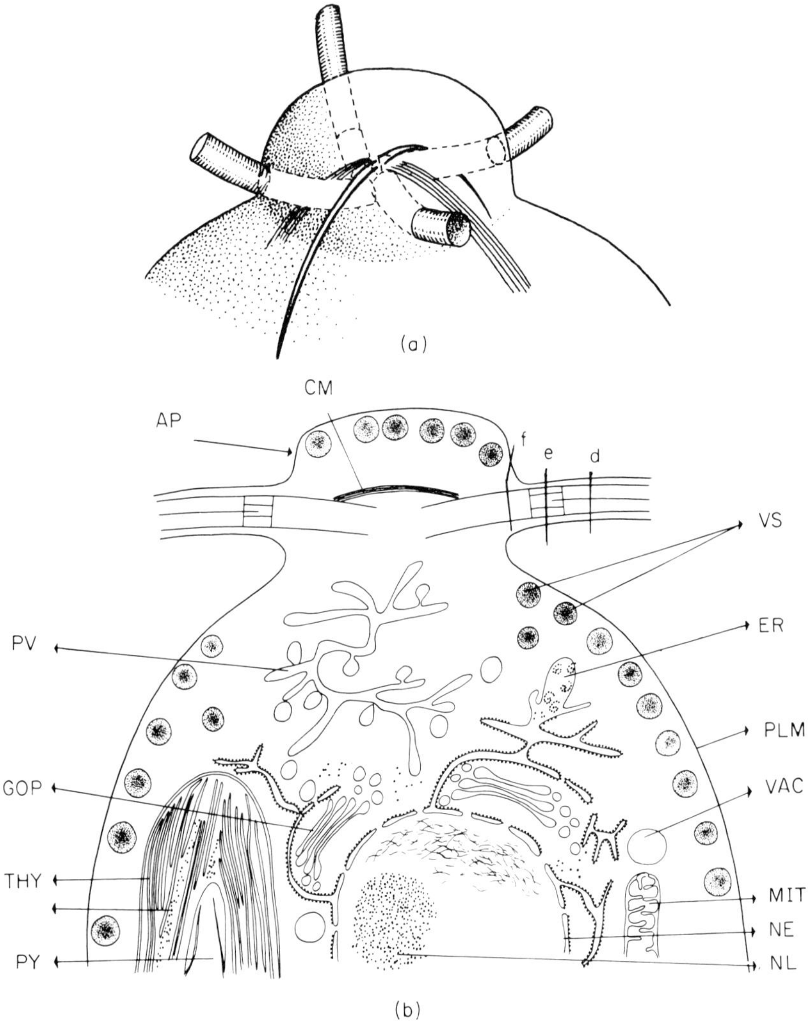

Fig. 3.2 (pp. 90–1) (a–i) Zoid of the cruciate type (naked quadriflagellate zoospore of *Stigeoclonium*). (a) Spatial diagram of flagellar apparatus showing anchorage of flagella in cell by cruciately arranged microtubular roots. (b) Longitudinal section (PY = pyrenoid; THY = thylakoid; GOP = perinuclear Golgibody; PV = pulsating vacuole; AP = apical papilla; CM = capping material; VS = vesicles with sticky material; PLM = plasmalemma; VAC = vacuole; MIT = mitochondrion; NE = nuclear envelope; NL = nucleolus). (c) Apical view on the flagellar apparatus (diagram), showing the basal parts of the four flagella and the four cruciately arranged microtubular roots. (d) Cross section of flagellum [for position, see (b) & (c)]. (e) Cross section of transition region showing star-shaped body [for position, see (b) & (c)]. (f) Cross section of basal body [for position see (b) & (c)]. (g, h) Cross sections of a 2-membered microtubular root, with associated transversely striated roots [for positions, see (c)]. (i) Cross section of 5-membered microtubular root [to the posterior end the number of microtubules decreases; for position, see (c)].

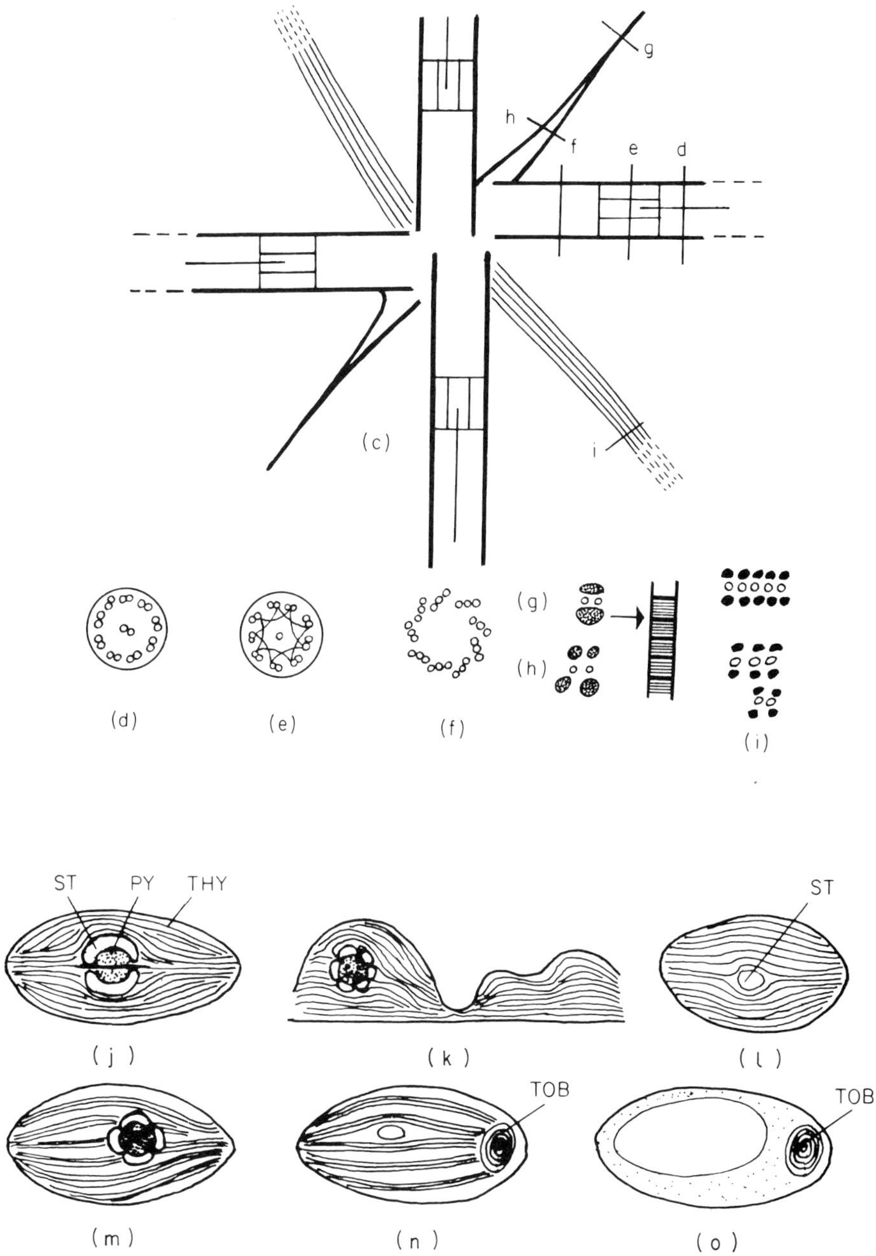

Fig. 3.2 (cont.) (j–n) Ultrastructure of chloroplasts and pyrenoids in siphonous chlorophytes (diagram). (j) In Cladophorales (PY = pyrenoids; ST = starch; THY = thylakoid); (k) in Acrosiphoniales; (l) in Dasycladales (ST = starch); (m) in Caulerpales (*Bryopsis* type); (n, o) in Caulerpales (*Halimeda* type) (o = amyloplast; TOB = thylakoid organising body). (= (a–j): based upon Manton 1964, after van den Hoek & Jahns 1978; (k–o) based upon Hori & Ueda 1967, 1975.)

The division Chlorophyta is here subdivided as follows:

Division Chlorophyta
  Class Chlorophyceae
    Order Volvocales* (incl. Tetrasporales)
      Chlorococcales
      Ulotrichales* (incl. Chlorosarcinales, Chaetophorales)
      Trentepohliales
      Klebsormidiales
      Coleochaetales
      Acrosiphoniales*
      Ulvales*
      Prasiolales*
      Sphaeropleales
      Cladophorales* (incl. Siphonocladales)
      Caulerpales*
      Dichotomosiphonales
      Dasycladales*
      Oedogoniales
      Zygnematales
  Class Charophyceae

Groups marked with an asterisk contain the marine representatives or seaweeds (most of the Volvocales, Ulotrichales and Prasiolales are freshwater or aerophytic).

2.1 Types of flagellate cell architecture

Within the here broadly conceived division of Chlorophyta, two distinct types of flagellate cell architecture can be distinguished, based on differences in the systems of flagellar roots, namely, the *cruciate type* and the *unilateral type* (Moestrup 1978, with further references).

The cruciate type of zoid

The flagella are anchored in the cell by four cruciately arranged microtubular roots, when the zoid is viewed on its anterior end. Two opposite roots each consist of two microtubules in most cases, and the two other opposite roots each consist of a varying number of microtubules (*c.* 3–8). Zoids of this type are approximately bilaterally symmetrical when viewed from the side (Fig. 3.2 a–i). Often one, two or four transversely striated roots are associated with microtubular roots. In other cases the basal bodies are connected by a striated band (e.g. *Chlamydomonas*). It is likely that the microtubular roots function as a cellular skeleton and at the same time anchor the flagella in the cell, whereas the striated roots function as cellular 'muscles' and are responsible for the flagellar beat (Salisbury 1978; Swanson *et al.* 1978; Salisbury & Floyd 1978; Robenek & Melkonian 1979). In many instances a broader microtubular root runs over an intraplastidial eyespot composed of plates of carotenoid globules clasped between thylakoids.

The cruciate type of zoid is the common type of green algal zoid, as it is represented by the unicellular biflagellate genus *Chlamydomonas* and by the biflagellate and quadriflagellate reproductive zoids in numerous species. Where possible, marine representatives will be stressed in the following discussions. However, much new fundamental knowledge on phylogenetic relationships among green algae has been gathered in studies on freshwater and aerophytic algae, and consequently these studies have to be given due attention.

More than 40 Chlorophycean species distributed over the orders Volvocales, Chlorococcales, Ulotrichales, Ulvales, Acrosiphoniales, Cladophorales, Caulerpales (Hori 1977), Dasycladales and Sphaeropleales are known to have the cruciate type of zoid, on the basis of electron microscopical observations (see Moestrup 1978 for a review). Transversely striated roots have been observed in the Volvocales (Manton & Parke 1965; Manton 1966; Lembi 1975), in the Ulotrichales (Manton 1964; Melkonian 1975), including *Ulothrix* (Sluiman *et al.* 1980a, b) and *Monostroma grevillei* (Jónsson & Chesnoy 1974), in the Ulvales, including *Ulva* (Micalef & Gayral 1972; Melkonian 1979) and *Enteromorpha* (Evans & Christie 1970; McArthur & Moss 1979), in the Acrosiphoniales, including *Urospora* (Kristiansen 1974), in the Caulerpales, including *Bryopsis* (Hori 1977) and in the Oedogoniales (Markowitz 1978; Moestrup 1978). Radially symmetrical, stephanokontan zoids (e.g. in *Bryopsis* and *Derbesia*) can probably be considered as derived from the cruciate type by repetition of the flagellar system.

THE UNILATERAL TYPE OF ZOID

The flagella are anchored in the cell by one broad band of many microtubules, running from the flagellar bases, along one side of the cell, towards the posterior cell-pole. Near the flagellar bases this microtubular band is part of a *multilayered structure* (MLS). MLS's show considerable differences regarding number of layers and microtubular number (Fig. 3.3).

Apart from the microtubular layer, two laminate layers which are each composed of differently oriented platelets, are characteristic. In the biflagellate representatives of this type the two flagella are inserted subapically and unilaterally, and they are directed backwards during swimming. An eyespot is always lacking. Biflagellate representatives of this type include the zoospores of members of the Coleochaetales (McBride 1971; Moestrup 1974), Klebsormidiales (Marchant *et al.* 1973; Pickett-Heaps 1975; Moestrup 1978; Rogers *et al.*

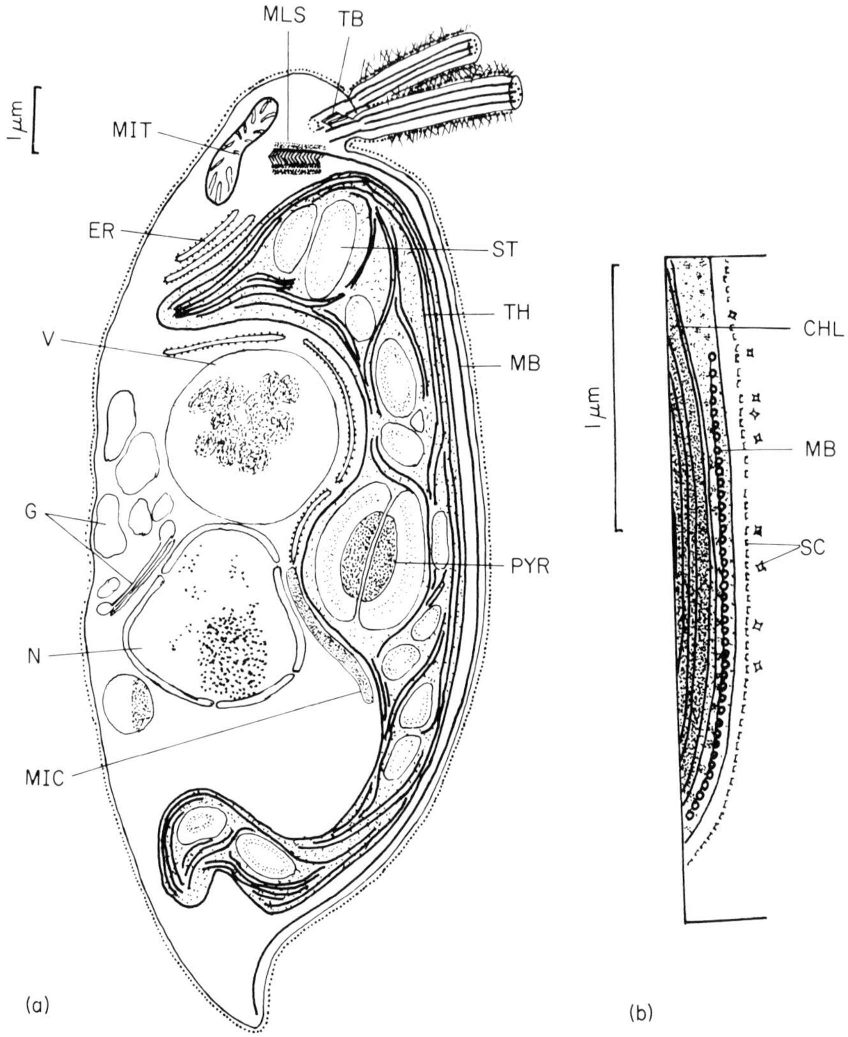

Fig. 3.3 Zoid of unilateral type (zoospore of *Chaetosphaeridium globosum*). (a) Longitudinal section; (b) transverse section of microtubular band. (MLS = multilayered structure; TB = transversely striated band connecting basal bodies; MIT = mitochondrion; G = Golgi-complex and pulsating vacuoles; N = nucleus; MIC = microbody; PYR = pyrenoid; MB = microtubular band; TH = thylakoid; ST = starch; CHL = chloroplast; SC = diamond-scales.) Based upon Moestrup 1974.

1978) and spermatozoids of higher plants, with multiflagellate spermatozoids of higher plants probably being derived from the biflagellate unilateral zoid.

Free-living chlorophytan flagellates of the unilateral type were unknown until recently. Only in multicellular freshwater and aerophytic chlorophytes had the unilateral zoid been observed so far. This suggested that, within the evolutionary lineage leading from freshwater to land chlorophytes, the unilateral reproductive zoid evolved from the cruciate reproductive zoid. However, recently the asymmetric, scaly green flagellate *Nephroselmis* was discovered to have a flagellar root system including three microtubular roots, one being multilayered. This points to an early phylogenetic divergence of the unilateral and cruciate zoid in the ancestral group of flagellate, scale-covered chlorophytes (Moestrup & Ettl 1979).

The intermediate type of zoid

Aerophytic filamentous algae in the Trentepohliales have (approximately) bilaterally symmetrical zoids with two roots which each contain a MLS-like structure (Graham & McBride 1975; Chapman & Madison 1976; Chapman 1977; Chapman & Good 1977). There are additional microtubular roots, and the root system as a whole shows a cruciate arrangement (Chapman 1979). Therefore, these zoids show features of both the cruciate and the unilateral zoid type and can be considered as belonging to an *intermediate* type.

2.2 Types of mitosis and cytokinesis

Recent ultrastructural investigations have led to the recognition of at least four types of mitosis-cytokinesis in the Chlorophyta (Fig. 3.4).

Type I

During anaphase the chromosomes move to the nuclear poles mainly by elongation of the interzonal spindle-microtubules; these are persistent during telophase and cytokinesis and keep the daughter nuclei widely apart. Cytokinesis takes place by a diaphragm-like furrow of the plasmalemma. In this furrow cell wall material is often deposited. Often the nuclear envelope is more or less broken down during mitosis (open mitosis).

Examples include members of the Volvocales (Pearson & Norris 1975; Mattox & Stewart 1977), Klebsormidiales and Ulotrichales (Floyd *et al.* 1972b; Stewart *et al.* 1973; Pickett-Heaps 1972a, 1974a, 1975), including *Ulothrix* (Lokhorst 1978).

The naked, asymmetric uniflagellate *Pedinomonas* (Pickett-Heaps & Ott 1974) has a persistent telophase spindle and a persistent nuclear envelope (closed mitosis), as have siphonous *Valonia* (Hori & Enomoto 1978) and *Cladophora* (Scott & Bullock 1976; McDonald & Pickett-Heaps 1976); these latter two genera, however, differ in other details, for instance in the presence of kinetochores and the fact that mitosis is not followed by cytokinesis. Filamentous aerophytic *Cephaleuros* has also a persistent telophase spindle and a closed mitosis during sporogenesis (Graham & McBride 1978) as have the secondary nuclei of *Batophora* (Dasycladales) (Liddle *et al.* 1976; Koop *et al.*

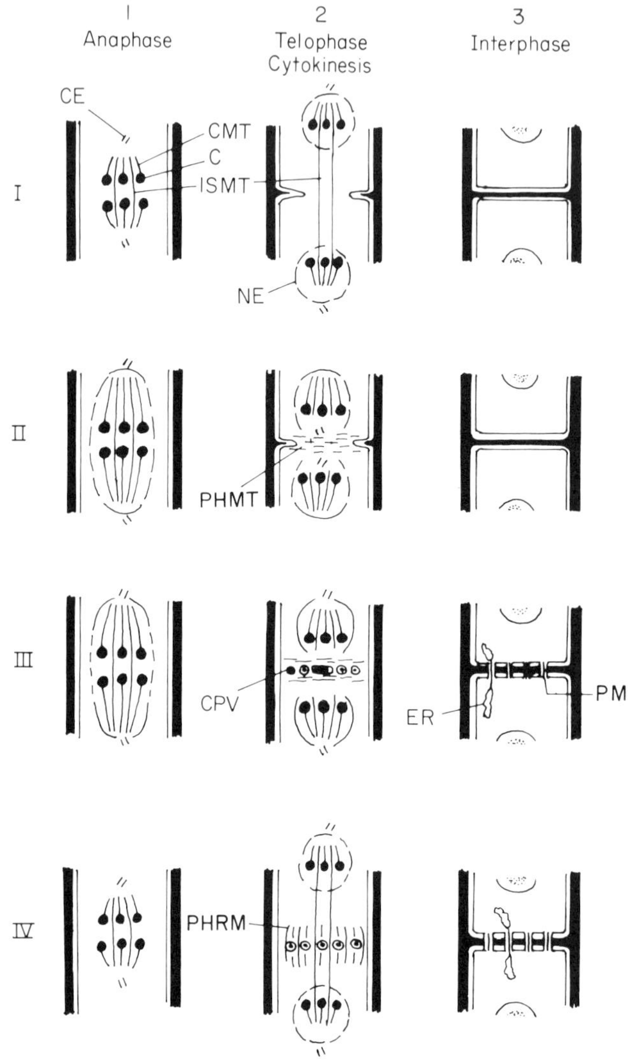

Fig. 3.4 Types I–IV of mitosis–cytokinesis (cf. section 2.2 of text). CE = centriole; CMT = chromosomal spindle-microtubule; C = chromosome; ISMT = interzonal spindle microtubule; NE = nuclear envelope; PHMT = phycoplast-microtubule; CPV = cell plate vesicle; PM = plasmodesma; PHRM = phragmoplast-microtubule (modified after van den Hoek & Jahns 1978).

1977). Light-microscopic observations suggest persistent telophase spindles and closed mitoses in *Bryopsis* and *Derbesia* (Caulerpales) (Rietema 1975). The occurrence of the persistent telophase spindle in such widely divergent organisms as vascular plants, green flagellates and Metazoa is thought to indicate its primitive nature (Pickett-Heaps 1975; Stewart & Mattox 1975a, b).

Type II

During anaphase the chromosomes move to the nuclear poles mainly by shortening of the chromosomal spindle-microtubules. During telophase the spindle soon collapses so that the daughter nuclei approach each other rapidly (non-persistent telophase spindle). The nuclear envelope remains intact during mitosis (closed mitosis) and has mostly polar openings (fenestrae). Cytokinesis takes place by invagination of the plasmalemma and fusion of a cell plate (a plate of vesicles) in a phycoplast (this is a plate of microtubules in the future cleavage plane, cf. Pickett-Heaps 1972b). Centrioles are often associated with the formation of the phycoplast. Examples include the naked salt water flagellate *Dunaliella* (Marano 1976) and other Volvocales (Johnson & Porter 1968; Stewart *et al.* 1974; Pickett-Heaps 1975; Floyd 1978); Chlorococcales (Pickett-Heaps 1970, 1975; Deason & O'Kelley 1979); sarcinoid (= packet-forming) Ulotrichales (Deason & O'Kelley 1979; Deason *et al.* 1979) and filamentous Ulotrichales (Pickett-Heaps 1973, 1975; Stewart *et al.* 1973). *Acrosiphonia* (Acrosiphoniales) seems to present a modified type II: the nuclei divide synchronously in the future division plane of the cell (Fig. 3.6d, e), and after that a common phycoplast-like structure which is limited to the ingrowing septum edge is formed between the two sets of daughter-nuclei (Jónsson 1962; Kornmann 1965; Hudson & Waaland 1974).

Type III

Type III resembles type II, but differs by the fact that plasmodesmata are left open during fusion of the cell-plate vesicles. These plasmodesmata resemble those of higher plants in having a dark core which is connected with the ER of the resultant cells. Examples are a number of freshwater Ulotrichales (McBride 1970; Floyd *et al.* 1972a; Stewart *et al.* 1973; Mattox *et al.* 1974; Stewart & Mattox 1975a). The assignment of *Ulothrix* to this type is probably the result of a misidentification (Lokhorst 1978 personal communication). Oedogoniales have a special elaboration of this type (Pickett-Heaps & Fowke 1969, 1970). So far, type III has not been found in green seaweeds.

Type IV

Type IV resembles type I in that the chromosomes move apart mainly by elongation of the interzonal spindle microtubules which are, here too, persistent during telophase. In addition, during the course of telophase a phragmoplast develops from the persistent spindle. A phragmoplast is a plate of microtubules in the cleavage plane which is parallel with the spindle axis. A cell plate (plate of vesicles with wall material) appears in the phragmoplast. Plasmodesmata with dark cores are left open during cytokinesis. Examples are the minute discoid multicellular freshwater alga *Coleochaete* (Marchant & Pickett-Heaps 1973), the stonewort *Chara* (Pickett-Heaps 1967, 1975), mosses, liverworts, and vascular plants. In the order Zygnematales *Spirogyra* (Fowke & Pickett-Heaps 1969a, b) has a cytokinesis which is intermediate between types I and IV, but that of *Mougeotia* (Bech-Hansen & Fowke 1972) conforms to type I. Until recently type IV was generally thought to be restricted to bryophytes and tracheophytes (see, however, Berghs 1906; Karling 1926; McAlister 1931). The discovery of the phragmoplast in green algae strengthens, of course, the theory that the ancestor of the higher plants was a multicellular green alga. So far, type IV has not been found in green seaweeds.

Mitosis–cytokinesis in perspective

The above typology of mitosis and cytokinesis is probably oversimplified and neglects various details whose taxonomic importance is difficult to estimate. For example, differences in the degree of persistence of the nuclear envelope, the presence or absence of perinuclear ER, the presence or absence of kinetochores and their degree of complexity, the involvement or not of centrioles or rhizoplasts (Mattox & Stewart 1977) and other details (Taylor 1976). *Ulva* (Løvlie & Bråten 1970), *Enteromorpha* (McArthur & Moss 1977) and three filamentous Ulotrichales (Mattox & Stewart 1974) have an early ingrowing cleavage furrow without a phycoplast and consequently conform to type I. However, their interzonal spindle was at first (Stewart & Mattox 1975) interpreted as 'essentially non-persistent' and consequently their mitosis–cytokinesis-type as closely related to type II; but later (in Sluiman *et al.* 1980) this interpretation was thought to be a 'misinterpretation' and their interzonal spindle to be persistent instead; this reversal of interpretation permitting the inclusion of the above mentioned five chlorophytes in a newly proposed class, Ulvaphyceae (characterized by type I mitosis–cytokinesis; the presence, in some species, of body-scales on the zoids; and several details of the flagellar root-system). This proposal is not accepted here, as it disregards differences in life-history and architecture (see Section 3 for general discussion).

The type of mitosis–cytokinesis appears to be an important character for the ordinal subdivision of the Chlorophyceae. However, many more species have to be screened for this character to permit sounder generalizations than are presently possible. This is true for relatively well-defined orders such as the Cladophorales, Dasycladales, Caulerpales, Prasiolales and Sphaeropleales, as well as the heterogeneous complex composed of the three traditional orders Ulotrichales, Chlorococcales and Volvocales as this character may possibly help to rearrange the genera in this complex into more natural orders (cf. Stewart & Mattox 1975a, b).

Diverging types of mitosis–cytokinesis are known for some closely related species. The flagellates *Nephroselmis* and *Platymonas* share an outer covering of stellate scales of a type unknown outside these two genera and *Prasinocladus* (Parke & Manton 1965); however, the former has a persistent telophase spindle and divides by simple invagination (type I) (Mattox & Stewart 1977) as *Heteromastix* (cf. Moestrup & Ettl 1979), whereas the latter has a collapsing telophase spindle and divides by a phycoplast (type II) (Stewart *et al.* 1974). This can be explained by assuming that these two flagellates are extant, comparatively little changed close relatives of the ancestral scale-covered, green flagellate in which type II arose from type I. Another explanation is that the various types of mitosis–cytokinesis arose more than once in the course of evolution.

2.3 Organizational level of the thallus

For several decades the organizational level of the thallus has been and continues to be an important criterion for the subdivision of the Chlorophyceae, as it is for the subdivision of various other algal classes. The following organizational levels can be distinguished (modified after Pascher 1931; Fott 1971; van den Hoek & Jahns 1978): 1, the flagellate unicellular level; 2, the flagellate colonial level; 3, the palmelloid level; 4, the coccoid level; 5, the sarcinoid (= packet-like) level (cf. Groover & Bold 1969); 6, the filamentous level; 7, the thallose level; 8, the siphonous level. The siphonocladous level (composed of multinucleate cells) can be considered as a special aspect of the siphonous level.

Seaweeds belong only to the palmelloid, the filamentous, the thallose, and the siphonous (including the siphonocladous) organizational levels. A few examples will be given.

Palmoclathrus stipitatus (Womersley 1971) (order: Volvocales) is a stalked, perforated rounded membrane composed of a stiff gelatinous matrix containing scattered cells. *Palmoclathrus* is one of two seaweed-genera with a *palmelloid organizational level* (Fig. 3.5a, b). *Ulothrix* (Fig. 3.5c–j), which is composed of unbranched, uniseriate filaments, and *Epicladia* (Fig. 3.5k, l), which consists of branched, uniseriate filaments, both belong to the *filamentous organizational level*. Both genera have uninucleate cells and parietal chloroplasts. The blade-like genera *Prasiola* (order Prasiolales) (Fig. 3.5m–p), *Monostroma* (order Ulotrichales) (Fig. 3.5q–v) and *Ulva* (order Ulvales) (Fig. 3.5c–h), and the intestine-like genus *Enteromorpha* (order Ulvales) (Fig. 3.5w–z, A, B) belong to the *thallose organizational level*. In *Prasiola* the uninucleate cells each contain a central, stellate chloroplast with a central pyrenoid (Fig. 3.5o, p); in *Monostroma, Ulva* and *Enteromorpha* the cells each contain one nucleus and a parietal chloroplast against one side of the cell (Fig. 3.5t, B, F, G).

To the *siphonocladous organizational level* (= filaments composed of multinucleate cells) belong unbranched *Urospora* (order: Acrosiphoniales) and *Chaetomorpha* (order: Cladophorales) (Fig. 3.6k–q), and branched *Acrosiphonia*

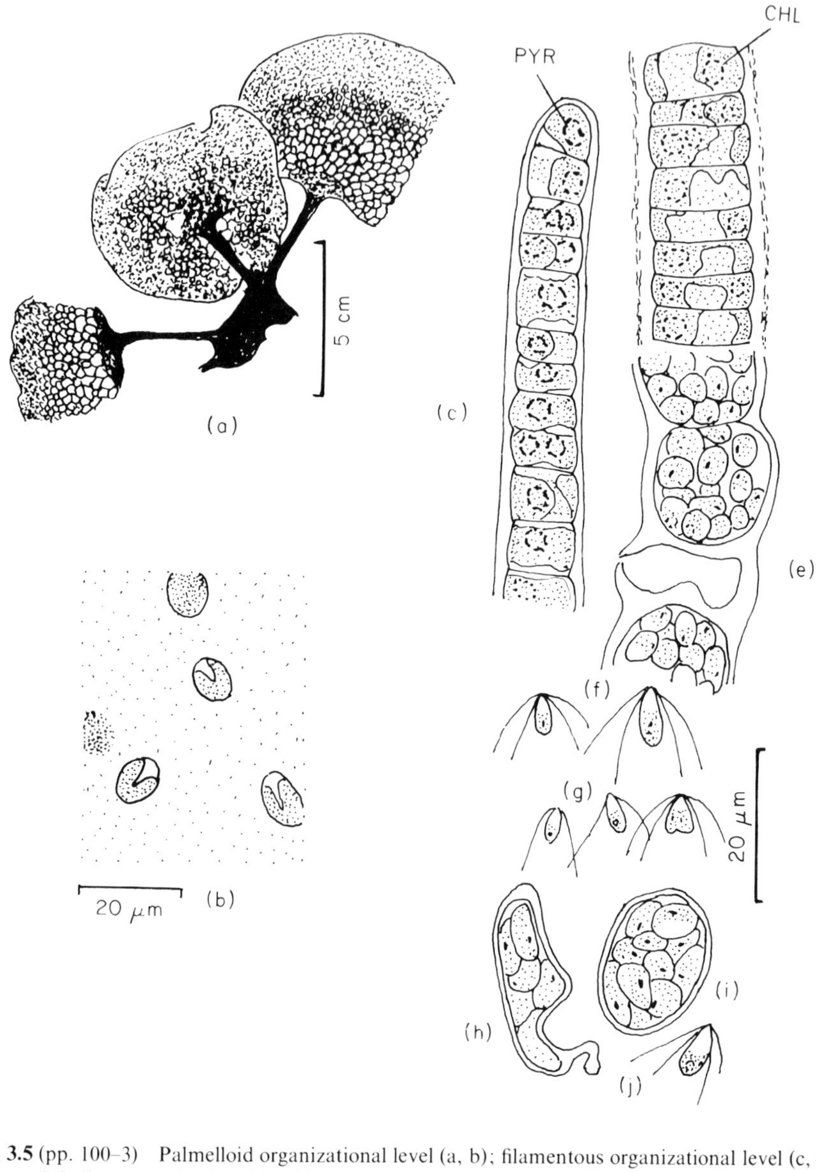

Fig. 3.5 (pp. 100–3) Palmelloid organizational level (a, b); filamentous organizational level (c, d, k, l); and thallose organizational level (m–p, q–t, w, A–H).

(a, b) *Palmoclathrus*; (a) habit, (b) cells dispersed in the gelatinous matrix (after Womersley, 1971).

(c–j) *Ulothrix flacca* (Dillw.) Thur. in le Jolis. (c, d) vegetative filament; (e) filament with zoosporangia, one zoosporangium emptied; (f) quadriflagellate zoospores produced by these sporangia; (g) two gametes and one planozygote; (h, i) stalked and unstalked hypnozygote, both their contents divided into zoospores; (j) quadriflagellate zoospore produced by hypnozygote (after Lokhorst 1978).

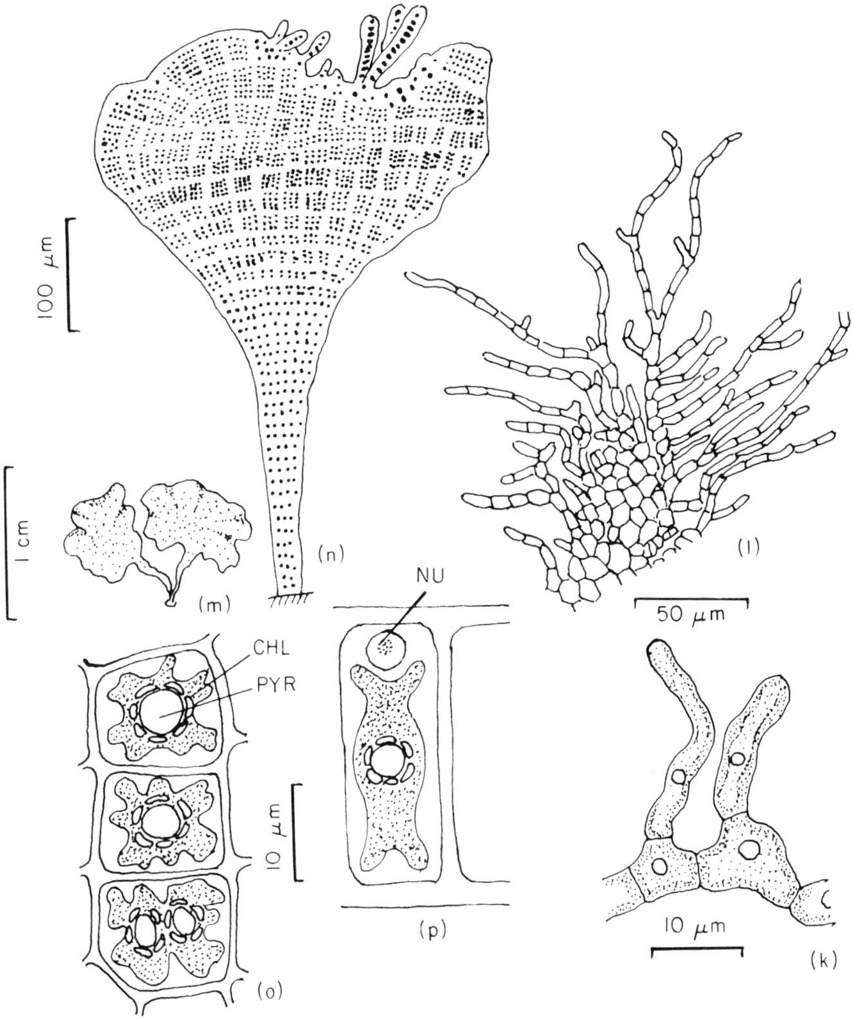

Fig. 3.5 (cont.) (k, l) *Epicladia flustrae* Reinke (based on Reinke 1889).
(m–p) *Prasiola stipitata* Suhr. (m, n) habit, see quadrangular groups of cells, note stellate central chloroplasts; (p) transverse section of thallus (after Hamel 1930, 1932 and Chadefaud 1960).
CHL = chloroplast; PYR = pyrenoid; NU nucleus; RHIZ = rhizoidal outgrowth of cell.

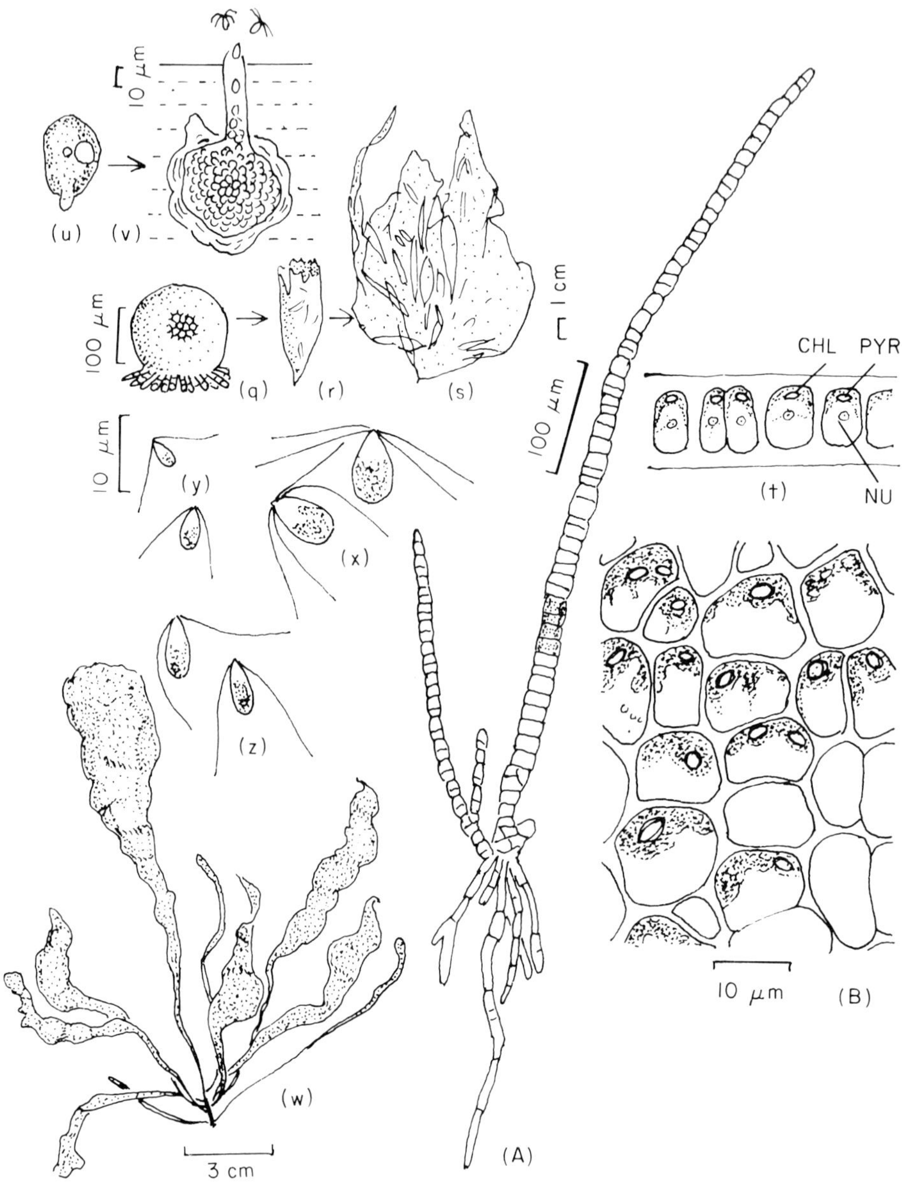

Fig. 3.5 (cont.) (q–v) *Monostroma grevillei* (Thur.) Wittr. (q–s) Development of monostromatic blade from sac-like young stage; (t) cross section of blade; (u) development of shell-boring hypnozygote from young, stalked zygote (based on Kornmann 1962).

(w–z, A, B) *Enteromorpha compressa* (L.) Grev. (w) habit (based on Bliding 1963); (x) quadriflagellate zoospores produced by sporophyte; (y, z) male and female gametes produced by gametophytes; (A) young, uniseriate filamentous plant; (B) typical 'cap-like' apically oriented chloroplasts (after Koeman unpublished).

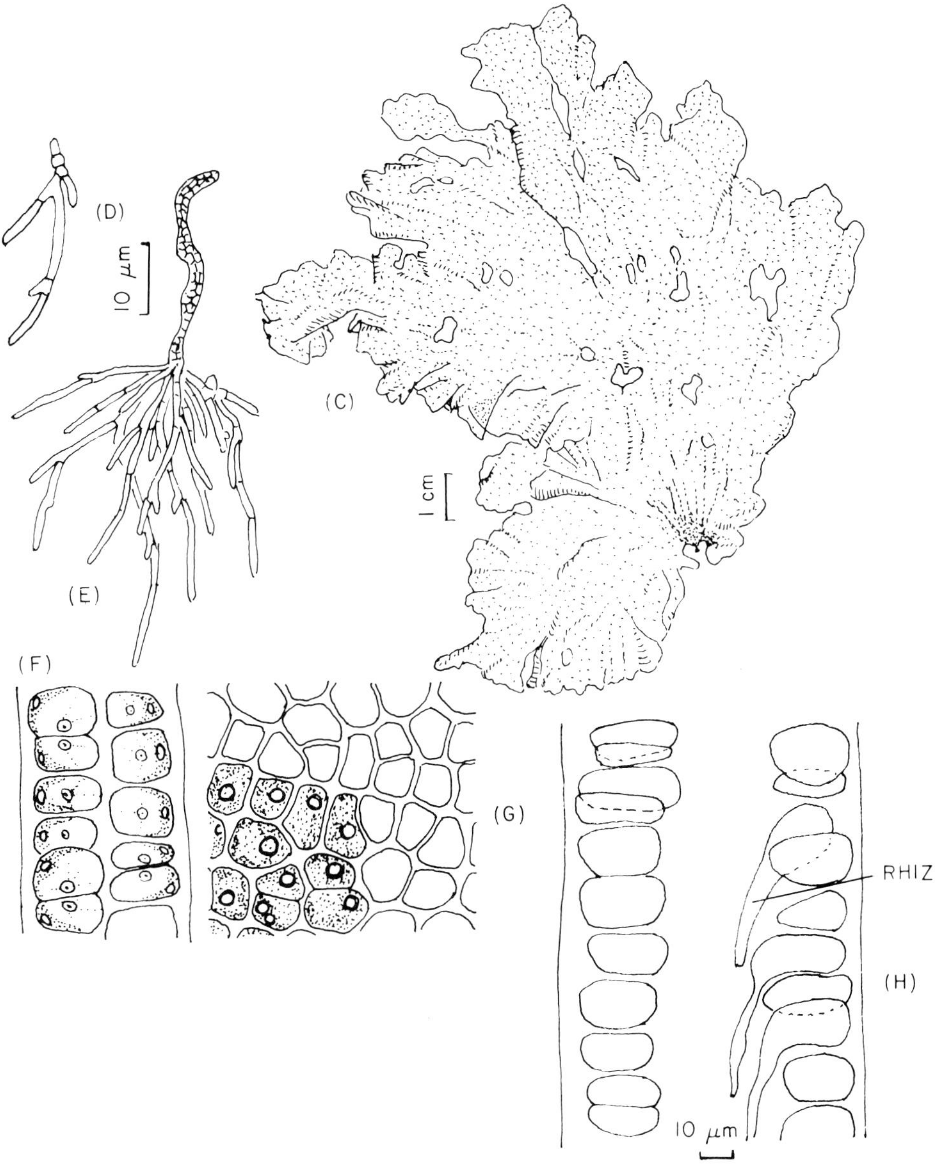

Fig. 3.5 (cont.) (C–H) *Ulva lactuca* L. (C) habit; (D) young, uniseriate filamentous stage; (E) young, tubular stage; (F) transverse section of blade; (G) surface view of blade; (H) transverse section of basal region showing rhizoidal cells (after Koeman & van den Hoek 1981).

(order: Acrosiphoniales) (Fig. 3.6a–e) and *Cladophora* (order: Cladophorales) (Fig. 3.6f–j). In the Acrosiphoniales each cell has a cylindrical, parietal, perforated chloroplast with many polypyramidal pyrenoids (Fig. 3.6d, e; see Section 2.5); in the Cladophorales each cell contains a parietal, often reticulate layer of small polygonal chloroplasts many of which contain bilenticular pyrenoids (Fig. 3.6j, see Section 2.5). Other representatives of the siphonocladous organizational level are the tropical genera *Struvea* (Fig. 3.6r, s), *Valonia* (Fig. 3.6t, u) and *Siphonocladus* (Fig. 3.6v, w). The latter three genera are often still ranged under the separate order Siphonocladales, mainly on the basis of a peculiar method of cell division, termed *segregative cell division* (cleavage of the protoplast into walled, rounded portions which later expand into new segments). However, this characteristic is shown by only two genera (*Dictyosphaeria* and *Siphonocladus*, see Fig. 3.6v, w). In other genera placed in the Siphonocladales, for instance *Valonia* (Fig. 3.6t, u) and *Struvea* (Fig. 3.6r), the modes of branching and cell division are variations of those in *Cladophora* (Fig. 3.6f). Similarity in cytology (a parietal reticulum of small polygonal chloroplasts with bilenticular pyrenoids) (cf. Section 2.5), life history (cf. Section 2.4), mode of mitosis (Type I mitosis in Fig. 3.4, taking place over the whole cell, see Fig. 3.6q) and the chemical composition of the cell wall (crystalline cellulose, cf. Section 2.6) strongly support the viewpoint that Siphonocladales and Cladophorales should be merged into one order, for which the correct name is Cladophorales Haeckel (Papenfuss & Chihara 1975; Valet 1976). *Struvea* (Fig. 3.6r) in its architecture much resembles some *Cladophora* species (van den Hoek 1963). Its terminal branch-systems form triangular blades in which the apical cells are attached to other cells by means of short hapteroid rhizoidal cells (Fig. 3.6s). Terminal hapteroid rhizoidal cells also occur in some *Cladophora* species (Fig. 3.6g).

In the *siphonous organizational level* each plant is in principle one multinucleate giant cell. Examples are *Bryopsis* (Fig. 3.7a–d) and *Caulerpa* (Fig. 3.7e) (order Caulerpales). In *Neomeris* (Fig. 3.7f–i) and other Dasycladales the thallus is also composed of one giant cell, which contains, however, only one large nucleus as long as the thallus is vegetative. Gametogenesis starts with the division of this nucleus (by meiosis) and the production of a large number of small nuclei (cf. Section 2.4).

In many Caulerpales the multinucleate siphons are used for the construction of more or less complex tissues. In *Rhipiliopsis* (van den Hoek 1978; Farghaly & Denizot 1979) this tissue is still rather simple and consists of branched, laterally fused siphons (by means of protuberances) which form a small, open reticulate blade-like frond (Fig. 3.7j–m) (compare with *Struvea*, Fig. 3.6r, which has a parallel construction on the basis of the siphonocladous organizational level). In the branched, finger-like fronds of *Codium* (Fig. 3.7n, o), the stipe of *Penicillus* (Fig. 3.7p) and in the strongly calcified segments of *Halimeda* (Fig. 3.7q, r) there is a plectenchymatous medulla of pale longitudinal siphons bearing a cortex of variously formed utricles in which the chloroplasts are concentrated.

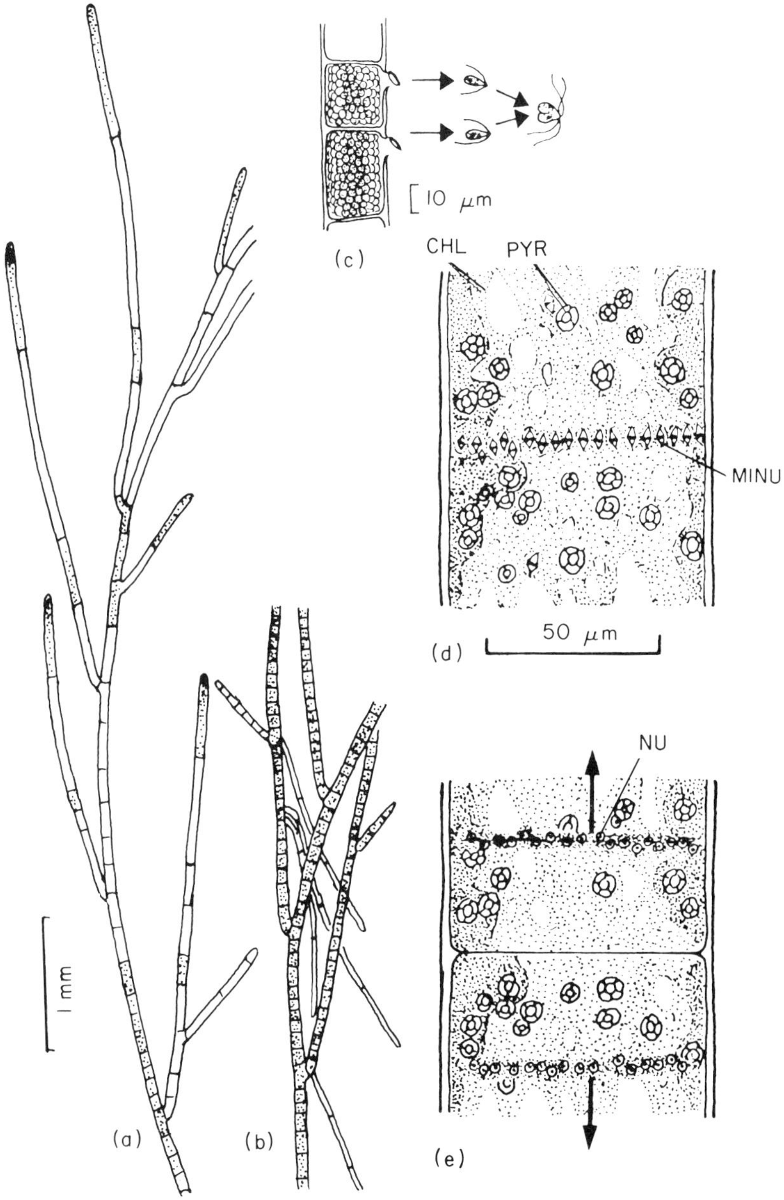

Fig. 3.6 (pp. 105–9) Siphonocladous organizational level.
(a–e) *Acrosiphonia*. (a, b) habit (based on Kornmann & Sahling 1977); (c) gametangia opening through lids (based on van den Hoek & Jahns 1978); (d) simultaneous mitoses in future plane of cleavage; (e) cleavage just finished, the daughter nuclei moving away from the plane of cleavage (based on Kornmann 1965).

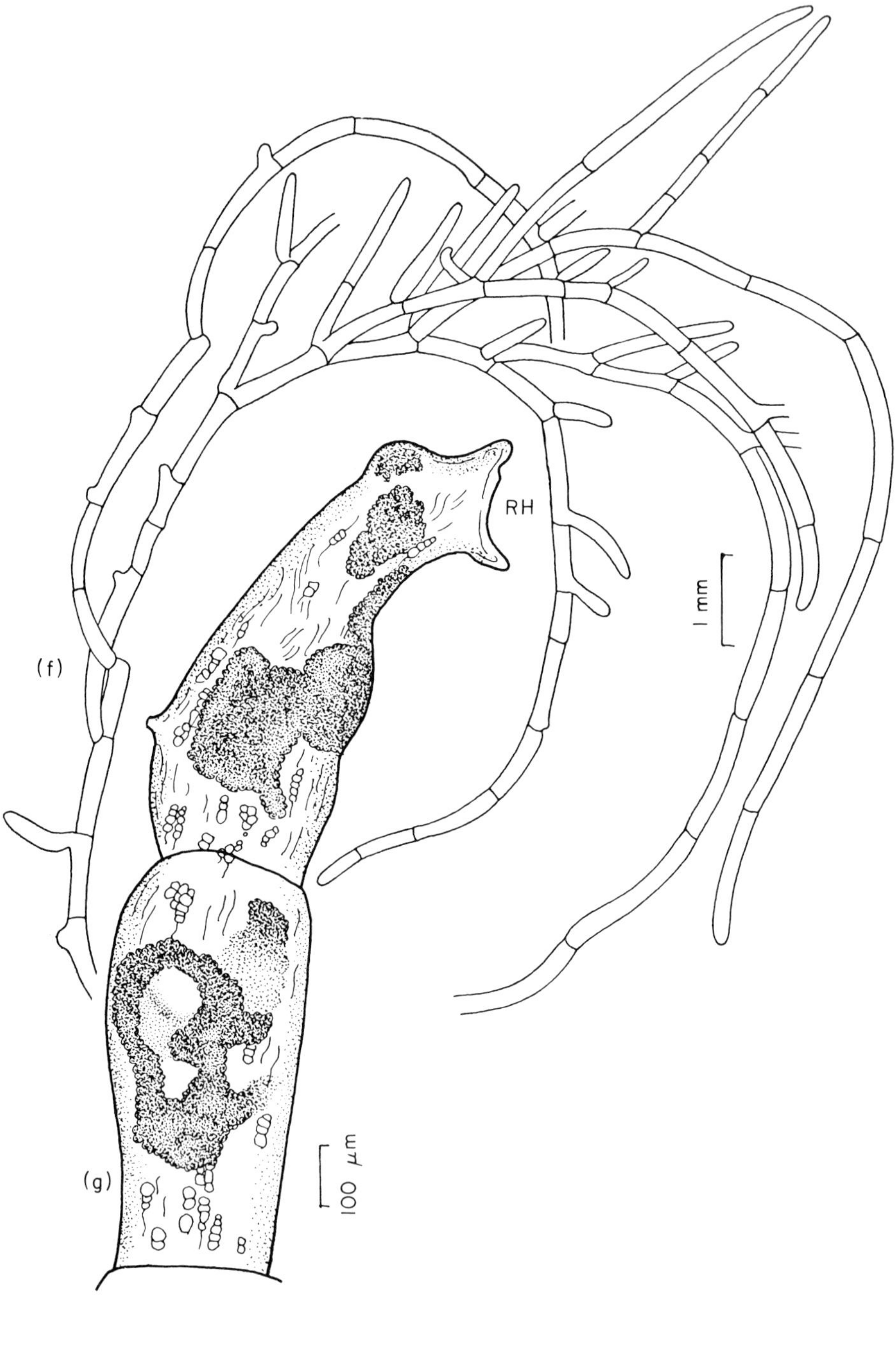
RH
1 mm
(f)
(g)
100 μm

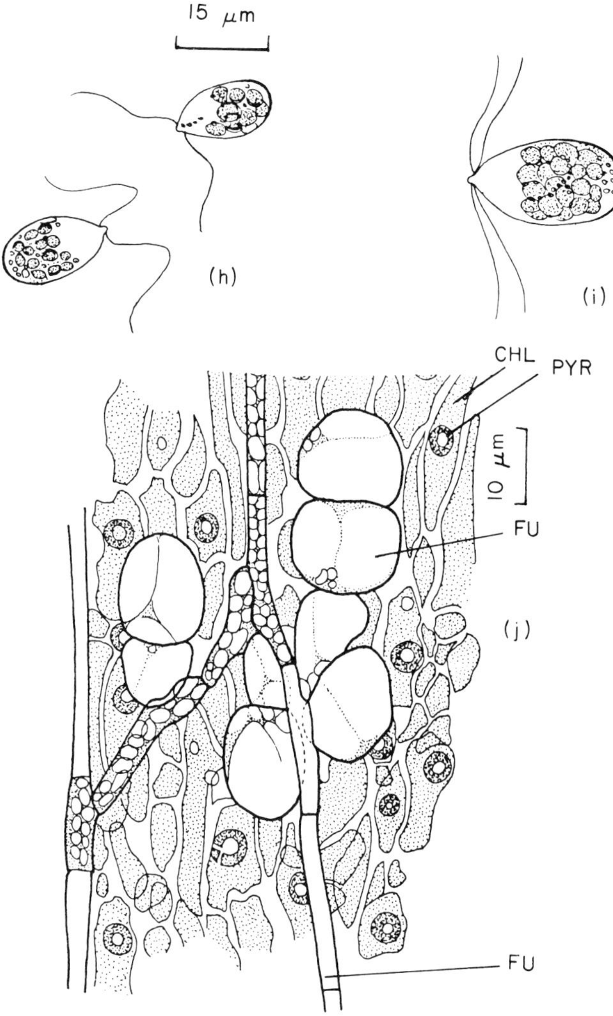

Fig. 3.6 (cont.) (f–j) *Cladophora catenata* (L.) Kütz. (= *Cl. fuliginosa* Kütz.). (f) Habit; (g) two terminal cells having divided their contents into a contracted, perforated cylinder of zoids, the apical cell ending in a hapteroid disc; (h) isogametes and (i) quadriflagellate zoospore (produced by separate plants); (j) surface view of cell showing polygonal chloroplasts with pyrenoids and fungus living in the cell wall (*Blodgettiomyces*) (original).

PYR = pyrenoid; CHL = chloroplast; FU = fungus growing in cell-wall; NU = nucleus; MINU = mitosing nucleus; RH = rhizoidal hapteron.

This construction resembles that of the multiaxial Rhodophyta (cf. Chapter 1). The development of these multiaxial structures shows considerable differences; for instance in *Penicillus* it starts as a discrete primordium formed by the swollen tip of a stolon (Fig. 3.7p) from which arise siphons growing upwards and downwards; whereas in *Codium* an undifferentiated tangle of siphons gives rise

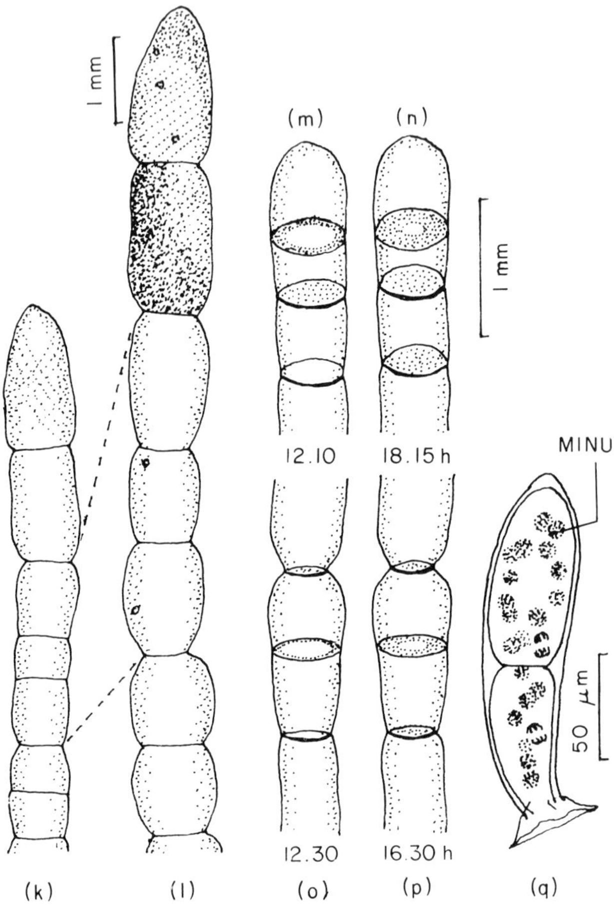

Fig. 3.6 (cont.) (k–q) *Chaetomorpha darwinii* (Hook.) Kütz. (k–l) Showing intercalary growth by cell enlargement; upper two cells in (l) are fertile; (m–p) growth by diaphragm-like divisions of apical and intercalary cells; (q) germling showing nuclei with approximately simultaneous divisions, spread over the whole cell (compare with (d, e)) (based on Kornmann 1969).

to mutually interwoven siphons forming the adult frond (Friedmann & Roth 1977).

The occurrence of analogous organizational levels in various algal classes is explained by assuming parallel evolution, within these classes, of the filamentous, thallose and siphonous levels from the primitive unicellular flagellate level, via the coccoid level. In Fott's (1970, 1971, 1974) opinion each level should correspond, in formal taxonomy, to an order. Accordingly, within the Chlorophyceae (Fott 1971) the flagellate level corresponds with the order Volvocales, the palmelloid level with the order Tetrasporales, the coccoid level with the order Chlorococcales, the filamentous level with the order Ulotrichales,

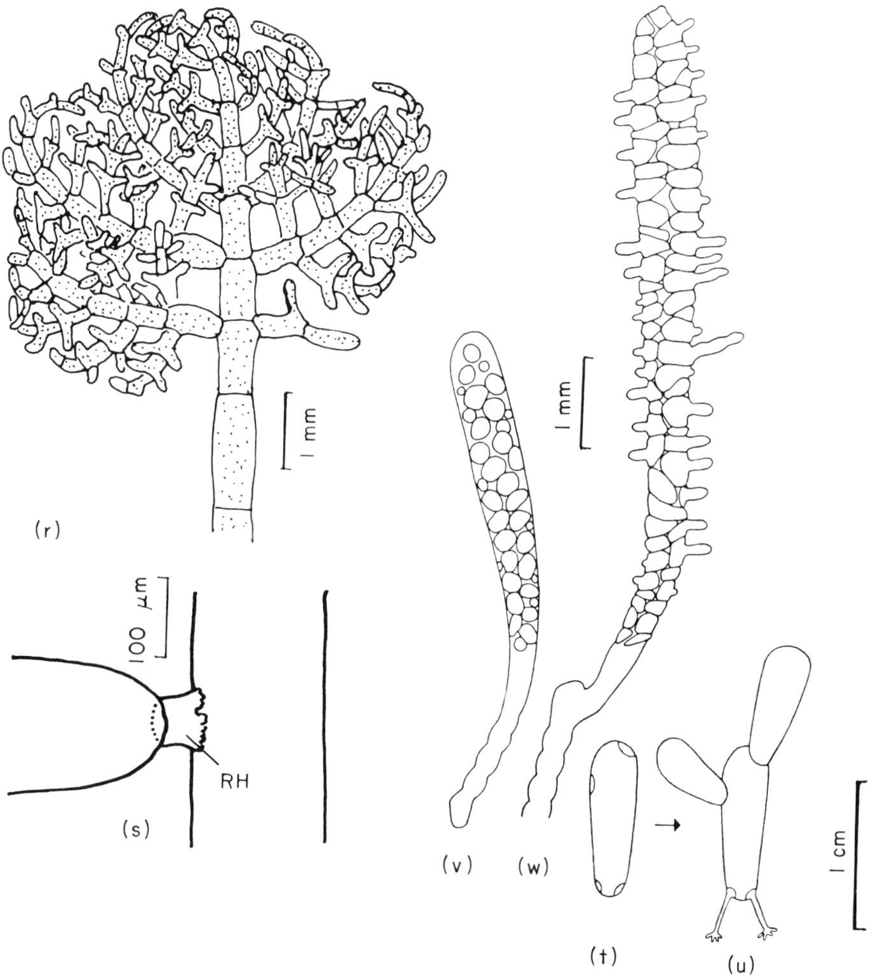

Fig. 3.6 (cont.) (r, s) *Struvea anastomosans* (Harv.) Picc. (r) Habit; (s) terminal hapteroid rhizoidal cell ((r) based on Vickers & Shaw 1908; (s) on Egerod 1952).

(t, u) *Valonia*, showing development of branches and rhizoids from small, cut off lenticular cells.

(v, w) *Siphonocladus tropicus* (Crouan) J. Ag., showing segregative cell division (after Egerod 1952).

the siphonous level with the order Siphonales (here: Caulerpales) and the siphonocladous level with the order Siphonocladales (here: Cladophorales). In addition, the Chlorosarcinales may contain the sarcinoid level (Groover & Bold 1969) (which is often placed in the Ulotrichales) and the Chaetophorales the

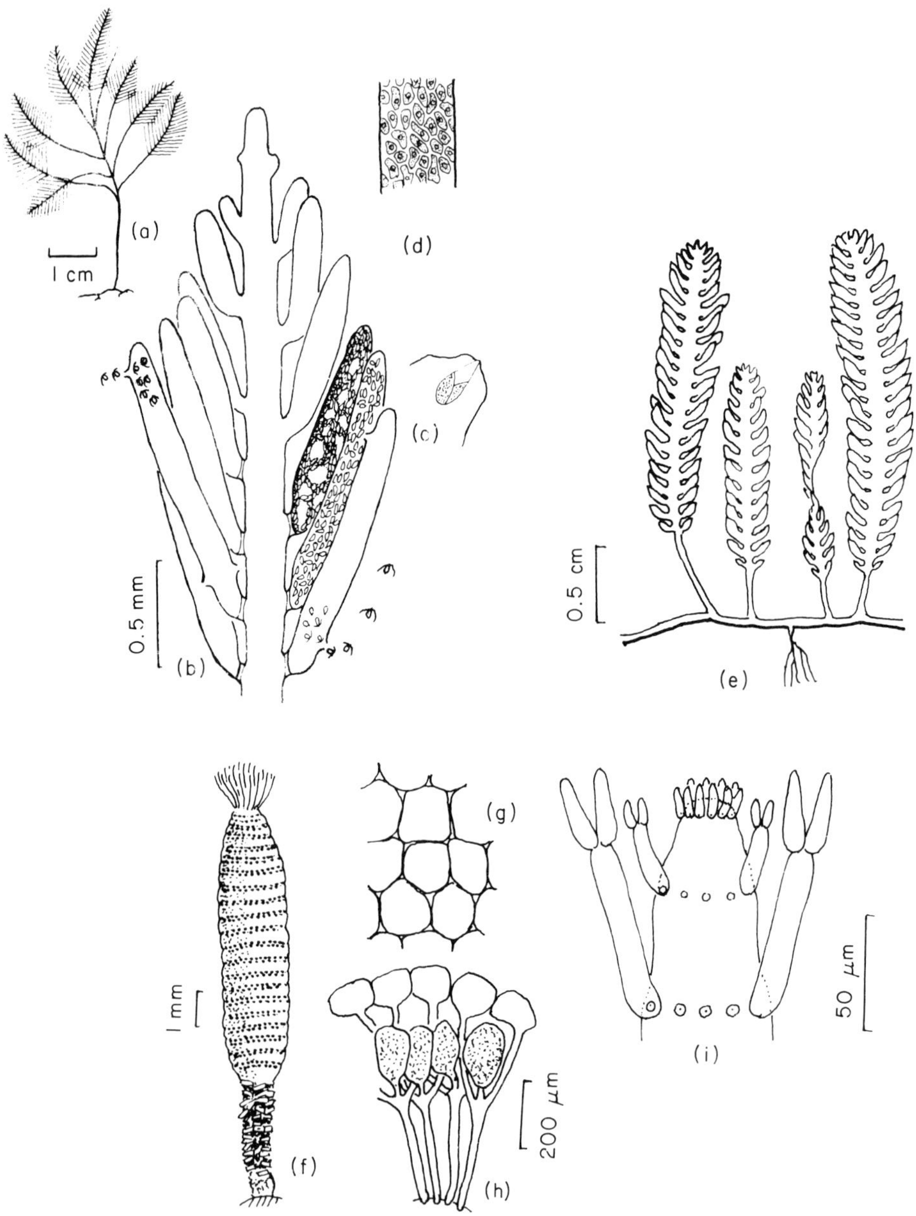

Fig. 3.7 (pp. 110–13) Siphonous organizational level.

(a–d) *Bryopsis plumosa* (Huds.) C.Ag. (a) habit; (b) laterals transformed into female gametangia; (c) female gamete; (d) chloroplasts with pyrenoids (after van den Hoek & Jahns 1978).

(e) *Caulerpa mexicana* (Sond.) J. Ag. (after Taylor 1960).

(f–i) *Neomeris annulata* Dickie. (f) habit; (g) surface view showing ends of radial laterals; (h) radial laterals bearing gametangia; (i) growing tip showing radial arrangement of laterals. ((f) after Taylor 1960; (g, h) after Joly 1967; (i) after Egerod 1952.)

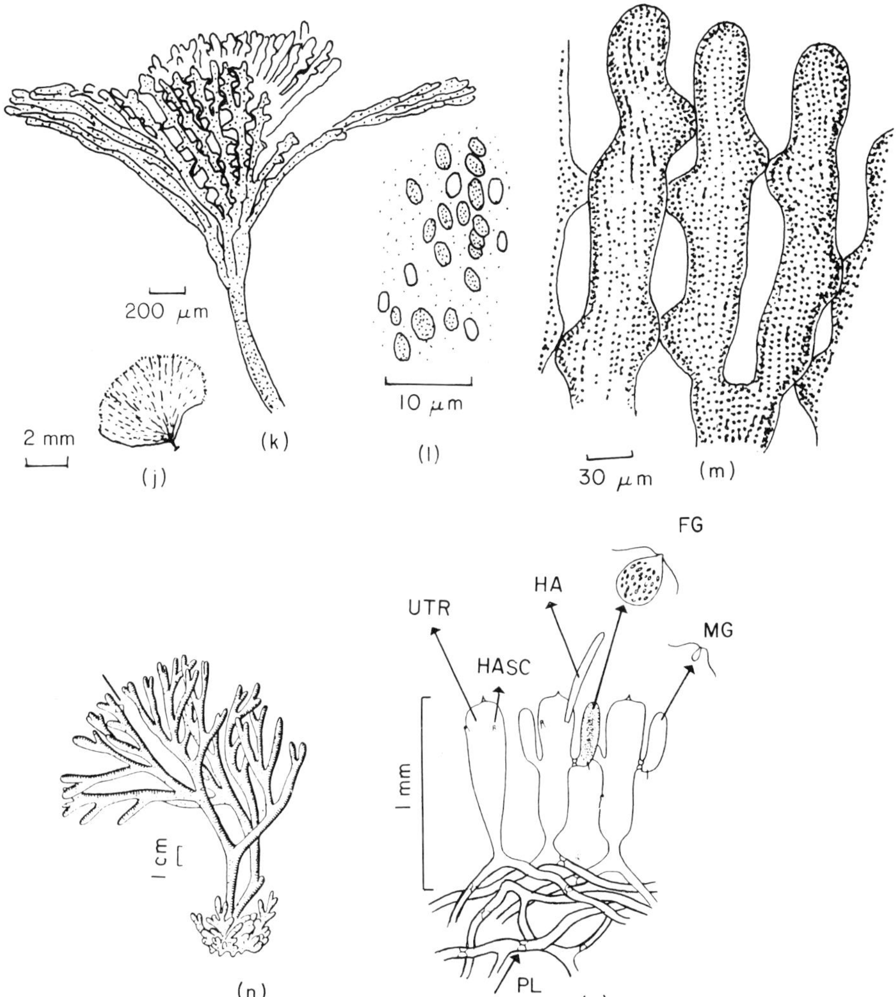

Fig. 3.7 (cont.) (j–m) *Rhipiliopsis reticulata* (van den Hoek) Farghaly & Denizot (1979). (j, k) Habit; (l) chloroplasts and amyloplasts; (m) detail of frond-structure (based on van den Hoek 1978).

(n–o) *Codium fragile* (Sur.) Hariot. (n) Habit; (o) structure showing medullary siphons and peripheral utricles; the utricles bearing hairs and female and male gametangia (after van den Hoek & Jahns 1978).

FG = female gamete; HA = hair; HASC = hair scar; MG = male gamete; PL = plug in siphon; UTR = utricle.

'branched filamentous' level (e.g. Bold & Wynne 1978). The above opinion, however, implies that each organizational level of one class should be internally homologous or, in other words, originated only once within each class. This is the reason why Fott (1974) placed the indubitably closely related 'coccoid' Desmidiaceae and 'filamentous' Zygnemataceae in a separate class, the

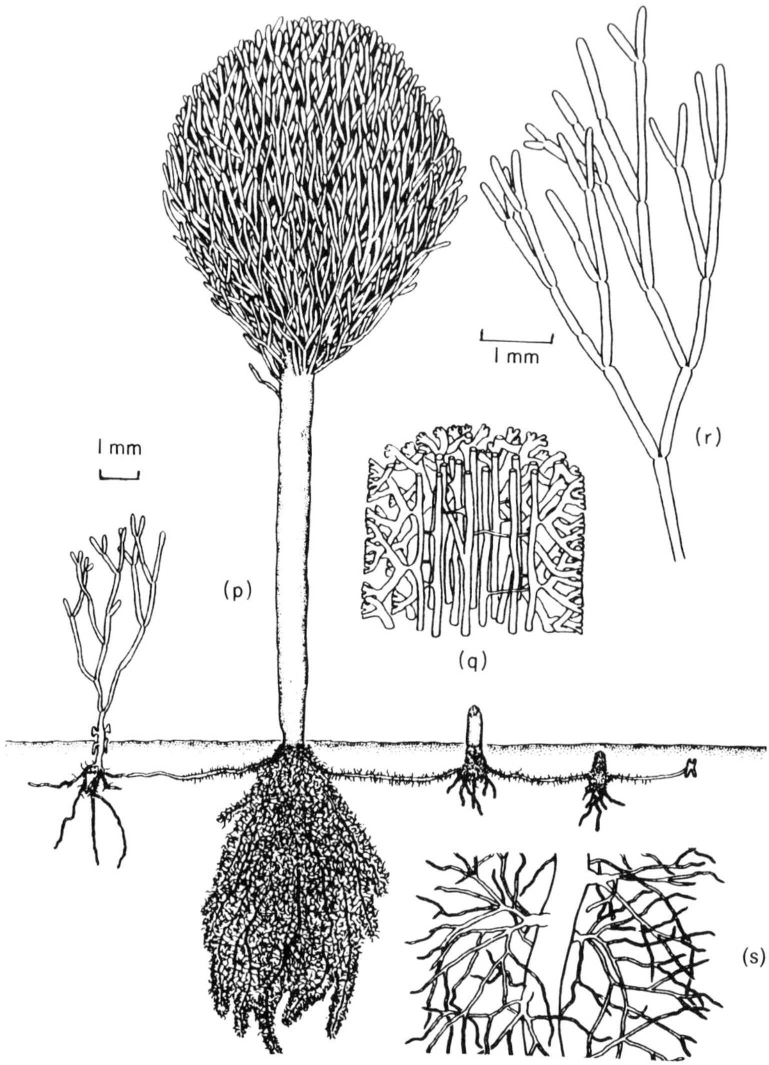

Fig. 3.7 (cont.) (p–s) *Penicillus capitatus* Lamarck. (p) Showing young plants and full-grown plant sprouting from stolon with rhizoids; (q) showing differentiation, in the stipe, of medullary filaments and cortex of utricles; (r) branched siphons in the 'brush'-like part (= capitulum); (s) branched rhizoids (after Friedmann & Roth 1977).

Conjugatophyceae, in which he thought the line of 'morphological advance' to have stopped with the unbranched filament. However, it is quite likely that these various organizational levels arose more than once within the Chlorophyceae, just as they arose more than once in all algal classes.

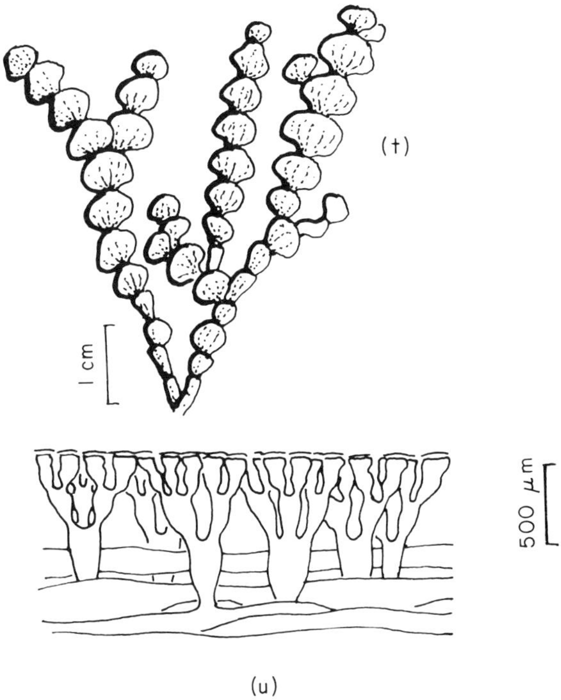

Fig. 3.7 (cont.) (t, u) *Halimeda*. (t) Habit of *Halimeda tuna* (Ell. & Sol.) Lamour; (u) longitudinal section of segment showing medullary filaments and cortex of utricles covered by calcareous layer (after Hillis 1959).

For instance 'siphonocladous' Acrosiphoniales and Cladophorales show a strikingly superficial resemblance and therefore were both placed until recently within the same order. However, they show considerable differences in details of their morphology, in chloroplast-structure, type of mitosis and cytokinesis, cell-wall structure and composition, and life-history, and it is reasonable to assume that they arose independently. Tiny siphonous *Characiosiphon* (a freshwater genus) clearly belongs to Chlorococcales and not Caulerpales, on the basis of its life-history (haplontic), type of mitosis–cytokinesis (phycoplast), and the absence of siphonein and siphonaxanthin (cf. Section 2.7) (Stewart *et al.* 1978). Such a 'polyphyletic' origin of the thallose level is suggested by the completely different ontogenies of various monostromatic blade-like Chlorophyceae ('*Monostroma*' *sensu lato*), if ontogenies are thought to contain vestiges of past phylogenies (Hirose & Yoshida 1964; Tatewaki 1969, 1972). The occurrence of the three types of zoid, three to four types of mitosis–cytokinesis, and three types of sexual fusion within the filamentous organizational level is a strong indication for the heterogeneity of this level and hence of the order Ulotrichales (including the Chaetophorales) (cf. Fig. 3.1). The filamentous level in the Chlorophyceae is therefore probably of multiple origin; this is paralleled by the multiple origin of this level in all algal classes regarded collectively.

The coccoid level is generally thought to be primitive with regard to the trichoid and thalloid level. There are, however, also arguments for the reverse possibility. For instance, in coccoid *Kirchneriella* transverse cell walls are formed in phycoplasts (Pickett-Heaps 1970). After that the cells separate by splitting along the transverse walls and are then transformed into autospores. Ontogenetically, in *Kirchneriella* the multicellular condition is therefore primitive, and the unicellular condition derived (van den Hoek & Jahns 1978). In other Chlorococcales cytokinesis much resembles that of *Kirchneriella* and takes place in a phycoplast (Pickett-Heaps 1975). There does not seem to be a fundamental difference between cell divisions leading to the formation of autospores and those giving rise to vegetative daughter cells ('eleutheroschisis' and 'desmoschisis' *sensu* Groover & Bold 1969). Therefore, although it is likely that the ancestral Chlorophyte was a unicellular flagellate, it is quite conceivable that 'regressive' evolution produced the unicellular coccoid level many times from the sarcinoid, filamentous and even the thallose organizational level.

The conclusion of this section is that the level of organization cannot be used as the sole or primary character separating chlorophycean orders, as organizational levels are likely to be heterogeneous. Several orders, however, are characterized by a very typical realization of the organizational level, for instance, the very characteristic architecture of the filamentous level in Oedogoniales and Zygnematales, and of the siphonocladous level in Acrosiphoniales and Cladophorales.

2.4 Types of life history and sexual fusion

Three types of life history occur in the Chlorophyta: the haplontic, the diplohaplontic and the diplontic life history (cf. Chapter 6). There are also three types of sexual fusion: iso-, aniso- and oogamy.

On the basis of karyological (Koop 1975; Liddle *et al.* 1976; Koop *et al.* 1977) and genetic evidence (Koop 1976), the Dasycladales is the only Chlorophycean order for which it is now reasonably certain that it is characterized by a diplontic life history. As long as the upright siphon-like plant remains vegetative it contains only one giant nucleus in the rhizoidal basal part. Gametogenesis starts with meiosis of this nucleus whose haploid daughter nuclei produce, via mitoses, a multitude of haploid nuclei some of which are in the end incorporated in gametangial cysts (one nucleus per cyst). After that the contents of the cyst divide into flagellate isogametes. These features are unique to the diplontic life history of the Dasycladales. The diplontic life history in the Prasiolales, for instance, is realized in a completely different way (see below).

As far as we know, in the order Caulerpales the *Bryopsis-Bryopsidella-Derbesia* complex is characterized by a diplohaplontic life history, which is nearly isomorphic in *Bryopsidella*, heteromorphic with a reduced gametophyte (the vesicular *Halicystis*-phase) in *Derbesia*, and heteromorphic with a reduced sporophyte in *Bryopsis* (see Fig. 6.4 in Chapter 6). The meiospores in this complex are stephanokontan (Rietema 1975). The life history of *Pseudobryopsis* much resembles that of *Bryopsis* (Tatewaki 1979). Both have reduced, creeping filamentous sporophytes with a giant nucleus and holocarpic production of stephanokontan zoospores. In different species or strains this filamentous stage gives rise directly to new *Bryopsis* plants (Rietema 1975), or *Pseudobryopsis* plants (Mayhoub 1974). Karyological evidence on the place of meiosis and the ploidy level of the phases is available for only *Derbesia* and *Bryopsidella* (Neumann 1969; Rietema 1972).

Other genera of Caulerpales are generally thought to have a diplontic life history (e.g. Chadefaud 1960; Fott 1971; Chapman & Chapman 1973), in which the anisogametes are meiotically formed by the diploid vegetative phase. Convincing evidence for the occurrence of meiosis during gametogenesis, however, is lacking. Zygotes of *Caulerpa* (Price 1972), *Udotea* (Meinesz 1969, 1972b), and *Halimeda* (Meinesz 1972a) have been shown to grow into new adult plants. In *Udotea* and *Halimeda* the zygote grows at first into a spherical uninucleate 'protosphere', which produces the initial of the adult plant. These data from cultures strongly suggest a diplontic life history, but possibly the 'protosphere' stage is comparable to the uninucleate filamentous stage of *Bryopsis* which is capable of either forming stephanokontan spores or of directly producing adult thalli. In that case the life history of *Udotea* and *Halimeda* could be heteromorphic diplohaplontic with a strongly reduced 'sporophyte' with vegetative meiosis. At present there are insufficient reasons to split the order Caulerpales into two or three siphonous orders either on basis of different types of life history (cf. Feldmann 1954; Chadefaud 1960; Chapman & Chapman 1973) or on the basis of the position of the nuclei outside or inside the layer of chloroplasts (Burr & West 1970; Rietema 1975). Heteroplastidic genera with holocarpic production of gametes (such as *Caulerpa, Halimeda, Udotea, Penicillus*) and homoplastidic genera with non-holocarpic production of gametes (*Bryopsis, Bryopsidella, Pseudobryopsis, Derbesia, Codium*) probably form two natural subgroups of the Caulerpales.

In at least one species of *Prasiola* (Prasiolales) spermatozoids and eggs are formed after meiosis of vegetative blade-cells followed by several mitoses; consequently the life history is diplontic. Each gamete is liberated separately from a small cell; therefore the gametangia are plurilocular, a unique feature in the Chlorophyceae (Friedmann 1959; Bravo 1965). In details of the oogamy and of the reproductive structures *Prasiola japonica* (Fujiyama 1955) and the closely related *Rosenvingiella constricta* (Kornmann & Sahling 1974) are very similar, but evidence points to a haplontic life history of the latter two species.

Stalked, thick-walled unicellular '*Codiolum*-like' phases occurring in the life histories of species of *Ulothrix, Monostroma, Acrosiphonia* and *Urospora* (Fig. 3.5h, i, u, v) can be interpreted as stalked hypnozygotes with attachment discs (Bold & Wynne 1978; van den Hoek & Jahns 1978). Consequently, the corresponding life histories, for example, those in the order Acrosiphoniales, can be interpreted as being haplontic, although evidence for the place of meiosis is lacking. Alternatively, the '*Codiolum*-like' phase can be interpreted as a strongly reduced, unicellular sporophytic phase (see Chapter 6) and the corresponding life history as heteromorphic diplohaplontic with a strongly reduced sporophyte, or 'nearly haplontic'. Actually, there are no sharp criteria (neither morphological, nor functional) for distinguishing unicellular sporophytes and hypnozygotes from one another, and therefore such a distinction is largely a matter of definition. Both are unicellular uninucleate diploid phases formed by fusion of gametes and permitting the survival over a period of unfavourable conditions.

The Cladophorales, which superficially resemble the Acrosiphoniales in thallus architecture, are characterized by an isomorphic diplohaplontic life history (Fig. 3.6h, i); this is supported by profuse karyological evidence (Föyn 1934a; Köhler 1956; Wik-Sjöstedt 1970; Patel 1971, 1972; Kornmann 1972; Verma & Verma 1976).

Within the thallose (= multicellular blade- and tube-like) organizational level a number of genera are characterized by an isomorphic diplohaplontic life history (Fig 3.5x–z) (for karyological evidence, see Föyn 1934b; Kapraun 1970, with further references; for genetic evidence, see Fjeld & Løvlie 1976), the development of the adult thallus from a juvenile monoseriate and subsequently tubular stage (Fig. 3.5 A, D, E), and the release of zoids through circular pores. This group is here regarded as the order Ulvales and encompasses the genera *Ulva, Enteromorpha, Ulvaria* (Bliding 1963, 1968; Gayral 1964, 1965), *Chloropelta* (Tanner 1980) and *Percursaria* (Kornmann 1956; Bliding 1963).

Other thallose chlorophytes are for the greater part monostromatic blades that often have (probably haplontic) life histories with hypnozygotes (or reduced unicellular sporophytes) and various types of early development and zoid release (e.g. Fig. 3.5q–v) (Kornmann 1962, 1964; Hirose & Yoshida 1964; Bliding 1968; Tatewaki 1969, 1972). These chlorophytes were until recently considered as species of the genus *Monostroma* but are presently dispersed over the genera *Monostroma s.s., Kornmannia* (Bliding 1968), *Gayralia* and *Protomonostroma* (Vinogradova 1969, 1974; Gayral 1971). These four genera are included here under the heterogeneous residual order Ulotrichales (Kornmann 1963; Parke & Dixon 1976). This is in contrast to other current systems which attribute them to the Ulvales (Bliding 1968; Vinogradova 1974, 1979).

Coleochaete (Coleochaetales) has a very characteristic type of oogamy and a haplontic life history. Meiosis takes place during germination of the hypnozygote (Allen 1905; Hopkins & McBride 1976).

The complex of the orders Ulotrichales, Chlorococcales and Volvocales seems to be characterized by haplontic life histories. Actually, it is the occurrence of hypnozygotes (or reduced unicellular sporophytes) which in the large majority of cases suggests the life histories to be haplontic (e.g. in *Ulothrix*, see Fig. 3.5c–j), but karyological and genetic evidence is comparatively rare, and it should be realized that hypnozygotes may also occur in diplontic life histories such as that of the watermould *Saprolegnia* (Bryant & Howard 1969; Howard & Moore 1970). Such evidence is available for a number of unicellular and colonial Volvocales (for review, see Bold & Wynne 1978), *Characiosiphon* (Chlorococcales) (Stewart *et al.* 1978), and several *Monostroma* species (for a review, see Tatewaki 1972). An alternation of isomorphic generations was postulated for certain marine *Ulothrix* populations (Perrot 1972), but this was not confirmed by later work and the only type of sexual cycle known for *Ulothrix* is an alternation between a gametophyte and a hypnozygote (or reduced sporophyte) generation (Lokhorst & Vroman 1972, 1974a, 1974b; Lokhorst 1978). Recently, an alternation of isomorphic phases was demonstrated in two marine species of the branched filamentous genus Entocladia (O'Kelly & Yarish 1981).

Iso-, aniso-, and oogamy occur in all three orders, in the order Volvocales and even within one genus, *Chlamydomonas*.

Most orders encompass numerous species for which only asexual reproduction is known (see for instance Yarish 1975). If life history type is considered as an important characteristic on the ordinal level, this implies that species without sexual reproduction are thought to have lost sexuality secondarily. An asexual mutant of *Derbesia marina* illustrates the possible derivation of purely asexual cycles from sexual cycles (Kornmann 1970).

The conclusion of this section is that types of life history and of sexual fusion as such can not be used as primary criteria for the distinction of chlorophycean orders any more than organizational level, because each type is heterogeneous and has probably arisen more than once in the course of evolution. However, the way in which the various life history types and sexual fusion are realized (e.g. by special features of the reproductive structures) is highly characteristic for the orders Cladophorales, Dasycladales, Acrosiphoniales, Zygnematales, Oedogoniales, Coleochaetales, and Ulvales. This is probably also valid for several other orders, but the available evidence is either incomplete (particularly with regard to the place of meiosis) or contradictory (e.g. for Prasiolales, and the heterogeneous order Ulotrichales).

2.5 Structure of the chloroplast and pyrenoids

Within the siphonous organizational level the orders are characterized by quite distinct types of chloroplast structure. The order Acrosiphoniales is characterized by the possession of one cylindrical, parietal reticulate chloroplast (Fig. 3.6d, e), which is filled with tightly appressed thylakoids and contains 'polypyramidal' pyrenoids (with radiating tubular thylakoid-extensions and covered by many radiating starch grains) (Fig. 3.2k); the order Cladophorales by the possession of numerous discoid chloroplasts united in a parietal reticulum (Fig. 3.6j) and which are filled with tightly appressed (often paired) thylakoids and usually contain 'bilenticular' pyrenoids (divided into two hemispheres by a single thylakoid, each hemisphere covered by a bowl-shaped starch grain) (Fig. 3.2j); the order Dasycladales by the possession of numerous discoid chloroplasts evenly filled with parallel thylakoids which may, or may not form granum-like structures and which lack pyrenoids (Fig. 3.2l); and the order Caulerpales by the possession of numerous discoid chloroplasts filled with lamellae (stacks) of various numbers of thylakoids. Within the latter order two groups can be distinguished: the homoplastidic group having only chloroplasts (with or without polypyramidal pyrenoids) (Figs 3.7d, 3.2m) and the heteroplastidic group (Fig. 3.7l) having amyloplasts in addition to chloroplasts, which characteristically contain a thylakoid-organizing body at one end of the chloroplast (Fig. 3.2n, o). Such a thylakoid-organizing body consists of concentric double membranes. The first group contains the marine genera *Derbesia, Bryopsis*, and multiaxial *Codium*, the second group contains *Caulerpa, Chlorodesmis*, and multiaxial *Halimeda, Udotea, Penicillus,* and *Avrainvillea* (Hori & Ueda 1967, 1975; Hori 1974; Borowitzka 1976; Colombo 1978; see also Jónsson 1962; van den Hoek 1963; Wik-Sjöstedt & Nordquist 1970; van den Driessche 1973; Chan *et al.* 1978). The heteroplastidic group is also characterized by holocarpic gamete-production (Meinesz 1972a, b, 1979).

Within the thallose organizational level the order Ulvales (as here conceived, see previous section) seems to be characterized by the pyrenoid being traversed by one to three thylakoids and being covered by two to several bowl-shaped starch plates. The monostromatic thalloid genera *Monostroma* (*sensu stricto*), *Kornmannia* and *Protomonostroma* (order Ulotrichales, see previous section) seem to have different characteristic pyrenoid-types (Hori 1972, 1973). Prasiolales are characterized by a central, stellate chloroplast with a central pyrenoid (Fig. 3.5o, p) (Friedmann 1959). However, within the trichoid genus *Ulothrix* at least four different pyrenoid-types can be distinguished, and these much resemble the types found in *Ulva, Enteromorpha* and *Monostroma* (Lokhorst 1978; Lokhorst & Star 1980).

Within the genus *Chlamydomonas* chloroplast- and pyrenoid structure vary widely, and this in contrast to other ultrastructural features (Ettl 1971), as it does in *Chlorococcum*-like Chlorococcales (Star 1955; Bold & Wynne 1978).

It can be concluded that the more derived chlorophycean orders (Cladophorales, Dasycladales, Caulerpales, Ulvales, Prasiolales, Oedogoniales, Zygnematales) are characterized by distinctive chloroplast- and/or pyrenoid-structure. Within the 'primitive' complex of Ulotrichales, Chlorococcales and

Volvocales, differences in chloroplast- and pyrenoid-structure may occur at the generic or even specific level (in *Chlamydomonas*).

2.6 Chemical composition and structure of the cell walls

Within the siphonous organizational level the structure and chemical composition of the cell wall seem to be characteristic at the ordinal level. In the Cladophorales the main wall polysaccharide is highly crystalline cellulose I, forming numerous lamellae of microfibrils oriented in a crossed fibrillar pattern. This is one important reason for merging the Siphonocladales with the Cladophorales (cf. Dawes 1969) (apart from similarities in thallus architecture, chloroplast and pyrenoid structure, and life-history types). The Charophyceae and higher plants also have crystalline cellulose walls. In the Acrosiphoniales mannan, xylan, glucan, and rhamnan constitute the cell wall in varying proportions (Bachman *et al.* 1976; Carlberg & Percival 1977). In the Dasycladales the main cell-wall polysaccharides are β-1,4-mannan and cellulose in varying proportions (for reviews, see Percival & McDowell 1967; Lewin 1974; Mackie & Preston 1974; Nisizawa & Sasaki 1975). The proportions vary not only with the genus but also with the morphological phase of the life history. For instance, in *Codium* mannan is the main polysaccharide; in *Bryopsis* and *Derbesia* xylan plus glucan are predominant in the walls of the gametophytic phase but mannan plus glucan predominate in the sporophytic phase (see Chapter 16). Undifferentiated *Codium*-filaments contain much glucan in addition to mannan in their walls, but differentiated *Codium* (multiaxial with utricles) has only traces of glucan (Huizing & Rietema 1975; Rietema 1975; Wutz & Zetsche 1976; Huizing *et al.* 1979). Mannan is the main constituent of the walls of vegetative cells in the Dasycladales, but cellulose I is dominant in the walls of gametangial cysts (Nisizawa *et al.* 1974; Herth *et al.* 1975). In the Acrosiphoniales (*Urospora*) mannan is the main wall polysaccharide of hypnozygotes, and glucan and xylan predominate in the vegetative phase (Bachman *et al.* 1976). Cell wall composition and structure should therefore be viewed as dynamic, changing with age and life history phase.

At present only very few species in the remaining green algae have been analysed critically with regard to the composition of the cell walls. Within the thallose level of organization xylomannan is mentioned for the Prasiolales and cellulose II for various Ulvales. Various filamentous Ulotrichales have cellulose II (reviews in Nisizawa & Sazaki 1975; van den Hoek & Jahns 1978).

Zoids covered with 0.05 μm large, diamond-shaped scales of the type that covers *Platymonas* and related green flagellates (at present often classified in the class Prasinophyceae; cf. van den Hoek & Jahns 1978) are now also known from widely divergent multicellular chlorophytes such as the unilateral spermatozoids of the Charophyceae, the unilateral zoids of Coleochaetales, and the cruciate zoids of *Ulothrix zonata* (Sluiman *et al.* 1980a, b), *Trichosarcina polymorpha, Pseudendoclonium basiliense*, and *Monostroma grevillei* (Ulotrichales) (Moestrup 1978). If these diamond-shaped scales are homologous, then their occurrence in such taxonomically distant chlorophytes can only be explained by considering them as vestiges of an ancestral, scale-covered green flagellate which were retained in some, but lost in most during a very early phylogenetic radiation in the chlorophytes. Probably scaly coverings were not phylogenetically transformed into microfibrillar polysaccharide-walls (as hypothesized by Stewart & Mattox 1978) but rather replaced by them, such as occurs ontogenetically in the developing zoids of *Monostroma grevillei*.

In this way the picture emerges of the hypothetical chlorophytan ancestor as an isokontan, scale-covered flagellate of the cruciate type, with a persistent telophase spindle, a closed mitosis, and cytokinesis by simple cleavage (Moestrup 1978; see also Sections 2.1 and 2.2). Apparently scale-covered green flagellates are part of the Chlorophyceae and should not be split off as the separate class Prasinophyceae (Stewart & Mattox 1975a; van den Hoek & Jahns 1978). Here they are considered as part of the order Volvocales. In *Chlamydomonas* and other non-scaly Volvocales the cell-wall consists of a crystalline glycoprotein-lattice (largely composed of hydroxyproline bound to arabinose and galactose) clasped between heteropolysaccharide-layers (Hills *et al.* 1973; Roberts 1974; Catt *et al.* 1976). This cell-wall type possibly replaced the scaly covering. In its turn, this cell-wall type is ontogenetically replaced by a layer of non-crystalline cellulose in the developing zoid of *Chlorococcum* (Miller 1978). This suggests how chlorophytes with cellulose walls can possibly be derived from scaly green flagellates.

The conclusion of this section is that a number of 'derived' orders are characterized by their cell-wall structure and composition (Cladophorales, Acrosiphoniales, Dasycladales, Caulerpales). Data are too incomplete to permit generalizations for the remaining orders.

2.7 The presence or absence of the xanthophylls siphonein and siphonaxanthin

The combined occurrence of the accessory photosynthetic pigments siphonein and siphonaxanthin has been determined in 36 species of the order Caulerpales (all species investigated); siphonaxanthin alone has been found in seven species and not in 27 species of the order Cladophorales (including Siphonocladales); and siphonein alone in two species of the order Dichotomosiphonales. Both pigments are lacking in Dasycladales, Acrosiphoniales, Sphaeropleales (for a review, see Goodwin 1974) and the tiny siphonous *Characiosiphon* (Chlorococcales) (Stewart *et al.* 1978). Within the siphonous organizational level the distribution of these two pigments therefore supports the division into orders.

Siphonaxanthin has also been discovered in *Microthamnion* (Weber 1973) and in *Ulva japonica* (but not in *U. pertusa* and two *Enteromorpha* species)

(Ikemori & Arasaki 1977; Kageyama & Yokohama 1977; Yokohama *et al.* 1977).

3 GENERAL DISCUSSION

Fig. 3.1 can be interpreted as a cross-section through an imaginary evolutionary tree of the chlorophytes in the plane of the present. The distances from the centre give an indication of the degree of derivation of each group (cf. Table 3.1). The broken lines can be interpreted as projections of past phylogenetic ramifications in the plane of the present. Figure 3.1 also illustrates that phylogenetic speculation about chlorophytes is, in all sobriety, a comparative morphological exercise with extant organisms.

Figure 3.1 illustrates the concept that organizational levels, types of life history, and types of sexual fusion are grades (increasingly advanced evolutionary levels) but that types of zoid and types of mitosis–cytokinesis represent clades (monophyletic units) (Huxley 1957). The derived organizational levels (e.g., the thallose, siphonous and 'conductive tissue' level) arose several times from the primitive ones (coccoid and/or filamentous levels). The derived life history types (the nearly diplontic, the diplohaplontic life histories) arose several times from the primitive haplontic life history. Oogamy arose many times from anisogamy. One branch with unilateral zoids and the almost completely concomitant phragmoplastic cytokinesis arose only once from a primitive chlorophyte with cytokinesis, by simple cleavage in a persistent telophase-spindle, and with the cruciate zoid-type; another branch with the cruciate zoid-type and with phycoplastic cytokinesis arose also only once from the same primitive ancestral chlorophytan group, which possibly also gave rise to the siphonous groups Dasycladales, Caulerpales and Cladophorales and the thallose group Ulvales without much changing the type of mitosis–cytokinesis.

In Fig. 3.1 the orders Oedogoniales, Zygnematales, Cladophorales, Dasycladales, Caulerpales, Acrosiphoniales, Ulvales, the class Charophyceae, and the divisions Bryophyta and Tracheophyta appear as well-defined groups with distinct sets of characters. This is emphasized by the fact that the organizational level and life history type of each group are realized by group-specific features (for more extensive recent descriptions, see Bold & Wynne 1978; van den Hoek & Jahns 1978).

The recently proposed orders Coleochaetales and Klebsormidiales (Stewart & Mattox 1975) also seem to be well-defined. However, for each order only two species are known, belonging to *Coleochaete* and *Chaetosphaeridium* in Coleochaetales, and to *Klebsormidium* and *Chlorokybus* in Klebsormidiales. The reason is that these orders are primarily characterized by ultrastructural features (zoid-type and type of mitosis–cytokinesis), which future EM-research will possibly discover in other algae now classified under the order Ulotrichales.

A recent redefinition of the class Charophyceae as including the orders

Zygnematales, Klebsormidiales, Coleochaetales and Charales (Stewart & Mattox 1975, 1978; Moestrup 1978) is not accepted here, as it disregards striking differences in thallus architecture and in reproductive features. Such a redefinition is not supported by Fig. 3.1.

Figure 3.1 clearly illustrates that the order Ulotrichales is heterogeneous, namely, with regard to the type of mitosis–cytokinesis, the type of life-history and sexual fusion, and the presence or absence of diamond-scales on the zoids. The order is here characterized by its organizational level (mainly filamentous) and life history type (which is supposedly mainly haplontic, but actual knowledge on this topic is still inadequate). It includes the orders Chaetophorales and Chlorosarcinales of other taxonomies. Possibly the order could be divided into one group with a persistent telophase spindle and simple cleavage (Ulotrichales) and another group with a collapsing telophase-spindle and a phycoplast with cleavage or cell-plate (Chaetophorales). The first group could consist of the filamentous descendants of the hypothetical ancestral, scale-covered green flagellate with type I mitosis–cytokinesis, and the second group of the filamentous descendants of the hypothetical scale-covered green flagellate in which type II mitosis–cytokinesis had evolved from type I (Fig. 3.4; cf. Section 2.2) (cf. Stewart & Mattox 1975a, b). However, the distinction of orders on the basis of the type of mitosis–cytokinesis alone, and the inadequacy of our knowledge of taxonomically important characters of many algae placed in Ulotrichales *sensu lato*, make such a division at present impracticable and undesirable. Therefore the order Ulotrichales is here maintained in a broad sense, but it is considered heterogeneous and residual, possibly containing unrecognized members of the newly recognized Klebsormidiales and Coleochaetales (Stewart & Mattox 1975).

The following orders seem to be well-founded on the basis of architecture of reproductive and vegetative structures, but their correct place in Fig. 3.1 is uncertain because information on type of life history and/or type of mitosis–cytokinesis is inadequate or lacking: Trentepohliales (cf. Bourrelly 1966; Chapman & Good 1977, 1978), Prasiolales, Sphaeropleales and Dichotomosiphonales. The Chlorococcales differ from Ulotrichales and Volvocales with cytokinesis by phycoplasts (type II, Fig. 3.4), only by the coccoid organizational level (no coccoid green algae are known to have type I mitosis–cytokinesis, query-mark in Fig. 3.1). This characteristic is insufficient for defining an order, particularly because there are species with intermediate morphologies and because there are reasons to believe that the coccoid level may have arisen more than once from the flagellate level, as well as from the sarcinoid and filamentous levels by 'regressive evolution' (cf. Section 2.3). In the future the taxonomic status of this order must be reconsidered in conjunction with a rearrangement of the Ulotrichales. Scaly green flagellates should not be placed in a separate class (Prasinophyceae) as zoids covered with similar scales are now known from various multicellular green chlorophytes (Fig. 3.1).

The heterogeneity of, and at the same time the apparent absence of sharp boundaries within, the Ulotrichales-Chlorococcales-Volvocales complex possibly reflects its relative primitiveness in that it contains many forms which have changed little since an early phylogenetic radiation in the chlorophytes. All three orders in this complex encompass marine, freshwater, as well as aerophytic species or genera (Fig. 3.1). Interestingly, the well-defined and more derived subgroups of the Chlorophyta are either entirely or predominantly marine and at the same time iso- to anisogamous; or they are limnetic (growing in freshwater) and/or aerophytic and at the same time oogamous (Fig. 3.1).

Of the marine chlorophytes the siphonous multiaxial, lime-encrusted, heteroplastidic, anisogamous, and possibly diplontic Caulerpales such as *Halimeda* and *Penicillus* (Fig. 3.7p–u) can be considered as most derived. Of the freshwater chlorophytes, the Charophyceae are most derived (siphonous with a highly specialized architecture, with unilateral zoids, phragmoplasts, oogamy, but rather primitively, a haplontic life history). It is perhaps not merely by chance that the Klebsormidiales, which seem to most resemble the ancestral group from which arose the aerophytic 'chlorophytes' with phragmoplasts and unilateral reproductive zoids (Bryophyta and Tracheophyta), contain a freshwater as well as an aerophytic species.

The green algae plus stoneworts, the mosses plus liverworts, and the vascular plants are here regarded as the mutually related divisions Chlorophyta, Bryophyta and Tracheophyta, respectively. An alternative possibility would be to include these three divisions in one division, the Chlorophyta, containing the subdivisions Chlorophytina (embracing the classes Chlorophyceae and Charophyceae), Bryophytina and Tracheophytina (van den Hoek & Jahns 1978). This classification would express the concept that green algae and higher plants are much more related with each other than with any other plant group.

Both of these classifications have the advantage of being practical by maintaining classes as they are generally conceived. Classes are, in algal systematics, the primary operational groups of high hierarchical level, and therefore their status should not be unnecessarily changed. The Chlorophyceae, as here conceived, conform largely to much traditional usage (e.g. Fritsch 1948; Chapman & Chapman 1973; Stewart 1974; Parke & Dixon 1976; Bold & Wynne 1978; van den Hoek & Jahns 1978) and a subdivision Tracheophytina as well as a division Tracheophyta can be subdivided into classes like Lycopsida, Coniferopsida, Ginkgopsida, Magnoliopsida, etc. (Kalkman 1972).

For the same practical reason it is undesirable to elevate chlorophycean orders to class-level (as proposed by Round 1971; see also Fott 1971, for Conjugatophyceae instead of Zygnematales, and Vinogradova 1979, for Siphonophyceae instead of Caulerpales). The choice of the hierarchical level of a group like Dasycladales is rather arbitrary in that orders, classes, and divisions cannot be really defined as phylogenetically equivalent entities; for instance, we are not capable of saying whether or not the order Rosales (Magnoliopsida) is phylogenetically equivalent to the chlorophycean orders Dasycladales and Volvocales.

ACKNOWLEDGEMENTS

The author wishes to thank Sibylla Nijhof and Jeannet Schröder-ter Avest for typing several versions of the manuscript.

4 REFERENCES

ALLEN C. E. (1905) Die Keimung der Zygote bei *Coleochaete*. *Ber. Dtsch. Bot. Ges.*, **23**, 285–92.

ANTIA N. J. (1977) A critical appraisal of Lewin's Prochlorophyta. *Br. Phycol. J.*, **12**, 271–6.

BACHMAN P., KORNMANN P. & ZETSCHE K. (1976) Regulation der Entwicklung und des Stoffwechsels der Grünalge *Urospora* durch die Temperatur. *Planta*, **128**, 241–5.

BECH-HANSEN C. W. & FOWKE L. C. (1972) Mitosis in *Mougeotia* sp. *Can. J. Bot.*, **50**, 1811–16.

BERGHS J. (1906) Le noyau et la cinèse chez le *Spirogyra*. *La Cellule*, **23**, 55–86.

BLIDING C. (1963) A critical survey of European taxa in Ulvales. Part I, *Capsosiphon, Percursaria, Blidingia, Enteromorpha*. *Opera botanica, Lund*, **8**, 1–160.

BLIDING C. (1968) A critical survey of European taxa in Ulvales. II. *Ulva, Ulvaria, Monostroma, Kornmannia*. *Bot. Not.*, **121**, 535–629.

BOLD H. C. & WYNNE M. J. (1978) *Introduction to the Algae. Structure and Reproduction*. Prentice-Hall, Englewood Cliffs, New Jersey.

BOROWITZKA M. A. (1976) Some unusual features of the ultrastructure of the chloroplasts of the green algal order Caulerpales and their development. *Protoplasma*, **89**, 129–47.

BOURRELLY P. (1966) *Les Algues d'Eau Douce. Initiation à la Systématique, Tome I.* Les Algues Vertes. N. Boubée, Paris.

BRAVO L. M. (1965) Studies on the life history of *Prasiola meridionalis*. *Phycologia*, **4**, 177–94.

BRYANT T. R. & HOWARD K. L. (1969) Meiosis in Oomycetes. I. A microspectrophotometric analysis of nuclear deoxyribonucleic acid in *Saprolegnia terrestris*. *Am. J. Bot.*, **56**, 1075–83.

BURR F. A. & WEST J. A. (1970) Light and electron microscope observations on the vegetative and reproductive structures of *Bryopsis hypnoides*. *Phycologia*, **9**, 17–37.

CARLBERG G. E. & PERCIVAL E. (1977) The carbohydrates of the green seaweeds *Urospora wormskioldii* and *Codiolum pusillum*. *Carbohydr. Res.*, **57**, 223–34.

CATT J. W., HILLS G. J. & ROBERTS K. (1976) A structural glycoprotein containing hydroxyproline, isolated from the cell wall of *Chlamydomonas reinhardii*. *Planta*, **131**, 165–71.

CHADEFAUD M. (1960) Les végétaux non vasculaires, cryptogamie. In *Traité de Botanique, Systématique*, vol. 1, eds. Chadefaud M. & Emberger L., p. 1018. Masson, Paris.

CHAN K.-Y, WONG S. L. L. & WONG M. H. (1978) Observations on *Chaetomorpha brachygona* Harv. (Chlorophyta, Cladophorales). I. Ultrastructure of the vegetative cells. *Phycologia*, **17**, 419–29.

CHAPMAN R. L. (1977) Scanning and transmission electron microscopic observations on zoosporogenesis in *Cephaleuros virescens* (Chlorophyta; Chroolepidaceae). *J. Phycol.*, **13** (suppl.), abstr. 57.

CHAPMAN R. L. (1979) Observations on the biflagellate gametes of *Cephaleuros* (Chroolepidaceae) and symmetry in green algal monads. *J. Phycol.*, **15** (suppl.), abstr. 19.

CHAPMAN R. L. & GOOD B. H. (1977) Some comparisons among *Cephaleuros, Phycopeltis, Trentepohlia* and other green algae. *J. Phycol.*, **13** (suppl.), abstr. 58.

CHAPMAN R. L. & MADISON S. W. (1976) Observations on the zoospores of *Cephaleuros virescens* (Chroolepidaceae). *J. Phycol.* **12** (suppl.), abstr. 70.

CHAPMAN V. J. & CHAPMAN D. J. (1973) *The Algae*. 2nd ed. MacMillan, London.

COLOMBO P. M. (1978) An ultrastructural study of thallus organization in *Udotea petiolata*. *Phycologia*, **17**, 227–35.

DAWES C. J. (1969) A study of the ultrastructure of a green alga, *Apjohnia laetevirens* Harvey with emphasis on cell wall structure. *Phycologia*, **8**, 77–84.

DEASON T. R. & O'KELLEY J. C. (1979) Mitosis and cleavage during zoosporogenesis in several coccoid green algae. *J. Phycol.*, **15**, 371–8.

DEASON T. R., RYALS P. E., O'KELLEY J. C. & BULLOCK K. W. (1979) Fine structure of mitosis and cleavage in *Friedmannia israelensis* (Chlorophyceae: Chlorosarcinaceae). *J. Phycol.*, **15**, 452–7.

EGEROD L. E. (1952) An analysis of the siphonous Chlorophycophyta. *Univ. of California Publ. Bot.*, **25**, 325–454.

ETTL H. (1971) *Chlamydomonas* als geeigneter Modellorganismus für vergleichende cytomorphologische Untersuchungen. *Arch. Hydrobiol.* **39**, *Algological Studies* 5, 259–300.

EVANS L. V. & CHRISTIE A. O. (1970) Studies on the shipfouling alga *Enteromorpha*. I. Aspects of the fine-structure and biochemistry of swimming and newly settled zoospores. *Ann. Bot.*, **34**, 451–66.

FARGHALY M. S. & DENIZOT M. (1979). Le genre *Rhipiliopsis*. Définition et place dans les Caulerpales. *Rev. Algol.*, N.S., **14**, 169–84.

FELDMANN J. (1954) Sur la classification des Chlorophycées siphonées. *VIII Congrès Int. Bot. Rapp. Comm. Sect.*, **17**, 96–8.

FJELD A. & LØVLIE A. (1976) Genetics of multicellular marine algae. In *The Genetics of Algae*, ed. Lewin R. A., pp. 219–35. Blackwell Scientific Publications, Oxford.

FLOYD G. L. (1978) Mitosis and cytokinesis in *Asteromonas gracilis*, a wall-less, green monad. *J. Phycol.*, **14**, 440–5.

FLOYD G. L., STEWART K. D. & MATTOX K. R. (1972a) Comparative cytology of *Ulothrix* and *Stigeoclonium*. *J. Phycol.*, **8**, 68–81.

FLOYD G. L., STEWART K. D. & MATTOX K. R. (1972b) Cellular organization, mitosis, and cytokinesis in the ulotrichalean alga, *Klebsormidium*. *J. Phycol.*, **8**, 176–84.

FOTT B. (1970) Taxonomische Einheiten in der Algologie. *Hydrobiologia*, **35**, 7–16.

FOTT B. (1971) *Algenkunde*. VEB Gustav Fischer Verlag, Jena.

FOTT B. (1974) The phylogeny of eukaryotic algae. *Taxon*, **23**, 446–61.

FOWKE L. C. & PICKETT-HEAPS J. D. (1969a) Cell-division in *Spirogyra*. I. Mitosis. *J. Phycol.*, **5**, 240–59.

FOWKE L. C. & PICKETT-HEAPS J. D. (1969b) Cell division in *Spirogyra*. II. Cytokinesis. *J. Phycol.*, **5**, 273–81.

FÖYN B. (1934a) Lebenszyklus, Cytologie und Sexualität der Chlorophycee *Cladophora suhriana* Kg. *Arch. Protistenk.*, **83**, 1–56.

FÖYN B. (1934b) Lebenszyklus und Sexualität der Chlorophycee *Ulva lactuca* L. *Arch. Protistenk.*, **83**, 154–77.

FRIEDMANN I. (1959) Structure, life-history and sex determination of *Prasiola stipitata* Suhr. *Ann. Bot.*, N.S., **23**, 571–94.

FRIEDMAN I. & ROTH W. C. (1977) Development of the siphonous green alga *Penicillus* and the *Espera* state. *Bot. J. Linn. Soc.*, **74**, 189–214.

FRITSCH F. E. (1948) *The Structure and Reproduction of the Algae*. vol. I. Cambridge University Press, Cambridge.

FUJIYAMA T. (1955) On the life-history of *Prasiola japonica* Yatabe. *J. Fac. Fish. Anim. Husb. Hiroshima Univer.*, **1**, 15–37.

GAYRAL P. (1964) Sur le démembrement de l'actuel genre *Monostroma* Thuret (Chlorophycées, Ulotricales). *C.R. Acad. Sci., Paris*, **258**, 2149–52.

GAYRAL P. (1965) *Monostroma* Thuret, *Ulvaria* Rupr. emend. Gayral, *Ulvopsis* Gayral (Chlorophycées, Ulotrichales). Structure, reproduction, cycles, position systématique. *Rev. Gén. Bot.*, **72**, 627–38.

GAYRAL P. (1971) Mise au point sur la systématique de l'ordre des Ulvales. *Bull. Soc. Phycol. Fr.*, **16**, 63–7.

GOODWIN T. W. (1974) Carotenoids and biliproteins. In *Algal Physiology and Biochemistry*, ed. Stewart W. D. P., pp. 176–205. Blackwell Scientific Publications, Oxford, England.

GRAHAM L. E. & MCBRIDE G. E. (1975) The ultrastructure of multilayered structures associated with flagellar bases in motile cells of *Trentepohlia aurea*. *J. Phycol.*, **11**, 86–96.

GRAHAM L. E. & MCBRIDE G. E. (1978) Mitosis and cytokinesis in sessile sporangium of *Trentepohlia aurea* (Chlorophyceae). *J. Phycol.*, **14**, 132–7.

GROOVER R. D. & BOLD H. C. (1969) *Phycological studies. VIII. The taxonomy and comparative physiology of the Chlorosarcinales and certain other edaphic algae*. Univ. Texas Publ., 6907, Austin.

HAMEL G. (1930) Chlorophycées des côtes françaises. *Rev. Algol.*, **5**, 1–54, 383–430.

HAMEL G. (1932) Chlorophycées des côtes françaises. *Rev. Algol.*, **6**, 9–73.

HERTH W., KUPPEL A. & FRANKE W. W. (1975) Cellulose in *Acetabularia* cyst walls. *J. Ultrastr. Res.*, **50**, 289–92.

HILLIS L. W. (1959) A revision of the genus *Halimeda* (order Siphonales). *Inst. Mar. Sci. Univ., Texas*, **6**, 321–403.

HILLS G. J., GUERNEY-SMITH M. & ROBERTS K. (1973) Structure, composition and morphogenesis of the cell wall of *Chlamydomonas reinhardii*. II. Electronmicroscopy and optical diffraction analysis. *J. Ultrastr. Res.* **43**, 79–192.

HIROSE H. & YOSHIDA K. (1964) A review of the life histories of the genus *Monostroma*. *Bull. Jap. Soc. Phycol.*, **12**, 19–31.

HOPKINS A. W. & MCBRIDE G. E. M. (1976) The life history of *Coleochaete scutata* (Chlorophyceae) studied by a Feulgen microspectrophotometric analysis of the DNA-cycle. *J. Phycol.*, **12**, 29–35.

HORI T. (1972) Ultrastructure of the pyrenoid of *Monostroma* (Chlorophyceae) and related genera. In *Contributions to the Systematics of Benthic Marine Algae of the North Pacific*, eds. Abbott I. A. & Kurogi M., pp. 17–32. Japanese Society of Phycology, Kobe, Japan.

HORI T. (1973) Comparative studies of pyrenoid ultrastructure in algae of the *Monostroma*-complex. *J. Phycol.*, **9**, 190–9.

HORI T. (1974) Electron microscope observations on the fine structure of the chloroplasts of algae. 2. The chloroplasts of *Caulerpa* (Chlorophyceae). *Int. Rev. Gesamt. Hydrobiol.*, **59**, 239–45.

HORI T. (1977) Electron microscope observations on the flagellar apparatus of *Bryopsis maxima* (Chlorophyceae)., *J. Phycol.*, **13**, 238–43.

HORI T. & ENOMOTO S. (1978) Electron microscope observations on the nuclear division in *Valonia ventricosa* (Chlorophyceae, Siphonoclades). *Phycologia*, **17**, 133–42.

HORI T. & UEDA R. (1975) The fine structure of algal chloroplasts and algal phylogeny. In *Advance of Phycology in Japan*, eds. Tokida J. & Hirose H., pp. 11–42. VEB Gustav Fischer Verlag, Jena.

HORI T. & UEDA R. (1976) Electron microscope studies on the fine structure of plastids in siphonous green algae with special reference to their phylogenetic relationships. *Sci. Rep. Tokyo Kyoiku Daigaku*, **12 (B)**, 225–44.

HOWARD K. L. & MOORE R. T. (1970) Ultrastructure of oogenesis in *Saprolegnia terrestris*. *Bot. Gaz.*, **131**, 311–36.

HUDSON P. R. & WAALAND J. R. (1974) Ultrastructure of mitosis and cytokinesis in the multinucleate green alga *Acrosiphonia*. *J. Cell Biol.*, **62**, 274–94.

HUIZING H. J. & RIETEMA H. (1975) Xylan and mannan as cell wall constituents of different stages in the life-histories of some siphoneous green algae. *Br. Phycol. J.*, **10**, 13–16.

HUIZING H. J., RIETEMA H. & SIETSMA J. H. (1979) Cell wall constituents of several siphoneous green algae in relation to morphology and taxonomy. *Br. Phycol. J.*, **14**, 25–32.

HUXLEY J. S. (1957) The three types of evolutionary process. *Nature* (*Lond.*), **180**, 454–5.

IKEMORI M. & ARASAKI S. (1977) Photosynthetic pigments in marine algae. I. Two-dimensional paper chromatographic separation of chlorophylls and carotenoids from green algae and sea grasses. *Bull. Jap. Soc. Phycol.*, **25**, 54–65.

JOHNSON U. G. & PORTER K. R. (1968) Fine structure of cell division in *Chlamydomonas reinhardii*. *J. Cell Biol.*, **38**, 403–25.

JOLY A. B. (1967) *Gêneros de Algas Marinhas da Costa Atlântica Latino-Americana*. University São Paulo.

JÓNSSON S. (1962) Recherches sur les Cladophoracées marines. *Ann. Sci. Nat. Bot.*, Sér. 12, **3**, 25–230.

JÓNSSON S. & CHESNOY L. (1974) Etude ultrastructurale de l'incorporation des axonèmes flagellaires dans les zygotes du *Monostroma grevillei* (Thuret) Wittr., chlorophycée marine. *C. R. Acad. Sci. Paris*, Sér. D, **278**, 1557–60.

KAGEYAMA A. & YOKOHAMA Y. (1977) Pigments and photosynthesis of deep-water green algae. *Bull. Jap. Soc. Phycol.*, **25**, 168–75.

KALKMAN C. (1972) *Mossen en Vaatplanten*. Oosthoek, Utrecht.

KAPRAUN D. F. (1970) Field and cultural studies of *Ulva* and *Enteromorpha* in the vicinity of Port Aransas, Texas. *Contrib. Mar. Sci.* **15**, 205–85.

KARLING J. S. (1926) Nuclear and cell division in *Nitella* and *Chara*. *Bull. Torrey Bot. Club*, **53**, 319–79.

KOEMAN R. T. P. & VAN DEN HOEK C. (1981) The taxonomy of *Ulva* (Chlorophyceae) in the Netherlands. *Br. Phyc. J.*, **16**, 9–53.

KÖHLER K. (1956) Entwicklungsgeschichte, Geschlechtsbestimmung und Befrüchtung bei *Chaetomorpha*. *Arch. Protistenk.*, **101**, 223–68.

KOOP H.-U. (1975) Über den Ort der Meiose bei *Acetabularia mediterranea*. *Protoplasma*, **85**, 109–14.

KOOP H.-U. (1976) Cross breeding experiments in *Acetabularia mediterranea*. *Ber. Dtsch. Bot. Ges.*, **89**, 499–505.

KOOP H.-U., HEUNERT H.-H. & SMID R. (1977) Division of the primary nucleus of *Acetabularia*. *Protoplasma*, **93**, 131–4.

KORNMANN P. (1956) Zur Morphologie und Entwicklung von *Percursaria percursa*. *Helgol. Wiss. Meeresunters.*, **5**, 259–72.

KORNMANN P. (1960) Die heterogene Gattung *Gomontia* II. Der fädige Anteil *Eugomontia sacculata nov. gen. nov. spec. Helgol. Wiss. Meeresunters.*, **7**, 59–71.

KORNMANN P. (1962) Die Entwicklung von *Monostroma grevillei*. *Helgol. Wiss. Meeresunters.*, **8**, 195–202.

KORNMANN P. (1963) Die Ulotrichales, neu geordnet auf der Grundlage entwicklungsgeschichtlicher Befunde. *Phycologia*, **3**, 60–8.

KORNMANN P. (1964) Über *Monostroma bullosum* (Roth) Thuret und *M. oxyspermum* (Kütz.) Doty. *Helgol. Wiss. Meeresunters.*, **11**, 13–21.

KORNMANN P. (1965) Zur Analyse des Wachstums und des Aufbaus von *Acrosiphonia*. *Helgol. Wiss. Meeresunters.*, **12**, 219–38.

KORNMANN P. (1969) Gesetzmässigkeiten des Wachstums und der Entwicklung von *Chaetomorpha darwinii* (Chlorophyta, Cladophorales). *Helgol. Wiss. Meeresunters.*, **19**, 335–54.

KORNMANN P. (1970) Eine Mutation bei der siphonalen Grünalge *Derbesia marina*. *Helgol. Wiss. Meeresunters.*, **21**, 1–8.

KORNMANN P. (1972) Ein Beitrag zur Taxonomie der Gattung *Chaetomorpha* (Cladophorales, Chlorophyta). *Helgol. Wiss. Meeresunters.*, **23**, 1–31.

KORNMANN P. & SAHLING P.-H. (1974) Prasiolales (Chlorophyta) von Helgoland. *Helgol. Wiss. Meeresunters.*, **26**, 99–133.

KORNMANN P. & SAHLING P.-H. (1977) Meeresalgen von Helgoland. *Helgol. Wiss. Meeresunters.*, **29**, 1–289.

KRISTIANSEN J. (1974) The fine structure of the zoospores of *Urospora penicilliformis*, with special reference to the flagellar apparatus. *Br. Phycol. J.*, **9**, 201–13.

LEEDALE G. T. (1974) How many are the kingdoms of organisms? *Taxon*, **23**, 261–70.

LEMBI C. A. (1975) A rhizoplast in *Carteria radiosa* (Chlorophyceae). *J. Phycol.*, **11**, 219–21.

LEWIN R. A. (1974) Biochemical taxonomy. In *Algal Physiology and Biochemistry*, ed. Stewart W. D. P., pp. 1–39. Blackwell Scientific Publications, Oxford.

LEWIN R. A. (1975) A marine *Synechocystis* (Cyanophyta, Chroococcales) epizoic on ascidians. *Phycologia*, **14**, 153–60.

LEWIN R. A. (1976) Prochlorophyta as a proposed new division of algae. *Nature* (*Lond.*), **261**, 697–8.

LEWIN R. A. (1977) *Prochloron*, type genus of the Prochlorophyta. *Phycologia*, **16**, 217.

LEWIN R. & WITHERS N. W. (1975) Extraordinary pigment composition of a Prokaryotic alga. *Nature (Lond.)*, **256**, 735–7.

LIDDLE L., BERGER S. & SCHWEIGER H.-G. (1976) Ultrastructure during development of the nucleus of *Batophora oerstedii* (Chlorophyta: Dasycladaceae). *J. Phycol.*, **12**, 261–72.

LOKHORST G. M. (1978) Taxonomic studies on the marine and brackish water species of *Ulothrix* (Ulotrichales, Chlorophyceae) in Western Europe. *Blumea*, **24**, 191–299.

LOKHORST G. M. & STAR W. (1980) Pyrenoid ultrastructure in *Ulothrix* (Chlorophyceae). *Acta Bot. Neerl.*, **29**, 1–15.

LOKHORST G. M. & VROMAN M. (1972) Taxonomic study on three freshwater *Ulothrix* species. *Acta Bot. Neerl.*, **21**, 449–80.

LOKHORST G. M. & VROMAN M. (1974a) Taxonomic studies on the genus *Ulothrix* (Ulotrichales, Chlorophyceae). II. *Acta Bot. Neerl.*, **23**, 369–98.

LOKHORST G. M. & VROMAN M. (1974b) Taxonomic studies on the genus *Ulothrix* (Ulotrichales, Chlorophyceae). III. *Acta Bot. Neerl.*, **23**, 561–602.

LØVLIE A. & BRÅTEN T. (1970) On mitosis in the multicellular green alga *Ulva mutabilis* Føyn. *J. Cell Sci.*, **6**, 109–29.

MCALISTER F. (1931) The formation of the achromatic figure in *Spirogyra setiformis*. *Am. J. Bot.*, **18**, 838–53.

MCARTHUR D. M. & MOSS B. L. (1977) The ultrastructure of cell walls in *Enteromorpha intestinalis* (L.) Link. *Br. Phycol. J.*, **12**, 359–68.

MCARTHUR D. M. & MOSS B. L. (1979) Gametogenesis and gamete structure of *Enteromorpha intestinalis* (L.) Link. *Br. Phycol. J.*, **14**, 43–57.

MCBRIDE G. E. (1970) Cytokinesis and ultrastructure in *Fritschiella tuberosa* Iyengar. *Arch. Protistenk.*, **112**, 365–75.

MCBRIDE G. E. (1971) The flagellar base in *Coleochaete* and its evolutionary significance. *J. Phycol.*, **7**, (suppl.) 13.

MCDONALD K. L. & PICKETT-HEAPS J. D. (1976) Ultrastructure and differentiation in *Cladophora glomerata*. I. Cell division. *Am. J. Bot.*, **63**, 592–601.

MACKIE W. & PRESTON R. D. (1974) Cell wall and intercellular region polysaccharides. In *Algal Physiology and Biochemistry* ed. Stewart W. D. P., pp. 40–85. Blackwell Scientific Publications, Oxford.

MANTON I. (1964) Observations on the fine structure of the zoospore and young germling of *Stigeoclonium*. *J. Exp. Bot.*, **15**, 399–411.

MANTON I. (1966) Observations on scale production in *Pyramimonas amylifera* Conrad. *J. Cell Sci.*, **1**, 429–38.

MANTON I. & PARKE M. (1965) Observations on the fine structure of two species of *Platymonas* with special reference to flagellar scales and the mode of origin of the theca. *J. Mar. Biol. Ass. U.K.*, **45**, 743–54.

MARANO F. (1976) Etude ultrastructurale de la division chez *Dunaliella*. *J. Microscopie Biol. Cell.*, **25**, 279–82.

MARCHANT H. J. & PICKETT-HEAPS J. D. (1973) Mitosis and cytokinesis in *Coleochaete scutata*. *J. Phycol.*, **9**, 461–71.

MARCHANT H. J., PICKETT-HEAPS J. D. & JACOBS K. (1973) An ultrastructural study of zoosporogenesis and the mature zoospore of *Klebsormidium flaccidum*. *Cytobios*, **8**, 95–107.

MARKOWITZ M. (1978) Fine structure of the zoospore of *Oedocladium carolinianum* (Chlorophyta) with special reference to the flagellar apparatus. *J. Phycol.*, **14**, 289–302.

MATTOX K. R. & STEWART K. D. (1974) A comparative study of cell division in *Trichosarcina polymorpha* and *Pseudendoclonium basiliense* (Chlorophyceae). *J. Phycol.*, **10**, 447–56.

MATTOX K. R. & STEWART K. D. (1977) Cell division in the scaly green flagellate *Heteromastix angulata* and its bearing on the origin of the Chlorophyceae. *Am. J. Bot.*, **64**, 931–45.

MATTOX K. R., STEWART K. D. & FLOYD G. L. (1974) The cytology and classification of *Schizomeris leibleinii*. I. The vegetative thallus. *Phycologia*, **13**, 63–70.

MAYHOUB H. (1974) Reproduction sexuée et cycle de développement de *Pseudobryopsis myura* (Ag.) Berthold (Chlorophycée, Codiale). *C. R. Acad. Sci. Paris,* Sér. D, **278**, 867–70.

MEINESZ A. (1969) Sur la reproduction sexuée de l'*Udotea petiolata* (Turra) Boerg. *C. R. Acad. Sci. Paris,* Sér. D, **269**, 1063–5.

MEINESZ, A. (1972a) Sur le cycle de l'*Halimeda tuna* (Ellis et Solander) Lamouroux (Udotéacée, Caulerpale). *C. R. Acad. Sci. Paris,* Sér. D, **275**, 1363–5.

MEINESZ A. (1972b) Sur le cycle de l'*Udotea petiolata* (Turra) Boergesen (Caulerpale, Udotéacée) *C. R. Acad. Sci. Paris,* Sér. D, **275**, 1975–7.

MEINESZ A. (1979) Contribution à l'étude de *Caulerpa prolifera* (Forsskål) Lamouroux (Chlorophycée, Caulerpale). *Bot. Mar.*, **22**, 117–21.

MELKONIAN M. (1975) The fine structure of the zoospores of *Fritschiella tuberosa* Iyeng. (Chaetophorineae, Chlorophyceae) with special reference to the flagellar apparatus. *Protoplasma*, **86**, 391–404.

MELKONIAN M. (1979) Structure and significance of cruciate flagellar root systems in green algae: zoospores of *Ulva lactuca* (Ulvales, Chlorophyceae). *Helgol. Wiss. Meeresunters.*, **32**, 425–35.

MICALEF H. & GAYRAL P. (1972) Quelques aspects de l'infrastructure des cellules végétatives et des cellules reproductrices d'*Ulva lactuca* L. (Chlorophycées). *J. Microsc.*, **13**, 417–28.

MILLER D. H. (1978) Cell wall chemistry and ultrastructure of *Chlorococcum oleofaciens* (Chlorophyceae). *J. Phycol.*, **14**, 189–94.

MOESTRUP Ø. (1974) Ultrastructure of the scale-covered zoospores of the green alga *Chaetosphaeridium*, a possible ancestor of the higher plants and bryophytes. *Biol. J. Linn. Soc.*, **6**, 111–25.

MOESTRUP Ø. (1978) On the phylogenetic validity of the flagellar apparatus in green algae and other chlorophyll a and b containing plants. *Biosystems*, **10**, 117–44.

MOESTRUP Ø. & ETTL H. (1979) A light and electron microscopical study of *Nephroselmis olivacea* (Prasinophyceae). *Opera Bot.*, **49**, 1–39.

NEUMANN K. (1969) Beitrag zur Cytologie und Entwicklung der siphonalen Grünalge *Derbesia marina*. *Helgol. Wiss. Meeresunters.*, **19**, 355–75.

NISIZAWA K., KURODA K., TOMITA Y. & SHIMAHARA H. (1974) Main cell wall constituents of the cysts of *Acetabularia*. *Bot. Mar.*, **17**, 16–19.

NISIZAWA K. & SASAKI S. F. (1975) Cell wall composition of algae from a phylogenetic point of view. In *Advance of Phycology in Japan*, eds. Tokida J. & Hirose H., pp. 42–52. VEB Gustav Fischer Verlag, Jena.

O'KELLY C. J. & YARISH C. (1981) Observations on marine Chaetophoraceae (Chlorophyta) II. On the circumscription of the genus *Entocladia* Reinke. *Phycologia*, **20**, 32–45.

PAPENFUSS G. F. & CHIHARA M. (1975) The morphology and systematic position of the green algae *Ernodesmis* and *Apjohnia*. *Phycologia*, **14**, 309–16.

PARKE M. & MANTON I. (1965) Preliminary observations on the fine structure of *Prasinocladus marinus*. *J. Mar. Biol. Ass. U.K.*, **45**, 525–36.

PARKE M. & DIXON P. (1976) Check-list of British marine algae. Third revision. *J. Mar. Biol. Ass. U.K.*, **56**, 527–94.

PASCHER A. (1931) Systematische Übersicht über die mit Flagellaten in Zusammenhang stehenden Algenreihen und Versuch einer Einreihung dieser Algenstämme in die Stämme des Pflanzenreiches. *Beih. Bot. Ztrbl. Abt.* 2, **48**, 317–32.

PATEL R. J. (1971) Cytotaxonomical studies of British marine species of *Chaetomorpha*. I. *Chaetomorpha linum* Kütz. and *Chaetomorpha aerea* Kütz. *Phykos*, **10**, 127–36.

PATEL R. J. (1972) Cytotaxonomical studies of British marine species of *Chaetomorpha*. II. *Chaetomorpha melagonium* Kütz. *Phykos*, **11**, 17–22.

PEARSON B. R. & NORRIS R. E. (1975) Fine structure of cell division in *Pyramimonas parkeae* Norris and Pearson (Chlorophyta, Prasinophyceae). *J. Phycol.*, **11**, 113–24.

PERCIVAL E. & MCDOWELL R. H. (1967) *Chemistry and Enzymology of Marine Algal Polysaccharides*. Academic Press, London.

PERROT Y. (1972) Les *Ulothrix* marins de Roscoff et le problème de leur cycle de reproduction. *Bull. Soc. Bot. France, Mém.*, **1972**, 67–74.

PICKETT-HEAPS J. D. (1967) Ultrastructure and differentiation of *Chara* sp. II. Mitosis. *Aust. J. Biol. Sci.*, **20**, 883–94.

PICKETT HEAPS J. D. (1970) Mitosis and autospore-formation in the green alga *Kirchneriella lunaris*. *Protoplasma*, **70**, 325–47.

PICKETT-HEAPS J. D. (1972a) Cell divison in *Klebsormidium subtilissimum* (formerly *Ulothrix subtilissima*) and its possible phylogenetic significance. *Cytobios*, **6**, 167–83.

PICKETT-HEAPS J. D. (1972b) Variation in mitosis and cytokinesis in plant cells: its significance in the phylogeny and evolution of ultrastructural systems. *Cytobios*, **5**, 59–77.

PICKETT-HEAPS J. D. (1973) Cell division and wall structure in *Microspora*. *New Phytol.*, **72**, 347–55.

PICKETT-HEAPS J. D. (1974a) Cell division in *Stichococcus*. *Br. Phycol. J.*, **9**, 63–73.

PICKETT-HEAPS J. D. (1974b) Ultrastructural morphology and cell division in *Pedinomonas*. *Cytobios*, **11**, 41–58.

PICKETT-HEAPS J. D. (1975) *Green Algae. Structure, Reproduction and Evolution in Selected Genera*. Sinauer, Sunderland, Mass.

PICKETT-HEAPS J. D. & FOWKE L. C. (1969) Cell division in *Oedogonium*. I. Mitosis, cytokinesis and cell elongation. *Aust. J. Biol. Sci.*, **22**, 857–94.

PICKETT-HEAPS J. D. & FOWKE L. C. (1970) Cell division in *Oedogonium*. II. Nuclear division in *O. cardiacum*. *Aust. J. Biol. Sci.*, **23**, 71–92.

PICKETT-HEAPS J. D. & OTT D. W. (1974) Ultrastructural morphology and cell division in *Pedinomonas*. *Cytobios*, **11**, 41–58.

PRICE I. A. (1972) Zygote development in *Caulerpa* (Chlorophyta, Caulerpales). *Phycologia*, **11**, 217–18.

REINKE J. (1889) *Atlas Deutscher Meeresalgen, 1. Heft*. Paul Parey, Berlin.

RIETEMA H. (1972) A morphological, developmental, and caryological study on the life-history of *Bryopsis halymeniae* (Chlorophyceae). *Neth. J. Sea Res.*, **5**, 445–57.

RIETEMA H. (1975) Comparative investigations on the life-histories and reproduction of some species in the siphoneous green algal genera Bryopsis and Derbesia. Thesis, University of Groningen.

ROBENEK H. & MELKONIAN M. (1979) Rhizoplast-membrane associations in the flagellate *Tetraselmis cordiformis* Stein (Chlorophyceae) revealed by freeze etching and thin sections. *Arch. Protistenk.*, **122**, 340–51.

ROBERTS K. (1974) Crystalline glycoprotein cell walls of algae: their structure, composition and assembly. *Phil. Trans. Roy. Soc. Lond.*, Ser B, **268** (891), 129–46.

ROGERS C. E., MATTOX K. R. & STEWART K. D. (1978) *Chlorokybus atmophyticus* Geitler, a chlorosarcinoid charophyte. *J. Phycol.*, **14** (suppl.), 32.

ROUND F. E. (1971) The taxonomy of the Chlorophyta. II. *Br. Phycol. J.*, **6**, 235–64.

SALISBURY J. L. (1978) The rhizoplast of *Platymonas subcordiformis* (Chlorophyta) is a contractile organelle. *J. Phycol.*, **14** (suppl.), abstr. 14.

SALISBURY J. L. & FLOYD G. L. (1978) Calcium-induced contraction of the rhizoplast of a quadriflagellate green alga. *Science*, **202**, 975–7.

SCOTT J. H., & BULLOCK K. W. (1976) Ultrastructure of cell division in *Cladophora*. Pregametangial cell division in the haploid generation of *Cladophora flexuosa*. *Can. J. Bot.*, **54**, 1546–60.

SCHULZ-BADES M. & LEWIN R. A. (1976) Fine structure of *Synechocystis didemni* (Cyanophyta, Chroococcales). *Phycologia*, **15**, 1–6.

SINGH R. N. (1942) Reproduction in *Draparnaldiopsis indica* Bharadwaja. *New Phytol.*, **41**, 262–73.

SINGH R. N. (1945) Nuclear phases and alternation of generations in *Draparnaldiopsis indica* Bharadwaja. *New Phytol.*, **44**, 118–29.

SINGH R. N. (1947) *Fritschiella tuberosa* Iyeng. *Ann. Bot.*, **11**, 159–64.

SLUIMAN H. J., MATTOX K. R. & STEWART K. D. (1980a) Moderne opvattingen over de fylogenie van groenwieren en landplanten. *Vakbl. Biol.*, **60**, 204–12.

SLUIMAN H. J., STEWART K. D. & MATTOX K. R. (1980b) The ultrastructure of the scaly zoospore of *Ulothrix zonata*: its taxonomic significance. *J. Phycol.*, **16** (suppl.), abstr. 40.

STARR R. C. (1955) A comparative study of *Chloroccum* Meneghini and other spherical zoospore-producing genera of the Chlorococcales. *Indiana Univ. Publ., Sci. Ser.*, **20**, 1–109.

STEWART J. K., STEWART J. R. & BOLD H. C. (1978) The morphology and life history of *Characiosiphon rivularis* Iyengar (Chlorophyta: Characiosiphonaceae): a light and electron-microscopic study. *Arch. Protistenk.*, **120**, 312–40.

STEWART K. D. & MATTOX K. R. (1975a) Comparative cytology, evolution and classification of the green algae with some consideration of the origin of other organisms with chlorophylls *a* and *b*. *Bot. Rev.*, **41**, 104–35.

STEWART K. D. & MATTOX K. R. (1975b) Some aspects of mitosis in primitive green algae, phylogeny and function. *BioSystems*, **7**, 310–15.

STEWART K. D. & MATTOX K. R. (1978) Structural evolution in the flagellated cells of green algae and land plants. *BioSystems*, **10**, 145–52.

STEWART K. D., MATTOX K. R. & CHANDLER C. D. (1974) Mitosis and cytokinesis in *Platymonas subcordiformis*, a scaly green monad. *J. Phycol.*, **10**, 65–79.

STEWART K. D., MATTOX K. R. & FLOYD G. L. (1973) Mitosis, cytokinesis, the distribution of plasmodesmata, and other cytological characteristics in the Ulotrichales, Ulvales, and Chaetophorales: phylogenetic and taxonomic considerations. *J. Phycol.*, **9**, 128–41.

STEWART W. D. P. (1974) A note on taxonomy. Appendix 1. In *Algal Physiology and Biochemistry*, ed. Stewart W. D. P., pp. 909–17. Blackwell Scientific Publications, Oxford.

SWANSON J. A., SALISBURY J. A., SALISBURY J. L. & FLOYD G. L. (1978) Ultrastructure of the flagellar apparatus of *Platymonas subcordiformis* (Chlorophyta) and a proposed function of the contractile rhizoplast. *J. Phycol.*, **14** (suppl.), abstr. 28.

TANNER C. E. (1980) *Chloropelta* gen. nov., an ulvaceous green alga with a different type of development. *J. Phycol.*, **16**, 128–37.

TATEWAKI M. (1969) Culture studies on the life history of some species of the genus *Monostroma*. *Scient. Pap. Inst. Algol. Res. Fac. Sci. Hokkaido Univ.*, **6**, 1–56.

TATEWAKI M. (1972) Life history and systematics in *Monostroma*. In *Contributions to the Systematics of Benthic Marine Algae of the North Pacific*, eds. Abbott I. A. & Kurogi M., pp. 1–15. Japanese Society Phycology, Kobe, Japan.

TATEWAKI M. (1979) Life history of *Pseudobryopsis* sp. (Codiales, Chlorophyta) Sôrui, *Jap. J. Phycol.*, **27**, 7–16.

TAYLOR F. J. R. (1976) Flagellate phylogeny: a study in conflicts. *J. Protozool.* **23**, 28–40.

TAYLOR W. R. (1960) *Marine Algae of the Eastern Tropical and Subtropical Coasts of the Americas*. Ann Arbor, Mich. Univ. Press, U.S.A.

THINH LUONG-VAN (1979) *Prochloron* (Prochlorophyta) associated with the ascidian *Trididemnum cyclops* Michaelsen. *Phycologia*, **18**, 77–82.

VALET G. (1976) Le *Struvea scoparia* (Chlorophycée) de Nouvelle Calédonie et sa position systématique. *Phycologia*, **15**, 421–4.

VAN DEN DRIESSCHE T. (1973) The chloroplasts of *Acetabularia*. The control of their multiplication activities. *Sub-Cell. Biochem.*, **2**, 33–67.

VAN DEN HOEK C. (1963) *Revision of the European Species of* Cladophora. Brill, Leiden. (Reprint by Koeltz, Koenigstein, 1976.)

VAN DEN HOEK C. (1978) Marine algae from the coral reef of Curaçao, Netherlands Antilles. I. Three new and one rarely observed species from the steep fore-reef slope. *Aquatic Bot.*, **5**, 47–61.

VAN DEN HOEK C. & JAHNS H. M. (1978) *Algen. Einführung in die Phykologie*. Georg Thieme Verlag, Stuttgart.

VAN DER VEER J. (1979) Pavlova and the taxonomy of flagellates, especially the Chrysomonads. Thesis, University of Groningen.

VERMA B. N. & VERMA S. K. (1976) A note on the cytology and life history of *Chaetomorpha gracilis* Kütz. *Phykos*, **15**, 84–94.

VICKERS A. & SHAW M. H. (1908) *Phycologia Barbadensis*. Paris.

VINOGRADOVA K. L. (1969) K sistematike poriadka Ulvales (Chlorophyta). *Botan. Žurn.*, **54**, 1347–55.

VINOGRADOVA K. L. (1974) *Ul'vovye vodorosli (Chlorophyta) mor'ej SSSR*. Izdatel'stvo 'Nauka', Leningrad, USSR.

VINOGRADOVA K. L. (1979) *Opredelitel' vodoroslej dal'nevostočnych morej SSSR, zel'enye vodorosli*. Izdatel'stvo 'Nauka', Leningrad, USSR.

WEBER A. (1973) Über die Chlorophylle und Carotenoide einiger Ulotrichales. *Mitt. Staatinst. Allg. Bot. Hamburg*, **14**, 25–9.

WHATLEY J. M. (1977) The fine structure of *Prochloron*. *New Phytol.*, **79**, 309–13.

WIK-SJÖSTEDT A. (1970) Cytogenetic investigations in *Cladophora*. *Hereditas*, **66**, 233–62.

WIK-SJÖSTEDT A. & NORDQUIST T. (1970) Preliminary observations on the ultrastructure of some *Cladophora* and *Spongomorpha* species. *Bot. Mar.*, **13**, 6–12.

WOMERSLEY, H. B. S. (1970) *Palmoclathrus*, a new deep water genus of Chlorophyta. *Phycologia*, **10**, 229–33.

WUTZ M. & ZETSCHE K. (1976) Zur Biochemie und Regulation des heteromorphen Generationswechsels der Grünalge *Derbesia-Halicystis*. *Planta*, **129**, 211–16.

YARISH C. (1975) A cultural assessment of the taxonomic criteria of selected marine Chaetophoraceae (Chlorophyta). *Nova Hedwigia*, **26**, 385–430.

YOKOHAMA Y., KAGEYAMA A., IKAWA T. & SHIMURA S. (1977) A carotenoid characteristic of chlorophycean seaweeds living in deep coastal waters. *Bot. Mar.*, **20**, 433–6.

CHAPTER 4
RHODOPHYTA: LIFE HISTORIES

JOHN A. WEST
Department of Botany, University of California,
Berkeley, California, U.S.A.

MAX H. HOMMERSAND
Department of Botany, University of North Carolina,
Chapel Hill, North Carolina, U.S.A.

1 INTRODUCTION

The past decade has seen a tremendous growth in our understanding of the life histories of red algae. This success is due to a very large extent to the adoption of enriched sea water media developed by von Stosch, Provasoli or McLachlan (see McLachlan 1973) that have made it possible to bring many species into a reproductive state in stationary culture under conditions of relatively high nutrient concentrations.

Prior to 1965 three types of life histories were recognized in the Florideophycidae: 1, a biphasic cycle in which a haploid gametophyte alternates with a haploid carposporophyte; 2, a biphasic cycle in which a haploid gametophyte alternates with a diploid carposporophyte; and 3, a triphasic cycle in which a haploid gametophyte produces a diploid carposporophyte which in turn gives rise to a diploid, free living tetrasporophyte (Drew 1955b; Kylin 1956).

The triphasic life cycle with isomorphic gametophytic and tetrasporophytic generations in which meiosis takes place in the tetrasporangium was first demonstrated in *Polysiphonia* in the classical study of Yamanouchi (1906). This work is exemplary in that not only were the stages of meiosis in the tetrasporangium documented but chromosome counts were made of nuclei undergoing mitosis at every critical stage in the life cycle. The male and female gametophytes with 20 chromosomes were clearly demonstrated to be haploid, and both the carposporophytic and tetrasporophytic generations with 40 chromosomes were shown to be diploid.

The distinction between biphasic and triphasic life histories provided the essential basis for separating the Nemalionales as a primitive order from the remaining orders of Florideophycidae. The first division of the zygote nucleus was found to be heterotypic in Nemalionales, with evident pairing of homologous chromosomes and with diakinesis figures that purported to show crossing over. The clearest examples of meiotic figures were published for *Scinaia* by Svedelius in 1915, for *Nemalion* by Kylin in 1916, for *Batrachospermum* by Kylin in 1917, for *Porphyra* by Dangeard in 1927, and for *Asparagopsis* and *Bonnemaisonia* by Svedelius in 1933 (see Kylin 1956).

By 1961, Magne (1952, 1960a,b, 1961a,b) had reinvestigated all of the critical genera in which it had been claimed that meiosis takes place immediately after fertilization. He argued that it was necessary to count the chromosomes using the Fuelgen technique in vegetative divisions in the gametophyte, the carposporophyte, and where possible the germinating carpospore. In each of the classical examples, Magne counted twice the number of chromosomes in the carposporophyte as in the gametophyte, and in some instances carpospore germlings were also found to be diploid. Magne (1964) went so far as to question whether haplontic (haplophasic) life histories even exist in red algae.

The implications of Magne's observations were clear. Members of the Nemalionales probably possessed triphasic life histories in which a heteromorphic tetrasporophyte had gone unrecognized. Most likely the missing phase resembled members of the *Acrochaetium-Rhodochorton* complex. The solution lay in carrying out life history studies in culture. At the meetings of the International Phycological Society in Naples in September 1963, Magne proposed to von Stosch that he undertake just such a study. Material of *Liagora farinosa* from the Mediterranean was procured. Von Stosch (1965) found that carpospores of *Liagora* gave rise to an *Acrochaetium*-like phase that reproduced by means of monospores and tetraspores. Tetrasporangia appeared to undergo meiosis producing four tetraspores. These produced filamentous protonemata that were virtually indistinguishable from the tetrasporophyte; however, buds arising on the presumably haploid protonemata developed into typical *Liagora* thalli. A complete life cycle of a member of the Nemalionales was first achieved by Ramus (1969) in *Pseudogloiophloea confusa*, a close relative of *Scinaia* from the west coast of North America.

The unequivocal demonstration of a triphasic life cycle with heteromorphic gametophytic and tetrasporophytic generations, coupled with cytological evidence that meiosis is not zygotic set the stage for a tempestuous rise in interest and research on the life histories of the red algae that has occupied phycologists during the past decade. The pioneering studies of Feldmann & Feldmann (1942) on *Bonnemaisonia* and *Asparagopsis* gained acceptance that was long overdue. The generalized *Liagora farinosa*-type life cycle was found to be widespread in marine species of Nemalionales. Freshwater red algae appeared to be an exception in that tetrasporangia are not known to be produced by the filamentous *Chantransia*-stage. Magne (1967b) provided the cytological evidence that led to a reinterpretation of the life histories of freshwater red algae with his elegant demonstration of meiosis during the transition from the diploid protonemal stage to the erect thallus in *Lemanea*. Recent observations by von Stosch & Theil (1979) have confirmed Magne's predictions of a *Lemanea*-type life cycle in *Batrachospermum* as well. Direct development of the gametophyte from a diploid protonemal phase without the formation of tetrasporangia has been proposed to account for the life histories of several marine red algae investigated in culture, but the site of meiosis has not, so far, been established cytologically in any of these studies.

At first it was thought that a heteromorphic life history of the *Liagora farinosa*- or *Lemanea*-type would prove to be diagnostic for the Nemalionales just as a biphasic, haplontic life cycle was formerly thought to characterize and set apart this order. The discovery of a heteromorphic life history in *Acrosymphyton* (Cryptonemiales) by Cortel-Breeman & van den Hoek (1970) that involves a *Hymenoclonium*-phase similar to that described for *Bonnemaisonia asparagoides* quickly set aside this notion.

A long list of genera belonging to the orders Cryptonemiales and Gigartinales have since been shown to possess prostrate filamentous systems or crustose stages that appear to be free-living, heteromorphic, presumably diploid stages that either bear tetrasporangia or give rise to gametophytes by direct development in culture. The list now includes some members of the Gloiosiphoniaceae, Dumontiaceae and Kallymeniaceae in the Cryptonemiales and the Calosiphoniaceae, Gymnophlaeaceae, Furcellariaceae, Phyllophoraceae and Gigartinaceae in the Gigartinales. In fact, only the Ceramiales and Rhodymeniales *sensu stricto* seem to be characterized diagnostically by an alternation of isomorphic generations.

There has not been active speculation recently concerning the evolution of the different types of life histories encountered in the Rhodophyta. The original concepts of Svedelius (1937) and Kylin (1945) that the biphasic, haplodiplontic (*Liagora tetrasporifera*-type) and triphasic, haplodiplontic (*Polysiphonia*-type) life cycles evolved through a 'delay' of meiosis seem to have been laid to rest.

It was Feldmann (1952) who first proposed that the triphasic (trigenetic) life cycle characterized by *Polysiphonia* is primitive in Florideophycidae. Feldmann postulated that the alternation of essentially isomorphic generations observed in most members of the family Acrochaetiaceae is primitive, along with the apparently simple organization of the filamentous thallus and reproductive structures. Feldmann argued that the *Polysiphonia*-type life history serves as a stem line that extends from the most primitive to the most specialized orders and families and that the variant life histories so frequently encountered in different families represent isolated deviations from the norm that occur almost randomly throughout the major taxonomic groups. This view was developed further by Feldmann (1972) and amplified and strongly defended by Magne (1972) who made hypotheses to account for the evolution of each of the types of life cycles then known to exist in the Florideophycidae.

In the last few years, as more and more studies have documented the variation that exists at every taxonomic level, the possibility of a phylogenetic interpretation of red algal life histories seems almost to be receding from us. It is not obvious in any of the more difficult groups whether an isomorphic or a heteromorphic life cycle is the more primitive. It is not even certain whether a biphasic or a triphasic cycle is primitive in extant genera. The recent discovery of a biphasic life history in *Palmaria palmata* (van der Meer & Todd 1980) is a case in point. The same is true of *Rhodochorton floridulum* (Stegenga 1978) and possibly *R. purpureum* (West 1969; Stegenga 1978) in which the tetrasporophyte can be interpreted as developing directly from the carpogonium. The primitive or advanced status of the biphasic life history of Bangiaceae is also not a settled matter.

If one argues, in agreement with Feldmann (1952), that the Acrochaetiaceae represent the primitive condition in Florideophycidae, then erect axes and whorled filaments as in *Batrachospermum* and primitive genera of most other orders must be secondarily derived. Given the restraint imposed by apical growth, the *Batrachospermum* mode of development and organization could have arisen several times leading to establishment of a number of independent phylogenetic lines. If this were true, then depending on circumstance the ancestors of a given line could have either an alternation of isomorphic or heteromorphic phases at inception. The ancestors of the Ceramiales, for example, may have possessed only isomorphic gametophytic and tetrasporophytic phases, or have passed through heteromorphic stages resembling those seen in taxa of Bonnemaisoniales. The erect multiaxial subfamilies of the Corallinaceae may also have exhibited an alternation of isomorphic phases from the time of their inception from prostrate progenitors. By contrast, the progenitors of the more primitive families of Gigartinales and Cryptonemiales may have begun with an alternation of heteromorphic phases very much like that exhibited by many modern representatives having simple thallus construction. There is no saying at present that the families of the orders Cryptonemiales and Gigartinales are monophyletic, and as the direction of evolution with respect to life histories of red algae continues to unfold, our perspective of the evolution and classification of the red algae will be profoundly affected.

The complexity of life histories encountered or thought to exist in red algae has bred an array of systems of terminology. As our concepts have evolved, so it seems has the number of terms in the literature. The history of terminology applied to life histories of red algae presents a fascinating story. Interested readers should consult Svedelius (1937), Kylin (1945), Feldmann (1952, 1972), Drew (1955b), Chapman & Chapman (1961), Magne (1972), Dixon (1970, 1973) and Umezaki (1977) in order to appreciate the kaleidoscope of changing perspectives.

2 SUBCLASS BANGIOPHYCIDAE

The subclass, Bangiophycidae, is generally subdivided into four or five orders (Dixon 1973; Kylin 1956). A sexual life history has been clearly demonstrated in two of these, Rhodochaetales and Bangiales.

2.1 Rhodochaetales

RHODOCHAETACEAE

This monotypic order contains only *Rhodochaete parvula*, a species that shares a number of features with both Bangiophycidae and Florideophycidae. Originally, *Rhodochaete* was thought to reproduce asexually by monospores cleaved out as in *Erythrotrichia*, but Magne (1960b) discovered that the thalli are, in fact, sexual and monoecious. Each male cell produces one spermatium that fertilizes an undifferentiated vegetative cell functioning as a carpogonium which forms a carposporangium releasing one carpospore. The fate of the presumably diploid carpospore is unknown. *Rhodochaete* is of particular interest because it exhibits the simplest form of sexual reproduction known in red algae.

2.2 Bangiales

BANGIACEAE

The life history of members of the Bangiales has intrigued phycologists for nearly a century, ever since Berthold (1882) first illustrated fertilization in *Porphyra, Bangia* and *Erythrotrichia*. Many early observations showed that aplanospores (monospores) released from thalli of *Porphyra* and *Bangia* exhibit bipolar germination and develop in a manner resembling the adult thallus, while carpospores germinate into fine, irregularly branched, procumbent filaments quite unlike the erect plant (Kylin 1956; Drew 1956). With the discovery that the filamentous alga *Conchocelis rosea* is linked with *Porphyra purpurea*, Drew (1949, 1954) established that the life history of *Porphyra* is essentially biphasic and involves an alternation between a macroscopic foliose phase and a microscopic filamentous phase. A complete life cycle was demonstrated for *Porphyra tenera* by Tseng & Chang (1955a).

In the absence of conclusive evidence of sexual reproduction Conway (1964) followed by Conway & Cole (1977) proposed that the terms α spore replace carpospore and β spore replace spermatium. The recent demonstration of fertilization in *Porphyra gardneri* by Hawkes (1978), using scanning and transmission electron microscopy, has finally resolved the controversy, and it is now generally accepted that the life history of *Porphyra* and *Bangia* is fundamentally sexual (Cole & Conway 1980). Even so, apomictic species are known that develop a foliose phase bearing spermatia and carpospores and a filamentous phase bearing conchospores in which the chromosome number is unchanged and fertilization and meiosis are presumably absent (Mumford & Cole 1977).

Figure 4.1 illustrates a generalized life cycle of a sexually reproducing species of *Porphyra*. Conchospores produced by the filamentous phase under short-day conditions late in the autumn or early winter germinate in a bipolar manner to produce the haploid *Porphyra* thallus. Young blades are recycled asexually by means of aplanospores which also exhibit bipolar germination. Later, under day-lengthening conditions and warming temperatures thalli produce spermatangia and carpogonia. Division of the carpogonium following fertilization is mitotic, forming packets of diploid carpospores which germinate initially in a unipolar fashion and ultimately develop the highly branched, prostrate filamentous system of the *Conchocelis*-phase. *Conchocelis* filaments may be recycled by monospores formed terminally in monosporangia, usually under long-day conditions. Under short days and falling temperatures terminal and intercalary cells of the *Conchocelis*-phase undergo meiosis and generate files of large, thick-walled cells, the conchosporangia. Alternatively, *Conchocelis* filaments may produce plantlets that release aplanospores or may develop directly into dwarf *Porphyra* thalli. The evidence for the generalized life cycle presented in Fig. 4.1 comes from several sources and has not been demonstrated in its entirety in any single species.

Fertilization has been seen and described by many workers since Berthold (1882). (For recent reviews see Hawkes 1978 and Cole & Conway 1980.) Spermatia attach to the wall above the scarcely differentiated carpogonium and penetrate by means of a fertilization canal. In agreement with Dangeard (1927) and Magne (1952), Hawkes (1978) has shown conclusively by light and electron microscopy that sperm nuclei enter the carpogonium; however, the fusion of the spermatial and carpogonial nuclei has not been reported. Division of the carpogonium is evidently mitotic since diploid chromosome numbers have been counted in carpospores and in cells of the *Conchocelis*-phase of many species since Magne's (1952) initial observations in *Porphyra linearis* (Giraud & Magne 1968; Chen *et al.* 1970b; Yabu 1975; Mumford & Cole 1977; Kito 1978; Hawkes 1978). Earlier observations (Dangeard 1927; Tseng & Chang 1955b) that meiosis takes place during the first divisions of the fertilized carpogonium have not received support. Giraud & Magne (1968) have illustrated stages of meiosis in terminal and intercalary cells of the *Conchocelis*-stage prior to formation of conchosporangia. Chromosome counts before and after conchospore formation

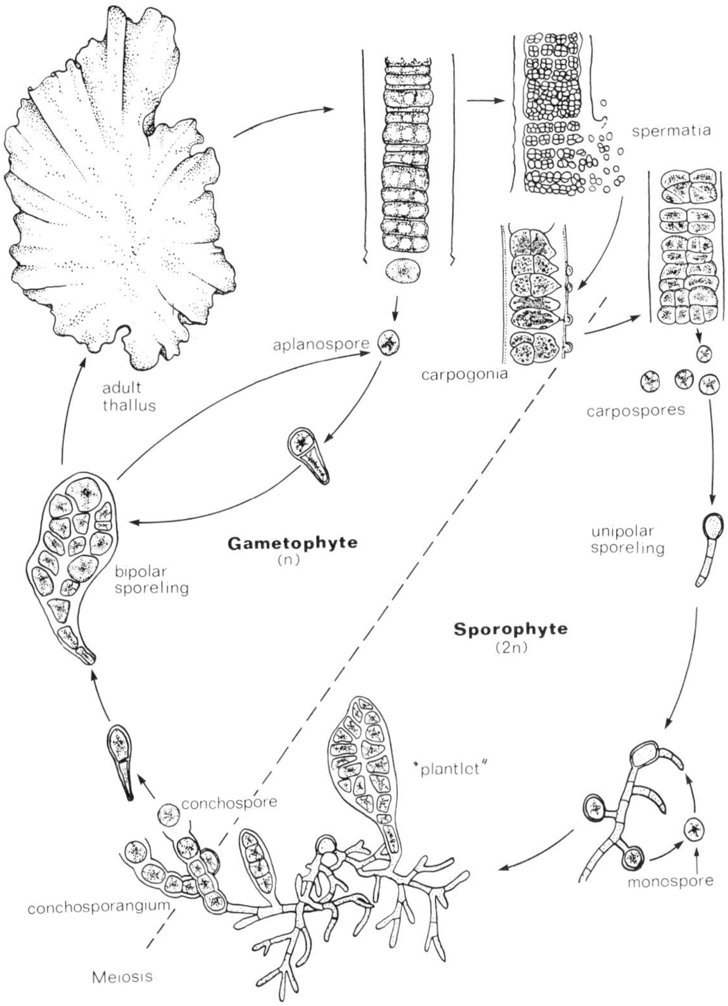

Fig. 4.1 Life history of *Porphyra*. (Based on Berthold, Thuret & Bornet, Tseng & Chang, Magne, and Cole & Conway.)

by Migita (1967) and Kito (1978) confirm this interpretation of the probable site of meiosis.

A number of interesting variations have been described in the life histories of both sexually reproducing and apomictic species of *Porphyra*. Asexual reproduction of the foliose phase by means of aplanospores is common in *Porphyra* species from the western Pacific Ocean and occurs in four of the six species of economic importance in Japan (Kurogi 1972; Miura 1975). A few species, including the economically important *P. tenera* and *P. yezoensis*, release aplanospores from young dwarf thalli from autumn to winter (Chang & Tseng 1962; Fukuhara 1968). The production of aplanospores greatly increases the set of foliose thalli on nets and extends the cultivation period of *Porphyra* grown commercially in Japan (Miura 1975).

It has been thought that most species of *Porphyra* from the northeastern Pacific Ocean do not reproduce by aplanospores (Conway *et al.* 1975), the major exception being *P. gardneri* (Hawkes 1980). The same was said to be true of *P. linearis* in Nova Scotia (Bird 1973). It now seems that most species in North America and Europe liberate simple protoplasts or aplanospores from the apical regions of young thalli in winter and that this may be a major means for the asexual multiplication of many species (Cole & Conway 1980). Aplanospores normally show bipolar germination and repeat the foliose phase. An interesting exception was seen in *P. rosengurtii* in which aplanospores can develop in a bipolar or a unipolar fashion to produce either *Porphyra* blades or *Conchocelis* filaments (Kapraun & Luster 1980). Not surprisingly *P. rosengurtii* is an apomictic species in which the chromosome number of both foliose and filamentous phases is the same.

All Japanese species so far investigated can produce the filamentous, shell-inhabiting *Conchocelis*-phase (Chihara 1975). In British Columbia and Washington only two species have been shown to lack a *Conchocelis*-phase (Conway & Cole 1977). One of these, *Porphyra subtumens*, reproduces solely by aplanospores, and in the other, *P. sanjuanensis*, the carpospore-like aplanospores show bipolar germination and develop into foliose thalli (Krishnamurthy 1972; Conway *et al.* 1975). The filamentous *Conchocelis*-phase can be maintained for long periods of time in culture and grows under a wide range of temperatures and photoperiods (Dring 1967; Conway & Cole 1977). It is probably a perennial, persistent stage in the life history of many *Porphyra* and *Bangia* species in nature as well. Though most species grow in shell, this is not a requirement. Asexual reproduction of the *Conchocelis*-phase is by fragmentation and the production of monospores in differentiated sporangia (Richardson 1972; Conway & Cole 1977; Cole & Conway 1980). Vegetative growth and monospore production are generally promoted by long days and summer temperatures (Iwasaki 1961; Conway & Cole 1977; Kapraun & Luster 1980) but there does not appear to be any clear evidence of photoperiodic control of monosporangia formation (Dring 1967a).

The formation of conchosporangia and the release of conchospores marks the transition from the filamentous to the foliose phase in both sexually reproducing and apomictic species of *Porphyra* and *Bangia*. Virtually all studies have confirmed the original observations of Kurogi (1959) that the production of conchospores is stimulated by short days (<12 LD) and inhibited by long days (>12 LD). (For reviews and literature citations see Dixon & Richardson 1970; Conway & Cole 1977; Kapraun & Luster 1980.) The photoperiodic nature of the short days response is discussed by Lüning in Chapter 9 of this volume.

The life history of *Bangia* exhibits the same variation between sexually reproducing and apomictic species as seen in strains of *Porphyra*. The freshwater form of *Bangia atropurpurea* reproduces entirely asexually by means of aplanospores (Nichols & Veith 1978; Sheath & Cole 1980). Likewise, some marine populations of *B. fuscopurpurea* that produce spermatia and carpospores have been shown to be completely apomictic with no evidence of sexual reproduction or change in chromosome number (Sommerfeld & Nichols 1970; Richardson 1970). The report by Yabu (1967) of fertilization and haploid chromosome numbers ($n = 3$ or 4) and diploid numbers ($2n = 6$ or 8) suggests that *Bangia* is also capable of a true sexual cycle as postulated earlier by Berthold (1882) and Dangeard (1927).

Erythropeltidaceae

Erythropeltidaceae produce both aplanospores (as in Bangiaceae) or monospores. Fertilization was illustrated by Berthold (1882) for *Erythrotrichia* and has been reported since in a number of genera, but Dixon (1973) considers the evidence unconvincing. Prostrate filamentous stages have been described in *Erythrotrichia* (Heerebout 1968), in *Smithora* (Richardson & Dixon 1969) and in *Porphyropsis* (Murray *et al.* 1972) which can reproduce asexually by monospores.

3 SUBCLASS FLORIDEOPHYCIDAE

3.1 Nemalionales

The Nemalionales as treated here includes the families Acrochaetiaceae, Batrachospermaceae, Lemaneaceae, Helminthocladiaceae, and Chaetangiaceae. The Bonnemaisoniales and Gelidiales are considered to be separate orders. Sexually reproducing members of the Acrochaetiaceae possesses either an isomorphic or heteromorphic, haplodiplontic life history. The remaining families, with the exception of the genus *Galaxaura*, all have a heteromorphic life history in which the diploid phase is filamentous and monosiphonous, resembling taxa in the Acrochaetiaceae.

Acrochaetiaceae

The Acrochaetiaceae is the only family presently placed in the order Nemalionales in which the gametophyte and tetrasporophyte generations are essentially isomorphic. It was partly for this reason that Feldmann (1962) regarded the family as representing a distinct order, the Acrochaetiales Chemin. Drew (1928), Woelkerling (1971), Dixon & Irvine (1977) and Garbary (1978) have argued that the species of this family cannot be assigned to separate genera. Following the single genus concept most species in this family have been formally transferred to *Audouinella*, the oldest available name based on *A. miniata* (= *A. hermanni*), a freshwater species. Most workers who have investigated life histories of Acrochaetiaceae in culture have tended to keep *Rhodochorton* separate from *Acrochaetium* or *Audouinella*. Recently Stegenga & Vroman (1977) have recognized *Chromastrum, Acrochaetium, Kylinia, Audouinella, Rhodochorton* and possibly *Liagoraphila* based on the presence of isomorphic or heteromorphic life histories, the mode of spore germination, postfertilization development, and chromatophore structure. This classification will be followed here.

An isomorphic life history was first recognized by Drew (1935) in the dioecious, freshwater species *Audouinella hermanni* (as *Rhodochorton violaceum*) and confirmed by Swale & Belcher (1963) in the monoecious species *A. investiens* (as *Rhodochorton investiens*). Morphologically similar tetrasporophytic and gametophytic plants were described for the marine species, *Acrochaetium botryocarpum* from Australia by Woelkerling (1970). Woelkerling (1971) later transferred this species to *Audouinella*. A slightly heteromorphic life history was seen in *Acrochaetium dasyae* from the Netherlands by Stegenga & Borsje (1976). In contrast, West (1968) found that *Acrochaetium pectinatum* has a heteromorphic life history. Tetrasporophytes 1–2 cm in height form tetrasporangia under short days (6–10 h). The tetraspores develop into dwarf gametophytes that are usually dioecious and less than 0.5 mm tall, except that some females reach 2–3 mm in length and produce spermatangia as well as carpogonia. Monospores give back the same phase as the parent. Because of the absence of a pyrenoid, Stegenga & van Wissen (1979) have proposed transferring this species to *Audouinella*.

The genus *Kylinia* was characterized by Feldmann (1958) as reproducing sexually with the spermatangia borne at the top of a long hyaline, hair-like cell (androphore). Recently Stegenga & von Wissen (1979) have shown that carpospores from *Kylinia rosulata* germinate in a unipolar fashion leaving an empty spore and producing a filament that branches into a prostrate system that in turn gives rise to a number of erect axes. These produce monosporangia under short days and tetrasporangia under long days. Monospores reproduce the tetrasporophyte while tetraspores give back the gametophyte. Stegenga & van Wissen equate the tetrasporophyte with *Acrochaetium strictum* from nature. They reject the conclusions of Boillot & Magne (1973) that *Acrochaetium pectinatum* is the tetrasporophyte of *Kylinia rosulata*.

Rhodochorton purpureum was shown by West (1969) to possess a distinctive life history in which the tetrasporophyte develops directly from clavate gonimoblast cells. This species produces normal tetrasporangia, bisporangia with two binucleate spores and other sporangial types under short days. Tetraspores commonly exhibit bipolar (occasionally unipolar) germination and develop into either male or female gametophytes. Similar behaviour was observed by Stegenga (1978) in cultures of *R. purpureum* from Holland and by Ohta & Kurogi (1979) in cultures from Muroran, Hokkaido. *Rhodochorton purpureum* is normally dioecious, but West (1970a) obtained a monoecious isolate from Chile, and Ohta & Kurogi (1979) grew a monoecious strain from Oshoro, Hokkaido. West (1972a) concluded from an examination of numerous isolates that strains of *R. purpureum* are genetically selected for temperature dependency of sporulation related to latitude. Plants from higher latitudes sporulate at lower temperatures. Vegetative growth is greatest under long days, while both gametogenesis and tetrasporogenesis are favoured by short days.

The life history of *Rhodochorton floridulum* is similar to *R. purpureum* except that the tetrasporophyte develops directly from the fertilized carpogonium with only one erect filament and one rhizoid. Recognizable gonimoblasts are lacking (Stegenga 1978). West (1970b, 1979) has described an epizoic species of *Rhodochorton* with a pseudoparenchymatous basal disc and an endozoic species with a distinctive basal system. Both species yielded only tetrasporangial plants in culture.

The genus *Chromastrum* was created by Papenfuss (1945) for species of Acrochaetiaceae whose cells contain one or more stellate chromatophores. Following Kylin's misidentification of a Swedish plant having stellate chromatophores with *Kylinia rosulata*, Papenfuss (1947) placed *Chromastrum* in synonymy under *Kylinia*. With the clarification of *Kylinia* by Feldmann (1958) the name *Chromastrum* again became available. Recently Stegenga & Mulder (1979) surveyed the species of *Chromastrum* and have modified the genus description to include species with stellate chromatophores in which the life history is heteromorphic. The gametophyte possesses a unicellular base and a small carposporophyte, while the tetrasporophyte develops a multicellular base from a septately germinating spore. Gametophytic and tetrasporophytic stages have been described from culture studies of six species previously identified as *Acrochaetium*. These include the type species *C. virgatulum/rhipidandrum* (Borsje 1973), *C. densum/catenulatum* (Stegenga & Vroman 1976), *C. polyblastum/hallandicum* (Stegenga & Borsje 1977), *C. reductum/kylinioides* (Stegenga & van Wissen 1979), *C. humile/moniliforme* (Stegenga & Mulder 1979), and *C. collopodum* (Stegenga & Mulder 1979). The geographical distribution of *Chromastrum* is primarily in temperate waters and especially in western Europe.

Batrachospermaceae

Magne (1967, 1972) postulated that *Batrachospermum* and other members of the freshwater family Batrachospermaceae would be found to have a *Lemanea*-type cycle in which the gametophyte buds directly from a diploid protonema. Free-living tetrasporophytes have never been seen in *Batrachospermum*. Hurdlebrink & Schwantes (1972) showed by Feulgen cytophotometry that the nuclei in cells of the erect thallus with whorled branches contain half as much DNA as cells of the prostate *Pseudochantransia*-phase. Supposedly meiosis takes place during the transition from the *Pseudochantransia* to the *Batrachospermum*-phase.

Recently von Stosch & Theil (1979) have discovered the site of meiosis in a species of *Batrachospermum* that lacks a typical *Pseudochantransia*-phase but instead forms erect axes directly from a crustose prothallus of richly branched filaments. The erect batrachospermoid axis is initially diploid. Meiosis is thought to take place in an apical cell that divides obliquely producing a short protuberance into which one daughter nucleus of the first meiotic division migrates and is separated as a polar body. The second division results in a somewhat larger polar body cut off by a wall. The viable meiospore continues as the apical initial of the now haploid axis bearing whorled lateral filaments. Only the two polar cells mark the point of transition from the diploid to the haploid axis. Cytological evidence for meiosis has recently been demonstrated for a species of *Batrachospermum* by Balakrishnan & Chaugule (1980). The life history is essentially like that of *Lemanea*.

Lemaneaceae

Magne (1967a,b) demonstrated that in *Lemanea* meiosis does not involve tetrasporangia but takes place instead in the apical cell of an erect branch. Thus, the transition from diploid to haploid phases takes place in conjunction with the direct development of the gametophyte from a diploid prothallus. Magne (1967b) described two types of behaviour in *Lemanea* which correspond to the subgenera *Lemanea* and *Paralemanea*. In both, the apical cell of an erect branch undergoes meiosis in which one of the daughter nuclei of both the first and second divisions is cut off as a polar body from the lower side of the apical cell. That meiosis has occurred is confirmed by chromosome counts in the diploid *Pseudochantransia*-phase and in the haploid, erect *Lemanea*-phase below and above the site of meiosis. In the subgenus *Lemanea* meiosis takes place after the erect axis bearing its whorled lateral filaments has reached 20–40 segments in length. In the subgenus *Paralemanea* meiosis precedes the cutting off of periaxial cells and the haploid axis is borne on a diploid monosiphonous pedicel of the prothallus or *Pseudochantransia*-phase.

Helminthocladiaceae

Nemalion, the problematic genus of so many classical investigations, has yielded mixed results in culture studies. Following the demonstration by Magne (1961a) that the carposporophyte and germling stage of the tetrasporophyte in *Nemalion helminthoides* is diploid ($2n = 16$), Fries (1967) was able to obtain filamentous *Acrochaetium*-like plants of *N. multifidum** in axenic culture that produced a few tetrasporangia. She later found tetrasporophytes in nature (Fries 1969). In contrast, Martin (1969) observed that the filamentous phase of *N. helminthoides*

* Dixon and Irvine (1977) regard *N. multifidum* as being conspecific with *N. helminthoides*.

developed from carpospores that overwinter in nature as thick-walled cells that appear to grow directly into gametophytes in the spring. Chen *et al.* (1978) reported that gametophytes of *N. helminthoides* from Nova Scotia need relatively high temperatures (15–20°C) and long days (16:8 LD) for normal growth. Carpospores develop into filamentous tetrasporophytes that require low temperatures (5°C) and short days (8:16 LD) to induce tetrasporogenesis. Monospores repeat the tetrasporophytic phase. Gametophytes are haploid ($n = 10$) and sporophytes diploid ($2n = 20$). Müller & Petronijevic (1979) find that the tetrasporophyte of *N. multifidum* from Kiel Bight survives and bears tetrasporangia only if *Calothrix parasitica* is present.

Following earlier observations by Umezaki (1967), Masuda & Umezaki (1977) showed that carpospores of *Nemalion vermiculare* from Japan germinate into filamentous tetrasporophytes that produce tetrasporangia under short days and 10–18°C. Liberated tetraspores first form filamentous plants similar to the tetrasporophyte which develop adult thalli under long days and 10–18°C reaching maturity in culture in four months. Surprisingly, monospores develop similarly to tetraspores and also produce erect *Nemalion* thalli.

The first clear demonstration of heteromorphic phases in the life history of a member of the Nemalionales was provided by von Stosch (1965) in *Liagora farinosa* from the western Mediterranean (see Fig. 4.2). Carpospores germinate into *Acrochaetium*-like plants consisting of branched, filamentous prostrate and erect systems that are recycled by monospores. Tetrasporangia, initiated under similar conditions to the monosporangia, appear to undergo meiosis and produce a gametophyte protonema that is virtually indistinguishable from the young sporophyte except for the secretion of mucilaginous material. Later, buds arise on the protonemal branches which grow into typical *Liagora* thalli that, in culture, remained uncalcified and did not fruit. Von Stosch (1965) concluded that, despite the morphological similarity between the tetrasporophytic phase and the protonemal stage of the gametophyte, they differ markedly in their physiological behavior and in their ability to develop buds.

The *Liagora farinosa*-type of life cycle has been confirmed in another species, *L. distenta*, by Couté (1971) and has been found in *Helminthocladia calvadosii* (Boillot 1971a, 1974) and *Helminthora divaricata* (Boillot 1971b, 1972). In *Helminthocladia calvadosii* carpospores germinated into diploid ($2n = 16$) filamentous sporophytes that bore cruciately divided tetrasporangia in the spring under conditions of natural illumination.

Boillot (1972) found that in *Helminthora divaricata* carpospores germinated into a diploid ($2n = 18$) filamentous tetrasporophyte that produced tetrasporangia in three months in winter or summer, not being dependent on photoperiod. Tetrasporangia underwent meiosis with nine bivalents seen at prometaphase I. The tetraspores germinated into filamentous protonemal stages that produced erect, multiaxial *Helminthora* thalli after three months.

In several species of *Liagora* and two species of *Helminthocladia* the carposporophyte bears terminal, cruciately divided carpotetrasporangia. These have long been thought to be the site of meiosis, and it generally has been held that the life history of such species is biphasic with only gametophytic and carposporophytic generations present. Bodard (1971) demonstrated cytologically that meiosis occurs at the time of formation of the carpotetraspores in *Helminthocladia senegalensis*, and Couté (1971) has shown that the carpospores of *Liagora tetrasporifera* produce a filamentous protonemal stage like that found in other species of *Liagora*. Tetrasporangia were never seen. Instead, small,

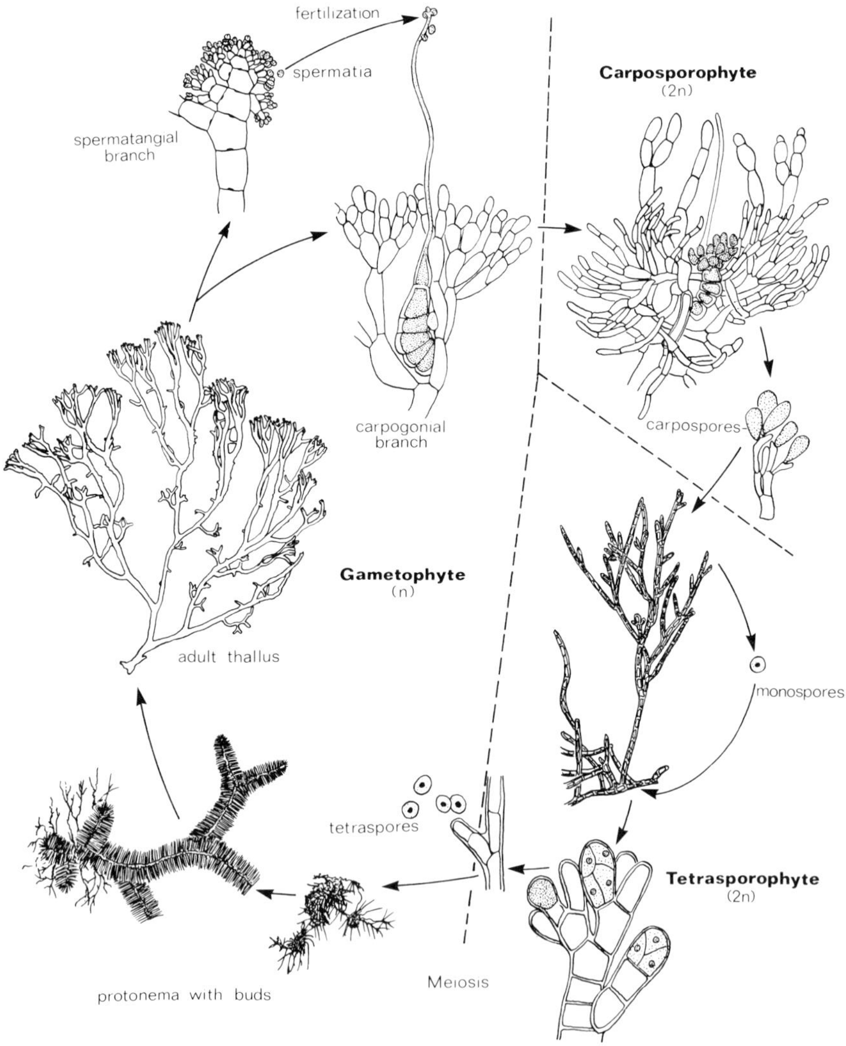

Fig. 4.2 Life history of *Liagora*. (Based on von Stosch and Desikachary.)

highly pigmented cells are formed which produce rhizoids and appear to develop directly into the young adult gametophyte.

CHAETANGIACEAE

Until fairly recently tetrasporangia were unknown in the majority of Chaetangiaceae except in *Galaxaura* and the closely related monotypic genus *Actinotrichia* (Kylin 1956). The gametophytes and tetrasporophytes of *Galaxaura* are anatomically distinct, and this has led to the creation of separate species for the two phases of the life history in most instances. Nonetheless, *Galaxaura* exhibits an essentially isomorphic alternation of generations. The taxonomy of gametophytic and tetrasporophytic phases of *Galaxaura* has been considered by Papenfuss & Chiang (1969). With the discovery that the life history of *Pseudogloiophloea* and *Scinaia* resembles *Liagora farinosa* in having *Acrochaetium*-like tetrasporophytic and protonemal gametophytic phases, the problem of the apparent absence of tetrasporangia in these genera has been resolved. *Chaetangium*, a genus of the southern hemisphere, remains to be investigated.

The complete life cycle of *Pseudogloiophloea confusa* was obtained in culture by Ramus (1969). A plant from Jalama Beach, California, liberated carpospores that developed into diploid ($2n = 8$–10) *Acrochaetium*-like tetrasporophytes that were recycled by monospores. Tetrasporangia are initiated with 16:8 LD at 2700 meter-candles illuminance and 15–20°C under conditions of nitrogen depletion. Tetraspores give rise to a filamentous protonema resembling the tetrasporophyte that also reproduces by monospores. Erect thalli develop from bud-like initials surrounded by mucilage when given 2200 meter-candles illuminance, 17°C and 3.0 mg/l phosphate under a variety of photoperiods. Thalli of the Jalama strain are monoecious and self-fertile, producing carposporophytes in culture. Carpospores from cultured material were used to repeat the life history.

The life history of *Scinaia forcellata* has been investigated by Boillot (1968, 1969, 1972). Carpospores germinate into small filamentous sporophytes that produce tetrasporangia. Monosporangia were not encountered. The tetraspores produce the protonema of the gametophyte which, in turn, initiates buds that initially grow uniaxially, later converting to multiaxial growth. Thalli did not reproduce in culture. Magne (1961b) has reported the chromosome numbers as $n = 8$ in the gametophyte and $2n = 16$ in the young tetrasporophyte. Boillot (1972) reports $2n = 12$–15 for the tetrasporophyte and $n = 6$ for the first division of the tetraspore.

Van den Hoek & Cortel-Breeman (1970b) investigated the life history of *Scinaia complanata*. Carpospores produce a tetrasporophyte that reproduces itself or gives rise to young gametophytes. Both processes seemed to be favoured by 17°C and long-day conditions. Only tetrasporangia were observed in culture, and it was suggested that tetraspores may possibly be responsible for recycling the tetrasporophyte as well as production of the gametophyte. Development of the adult phase of the gametophyte could be ontogenetically uniaxial or multiaxial, involving several remote initials.

The life history of *Scinaia turgida* was investigated by Boillot (1971c, 1972) cytologically in culture. Tetrasporophytes were diploid ($2n = 12$) yielding six bivalents in prophase I at the end of diakinesis. Six chromosomes are also counted in the prophase of the second division. Development of the gametophytes was not followed in culture.

3.2 Palmariales

Palmariaceae

The Palmariaceae has been elevated to ordinal status by Guiry and Irvine (in Guiry 1978a) and includes four genera: *Palmaria, Halosaccion, Leptosarca* and *Neohalosacciocolax* (Guiry 1974; Lee & Kurogi 1978). The Palmariales is presently distinguished from other orders of red algae on the basis of tetrasporangial stalk cells and the unusual life history in *Palmaria* and *Halosaccion*. *Palmaria* and *Halosaccion* were formerly regarded as being strictly asexual with tetrasporophytes reproducing by apomeiotic tetrasporangia spores germinating to yield only male gametophytes and tetrasporophytes. Magne (1964) concluded that the chromosome number in *Palmaria palmata* was 14 in all phases. Yabu (1976) observed meiosis in the tetrasporangia of *P. palmata* ($n = 21$–26, $2n = 40$–50) and in *Halosaccion saccatum* ($n = 17$, $2n = c.\ 30$). Van der Meer (1976) verified Yabu's observations with approximate chromosome numbers (38–48, 18–23) for pre- and post-division tetrasporangial nuclei of *P. palmata*. Sparling (1961), Yabu (1976) and others cultured tetraspores of *P. palmata* but none of the progeny reproduced. More recently, van der Meer & Todd (1980) and van der Meer & Chen (1979) presented cultural and cytological data that indicate that the life histories of both *P. palmata* (see Fig. 4.3) and *H. ramentaceum* are sexual. Spore tetrads isolated from field-collected tetrasporophytes showed a 2:2 segregation of large (30–50 cm long) haploid male plants with flat fronds, which produced spermatangia, and small haploid female plants, either simple discs (0.1 mm diam.) or discs with short, slender blades (1 mm long). The diminutive female produced numerous carpogonia with elongated trichogynes. After fertilization the diploid tetrasporophyte blades developed directly from the carpogonia. The tetrasporangia formed on these diploid blades are of normal morphology and produce viable tetraspores. On the basis of the published data, it can be concluded that a sexual life history occurs in the Palmariaceae involving vegetatively dimorphic gametophytes and the complete absence of a carposporophytic phase (Fig. 4.3) in a fashion similar to that of *Rhodochorton purpureum* (West 1969). The evolutionary basis for this behaviour may prove to be very interesting.

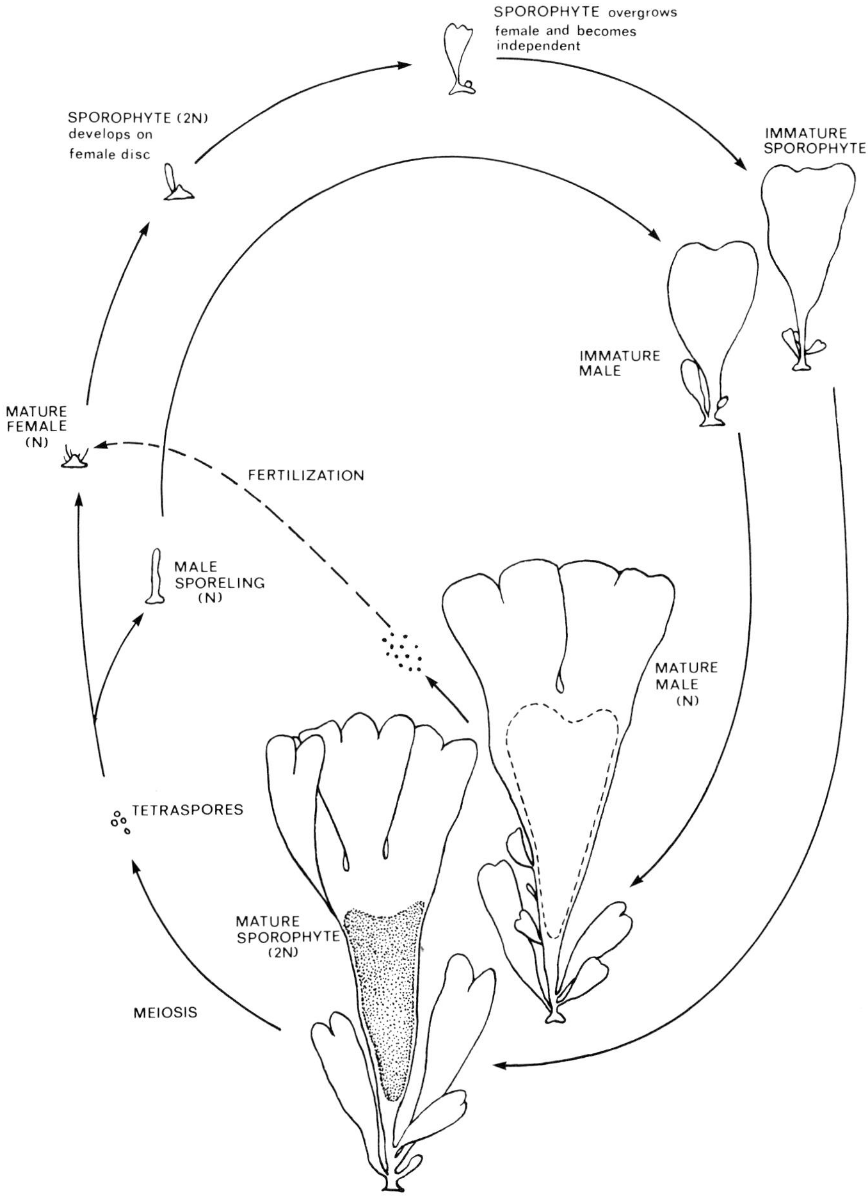

Fig. 4.3 Life history of *Palmaria palmata*. (Reproduced from van der Meer & Todd.)

Rhodophysema and related genera of the Cryptonemiales which are known to possess only tetrasporophytes and male gametophytes are being reinvestigated in view of the life histories reported in Palmariales (see p. 160).

3.3 Gelidiales

Gelidiaceae

Members of the Gelidiales either possess a *Polysiphonia*-type life history with isomorphic gametophytes and tetrasporophytes or are known only from tetrasporic specimens. Haploid gametophytes and diploid tetrasporophytes have been recoded in *Gelidium* and *Acanthopeltis* (Kaneko 1968; Yabu 1975). The basic chromosome number appears to be five (Kaneko 1968). Field and laboratory studies on the ecology and reproduction of gelidioid algae have been reviewed by Santelices (1974). It appears as though no one has raised sporelings of any member of the Gelidiales to maturity in culture. The mode of germination of both carpospores and tetraspores is unique and characteristic of the order Gelidiales as a whole. It has been described in detail for *Gelidium* and *Pterocladia* (Katada 1955; Boillot 1963; Kaneko 1966) and for *Gelidiella* (Chihara & Kamura 1963). Division of the spore nucleus may or may not take place before formation of the germ tube. Most of the spore contents pass into the germ tube leaving a largely evacuated spore which is cut off (Kaneko 1966). The initial then divides obliquely by a curved wall into cells of unequal size. One, two or more rhizoids are initiated from the distal end, and an apical initial is subsequently formed from the proximal end near the empty spore wall (Chihara & Kamura 1963; Kaneko 1966).

3.4 Bonnemaisoniales

The reproduction, life history and systematics of the Bonnemaisoniaceae have been reviewed by Chihara & Yoshizaki (1972) who argue in agreement with Feldmann & Feldmann (1942) and J. Feldmann (1952) that the Naccariaceae and Bonnemaisoniaceae be placed in a distinct order, the Bonnemaisoniales. Both families exhibit life histories involving an alternation of heteromorphic generations. The free-living, presumably diploid generation is either a branched, prostrate filamentous plant of the *Hymenoclonium serpens*-type, or a monosiphonous or polysiphonous creeping or erect thallus that normally expresses some of the features of the gametophytic generation.

Naccariaceae

The family Naccariaceae at present includes only two genera, *Atractophora* and *Naccaria*. The life history of *Atractophora hypnoides* has been investigated by Boillot (1967) and that of *Naccaria wiggii* by Boillot (1967), Jones & Smith (1970) and Boillot & L'Hardy-Halos (1975). Initially Boillot (1967) observed that in both *Atractophora* and *Naccaria* the prostrate protonemal stage formed from the carpospores gives rise by direct development to the erect axes of the adult gametophyte. Subsequently, Boillot & L'Hardy-Halos (1975) obtained tetrasporophytes of *Naccaria* in culture that released tetraspores which germinated into the prothallial stage of the gametophyte and, in turn, produced the erect axis of the adult. Jones & Smith (1970) obtained tetrasporangia in slide cultures from carpospores of *Naccaria* that had been cultured in seawater in the laboratory and subsequently transferred to the natural environment.

Boillot & L'Hardy-Halos (1975) observed that if carpospores of *Naccaria wiggii* are thinly dispersed they first divide one to two times perpendicular to the substrate and produce prostrate filaments that grow side by side in unistratose seriate fascicles. The filaments exhibit negative phototropism under lateral illumination and resemble the *Hymenoclonium*-phase of *Acrosymphyton* seen by Cortel-Breeman & van den Hoek (1970). Tetrasporangia are produced in late February, but are suppressed in longer daylengths after mid-March. The tetraspores germinate initially in the same way as the carpospores to produce a stellately branched protonema from the central spore area. Soon, however, clumps are formed and erect axes arise from individual cells and develop into adult thalli of *Naccaria*. Only male plants were obtained in culture.

In the earlier study (Boillot 1967) the germlings of *Atractophora* and *Naccaria* both developed similarly to the gametophytic prothallus and produced erect axes directly. Boillot & L'Hardy-Halos (1975) speculated that the difference in the two results could be due: 1, to rapid production and dispersal of tetraspores in the earlier culture; 2, to *in situ* germination of tetraspores with the suppression of the gametophytic prothallus as in some *Acrosymphyton* isolates studied by Cortel-Breeman & van den Hoek (1970); or 3, to the fact that the two sets of observations are correct and that variability exists in the life cycle of members of the family Naccariaceae.

Bonnemaisoniaceae

Several versions of a heteromorphic life history are encountered in the Bonnemaisoniaceae. Chihara & Yoshizaki (1972) observed two major patterns in the development of the carpospore, the 'direct disc type' in bilaterally branched species and the 'bipolar erect type' in radially branched species. In *Bonnemaisonia asparagoides* (Feldmann & Mazoyer 1937; Feldmann & Feldmann 1942; Kylin 1945; Feldmann G. 1966) and *Bonnemaisonia clavata* (Feldmann & Feldmann 1939b) from Europe, and in *Delisea* (including *Ptilonia*) from Japan (Chihara 1962; Chihara & Yoshizaki 1978), the carpospores, which measure 50–80 μm in diameter, divide vertically in half and then subdivide to form a disc that later produces a prostrate system bearing gland cells that resembles the genus *Hymenoclonium*. In both *Bonnemaisonia asparagoides* and *Delisea*, the *Hymenoclonium*-stage gives rise directly to erect thalli that resemble the gametophyte. Earlier speculations by Feldmann & Feldmann (1942) that the *Hymenoclonium*-stage bears tetrasporangia was refuted by Kylin (1945).

What had earlier been regarded as tetraspores were later interpreted as irregular divisions into four cells that develop into small plantlets of the *Bonnemaisonia* gametophyte (Feldmann G. 1966). Figure 4.4. depicts the life history of *B. asparagoides* as recorded by G. Feldmann. None of these species has been grown to maturity in culture, and the presumption that the erect thalli are gametophytes rests on the apparent absence of erect tetrasporophytes in nature. The site of meiosis may well take place during initiation of a bud in the *Hymenoclonium* stage, but this has not been established cytologically. More

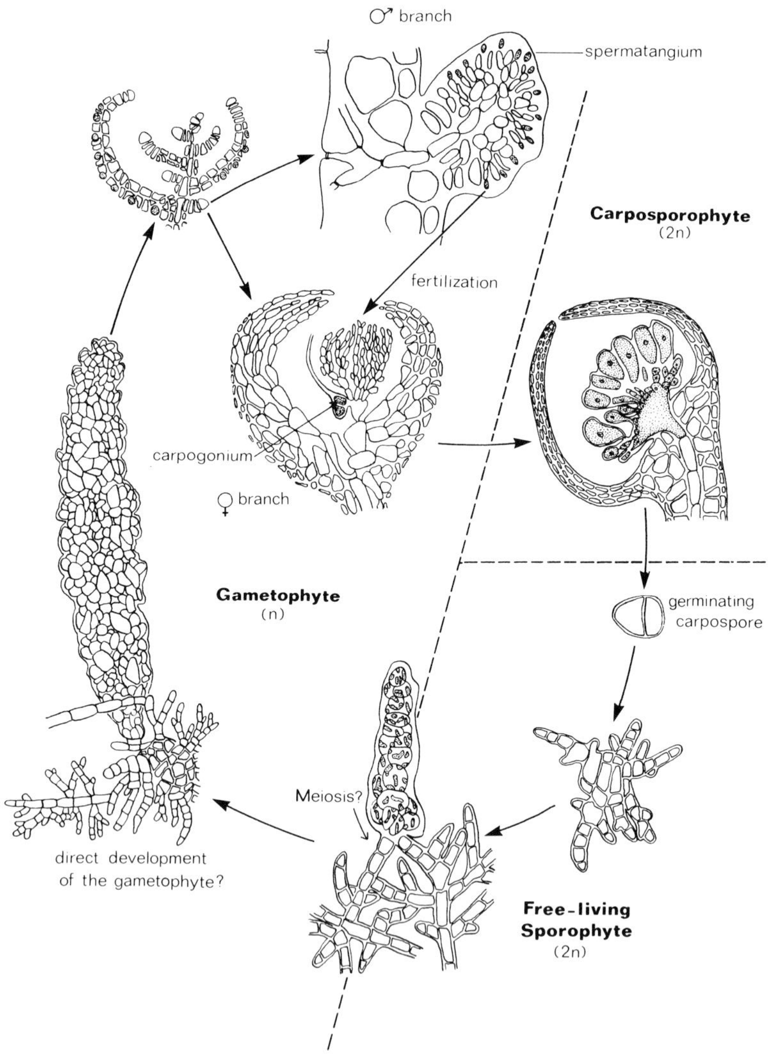

Fig. 4.4 Life history of *Bonnemaisonia asparagoides*. (Based on G. Feldmann and Svedelius.)

recently Cortel-Breeman (1975) has claimed that in her cultures *Bonnemaisonia* thalli definitely did not grow directly on *Hymenoclonium* plants obtained from carpospores in culture, but arose from some kind of spores, probably tetraspores.

The life history of another species of *Bonnemaisonia, B. geniculata* from central California, has been investigated recently by Shevlin & Polanshek (1978). In its vegetative and reproductive development *B. geniculata* stands closest to *B. asparagoides* (Hudson & Wynne 1969). The large carpospores (85–105 μm) of *B. geniculata* first divide vertically into two cells each of which develops both an erect and a rhizoidal axis (double bipolar germination). The erect axes develop into plantlets that resemble tetrasporophytes found in the field from February to June, in which the tetrasporangia are borne in determinate lateral stichidia. Tetraspores from field material showed bipolar germination and gave rise to male or female plants in a 1 : 1 ratio. When placed together on a rotary shaker, fertilization takes place and the carposporophytes that develop bear carposporangia that germinate normally. The gametophytes are epiphytic while the tetrasporophytes form an epilithic turf that probably perennates as a basal rhizoidal system.

Levring reported earlier (1953, 1955) that *Leptophyllis conferta* and two species of *Delisea* from Australia and New Zealand possess free living tetrasporophytes that are isomorphic with their gametophytes. Also, Wynne (1970) described erect tetrasporic plants with compound tetrasporangial stichidia in a new bilaterally branched genus, *Pleuroblepharidella*, from the western North Pacific. The life history of this monotypic genus is unknown.

The radially branched Bonnemaisoniaceae include *Bonnemaisonia hamifera* and *B. nootkana* in both of which the tetrasporophytic phase has long been known as the genus *Trailliella*, and *Asparagopsis armata* and the closely related species *A. taxiformis*, in both of which the tetrasporophyte is known as *Falkenbergia*. In both species complexes carpospores exhibit bipolar germination to produce a free-living tetrasporophyte that is morphologically distinct from the gametophyte. The life history of *Bonnemaisonia hamifera* and *Trailliella intricata* has been investigated in culture in Europe by Feldmann & Feldmann (1942), Harder (1948), Harder & Koch (1949), Koch (1950), and Kornmann & Sahling (1962), from Atlantic North America by Chen *et al.* (1969, 1970a), and from Japan by Chihara (1961). The carpospores of *B. nootkana* from Pacific North America also develop into *Trailliella intricata* (Chihara 1965). The complete life cycle of *Bonnemaisonia hamifera* has not been obtained in culture in the laboratory. Tetraspores from *Trailliella intricata* have given rise to the gametophytic phase of *B. hamifera* in cultures from Europe (Harder & Koch 1949; Kornmann & Sahling 1962), Atlantic North America (Chen *et al.* 1969), and Japan (Chihara 1961). Tetraspores are seen only rarely in nature in the autumn (Dixon & Irvine 1977) and have been produced in culture only under autumn conditions (Koch 1950).

Sexual plants of *Bonnemaisonia hamifera* are well known in Japan where the species may have originated; however, in the North Atlantic and in southern California collectors often encounter only sterile plants, isolated male gametophytes, or female gametophytes with sterile cystocarps. Tetraspores from a single tetrad have been known to produce only male gametophytes in culture (Chen *et al.* 1969) or only female gametophytes (Koch 1950), suggesting that sexuality is phenotypically determined. More recently, however, male and female plants with carpospores have been seen in Europe (Kornmann & Sahling 1962; Bichard-Breaud & Floc'h 1966). Many workers have commented on the sporadic occurrence and discontinuity in the distribution of asexual and sexual phases of *B. hamifera*. For reviews of this topic see Dixon (1970) and Dixon & Irvine (1977).

The life history of *Asparagopsis armata* and of the closely related species *A. taxiformis* has presented many of the same problems as *Bonnemaisonia hamifera*. The European *Asparagopsis armata* is thought to have originated in Australia and become naturalized in the south of France during the first quarter of this century from where it spread throughout the temperate waters of Atlantic Europe, North Africa and the western Mediterranean (see however, Dixon 1965).

Feldmann & Feldmann (1939a, 1942) demonstrated that the carpospores of *Asparagopsis armata* give rise to tetrasporophytes that correspond to *Falkenbergia rufolanosa*. Germinated tetraspores of *Falkenbergia* were not shown conclusively to produce *Asparagopsis* plants in these early studies. Subsequently, von Stosch (1964) obtained plantlets identifiable as *Asparagopsis* from tetraspores of *Falkenbergia rufolanosa* from Atlantic Europe. Upon transferring her material to von Stosch's medium, G. Feldmann (1965) was able to obtain young plants identifiable as *Asparagopsis armata* from spores released from *Falkenbergia rufolanosa* in late autumn. Cytological studies were attempted, but reliable chromosome counts could not be obtained. The site of meiosis has not been established with certainty for any member of the Bonnemaisoniaceae. Chihara (1961) was able to demonstrate that carpospores of the pan-tropical species *Asparagopsis taxiformis* give rise to germlings identical with *Falkenbergia hillebrandii*. Tetraspores from field-collected material of *F. hillebrandii* initially produce a stage resembling *Falkenbergia*; however, the germling ultimately shifts from an alternate-distichous branching pattern to the triradiate pattern characteristic of adult plants of *Asparagopsis*.

3.5 Cryptonemiales

The Cryptonemiales, as circumscribed by Kylin (1956), consists of a central group of families including the Dumontiaceae, Gloiosiphoniaceae, Endocladiaceae and Tichocarpaceae, Halymeniaceae and Kallymeniaceae in which the carpogonia and auxiliary cells are borne either in the same or in separate special branch systems within the vegetative thallus. In a second cluster of families including the Rhizophyllidaceae*, Polyideaceae†, Peyssonneliaceae, Hildenbrandiaceae and Corallinaceae, spermatangia, carpogonia and tetrasporangia are normally borne in separate, superficial nemathecia or in conceptacles. This second group does not fit well either within Cryptonemiales or Gigartinales and probably comprises one or more natural orders. The anomalous parasitic family Choreocolacaceae is of uncertain position.

DUMONTIACEAE, WEEKSIACEAE

The Dumontiaceae is well investigated with respect to life histories in culture. *Dumontia* and the majority of genera belonging in the family have a *Polysiphonia*-type life history. Other genera (*Pikea, Farlowia, Acrosymphyton* and *Thuretellopsis*) with heteromorphic phases should, perhaps, be assigned to a different family. Those with heteromorphic life histories are discussed first.

Farlowia, morphologically similar to *Pikea*, contains four species that are restricted to the North Pacific. Tetrasporophytes are unknown except in *F. conferta* in which Abbott (1968) has indicated that tetrasporangial sori occur on branch tips. This observation needs to be verified. DeCew & West (1981a) cultured carpospores from *F. conferta, F. compressa* and *F. mollis*. All developed into crustose plants resembling *Haematocelis* in vegetative structure. *Haematocelis*, with terminal zonately divided tetrasporangia, is also considered the tetrasporophytic phase of *Schizymenia* (see pp. 167–8).

Thuretellopsis, a monotypic genus endemic to Washington, is known in nature only in the erect gametophytic phase. Richardson & Dixon (1970) observed that carpospores germinated in culture, forming prostrate filamentous plants in 9:15 LD and radial crusts in 16:8 LD. The tetrasporophytes are recycled by monospores. Tetrasporangia produced in 9:15 LD released tetraspores that gave rise to juvenile upright gametophytes; however, these failed to produce spermatia or carpogonial branches and bore only monosporangia. The tetrasporophyte crust obtained in culture somewhat resembles *Rhododiscus pulcherrimus* with terminal cruciate-decussate tetrasporangia according to Parke & Dixon (1976).

Pikea is restricted to the North Pacific and tetrasporophytes are unknown in field plants. Chihara (1972) cultured carpospores of two isolates of *P. californica* from Japan, both of which gave rise directly to basal discs that bore erect shoots that were anatomically similar to those of the field collected gametophytes but did not reproduce. By contrast, Scott & Dixon (1971) obtained a crustose phase from California isolates that bore tetrasporangia and monosporangia. Isolated tetraspores produced basal discs with erect *Pikea*-like fronds that did not form gametangia. The gametophyte basal discs and tetrasporophyte crusts were anatomically similar.

* Removed to the Gigartinales by Wiseman (1975).

† Removed to the Gigartinales by Papenfuss (1966)

Several explanations may be offered for the apparent discrepancies in these results. Dixon *et al.* (1972) suggest two interpretations of Chihara's data: 1, the Japanese isolates formed tetrasporangia and spores germinated *in situ* to form the gametophytic phase or 2, a *Lemanea*-type life history occurred involving meiosis in the apical cell of an erect shoot primordium. A third possibility is that the carpospores were derived by apomixis without fertilization and germinated directly to form a new gametophyte. No conclusion is possible at this time in the absence of cytological studies.

By far the most extensive and careful investigations of any Dumontiaceae is the work on *Acrosymphyton purpuriferum* (see Fig. 4.5) by Cortel-Breeman & van den Hoek (1970), Cortel-Breeman (1975), Cortel-Breeman & Ten Hoopen (1978) and Breeman (1979). The gametophyte is an erect, cylindrical, branched monoecious plant. Tetrasporophytes were previously unknown. In culture, the carpospores gave rise to crustose plants resembling *Hymenoclonium serpens* which formed tetrahedral tetrasporangia in short days (8:16 LD). Light breaks in the dark period did not influence sporulation, in contrast to typical phytochrome mediated systems. The tetrasporelings developed into typical gametophytes which reproduced sexually in culture. *Hymenoclonium* plants isolated from nature, sporulated in culture and gave rise either to *Acrosymphyton* or *Bonnemaisonia asparagoides*, indicating the morphological similarity between the tetrasporophytes of these two genera. DNA microspectrophotometry indicated that the gametophytes were haploid in *Acrosymphyton* and the sporophytes diploid. Infrequently *Acrosymphyton* tetrasporophytes reproduced asexually by apomeiotic propagules that repeated the tetrasporophytic phase. No apomictic reproduction was seen in the gametophytic phase.

Hyalosiphonia, a monotypic genus from China, Japan and Korea that is related to *Cryptosiphonia* from Pacific North America, has a *Polysiphonia*-type life history as based on observations of field material (Chihara & Yoshizaki 1971). Umezaki (1972) cultured carpospores from field-collected specimens of *H. caespitosa* which formed basal discs and upright uniaxial shoots. These produced tetrasporangia three months after the plants were placed in 7:17 LD. Erect plants differentiated from the tetraspores but failed to reproduce.

Neodilsea integra from Hokkaido, Japan, investigated in culture by Masuda (1973b), has a *Polysiphonia*-type life history. Tetrasporangia developed in both 14:10 LD and 10:14 LD. Tetraspores and carpospores exhibited similar development patterns forming numerous colourless unicellular hairs on the discs and erect shoots. Crossing of male and female gametophytes resulted in viable carposporophytes. Females isolated from males produced no cystocarps indicating that apomixis does not occur. Chihara (1967) and Zinova (1961) point out that specimens of *N. integra* from Arctic Seas do not form normal tetrasporangia but monosporangia-like structures. It is possible that the lower temperature results in a loss of sexual reproduction as suggested by Dixon (1973). Masuda (1973a) obtained discs and young adult thalli in culture from germinated tetraspores of three additional species of *Neodilsea*.

The Weeksiaceae, established by Abbott (1968) on the basis of post-fertilization patterns, includes the multiaxial genera *Weeksia* and *Constantinea* and the uniaxial genus *Leptocladia*, all formerly in the Dumontiaceae. All presumably have isomorphic phases, but none have been carried through their complete life histories in culture.

GLOIOSIPHONIACEAE

Gloiosiphonia, widely distributed in the Atlantic and Pacific, contains at least two species. Isomorphic tetrasporophytes have been very rarely recorded. According to Morohoshi & Masuda (1980) only two records are known of tetrasporangia in erect thalli of *G. capillaris*. None is reported for *G. verticillaris*.

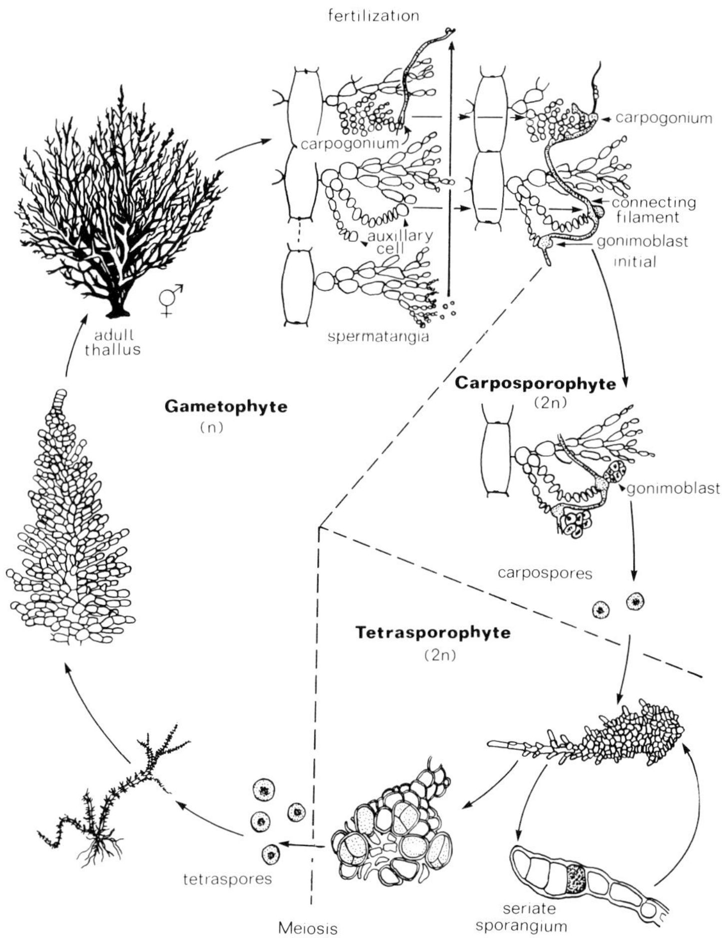

Fig. 4.5 Life history of *Acrosymphyton purpuriferum*. (Based on Cortel-Breeman.)

Edelstein (1970) and Edelstein & McLachlan (1971) cultured carpospores of *G. capillaris* from Nova Scotia that developed into tetrasporangiate crusts that resemble *Cruoriopsis hauckii*. Tetrasporelings differentiate into typical monoecious gametophytes. Morohoshi & Masuda (1980), culturing *G. capillaris* from Hokkaido, Japan, obtained very different results. Carpospores from field-collected gametophytes differentiated directly into basal discs and erect cylindrical monoecious fronds that produced cystocarps. The carpospores repeated the cycle, indicating the complete absence of tetrasporophytes. It was not established whether syngamy was necessary to produce cystocarps.

Isolates of *Gloiosiphonia verticillaris* from California show a standard sexual cycle (DeCew & West unpublished). The crustose *Cruoriopsis*-type tetrasporophytes sporulate in 16°C, 8:16 LD but not in 15°C, 16:8 LD. The tetraspore germlings develop into self-fertile monoecious gametophytes. Isolates of plants different in vegetative development from both *G. verticillaris* and *G. capillaris* that were obtained from California and Baja California are possibly equivalent to *G. californica*. The *Cruoriopsis*-type tetrasporophytes sporulated vigorously in low nitrate medium (8 µg–at N/l). The tetraspores germinated readily yielding upright plants with direct cortication rather than descending cortication characteristic of *G. verticillaris*. These then differentiated into monoecious self-compatible gametophytes (DeCew & West unpublished). One isolate of '*G. californica*' from Stillwater Cove, Monterey County, California, is apparently apogamic, carpospores giving rise to gametophytes directly.

Chihara (1972) cultured carpospores of *Schimmelmannia plumosa* from Japan that gave rise to erect uniaxial thalli similar to gametophytes from the field but these did not reproduce; thus the tetrasporophytes remain unknown. Tetrasporophytes are also unrecorded for *Thuretella* from Atlantic Europe, and *Gloeophycus* from Korea and Hokkaido (Lee & Yoo 1980; Kaneko *et al.* 1980).

Endocladiaceae and Tichocarpaceae

The *Polysiphonia*-type life history is apparently exhibited in the Endocladiaceae (*Endocladia* and *Gloiopeltis*) and the monogeneric multiaxial family Tichocarpaceae (*Tichocarpus*) based on field observations.

Halymeniaceae

Halymenia floresia (van den Hoek & Cortel-Breeman 1970) is the only member of the Halymeniaceae for which a complete life history has been attempted in culture. Regrettably, it falls short in that achievement. Carpospores of *Halymenia floresia* developed into acrochaetioid plants forming only monosporangia. Monospores germinated into small foliose *Halymenia* plants that failed to reproduce. Several attempts were made to induce tetrasporogenesis in the filamentous stage by varying light and temperature conditions, but without success. The authors state that *H. floresia* reverts to an acrochaetioid condition during the low temperature periods of the winter.

Kallymeniaceae

Murray & Dixon (1972) cultured carpospores of *Callophyllis firma* from Monterey, California. Fertile tetrasporophytes developed in two months, and tetraspores germinated forming gametophytes. Cystocarps formed and released viable carpospores. The entire life history was completed in six months. Both phases were morphologically identical. This life history pattern is probably representative of most members of the Kallymeniaceae. Codomier (1972) germinated carpospores of three species of *Kallymenia* which produced monostromatic creeping discoid thalli in which fusions frequently take place between adjacent cells. The prostrate stage of *K. microphylla* was equated by Codomier (1973) with *Rhodochorton hauckii*. Hansen (1977) obtained prostrate germlings from carpospores of *Cirrulicarpus carolinensis* from North Carolina that had the general appearance of *Hymenoclonium serpens* and also exhibited fusions between adjacent cells similar to those seen in *Kallymenia*. The monostromatic prostrate system developed into a layered crust that subsequently gave rise to erect multiaxial gametophytic blades. Tetrasporophytes were not seen in field or cultured material and the life history may be apomictic.

Choreocolacaceae

The Choreocolacaceae with three parasitic genera, *Choreocolax, Harveyella* and *Holmsella*, has been a subject of recent investigations on physiology, fine structure and ecology. *Harveyella mirabilis* (Goff & Cole 1976a), on the basis of field investigations of phenology, probably exhibits a typical *Polysiphonia*-type life history. Initial spore germination occurs in host wounds inflicted primarily by grazing isopods or amphipods (Goff & Cole 1976b). Rhizoidal cells establish secondary pit connections with host cells and subsequently develop a colourless reproductive pustule. *Holmsella* (Callow *et al.* 1979) and *Choreocolax* (Kugrens & West 1973) also were investigated for physiology and fine structure, but no critical work on reproduction and life histories has been undertaken.

Rhizophyllidaceae, Polyideaceae, Peyssoneliaceae

These three families with carpogonial branches and auxiliary cells borne in special nemathecia were clustered by Kylin (1956) in the Cryptonemiales. Papenfuss (1966) transferred the Polyideaceae to Gigartinales and Wiseman (1975) similarly placed the Rhizophyllidaceae in Gigartinales. There is no major evidence to suggest that this group of families is closely related to either order. All appear to possess isomorphic life histories based on field observations. Both zonately and cruciately divided tetrasporangia are recorded in members of all three families. For details see Wiseman (1977) (*Ochtodes*), Dixon & Irvine (1977) (*Polyides*), Nozawa (1970) (*Rhodopeltis*), Denizot (1968) and Nozawa (1969, 1972) (*Peyssonnelia, Cruoriella, Polystrata*).

Genera removed from Peyssonneliaceae

The genera *Rhodophysema* (Fletcher 1975, 1977; Ganesan & West 1975; Masuda & Ohta 1975; South & Whittick 1976), *Rhodophysemopsis* (Masuda 1976a) and *Pseudorhododiscus* (Masuda 1976b) sometimes placed in Peyssonneliaceae* are reproductively distinct and have in the past been thought to lack female gametophytes. Usually, only tetrasporophytes and sometimes male gametophytes are seen in field-collected specimens and in laboratory cultures. Tetrasporangia have been interpreted as apomeiotic in species that have been investigated because only successive generations of tetrasporophytes normally occur in culture. South & Whittick (1976) counted 30–35 chromosomes in nuclei of successive tetrasporophytic generations of *Rhodophysema elegans* from culture. Masuda & Ohta (1975) have observed male plants and male tetrasporangial plants of *Rhodophysema georgii* in culture. Reinvestigation by DeCew of the same isolate of *R. elegans* studied by Ganesan & West (1975) strongly suggest the occurrence of a distinct sexual-type of life history in this species. Probable spermatangia and carpogonia with elongate trichogynes are present in young monoecious crusts. These structures were interpreted as hairs by the earlier authors. Tetrasporangia appear to develop directly from a generative stalk cell that is cut off by a 'fertilized' carpogonium. It is likely that the tetrasporophytic phase is reduced to a single generative stalk cell which produces tetrasporangia successively. This turn of events suggests that the life history of all other species of *Rhodophysema* as well as *Rhodophysemopsis* and *Pseudorhododiscus* should be re-examined carefully.

Pseudorhododiscus (Masuda 1976b) is similar in reproductive patterns to *Rhodophysema georgii* in that tetraspores from tetrasporophytes gave rise to successive generations of male plants and tetrasporophytes in culture. *Rhodophysemopsis* has not been cultured, but Masuda (1976a) reported only males and tetrasporophytes in field-collected plants. *Porphyrodiscus*, with zonately divided tetrasporangia borne in nemathecia, is the tetrasporophyte of *Ahnfeltia plicata* (Farnham & Fletcher 1976; Chen 1977). The tetrasporangia of *Pulvinia* (Hollenberg 1970) are borne on stalk cells similar to those of the Palmariaceae. The biology of this genus needs to be investigated in culture.

Although Kylin placed *Erythrodermis* in the Peyssonneliaceae, it is more closely allied to *Petrocelis* and at least one species is the tetrasporophyte of *Gymnogongrus flabelliformis* (Masuda *et al.* 1979) and *G. leptophyllus* (DeCew & West 1981).

* Denizot (1968) relegates *Rhodophysema, Coriophyllum* and other genera to *Incertae Sedis*.

Coriophyllum, a genus endemic to Pacific North America, is known only as a tetrasporophyte in field collections; however, culture studies (DeCew unpublished) have shown that crustose male and female plants are derived from tetraspores. The 'fertilized' carpogonial cell probably gives rise directly to gonimoblast tissue that is intrusive in the gametophyte. At maturity it reproduces nemathecia bearing zonately divided tetrasporangia. The taxonomic position of this genus is uncertain.

HILDENBRANDIACEAE

The Hildenbrandiaceae, a monogeneric family, may be uniform in its reproductive patterns. The nine species of *Hildenbrandia* are reported to be exclusively tetrasporangiate, and this is borne out by culture investigations of *H. occidentalis* and *H. prototypus* (DeCew & West 1977). Successive tetrasporangiate generations developed in each species. Careful re-examination is needed, however, because there appear to be small carpogonia with trichogynes present that were previously regarded as hair cells (DeCew unpublished).

CORALLINACEAE

Among the red algae the Corallinaceae is unquestionably the most widely distributed family with greatest total biomass and the greatest number of taxa–approximately than 50 genera and 400 species.

Coralline algae generally have a *Polysiphonia*-type life history with isomorphic phases except in a few instances. Foslie (1905) first noted the high frequency of bisporic sporangia in crustose corallines of colder regions, and concluded that the frequency of meiosis was reduced, eliminating the normal sexual cycle. This was verified in the careful cytological studies on *Lithophyllum* bispores by Suneson (1950). More recently, Gittins (1977) has noted a similar pattern in the articulated coralline *Lithothrix aspergillum* from the Pacific coast of North America. In cold upwelling areas bisporic plants predominate.

Development of non-articulated crustose coralline sporelings has been investigated in the field (Adey & Vassar 1975), but optimal requirements for growth and reproduction in culture remain almost totally unknown. In the laboratory virtually all work is done with unsupplemented natural seawater requiring frequent changing. Notoya (1976) investigated the influence of temperature, light intensity, salinity and pH on sporeling development in crustose corallines and Chihara (1974) has described the germination patterns of a number of taxa. No one has obtained a complete life history of a species of crustose corallines in culture.

The only successful culture work on coralline life histories is the brief but elegant paper by von Stosch (1969). Carpospores from field-collected female gametophytes of *Corallina officinalis* germinated forming expanded basal discs from which erect articulated fronds arose. Tetrasporangial conceptacles developed and the viable tetraspores produced male and female gametophytes in culture in a 1 : 1 ratio. Fertilization resulted in formation of viable carpospores that germinated ultimately giving rise to erect articulated plants that did not, however, become reproductive. *Jania rubens*, also cultured by von Stosch, grew best in 12 μg-at./l phosphate (a quarter of the normal 50 μg-at./l level). Low phosphate stimulated calcification, and high phosphate inhibited the process. After four years in culture tetrasporophytes formed conceptacles when the iron concentration was increased from 1 μg-at./l to 11 μg-at./l and released viable tetraspores that developed into monoecious gametophytes. (Usually the coralline algae are considered to be dioecious.) Normal spermatangia and carpogonial branches developed in the male and female conceptacles, and fertilization did occur after several years. Carposporelings developed into plants bearing both male and tetrasporangial conceptacles. The tetraspores did give rise to normal gametophytes. Von Stosch also undertook cytological investigations with *Jania* and observed that the gametophytes were haploid and the sporophytes diploid. One large chromosome was evident in each haploid set suggesting the presence of sex chromosomes.

Before extensive culture work on coralline life histories can be undertaken, it is clear we need to know much more about the nutritional physiology of these algae and to develop more suitable media than any currently available.

3.6 Gigartinales

The order Gigartinales contains a heterogenous assemblage of families in which the auxiliary cell is said to be derived from an ordinary vegetative cell (Kylin 1956). In this treatment families that appear to be closely related are clustered or arranged in sequence and families of less certain position are appended at the end of the section.

Chondriellaceae and Gigartinaceae

The monotypic family Chondriellaceae, known only from the Juan Fernandez Islands, is distinguished from the Gigartinaceae by its tetrasporangia forming terminally on cortical filaments. The life history is presumed to be isomorphic.

The Gigartinaceae consists of four widely distributed genera (*Chondrus, Gigartina, Iridaea* and *Rhodoglossum*) of major economic importance. At present, generic limits are not clear and there are marked differences of opinion on this matter.

Chondrus crispus, best known for its carrageenan production, has been cultured extensively. Chen & McLachlan (1972) provided the first complete culture study on the life history which is a normal isomorphic type. Of the six species of *Chondrus* known from Japan only *C. ocellatus* has been cultured through its life history (West & Lee unpublished). Presently there is no evidence to indicate that apomixis occurs in these or any other species. One strain of *Rhodoglossum affine* from California cultured by Chen & McLachlan (1980) shows a standard isomorphic life history. West & Guiry (unpublished) cultured *R. californicum* from California which also exhibits an isomorphic life history. Kim (1976) cultured two isolates of *R. roseum*, both of which reproduced sexually.

Species from Pacific North America placed in *Iridaea* appear to have a standard *Polysiphonia*-type life history, but it is also reasonable to assume that some populations are apogamic or apomeiotic. Kim (1976) demonstrated that tetraspores of an isolate of *I. cordata* var. *splendens* from Washington germinated to form other tetrasporophytes, while an isolate of *I. cordata* var. *cordata* maintained a normal sexual life history. Kim (1976) also found that one isolate of *I. cornucopiae* from Washington was apomeiotic giving rise to three successive tetrasporophytic generations in culture, whereas another strain from British Columbia reproduced sexually. Two isolates of *Iridaea heterocarpa* from Washington reproduced sexually. The frequency of gametophytes and tetrasporophytes may not be equal. *Iridaea cordata* exhibits a higher frequency of tetrasporophytes than gametophytes (Hansen & Doyle 1976). By contrast *Iridaea flaccida* (Abbott 1980) has a very low frequency (20%) of tetrasporophytes. Phase frequency differences may result from many causes unrelated to life history differences. Differential rates of metabolism, growth or grazing, vegetative propagation, spore production and spore germination, are only a few examples. Wide variations in phase frequency within populations of Gigartinaceae are probably common as, for example, in *Rhodoglossum affine, Gigartina leptorhynchos, G. canaliculata, G. corymbifera* (Abbott 1980), and *Chondrus crispus* (Craigie & Pringle 1978; Penncavage *et al.* 1979).

Data based on reproductive patterns of *Gigartina* in culture are limited. All phases are recorded from field collections of most species of *Gigartina* but it is questionable whether or not sexual reproduction is a consistent phenomenon in all populations of each species (see Abbott 1980). Several *Gigartina* species have been grown successfully through complete sexual life histories, including isolates of *G. canaliculata, G. leptorhynchos* and *G. ornithorhynchos* (*G. spinosa*) all from Monterey, California, as well as *G. teedii* from England and Brasil (West & Guiry unpublished), and *G. intermedia* from Korea (West & Lee unpublished).

Four isolates of *G. johnstonii* from Puerto Peñasco, Sonora, Mexico, reproduced in normal sexual patterns, but two third-generation plants of one isolate developed mixed-phase reproduction (West & Guiry unpublished). A female plant that had produced only procarps for nine months then produced both spermatangia and procarps and later developed cystocarps. Viable carpospores formed tetrasporophytes that released viable tetraspores but these were not cultured to maturity. One male plant produced normal spermatangia for three months, and then one branch produced tetrasporangial sori as well as spermatangia. Tetraspore germlings developed into dioecious male and female plants. No cytological investigations were done.

In *Gigartina* subgenus *Mastocarpus*, which includes about five widely distributed species, extensive culture studies on the reproductive patterns have been undertaken within the last ten years. Tetrasporophytes were unknown until West (1972b) reported that tetraspores of *Petrocelis* developed into dioecious gametophytes resembling both *G. papillata* and *G. jardinii*. Female plants bore cystocarps releasing carpospores that germinated into crustose *Petrocelis* plants which often formed and released tetraspores under short day (8:16 LD) conditions. Subsequent culture work on numerous isolates of *G. papillata* ranging from Alaska to Mexico revealed that two reproductive processes are evident. In addition to the normal sexual cycle some strains appear to be obligately apogamic. The carpospores form upright female gametophytes that develop cystocarps, apparently without fertilization. Successive generations remain apogamic. Approximately 60% of the isolates are apogamic and 40% sexual. The frequency of sexual reproduction increases in the southern range of distribution. Sexual isolates within this pleomorphic species complex are not all interfertile (Polanshek & West 1975, 1977).

In *G. jardinii* the frequency of sexual reproduction is higher (75%) than in *G. papillata*, and there appears to be no geographic pattern. Although all the cultured gametophytes derived from *G. jardinii* blade tips are completely interfertile throughout the range, none of these are compatible with gametophytes that develop from *Petrocelis* tetraspores. Carpospores from all sexual plants of *G. jardinii* develop into *Petrocelis*-like crusts which, however, do not sporulate in culture.

Gigartina stellata from the North Atlantic shows a similar pattern. *Petrocelis cruenta* tetraspores develop into *G. stellata* gametophytes. Many isolates of *G. stellata*, however, appear apogamic, the frequency being higher in northern regions. At the southern end of the range of the species all isolates are sexual (West *et al.* 1977; West & Guiry unpublished). Chen *et al.* (1974), Rueness (1978), Dion & Delépine (1979) have also noted the occurrence of putative apomicts in *G. stellata*. Isolates of *Gigartina ochotensis, G. pacifica* (Masuda & Uchida 1976; Masuda *et al.* unpublished) and *G. mammilosa* (Masuda & Ohno unpublished) also appear to be either apogamic or sexual.

Recently Chen and Craigie (1981) and Guiry (unpublished) have reinvestigated the carrageenan chemistry of the developmental stages of *G. stellata*. The basal discs of a strain reported to be apomictic by Chen *et al.* (1974) produce a lambda carrageenan, whereas the upright axes contain a kappa-type carrageenan (Chen and Craigie 1981). Guiry (unpublished) has observed a similar phenomenon in other 'apomictic strains.' If carrageenan type is truly genetically linked, with the haploid phase producing kappa and the diploid phase producing lambda types, then the formation of carposporophytes by an apomictic process is very doubtful. It is possible that special types of somatic fertilization and meiosis occur which were not evident previously. Considerable additional work is needed.

It has been proposed (Kim 1976) that the subgenus *Mastocarpus* of *Gigartina* be elevated to familial status (Mastocarpaceae) on the basis of reproductive morphology and life history patterns; however, additional investigations are necessary. This subgenus also shows affinities with some genera of the Phyllophoraceae with respect to reproduction and life history.

PHYLLOPHORACEAE

The Phyllophoraceae* possesses a variety of life histories. Species of the genus *Ahnfeltia* all probably have crustose tetrasporophytes. The tetrasporophyte of the type species, *A. plicata*, has been linked with *Porphyrodiscus simulans* (Farnham & Fletcher 1976; Chen 1977). The location and development of carpogonia and spermatangia are not yet understood. Magruder (1977) cultured the male and female gametophytes and the crustose (*Erythrodermis*-like) tetrasporophytes of *A. concinna. Ahnfeltia paradoxa* in Japan, which is morphologically similar to *Gymnogongrus furcellatus* from South America, is currently under investigation by Masuda. Preliminary observations indicate that it has an *Erythrodermis*-like crustose tetrasporophyte similar to that reported for *G. furcellatus* from Chile by Candia & Kim (1977). *Ahnfeltia gigartinoides* and a new species from Pacific Mexico also have an *Erythrodermis*-like tetrasporophyte (DeCew & West 1977b). The two species belong in a new genus.

Gymnogongrus is a large genus exhibiting a great diversity of life history patterns. *Gymnogongrus griffithsiae*, the type species, and *G. crenulatus* are considered to be tetrasporoblastic, the diploid tetrasporophyte developing directly from the carposporophyte attached to the female gametophyte (Schotter 1968; Ardré 1977b, 1978). Cordeiro-Marino & Poza (1980) have verified this through culture and cytological studies on *G. griffithsiae* in Brazil. Several other species also appear to be of the tetrasporoblastic type. A second and larger group of *Gymnogongrus* species thought to lack a tetrasporophyte have been investigated through culture studies and are now known to have an *Erythrodermis*-type tetrasporophyte: *G. leptophyllus* and *G. linearis* (DeCew & West 1981b), *G. flabelliformis* (Masuda *et al.* 1979), *G. furcellatus* (Candia & Kim 1977), *G. martinensis* (West & DeCew unpublished). There is also an indication that in some localities *G. leptophyllus* bypasses normal tetrasporophytes, the carpospores developing directly into new upright female gametophytes.

Ceratocolax, the parasite of *Phyllophora*, is tetrasporoblastic judging from field observations. *Phyllophora* (Newroth 1971, 1972), like *Gymnogongrus*,

* Guiry (1978b) proposed that the name Tylocarpaceae Kuetzing, which has nomenclatural priority over the Phyllophoraceae, be adopted pending revision of the familial structure in the Gigartinales.

appears to have both tetrasporoblastic (*P. truncata*) and isomorphic (*P. pseudoceranoides*) life history patterns. Tetrasporophytes are not recorded for some species of *Petroglossum* and *Ozophora* but, in general, the evidence suggests the two genera are isomorphic. In *Schottera* and *Stenogramme* tetrasporophytes and gametophytes are considered to be isomorphic, but no definitive evidence is available from culture studies.

Besa has been shifted between the Gigartinaceae and Phyllophoraceae by various authors, principally because of vegetative anatomy that is intermediate between the two families. *Besa papillaeformis*, the type species, has an isomorphic life history according to DeCew & Silva (unpublished). Hollenberg (1969) speculated that *Erythrodermis pacifica* might be the tetrasporophyle of *B. stipitata* The familial assignment of *Ahnfeltia gracilis*, a species previously placed in *Besa* (see Mikami 1965), is still open to question. The interpretations of *Besa* listed here have not yet been verified by studies with cultures.

Cruoriaceae

All genera assigned to the Cruoriaceae (*Petrocelis, Cruoria, Cruoriopsis* and *Haematocelis*) by Kylin (1956) apparently are crustose tetrasporophytes of upright foliose genera of other families in the Cryptonemiales (see pp. 155–8) and Gigartinales (see pp. 162–6; 166–9). The scattered records of carposporophytes in *Cruoria, Cruoriopsis* and *Petrocelis* (see Kylin 1956) are unclear. *Petrocelis hennedyi* is the only species of the genus reported to have carposporophytes (Rosenvinge 1917) but this has never been re-examined, nor has *P. hennedyi* been cultured to maturity. Denizot (1968) relegated *Cruoriopsis* to the Peyssonneliaceae without explanation. All the foliose genera in other families seem to have nomenclatural priority; so both the family Cruoriaceae and the crustose genera may, perhaps, be eliminated or designated as form-genera.

Calosiphoniaceae

The Calosiphoniaceae with two genera, *Schmitzia* and *Calosiphonia*, bears a resemblance to the Gloiosiphoniaceae (Cryptonemiales). Tetrasporophytes are unknown in the field for either genus. Mayhoub (1973, 1975) and Mayhoub *et al.* (1976) cultured carpospores of *C. vermicularis* from Syria. Two types of development and reproduction were observed, one sexual and the other apparently apomictic. In the first, carpospore germlings developed into crustose tetrasporophytes that produced zonately divided tetrasporangia in short day ($<$ 8:16 LD) conditions and remained vegetative or produced monosporangia under long day ($>$12:12 LD) conditions. Mayhoub did not consider the taxonomic placement of the tetrasporophytic phase, but the vegetative anatomy of the crust and terminal zonately divided tetrasporangia indicate that it is morphologically equivalent to *Haematocelis*, the crustose tetrasporophyte of *Schizymenia dubyi* (Gigartinales) and *Farlowia* (Cryptonemiales). Tetraspore germlings then developed into erect gametophytes with the basal discs forming monospores. Mayhoub (1975) provided no information about sexual reproduction in these gametophytes. The second type of development appears to be apomictic. Carpospores from field-collected plants germinated to form small discs bearing upright fronds. These are monoecious, forming carpogonial branches and spermatangia. Mayhoub (1973) indicated that fertilization occurs to form cystocarps, equating this to the *Lemanea*-type development (Magne 1976b), but later Mayhoub (1975) stated that the cystocarps developed without fertilization in this instance. One may assume that *Calosiphonia* exhibits either sexual or nonsexual reproductive patterns similar to those of *Gigartina*, subgenus *Mastocarpus*, but the published evidence is incomplete.

Feldmann (1954) cultured the carpospores of *Schmitzia neapolitana* (as *Bertholdia neapolitana*) which formed a crustose stage anatomically similar to that of *Calosiphonia* but these did not reproduce.

Gymnophlaeaceae (formerly Nemastomaceae), Sebdeniaceae

The Gymnophlaeaceae show a mixture of isomorphic and heteromorphic life history types. No tetrasporophytes are recorded for *Adelophyton* (Kraft 1975). All species of *Nemastoma* are believed to be isomorphic. Umezaki (1974) cultured tetraspores and carpospores of field-collected *N. nakamurae* from Japan. Upright fronds were formed in each phase, but they did not reproduce in culture. In *Platoma*, only *P. bairdii* is reported as being isomorphic: the other seven species including the type, *P. cyclocolpa*, have no recorded tetrasporophytes (Norris & Bucher 1977). Tetrasporophytes were not known in *Predaea* until Lemus & Ganesan (1981) cultured carpospores of *P. feldmannii* and *P. pusilla* from Venezuela and observed the development of an acrochaetioid phase bearing tetrasporangia. Release and germination of tetraspores were not observed, but the study provides the first evidence for a heteromorphic life history in the Gymnophlaeaceae. *Titanophora* with seven species is widely distributed in tropical waters. No tetrasporophytes are recorded, nor have culture studies been attempted.

Species of *Schizymenia* are restricted to temperate and cold waters and constitute a heterogeneous assemblage with respect to life history patterns. Plants with tetrasporoangia have not been obtained in cultures in the type species *S. dubyi*, but Ardré (1977a) believes that *Haematocelis rubens* may be the tetrasporophytic phase because it is frequently associated with the base of *S. dubyi* obtained in a number of localities in the North Atlantic. Sciuto *et al.* (1979) indicate that *S. dubyi* produces a distinctive amino acid, pyrrolidine-2,5-

dicarboxylic acid, that is also synthesized by *H. rubens*. None of the 35 other red algal genera analysed contained this compound. On the other hand, carpospores of *S. dubyi* cultured by Codomier (1974) gave rise to discs with upright multiaxial blades that did not reproduce. DeCew & Moe (unpublished) observed that tetraspores of *Haematocelis* from Baja California, Mexico and California gave rise to multiaxial foliose plants bearing gland cells and carpogonial branches characteristic of *Schizymenia*; however, carpospores derived from *Schizymenia pacifica* (Monterey, California) germinated to become upright foliose blades with vesiculate cells (Moe unpublished). All *Schizymenia* species recorded for the northeastern Pacific are presumed to have isomorphic life histories (Abbott & Hollenberg 1976), but this is doubtful. Specimens recorded as *Schizymenia pacifica* tetrasporophytes (Abbott 1967) all lack vesiculate cells and are probably not *Schizymenia* (DeCew & Moe unpublished). It is possible that *Schizymenia* does not belong in the Gymnophlaeaceae.

The monogeneric family Sebdeniaceae appears to be allied to the Gymnophlaeaceae. *Sebdenia* probably exhibits a *Polyisphonia*-type life history but culture studies have not been undertaken.

Furcellariaceae

In the Furcellariaceae, a standard isomorphic life history is evident only in *Furcellaria* on the basis of cytological evidence (Austin 1960a,b). *Halarachnion* (Boillot 1965) and *Neurocaulon* (Codomier 1969) are heteromorphic with a crustose, *Cruoria*-like tetrasporophyte. Based on the vegetative morphology, postfertilization development, zonate tetrasporangia and heteromorphic life histories, both *Neurocaulon* and *Halarachnion* should be placed in a separate family, retaining only *Furcellaria* in the Furcellariaceae.

Solieriaceae (including Rhabdoniaceae), Rissoellaceae

Gabrielson (1980) has proposed the transfer of *Rhabdonia* and three other Australian genera to Solieriaceae. The Solieriaceae is of widespread temperate and tropical distribution.

In the majority of genera belonging to the Solieriaceae the life history is presumably of the *Polysiphonia*-type. Polne *et al.* (1980) have described an isomorphic life history for *Eucheuma uncinatum* (Gulf of California, Mexico) utilizing a running seawater culture system. The reproductive patterns of some other species of *Eucheuma* may be variable. For example, field observations of *E. isiforme* and *E.* nov. sp. (tropical Atlantic) indicate a scarcity or absence of male plants (Dawes *et al.* 1974). *Agardhiella subulata* normally exhibits isomorphic phases, however, dwarf male gametophytes commonly develop *in situ* from tetrasporangia on tetrasporophytes (Gabrielson & Hommersand

1981). In *Gardneriella*, a parasite of *Agardhiella gaudichaudii* in California, Goff (1980) reports the meiotic formation of binucleate bispores. In culture each spore gave rise to one male and one female gametophyte. These are intercompatible forming carposporophytes with viable diploid carpospores. The Australian genera of Solieriaceae and Rhabdoniaceae investigated by Min-Thein & Womersley (1976) generally possess isomorphic gametophytes and tetrasporophytes as do most other genera in which field material has been examined (Kylin 1956; Papenfuss & Edelstein 1974).

Two exceptional genera, *Turnerella* and *Opuntiella*, most likely possess heteromorphic life histories and probably do not belong in the Solieriaceae. Carpospores of *Turnerella pennyi* gave rise in culture to crustose tetrasporophytes resembling *Cruoria rosea* (South *et al.* 1972). Vegetative cells of a female gametophyte were haploid (n = 18–20) and carposporophyte cells were diploid ($2n$ = 36–40). *Turnerella mertensiana* in Japan, recently investigated by Kasahara (1980), has a crustose tetrasporophyte somewhat similar to *Cruoria* and produces viable tetraspores that develop into foliose gametophytes.

Opuntiella is morphologically similar to *Turnerella*. Both genera have numerous vesiculate cells in the cortical layer. One never sees tetrasporangia in the blades of *Opuntiella*, even though they were reported by some earlier workers. Moe & DeCew (unpublished) believe that the tetrasporangia reported were merely fractured vesiculate cells observed in dried specimens. They believe that *Opuntiella* has a heteromorphic life history with a crustose tetrasporophyte resembling *Cruoria profunda*, a species that also has conspicuous vesiculate cells.

The monogeneric family Rissoellaceae with but one species, *Rissoella verruculosa* from the Mediterranean Sea, apparently has an isomorphic life history (Kylin 1956).

Caulacanthaceae, Cystocloniaceae (formerly Rhodophyllidaceae), Hypneaceae

The Caulacanthaceae includes widely distributed genera (see Gabrielson 1980). None of these have been cultured, although their life histories are presumed to be isomorphic (Searles 1968).

Min-Thein & Womersley (1976) provide the most recent summary of reproductive characters of Cystocloniaceae and a discussion of species endemic to Australia. *Calliblepharis, Rhodophyllis* and *Cystoclonium* are more widely distributed (Kylin 1956). All evidently have isomorphic life histories. Mixed phase reproduction has been recorded once in a plant of *Cystoclonium purpureum* with tetrasporangia and cystocarps and a plant with tetrasporangia and spermatangia from field collections (Edelstein *et al.* 1974).

The Hypneaceae includes *Hypnea*, an ubiquitous genus of tropical seas,

Rhododactylis from Australia *Hypneocolax* a parasite on *Hypnea*. Complete life histories are not known. Certain species in some localities have a high frequency of tetrasporophytes, suggesting that the principal reproductive mode is by apomeiosis. Børgesen (1920) reported bispores and neutral spores in *Hypneocolax*, while Fan (1961) recorded zonate tetrasporangia. Tetrasporangia are unrecorded in *Rhododactylis*.

Mychodeaceae, Mychodeophyllaceae, Acrotylaceae, Dicranemaceae

The vegetative development and reproduction of these four families have been investigated by Kraft (1977a,b, 1978). All species are considered to have isomorphic life histories, but none has been examined in culture.

Sarcodiaceae

Within the Sarcodiaceae two genera (*Trematocarpus* and *Sarcodia*) are believed to be isomorphic on the basis of field-collected material. Mshigeni (1976) cultured carpospores of *S. montagneana* in Tanzania, but the life history was not completed. In a third genus *Chondrymenia*, which is monotypic, tetrasporangia are not recorded (Kylin 1956).

Plocamiaceae

The Plocamiaceae comprises two genera, the widely distributed *Plocamium* and its parasite *Plocamiocolax*. Both taxa are isomorphic according to most phenological information (Womersley 1971b; Abbott & Hollenberg 1976; South & Adams 1979).

Gracilariaceae

Members of the Gracilariaceae are generally thought to have the standard *Polysiphonia*-type life history. The complete life history of *Gracilaria verrucosa* in Japan was fully investigated in culture by Ogata *et al.* (1972). Subsequently, a number of other species have been studied, particularly by the research group of the Atlantic Research Laboratory in Halifax, Nova Scotia. Hybridization experiments between *G. foliifera* × *G. tikvahiae* (referred to as *G.* sp.) resulted in cystocarp-like swellings, but no viable carpospores were released (McLachlan *et al.* 1977). Three isolates of *G. tikvahiae* interbred freely, all having identical chromosome numbers ($n = 24$). Most of these studies (van der Meer 1977, 1978, 1979a,b; van der Meer & Bird 1977; van der Meer & Todd 1977) centre on Mendelian and non-Mendelian inheritance of phycobiliprotein pigment mutants but also focus on the nature of mixed phase reproduction involving diploid tetrasporophytes bearing diploid spermatangia or diploid carpogonia as well as tetrasporangia. Viable triploid or polyploid carpospores were formed by crossing isolates of different ploidy levels. Many carpospore germlings failed to reach reproductive maturity, but in those forming tetrasporangia nuclear divisions were apparently meiotic with as many as 18 trivalents. In other experiments mitotic recombination (van der Meer & Todd 1977) gave rise to genetically homozygous tissue on heterozygous hybrid tetrasporophytes. As they point out, mitotic recombination may be the cause of mixed phase reproduction in numerous red algae although earlier investigations did not involve karyological analysis. Although mixed-phase reproduction was common in the *Gracilaria* isolates investigated by van der Meer & Todd (1977), this behaviour is not reported by other investigators except Cabioch (1972) for either field-collected or cultured material.

Sphaerococcaceae, Phacelocarpaceae, Nizymeniaceae

There is some question about the ordinal position of these three families. The Sphaerococcaceae is presently a monogeneric family (Searles 1968) with *Sphaerococcus* as the monotypic genus. Although it is common in Atlantic Europe and the Mediterranean Sea, it has not been cultured. The life history is assumed to be of the *Polysiphonia*-type; however, tetrasporangiate plants have not been seen with certainty. They are 'reputedly zonate' (Dixon & Irvine 1977). The Phacelocarpaceae, established by Searles (1968) now includes only the genus *Phacelocarpus*, since *Nizymenia* and *Stenocladia* have been removed to the Nizymeniaceae (Womersley 1971a), partly on the basis of their having cruciate rather than zonate tetrasporangia. All have a *Polysiphonia*-type life history based on examination of field-collected plants.

3.7 Rhodymeniales

With the removal of the Palmariaceae to the Palmariales (Guiry 1978a). The Rhodymeniales forms a natural order of three families: Rhodymeniaceae, Lomentariaceae, and Champiaceae (Lee 1978). The Rhodymeniales basically exhibits a *Polysiphonia*-type life history. Most present information is based on observations of field-collected specimens. The few laboratory culture studies are discussed below.

Rhodymeniaceae

Minium, a new monotypic crustose genus epiphytic on articulate corallines, is known only from California (Moe 1979). Observations on plants in culture and from the field indicate that this genus shows a typical isomorphic life history. Carpospores germinated to form tetrasporophytes that developed tetrasporangia. Tetraspores gave rise to normal male and female gametophytes, but fertilization did not occur in culture although numerous cystocarps were evident in field collections. It is possible that isolates of *Minium* are self incompatible.

The life history of *Fauchea laciniata* differs in different isolates. In one isolate carpospores germinated forming viable tetrasporophytes (Sparling 1961); however, in another from the same region (West & Norris 1966), tetraspores germinated to form a second generation of tetrasporophytes in culture, suggesting that tetrasporangia were apomeiotic. As is common for other species of red algae, different populations may exhibit either sexual or apomictic reproductive patterns.

Lomentariaceae, Champiaceae

Lomentaria, widely distributed in both hemispheres, has infrequently been cultured in the laboratory. Lee & West (1980b) completed the life history of isolates of *L. hakodatensis* from Korea. *Binghamiopsis caespitosa*, a new genus and species from Pacific North America (Lee *et al.* 1981) closely related to *Binghamia* and *Lomentaria*, also has been cultured and shows the normal *Polysiphonia*-type life history although, as with *Minium*, the male and female plants from a single tetrasporophyte appear to be incompatible. *Gastroclonium coulteri, Botryocladia pseudodichotoma* (Lee & West unpublished) and *Champia parvula* from Rhode Island (R. Steele personal communication) and from Korea (Lee & West unpublished) exhibit *Polysiphonia*-type patterns. At least one species of *Lomentaria* (*L. orcadensis*) is apomeiotic, recycling successive tetrasporophyte generations, according to the cytological evidence of Svedelius (1937). Bird (1978) noted a complete absence of gametophytes in *L. orcadensis* from Nova Scotia. Both *Champia* and *Lomentaria* contain numerous rather similar species easily grown in the laboratory and would be good candidates for biosystematic studies involving culture and hybridization.

3.8 Ceramiales

The four families of the Ceramiales characteristically possess a *Polysiphonia*-type life history with isomorphic gametophytic and tetrasporophytic generations. Both carpospores and tetraspores normally germinate in a bipolar manner to produce a primary rhizoid and an erect axis. Variations in the life history that have been encountered both in nature and under laboratory conditions involve only secondary modifications of the *Polysiphonia*-type life cycle.

Ceramiaceae

Species of Ceramiaceae are easily maintained in the laboratory and many have been carried through a complete life history in culture. Species may be regularly monoecious or dioecious, with dioecious species sometimes producing both sexes on the same plant as an aberrant behaviour. More common is the occurrence of mixed phases with both sexual stages and tetrasporangia occurring on the same plant. The occurrence of mixed phases in nature has been reviewed by Knaggs (1969). In addition, there are species in which tetrasporangia are replaced by polysporangia that are known to undergo meiosis, and there are forms in which reproduction is apomeiotic or essentially asexual by means of tetrasporangia, bisporangia, monosporangia, seirosporangia, parasporangia, pseudosporangia, gemmae, propagules or fragmentation. Guiry (1978a) has reviewed the classification of sporangial types in Florideophycidae including Ceramiales. Most life history studies of Ceramiaceae carried out in culture have not incorporated either cytological observations or genetic investigations of key features, and the unusual phenomena seen in Ceramiaceae, such as mixed phases, are largely unexplained.

The life history of *Antithamnion* has been investigated in culture in *A. cruciatum* (Kapraun 1977a; Whittick & Hooper 1977), *A. plumula** (Sundene 1959, 1975; L'Hardy-Halos 1968; Rueness & Rueness 1975), *A. boreale* (Sundene 1962), *A. tenuissimum* (Sundene 1965a; Rueness & Rueness 1973), *A. occidentale*† and *A. defectum* (West & Norris 1966), *A spirographidis*‡ and *A. sarniense*‡ (Drew 1955a; Sundene 1965b; L'Hardy-Halos 1968) and *A. nipponicum* (Lee & West 1980a).

Particularly interesting results have been obtained in hybridization studies between varieties and strains of *Pterothamnion plumula* (as *Antithamnion plumula*). Sundene (1975) attempted to cross 13 strains of *A. plumula* including var. *plumula*, var. *bebbii* and var. *crispum*. Most of the crosses were intersterile, while a few were interfertile. Crosses between the distichously branched var. *plumula* and the verticillate var. *bebbii* gave positive results in some instances, unlike earlier reports (Sundene 1959). The problem posed by Sundene appears to have been resolved by Rueness & Rueness (1975) who crossed interfertile strains of var. *plumula* from Plymouth, England, with var. *bebbii* from Kristineberg, Sweden, and analysed the variation in the F_1 and F_2 hybrids both by tetrad analysis and random sampling of spore germlings. There was no reduction in the viability of carpospores or tetraspores from the crosses. The F_1 always expressed the var. *plumula* phenotype. Following meiosis in the tetrasporangia the two morphological types, var. *plumula* and var. *bebbii* segregated 1:1 in the F_2. The male and female sexes also segregated 1:1 and appeared to be unlinked with the morphogenetic factor. The results suggest that branching of the pinnae is controlled by a single gene represented by a dominant allele in var. *plumula* and a recessive allele in var. *bebbii*.

* Wollaston (1979) and Moe & Silva (1980) recognize *Pterothamnion* Naegeli (type species of *P. plumula* [Ellis] Naegeli) as distinct from *Antithamnion*.

† Wollaston (1971) has transferred *A. occidentale* to a new genus, *Scagelia*.

‡ Both species are placed in *Antithamnionella* by Wollaston (1968).

Antithamnionella sarniensis and *A. spirographidis* (both as *Antithamnion*) have been regarded as belonging to separate, but closely related, species (L'Hardy-Halos 1968) or as conspecific (Sundene 1965b). Sexual and asexual phases of *A. spirographidis* were grown in culture by Drew (1955a), who obtained evidence of segregation for sex at meiosis but noted one fertile plant bearing both spermatangia and carpogonia. Sundene (1965b) germinated tetrasporangia from plants identified as *A. sarniense* from the Channel Islands and obtained male and female gametophytes in culture which agreed with the description for the sexual phase of *A. spirographidis*.

The life history of the type species of *Antithamnion, A. cruciatum*, has been only partially elucidated in culture. Sexual plants are known from Europe (L'Hardy-Halos 1968) and North Carolina (Kapraun 1977a) suggesting the presence of a *Polysiphonia*-type life history; however, sexual plants appear to be absent throughout the northern range of this species. Whittick & Hooper (1977) counted around 90–100 chromosomes in a strain of *A. cruciatum* from Newfoundland that forms structures resembling zonately divided tetrasporangia. The cytological data suggest that the Newfoundland plants may represent an asexual, polyploid population.

Mixed phases have been seen and described in *Antithamnion occidentale* (West & Norris 1966), *A. defectum* (as *A. pygmaeum*) (West & Norris 1966; Lee & West 1980a), and *A. tenuissimum* (Rueness & Rueness 1973). In *A. occidentale* West & Norris (1966) observed that tetraspores formed on male plants give rise to non-sporangiate male plants. Both male and male/tetrasporangial plants crossed with female plants and resulting carpospores developed into tetrasporophytes. In *A. defectum* tetraspores from female/tetrasporangial thalli grew into functional female plants.

Sundene (1965a) had reported that *A. tenuissimum* is a dioecious species with a typical alternation of isomorphic generations. Mixed phases were seen in only one plant. In contrast, Rueness & Rueness (1973) obtained plants in culture that bore tetrasporangia and spermatangia on the same thallus. These were shown to be haploid ($n \simeq 32$) and to produce functional spermatia. Carpospores resulting from crosses between male/tetrasporangial plants and normal females develop into normal tetrasporophytes ($2n \simeq 64$). The tetraspores give rise to females, males and male/tetrasporangial plants. Tetrasporangia on the male/tetrasporangial plants appear to develop apomeiotically and give rise to non-sporangiate male and female plants. The formation of mixed phases with tetrasporangia borne on haploid plants in *Antithamnion* stands in contrast to the production of gametangia and tetrasporangia on diploid plants in *Callithamnion* (Hassinger-Huizinga 1952) and *Spermothamnion* (Drew 1934).

A *Polysiphonia*-type life history was demonstrated in *Antithamnion nipponicum* (Lee & West 1980a) with neither mixed phase reproduction nor apomixis.

A relatively stable *Polysiphonia*-type life history has been described in *Callithamnion byssoides* from Texas (Edwards 1969a, 1971), North Carolina

(Kapraun 1978) and Norway (Rueness & Rueness 1980), and in *Callithamnion baileyi* (Whittick & West 1979), *C. bipinnatum* (Rueness & Rueness 1980), *C. hookeri* (Edwards 1977; Rueness & Rueness 1978) and *C. corymbosum* (Hassinger-Huizinga 1952). Irregularities, with gametangia and sporangia borne on the same thallus, have been reported in ten species from nature (Knaggs 1969) and were seen in the majority of the species listed above.

The life history of *Callithamnion byssoides* from Texas was completed in culture in about one month (Edwards 1969a). Kapraun (1978a) showed that the North Carolina and Texas isolates of *C. byssoides* demonstrate a differential sensitivity to light intensity and temperature and Rueness & Rueness (1980) propose that both the North Carolina and Texas strains are morphologically distinct from the European *C. byssoides* and do not belong in that species. The occasional tetrasporangia that develop on male plants of *C. byssoides* from Norway are thought to be haploid since the cultures were derived from tetraspores of normal tetrasporophytes (Rueness & Rueness 1980).

A form of *Callithamnion hookeri* from Sweden that reproduces solely by means of parasporangia borne adaxially on branchlets in lieu of tetrasporangia has been cultured through four successive generations (Rueness & Rueness 1978). *Callithamnion baileyi* from Newfoundland and Massachusetts is a monoecious species studied by Whittick & West (1979) that completes its triphasic life cycle in less than five months depending on temperature and light intensity. Mixed phases were encountered with spermatangia and occasionally carpogonia borne on tetrasporangial plants. Tetraspores from these plants produce monoecious gametophytes which bear numerous carposporophytes. Presumably the mixed-phase tetrasporophytes are diploid.

In the very detailed study of *Callithamnion corymbosum* by Hassinger-Huizinga (1952) isolates of this dioecious species were cultured through nine generations. Tetrad analysis normally showed two male and two female plants per sporangium. Male/tetrasporangial and female/tetrasporangial plants were encountered infrequently and were diploid and carried one or two extra chromosomes ($2n+2 = 62$ or $2n+4 = 64$). Hassinger-Huizinga undertook a genetic analysis of the process of sex determination in this species but her interpretation has not been generally accepted.

A study of the distribution and ecology of British species of *Callithamnion* involving field transplantations, observations of seasonal populations and culture studies (Price 1978) demonstrates that many characters commonly relied upon for distinguishing species vary with the environment. Robustness of growth, cortication, axial cell form and size, branching patterns and hair formation are particularly affected. Species circumscription will need to take these factors into consideration.

The life history of a dioecious strain of *Spermothamnion repens* from Norway has been followed in culture for two successive generations by Rueness (1971). Tetraspore germlings give rise to male and female gametophytes and carpospore germlings produce mature tetrasporophytes within three to eight months. Mixed phases were not seen, and there were no deviations from the *Polysiphonia*-type life history.

In an earlier study Drew (1934) reported that diploid plants ($2n = 60$) of *Spermothamnion repens* (as *S. turneri*) from Woods Hole, Massachusetts, produce male and female sex organs in addition to tetrasporangia. The tetrasporangia evidently undergo meiosis producing haploid tetraspores ($n = 30$) that germinate readily. Cystocarps that matured following fertilization were tetraploid ($4n = 120$) with one exception that was triploid. In subsequent collections Drew (1943) found triploid plants ($n = 90$), mainly in April, which were sterile. Some filaments, also collected in April, possessed 45–50 chromosomes, half the triploid number.

In Britain, cytological investigations of *Spermothamnion repens* (as *S. turneri*) by Drew (1943) revealed haploid, diploid and triploid individuals with diploid plants being most common. Diploid plants ($2n = 60$) also bore carpogonial branches, but neither cystocarps nor spermatangial branches were seen on diploid individuals. Haploid plants ($n = 30$) collected in March or April bear spermatangia and carpogonia and produce normal cystocarps. Germinating carpospores were counted with triploid chromosome numbers ($3n = 90$). Triploid plants are common in nature and, as at Woods Hole, they are sterile. Individuals with 45–50 chromosomes, presumably derived from triploids, do, however, occur in nature and sporangial mother cells are occasionally seen on haploid plants. The data provided by Drew for *Spermothamnion repens* suggests a mechanism by which polyploid series may have evolved in Ceramiaceae.

Drew (1937) established that *Tiffaniella snyderae* (as *Spermothamnion snyderae*) from California is dioecious with typical sexually reproducing haploid gametophytes ($n = 32$). Diploid plants ($2n = 64$) bear polysporangia that contain from two to nine nuclei, all of which undergo meiosis. The polysporangia of *T. snyderae* are therefore, homologous with normal tetrasporangia.

Rueness (1968) found that paraspores from *Plumaria elegans* give rise only to parasporangial plants in culture. Parasporangia-bearing plants have a wider and more northern distribution than sexual strains. Drew (1939) had earlier established that plants bearing parasporangia are triploid ($3n = 91$–93) and reproduce entirely asexually. Sexual plants of *Plumaria elegans* from England have a normal *Polysiphonia*-type life cycle with male, female and tetrasporic plants ($n = 31$, $2n = 62$–64).

In the tribe Griffithsieae sexual plants are thought to be haploid, the sexes separate, and mixed phases are uncommon. Recently Waaland (1978) has achieved somatic cell fusion between vegetative cells of a male and female isolate of *Anotrichium tenue* (as *Griffithsia tenuis*) that have regenerated new plants. At first, the hybrids generally produce tetrasporangial branches, characteristic of the diploid phase and later either male or female reproductive structures. Waaland suggests that nuclear fusion is not required, and that male and female nuclei act in concert in a heterokaryon to modify the morphological expression that results in tetrasporangial branches. Waaland (1975) and Waaland & Watson (1980) find that when two cells of *Griffithsia pacifica* are held together a descending repair rhizoid produces a hormone (rhodomorphin) that induces formation of an ascending repair shoot cell leading to their fusion (see Chapter 18). The hormones of three species (*G. pacifica, G. globulifera* and *G. tenuis*) do not crossreact to induce or attract repair shoots of the other species and there are no interspecific cell fusions. Parasexual hybrids between three strains of *G. pacifica* led to successful fusions; however, there was evidence of cytoplasmic incompatibility in hybrid cells of some crosses.

Rueness (1978b) has investigated the biosystematics of the *Ceramium diphanum-strictum* species complex experimentally with crosses between several isolates from Scandinavia and New Jersey. Crosses between *Ceramium strictum* and *C. tenuissimum* from Norway were intersterile. Interestingly, seemingly identical allopatric strains of *C. strictum* from Norway and New Jersey were also completely sterile in attempted crosses. Crosses between *C. strictum* from Oslofjord and the very different looking *C. tenuicorne* from the Baltic Sea were fully fertile, and no reduced viability or fertility was observed in F_1 or F_2 hybrids or the backcross hybrids. Hybrids were morphologically of the *C. tenuicorne*-type while growth responses and physiological tolerances of the hybrid were intermediate relative to the two parents. Whereas *C. tenuicorne* grew best with less than 10‰ salinity and *C. strictum* with 30‰ salinity, the optimum salinity for the hybrid was intermediate (10–20‰). A triploid strain ($3n = 60$) of *C. strictum* from Oslofjord (Rueness 1973a) reproduced solely asexually by means of parasporangia. This finding is consistent with Drew's (1939) observations on *Plumaria* and suggests the occurrence of a polyploid series. Rao *et al.* (1978) have investigated meiosis in tetraspore mother cells of three species of *Ceramium* ($2n \simeq 42$; 84) from India. They conclude that hybridization and polyploidy play a role in the evolution of these species.

Edwards (1973) demonstrated a *Polysiphonia*-type life history in culture for three British species of *Ceramium*: *C. shuttleworthianum*, *C. rubrum* and *C. pedicellatum*. The reported absence of sexual plants in northern localities in Scotland was attributed to inhibition of gametangia production by the environment.

Delesseriaceae

The Delesseriaceae have not received the attention given to the Ceramiaceae. Field observations suggest that most species have an isomorphic life history. Only two species, *Membranoptera platyphylla* (West & Norris 1966) and *Membranoptera multiramosa* (Waaland & Kemp 1972) have been investigated in culture. *Membranoptera multiramosa* was monoecious in culture. Mixed phases

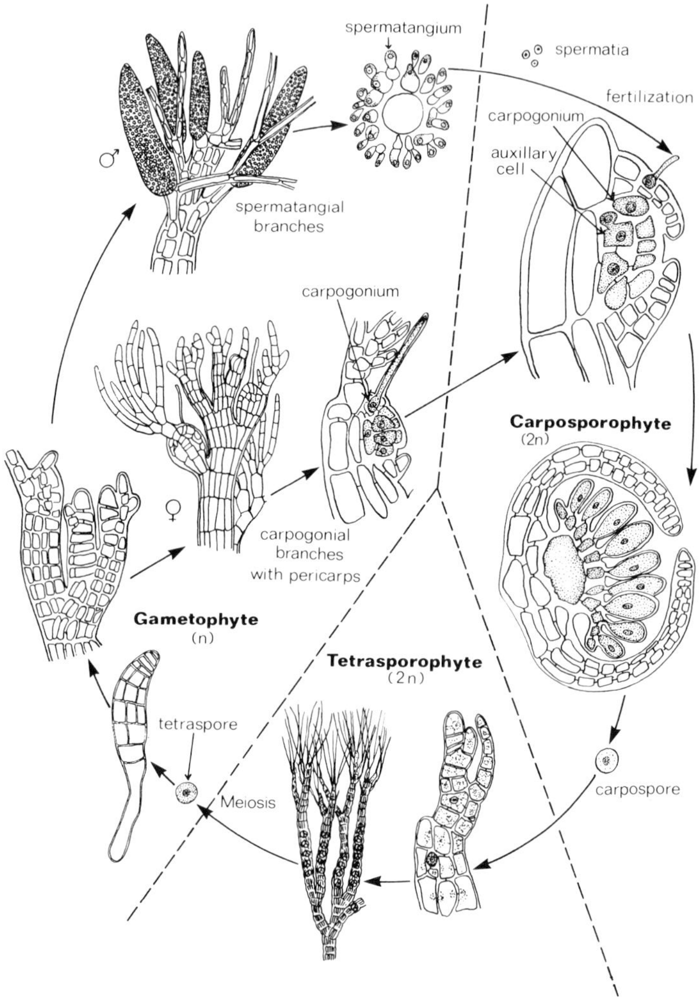

Fig. 4.6 Life history of *Polysiphonia*. (Based on Falkenberg, Kylin and Iyengar & Balakrishnan.)

with cystocarps and tetrasporangia have been reported on the same thallus in *M. alata* in nature (Edelstein & McLachlan 1967).

DASYACEAE

The life histories of a few Dasyaceae have been investigated in culture. Parsons (1975) observed that *Dasya clavigera* possesses a typical *Polysiphonia*-type life history, and Lee & West (1979) found the same to be true of *Dasysiphonia chejuensis*; however, they also encountered female/tetrasporangia mixed-phase reproduction. In contrast, West (1970c) showed that *Heterosiphonia japonica* (as

H. densiuscula and *H. asymmetrica*) is obligately apomeiotic. Sporeling stages of numerous species of Dasyaceae have been followed in culture (see Parsons 1975).

RHODOMELACEAE

It is generally assumed that most Rhodomelaceae possess a typical *Polysiphonia*-type life history (see Fig. 4.6) although very few species have actually been investigated in culture. Isomorphic life histories and ecological adaptations of several species of *Polysiphonia* have been followed in culture, often in a crossed-gradient apparatus, including *P. hemispherica* var. *boldii* (as *P. boldii*) (Edwards 1969b, 1970b), *P. denudata* (Edwards 1970a; Kapraun 1978b), *P. hemispherica* (Rueness 1971), *P. ferulacea* (Kapraun 1977b, 1978b) and *P. urceolata* (Kapraun 1978b, 1979). Asexual reproductive propagules resembling spermatangial branches were found on male, female and tetrasporic plants of *Polysiphonia ferulacea* (Kapraun 1977b), and male/tetrasporangia and female/tetrasporangia mixed-phase reproduction has been recorded for *Polysiphonia urceolata* (see Kapraun 1979).

Rueness (1973b) carried out hybridization experiments between *P. hemispherica* from Scandinavia and *P. boldii* from Texas. Cross-fertilization between these widely separated populations yielded fertile carposporophytes. Tetraspores produced from the interspecific crosses exhibited greatly reduced viability compared to tetraspores derived from intraspecific crosses. It was proposed that *P. boldii* be reduced to varietal status as *P. hemispherica* var. *boldii*.

ACKNOWLEDGEMENTS

Sincere thanks go to Francia Hommersand for her help in preparing and editing the manuscript and to Patricia Wyman and Susan Sizemore for their contributions to the design and preparation of the life history diagrams.

4 REFERENCES

ABBOTT I. A. (1967) Studies in the foliose red algae of the Pacific Coast II. *Schizymenia*. *Bull. So. Calif. Acad. Sci.*, **66**, 161–74.

ABBOTT I. A. (1968) Studies in some foliose red algae of the Pacific coast. III. Dumontiaceae, Weeksiaceae, Kallymeniaceae. *J. Phycol.*, **4**, 180–98.

ABBOTT I. A. (1980) Seasonal population biology of some carrageenophytes and agarophytes. In *Pacific Seaweed Aquaculture*, eds. Abbott I. A., Foster M. S. & Eklund L. F., pp. 45–53. California Sea Grant Program, La Jolla.

ABBOTT I. A. & HOLLENBERG G. (1976) *Marine Algae of California*. Stanford University Press, California.

ADEY W. H. & VASSAR J. M. (1975) Colonization, succession and growth rates of tropical crustose coralline algae. *Phycologia*, **14**, 55–69.

ARDRÉ F. (1977a) Sur le cycle du *Schizymenia dubyi* (Chauvin ex Duby) J. Agardh (Némastomacée, Gigartinale). *Rev. Algol.* N.S., **12**, 73–86.

ARDRÉ F. (1977b) Sur le cycle morphologique du *Gymnogongrus crenulatus* (Turn.) J. Ag. (Gigartinale, Rhodophycée) en culture. *C. R. Acad. Sci. Paris*, **285D**, 513–14.

ARDRÉ F. (1978) Sur les cycles morphologiques du *Gymnogongrus crenulatus* (Turn.) J. Ag. et du *Gymnogongrus devoniensis* (Grev.) Schott. (Gigartinales, Phyllophoracée) en culture. *Rev. Algol.* N.S., **13**, 151–76.

AUSTIN A. (1960a) Life history and reproduction of *Furcellaria fastigiata* (L.) Lam. 1. The haploid plants and the development of the carposporophyte. *Ann. Bot.* N.S., **24**, 257–74.

AUSTIN A. (1960b) Life history and reproduction of *Furcellaria fastigiata* (L.) Lam. 2. The tetrasporophyte and reduction division in the tetrasporangium. *Ann. Bot.* N.S., **24**, 296–310.

BALAKRISHNAN M. S. & CHAUGULE B. B. (1980) Cytology and life history of *Batrachospermum mahabaleshwarensis*. *Crytogamie: Algologie*, **1**, 83–97.

BERTHOLD G. (1882) Die Bangiaceen des Golfes von Neapel und der angrenzenden Meeresabschnitte. *Fauna und Flora des Golfes von Neapel*, **8**, 1–28.

BICHARD-BREAUD J. and FLOC'H J.-Y. (1966) Presence des gamétophytes males de *Bonnemaisonia hamifera* Hariot (Floridée, Bonnemaisoniacée) sur la côte atlantique française. *C. R. Acad. Sci. Paris, Ser.*, **262D**, 1949–51.

BIRD C. J. (1973) Aspects of the life history and ecology of *Porphyra linearis* (Bangiales, Rhodophyceae). *Can. J. Bot.*, **50**, 1859–63.

BIRD C. J. (1978) Investigations of the marine algae of Nova Scotia. XV. *Lomentaria orcadensis* (Rhodophyta, Champiaceae). *Proc. N.S. Inst. Sci.*, **28**, 189–90.

BODARD M. (1971) Étude morphologique et cytologique d'*Helminthocladia senegalensis* (Rhodophycées), Némalionale nouvelle à carpotétraspores et à cycle haplodiplophasique. *Phycologia*, **10**, 361–74.

BOILLOT A. (1963) Recherches sur le mode de développement des spores du genre *Gélidium* (Rhodophycées, Gelidiales). *Rev. Gén. Bot.*, **70**, 131–7.

BOILLOT A. (1965) Sur l'alternance de générations hétéromorphes d'une Rhodophycée, *Halarachnion ligulatum* (Woodward) Kützing (Gigartinales, Furcellariacées). *C. R. Acad. Sci. Paris*, **261**, 4191–3.

BOILLOT A. (1967) Sur le développement des carpospores de *Naccaria wiggii* (Turner) Endlicher et d'*Atractophora hypnoides* Crouan (Naccariacées, Bonnemaisoniales). *C. R. Acad. Sci. Paris*, **264D**, 257–60.

BOILLOT A. (1968) Sur l'existence d'un tétrasporophyte dans le cycle de *Scinaia furcellata* (Turner) Bivona, Némalionales. *C. R. Acad. Sci. Paris*, **266D**, 1831–2.

BOILLOT A. (1969) Sur le développement des tétraspores et l'édification du gametophyte chez *Scinaia furcellata* (Turner) Bivona, (Rhodophycées Némalionales). *C. R. Acad. Sci. Paris*, **268D**, 273–5.

BOILLOT A. (1971a) Sur le cycle d'*Helminthocladia calvadosii* (Lamouroux) Setchell. *Bull. Soc. Phycol. France*, **16**, 106–10.

BOILLOT A. (1971b) Sur la présence de tétraspores dans le cycle d'*Helminthora divaricata* (C. Agardh) J. Agardh (Rhodophycées, Némalionales). *C. R. Acad. Sci. Paris*, **272D**, 932–3.

BOILLOT A. (1971c) Sur le cycle de *Scinaia turgida*, Chemin. *Bull. Soc. Phycol. France*, **16**, 68–9.

BOILLOT A. (1972) Cycle biologique de quelques Némalionales. *Botaniste*, **55**, 207–50.

BOILLOT A. (1974) Le sporophyte de l'*Helminthocladia calvadosii* (Lamouroux) Setchell (Rhodophycée, Némalionales). *Botaniste*, **56**, 187–92.

BOILLOT A. & L'HARDY-HALOS M.-Th. (1975) Observations en culture d'une Rhodophycée Bonnemaisoniale: le *Naccaria wiggii* (Turner), Endlicher. *Bull. Soc. Phycol. France*, **20**, 30–5.

BOILLOT A. & MAGNE F. (1973) Le cycle biologique de *Kylinia rosulata* Rosenvinge (Rhodophycées, Acrochaetiales). *Bull. Soc. Phycol. France*, **18**, 47–53.

BØRGESEN F. (1920) The Marine Algae of the Danish West Indies. Part 3: Rhodophyceae (6). *Dansk Bot. Arkiv*, **3**, 369–504.

BORSJE W. J. (1973) The life history of *Acrochaetium virgatulum* (Harv.) J. Ag. in culture. *Br. Phycol. J.*, **8**, 205.

BREEMAN A. M. (1979) The caryological phases in the life history of *Acrosymphyton purpuriferum* (J. Ag.) Sjöst. (Rhodophyceae, Cryptonemiales). *Phycologia*, **18**, 146–8.
CABIOCH J. (1972) Un nouveau cas d'anomalie du cycle des Gigartinales (Algues Floridées). *C. R. Acad. Sci. Paris*, **275**, 1979–81.
CALLOW J., CALLOW M. & EVANS L. (1979) Nutritional studies on the parasitic red alga *Choreocolax polysiphoniae*. *New Phytol.*, **83**, 451–62.
CANDIA A. I. & KIM D. H. (1977) Resultados preliminares de los estudios de ciclo de vida de *Gymnogongrus furcellatus* (C. Agardh) J. Agardh (Phyllophoraceae, Gigartinales). *Gayana Miscelanea. No. 5. Primeras Jornadas Nacionales de Agricultura*, 77–78.
CHANG T. J. & TSENG C. K. (1962) The Chinese *Porphyra* and their geographical distribution. *Oceano. Limn. Sin.*, **4**, 183–8.
CHAPMAN D. J. & CHAPMAN V. J. (1961) Life histories in the algae. *Ann. Bot.* N.S., **25**, 547–61.
CHEN L. C-M. (1977) The sporophyte of *Ahnfeltia plicata* (Huds.) Fries (Rhodophyceae, Gigartinales) in culture. *Phycologia*, **16**, 163–8.
CHEN L. C-M. & CRAIGIE J. S. (1981) Carrageenan analysis in apomictic *Gigartina stellata*—new puzzles. *Proc. Int. Seaweed Symp.*, **10**, (in press).
CHEN L. C-M., EDELSTEIN T., BIRD C. & YABU H. (1978) A culture and cytological study of the life history of *Nemalion helminthoides* (Rhodophyta, Nemaliales). *Proc. N. S. Inst. Sci.*, **28**, 191–9.
CHEN L. C-M., EDELSTEIN T. & MCLACHLAN J. (1969) *Bonnemaisonia hamifera* Hariot in nature and in culture. *J. Phycol.*, **5**, 211–20.
CHEN L. C-M., EDELSTEIN T. & MCLACHLAN J. (1970a) Vegetative development of the gametophyte of *Bonnemaisonia hamifera* from a filamentous state. *Can. J. Bot.*, **48**, 523–5.
CHEN L. C-M., EDELSTEIN T. & MCLACHLAN J. (1974) The life history of *Gigartina stellata* (Stackh.) Batt. (Rhodophyceae, Gigartinales) in culture. *Phycologia*, **13**, 287–94.
CHEN L. C-M., EDELSTEIN T., OGATA E. & MCLACHLAN J. (1970b) The life history of *Porphyra miniata*. *Can. J. Bot.*, **48**, 385–9.
CHEN L. C-M. & MCLACHLAN J. (1972) The life history of *Chondrus crispus*. *Can. J. Bot.*, **50**, 1055–60.
CHEN L. C-M. & MCLACHLAN J. (1980) *Rhodoglossum affine* (Harv.) Kylin (Gigartinaceae, Rhodophyta) in culture. *Syesis*, **12**, 113–16.
CHIHARA M. (1961) Life cycle of the bonnemaisoniaceous algae in Japan (1). *Sci. Rep. Tokyo Kyoiku Daigaku Sect.*, **10B**, 121–53.
CHIHARA M. (1962) Life cycle of the bonnemaisoniaceous algae in Japan (2). *Sci. Rep. Tokyo Kyoiku Daigaku Sect.*, **11B**, 27–53.
CHIHARA M. (1965) Germination of the carpospores of *Bonnemaisonia nootkana*, with special reference to the life cycle. *Phycologia*, **5**, 71–9.
CHIHARA M. (1967) Some marine algae collected at Cape Thompson of the Alaskan Arctic. *Bull. Nat. Sci. Mus. Tokyo*, **10**, 183–200.
CHIHARA M. (1972) Germination of carpospores of *Pikea californica* and *Schimmelmannia plumosa* as found in Japan with special reference to their life history. *Soc. Bot. France Mém.*, **1972**, 313–22.
CHIHARA M. (1974) The significance of reproductive and spore germination characteristics to the systematics of the Corallinaceae: nonarticulated coralline algae. *J. Phycol.*, **10**, 266–74.
CHIHARA M. (1975) Life history of algae: Rhodophyta, their life histories. In *Advance in Phycology in Japan*, eds. Tokida J. & Hirose H., pp. 137–50. W. Junk, The Hague.
CHIHARA M. & KAMURA S. (1963) On the germination of tetraspores of *Gelidiella acerosa*. *Phycologia*, **3**, 71–4.
CHIHARA M. & YOSHIZAKI M. (1971) The thallus structure and reproductive system of *Hyalosiphonia caespitosa* (Cryptonemiales, Rhodophyta). *Bot. Mag. Tokyo*, **84**, 319–25.
CHIHARA M. & YOSHIZAKI M. (1972) Bonnemaisoniaceae: Their gonimoblast development, life history and systematics. In *Contributions to the Systematics of Benthic Marine Algae of the North Pacific*, eds. Abbott I. A. & Kurogi M., Jap. Soc. Phycol., Kobe, Japan.
CHIHARA M. & YOSHIZAKI M. (1978) Anatomical and reproductive features of *Ptilonia okadai* (Rhodophyta, Bonnemaisoniaceae). *Phycologia*, **17**, 382–7.

CODOMIER L. (1969) Sur l'alternance de générations hétéromorphes du *Neurocaulon grandifolium* (Rhodophycée, Gigartinale). *C. R. Acad. Sci. Paris*, **269**, 1060–2.

CODOMIER L. (1972) Sur le développement des spores et sur l'origine des cellules étoilées médullaires des *Kallymenia* (Rhodophycées, Cryptonemiales). *C. R. Acad. Sci. Paris*, **274D**, 369–71.

CODOMIER L. (1973) Sur le développement des spores et la formation du thalle rampant de *Kallymenia microphylla* J. Ag. (Rhodophycées, Cryptonemiales). *Giorn. Bot. Ital.*, **107**, 269–80.

CODOMIER L. (1974) On the development of a foliose rhodophyte, *Schizymenia dubyi* J. Ag. *8th Int. Seaweed Symposium*, **1974** (abst.).

COLE K. & CONWAY E. (1980) Studies in the Bangiaceae: Reproductive modes. *Bot. Mar.*, **23**, 545–53.

CONWAY E. (1964) Autecological studies of the genus *Porphyra* II. *Porphyra umbilicalis* (L.) J. Ag. *Brit. Phycol. Bull.*, **2**, 349–63.

CONWAY E. & COLE K. (1977) Studies in the Bangiaceae: structure and reproduction of the conchocelis of *Porphyra* and *Bangia* in culture (Bangiales, Rhodophyceae). *Phycologia*, **16**, 205–16.

CONWAY E., MUMFORD JR., T. & SCAGEL R. F. (1975) The genus *Porphyra* in British Columbia and Washington. *Syesis*, **8**, 185–244.

CORDEIRO-MARINO M. & POZA A. (1981) Life history of *Gymnogongrus griffithsiae* (C. Agardh) J. Agardh (Phyllophoraceae, Gigartinales). *Proc. Int. Seaweed Symp.*, **10** (in press).

CORTEL-BREEMAN A. M. (1975) The life history of *Acrosymphyton purpuriferum* (J. Ag.) Sjöst. (Rhodophyceae, Cryptonemiales). Isolation of tetrasporophytes. With some remarks on the tetrasporophyte of *Bonnemaisonia asparagoides* (Woodw.) C. Ag. (Nemalionales). *Acta Bot. Neerl.*, **24**, 111–27.

CORTEL-BREEMAN A. M. & VAN DEN HOEK C. (1970) Life history studies on Rhodophyceae I. *Acrosymphyton purpuriferum* (J. Ag.) Kyl. *Acta Bot. Neerl.*, **19**, 265–84.

CORTEL-BREEMAN A. M. & TEN HOOPEN A. (1978) The short day response in *Acrosymphyton purpuriferum* (J. Ag.) Sjöst. (Rhodophyceae, Cryptonemiales). *Phycologia*, **17**, 125–32.

COUTÉ A. (1971) Sur le cycle morphologique du *Liagora tetrasporifera* comparé à celui du *Liagora distenta* (Rhodophycées, Nemalionales, Helminthocladiacées). *C. R. Acad. Sci. Paris*, **273**, 626–9.

CRAIGIE J. and PRINGLE J. (1978) Spatial distribution of tetrasporophytes and gametophytes in four maritime populations of *Chondrus crispus*. *Can. J. Bot.*, **56**, 2910–14.

DANGEARD P. (1927) Recherches sur les *Bangia* et les *Porphyra*. *Botaniste*, **18**, 183–244.

DAWES C., MATHIESON A. & CHENEY D. (1974) Ecological studies of Floridian *Eucheuma* (Rhodophyta, Gigartinales). I. Seasonal growth and reproduction. *Bull. Mar. Sci.*, **24**, 235–73.

DECEW T. & WEST J. A. (1977) Culture studies on the marine red algae *Hildenbrandia occidentalis* and *H. prototypus* (Cryptonemiales, Hildenbrandiaceae). *Bull. Jap. Soc. Phycol.* **25** suppl. (Mem. Iss. Yamada), 31–41.

DECEW T. C. & WEST J. A. (1981a) Investigations on the life histories of three species of *Farlowia* (Rhodophyta, Cryptonemiales, Dumontiaceae) from the Pacific Coast of North America. *Phycologia*, **20b**, (in press).

DECEW T. C. & WEST J. A. (1981b) Life histories in the Phyllophoraceae (Rhodophyceae, Gigartinales) from the Pacific coast of North America. I. *Gymnogongrus linearis* and *G. leptophyllis*. *J. Phycol.*, **17** (in press).

DENIZOT M. (1968) Les Algues Floridées Encroutantes (à L'exclusion des Corallinacées). Mus. Nat. Hist. Natur. Paris.

DION P. & DELÉPINE R. (1979) Cycles de développement de *Gigartina stellata* et *Petrocelis cruenta* (Rhodophyceae, Gigartinales) étudiés *in situ* à Roscoff. *Rev. Algol.* N.S., **14**, 327–41.

DIXON P. S. (1965) Perennation, vegetative propagation and algal life histories, with special reference to *Asparagopsis* and other Rhodophyta. *Bot. Gothoburg*, **3**, 67–74.

DIXON P. S. (1970) The Rhodophyta: some aspects of their biology, II. *Oceanogr. Mar. Biol. Ann. Rev.*, **8**, 307–52.

DIXON P. S. (1973) *Biology of the Rhodophyta*. Hafner Press, New York.

DIXON P. S. & IRVINE L. (1977) *Seaweeds of the British Isles, Volume I, Rhodophyta. Part I, Introduction,* Nemaliales, *Gigartinales*. British Museum (Natural History).

DIXON P. S., MURRAY S., RICHARDSON W. & SCOTT J. (1972) Life history studies in genera of the Cryptonemiales. *Soc. Bot. France Mém.*, **1972**, 323–32.

DIXON P. S. & RICHARDSON W. N. (1970) Growth and reproduction in red algae in relation to light and dark cycles. *Ann. N.Y. Acad. Sci.*, **175**, 764–77.

DREW K. M. (1928) A revision of the genera *Chantransia, Rhodochorton*, and *Acrochaetium . . . Univ. Calif. Publ. Bot.*, **14**, 139–224.

DREW K. M. (1934) Contributions to the cytology of *Spermothamnion turneri* (Mert.) Aresch. I. The diploid generation. *Ann. Bot.*, **48**, 549–73.

DREW K. M. (1935) The life history of *Rhodochorton violaceum* (Kuetz.) comb. nov. (*Chantransia violacea* (Kuetz.)). *Ann. Bot.*, **49**, 439–50.

DREW K. M. (1937) *Spermothamnion snyderae* Farlow, a Floridean alga bearing polysporangia. *Ann. Bot.*, N.S., **1**, 463–76.

DREW K. M. (1939) An investigation of *Plumaria elegans* (Bonnem.) Schmitz with special reference to triploid plants bearing parasporangia. *Ann. Bot.* N.S., **3**, 347–67.

DREW K. M. (1943) Contributions to the cytology of *Spermothamnion turneri* (Mert.) Aresch. II. The haploid and triploid generations. *Ann. Bot.* N.S., **7**, 23–70.

DREW K. M. (1949) *Conchocelis*-phase in the life history of *Porphyra umbilicalis* (L.) Kuetz. *Nature (Lond.)*, **164**, 748–9.

DREW K. M. (1954). Studies in the Bangioideae. III. The life history of *Porphyra umbilicalis* (L.) Kütz. var. *laciniata* (Lightf.) J. Ag. *Ann. Bot.*, N.S., **18**, 183–211.

DREW K. M. (1955a) Sequence of sexual and asexual phases in *Antithamnion spirographidis* Schiffner. *Nature*, (*Lond.*) **175**, 813–14.

DREW K. M. (1955b) Life histories in the algae with special reference to the Chlorophyta, Phaeophyta and Rhodophyta. *Biol. Rev.*, **30**, 343–90.

DREW K. M. (1956) Reproduction in the Bangiophycidae. *Bot. Rev.*, **22**, 553–611.

DRING M. J. (1967) Effects of daylength on growth and reproduction of the conchocelis-phase of *Porphyra tenera*. *J. Mar. Biol. Assoc. U.K.*, **47**, 501–10.

EDELSTEIN T. (1970) The life history of *Gloiosiphonia capillaris* (Hudson) Carmichael. *Phycologia*, **9**, 55–9.

EDELSTEIN T., BIRD C. J. & MCLACHLAN J. (1974) Tetrasporangia and gametangia on the same thallus in the red algae *Cystoclonium purpureum* (Huds.) Batt. and *Chondria baileyana* (Mont.) Harv. *Br. Phycol. J.*, **9**, 247–50.

EDELSTEIN T. & MCLACHLAN J. (1967) Cystocarps and tetrasporangia on the same thallus in *Membranoptera alata* and *Polysiphonia urceolata*. *Br. Phycol. Bull.*, **3**, 185–7.

EDELSTEIN T. & MCLACHLAN J. (1971) Further observations on *Gloiosiphonia capillaris* (Hudson) Carmichael in culture. *Phycologia*, **10**, 215–19.

EDWARDS P. (1969a) The life history of *Callithamnion byssoides* in culture *J. Phycol.*, **5**, 266–8.

EDWARDS P. (1969b) Field and cultural studies on the seasonal periodicity of growth and reproduction of selected Texas benthic marine algae. *Cont. Mar. Sci. Univ. Tex.*, **14**, 59–114.

EDWARDS P. (1970a) Field and cultural observations on the growth and reproduction of *Polysiphonia denudata* from Texas. *Br. Phycol. J.*, **5**, 145–53.

EDWARDS P. (1970b) Attempted hybridization in the red algal genus *Polysiphonia*. *Nature*, (*Lond.*), **226**, 467–8.

EDWARDS P. (1971) Effects of light intensity, daylength, and temperature on growth and reproduction of *Callithamnion byssoides*. In *Contributions in Phycology*, eds. Parker B. C. & Brown R. M., pp. 163–74. Allen Press, Lawrence, Kansas.

EDWARDS P. (1973) Life history studies of selected British *Ceramium* species. *J. Phycol.*, **9**, 181–4.

EDWARDS P. (1977) An analysis of the pattern and rate of cell division, and morphogenesis of sporelings of *Callithamnion hookeri* (Dillw.) S. F. Gray (Rhodophyta, Ceramiales). *Phycologia*, **16**, 189–96.

FAN K. (1961) Studies on *Hypneocolax*, with a discussion on the origin of parasitic red algae. *Nova Hedwigia*, **3**, 119–28.

FARNHAM W. F. & FLETCHER R. L. (1976) The occurrence of a *Porphyrodiscus simulans* Batt. phase in the life history of *Ahnfeltia plicata* (Huds.) Fries. *Br. Phycol.*, **11**, 183–90.

FELDMANN G. (1965) Le développement des tétraspores de *Falkenbergia rufolanosa* et le cycle des Bonnemaisoniales. *Rev. Gén. Bot.*, **72**, 621–6.

FELDMANN G. (1966) Sur le cycle haplobiontique du *Bonnemaisonia asparagoides* (Woodw.) Ag. *C. R. Acad. Sci. Paris*, **262**, 1695–8.

FELDMANN J. (1952) Les cycles de reproduction des Algues et leurs rapports avec la phylogénie. *Rev. Cytol. Biol. Végét.*, **13**, 1–49.

FELDMANN J. (1954) Recherches sur la structure et le développement des Calosiphoniacées, (Rhodophycées-Gigartinales). *Rev. Gén. Bot.*, **61**, 453–99.

FELDMANN J. (1958) Le genre *Kylinia* Rosenvinge (Acrochaetiales) et sa reproduction. *Bull. Soc. Bot. Fr.*, **105**, 493–500.

FELDMANN J. (1962) The Rhodophyta order Acrochaetiales and its classification. *Proc. 9th Pacif. Sci. Congr.*, **1957 (4)**, 219–21.

FELDMANN J. (1972) Les problèmes actuels de l'alternance de génerations chez les Algues. *Soc. Bot. Fr. Mém.*, **1972**, 7–38.

FELDMANN J. & FELDMANN G. (1939a) Sur le développement des carpospores et l'alternance de générations de l'*Asparagopsis armata* Harvey. *C. R. Acad. Sci. Paris*, **208**, 1240–2.

FELDMANN J. & FELDMANN G. (1939b) Sur l'alternance de générations chez les Bonnemaisoniacées. *C. R. Acad. Sci. Paris*, **208**, 1425–7.

FELDMANN J. & FELDMANN G. (1942) Recherches sur les Bonnemaisoniacées et leur alternance de générations. *Ann. Sci. Nat. Bot.* Sér., **11**, 75–175.

FELDMANN J. & MAZOYER G. (1937) Sur l'identité de l'*Hymenoclonium serpens* (Crouan) Batters avec le protenéma du *Bonnemaisonia asparagoides* (Woodw.) C. Ag. *C. R. Acad. Sci. Paris*, **205**, 1084–5.

FLETCHER R. (1975) The life history of *Rhodophysema georgii* in laboratory culture. *Mar. Biol.*, **31**, 299–304.

FLETCHER R. (1977) Studies on the life history of *Rhodophysema elegans* in laboratory culture. *Mar. Biol.*, **40**, 291–7.

FOSLIE M. H. (1905) New *Lithothamnia* and systematical remarks. *Norske Vid. Selsk. Skrift.*, **1905**, 1–38.

FRIES L. (1967) The sporophyte of *Nemalion multifidum* (Weber & Mohr) J. Ag. *Svensk. Bot. Tidskr.*, **61**, 457–62.

FRIES L. (1969) The sporophyte of *Nemalion multifidum* (Weber & Mohr) J. Ag. found on the Swedish West Coast. *Svensk. Bot. Tidskr.*, **63**, 139–41.

FUKUHARA E. (1968) Studies on the taxonomy and ecology of *Porphyra* of Hokkaido and its adjacent waters. *Bull. Hokkaido reg. Fish. Res. Lab.*, **34**, 40–99.

GABRIELSON P. W. (1980) Phylogenetic relationships of selected genera in the family Solieriaceae and taxonomic implications. *J. Phycol.*, **16**, (suppl), abst. 11.

GABRIELSON P. & HOMMERSAND M. H. (1981) The vegetative and reproductive morphology of *Agardhiella subulata* (C. Agardh) Kraft & Wynne (Solieriaceae, Gigartinales). *J. Phycol.* (in press).

GANESAN E. K. & WEST J. A. (1975) Culture studies on the marine red alga *Rhodophysema elegans* (Cryptonemiales, Peyssonneliaceae). *Phycologia*, **14**, 161–6.

GARBARY D. (1978) On the phylogenetic relationships of the Acrochaetiaceae (Rhodophyta). *Br. Phycol. J.*, **13**, 247–54.

GIRAUD A. & MAGNE F. (1968) La place de la méoise dans le cycle de développement de *Porphyra umbilicalis*. *C. R. Acad. Sci. Paris.*, **267**, 86–588.

GITTINS B. T. (1977) Contributions to the biology of *Lithothrix aspergillum* J. E. Gray (Corallinaceae, Rhodophyceae). *J. Phycol.*, **13** (Suppl.), abst. 468.

GOFF L. (1980) The role of binucleate bispores in the parasitic red alga *Gardneriella tuberifera*. *J. Phycol.*, **16** (Suppl.), abst. 14.

GOFF L. & COLE K. (1976a) The biology of *Harveyella mirabilis* (Cryptonemiales, Rhodophyceae). III. Spore germination and subsequent development within the host *Odonthallia floccosa* (Ceramiales, Rhodophyceae). *Can. J. Bot.*, **54**, 268–80.

GOFF L. J. & COLE K. (1976b) The biology of *Harveyella mirabilis* (Cryptonemiales, Rhodophyceae). IV. Life history and phenology. *Can. J. Bot.*, **54**, 281–92.

GUIRY M. (1974) A preliminary consideration of the taxonomic position of *Palmaria palmata* (Linnaeus) Stackhouse = *Rhodymenia palmata* (Linnaeus) Greville. *J. Mar. Biol. Ass. U.K.*, **54**, 509–28.

GUIRY M. (1978a) The importance of sporangia in the classification of the Florideophyceae. In *Modern Approaches to the Taxonomy of Red and Brown Algae*, eds. Irvine D. E. G. & Price J. H., Academic Press, pp. 111–44, London & New York.

GUIRY M. (1978a) Notes on some family names of Florideophyceae (Rhodophyta). *Taxon*, **27**, 191–5.

HANSEN G. I. (1977) *Cirrulicarpus carolinensis*, a new species in the Kallymeniaceae (Rhodophyta). *Occasional Pap. Farlow Herb. Harvard Univ.*, **12**, 1–22.

HANSEN J. & DOYLE W. (1976) Ecology and natural history of *Iridaea cordata* (Gigartinaceae): Population structure. *J. Phycol.*, **12**, 273–8.

HARDER R. (1948) Einordnung von *Trailliella intricata* in Generationswechsel der Bonnemaisoniaceae. *Nachr. Akad. Wiss. Göttingen*, **1948**, 24–7.

HARDER R. & KOCH W. (1949) Life history of *Bonnemaisonia hamifera* (*Trailliella intricata*). *Nature (Lond.)*, **163**, 106.

HASSINGER-HUIZINGA, H. (1952) Generationswechsel und Geschlechtsbestimmung bei *Callithamnion corymbosum* (Sm.) Lyngb. *Arch. Protistenk.*, **98**, 91–124.

HAWKES M. W. (1978) Sexual reproduction in *Porphyra gardneri* (Smith and Hollenberg) Hawkes (Bangiales, Rhodophyta). *Phycologia*, **17**, 326–50.

HAWKES M. W. (1980) Ultrastructure characteristics of monospore formation in *Porphyra gardneri* (Rhodophyta). *J. Phycol.*, **16**, 192–6.

HEEREBOUT G. R. (1968) Studies on the Erythropeltidaceae (Rhodophyceae-Bangiophycidae). *Blumea*, **16**, 139–57.

HOLLENBERG G. J. (1969) New species of marine algae from Washington, USA. *Syesis*, **2**, 163–9.

HOLLENBERG G. J. (1970) Phycological notes. IV. Including new marine algae and new records for California. *Phycologia*, **9**, 61–72.

HUDSON P. R. & WYNNE M. J. (1969) Sexual plants of *Bonnemaisonia geniculata* (Nemaliales). *Phycologia*, **8**, 207–13.

HURDLEBRINK L. & SCHWANTES H. O. (1972) Sur le cycle de développement de *Batrachospermum*. *Soc. Bot. Fr. Mém.*, **1972**, 269–74.

IWASAKI H. (1961) The life cycle of *Porphyra tenera* in vitro. *Biol. Bull.*, **121**, 173–87.

JONES W. E. & SMITH R. M. (1970) The occurrence of tetraspores in the life history of *Naccaria wiggii* (Turn.) Endl. *Br. Phycol. J.*, **5**, 91–5.

KANEKO T. (1966) Morphological and developmental studies of Gelidiales. I. Behaviour of the nucleus in early stages of tetraspore germination in *Gelidium vagum* Okamura. *Bull. Jap. Soc. Phycol.*, **14**, 62–70.

KANEKO T. (1968) Morphological and developmental studies of Gelidiales. II. On *Acanthopeltis japonica* Okamura. *Bull. Fac. Fish. Hokkaido Univ.*, **19**, 165–72.

KANEKO T., MATSUYAMA K. & YAMADA I. (1980) On *Gloeophycus koreanum* I. K. Lee and S. A. Yoo (Rhodophyta, Gloiosiphoniaceae) in Hokkaido. *Jap. J. Phycol.*, **28**, 97–104.

KAPRAUN D. F. (1977a) Studies on the growth and reproduction of *Antithamnion cruciatum* (Rhodophyta, Ceramiales) in North Carolina. *Norw. J. Bot.*, **24**, 269–74.

KAPRAUN D. F. (1977b) Asexual propagules in the life history of *Polysiphonia ferulacea* (Rhodophyta, Ceramiales). *Phycologia*, **16**, 417–26.

KAPRAUN D. F. (1978a) Field and culture studies on growth and reproduction of *Callithamnion byssoides* (Rhodophyta, Ceramiales) in North Carolina. *J. Phycol.*, **14**, 21–4.

KAPRAUN D. F. (1978b) Field and culture studies on selected North Carolina *Polysiphonia* species. *Bot. Mar.*, **21**, 143–53.

KAPRAUN D. F. (1979) Comparative studies of *Polysiphonia urceolata* from three North Atlantic sites. *Norw. J. Bot.*, **26**, 269–76.

KAPRAUN D. F. & LUSTER D. G. (1980) Field and culture studies of *Porphyra rosengurtii* Coll. et Cox (Rhodophyta, Bangiales) from North Carolina. *Bot. Mar.*, **23**, 449–57.

KASAHARA K. (1980) On the life history of *Turnerella mertensiana* (Rhodophyta, Gigartinales). *Bot. Mag. Tokyo*, **93**, 117–23.

KATADA M. (1955) Fundamental studies on the propagation of Gelidiaceous algae with special reference to shedding and adhesion of the spores, germination, growth and vegetative reproduction. *J. Shimonoseki Coll. Fish.*, **5**, 1–87.

KIM D. (1976) A study of the development of cystocarps and tetrasporangial sori in Gigartinaceae. *Nova Hedwigia*, **27**, 1–146.

KITO H. (1978) Cytological studies on genus *Porphyra*. *Bull. Tohoku Reg. Fish. Res. Lab.*, **39**, 29–84.

KNAGGS F. (1969) A review of Florideophycidean life histories and the culture techniques employed in their investigation. *Nova Hedwigia*, **18**, 293–330.

KOCH W. (1950) Entwicklungsgeschichtliche und physiologische Untersuchungen an Laboratoriumskulturen der Rotalge *Trailliella intricata* Batters (Bonnemaisoniaceae). *Arch. Mikrobiol.*, **14**, 635–60.

KORNMANN P. & SAHLING P-H. (1962) Geschlechtspflanzen von *Bonnemaisonia hamifera* Hariot bei Helgoland. *Helgol. Wiss. Meeresunters.*, **8**, 298–301.

KRAFT G. (1975) Considerations of the order Cryptonemiales and the families Nemastomaceae and Furcellariaceae (Gigartinales, Rhodophyta) in the light of the morphology of *Adelophyton corneum* (J. Agardh) gen. et comb. nov. from Southern Australia. *Br. Phycol. J.*, **10**, 279–90.

KRAFT G. (1977a) Studies of marine algae in the lesser-known families of the Gigartinales (Rhodophyta) I. The Acrotylaceae. *Aust. J. Bot.*, **25**, 97–140.

KRAFT G. (1977b) Studies of the marine algae in the lesser-known families of the Gigartinales (Rhodophyta) II. The Dicranemaceae. *Aust. J. Bot.*, **25**, 219–67.

KRAFT G. (1978) Studies of marine algae in the lesser-known families of the Gigartinales (Rhodophyta) III. The Mychodeaceae and Mychodeophyllaceae. *Aust. J. Bot.*, **26**, 515–610.

KRISHNAMURTHY V. (1972) A revision of the species of the algal genus *Porphyra* occurring on the Pacific coast of North America. *Pac. Sci.*, **26**, 24–49.

KUGRENS P. & WEST J. (1973) Ultrastructure of an alloparasitic red alga *Choreocolax polysiphoniae*. *Phycologia*, **12**, 175–86.

KUROGI M. (1959) Influence of light on the growth and maturation of conchocelis-thallus of *Porphyra*. I. Effects of photoperiod on the formation of monosporangia and liberation of monospores. *Bull. Tohoku Reg. Fish. Res. Lab.*, **15**, 33–42.

KUROGI M. (1972) Systematics of *Porphyra* in Japan. In *Contributions to the Systematics of Benthic Marine Algae of the North Pacific*, eds. Abbott I. A. & Kurogi M., pp. 167–91. Jap. Soc. Phycol. Kobe, Japan.

KYLIN H. (1935) Remarks on the life-history of the Rhodophyceae. *Bot. Rev.*, **1**, 138–48.

KYLIN H. (1945) Über den Generationswechsel von *Bonnemaisonia asparagoides*. *K. Fysiolgr. Sällsk. Lund Förh.*, **15**, 205–10.

KYLIN H. (1956) *Die Gattungen der Rhodophyceen*. Gleerup, Lund.

LAWSON R. P. & RUSSELL G. (1967) Simultaneous occurrence of carposporophytes and tetrasporangia in *Polysiphonia urceolata*. *Br. Phycol. Bull.*, **3**, 249–50.

LEE I. K. (1978) Studies on Rhodymeniales from Hokkaido. *J. Fac. Sci. Hokkaido Univ.* Series V (Bot.), **11**, 1–194.

LEE I. K. & KUROGI M. (1978) *Neohalosacciocolax aleutica* gen. et sp. nov. (Rhodophyta), parasitic on *Halosaccion minjaii* I. K. Lee from North Pacific. *Br. Phycol. J.*, **13**, 131–9.

LEE I. K. & WEST J. A. (1979) *Dasysiphonia chejuensis* gen. et sp. nov. (Rhodophyta, Dasyaceae) from Korea. *Syst. Bot.*, **4**, 115–29.

LEE I. K. & WEST J. A. (1980a) *Antithamnion nipponicum* Yamada et Inagaki (Rhodophyta, Ceramiales) in culture. *Jap. J. Phycol.*, **28**, 19–27.

LEE I. K. & WEST J. A. (1980b) A life history of *Lomentaria hakodatensis* Yendo (Rhodophyta, Lomentariaceae) in culture. *Bot. Mar.*, **23**, 419–23.

LEE I. K., WEST J. A. & HOMMERSAND M. (1981) *Binghamiopsis caespitosa* gen. et sp. nov. (Lomentariaceae, Rhodophyta) from the eastern Pacific. *Phycologia* (in press).

LEE I. K. & YOO S. A. (1980) *Gloeophycus koreanum* gen. et sp. nov. (Rhodophyta, Gloiosiphoniaceae) from Korea. *Phycologia*, **18**, 347–54.

LEMUS A. & GANESAN E. K. (1981) Morphological and culture studies in two species of *Predaea* De Toni (Rhodophyta, Gymnophlaeaceae). *Bol. Inst. Univ. Oriente*, **15** (in press).

LEVRING T. (1953) The marine algae of Australia. I. Rhodophyta: Goniotrichales, Bangiales and Nemalionales. *Arkiv f. Bot.*, **2**, 457–566.

LEVRING T. (1955) Contributions to the marine algae of New Zealand. *Arkiv f. Bot.*, **3**, 407–32.

L'HARDY-HALOS M. T. (1968) Les Ceramiaceae (Rhodophyceae, Florideae) des côtes de Bretagne: I. Le Genre *Antithamnion* Nägeli. *Rev. Algol.*, **9**, 152–83.

MCLACHLAN J. (1973) Growth media-marine. In *Handbook of Phycological Methods. Culture Methods and Growth Measurements*, ed. Stein J. R., pp. 25–51. Cambridge University Press, Cambridge.

MCLACHLAN J., VAN DER MEER J. P. & BIRD N. L. (1977) Chromosome numbers of *Gracilaria foliifera* and *Gracilaria* sp. (Rhodophyta) and attempted hybridization. *J. Mar. Biol. Ass. U.K.*, **57**, 1137–41.

MAGNE F. (1952) La structure du noyau et le cycle nucléaire chez le *Porphyra linearis* Greville. *C. R. Acad. Sci. Paris*, **234**, 986–8.

MAGNE F. (1960a) Sur le lieu de la méiose chez le *Bonnemaisonia asparagoides* (Woodw.) *C. R. Acad. Sci. Paris*, **250**, 2742–4.

MAGNE F. (1960b) Le *Rhodochaete parvula* Thuret (Bangioidée) et sa reproduction sexuée. *Cah. Biol. Mar.*, **1**, 407–20.

MAGNE F. (1961a) Sur le cycle cytologique du *Nemalion helminthoides* (Velley) Batters. *C. R. Acad. Sci. Paris*, **252**, 157–9.

MAGNE F. (1961b) Sur la caryologie de deux Rhodophycées considerées jusqu'ici comme à cycle cytologique entièrement haplophasique. *C. R. Acad. Sci. Paris*, **252**, 4023–4.

MAGNE F. (1964) Recherches caryologiques chez les Floridées (Rhodophycées). *Cah. Biol. Mar.*, **5**, 461–671.

MAGNE F. (1967a) Sur l'existence, chez les *Lemanea* (Rhodophycées, Némalionales), d'un type de cycle de développement encore inconnu chez les Algues rouges. *C. R. Acad. Sci. Paris*, **264D**, 2632–3.

MAGNE F. (1967b) Sur le déroulement et le lieu de la méiose chez Lémanéacées (Rhodophycées, Némalionales). *C. R. Acad. Sci. Paris*, **265D**, 670–3.

MAGNE F. (1972) Le cycle de développement des Rhodophycées et son evolution. *Soc. Bot. Fr. Mém.*, **1972**, 247–68.

MAGRUDER W. H. (1977) The life history of the red alga *Ahnfeltia concinna* (Rhodophyta, Gigartinales). *Phycologia*, **16**, 197–203.

MARTIN M. T. (1969) A review of life histories in the Nemalionales and some allied genera. *Br. Phycol. J.*, **4**, 145–58.

MASUDA M. (1973a) *Neodilsea crispata*, a new species of red algae (Cryptonemiales, Rhodophyta). *J. Jap. Bot.*, **48**, 35–48.

MASUDA M. (1973b) Taxonomy and life history of *Neodilsea integra* (Kjellm.) A. Zinova var. *longissima* var. nov. (Rhodophyta, Cryptonemiales). *Bull. Nat. Sci. Mus. Tokyo*, **16**, 459–72.

MASUDA M. (1976a) Taxonomic notes on *Rhodophysemopsis* gen. nov. (Rhodophyta). *J. Jap. Bot.*, **51**, 175–87.

MASUDA M. (1976b) On a new red algal genus *Pseudorhododiscus*. *Acta Phytotax. Geobot.*, **27**, 123–32.

MASUDA M., DECEW T. & WEST J. (1979) The tetrasporophyte of *Gymnogongrus flabelliformis* Harvey (Gigartinales, Phyllophoraceae). *Jap. J. Phycol.*, **27**, 63–73.

MASUDA M. & OHTA M. (1975) The life history of *Rhodophysema geogii* Batters (Rhodophyta, Cryptonemiales). *J. Jap. Bot.*, **50**, 1–10.

MASUDA M. & UCHIDA T. (1976) On the life history of *Gigartina ochotensis* (Rupr.) Rupr. from Muroran, Hokkaido. *Bull. Jap. Soc. Phycol.*, **24**, 41–7.

MASUDA M. & UMEZAKI I. (1977) On the life history of *Nemalion vermiculare* Suringar (Rhodophyta) in culture. *Bull. Jap. Soc. Phycol.*, **25** (suppl.), Mem. Iss. Yamada, 129–36.

MAYHOUB H. (1973) Cycle du développement de *Calosiphonia vermicularis* (J. Agardh) Schmitz (Rhodophycée, Gigartinale). *C. R. Acad. Sci. Paris*, **277**, 1137–40.

MAYHOUB H. (1975) Nouvelles observations sur le cycle de développement du *Calosiphonia vermicularis* (J. Ag.) Sch. (Rhodophycée, Gigartinale). *C. R. Acad. Sci. Paris*, **280**, 2441–3.

MAYHOUB H., GAYRAL P. & JACQUES R. (1976) Action de la composition spectrale de la lumière sur la croissance et la reproduction de *Calosiphonia vermicularis* (J. Agardh) Schmitz (Rhodophycées, Gigartinales). *C. R. Acad. Sci. Paris*, **283**, 1041–4.

MIGITA S. (1967) Cytological studies on *Porphyra yezoensis* Ueda. *Bull. Fac. Fish. Nagasaki Univ.*, **24**, 55–64.

MIKAMI H. (1965) A systematic study of the Phyllophoraceae and Gigartinaceae from Japan and its vicinity. *Sci. Pap. Inst. Algol. Res. Fac. Sci. Hokkaido Univ.*, **5**, 181–285.

MIN-THEIN U. & WOMERSLEY H. (1976) Studies on southern Australian taxa of Solieriaceae, Rhabdoniaceae and Rhodophyllidaceae (Rhodophyta). *Aust. J. Bot.*, **24**, 1–166.

MIURA A. (1975) *Porphyra* cultivation in Japan. In *Advance of Phycology in Japan*, eds. Tokida J. & Hirose H., pp. 273–304. W. Junk, The Hague.

MOE R. L. (1979) *Minium parvum* gen. et sp. nov., a crustose member of the Rhodymeniales (Rhodophyta). *Phycologia*, **18**, 38–46.

MOE R. L. & SILVA P. C. (1980) Morphological and taxonomic studies on Antarctic Ceramiaceae (Rhodophyceae). II. *Pterothamnion antarcticum* (Kylin) comb. nov. (*Antithamnion antarcticum* Kylin). *Br. Phycol. J.*, **15**, 1–17.

MOROHOSHI H. & MASUDA M. (1980) The life history of *Gloiosiphonia capillaris* (Hudson) Carmichael (Rhodophyceae, Cryptonemiales). *Jap. J. Phycol.*, **28**, 81–92.

MSHIGENI K. (1976a) Contributions on the developmental biology of Tanzanian Rhodophyta. I. Spore germination in *Sarcodia montagneana* (J. Agardh). *Nova Hedwigia*, **28**, 649–55.

MSHIGENI K. (1976b) Contributions on the developmental biology of Tanzanian Rhodophyta. III. Spore germination in *Sarconema filiforme* (Sonder) Kylin. *Nova Hedwigia*, **28**, 661–5.

MÜLLER K. & PETRONIJEVIC E. (1979) Untersuchungen zum Generationswechsel von *Nemalion multifidum* (Weber et Mohr) J. Ag. *Nova Hedwigia*, **31**, 293–318.

MUMFORD JR., T. F. & COLE K. (1977) Chromosome numbers for fifteen species in the genus *Porphyra* (Bangiales, Rhodophyta) from the west coast of North America. *Phycologia*, **16**, 373–7.

MURRAY S. N. & DIXON P. (1973) The life history of *Callophyllis firma* (Kylin) Norris in laboratory culture. *Br. Phycol. J.*, **7**, 165–8.

MURRAY S., DIXON P. & SCOTT J. L. (1972) The life history of *Porphyropsis coccinea* var. *dawsonii* in culture. *Br. Phycol. J.*, **7**, 323–33.

NEWROTH P. (1971) Studies on life histories in the Phyllophoraceae. I. *Phyllophora truncata* (Rhodophyceae, Gigarintales). *Phycologia*, **10**, 345–54.

NEWROTH P. (1972) Studies on life histories in the Phyllophoraceae. II. *Phyllophora pseudoceranoides* and notes on *P. crispa* and *P. heredia* (Rhodophyta, Gigartinales). *Phycologia*, **11**, 99–107.

NICHOLS H. W. & VEITH G. M. (1978) Development of *Bangia atropurpurea*. *Phytomorphology*, **28**, 322–8.

NORRIS J. & BUCHER K. (1977) The genus *Platoma* (Gigartinales, Rhodophyta) with a description of *P. abbottiana* sp. nov. *J. Phycol.*, **13**, 155–62.

NOTOYA M. (1976) On the influence of various culture conditions on the early development of spore germination in three species of the crustose corallines (Rhodophyta). (Preliminary report.) *Bull. Jap. Soc. Phycol.*, **24**, 137–42.

NOZAWA Y. (1969) Systematic anatomy of the Squamariaceae in the southern islands of Japan (2): *Cruoriella fissurata* Dawson. *Bull. Jap. Soc. Phycol.*, **17**, 19–24.

NOZAWA Y. (1970) Systematic anatomy of the red algal genus *Rhodopeltis*. *Pacific Sci.*, **24**, 99–133.

NOZAWA Y. (1972) Systematic anatomy of the Squamariaceae in the southern islands of Japan (4): *Peyssonnelia conchicola* Picc. & Grun., *P. rubra* var. *orientalis* Weber van Bosse. *Bull. Jap. Soc. Phycol.*, **20**, 41–7.

OGATA E., MATSUI T. & NAKAMURA H. (1972) The life cycle of *Gracilaria verrucosa* (Rhodophyceae, Gigartinales) *in vitro*. *Phycologia*, **11**, 75–80.

OHTA M. & KUROGI M. (1979) On the life history of *Rhodochorton purpureum* (Lightf.) Rosenvinge from Hokkaido in culture. *Jap. J. Phycol.*, **27**, 161–7.

PAPENFUSS G. F. (1945) Review of the *Acrochaetium-Rhodochorton* complex of the red algae. *Univ. Calif. Publ. Bot.*, **18**, 299–334.

PAPENFUSS G. F. (1947) Further contributions toward an understanding of the *Acrochaetium-Rhodochorton* complex of the red algae. *Univ. Calif. Publ. Bot.*, **18**, 433–47.

PAPENFUSS G. F. (1966) A review of the present system of classification of the Florideophycidae. *Phycologia*, **5**, 247–55.

PAPENFUSS G. F. & CHIANG Y. M. (1969) Remarks on the taxonomy of *Galaxaura* (Nemaliales, Chaetangiaceae). *Proc. Int. Seaweed Symp.*, **6**, 303–14.

PAPENFUSS G. F. & EDELSTEIN T. (1974) The morphology and taxonomy of the red alga *Sarconema* (Gigartinales: Solieriaceae). *Phycologia*, **13**, 31–44.

PARKE M. & DIXON P. (1976) Check list of British marine algae—third revision. *J. Mar. Biol. Ass. U.K.*, **56**, 527–94.

PARSONS M. J. (1975) Morphology and taxonomy of the Dasyaceae and the Lophothalieae (Rhodophyceae) of the Rhodophyta. *Aust. J. Bot.*, **23**, 549–713.

PENNCAVAGE N., SCHATZ S. S., MCCANDLESS E. & FIORE J. (1979) Distribution of gametophytes and sporophytes of *Chondrus crispus* in the vicinity of Woods Hole. *Biol. Bull.*, **157**, 387–8.

POLANSHEK A. & WEST J. (1975) Culture and hybridization studies on *Petrocelis* (Rhodophyta) from Alaska and California. J. Phycol., **11**, 434–9.

POLANSHEK A. & WEST J. (1977) Culture and hybridization studies on *Gigartina papillata* (Rhodophyta). *J. Phycol.*, **13**, 141–9.

POLNE M., NEUSHUL M. & GIBOR A. (1980) Growing *Eucheuma uncinatum* in culture: the domestication of a marine crop plant. In *Pacific Seaweed Aquaculture*, eds. Abbott I. A., Eklund L. F. & Foster M. S, pp. 115–22. California Sea Grant Program, La Jolla.

PRICE J. H. (1978) Ecological determination of adult form in *Callithamnion*: its taxonomic implications. In *Modern Approaches to the Taxonomy of Red and Brown Algae*, eds. Irvine D. E. G. & Price J. H., pp. 263–300. Academic Press, London.

RAMUS J. (1969) The development sequence of the marine red alga *Pseudogloiophloea* in culture. *Univ. Calif. Publ. Bot.*, **52**, 1–42.

RAO B. G. S., SUNDARI M. & RAO M. V. (1978) Chromosome behaviour at meiosis and its bearing on the cytotaxonomy of *Ceramium* species. *Bot. Mar.*, **21**, 123–9.

RICHARDSON N. (1970) Studies on the photobiology of *Bangia fuscopurpurea*. *J. Phycol.*, **6**, 215–19.

RICHARDSON N. (1972) Spore classification in the genera *Bangia* and *Porphyra*. *Br. Phycol. J.*, **7**, 49–51.

RICHARDSON N. & DIXON P. (1969) The Conchocelis phase of *Smithora naiadum* (Anders.) Hollenb. *Br. Phycol. J.*, **4**, 181–3.

RICHARDSON N. & DIXON P. (1970) Culture studies on *Thuretellopsis peggiana* Kylin. *J. Phycol.*, **6**, 154–9.

ROSENVINGE L. K. (1971) Marine algae of Denmark. Part II. Rhodophyceae II. (Cryptonemiales). *K. danske Vidensk. Selsk. Skr. 7 Raekke Naturv. Mat. Afd.*, **7**, 155–283.

RUENESS J. (1968) Paraspores from *Plumaria elegans* (Hornem.) Schmitz in culture. *Norw. J. Bot.*, **15**, 220–4.

RUENESS J. (1971a) *Polysiphonia hemispherica* Aresch. in Scandinavia. *Norw. J. Bot.*, **18**, 65–74.

RUENESS J. (1971b) The life history of *Spermothamnion repens* (Dillw.) Rosenv. in culture. *Norw. J. Bot.*, **18**, 93–5.

RUENESS J. (1973a) Culture and field observations on growth and reproduction of *Ceramium strictum* Harv. from the Oslofjord, Norway. *Norw. J. Bot.*, **20**, 61–5.

RUENESS J. (1973b) Speciation in *Polysiphonia* (Rhodophyceae, Ceramiales) in view of hybridization experiments: *P. hemisphaerica* and *P. boldii*. *Phycologia*, **12**, 107–9.

RUENESS J. (1978a) A note on development and reproduction in *Gigartina stellata* (Rhodophyceae, Gigartinales) from Norway. *Br. Phycol. J.*, **13**, 87–90.

RUENESS J. (1978b) Hybridization in red algae. In *Modern Approaches to the Taxonomy of Red and Brown Algae*, eds. Irvine D. E. G. & Price J. H., pp. 247–62. System. Assoc. Sp. Vol. No. 10, Academic Press, London & New York.

RUENESS J. & RUENESS M. (1973) Life history and nuclear phases of *Antithamnion tenuissimum*, with special reference to plants bearing both tetrasporangia and spermatangia. *Norw. J. Bot.*, **20**, 205–10.

RUENESS J. & RUENESS M. (1975) Genetic control of morphogenesis in two varieties of *Antithamnion plumula* (Rhodophyceae, Ceramiales). *Phycologia*, **14**, 81–5.

RUENESS M. & RUENESS J. (1978) A parasporangium-bearing strain of *Callithamnion hookeri* (Rhodophyceae, Ceramiales) in culture. *Norw. J. Bot.*, **25**, 201–5.

RUENESS J. & RUENESS M. (1980) Culture and field observations on *Callithamnion bipinnatum* and *C. byssoides* (Rhodophyta, Ceramiales) from Norway. *Sarsia*, **65**, 29–34.

SANTELICES B. (1974) Gelidioid Algae. A brief resume of the pertinent literature. *Mar. Agron., U.S. Sea Grant Prog., Univ. Hawaii, Tech. Rep.*, **1**, 1–111.

SCHOTTER G. (1968) Recherches sur les Phyllophoracées. *Bull. Institut Oceanogr., Monaco*, **67**, (no. 1383).

SCIUTO S., PIATELLI M., CHILLEMI R., FURNARI G. & CORMACI M. (1979) The implications of *Haematocelis rubens* J. Agardh in the life history of *Schizymenia dubyi* (Chauvin) J. Agardh (Rhodophyta, Gigartinales): A chemical study. *Phycologia*, **18**, 96–9.

SCOTT J. & DIXON P. (1971) The life history of *Pikea californica* Harv. *J. Phycol.*, **7**, 295–300.

SEARLES R. B. (1968) Morphological studies of red algae of the order Gigartinales. *Univ. Calif. Publ. Bot.*, **43**, 1–86.

SHEATH R. G. & COLE K. M. (1980) Distribution and salinity adaptations of *Bangia atropurpurea* (Rhodophyta), a putative migrant into the Laurentian Great Lakes. *J. Phycol.*, **16**, 412–20.

SHEVLIN D. & POLANSHEK A. (1978) Life history of *Bonnemaisonia geniculata* (Rhodophyta): laboratory and field study. *J. Phycol.*, **14**, 282–9.

SOMMERFELD M. R. & NICHOLS H. W. (1970) Developmental and cytological studies in *Bangia fuscopurpurea* in culture. *Am. J. Bot.*, **57**, 640–5.

SOUTH G. & ADAMS N. (1979) A revision of the genus *Plocamium* Lamouroux (Rhodophyta, Gigartinales) in New Zealand. *Phycologia*, **18**, 120–32.

SOUTH G., HOOPER R. & IRVINE L. (1972) The life history of *Turnerella pennyi*. *Br. Phycol. J.*, **7**, 221–33.

SOUTH G. R. & WHITTICK A. (1976) Aspects of the life history of *Rhodophysema elegans* (Rhodophyta, Peyssonneliaceae). *Br. Phycol. J.*, **11**, 349–54.

SPARLING S. R. (1961) A report on the culture of some species of *Halosaccion*, *Rhodymenia* and *Fauchea*. *Amer. J. Bot.*, **48**, 493–9.

STEGENGA H. (1978) The life histories of *Rhodochorton purpureum* and *Rhodochorton floridulum* (Rhodophyta, Nemaliales) in culture. *Br. Phycol. J.*, **13**, 279–89.

STEGENGA H. & BORSJE W. J. (1976) The morphology and life history of *Acrochaetium dasyae* Collins (Rhodophyta, Nemaliales). *Acta Bot. Neerl.*, **25 (1)**, 15–29.

STEGENGA H. & BORSJE W. J. (1977) The morphology and life history of *Acrochaetium polyblastum* (Rosenv.) Borg. and *Acrochaetium hallandicum* (Kylin) Hamel (Rhodophyta, Nemaliales). *Acta Bot. Neerl.*, **26**, 451–70.

STEGENGA H. & MULDER A. S. (1979) Remarks on the *Audouinella microscopica* (Näg.) Woelkerling complex with a brief survey of the genus *Chromastrum* Papenfuss (Rhodophyta, Nemaliales). *Acta Bot. Neerl.*, **28**, 289–311.

STEGENGA H. & VROMAN M. (1976) The morphology and life history of *Acrochaetium densum* (Drew) Papenfuss (Rhodophyta, Nemaliales). *Acta Bot. Neerl.*, **25**, 257–80.

STEGENGA H. & VROMAN M. (1977) The significance of life history characters for the systematics of the Acrochaetiaceae (abstr.) *J. Phycol.*, **13** (suppl.), 64.

STEGENGA H. & VAN WISSEN M. J. (1979) Remarks on the life histories of three Acrochaetioid algae (Rhodophyta, Nemaliales). *Acta Bot. Neerl.*, **28**, 97–115.

SUNDENE O. (1959) Form variation in *Antithamnion plumula*. Experiments on Plymouth and Oslofjord strains in culture. *Nytt Mag. Bot.*, **7**, 181–7.

SUNDENE O. (1962) Reproduction and morphology in strains of *Antithamnion boreale* originating from Spitsbergen and Scandinavia. *Skr. Norsk. Vidensk. Akad. Skr. Mat.-Nat. Fk.*, N.S., **5**, 1–19.

SUNDENE O. (1965a) *Antithamnion tenuissimum* (Hauck) Schiffner in culture. *Nytt Mag. Bot.*, **11**, 5–10.

SUNDENE O. (1965b) The conspecificity of *Antithamnion sarniense* and *A. spirographidis* in view of culture experiments. *Nytt Mag. Bot.*, **11**, 35–42.

SUNDENE O. (1975) Experimental studies on form variation in *Antithamnion plumula* (Rhodophyceae). *Norw. J. Bot.*, **22**, 35–42.

SUNESON S. (1950) The cytology of bispore formation in two species of *Lithophyllum* and the significance of the bispores in the Corallinaceae. *Bot. Not.*, **1950**, 429–50.

SVEDELIUS N. (1931) Nuclear phases and alternation in the Rhodophyceae. *Bot. Zbl.*, **48**, 38–59.

SVEDELIUS N. (1937) The apomeiotic tetrad division in *Lomentaria rosea* in comparison with the normal development in *Lomentaria clavellosa*. *Symb. Bot. Upsal.*, **2**, 1–53.

SWALE E. & BELCHER J. (1963) Morphological observations on wild and cultured material of *Rhodochorton investiens* (Lenormand) comb. nov. (*Balbiania investiens* (Lenormand) Sirodot). *Ann. Bot.* N.S., **27**, 281–90.

TSENG C. K. & CHANG T. J. (1955a) Studies on the life history of *Porphyra ternera* Kjellm. *Scientia Sinica*, **4**, 375–98. (English transl. Combined from *Acta Bot. Sinica*, **3**, 287–302 (1954) and **4**, 27–46 (1955).)

TSENG C. K. & CHANG T. J. (1955b) Studies on *Porphyra* III. Sexual reproduction of *Porphyra*. *Acta Bot. Sin.*, **4**, 153–66.

UMEZAKI I. (1967) The tetrasporophyte of *Nemalion vermiculare* Sur. *Rev. Algol.*, **9**, 19–24.

UMEZAKI I. (1972) The life history of *Hyalosiphonia caespitosa* (Dumontiaceae, Rhodophyta). *J. Jap. Bot.*, **47**, 278–88.

UMEZAKI I. (1974) The life history of *Nemastoma nakamurae* Yendo (Nemastomataceae, Rhodophyta) in culture. *J. Jap. Bot.*, **49**, 346–52.

UMEZAKI I. (1977) Life histories in the Florideophyceae and their evolution. *Acta Phytotax. Geobot.*, **28**, 1–18.

VAN DEN HOEK C. & CORTEL-BREEMAN A. M. (1970a) Life history studies on Rhodophyceae. II. *Halymenia floresia* (Clem.) Ag. *Acta Bot. Neerl.*, **19**, 341–62.

VAN DEN HOEK C. & CORTEL-BREEMAN A. M. (1970b) Life history studies on Rhodophyceae. III. *Scinaia complanata* (Collins) Cotton. *Acta Bot. Neerl.*, **19**, 457–67.

VAN DER MEER J. P. (1976) A contribution towards elucidating the life history of *Palmaria palmata* (= *Rhodymenia palmata*). *Can. J. Bot.*, **54**, 2903–6.

VAN DER MEER J. P. (1977) Genetics of *Gracilaria* sp. (Rhodophyceae, Gigartinales). II. The life history and genetic implications of cytokinetic failure during tetraspore formation. *Phycologia*, **16**, 367–71.

VAN DER MEER J. P. (1978) Genetics of *Gracilaria* sp. (Rhodophyceae, Gigartinales). III. Non-Mendelian gene transmission. *Phycologia*, **17**, 314–18.

VAN DER MEER J. P. (1979a) Genetics of *Gracilaria* sp. (Rhodophyceae, Gigartinales). V. Isolation and characterization of mutant strains. *Phycologia*, **18**, 47–54.

VAN DER MEER J. P. (1979b) Genetics of *Gracilaria tikvahiae* (Rhodophyceae). VI. Complementation and linkage analysis of pigmentation mutants. *Can. J. Bot.*, **57**, 64–8.

VAN DER MEER J. P. & BIRD N. L. (1977) Genetics of *Gracilaria* sp. (Rhodophyceae, Gigartinales). I. Mendelian inheritance of two spontaneous green variants. *Phycologia*, **16**, 159–61.

VAN DER MEER J. P. & CHEN L. C.-M. (1979) Evidence for sexual reproduction in the red algae *Palmaria palmata* and *Halosaccion ramentaceum*. *Can. J. Bot.*, **57**, 2452–9.

VAN DER MEER J. P. & TODD E. R. (1977) Genetics of *Gracilaria* sp. (Rhodophyceae, Gigartinales). IV. Mitotic recombination and its relationship to mixed phases in the life history. *Can. J. Bot.*, **55**, 2810–17.

VAN DER MEER J. P. & TODD E. R. (1980) The life history of *Palmaria palmata* in culture. A new type for the Rhodophyceae. *Can. J. Bot.*, **58**, 1250–6.

VON STOSCH H. A. (1964) Wirkungen von Jod und Arsenit auf Meeresalgen in Kultur. *4e Cong. Int. Alg. Mar. Biarritz*, **1961**, 142–50.

VON STOSCH H. A. (1965) The sporophyte of *Liagora farinosa* Lamour. *Br. Phycol. Bull.*, **2**, 486–96.

VON STOSCH H. A. (1969) Observations on *Corallina, Jania* and other red algae in culture. *Proc. Int. Seaweed Symp.*, **6**, 389–99.

VON STOSCH H. A. & THEIL G. (1979) A new mode of life history in the freshwater red algal genus *Batrachospermum*. *Am. J. Bot.*, **66**, 105–7.

Waaland J. R. & Kemp C. I. (1972) Observations on the life history of *Membranoptera multiramosa* Gardner (Rhodophyceae, Ceramiales) in culture. *Phycologia*, **11**, 15–18.

Waaland S. D. (1975) Evidence for a species-specific cell fusion hormone in red algae. *Protoplasma*, **86**, 253–61.

Waaland S. D. (1978) Parasexually produced hybrids between female and male plants of *Griffithsia tenuis* C. Agardh, a red alga. *Planta*, **138**, 65–8.

Waaland S. D. & Watson B. A. (1980) Isolation of a cell-fusion hormone from *Griffithsia pacifica* Kylin, a red alga. *Planta*, **149**, 493–7.

West J. A. (1968) Morphology and reproduction of the red alga *Acrochaetium pectinatum* in culture. *J. Phycol.*, **4**, 89–99.

West J. A. (1969) The life histories of *Rhodochorton purpureum* and *R. tenue* in culture. *J. Phycol.*, **5**, 12–21.

West J. A. (1970a) A monoecious isolate of *Rhodochorton purpureum*. *J. Phycol.*, **6**, 368–70.

West J. A. (1970b) The life history of *Rhodochorton concrescens* in culture. *Br. Phycol. J.*, **5**, 179–86.

West J. A. (1970c) The conspecificity of *Heterosiphonia asymmetrica* and *H. densiuscula* and their life histories in culture. *Madroño*, **20**, 313–19.

West J. A. (1972a) Environmental regulation of reproduction in *Rhodochorton purpureum*. In *Contributions to the Systematics of the Benthic Marine Algae of the North Pacific*, eds. Abbott I. A. & Kurogi M., pp. 213–30. Jap. Soc. Phycol. Kobe, Japan.

West J. A. (1972b) The life history of *Petrocelis franciscana*. *Br. Phycol. J.*, **7**, 299–308.

West J. A. (1979) The life history of *Rhodochorton membranaceum*, an endozoic red alga. *Bot. Mar.*, **22**, 111–15.

West J. A. & Norris R. E. (1966) Unusual phenomena in the life histories of Florideae in culture. *J. Phycol.*, **2**, 54–7.

West J. A., Polanshek A. & Guiry M. (1977) The life history in culture of *Petrocelis cruenta* J. Agardh (Rhodophyta) from Ireland. *Br. Phycol. J.*, **12**, 45–53.

West J. A., Polanshek A. & Shevlin D. (1978) Field and culture studies on *Gigartina agardhii* (Rhodophyta). *J. Phycol.*, **14**, 416–26.

Whittick A. & Hooper R. G. (1977) The reproduction and phenology of *Antithamion cruciatum* (Rhodophyta: Ceramiaceae) in insular Newfoundland. *Can. J. Bot.*, **55**, 520–4.

Whittick A. & West J. A. (1979) The life history of a monoecious species of *Callithamnion* (Rhodophyta, Ceramiaceae) in culture. *Phycologia*, **18**, 30–7.

Wiseman D. R. (1975) On the status of the red algal family, the Rhizophyllidaceae (Gigartinales). *Taxon*, **24**, 489–90.

Wiseman D. R. (1977) Observations of the reproductive morphology of the red algal genus *Ochtodes* J. Agardh (Rhizophyllidaceae, Gigartinales). *Phycologia*, **16**, 1–8.

Woelkerling W. J. (1970) *Acrochaetium botryocarpum* (Harvey) J. Ag. in Southern Australia. *Br. Phycol. J.*, **5**, 159–71.

Woelkerling W. J. (1971) Morphology and taxonomy of the *Audouinella* complex (Rhodophyta) in southern Australia. *Aust. J. Bot.*, Suppl. **1**.

Wollaston E. M. (1968) Morphology and taxonomy of southern Australian genera of Crouanieae Schmitz (Ceramiaceae, Rhodophyta). *Aust. J. Bot.*, **16**, 217–417.

Wollaston E. M. (1971) *Antithamnion* and related genera occurring on the Pacific coast of North America. *Syesis*, **4**, 73–92.

Wollaston E. M. (1979) Recognition of *Pterothamnion* Naegeli with taxonomic notes on *P. simile* (Hooker and Harvey) Naegeli and *Platythamnion nodiferum* (J. Agardh) Wollaston (Rhodophyta, Ceramiaceae). *Trans. Roy. Soc. S. Aust.*, **103**, 191–6.

Womersley H. B. S. (1971a) The relationships of *Nizymenia* and *Stenocladia* (Gigartinales, Rhodophyta). *Phycologia*, **10**, 199–203.

Womersley H. B. S. (1971b) The genus *Plocamium* (Rhodophyta) in Southern Australia. *Trans. Roy. Soc. S. Aust.*, **95**, 9–27.

Wynne M. J. (1970) Marine algae of Amchitka Island (Aleutian Islands). II. Bonnemaisoniaceae. *Pacific Sci.*, **24**, 433–8.

Yabu H. (1967) Nuclear division in *Bangia fuscopurpurea* (Dillwyn) Lyngbye. *Bull. Fac. Fish. Hokkaido Univ.*, **17**, 163–4.

YABU H. (1975) Cytological studies of the Rhodophyta and Chlorophyta. *In Advance of Phycology in Japan*, eds. Tokida J. & Hirose H., pp. 125–35. W. Junk, The Hague.
YABU H. (1976) A report on the cytology of *Rhodymenia palmata, R. pertusa* and *Halosaccion saccatum* (Rhodophyta). *Bull. Fac. Fish. Hokkaido Univ.*, **27**, 51–62.
YAMANOUCHI S. (1906) The life history of *Polysiphonia violacea*. *Bot. Gaz.*, **42**, 401–49.
ZINOVA A. (1961) De alga rubra *Dilsea integra* (Kjellm.) Rosenv. *Not. Bot. Mater. Spor. Rast. Bot. Inst. Akad. Nauk SSSR.*, **14**, 82–6.

CHAPTER 5
PHAEOPHYTA: LIFE HISTORIES

P. M. PEDERSEN
University of Copenhagen, Institut for Sporeplanter,
Ø. Farimagsgade 2D, DK-1353 Copenhagen K., Denmark

1 INTRODUCTION

Investigations of the life histories of brown algae (Phaeophyceae) have been carried out for over seventy years. A considerable amount of literature has accumulated, especially from the early studies of the French phycologist Camille Sauvageau, who must be considered as the pioneer of this subject. The results obtained by him and contemporaries were remarkable considering the limited culturing facilities and the often serious problem of diatom contamination in the cultures. Many of the results obtained during that period, still contribute to the classification of the brown algae and to our knowledge of their life histories.

At the present time, the culturing techniques are much more refined with improved media, controlled environmental conditions, and the use of germanium dioxide to prevent the growth of diatoms. This has added a new aspect to our understanding of the biology of species; however, even though many valuable papers have been published, questions still exist. The recent literature has been excellently reviewed by Scagel (1966), Russell (1973) and Wynne & Loiseaux (1976). The primary purpose of this chapter, therefore, is not to review the more recent papers, but to throw light on the gaps in our knowledge, especially with regard to aspects such as intraspecific variation, the influence of environmental conditions, and the role of the microthalli; hopefully this will stimulate further investigations.

The initial phase in the development of most brown algae is called the microthallus or prostrate system in this chapter and it is mentioned if this phase is reported to be gametophytic.

2 LIFE HISTORIES IN GENERAL

2.1 Germination of zoospores

Reproduction by zoospores is probably the most common feature in the life histories of brown algae belonging to the less advanced orders which in particular have been studied by the present author. Germination of zygotes, monospores, and other more sophisticated reproductive structures has been omitted from this section but different patterns in the development of fucoid embryos are described in Chapter 2 p. 65. Upon settlement brown algal zoospores from unilocular and plurilocular sporangia round up and form embryospores. From this point, however, the developmental sequence may differ, with two main types of germination patterns described: unipolar and stellate.

Unipolar germination

A During germination the embryospore derived from a zoospore from a plurilocular sporangium develops a 'germ tube' which is later closed by a cross-wall. Both cells contain chloroplast(s) and a nucleus. Further divisions take place in the youngest cell and a filamentous (Fig. 5.1A) or discoid germling is formed. The germling later grows into a more or less well-developed prostrate system on which erect thalli later occur, however, the cell corresponding to the embryospore does not directly take part in the formation of an erect part (see below). This type of germination is very common among plants of several orders, e.g. *Delamarea attenuata* (Dictyosiphonales), *Ectocarpus siliculosus* (Ectocarpales), *Scytosiphon lomentaria* (Scytosiphonales) and many others. A modification of this type is found in *Pilayella gardneri* (Wynne 1969). Immediately after settling, the embryospores form small rhizoidal outgrowths ('amoeboid' stage).

In at least two cases, *Litosiphon pusillus* and *Myriotrichia clavaeformis* (Dictyosiphonales), the embryospores may differentiate very rapidly. The few-celled germling does not always develop into a branched, prostrate system. Instead, the germination tube develops into a rhizoidal filament representing the initial attachment system, whereas the oldest cell, corresponding to the embryospore, develops directly into an uniseriate, initial, erect filament (Kylin 1933; Pedersen 1978).

B During the germination process the embryospore (e) similarly forms a 'germ tube', but in this case the cell contents migrate into the terminal part of the tube, which swells (ts). This swelling is closed by a wall and only a small cytoplasmic body (c) is left in the proximal part of the 'germ tube'. The apical cell divides and develops into a filamentous or discoid germling like type A with the empty embryospore and a part of the 'germ tube' as an inactive appendix (Fig. 5.1B–

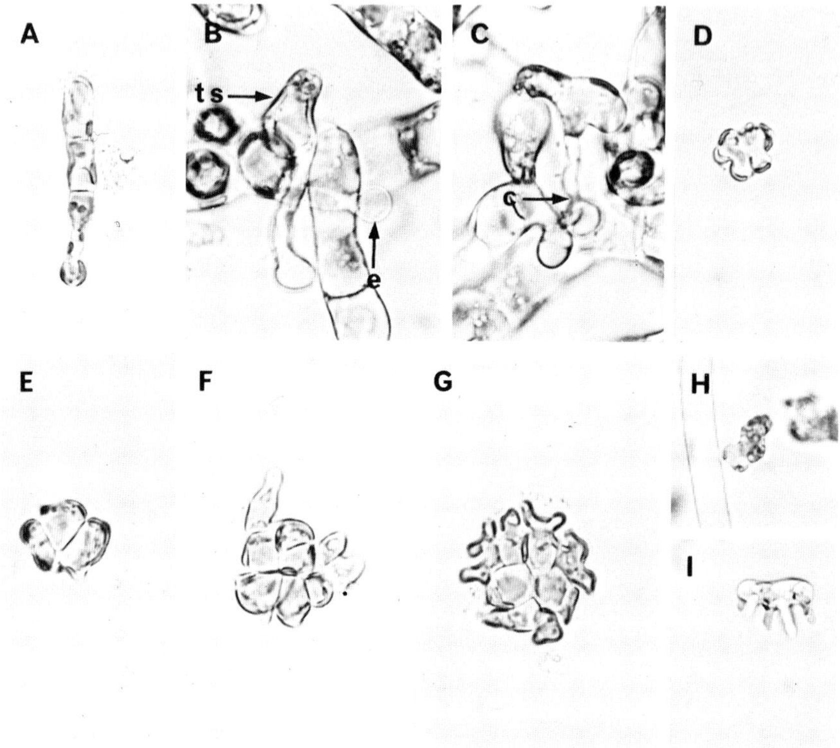

Fig 5.1 Germination of zoospores from plurilocular sporangia. (A) Unipolar germination type A in *Phaeostroma longisetum*. The embryospore is not empty. (B), (C). Unipolar germination type B in *Coelocladia arctica*. The embryospore (e, arrow) is left empty together with a part of the germination tube except for a small body of cytoplasm (c, arrow). The cell content migrates to a terminal swelling (ts, arrow). (D-G) Stellate germination in *Myrionema strangulans*. (H, I) Unipolar germination in the same species with later development of rhizoidal outgrowths (all × 600).

C). This type of germination is found, for example, in *Coelocladia arctica* (Dictyosiphonales).

A modification of this type is found in *Phaeostroma pustulosum* (Ectocarpales) (Fig. 5.2A–C). In this species a nucleus is obviously left in the embryospore together with a small cytoplasmic body, as the embryospore (e) is able to form a juvenile hair (h), i.e. a hair consisting of only long cells without a basal meristem. Formation of hairs often takes place in crowded cultures. Also small rhizoidal outgrowths are sometimes formed from the embryospore.

Stellate germination

This type of germination seems to represent an adaptation to an epiphytic habitat. The embryospore forms three to five lobes, which are later closed by

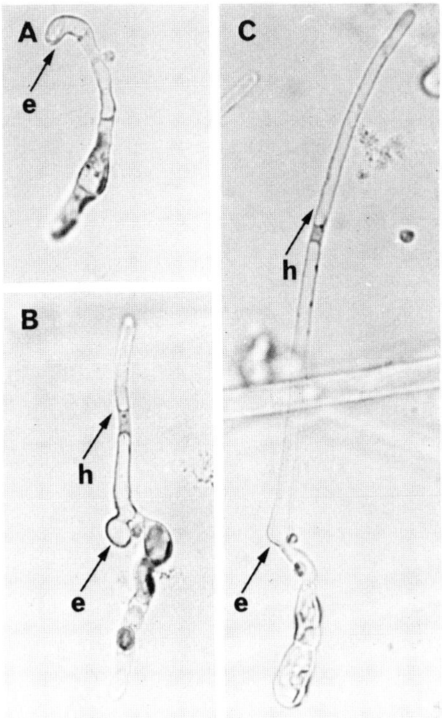

Fig. 5.2 Germination of zoospores from plurilocular sporangia of *Phaeostroma pustulosum*. (A) Unipolar germination type B. (B) Initial development of a hair (h, arrow) from the embryospore (e, arrow). (C) Very long initial hair from the embryospore; a few cross walls may later develop (all × 600).

oblique walls. This few-celled stage develops into a disc by the radiate formation of filaments often with a characteristic Y-shaped branching pattern along the margin (Fig. 5.1D–G). This type of germination occurs in, e.g. *Myrionema strangulans* and related species (Ectocarpales), *Hummia onusta* (Dictyosiphonales), and *Cladosiphon zosterae* (Chordariales, Chordariaceae). Many plants, however, which are considered as epiphytes, do not have this type of germination, e.g. *Halothrix lumbricalis*.

Embryospores formed of zoospores from unilocular sporangia show the same germination pattern as those formed of zoospores from plurilocular sporangia, for example, unipolar type A in *Striaria attenuata* (Dictyosiphonales), type B in *Ralfsia confusa* (Hollenberg 1969), and stellate in *Hecatonema streblonematoides* (Loiseaux 1970b). It is noteworthy that the same apparent species, *Ralfsia clavata,* shows both type A (Loiseaux 1968) and type B (Kylin 1934) germination pattern. This may be related to the fact that *R. clavata* in some cases is a stage in the life history of *Petalonia fascia* (Edelstein *et al.* 1970) while in other cases it behaves in culture like an independent species (Loiseaux 1968).

In many Myrionemataceae, e.g. *Myrionema strangulans,* some embryospores, even those formed by zoospores from the same sporangium, may have unipolar germination, whilst others have stellate germination. This phenomenon is known as heteroblasty, which seems to be an unexplained factor in brown algal life histories. Stellate germination is certainly genetically determined, but in some cases it seems to be triggered by contact with the substratum. I have observed in cultures of *M. strangulans* that some zoospores germinate unipolarly and form few-celled filamentous germlings but these then later form small outgrowths like the subdivided embryospore (Fig. 5.1H, I). Therefore, the difference among the swarmers is not always real but merely a morphological response to various degrees of substratum contact. Transition from a filamentous to a disc-shaped prostrate system has also been observed in *Scytosiphon lomentaria*.

The germination pattern of the embryospores of *Cladosiphon okamuranus*, described by Shinmura (1974) is an example which supports my assumption that heteroblasty is related to the degree of substratum contact. Although most of the embryospores germinated stellately, a few germinated unipolarly into filamentous germlings which secondarily formed discs (cf. Shinmura 1974, Pl. 2, Fig. 27). Such species are well adapted to grow on either firm or loose substrata.

2.2 Heteromorphic, biphasic life histories

Heteromorphic, biphasic life histories have been known since the early part of the century. The basic pattern of this type of life history is as shown by Sauvageau (1915) with respect to *Saccorhiza polyschides*: meiosis occurs in the unilocular sporangia on the macrothalli, the embryospores germinate and develop into haploid microthalli (gametophytes), and the diploid phase is re-established by sexual reproduction, in this case by oogamy. The sex of the gametophytes is now known to be genetically determined since a large X-chromosome pairs with a smaller Y-chromosome during meiosis as observed by Evans (1965) in *S. polyschides*. This type of life history is uniformly present in the order Laminariales.

Deviations from this rigid scheme seem to exist; for example, Nakahara & Yamada (1974) reported that a few unfertilized eggs of *Agarum cribrosum* developed into normal-looking sporophytes. The biological significance of these 'sporophytes' is not known.

The very precise knowledge obtained recently on the temperature and light demands for growth and fertilization of laminarian gametophytes (cf. e.g. Lüning & Neushul 1978) has made it possible to make crossing experiments between populations. Such crossing experiments are important with regard to our understanding of the process of speciation in the sea and to our knowledge of individual taxa. Thus Nakahara & Yamada (1974) showed interfertility between four local forms of *Agarum cribrosum*. In earlier literature several species of *Laminaria* (section Simplices) have been reported from the North Atlantic, separated by such characters as the presence or absence of mucilage ducts, hollow stipes, and bullose blades. These various taxa, except *L. solidungula*, are now generally reduced to one or two species, *L. saccharina* and *L. longicruris*. Lüning *et al.* (1978) showed interfertility between *L. longicruris* from Canada with *L. saccharina* in Europe, but the F_1 generation was not followed to fertility. It is rather surprising that no reproductive isolating mechanisms have evolved in this case, which seems to be an example of allopatric speciation in benthic, marine algae with the Gulf Stream as a strong barrier.

Similar life histories exist in the orders Sporochnales and Desmarestiales. The gametophytes are either monoecious or dioecious, but there is not cytological evidence to substantiate that dioecism is genetically determined. In *Sporochnus pedunculatus* the possible determination must be vague as the gametophytes can carry either oogonia or antheridia or both depending on the geographic origin of the species. Thus plants from the English Channel seem to be always monoecious and plants from the Mediterranean always dioecious (Caram 1965).

Heteromorphic life histories are also common in the order Chordariales, which contains plants with haplostichous erect thalli. The basic pattern of the life history is in agreement with that of *Saccorhiza polyschides*, but the diploid generation is re-established by isogamous/anisogamous copulation between gametes from plurilocular gametangia on the gametophytes. This type of life history has been shown for several species of Chordariales, in some cases supported by cytological data, e.g. in *Sphaerotrichia divaricata* by Ajisaka & Umezaki (1978). The gametophytes were clearly streblonematoid and carried uniseriate, plurilocular sporangia and hairs. Fertility and sexual fusion (isogamous or anisogamous) depended on environmental conditions.

Although many chordarialean species have only unilocular sporangia on the macrothalli, some have both uni- and plurilocular sporangia. The fate of the swarmers from the plurilocular sporangia on macrothalli of *Cladosiphon okamuranus* was reported by Shinmura (1974). They germinated into *Myrionema*-like discs which later carried plurilocular sporangia. New macrothalli developed from these discs under certain culture conditions, but under other more unfavourable environmental conditions the microthalli existed alone. In all other respects *C. okamuranus* showed a typical chordarialean life history (Shinmura 1977) which in this case, was supported by cytological data showing eight and 16 chromosomes in gametophytes and sporophytes, respectively. The basic pattern in chordarialean life histories certainly has several modifications. For example, Kornmann (1962b) has shown that new macrothalli of *Chordaria flagelliformis* develop directly from swarmers of the unilocular sporangia. Kornmann also observed relatively few microthalli in the cultures. These became extensively fertile at low temperatures, but even though new macrothalli developed, copulation between swarmers was never observed. He, therefore, concluded that the life history of *Chordaria flagelliformis* was a heteromorphic, monophasic alternation of generations and that the 'gametophytes' reported by Caram (1955) represented contaminants.

This heteromorphic, monophasic type of life history seems to be rather common among brown algae, especially in the less advanced orders, Dictyosiphonales, Chordariales and Scytosiphonales. The assumption that meiosis invariably occurs in the unilocular sporangium has resulted in rather speculative interpretations of culture results. Caram (1965) suggested that the swarmers from the unilocular sporangia of the chordarialean species *Sauvageaugloia chordariaeformis* copulated, in an attempt to explain why these swarmers developed directly into new macrothalli. Later Caram (1972) listed a number of species belonging to different orders in which this phenomenon seemingly took place or was reported to do so. This was seriously questioned by Müller (1975), and I must share his scepticism and agree with him that cytological data are imperative to support such reports. I am quite sure that many such reports have little foundation.

Asensi (1978) reported on an unusual life history in *Chordaria linearis*. The swarmers from the unilocular sporangia germinated into uniseriate, prostrate systems from which new erect thalli directly developed. These erect thalli were initially uniseriate with a basally sheathed, terminal true hair but later became pluriseriate by longitudinal divisions, mainly in one plane, to form flat parenchymatous thalli. Sporangia were not observed on the thalli developed in culture. This report must, however, be a regretable misunderstanding. *Chordaria linearis* is a true chordarialean species and, therefore, a satisfactory explanation must be found to explain the alleged metamorphosis of a parenchyma into uniseriate filaments. This seems, however, impossible.

In the order Dictyosiphonales heteromorphic life histories have been known for a long time. Sauvageau (1917) showed that small plants with plurilocular sporangia (microthalli) developed from swarmers of unilocular sporangia on the macrothallus of *Dictyosiphon foeniculaceus,* these initial results being later substantiated by him (Sauvageau 1929). Probably with his earlier observations on the life history of *Saccorhiza* and his find of gametophytes in the Sporochnales in mind, his interpretation of the *Dictyosiphon* microthalli as gametophytes was to be expected. Sauvageau, however, never observed copulation between swarmers from the supposed gametophytes. In fact the swarmers 'se meuvent pendant une heure environ sans paraître s'attirer mutuellement'. That copulation had taken place in some cases was concluded by the presence of 'zygotes', cells with two chloroplasts and two eye-spots. The presence of such cells may, of course, indicate that copulation has occurred, but it is insufficient evidence, as it is well-known that some swarmers liberated from a sporangium may only be partially cleaved. Such incompletely separated cells containing two chloroplasts and two eye-spots may round off and form embryospores as for normal swarmers.

The interpretation that the small plants with plurilocular sporangia in the life history of *Dictyosiphon foeniculaceus* represented gametophytes, which by isogamous copulation of the swarmers formed the sporophytic generation, was indirectly supported by the observations of Knight (1929) on *Ectocarpus siliculosus*, who showed cytologically that meiosis took place in unilocular sporangia. The observations of Sauvageau and Knight are probably the main reason why it has never been questioned that the basis for the heteromorphic life history was a change in the ploidy level. This is very important as this type of life history is one of the characters which separate the Dictyosiphonales from other parenchymatous brown algae and has resulted in a placing of this order close to very specialized orders, e.g. Laminariales, in the taxonomic hierarchy. Subsequent culture studies have not clarified this matter. If microthalli are present at all in the life history of the species studied, they may be physically separated from the erect phase (Fig. 5.3A). This separation may be a question of time and/or environmental conditions, but in some cases the separation seems genetically determined and the new erect phase develops only on prostrate systems derived from swarmers of plurilocular sporangia (Fig. 5.3B). Even in this situation the microthalli do not seem to function as gametophytes, but merely as a phase which is able to both survive unfavourable conditions and rapidly reproduce by sporangia formation when suitable conditions return. Gametophytes are stated to be absent in *Dictyosiphon chordaria*, a species rather similar to *Dictyosiphon foeniculaceus*. Føyn (1934) showed that the swarmers from the unilocular sporangia developed into sterile, uniseriate, branched prostrate systems from which new parenchymatous thalli are developed. This was repeated through four successive generations. This study is, however, inconclusive as it may well be that the culture conditions used were just not suitable for the development of sporangia on the microthalli.

The above mentioned examples show that very accurate information on sexual reproduction, preferably supported by cytological data, must be present before a species is placed among other algae with this type of life history. Further investigations will probably show that several species in fact have a monophasic type of life history, which is discussed in the following section.

2.3 Direct, monophasic life histories

Absence of sexual reproduction (i.e. direct development of the swarmers) separates this type of life history from the biphasic type. The life history may consist of successive generations of identical plants or it may be heteromorphic,

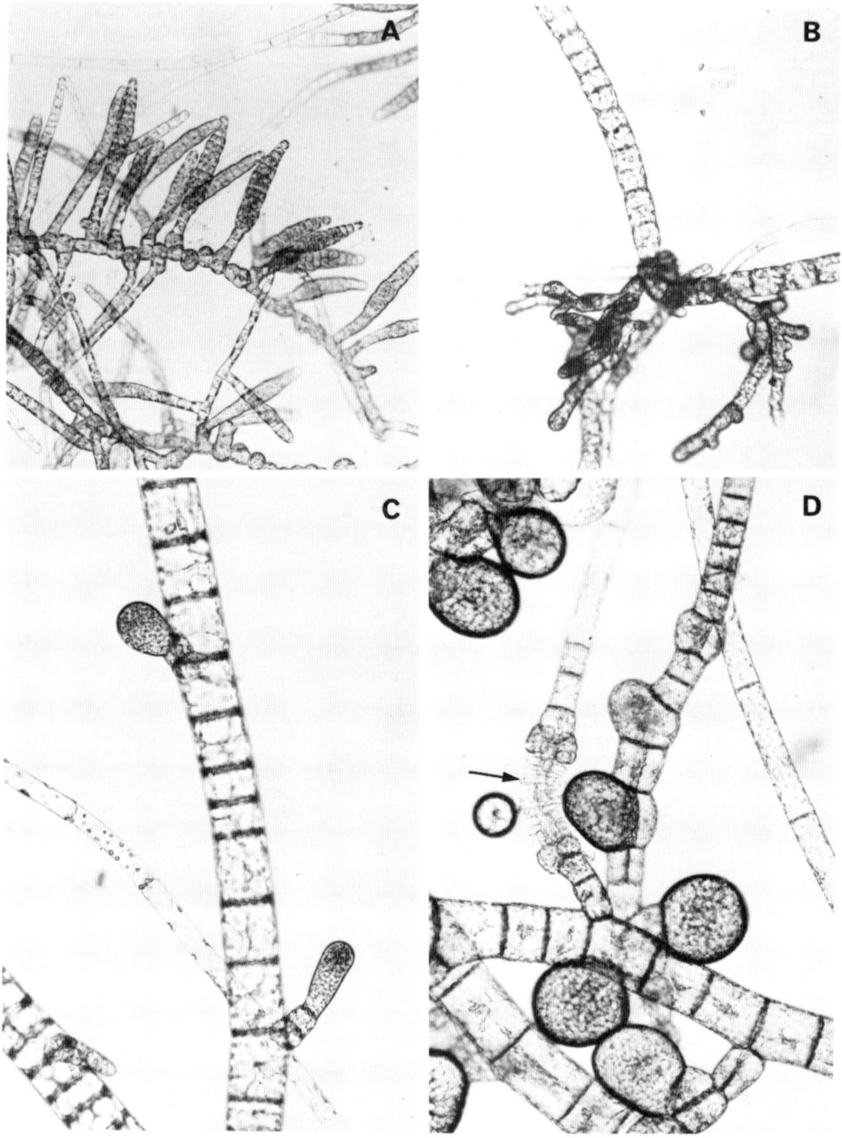

Fig. 5.3 Heteromorphic (A, B) and isomorphic (C, D) life histories. (A) Microthallus of *Asperococcus fistulosus*; note the hecatonematoid morphology of the plant (× 150). (B) Young, erect filament of *Delamarea attenuata* from sterile, prostrate system (× 150). (C) Sporophyte of *Haplospora globosa* (× 150). (D) The gametophyte of *Haplospora globosa* with empty antheridia (arrow) and monospores (× 150).

i.e. microthalli, which have the same ploidy level as the parent generation, are included.

This type of life history is very common among microthallus-like species which may carry either plurilocular or unilocular sporangia. The swarmers from these sporangia develop directly into plants similar to the parent generation, e.g. in *Mikrosyphar* species and in *Compsonema saxicola* (see p. 214). The reality of the absence of microthalli in plants forming erect thalli may of course always be questioned, but it is a fact that many members of the Ectocarpales have only one type of sporangium, often plurilocular, on the erect parts. The embryospores formed by the swarmers from these sporangia germinate and grow into sterile prostrate systems from which identical erect thalli later develop. In principle, this may continue in an endless series of generations if the environmental conditions are continuously favourable. This type of life history occurs in many species of *Giffordia,* species of *Acinetospora, Polytretus reinboldii, Kuckuckia spinosa* and *Sorocarpus micromorus.*

Whether these species are haploid or diploid has not been investigated to any considerable extent. Chromosomes have been counted in *Sorocarpus micromorus* by Pedersen (1974), and these counts showed nine chromosomes, a number quite close to the commonly accepted haploid number of eight in the Ectocarpaceae (Cole 1967). *Sorocarpus micromorus* is therefore considered to be a haploid, asexual species in the Atlantic, and it seems to follow a similar pattern in Australia (Clayton 1974). The only report of unilocular sporangia in this species (Takamatsu 1936) obviously needs confirmation as these sporangia have not been observed subsequently.

With regard to the presence of unilocular sporangia a similar situation seems to exist in *Polytretus reinboldii.* Unilocular sporangia were not observed on Danish material in culture (Pedersen 1977), and some cells which resembled unilocular sporangia in material from nature were interpreted as ascocysts by Rosenvinge & Lund (1941). Similar cells were reported by Kurogi (1978) who illustrated cells with a homogenous content, some of them empty, but this author was also uncertain about their nature. According to Jenneborg (1977) the formation of these cells is probably caused by the phycomycete *Eurychasma dicksonii.*

The widely distributed alga *Elachista fucicola* is an example of a species with a direct type of life history, although an alternation between gametophytes and sporophytes could be expected by the presence of only unilocular sporangia at the bases of the assimilating filaments. Kylin (1937) showed that the swarmers from these sporangia developed directly into new macrothalli. These results were explained by the absence of meiosis in the unilocular sporangia, but the macrothalli were still assumed to be diploid. Kornmann (1962a) showed that plurilocular sporangia developed on the prostrate systems at low temperature, and this report was later confirmed by Koeman & Cortel-Breeman (1976) working on Dutch material. The last mentioned authors supported their observations with cytological data and showed that all developmental stages had the same chromosome number (20–24) except for a few germlings which developed from a unilocular sporangium and were found to have half the number present in the cells of their parent plants. Similar results were obtained by Pedersen (1979a) on Greenland strains. These strains also had plurilocular sporangia on the prostrate systems, but these sporangia were morphologically different from those of the European strains. With regard to development and fertility, the Greenland strains developed only macrothalli and unilocular sporangia at low temperature (4° C) which was exactly the opposite to that recorded for European strains. This difference was expected as the Greenland plants are adapted to constant low temperatures. The results further suggested a physiological difference which later was shown to be genetically determined. The microthalli and the macrothalli were physically connected, and all developmental stages had about eight chromosomes. Based on these studies it seems very unlikely that the microthalli have a sexual function in the life history of *Elachista fucicola,* and it is also premature to state that a plant is diploid because it carries unilocular sporangia.

On the other hand studies on life histories and chromosomes of algae over their geographical range open new aspects in our understanding of individual species. Data are accumulating which suggest that a brown alga is not always haploid or diploid or shifts between the two ploidy levels, but strains of higher ploidy levels may also occur. The Dictyosiphonalean species *Isthmoplea sphaerophora* shows a similar variation in chromosome number as *Elachista fucicola* within its distribution area. The plants from the Oslofjord, Norway, which must be considered as a marginal area of its distribution, had 24 chromosomes (Rueness 1974), while plants from Greenland, where it occurs commonly, had only eight chromosomes (Pedersen 1975). Therefore, a species must not be considered as fully understood when it is studied in only one part of its distribution area. The cosmopolitan species especially must consist of a wide range of physiological or genetic strains as they are adapted to a wide range of environmental conditions.

The monophasic, direct type of life history is also very common in the Dictyosiphonales, and swarmers from both unilocular and plurilocular sporangia may show this type of development. *Punctaria plantaginea* has only unilocular sporangia, formed by subdivisions of surface cells of the parenchymatous thallus. One strain from Greenland was studied in culture. Liberation of swarmers from the unilocular sporangia was often observed but no sexual activity occurred and they germinated directly into uniseriate, branched, prostrate systems which were sterile under all conditions tested. New erect thalli, again bearing unilocular sporangia, developed directly from the prostrate systems. The cytological basis for this life history is still unknown.

Desmotrichum undulatum has only plurilocular sporangia and shows the same type of development in both the U.S.A. (Rhodes 1970) and Denmark (Pedersen unpublished).The swarmers from the plurilocular sporangia on the parenchymatous thalli germinate into prostrate systems which later carry siliquose plurilocular sporangia. New erect thalli develop from these prostrate systems, in the Danish strain both at 4° C and 15° C, in the U.S.A. strain at 6° C, but not at 21° C. The fertile prostrate systems clearly resemble the genus *Hecatonema* (Ectocarpaceae) and other species related to *Desmotrichum,* e.g. *Punctaria latifolia* (Clayton & Ducker 1970), *Asperococcus fistulosus* and *A. turneri* (Knight *et al.* 1935) are also shown to have hecatonematoid, prostrate systems. The hecatonematoid phase of the above-named species may be found unconnected to a parenchymatous phase in nature and might therefore be identified as a separate species, e.g. *Hecatonema maculans.* The latter species, however, in some cases can have an independent existence (Clayton 1974), and this 'taxon' must, consequently, be considered composed of various genotypes which are indistinguishable on morphological characteristics. From a phylogenetic point of view, *Hecatonema* may be considered derived from extinct or living members of the Punctariaceae (including the Asperococcaceae) by natural selection of microthalli which have lost the ability to form erect, parenchymatous thalli. Therefore, the placing of *Hecatonema* in the Punctariaceae rather than the Ectocarpaceae would be more natural.

2.4 Isomorphic life histories

The classic example of this type of life history is found in the Dictyotales. The gametophytes carry oogonia and antheridia, often on separate identical plants. The resultant zygote develops into a sporophyte which is morphologically indistinguishable from the gametophyte in the sterile condition. At fertility the sporophyte carries tetrasporangia and each tetraspore, formed by meiotic divisions, develops into either a male or female gametophyte. Deviations from this cyclic pattern, however, do exist. Gaillard (1972) has for example observed that tetraspores from *Padina pavonica* develop directly into new sporophytes, suggesting that the tetraspores are formed by apomeiosis.

Ectocarpus siliculosus is another well-known example of a plant with an isomorphic alternation type life history. It is, however, not always a relatively simple alternation between a gametophytic and sporophytic generation with the added asexual development of the swarmers from plurilocular sporangia, as shown by Papenfuss (1935). Müller (1967) has shown, based on extensive studies on this species, that its life history is much more complex, including male and female haploid sporophytes (asexual plants bearing a sex factor). These plants carry unilocular sporangia and/or plurilocular sporangia. Tetraploid plants with both types of sporangia are also known. This complex life history represents, of course, the total potential development of the species, but this is not necessarily realized in all parts of its widespread geographical distribution. In Greenland, for example, *Ectocarpus siliculosus* has, in nature, only plurilocular sporangia.

Under culture conditions which correspond to the summer conditions in Greenland, the life history is direct as the swarmers from the plurilocular sporangia germinate into new plants with similar reproductive structures. It is unknown whether these plants are gametophytes, sporophytes, or plants at a higher ploidy level.

An advanced type of isomorphic alternation is possibly present in the genera *Haplospora* and *Tilopteris* (Tilopteridales). These genera carry plurilocular sporangia and/or monosporangia, the latter a reproductive structure which seldom occurs in the brown algae. South (1975) has proposed the term parthenospores for the monospores in *Tilopteris,* meiospores for the quadrinucleate spores in *Haplospora*. It was suspected before the turn of the century that plants of *Tilopteris* and *Haplospora* with both plurilocular sporangia and monosporangia represented gametophytes (Fig. 5.3D), while plants with only monosporangia (quadrinucleate in *Haplospora*) were sporophytes (Fig. 5.3C) formed by oogamy. This assumption has, however, been difficult to support, especially for *Tilopteris* (cf. Dammann 1930; South 1972). The observations by Sundene (1966) on *Haplospora* seem to support the original assumption, but the oogamy is described in rather vague terms: 'zoids were liberated and a great many of them were observed congregating around the nonmotile cell, an indication that they were attracted by the latter'. This observation obviously needs cytological support.

2.5 Scytosiphonalean life histories

In the heteromorphic, biphasic life histories previously described the gametophytic generation is reduced in size compared with the sporophytic generation. Among the Scytosiphonales where sexuality is reported the gametophytes are large plants, whereas the sporophytes are reduced. It has been known for some years that a crustose phase is included in the life history of some Scytosiphonales. The role of the so-called ralfsioid stages in the life history seems to vary, likewise the influence of environmental conditions, especially the photoperiod, on the formation of erect thalli. This will be exemplified in the following discussion.

The cosmopolitan species *Scytosiphon lomentaria* has been studied extensively in culture from various parts of its distributional range. The results indicate that a very plastic life history is present with the occurrence of a wide range of genotypes which may be given taxonomic status at the subspecies or variety level.

Asexual strains

Many authors report that swarmers from the plurilocular sporangia act as zoospores and germinate into prostrate systems with quite variable morphologies. The prostrate systems are described as filamentous, 'knot' filamentous, ralfsioid, *Microspongium*-like (Lund 1966), and *Streblonema–Compsonema*-like

(Loiseaux 1970a). Transitions from filamentous to 'knot' filamentous phases occur depending on the degree of contact with the substratum (Kristiansen & Pedersen 1979). This has also been observed in *Petalonia fascia* by Edelstein *et al.* (1970). The *Ralfsia*-like crusts are often sterile in culture, but they may become fertile with unilocular sporangia, thus resembling *Ralfsia clavata* or *Microspongium gelatinosum.* Clayton (1978) has reported cushion-shaped microthalli with plurilocular sporangia. Whether fertility or sterility occurs seems to depend on the strains studied. Wynne (1969) has reported that only one strain of *Scytosiphon lomentaria* had fertile crusts developed from cylindrical thalli. Also only one strain in a study by Kristiansen & Pedersen (1979) had fertile crusts, although 48 strains were studied and many of these were cultivated under similar conditions. The crusts with unilocular sporangia developed from the seventh generation of cylindrical thalli under constant culturing conditions (15° C, 16-h photoperiods). Another strain had very small, lumpy, prostrate systems which never developed into crusts or 'knot' filaments in any conditions of photoperiod and temperature. *Compsonema–Streblonema*-like phases (Loiseaux 1970a) were not observed in the Danish strains of *Scytosiphon,* and the life history of *Compsonema saxicola* (see p. 214) did not include an erect parenchymatous phase.

Based on nomenclatural considerations and studies on *Microspongium globosum* in culture (Pedersen unpublished), Kristiansen & Pedersen (1979) proposed to separate *Microspongium gelatinosum sensu* Reinke (1889, cf. pl. 8) with unilocular sporangia from *M. gelatinosum* (Reinke 1888) in its original sense as a plant with only plurilocular sporangia. Therefore, crustose plants of *Scytosiphon lomentaria,* previously identified as *M. gelatinosum,* may now be called *Ralfsia*-like in all parts of its distribution area.

Formation of erect thalli depends in some cases on environmental conditions. Wynne (1969) showed that photoperiod and temperature played a role, and Dring & Lüning (1975) demonstrated a true photoperiodic response in that erect thalli developed only under SD conditions, the response being inhibited by a short light break in the dark period. The Danish strains showed no response to daylength as erect thalli were formed under both cold (4° C) and warm (15° C) LD conditions as well as under cold (*c.* 5° C) SD conditions. Similar results were obtained by Clayton (1978). These results are in agreement with the seasonal occurrence of cylindrical thalli in nature in Denmark.

A further complication in our understanding of *Scytosiphon lomentaria* is the occurrence of forms with complanate, erect thalli. These thalli may resemble a narrow *Petalonia*, but they are hollow as in *Scytosiphon.* Clayton (1976) studied such plants from Australia in culture and found that the F_1 generation of erect thalli from wild, complanate thalli was either cylindrical or complanate, but how is it possible to distinguish genetically or phenetically flat thalli in nature?

The complanate variety of *Scytosiphon lomentaria* was first described by Rosenvinge (1893). Observations on cultures of this form from Greenland have not shown cylindrical thalli in the life history under cold, LD and SD conditions, nor under warm, LD conditions, and the absence of paraphyses seems to be a reliable character, which separates this plant from phenetically flat plants. The prostrate systems are 'knot' filamentous or ralfsioid but have remained sterile.

Sexual strains

Nakamura & Tatewaki (1975) confirmed Tatewaki's (1966) earlier report of a heteromorphic life history for *Scytosiphon lomentaria* in Japan. The diploid phase was *Ralfsia*-like and bore unilocular sporangia, but similar haploid plants also occurred. The cylindrical thalli were heterothallic. These authors also reported on sexual reproduction in other members of the Scytosiphonales, namely, *Petalonia zosterifolia, Colpomenia bullosa* and *Endarachne binghamiae*.

Recently Clayton (1979) reported on sexuality in *Colpomenia peregrina* from Australia. The number of functional gametophytes found in nature was extremely low, and they seemed to be seasonally distributed. The zygotes were formed by anisogamous fusion, and the zygotes and parthenogametes developed into cushion-shaped diploid or haploid plants, respectively, bearing unilocular sporangia. Several strains, however, reproduced new saccate thalli asexually.

3 MORPHOLOGICAL PLASTICITY—PLEIOMORPHISM

3.1 Introduction

During the last 20 years, evidence has accumulated to show that environmental conditions play a much greater role in influencing morphological expression of freshwater and marine algae than previously believed. This certainly equally applies to the marine brown algae as improved culture techniques and facilities have made it possible to study the effects of combined environmental factors on both the morphology and development of plants. The results obtained by the study of unialgal cultures must preferably be compared with observations in nature, because results from culture studies can always be refuted as being products of artificial conditions.

In certain geographic areas the marine vegetation has been thoroughly investigated throughout the last 200 years. In such areas it is often sufficient to compare the results from culture with those available in the literature, which can contain valuable information on the morphology, reproduction, and seasonality of the various species.

Investigations of species whose life histories are unknown, are still important but should no longer be considered the only purpose of the investigation. The study may, with little further effort, be transformed into a combined ecological/biological study. Studies of this type have shown, especially within the last ten years, that taxa at various levels in the traditional taxonomic hierarchy, even belonging to different orders, may be developmental stages or morphological variations of the same species. Such studies may be the gateway to a better understanding of the phylogenetic pathways in the brown algae and, consequently, to a more natural taxonomic system. Some examples of morphological plasticity in the brown algae governed by environmental conditions are presented below.

3.2 Feldmannia-Acinetospora

The genus *Feldmannia* was erected and placed in the Ectocarpaceae by Hamel (1939) to receive plants with erect filaments (*Ectocarpus*-like) possessing a basal meristem and hypomeristematic branching. *Acinetospora* is an older genus erected by Bornet (1891). This genus is sometimes placed (Fritsch 1945) in the Tilopteridales close to *Tilopteris* and *Haplospora* obviously due to the presence of monospores, which are of rare occurrence within the brown algae.

It was already suspected by Buffham (1893) that *Acinetospora* might be the sporophyte of *Feldmannia*. This opinion was supported by Kornmann (1953) based on culture studies of *Acinetospora*. The gametophytes obtained were identifiable as both *Feldmannia lebelii* and *F. padinae*. This report was, however, not supported by cytological data.

Knoepffler-Péguy (1972, 1977) also showed a connection between the two genera, this being a function of changes in the environmental conditions rather than reflecting a change in ploidy level. She studied plants from Sweden and the Mediterranean, and the plants passed into *Acinetospora*-morphology during the cold season and back into *Feldmannia*-morphology during the warm season. These changes were not related to changes in the ploidy level as reported by Kornmann. The temperature, therefore, seems decisive in determining the morphology of the species in question, although this factor often acts in combination with salinity in controlling a combination of traits. Knoepffler-Péguy concluded that it is no longer possible to maintain *Acinetospora* and *Feldmannia* as distinct genera.

3.3 Giffordia ovata

Giffordia ovata is an example of a species which shows considerable phenotypic plasticity depending on environmental conditions. Several varieties of this species have been described, and one of these, *Ectocarpus ovatus* var. *intermedius*, was later given specific rank by Lund (1959).

The life history of *G. ovata* (as *G. fuscata*) was described by Kornmann (1954). He reported a monophasic, heteromorphic life history, and this was, in most details, confirmed by Pedersen (1979b) on material from Greenland. The strain from Greenland was additionally exposed to various sets of culture conditions to investigate the effects of temperature, salinity, light intensity, and photoperiod on the development and morphology. Temperatures between 4° C and 15° C (no intermediate temperatures were investigated) had no effect on the development and morphology of the erect thalli. The thalli had the morphologic characteristics of the species with opposite branches, or a branch opposite a sporangium, or opposite sporangia scattered along the axis at shorter or longer intervals. The age of the plants, however, strongly influenced their morphology. The plants with the above described morphology looked different two weeks later, all or nearly all cells of the axes being then beset with sporangia, often in whorls of four.

Reduced light intensity played a role under cold, LD conditions on the morphology of the erect filaments. These were now short, almost unbranched, and the sporangia often occurred singly. Such filaments resemble *Ectocarpus holmii* Rosenvinge (1893), later reduced to a form of *E. ovatus* by Rosenvinge (1898). The suboptimal light intensity also resulted in fertile, prostrate systems.

In crowded cultures, probably due to nutritional deficiency, the plants responded morphologically by forming pseudohairs. Apical or subapical cells of the laterals became meristematic and formed long, hyaline cells outwards. The laterals thus consisted of a few normal cells at the base, a meristem, and long hyaline cells, sometimes ending in vegetative cells depending on where the meristem was inserted. Plants with such laterals are in agreement with *Giffordia intermedia,* and there is consequently no reason to maintain the two taxa as distinct.

The width of the erect filaments, and length and width of the plurilocular sporangia are features which have been used to distinguish varieties. These characters also show considerable variability in the strain from Greenland and overlap the plants from Scandinavia from which the plants from Greenland have been separated as a variety.

The salinity also played an important role in the development of *G. ovata.* Plants cultivated under cold, LD conditions at 10‰ lost their viability, and erect filaments and sporangia never developed. This factor is, therefore, decisive in relation to the geographical distribution.

3.4 The basal systems of Litosiphon pusillus

Litosiphon pusillus (Carm. ex Hook.) Harv. has been extensively studied in culture from various parts of its distributional range by Sauvageau (1933), Kylin (1933), Dangeard (1965) and Nygren (1975). It is, therefore, well-known that it has fertile, prostrate systems (microthalli) with one type of plurilocular sporangium (Nygren 1975) or two types of plurilocular sporangia together with unilocular sporangia (Sauvageau 1933). A Danish strain (Pedersen unpublished) has a similar stage in its life history which forms under some culture conditions, plurilocular sporangia with large loculi (type 1) (Fig. 5.4A) and, under other conditions, sporangia with small loculi (type 2) (Fig. 5.4B, C) and oval to oblong unilocular sporangia. Under optimal culture conditions physical connections are present between the macrothalli and the microthalli.

The microthalli may exist alone under various suboptimal conditions. Their morphology varies depending on the conditions but can be characterized as streblonematoid. Using reduced salinity, for example 10‰ instead of 30‰, or reduced illuminance, 700 lx instead of *c.* 2500 lx, the formation of erect thalli is inhibited, but the reproductive potential of the microthalli remains unaffected.

Some drastic changes occur when the cultures are exposed to low salinity (10‰) combined with low illuminance (700 lx). Erect thalli occur rarely, and if present only as short, uniseriate filaments with terminal hairs. Unilocular sporangia and plurilocular sporangia (type 1) are not formed on the microthalli. The plurilocular sporangia (type 2) remain unaffected (Fig. 5.4D, E) with regard to frequency but a reduction in size takes place, this being more enhanced if the light intensity is further reduced.

Litosiphon pusillus is able to grow in culture at salinities as low as 5‰ but exists only as microthalli with plurilocular sporangia of type 2. Such thalli are, of course, not identified as *Litosiphon pusillus* but merely as a species of *Streblonema*. It is, in fact possible to find descriptions of plants in the literature which are probably microthalli of *Litosiphon pusillus*. *Streblonema thuretii* Sauvageau, *S. volubile* (Crouan frat.) Thur. in Le Jolis and *S. danicum* Kylin are all species which have unilocular sporangia with a morphology similar to those of *Litosiphon* microthalli. When the microthalli only carry plurilocular sporangia type 2, these may be identified as *Streblonema oligosporum* Strömfelt, a species described from the Baltic.

The observations on plants in culture are in agreement with observations in nature. The three first mentioned species of *Streblonema* are often found as endophytes, growing in a habitat where competition for light occurs, and this probably has a similar effect on the erect thalli as described above. In Denmark the erect thalli of *Litosiphon pusillus* are only reported from areas with relatively high salinities (more than 10‰) and disappear in areas with constant low salinities, i.e., towards the Baltic. Based on the results of the culture studies it is concluded that *Litosiphon pusillus* is able to penetrate far into the Baltic as microthalli with small plurilocular sporangia of type 2. These microthalli are in agreement with *Streblonema oligosporum* which occurs in the Öregrund archipelago (Wærn 1952) near Stockholm, Sweden. It would be very interesting to cultivate *S. oligosporum* from this area in order to investigate the fate of the swarmers from the plurilocular sporangia. Do they immediately develop into microthalli from which macrothalli develop under optimal culture conditions (as I am inclined to expect) or has there since the glaciation been a selection of plants which are unable to form erect thalli?

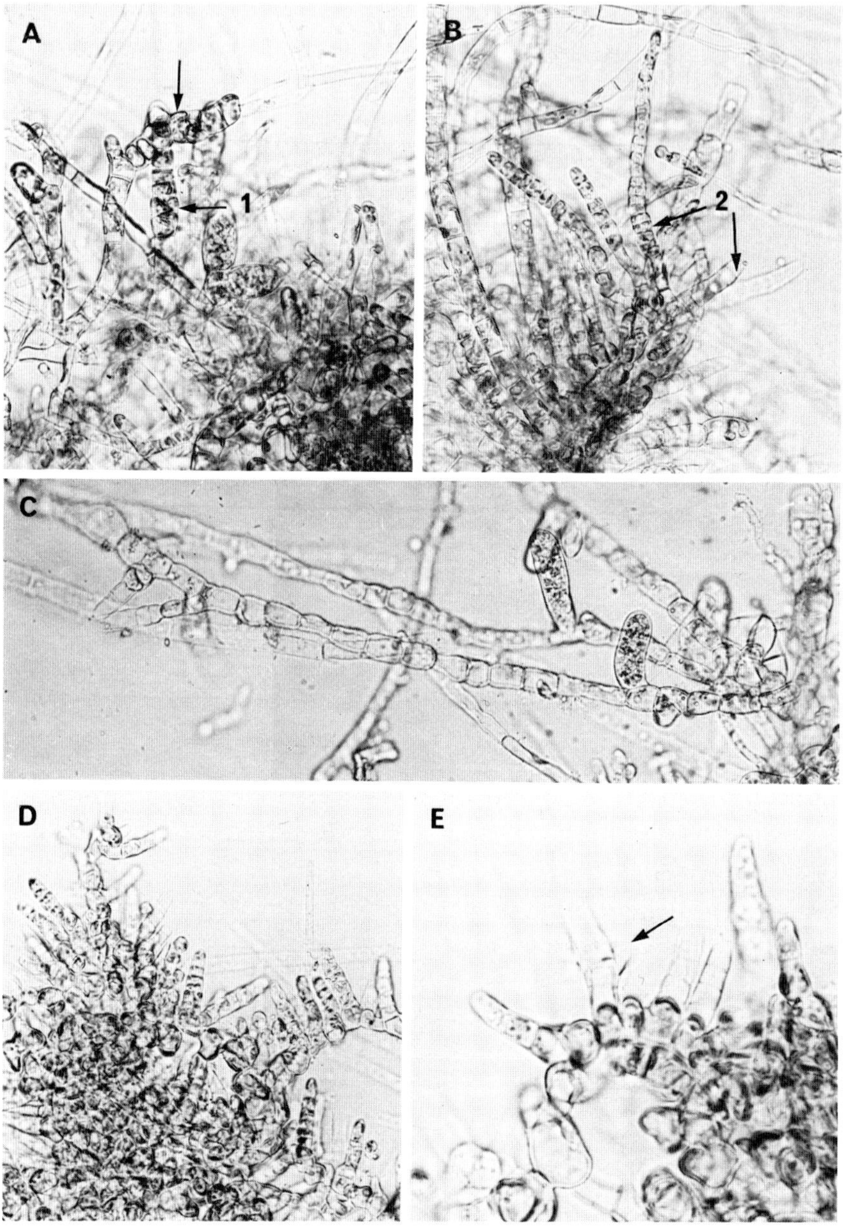

Fig. 5.4 Microthalli of *Litosiphon pusillus* under various culture conditions. (A) Plurilocular sporangium type 1 (arrow 1) with large loculi. The plant also carries unilocular sporangia. Note that the type 1 sporangium has loculi of about the same size as the loculi in the fertile, juvenile, erect thallus (arrow) (× 300). (B) Mature and empty plurilocular sporangia type 2 (arrow, 2) with smaller loculi (× 300). (C) Empty plurilocular sporangia type 2 on the same filament as a sessile, unilocular sporangium (× 300). (D) Disc-shaped plant with plurilocular sporangia type 2, exclusively (× 300). (E) Detail of plant with empty plurilocular sporangia type 2, one is branched (arrow) (× 600).

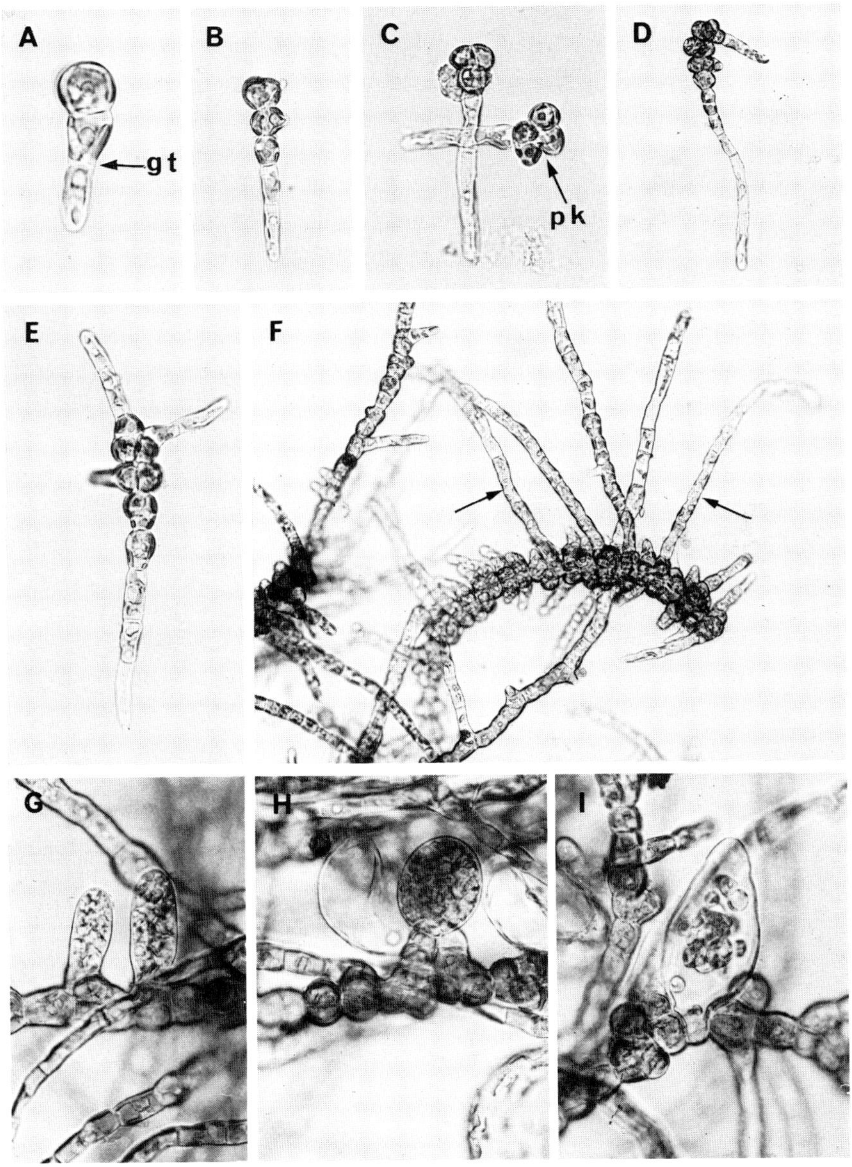

Fig. 5.5 *Compsonema saxicola* in culture under cold SD conditions. (A) Unipolar germination with formation of a germination tube (gt) (×600). (B, C) Further development of the germling with longitudinal divisions in the oldest part forming a parenchymatous knot (pk). The germination tube develops into a rhizoidal filament (×300). (D, E) Development of laterals from parenchymatous parts (×150, ×300). (F) Well-developed parenchymatous part with uniserate laterals (arrows) (×150). (G–I) Development of unilocular sporangia from parenchymatous parts of the thalli. The sporangia are sessile (G, I) or terminal on a short stalk (H) (×300).

3.5 The life history of Compsonema saxicola

Compsonema saxicola (Kuck.) Kuck. was described as *Myrionema saxicola* (?) by Kuckuck (1897) and transferred to *Compsonema* Kuckuck (1899) by Kuckuck (1953). Obviously due to difficulties in placing the species at the generic level, it has been placed in two different families in the Ectocarpales, initially in the Myrionemataceae by Kuckuck (1899), as followed by Parke & Dixon (1976), later in the Ectocarpaceae close to *Protectocarpus speciosus* (Børg.) Kuck. by Kuckuck (1953).

Only unilocular sporangia are found on *Compsonema saxicola* in nature. The life history of the species is unknown, but a monophasic, direct type is suspected. The very similar species *Compsonema sporangiiferum* Setchell & Gardner shows connection with a minute *Scytosiphon* (Loiseaux 1970a) which made it further necessary to study *Compsonema saxicola*.

Danish material of *Compsonema saxicola* was studied in culture under various conditions. The swarmers from the unilocular sporangia round up after settling, and the embryospores germinate unipolarly and form unbranched germlings. The uniseriate stage is ephemeral as longitudinal divisions occur in the oldest cells and a few-celled parenchymatous knot (pk) is formed while the youngest cell of the germination tube (gt) develops into a long-celled rhizoidal filament (Fig. 5.5A–C). New, uniseriate laterals develop from the parenchymatous part while longitudinal divisions occur continuously and form larger parenchymatous parts (Fig. 5.5D–F).

Under cold, SD conditions further generations develop unilocular sporangia extensively, but these are also formed under cold, LD conditions. The sporangia always occur on the parenchymatous part of the plants like in nature, and they are either sessile or terminal on a short stalk (Fig. 5.5G–I). This monophasic, direct type of life history has been repeated through six generations without formation of plurilocular sporangia and erect thalli.

The placing of *Compsonema saxicola* in either the Myrionemataceae or in the Ectocarpaceae is in my opinion rather unnatural. The parenchymatous nature of the plants excludes placing it in either family, but this characteristic together with the presence of only one chloroplast per cell suggests its being placed in the Scytosiphonaceae.

ACKNOWLEDGEMENTS

The unpublished results referred to have been financially supported by the Danish Natural Science Research Council and the University of Copenhagen. Tyge Christensen has kindly read the manuscript and has suggested very useful improvements. Lene Christiansen is thanked for technical assistance.

4 REFERENCES

AJISAKA T. & UMEZAKI I. (1978) The life history of *Sphaerotrichia divaricata* (Ag.) Kylin (Phaeophyta, Chordariales) in culture. *Jap. J. Phycol.*, **26**, 53–9.

ASENSI A. O. (1978) Observaciones sobre *Chordaria linearis* (Hook. et Harv.) Cotton (Phaeophyta) y su desarrollo en cultivo. *Darwiniana*, **21**, 331–42.

BORNET E. (1891) Note sur quelques *Ectocarpus*. *Bull. Soc. Bot. Fr.*, **38**, 353–72.

BUFFHAM T. H. (1893) Algological notes. *Grevillea*, **21**, 86–93.

CARAM B. (1955) Sur l'alternance de générations chez *Chordaria flagelliformis*. *Bot. Tidsskr.*, **52**, 18–36.

CARAM B. (1965) Recherches sur la reproduction et le cycle sexué de quelques Phéophycées. *Vie Milieu*, **16**, 21–221.

CARAM B. (1972) Le cycle de reproduction des Phéophycées—Phéosporées et ses modifications. *Bull. Soc. Bot. Fr.*, Mém. **1972**, 151–60.

CLAYTON M. (1974) Studies on the development, life history and taxonomy of the Ectocarpales (Phaeophyta) in southern Australia. *Aust. J. Bot.*, **22**, 743–813.

CLAYTON M. (1976) The morphology, anatomy and life history of a complanate form of *Scytosiphon lomentaria* (Scytosiphonales, Phaeophyta) from southern Australia. *Mar. Biol.*, **38**, 201–8.

CLAYTON M. (1978) Morphological variation and life history in cylindrical forms of *Scytosiphon lomentaria* (Scytosiphonaceae: Phaeophyta) from southern Australia. *Mar. Biol.*, **47**, 349–57.

CLAYTON M. (1979) The life history and sexual reproduction of *Colpomenia peregrina* (Scytosiphonales, Phaeophyta) in Australia. *Br. Phycol. J.*, **14**, 1–10.

CLAYTON M. & DUCKER S. C. (1970) The life history of *Punctaria latifolia* Greville (Phaeophyta) in southern Australia. *Aust. J. Bot.*, **18**, 293–300.

COLE K. (1967) Chromosome numbers in the Phaeophyceae. *Can. J. Genet. Cytol.*, **9**, 519–30.

DAMMANN H. (1930) Entwicklungsgeschichtliche und cytologische Untersuchungen an helgoländer Meeresalgen. *Wiss. Meeresunters. Abt. Helgoland*, N. F., **18**(4), 1–36.

DANGEARD P. (1965) Sur le cycle évolutif de *Litosiphon pusillus* (Carm.) Harvey. *Botaniste*, **49**, 47–63.

DRING M. J. & LÜNING K. (1975) A photoperiodic response mediated by blue light in the brown alga *Scytosiphon lomentaria*. *Planta*, **125**, 25–32.

EDELSTEIN T., CHEN L. C-M. & MCLACHLAN J. (1970) The life cycle of *Ralfsia clavata* and *R. bornetii*. *Can. J. Bot.*, **48**, 527–31.

EVANS L. V. (1965) Cytological studies in the Laminariales. *Ann. Bot.*, N.S. **29**, 541–62.

FRITSCH F. E. (1945) *The Structure and Reproduction of the Algae*, vol. 2. Cambridge University Press, Cambridge.

FØYN B. R. (1934) Über den Lebenscyklus einiger Braunalgen. *Bergens Mus. Årb*, **1934**, 1–9.

GAILLARD J. (1972) Quelques remarques sur le cycle reproducteur des Dictyotales et sur ses variations. *Bull. Soc. Bot. Fr., Mém.* **1972**, 145–50.

HAMEL G. (1939) Sur la classification des Ectocarpales. *Bot. Notiser.*, **1939**, 65–70.

HOLLENBERG G. J. (1969) An account on the Ralfsiaceae (Phaeophyta) of California. *J. Phycol.*, **5**, 290–301.

JENNEBORG L. H. (1977) *Eurychasma*-infection of marine algae. Changes in algal morphology and taxonomical consequences. *Bot. Mar.* **20**, 499–507.

KNIGHT M. (1929) Studies in the Ectocarpaceae. II. The life history and cytology of *Ectocarpus siliculosus* Dillw. *Trans. Roy. Soc. Edinb.*, **56**, 307–32.

KNIGHT M., BLACKLER M. C. H. & PARKE M. W. (1935) Notes on the life-cycle of species of *Asperococcus*. *Trans. Liverpool Biol. Soc.*, **1935**, 79–97.

KNOEPFFLER-PÉGUY M. (1972) Comportement de deux espéces suedoises de *Feldmannia* cultivées en diverses conditions de temperature et de salinité. *Bull. Soc. Bot. Fr.* Mém., **1972**, 101–4.

KNOEPFFLER-PÉGUY M. (1977) Polymorphisme et environment chez les *Feldmannia* (Ectocarpacées). *Rev. Algol.*, N.S. **12**, 111–28.

KOEMAN R. P. T. & CORTEL-BREEMAN A. M. (1976) Observations on the life history of *Elachista fucicola* (Vell.) Aresch. (Phaeophyceae) in culture. *Phycologia*, **15**, 107–17.

KORNMANN P. (1953) Der Formenkreis von *Acinetospora crinata* (Carm.) nov. comb. *Helgol. Wiss. Meeresunters.*, **4**, 203–24.

KORNMANN P. (1954) *Giffordia fuscata* (Zan.) Kuck. nov. comb., eine Ectocarpacee mit heteromorphen, homophasischen Generationen. *Helgol. Wiss. Meeresunters.*, **5**, 41–52.

KORNMANN P. (1962a) Plurilokuläre Sporangien bei *Elachista fucicola*. *Helgol. Wiss. Meeresunters.*, **8**, 293–7.

KORNMANN P. (1962b) Die Entwicklung von *Chordaria flagelliformis*. *Helgol. Wiss. Meeresunters.*, **8**, 276–9.

KRISTIANSEN AA. & PEDERSEN P. M. (1979) Studies on the life history and seasonal variation of *Scytosiphon lomentaria* (Fucophyceae, Scytosiphonales). *Bot. Tidsskr.*, **74**, 31–56.

KUCKUCK P. (1897) Bemerkungen zur marinen Algenvegetation von Helgoland II. *Wiss. Meeresunters., Abt. Helgoland*, **2**, 373–400.

KUCKUCK P. (1899) Beiträge zur Kenntnis der Meeresalgen. 8. *Compsonema*, ein neues Genus der Phaeosporeen. *Wiss. Meeresunters., Abt. Helgoland*, **3**, 56–60.

KUCKUCK P. ed. KORNMANN P. (1953) In Ectocarpaceen-Studien I. *Hecatonema, Chilionema, Compsonema*. *Helgol. Wiss. Meeresunters.*, **4**, 316–52.

KUROGI M. (1978) The genus *Polytretus* (Ectocarpaceae, brown algae) in Japan. *J. Fac. Sci. Hokkaido Univ.*, Ser. V, **11**, 237–48.

KYLIN H. (1933) Über die Entwicklungsgeschichte der Phaeophyceen. *Acta Univ. Lund.*, **29**(7).

KYLIN H. (1934) Zur Kenntnis der Entwicklungsgeschichte einiger Phaeophyceen. *Acta Univ. Lund.*, **30**(9).

KYLIN H. (1937) Bemerkungen über die Entwicklungsgeschichte einiger Phaeophyceen. *Acta Univ. lund.*, **33**(1).

LOISEAUX S. (1968) Recherches sur les cycles de développement des Myrionématacées (Phéophycées) III, IV. *Rev. Gen. Bot.*, **75**, 295–318.

LOISEAUX S. (1970a) *Streblonema anomalum* S. et G. and *Compsonema sporangiiferum* S. et G., stages in the life history of a minute *Scytosiphon*. *Phycologia*, **9**, 185–91.

LOISEAUX S. (1970b) Notes on several Myrionemataceae from California using culture studies. *J. Phycol.*, **6**, 248–60.

LUND S. (1959) The marine algae of East Greenland I. Taxonomical part. *Meddr Grønland*, **156**(1), 1–247.

LUND S. (1966) On a sporangia-bearing microthallus of *Scytosiphon lomentaria* from nature. *Phycologia*, **6**, 67–78.

LÜNING K., CHAPMAN A. R. O. & MANN K. H. (1978) Crossing experiments in the non-digitate complex of *Laminaria* from both sides of the Atlantic. *Phycologia*, **17**, 293–8.

LÜNING K. & NEUSHUL M. (1978) Light and temperature demands for growth and reproduction of laminarian gametophytes in southern and central California. *Mar. Biol.*, **45**, 297–309.

MÜLLER D. G. (1967) Generationswechsel, Kernphasenwechsel und Sexualität der Braunalge *Ectocarpus siliculosus* im Kulturversuch. *Planta*, **75**, 39–54.

MÜLLER D. G. (1975) Experimental evidence against sexual fusions of spores from unilocular sporangia of *Ectocarpus siliculosus* (Phaeophyta). *Br. Phycol. J.*, **10**, 215–321.

NAKAHARA H. & YAMADA I. (1974) Crossing experiments between four local forms of *Agarum cribrosum* Bory (Phaeophyta) from Hokkaido, northern Japan. *J. Fac. Sci. Hokkaido Univ.*, Ser. V, **10**, 49–54.

NAKAMURA Y. & TATEWAKI M. (1975) The life histories of some species of the Scytosiphonales. *Scient. Pap. Inst. Algol. Res. Hokkaido Univ.*, **6**, 57–93.

NYGREN S. (1975) Life history of some Phaeophyceae from Sweden. *Bot. Mar.*, **18**, 131–41.

PAPENFUSS G. F. (1935) Alternations of generations in *Ectocarpus siliculosus*. *Bot. Gaz.*, **96**, 421–46.

PARKE M. & DIXON P. S. (1976) Check-list of British marine algae—third revision. *J. Mar. Biol. Ass. U.K.*, **56**, 527–94.

PEDERSEN P. M. (1974) The life history of *Sorocarpus micromorus* (Phaeophyceae, Ectocarpaceae) in culture. *Br. Phycol. J.*, **9**, 57–61.

PEDERSEN P. M. (1975) Culture studies on marine algae from West Greenland I. Chromosomal information relating to the life history of *Isthmoplea sphaerophora* (Phaeophyceae, Dictyosiphonales). *Br. Phycol. J.*, **10**, 165–8.

Pedersen P. M. (1977) *Polytretus reinboldii*, a rare brown alga in culture (Ectocarpales, Sorocarpaceae fam. nov.) *Bot. Notiser*, **130**, 35–40.

Pedersen P. M. (1978) Culture studies in the pleomorphic brown alga *Myriotrichia clavaeformis* (Dictyosiphonales, Myriotrichiaceae). *Norw. J. Bot.*, **25**, 281–91.

Pedersen P. M. (1979a) Culture studies on the brown algae *Halothrix lumbricalis* and *Elachista fucicola*. *Bot. Notiser*, **132**, 151–9.

Pedersen P. M. (1979b) Culture studies on marine algae from West Greenland IV. *Giffordia ovata* (Fucophyceae, Ectocarpales), life history and morphological expressions. *Bot. Tidsskr.*, **74**, 57–65.

Reinke J. (1888) Die braunen Algen (Fucaceen und Phaeosporeen) der Kieler Bucht. *Ber. Dtsch. Bot. Ges.*, **6**, 14–20.

Reinke J. (1889) *Atlas deutscher Meeresalgen* I. P. Parey, Berlin.

Rhodes R. G. (1970) Relation of temperature to development of the macrothallus of *Desmotrichum undulatum*. *J. Phycol.*, **6**, 312–4.

Rosenvinge L. K. (1893) Grønlands havalger. *Meddr. Grønland*, **3**, 765–981.

Rosenvinge L. K. (1898) Deuxième mémoire sur les algues marines du Groenland. *Meddr. Grønland*, **20**, 1–125.

Rosenvinge L. K. & Lund S. (1941) The marine algae of Denmark. II. Phaeophyceae. I. Ectocarpaceae and Acinetosporaceae. *Biol. Skr.*, **1(4)**, 1–79.

Rueness J. (1974) Life history in culture and chromosome number in *Isthmoplea sphaerophora* (Phaeophyceae) from southern Scandinavia. *Phycologia*, **13**, 323–8.

Russell G. (1973) The Phaeophyta: a synopsis of some recent developments. *Oceanog. Mar. Biol. Ann. Rev.*, **11**, 45–88.

Sauvageau C. (1915) Sur la sexualité heterogamique d'une Laminaire. *C. R. Acad. Sci., Paris*, **161**, 769–99.

Sauvageau C. (1917) Sur un nouveau type d'alternance des générations chez les algues brunes (*Dictyosiphon foeniculaceus*). *C. R. Acad. Sci., Paris*, **164**, 829–31.

Sauvageau C. (1929) Sur le développement de quelques Phéosporées I. *Dictyosiphon* Grev. *Bull. Stn Biol. Arcachon*, **26**, 253–64.

Sauvageau C. (1933) Sur quelques algues Phéosporées de Guéthary (Basses-Pyrénées). *Bull. Stn biol. Arcachon*, **30**, 1–128.

Scagel R. G. (1966) The Phaeophyceae in perspective. *Oceanog. Mar. Biol. Ann. Rev.*, **4**, 123–94.

Shinmura I. (1974) Studies on the cultivation of an edible brown alga, *Cladosiphon okamuranus* III. Development of zoospore from plurilocular sporangium. *Bull. Jap. Soc. Scient. Fish.*, **40**, 1213–22.

Shinmura I. (1977) Life-history of *Cladosiphon okamuranus* Tokida from southern Japan. *Bull. Jap. Soc. Phycol.*, **25**, 333–40.

South G. R. (1972) On the life history of *Tilopteris mertensii* (Turn. in Sm.) Kütz. *Proc. Seaweed Symp.* **7**, 83–9.

South G. R. (1975) Contributions to the flora of marine algae of eastern Canada III. Order Tilopteridales. *Naturaliste Can.*, **102**, 693–702.

Sundene O. (1966) *Haplospora globosa* Kjellman and *Scaphospora speciosa* Kjellman in culture. *Nature (Lond.)* **209**, 937–8.

Takamatsu M. (1936) *Sorocarpus* aus der Matsushima-Bucht (Prov. Miyagi, Japan), auch morphologisch untersucht. *Res. Bull. Saito Ho-on Kai Mus.*, **8**, 71–99.

Tatewaki M. (1966) Formation of a crustose sporophyte with unilocular sporangia in *Scytosiphon lomentaria*. *Phycologia*, **6**, 62–6.

Wærn M. (1952) Rocky-shore algae in the Öregrund archipelago. *Acta Phytogeogr. Suec.*, **30**, 1–298.

Wynne M. J. (1969) Life history and systematic studies of some Pacific North American Phaeophyceae (brown algae). *Univ. Calif. Publs. Bot.*, **50**, 1–88.

Wynne M. J. & Loiseaux S. (1976) Recent advances in life history studies of the Phaeophyta. *Phycologia*, **15**, 435–52.

CHAPTER 6
CHLOROPHYTA: LIFE HISTORIES

CHRISTOPHER E. TANNER
Division of Natural Science and Mathematics, St. Mary's College of Maryland, St. Mary's City, Maryland 20686, U.S.A.

1 INTRODUCTION

The term life history was probably best defined by Russell (1973) as the sum of an organism's 'known morphological, cytological and reproductive phases'. Unlike higher plants the life histories of algae do not always include an alternation of generations or even an obligatory life cycle (Drew 1955; Chapman & Chapman 1961). Discussion of algal life histories is further complicated by the flexible relationship between morphological phases and cytological events (Feldmann 1972; Fjeld & Løvlie 1976). The green algae are probably the most morphologically diverse of all divisions or phyla of organisms. This diversity is reflected by their life histories which rival the brown algae for complexity and variability. However, unlike the brown algae (Russell 1973; Wynne & Loiseaux 1976) or red algae (Dixon 1970, 1973), there are no recent reviews of life histories of the green algae except for a brief article by Vinogradova (1976). The purpose, then, of this chapter is to review advances in our knowledge of life histories of marine green macroalgae primarily since Chapman's (1964) review of the Chlorophyta.

Although publications describing life histories of the green algae appeared prior to the turn of the century, up to 1921 all green algae were thought to be haploid, producing gametes that fused to form diploid resting zygotes (Smith 1938; Bold & Wynne 1978). Resting zygotes were thought to eventually return to the haploid vegetative phase through the production of meiospores. This type of life history was later found to be common among freshwater green algae, but early investigations of marine macroalgae demonstrated the occurrence of diploid green algae which reproduced solely by haploid gametes (Williams 1925; Schussnig 1930) and green algae which demonstrated alternations between cytologically different isomorphic phases (Hartmann 1929; Föyn 1929, 1934) and heteromorphic phases (Kornmann 1938). In fact subsequent studies indicate that true resting zygotes are exceedingly rare, if not entirely absent, among the green seaweeds.

Our understanding of green algal life histories has increased considerably in the last couple of decades, stimulated by the inclusion of reproductive and life history characteristics into classification schemes previously based on morphological criteria. Examples of the use of life histories in the taxonomy of green algae are the division of ulvaceous algae between the Ulotrichales and Ulvales (see Sections 3.2 and 3.3) and the establishment of the Acrosiphoniales (see Section 3.5). Yet for a large number of species information about life histories is insufficient or entirely lacking. As pointed out by Wynne & Loiseaux (1976) knowledge of the ploidy levels of the morphological phases is essential for the interpretation of life histories. For the marine green algae unequivocal cytological evidence is available for only a few species, and even less is known about the genetic control of life cycles. These areas demand further study before many more changes based on life histories are made in the classification of green algae. Recent advances have been made in the investigation of interactions between life histories and environmental factors. This area of study promises to be valuable for understanding the distribution of algae and the variation in their life histories and vegetative form (see Chapters 8 and 12).

2 TERMINOLOGY

Two types of system for the classification of life histories have been used. Svedelius (1931), Feldmann (1952), Chadefaud (1952), Drew (1955) and Chapman (1964) proposed complex terminologies and symbols. Kylin (1938), Scagel (1966a) and Dixon (1963, 1970) felt that the terminologies employed were too complex and supported the use of generic types for the classification of life histories. The diversity of the life histories found to date in the green algae and our insufficient knowledge of the subtle relationships between morphological phases and cytological events (see Fjeld & Løvlie 1976) make the first type of classification extremely cumbersome and both types deceptive. Interpretations of life histories are often influenced by pre-existing models and the associated terminologies (van den Hoek *et al.* 1972). Russell (1973) notes that life histories are often condensed into formulas that eventually become accepted as truths. Neither type of system is flexible enough without constant major revision to incorporate new information concerning variations in life histories.

Rather than use preconceived categories or types for the classification of life histories, the terminology used in this review refers to specific events or phases. *Somatic* is used interchangeably with vegetative. Somatic phases which produce gametes are referred to as *gametophytes*, whereas somatic phases that reproduce solely through spores or by vegetative means are referred to as *sporophytes*. Gametophytes that produce two types of gametes on the same thallus are referred to as *bisexual*, whereas gametophytes producing one type are called *unisexual*. *Isomorphic* and *heteromorphic* are used to indicate that two or more somatic phases are either morphologically similar or dissimilar. The location of meiosis is specified by the terms *zygotic, sporic, gametic* and *somatic*. Zygotic meiosis occurs in the zygote at germination. Sporic meiosis occurs in the production of spores and gametic meiosis occurs in the production of gametes. Somatic meiosis occurs in vegetative cells and is not followed immediately by the production of reproductive cells.

3 TAXONOMIC TREATMENT

3.1 Prasiolales

Although life histories and reproduction in members of this small order have received considerable attention (Fujiyama 1955; Friedmann 1959, 1963, 1969; Friedmann & Manton 1960; Manton & Friedmann 1960; Bravo 1962, 1965; Cole & Akintobe 1963; Kornmann & Sahling 1974; Edwards 1975; Hanic 1979), a number of questions remain unresolved. The relationship between morphological, reproductive and cytological events appears to be flexible with reports of both somatic meiosis (Friedmann 1959; Cole & Akintobe 1963) and zygotic meiosis (Fujiyama 1955; Kornmann & Sahling 1974), although the latter require further substantiation.

The life history of species with somatic meiosis (Fig. 6.1; *Prasiola*

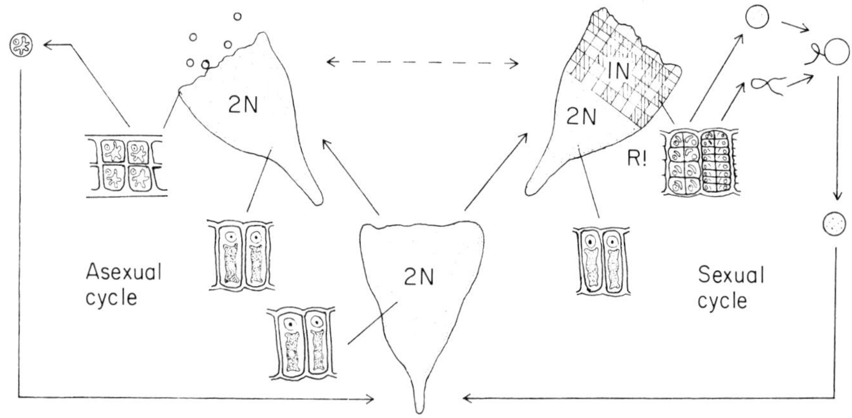

Fig 6.1 Life history of *Prasiola* species with somatic meiosis (modified from Friedmann 1959; and van den Hoek & Jahns 1978). Vegetative areas of the thallus are monostromatic, whereas sporogenic areas are distromatic, and gametogenic areas are polystromatic. R! = meiosis.

meridionalis, P. stipitata) has been interpreted as a modified alternation between a diploid sporophyte and a haploid gametophyte (Friedmann 1959; Cole & Akintobe 1963) even though meiosis and gametogenesis appear not to be obligatory but rather are induced by a combination of environmental factors (Friedmann 1959, 1963, 1969). Asexual reproduction is by diploid aplanospores that are released through degeneration of the sporangial cell walls and intercellular matrix (Friedmann 1959; Bravo 1965). Humidity, water temperature, photoperiodism, and culture medium have all been suggested as possible factors that control the induction of sexuality (Friedmann 1963, 1969; Bravo 1965). Sexuality is preceded by meiotic divisions in cells of the upper portion of the blade, followed by several mitotic divisions. The result is a mosaic pattern of paired polystromatic macrogametangial and microgametangial areas (Fig. 6.1). Friedmann (1959) and Cole & Akintobe (1963) interpreted the mosaic pattern of macrogametangia and microgametangia as the result of meiotic sex determination, whereas Bravo (1965) stated that in *P. meridionalis* all of the cells resulting from a single meiosis developed into either macrogametangia or microgametangia but not both. Bravo's interpretation is supported by Friedmann's (1959, text Fig. 2) observation that meiotic divisions occurred in the plane of the thallus resulting in a polystromatic blade. If sex segregation occurred during meiosis then the areas of macrogametangia and microgametangia would overlap instead of being in distinct sarcinoid packets (Fig. 6.1). As in the release of spores, gametes are released through the rupture and disintegration of the gametangial tissue. Friedmann (1959) and Cole & Akintobe (1963) reported that the inner walls of the gametangia disintegrate initially resulting in a bladder containing nonmotile macrogametes and biflagellate microgametes. Syngamy is unusual in these species in that one of the microgamete flagella is absorbed into the cytoplasm of the macrogamete, and the resulting zygote propels itself by means of the remaining posteriorly positioned flagellum (Friedmann & Manton 1960; Manton & Friedmann 1960; Cole & Akintobe 1963). The ultrastructure of the gametes and zygote of *P. stipitata* was described by Manton & Friedmann (1960).

Life histories that differ from the above have been reported by Fujiyama (1955), and Kornmann and Sahling (1974). Fujiyama described the freshwater species *Prasiola japonica* as haploid with meiosis occurring in the zygote. As in the previously discussed species, *P. japonica* produced macrogametangia and microgametangia in a mosaic pattern. However, his chromosome counts need to be confirmed. Kornmann & Sahling (1974) reported a variety of life histories in species of *Prasiola* and *Rosenvingiella* from Helgoland. *Prasiola calophylla* was described as having an alternation of heteromorphic asexual phases in which released aplanospores developed into aplanosporangia that released more aplanospores. Note should be made of the similarities between this report and the observation of Bravo (1965) of sexual *Rosenvingiella*-like thalli which released round cells that eventually produced gametes. Kornmann & Sahling

(1974) reported that *Rosenvingiella constricta* was unisexual with meiosis occurring in the zygote. They based their interpretation on the observation that each zygote germinated to produce four plants, two of each sex. However, these observations need to be substantiated with chromosome counts.

Several authors have suggested that *Rosenvingiella* is a developmental stage in the life history of species of *Prasiola* (Friedmann 1959, 1969; Bravo 1965; Edwards 1975). This interpretation has been challenged by Kornmann & Sahling (1974) and Hanic (1979). In both of these investigations species of *Rosenvingiella* and *Prasiola* reproduced true to type. Also, if *R. constricta* is unisexual and has zygotic meiosis as reported by Kornmann & Sahling (1974) a connection between this species and *Prasiola* is automatically ruled out.

3.2 Ulotrichales

The concept of this order is presently in a state of fluctuation as a result of both life history studies (Kornmann 1963a, 1965b, 1973) and cytological studies (Stewart *et al.* 1973; Stewart & Mattox 1975). Kornmann (1963a, 1965b) defined the Ulotrichales as containing filamentous (*Ulothrix*), discoid (*Gomontia*) and thallose (*Monostroma*) green algae characterized by an alternation between multicellular gametophytes and unicellular *Codiolum*-like sporophytes. He later (1970b, 1973) divided these algae among several orders and placed them along with the Acrosiphoniales into the new class Codiolophyceae, characterized by a *Codiolum*-like sporophyte. Although not widely followed at this time (Parke & Dixon 1976; Chapter 3), his concept of the Codiolophyceae indicates the common occurrence of this type of life history in green algae.

Among the marine uninucleate green algae, *Codiolum*-like stages have been observed in species of *Ulothrix* (Kornmann 1963a, 1963b; Perrot 1968, 1970, 1971, 1972; Ottgai & Fujiyama 1977; Lokhorst 1978), *Monostroma* (Kornmann & Sahling 1962; Kornmann 1962, 1964a; Hirose & Yoshida 1964; Gayral 1965; Kida 1966; Bliding 1968; Tatewaki 1972), *Gomontia* (Kornmann 1973; Wilkinson & Burrows 1972, as *Codiolum polyrhizum*), and *Collinsiella* (Chihara 1958, 1959a, 1960). The *Codiolum*-like stage of these algae results from sexual fusion or parthenogenetic development of gametes and is regarded as a distinct sporophytic phase, since it increases several times in size after settling and often develops a well defined stalk that attaches to the substrate (Lokhorst 1978). The morphology of the sporophyte varies according to species (Lokhorst 1978) and is often modified by the substrate in shell-boring forms (Kornmann 1959; Chihara 1962; Wilkinson & Burrows 1972; Wilkinson 1975). Cytological aspects of the life histories of these species have not been clarified; however, it is assumed that meiosis occurs during the first two divisions of the uninucleate sporophyte, followed by mitosis to produce zoospores (Bold & Wynne 1978;

Lokhorst 1978). Although these genera show similarities in their general life histories, considerable variation in specific life history events may occur.

Freshwater, estuarine, and marine species of *Ulothrix* usually demonstrate a heteromorphic life history with a *Codiolum*-like sporophyte (Kornmann 1963b; Lokhorst & Vroman 1972, 1974a, 1974b; Lokhorst 1978). However, Perrot (1968, 1970, 1972) reported an alternation of isomorphic filamentous gametophytic and sporophytic phases for isolates of *U. flacca* and *U. subflaccida*. Lokhorst (1978) studied numerous isolates of these two species and observed no evidence of an isomorphic life history. Species of *Ulothrix* can be either unisexual or bisexual (Perrot 1972; Lokhorst 1978) and in a few instances possibly both (*U. speciosa* in Kornmann 1964b; *U. subflaccida* in Lokhorst 1978). Asexual reproduction by zoospores in the filamentous gametophyte is a common occurrence (Kornmann 1963b; Perrot 1972; Lokhorst 1978), and is induced by short photoperiods in most species (Lokhorst 1978). One species, *U. acrorhiza*, has only been observed to reproduce by zoospores in culture (Kornmann 1963b). Other forms of asexual reproduction include aplanospore formation and fragmentation (Lokhorst 1978). Both zoospores and gametes are released from gametophytes through irregular holes, either individually or together in a hyaline mucilaginous envelope (Lokhorst & Vroman 1974b; Lokhorst 1978). Also of interest are the differences in development of the filamentous gametophytes between 'freshwater' and 'marine' species (Lokhorst 1978). In predominantly freshwater species, each zoospore gives rise to only one filament. However, in marine species when the germling is three to ten cells long, longitudinal divisions (referring to the plane of the new cell plate) result in two to several upright filaments.

A variety of reproductive details, life histories, and developmental patterns have been reported for monostromatic thallose species previously assigned to *Monostroma* (Gayral 1962, 1965, 1971; Kornmann 1962, 1964a, Kornmann & Sahling 1962; Hirose & Yoshida 1964; Kida 1966; Dube 1967; Bliding 1968; Tatewaki 1969, 1972; Kapraun & Flynn 1973). It is now apparent that this group contains a number of superficially similar genera (Kornmann 1964a; Gayral 1964, 1965; Bliding 1968; Vinogradova 1969, 1974). However, as yet there is little agreement among authors regarding the significance of the life history differences. Table 6.1 lists most of the genera that are currently recognized by one or more authors and their life history characteristics. Of these genera, *Ulvaria* and occasionally *Kornmannia* are placed with the Ulvaceae because of similarities in life histories and development (Bliding 1968; Vinogradova 1974; Kornmann & Sahling 1977). The heteromorphic species are often placed in their own family, the Monostromataceae, either within the Ulvales (Bliding 1968; Vinogradova 1974) or within the Ulotrichales (Round 1974; Parke & Dixon 1976). In species with an alternation between isomorphic gametophytic and sporophytic phases, meiosis is known to occur during sporogenesis (Dube 1967). In heteromorphic species with *Codiolum*-like

Table 6.1 Life histories, reproduction and development in monostromatic species previously placed in *Monostroma*.

Genus & species	Life history	Reproduction	Method of swarmer release from leafy thallus	Development	References
Ulvaria Ruprecht *U. obscura* (Kützing) Gayral (*U. fusca* (P. et R.) Ruprecht) *U. splendens* (Ruprecht)	Isomorphic leafy gametophyte leafy sporophyte	Unisexual biflag. gametes quadriflag. spores	One at a time through a pore	Erect fil. → saccate germling → blade	Gayral (1962, 1965) Dube (1967) Bliding (1968)
Kornmannia Bliding *K. zostericola* (Tilden) Bliding	Heteromorphic discoidal gametophyte leafy sporophyte	Bisexual biflag. gametes quadriflag. spores	One at a time through a pore	Prostrate disc → upheaval → saccate frond→ blade or prostrate disc → erect fil → blade	Yamada & Kanda (1941) Yamada & Tatewaki (1965) Tatewaki (1972)
K. leptoderma (Kjell.) Bliding	No alternation leafy thallus	Quadriflag. swarmers	One at a time through a pore	Prostrate disc → upheaval → saccate frond → blade	Kornmann & Sahling (1962) Hirose & Yoshida (1964) Bliding (1968) Tatewaki (1972)
Gayralia Vinogradova *G. oxysperma* (Kützing) Vinogr.	No alternation leafy thallus	Biflag. swarmers (quadriflag. swarmers)	Disintegration of cell wall	Erect fil. → saccate germling →blade	Kornmann (1964a) Gayral (1965) Bliding (1968) Tatewaki (1972) Kapraun & Flynn (1973)
Protomonostroma Vinogradova *P. undulatum* (Wittr.) Vinogr.	Heteromorphic asexual leafy and *Codiolum*-like phases	Quadriflag. swarmers	Simultaneously after disintegration of cell wall	Erect fil. → blade	Kornmann & Sahling (1962) Tatewaki (1972)
Capsosiphon Gobi *C. groenlandicum* (J. Ag) Vinogr.	Heteromorphic leafy gametophyte *Codiolum*-like sporophyte	Bisexual biflag. gametes aplanospores	In a hyaline envelope through a slit	Erect fil. → saccate germling → tubular frond	Tatewaki (1972)
Monostroma Thuret *M. bullosum* (Roth) Thuret *M. angicava* Kjellman *M. grevillei* (Thuret) Wittr.	Heteromorphic leafy gametophyte *Codiolum*-like sporophyte	Unisexual biflag. gametes quadriflag. spores	Simultaneously through an irregular opening	Prostrate disc → upheaval → saccate frond → blade	Suneson (1947) Kornmann (1964a) Gayral (1965) Bliding (1968) Tatewaki (1972)

sporophytes meiosis is assumed to occur during sporogenesis, but as yet cytological proof of this is lacking. *Gomontia* (Kornmann 1965b; Parke & Dixon 1976) and *Collinsiella* (Kornmann 1965b) are sometimes placed in the Monostromataceae because of developmental and life history similarities to *Monostroma*. In both of these genera discoidal gametophytes, that are bisexual in *Gomontia* (Kornmann 1973) and unisexual in *Collinsiella* (Chihara 1960), alternate with a shell-boring *Codiolum*-phase. Gametophytes of *C. tuberculata* from the northeast Pacific are also capable of reproducing asexually by quadriflagellate zoospores (personal observation). *Collinsiellopsis*, which resembles the gametophytic phase of *Collinsiella*, has only been observed to reproduce by quadriflagellate zoospores (Chihara 1967).

Few of the life histories of marine chaetophoracean algae are completely known. This is a particularly difficult group to work with because of confusion over generic concepts (Papenfuss 1962; Yarish 1975) and the considerable phenotypic variation observed in some genera (Yarish 1976). Asexual reproduction predominates under most culture conditions, although gametes and syngamy have been observed on a few occasions. Of the genera studied in culture *Phaeophila* (Nielsen 1972), *Pseudendoclonium* (Yarish 1975) and *Syncoryne* (Nielsen & Pedersen 1977) have only been observed to release quadriflagellate zoospores. *Acrochaete* (Nielsen 1979), *Bolbocoleon* (Yarish 1975), *Ectochaete* (Moewus 1949), *Entocladia* (O'Kelly & Yarish 1981), *Eugomontia* (Kornmann 1960), *Ochlochaete* (Nielsen 1978) and *Ulvella* (Papenfuss 1962; Nielsen 1977) have been observed to release quadriflagellate and biflagellate swarmers, although syngamy has only been observed in *Ectochaete, Entocladia* and *Eugomontia*. O'Kelly and Yarish (1981) have completed the life history of *Entocladia* which demonstrates an alternation between isomorphic filamentous gametophytic and sporophytic phases. This life history study and recent cytological investigations (Mattox & Stewart 1973; Floyd & Yarish 1978) suggest that at least some of the chaetophoracean genera have close affinities with the Ulvales. *Eugomontia* also has an alternation of isomorphic generations (Kornmann 1960); however, sporangia on the sporophytic phase are often liberated and develop into stages that are indistinguishable from the *Codiolum*-phase of heteromorphic ulotrichalean algae (Wilkinson & Burrows 1970).

3.3 Ulvales

Although members of this order are often placed in the Ulotrichales, some authors, including this one, are of the opinion that there are sufficient differences in life histories and reproduction to separate them from the primarily heteromorphic ulotrichalean algae (Round 1971; Parke & Dixon 1976). The separation is further supported by recent cytological and ultrastructural evidence that suggests that *Ulva, Enteromorpha, Percursaria* and related freshwater genera represent a separate line of evolution from other chlorophycean algae (Stewart & Mattox 1975, 1978; see Chapter 3).

The genera *Percursaria, Enteromorpha, Ulva* and *Ulvaria* have been shown to have a basic life history of an alternation of isomorphic diploid sporophytic and haploid gametophytic phases (Kornmann 1956; Föyn 1959; Bliding 1963, 1968; Dube 1967). The reproductive, culture and cytological evidence for this type of life history in *Enteromorpha* and *Ulva* is overwhelming. However, detailed studies of the species, *Ulva mutabilis*, have shown that morphological and cytological cycles do not always coincide and that there are 12 morphologically identical phases possible which differ in their cytological and reproductive details (Fig. 6.2; Föyn 1958, 1962; Løvlie 1964; Hoxmark & Nordby 1974; Hoxmark 1975; Fjeld & Løvlie 1976; Løvlie & Bryhni 1978). In this life history diploid sporophytes produce equal numbers of zoospores of two mating types (mt^+ and mt^-) through segregation of chromosomes during meiosis. Mating types appear to be carried as a single allelic pair in the sporophytes (Fjeld & Løvlie 1976). Haploid gametophytes are always of one mating type (unisexual) and produce gametes by mitosis which, following copulation with gametes of the opposite mating strain, return to the sporophytic phase. Gametes can also germinate parthenogenetically. In *U. mutabilis* most of the parthenogenetic gametes develop into parthenosporophytes, although 1–2% develop into gametophytes of the same mating type (Fig. 6.2; Hoxmark 1975; Fjeld & Løvlie 1976). Parthenosporophytes remain haploid or partially or completely undergo a process called 'diploidization' (Hoxmark & Nordby 1974) whereby the chromosome number is doubled within somatic cells. The two mating strains differ in their abilities to undergo diploidization (Fig. 6.2; Hoxmark 1975; Fjeld & Løvlie 1976). Both diploid and haploid cells in parthenosporophytes undergo meiosis to form zoospores. During the first division of meiosis in haploid cells chromatid pairs segregate randomly between daughter cells resulting in a low frequency of viable spores (Hoxmark & Nordby 1974). Zoospores from parthenosporophytes are always of one mating strain, indicating the allelic nature of sex determination. Although zoospores have never been observed to produce sporophytes in *U. mutabilis*, diploid zoospores (0.1%) have been observed that developed into diploid gametophytes (Fig. 6.2; Föyn 1962; Fjeld & Løvlie 1976). These produced neutral or sexual diploid gametes which returned to the sporophytic phase.

Parthenogenetic development of gametes into gametophytic or sporophytic phases has also been observed in other species of *Ulva* (Bliding 1968; Rhyne 1973; Tanner 1979), species of *Enteromorpha* (Kapraun 1970) and in *Percursaria percursa* (Kornmann 1956), indicating that some of the life history variations described for *U. mutabilis* may be of widespread occurrence.

The life history variations observed in *Ulva mutabilis* could explain the occurrence of asexual strains of algae in populations at the border of the geographic distribution of a given species. In *U. mutabilis* Løvlie and Bryhni

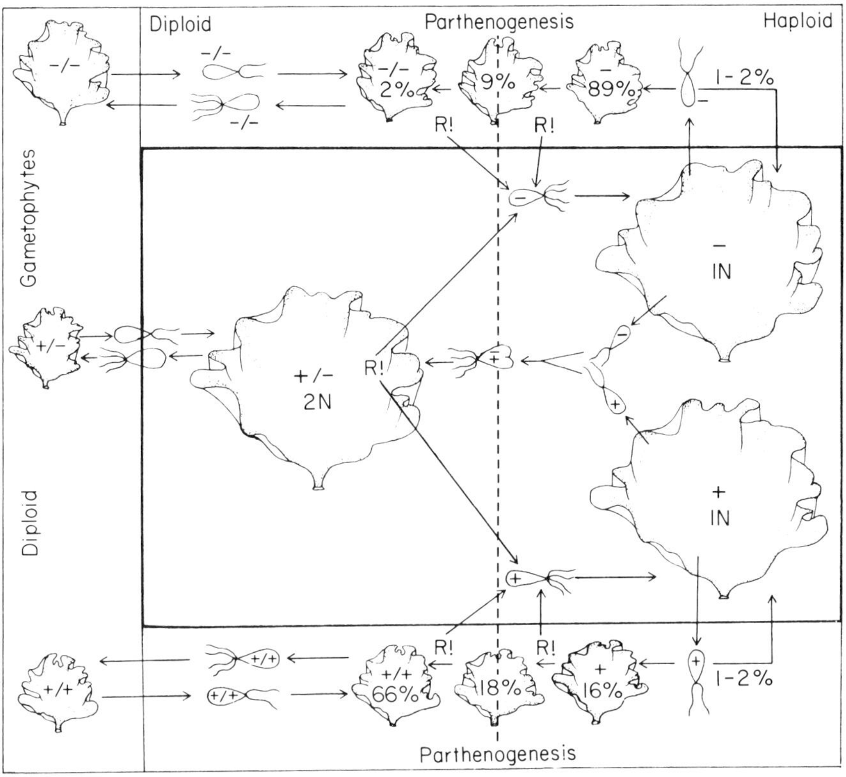

Fig. 6.2 Life history of *Ulva*. The inner box illustrates the basic life history of an alternation between isomorphic diploid sporophytic and haploid gametophytic phases. The marginal boxes represent alternative life history events. Gametes develop parthenogenetically into gametophytes or sporophytes. Parthenosporophytes are haploid, diploid or partially haploid and diploid. Percentages are for *Ulva mutabilis* (Fjeld & Løvlie 1976). R! = meiosis.

(1978) predicted that in border populations where one might expect low population densities and temperatures below the optimal for gamete fusion, the equilibrium between sexual and parthenogenetic reproduction would be disturbed resulting in a predominance of the latter. This could possibly explain the rarity of sporophytes in populations of *U. lactuca* on the Swedish west coast (Kylin 1949; Levring 1955) and the existence of strains of *Ulva, Enteromorpha* and *Ulvaria* that reproduce strictly by biflagellate neutral swarmers (Bliding 1963, 1968; Kapraun 1970).

Asexual strains which reproduce solely by quadriflagellate zoospores have not been observed in *Ulva*; however, in *Blidingia* and several taxa of *Enteromorpha* quadriflagellate swarmers are the only observed means of reproduction other than vegetative methods (Bliding 1963). In a few isolates of *Ulva* (Föyn 1958; Bliding 1968; Hoxmark 1975) and in *Percursaria percursa* (Kornmann 1956) thalli have been observed to release both biflagellate and quadriflagellate swarmers from the same thallus. In *U. mutabilis* these were released from parthenosporophytes, suggesting yet another possible variation to this complex life history (Hoxmark 1975).

In addition to reproduction by biflagellate and quadriflagellate swarmers, genera in the Ulvales may reproduce vegetatively through the resumption of totipotency of vegetative cells in the rhizoidal holdfast (Løvlie 1964; Dube 1967; Chihara 1968; Tanner 1980) or in the blade (Bonneau 1978). Another variation that has been observed in cultures of *Ulva lactuca* (Föyn 1934; Bonneau 1978) and *U. fasciata* (Tanner 1979) is the occurrence of diminuitive, floating, monostromatic, globose plants. These developed vegetatively either from germlings breaking away from the substrate (Föyn 1934) or through the sloughing off of vegetative cells which then acted as aplanospores, developing in a pattern different from that of motile swarmers (Bonneau 1978; Tanner 1979). In these examples the globose plants released motile swarmers that in the latter two studies developed normally. The significance of this floating globose stage in terms of dispersal is discussed by Bonneau (1978) and Tanner (1979).

3.4 Cladophorales

With few exceptions the members of this order that have been investigated demonstrate an alternation between isomorphic sporophytic and gametophytic phases with meiosis occurring during sporogenesis (Köhler 1956; Bliding 1957; Sinha 1958; Jónsson 1962b; van den Hoek 1963; Migita 1967; Kornmann 1969, 1972a; Wik-Sjöstedt 1970; Nienhuis 1974, 1975; Koshizaka 1978). Gametophytes are generally unisexual and produce isogametes. Parthenogenetic development of gametes is a common phenomenon (Köhler 1956; Migita 1967; Wik-Sjöstedt 1970; Koshizaka 1978) and somatic nuclei in parthenogenetic thalli have been reported to undergo 'diploidization' (Wik-Sjöstedt 1970). Some species reproduce almost entirely through fragmentation (Chapman & Chapman 1973; Nienhuis 1974, 1975; Bold & Wynne 1978). Reproduction by biflagellate zoospores has been reported for species of *Cladophora* (Wik-Sjöstedt 1970) and *Rhizoclonium* (Russell 1971; Nienhuis 1974, 1975). In *Rhizoclonium* biflagellate spores are produced more commonly in estuarine conditions and appear to be a means of propagating the sporophytic phase (Nienhuis 1974, 1975). Nienhuis has suggested that the alternation between asexual and sexual phases is inhibited by the relatively unstable conditions in estuaries, resulting in sterile populations reproducing by fragmentation. The suppression of reproductive capacity in *Urospora* occurs primarily in soft substrate areas and may be related to a decrease in salinity or an increased level of available nitrogen (Nienhuis 1974).

Life histories that deviate from that discussed above have been reported for a few species of *Cladophora*. Archer and Burrows (1960) reported that *C.*

rupestris alternated with a *Codiolum*-like phase. However this report has not been supported by subsequent studies which have demonstrated an alternation between isomorphic sporophytic and gametophytic phases for this species (Wik-Sjöstedt 1970; Koshizaka 1978). Siddique and Faridi (1977) have recently reported that the freshwater species, *Cladophora crispata*, has zygotic meiosis.

3.5 Acrosiphoniales

Life history was one of the criteria used to separate genera in this order from cladophoracean algae (Jónsson 1959b, 1962b; Kornmann 1973). As in the Ulotrichales the life history of this group is characterized by an alternation between macrophytic gametophytes and a *Codiolum*-like or *Chlorochytrium*-like sporophytic phase (Fig. 6.3b). Though cytological evidence is meagre, meiosis is assumed to occur in the sporophyte during zoosporogenesis. Species are either unisexual or bisexual and produce isogametes or anisogametes. Parthenogenesis is a relatively common phenomenon, resulting in the propagation of either the sporophytic or gametophytic phase (Jónsson 1957, 1967; Kornmann 1963a; Hanic 1965). The gametophytic phase of *Urospora* is also capable of reproducing itself through quadriflagellate or biflagellate mitospores, particularly at low temperatures (Kornmann 1961a, 1961c, 1963a; Hanic 1965).

A number of interesting variations of the observed life history outlined above have been found for species of *Urospora* and *Acrosiphonia* (Fig. 6.3). In *U. speciosa, U. vancouveriana* and *U. wormskioldii* a multicellular dwarf phase develops from zoospores released either from a filamentous gametophyte or from the *Codiolum* phase (Fig. 6.3a; Kornmann 1961a, 1961c, 1963a; Hanic 1965). Dwarf plants produce biflagellate or quadriflagellate swarmers that germinate into the *Codiolum* phase. The type of swarmers and morphological phase produced in culture for these species is temperature dependent, sexual reproduction being suppressed at low temperatures (Hanic 1965).

Variations in the life history of *Acrosiphonia* have been observed which demonstrate a gradation between a heteromorphic life history with sporic meiosis and a life history with zygotic meiosis and without an alternation of morphological phases (Kornmann 1964c, 1965a, 1970b). *Acrosiphonia spinescens* has a typical heteromorphic life history as described for the order (Fig. 6.3b; Jónsson 1957, 1959a; Kornmann 1970c, 1973). In *A. grandis* a free-living *Codiolum* phase is produced from zygotes or parthenogenetic gametes, but this phase develops directly into the filamentous gametophyte (Fig. 6.3b; Kornmann 1970c, 1970d). Although the nucleus of the *Codiolum* phase undergoes a number of simultaneous divisions prior to germination, it is not known whether meiosis occurs at this stage. In *A. arcta* from Helgoland the zygote undergoes two nuclear divisions, possibly representing meiosis, and develops into the

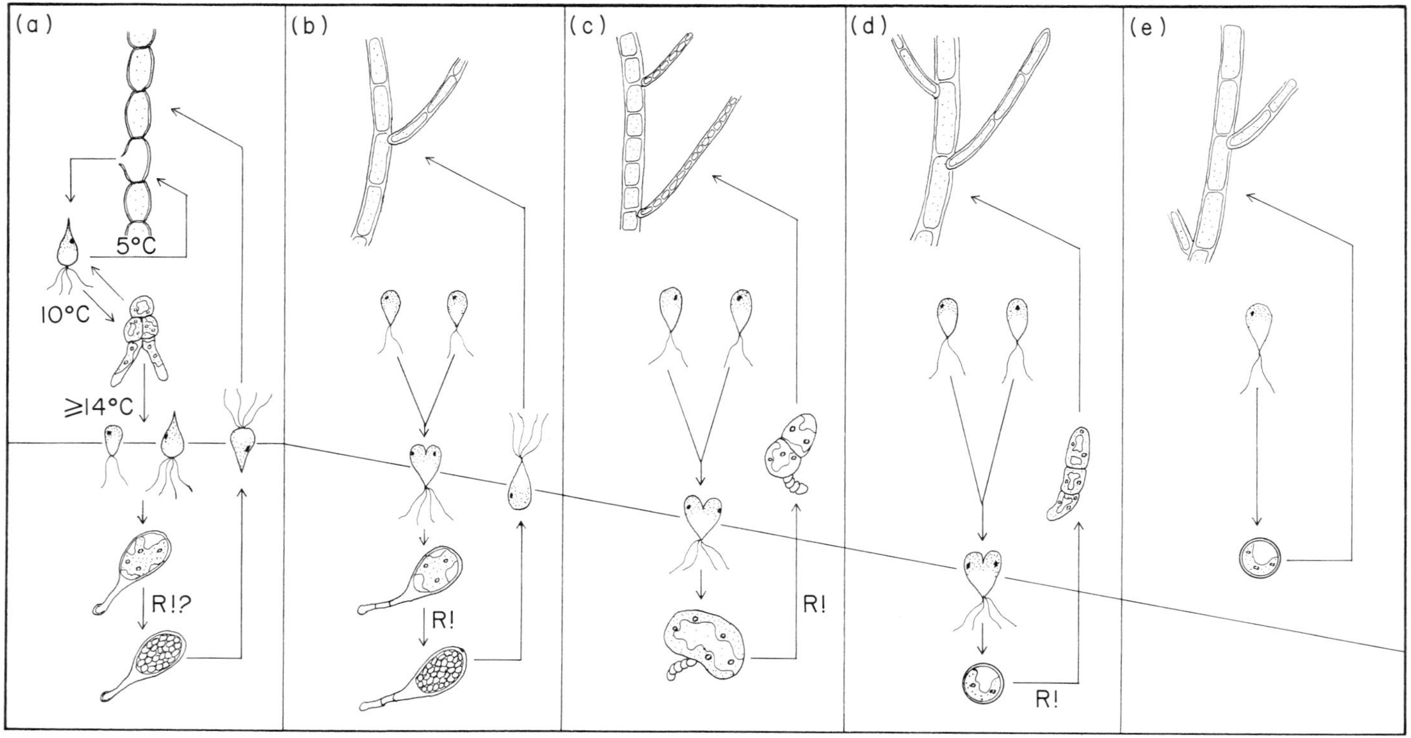

Fig. 6.3 Life histories in the Acrosiphoniales. (a) *Urospora wormskioldii* (modified from Kornmann 1961c; Bachman *et al.* 1976). (b) *Acrosiphonia spinescens*—a filamentous gametophytic phase alternates with a *Codiolum*-like sporophytic phase. (c) *Acrosiphonia grandis*—the *Codiolum*-like phase develops directly into the filamentous gametophyte. (d) *Acrosiphonia arcta*—zygotic meiosis. (e) *Acrosiphonia sonderi*—asexual reproduction only. (b–e) Modified from Kornmann (1970c). R! = meiosis.

filamentous gametophyte (Fig. 6.3d; Kornmann 1965a, 1970c). Gametes also develop parthenogenetically into the filamentous phase. In this species the *Codiolum* phase is entirely lacking, and meiosis is assumed to be zygotic.

In some isolates of *Acrosiphonia sonderi* sexual reproduction is suppressed or absent, and the filamentous phase reproduces solely by means of biflagellate swarmers. Jónsson (1967) studied sexual isolates of *A. sonderi* in which the zygote either developed directly into the filamentous gametophytic phase or developed into a *Codiolum* phase that propagated itself by means of biflagellate swarmers (Fig. 6.3e). After following the nuclear events ensuing sexual fusion, he claimed that in the zygotes which developed directly the nuclei failed to fuse resulting in a n+n condition in the zygote. Jónsson suggested that this was an example of facultative karyogamy and may have led to the development of asexual strains.

Though *Codiolum* (Hollenberg 1958; Fan 1959; Hanic 1965; Kornmann 1961b, 1964d) and *Chlorochytrium* (Jónsson 1966; Chihara 1969; Kornmann 1964d; Miyaji & Kurogi 1976) have both been reported as the sporophytic phase in this order, it is now thought by several researchers that the phenotypic differences between these thalli are caused by environmental conditions (Chihara 1962; Kornmann 1972b, 1973).

3.6 Caulerpales

With the exception of the Bryopsidaceae all the members of this morphologically diverse order are assumed to have gametic meiosis with a single diploid vegetative phase (Chapman 1964). Unfortunately, cytological evidence is lacking for all of the genera with gametic meiosis except *Codium* (Williams 1925; Schussnig 1930, 1950) and *Caulerpa* (Schussnig 1939), and the evidence for meiosis in the latter has been questioned (Goldstein & Morrall 1970). However, the development of zygotes directly or through a protonemal stage into the gametophyte has been cited as evidence for gametic meiosis (Price 1972). All the species studied produced anisogametes except a few isogamous species reported for *Halimeda* (Chapman & Chapman 1973). Some genera have simple differentiated gametangia that are cut off by a septum or plug from the rest of the siphonous thallus as in *Codium* (Arasaki *et al.* 1955; Borden & Stein 1969), whereas others have compound gametangia that appear to be continuous with the siphonous thallus as in *Halimeda* (Graham 1975) and *Chlorodesmis* (Ducker 1965). However, in *Halimeda* the sudden eruption of the gametangial contents into the gametophore suggest some sort of barrier (Graham 1975). In *Udotea* (Meinesz 1969) and *Caulerpa* (Iyengar 1940; Goldstein & Morrall 1970) differentiated gametangia are not produced; rather, gametes are formed by a conversion of the somatic protoplasm into gametes by nuclear divisions and cytoplasmic cleavage. Both unisexual and bisexual species have been reported for the genera with gametic meiosis in this order. *Codium fragile* can either be unisexual or bisexual, the bisexual thalli possibly resulting from the interweaving of filaments from two or more zygotes (Arasaki *et al.* 1955). In the northeast Atlantic populations of *C. fragile* subspecies *tomentosoides* have been observed to release biflagellate swarmers that germinated directly (Churchill & Moeller 1972; Ramus 1972). In some genera vegetative reproduction may be important (Borden & Stein 1969; Fralick & Mathieson 1972). Life histories of most of the species and genera in this group are either unknown or incompletely studied. It is essential that future studies include information regarding the development of zygotes and detailed cytological evidence.

Early work on *Bryopsis* (Schussnig 1932; Zinnecker 1935) indicated that the vegetative thallus was diploid with gametic meiosis. This interpretation was adopted for both *Bryopsis* (Fritsch 1935; Smith 1955) and *Halicystis* (Fritsch 1935). However, the discovery of alternation between *Halicystis ovalis* and *Derbesia marina* (Kornmann 1938) and between *Bryopsis halymeniae* and *Derbesia neglecta* (Hustede 1960, 1964) led more recent investigators to question the earlier studies (Rietema 1970, 1972; van den Hoek *et al.* 1972). These recent investigations show that a variety of life histories is possible even within a single species (Fig. 6.4; Rietema 1970).

Most of these investigations suggest that an alternation of heteromorphic haploid gametophytic and diploid sporophytic phases is usually obligatory (Fig. 6.4c–g; Hustede 1960, 1964; Ziegler & Kingsbury 1964; Neumann 1967, 1969a; Rietema 1969, 1971a, 1972; Kornmann 1970a; Kornmann & Sahling 1976; Mayhoub 1976; Kobara & Chihara 1978b; Okuda *et al.* 1979), although cytological evidence is available in only a few instances (Neumann 1967, 1969a; Rietema 1972). Gametes are always of two sizes and are either produced in differentiated gametangia that bud off from the vegetative thallus as in *Pseudobryopsis* (Feldmann 1969; Kobara & Chihara 1978a), in vegetative pinnules that are converted into gametangia as in *Bryopsis* (Burr & West 1970; Rietema 1970; Diaz-Piferrer & Burrows 1974) or through conversion of part of the vegetative protoplasm into a mass of gametes as in *Halicystis* (Hollenberg 1935). The stephanokontan spores produced by the sporophytic phases indicate a close relationship between these genera. Stephanokontan spores of *Pseudobryopsis* differ morphologically from those of other genera by abruptly tapering posteriorly (Fig. 6.4e; Kobara & Chihara 1978b; Okuda *et al.* 1979).

In some isolates of *Bryopsis* the zygote develops into a protonemal stage that may develop directly into the gametophyte or may produce spores (Fig. 6.4g; Neumann 1969b; Rietema 1970, 1971b; Bartlett & South 1973; Diaz-Piferrer & Burrows 1974). Neumann (1969b) reported that meiosis was gametic in an isolate of *Bryopsis hypnoides* with direct development, although he did not produce any cytological evidence. Rietema (1971b, 1972) challenged Neumann's interpretation and suggested that in northern latitudes the sporophyte of

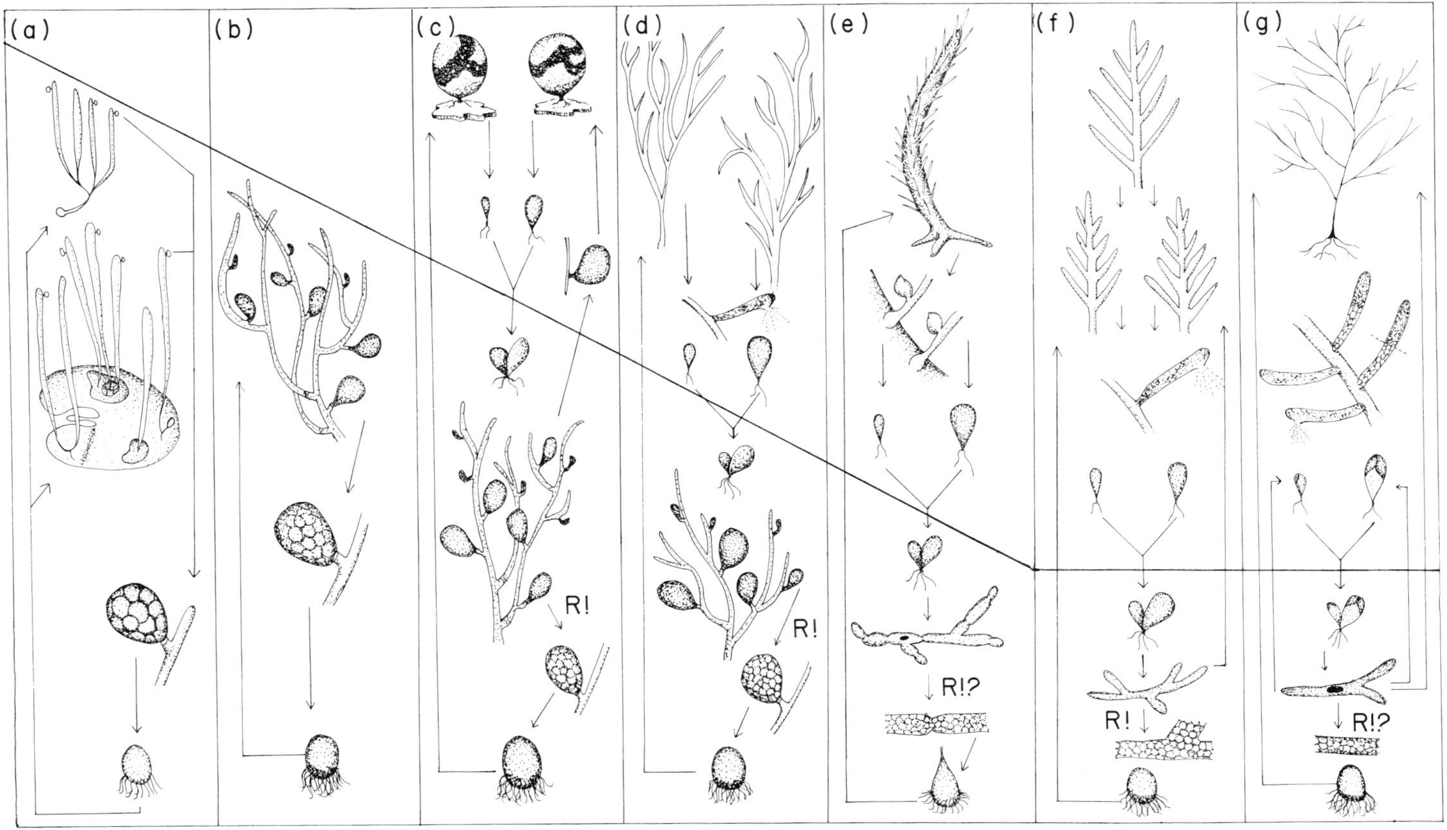

Fig. 6.4 Life histories in the Bryopsidaceae. (a) *Pedobesia clavaeformis* (modified from MacRaild & Womersley 1974). (b) Asexual isolate of *Derbesia marina* (Kornmann 1966, 1970a). (c) Alternation between a sporophytic *Derbesia* phase and a gametophytic *Halicystis* phase. (d) *Bryopsidiella neglecta* (modified from van den Hoek *et al*. 1972. (e) *Pseudobryopsis* (modified from Kobara & Chihara 1978b). (f) *Bryopsis* species (modified from van den Hoek *et al*. 1972). (g) *Bryopsis hypnoides* (modified from Rietema 1971b). R! = meiosis.

Bryopsis lost the capacity to produce spores and that meiosis occurred somatically within the sporophyte (Rietema 1970, 1971b). In support of Rietema is the investigation of *Derbesia marina* by Sears and Wilce (1970) in which sporangia of the sporophytic phase failed to cleave into spores but rather developed directly into gametophytes. Meiosis has been shown to occur within the sporangium in *Derbesia marina* (Neumann 1969a) and *D. neglecta* (Rietema 1972). An alternative explanation would be the development of parthenogenetic gametes into a haploid protonemal stage that develops directly into the gametophytic phase as has been hypothesized to occur in border populations of diplohaplontic algae (Løvlie & Bryhni 1978). Parthenogenetic development has been suggested for *Halicystis* (Ziegler & Kingsbury 1964; Wutz & Zetsche 1976), although Rietema (1969) reports that neither macrogametes nor microgametes of *Bryopsis plumosa* develop parthenogenetically. An interesting variation is the report by Bartlett and South (1973) that the protonemal stage of *Bryopsis hypnoides* may develop directly into the gametophyte or may reproduce sexually (Fig. 6.4g). Rietema (1971b) and Diaz-Piferrer and Burrows (1974) report that the protonemal stage of this species can either develop directly or reproduce by stephanokontan spores.

In other studies an absence of sexual reproduction was demonstrated for *Derbesia* and *Pedobesia* (Fig. 6.4a, b; Kornmann 1966, 1970a; Neumann 1969a; Sears & Wilce 1970; MacRaild & Womersley 1974). Bernatowicz (1958) suggested that this could be caused by the gametophytic and sporophytic phases having different temperature tolerances. In a study of a heteromorphic isolate of *Derbesia marina* by Kornmann (1966, 1970a) a mutation occurred which resulted in an asexual haploid isolate of *Derbesia* that reproduced solely by stephanokontan spores.

The protonemal phases of *Bryopsis hypnoides* (Neumann 1969b), *B. monoica* (Neumann 1970), and *Pseudobryopsis* (Neumann 1970; Okuda *et al.* 1979) are reported to have a single giant nucleus that resembles the primary nucleus of *Acetabularia*. Burr and West (1971) investigated the ultrastructure of the primary nucleus in *Bryopsis, Acetabularia* and *Batophora* and observed that the primary nucleus in *Bryopsis* differed from those of the other two genera by having an extensive peripheral reticulum composed of microfibrils.

Although numerous culture studies have now been made of this family our understanding of the life histories has been hindered by the almost total absence of cytological information and the influence of pre-existing models (Rietema 1971b; van den Hoek *et al.* 1972). It is possible that cytological and morphological cycles are independent and that meiosis can occur in either the gametophytic or sporophytic phases. However, on the basis of the evidence at hand, it seems more likely that meiosis is restricted to the sporophyte and that the deviant life histories are caused by meiosis in the sporophyte not being followed by cytokinesis or by the parthenogenetic development of gametes.

3.7 Siphonocladales

Our knowledge of the life histories in the Valoniaceae has essentially remained unchanged since Chapman's (1964) review of the Chlorophyta. Chihara (1953, 1959b) observed isomorphic spore-producing and unisexual gamete-producing thalli of *Valonia macrophysa*. Van den Hoek & Jahns (1978) interpreted this as an alternation between isomorphic diploid sporophytic and haploid gametophytic phases. However, Schechner-Fries (1934) and Schussnig (1938) reported that meiosis was gametic in *Valonia utricularis*, and Chihara (1959b) proposed that *Valonia* was diploid and could either reproduce gametes by meiosis or produce mitospores for multiplication of the diploid gametophyte. This interpretation has been accepted by Chapman (1964) and Bold & Wynne (1978).

In contrast to the Valoniaceae, there is good evidence for an alternation of isomorphic diploid sporophytic and haploid gametophytic phases in *Anadyomene* (Enomoto & Hirose 1970; Mayhoub 1975) and *Microdictyon* (Iyengar & Ramanathan 1941), both members of the Anadyomenaceae. Enomoto & Hirose (1970) clearly demonstrated that in *Anadyomene* zoospores were haploid, whereas zygotes that developed into sporophytes were diploid. Production of swarmers in culture is regulated by temperature, with warmer temperatures (23–25°C) promoting fertility (Jónsson 1962a; Enomoto & Hirose 1970). The Anadyomenaceae are often aligned with the Cladophorales because of the absence of segregative cell division (Papenfuss 1955), similarities in life histories (Enomoto & Hirose 1970), and the similarities in chloroplast structure (Hori & Ueda 1967). However, there is disagreement as to whether the Anadyomenaceae should be placed in the Cladophorales (Papenfuss 1955) or whether the Cladophoraceae should be placed in the Siphonocladales (Vinogradova 1972; Chapman & Chapman 1973). Round (1971) is of the opinion that the morphological differences are sufficient to keep the Siphonocladales, including the Anadyomenaceae, separate from the Cladophorales. Further investigations into the life histories of the Valoniaceae may help to resolve these taxonomic problems.

3.8 Dasycladales

Although the morphological and reproductive cycles in this order have long been known (Fig. 6.5; Bonotto 1975) the position of meiosis and the cytological cycle were largely a matter of speculation until recently. Schulze (1939) suggested that meiosis occurred during gametogenesis, and this was supported by evidence by Woodcock & Miller (1973a) for a nuclear arrangement in the shape of a tetrad in maturing cysts. Puiseux-Dao (1970) questioned Schulze's meiotic figures, and Green (1973, 1976b) suggested that meiosis occurred prior to gametogenesis because cysts which produced gametes were unisexual. Green is supported by Koop's (1975, 1977) recent microspectrophotometric analyses of DNA in *Acetabularia*. Apparently cyst nuclei have the same amount of DNA as

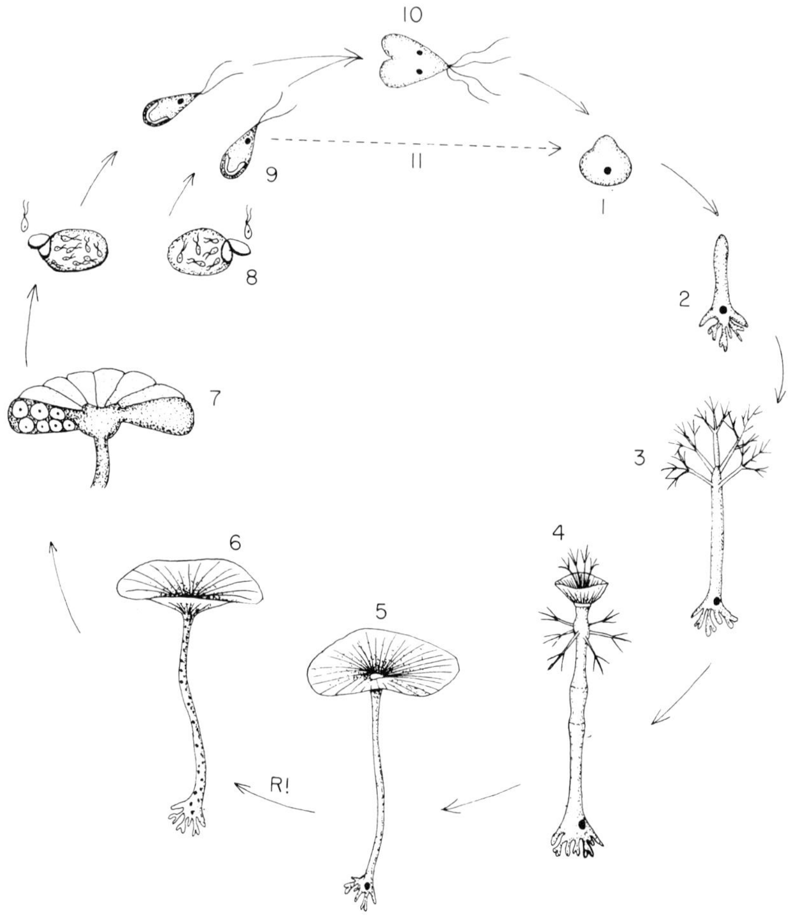

Fig. 6.5 Life history in *Acetabularia mediterranea* (modified from Bonotto 1975). (1) Germinating zygote or parthenogenetic gamete. (2) Development of stalk and rhizoids. (3) Whorl formation. (4) Initiation of cap. (5) Growth of cap and meiotic division of primary nucleus. (6) Migration of secondary nuclei into cap rays. (7) Development of cysts in rays. (8) Release of gametes from cysts. (9) Isogametes. (10) Syngamy. (11) Parthenogenesis.

gametes and half the amount of DNA found in zygotes, suggesting that meiosis occurs prior to cyst formation and probably during the breakdown of the primary nucleus (Fig. 6.5; Koop 1977). Koop (1979) reviews the evidence for meiosis in the primary nucleus.

In addition, zoospores released from cysts have been reported to develop directly into new plants (Puiseux-Dao 1962, 1970). Green (1976a) reported parthenogenetic development in *Acetabularia cliftonii*, though the parthenogenetic thalli failed to form normal apices. New thalli can also develop vegetatively from basal rhizoids (Bonotto & Kirchman 1970; Bonotto 1975). The primary nucleus, cyst formation and gametogenesis have all been studied extensively with the electron microscope (Crawley 1966; Burr & West 1971; Woodcock & Miller 1973a, 1973b).

4 DISCUSSION

Although our knowledge of life histories in the green algae has increased considerably during the last two decades, all the possible combinations of life history events have been analysed for relatively few species. From the information that is available it is clear that there are a number of possible cycles within the life histories of green algae that do not always coincide. These include morphological, reproductive, cytological and biochemical cycles. The complex interrelations between the first three cycles are illustrated by *Ulva mutabilis*, in which there is only one morphological form but twelve different cytological and reproductive combinations (Fig. 6.2). In *U. mutabilis* the position of meiosis is fixed in relation to the type of swarmer produced, whereas the ploidy level is independent of both meiosis and the reproductive phase of the thallus. A close relationship between meiosis and the reproductive cycle appears to hold true for most green algae and is generally consistent within closely related genera. The most obvious exception to this is seen in the Prasiolales in which both somatic and zygotic meiosis are apparently possible within the same genus.

In addition to morphological, reproductive and cytological cycles, genera belonging to the Acrosiphoniales (Bachmann *et al.* 1976) and the Caulerpales (Huizing & Rietema 1975; Wutz & Zetsche 1976; Huizing *et al.* 1979) demonstrate a regular alternation in the chemistry of their cell walls. The change in cell wall chemistry is generally associated with changes in the morphology of the vegetative phases. Considering that morphogenesis in green algae is regulated by the structure of the wall (Bryhni 1974, 1978), it is not really surprising that changes in morphology are associated with chemical differences in the cell wall. The influence of wall chemistry on morphology is supported by the relatively small difference in wall chemistry between the almost isomorphic gametophytic and sporophytic phases of *Bryopsidiella neglecta* (Fig. 6.4d; Huizing *et al.* 1979).

A number of studies have indicated that environmental factors influence or control life histories in some species of green algae. Some of the factors that have been discussed are temperature, light quality and period, exposure, salinity and levels of nutrients or growth factors (Dring 1974). In species of *Ulothrix* and *Urospora* the alternation between morphological phases has been shown to be regulated in culture by temperature (Fig. 6.3a; Kornmann 1961a, 1961c; Hanic 1965; Perrot 1972; Bachmann *et al.* 1976; Lokhorst 1978). Temperature has also been suspected as the regulating factor for a number of green algae, such as *Collinsiella* (Chihara 1960) and *Prasiola* (Bravo 1965), that show a regular seasonal alternation in their life histories. Photoperiodic phenomena involving a phytochrome-like system have been demonstrated for both red and brown algae (Dring 1970, 1974; Chapter 9). Although photoperiodic control of life histories has been suggested for *Bryopsis* (Rietema 1972), *Derbesia* (Rietema 1973), *Prasiola* (Friedmann 1963, 1969) and other green algae, the presence of a phytochrome or similar pigment system has yet to be demonstrated. In some green algae light quality influences reproduction (Chapter 9). Dring (1974) is of the opinion that blue light related phenomena may explain the induction of reproduction by transfer from dark to light or vice versa observed in a number of marine algae. In *Prasiola stipitata* the degree of wetting appears to influence the percentage of plants that undergoes meiosis and sexual reproduction (Friedmann 1963, 1969). In some filamentous green algae the degree of sexuality appears to be related to salinity or stability of the environment (Russell 1971; Nienhuis 1974; Lokhorst 1978).

Changes in temperature (Sears & Wilce 1970), light intensity and photoperiod (Dring 1974; Chapter 9), culture medium (Nilsen & Nordby 1975) or combinations of these factors (Nordby & Hoxmark 1972) induce reproduction in many species of marine green algae and can act as an entraining stimulus for endogenous rhythms (Page & Kingsbury 1968; Page & Sweeney 1968). In nature, reproduction is often periodic, generally in relation to tidal cycles. Release of swarmers usually occurs at sunrise during either spring tides (Smith 1947; Chihara 1959b; Christie & Evans 1962) or neap tides (Ohno 1972; Sawada 1972; Sawada & Watanabe 1974; Okuda 1975). It is unclear as to what environmental factors cause the synchrony between reproductive and tidal cycles; however, it is likely that the reproductive periodicity is caused by endogenous rhythms, perhaps entrained by a number of different environmental factors. Such endogenous rhythms have been demonstrated in cultures of *Derbesia* (Page & Kingsbury 1968; Page & Sweeney 1968).

Much attention has been paid to the phylogeny of algal groups based upon the hypothetical relationships between different types of life histories. Although there is general agreement that a *Chlamydomonas*-like life history with zygotic meiosis is the most primitive (Roe 1975), further speculation regarding the evolution of life histories in the green algae is not warranted until the present classification schemes have been exhaustively tested cytologically as has been done already for a number of green algae (Pickett-Heaps & Marchant 1972; Stewart & Mattox 1975, 1978). It is probable that similarities in life histories in many instances do reflect phylogenetic relationships; however, we do not yet fully understand the genetic mechanisms for the regulation of life histories, or even the degree of stability of various life history events.

5 REFERENCES

ARASAKI S., TOKUDA H. & FUJIYAMA K. (1955) The reproduction and morphogeny in *Codium fragile* (Japanese with English summary). *Bot. Mag., Tokyo*, **69**, 39–45.

ARCHER A. & BURROWS E. M. (1960) Heteromorphic life-histories as a family criterion in the Cladophorales. *Br. Phycol. Bull.*, **2**, 31–3.

BACHMANN P., KORNMANN P. & ZETSCHE K. (1976) Regulation der Entwicklung und des Stoffwechsels der Grünalge *Urospora* durch die Temperatur. *Planta*, **128**, 241–5.

BARTLETT R. B. & SOUTH G. R. (1973) Observations on the life history of *Bryopsis hypnoides* Lamour. from Newfoundland. *Acta. Bot. Meer.*, **22**, 1–15.

BERNATOWICZ A. J. (1958) Ecological isolation and independent speciation of the alternate generations of plants. *Biol. Bull.*, **115**, 323 (abstr.).

BLIDING C. (1957) Studies on *Rhizoclonium* I. Life history of two species, *Bot. Not.*, **110**, 271–5.

BLIDING C. (1963) A critical survey of European taxa in Ulvales. Part I. *Capsosiphon, Percursaria, Blidingia, Enteromorpha. Opera Bot.*, **8** (3), 1–160.

BLIDING C. (1968) A critical survey of European taxa in Ulvales. Part II. *Ulva, Ulvaria, Monostroma, Kornmannia. Bot. Not.*, **121**, 535–629.

BOLD H. C. & WYNNE M. J. (1978) *Introduction to the Algae—Structure and Reproduction.* Prentice-Hall, Englewood Cliffs, New Jersey.

BONNEAU E. R. (1978) Asexual reproduction capabilities in *Ulva lactuca* L. (Chlorophyceae). *Bot. Mar.*, **21**, 117–21.

BONOTTO S. (1975) Morphogenesis in normal, branched and irradiated *Acetabularia mediterranea. Pubbl. Staz. Zool. Napoli*, **39**, 96–107.

BONOTTO S. & KIRCHMANN R. (1970). Sur les processus morphogénétiques d'*Acetabularia mediterranea. Bull. Soc. Roy. Bot. Belgique*, **103**, 255–72.

BORDEN C. A. & STEIN J. R. (1969) Reproduction and early development in *Codium fragile* (Suringar) Hariot: Chlorophyceae. *Phycologia*, **8**, 91–9.

BRAVO L. (1962) A contribution to the knowledge of the life history of *Prasiola meridionalis. Phycologia*, **2**, 17–23.

BRAVO L. (1965) Studies on the life history of *Prasiola meridionalis. Phycologia*, **4**, 177–94.

BRYHNI E. (1974) Genetic control of morphogenesis in the multicellular alga *Ulva mutabilis*. Defect in cell wall production. *Dev. Biol.*, **37**, 273–9.

BRYHNI E. (1978) Quantitative differences between polysaccharide compositions in normal differentiated *Ulva mutabilis* and the undifferentiated mutant *lumpy. Phycologia*, **17**, 119–24.

BURR F. A. & WEST J. A. (1970) Light and electron microscope observations on the vegetative and reproductive structures of *Bryopsis hypnoides. Phycologia*, **9**, 17–37.

BURR F. A. & WEST J. A. (1971) Comparative ultrastructure of the primary nucleus in *Bryopsis* and *Acetabularia. J. Phycol.*, **7**, 108–13.

CHADEFAUD M. (1952) Sur le cycle sexuel des organismes et son évolution. *Rev. Sci., Paris*, **90**, 49–57.

CHAPMAN D. J. & CHAPMAN V. J. (1961) Life histories in the algae. *Ann. Bot.* N.S., **25**, 547–61.

CHAPMAN V. J. (1964) The Chlorophyta. *Oceanogr. Mar. Biol. Ann. Rev.*, **2**, 193–228.

CHAPMAN V. J. & CHAPMAN D. J. (1973) *The Algae*, 2nd edn. Macmillan, London & New York.

CHIHARA M. (1953) Studies on the life history of the green algae in the warm seas around Japan (2). On the life-history of *Valonia macrophysa* Kütz. (Japanese with English Summary). *J. Jap. Bot.*, **28**, 353–61.

CHIHARA M. (1958) Studies on the life-history of the green algae in the warm seas around Japan (7). On the sexual reproduction in *Collinsiella* (Japanese with English summary). *J. Jap. Bot.*, **33**, 307–13.

CHIHARA M. (1959a) Studies on the life history of the green algae in the warm seas around Japan (8). On the life-history of *Collinsiella cava* (Yendo) Printz. (Japanese with English summary). *J. Jap. Bot.*, **32**, 193–201.

CHIHARA M. (1959b) Studies on the life-history of the green algae in the warm seas around Japan (9). Supplementary note on the life-history of *Valonia macrophysa* Kütz. *J. Jap. Bot.*, **34**, 257–66.

CHIHARA M. (1960) *Collinsiella* in Japan, with special reference to the life-history. *Sc. Rep. Tokyo Kyoiku Daigaku*, **9B**, 181–98.

CHIHARA M. (1962) Occurrence of the *Gomontia*-like phase in the life history of certain species belonging to *Collinsiella* and *Monostroma* (a preliminary note). *J. Jap. Bot.*, **37**, 44–5.

CHIHARA M. (1967) *Collinsiellopsis*, a new genus of green algae. *Phycologia*, **6**, 87–95.

CHIHARA M. (1968) Field, culturing and taxonomic studies of *Ulva fenestrata* P. & R. and *Ulva scagelii* sp. nov. (Chlorophyceae) in British Columbia and northern Washington. *Syesis*, **1**, 87–102.

CHIHARA M. (1969) Culture study of *Chlorochytrium inclusum* from the northeast Pacific. *Phycologia*, **8**, 127–33.

CHRISTIE A. O. & EVANS L. V. (1962) Periodicity in the liberation of gametes and zoospores of *Enteromorpha intestinalis* Link. *Nature* (*Lond.*), **193**, 193–4.

CHURCHILL A. C. & MOELLER H. W. (1972) Seasonal patterns of reproduction in New York populations of *Codium fragile* (Sur.) Hariot subsp. *tomentosoides* (van Goor) Silva. *J. Phycol.*, **8**, 147–52.

COLE K. & AKINTOBE S. (1963) The life cycle of *Prasiola meridionalis* Setchell and Gardner. *Can. J. Bot.*, **41**, 661–8.

CRAWLEY J. C. W. (1966) Some observations on the fine structure of the gametes and zygotes of *Acetabularia*. *Planta*, **69**, 365–76.

DIAZ-PIFERRER M. & BURROWS E. M. (1974) The life history of *Bryopsis hypnoides* Lamour. from Anglesey, North Wales and from the Caribbean. *J. Mar. Biol. Ass. U.K.*, **54**, 529–38.

DIXON P. S. (1963) Terminology and algal life histories, with particular reference to the Rhodophyta. *Ann. Bot.* N.S., **27**, 353–5.

DIXON P. S. (1970) The Rhodophyta: some aspects of their biology. II. *Oceanogr. Mar. Biol. Ann. Rev.*, **8**, 307–52.

DIXON P. S. (1973) *Biology of the Rhodophyta*. Hafner Press, New York.

DREW K. M. (1955) Life histories in the algae with special reference to the Chlorophyta, Phaeophyta and Rhodophyta. *Biol. Rev.*, **30**, 343–90.

DRING M. J. (1970) Photoperiodic effects in microorganisms. In *Photobiology of Microorganisms*, ed. Halldal P., pp. 345–68. Wiley-Interscience, New York.

DRING M. J. (1974) Reproduction. In *Algal Physiology and Biochemistry*, ed. Stewart W. D. P., pp. 814–37. University of California Press, Berkeley, California.

DUBE M. A. (1967) On the life history of *Monostroma fuscum* (Postels *et* Ruprecht) Wittrock. *J. Phycol.*, **3**, 64–73.

DUCKER S. C. (1965) The structure and reproduction of the green alga *Chlorodesmis bulbosa*. *Phycologia*, **4**, 149–62.

EDWARDS P. (1975) Evidence for a relationship between the genera *Rosenvingiella* and *Prasiola* (Chlorophyta). *Br. Phycol. J.*, **10**, 291–7.

ENOMOTO S. & HIROSE H. (1970) On the life history of *Anadyomene wrightii* with special reference to the reproduction, development, and cytological sequence. *Bot. Mag. Tokyo*, **83**, 270–80.

FAN K. C. (1959) Studies on the life histories of marine algae I. *Codiolum petrocelidis* and *Spongomorpha coalita*. *Bull. Torrey Bot. Club*, **86**, 1–12.

FELDMANN J. (1952) Les cycles de reproduction des Algues et leurs rapports avec la phylogénie. *Rev. Cytol.*, **13**, 1–49.

FELDMANN J. (1969) *Pseudobryopsis myura* and its reproduction. *Am. J. Bot.*, **56**, 691–5.

FELDMANN J. (1972) Les problémes actuels de l'alternance de générátions chez les algues. *Soc. Bot. France, Mém.*, **1972**, 7–38.

FJELD A. & LØVLIE A. (1976) Genetics of multicellular marine algae. In *The Genetics of Algae*, ed. Lewin R. A. pp. 219–35. University of California Press, Berkeley, California.

FLOYD G. L. & YARISH C. (1978) Comparative ultrastructure of the vegetative cells of selected marine Chaetophoraceae (Chlorophyta), *J. Phycol.*, **14**, (suppl.) abstr. 32.

FÖYN B. (1929) Untersuchungen über die Sexualität und Entwicklung von Algen. IV. Vorläufig Mitteillung über die Sexualität und den Generationswechsel von *Cladophora* und *Ulva*. *Ber. Dtsch. Bot. Ges.*, **47**, 495–506.

FÖYN B. (1934) Lebenszyklus und Sexualität der Chlorophycee *Ulva lactuca* L. *Arch. Protistenk.*, **83**, 154–77.

FÖYN B. (1958) Über die Sexualität und den Generationswechsel von *Ulva mutabilis* sp. n. *Arch. Protistenk.*, **102**, 473–80.

FÖYN B. (1959) Geschlechtkontrollierte Vererbung bei der marinen Grünalge *Ulva mutabilis*. *Arch. Protistenk*, **104**, 236–53.

FÖYN B. (1962) Diploid gametes in *Ulva*. *Nature* (*Lond.*), **193**, 300–1.

FRALICK R. A. & MATHIESON A. C. (1972) Winter fragmentation of *Codium fragile* (Suringar) Hariot ssp. *tomentosoides* (van Goor) Silva (Chlorophyceae, Siphonales) in New England. *Phycologia*, **11**, 67–70.

FRIEDMANN I. (1959) Structure, life-history, and sex determination of *Prasiola stipitata* Suhr. *Ann. Bot.*, N.S., **23**, 571–94.

FRIEDMANN I. (1963) Ecological aspects of the occurrence of meiosis in *Prasiola stipitata* Suhr. *Proc. Int. Seaweed Symp.*, **4**, 186–90.

FRIEDMANN I. (1969) Geographic and environmental factors controlling life history and morphology in *Prasiola stipitata* Suhr. *Öst. Bot. Z.*, **116**, 203–25.

FRIEDMANN I. & MANTON I. (1960) Gametes, fertilization and zygote development in *Prasiola stipitata* Suhr. I. Light microscopy, *Nova Hedwigia*, **1**, 333–44.

FRITSCH F. E. (1935) *The Structure and Reproduction of the Algae*, vol. I. Cambridge University Press, Cambridge.

FUJIYAMA T. (1955) On the life-history of *Prasiola japonica* Yatabe. *J. Fac. Fish. Anim. Husb. Hiroshima Univer.*, **1**, 15–45.

GAYRAL P. (1962) Reproduction et développement de *Monostroma obscurum* (Kütz). J. Ag. *Soc. Bot. France Bull.*, **109**, 53–9.

GAYRAL P. (1964) Sur le démembrement de l'actuel genre *Monostroma* Thuret (Chlorophycées, Ulotrichales s.l). *C. R. Acad. Sci. Paris*, Ser. D., **258**, 2149–52.

GAYRAL P. (1965) *Monostroma* Thuret, *Ulvaria* Rupr. emend. Gayral, *Ulvopsis* Gayral (Chlorophycées, Ulotrichales): structure, reproduction, cycles, position systématique. *Rev. Gen. Bot.*, **72**, 627–38.

GAYRAL P. (1971) Mise au point sur la systématique de l'ordre des Ulvales. *Soc. Phycol. France Bull.*, **16**, 63–7.

GOLDSTEIN M. & MORRALL S. (1970) Gametogenesis and fertilization in *Caulerpa*. *Ann. N.Y. Acad. Sci.*, **175**, 660–72.

GRAHAM E. (1975) Fruiting in *Halimeda* (Order Siphonales) 1. *Halimeda cryptica* Colinvaux and Graham. *Bull. Mar. Sci.*, **25**, 130–3.

GREEN B. R. (1973) Evidence for the occurrence of meiosis before cyst formation in *Acetabularia mediterranea* (Chlorophyceae, Siphonales). *Phycologia*, **12**, 233–5.

GREEN B. R. (1976a) Abnormal cells resulting from asexual reproduction in *Acetabularia* (Chlorophyceae, Siphonales). *Phycologia*, **15**, 161–4.

GREEN B. R. (1976b) Approaches to the genetics of *Acetabularia*. In *The Genetics of Algae*, ed. Lewin R. A., pp. 236–56. University of California Press, Berkeley, California.

HANIC L. A. (1965) Life History Studies of Urospora and Codiolum from Southern British Columbia. Ph.D. Dissertation, University of British Columbia.

HANIC L. A. (1979) Observations on *Prasiola meridionalis* (S. & G.) and *Rosenvingiella constricta* (S. & G.) Silva (Chlorophyta, Prasiolales) from Galiano Island, British Columbia. *Phycologia*, **18**, 71–6.

HARTMANN M. (1929) Untersuchungen über die Sexualität und Entwicklung von Algen. III. Über die Sexualität und den Generationswechsel von *Chaetomoprha* und *Enteromorpha*. *Ber. Dtsch. Bot. Ges.*, **47**, 485–94.

HIROSE H. & YOSHIDA K. (1964) Review of the life history of the genus *Monostroma* (Japanese with English summary). *Bull. Jap. Soc. Phycol*, **12**, 19–31.

HOLLENBERG G. J. (1935) A study of *Halicystis ovalis*. I. Morphology and reproduction. *Am. J. Bot.*, **22**, 782–812.

HOLLENBERG G. J. (1958) Observations concerning the life cycle of *Spongomorpha coalita* (Ruprecht) Collins. *Madroño*, **14**, 249–51.

HORI T. & UEDA R. (1967) Electron microscope studies on the fine structure of plastids in siphonous green algae with special reference to their phylogenetic relationships. *Sci. Rept. Tokyo Kyoiku Daigoku*, Sec. B., **12**, 225–44.

HOXMARK R. C. (1975) Experimental analysis of the life cycle of *Ulva mutabilis*. *Bot. Mar.*, **18**, 123–9.

HOXMARK R. C. & NORDBY Ø. (1974) Haploid meiosis as a regular phenomenon in the life cycle of *Ulva mutabilis*. *Hereditas*, **76**, 239–50.

HUIZING H. J. & RIETEMA H. (1975) Xylan and mannan as cell wall constituents of different stages in the life-histories of some siphonous green algae. *Br. Phycol. J.*, **10**, 13–16.

HUIZING H. J., RIETEMA H. & SIETSMA J. H. (1979) Cell wall constituents of several siphoneous green algae in relation to morphology and taxonomy. *Br. Phycol. J.*, **14**, 25–32.

HUSTEDE H. (1960) Über den Generationswechsel zwischen *Derbesia neglecta* Berth und *Bryopsis halymeniae* Berth. *Naturwissenschaften*, **47**, 19.

HUSTEDE H. (1964) Entwicklungsphysiologische Untersuchungen über den Generationswechsel zwischen *Derbesia neglecta* Berth. und *Bryopsis halymeniae* Berth. *Bot. Mar.*, **6**, 134–42.

IYENGAR M. O. P. (1940) On the formation of gametes in *Caulerpa*. *J. Ind. Bot. Soc.*, **18**, 191–4.

IYENGAR M. O. P. & RAMANATHAN K. R. (1941) On the life-history and cytology of *Microdictyon tenius* (Ag.) Decsne. *New Phytol.*, **20**, 157–9.

JÓNSSON S. (1957) Sur la reproduction et la sexualité de l'*Acrosiphonia spinescens* (Kütz) Kjellm. *C.R. Acad. Sci. Paris*, Ser. D., **244**, 1251–4.

JÓNSSON S. (1959a) L'existence de l'alternance hétéromorphe de générations entre l'*Acrosiphonia spinescens* Kjellm. et le *Codiolum petrocelidis* Kuck. *C.R. Acad. Sci. Paris*, Ser. D, **248**, 835–7.

JÓNSSON S. (1959b) Le cycle de développement de *Spongomorpha lanosa* (Roth) Kütz. et la nouvelle famille des Acrosiphoniacées. *C.R. Acad. Sci. Paris*, Ser. D, **248**, 1565–7.

JÓNSSON S. (1962a) Sur la reproduction de l'*Anadyomene stellata* (Wulf.) Ag. de la Mediterranée. *C.R. Acad. Sci. Paris*, Ser. D, **255**, 1983–5.

JÓNSSON S. (1962b) Recherches sur des Cladophoracées marines (structure, reproduction, cycles comparés, conséquences systématiques). *Ann. Sci. Nat. Bot.* Ser. **12**, 25–230.

JÓNSSON S. (1966) Sur l'identification du sporophyte de *Spongomorpha lanosa* (Roth) Kütz. (Acrosiphoniacées). *C.R. Acad. Sci. Paris*, Ser. D, **262**, 626–9.

JÓNSSON S. (1967) Sur l'existence de générations mictohaploïdes issues de faux zygotes codioïales chez l'*Acrosiphonia sonderi* (Kütz) Kornm. *C.R. Acad. Sci. Paris*, Ser. D, **264**, 1009–12.

KAPRAUN D. F. (1970) Field and cultural studies of *Ulva* and *Enteromorpha* in the vicinity of Port Aransas, Texas. *Contrib. Mar. Sci.*, **15**, 205–85.

KAPRAUN D. F. & FLYNN E. H. (1973) Culture studies of *Enteromorpha linza* (L.) J. Ag. and *Ulvaria oxysperma* (Kützing) Bliding (Chlorophyceae, Ulvales) from Central America. *Phycologia*, **12**, 145–52.

KIDA W. (1966) Studies on the morphology and ecology of *Monostroma* in Ise Bay and vicinity, Japan. *J. Fac. Fish., Pref. Univ. Mie*, **7**, 82–159.

KOBARA T. & CHIHARA M. (1978a) On the taxonomy and sexual reproduction of the siphonous green alga *Pseudobryopsis hainanensis* Tseng. *J. Jap. Bot.*, **53**, 341–52.

KOBARA T. & CHIHARA M. (1978b) On the life history of *Pseudobryopsis hainanensis* (Chlorophyceae). *J. Jap. Bot.*, **53**, 353–60.

KOHLER K. (1956) Entwicklungsgeschichte, Geschlechtsbestimmung, und die Fruchtung bei *Chaetomorpha*. *Arch. Protistenk.*, **101**, 223–68.

KOOP H.-U. (1975) Über den Ort der Meiosis bei *Acetabularia mediterranea*. *Protoplasma*, **84**, 109–14.

KOOP H.-U. (1977) Genetic aspects of *Acetabularia mediterranea*. In *Progress in Acetabularia Research*, ed Woodcock C. L. F., pp. 7–18. Academic Press, New York.

KOOP H.-U. (1979) The life cycle of *Acetabularia* (Dasycladales, Chlorophyceae): a compilation of evidence for meiosis in the primary nucleus. *Protoplasma*, **100**, 353–66.

KORNMANN P. (1938) Zur Entwicklungsgeschichte von *Derbesia* und *Halicystis*. *Planta*, **28**, 464–70.

KORNMANN P. (1956) Zur Morphologie und Entwicklung von *Percursaria percursa*. *Helgol. Wiss. Meeresunters.*, **5**, 259–72.

KORNMANN P. (1959) Die heterogene Gattung *Gomontia*. I. Der sporangiale Anteil, *Codiolum polyrhizum*. *Helgol. Wiss. Meeresunters.*, **6**, 229–38.

KORNMANN P. (1960) Die heterogene Gattung *Gomontia*. II. Der fädige Anteil *Eugomontia sacculata* nov. gen. nov. spec. *Helgol. Wiss. Meeresunters.*, **7**, 59–71.

KORNMANN P. (1961a) Die Entwicklung von *Codiolum gregarium* A. Braun. *Helgol. Wiss. Meeresunters.*, **7**, 252–9.

KORNMANN P. (1961b) Über *Spongomorpha lanosa* und ihre Sporophytenformen. *Helgol. Wiss. Meeresunters.*, **7**, 195–205.

KORNMANN P. (1961c) Über *Codiolum* und *Urospora*. *Helgol. Wiss. Meeresunters.*, **8**, 42–57.

KORNMANN P.(1962) Die Entwicklung von *Monostroma grevillei*. *Helgol. Wiss. Meeresunters.*, **8**, 195–202.

KORNMANN P. (1963a) Die Ulotrichales, neu geordnet auf der Grundlage entwicklungsgeschichtlicher Befunde. *Phycologia*, **3**, 60–8.

KORNMANN P. (1963b) Der Lebenszyklus einer marinen *Ulothrix*-Art. *Helgol. Wiss. Meersunters.*, **8**, 357–60.

KORNMANN P. (1964a) Über *Monostroma bullosum* (Roth) Thuret und *M. oxyspermum* (Kütz) Doty. *Helgol. Wiss. Meeresunters.*, **11**, 13–21.

KORNMANN P. (1964b) Die *Ulothrix*-Arten von Helgoland. I. *Helgol. Wiss. Meeresunters.*, **11**, 27–38.

KORNMANN P. (1964c) Der Lebenszyklus von *Acrosiphonia arcta*. *Helgol. Wiss. Meeresunters.*, **11**, 110–17.

KORNMANN P. (1964d) Zur Biologie von *Spongomorpha aeruginosa* (Linnaeus) van den Hoek. *Helgol. Wiss. Meeresunters.*, **11**, 200–8.

KORNMANN P. (1965a) Was ist *Acrosiphonia arcta*? *Helgol. Wiss. Meeresunters.*, **12**, 40–51.

KORNMANN P. (1965b) Ontogenie und Lebenszyklus der Ulotrichales in phylogenetischer Sicht. *Phycologia*, **4**, 163–72.

KORNMANN P. (1966) Eine erbliche Variante von *Derbesia marina*. *Naturwissenschaften*, **53**, 161.

KORNMANN P. (1969) Gesetzmassigkeiten des Wachstums und der Entwicklung von *Chaetomorpha darwinii* (Chlorophyta, Cladophorales). *Helgol. Wiss Meeresunters.*, **19**, 335–54.

KORNMANN P. (1970a) Mutation in the siphonaceous green alga *Derbesia marina*. *Helgol. Wiss. Meeresunters.*, **21**, 1–8.

KORNMANN P. (1970b) Advances in marine phycology on the basis of cultivation. *Helgol. Wiss. Meeresunters.*, **20**, 39–61.

KORNMANN P. (1970c) Phylogenetische Beziehungen in der Grünalgengattung *Acrosiphonia*. *Helgol. Wiss. Meeresunters.*, **21**, 292–304.

KORNMANN P. (1970d) Der Lebenszyklus von *Acrosiphonia grandis* (Acrosiphoniales, Chlorophyta). *Mar. Biol.*, **7**, 324–31.

KORNMANN P. (1972a) Ein Beitrag zur Taxonomie der Gattung *Chaetomorpha* (Cladophorales, Chlorophyta). *Helgol. Wiss. Meeresunters.*, **23**, 1–31.

KORNMANN P. (1972b) Les sporophytes vivant en endophyte de quelques Acrosiphoniacées et leur rapports biologiques et taxonomiques. *Soc. Bot. Fr., Mém.*, **1972**, 75–86.

KORNMANN P. (1973) Codiolophyceae, a new class of Chlorophyta. *Helgol. Wiss. Meeresunters.*, **25**, 1–13.

KORNMANN P. & SAHLING P.-H. (1962) Zur Taxonomie und Entwicklung der *Monostroma*-Arten von Helgoland. *Helgol. Wiss. Meeresunters.*, **8**, 302–20.

KORNMANN P. & SAHLING P.-H. (1974) Prasiolales (Chlorophyta) von Helgoland. *Helgol. Wiss. Meeresunters.*, **26**, 99–133.

KORNMANN P. & SAHLING P.-H. (1976) Wiedereinführung von *Bryopsis lyngbyei* (Bryopsidales, Chlorophyta) als selbständige Art. *Helgol. Wiss. Meeresunters.*, **28**, 217–25.

KORNMANN P. & SAHLING P.-H. (1977) Meeresalgen von Helgoland. *Helgol. Wiss. Meeresunters.*, **29**, 1–289.

KOSHIZAKA M. (1978) On the life history of eight species of *Cladophora* (Chlorophyta, Cladophoraceae) (Japanese with English abstract). *Bull. Fac. Fish., Hokkaido Univ.*, **29**, 325–37.

KYLIN H. (1938) Beziehungen zwischen Generationswechsel und Phylogenie. *Arch. Protistenk.*, **90**, 432–47.

KYLIN H. (1949) Die Chlorophyceen der Schwedischen Westküste. *Lunds. Univ. Årsskrift N.F. Avd.*, **45**, 1–79.

LEVRING T. (1955) Some remarks on the structure of the gametes and the reproduction of *Ulva lactuca*. *Bot. Not.*, **108**, 40–5.

LOKHORST G. M. (1978) Taxonomic studies on the marine and brackish-water species of *Ulothrix* (Ulotrichales, Chlorophyceae) in western Europe. *Blumea*, **24**, 191–299.

LOKHORST G. M. & VROMAN M. (1972) Taxonomic study on three freshwater *Ulothrix* species. *Acta. Bot. Neerl.*, **21**, 449–80.

LOKHORST G. M. & VROMAN M. (1974a) Taxonomic studies on the genus *Ulothrix* (Ulotrichales, Chlorophyceae) II. *Acta. Bot. Neerl.*, **23**, 369–98.

LOKHORST G. M. & VROMAN M. (1974b) Taxonomic studies on the genus *Ulothrix* (Ulotrichales, Chlorophyceae) III. *Acta. Bot. Neerl.*, **23**, 561–602.

LØVLIE A. (1964) Genetic control of division rate and morphogenesis in *Ulva mutabilis* Föyn. *C.R. Trav. Lab. Carlsberg*, **34**, 77–168.

LØVLIE A. & BRYHNI E. (1978) On the relation between sexual and parthenogenetic reproduction in haplo-diplontic algae. *Bot. Mar.*, **21**, 155–63.

MACRAILD G. N. & WOMERSLEY H. B. S. (1974) The morphology and reproduction of *Derbesia clavaeformis* (J. Agardh) De Toni (Chlorophyta). *Phycologia*, **13**, 83–93.

MANTON I. & FREIDMANN I. (1960) Gametes, fertilization and zygote development in *Prasiola stipitata* Suhr. II. Electron microscopy. *Nova Hedwigia*, **1**, 443–62.

MATTOX K. R. & STEWART K. D. (1973) Observations on the zoospores of *Pseudendoclonium basiliense* and *Trichosarcina polymorpha* (Chlorophyceae). *Can. J. Bot.*, **51**, 1425–30.

MAYHOUB H. (1975) Reproduction sexuée et cycle du développement de l'*Anadyomene stellata* (Wulf.) Ag. de la Mediterranée orientale. *C. R. Acad. Sci. Paris*, Ser. D, **280**, 587–90.

MAYHOUB H. (1976) Sur l'existence d'un *Derbesia* dans le cycle de l'*Halicystis boergesenii* Iyengar et Raman (Chlorophycées, Derbésiales). *C. R. Acad. Sci. Paris*, Ser. D, **282**, 707–10.

MEINESZ A. (1969) Sur la reproduction sexuée de l' *Udotea petiolata* (Turra) Boerg. *C. R. Acad. Sci. Paris*, Ser. D, **269**, 1063–5.

MIGITA S. (1967) On the structure and life history of *Rhizoclonium riparium* Kütz. from Kyushu. *Bull. Jap. Soc. Phycol.*, **15**, 9–17.

MIYAJI K. & KUROGI M. (1976) On the development of zoospores of the green alga *Chlorochytrium inclusum* (Japanese with English summary). *Bull. Jap. Soc. Phycol.*, **24**, 121–9.

MOEWUS L. (1949) Zur Biologie und Systematik der Gattung *Ectochaete* (*E. polymorpha* und *E. ramulosa*). *Bot. Not.*, **1949**, 283–312.

NEUMANN K. (1967) Der Ort der Meiosis und die Sporenbildung bei der siphonalen Grünalge *Derbesia marina*. *Naturwissenschaften*, **54**, 121.

NEUMANN K. (1969a) Beitrag zur Cytologie und Entwicklung der siphonalen Grünalge *Derbesia marina*. *Helgol. Wiss. Meeresunters.*, **19**, 355–75.

NEUMANN K. (1969b) Protonema mit Riesenkern bei der siphonalen Grünalge *Bryopsis hypnoides* und weitere cytologische Befunde. *Helgol. Wiss. Meeresunters.*, **19**, 45–57.

NEUMANN K. (1970) Einkerniges Protonema bei *Bryopsis* und *Pseudobryopsis myura*. *Helgol. Wiss Meeresunters.*, **20**, 213–5.

NIELSEN R. (1972) A study of the shell-boring marine algae around the Danish island Laesø. *Bot. Tiddskr.*, **67**, 245–69.

NIELSEN R. (1977) Culture studies on *Ulvella lens* and *Ulvella setchellii*. *Br. Phycol., J.*, **12**, 1–5.

NIELSEN R. (1978) Variation in *Ochlochaete hystrix* (Chaetophorales, Chlorophyceae) studied in culture. *J. Phycol.*, **14**, 127–31.

NIELSEN R. (1979) Culture studies on the type species of *Acrochaete*, *Bolbocoleon* and *Entocladia* (Chaetophoraceae, Chlorophyceae). *Bot. Not.*, **132**, 441–9

NIELSEN R. & PEDERSEN P. M. (1977) Separation of *Syncoryne reinkei* nov. gen., nov. sp. from *Pringsheimiella scutata* (Chlorophyceae, Chaetophoraceae). *Phycologia*, **16**, 411–6.

NIENHUIS P. H. (1974) Variability in the life cycle of *Rhizoclonium riparium* (Roth) Harv. (Chlorophyceae: Cladophorales) under Dutch estuarine conditions. *Hydrobiol. Bull.*, **8**, 172–8.

NIENHUIS P. H. (1975) *Biosystematics and ecology of* Rhizoclonium riparium *(Roth) Harv. (Chlorophyceae, Cladophorales) in the estuarine area of the rivers Rhine, Meuse and Scheldt.* Bronder Offset B.V., Rotterdam.

NILSEN G. & NORDBY Ø. (1975) A sporulation-inhibiting substance from vegetative thalli of the green alga *Ulva mutabilis* Föyn. *Planta*, **125**, 127–39.

NORDBY Ø. & HOXMARK R. C. (1972) Changes in cellular parameters during synchronous meiosis in *Ulva mutabilis* Föyn. *Exp. Cell Res.*, **75**, 321–8.

OHNO M. (1972) The periodicity of gamete liberation in *Monostroma*. *Proc. Int. Seaweed Symp.*, **7**, 405–9.

O'KELLY C. J. & YARISH C. (1981) Observations on marine Chaetophoraceae (Chlorophyta) II. On the circumscription of the genus *Entocladia* Reinke. *Phycologia*, **20**, 32–45.

OKUDA K., ENOMOTO S., & TATEWAKI M. (1979) Life history of *Pseudobryopsis* sp. (Codiales, Chlorophyta). *Jap. J. Phycol.*, **27**, 7–16.

OKUDA T. (1975) Reproduction of ulvaceous algae with special reference to the periodic fruiting II. Fruiting of *Ulva pertusa* during the neap tides in Okayama. *J. Fac. Agr., Kyushu Univ.*, **19**, 149–57.

OTTGAI M. & FUJIYAMA T. (1977) Studies on the life-history of *Ulothrix* II. On the sexual reproduction and life-history of *Ulothrix implexa*, Kützing prox. (Japanese with English summary). *J. Shimonseki Univ. Fish.*, **24**, 337–42.

PAGE J. Z. & KINGSBURY J. M. (1968) Culture studies on the marine green alga *Halicystis parvula-Derbesia tenuissima*. II. Synchrony and periodicity in gamete formation and release. *Am. J. Bot.*, **55**, 1–11.

PAGE J. Z. & SWEENEY B. M. (1968) Culture studies on the marine green alga *Halicystis parvula-Derbesia tenuissima*. III. Control of gamete formation by an endogenous rhythm. *J. Phycol.*, **4**, 253–60.

PAPENFUSS G. F. (1955) Classification of the algae. In *A Century of Progress in the Natural Sciences, 1853–1953*, pp. 115–224. California Academy Science, San Francisco, California.

PAPENFUSS G. F. (1962) On the circumscription of the green algal genera *Ulvella* and *Pilinia*. *Phykos*, **1**, 6–12.

PARKE M. & DIXON P. S. (1976) Check-list of British marine algae—third revision. *J. Mar Biol. Ass. U.K.*, **56**, 527–94.

PERROT Y. (1968) Sur le cycle de deux formes d'*Ulothrix flacca* (Dillw.) Thuret de la région de Roscoff. *C. R. Acad. Sci. Paris*, Ser. D, **266**, 1953–5.

PERROT Y. (1970) Sur la spécifité et le cycle de l'*Ulothrix subflaccida* (Wille) de la région de Roscoff. *C. R. Acad. Sci. Paris*, Ser. D, **270**, 932–3.

PERROT Y. (1971) Sur le cycle de reproduction de l'*Ulothrix pseudoflacca* Wille de la région de Roscoff. *C. R. Acad. Sci. Paris*, Ser. D, **273**, 858–9.

PERROT Y. (1972) Les *Ulothrix* marins de Roscoff et le problème de leur cycle de reproduction. *Soc. Bot. Fr., Mém.*, **1972**, 67–74.

PICKETT-HEAPS J. D. & MARCHANT H. J. (1972) The phylogeny of the green algae: a new proposal. *Cytobios*, **6**, 255–64.

PRICE I. R. (1972) Zygote development in *Caulerpa* (Chlorophyta, Caulerpales). *Phycologia*, **11**, 217–8.

PUISEUX-DAO S. (1962) Recherches biologiques et physiologiques sur quelques Dasycladacées. *Rev. Gén. Bot.*, 819, 409–503.

PUISEUX-DAO S. (1970) *Acetabularia and Cell Biology*. Springer, New York.

RAMUS J. (1972) Differentiation of the green alga *Codium fragile*. *Am. J. Bot.*, **59**, 478–82.

RHYNE C. F. (1973) Field and experimental studies on the systematics and ecology of *Ulva curvata* and *Ulva rotundata*. *Sea Grant Publ.* UNC–SG–73–09.

RIETEMA H. (1969) A new type of life history in *Bryopsis*, Chlorophyceae, Caulerpales. *Acta Bot. Neerl.*, **18**, 615–9.

RIETEMA H. (1970) Life histories of *Bryopsis plumosa*, Chlorophyceae, Caulerpales, from European coasts. *Acta Bot. Neerl.*, **19**, 859–66.

RIETEMA H. (1971a) Life history studies in the genus *Bryopsis*, Chlorophyceae. Part 3. The life history of *Bryopsis monoica*. *Acta Bot. Neerl.*, **20**, 205–10.

RIETEMA H. (1971b) Life history studies in the genus *Bryopsis*, Chlorophyceae. Part 4. Life histories in *Bryopsis hypnoides* from different points along the European coasts. *Acta. Bot. Neerl.*, **20**, 291–8.

RIETEMA H. (1972) A morphological, developmental, and caryological study on the life-history of *Bryopsis halymeniae* (Chlorophyceae). *Neth. J. Sea Res.*, **5**, 445–57.

RIETEMA H. (1973) The influence of daylength on the morphology of the *Halicystis parvula* phase of *Derbesia tenuissima* (De Not.) Crn. (Chlorophyceae, Caulerpales). *Phycologia*, **12**, 11–6.

ROE K. E. (1975) Origin of the alternation of generations in plants: reconsideration of the traditional theories. *The Biologist*, **57**, 1–13.

ROUND F. E. (1971) The taxonomy of the Chlorophyta II. *Br. Phycol. J.*, **6**, 235–64.

RUSSELL G. (1971) Marine algal reproduction in two British estuaries. *Vie et Milieu*, **22**, (suppl.) 219–32.

RUSSELL G. (1973) The Phaeophyta: a synopsis of some recent developments. *Oceanogr. Mar. Biol. Ann. Rev.*, **11**, 45–88.

SAWADA T. (1972) Periodic fruiting of *Ulva pertusa* at three localities in Japan. *Proc. Int. Seaweed Symp.*, **7**, 229–30.

SAWADA T. & WATANBE T. (1974) Reproduction of ulvaceous algae with special references to the periodic fruiting I. Semilunar rhythmicity of *Ulva pertusa* in northern Kyushu. *J. Fac. Agr., Kyushu Univ.*, **18**, 81–8.

SCAGEL R. F. (1966a) The Phaeophyceae in perspective. *Oceanogr. Mar. Biol. Ann. Rev.*, **4**, 123–94.

SCAGEL R. F. (1966b) Marine algae of British Columbia and northern Washington, Part I: Chlorophyceae (green algae). *Nat. Mus. Canada Bull.*, **207**, 1–257.

SCHECHNER-FRIES M. (1934) Beiträge zur Entwicklungsgeschichte der Protophyten. XI. Der Phasenwechesel von *Valonia utricularis* (Roth) Ag. *Oesterr. Bot. Zeitschr.*, **83**, 241–54.

SCHULZE K. L. (1939) Cytologische Untersuchungen an *Acetabularia mediterranea* und *Acetabularia wettsteinii*. *Arch. Protistenk.*, **92**, 179–225.

SCHUSSNIG B. (1930) Der Generations- und Phasenwechsel bei den Chlorophyceen. *Oesterr. Bot. Zeitschr.*, **79**, 58–77.

SCHUSSNIG B. (1932) Der Generations- und Phasenwechsel bei den Chlorophyceen III. *Oesterr. Bot. Zeitschr.*, **81**, 296–8.

SCHUSSNIG B. (1938) Der Kernphasenwechsel von *Valonia utricularis* (Roth) Ag. *Planta*, **28**, 43–59.

SCHUSSNIG B.(1939) Ein Beitrag zur Entwicklungsgeschichte von *Caulerpa prolifera*. *Bot. Not.*, **92**, 75–96.

SCHUSSNIG B. (1950) Die Gametogenesis von *Codium decorticatum* (Woodw.) Howe. *Sv. Bot. Tidskr.*, **44**, 55–71.

SEARS J. R. & WILCE R. T. (1970) Reproduction and systematics of the marine alga *Derbesia* (Chlorophyceae) in New England. *J. Phycol.*, **6**, 381–92.

SIDDIQUE M. & FARIDI M. A. F. (1977) The life history of *Cladophora crispata* (Roth) Ag. *Pak. J. Bot.*, **9**, 159–62.

SINHA J. P. (1958) Chromosome numbers and life cycles in members of Cladophorales. *Br. Phycol. J.*, **1**, 24–7.

SMITH G. M. (1938) Nuclear phases and alternation of generations in the Chlorophyceae. *Bot. Rev.*, **4**, 132–9.

SMITH G. M. (1947) On the reproduction of some Pacific coast species of *Ulva*. *Am. J. Bot.*, **34**, 80–7.

SMITH G. M. (1955) *Cryptogamic Botany. Vol. I. Algae and Fungi*, 2nd edn. McGraw Hill, New York.

STEWART K. D. & MATTOX K. R. (1975) Comparative cytology, evolution and classification of the green algae with some consideration of the origin of other organisms with chlorophylls *a* and *b*. *Bot. Rev.*, **41**, 104–35.

STEWART K. D. & MATTOX K. R. (1978) Structural evolution in the flagellated cells of green algae and land plants. *BioSystems*, **10**, 145–52.

STEWART K. D., MATTOX K. R. & FLOYD G. L. (1973) Mitosis, cytokinesis, the distribution of plasmodesmata, and other cytological characteristics in the Ulotrichales, Ulvales, and Chaetophorales: phylogenetic and taxonomic considerations. *J. Phycol.*, **9**, 128–41.

SUNESON A. (1947) Notes on the life history of *Monostroma*. *Sv. Bot. Tidskr.*, **41**, 235–46.

SVEDELIUS N. (1931) Nuclear phases and alternation in the Rhodophyceae. *Beih. Bot. Centralbl.*, **48**, 38–59.

TANNER C. E. (1979) The Taxonomy and Morphological Variation of Distromatic Ulvaceous Algae (Chlorophyta) from the Northeast Pacific. Ph.D. Dissertation, University of British Columbia, Vancouver.

TANNER C. E. (1980) *Chloropelta* gen. nov., an ulvaceous green alga with a different type of development. *J. Phycol.*, **16**, 128–37.

TATEWAKI M. (1969) Culture studies on the life history of some species of the genus *Monostroma*. *Mem. Fac. Fish., Hokkaido Univ.*, **2**, 1–264.

TATEWAKI M. (1972) Life history and systematics in *Monostroma*. In *Contributions to the Systematics of Benthic Marine Algae of the North Pacific*, eds. Abbott I. A. & Kurogi M., pp. 1–15. Japanese Society of Phycology, Kobe.

VAN DEN HOEK C. (1963) *Revision of the European Species of* Cladophora. Brill, Leiden.

VAN DEN HOEK C., CORTEL-BREEMAN A. M., RIETEMA H. & WANDERS J. B. W. (1972) L'interprétation des données obtenues, par des cultures unialgales, sur les cycles évolutifs des algues. Quelques exemples tirés des recherches conduites au laboratoire de Gronique. *Soc. Bot. Fr., Mém.*, **1972**, 45–66.

VAN DEN HOEK C. & JAHNS H. M. (1978) *Algen, Einführung in die Phykologie*. Georg Thieme Verlag, Stuttgart, Germany.

VINOGRADOVA K. L. (1969) K sistematike porjadka Ulvales (Chlorophyta), s.l. Risunkom. *Botaniceskij Zhurn., Leningrad*, **54**, 1347–55.

VINOGRADOVA K. L. (1972) On the taxonomic position and phylogeny of the order Siphonocladales (Chlorophyta) (Russian with English Summary). *Botaniceskij Zhurn. (Leningrad)*, **57**, 1232–9.

VINOGRADOVA K. L. (1974) *Ulvales (Chlorophyta) of the USSR Seas* (Russian). Nauka, Leningrad.

VINOGRADOVA K. L. (1976) Life cycles of the Chlorophyta and some aspects of their evolution (Russian with English summary). *Botaniceskij Zhurn. (Leningrad)*, **61**, 1041–8.

WIK-SJØSTEDT A. (1970) Cytogenetic investigations in *Cladophora*. *Hereditas*, **66**, 233–62.

WILKINSON M. (1975) The occurrence of shell-boring *Phaeophila* species in Britain. *Br. Phycol. J.*, **10**, 235–40.

WILKINSON M. & BURROWS E. M. (1970) *Eugomontia sacculata* Kornm. in Britain and North America. *Br. Phycol., J.*, **5**, 235–8.

WILKINSON M. & BURROWS E. M. (1972) An experimental taxonomic study of the alga confused under the name *Gomontia polyrhiza*. *J. Mar. Biol. Ass. U.K.*, **52**, 49–57.

WILLIAMS M. M. (1925) Cytology of the gametangia of *Codium tomentosum* (Stackh.). *Proc. Linn. Soc. N.S.W.*, **50**, 98–111.

WOODCOCK C. L. F. & MILLER G. J. (1973a) Ultrastructural features of the life cycle of *Acetabularia mediterranea*. I. Gametogenesis. *Protoplasma*, **77**, 313–9.

WOODCOCK C. L. F. & MILLER G. J. (1973b) Ultrastructural features of the life cycle of *Acetabularia mediterranea*. II. Events associated with the division of the primary nucleus and the formation of cysts. *Protoplasma*, **77**, 331–41.

WUTZ M. & ZETSCHE K. (1976) Zur Biochemie und Regulation des heteromorphen Generationswechsels der Grünalge *Derbesia-Halicystis*. *Planta*, **129**, 211–16.

WYNNE M. J. & LOISEAUX S. (1976) Recent advances in life history studies of the Phaeophyta. *Phycologia*, **15**, 435–52.

YAMADA Y. & KANADA T. (1941) On the culture experiment of *Monostroma zostericola* and *Enteromorpha nana* var. *minima*. *Sci. Pap. Inst. Algol. Res., Fac. Sci., Hokkaido Imp. Univ.*, **2**, 217–26.

YAMADA Y. & SAITO E. (1938) On some culture experiments with the swarmers of certain species belonging to the Ulvaceae. *Sci. Pap. Inst. Algol. Res., Fac. Sci., Hokkaido Imp. Univ.*, **2**, 35–51.

YAMADA Y. & TATEWAKI M. (1965) New findings on the life history of *Monostroma zostericola* Tilden. *Sci. Pap. Inst. Algol. Res., Fac. Sci., Hokkaido Imp. Univ.*, **5**, 105–17.

YARISH C. (1975) A cultural assessment of the taxonomic criteria of selected marine Chaetophoraceae (Chlorophyta). *Nova Hedwigia*, **26**, 385–430.

YARISH C. (1976) Polymorphism of selected marine Chaetophoraceae (Chlorophyta). *Br. Phycol. J.*, **11**, 29–38.

ZIEGLER J. R. & KINGSBURY J. M. (1964) Cultural studies on the marine green alga *Halicystis parvula-Derbesia tenuissima*. I. Normal and abnormal sexual and asexual reproduction. *Phycologia*, **4**, 105–16.

ZINNECKER E. (1935) Reduktionsteilung, Kernphasenwechsel und Geschlechtsbestimmung bei *Bryopsis plumosa* (Huds.) Ag. *Oesterr. Bot. Zeitschr.*, **84**, 53–72.

CHAPTER 7
CYTOLOGY AND ULTRASTRUCTURE

SUSAN H. BRAWLEY
Marine Systems Laboratory, Department of Paleobiology,
Smithsonian Institution, Washington, D.C. 20560, U.S.A.

RICHARD WETHERBEE
School of Botany, University of Melbourne,
Parkville 3052, Australia

1 INTRODUCTION

Algal cells have interesting, unique features, but they are similar in many respects to other eukaryotic cells. Little tissue development is observed however in most algae compared with that present in higher plants and animals. As a result, the algae have frequently provided superior systems for understanding processes such as secretion, differentiation, and gene expression, which are common to all eukaryotic cells.

It is helpful as we examine algal structure to remember that we are observing an evolutionary continuum. For example, the endosymbiotic theory for the origin of plastids and mitochondria from prokaryotic organisms (i.e. blue-green algae and/or *Prochloron*-type cells and bacteria) has gained increasing credence

(Margulis 1970; Stanier & Cohen-Bazire 1977; Giddings *et al.* 1980). The different types of organisms 'captured' by the progenitors of each algal group may explain the extra membranes associated with the plastids in some groups (Section 4.1; Lee 1977; Gibbs 1978), and this event might also have been a key factor in suppressing tissue development and specialization. For example, tissue specialization resulting in improved locomotion and food capture was unlikely to evolve after chloroplasts were present within the cell. The rhizoplasts (Section 4.5) observed in some algal motile cells may be instructive in this regard since they resemble myofibrils in the muscular tissue of animals. Certainly the ability of the ancestral algal cells to capture and retain a 'food factory', the plastid, was significant and must have had a profound effect on evolution at the cellular and tissue levels.

We have emphasized recent literature in our review of cellular structure within the marine Rhodophyta, Phaeophyta, and Chlorophyta. Additional information, including many references to the classical literature, may be found in Dodge (1973). In some cases, we mention common and important organelles (e.g. the lysosome) briefly because little is known about them in the marine algae. In addition to reviewing general information on the common algal organelles, we discuss a number of special topics to emphasize our opinion that cellular structure is most meaningful when it is viewed in a functional context. In fact, such an orientation is essential to understanding cellular development and evolutionary trends within the algae. We have assumed some basic knowledge from the reader about the various organelles, but detailed information on organelle structure and function may be found in a number of recent books and reviews (e.g. Gunning & Steer 1975; Novikoff 1976; Brinkley & Porter 1977; Smith 1977; Dustin 1978; Gillham 1978; Beevers 1979; Raff 1979; Roland & Vian 1979; Whaley & Dauwalder 1979).

2 RHODOPHYTA

2.1 Introduction

The red algae are an ancient assemblage, and although members of this division may exhibit extreme complexity in their reproductive development and life histories, they also possess several primitive features. These include the presence of phycobilin pigments and lack of flagellated cells, both features of the prokaryotic blue-green algae. The red algae are therefore believed to have diverged from very primitive stock and are often postulated to be the oldest of eukaryotic cells.

Fine structural and cytological studies on the red algae have confirmed many of their unique and interesting features. A great majority of the early work has been descriptive, though recent investigations have concentrated on the mechanisms and controls of development. The excellent work on the morphogenesis of *Griffithsia* by S. D. Waaland and colleagues (e.g. Waaland 1978) has helped demonstrate the potential for using some types of red algal cells for experimental studies. Other areas of increasing interest include ultrastructural studies on cell division and differentiation of reproductive structures (e.g. Scott *et al.* 1980), the development and nutrition of red algal parasites (e.g. Goff 1979a, b) and the structure, composition and function of pit connections (e.g. Pueschel 1980; Wetherbee & Kraft 1981). These and other select areas of investigation are discussed here, though other areas of interest are only discussed briefly, as another review on red algal cytology and ultrastructure (Duckett & Peel 1978) has recently appeared.

2.2 Chloroplasts, pyrenoid and mitochondria

Fine-structural aspects of the chloroplast, in addition to pigment composition and other biochemical criteria, are of considerable significance as defining features of the Rhodophyta (Dixon 1973; Duckett & Peel 1978). Chloroplasts are surrounded by an envelope composed of two membranes and contain single, unstacked thylakoids (Fig. 7.1). One or more peripheral thylakoids (or 'inner limiting discs', Bouck 1962) may encircle the plastid (Bisalputra 1974; Hara & Chihara 1974). The water-soluble accessory pigments, allophycocyanin, R-phycocyanin and phycoerythrin, are assembled into structures known as phycobilisomes which are distributed in a distinct pattern on the thylakoid surface (Gantt & Conti 1965; Neushul 1970). Several different cytochemical techniques, including immunoelectron microscopy, have been used to generate a model for the structure of phycobilisomes in *Porphyridium* (Gantt & Lipschultz 1977; Gantt 1977). They consist of an inner layer of allophycocyanin which abutts the photosynthetic membrane and is surrounded by a hemispherical layer of R-phycocyanin and covered by a peripheral layer of phycoerythrin. Most work in this area has been done on unicellular soil algae and has been reviewed by Gantt (1977).

Genophores, or electron-transparent regions presumed to contain DNA, are typically scattered in the stroma. The DNA fibrils are attached to the thylakoid membranes, which may be involved in their replication and segregation during chloroplast division (Bisalputra & Bisalputra 1967; Bisalputra 1974). Small electron-dense 'globules' ('lipid granules' or 'plastoglobuli') (Fig. 7.1) are also observed in the stroma (Bisalputra 1974). They have not been characterized in the red algae, though work on analagous structures in higher plants suggests they may function in lipid synthesis and storage (Sprey & Lichtenthaler 1966). Floridean starch is not found within red algal chloroplasts, but accumulates within the cytoplasm, apparently in close association with the ER (Borowitzka 1978; Pueschel 1979).

Pyrenoids with a granular, or in the case of *Smithora* (McBride & Cole 1972), crystalline, matrix have been observed in some red algae, particularly in the

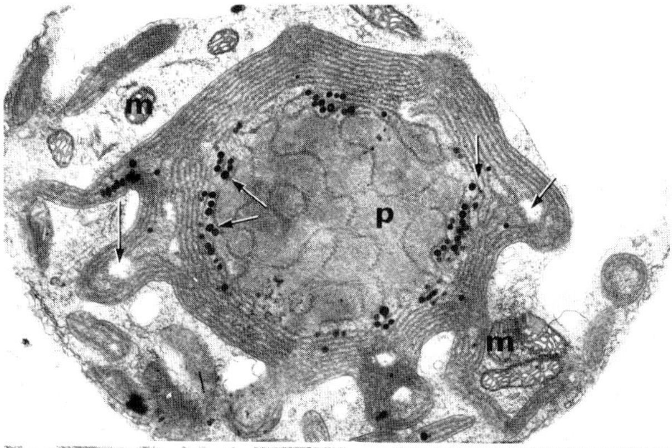

7.1

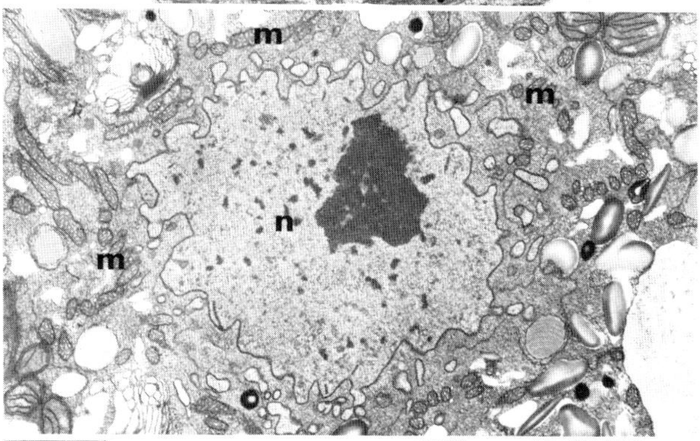

7.2

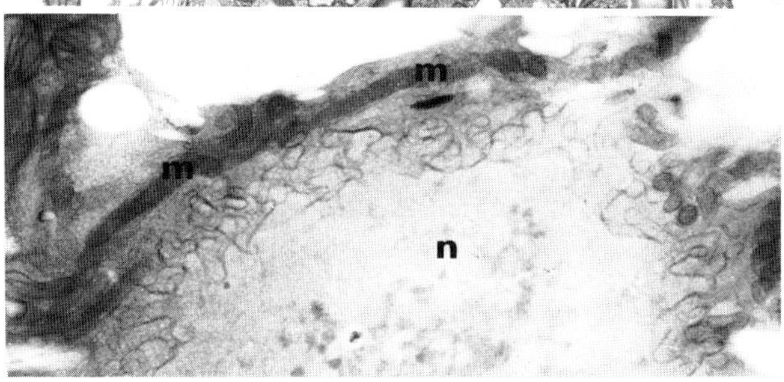

7.3

Fig. 7.1 Stellate chloroplast of *Nemalion*. Note the central pyrenoid (p), single, unstacked thylakoids, electron dense 'plastoglobuli' (small arrows) and electron-transparent genophores (large arrows). Mitochondria (m) are also present. (Courtesy Sue Ramm-Anderson.) (× 10 000.)

Fig. 7.2 Highly evaginated nucleus (n) during early carpospore development in *Polysiphonia*. Profiles of mitochondria are prominent around the nucleus and other organelles and storage products are also observed. (Wetherbee & West 1977.) (× 4960.)

Fig. 7.3 High voltage electron micrograph of a one μmol thick section of a carpospore in the same stage of development as seen in Fig. 7.2. The highly evaginated surface of the nucleus is surrounded by extensive profiles of mitochondria which would suggest that they are much larger in size, and fewer in number, than indicated by thin sections. (Wetherbee unpublished.) (× 12 560.)

more 'primitive' forms (Bangiophycidae and the Nemalionales) (McBride & Cole 1972; Duckett *et al.* 1974; Lin *et al.* 1977; Hawkes 1978). Three different types of pyrenoids have been described (see Dodge 1973) and may be of systematic importance (Hara & Chihara 1974; Duckett & Peel 1978). Some work on the chemical composition of brown algal pyrenoids has been undertaken (see Section 3.1), but to our knowledge, red algal pyrenoids have yet to be isolated and characterized.

Although the major structural and biochemical features of the red algal chloroplast are consistent throughout the division, there are distinct morphological differences which may be of systematic significance within the group. The structure of pyrenoids, if present, and the presence or absence of one or more peripheral thylakoids, has received the most attention to date. These criteria have been extensively reviewed by Duckett and Peel (1978).

Proplastids are common in the meristematic cells of many red algae and typically consist only of an outer envelope, a genophore and an occasional thylakoid or thylakoid fragment (Tripodi 1974; Bisalputra 1974; Borowitzka 1978; Wetherbee 1980). A peripheral thylakoid is present in some species. The structure and continuity of red algal chloroplasts, their development from proplastids, and their subsequent replication has been reviewed by Bisalputra (1974). However, to date, only straight ultrastructural observations have been made, and the precise manner of thylakoid ontogeny is not known for any red alga. There is evidence to suggest that the peripheral thylakoid, if present, invaginates to form a few thylakoids which then bifurcate to duplicate the complete complement (Brown & Weier 1968; Bisalputra 1974; Honsell *et al.* 1978). In plastids without peripheral thylakoids the evidence is unclear, but thylakoids may arise from blebs of the envelope or from pre-existing thylakoids or thylakoid fragments. Borowitzka (1978) has described a 'pro-lamellar body' during thylakoid development in the coralline red alga *Lithothrix*. However, the micrographs are inconclusive and further cytochemical and structural work is required in order to establish the mechanism of chloroplast ontogeny in this and other red algal groups.

The structure of the red algal mitochondrion is similar to that of most algae, consisting of a double membrane envelope, tubular cristae, and electron-transparent regions with presumed DNA fibrils (Dodge 1973) (Figs 7.1, 7.2). Electron dense, spherical inclusions of unknown composition have been observed in some species (e.g. *Smithora*, McBride & Cole 1971). The density of mitochondrial profiles increases markedly in rapidly differentiating cells, particularly during the formation of spores (McBride & Cole 1971; Scott & Dixon 1973a; Pueschel 1979). The exact number of mitochondria in any one cell has not been established, and may be far less than assumed from single thin sections (Fig. 7.3). High concentrations are most conspicuous in the cytoplasm adjacent to nuclei (Figs 7.2, 7.3), and in close association with the forming face of dictyosomes (Figs 7.5–7.7) (see Section 2.3). Connections between the outer

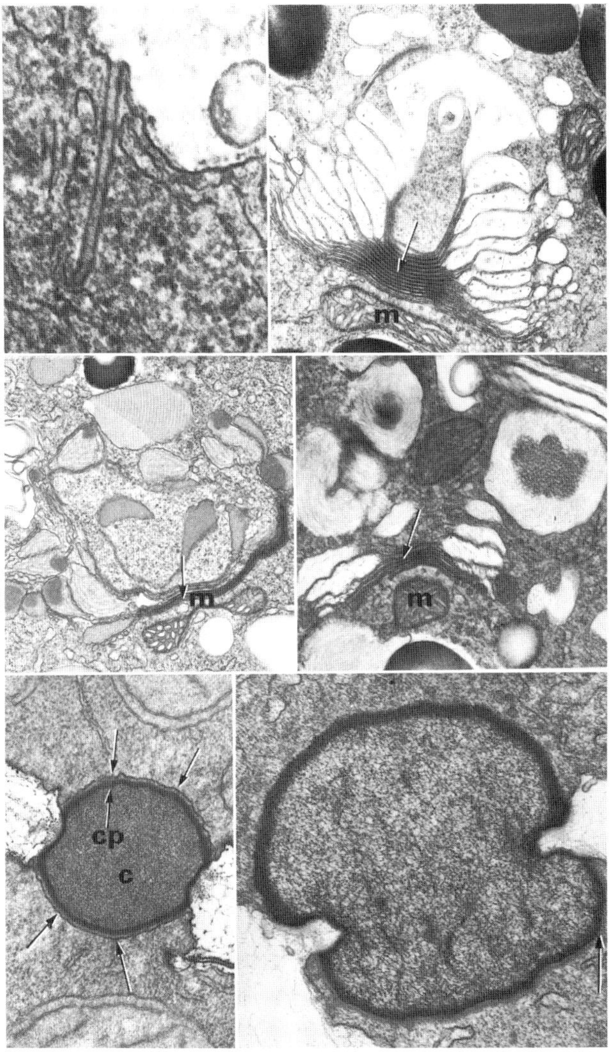

7.4 7.5

7.6 7.7

7.8 7.9

Fig. 7.4 Tubular extension of the plasma membrane observed during late carpospore development in *Polysiphonia*. (Wetherbee 1978.) (× 33 600.)

Figs 7.5–7.7 Three stages in Golgi body activity during carposporogenesis in *Polysiphonia*. Note the three distinct products and the presence of a mitochondrion at the forming face. The cisternae membranes are closely appressed in the central region (arrows). (Wetherbee unpublished.) (7.5— × 17 600, 7.6— × 11 200, 7.7— × 14 400.)

Fig. 7.8 Pit connection between two vegetative cells of *Polysiphonia*. The compact core (c) and two cap layers (cp) are overlain by the cap membranes (arrows). The plasma membrane is continuous between cells, and with the cap membranes, therefore making the pit connection extracellular. (Wetherbee 1979.) (× 20 800.)

Fig. 7.9 Intracellular 'transfer connection' between cells of the carposporophyte in *Polysiphonia*. The core is less compact and cap membranes are not present. The outer layer of the caps appears striated (arrow) when the section is oriented correctly. (Wetherbee 1979.) (× 40 000.)

mitochondrial envelope and the ER have been observed in developing spores of *Ceramium* (Chamberlain & Evans 1971). Mitochondria of unusual morphology have been observed in the fusion cell of *Erythrocystis* and *Polysiphonia* (Tripodi *et al.* 1972; Tripodi & de Masi 1977) where they contain cristae of atypical disposition (see Osborn *et al.* 1979). As the fusion cell is a site of cytoplasmic breakdown (see Section 2.7) this may account for their unusual structure.

2.3 Endomembranes (the Golgi complex, endoplasmic reticulum, plasma membrane)

With the exception of chloroplasts, more is known about the structure, differentiation and function of the Golgi complex than of any other organelle of the red algae. Some of the best work has been undertaken on the unicellular soil alga *Porphyridium* (e.g. Ramus 1972; Ramus & Robins 1975), which is beyond the scope of this review. During periods of cellular differentiation the Golgi complex and Golgi-derived vesicles often undergo distinct morphological changes which are associated with the formation of diverse products (Figs 7.5–7.7) (Peyrière 1970; Kugrens & West 1972c, 1973a, 1974; Chamberlain & Evans 1973; Scott & Dixon 1973a; Konrad Hawkins 1974a, b; Wetherbee & West 1977; Wetherbee 1978). Golgi bodies (='dictyosomes') typically possess six to ten cisternae and vary greatly in size according to the developmental state of the cell. Cisternae are normally separated from one another by a space of approximately 10 nm. However, when the size and activity of the Golgi body is at its maximum, this intercisternal space disappears and adjacent cisternae become closely appressed (Figs 7.5–7.7) (Alley & Scott 1977; Duckett & Peel 1978). This feature, which apparently maintains the overall shape and cohesiveness of the stack during intense activity, has not been observed in other plant or animal cells. Duckett and Peel (1978) have suggested that this feature might be of systematic value, and restricted to the 'higher' Florideophycideae, and absent in the Nemalionales and Bangiophycideae. However, Golgi bodies of similar morphology have recently been observed in developing carpospores of *Nemalion* (Ramm-Anderson & Wetherbee unpublished) and also appears to be present in *Smithora* (McBride & Cole 1971, see Fig. 8).

The manner in which secretory vesicles form at the distal face of the Golgi often changes during development (Scott & Dixon 1973a; Alley & Scott 1977; Wetherbee & West 1977; Wetherbee 1978). Vesicles may arise by vesiculation at the periphery of cisternae, or alternatively, entire cisternae may detach and condense into single vesicles. Mitochondria have been consistently found at the forming face of Golgi bodies in all red algae of the subclass Florideophycideae studied to date (Figs 7.5–7.7) (Peyrière 1970, 1975; Konrad Hawkins 1974a, b; Alley & Scott 1977). McBride & Cole (1972) reported that there is no strict association between mitochondria and Golgi bodies in the Bangiophycean alga *Smithora*, though Hawkes (personal communication) has observed the relationship in *Porphyra*. In developing tetraspores of *Palmaria*, the mitochondrion-Golgi complex appears associated with chloroplasts (Pueschel 1979).

There appears to be a uniform sequence of events whereby the Golgi apparatus secretes extracellular products during the differentiation of spores in several red algae. Golgi-derived vesicles are first active in mucilage formation followed later by the production of cored vesicles which are composed of a glycoprotein (Chamberlain & Evans 1973), and appear to function in spore protection and/or adhesion following release (Peyrière 1970; Chamberlain & Evans 1973; Pueschel 1979). An additional developmental stage has been observed during carposporogenesis in *Polysiphonia* (Wetherbee & West 1976, 1977). Following initial wall formation, the Golgi apparatus differentiates to become active in the production of striated vesicles which remain with the cytoplasm. This change in Golgi activity coincides with the breakdown of proteinaceous vesicles in the cytoplasm and a role in protein transformation and turnover is suggested. The continuity of the Golgi apparatus was observed throughout all developmental stages (Wetherbee 1980).

During monospore formation in *Smithora* (McBride & Cole 1972), populations of Golgi bodies appear active in both cell wall and vacuole formation, and in the gland cells of *Botryocladia*, the Golgi plays an active role in the production of an inner slime (Young 1978). Fibrous vesicles in the developing spermatia of several red algae are derived, in part, from Golgi vesicles, and appear to have a role in the liberation of spermatia (see Section 2.8).

The ER is structurally distinctive when surrounding postmeiotic nuclei in *Corallina* (Peel *et al.* 1973), as crystalline arrays in the fusion cell of *Erythrocystis* (Tripodi & de Masi 1977) and during the formation of cytoplasmic annulate lamellae (Wetherbee *et al.* 1974; Tripodi 1974). Similar to the Golgi complex, the ER is most conspicuous in cells undergoing differentiation, and has been implicated in the packaging and deposition of cell wall materials (Ramus 1969a; McBride & Cole 1972; Chamberlain & Evans 1973), in the formation of floridean starch (Borowitzka 1978; Pueschel 1979) and in the production of various types of cytoplasmic vesicles and vacuoles (Kugrens & West 1972, 1973, 1974; Scott & Dixon 1973b; Wetherbee & West 1977; Hawkes 1978; Young 1977, 1979a, b, c). ER-derived vesicles often accumulate only during precise developmental stages, and appear to play a role in the formation and isolation of various components which are necessary for subsequent stages in development (e.g. Kugrens & West 1972, 1973, 1974; Wetherbee & West 1977). A role in the formation of storage products has been observed in developing gland cells of *Opuntiella* and *Antithamnion* (Young 1979b, c) and also in the vegetative cells of *Antithamnion* (Young 1979a).

Recent studies have shown the surface area of the plasma membrane to increase markedly in some cells. Extensive invaginations of the plasma membrane in epidermal cells of the calcified red algae *Corallina* and *Lithothrix* suggest a role in the uptake of solutes (Borowitzka & Vesk 1978, 1979). Termed 'transfer cells', these cells are morphologically similar to those described in higher plants. In other red algal cells, membranous tubules of uniform diameter ($\simeq$50 nm) have been observed within the cytoplasm (Fig. 7.4) (Peyrière 1969; Kugrens & West 1973; Wetherbee 1978). Apparently derived from the ER, these tubules have been observed to fuse with the plasma membrane in carpospores of *Polysiphonia* (Wetherbee 1978). This vast increase in surface area of the plasma membrane might be associated with absorption and/or secretion. Experimental studies are needed to confirm these suspicions.

Lomasomes ('multivesicular bodies' or 'paramural bodies') often appear associated with the ER and plasma membrane, and have been implicated in cell wall formation (e.g. Ramus 1969a; McBride & Cole 1972; Chamberlain & Evans 1973; Goff 1979b). Cytoplasmic extensions enclosed by the plasma membrane and penetrating within the cell wall (plasmalemmavilli, Cottler 1971; Young 1979b; Goff 1979b) have likewise been implicated in either secretion (Cottler 1971) or absorption (Goff 1979b). However, there is considerable doubt, in most cases, whether these structures are real or artifacts or fixation. Reliable judgements as to the function of lomasomes and plasmalemmavilli in red algal cells are therefore not possible at this time.

2.4 The cell wall

Electron micrographs of red algal cell walls reveal a layered fibrillar appearance, though the thickness and general orientation of the fibrillar component varies considerably in the different areas of the same plant, between different plants and according to the nature and length of fixation (Gordon & McCandless 1973; Duckett & Peel 1978; Wetherbee unpublished). The randomly oriented microfibrils of the inner layer are composed of cellulose in the Florideophycidae (Myers *et al.* 1956) and mostly xylan in the Bangiophycidae (Frei & Preston 1964). Gretz *et al.* (1980) have obtained evidence for cellulose in the *Conchocelis* phase of *Bangia*. As cellulose was reported absent for the *Bangia* phase (Frei & Preston 1964), the wall composition appears to differ for alternate phases in the life history of the same organism. It is suggested that the *Conchocelis* phase may represent a structural and biochemical link between the two classes of red algae.

The wall matrix and outer mucilaginous layer are generally composed of a sulphated polysaccharide (for review see Mackie & Preston 1974). Gordon-Mills & McCandless (1975) used fluorescent antibody techniques to locate the commercially important carrageenans within the immediate cell walls and matrix of *Chondrus*. Specific carrageenans characterize different phases in the life history and are observed in the first cells (carpospores and tetraspores) formed by each generation (McCandless *et al.* 1973; Gordon-Mills & McCandless 1975). A multilayered cuticle is observed in some species (e.g. Scott

& Dixon 1973a) and has been demonstrated to be proteinaceous in *Porphyra* (Hanic & Craigie 1969). The involvement of the endomembrane system in cell wall formation, as well as the composition and apparent role of the different wall layers during the formation and release of spores, is discussed elsewhere in this chapter (see Sections 2.3 and 2.8).

$CaCO_3$ crystals (calcite or aragonite) are observed in the cell walls of several diverse taxa of the red algae, though little is known of the mechanism of calcification (Borowitzka 1977). Crystals are always arranged within an organic matrix, and are absent from the thin layer of wall material adjacent to the plasma membrane (Borowitzka & Vesk 1979). In *Corallina* the calcite crystals nearest the cell are small and oriented at right angles to the plasma membrane, while further away they increase in size and become more randomly oriented (Borowitzka & Vesk 1978). Secondary deposits of $CaCO_3$ were reported in *Lithothrix*, indicating that at least two mechanisms of calcification are involved (Borowitzka & Vesk 1979). Some brown and green algae also have calcified cell walls, and this carbonate production makes an important contribution to the sediment of shallow, tropical marine waters (Wefer 1980).

2.5 The nucleus (mitosis and meiosis)

The physical relationship between the nucleus and cytoplasm during cell differentiation in many red algae far exceeds that normally observed in plant cells. Profiles of nuclei are often highly evaginated. The increase in both surface area and frequency of annular pores may facilitate the movement of materials between the nucleoplasm and the cytoplasm during development (Tripodi 1971; Kugrens & West 1972; Scott & Dixon 1973; Chamberlain & Evans 1973; Wetherbee & Wynne 1973; Pueschel 1979). Although nuclear fragments often appear dissociated from the highly evaginated surface, serial sections usually show them to be continuous. However, in developing carpospores of *Polysiphonia* (Tripodi 1974), blebs surrounded by a double envelope were observed, but do not migrate far into the cytoplasm before breaking down. In modified vegetative cells of *Corallina* evaginations of the outer membrane appear to pinch off and accumulate in the cytoplasm (Peel & Duckett 1975). During carposporogenesis in *Polysiphonia* (Wetherbee & Wynne 1973), stacks of lamellae accumulate within the nucleoplasm beneath regions of envelope devoid of pores. Cytoplasmic products located within the nucleoplasm may be associated with these regions. Annulate lamellae, which are characterized by the presence of pore complexes similar to those of the nuclear envelope, have been reported within the nucleoplasm (Peel *et al.* 1973; Wetherbee *et al.* 1974). A complex of precisely oriented ER and electron-dense ground plasm in contact with the nuclear envelope surrounds postmeiotic cells in *Corallina* (Peel *et al.* 1973) and may be important in subsequent cytoplasmic differentiation.

McDonald's (1972) ultrastructural description of mitosis in *Membranoptera*

7.10 7.11 7.12 7.13 7.14

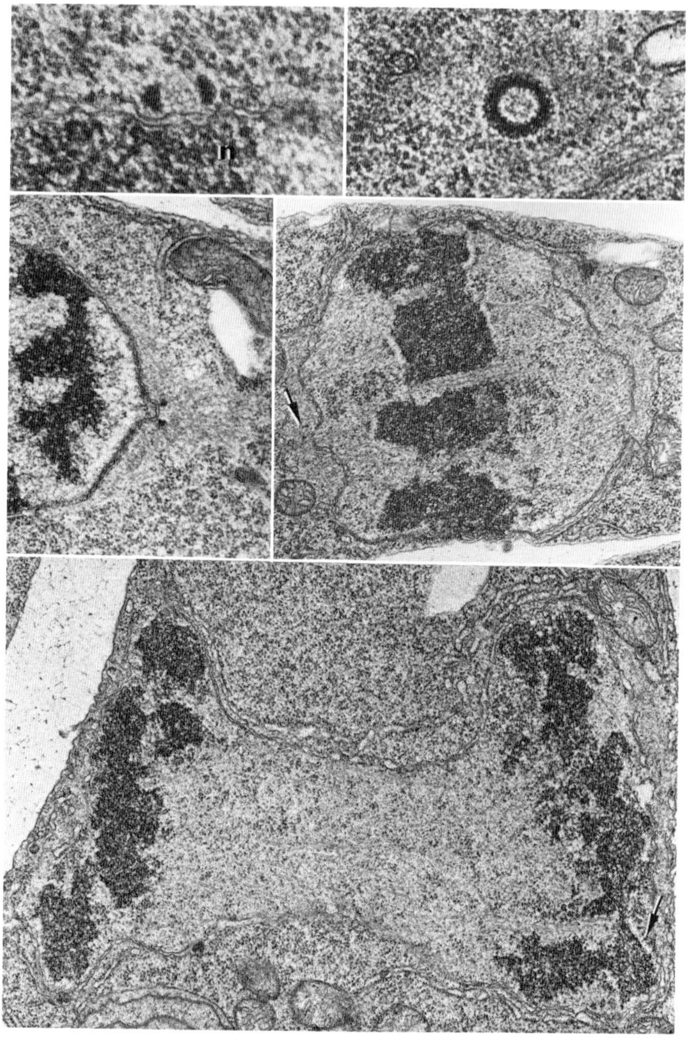

Figs 7.10–7.14 *Polysiphonia* male reproductive branch cells. (Courtesy Dr. Joe Scott.)

Fig. 7.10 *Polysiphonia* interphase nucleus with one of a pair of longitudinally sectioned polar rings closely associated with the nuclear envelope. Microtubules and a ribosome-free zone of exclusion are absent at this time. Note fibrous center of polar ring nucleus (n). (× 68 800.)

Fig. 7.11 Transversely sectioned interphase polar ring. An outer layer of material showing a periodic structuring is evident as well as the ring-like organization of this 'nucleus-associated organelle'. (× 60 800.)

Fig. 7.12 Late prophase nucleus with smooth surfaced 'nuclear envelope protrusion' subjacent to an attached polar ring. A microtubule-filled exclusion zone is well developed at this stage of mitosis. (× 19 200.)

Fig. 7.13 Metaphase. Opposed kinetochores on chromosomes in a typical equatorial plate arrangement, nucleolar fragments, and chromosomal and non-chromosomal spindle microtubules can be seen in the nucleoplasm. A layer of perinuclear endoplasmic reticulum surrounds most of the nucleus, whose envelope remains intact except at the fenestrated poles. One of the two dissociated polar ring portions (arrow) can be seen at one of the flattened division poles. (× 16 400.)

has recently been supplemented by extensive work on other ceramialean red algae (Figs 7.10–7.14), including *Polysiphonia* (Scott *et al.* 1981) and *Dasya* (Phillips & Scott 1981). These genera are characterized by a closed, fenestrated spindle, one or more layers of perinuclear endoplasmic reticulum, layered kinetochores, and chromosomal and non-chromosomal microtubules. Unique organelles, termed the 'polar rings' (Figs 7.10–7.13) were found to arise during interphase–early prophase and remain throughout mitosis. Polar rings may function as 'microtubule organizing centres' (McDonald 1972; Scott *et al.* 1981), though spindle microtubules have also been reported to originate from diffuse MTOC-like material found along the inner membrane of the nuclear envelope (Scott *et al.* 1981). Although the mechanism of mitosis appears to be similar within the Ceramiales, work has not been undertaken on other representatives of the Florideophycidae. However, there is evidence to suggest that differences do exist between the two classes. Several structural aspects of mitosis in the bangiophycidean red alga *Porphyridium* (Bronchart & Demoulin 1977; Schornstein & Scott 1980) differ from those described for the Ceramiales. This work appears promising, though it is not known at this time whether the features of cell division in the red algae will be of systematic value.

Ultrastructural studies on meiosis are scarce and incomplete. Synaptonemal complexes, structures indicative of meiosis, have been described in several red algae (Kugrens & West 1972a; Pueschel 1979), and found to be similar in ultrastructure to those observed in other organisms. Studies on late telophase in some members of the Florideophycideae (Scott & Thomas 1975) showed the four nuclei to be interconnected, and suggested a unique type of uninucleate meiosis where both divisions occurred within the nuclear envelope of the mother cell. However, Scott (personal communication) now suspects this information may be inaccurate, as the four nuclei formed in *Dasya* appear to separate first and then spring back together and fuse with one another before once again separating as cleavage proceeds. The process of meiosis appears to be both unique and interesting, and much work is needed in this area.

2.6 Pit connections (septal plugs)

Red algal pit connections are a common feature in the sub-class Florideophycidae (Figs 7.8, 7.9) and in a few species of the Bangiophycidae (e.g. the conchocelis stage of *Porphyra* and *Bangia*, Lee & Fultz 1970; Bourne *et al.* 1970; Lee 1971).

Fig. 7.14 Late anaphase is characterized by an interzonal spindle region whose length varies according to the cell type in which it occurs. In small vegetative cells, such as is shown in this micrograph, the two chromatin masses are separated by a relatively short distance. The chromosomes are closely associated with the nuclear-envelope at their kinetochore regions (arrow) and spindle mts are still visible in the interzonal spindle region. (×17 600.)

A relatively undifferentiated pit connection has recently been observed between cells of *Rhodochaete* (Bangiophycideae) (Boillot 1978). With this exception, the same fundamental structure of pit connections is observed throughout the division and is characterized by a central core (or plug) plus one or more electron-opaque caps (e.g. Bouck 1962; Ramus 1969a, b; Lee 1971; Pueschel 1976). The plasma membrane is continuous between cells, and in addition, may become continuous with a cap membrane which isolates the pit connection from the cytoplasm of the cell which produces it (Fig. 7.8). When cap membranes are formed by both cells, the pit connection is totally surrounded by the plasma membrane, and is therefore extracellular (Ramus 1969a, b; Pueschel 1976). Pit connections continue to differentiate as the cells they connect mature, increasing in size and changing shape throughout development (Bouck 1962; Ramus 1969b; Konrad Hawkins 1972; Duckett *et al.* 1974; Wetherbee 1978, 1980; Wetherbee & Kraft 1981). Variations in pit connection morphology have been noted between different phases of the life history of the same organism (Ramus 1969b; Duckett *et al.* 1974; Peyrière 1977a; Wetherbee 1978, 1980), and between different species (e.g. Feldmann *et al.* 1977).

ER cisternae occupy the septal aperture between cells after incomplete cytokinesis in *Pseudogloiophloea* (Ramus 1969a, b), and material in these cisternae is deposited to form the core of the pit connection. Ramus (1971) reported that pit connections isolated from *Griffithsia* are composed of an acidic polysaccharide-protein complex. Pueschel (1981) has studied the cytochemical properties of several florideophycidean and bangiophycidean pit connections and finds the cores to be composed entirely of protein, while the outer cap layer, when present, is reported to be at least partly polysaccharide.

Since the cytoplasm of adjacent cells connected by pit connections is interrupted rather than continuous, the term 'septal plug' is preferred by some authors (Dixon 1973; Marchant 1976), though this designation implies these structures may be barriers to transport and communication. Translocation along files of red algal cells (Hartmann & Eschrich 1969) and between photosynthetic red algae and their colourless parasites (Evans *et al.* 1973, 1978; Goff 1979a) has been demonstrated using radioactive tracers. However, the exact avenue of translocation has not been ascertained, but seemingly would be through either the cell walls or the pit connections. Evidence for intercellular transport via pit connections is circumstantial. Significant changes in the morphology of pit connections have been noted where transport between cells would be expected. The absence of one, or both, cap membranes has received recent attention. If both are absent, the pit connections are intracellular rather than extracellular, and this feature may be advantageous if transport occurs by this avenue (Peyriere 1977a; Wetherbee 1978, 1980; Wetherbee & Scott 1980; Quirk & Wetherbee 1980) (Fig. 7.9). In a species of *Cryptonemia* with extremely thick stalks, pit connections become progressively more convoluted with increasing distance from the stalk surface (Wetherbee & Kraft 1981). The core is electron transparent while cap membranes are retained and follow the contours of the evaginations. It is hypothesized that the increase in surface area may aid in the transport of solutes towards the deeply-buried layers of living cells. The term 'transfer connection' (Wetherbee 1979) has been proposed to include those red algal pit connections whose morphology or position (or both) within a thallus would suggest a role in enhancing intercellular transport. Alternatively, Goff (1979b) has found the host-penetrating cells of the red algal parasite *Harveyella* to have an extensive endomembrane and vacuolar system which she believes is involved in the uptake and processing of materials derived from host cells following diffusion through the intervening cell walls. More work is needed before the role of pit connections in red algal development is understood.

2.7 Structure and development of the carposporophyte

Few fine structural studies have been conducted on the development and interrelationship of the various components of the carposporophyte. Even so, fundamental differences in interpretation exist, and are concerned with the exact location of the meristematic cell (or cells) responsible for carposporophyte formation. The carposporophyte of *Polysiphonia* was found to differentiate into distinct regions (Wetherbee 1980). The meristematic region is believed to be located on the surface, and consists of actively dividing gonimoblast cells plus developing carpospores. Gonimoblast cells which have ceased dividing, plus certain cells of the haploid gametophyte appear to degenerate and are eventually incorporated into the fusion cell which expands outward during development. Kugrens and West (1974) observed a similar developmental pattern in the carposporophyte of *Erythrocystis*. However, in Tripodi's (1974, 1977) interpretation of carposporophyte development in these same two genera (*Polysiphonia* and *Erythrocystis*), the fusion cell is described as meristematic, continually budding-off cells which eventually develop into carpospores. Studies on postfertilization development in *Nemalion* are similarly contradictory. Ramm-Anderson (1980) described carposporangium formation as occurring by proliferation from meristematic cells located near the surface of the carposporophyte, while Duckett & Peel (1978) consider the fusion cell itself to be meristematic in *Nemalion*. More attention must be directed to an understanding of the classical literature (which stresses the apical growth of gonimoblast filaments), and to the morphology of material prepared for light microscopy. Postfertilization development in the Rhodophyta is notoriously complex, and it is understandably difficult to establish developmental patterns using only thin sections of fixed material. Thus, the proposition that the mature fusion cell of *Polysiphonia* and *Erythrocystis* is meristematic needs to be substantiated by developmental studies. The belief that mitochondria (Tripodi 1977) or other cellular components (e.g. pyrenoids; Duckett & Peel 1978) arise *de novo* within gonimoblast cells produced by the fusion cell requires verification, since in our opinion the fusion cell of those red algae studied to date is involved only in absorbing cells, not in generating them.

In gonimoblast filaments of the coralline alga *Lithothrix* (Borowitzka 1978), sequential stages in carpospore formation have been observed with the electron microscope. The least differentiated cell is at the base of the filament and adjacent to the fusion cell. Which cell (if any) remains meristematic after initial formation is not clear. The basal cell could be meristematic itself, or conceivably be continually replaced by a meristematic fusion cell. As with most red algae, studies on carposporophyte formation in *Lithothrix* are incomplete, and more information is required. The anatomy of postfertilization development differs greatly among the various taxonomic groups, and the components of the carposporophyte may function differently as well. The manner and avenue of providing nutrients to the different regions of the carposporophyte also need to be elucidated. In some groups, the outer meristematic regions contain only proplastids, and nutritive assistance would appear vital to the developing gonimoblast (Wetherbee 1978, 1980). Certain cells of the gametophyte and carposporophyte appear to be commonly modified to provide this assistance, though no direct evidence for translocation towards, or within, the carposporophyte generation is presently available.

2.8 Differentiation of gametes and spores

In contrast to the published literature available on the formation of gametes and subsequent fertilization in the green and brown algae, few reports are available on the Rhodophyta. The structure of the carpogonium and spermatium in *Porphyra* (Bangiales), as well as the mechanism of fertilization, has been described by Hawkes (1978). The carpogonium of this relatively simple red alga is a slightly modified vegetative cell. The spermatium attaches to a protrichogyne, and the contents are transferred to the carpogonium via a fertilization canal. Parallel studies are needed for other red algae.

The fine structural features which characterize spermatia in all red algae studied to date are basically similar (Hawkes 1978; Duckett & Peel 1978), and are distinguished from those of the other benthic seaweeds by the lack of flagellated cells. Flagellar roots have been reported in the spermatia of *Bonnemaisonia hamifera* (Simon-Bichard-Bréaud 1971, 1972), but these findings have not been verified (Young 1977), and are generally discounted by most workers. Red algal spermatia appear colourless, and plastids are entirely absent, or greatly reduced in size and complexity (for review see Duckett & Peel 1978; Hawkes 1978). One, or occasionally two, large fibrous vesicles ('spermatial vesicles') are formed by the ER and Golgi complex, and are believed to contain a mucopolysaccharide (Peyrière 1971; Scott & Dixon 1973b). Release of the spermatial vesicle may cause an increase in osmotic pressure resulting in the rupture of the spermatangial wall and the escape of the spermatia (Kugrens &

West 1972; Scott & Dixon 1973b; Young 1977). Other characteristic morphological features of red algal spermatia were recently reviewed (Duckett & Peel 1978).

Several ultrastructural and cytological investigations have dealt with aspects of carpospore formation (Tripodi 1971, 1972, 1974; Kugrens & West 1973a, 1974; Wetherbee & Wynne 1973; Triemer & Vasconcelos 1977; Wetherbee & West 1977; Wetherbee 1978, 1980) and tetraspore formation (Peyrière 1969, 1970; Kugrens & West 1972a, c; Scott & Dixon 1973a; Chamberlain & Evans 1973; Peel *et al.* 1973; Konrad Hawkins 1974a; Alley & Scott 1977; Pueschel 1979). As discussed above, the endomembrane system, especially the Golgi complex, plays an important role in spore development (see Sections 2.3 and 2.4). Some features of spore production are observed in all species studied to date, including the accumulation of floridean starch and the formation of cored vesicles. Other stages in development appear less widespread, such as the formation of 'striped vesicles' in *Erythrocystis* (Kugrens & West 1974). Cytoplasmic differentiation appears to be essentially similar for both spore types and is concerned with the formation of storage products and various vesicles responsible for spore protection and subsequent adhesion.

2.9 Parasitic red algae

Parasitic red algae occur in four orders of the Florideophycideae but are not found in the Bangiophycideae. The basic morphology and taxonomic distribution of these organisms has recently been reviewed (Evans *et al.* 1978). Evidence for their parasitic nature—such as their reduced size, lack of pigmentation and penetration of the host tissue by a network of rhizoids, has been confirmed with some species by experiments using the radioactive tracer ^{14}C. Evans *et al.* (1973) demonstrated that ^{14}C assimilated by the host *Gracilaria verrucosa* during photosynthesis was incorporated into the alloparasite *Holmsella pachyderma*. They obtained similar results for the *Polysiphonia lanosa* (host)—*Choreocolax polysiphoniae* (parasite) association (Evans *et al.* 1978). Goff (1976, 1979a) likewise demonstrated the transfer of photo-assimilated ^{14}C bicarbonate to the alloparasite *Harveyella mirabilis* from its host *Odonthalia floccosa*. Greatly reduced plastids have been observed in *Holmsella* (Peyrière 1976), *Harveyella* (Peyrière 1976; Goff 1976) and *Choreocolax* (Kugrens & West 1973b); they consist of an outer double membrane envelope and no thylakoids. An occasional thylakoid is present in the plastids of *Levringiella* (Kugrens & West 1973a), while in other so-called 'parasites', plastid development and pigmentation can be more extensive (for review see Evans *et al.* 1978).

Secondary pit connections between host and parasite have been demonstrated in several associations. The fine structure of the host–parasite pit connection is generally reported to be similar to that between two parasite cells (Kugrens & West 1973; Goff 1976; Peyrière 1977b). However, in the case of the *Holmsella–Gracilaria* association (Quirk & Wetherbee 1980), the surface of the pit connection adjoining the cytoplasm of the parasite cell is morphologically identical to a parasite–parasite pit connection, while the surface adjoining the host cell more closely resembles that of a host–host connection. Kugrens and West (1973b) have reported that host–parasite pit connections are initiated by the host, which develops a cytoplasmic protuberance towards the parasite. The parasite cell then buds off a daughter cell which fuses with the protuberance. The parasite–parasite pit connection formed between the parent cell and its budded daughter cell is now a host–parasite pit connection. Goff (1976) likewise observed a protuberance by the host cell towards *Harveyella*, but feels that the host–parasite pit connections are formed following the fusion of the host protuberance with the parasite. Peyrière (1977b) observed secondary pit connection formation between *Harveyella* and a different host (*Rhodomela*), but protuberances were not observed and the *Harveyella* cells appeared entirely responsible for establishing the connection.

While translocation between host and parasite has been demonstrated, the exact avenue has not. Goff (1979b) has proposed that nutrient transport takes place through the host cell wall into the parasite rhizoidal cells, and that these cells are modified to accomplish this role. Alternatively, nutritional materials could be transported via pit connections. In *Holmsella* parasite–parasite pit connections lack cap membranes and are intracellular (Quirk & Wetherbee 1980), while host–parasite pit connections normally have a cap membrane only on the host side. Similar relationships between pit connections and their associated membranes have been observed where transport between cells would be expected (Peyriere 1977a; Wetherbee 1978, 1980; Wetherbee & Scott 1980).

2.10 Morphogenesis in *Griffithsia*

Several interesting studies on factors involved in growth and regeneration of red algal cells have been made by Waaland and co-workers. *Griffithsia pacifica* can regenerate from single, isolated cells in a relatively rapid and predictable pattern (Duffield *et al.* 1972). Apical cell growth is insensitive to the effect of light intensity and photoperiod (provided light is present), while division of nodal cells, and thus algal branching, is promoted by both increased light intensity and duration (Waaland & Cleland 1972).

A previously undescribed pattern of cell elongation—bipolar band growth—has been observed in several red algal species (Waaland *et al.* 1972; Waaland & Waaland 1975). Growth of intercalary cells of *Griffithsia pacifica* is confined to two bands of about 20 μm in length, one at each end of the cell. Elongation occurs by the addition of new cell wall material with the microfibrils being arranged randomly. The number (whether one or two) and the position of bands of elongation are constant within a species but vary from species to species (Waaland & Waaland 1975).

The process of cell repair in *G. pacifica* has also been investigated (Waaland

& Cleland 1974). An atypical shoot cell and a rhizoid cell are cut off on the appropriate side of a dead intercalary cell, grow toward each other through the lumen of the dead cell (Fig. 18.6, p. 646), and finally fuse to form a new cell. Intercellular communication between the repair shoot and rhizoidal cells appears mediated by the hormone 'rhodomorphin' (Waaland 1975). Cell repair can be induced between two plants of the same species, but not between two different species (Waaland 1975).

Isolated hybrid cells resulting from somatic fusion between the vegetative cells of male and female strains of *Griffithsia tenuis* have regenerated new plants (Waaland 1978). The hybrid filaments first produce reproductive structures characteristic of the diploid, tetrasporic phase, suggesting that the unfused male and female nuclei act in concert to produce these structures. As the filament continues growing, sexual stages (male or female) are formed, and may result from the nuclei of one sex being left behind during cell divisions, the remaining nuclei controlling production of the appropriate reproductive structure. Determining what controls the formation of reproductive structures is a difficult problem, and the excellent system developed by Waaland and colleagues has considerable potential for unravelling the control mechanisms responsible for morphogenesis in the red algae. (See also Chapter 18.)

3 PHAEOPHYTA

Despite the morphological diversity within the Phaeophyta, continued investigation of members of the brown algae has confirmed the distinctive cytological features of the group (Dodge 1973). After a general discussion of the organelles of brown algal cells, we conclude this section on the Phaeophyta with a consideration of several topics which have received particular attention in recent investigations: 1, gametogenesis and zoosporogenesis; 2, fertilization, and 3, cellular polarization in brown algal embryos. Tissue specialization in the Laminariales is discussed in Chapter 15.

3.1 Chloroplasts, pyrenoid and mitochondria

The plastids of brown algal cells occur in many patterns—as discs, parietal plates, or in stellate configurations and, depending on the species, one to many plastids per cell are present. The plastid is enveloped by at least two unit membranes (i.e. the inner and outer chloroplast membranes), and two additional membranes often surround the plastid. These form the chloroplast endoplasmic reticulum (Gibbs 1962; Bouck 1965), which is continuous with the outer membrane of the nuclear envelope (Figs 7.15, 7.16). At one time, the chloroplast ER-nuclear envelope association was believed to be of phylogenetic value, since it is commonly observed in less specialized groups such as the Ectocarpales but rarely in more complex orders. However, observations of *Macrocystis* (Chi & Neushul 1972) and *Chorda* (Toth 1976) indicate that the chloroplast-ER nuclear

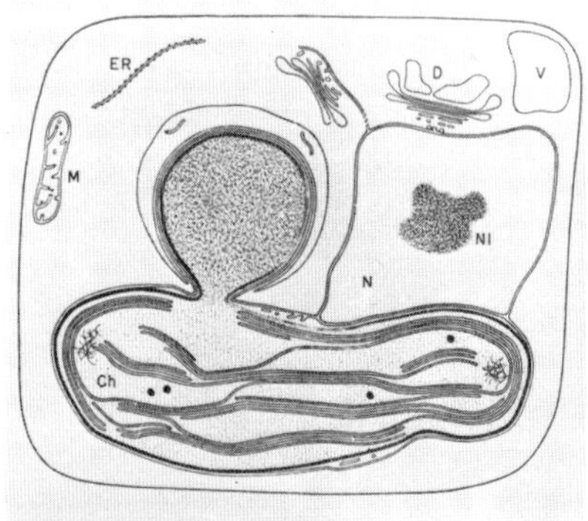

7.15

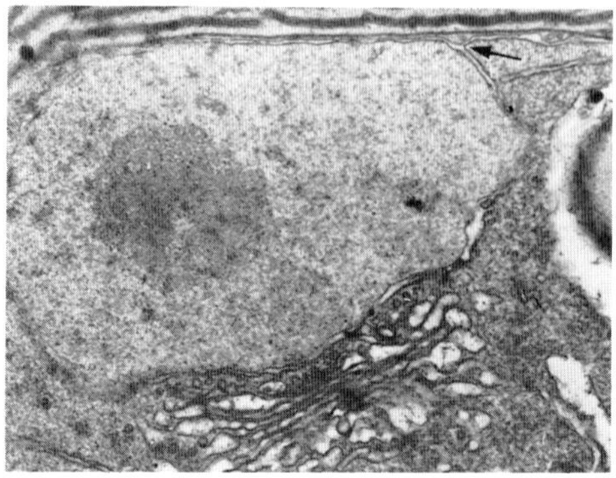

7.1

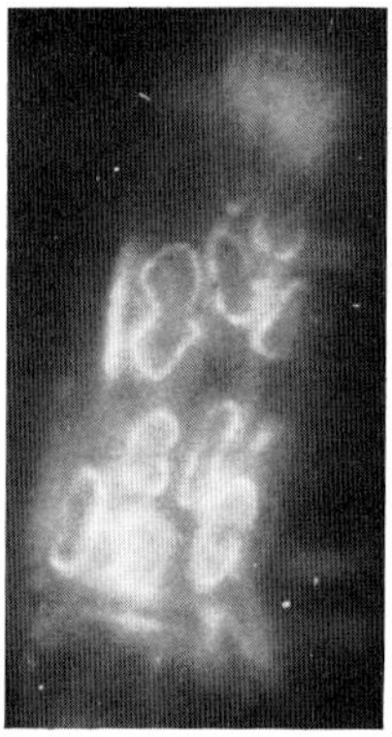

7.17

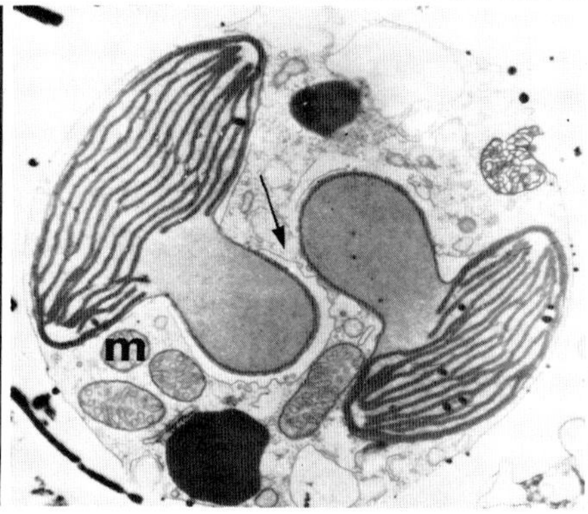

7.18

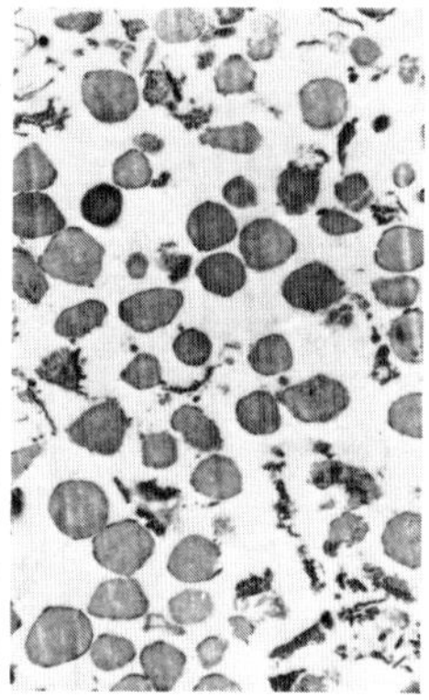

7.19

Fig. 7.15 Generalized depiction of a brown algal cell. The nuclear envelope is continuous with the endoplasmic reticulum (ER) including the chloroplast ER. Mitochondria (M), chloroplast (Ch), small vacuole (V), Golgi body (D = dictyosome). Microtubules and microfilaments were unknown constituents of brown algal cells when this drawing was made. (Bouck 1965.)

Fig. 7.16 The interphase nucleus of an assimilatory cell of *Chorda filum*. The nuclear envelope clearly extends around the plastid (arrow) to form the chloroplast endoplasmic reticulum. Also note the perinuclear Golgi body. (Bouck 1965.) (× 13 250.)

Fig. 7.17 The genophore of the brown algal plastid forms a continuous peripheral ring as clearly observed in this DAPI-stained *Sphacelaria* filament. (Coleman 1979.) (× 2000.)

Fig. 7.18 A *Pilayella* zoospore with many of the common organelles of a brown algal cell. Note the tubular cristae in the mitochondria (m). The stalked pyrenoid is surrounded by a sheath of polysaccharide (arrow). The osmiophilic inclusions are probably physodes. (Courtesy of Dr. Len Evans.) (× 13 000.)

Fig. 7.19 Pyrenoids isolated from *Pilayella* cells following treatment of cells with 0.1% $HgCl_2$ and fractionation on silica sol gradients. The pyrenoids retain some of the membranes of the chloroplast ER and chloroplast envelope. (Kerby & Evans 1978.) (× 2300.)

envelope association is present in young germlings of the highly specialized Laminariales. This character may be present in most brown algae during some stage of their life cycles, but it appears to become infrequent in cells containing many plastids and undergoing extensive subcellular differentiation.

Phaeophycean plastids typically have three thylakoids per lamella, and a girdle lamella is found near the chloroplast envelope. The lamellar arrangement is more irregular in dividing plastids. Both longitudinal and transverse divisions of the plastids occur within the same cell (Berkaloff & Rousseau 1979). The chloroplast matrix in many species of brown algae contains small electron-opaque globuli, most of which are presumed to consist of lipid or phenolic material (Pellegrini 1980).

The genophore of the brown algal plastid is arranged in a planar ring near the chloroplast envelope. This arrangement of chDNA was first demonstrated in *Sphacelaria* with electron microscopy (Bisalputra & Bisalputra 1969) and has been verified by fluorescence microscopy (Fig. 7.17) (Coleman 1979).

The pyrenoid has a homogeneous matrix and usually lies on a short stalk at the periphery of the chloroplast (Fig 7.18). Kerby & Evans (1978) isolated pyrenoids from *Pilayella littoralis* following pretreatment of the alga in 0.1% $HgCl_2$ in sea water (Fig. 7.19). Their analysis of the matrix material indicated that two major polypeptides were present, and these appeared to be subunits of ribulose-1, 5-bisphosphate carboxylase (RuBPCase). This biochemical evidence linking the pyrenoid with polysaccharide synthesis supports earlier work on green algal pyrenoids (Holdsworth 1971).

Pyrenoids are common in plastids of some orders e.g., Ectocarpales (Evans 1966; Lofthouse & Capon 1975) but are poorly developed in others e.g., the eggs of some Fucales (Evans 1968) and in vegetative cells of Sphacelariales (Hori 1972). Pyrenoids are observed in some members of the Dictyotales and Laminariales (Chi 1971) and have been reported in *Cutleria* (Hori 1972; La Claire & West 1978).

Algal mitochondria typically have tubular cristae, and in brown algal cells the cristae are usually perpendicular or irregularly arranged with respect to the longitudinal axis of the organelle (Schmitz & Srivastava 1974; Markey & Wilce 1976a; Brawley *et al.* 1976a). Sperm mitochondria in *Fucus*, however, have longitudinally-oriented cristae (Brawley *et al.* 1976a). The calcium environment of the mitochondrion may determine the shape and orientation of cristae (Osborn *et al.* 1979).

Intracristal material, which is sometimes tubular (Pollock & Cassell 1977), is found within the mitochondria of *Fucus* sperm (Brawley *et al.* 1976a; Pollock & Cassell 1977) and *Pilayella* zoospores (Markey & Wilce 1976a). In other organisms, intracristal material frequently occurs in pathological conditions (refs. in Brawley *et al.* 1976a). Since intracristal material is present during the differentiation of the gametes and zoospores of many brown algae, it is unlikely to be indicative of abnormal or degenerate organelle structure. Functional studies of these mitochondria should prove interesting.

Tissues with high energy requirements in more complex organisms are frequently packed with mitochondria. While mitochondria are numerous in most algal cells, their intracellular distribution is rarely notable; however, cellular differentiation in the brown algae includes several cases of mitochondrial localization. Mitochondria are especially numerous in sieve elements of the Laminariales (Schmitz & Srivastava 1975; Sideman & Scheirer 1977), reflecting their high metabolic rates. The localization of mitochondria in the fertilized egg of *Fucus* prior to pronuclear fusion may be related to the weak polar axis associated with fertilization (Section 3.8).

3.2 Endomembranes (the Golgi complex, endoplasmic reticulum, vesicles)

The Golgi complex is generally perinuclear in brown algal cells, including specialized cells of the Laminariales (e.g. sieve elements of *Laminaria*, Schmitz & Srivastava 1974), though Golgi bodies (= dictyosomes) are present throughout the cytoplasm of many fucalean cells (Bouck 1965; Brawley *et al.* 1977). The Golgi body typically has six to eight cisternae which become horse-shoe shaped during vesicle production (Brawley *et al.* 1976b, 1977; Pellegrini 1976). The Golgi complex in gametangial cells of *Cutleria* (La Claire & West 1978) is composed of 20 or more cisternae. Many ultrastructural observations suggest that the Golgi complex may be formed from vesiculation of the nuclear envelope (Bouck 1965; review by Whaley & Dauwalder 1979) but cytochemical support for this proposal is still lacking.

In the brown algae, the Golgi complex has been identified as the site of polysaccharide sulphation (Evans & Callow 1974; Callow *et al.* 1978) and may be involved in mastigoneme production (Loiseaux 1973; Markey & Wilce 1976a). Coughlan & Evans (1978) successfully isolated a Golgi fraction from *Fucus* embryos which contained thiamine pyrophosphatase, inosine diphosphatase and galactosyl transferases in an active state. The isolated Golgi cisternae were able to transfer galactose to fucoidan and N-acetyl glucosamine from UDP-galactose. Assessments of the function of the Golgi complex in algal cells should consider the hypothesis (Palade 1975; Whaley & Dauwalder 1979) that the Golgi complex may be important in cell recognition through the addition to the plasmalemma of membrane that contains specialized receptors, in addition to the assembly and secretion role attributed to the Golgi complex classically.

The endoplasmic reticulum is common throughout the cytoplasm of most brown algal cells. Few distinctive features are reported compared with other cells. An exception is the close association of the ER with the plastids in many brown algae (chloroplast ER, Section 3.1). The ER is involved in at least some aspect of mastigoneme production in the brown algae (Markey & Wilce 1976a, b). Highly dilated cisternae of RER are observed in some cells (e.g. the thallus cell in the *Fucus* embryo; Brawley *et al.* 1977), and RER proliferates markedly in *Fucus* cells treated with colchicine (Brawley unpublished).

Oliveira & Bisalputra (1977) have described vacuole formation in *Ectocarpus* cells through the fusion of vesicles presumed to be of ER origin. Pellegrini (1979) examined vacuole formation in *Cystoseira* cells. She found that the provacuoles were acid phosphatase positive and concluded that vacuole formation occurred through the merging of vesicles produced by the Golgi complex or GERL (*G*olgi related smooth *ER* which produces lysosomes).

A variety of vesicle types has been identified in the cytoplasm of brown algal cells (Brawley *et al.* 1976b). Unfortunately, the contents of few vesicles are known, resulting in ambiguities about their functions. Vesicles with 'granular-cored' contents are secreted by *Cutleria* gametangia shortly before the gametes are released and may be involved in pore formation (La Claire & West 1978). The best characterized vesicles are probably the osmiophilic physodes which contain polyphenolics and are turquoise after toluidine blue staining (McCully 1968). The physodes are closely associated with the plastids of some species (e.g. *Dictyota*, Evans & Holligan 1972; *Cystoseira*, Pellegrini 1980), but there is no direct evidence of their production by the plastids. Pellegrini (1980) has observed the fusion of small acid phosphatase-positive vesicles with the physodes and compared the resulting vesicles to secondary lysosomes.

3.3 Cell wall

The structural organization of cell wall microfibrils was a favourite subject of early ultrastructural studies (see Dodge 1973; Mackie & Preston 1974) because cleaned and shadowed preparations of the cell wall were easy to prepare at a time when the use of cytoplasmic fixatives was a rudimentary art. The architecture of the cell wall is well-known only for the walls of a few higher plants (e.g. Talmadge *et al.* 1973). In the brown algae, cell wall formation has been best studied in the Fucales.

Wall formation in the Fucales follows fertilization, and the mature wall has three distinctive layers (Vreugdenhil *et al.* 1976): an inner fibrillar layer (sulphated fucan and alginic acid), an outer fibrillar layer (alginic acid and cellulose) and an exterior amorphous layer (sulphated fucan and alginic acid) (Fig. 7.20). The exterior amorphous layer is thickened at the rhizoid tip. The sulphated fucan (fucoidan) in this outer layer is the 'glue' which holds the young embryo in place in the wave-swept intertidal zone. The permanent cell wall in *Halidrys* is not formed until about 24 h after fertilization but is preceded by a fibrillar zygotic wall whose outer layer is mucilaginous (Hardy & Moss 1978).

Stevens & Quatrano (1978) have demonstrated that cellulose is synthesized after fertilization in *Fucus* without a requirement for postfertilization RNA or protein synthesis. Cellulose is present in the wall within 20 min after fertilization. Peng & Jaffe (1976) observed strings of membrane particles on the plasma membrane of *Pelvetia* (Fucales) eggs following fertilization (Fig. 7.22), and they suggested that these particle strings might be involved in cellulose microfibril orientation (Figs 7.21, 7.22). Membrane particles were evenly distributed on the plasma membrane before fertilization but particle strings were observed in membranes of some vesicles both before and after fertilization. Peng & Jaffe (1976) calculated that the vesicle particle strings were insufficient in number to

7.20 7.21 7.23

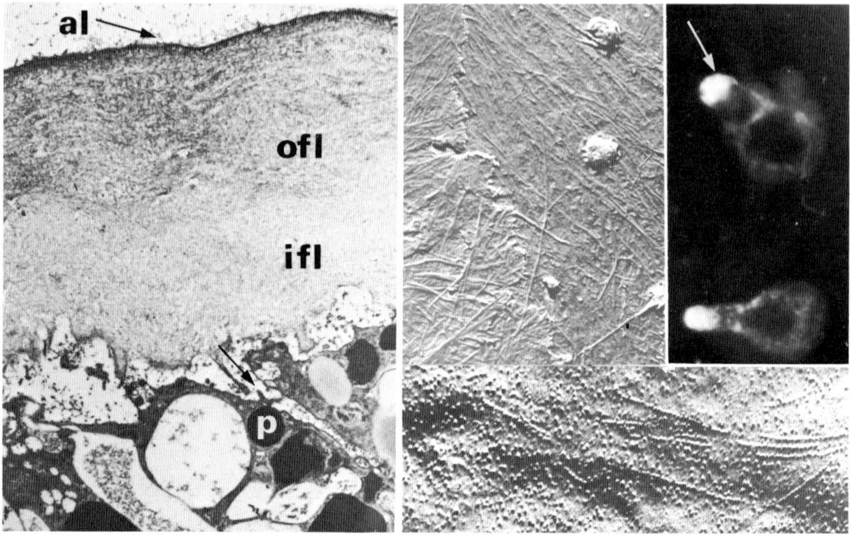

7.22

Fig. 7.20 The wall of a *Fucus* embryo with three distinctive layers: the inner fibrillar layer (ifl), outer fibrillar layer (ofl) and amorphous layer (al). These consist respectively of: sulphated fucan and alginic acid, alginic acid and cellulose, and sulphated fucan and alginic acid. The cytoplasm contains physodes (p) and other polysaccharide containing vesicles are observed in the cytoplasm. Membrane furrowing is observed as part of cytokinesis (arrow). (Brawley 1978.) (× 16 700.)

Fig. 7.21 Two lamellae of microfibrils in the inner fibrillar layer of a cross-fractured *Fucus* wall. (Peng & Jaffe 1976.) (× 21 000.)

Fig. 7.22 Particle strings on the P face of the plasma membrane of a 0.5 h old *Fucus* embryo after freeze fracture. These particle strings may be involved in the orientation of cellulose microfibrils in the cell wall. (Peng & Jaffe 1976.) (× 73 600.)

Fig. 7.23 Two-celled *Fucus* embryos stained with fluorescein-conjugated ricin, a lectin specific for sulphated and unsulphated fucoidan. In normal embryos fucoidan is localized in the cell wall at the rhizoid tip (arrow). This polysaccharide acts as a glue. (Hogsett & Quatrano 1978.) (× 200.)

account for those in the plasmalemma by vesicle fusion alone following fertilization. They hypothesized that particle aggregation within the plasma membrane also occurred.

Identification of specific enzymes to lyse wall polymers would greatly advance our understanding of its composition. Antibodies were developed against alginate by Vreeland (1972) to study wall biogenesis, and lectins were used by Hogsett & Quatrano (1978) to study cell wall fucoidan. The lectin ricin binds to the galactose residues of fucoidan (Fig. 7.23) and is a specific probe for *in situ* studies of cell wall fucoidan (Section 3.9).

3.4 The nucleus and centroioles, mitosis, and meiosis

There are few comprehensive studies of mitosis in the brown algae. Some stages of mitosis have been observed in *Fucus* (Bouck, in Leedale 1970; Brawley *et al.*

1977; Berkaloff & Rousseau 1979), *Chorda* (Toth 1974), *Cutleria* (La Claire & West 1979), *Hormosira* (Forbes & Hallam 1979), *Pilayella* (Markey & Wilce 1975), and *Zonaria* (Neushul & Dahl 1972).

The paired centrioles occupy a cleft in the nuclear envelope during interphase (Markey & Wilce 1975; Brawley 1978). They replicate prior to mitosis and migrate to opposite poles of the nucleus during early prophase. A cleft in the nuclear envelope remains associated with the migrating pair of centrioles in *Pilayella* (Markey & Wilce 1975). The nucleolus disappears in late prophase and reappears in telophase (Toth 1974; Markey & Wilce 1975; Brawley *et al.* 1977; La Claire & West 1979). Paired centrioles are observed at the spindle poles, and microtubules often emanate from the electron-opaque pericentriolar material (e.g. *Pilayella*, Markey & Wilce 1975). Investigations of the pericentriolar material in animal cells suggest that it may be more important than the associated centrioles as a microtubule organizing centre (MTOC) (Berns & Richardson 1977; Telzer & Rosenbaum 1979).

Through metaphase the nuclear envelope is intact except for polar fenestrae, but it breaks down during anaphase (Figs 7.24, 7.25) (Toth 1974; Brawley *et al.* 1977; La Claire & West 1979). Technical problems are probably responsible for conflicting reports in the literature regarding the degree to which the nuclear envelope disappears during mitosis. Anaphase is apparently brief, and many of the reports on mitosis in the brown algae are based solely upon observations of metaphase. Serial sectioning is essential for observing details of centriolar position during mitosis and for a more complete documentation of all mitotic stages.

Some stages of meiosis have been described in *Pilayella* (Toth & Markey 1973) and *Fucus* (Berkaloff & Rousseau 1979). In both cases synaptonemal complexes were observed, and the early stages appear quite similar to observations on mitotic nuclei.

3.5 Microtubules and microfilaments

Microtubules sometimes occur near the site of cytokinesis (Markey & Wilce 1975), but their involvement in cytokinesis or wall formation has not been demonstrated in the brown algae. Microtubules are rarely observed in the cells of *Fucus* except during mitosis (Brawley *et al.* 1977), and wall formation is not affected in embryos treated with colchicine (Stevens 1975). However, an abnormal position of the egg nucleus is observed in *Fucus* eggs after colchicine treatment, and pronuclear migration following fertilization is completely blocked by this drug (Brawley & Quatrano 1979a) (Fig. 7.27).

Cytochalasins B and D strongly inhibit the movement of Golgi vesicles, suggesting that vesicle movement is mediated by actin microfilaments (Brawley & Quatrano 1979b). Microfilaments are also suspected to have a role in axis fixation in *Fucus* embryos (Brawley & Quatrano 1979b; Quatrano *et al.* 1979), but their function in this process was not directly demonstrated. Furrowing has often been associated with actin in animal cells (e.g. Schroeder 1979), and it is blocked by cytochalasin treatment. Furrowing still continues in *Fucus* embryos undergoing cytokinesis in the presence of cytochalasin, but is slower than normal (Brawley & Quatrano 1979b).

7.24 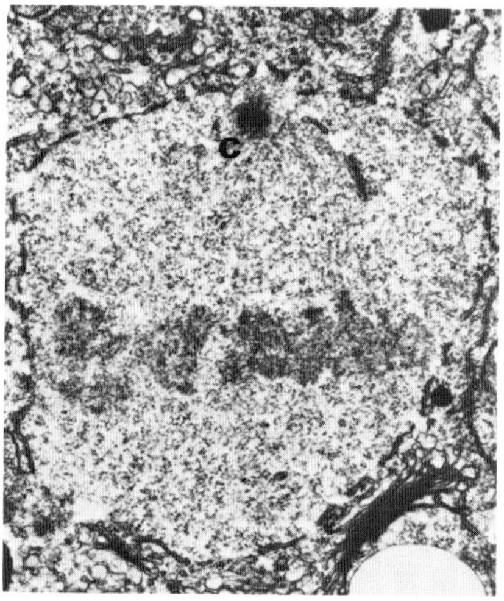

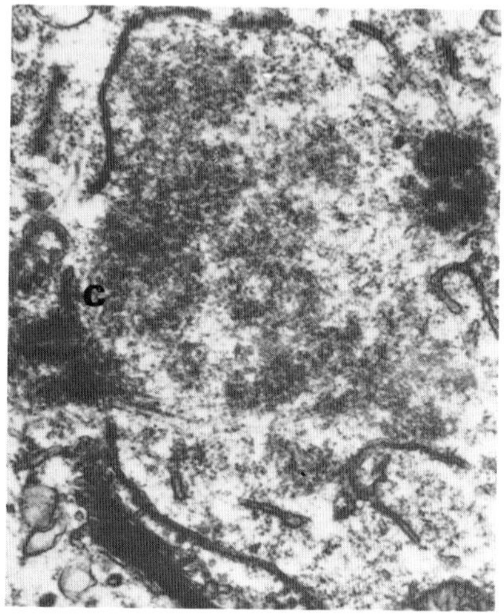 7.25

Figs 7.24, 7.25 Mitosis in *Cutleria* vegetative cells. In metaphase (**7.24**) the nuclear envelope fragments and breaks down during anaphase (× 13 600) (**7.25**). Note the position of the centrioles (c) (× 18 400). (Courtesy of Dr. John La Claire II.)

Chloroplast movement in response to changes in light intensity is common in many brown algal cells (Rüffer *et al.* 1978), but whether this movement is mediated by microtubules or microfilaments is unknown. Chloroplast movements associated with fertilization and mitosis in *Fucus* (Brawley *et al.* 1977; Brawley 1978) are unaffected by cytochalasin treatment.

3.6 Flagella

Recent studies have added little to our knowledge of flagella in the brown algae (see Dodge 1973, for summary of the work of Manton and others), which was the subject of extensive study when algae were first examined with transmission electron microscopy. Mastigonemes (Bouck 1969) decorate the anterior flagellum in the normally heterokont flagella (Fig. 7.26) of reproductive cells and are initially localized on one side at the flagellar base (Markey & Wilce 1976a; La Claire & West 1978). Developing mastigonemes are first observed in the nuclear envelope (Bouck 1969; Markey & Wilce 1976a; Berkaloff & Rousseau 1979), and may be associated with the ER and/or Golgi complex before deposition onto the anterior flagellum.

The anterior flagellum is important in gamete recognition in the brown algae (Jaenicke & Müller, in Jaenicke 1974; Callow *et al.* 1978) and may have a tightly coiled apical region which is undecorated (e.g. Toth 1976). The posterior flagellum is smooth, and one of its most important characters is a more electron-opaque region of plasma membrane near its base, which faces a similar

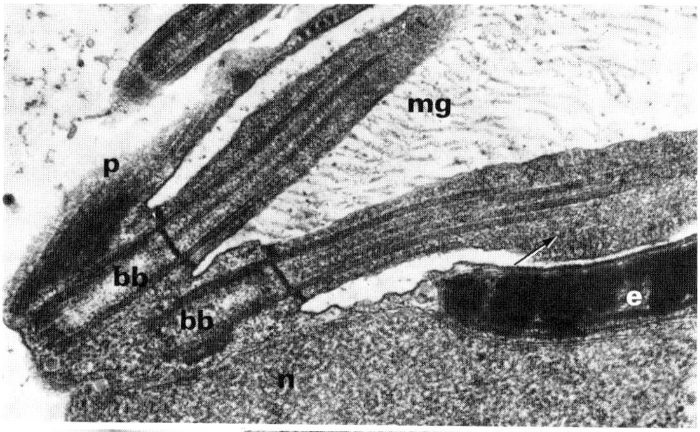

7.26

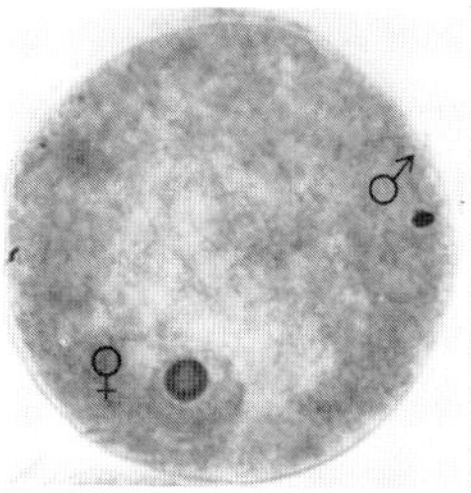

7.27

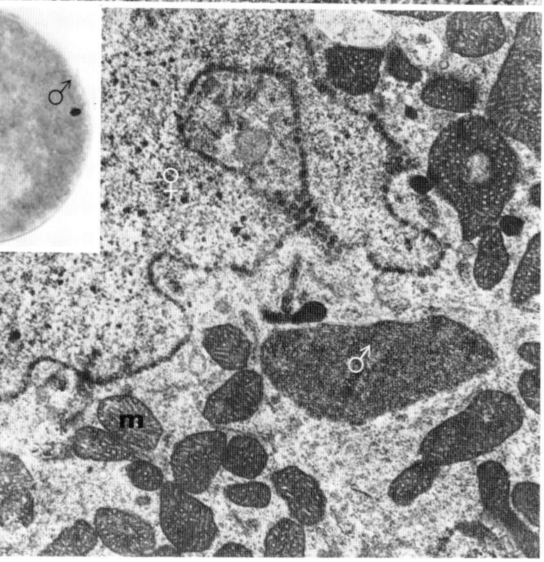

7.28

Fig. 7.26 Longitudinal section of the flagella and basal bodies (bb) in a *Fucus* sperm cell. Mastigonemes (mg) decorate the anterior flagellum. A basal swelling (arrow) is observed near the base of the posterior flagellum near the plasma membrane overlying the eyespot (e). Proboscis (p), Nucleus (n). (Berkaloff & Rousseau 1979.) (× 36 000.)

Fig. 7.27 Fertilized *Fucus* egg in which pronuclear migration and fusion were blocked by incubation in colchicine. Note the large nucleolus in the egg nucleus (♀). Sperm nucleus (♂). (Brawley & Quatrano 1979a.) (× 560.)

Fig. 7.28 The sperm nucleus (♂) and egg nucleus (♀) prior to karyogamy. Note the asymmetric concentration of egg mitochondria (m) in the vicinity of the sperm nucleus before pronuclear fusion. (Brawley *et al.* 1976a.) (× 8400.)

region on the adjacent plasma membrane of the cell (Toth 1974; Caram 1975; Markey & Wilce 1976a; La Claire & West 1978; Berkaloff & Rousseau 1979). The diameter of this region near the base of the posterior flagellum is often larger than that through the rest of the sheath surrounding the flagellar microtubules.

Centrioles move to the cell surface prior to formation of the flagella and serve as basal bodies (Toth 1976; La Claire & West 1978; Berkaloff & Rousseau 1979). The plasma membrane extends to become the flagellar sheath as microtubule nucleation occurs from the centriolar complex. Reports of flagellar formation within cytoplasmic vesicles in some brown algae (Loiseaux 1973; Markey & Wilce 1976a,b) should be verified with serial sectioning or observations of earlier stages of developing flagella (cf. Sorokin 1962).

Many brown algal gametes are negatively phototactic but the structural details and mechanism by which light is perceived and flagellar motion affected are poorly understood. A modified chloroplast, the eyespot (Dodge 1973), lies beneath the plasma membrane near the posterior flagellum, and flagellar rootlet microtubules are frequently associated with this poorly understood organelle (Bouck 1970; La Claire & West 1978; Section 4.5).

3.7 Gametogenesis and zoosporogenesis

Essential features of gametogenesis and zoosporogenesis are similar: formation of a reproductive structure and subcompartmentalization of its contents preceded either by mitotic, or both meiotic and mitotic, nuclear divisions, to form gametes or zoospores. Formation of flagella in the gametes or zoospores occurs at the end of this cleavage process. Recent investigations of zoospore formation in the marine Phaeophyta include those of Toth (1974, 1976) on *Chorda*, Lofthouse & Capon (1975) on *Ectocarpus* and Markey & Wilce (1976a) on *Pilayella*. Gametogenesis has been examined cytologically in *Cutleria* by La Claire & West (1978, 1979), in *Pilayella* by Markey & Wilce (1975, 1976b) and in *Fucus* by Berkaloff & Rousseau (1979).

The Golgi complex is typically active in some stage of gametogenesis and zoosporogenesis, and Golgi-derived vesicles are usually important in cleavage. Cleavage occurs through a combination of furrowing and internal vesicle fusion generally (Toth 1974; Markey & Wilce 1975, 1976a, b; La Claire & West 1979); however, cleavage of the mature gametangium into 64 individual sperm in *Fucus* occurs entirely by vesicle fusion (Berkaloff & Rousseau 1979).

Characteristic organelle associations are observed as *Chorda* zoospores differentiate. A chloroplast is closely associated with each nucleus in the sporangium, and each zoospore ultimately has a single chloroplast which develops an eyespot (Toth 1974). Toth (1976) has also followed zoospore settlement and early embryogenesis in *Chorda*. The zoospores retract their flagella, and a germ tube forms. Most of the cytoplasm migrates into the germ tube, but a nucleus and some other organelles still occupy a small area of the basal cell after the first mitosis. Chloroplast replication occurs concurrent with the first mitosis, and a single eyespot remains in one of the new plastids.

Differentiation of the paraphyses and zoosporangia from the meristoderm in *Chorda* was described using light microscopy by Toth (1976). A thick polysaccharide layer which covers the paraphyses and sporangia can be observed when ruthenium red is included in the fixative. The sporangial apices have a distinctive cap which is less compact than the wall surrounding the rest of the cell.

Osmiophilic material is polarized exterior to the chloroplasts near the plasma membrane in female gametes of *Cutleria* (La Claire & West 1978). This material, presumed to be polyphenolic, is secreted prior to release of the gametes from the gametangium. Mature female gametes have a centrallylocated nucleus, and numerous plastids lie in the posterior half of the cell. Mitochondria, Golgi bodies, and numerous vesicles are located near the basal bodies, and a plastid associated with the posterior flagellum develops an eyespot. The posterior flagellum has the characteristic basal swelling and differentiated plasma membrane (Section 3.6).

3.8 Fertilization

Recent studies of fertilization have focused on several members of the Fucales. Fertilization in this order involves fusion of a large non-motile egg and a small biflagellate sperm. The egg surface in *Fucus* (Callow *et al.* 1978) and *Sargassum* (Smith 1976) appears irregular when observed by scanning electron microscopy. Presumably, this is due to the large number of vesicles at the egg surface prior to fertilization, but the actual irregularity may be artifactual (Shelton & Mowczko 1978).

A pheromone released by the egg attracts numerous sperm to each egg; superfluous sperm become embedded in the young wall at the egg surface (Callow *et al.* 1978; Hardy & Moss 1978). The cellular site of pheromone production and its mode of release from the egg remain unknown but the pheromone has been chemically characterized as a trans, cis-1,3,5-octatriene (Müller & Jaenicke 1973; Jaenicke 1974; see Chapter 19). Autoradiographic evidence indicates that the pheromone released by the female gamete of *Ectocarpus* is initially bound by the anterior flagellum of the male gamete (Jaenicke & Müller in Jaenicke 1974). Membrane fusion in *Fucus* gametes occurs between the anterior sperm flagellum and the egg plasmalemma (Callow *et al.* 1978). Gamete recognition involves membrane receptors and has been investigated using lectins (Bolwell *et al.* 1977, 1979).

The egg surface is smooth following fertilization (Callow *et al.* 1978). Contents of the vesicles released following fertilization are involved in wall formation, which begins immediately after fertilization. These vesicles have been termed 'cortical vesicles' (Schröter *et al.* 1976), but this is a confusing term since they are very different in composition from cortical vesicles in animal eggs. The rapid secretion of wall material probably serves as one block to polyspermy, which is rare in *Fucus* eggs (Brawley & Quatrano 1979a). Polyspermy occurs in *F. serratus* when eggs are exposed to extremely high concentrations of sperm (Callow *et al.* 1978).

The ultrastructure of fertilization and pronuclear migration and fusion has been described in *Fucus* by Brawley *et al.* (1976a). The nuclear envelope surrounding the sperm nucleus breaks down following fertilization, but it reforms prior to fusion of the egg and sperm nuclei. The eyespot and sperm mitochondria trail the sperm nucleus as it migrates toward the egg nucleus, but the flagellar microtubules remain in the cortical cytoplasm. Pronuclear migration is blocked by incubation of fertilized eggs in colchicine but not by incubation in cytochalasin B (Brawley & Quatrano 1979a); migration therefore appears to be microtubule mediated.

A distinctive feature of pronuclear fusion in *Fucus* compared to other organisms having similar gametes and fertilization processes (i.e. marine invertebrates) is the very large and asymmetric concentration of egg mitochondria around the pronuclei just before their fusion (Brawley *et al.* 1976a) (Fig. 7.28). The localization of mitochondria in the fertilized egg of *Fucus* may be related to the weak cell polarization associated with fertilization and the migration path of the sperm nucleus in the egg cytoplasm (Knapp 1931). Intracellular calcium levels increase rapidly in many eggs following fertilization (e.g. Ridgeway *et al.* 1977; Gilkey *et al.* 1978). Since mitochondria can be involved in the regulation of intracellular calcium (Lehninger 1970; Osborn *et al.* 1979), and calcium-carried current is associated with axis fixation in embryos of the Fucales (Section 3.9), the asymmetric concentration of mitochondria in the *Fucus* egg may play a role in fixation of the weak, fertilization-associated polar axis. (Also see Chapter 18.)

If fertilized eggs have been incubated in colchicine to block pronuclear migration, chromosome condensation and mitosis occur in the egg nucleus after drug removal, but a spindle is not formed in association with the sperm chromosomes after their condensation (Brawley & Quatrano 1979a). This suggests that the egg centrioles (or the MTOC material associated with them), and not those of the sperm, are important in normal embryonic mitosis. The fate of the sperm centrioles remains an interesting question since they were not observed migrating with other sperm organelles (Brawley *et al.* 1976a). If sperm centrioles are not involved in mitosis in the embryo, Yamanouchi's observation (1909) of multipolar spindle formation was probably due to an abnormal mitosis, not polyspermy. Callow *et al.* (1978) did not report on embryological development in batches of eggs in which they induced polyspermy.

Comparative cytological studies of fertilization in brown algae in which both gametes are motile and in the Laminariales where the plants are also oogamous, as in the Fucales, will be of particular interest. Cytological studies of *Ectocarpus* are especially needed to complement our biochemical knowledge of gamete attraction in this genus.

3.9 Polar axis fixation

Aspects of axis fixation in the Fucaceae have interested a large number of workers (Quatrano 1973; Novotny & Forman 1974; Nuccitelli & Jaffe 1975; Ohmori & Miyazaki 1976; Ogawa 1977; Callow *et al.* 1978; Hogsett & Quatrano 1978; Nuccitelli 1978; Schröter 1978; Brawley & Quatrano 1979b) and is best investigated in *Fucus* and *Pelvetia*. Light is the strongest external stimulus determining axis fixation, the rhizoid forming on the shaded side of a unilaterally illuminated zygote. The axis is labile until a few hours before germination (Jaffe 1968). Recent studies have attempted to identify the intracellular basis for axis fixation. Nuccitelli & Jaffe (1975) and Nuccitelli (1978) have identified a current entering the prospective rhizoid tip area which is believed to be associated with axis induction. Axis fixation is also closely timed to the polar secretion of the polysaccharide fucoidan into the wall at the rhizoid tip (Novotny & Forman 1974).

Several investigators (Jaffe *et al.* 1974; Nuccitelli 1978) speculated that vesicle movement occurred by 'self-electrophoresis' to the rhizoid tip, through which the calcium-carried current enters, due to the negative charge of the fucoidan after sulphation. This hypothesis stimulated investigations aimed at identifying the effect of sulphate deletion from the growth medium on rhizoid formation under conditions not inhibitory to protein synthesis e.g., after addition of methionine (Crayton *et al.* 1974). Rhizoids formed when sulphate was deleted under these conditions.

Embryos grown in sulphate-deleted medium containing methionine have been termed met-embryos (Brawley & Quatrano 1979b). Hogsett & Quatrano (1978) found no fucoidan in the cell walls of met-embryos using the lectin ricin which binds specifically to both sulphated and unsulphated fucoidan (Section 3.3). When met-embryos were viewed ultrastructurally, similar organelle distributions were found in met- and normal embryos except that there were markedly fewer free Golgi vesicles in the cytoplasm of met-embryos, and the Golgi bodies appeared inactive (no horseshoe-shaped cisternae) (Brawley & Quatrano 1979b). Fucoidan synthesis is apparently shut off, without affecting axis fixation, when sulphate is absent. Autoradiographic work on the *Fucus* embryo provides strong evidence for rejecting the self-electrophoretic hypothesis since ^{35}S-fucoidan is secreted around the entire cell wall of the normal l-celled embryo at a time when the axis is fixed and the current localized (Brawley & Quatrano 1979b).

Cytochalasin B (CB) reversibly inhibits axis fixation but not mitosis

(Quatrano 1973). This fact has led to speculation that actin microfilaments are required for axis fixation in the embryo. Actin has not been directly demonstrated in the embryo, but cytochalasin D (CD), a highly specific actin inhibitor, also prevents axis fixation (Brawley & Quatrano 1979b). CB and CD also inhibit vesicle movement in *Fucus* without affecting fucoidan sulphation or Golgi body morphology. The initially random secretion of ^{35}S-fucoidan in both 1- and 2-celled met-embryos chased with Na_2SO_4 is of especial interest. Brawley & Quatrano (1979b) speculate that this is due to a loose actin network resulting from an internal pH change in cells grown without sulphate. Actin could be responsible for the current associated with axis formation through its localized association with the plasmalemma.

Further progress in investigations of axis fixation requires study of the effects of incubation in CB and sulphateless medium on the current patterns observed in *Fucus* embryos and the direct identification of actin. Ultrastructural studies of *Sargassum* and *Coccophora* would also be important because these genera reportedly have fixed axes prior to fertilization (Nakazawa 1950, 1951). Whether there is an asymmetric mitochondrial localization or other cytoplasmic gradient in the eggs of *Sargassum* and *Coccophora* is unknown.

4 CHLOROPHYTA

The green algae have received extensive study from microscopists, but much of the existing information concerns freshwater members of the division (Pickett-Heaps 1975). Several freshwater Chlorophyta (e.g. *Chlamydomonas*) have been used extensively in cellular and developmental studies.

Characters such as whether the mitotic spindle is open or closed and whether cell division is associated with a 'phycoplast' or 'phragmoplast' have been used in recent phylogenetic schemes for the green algae (Pickett-Heaps 1972, 1979; Stewart & Mattox 1975). These concepts are less helpful for understanding relationships among the marine green algae studied so far. In algae representing more than five orders of marine Chlorophyta, the spindle is closed or slightly perforated, and cytokinesis occurs by cleavage without a phycoplast or phragmoplast. Little information is available for the marine Chlorophyta concerning other important characters such as glycolate enzymes and the microtubular cytoskeleton of reproductive cells. This information may be useful in understanding phylogenetic relationships within these algae (see Chapter 3).

We mention microtubules and microfilaments only briefly in the following discussion on organelles in the marine Chlorophyta but they are known to affect cell development and influence wall formation in a number of the freshwater members (Marchant & Pickett-Heaps 1974; Palevitz & Hepler 1975; Marchant 1977; Robinson 1977; Quader *et al.* 1978). Their roles in cells of the marine green algae should be investigated.

The Dasycladales, especially *Acetabularia*, have interested a wide range of biologists, and some aspects of their development are discussed below (Sections 4.1, 4.4, 4.6). We also review recent work on the production of protoplasts in the marine Chlorophyta.

4.1 Chloroplasts, pyrenoid and mitochondria

When viewed with the light microscope, green algal plastids are observed in a variety of shapes, and these are useful in algal identification. The plastids may be disc-shaped, form a reticulum or a parietal, cup-shaped plastid may be observed.

The thylakoids of green algal plastids are arranged as two to many per lamella. In some, an irregular stacking results in a thylakoid arrangement reminiscent of the grana and intergrana regions of the plastids in the Charophyceae and higher plants.

Starch is synthesized and stored within the chloroplasts of most green algae, but plastids differentiate from a common stem organelle into chloroplasts and amyloplasts in some of the Caulerpales (Udoteaceae and Caulerpaceae), a condition known as heteroplasty. The chloroplasts are rather typical green algal plastids, whereas the amyloplasts have few thylakoids and are differentiated as starch-storing organelles (Borowitzka & Larkum 1974). In contrast to the green algae, polysaccharide storage occurs in cytoplasmic vesicles in the red and brown algae.

Many of the recent studies of green algal chloroplasts have investigated special features of plastids of the Caulerpales. Many chloroplasts in this coenocytic order of green algae contain structures termed the 'concentric lamellar system' (CLS) or 'polar thylakoid organizing body' (Hori & Ueda 1967; Borowitzka & Larkum 1974; Calvert *et al.* 1976; Colombo 1978). The CLS (Fig. 7.29) does not contain chlorophyll and appears to be continuous with the inner membrane of the chloroplast envelope (Borowitzka 1976). It is composed of concentric lamellae and may range from 3–25 pairs of thylakoids. The CLS is present in both chloroplasts and amyloplasts. It is postulated (Borowitzka & Larkum 1974) to be a thylakoid organizing centre but biochemical and developmental evidence is needed to confirm this hypothesis. Girdle bands of thylakoids are also present in some Caulerpalean plastids (Calvert *et al.* 1976).

Wright and Grant (1978) have suggested that the extra membrane systems in Caulerpalean plastids are responsible for greater resistance of *Caulerpa* chloroplasts to detergent disruption (i.e. compared to those in higher plants). If these membrane specializations do confer greater osmotic and chemical stability, they might be important in the evolution of an interesting plant–animal interaction (Giles & Sarafis 1974). For example, plastids of *Caulerpa*, *Codium* and several other coenocytic green algae are known to remain functional for weeks following their phagocytosis by cells of saccoglossan molluscs and

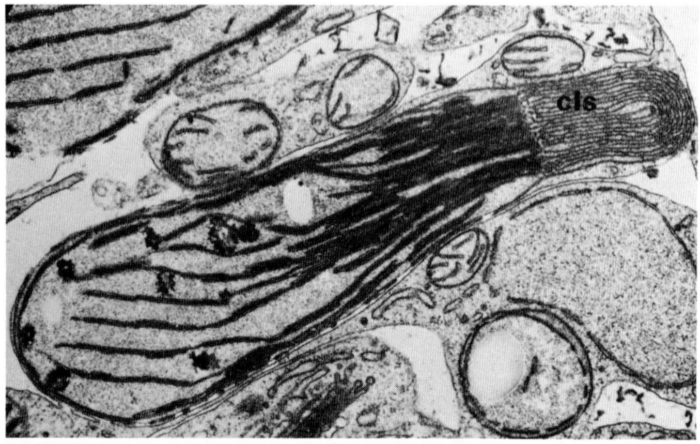

7.29

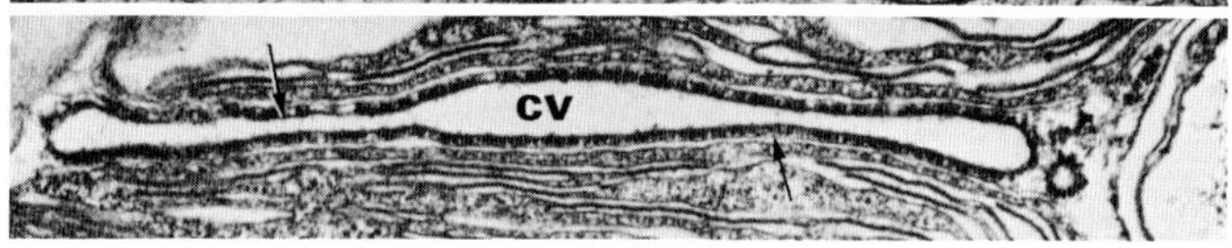

7.30

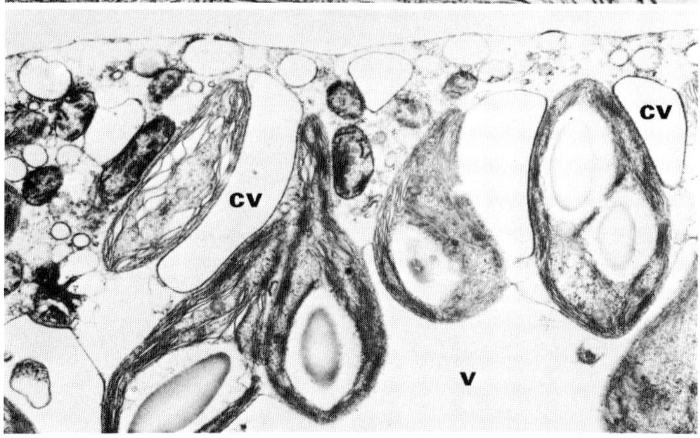

7.31

Fig. 7.29 A chloroplast of *Caulerpa paspaloides*. Note the concentric lamellar system (cls). The chloroplast lamellae have a varying number of thylakoids (Calvert *et al.* 1976.) (× 17 600)
Figs 7.30, 7.31 Cleavage of the multinucleate cap of *Acetabularia mediterranea* occurs through fusion of cytotomic vesicles (cv) with smaller coated vesicles and with the plasma membrane. These vesicles are associated through intramembranous cross-bridges (arrows) with the plastids and mitochondria (**7.30**). Vacuole (v). (Fig. 7.30 Franke *et al.* 1977 [× 31 200]) (Fig. 7.31 Franke *et al.* 1976 [× 10 880])

make an important contribution to the animal's nutrition (Trench & Olhorst 1976).

Plastid ultrastructure has been used to support previous schemes (Svedelius 1906; Børgesen 1907) for the direction of speciation within the genus *Caulerpa*. Primitive characters (Calvert *et al.* 1976) include the presence of a pyrenoid, a complex CLS, starch formation in the chloroplast, and large plastid size. The pyrenoid may be a well defined region within the plastid and is surrounded by a solid band of starch (e.g. *C. microphysa*), a region of greater stromal density traversed by thylakoid bands (e.g. *C. elongata*), or entirely is absent (e.g. *C. racemosa*).

Coleman (1979) observed chloroplast DNA *in situ* in *Acetabularia*, *Batophora* (Dasycladales) and *Codium* (Caulerpales) using the DNA-specific, fluorescent dye, 4,6-diamidino-2-phenylindole (DAPI). Her observations of chDNA within green algal plastids show an arrangement not dissimilar to the peripheral ring of chDNA in the brown algal plastid, although it is less planar. Although large nucleoids of chDNA are observed in green algal plastids, DAPI staining shows that they are interconnected in a continuous 'necklace' of chDNA. Coleman has confirmed an earlier observation (Woodcock & Bogorad 1970) that plastids in *Acetabularia* have a varying amount of DNA. She observed a variation in size of the nucleoids within the chloroplasts of vegetative cells and only *c.* 25% of *Acetabularia* and *Batophora* chloroplasts contained any DNA. Chloroplasts in *Acetabularia* cysts, however, contained as many as nine nucleoids/plastid, and 98% of all plastids contained chDNA.

Mitochondria in green algal cells resemble those in other eukaryotic cells. The inner membrane of the mitochondrial envelope enfolds to form tubular cristae. Most cells contain many mitochondria, but male gametes of *Bryopsis* have a single large mitochondrion (Burr & West 1970). Several unusual mitochondrial conformations have been observed. Scott & Bullock (1976) observed an unusual mitochondrial morphogenesis in *Cladophora flexuosa* during the late stages in gamete formation. The mitochondria were often highly contorted, and regions of the mitochondria containing DNA were separated from the rest of the mitochondrial matrix by a double membrane. Mitochondria outside the differentiated gametangial area in *Derbesia* (*Halicystis* phase) often have tubular paracrystalline inclusions in the matrix and concentrically arranged cristae (Wheeler & Page 1974; cf. Osborn *et al.* 1979).

4.2 Endomembranes (the Golgi complex, vesicles, endoplasmic reticulum, microbodies)

Golgi bodies (=dictyosomes) of green algae appear similar to those described in other plant cells. Very large vesicles form at the ends of Golgi cisternae (Franke *et al.* 1976, 1977; Hori 1977), but there are no other conformational changes associated with increased Golgi activity (cf. Section 3.2). The typical Golgi body has *c.* seven to ten cisternae.

Franke *et al.* (1977) have described a process of cytoplasmic cleavage in *Acetabularia* involving coated vesicles that they hypothesize are produced by the Golgi body (Figs 7.30, 7.31). Cysts form in the cap through the alignment and fusion of large coated vesicles (Franke *et al.* 1976, 1977). Microtubules are associated with these vesicles and may be responsible for vesicle orientation.

The 'cytotomic' and smaller coated vesicles align near invaginations of the vacuole and plasma membrane. The large, cytotomic vesicles are associated along their flattened surfaces with mitochondria and chloroplasts through regularly-spaced, electron-opaque, crossbridging elements. Fusion of cytotomic vesicles with other vesicles occurs at their free ends which have typical bristle coats. Franke *et al.* (1976) believed that these vesicles are aligned and associated with plastids and mitochondria in a way that assures adequate distribution of these organelles to each cyst.

Many cells of the green algae are highly vacuolated. This is especially true in coenocytic algae where the response to wounding (e.g. *Caulerpa*, Dreher *et al.* 1978) involves gel formation by the vacuolar matrix. In *Caulerpa* the external plug is reinforced by an internal wound plug. This is secreted for up to 11 h after wounding, by which time wall formation begins. A number of protoplasts form near the wound site as the cytoplasm retracts, but these coalesce as the cytoplasm moves back toward the repaired area. The external plug and internal plug are of different composition. Sulphated glycoprotein or a mixture of protein and sulphated polysaccharides forms the internal wound plug.

Protein bodies formed by the rough ER appear to be important in formation of a wound plug in *Bryopsis* (Burr & West 1971). They break down into rod-like and then granulo-fibrillar material when secreted into the vacuole. The association of microtubules with the protein bodies as they disperse is particularly interesting. Large arrays of microtubules are also observed inside the vacuole of *Penicillus*, associated with calcium oxalate crystals (Turner & Friedmann 1974). The vacuole of *Penicillus* contains other electron-opaque bodies and a fibrillo-granular matrix which probably functions in a fashion similar to the contents of *Bryopsis* and *Caulerpa* vacuoles following wounding.

Microbodies are spherical, electron-opaque bodies with a finely granular matrix. They are surrounded by a single membrane. Microbodies may be closely associated with mitochondria, ER, lipid bodies and chloroplasts, and they are important in specific metabolic activities, e.g., glycolate metabolism. This organelle has been identified morphologically in a number of green algae including *Enteromorpha*, *Ulva*, *Ulothrix* and *Cladophora* (Silverberg 1975). Menzel (1976) has identified microbodies in *Acetabularia* using the DAB technique to identify catalase, a marker enzyme for microbodies.

4.3 Cell wall

The composition of cell wall material and the microfibrillar architecture of its various layers are known in many marine Chlorophyta from the work of Preston and others (see Dodge 1973; Mackie & Preston 1974). Some algae, such as *Valonia*, have highly cellulosic walls and microfibrils form well-ordered arrays. Mannan and xylan are important wall constituents in many species belonging to the Caulerpales. Cytochemical data are available on the early zygote and rhizoid cell walls of *Ulva* (Bråten 1975) and *Enteromorpha* (Callow & Evans 1974), algae well known as ship-fouling organisms, and cell wall proteins have been studied in *Acetabularia* (Göke *et al.* 1974; Paradies & Werz 1977a, b). Hanic & Craigie (1969) analysed the outer cuticle of the cell wall in several marine Chlorophyta and found that it was highly proteinaceous.

The alternating gametophytic and sporophytic generations of an algal life cycle often involve two morphologically dissimilar phases. Changes in organelle structure and function from the gametophytic to the sporophytic phase of a single species are of particular interest. They offer excellent possibilities for studying the developmental consequences of evolutionary selection for different plant morphologies. The differences in wall composition between the gametophyte and sporophyte of several green algae present such a case. The sporophytes of *Derbesia* and *Bryopsis* contain mannan as the main cell wall polysaccharide, whereas the gametophytic plants have xylan and cellulose as the main wall constituents (Huizing & Rietema 1975). A similar change in wall composition is observed in *Acetabularia*, in which the cyst walls are composed of cellulose (Herth *et al.* 1975) but the vegetative plant wall is composed of β-1,4-mannan.

A septal plug occurs in the wall between cells of one marine green alga, the first report of pit connections in an organism other than red algae and fungi (Brawley & Sears 1980). Cytoplasmic continuity is maintained in some green and brown algae by plasmodesmata which traverse the crosswall between cells, but whether the septal plug is a temporary or permanent barrier to intercellular communication is unknown.

4.4 Nucleus, mitosis, and cytokinesis

Mitosis and cytokinesis in the green algae have received extensive study. Although most of these studies treat freshwater species (Pickett-Heaps 1975, 1979), an increasing amount of information is accumulating on marine members of the division.

The interphase nucleus in cells of most of the marine green algae is characterized by large blocks of heterochromatin (Scott & Bullock 1976; Hori & Enamoto 1978a, c). The nuclear envelope remains completely intact (Fig. 7.32) during mitosis in many species (*Cladophora*, Scott & Bullock 1976; *Dictyosphaeria*, Hori & Enamoto 1978b; *Valonia*, Hori & Enamoto 1978c; secondary nuclei of *Batophora*, Liddle *et al.* 1976, and *Acetabularia*, Godward *et al.* 1979). In *Bryopsis* (Burr & West 1970) and *Ulva* (Løvlie & Bråten 1970), the nuclear envelope remains intact except for polar fenestrae, while in *Acrosiphonia* (Hudson & Waaland 1976), the entire nuclear envelope is distinctly perforated by anaphase.

Centrioles may be displaced slightly to the side of the mitotic axis during division (Løvlie & Bråten 1970; Scott & Bullock 1976; Hori & Enamoto 1978c), but they were not observed during mitosis in the Dasycladales (Liddle *et al.*

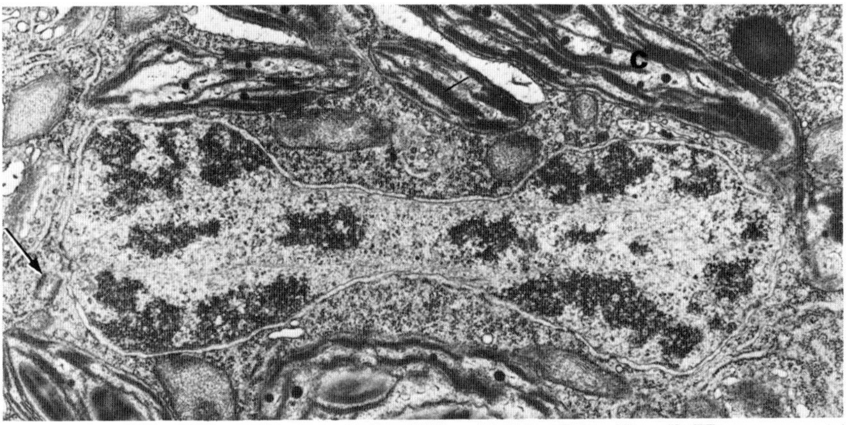

7.32

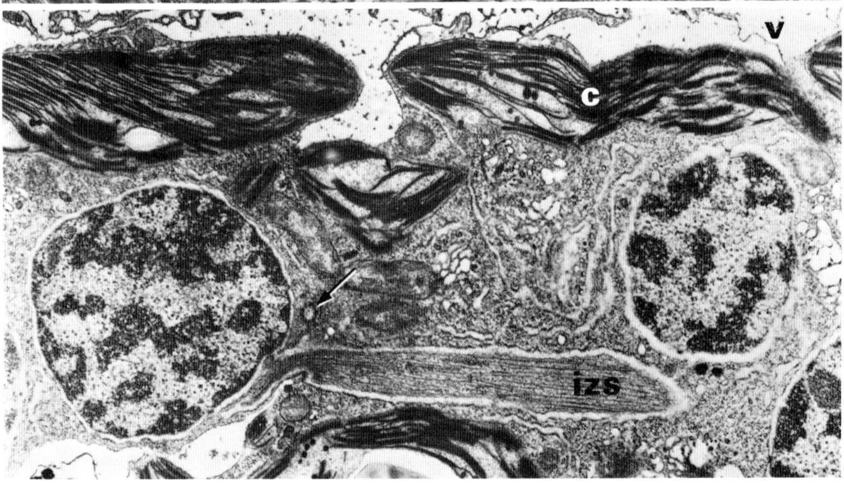

7.33

Figs 7.32, 7.33 Mitosis in *Cladophora* occurs within a completely closed spindle. (**7.32**) Mid-anaphase (× 27 500); (**7.33**) late-telophase (× 11 000). A residual interzonal spindle (izs) is observed in late telophase. Note centrioles (arrow) and chloroplasts (c). Rough ER and free ribosomes are abundant. Vacuole (v). (Scott & Bullock 1976.)

1976) even when nuclei were serially sectioned (Godward *et al.* 1979). Polar microtubules are observed in late prophase in *Cladophora* while shorter microtubules are found attached to the condensed chromosomes at well-defined kinetochores during metaphase and anaphase. Daughter nuclei frequently pinch off from a residual, interzonal spindle in late telophase (Scott & Bullock 1976; Hori & Enamoto 1978b, c) (Fig. 7.33). The microtubules in the interzonal spindle are localized near the nuclear envelope.

Two or more cisternae of endoplasmic reticulum closely surround the nucleus during mitosis in the white spot nuclei of *Acetabularia* (Godward *et al.* 1979). These cisternae seem to be assembled by a body similar to the prolamellar body observed in proplastids of higher plants.

The nucleolus disappears by metaphase in *Acrosiphonia* (Hudson & Waaland 1976), and is extruded during anaphase in white spot nuclei of *Acetabularia* (Godward *et al.* 1979). More typically (e.g. Scott & Bullock 1976; Hori & Enamoto 1978a, b), the nucleolus remains essentially intact through anaphase and nucleolar fragments then move to both poles.

Cytokinesis follows mitosis in *Ulva*, and the division furrow passes between the centrioles which lie close together following nuclear rotation during telophase (Bråten & Nordby 1973). In *Cladophora*, *Dictyosphaeria* and *Valonia*, nuclear division is asynchronous and independent of cytokinesis. Microtubules are not associated with cytokinesis in these species, and cleavage occurs through membrane furrowing (Scott & Bullock 1976; Hori & Enamoto 1978a, b). A cleavage furrow is also responsible for cytokinesis in *Enteromorpha* (McArthur & Moss 1978) and *Acrosiphonia* (Hudson & Waaland 1976). A few microtubules lie in a ring near the furrow in *Acrosiphonia*.

Mitosis and cytokinesis are especially interesting in *Acrosiphonia* (Figs 7.34, 7.35). The synchrony of nuclear and cytoplasmic division in this alga varies as the plant matures (Hudson & Waaland 1976). Mitosis is signaled in a cell when nuclei migrate to form a band in the cell. Mitosis begins with centriole replication and their migration to the spindle poles. It is interesting that not all of the nuclei in the mitotic area migrate into the band and undergo mitosis. Those that do are associated with centrioles and microtubules. When mitosis occurs at the filament tip, the apical cell always receives many more nuclei than the other daughter cell because the cytokinetic furrow divides the mitotic area unevenly, and the band is several nuclei thick. Intercalary cell divisions, however, result in an equal distribution of nuclei (Hudson & Waaland 1976).

Meiosis during formation of quadriflagellate zoospores (Bråten & Nordby 1973; Nordby 1974) is similar in cytological detail to mitosis (Løvlie & Bråten 1970). A final duplication of the centrioles occurs in meiosis without further nuclear or cell division (Bråten & Nordby 1973). The nucleus migrates away from its position near the chloroplast before both mitosis and meiosis in *Ulva* (Bråten & Nordby 1973; Nordby 1974).

The site of meiosis in thalli of the Dasycladales was unknown until recently when Green (1973), Koop (1975, 1979), and Spring *et al.* (1975) provided evidence that meiosis in *Acetabularia* occurs in the primary nucleus. Liddle *et al.* (1976) observed synaptonemal complexes during meiosis in the primary nucleus of *Batophora*.

The association between the cytoplasm and primary nucleus in the Dasycladales has been the subject of much investigation. The ease of enucleation and nuclear transplantation in these algae has made them important systems for studies of cell regeneration and nucleo-cytoplasmic interactions. This work is extensively reviewed (Hämmerling 1963; Puiseux-Dao 1970; Quatrano 1974; Woodcock 1977). Fine structural studies were made of nuclear morphogenesis during aging (Berger & Schweiger 1975; Zerban & Werz 1975; Liddle *et al.*

7.34 7.35 7.36

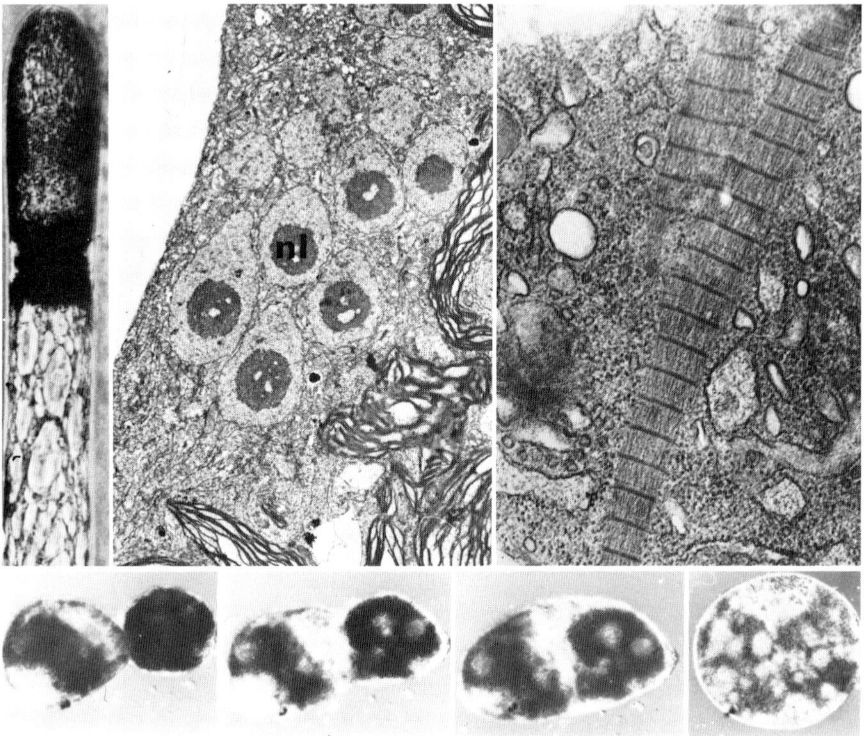

7.37 7.38 7.39 7.40

Fig. 7.34 Mitosis and cytokinesis in a living tip cell of *Acrosiphonia*. The dark, banded area is where mitotic nuclei are localized. Cytokinesis is accomplished by furrowing. Note the reticulate chloroplast. (Hudson & Waaland 1976.) (× 100.)

Fig. 7.35 Longitudinal section through the mitotic region of an *Acrosiphonia* filament (cf. Fig. 7.34). Note the large nucleolus (nl) in each nucleus. (Hudson & Waaland 1976.) (× 2700.)

Fig. 7.36 Portions of the two rhizoplasts associated with the flagellar bases (located near the upper right of the micrograph) of a *Urospora* zoospore. Note the periodic banding of the fibrils in the rhizoplast. (Kristiansen 1974.) (× 26 900.)

Figs 7.37–7.40 (× 70) Successive stages of protoplast fusion in *Acetabularia*. In Fig. 7.37 two protoplasts prepared from cysts have been pushed together, and fusion is complete by Fig. 7.40. (Primke *et al.* 1978.)

1976) and during fragmentation of the primary nucleus to produce secondary nuclei (Woodcock & Miller 1973; Liddle *et al.* 1976). Cairns *et al.* (1978) have also used *Acetabularia* to translate animal virus RNA.

4.5 Flagella, rhizoplast, and phototaxis

The gametes of green seaweeds are biflagellate, whereas zoospores are bi-, quadri-, or multiflagellate. The flagellar surface is smooth, and insertion is typically apical. The transition zone between the 9+2 axoneme and the basal body in the flagella of some green algae has a stellate character (Ringo 1967; also see Dodge 1973), but in zoospores of *Urospora* this region has 'wings' attached to one tubule in each of the doublets (Kristiansen 1974). *Urospora* zoospores also have prominent striated fibrillar rootlets called rhizoplasts (= fibrous roots) (Fig. 7.36) and microtubular rootlets with nine parallel microtubules per rootlet (Kristiansen 1974). Rhizoplasts are clearly plant 'muscles', contractile organelles intimately concerned with flagellar movement. Salisbury & Floyd (1978) demonstrated their contraction in response of the unicellular green alga *Platymonas* to incubation in 2 mM $CaCl_2$. Salisbury & Floyd (1978) concluded that motile cells with large apical pits might be expected to have particularly large rhizoplasts to overcome the physical constraint of the pit on flagellar movement. Rhizoplasts are also part of the flagellar apparatus in male gametes of *Bryopsis* (Hori 1977).

The chloroplast in many zoospores and gametes of the marine Chlorophyta has a region differentiated as an eyespot. Melkonian & Robenck (1979) have investigated membranes adjacent to the eyespot in the chloroplast envelope and plasma membrane in *Tetraselmis* (a freshwater green flagellate) with freeze fracture techniques. They suggested that the photoreceptor is located in the outer membrane of the chloroplast envelope because of the increased number of intramembraneous particles found in this membrane in the eyespot region. Melkonian & Robenck (1979) hypothesized that changes in electrical potential affect locomotion after photoreception occurs, and that a region analogous to a gap junction exists between the outer chloroplast envelope and the plasma membrane. However, much remains to be discovered about the relationship between the photoreception event and its transduction to the flagella and rhizoplast to control locomotion. A constant distance is maintained between the eyespot and the attachment site of one of the rhizoplasts to the plasmalemma, and this is responsible for the apparently varying position of the eyespot within the cell in *Tetraselmis*.

4.6 Protoplast formation and fusion

One of the most interesting and promising areas of recent research on the marine green algae involves protoplast preparation and fusion. Protoplasts of the coenocytic alga *Bryopsis* were made by cutting the thallus and squeezing cytoplasm out of the plant (Tatewaki & Nagata 1970; Kobayashi & Kanaizuka 1977). Some of these protoplasts regenerated whole thalli. Kobayashi & Kanaizuka (1977) separated the cytoplasm into plastid and non-plastid fractions before centrifuging the two fractions together to make protoplasts.

Primke *et al.* (1978) produced protoplasts by treating *Acetabularia* cysts with papain or proteinase K (Figs 7.37–7.40). Protoplast fusion resulted when two of these cyst-derived protoplasts were pushed together, and a protoplast could also be fused with a cytoplast. It is interesting that protoplasts made from cells in the same cyst (i.e. same mating type) could be fused more successfully than those from different cysts (i.e. different mating type). Fused protoplasts differentiated some stalk and whorl-like structures during the first 24 h following their production but then dedifferentiated to spherical protoplasts which remained viable for at least three weeks.

Millner *et al.* (1979) used enzymatic digestion in a medium with high osmolarity to produce protoplasts from *Enteromorpha intestinalis*. Only 10% of these protoplasts were viable after 24 h, but Millner *et al.*'s techniques may be useful in isolating protoplasts from other multicellular marine algae.

The investigation of many cellular activities is facilitated by microinjection or organelle isolation. Since the plant cell wall makes application of these techniques difficult, the use of protoplasts may be of great value in cytological and developmental studies of the algae.

5 REFERENCES

ALLEY C. D. & SCOTT J. L. (1977) Unusual dictyosome morphology and vesicle formation in tetrasporangia of the marine red alga *Polysiphonia denudata*. *J. Ultrastr. Res.*, **58**, 289–98.

BEEVERS H. (1979) Microbodies in higher plants. *Ann. Rev. Plant Physiol.*, **30**, 159–93.

BERGER S. & SCHWEIGER H. G. (1975) Cytoplasmic induction of changes in the ultrastructure of the *Acetabularia* nucleus and perinuclear cytoplasm. *J. Cell Sci.*, **17**, 517–29.

BERKALOFF C. & ROUSSEAU B. (1979) Ultrastructure of male gametogenesis in *Fucus serratus* (Phaeophyceae). *J. Phycol.*, **15**, 163–73.

BERNS B. W. & RICHARDSON S. M. (1977) Continuation of mitosis after selective laser microbeam destruction of the centriolar region. *J. Cell Biol.*, **75**, 977–82.

BISALPUTRA T. (1974) Plastids. In *Algal Physiology and Biochemistry*, ed. Stewart W. D. P., pp. 124–60. Blackwell Scientific Publications, Oxford.

BISALPUTRA T. & BISALPUTRA A. A. (1967) The occurrence of DNA fibrils in chloroplasts of *Laurencia spectabilis*. *J. Ultrastr. Res.*, **17**, 14–22.

BISALPUTRA T. & BISALPUTRA A. A. (1969) The ultrastructure of the chloroplast of a brown alga *Sphacelaria* sp. I. Plastid DNA configuration—The chloroplast genophore. *J. Ultrastr. Res.*, **29**, 151–70.

BOILLOT A. (1978) Les ponctuations intercellulaires du *Rhodochaete parvula* Thuret (Rhodophycées, Bangiophycidées). *Rev. Algol.*, **13**, 251–8.

BOLWELL G. P., CALLOW J. A., CALLOW M. E. & EVANS L. V. (1977) Cross-fertilization in fucoid seaweeds. *Nature (Lond.)*, **268**, 626–7.

BOLWELL G. P., CALLOW J. A., CALLOW M. E. & EVANS L. V. (1979) Fertilization in brown algae. II. Evidence for lectin-sensitive complementary receptors involved in gamete recognition in *Fucus serratus*. *J. Cell Sci.*, **36**, 19–30.

BØRGESEN F. (1907) An ecological and systematic account of the *Caulerpas* of the Danish West Indies. *K. Danske Vidensk. Selsk. Skrifter. Ser. VII. Naturvid. og Math. Afd.*, **4**, 337–92.

BOROWITZKA M. A. (1976) Some unusual features of the ultrastructure of the chloroplasts of the green algal order Caulerpales and their development. *Protoplasma*, **89**, 129–47.

BOROWITZKA M. A. (1977) Algal calcification. *Oceanogr. Mar. Biol. Ann. Rev.*, **15**, 189–223.

BOROWITZKA M. A. (1978) Plastid development and floridean starch grain formation during carposporogenesis in the coralline red alga *Lithothrix aspergillum* Gray. *Protoplasma*, **95**, 217–28.

BOROWITZKA M. A. & LARKUM A. W. D. (1974) Chloroplast development in the Caulerpalean alga *Halimeda*. *Protoplasma*, **81**, 131–44.

BOROWITZKA M. A. & VESK M. (1978) Ultrastructure of the Corallinaceae. I. The vegetative cells of *Corallina officinalis* and *C. cuvierii*. *Mar. Biol.*, **46**, 295–304.

BOROWITZKA M. A. & VESK M. (1979) Ultrastructure of the Corallinaceae (Rhodophyta). II. Vegetative cells of *Lithothrix aspergillum*. *J. Phycol.*, **15**, 146–53.

BOUCK G. B. (1962) Chromatophore development, pits, and other fine structure in the red alga, *Lomentaria baileyana* (Harv.) Farlow, *J. Cell Biol.*, **12**, 553–69.

BOUCK G. B. (1965) Fine structure and organelle associations in brown algae. *J. Cell Biol.*, **26**, 523–37.

BOUCK G. B. (1969) Extracellular microtubules. The origin, structure, and attachment of flagellar hairs in *Fucus* and *Ascophyllum* antherozoids. *J. Cell Biol.*, **40**, 446–60.

BOUCK G. B. (1970) The development and postfertilization fate of the eyespot and the apparent photoreceptor in *Fucus* sperm. *Ann. N.Y. Acad. Sci.*, **175**, 673–85.

BOURNE V. L., CONWAY E. & COLE K. (1970) On the ultrastructure of pit connections in the conchocelis phase of the red alga *Porphyra perforata*. J. Ag. *Phycologia*, **9**, 79–81.

BRÅTEN T. (1975) Observations on mechanisms of attachment in the green alga *Ulva mutabilis* Føyn. An ultrastructural and light microscopical study of zygotes and rhizoids. *Protoplasma*, **84**, 161–73.

BRÅTEN T. & NORDBY Ø. (1973) Ultrastructure of meiosis and centriole behaviour in *Ulva mutabilis* Føyn. *J. Cell Sci.*, **13**, 69–81.

BRAWLEY S. H. (1978) Cytological Studies of Embryogenesis in the Brown Alga Fucus: Fertilization and the Formation of a Polarized Embryo. Ph.D. dissertation, University of California, Berkeley.

BRAWLEY S. H. & QUATRANO R. S. (1979a) Effects of microtubule inhibitors on pronuclear migration and embryogenesis in *Fucus distichus* (Phaeophyta). *J. Phycol.*, **15**, 266–72.

BRAWLEY S. H. & QUATRANO R. S. (1979b) Sulfation of fucoidin in *Fucus* embryos. IV. Autoradiographic investigations of fucoidin sulfation and secretion during differentiation and the effects of cytochalasin treatment. *Develop. Biol.*, **73**, 193–205.

BRAWLEY S. H., QUATRANO R. S. & WETHERBEE R. (1977) Fine-structural studies of the gametes and embryo of *Fucus vesiculosus* L. (Phaeophyta). III. Cytokinesis and the multicellular embryo. *J. Cell Sci.*, **24**, 275–94.

BRAWLEY S. H. & SEARS J. R. (1980) Septal plugs in a marine green alga. *J. Cell Biol.*, **87**, 62a.

BRAWLEY S. H., WETHERBEE R. & QUATRANO R. S. (1976a) Fine-structural studies of the gametes and embryo of *Fucus vesiculosus* L. (Phaeophyta). I. Fertilization and pronuclear fusion. *J. Cell Sci.*, **20**, 233–54.

BRAWLEY S. H., WETHERBEE R. & QUATRANO R. S. (1976b) Fine-structural studies of the gametes and embryo of *Fucus vesiculosus* L. II. The cytoplasm of the egg and young zygote. *J. Cell Sci.*, **20**, 255–71.

BRINKLEY B. R. & PORTER K. R. (1977) *International Cell Biology. 1976–1977*. The Rockefeller University Press, New York.

BRONCHART R. & DEMOULIN V. (1977) Unusual mitosis in the red alga *Porphyridium purpureum*. *Nature* (*Lond.*), **268**, 80–1.

BROWN D. L. & WEIER T. E. (1968) Chloroplast development and ultrastructure in the freshwater red alga *Batrachospermum*. *J. Phycol.*, **4**, 199–206.

BURR F. A. & WEST J. A. (1970) Light and electron microscope observations on the vegetative and reproductive structures of *Bryopsis hypnoides*. *Phycologia*, **9**, 17–38.

BURR F. A. & WEST J. A. (1971) Protein bodies in *Bryopsis hypnoides*: Their relationship to wound-healing and branch septum development. *J. Ultrastr. Res.*, **35**, 476–98.

CAIRNS E., GESSCHWENDER H. H., PRIMKE M., YAMAKAWA M., TRAUB P. & SCHWEIGER H. G. (1978) Translation of animal virus RNA in the cytoplasm of a plant cell. *P.N.A.S.* (*USA*), **75**, 5557–9.

CALLOW M. E., COUGHLAN S. H. & EVANS L. V. (1978) The role of Golgi bodies in polysaccharide sulphation in *Fucus* zygotes. *J. Cell Sci.*, **32**, 337–56.

CALLOW M. E. & EVANS L. V. (1974) Studies on the ship-fouling alga *Enteromorpha*. III. Cytochemistry and autoradiography of adhesive production. *Protoplasma*, **80**, 15–27.

CALLOW M. E., EVANS L. V., BOLWELL G. P. & CALLOW J. A. (1978) Fertilization in brown algae. I. SEM and other observations on *Fucus serratus*. *J. Cell Sci.*, **32**, 45–54.

CALVERT H. E., DAWES C. H. & BOROWITZKA M. A. (1976) Phylogenetic relationships of *Caulerpa* (Chlorophyta) based on comparative chloroplast ultrastructure. *J. Phycol.*, **12**, 149–62.

CARAM B. (1975) Aspects ultrastructuraux de la spermatogenèse chez le *Cutleria adspersa* (Phéophycées, Cutlériales) de la côte méditerranéenne française. *C.R. Acad. Sci. Paris*, **281** (D), 1089–92.

CHAMBERLAIN A. H. L. & EVANS L. V. (1973) Aspects of spore production in the red alga *Ceramium*. *Protoplasma*, **76**, 139–59.

CHI E. Y. (1971) Brown algal pyrenoids. *Protoplasma*, **72**, 101–4.

CHI E. Y. & NEUSHUL M. (1972) Electron-microscopic studies of sporogenesis in *Macrocystis*. *Proc. Int. Seaweed Symp.*, **7**, 181–7.

COLEMAN A. W. (1979) Use of the fluorochrome 4,6-diamidino-2-phenylindole in genetic and developmental studies of chloroplast DNA. *J. Cell Biol.*, **82**, 299–305.

COLOMBO P. M. (1978) An ultrastructural study of thallus organization in *Udotea petiolata*. *Phycologia*, **17**, 227–35.

COLOMBO P. M. & POSTAI E. (1979) The process of wound repair in *Udotea petiolata* (Siphonales). *Cytobios*, **23**, 7–16.

COTTLER M. H. (1971) Plasmalemmal extensions in *Chondrus crispus* (L.) Stackh. *J. Ultrastr. Res.*, **37**, 31–6.

COUGHLAN S. & EVANS L. V. (1978) Isolation and characterization of Golgi bodies from vegetative tissue of the brown alga *Fucus serratus*. *J. Exp. Bot.*, **29**, 55–68.

CRAYTON M. A., WILSON E. & QUATRANO R. S. (1974) Sulfation of fucoidin in *Fucus* embryos. II. Separation from initiation of polar growth. *Develop. Biol.*, **39**, 164–7.

DIXON P. S. (1973) *Biology of the Rhodophyta*. Oliver & Boyd, Edinburgh.

DODGE, J. D. (1973) *The Fine Structure of Algal Cells*. Academic Press, London.

DREHER T. W., GRANT B. R. & WETHERBEE R. (1978) The wound response in the siphonous alga *Caulerpa simpliciuscula* C. Ag.: Fine structure and cytology. *Protoplasma*, **96**, 189–203.

DUCKETT J. G., BUCHANAN J. S., PEEL M. C. & MARTIN M. T. (1974) An ultrastructural study of pit connections and percurrent proliferations in the red alga *Nemalion helminthoides* (vell. in with) batt. *New Phytol.*, **73**, 497–507.

DUCKETT J. G. & PEEL M. C. (1978) The role of transmission electron microscopy in elucidating the taxonomy and phylogeny of the Rhodophyta. In *The Systematics Association Special Volume No. 10—Modern Approaches to the Taxonomy of Red and Brown Algae*, eds. Irvine D. E. G. & Price J. H., pp. 157–204. Academic Press, London.

DUFFIELD E. C., WAALAND S. D. & CLELAND R. (1972) Morphogenesis in the red alga, *Griffithsia pacifica*: Regeneration from single cells. *Planta*, **105**, 185–95.

DUSTIN P. (1978) *Microtubules*. Springer-Verlag, Berlin.

EVANS L. V. (1966) Distributions of pyrenoids among some brown algae. *J. Cell Sci.*, **1**, 449–54.

EVANS L. V. (1968) Chloroplast morphology and fine structure in British Fucoids. *New Phytol.*, **67**, 173–8.

EVANS L. V. & CALLOW M. E. (1974) Polysaccharide sulphation in *Laminaria*. *Planta*, **117**, 93–95.

EVANS L. V., CALLOW J. A. & CALLOW M. E. (1973) Structural and physiological studies on the parasitic red alga *Holmsella*. *New Phytol.*, **72**, 393–402.

EVANS L. V., CALLOW J. A. & CALLOW M. E. (1978) Parasitic red algae: an appraisal. In *The Systematics Association Special Volume No. 10—Modern Approaches to the Taxonomy of Red and Brown Algae*, eds. Irvine D. E. G. & Price J. H., pp. 87–110. Academic Press, London.

EVANS L. V. & HOLLIGAN M. S. (1972) Correlated light and electron microscope studies on brown algae. II. Physode production in *Dictyota*. *New Phytol.*, **71**, 1173–80.

FELDMANN J., FELDMANN G. & GUGLIELMI G. (1977) Nouvelles observations sur l'ultrastructure des synapses des Rhodophycées. *Rev. Algol.* N.S., **7** (1–2), 11–30.

FORBES M. A. & HALLAM N. D. (1979) Embryogenesis and substratum adhesion in the brown alga *Hormosira banksii* (Turner) Decaisne. *Br. Phycol. J.*, **14**, 69–81.

FRANKE W. W., KARTENBECK J. & SPRING H. (1976) Involvement of bristle coat structures in surface membrane interactions during coenocytotomic cleavage in caps of *Acetabularia mediterranea*. *J. Cell Biol.*, **71**, 196–206.

FRANKE W. W., SPRING H., KARTENBECK J. & FALK H. (1977) Cyst formation in some Dasycladacean green algae. I. Vesicle formation during coenocytotomy in *Acetabularia mediterranea*. *Cytobiologie*, **14**, 229–52.

FREI E. & PRESTON R. D. (1964) Non-cellulosic structural polysaccharides in algal cell walls. II. Association of xylan and mannan in *Porphyra umbilicalis*. *Proc. Roy. Soc.* (*London*), **160**, 314–27.

GANTT E. (1977) Recent contributions in phycobiliproteins and phycobilisomes. *Photochem. Photobiol.*, **26**, 685–9.

GANTT E. & CONTI S. F. (1965) The ultrastructure of *Porphyridium cruentum*. *J. Cell Biol.*, **26**, 305–38.

GANTT E. & LIPSCHULTZ C. A. (1977) Probing phycobilisome structure by immuno-electron microscopy. *J. Phycol.*, **13**, 185–92.

GIBBS S. (1962) Nuclear envelope chloroplast relationships in algae. *J. Cell Biol.*, **14**, 433–44.

GIBBS S. (1978) The chloroplasts of *Euglena* may have evolved from symbiotic green algae. *Can. J. Bot.*, **56**, 2883–9.

GIDDINGS T. H. JR., WITHERS N. W. & STAEHELIN A. (1980) Supramolecular structure of stacked and unstacked regions of the photosynthetic membranes of *Prochloron* sp., a prokaryote. *P.N.A.S.* (*U.S.A.*), **77**, 352–6.

GILES K. L. & SARAFIS V. (1974) Implications of rigescent integuments as a new structural feature of some algal chloroplasts. *Nature* (*Lond.*), **248**, 512–13.

GILKEY J. C., JAFFE L. F., RIDGEWAY E. B. & REYNOLDS G. T. (1978) A free calcium wave traverses the activating egg of the medaka, *Oryzias latipes*. *J. Cell Biol.*, **76**, 448–66.

GILLHAM N. W. (1978) *Organelle Heredity*. Raven Press, New York.

GODWARD M. B. E., BETH K. & PACEY J. (1979) Nuclear division in the cyst, and white spot nuclei preceding cyst formation, in *Acetabularia wettsteinii*. *Protoplasma*, **101**, 37–46.

GOFF L. J. (1976) The biology of *Harveyella mirabilis* (Cryptonemiales; Rhodophyceae). V. Host responses to parasite infection. *J. Phycol.*, **12**, 313–28.

GOFF L. J. (1979a) The biology of *Harveyella mirabilis* (Cryptonemiales, Rhodophyceae). VI. Translocation of photoassimilated ^{14}C. *J. Phycol.*, **15**, 82–7.

GOFF L. J. (1979b) The biology of *Harveyella mirabilis* (Cryptonemiales, Rhodophyceae). VII. Structure and proposed function of host-penetrating cells. *J. Phycol.*, **15**, 87–100.

GÖKE L., PARADIES H. H. & WERZ G. (1974) Cell wall polypeptides of *Polyphysa* (*Acetabularia*) *cliftonii*: Amino acid composition of stalk and cap cell wall polypeptides. *Biochem. Biophys. Res. Comm.*, **60**, 22–7.

GORDON E. M. & MCCANDLESS E. L. (1973) Ultrastructure and histochemistry of *Chondrus crispus* Stackhouse. *Proc. Nova Scotia Inst. Sci.*, **27** (Suppl. 1), 11–133.

GORDON-MILLS E. M. & MCCANDLESS E. L. (1975) Carrageenans in the cell walls of *Chondrus crispus* Stack. (Rhodophyceae, Gigartinales). I. Localization with fluorescent antibody. *Phycologia*, **14**, 275–81.

GREEN B. R. (1973) Evidence for the occurrence of meiosis before cyst formation in *Acetabularia mediterranea* (Chlorophyceae, Siphonales). *Phycologia*, **12**, 233–5.

GRETZ M. R., ARONSON J. M. & SOMMERFELD M. R. (1980) Cellulose in the cell walls of the Bangiophyceae (Rhodophyta). *Science*, **20**, 779–81.

GUNNING B. E. S. & STEER M. W. (1975) *Ultrastructure and the Biology of Plant Cells*. Edward Arnold, London.

HÄMMERLING J. (1963) Nucleo-cytoplasmic interactions in *Acetabularia* and other cells. *Ann. Rev. Plant Physiol.*, **14**, 65–92.

HANIC L. A. & CRAIGIE J. S. (1969) Studies on the algal cuticle. *J. Phycol.*, **5**, 89–102.

HARA Y. & CHIHARA M. (1974) Comparative studies on the chloroplast ultrastructure in the Rhodophyta, with special reference to their taxonomic significance. *Sci. Rep. Tokyo Kyoiku Daiguku*, Sec. B., **15**, 209–35.

HARDY F. G. & MOSS B. L. (1978) The attachment of zygotes and germlings of *Halidrys siliquosa* (L.) Lyngb. (Phaeophyceae, Fucales). *Phycologia*, **17**, 69–78.

HARTMANN T. & ESCHRICH W. (1969) Stofftranspost in Rotalgen. *Planta*, **85**, 303–12.

HAWKES M. W. (1978) Sexual reproduction in *Porphyra gardneri* (Smith et Hollenberg) Hawkes (Bangiales, Rhodophyta). *Phycologia*, **17**, 329–53.

HERTH W., KUPPER A. & FRANKE W. W. (1975) Cellulose in *Acetabularia*. *J. Ultrastr. Res.*, **50**, 289–92.

HOGSETT W. E. & QUATRANO R. S. (1978) Sulphation of fucoidin in *Fucus* embryos III. Required for localization in the rhizoid wall. *J. Cell Biol.*, **78**, 866–73.

HOLDSWORTH R. H. (1971) The isolation and partial characterization of the pyrenoid protein of *Eremosphaera viridis*. *J. Cell Biol.*, **15**, 499–513.

HONSELL E., AVANZINI A. & GHIRARDELLI L. A. (1978) Two ways of chloroplast development in vegetative cells of *Nitophyllum punctatum* (Rhodophyta). *J. Submicrosc. Cytol.*, **10 (2)**, 227–37.

HORI T. (1972) Survey of pyrenoid distribution in the vegetative cells of brown algae. *Proc. Int. Seaweed Symp.*, **7**, 165–71.

HORI T. (1977) Electron microscope observations on the flagellar apparatus of *Bryopsis maxima* (Chlorophyceae) *J. Phycol.*, **13**, 238–43.

HORI T. & ENAMOTO S. (1978a) Developmental cytology of *Dictyosphaeria cavernosa*. I. Light and electron microscope observations on cytoplasmic cleavage in zooid formation. *Bot. Mar.*, **21**, 401–8.

HORI T. & ENAMOTO S. (1978b) Developmental cytology of *Dictyosphaeria cavernosa*. II. Nuclear division during zooid formation. *Bot. Mar.*, **21**, 477–81.

HORI T. & ENAMOTO S. (1978c) Electron microscope observations on the nuclear division in *Valonia ventricosa* (Chlorophyceae, Siphonocladales). *Phycologia*, **17**, 133–42.

HORI T. & UEDA R. (1967) Electron microscope studies on the fine structure of plastids in siphonous green algae with special reference to their phylogenetic relationships. *Sci. Rept. Tokyo Kyoiku Daigoku*, Sec. B., **12**, 225–44.

HUDSON P. R. & WAALAND J. R. (1976) Ultrastructure of mitosis and cytokinesis in the multinucleate green alga *Acrosiphonia*. *J. Cell Biol.*, **62**, 274–94.

HUIZING H. R. & RIETEMA H. (1975) Xylan and mannan as cell wall constituents of different stages in the life-histories of some siphonous green algae. *Br. Phycol. J.*, **10**, 13–16.

JAENICKE J. (1974) Chemical signal transmission by gamete attractants in brown algae. In *Biochemistry of Sensory Functions*, ed. Jaenicke L., pp. 307–9. Springer-Verlag, Berlin.

JAFFE L. F. (1968) Localization in the developing *Fucus* egg and the general role of localizing currents. *Adv. Morphog.*, **7**, 295–328.

JAFFE L. F., ROBINSON K. R. & NUCCITELLI R. (1974) Local cation entry and self-electrophoresis as an intracellular localization mechanism. *Ann. N.Y. Acad. Sci.*, **238**, 372–89.

KERBY N. W. & EVANS L. V. (1978) Isolation and partial characterization of pyrenoids from the brown alga *Pilayella littoralis* (L.) Kjellm. *Planta*, **142**, 91–5.

KNAPP E. (1931) Entwicklungsphysiologische Untersuchungen an Fucacean Eiern. *Planta*, **14**, 731–51.

KOBAYASHI K. & KANAIZUKA Y. (1977) Reassembly of living cells from dissociated components in *Bryopsis*. *Plant Cell Physiol.*, **18**, 1373–7.

KONRAD HAWKINS E. (1974a) Golgi vesicles of uncommon morphology and wall formation in the red alga, *Polysiphonia*. *Protoplasma*, **80**, 1–14.

KONRAD HAWKINS E. (1974b) Growth and differentiation of the Golgi apparatus in the red alga, *Callithamnion roseum*. *J. Cell Sci.*, **14**, 633–55.

KOOP H-U (1975) Über den Ort der Meiose bei *Acetabularia mediterranea*. *Protoplasma*, **85**, 109–14.

KOOP H-U (1979) The life cycle of *Acetabularia* (Dasycladales, Chlorophyceae): A compilation of evidence for meiosis in the primary nucleus. *Protoplasma*, **100**, 353–66.

KRAFT G. T. (1977) Transfer of the New Zealand red alga *Tylotus proliferus* (Gracilariaceae, Gigartinales) to the genus *Gracilaria*. *N.Z. J. Bot.*, **15**, 495–502.

KRISTIANSEN J. (1974) The fine structure of the zoospores of *Urospora penicilliformis*, with special reference to the flagellar apparatus. *Br. Phycol. J.*, **9**, 201–13.

KUGRENS P. & WEST J. A. (1972a) Synaptonemal complexes in red algae. *J. Phycol.*, **8**, 187–91.

KUGRENS P. & WEST J. A. (1972b) Ultrastructure of spermatial development in the parasitic red algae *Levringiella gardneri* and *Erythrocystis saccata*. *J. Phycol.*, **8**, 331–43.

KUGRENS P. & WEST J. A. (1972c) Ultrastructure of tetrasporogenesis in the parasitic red alga *Levringiella gardneri* (Setchell) Kylin. *J. Phycol.*, **8**, 370–83.

KUGRENS P. & WEST J. A. (1973a) The ultrastructure of carpospore differentiation in the parasitic red alga *Levringiella gardneri* (Setchell) Kylin. *Phycologia*, **12**, 163–73.

KUGRENS P. & WEST J. A. (1973b) The ultrastructure of an alloparasitic red alga *Choreocolax polysiphoniae*. *Phycologia*, **12**, 175–86.

KUGRENS P. & WEST J. A. (1974) The ultrastructure of carposporogenesis in the marine hemiparasitic red alga *Erythrocystis saccata*. *J. Phycol.*, **10**, 139–47.

LA CLAIRE J. W. & WEST J. A. (1978) Light and electron microscopic studies of growth and reproduction in *Cutleria* (Phaeophyta). I. Gametogenesis in the female plant of *C. hancockii*. *Protoplasma*, **97**, 93–110.

LA CLAIRE J. W. & WEST J. A. (1979) Light- and electron microscopic studies of growth and reproduction in *Cutleria* (Phaeophyta). II. Gametogenesis in the male plant of *C. hancockii*. *Protoplasma*, **101**, 247–67.

LEE R. E. (1971) The pit connections of some lower red algae: Ultrastructure and phylogenetic significance. *Br. Phycol. J.*, **6**, 29–38.

LEE R. E. (1977) Evolution of algal flagellates with chloroplast ER from the ciliates. *S. Afr. J. Sci.*, **73**, 179–82.

LEE R. E. & FULTZ S. A. (1970) Ultrastructure of the *Conchocelis* stage of the marine red alga *Porphyra leucosticta*. *J. Phycol.*, **6**, 22–8.

LEEDALE G. F. (1970) Phylogenetic aspects of nuclear cytology in the algae. *Ann. N.Y. Acad. Sci.*, **175**, 429–51.

LEHNINGER A. L. (1970) Mitochondria and calcium ion transport. *Biochem. J.*, **119**, 129–38.

LIDDLE L., BERGER S. & SCHWEIGER H. G. (1976) Ultrastructure during development of the nucleus of *Batophora oerstedii* (Chlorophyta, Dasycladaceae). *J. Phycol.*, **12**, 261–72.

LIN H-P, SWAFFORD J. R. & SOMMERFELD M. R. (1977) Comparative ultrastructure of the alternating vegetative phases of *Bangia fuscopurpurea* (Bangiophycidae, Rhodophyta). *Bot. Mar.*, **20**, 339–44.

LOFTHOUSE P. F. & CAPON B. (1975) Ultrastructural changes accompanying mitosporogenesis in *Ectocarpus* (Phaeophyta, Ectocarpales). *Phycologia*, **16**, 235–43.

LOISEAUX S. (1973) Ultrastructure of zoidogenesis in unilocular zoidocysts of several brown algae. *J. Phycol.*, **9**, 277–89.

LØVLIE A. & BRÅTEN T. (1970) On mitosis in the multicellular alga *Ulva mutabilis* Føyn. *J. Cell Sci.*, **6**, 109–29.

MCARTHUR D. M. & MOSS B. L. (1978) Ultrastructural studies of vegetative cells, mitosis, and cell division of *Enteromorpha intestinalis* (L.) Link. *Br. Phycol. J.*, **13**, 255–67.

MCBRIDE D. L. & COLE K. (1969) Ultrastructural characteristics of the vegetative cell of *Smithora naiadum*. *Phycologia*, **8**, 177–86.

MCBRIDE D. L. & COLE K. (1971) Electron microscopic observations on the differentiation and release of monospores in the marine red alga *Smithora naiadum*. *Phycologia*, **10**, 49–61.

MCBRIDE D. L. & COLE K. (1972) Ultrastructural observations on germinating monospores in *Smithora naiadum* (Rhodophyceae, Bangiophycidae). *Phycologia*, **11**, 181–91.

MCCANDLESS E. L., CRAIGIE J. S. & WALTER J. A. (1973) Carrageenans in the gametophyte and sporophyte stages of *Chondrus crispus*. *Planta*, **112**, 201–12.

MCCULLY M. E. (1968) Histological studies on the genus *Fucus*. II. Histology of the reproductive tissues. *Protoplasma*, **66**, 205–30.

McDonald K. (1972) The ultrastructure of mitosis in the marine red alga *Membranoptera platyphylla*. *J. Phycol.*, **8**, 156–66.

Mackie M. & Preston R. D. (1974) Cell wall and intercellular region polysaccharides. In *Algal Physiology and Biochemistry*, ed. Stewart W. D. P., pp. 40–85. Blackwell Scientific Publications, Oxford.

Marchant H. J. (1976) Plasmodesmata in algae and fungi. In *Intercellular Communication in Plants: Studies of Plasmodesmata*, eds. Gunning B. E. S. & Robards A. W., pp. 59–78. Springer-Verlag, New York.

Marchant H. J. (1977) Colony formation and inversion in the green alga *Eudorina elegans*. *Protoplasma*, **93**, 325–39.

Marchant H. J. & Pickett-Heaps J. D. (1974) The effect of colchicine on colony formation in the algae *Hydrodictyon, Pediastrum* and *Sorastrum*. *Planta*, **116**, 291–300.

Margulis L. (1970) *Origin of Eucaryotic Cells*. Yale University Press, New Haven.

Markey D. R. & Wilce R. T. (1975) The ultrastructure of reproduction in the brown alga *Pylaiella littoralis*. I. Mitosis and cytokinesis in the plurilocular gametangia. *Protoplasma*, **85**, 219–41.

Markey D. R. & Wilce R. T. (1976a) The ultrastructure of reproduction in the brown alga *Pylaiella littoralis*. II. Zoosporogenesis in the unilocular sporangia. *Protoplasma*, **88**, 147–73.

Markey D. R. & Wilce R. T. (1976b) The ultrastructure of reproduction in the brown alga *Pylaiella littoralis*. III. Later stages of gametogenesis in the plurilocular gametangia. *Protoplasma*, **88**, 175–86.

Melkonian M. & Robenek H. (1979) The eyespot of the flagellate *Tetraselmis cordiformis* Stein (Chlorophyceae): Structural specialization of the outer chloroplast membrane and its possible significance in phototaxis of green algae. *Protoplasma*, **100**, 183–97.

Menzel D. (1976) Cytochemischer Nachweis von Katalase in Microbodies bei *Acetabularia mediterranea*. *Planta*, **130**, 181–4.

Millner P. A., Callow M. E. & Evans L. V. (1979) Preparation of protoplasts from the green alga *Enteromorpha intestinalis* (L.) Link. *Planta*, **147**, 174–7.

Müller D. G. & Jaenicke L. (1973) Fucoserraten, the female sex attractant of *Fucus serratus* L. (Phaeophyta.) *F.E.B.S. Lett.*, **30**, 137–9.

Myers A., Preston R. D. & Ripley G. W. (1956) Fine structure in the red algae. I. X-ray and electron microscope investigation of *Griffithsia flosculosa*. *Proc. Roy. Soc.* (*London*), Ser. B. Biol. Sci., **144**, 450.

Nakazawa S. (1950) Origin of polarity in the eggs of *Sargassum confusum*. *Sci. Rept. Tohoku Univ.*, (4th Ser.), **18**, 424–33.

Nakazawa S. (1951) Invalid stratification to the egg polarity in *Coccophora* and *Sargassum*. *Sci. Repts. Tohoku Univ.*, (4th Ser.), **19**, 73–8.

Neushul M. (1970) A freeze-etching study of the red alga *Porphyridium*. *Am. J. Bot.*, **57**, 1231–9.

Neushul M. & Dahl A. L. (1972) Ultrastructural studies of brown algal nuclei. *Am. J. Bot.*, **59**, 401–10.

Nordby Ø. (1974) Light microscopy of meiotic zoosporogenesis in *Ulva mutabilis* Føyn. *J. Cell Sci.*, **15**, 443–55.

Novikoff A. (1967) The endoplasmic reticulum: a cytochemist's view (a review). *P.N.A.S.* (*U.S.A.*), **73**, 2781–7.

Novotny A. M. & Forman M. (1974) The relationship between changes in cell wall composition and the establishment of polarity in *Fucus* embryos. *Develop. Biol.*, **40**, 162–93.

Nuccitelli R. (1978) Öoplasmic segregation and secretion in the *Pelvetia* egg is accompanied by a membrane-generated electrical current. *Develop. Biol.*, **62**, 13–33.

Nuccitelli R. & Jaffe L. F. (1975) The pulse current pattern generated by developing fucoid eggs. *J. Cell Biol.*, **64**, 636–43.

Ogawa H. (1977) Sexuality in *Sargassum sagamianum* var. *yezoense* and development of its fertilized egg. *Bull. Jap. Soc. Phycol.*, **24**, 20–4.

Ohmori T. & Miyazaki S. (1976) An analysis of embryogeny in *Sargassum thunbergii*. II. On the segmentation of embryo and the formation of rhizoid cell. *Bull. Jap. Soc. Phycol.*, **23**, 93–8.

Oliveira L. & Bisalputra T. (1977) Ultrastructural and cytochemical studies on the nature and origin of the cytoplasmic inclusions of aging cells of *Ectocarpus* (Phaeophyta, Ectocarpales). *Phycologia*, **16**, 235–44.

OSBORN J. C., DUNCAN D. J. & SMITH J. L. (1979) Role of calcium ions in the control of embryogenesis of *Xenopus*. *J. Cell Biol.*, **80**, 589–604.

PALADE G. E. (1975) Intracellular aspects of the process of protein secretion. *Science* (*Wash. D.C.*), **189**, 347–58.

PALEVITZ B. A. & HEPLER P. K. (1975) Identification of actin *in situ* at the ectoplasm–endoplasm interface of *Nitella*: microfilament-chloroplast association. *J. Cell Biol.*, **65**, 29–38.

PARADIES H. H. & WERZ G. (1977a) Solution properties of a cell wall protein from *Acetabularia* (*Polyphysa*) *cliftonii*. I. α-precipitation of the protein. *Protoplasma*, **93**, 443–62.

PARADIES H. H. & WERZ G. (1977b) Solution properties of a cell wall protein from *Acetabularia* (*Polyphysa*) *cliftonii*. II. β-precipitation of the protein. *Protoplasma*, **93**, 463–72.

PEEL M. C. & DUCKETT J. G. (1975) Studies of spermatogenesis in the Rhodophyta. In *The Biology of the Male Gamete*, eds. Duckett J. G. & Racey P. A. *J. Linn. Soc.* (*Biol.*), Supp. No. 1, **7**, 1–13.

PEEL M. C., LUCAS I. A. N., DUCKETT J. G. & GREENWOOD A. D. (1973) Studies of sporogenesis in the Rhodophyta. I. An association of the nuclei with endoplasmic reticulum in post-meiotic tetraspore mother cells of *Corallina officinalis* L. *Z. Zellforsch.*, **147**, 59–74.

PELLEGRINI L. (1976) Variations ultrastructurales de l'appareil de Golgi chez la *Cystoseira stricta* Sauvageau (Pheophycée, Fucale), *Protoplasma*, **90**, 204–28.

PELLEGRINI L. (1979) On the origin and development of vacuoles in promeristematic cells of *Cystoseira stricta* Sauvageau (Phaeophyta, Fucales). *Protoplasma*, **101**, 89–102.

PELLEGRINI L. (1980) Cytological studies on physodes in the vegetative cells of *Cystoseira stricta* Sauvageau (Phaeophyta, Fucales). *J. Cell Sci.*, **41**, 209–32.

PENG H. B. & JAFFE L. F. (1976) Cell wall formation in *Pelvetia* embryos. A freeze-fracture study. *Planta*, **133**, 57–71.

PEYRIÈRE M. (1969) Infrastructure cytoplasmique du tetrasporocyste de *Griffithsia flosculosa* (Rhodophycée, Ceramicée) pendant la prophase meiotique. *C.R. Acad. Sci. Paris*, **269**, 2332–4.

PEYRIÈRE M. (1970) Evolution de l'appareil de Golgi de la tetrasporogenèse de *Griffithsia flosculosa* (Rhodophycée, Ceramiacèe). *C.R. Acad. Sci. Paris*, **270**, 2071–4.

PEYRIÈRE M. (1971) Etude infrastructurale des spermatocystes du *Griffithsia flosculosa* (Rhodophycée). *C.R. Acad. Sci. Paris*, Ser. D., **273**, 2071–4.

PEYRIÈRE M. (1975) A propos des relations dictyosomes—mitochondries observées chez les rhodophycées floridées. *C.R. Acad. Sci. Paris*, Ser. D., **281**, 1579–82.

PEYRIÈRE M. (1976) Etude infrastructurale du plastidome de deux Rhodophycees parasites *Harveyella mirabilis* (Reinsch) Schmitz et Reinke et *Holmsella pachyderma* (Reinsch) Sturch. *C.R. Acad. Sci. Paris*, Ser. D., **283**, 1169–71.

PEYRIÈRE M. (1977a) Infrastructure des synapses du *Griffithsia flosculosa* (Ellis) Batters et de quelques autres Rhodophycées floridées. *Rev. Algol.* N.S., **7 (1–2)**, 31–43.

PEYRIÈRE M. (1977b) Ultrastructure d'*Harveyella mirabilis* (Cryptonemiales, Rhodophycée) parasite de *Rhodomela confervoides* (Ceramiale, Rhodophycée): origine des synapses secondaires entre cellules de l'hôte et du parasite et entre cellules du parasite. *C.R. Acad. Sci. Paris*, Ser. D., **285**, 965–8.

PHILLIPS D. & SCOTT J. (1981) Ultrastructure of cell division and reproductive differentiation of male plants in the Florideophycidae (Rhodophyta). Mitosis in *Dasya baillouviana*. *Protoplasma*, (in press).

PICKETT-HEAPS J. D. (1972) Variation in mitosis and cytokinesis in plant cells; its significance in phylogeny and evolution of ultrastructural systems. *Cytobios*, **5**, 59–77.

PICKETT-HEAPS J. D. (1975) *Green Algae: Structure, Reproduction and Evolution in Selected Genera*. Sinauer, Stamford.

PICKETT-HEAPS J. D. (1979) Electron microscopy and the phylogeny of green algae and land plants. *Am. Zool.*, **19**, 545–54.

POLLOCK E. G. & CASSELL R. Z. (1977) An intracristal component of *Fucus* sperm mitochondria. *J. Ultrastr. Res.*, **58**, 172–7.

PRIMKE M., BERGER S. & SCHWEIGER H-G. (1978) Protoplasts from *Acetabularia*: isolation and fusion. *Cytobiologie*, **16**, 375–80.

PUESCHEL C. M. (1976) A freeze-etch study of the ultrastructure of red algal pit plugs. *Protoplasma*, **91**, 15–30.

PUESCHEL C. M. (1979) Ultrastructure of tetrasposogenesis in *Palmaria palmata* (Rhodophyta) *J. Phycol.*, **15**, 409–24.

PUESCHEL C. M. (1981) A reappraisal of the cytochemical properties of Rhodophycean pit plugs. *Phycologia*, **19**, 210–17.

PUISEUX-DAO S. (1970) *Acetabularia and Cell Biology*. Logos Press Ltd., London.

QUADER H., WAGENBRETH I. & ROBINSON D. G. (1978) Structure, synthesis and orientation of microfibrils. V. On the recovery of *Oocystis solitaria* from microtubule inhibitor treatment. *Cytobiologie*, **18**, 39–51.

QUATRANO R. S. (1973) Separation of processes associated with differentiation of two-celled *Fucus* embryos. *Develop. Biol.*, **30**, 209–13.

QUATRANO R. S. (1974) Developmental biology: Development in marine organisms. In *Experimental Marine Biology*, ed. Mariscal R. N., pp. 303–45. Academic Press, London & New York.

QUATRANO R. S., BRAWLEY S. H. & HOGSETT W. E. (1979) The control of the polar deposition of a sulfated polysaccharide in *Fucus* zygotes. In *Determinants of Spatial Organization*, eds. Subtelny S. & Konigsberg I. R., pp. 77–96. Academic Press, London & New York.

QUIRK H. M. & WETHERBEE R. (1980) Structural studies on the host-parasite relationship between the red algae *Holmsella* and *Gracilaria*. *Micron*, **11**, 511–12.

RAFF E. C. (1979) The control of microtubule assembly *in vivo*. *Int. Rev. Cytol.*, **59**, 1–96.

RAMM-ANDERSON S. M. (1980) Comparative fine structural aspects of postfertilization development in the Nemalionales (Rhodophyta) *Micron*, **11**, 517–18.

RAMUS J. (1969a) Pit connection formation in the red alga *Pseudogloiophloea*. *J. Phycol.*, **5**, 57–63.

RAMUS J. (1969b) Dimorphic pit connections in the red alga *Pseudogloiophloea*. *J. Cell Biol.*, **41**, 340–5.

RAMUS J. (1971) Properties of septal plugs from the red alga *Griffithsia pacifica*. *Phycologia*, **10**, 99–103.

RAMUS J. (1972) The production of extracellular polysaccharide by the unicellular red alga *Porphyridium aerugineum*. *J. Phycol.*, **8**, 97–111.

RAMUS J. & ROBINS D. M. (1975) The correlation of Golgi activity and polysaccharide secretion in *Porphyridium*. *J. Phycol.*, **11**, 70–4.

RIDGEWAY E. B., GILKEY J. C. & JAFFE L. F. (1977) Free calcium increases explosively in activating medaka eggs. *P.N.A.S.* (*U.S.A.*), **74**, 623–7.

RINGO D. L. (1967) Flagellar motion and fine structure of the flagellar apparatus in *Chlamydomonas*. *J. Cell Biol.*, **33**, 543–71.

ROBINSON D. G. (1977) Plant cell wall synthesis. *Adv. Bot. Res.*, **5**, 89–151.

ROLAND J. C. & VIAN B. (1979) The wall of the growing plant cell: Its three dimensional organization. *Int. Rev. Cytol.*, **61**, 129–66.

RÜFFER U., NULTSCH W. & PFAU J. (1978) Untersuchungen zur Lichtinduzierten Chromatophoren-verlagerung bei *Fucus vesiculosus*. *Helgol Wiss. Meeresunters.*, **31**, 333–46.

SALISBURY J. L. & FLOYD G. L. (1978) Calcium-induced contraction of the rhizoplast of a quadriflagellate green alga. *Science*, **202**, 975–7.

SCHMITZ K. & SRIVASTAVA L. M. (1974) Fine structure and development of sieve tubes in *Laminaria groenlandica* Rosenv. *Cytobiologie*, **10**, 66–78.

SCHMITZ K. & SRIVASTAVA L. M. (1975) On the fine structure of sieve tubes and the physiology of assimilate transport in *Alaria marginata*. *Can. J. Bot.*, **53**, 861–76.

SCHORNSTEIN K. L. & SCOTT J. L. (1980) Reevaluation of mitosis in the red alga *Porphyridium purpureum*. *Nature* (*Lond.*), **283**, 409–10.

SCHROEDER T. E. (1970) The contractile ring. Fine structure of dividing mammalian (HeLa) cells and the effects of cytochalasin B. *Z. Zellforsch Microsk. Anat.*, **109**, 431–49.

SCHRÖTER K. (1978) Asymmetrical jelly secretion of zygotes of *Pelvetia* and *Fucus*. An early polarization event. *Planta*, **140**, 69–73.

SCHRÖTER K., PENG H. B. & JAFFE L. F. (1976) Rapid cell wall formation by cortical vesicle fusion after fertilization of fucoid oocytes. *J. Cell Biol.*, **70**, 51a.

SCOTT J. L., BOSCO C., SCHORNSTEIN K. & THOMAS J. (1981) Ultrastructure of cell division and reproductive differentiation of male plants in the Florideophyceae (Rhodophyta): cell division in *Polysiphonia*. *J. Phycol.*, **16**, 507–24.

SCOTT J. L. & BULLOCK K. W. (1976) Ultrastructure of cell division in *Cladophora*. Pregametangial cell division in the haploid generation of *Cladophora flexuosa*. *Can. J. Bot.*, **54**, 1546–60.

SCOTT J. L. & DIXON P. S. (1973a) Ultrastructure of tetrasporogenesis in the marine red alga *Ptilota hypnoides*. *J. Phycol.*, **9**, 29–46.

SCOTT J. L. & DIXON P. S. (1973b) Ultrastructure of spermatium liberation in the marine red alga *Ptilota densa*. *J. Phycol.*, **9**, 85–91.

SCOTT J. L. & THOMAS J. P. (1975) Electron microscope observations of telophase II in the Florideophyceae. *J. Phycol.*, **11**, 474–6.

SHELTON E. & MOWCZKO W. E. (1978) Membrane blisters: A fixation artifact. A study in fixation for SEM. *Scanning*, **1**, 166–73.

SIDEMAN E. H. & SCHEIRER D. C. (1977) Some fine structural observations on developing and mature sieve elements in the brown alga *Laminaria saccharina*. *Am. J. Bot.*, **64**, 649–57.

SILVERBERG B. A. (1975) An ultrastructural and cytochemical characterization of microbodies in the green algae. *Protoplasma*, **83**, 269–95.

SIMON-BICHARD-BRÉAUD J. (1971) Un appareil cinétique dans les gametocystes mâles d'une Rhodophycée: *Bonnemaisonia hamifera* Hariot. *C.R. Acad. Sci. Paris*, **273**, 1272–5.

SIMON-BICHARD-BRÉAUD J. (1972) Formation de la crypte flagellaire et evolution de son contenu au cours de la gamétogenèse mâle chez *Bonnemaisonia hamifera* Hariot. (Rhodophycée) *C.R. Acad. Sci. Paris*, **274**, 1796–9.

SMITH H. (ed.) (1977) *The Molecular Biology of Plant Cells*. Univ. California Press, Berkeley.

SMITH M. T. (1976) Gamete release in *Sargassum heterophyllum* (Phaeophyceae). *Proc. E.M. Soc. So. Afr.*, **6**, 61–2.

SOROKIN S. (1962) Centrioles and the formation of rudimentary cilia by fibroblasts and smooth muscle cells. *J. Cell Biol.*, **15**, 363–77.

SPREY B. & LICHTENTHALER H. (1966) Zur Frage de Beziehungen zwischen Plastoglobuli und Thylakoid Genese in Gerstenkeim Lingen. *Z. Naturf.*, **21b**, 697–9.

SPRING H., SCHEER U., FRANKE W. W. & TRENDELENBURG M. F. (1975) Lampbrush-type chromosomes in the primary nucleus of the green alga *Acetabularia mediterranea*. *Chromosoma*, **50**, 25–43.

STANIER R. Y. & COHEN-BAZIRE G. (1977) Phototrophic prokaryotes: the Cyanobacteria. *Ann. Rev. Microbiol.*, **31**, 225–74.

STEVENS P. (1975) Cell wall formation in *Fucus* zygotes: cellulose synthesis and deposition after fertilization. Ph.D. Dissertation. Oregon State University, Corvallis.

STEVENS P. & QUATRANO R. S. (1978) Cell wall assembly in *Fucus* zygotes. II. Cellulose synthesis and deposition is controlled at the post-translational level. *Develop. Biol.*, **62**, 518–25.

STEWART K. D. & MATTOX K. R. (1975) Comparative cytology, evolution and classification of the green algae with some consideration of the origin of other organisms with chlorophylls *a* and *b*. *Bot. Rev.*, **41**, 104–35.

SVEDELIUS S. N. (1906) Reports on the marine algae of Ceylon. Ecological and systematic studies of the Ceylon species of *Caulerpa*. *Ceylon Mar. Biol. Rep.*, **4**, 81–144.

TALMADGE K. W., KEEGSTRA K., BAUER W. D. & ALBERSHEIM P. (1973) The structure of plant cell walls. I. The macromolecular components of the walls of suspension-cultured sycamore cells with a detailed analysis of the pectic polysaccharides. *Plant Physiol.*, **51**, 158–73.

TATEWAKI M. & NAGATA K. (1970) Surviving protoplasts *in vitro* and their development in *Bryopsis*. *J. Phycol.*, **6**, 401–3.

TELZER B. R. & ROSENBAUM J. L. (1979) Cell cycle-dependent, *in vitro* assembly of microtubules onto the pericentriolar material of HeLa cells. *J. Cell Biol.*, **81**, 484–8.

TOTH R. (1974) Sporangial structure and zoosporogenesis in *Chorda tomentosa* (Laminariales). *J. Phycol.*, **10**, 170–85.

TOTH R. (1976) The release, settlement, and germination of zoospores in *Chorda tomentosa* (Laminariales). *J. Phycol.*, **12**, 222–33.

TOTH R. & MARKEY D. R. (1973) Synaptonemal complexes in brown algae. *Nature* (*Lond.*), **243**, 236–7.

TRENCH R. K. & OHLHORST S. (1976) The stability of chloroplasts of siphonaceous algae in symbiosis with marine slugs. *New Phytol.*, **76**, 99–109.

TRIEMER R. E., AUREA C. & VASCONCELOS A. C. (1977) The ultrastructure of carposporogenesis in *Caloglossa leprieurii* (Delesseriaceae, Ceramiales). *Amer. J. Bot.*, **64**, 825–34.

TRIPODI G. (1971) The fine structure of the cystocarp in the red alga *Polysiphonia sertularioides* (Grat.) J. Ag. *J. Submicr. Cytol.*, **3**, 71–9.

TRIPODI G. (1974) Ultrastructural changes during carpospore formation in the red alga *Polysiphonia*. *J. Submicr. Cytol.*, **6**, 275–86.

TRIPODI G. & DE MASI F. (1977) The post-fertilization stages of red algae: the fine structure of the fusion cell of *Erythrocystis*. *J. Submicr. Cytol.*, **9 (4)**, 389–401.

TRIPODI G., PIZZOLORIGO P. & GIANNATTASIO M. (1972) A DN-ase sensitive twisted structure in the mitochondrial matrix of *Polysiphonia* (Rhodophyta). *J. Cell Biol.*, **55**, 530–2.

TURNER B. & FRIEDMANN E. I. (1974) Fine structure of capitular filaments in the coenocytic green alga *Penicillus*. *J. Phycol.*, **10**, 125–33.

VREELAND V. (1972) Immunocytochemical localization of the extracellular polysaccharide alginic acid in the brown seaweed, *Fucus distichus*. *J. Histochem. Cytochem.*, **20**, 358–67.

VREUGDENHIL D., DIJKSTRA M. C. & LEBBENGA K. K. (1976) The ultrastructure of the cell wall of normal and apolar embryos of *Fucus vesiculosus*. *Protoplasma*, **88**, 305–13.

WAALAND S. D. (1975) Evidence for a species-specific cell fusion hormone in red algae. *Protoplasma*, **86**, 253–61.

WAALAND S. D. (1978) Parasexually produced hybrids between female and male plants of *Griffithsia tenuis* C. Agardh, a red alga. *Planta*, **138**, 65–8.

WAALAND S. D. & CLELAND R. (1972) Development in the red alga, *Griffithsia pacifica*: Control by internal and external factors. *Planta*, **105**, 196–204.

WAALAND S. D. & CLELAND R. (1974) Cell repair through cell fusion in the red alga *Griffithsia pacifica*. *Protoplasma*, **79**, 185–96.

WAALAND S. D. & WAALAND J. R. (1975) Analysis of cell elongation in red algae by fluorescent labelling. *Planta*, **126**, 127–38.

WAALAND S. D., WAALAND J. R. & CLELAND R. (1972) A new pattern of plant cell elongation: bipolar band growth. *J. Cell Biol.*, **54**, 184–90.

WEFER G. (1980) Carbonate production by algae *Halimeda*, *Penicillus* and *Padina*. *Nature* (*Lond.*), **285**, 323–4.

WETHERBEE R. (1978a) Differentiation and continuity of the Golgi apparatus during carposporogenesis in *Polysiphonia* (Rhodophyta). *Protoplasma*, **95**, 347–60.

WETHERBEE R. (1978b) The presence of the tubular plasmalemmal structure during carposporogenesis in the red alga *Polysiphonia*. *Protoplasma*, **94**, 341–5.

WETHERBEE R. (1979) Transfer connections: specialized pathways for nutrient translocation in a red alga? *Science*, **204**, 858–9.

WETHERBEE R. (1980) Post-fertilization development in the red alga *Polysiphonia*. I. Proliferation of the carposporophyte. *J. Ultrastr. Res.*, **70**, 259–74.

WETHERBEE R. & KRAFT G. T. (1981) Morphological and fine structural features of pit connections in *Cryptonemia* sp., a highly differentiated marine red alga from Australia. *Protoplasma*, **106**, 167–72.

WETHERBEE R. & SCOTT F. J. (1980) The fine structure and distribution of 'transfer connections' in the red alga *Polysiphonia*. *Micron*, **11**, 509–10.

WETHERBEE R. & WEST J. A. (1976) Unique Golgi apparatus and vesicle formation in a red alga. *Nature* (*Lond.*), **259**, 566–7.

WETHERBEE R. & WEST J. A. (1977) Golgi apparatus of unique morphology during early carposporogenesis in a red alga. *J. Ultrastr. Res.*, **58**, 119–33.

WETHERBEE R., WEST J. A. & WYNNE M. J. (1974) Annulate lamellae in post-fertilization development in *Polysiphonia novae-angliae* (Rhodophyta). *J. Ultrastr. Res.*, **49**, 401–4.

WETHERBEE R. & WYNNE M. J. (1973) The fine structure of the nucleus and nuclear associations of developing carposporangia in *Polysiphonia novae-angliae* (Rhodophyta) *J. Phycol.*, **9**, 402–7.

WHALEY W. G. & DAUWALDER M. (1979) The Golgi apparatus, the plasma membrane and functional integration. *Int. Rev. Cytol.*, **58**, 199–246.

WHEELER A. E. & PAGE J. Z. (1974) The ultrastructure of *Derbesia tenuissima* (De Notaris) Crouan. I. Organization of the gametophyte protoplast, gametangium, and gametangial pore. *J. Phycol.*, **10**, 336–51.

WOODCOCK C. L. F. ed. (1977) *Progress in Acetabularia Research*. Academic Press, London & New York.

WOODCOCK C. L. F. & BOGORAD L. (1970) Evidence for variation in the quantity of DNA among plastids of *Acetabularia*. *J. Cell Biol.*, **44**, 361–75.

WOODCOCK C. L. F. & MILLER G. J. (1973) Ultrastructural features of the life cycle of *Acetabularia mediterranea*. II. Events associated with the division of the primary nucleus and the formation of cysts. *Protoplasma*, **77**, 331–41.

WRIGHT S. W. & GRANT B. R. (1978) Properties of chloroplasts isolated from siphonous algae. Effects of osmotic shock and detergent treatment on intactness. *Plant Physiol.*, **61**, 768–71.

YAMANOUCHI S. (1909) Mitosis in *Fucus*. *Bot. Gaz.*, **47**, 173–97.

YOUNG D. N. (1977) A note on the absence of flagellar structures in spermatia of *Bonnemaisonia*. *Phycologia*, **16 (2)**, 219–22.

YOUNG D. N. (1978) Ultrastructural evidence for a secretory function in the gland cells of the marine red alga *Botryocladia pseudodichotoma* (Rhodymeniaceae). *Protoplasma*, **94**, 109–26.

YOUNG D. N. (1979a) Ontogeny, histochemistry and fine structure of cellular inclusions in vegetative cells of *Antithamnion defectum* (Ceramiaceae, Rhodophyta). *J. Phycol.*, **15**, 42–8.

YOUNG D. N. (1979b) Fine structure and histochemistry of vesicle cells of the red alga *Antithamnion defectum* (Ceramiaceae). *J. Phycol.*, **15**, 49–57.

YOUNG D. N. (1979c) Fine structure of the 'gland cells' of the red alga, *Opuntiella californica* (Solieriaceae, Gigartinales) *Phycologia*, **18**, 288–95.

ZERBAN H. & WERZ G. (1975) Changes in frequency and total number of nuclear pores in the life cycle of *Acetabularia*. *Exp. Cell Res.*, **93**, 472–7.

SECTION II
ECOLOGY

INTRODUCTION

CHRISTOPHER S. LOBBAN

The separation of Chapters 8–19 into a section on ecology and a section on physiology and biochemistry is to a certain extent artificial. Ecology includes the study of responses of populations and individuals to their environment. Physiology includes the study of the responses to their environment of the cells making up individuals. Thus physiology can be regarded as ecology at the cellular level, while biochemistry is the view from the molecular level (the continuity is particularly evident in unicellular individuals). Nevertheless, a distinction between ecology and physiology–biochemistry can be seen not only in levels of organization, but also in the question asked: what are the environmental variables which impinge upon organisms? and, what are the processes taking place inside cells? It is chiefly on the basis of the question asked that we have categorized the subject matter of these twelve chapters.

Very few seaweeds have been studied in detail in a variety of conditions representing the range of their habitat, and this is cause for us to draw generalizations cautiously, and to interpret them with suspicion. The diversity of seaweeds—three Divisions—and of their habitats means that there may be very few universal truths about seaweeds beyond the basic cell biochemistry common to all eukaryotes.

Three problems of a general nature need to be borne in mind in carrying out and interpreting ecological studies: plants are different from animals; size is important; and environmental factors are multi-faceted.

Plants are different from animals, a fact which we anthropocentric animals sometimes forget when trying to understand plants (Bidwell 1979). Certainly there is great conceptual difficulty for us to visualize an environment as it impinges on a plant. The difficulty is compounded when we try to think of seaweeds, for these are frequently in an aqueous environment of which even those of us who are divers have had little experience. Several differences between plants and animals may be mentioned here for their ecological implications. 1, Animals must operate within narrow limits of body size and structure; the number and position of limbs is especially important. On the other hand multicellular plants, being non-motile, are not constrained by mechanical considerations, and may add or lose limbs (branches) continuously within limits imposed by certain stresses (wave action, for example, in the case of seaweeds). 2, Their immobility also means that plants have a very limited ability to grow through their environment in search of essentials and that they cannot take shelter from adverse conditions; again this contrasts with man and the animals with which he is most familiar. Once a spore or zygote settles in a particular spot it must survive whatever conditions prevail there—or die. 3, Whereas animals undergo an initial period of development followed by a more or less steady state (adulthood), plants develop throughout their lifetime. Various parts of a plant may develop, mature and die largely independent of the rest of the plant, particularly in perennials, but animals function as single units. The upshot of these and other differences has been succinctly stated by Evans (1972, p. 10): 'Complex intellectual processes are needed to attain even an imperfect understanding of the life of a plant, and every step of the argument must be checked by observation and experiment, as organisms so alien to us are full of surprises—only too often the response of a plant to some change of circumstances is anything but what a human being would expect.'

There are important consequences of the size difference between man and algae (Allen 1977). Individual algae, whether phytoplankton or seaweeds, respond only to conditions at their surfaces, not to average conditions. As Allen (1977, p. 255) says, '. . . just as mean soil moisture for a county gives little insight into the particular water relations of a tree, nutrients in the whole bottle do not have a simple relationship with culture organisms.' The difficulty becomes acute in situations where our measuring probes are very much larger than the volume to be measured—imagine trying to measure the pH immediately adjacent to a zoospore—or where the size of the probe is sufficient to alter the parameter it is supposed to measure—as in trying to measure water velocities in the boundary layer over a seaweed. Compared to the size of a zoospore even microclimate instrumentation (e.g. Forstner & Rützler 1970) is several orders of magnitude too large. (This applies equally to measuring the microenvironment of animals.) Any localized change caused by an alga—in nutrient concentrations, pH or temperature, for example—will be below the scale at which we can detect heterogeneity. Only under conditions of limited water movement, such as in tide pools or culture flasks, may these changes accumulate to the extent of significantly altering the average for the water mass. However, the 0.3 μm microelectrode technique of Nichols & Rikmenspoel (1977) perhaps gives an inkling of the sophistication which may one day be achieved in environmental studies.

Major difficulties in untangling the web of cause and effect in ecological study arise because environmental variables have many facets and because factors interact. Light, for example, has components of quality, flux rate, total daily flux, photoperiod, and so forth; and bright sunshine not only delivers a high light flux but may also cause a rise in temperature and in desiccation rate of an emersed seaweed. The difficulties stem in part from our need to consider the environment as if it consisted of discrete and measurable factors, each having some specific and separate effect on the plant. However, as Evans (1972, p. 34) warns us, 'Environmental "factors" are products of the human mind, convenient abstractions devised in order to make it possible to analyse and discuss complex problems: and like all such convenient abstractions they are separated by a gulf from things as they are.' Further, the effects of environmental variables are complicated by other factors which include: genetic differences, which confound our interpretation of the effect on the individual and force us to work with population means; non-linearity of plant response; effects of the plant on the environment (e.g. self-shading); ontogenetic drift; and persisting effects of past environments (Evans 1972). The last of these is a consequence of the plant continuing development throughout its lifetime; plants integrate their environmental history into their form and thus into their overall response to the current environment. Perhaps the most awkward complicating factor from an experimental view-point is competition which, while it has received little study so far, 'seems to sharpen the sensitivity of a species to its environmental conditions and to reduce the amplitude of those conditions in which the species is vigorous' (Russell & Fielding 1974, p. 696). This factor probably causes the greatest difference between laboratory and field responses (Bannister 1976).

Studies on the tolerance or response of an organism to specific environmental factors are the first step in the elucidation of the totality of the organism's response to its environment. Having separated the factors into discrete, measurable packets for purposes of reasoning and experimentation, we must then put everything together again, that is, to attempt multivariable modelling. As a beginning we may construct three- or multi-dimensional contour maps, showing 'response surfaces' which define an organism's tolerance zone (Alderdice 1972). Lehnberg (1978) has made some maps of this type for *Delesseria sanguinea*. The principal challenge to seaweed ecologists seems to be to amass enough experimental data, with enough of the non-experimental factors accounted for, to be able to begin such grand syntheses.

REFERENCES

ALDERDICE D. J. (1972) Factor combinations. Responses of marine poikilotherms to environmental factors acting in concert. In *Marine Ecology*, vol. I, part 3, ed. Kinne O., pp. 1659–722. John Wiley & Sons, New York.

ALLEN T. F. H. (1977) Scale in microscopic algal ecology: a neglected dimension. *Phycologia*, **16**, 253–7.

BANNISTER P. (1976) *Introduction to Physiological Plant Ecology*. Halstead/Wiley, New York.

BIDWELL R. G. S. (1979) *Plant Physiology*, 2nd edn. Macmillan, New York.

EVANS G. C. (1972) *The Quantitative Analysis of Plant Growth*. Blackwell Scientific Publications, Oxford.

FORSTNER M. & RÜTZLER K. (1970) Measurements of the microclimate in littoral marine habitats. *Oceanogr. Mar. Biol. Ann. Rev.*, **8**, 225–49.

LEHNBERG W. (1978) Die Wirkung eines Licht-Temperatur-Salzgehalt Komplexes auf den Gaswechsel von *Delesseria sanguinea* (Rhodophyta) aus der westlichen Ostsee. *Bot. Mar.*, **21**, 485–97.

NICHOLS K. M. & RIKMENSPOEL R. (1977) Mg^{2+}—dependent electrical control of flagellar activity in Euglena. *J. Cell Sci.*, **23**, 211–25.

RUSSELL G. & FIELDING A. H. (1974) The competitive properties of marine algae in culture. *J. Ecol.*, **62**, 689–98.

CHAPTER 8
GEOGRAPHICAL DISTRIBUTION

LOUIS D. DRUEHL
Department of Biological Sciences, Simon Fraser University,
Burnaby, British Columbia, Canada

1 INTRODUCTION

'The field of phytogeography is vast, the literature inexhaustible, its data capable of a variety of interpretation, extraordinary manipulation, and distortion. It has the fascination of a chess game. It is a valid, though treacherous, field of investigation' (Silva 1962). This is an appropriate evaluation of our present understanding of phytogeography. Four problems contribute to this state of affairs: 1, sparseness of records—the floras of very few areas have been thoroughly documented (often, floristic changes with season and depth are not known); 2, proof of absence—any phycologist who has completed a seaweed survey must question the credibility of discontinuous distributions (the problem lies in distinguishing between the proof of absence and the absence of proof); 3, uniformity of taxonomic criteria—the use of different taxonomic criteria for the same taxa results in an incorrect picture of seaweed distributions—for example, species of *Laminaria* are generally distinguished by anatomical features in the Pacific, whereas these features are not employed in the northwest Atlantic; further, the shape of *Laminaria* blades is used in distinguishing species of *Laminaria* only in Japan and thus, the number of species encountered in an area may in part be a function of criteria used to distinguish them; 4, cause and effect—many recent studies have attempted to explain seaweed distributions in terms of prevailing environmental conditions. A major assumption in these studies is that seaweed distributions are static. However, it is probable that present distributions reflect adjustment to environmental changes which occurred a long time ago (e.g. glacial recession).

In spite of these difficulties, we need to intensify our studies on seaweed distributions. As our knowledge of biogeography grows, we will be better able to interpret the effects of environmental change (natural and man-induced) on distribution patterns. However, the growth of biogeography must not outstrip advances in supporting disciplines, namely taxonomy, eco-physiology, paleoclimatology, and paleobotany.

2 SEAWEEDS AS ENVIRONMENTAL MONITORS

Seaweeds exist for the greatest part of their lives as attached organisms, monitoring the environment of passing currents and waves. During a small part of their lives, they exist as planktonic spores, gametes or free-floating zygotes. Here they move with a set of environmental parameters, particularly those associated with the drop of water in which they are contained. To appreciate forces influencing seaweed distributions, it is important to understand both their planktonic and benthic stages.

2.1 Planktonic stages

The problem of seaweed dispersal cannot be properly investigated without first understanding the biology of planktonic life history stages. The following paragraphs explore this relatively unknown area.

How long can a spore remain suspended and is it capable of meeting its energy requirements through photosynthesis? In a unique study, Kain (1964) found that the meiospores of *Laminaria hyperborea* lost their motility after one day, but remained suspended for several days. Kain also discovered that the meiospores were just capable photosynthetically of meeting their respiration requirements under saturating light conditions.

How close to each other must dioecious gametophytes settle to ensure subsequent fertilization? The first clue to this question came from an inadvertent discovery. Sundene (1962) observed a new generation of *Alaria esculenta* restricted to *c.* 10 m of a transplant site where he had introduced the parent generation. Anderson & North (1965) determined that the concentration of new sporophytes resulting from meiospores of a single *Macrocystis pyrifera* decreased inversely as the cube of the distance from the parent plant. They also noted that there was no decrease in the density of new sporophytes for up to 30 m from a concentration of mature sporophytes. In a similar study, Dayton (1973) found that the intertidal alga *Postelsia palmaeformis* had an effective meiospore dispersal range of up to 3 m.

Evidence from colonization studies of the island of Surtsey suggests that many seaweeds have an effective dispersal range of kilometres rather than metres (Jónsson 1972). Surtsey originated from a series of submarine eruptions starting 14 November 1963, approximately 30 km from Iceland and 5 km from the nearest point of land. The first seaweed (*Urospora*) was noted in 1965. By 1966, *Alaria, Porphyra, Petalonia* and other species were established. The Surtsey flora appeared to stabilize by 1968 when *Laminaria, Phycodrys, Desmarestia* and others became established (Table 8.1). One cannot discern from Jónsson's study whether spores and gametes, or drift plants inoculated the island.

Table 8.1 Numbers of species observed on Surtsey subsequent to its origin in 1963 (from Jónsson 1972).

	1964	1965	1966	1967	1968	1969	1970
Rhodophyta	0	0	1	1	7	5	7
Chlorophyta	0	1	5	6	10	9	9
Phaeophyta	0	0	6	7	10	10	9

Also relevant to understanding the importance of proximity of gametophytes are the recent findings of Müller *et al.* (see Chapter 19) on sex attractants in seaweeds, and Fetter (1977) on red algal spermatia dispersal. The red algae have been considered to be relatively inefficient in their sexual reproduction due to the lack of flagellated sperm. However, Fetter (1977) has shown that spermatia of *Tiffaniella snyderae* are dispersed in long spiralling slime strands, indicating that the red algae have effective mechanisms for gamete dispersal other than the more conventional method of swimming.

2.2 Benthic stages

During its benthic stage, a seaweed may exist for varying lengths of time as different morphologies and ploidy levels, each having different physiological tolerances and requirements. These features dictate a seaweed's response to its environment. The following paragraphs briefly explore these biological parameters. Chapters 4–6, 11 and 12 are also relevant.

The longer a seaweed persists, the greater the environmental variability that it will encounter in any given area. The question is—how long does a plant normally persist, and does any given species have a maximum life span? Generally, there is a tendency to classify seaweeds according to their longevity (Garbary 1976; see also Chapter 11). One scheme classifies seaweeds as ephemeral (short-lived, and often having several generations within one year); annual (persisting for the greater part of a year); or perennial (persisting for some years). However, some examples indicate that these terms, which were borrowed from flowering plant terminology, may not be strictly applicable to seaweeds. For example, *Costaria costata* is perennial subtidally and ephemeral in the sheltered, intertidal waters of British Columbia, Canada (Druehl & Hsaio

1977). Conversely, the intertidal and ephemeral *Laminaria ephemera* can persist subtidally for at least one year (Druehl 1968). These examples indicate that extrinsic and not intrinsic factors may determine the longevity of seaweeds. A second source of error in interpreting seaweed longevities comes from the assumption that when the fleshy portion of a seaweed 'disappears' the entire plant is gone. Dixon (1965) has demonstrated that several so-called annuals are actually perennial, in some cases perennating from a few cells.

Among the perennials there are reported cases of extended longevity: *Laminaria hyperborea,* 12 years (Kain 1977); *Pterygophora californica, c.* 25 years (Frye 1918); and *Pterosiphonia complanata,* more than 40 years (Dixon 1965).

We know the life spans of few seaweeds. Thus when attempting to correlate seaweed distributions with environmental parameters we do not know the seasonal range of their exposure. The alternation of independent haploid and diploid generations, and the common occurrence of asexual reproduction in marine algae allow for possible environmental selection of an optimal ploidy phase, which implies optimal morphology and physiology.

Numerous examples of apparently disjunct distributions of ploidy levels in red algae exist (Feldmann 1957; Dixon 1965; McLachan *et al.* 1969). Generally, the diploid stage (tetrasporophyte) of red algae has the wider distribution, indicating that this stage has a broader environmental tolerance than does the haploid gametophyte. In red algae having an alternation of heteromorphic generations (e.g. *Bonnemaisonia*) this differential distribution is relatively easy to determine. However, in those species having an alternation of isomorphic generations (e.g. *Polysiphonia*) this is more difficult; here one must at least observe sporangia or gametangia, or preferably—in view of the complexity of brown algal life histories (Chapter 5)—do a chromosome count. In this latter case it is possible that both ploidy levels are equally distributed but only one generation is capable of sexual expression (meiosis or syngamy) (Dixon 1965). Figure 8.1 illustrates possible life history responses of seaweeds to increasing environmental stress. Seaweeds with alternation of heteromorphic generations may persist in a locality by existing in one season as a gametophyte and in another season as a sporophyte. For example, mid-intertidal *Laminaria saccharina* persists as gametophytes through the summer and autumn months, and as sporophytes the remainder of the year in a British Columbia inlet (Hsaio & Druehl 1973). Apparently the microscopic gametophyte can survive the prolonged periods of daytime emersion whereas the macroscopic sporophyte cannot.

An analysis of the distribution of isomorphic and heteromorphic life history stages of the brown and green seaweeds of Japan supports the hypothesis that having alternate life forms better adapts a seaweed to environmental stress. Nakahara & Masuda (1971) have shown that the relative number of species with heteromorphic life cycles decreases with the mean annual seawater temperature and with the annual range of temperatures (Fig. 8.2).

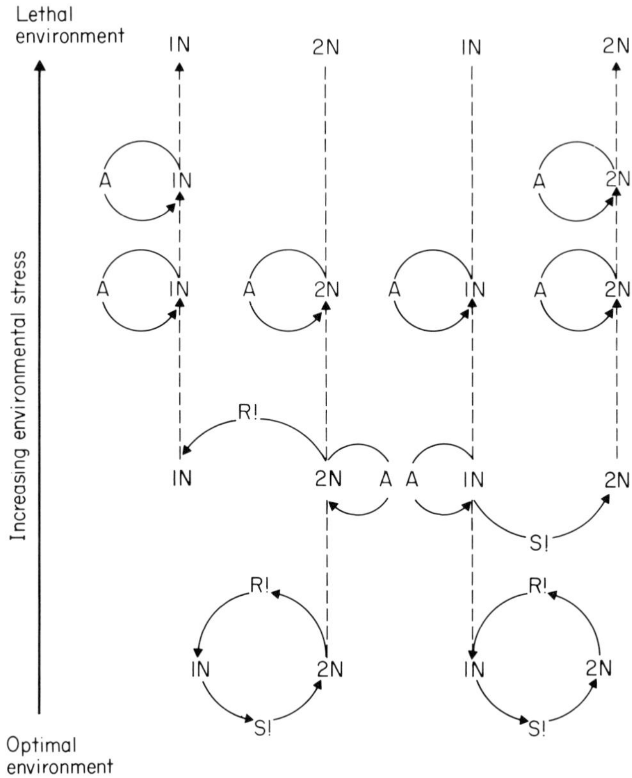

Fig. 8.1 Life cycle responses to an increasingly adverse environment. R! = meiosis; S! = syngamy; A = asexual reproduction; 1N = haploid; 2N = diploid; broken line = mode of dispersal.

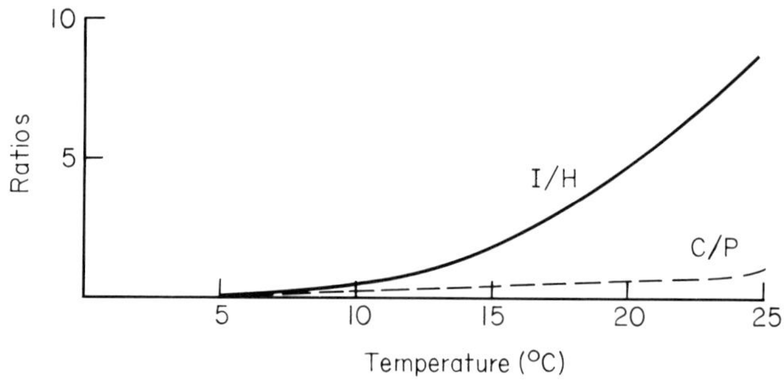

Fig. 8.2 The ratio of Chlorophyta and Phaeophyta isomorphic/heteromorphic life cycles (I/H) and the ratio of Chlorophyta and Phaeophyta (C/P) as a function of mean annual sea-water temperature in Japan (after Nakahara & Masuda 1971).

3 FACTORS INFLUENCING HORIZONTAL DISTRIBUTIONS

Factors influencing seaweed distributions can be divided into two categories. Factors influencing a plant's well-being, e.g., temperature, salinity, water motion, nutrients, and biological competition, dictate whether a plant can exist in a particular place, whereas physical factors influencing dispersal, e.g. land masses, ocean currents, and substrate type, dictate whether or not a plant can be introduced to a locality. Following is a discussion of how patterns of seaweed distribution may reflect prevailing environmental conditions, characterized by temperature, salinity, and water motion. The effect of biological competition, to the extent that it is understood, is also discussed. (Broad reviews covering all aspects of temperature and salinity in relation to marine plants have been published by Gessner 1970 and Gessner & Schramm 1971 respectively.)

It cannot be overemphasized that the following examples of distribution relative to environmental parameters are only correlations. Cause and effect have for the most part not been established, and at best there is only *prima facie* evidence supporting the role of selected environmental parameters.

3.1 Temperature

Since Alexander von Humbolt introduced in 1805 the concept of environmental factors governing biotic distributions, temperature has been generally regarded as a leading factor. One of the earliest attempts to relate water temperatures to seaweed distributions was by Setchell (1893). Setchell (1920) applied the concepts of stenothermy (narrow temperature tolerance) and eurythermy (wide temperature tolerance), as presented by M. Moebuis in 1897, to a scheme of climatic evaluation proposed by C. Hart Merriam in 1894 to delineate crop zones in the United States. Merriam's scheme employed the mean temperature for the warmest six weeks (isothere) and the mean temperature for the coldest six weeks (isocryme) of the year. Setchell proposed a series of coastal zones separated by 5° C isotheres and isocrymes. Seaweed species restricted to one zone were considered stenothermal, and those occupying more than one zone eurythermal. Often, a species occupying more than one zone reproduces in the winter in the more equatorward zone and in the summer in the more poleward zone. An example of this is *Ascophyllum nodosum* which reproduces during the summer in Greenland and late winter in Long Island Sound (Setchell 1920). Isocrymes were generally thought to limit poleward distributions and isotheres equatorward distributions. Figure 8.3 depicts potential responses of seaweeds to temperature extremes according to Setchell's hypotheses. However, Hutchins (1947) proposed that either isocrymes or isotheres could limit both poleward and equatorward distributions.

Temperature has been employed in assessing seaweed distribution patterns

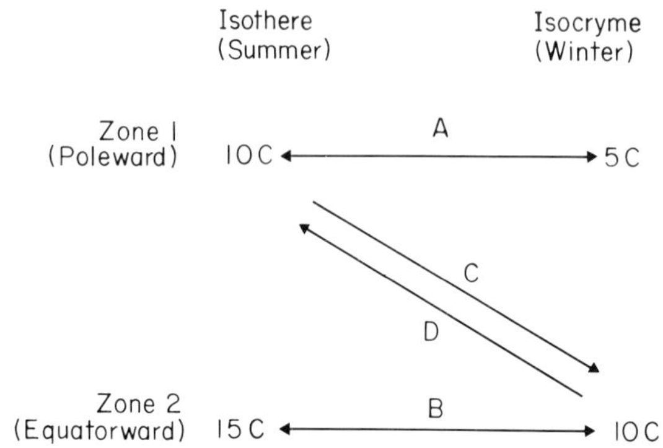

Fig. 8.3 Hypothetical seaweed response to different temperature regimes. Stenothermal species A and B reproduce within their respective temperature zones. The equatorward distribution of species A is blocked by the zone 2 isothere and the poleward distribution of species B is blocked by the zone 1 isocryme. Eurythermal species C and D occupy two zones. Species C reproduces during the summer in zone 1 and during the winter in zone 2 and species D does the opposite.

in the northeast Pacific. Scagel (1963) described an abrupt transition in the seaweed flora of the northeast Pacific at Point Conception, California, and a gradual transition in species composition from Northern California to Sitka, Alaska (Fig. 8.4). The Point Conception transition coincides with a region of rapid temperature change, whereas the more northern shores have relatively uniform temperatures (Fig. 8.5).

Temperature regimes have been employed in delineating phytogeographic provinces and seaweed distributions on a global scale. Michanek (1979) provides a comprehensive review of this development. Figure 8.6 represents Michanek's interpretation of climatic regions and prevailing temperature regimes.

Some discontinuities in seaweed distributions may be accredited to irregular temperature distributions. Dawson (1951) demonstrated the presence of cool northern seaweed species in areas of subtropical Baja California, Mexico, subjected to upwelling. Further, the presence of two *Laminaria* species as far equatorward as 19° S latitude in Brazil appear to be related to an upwelling system. These *Laminaria* are found only in waters deeper than 40 m (Oliveira & Quége 1978).

3.2 Salinity

Salinity is essentially an expression of the total salt per kilogram of water. Values are given as parts per thousand (‰). This expression tells us little about how salinity might affect a seaweed and most workers agree that plant response cannot be attributed to salinity *per se*. Generally, salinity has been considered only in assessing local distributions.

Care must be taken in interpreting salinity data, particularly in areas

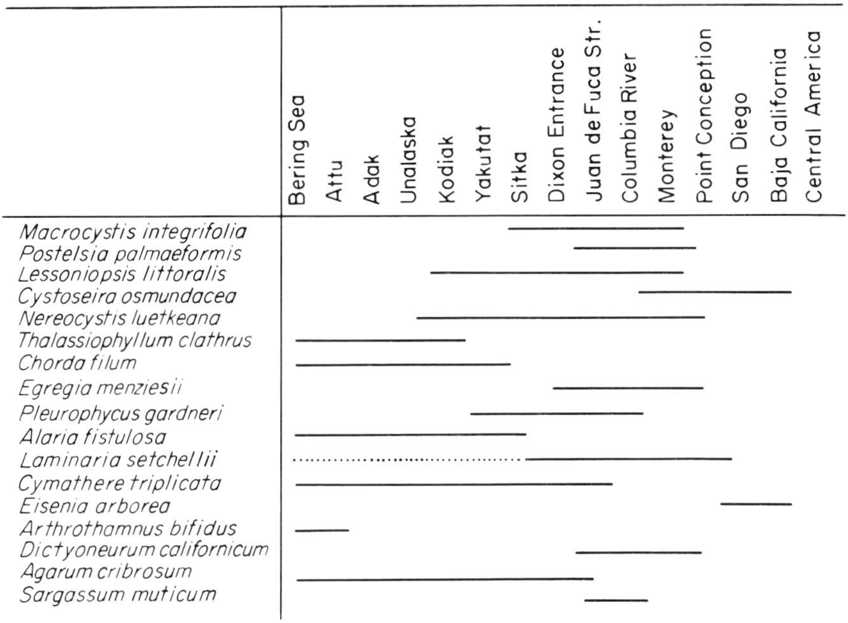

Fig. 8.4 The northeast Pacific distributions of selected seaweeds (from Scagel 1963).

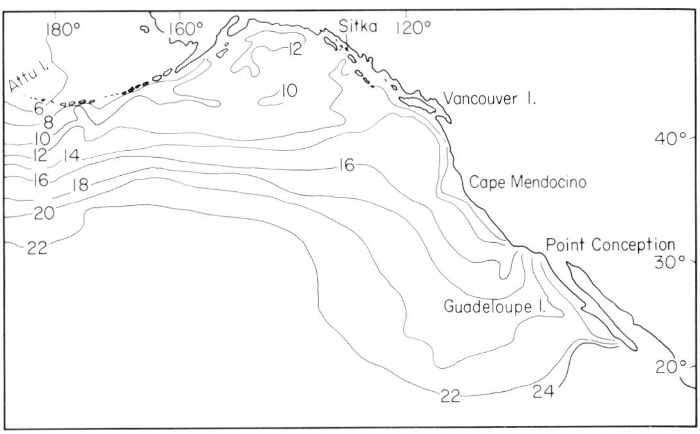

Fig. 8.5 Seawater temperature (°C) at 10 m depth for the northeast Pacific (from Scagel 1963).

subjected to tidal flushing. Here, daily or weekly values may fail to adequately characterize the salinity environment. Figure 8.7 depicts mean monthly values of salinity for three years and salinity fluctuations during two days for a British Columbia inlet (Hsaio 1972). The range of salinity in one day (14–27‰) in June approaches the mean monthly range as observed over the three years (19–28‰).

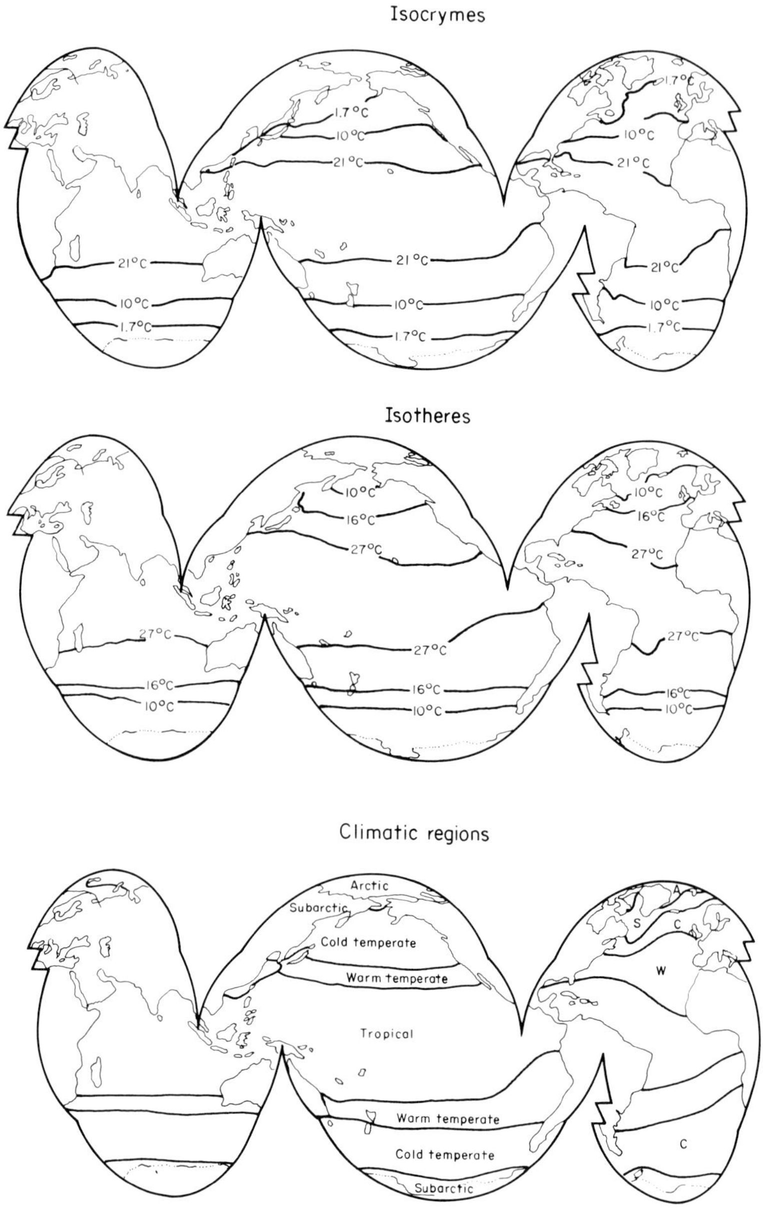

Fig. 8.6 Oceanic isocrymes, isotheres and climatic regions (from Michanek 1979).

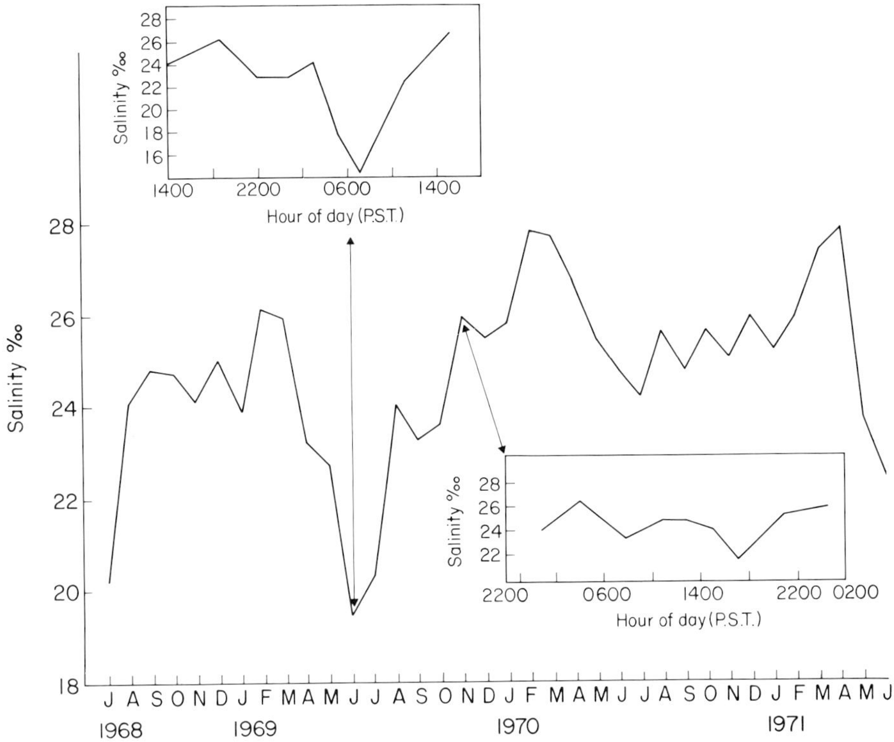

Fig. 8.7 Seasonal variation in salinity (monthly means) in a British Columbia inlet as measured weekly at low tide. Inserts illustrate summer and winter variation over 24 hours (after Hsiao 1972).

A second complication in evaluating salinity conditions arises from the vertical distribution of salinity in the water column. Often a steep gradient of salinity can exist in the upper few meters with low salinity water overlying that of higher salinity. Under these conditions an attached seaweed may be subjected to considerable salinity variation as the tide ebbs and floods.

Several descriptive systems for classifying ranges of salinities have been proposed, but these are often confusing and contradictory. The best practice is to simply state the salinity (‰). As information on algal distributions in relation to salinity accumulates, individual algal species or groups of species may ultimately be used to define salinity regimes on a more meaningful biological basis.

Seaweeds can be classified by their response to salinity. A euryhaline species is one tolerant to a wide range of salinities, and a stenohaline species one tolerant to a narrow range of salinities. This seemingly straightforward classification can be deceiving, however, as shown by the following examples. *Laminaria saccharina* in British Columbia has been shown to have a wide salinity tolerance, based on individuals of a single ecotype (Druehl 1967a). On the other hand, Yarish *et al.* (1979), studying salinity responses in two Rhodophyta species, found that the apparent euryhalinity in both species was a reflection of the overlap of tolerance ranges of a series of ecotypes.

Floristic changes as a function of salinity have been catalogued by Munda (1978) for four Icelandic fjords. Two trends are apparent (Table 8.2). First, the total number of species increased from the lowest (0–5‰) to the highest salinities

Table 8.2 Species composition of two Icelandic fjords as a function of salinity (after Munda 1978).

	Salinity‰			
	0–5	5–10	10–15	15–20
	Borgarfjördur			
Chlorophyta	5	12	11	20
Rhodophyta	0	1	1	22
Phaeophyta	2	10	16	25
Total	7	23	28	67
	Dýrafjördur			
Chlorophyta	11	14	17	21
Rhodophyta	2	2	8	15
Phaeophyta	3	11	17	21
Total	16	27	42	57

(15–20‰) encountered. Second, the Rhodophyta were least tolerant of the lower salinities and Chlorophyta most tolerant. Only two species were found in all salinities in the four fjord systems: *Enteromorpha intestinalis* and *Pilayella littoralis*.

The intertidal distribution of any particular species may end along a salinity gradient, but continue subtidally into waters supporting lower surface salinities. *Laminaria saccharina, Costaria costata* and *Constantinea subulifera* are known to grow in the lower intertidal and shallow subtidal regions in areas of the northeast Pacific having surface salinities in excess of 28‰. However, these species are known only from the subtidal region in an inlet having surface salinities of 12–19‰, where their upper limit is restricted by the 24–25‰ isohaline (Druehl 1967b).

3.3 Temperature/salinity

Increasing evidence supports the contention that combinations of environmental factors, rather than individual factors, may determine seaweed distributions. Scagel (1961) outlines the problems of viewing the total environment, and proposes a pragmatic approach based upon laboratory studies to focus on potentially important parameters. As he states, this approach has inherent dangers of oversimplification, but is necessary to reduce the variables to a manageable number. For some seaweeds, the adverse effects of high temperatures may be ameliorated by high salinity, and the adverse effects of low salinity offset by low temperatures. This has been demonstrated under laboratory conditions (Biebl 1972; Druehl 1967a).

The distribution of *Macrocystis integrifolia* in British Columbia has been correlated with temperature/salinity regimes (Druehl 1978). This species is present in Nootka Sound but absent from Entrance Island. Both locations experience similar temperature and salinity ranges. However, they differ in that when the Nootka plants are subjected to the higher temperatures, the salinity is high. At Entrance Island the higher temperatures are associated with low salinities (Fig. 8.8).

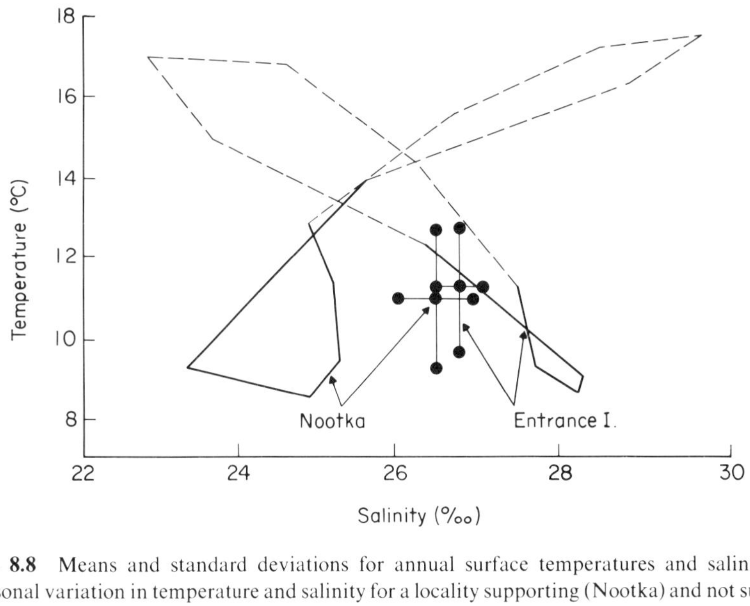

Fig. 8.8 Means and standard deviations for annual surface temperatures and salinities, and seasonal variation in temperature and salinity for a locality supporting (Nootka) and not supporting (Entrance Island) *Macrocystis integrifolia* populations. Solid lines depict winter conditions and broken lines summer conditions (after Druehl 1978).

A similar mechanism may help explain latitudinal distributions which have previously been related solely to temperature. For example, the southern extremes of Laminariales distribution along the west coast of North America are related to an increase in temperature and salinity (Figs. 8.9, 8.10). Possibly the salinity increase allows for a higher temperature tolerance and consequent extended equatorward distribution of the Laminariales.

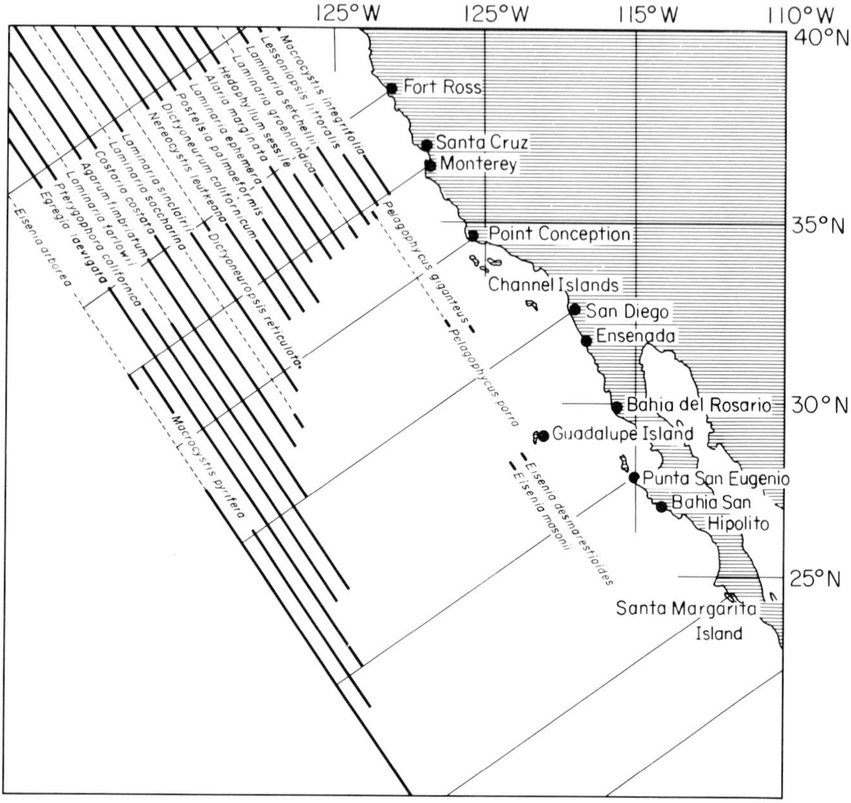

Fig. 8.9 General distribution of the Laminariales near the southern extreme of distribution in the northeast Pacific (from Druehl 1970).

3.4 Water motion

Waves and currents are two forms of water motion which may influence local seaweed distributions. These operate through a variety of mechanisms: force, abrasion, and movement *per se* may occlude or remove seaweeds from a particular habitat; waves and spray may offset adverse desiccation effects, and water motion may replenish nutrients and disrupt thermo- and haloclines.

Water motion has been attributed with influencing the morphology of seaweeds (see Chapter 12), the composition of seaweed communities (Sundene 1953; Chapter 11) and the relative biomass of seaweeds (Conover 1968).

3.5 Biological control

Biological control of horizontal distributions has received little theoretical consideration and no experimental effort. Pielou (1977) studied the latitudinal

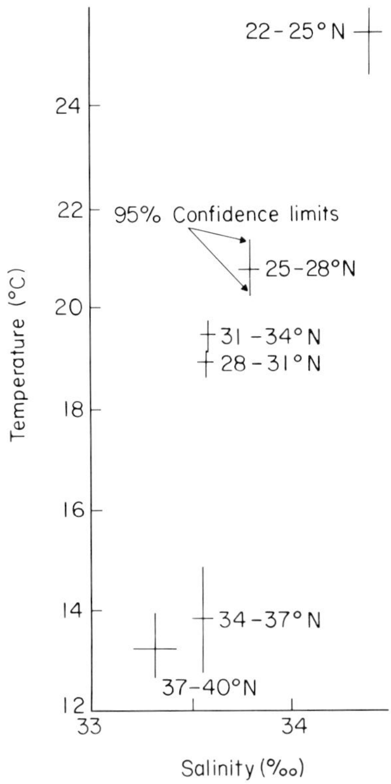

Fig. 8.10 August surface temperatures and salinities for different sectors of coast coinciding with the southern extreme of northeast Pacific Laminariales distribution (see Fig. 8.9). 22–25° N represents a region of no known Laminariales (Druehl unpublished).

extension of seaweeds along the eastern coast of the Western hemisphere. She concluded that the horizontal distributions of congeneric species were not influenced by competition.

4 COMPARATIVE FLORISTICS

Initially, phycologists compared species lists to establish regional relationships. From these early studies it became apparent that widely separated areas often have similar floristic elements (e.g. Japan–California, east–west North Atlantic, Brazil–Caribbean). Latitudinal changes in floristic composition were also noted. For example, tropical floras were reported to have a Rhodophyta/Phaeophyta (R/P) ratio of 4 or greater (Feldmann 1937). The species diversity and R/P ratio as a function of latitude are shown for the western Atlantic in Fig. 8.11, where it can be seen that the R/P ratio only falls below 4 in the northwest Atlantic. In the southwest Atlantic R/P values never fall below 4.2. Clearly, these earlier established relationships require re-investigation.

Some seaweeds have large latitudinal distributions and others are geographically restricted. The relative frequency of narrow and wide distributions are similar for the Rhodophyta, Phaeophyta and Chlorophyta (Fig. 8.12; Pielou 1977). The most common span is less than 10 degrees of latitude and the second most common span is 40–50 degrees of latitude.

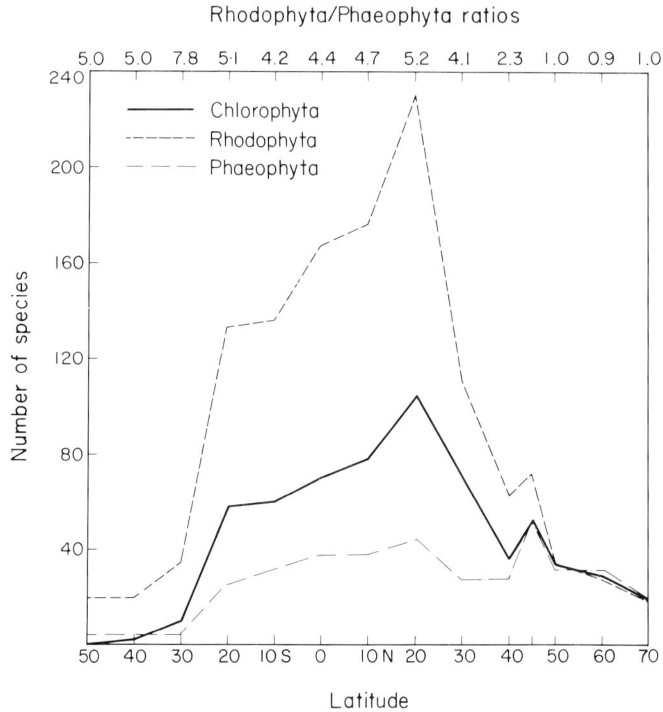

Fig. 8.11 An estimation of seaweed diversity and the Rhodophyta/Phaeophyta ratio as a function of latitude in the western Atlantic (after Pielou 1977).

Discontinuities in seaweed distributions have long intrigued phycologists. Continental drift, climatic change and dispersal/transport systems have all been suggested as mechanisms which could create discontinuities (Silva 1962; Hughes 1975; van den Hoek 1975). Large discontinuities, such as the bipolar distribution of *Macrocystis pyrifera,* probably resulted from causes of great antiquity (continental drift?). Smaller discontinuities, such as the discontinuity from British Columbia to central California of *Eisenia arborea* are likely the result of more recent causes (glaciation?).

The biogeography of the Atlantic seaweed flora has been intensely investigated in recent years (Pielou 1977; Lawson 1978; Kühnemann 1972; van den Hoek 1975). Van den Hoek (1975), using cluster analysis, divided the north Atlantic into five phytogeographic regions. By using ordination techniques, Lawson (1978) was able to suggest seven seaweed floras for the tropical and subtropical Atlantic (Fig. 8.13). Lawson (1978) was critical of the cluster analysis employed by van den Hoek (1975); however, Lawson's results in the region where their two studies overlap complement van den Hoek's findings.

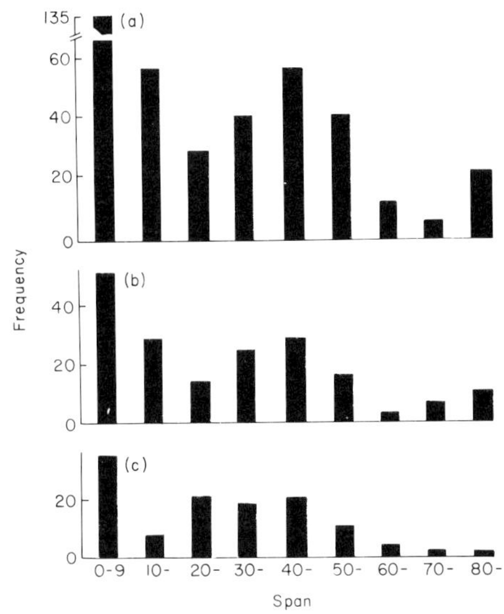

Fig. 8.12 The frequency distributions of spans of Rhodophyta (a), Chlorophyta (b) and Phaeophyta (c) occurring in the western Atlantic. Spans are in degrees of latitude (from Pielou 1977).

5 RECENT FLORISTIC CHANGES

Recent floristic changes can be attributed to intentional and unintentional introductions by man, and to environmental change.

Some known introductions are shown in Fig. 8.14. Some of these were introduced for maricultural purposes (e.g. *Eucheuma*), and others were accidentally introduced (e.g. *Sargassum*). The success of some species in establishing themselves in geographic regions distant from their normal areas of distribution suggests that inefficient long-range dispersal mechanisms, and not abiotic factors, may be limiting seaweed distributions.

Local floristic changes as a response to a man's alteration of the environment have been recorded. These include algal response to thermal pollution (Abbott

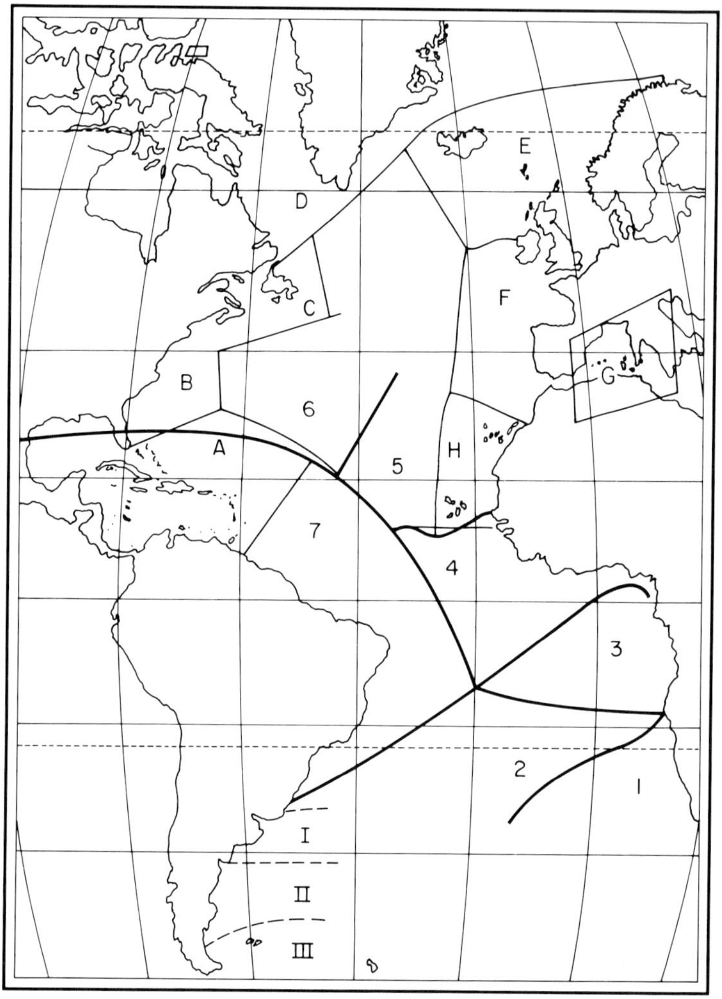

Fig. 8.13 The distribution of distinct seaweed floras in the Atlantic Ocean. (A–H from van den Hoek (1975), 1–7 from Lawson (1978), and I–III from Kühnemann (1972).)

& North 1972) and industrial/domestic pollution (Bokn & Lein 1978; Thom & Widdowson 1977). Generally, the floras responded to pollution by showing decreased species diversity, an enhanced Chlorophyta component, and selection for small, often turf-like plants. The waters of Mitajire Bay, Japan, have been partially de-polluted, and its seaweed flora has shown reversals in such trends (Teramoto & Kawamori 1977).

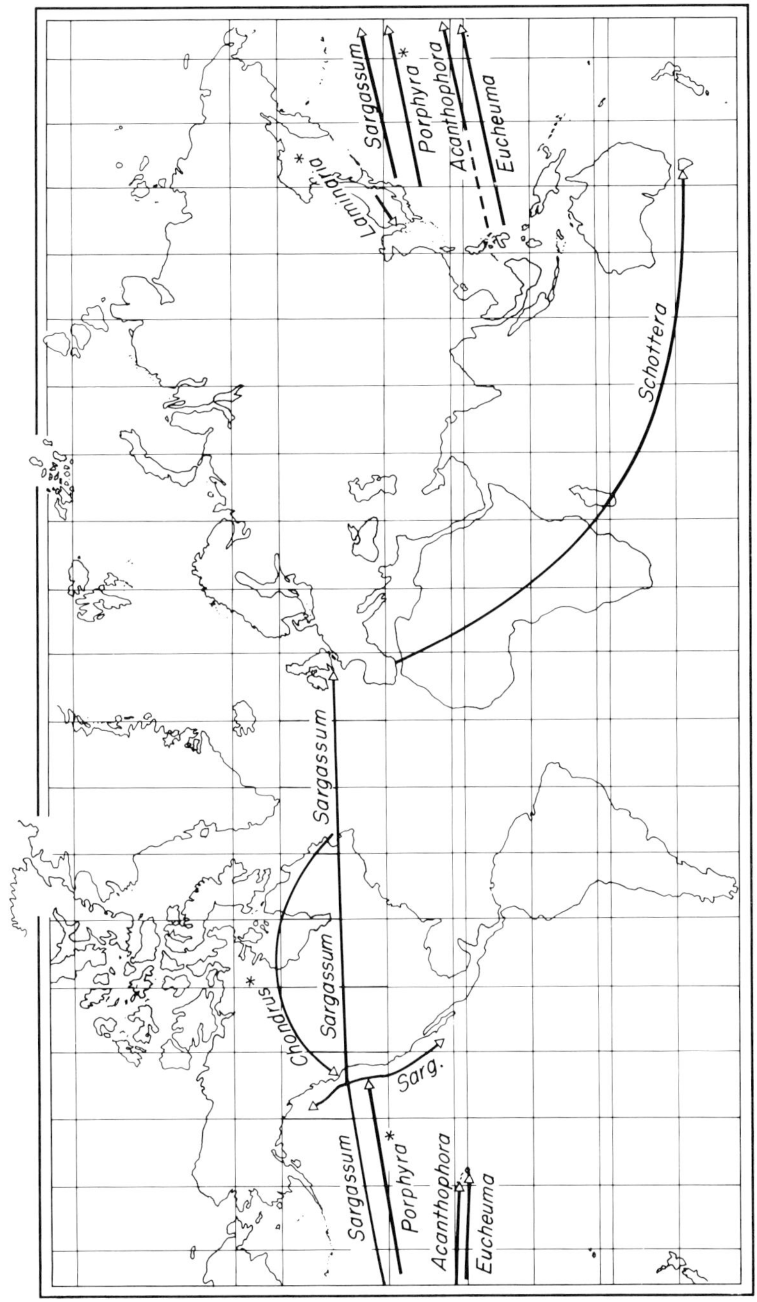

Fig. 8.14 Examples of seaweeds whose distributions have been altered by man. *are cases where the seaweed has been moved but natural populations are not known to be established.

6 REFERENCES

ABBOTT I. A. & NORTH W. J. (1972) Temperature influences on floral composition in California coastal waters. *Proc. Int. Seaweed Symp.*, **7**, 72–9.

ANDERSON E. K. & NORTH W. J. (1966) *In situ* studies of spore production and dispersal in the giant kelp *Macrocystis*. *Proc. Int. Seaweed Symp.*, **5**, 73–86.

BIEBL R. (1972) Temperature resistance of marine algae. *Proc. Int. Seaweed Symp.*, **7**, 23–8.

BOKN T. & LEIN T. E. (1978) Long-term changes in fucoid association of the inner Oslofjord, Norway. *Norw. J. Bot.*, **25**, 9–14.

CAREFOOT T. H. (1973) Feeding, food preference, and the uptake of food energy by the supralittoral isopod *Ligia pallasii*. *Mar. Biol.*, **18**, 228–36.

CHAPMAN A. R. O. (1973) A critique of prevailing attitudes towards the control of seaweed zonation on the sea shore. *Bot. Mar.*, **16**, 80–2.

CONOVER J. T. (1968) The importance of natural diffusion gradients and transport of substances related to benthic marine plant metabolism. *Bot. Mar.*, **11**, 1–9.

DAWSON E. Y. (1951) A further study of upwelling and associated vegetation along Pacific Baja California, Mexico. *J. Mar. Res.*, **10**, 39–58.

DAYTON P. K. (1971) Competition, disturbance, and community organization: The provision and subsequent utilization of space in a rocky intertidal community. *Ecol. Monogr.*, **41**, 351–89.

DAYTON P. K. (1973) Dispersion, dispersal, and persistence of the annual intertidal alga, *Postelsia palmaeformis* Ruprecht. *Ecology*, **54**, 433–8.

DAYTON P. K. (1975) Experimental evaluation of ecological dominance in a rocky intertidal algal community. *Ecol. Monogr.*, **45**, 137–59.

DIXON P. S. (1965) Perennation, vegetative propagation and algal life histories, with special reference to *Asparagopsis* and other Rhodophyta. *Bot. Gothoburg.*, **3**, 67–74.

DOTY M. S. (1946) Critical tide factors that are correlated with vertical distribution of marine algae along the Pacific Coast. *Ecology*, **27**, 315–28.

DRUEHL L. D. (1967a) Distribution of two species of *Laminaria* as related to some environmental factors. *J. Phycol.*, **3**, 103–8.

DRUEHL L. D. (1967b) Vertical distribution of some benthic marine algae in a British Columbia inlet, as related to some environmental factors. *J. Fish. Res. Bd. Canada*, **24**, 33–46.

DRUEHL L. D. (1968) Taxonomy and distribution of northeast Pacific species of *Laminaria*. *Can. J. Bot.*, **46**, 539–47.

DRUEHL L. D. (1970) The pattern of Laminariales distribution in the northeast Pacific. *Phycologia*, **9**, 237–47.

DRUEHL L. D. (1978) The distribution of *Macrocystis integrifolia* in British Columbia as related to environmental parameters. *Can. J. Bot.*, **56**, 69–79.

DRUEHL L. D. & HSIAO S. I. C. (1977) Intertidal kelp response to seasonal environmental changes in a British Columbia inlet. *J. Fish. Res. Bd. Canada*, **34**, 1207–11.

FELDMANN J. (1938) Recherches sur la végétation marine de la Méditeranee. La côte des Albères. *Rev. Algol.*, **10**, 1–339.

FELDMANN J. (1957) La reproduction des algues marines dans ses rapports avec leur situation géographique. *Coll. Int. Biol. Mar. Sta. Roscoff, Ann. Biol.*, **33**, 49–56.

FETTER R. (1977) Red algae reproduction via spermatia containing slime strands. *J. Phycol.*, **13** (suppl.), abstr. 466.

FRYE T. C. (1918) The age of *Pterygophora californica*. *Publ. Puget Sound Biol. Sta.*, **2**, 65–71.

GARBARY D. (1976) Life-forms of algae and their distribution. *Bot. Mar.*, **19**, 97–106.

GESSNER F. (1970) Temperature: plants. In *Marine Ecology*, vol. I, part 1, ed. Kinne O., pp. 363–406. John Wiley & Sons, New York.

GESSNER F. & SCHRAMM W. (1971) Salinity: plants. In *Marine Ecology*, vol. I, part 2, ed. Kinne O., pp. 705–820. John Wiley & Sons, New York.

HSIAO S. I. C. (1972) Nutritional requirements for gametogenesis in *Laminaria saccharina* (L.) Lamour. Ph. D. Thesis, Simon Fraser University, Burnaby, British Columbia, Canada.

HSIAO S. I. C. & DRUEHL L. D. (1973) Environmental control of gametogenesis in *Laminaria saccharina*. IV. *In situ* gametogenesis and microsporophyte production. *J. Phycol.*, **9**, 160–4.

HUGHES T. (1975) The case for creation of the North Pacific Ocean during the Mesozoic Era. *Paleogeogr. Paleoclimatol. Paleoecol.*, **18**, 1–43.

HUTCHINS L. W. (1947) The basis for temperature zonation in geographical distributions. *Ecol. Monogr.*, **17**, 325–35.

JÓNSSON S. (1972) Marine benthic algae recorded in Surtsey during the field seasons of 1969 and 1970. *Surtsey Res. Prog. Rep.*, **6**, 75–6.

KAIN J. M. (1964) Aspects of the biology of *Laminaria hyperborea*. III. Survival and growth of gametophytes. *J. Mar. Biol. Ass. U.K.*, **44**, 415–33.

KAIN J. M. (1977) The biology of *Laminaria hyperborea*. X. The effects of death on some populations. *J. Mar. Biol. Ass. U.K.*, **57**, 587–607.

KÜHNEMANN O. (1972) Bosquejo fitogeográfico de la vegetación marina del litoral argentino. *Physis (Buenos Aires)*, **31**, 117–42 and 295–325.

LAWSON G. W. (1978) The distribution of seaweed in the tropical and subtropical Atlantic Ocean: a quantitative approach. *Bot. J. Linn. Soc.*, **76**, 177–93.

MCLACHLAN J., CHEN L. C. M. & EDELSTEIN T. (1969) Distribution and life history of *Bonnemaisonia hamifera* Hariot. *Proc. Int. Seaweed Symp.*, **6**, 245–9.

MICHANEK G. (1979) Phytogeographical provinces and seaweed distribution. *Bot. Mar.*, **22**, 375–91.

MUNDA I. M. (1978) Salinity dependent distribution of benthic algae in estuarine areas of Icelandic fjords. *Bot. Mar.*, **21**, 451–68.

NAKAHARA H. & MASUDA M. (1971) Type of life cycle and geographical distribution of marine green and brown algae in Japan. *Mar. Sci. Mon.*, **3, (III)**, 24–6 (Japanese with English summary).

OLIVEIRA E. C. & QUÉGE N. (1978) O gênero *Laminaria* (Phaeophyta) no Brazil, ocorrência e potencialidade econômica. *Pesquisa o Desenuoluimento*, **1**. Instituto de Pesquisa Tecnólogicas do Estado de São Paulo S/A–IPT.

PIELOU E. C. (1977) The latitudinal spans of seaweed species and their patterns of overlap. *J. Biogeogr.*, **4**, 299–311.

SCAGEL R. F. (1961) Ecology of marine algae. A synthetic approach to some problems in marine algal ecology, In *Recent Advances in Botany*, pp. 175–180. University Toronto Press.

SCAGEL R. F. (1963) Distribution of attached marine algae in relation to oceanographic conditions in the northeast Pacific. In *Marine Distributions*, ed. Dunbar M. J., pp. 37–50. University Toronto Press, Roy. Soc. Canada Spec. Publ. No. 5.

SETCHELL W. A. (1893) On the classification and geographic distribution of the Laminariaceae. *Trans. Conn. Acad.*, **9**, 333–75.

SETCHELL W. A. (1920) Stenothermy and zone-invasion. *Am. Nat.*, **54**, 385–97.

SILVA P. C. (1962) Comparison of algal floristic patterns in the Pacific with those in the Atlantic and Indian Oceans, with special reference to *Codium*. *Proc. 9th Pac. Sci. Congr.*, **4**, 201–16.

SUNDENE O. (1953) The algal vegetation of Oslofjord. *Norske Vidensk. Akad. Oslo, Mat.-Naturv. Klasse 1953*, No. 2.

SUNDENE O. (1962) The implications of transplant and culture experiments on the growth and distribution of *Alaria esculenta*. *Nytt Mag. Bot.*, **9**, 155–74.

TERAMOTO K & KAWAMORI Y. (1977) Changes in marine algal flora related with the improvement of the water quality in Mitajiri Bay, Yamaguchi Prefecture. *Bull. Jap. Soc. Phycol.*, **25**, 210–16.

THOM R. M. & WIDDOWSON T. B. (1977) A resurvey of E. Yale Dawson's 42 intertidal transects on the southern California mainland after 15 years. *Bull. So. Calif. Acad. Sci.*, **77**, 1–13.

VAN DEN HOEK C. (1975) Phytogeographic provinces along the coasts of the northern Atlantic Ocean. *Phycologia*, **14**, 317–30.

YARISH C., EDWARDS P. & CASEY S. (1979) A cultural study of salinity responses in ecotypes of two estuarine red algae. *J. Phycol.*, **15**, 341–6.

CHAPTER 9
LIGHT

K. LÜNING
Biologische Anstalt Helgoland, Palmaille 9, 2000 Hamburg 50,
Federal Republic of Germany

1 INTRODUCTION

Along the margins of the oceans the seaweeds have colonized a zone which reaches from about high-water mark down to a maximum depth of about 200 m. Light quantity, and light quality undergo dramatic changes within this region. For the ecologist and the physiologist dealing with seaweeds, it is of particular interest to obtain detailed knowledge, first about the seasonal course of total and spectral irradiance within the seaweed zone, and second, about the critical light levels for a particular alga. The first objective requires regular light measurements in the sea, and the second detailed analyses of saturating and minimum irradiances for photosynthesis and growth of any particular seaweed species, and where appropriate, lethal irradiances.

In all photosynthetic plants light is absorbed by pigments and used as energy in the process of photosynthesis (Chapter 13). In addition, light is absorbed in many plants by non-photosynthetic pigments and utilized as an environmental signal in the processes of photomorphogenesis. Such processes may synchronize the life cycle of a plant to the seasonal cycle of environmental factors. Our knowledge about photomorphogenetic reactions of seaweeds is still in its infancy compared to what is known about flowering plants. However, there is evidence that such responses exist among seaweeds.

In the past, experimental approaches to these objectives have been largely hampered by the lack of adequate measuring instruments, most phycological laboratories being equipped with at best a light meter calibrated in lux or foot-candles. In recent years the phycologist has been provided with sensitive and convenient electronic light meters, as well as detailed recommendations by physical oceanographers on how to measure light under water.

2 LIGHT MEASUREMENT

Detailed recommendations for the measurement of light under water are given by Jerlov (1968, 1976), Tyler (1973a, 1975, 1977), Tyler & Smith (1966, 1970), and Smith & Tyler (1976). Older reviews of the field by Strickland (1958), Sauberer (1962) and Westlake (1965) should also be consulted, and also reviews on light measurement above water (e.g. Gates 1965; Gaastra 1968; Szeicz 1968). Discussion is restricted in the present chapter to some of the more general and recent ideas which are relevant to the practical work of the phycologist.

2.1 Terminology, units and conversion factors

The spectral range of the solar radiant energy which reaches the surface of the oceans includes the near ultraviolet (290–380 nm), radiation which is visible to the human eye ('light', 380–760 nm), and the infra-red (760–3000 nm). Fifty per cent of the total energy is included in the waveband 300–700 nm (Szeicz 1974), and 45% in the waveband 400–700 nm (Anderson 1967). Photosynthesis of seaweeds can occur at wavelengths as short as 300 nm (e.g. in *Ulva lactuca*, Halldal 1964) and continue to about 700 nm (Govindjee & Braun 1974). If phytochrome is involved in the photomorphogenesis of a plant, one has to consider far-red (700–760 nm) and part of the infra-red (up to 800 nm) as biologically active (Smith 1975). Hence, when one uses the term 'light' in describing photobiological reactions of plants, one has to bear in mind that the plant uses a broader range of wavelengths than the human eye and also has a quite different spectral sensitivity. There can be no justification, therefore, for the plant biologist to use a light meter calibrated in lux or foot-candles and designed for photometric, rather than radiometric measurements (Bickford & Dunn 1972; McCree 1973; Tyler 1973b).

Of the large number of fundamental quantities recommended by the Committee on Radiant Energy in the Sea for use in optical oceanography (see Jerlov 1976) the plant biologist is normally just concerned with the measurement of 'irradiance', i.e. the radiant energy which is incident upon unit area per unit time. According to the Système International d'Unitès (SI) only the following units (or multiples of them) are permitted (see Incoll *et al.* 1977).

$$1\ \mathrm{W\ m^{-2}} = 1\ \mathrm{J\ m^{-2}\ s^{-1}}. \qquad (1)$$

For conversion of 'calories per unit area per unit time', a unit which has been widely used in meteorology, the following equations can be used:

$$1 \text{ cal cm}^{-2} \text{ min}^{-1} = 4.19 \text{ J cm}^{-2} \text{ min}^{-1} = 698 \text{ W m}^{-2} \qquad (2)$$

$$1 \text{ kcal cm}^{-2} \text{ year}^{-1} = 4.19 \text{ kJ cm}^{-2} \text{ year}^{-1} = 41.9 \text{ MJ m}^{-2} \text{ year}^{-1}. \qquad (3)$$

Instead of measuring irradiance in terms of energy one can also record the number of quanta (called 'photons' in the visible range) received per unit area per unit time. This is a logical approach, since the primary act in photochemical reactions depends on the number of quanta absorbed, and not on their energy content (see Chapter 13). Neither term nor unit has been prescribed so far for this type of measurement. Regarding terms one may still choose between 'photon irradiance' (Muray *et al.* 1971), 'quantum irradiance' (Tyler 1975), 'quanta irradiance' (Jerlov 1976), 'photon flux density' (Incoll *et al.* 1977) or 'photon fluence rate' or 'photon flux' (see Hartmann 1977; Mohr & Schopfer 1978). In this paper 'photon flux density' will be used. As to units, the SCOR Working Group 15 on Photosynthetic Radiant Energy (Tyler 1975) has recommended that 'quanta per unit area per sec' be used as the unit in biological oceanography. Many authors now favour the unit 'microeinsteins per m^2 per s', probably because, according to Incoll *et al.* (1977), this unit was used by the designers and manufacturers of one of the first commercially available versions of the quanta meter, who in this way have 'set the standard':

$$1 \ \mu E \text{ m}^{-2} \text{ s}^{-1} = 6.02 \times 10^{17} \text{ quanta m}^{-2} \text{ s}^{-1} = (119.7/\lambda) \text{ W m}^{-2} \qquad (4)$$

where $E = 6.02 \times 10^{23}$ quanta (Avogadro's number), and λ is measured in nm. The factor for conversion of photon flux density into irradiance, as given in equation (4) results from the energetical definition of the quantum:

$$1 \text{ quantum s}^{-1} = h\,c/\lambda = (1987/\lambda)\ 10^{-19} \text{ W} \qquad (5)$$

where $h = 6.62 \times 10^{-34}$ W s^2 (Planck's constant), $c = 3 \times 10^8$ m s^{-1} (velocity of light), and λ is measured in nm. Objections have been raised to the use of 'Einstein (E)' in this context by Incoll *et al.* (1977), because this is not an SI unit. These authors recommend the use of 'mol' instead, which would not cause any re-calibration problems, since

$$1 \ \mu\text{mol m}^{-2} \text{ s}^{-1} = 1 \ \mu E \text{ m}^{-2} \text{ s}^{-1}. \qquad (6)$$

The only disadvantage might then be seen in the fact that the last verbal hint has disappeared to indicate that a quantity having to do with light was being measured.

The commonly used term 'light intensity' is incorrect, since the word 'intensity' is incorporated in the fundamental term 'radiant intensity', and this is concerned with radiant flux emitted by a source (see Jerlov 1976), and not with that received by unit area per unit time (irradiance). Therefore, when specific light measurements are discussed, one should use the term appropriate to the units (irradiance, photon flux density, or, when referring to old measurements performed in lux or foot-candles, illuminance). The term 'light intensity', which is 'deeply entrenched in scientific jargon' (Incoll *et al.* 1977), should only be used as a general phrase, when detailed measurements are not discussed, and even here the term 'light quantity' would be preferable, since it does not clash with the use of the word 'intensity' in physical definitions.

Since units of irradiance ($W\ m^{-2}$) and of photon flux density ($\mu E\ m^{-2}\ s^{-1}$) are used side by side in plant physiology and ecology, it is important to realize the possibilities, and the limitations, of interconverting these two quantities. For monochromatic light (half-band width <20 nm; see Hartmann 1977) conversion is straightforward by using equation (4). For broad-band field conversion it is as accurate to measure these fields stepwise, say in intervals of 10 nm, by means of a spectroradiometer, converting the intervals separately and integrating over the whole spectrum measured.

If for broad-band fields only a rough conversion is needed, appropriate conversion factors, for natural as well as for the commonly used artificial light fields, can be found in the literature (Table 9.1). In the sea the total variation of the ratio Q/E (total quanta/total energy, waveband 400–700 nm) varies by no more than $\pm$ 10%, as was determined by Jerlov & Nygard (1969) and Morel & Smith (1974). According to the latter authors the following equation can be used for conversion:

$$1\ W\ m^{-2} = 2.50 \times 10^{18}\ \text{quanta}\ m^{-2}\ s^{-1} = 4.2\ \mu E\ m^{-2}\ s^{-1}. \tag{7}$$

Table 9.1 Conversion factors from units of irradiance ($1\ W\ m^{-2}$ in each case) to quantum irradiance and illuminance in the waveband 400–700 nm.

Light source	Quantum irradiance ($\mu E\ m^{-2}\ s^{-1}$)	Illuminance (lux)	Reference
Daylight above water surface	4.6	~250	Morel & Smith (1974), McCree (1972)
Daylight below water surface	4.2	—	Morel & Smith (1974)
White fluorescent tubes			
'Daylight 5000 De Luxe'[1]	4.6	~250	This paper
'Cool white'[2]	4.6	~360	McCree (1972)
'Warm white'[3]	4.7	~390	McCree (1972)
Quartz iodine lamp[4]	5.0	~250	McCree (1972)
Tungsten (100 W)	—	~240	Gaastra (1959)

[1] Osram 40 W/19, as measured with spectroradiometer (EG & G Inc., Salem, Mass.).
[2] Sylvania FR96T12-CW-VHO-135.
[3] Westinghouse F96T12/WW/SHO.
[4] Sylvania 250W-T-4Q-CL.

For conversion of yearly integrals the following equation may then be used (1 Megajoule = 10^6 Joule):

$$1 \text{ MJ m}^{-2} \text{ year}^{-1} = 4.2\, E \text{ m}^{-2} \text{ year}^{-1}. \quad (8)$$

Morel & Smith (1974) determined the ratio Q/E for 67 different spectral distributions of irradiance above water (400–700 nm), for sun altitudes ranging from 22–90°, for clear blue skies to completely overcast skies, and found it to be even more constant ($\pm$ 6%):

$$1 \text{ W m}^{-2} = 2.77 \times 10^{18} \text{ quanta m}^{-2} \text{ s}^{-1} = 4.6\, \mu E \text{ m}^{-2} \text{ s}^{-1}. \quad (9)$$

White fluorescent tubes and quartz iodide lamps are nowadays often used for laboratory cultivation of seaweeds. As can be seen from Table 9.1, 4.6–5.0 μE m^{-2} s^{-1} are equivalent to 1 W m^{-2} in these light fields. Although it is to be hoped that no more measurements will be performed in 'lux' in plant biology, one still has to deal with such measurements in the older literature. Rough conversion factors are, therefore, also included in Table 9.1. Finally, a convenient rule-of-thumb conversion factor emerges from these comparisons, for daylight and for several artificial light fields:

$$1 \text{ W m}^{-2} \simeq 5\ \mu E \text{ m}^{-2} \text{ s}^{-1} \simeq 250 \text{ lux}. \quad (10)$$

2.2 Measuring instruments

From the foregoing it is clear that the spectral response of a light meter for biological purposes, at any wavelength to be measured, should be proportional either to the incident energy ('energy meter', 'irradiance meter') or to the number of incident quanta ('quanta meter'), as shown in Fig. 9.1. The limits of spectral sensitivity have to be set to some standardized range. For productivity work in biological oceanography the range 350–700 nm (for photon flux density) has been recommended (Tyler 1975), whereas higher plant ecologists tend to define the range 400–700 nm (again for photon flux density) as 'photosynthetically active radiation' (PAR; McCree 1972; Shibles 1976). Irradiance meters and quanta meters must be equipped with a cosine collecting surface in order to receive flux as the cosine of the incident angle (see Smith 1969; Jerlov 1976), and for underwater measurements the immersion effect, which causes a decrease of the meter's sensitivity in underwater use, must be recognized (see Westlake 1965; Smith 1969).

Different photoelectric detectors (combined with filters correcting for the selective spectral response of the detector) have been used for the construction of quanta meters, for instance the selenium photovoltaic cell in the Danish quanta meter (Jerlov & Nygard 1969) or the silicon photodiode in the lambda quanta meter (manufactured by Li-Cor, Inc., Lincoln, Nebraska; see Biggs *et al.* 1971).

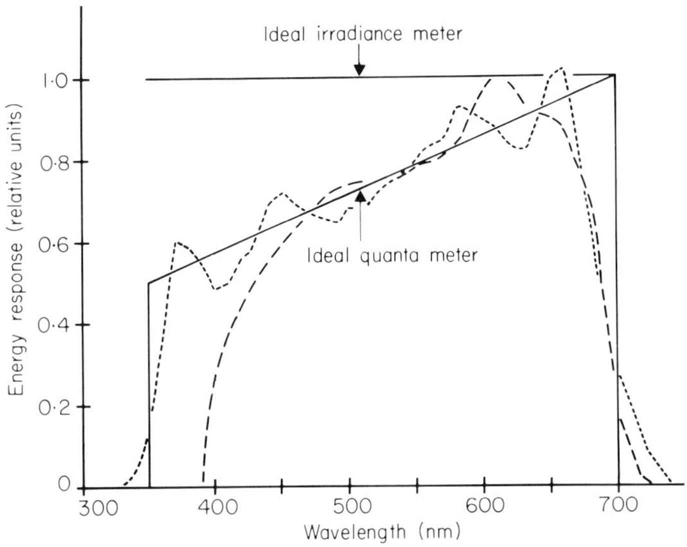

Fig. 9.1 Energy sensitivity of ideal irradiance meter, ideal quanta meter (solid lines), Danish quanta meter (Jerlov & Nygard 1969; dotted line), and of quanta meter designed by Biggs *et al.* (1971; broken line).

The thermopile, a classical 'black receiver', with an equal energy response from the ultraviolet to the middle infra-red (Withrow & Withrow 1956), is appropriate as an energy meter for laboratory work, but it is of little use for underwater measurements due to its low sensitivity, although occasional trials have been made in this direction (e.g. Stephens & Strickland 1962; Platt *et al.* 1970). A simple energy meter for underwater use has been described by Jerlov & Nygard (1969).

Recording spectroradiometers for underwater use, equipped with a silicon photodiode and interference filters, have been described e.g. by Burr & Duncan (1972) and Uphoff & Hergenrader (1976). The construction of the commercially available quanta spectrometer QSM-2500 (Techtum Instrument, Umea, Sweden) was originally initiated by Halldal (see Halldal & Halldal 1973). More sophisticated measuring instruments, e.g. equipped with a double-grating monochromator (Tyler & Smith 1966), are beyond the normal reach of the aquatic ecologist.

3 LIGHT AND ALGAL DISTRIBUTION

Physical oceanographers have described the changes that occur as natural daylight penetrates the ocean, and have provided biologists with a scheme of optical water types (Jerlov 1951, 1976), as well as with fundamental measurements of irradiance, photon flux density and spectral irradiance (e.g.

Tyler & Smith 1967; Morel & Caloumenos 1973; Jerlov 1976). Most of these measurements have been performed in the open ocean, and so the phycologist still has to make his own measurements if he wants information about seasonal light climate in the seaweed zone of most coastal sites. For regions where spectral transmittance remains fairly constant over long periods, the yearly cycle of underwater irradiance may be calculated from meteorological irradiance measurements performed above water and from occasional measurements of spectral transmittance (Weinberg 1976; Weinberg & Cortel-Breeman 1978). In other regions, continuous underwater measurements are required, although these have only rarely been obtained so far (e.g. Kain 1971; Lüning 1971; Chapman *et al.* 1976; Hatcher *et al.* 1977; Lüning & Dring 1979).

In the present chapter the basic facts about the changes of light quality and quantity in the sea are briefly reviewed (for more detailed information, see Jerlov 1976), and the implications of these physical parameters are discussed with particular reference to the conspicuous biological boundaries within the seaweed zone.

3.1 Light quality

The spectral distribution of the global radiation arriving at the ocean's surface consists of two components: direct sunlight, and an indirect component, which may arrive from a completely 'blue' sky, or at the other extreme, from a totally overcast sky (Gates 1965; Gaastra 1968; Szeicz 1974). Spectral distribution curves for daylight vary around a peak at about 560 nm, so that the main differences lie in the proportion of red and blue light. This was pointed out by Morel & Smith (1974) on the basis of 18 measurements performed by Morel & Caloumenos (1973, 1974) over a wide range of solar elevations (37–84°) and irradiances (400–700 nm; 15–486 W m^{-2}) and in conditions ranging from completely open to completely covered skies.

In the sea, light is attenuated due to absorption and scattering. Sea salts have little influence (Morel 1974) and so the absorption of light in pure seawater, as well as in distilled water, is minimal in the blue and maximal in the red region. In coastal seawater, however, shorter wavelengths are absorbed strongly by dissolved organic compounds ('Gelbstoff'), which may either be decay products brought into the sea by rivers, or may be liberated by brown algae (Craigie & McLachlan 1964; Sieburth & Jensen 1968) or phytoplankton (Fogg 1966). Scattering does not in itself attenuate the light, but increases its optical path length and thereby the absorption. Scattering by small particles, such as water molecules, increases towards shorter wavelengths (in proportion to the reciprocal of the fourth power of wavelength), but is not as important as scattering by larger particles in the sea, which is virtually independent of wavelength. However, absorption by particulate matter is again strongest at shorter wavelengths (Jerlov 1976).

The main result in relation to spectral transmittance is that blue and ultraviolet are attenuated much faster with depth in coastal waters (rich in 'yellow substance' and particulate matter) than in oceanic waters. Using common-sense terms, one may, according to Morel & Smith (1974), speak of 'green' (coastal) and 'blue' (oceanic) waters (for examples of underwater spectral distribution see e.g. Jerlov 1976; Tyler & Smith 1970; Morel & Caloumenos 1973, 1974). In consequence, marine waters have been classified by Jerlov (1951) into the oceanic water types I, II, III (IA and IB added by Jerlov 1964) and the coastal water types 1–9 (Fig. 9.2).

In oceanic water type I the maximum transmittance (per m) occurs at 475 nm (if 25 nm-wavebands are considered) and amounts to 98.2%. Even at 310 nm it is still 86%, so that ultraviolet plays a considerable part in clear water. On the other hand, in coastal water type 9 the maximum transmittance (per m) occurs at 575 nm, and is only 56%, whereas transmittance is as low as 5% at 375 nm. Transmittance in the red, at 675 nm ranges from 66% (type I) to 40% (type 9). Infra-red (> 1300 nm) is totally absorbed by the uppermost cm of water (Jerlov 1976), but far-red might still play an ecological role for photomorphogenetic reactions in the upper metres (Dring 1971), since transmittance at 750 nm by distilled water is still as high as $10\%\,m^{-1}$ (see Dring 1971 for references) and down-welling spectral irradiance could still be measured at 740 nm at about 6 m depth in clear water (Smith *et al.* 1973).

The system of water types provides the biologist with a powerful tool for assessing the performance of photobiological reactions in the sea. The

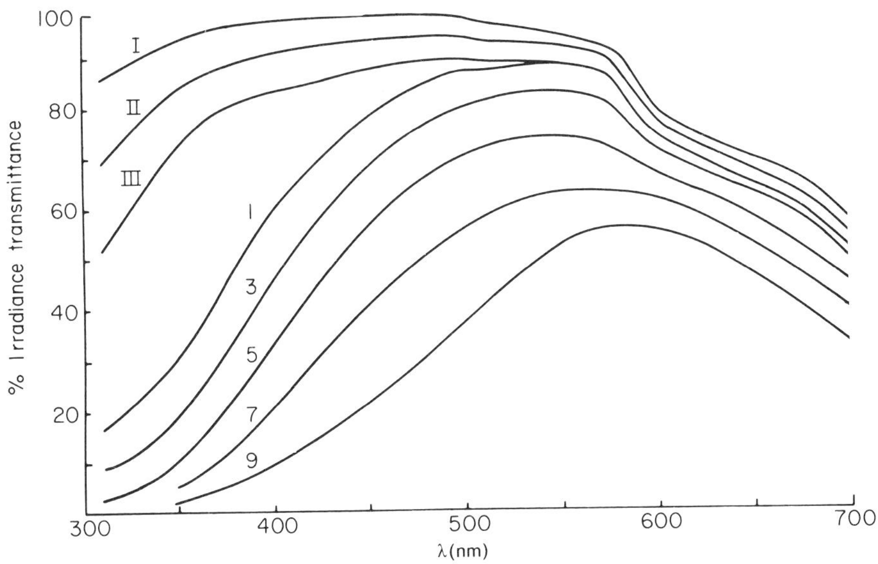

Fig. 9.2 Transmittance per meter of downward irradiance for Jerlov's optical water types (from Jerlov 1976).

application of the system to a particular ecological site is facilitated by the fact that the current water type in an area can be estimated from transmittance in a narrow waveband (preferably about 465 nm, Jerlov 1974) and then the spectral distribution of the light at any depth can be calculated, under certain precautions (see below), from the transmittance data given by Jerlov (1976).

The geographical distribution of different water types has been determined for a number of ocean areas (Jerlov 1976). Type I is represented at many stations measured in the Pacific, and even clearer water than indicated by type I has been found in the Sargasso Sea, whereas type III prevails in upwelling areas. For most of the areas where seaweeds grow, however, phycologists still have to determine the prevailing water type and to assess the seasonal fluctuations, which have been shown to occur e.g. near Helgoland, North Sea, by Lüning & Dring (1979), who also listed some examples of the water types in areas where the seaweed flora has been studied.

The characteristic biological boundaries within the seaweed zone, such as the lower limit of photophilous algae, of the massive brown algae, or the deepest-growing seaweeds have often been largely attributed to the changes in light quality with depth (see Gessner 1955; Levring 1947, 1961, 1966, 1968). Since light quality and light quantity change concurrently in the sea, it is difficult to disentangle their effects. This has recently been pointed out by Dring (1980), who also emphasized that the experimental evidence for the validity of the so-called 'theory of chromatic adaption', which interprets vertical zonation in terms of light quality, is almost nil (see Chapter 13).

3.2 Light quantity

The total solar irradiance which arrives at the surface of the earth's atmosphere has been measured in satellite and rocket experiments as 1400 W m^{-2} (the 'solar constant'; Laue & Drummond 1968), 99% of which falls in the region 200–4000 nm. This irradiance is reduced by absorption and scattering during passage through the atmosphere, and a maximum of about 70% of the solar constant (1000 W m^{-2}) can be measured at the surface of the earth in the tropics, but less than 50% in more cloudy temperate regions (Gates 1962). Of this solar radiation, only 45% is in the range of 400–700 nm (PAR; Anderson 1967). Temperate regions receive about half, polar regions about one third of the maximum values which are recorded for high-pressure tropical regions, where clouds are few (Geiger 1965).

How much of the yearly integral of PAR reaches each of the biological boundaries within the seaweed zone? Although few direct measurements are available, it is possible to estimate the magnitudes to be expected. At Helgoland, a rocky island in the North Sea, the lower limit of algal vegetation (crustose corallines) occurs at 15 m water depth. At this location, the lower algal limit is clearly determined by lack of light, since the rocky substrate extends deeper

(Lüning 1970). Underwater light was measured every 20 min for a whole year (Lüning & Dring 1979), and the yearly integral of irradiance at the lower algal limit was 1.3 MJ m^{-2} $year^{-1}$, corresponding to a photon flux density of 6 E m^{-2} $year^{-1}$ (Table 9.2). The prevailing water type was type 7.

The lower limit of crustose corallines at 120 m depth off Corsica, provides a contrasting example from clear, oceanic waters. Here water type IA prevails (Ivanoff 1957) and hence about 0.1% of surface light penetrates to 120 m (Jerlov 1970, 1976). Since the yearly integral of surface light (400–700 nm) in the Mediterranean is about 3000 MJ m^{-2} $year^{-1}$ (Geiger 1965), the deepest algae receive an irradiance of about 3 MJ m^{-2} $year^{-1}$, or a photon flux density of 12.6 E m^{-2} $year^{-1}$ (using equation (8) for conversion). These values are only double those measured in Helgoland, even though there is an eightfold difference in the depths considered. This result, and those of similar calculations based on the limited data available (see Lüning & Dring 1979 for further examples), suggest that it should be possible to arrive at 'annual quantum requirements' for the major algal zones.

If the lower limit of the perennial benthic algal vegetation (termed 'extinction depth' by Sears & Cooper 1978) can be characterized by an annual input in the order of 10 E m^{-2} $year^{-1}$, the deepest growing perennial kelps (Laminariales) seem to require at least 70 E m^{-2} $year^{-1}$ to build up their thallus (see Table 9.2, 8 m water depth). Like other species of the deeper sublittoral, many kelp species are well-adapted to survive periods of little, even no light, as in subpolar and polar regions, due to their strategy of growth and building-up reserve materials. In contrast, species of the upper sublittoral probably allocate more of the carbon fixed during the year to growth and reproduction than to reserve materials (Lüning 1979). Consequently we may expect the 'annual quantum requirement' of perennial algae inhabiting the upper sublittoral to be higher than that of the deepest kelp species. (For discussion of heterotrophy in seaweeds see Kremer, Chapter 14.)

An alternative approach to the definition of light-determined biological boundaries in the sea is to use the depth at which the irradiance has been reduced to a certain percentage of the irradiance above the surface. For

Table 9.2 Integrals of irradiance and quantum irradiance (both 400–700 nm) at different water depths (below M.L.W.S.) near Helgoland (North Sea), according to Lüning & Dring (1979). Lower limit of algal vegetation at 15 m, of *Laminaria hyperborea* (deepest kelp) at 8 m (Lüning 1970).

Water depth (m)	Irradiance (MJ m^{-2} $year^{-1}$)	Quantum irradiance (E m^{-2} $year^{-1}$)	Percentage of surface irradiance
2	227.2	1037.2	11
4	84.6	387.7	4
8	15.6	71.2	0.7
10	7.3	33.4	0.3
15	1.3	6.0	0.05

instance, the euphotic zone for phytoplankton has generally been defined as waters down to the 1% depth of surface irradiance, which is about 110 m in the clearest water (water type I) or 6 m in coastal water type 9 (Jerlov 1976; Steemann Nielsen 1974, 1975). When the percentage of surface irradiance is calculated for the available records of the deepest multicellular algae from different geographical locations with different prevailing water types, the lower limit of multicellular algae is found to occur at 0.05–0.1% of surface light (see Table 9.2 and, for further examples, Lüning & Dring 1979). This limit is 10–20 times lower than that generally accepted for phytoplankton, and the explanation for this difference probably lies in the greater ability of benthic plants to adjust their metabolism to extremely low irradiances (low dark respiration, allocation of reserves; see Boardman 1977), since they are exposed to a more constant and predictable light climate than phytoplankton. One may infer the light-determined depth of the lower algal limit—in the absence of grazing pressure (see e.g. Foster 1975)—from Fig. 9.3 which is based on a diagram given by Jerlov (1976), but extrapolated to 200 m for oceanic and to 40 m for coastal water types. For example, crustose corallines and green algae were found at 175 m water depth on deep forereefs off Jamaica and the Bahamas (Lang 1974). At this depth one can expect 0.05% of surface irradiance in water type I (see Fig. 9.3). The major limitations of this approach lie in the possible occurrence of stratification and of heterogenous waters, whose transmittance changes with depth. This danger was emphasized by Jerlov (1976), and the curves in Fig. 9.3 should only be used as a rough guide to the magnitudes to be expected. Another limitation, of course, lies in the geographical variation in the yearly integral of global irradiation above the water surface, which leads to a threefold difference in magnitude between the tropics and the poles (Geiger 1965).

As indicated in Fig. 9.3, another conspicuous biological boundary within the seaweed zone, the lower limit of massive brown algae (Laminariales in the Northern Hemisphere), appears to occur at the depth at which surface irradiance has been reduced to 0.7–1.4% (for examples see Lüning & Dring 1979), which is roughly the 1% light limit used to characterize the lower boundary of the euphotic zone for phytoplankton also. In absolute terms, this depth is located at 95 m for *Laminaria rodriguezii* near Corsica (Mediterranean, water type IA; Fredj 1972) or at 25 m for *Laminaria hyperborea* near Roscoff (Atlantic, water type III; Boutler *et al*, 1974) and, also for the latter species, at 8 m depth near Helgoland (North Sea, water type 7; see Table 9.2).

4 CRITICAL LIGHT LEVELS FOR PHOTOSYNTHESIS AND GROWTH

A complete understanding of the ways in which different stages of the life-cycles of any seaweeds are related to various aspects of the light climate in its natural environment will require detailed quantitative analysis of all its responses to

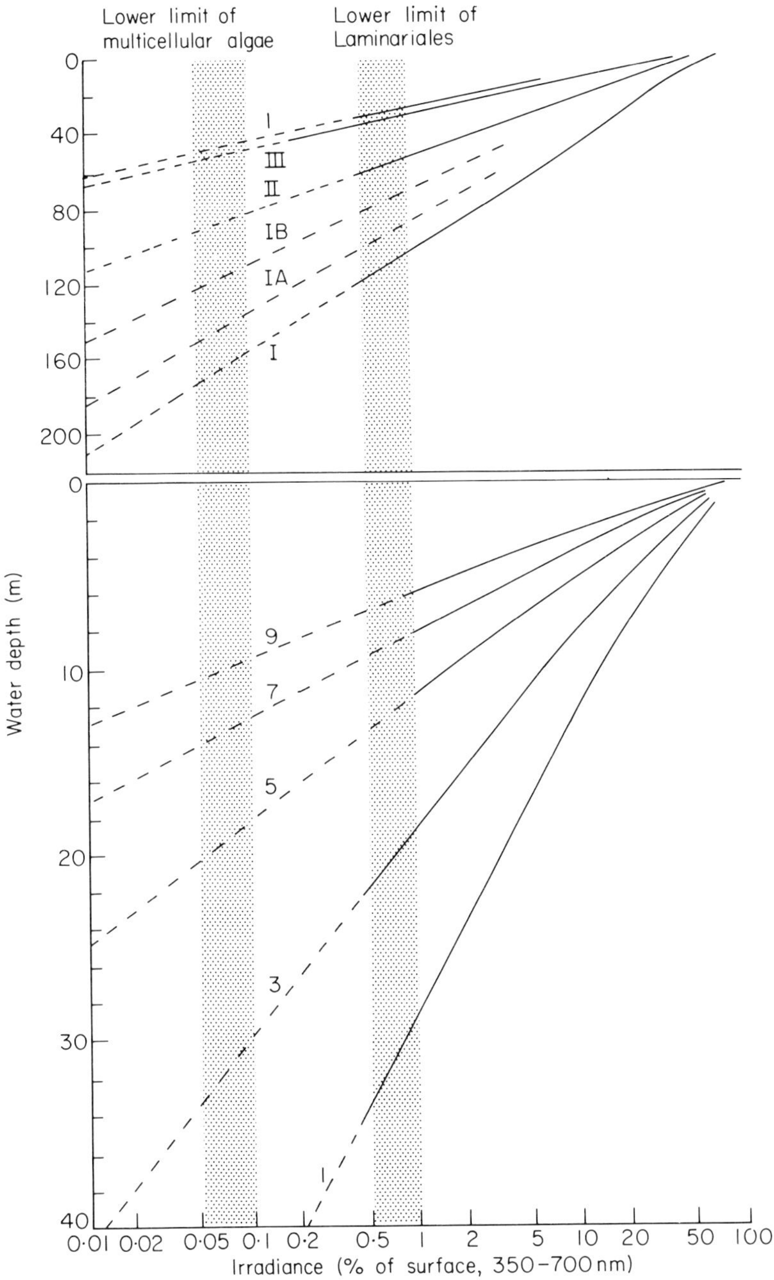

Fig. 9.3 Percentage of surface downward irradiance (350–700 nm) in relation to water depth and Jerlov water type. (Range 100–0.5% (solid lines) from Jerlov (1970, 1976), lower range (dotted lines) obtained by extrapolation from Lüning & Dring 1979).

light. At present there are no species for which we can even approach this goal, and the best that we can do is put values on the maximum and minimum light levels defining the limits within which the major physiological processes can operate. Attempting to do this will at least point out the major gaps in our knowledge and hopefully stimulate seaweed ecologists to undertake more ecophysiological work. This would help not only to close the gap between our knowledge of the ecophysiology of seaweeds and that of phytoplankton and higher plants, but also to enrich our understanding of plant—environment relations as a whole, since seaweeds are subjected to a number of environmental stresses that no other plants experience. Much of the older, largely descriptive literature has been reviewed by Gessner (1955) and Hellebust (1970).

4.1 Inhibiting levels of UV-C and UV-B

It is well-known that electro-magnetic radiation at wavelengths shorter than about 310 nm ('far UV', Halldal 1967; or 'UV-C' at <280 nm and 'UV-B' at 280–315 nm, Caldwell 1971) has damaging effects on all organisms. Nucleic acids absorb maximally at 260 nm, and proteins at 280 nm, and the effects of UV radiation, therefore, result from disruption of DNA replication and of protein synthesis. (Caldwell 1971). Seaweeds are as vulnerable as other organisms, and algae from both the intertidal (e.g. *Ulva lactuca, Ceramium ciliatum*) and the sublittoral (e.g. *Delesseria sanguinea, Brongniartella byssoides*) were killed by exposures of 1–4 min to UV-C (irradiance not stated; Biebl 1952), and similar results were obtained with the epidermal tissue of higher plants.

The shortest wavelengths that reach the earth's surface in solar radiation are at about 290 nm (although somewhat shorter wavelengths have been recorded at high altitudes, Caldwell 1971), so that UV-C does not occur in any natural habitat, although it can be readily obtained in low-pressure mercury vapour radiation sources, with a maximum emission at 253.7 nm ('germicidal lamps'). UV-B (280–315 nm), at least in its long-tail part, does occur in natural habitats, however, and, since clear waters have a high transmittance at 310 nm (86% m^{-1} in water type I; Fig. 9.2), it also occurs in the upper part of the sublittoral (Calkins 1975). Nucleic acids and proteins absorb wavelengths up to 310 nm, and UV-B irradiation of higher plants has resulted in various deleterious effects, such as structural damage to chloroplasts and other organelles (Krizek 1975; Brandle *et al.* 1977; Halldal & Taube 1972). UV-B radiation could, therefore, be involved in the control of the upper limit of seaweeds, and its effects deserve investigation.

4.2 Inhibiting levels of UV-A and light

The plant does not discriminate between light and near UV (310–390 nm, Halldal 1967; or 'UV-A' at 315–400 nm, Caldwell 1971), since the absorption spectra of the major plant pigments (chlorophylls, carotenoids, flavines) cover both regions, and both are used for photochemical work. Under high irradiances, photosensitized oxidations of the excited chlorophyll *a* may occur, and this may contribute to the lethal effects of high light that have been observed in some seaweeds. One to two hour's exposure to about 100 000 lux (400 W m^{-2}) caused extensive cytomorphological damage and death in sublittoral red algae, although intertidal species (*Ulva, Cladophora, Porphyra*) survived 5 h at the same illuminance without any damage being detectable 24 h later (Biebl 1956).

A more detailed study of the lethal effects of light on seaweeds has been carried out by Lüning (1980a) with spores of *Laminaria* spp. Thirty min exposure to December sunlight at Helgoland (340–400 μE m^{-2} s^{-1}) killed over 90% of the spores of all three species (*L. digitata, L. saccharina* and *L. hyperborea*), and laboratory experiments using photon flux densities of 400–2360 μE m^{-2} s^{-1} from a xenon lamp showed that 50% of the spores were killed by a quantum dose of about 0.3 E m^{-2} for *L. digitata* and *L. hyperborea* and of 0.5 E m^{-2} for *L. saccharina*. Clear reciprocity was observed over this range of irradiances (i.e. the response was proportional to the total quantum dose and was independent of the photon flux density and the length of the treatment), but when the spores were exposed to much larger quantum doses over long periods in lower irradiances of fluorescent light (e.g. 100 μE m^{-2} s^{-1} for 1 week = quantum dose of 60 E m^{-2}), almost complete survival was observed. Although the improved survival of spores in fluorescent light may be due, in part, to the practical absence of UV-B in such light compared with sunlight or xenon sources, it seems clear that there must be a threshold photon flux density below which no serious or irreversible damage occurs, and it is only above this value that reciprocity starts to operate. The mechanism of this lethal effect of high light has still to be worked out, but the reduced sensitivity of the spores in blue light compared with red and green light (Lüning 1980a) suggests that carotenoids may have a protective function, and this, in turn, suggests a parallel with the photoinhibition of photosynthesis (Krinsky 1968).

The inhibition of photosynthesis in high irradiances has been extensively reported and studied in phytoplankton (see Harris 1978), but has received relatively little attention from seaweed physiologists and ecologists (see also Chapter 13). Montfort (1929, 1933, 1936) found that the photosynthetic rates of sublittoral seaweeds decreased after shorter exposures to full daylight than did those of intertidal species, but it is difficult to derive reliable absolute values from these early results. The only recent study which has also used natural daylight compared two species of *Laminaria*, and showed that the photosynthetic rate of *L. hyperborea* (deep sublittoral) fell by 50% after 5 h at 250–300 W m^{-2}, whereas that of *L. digitata* (upper sublittoral) was unchanged (Drew 1974). A somewhat confused picture emerges from the results of experiments using high irradiances from artificial light sources (tungsten or tungsten-halogen lamps). King & Schramm (1976) reported that the photosynthesis of *Laminaria saccharina* was inhibited above 30 W m^{-2} in winter, but did not decline even at 200 W m^{-2} in summer, whereas Stocker & Holdheide (1937) had found that the photosynthesis of this species in summer declined after only 4–7 min above 4000 lux (16 W m^{-2}). Other sublittoral species (*Delesseria sanguinea, Phycodrys sinuosa*) were inhibited above 100 W m^{-2} in summer, whereas several intertidal species (e.g. *Enteromorpha intestinalis, Ectocarpus* sp.) showed inhibition at higher irradiances, and *Fucus* spp. showed no inhibition even at 600 W m^{-2} (King & Schramm 1976). It is clear that more work is needed on this topic, and the ecological interpretation of the results would be facilitated if natural light sources were used, or artificial light fields which simulate the spectral distribution of daylight above and below water. Work with phytoplankton species has emphasized that photoinhibition of photosynthesis does not set in immediately the plant is exposed to high irradiances, and that different species can withstand different lengths of exposure before inhibition occurs (Harris 1978). It is important, therefore, to measure photosynthesis after different periods of exposure to high irradiances, and to follow the example of King & Schramm (1976) in measuring the response of a single species in different seasons. It would also be helpful to measure—or exclude—the UV-B (or even UV-C) contained in a light source, so that the effects of these wavelengths can be identified and distinguished from the photoinhibition of photosynthesis by photosynthetically active radiation.

The only investigation of the effects of high irradiances on the growth rate of seaweeds is the recent work by Strömgren (1977), using halogen mercury lamps and a very sensitive laser diffraction method to measure growth. The maximum irradiance used was 130 W m^{-2}, but no inhibitory effects were observed on the growth of any of the five intertidal species of fucoids studied. It is hoped that this technique will be used in further studies of both intertidal and sublittoral species.

4.3 Saturating and compensation light levels for photosynthesis

The irradiances which were found to saturate photosynthesis in various recent investigations of seaweeds are listed in Table 9.3, and a rather clear pattern emerges from these results. *Fucus* and other eulittoral species are light-saturated at about 500 μE m^{-2} s^{-1}, laminarians and other species from the middle sublittoral are saturated by about 200 μE m^{-2} s^{-1}, whereas deep-water red algae require only about 100 μE m^{-2} s^{-1} for saturation. In white light, these values correspond roughly to irradiances of 100, 40 and 20 W m^{-2}, or to illuminances of 25, 10 and 5 klux. These values are similar to those reported as early as 1937 by Stocker & Holdheide. It is not always easy to read off values for the saturating irradiance from P versus I curves, especially for species with thick thalli, since the approach to the saturation plateau is often rather gradual, and, in some species, complete saturation may never be achieved. In *Laminaria* spp., for

Table 9.3 Light levels for saturation of photosynthesis (gross or net; artificial light) in eulittoral and sublittoral seaweeds. Values in brackets have been converted from original data using equation (10). Where ranges were stated in original publication, lower limit of range has been given here.

	Quantum irradiance ($\mu E\ m^{-2}\ s^{-1}$)	Irradiance ($W\ m^{-2}$)	Illuminance (lux)
(A) Eulittoral species			
Fucus vesiculosus[1]	(600)	(120)	30 000
Fucus serratus[2]	500	(100)	(25 000)
Codium fragile[3]	(500)	100	(25 000)
Monostroma nitidum[4]	(500)	(100)	25 000
Gigartina stellata[5]	(460)	(92)	23 000
Porphyra tenera[6]	(400)	(80)	20 000
(B) Upper and medium sublittoral species			
Laminaria spp.[2]	150	(30)	(7 500)
Laminaria hyperborea[7]	(200)	40	(10 000)
Laminaria japonica[8]	(200)	(40)	10 000
Macrocystis pyrifera[9]	(220)	(44)	11 000
Undaria pinnatifida[10]	(200)	40	(10 000)
Chondrus crispus[11]	180	(36)	(9 000)
Palmaria palmata[12]	210	(42)	(10 500)
Phyllophora antarctica[13]	(160)	(32)	8 000
(C) Deep-water red algae			
Delesseria sanguinea[2]	60	(12)	(3 000)
Ptilota serrata[14]	70	(14)	(3 500)
Plocamium telfairiae[15]	(60)	(12)	3 000

[1] Kanwisher 1966
[2] Lüning 1979
[3] Ramus 1978
[4] Yokohama 1973a
[5] Mathieson & Burns 1971
[6] Ogata & Matsui 1965
[7] Kain, Drew & Jupp 1976
[8] Niihara 1975
[9] Clendenning 1971
[10] Kageyama & Yokohama 1974
[11] Mathieson & Norall 1975a
[12] Robbins 1978
[13] Ohno 1976
[14] Mathieson & Norall 1975b (Fig. 1)
[15] Yokohama 1973b

example, this is almost certainly due to the gradual increase in the light penetrating to the shaded side of the thallus (Lüning 1979). Indeed, photoinhibition may occur at the upper side of a unilaterally-illuminated flat thallus long before the photosynthetic layers on the lower side have reached saturation (see also Chapter 13).

Most of the values listed in Table 9.3 were measured after relatively short exposures (i.e. minutes) to high irradiances, and further work is therefore needed to establish whether these irradiances are inhibitory over a longer time period (see Section 4.2). The irradiances at which saturation occurs may also vary according to season (e.g. Kanwisher 1966; Lüning 1971; Mathieson & Norall 1975b; King & Schramm 1976) and the extent of these variations needs to be established in a wider range of species. However, the results listed in Table

9.3 were obtained with a large number of species from a variety of geographical locations, using different techniques, and it seems unlikely that the general magnitudes apparent from these results will be greatly changed by subsequent work, or that any deep sublittoral alga will ever be found that is saturated at the same levels as *Fucus* spp.

Intertidal algae can clearly be regarded as 'sun plants', and the sublittoral species as 'shade plants' since the saturating light levels that have been found correspond quite closely with those reported for higher C-3 plants: 20–25 klux for sun plants, and 4–5, or at most 10, klux for shade plants (Böhning & Burnside 1956; Boardman 1977). Surprisingly, this classification does not fit that for marine phytoplankton (Steemann Nielsen 1975) quite so neatly. Surface phytoplankton are saturated at similar levels (500 $\mu E\ m^{-2}\ s^{-1}$) to intertidal seaweeds, even though the plankton are, strictly speaking, sublittoral; the shade phytoplankton from the bottom of the photic zone (1% of surface green light) correspond to the middle sublittoral benthic plants (saturation at 200 $\mu E\ m^{-2}\ s^{-1}$). There are some recent observations that deep-water phytoplankton may be saturated at irradiances as low as those for deep sublittoral red algae (Shimura & Ichimura 1973; Savidge 1979), but, in general, it seems that phytoplankton do not become as strongly shade-adapted as benthic marine plants, and this agrees with the deeper penetration of benthic plants into the sea (see Section 3).

From the existing data on photosynthetic compensation points in seaweeds not much can be extracted besides magnitudes and a seasonal rise with temperatures (Table 9.4). The range (1.1–2.3 W m^{-2} in winter, equivalent to 300–600 lux; 3–4.5 W m^{-2} in summer, equivalent to 750–1100 lux) compares with the magnitudes known from higher plants (200–500 lux in non-woody shade plants, 1000–2000 lux in non-woody sun plants; Larcher 1975). A clear influence of dark respiration due to differential growth activity is provided by the example of *Laminaria hyperborea* (Table 9.4) in which the old frond respires at only half the rate of the new frond (Lüning 1971).

4.4 Saturating and minimum light levels for growth

The light requirements for growth have been determined less often than those for photosynthesis and, for obvious reasons, most of this work has been done with smaller seaweeds or algal germlings. In general, growth appears to be light-saturated at considerably lower irradiances than photosynthesis. Adult thalli of eulittoral species (e.g. Fucales) are light-saturated at 150–250 $\mu E\ m^{-2}\ s^{-1}$, whereas the more bulky species of the upper sublittoral (*Laminaria, Chondrus, Codium*) are saturated at 30–100 $\mu E\ m^{-2}\ s^{-1}$, and the optimum photon flux density for deep-water red algae seems to be around 10 $\mu E\ m^{-2}\ s^{-1}$ (Table 9.5). The reasons for this discrepancy between the optima for growth and photosynthesis are not precisely known, but irradiances which are below saturation in short-term experiments, may become saturating over the longer

Table 9.4. Photosynthetic compensation irradiance for macroscopic thalli of various seaweeds, as determined in artificial light. W = winter, Sp = spring, S = summer, A = autumn, NS = no season stated, EL = eulittoral, SL = sublittoral. Where necessary, conversions to irradiance were performed using equation (10).

Species	Habitat	Compensation point (W m^{-2})				
		W	SP	S	A	NS
(A) Green algae						
Acrosiphonia centralis[1]	EL		2			
Cladophora glomerata[1]	EL			3		
Enteromorpha intestinalis[1]	EL	1.1		4.5		
Enteromorpha intestinalis[4]	EL					3–4
Chaetomorpha darwinii[5]	SL					0.5–1
(B) Brown algae						
Ectocarpus confervoides[1]	EL		1.4			
Scytosiphon lomentaria[1]	EL		2.5			
Fucus serratus[1]	EL	1.5	1.5		3.5	
Laminaria digitata[1]	SL	1.2	1.6		3.1	
Laminaria hyperborea[2]	SL					
Old frond		1.6	1.3			
New frond		2.3	2.6	3.0	2.6	
(C) Red algae						
Dumontia incrassata[1]	EL	1.7	2.3			
Porphyra leucosticta[1]	EL	1.2	2.3			
Porphyra tenera[3]	EL					1.4
Delesseria sanguinea[1]	SL	1.3	1.3	3.0	3.5	
Phycodrys rubens[1]	SL	1.7	1.3	3.8	3.8	
Furcellaria lumbricalis[1]	SL	1.6	1.4		4.0	
Palmaria palmata[6]	SL					1.3

[1] King & Schramm (1976).
[2] Lüning (1971).
[3] Ogata & Matsui (1965; value for 25° C).
[4] Raven & Smith (1977; value for 15° C).
[5] Raven *et al.* (1979; value for 15–25° C).
[6] Robbins (1978; value for 4–8° C).

term that is required to measure growth. It is becoming clear from work on phytoplankton that photosynthesis is only one of many processes involved in growth, and that quite often the growth rate will be unable to keep up with high photosynthetic rates, since some other process (e.g. availability of nutrients and carbon dioxide, inhibition by high oxygen tensions; Harris 1978) is limiting. In addition, growth depends on a variety of non-photosynthetic but light-dependent processes, which may become inhibited at lower irradiances than those required to saturate photosynthesis.

During the ontogenetic development of many species, the proportion of non-photosynthetic, internal tissues increases, and, therefore, the light requirements of mature plants are often greater than those of sporeling stages. This is well illustrated by the results for *Ascophyllum nodosum* and for *Laminaria* spp. (Table 9.5) since the saturating irradiances for the growth of the adult plants are between five and eight times higher than the corresponding values for the

Table 9.5 Light levels for saturation of growth rates in eulittoral and sublittoral seaweeds, as grown in the laboratory in artificial light. Values in brackets have been converted from original data using equation (10).

	Quantum irradiance ($\mu E\ m^{-2}\ s^{-1}$)	Irradiance ($W\ m^{-2}$)	Illuminance (lux)
(A1) Eulittoral species: adult thalli			
Ascophyllum nodosum, Pelvetia canaliculata[1]	(250)	50	(12 500)
Gigartina stellata[2]	(170)	(34)	>8 600
Fucus edentatus, F. distichus[3]	(150)	30	(7 500)
Fucus spiralis, F. vesiculosus, F. serratus[1]	(150)	30	(7 500)
(A2) Eulittoral species: juvenile stages			
Dumontia incrassata (germlings)[4]	(100)	(20)	5 000
Scytosiphon lomentaria (germlings)[5]	(60)	(12)	3 000
Ascophyllum nodosum (germlings)[6]	(30)	(6)	1 500
(B1) Sublittoral species: adult thalli			
Chondrus crispus[2]	(94)	(19)	4 700
Laminaria saccharina[7]	70	(14)	(3 500)
Codium fragile[8]	28	(6)	(1 400)
Pleonosporium squarrulosum[9]	(11)	(2)	540
(B2) Sublittoral species: juvenile stages			
Laminaria spp. (early sporophytes)[10]	(15)	3	(750)
Laminaria digitata (gametophytes)[11]	(8)	1.5	(375)
Halarachnion ligulatum (germlings)[4]	(10)	(2)	500
Antithamnion plumula (germlings)[12]	(9)	1.7	(425)
Plumaria elegans (germlings)[13]	(5)	1	(250)
Brongniartella byssoides (germlings)[12]	(4)	0.7	(175)

[1] Strömgren 1977
[2] Burns & Mathieson 1972
[3] McLachlan 1974
[4] Jones & Dent 1971
[5] Ohno 1969
[6] Sheader & Moss 1975
[7] Fortes & Lüning (unpublished)
[8] Hanisak 1979
[9] Murray & Dixon 1973
[10] Kain 1969
[11] Cosson 1975
[12] Boney & Corner 1963
[13] Boney & Corner 1962

microscopic stages. Germlings of deep-water red algae achieve their maximum growth rates at irradiances as low as 4–5 $\mu E\ m^{-2}\ s^{-1}$.

The lowest irradiance at which microscopic stages of seaweeds can grow, and this means an irradiance somewhat above the photosynthetic compensation irradiance, has been investigated for gametophytes and young sporophytes of *Laminaria* spp. It was found to be around 0.05 $W\ m^{-2}$ (Kain 1966, 1969). Similar values were obtained by Cosson (1975) for *L. digitata* (0.1 $W\ m^{-2}$), and Burrows (1971) found that young sporophytes of *L. saccharina* could even grow at 0.03 $W\ m^{-2}$.

Our knowledge of the relationship between growth and irradiance can be described, at best, as scanty, but even less is known about growth rates of seaweeds exposed to different spectral regions or to simulated underwater light fields. We cannot hope to obtain a clearer picture of the factors controlling growth in the sea until more experimental work has been done.

5 PHOTOMORPHOGENESIS

The term photomorphogenesis covers all those responses in which light acts as a signal that triggers a change in the pattern of metabolism, and often also in morphological development. The light which the plant uses as an environmental signal is absorbed by sensor pigments which are not involved in photosynthesis. The energy for these reactions does not come from the light signal, so that photomorphogenetic receptors are usually light-saturated at much lower irradiances than photosynthetic receptors. The search for photomorphogenetic responses in seaweeds has started relatively recently (for reviews see Dring 1970, 1971, 1974), and this section will review the progress in studies of the photoperiodic control of seaweed life cycles, and the morphological development in narrow spectral regions.

5.1 Photoperiodic reactions

Hundreds of photoperiodic responses are known for higher plants (Vince-Prue 1975), but for seaweeds it has long been thought that the regular yearly change in sea water temperature might be an adequate seasonal indicator, and that, therefore, marine organisms would not need to resort to photoperiodic responses (Bünning 1964). There is now adequate evidence from all three major groups that photoperiodic responses do exist in seaweeds.

Red algae

The first hints that photoperiodic reactions might trigger the seasonal development of seaweeds came from two Japanese species whose reproduction was restricted to, or enhanced in, the autumn. Photoperiodic control was suggested for both conchosporangia production in the *Conchocelis*-phase of *Porphyra tenera* (Kurogi 1959), and tetrasporangia production in the *Trailliella*-phase of *Bonnemaisonia hamifera* (Chihara 1961). An important criterion for a genuine photoperiodic response is that an interrupted-night regime (e.g. $8{:}\overline{7.5}{:}1{:}\overline{7.5}$) should have the same effect as a long day (e.g. $16{:}\overline{8}$; Vince-Prue 1975). This was shown to be true for the *Conchocelis* phase of *Porphyra tenera* by Dring (1967a, 1967b) and, independently, by Rentschler (1967), and these authors also showed that red light was most effective as a light break, and that the effects of red could be reversed by subsequent exposure to far-red light. Similar evidence was subsequently obtained for the *Conchocelis*-phase of *Bangia fuscopurpurea* (Richardson 1970). The implication of these results is that a pigment similar to phytochrome of higher plants acts as the photoreceptor in these algae. This view seems to be supported by work with crude extracts from *Acrochaetium daviesii* (van der Welde *et al.* 1978). It should be emphasized that the evidence existing so far is too meagre to clearly demonstrate the identity of algal 'phytochrome' with the phytochrome of higher plants. More recently, the response of the *Trailliella*-phase of *Bonnemaisonia hamifera* to short days has also been shown to be inhibited by a light break (Lüning 1980b), but an action spectrum is still lacking for this effect.

As far as can be judged from the experimental data available, the ecological significance of the photoperiodic behaviour of *Porphyra* and *Bonnemaisonia* can be interpreted as follows. The critical daylength, at which more than 50% of the plants form sporangia, is 10–12 h in both the *Conchocelis*-phase of *Porphyra tenera* (Kurogi & Sato 1962; Dring 1970; Rentschler 1967) and the *Trailliella*-phase of *Bonnemaisonia hamifera* (Lüning, unpublished). Very few sporangia are formed at daylengths exceeding 12 h light per day, so that in the original habitat of these plants (Hondo, Japan) induction of sporangium formation cannot occur before October (Iwasaki 1961, Chihara 1961). In both species, sporangium formation can be induced in a short-day regime at 15° C, but not at 10° C or less (Kurogi & Akiyama 1966, Lüning 1980c). Since the temperature in early October in Matsushima Bay is 18° C but falls to 10° C by Mid-November (Iwasaki 1961), sporangium production will be restricted to about six weeks in autumn, and no further reproduction will occur in the spring since the temperature does not reach 10° C until mid-April, when the daylength is 13 h. These combined daylength and temperature requirements may also explain the distribution of *Bonnemaisonia hamifera* in Europe. This species was introduced into Europe near the end of the last century, but only the tetrasporophytic *Trailliella*-phase can be found on Scandinavian shores, whereas on British and French coasts the gametophytic *Bonnemaisonia* phase can also be found (Feldmann 1956). The reason may simply be that the *Trailliella*-phase can never produce sporangia at northern latitudes because the required combination of relatively high temperatures with short daylengths never occurs.

All of the algal species mentioned so far as exhibiting a genuine photoperiodic response are characterized by a heteromorphic life cycle, and other examples are available in this context. In the Mediterranean *Calosiphonia vermicularis* (Gigartinales) the tetrasporophytic *Hymenoclonium* stage also forms sporangia only under short-day conditions, (night-break effective, Mayhoub 1975, 1976). The same was found in the *Falkenbergia* phase of *Asparagopsis armata* (effect of night-break not investigated, Oza 1976), as well as in the crustose tetrasporophyte of *Acrosymphyton purpuriferum* (night-break apparently not effective, Cortel-Breeman & ten Hoopen 1978). The most likely hypothesis to explain this coincidence between photoperiodic behaviour and heteromorphic life cycle may be that one of the two phases can only survive part of the year, so that the initiation of this phase has to be timed very accurately. Detailed physiological analyses, especially on the temperature resistance of the different phases, are required to test the validity of this hypothesis.

There may be different photoperiodic races or ecotypes within a species. West (1972) investigated different strains of *Rhodochorton purpureum*, which all sporulated at daylengths up to 14 h, but not from 16 h onwards, and observed maximum sporulation at 10° C in strains from Alaska and Washington, but 15° C in a Californian strain. More recently these strains have been shown to differ in critical daylength (12 h in Californian strain, 14 h in Alaskan and Washington strains), and in contrast to the earlier results (West 1972) a night-break was found to be effective in this species, provided that it was combined with a 10 h photoperiod (Dring & West, unpublished).

Brown algae

The photoperiodic responses discovered so far among brown algae are restricted to the Scytosiphonaceae, although receptacle formation in *Fucus distichus* might also be under photoperiodic control (Bird & McLachlan 1976). The swarmers of *Scytosiphon lomentaria* form a prostrate system, and this continues to grow as a crust in long days, whereas 10–12 short days will induce 90% of the crusts to form erect thalli (Dring & Lüning 1975a). This species is extremely sensitive to light given in the middle of a long dark period, and 1 min or less will completely inhibit the response to short days. In contrast to the red algae investigated so far and most higher plants, phytochrome is not the photoreceptor for this response, since only blue light and near UV is effective as a night-break. The quantum dose required for inhibition is only 20 $\mu E\,m^{-2}$ of 449 nm light (e.g. 0.33 $\mu E\,m^{-2}\,s^{-1}$ for 1 min) so that the pigment involved—whatever it is—is just as sensitive as phytochrome in higher plants.

Photoperiodic ecotypes also exist in *Scytosiphon lomentaria*. Critical daylength increased with latitude in a series of isolates from 32 to 69° N (Lüning 1980b), and this response is similar to that in higher plants (McMillan 1974). Furthermore, isolates from different latitudes show photoperiodic induction over different temperature ranges. The two extremes were the production of erect thalli in short days at 5–20° C in the isolates from 32–48° N, but only at 5–10° C in an isolate from northern Norway (69° N). These different, genetically fixed patterns mean that *Scytosiphon lomentaria* produces its erect thalli near its southern limit in the winter but at high latitudes in late summer. The ecological significance of this contrast may lie in the inability of the erect thalli to survive high summer temperatures in subtropical regions, and the necessity to wait for the ice to break up, before the development of erect thalli can proceed in subarctic regions. It is clear that different critical daylengths are required to match these conditions, but more investigations are needed to understand the whole complex of factors which synchronize this brown alga with its environment.

A similar response to short days has been found in the related species *Petalonia fascia* and *P. zosterifolia* (Lüning 1980b), but the fact that almost no photoperiodic responses have so far been discovered in brown algae outside the Scytosiphonaceae calls for further investigations. The onset of fertility in the gametophytes of several laminarian species, as well as of *Desmarestia aculeata, D. viridis* and *Chorda filum* is definitely not under photoperiodic control (Lüning 1980c), but it is tempting to speculate from analogy with heteromorphic life cycles among red algae that the reproduction of laminarian sporophytes may well be controlled by photoperiod.

Green algae

Since the green alga *Mougeotia* sp. exhibits phytochrome-controlled chloroplast movements (Haupt 1959) and has long been a favourite object of phytochrome research, green algae might seem to be good candidates for phytochrome-controlled photoperiodic responses similar to those in higher plants. The only genuine response (i.e. night-break effective) so far established, however, is the induction of sporulation by short days in the *Codiolum*-stage of *Monostroma grevillei* (Lüning 1980b), and the nature of the photoreceptor has still to be investigated. Once again this response is associated with a heteromorphic life cycle, and there is a clear incentive to investigate other such species to establish the extent of photoperiodic control in complex algal life histories.

5.2 Other photomorphogenetic reactions

Few seaweed species have so far been cultivated in pure blue, green or red light, and investigations of this type would not only provide urgently needed action spectra for growth, which would be a useful supplement to what we know about action spectra of photosynthesis, but possibly also reveal that certain developmental steps can only be performed in particular spectral regions. In other words, it seems likely that many photomorphogenetic responses in seaweeds are still waiting to be discovered.

The first examples of such responses include the observations that growth ceases completely in *Acetabularia mediterranea* after two to three weeks' cultivation at wavelengths longer than 520 nm (Clauss 1968) and is drastically reduced in *Dictyota dichotoma* under the same conditions (Müller & Clauss 1976). These responses can be regarded as morphological expressions of the complex and poorly-understood regulation of carbon and nitrogen metabolism by light. The balance between protein and carbohydrate synthesis seems to be tilted towards proteins by blue light and towards carbohydrates by red light (Kowallik 1970; Clauss 1972; Andersag & Pirson 1976).

Several distinct morphological effects of blue light have been reported for brown algae. Germlings of *Scytosiphon lomentaria* form hairs and two-dimensional crusts (this effect is similar to induction of two-dimensional prothallia in fern gametophytes) only in blue light, whereas they continue to grow as hairless filaments in green or red light (Dring & Lüning 1975b), and hair formation is also enhanced by blue light in *Dictyota dichotoma* (Müller & Clauss

1976). Blue light produces a different response in the gametophytes of many Laminariales in that egg production is induced, and the quantum dose required for induction increases with temperature (Lüning 1980a). In red and green light, the gametophytes continue to grow vegetatively (Lüning & Dring 1975b; Lüning & Neushul 1978), but in *Dictyota dichotoma*, it has been claimed that it is red light above 600 nm that induces sorus formation (Müller & Clauss 1976). Duncan (1973) discussed the possibility that stipe elongation in *Nereocystis luetkeana* might be triggered by a phytochrome-like system.

The light is apparently used as a signal that conditions are suitable for the morphological development to take place, and usually the signal seems to mean simply that light is present. Since natural light is polychromatic in most habitats, there may be little ecological significance in the fact that different species utilize different spectral regions to trigger similar morphological responses. The exception for this might be in deeper water, where one might expect the absorption properties of the photoreceptor to match the spectral distribution of the prevailing light, and the various blue light-mediated responses found so far among brown algae seem to support this idea.

ACKNOWLEDGEMENTS

I want to thank Dr. M. J. Dring for stimulating discussion and linguistic revision of the manuscript, and also Dr. M. J. Duncan for helpful comments.

6 REFERENCES

Andersag R. & Pirson A. (1976) Verwertung von Glukose in *Chlorella*kulturen bei Blau- und Rotlichtbestrahlung. *Biochem. Physiol. Pflanzen*, **169**, 71–85.

Anderson M. C. (1967) Photon flux, chlorophyll content, and photosynthesis under natural conditions. *Ecology*, **48**, 1050–3.

Bickford E. & Dunn S. (1972) *Lighting for Plant Growth*. Kent State University Press, Kent, Ohio.

Biebl R. (1952) Resistenz der Meeresalgen gegen sichtbares Licht und gegen kurzwellige UV-Strahlen. *Protoplasma*, **41**, 353–77.

Biebl R. (1956) Lichtresistenz von Meeresalgen. *Protoplasma*, **46**, 63–89.

Biggs W. W., Edison A. R., Eastin J. D., Brown K. W., Maranville J. W. & Clegg M. D. (1971) Photosynthesis light sensor and meter. *Ecology*, **52**, 125–31.

Bird N. L. & McLachlan J. (1976) Control of formation of receptacles in *Fucus distichus* L. subsp. *distichus* (Phaeophyceae, Fucales). *Phycologia*, **15**, 79–84.

Boardman N. K. (1977) Comparative photosynthesis of sun and shade plants. *Ann. Rev. Plant Physiol.*, **28**, 355–77.

Böhning R. H. & Burnside C. A. (1956) The effect of light intensity on rate of apparent photosynthesis in leaves of sun- and shade plants. *Am. J. Bot.*, **43**, 557–61.

Boney A. D. & Corner D. S. (1962) The effect of light on the growth of sporelings of the intertidal red alga *Plumaria elegans* (Bonnem.) Schm. *J. Mar. Biol. Ass. U.K.*, **42**, 65–92.

Boney A. D. & Corner D. S. (1963) The effect of light on the growth of sporelings of the red algae *Antithamnion plumula* and *Brongniartella byssoides*. *J. Mar. Biol. Ass. U.K.*, **43**, 319–25.

Boutler J., Cabioch L. & Grall J.-R. (1974) Quelques observations sur la pénétration de la lumière dans les eaux marines au voisinage de Roscoff et ses conséquences écologiques. *Bull. Soc. Phycol. Fr.*, **19**, 129–40.

BRANDLE J. R., CAMPBELL W. F., SISSON W. B. & CALDWELL M. M. (1977) Net photosynthesis, electron transport capacity, and ultrastructure of *Pisum sativum* exposed to ultraviolet-B radiation. *Plant. Physiol.*, **60**, 165–9.

BÜNNING E. (1964) *The Physiological Clock*. Springer-Verlag, Berlin.

BURNS R. L. & MATHIESON A. C. (1972) Ecological studies of economic red algae. II. Culture studies of *Chondrus crispus* (Stackhouse) Batters. *J. Exp. Mar. Biol. Ecol.*, **8**, 1–6.

BURR A. H. & DUNCAN M. J. (1972) Portable spectroradiometer for underwater environments. *Limnol. Oceanogr.*, **17**, 466–74.

BURROWS E. M. (1971) Assessment of pollution effects by the use of algae. *Proc. Roy. Soc. Lond. B*, **177**, 295–306.

CALDWELL M. M. (1971) Solar UV irradiation and the growth and development of higher plants. In *Photophysiology* **6**, ed. Giese A. C., pp. 131–77. Academic Press, New York and London.

CALKINS J. (1975) Measurements of the penetration of solar UV-B into various natural waters. In *Climatic Impact Assessment Program, Monograph* **5**, ed. U.S. Dept. of Transportation, var. pp. Washington, D.C.

CHAPMAN A. R. O., HOYT B. & PATON B. E. (1976) A self-contained instrument for logarithmic recording of submarine quantum irradiance. *Mar. Biol.*, **38**, 91–4.

CHIHARA M. (1961) Life cycle of the Bonnemaisoniaceous algae in Japan (1). *Science reports of Tokyo Kyoiku Daigaku*, **10**, 121–54.

CLAUSS H. (1968) Beeinflussung der Morphogenese, Substanzproduktion und Proteinvermehrung von *Acetabularia mediterranea* durch sichtbare Strahlung. *Protoplasma*, **65**, 49–80.

CLAUSS H. (1972) Der Einfluß von Rot- und Blaulicht auf die Photosynthese von *Acetabularia mediterranea* und auf die Verteilung des assimilierten Kohlenstoffs. *Protoplasma*, **74**, 357–79.

CLENDENNING K. A. (1971) Photosynthesis and general development in *Macrocystis*. *Nova Hedwigia (Beih.)*, **32**, 169–90.

CORTEL-BREEMAN A. M. & TEN HOOPEN A. (1978) The short day response in *Acrosymphyton purpuriferum* (J.Ag.) Sjöst. (Rhodophyceae, Crytonemiales). *Phycologia*, **17**, 125–32.

COSSON J. (1975) Action des conditions d'éclairement sur la croissance des gamétophytes de *Laminaria digitata* (L.) Lamouroux (Phéophycée, Laminariale). *Bull. Soc. Phycol. France*, **20**, 50–4.

CRAIGIE J. S. & MCLACHLAN J. (1964) Excretion of coloured ultraviolet-absorbing substances by marine algae. *Can. J. Bot.*, **42**, 23–33.

DREW E. A. (1974) Light inhibition of photosynthesis in macro-algae. *Br. Phyc. J.*, **9**, 217–18.

DRING M. J. (1967a) Effects of daylength on growth and reproduction of the *Conchocelis*-phase of *Porphyra tenera*. *J. Mar. Biol. Ass. U.K.*, **47**, 501–10.

DRING M. J. (1967b) Phytochrome in red alga, *Porphyra tenera*., *Nature (London)*, **215**, 1411–12.

DRING M. J. (1970) Photoperiodic effects in microorganisms. In *Photobiology of Microorganisms*, ed. Halldal P., pp. 345–68. Wiley & Sons, London.

DRING M. J. (1971) Light quality and the photomorphogenesis of algae in marine environments. In *Fourth European Marine Biology Symposium*, ed. Crisp D. J., pp. 375–92. University Press, Cambridge.

DRING M. J. (1974) Reproduction. In *Algal Physiology and Biochemistry*, ed. Stewart W. D. P., pp. 814–37. Blackwell Scientific Publications, Oxford.

DRING M. J. (1981) Chromatic adaptation of photosynthesis in benthic marine algae: an examination of its ecological significance using a theoretical model. *Limn. Oceanogr.*, **26**, 271–84.

DRING M. J. & LÜNING K. (1975a) A photoperiodic response mediated by blue light in the brown alga *Scytosiphon lomentaria*. *Planta*, **125**, 25–32.

DRING M. J. & LÜNING K. (1975b) Induction of two-dimensional growth and hair formation by blue light in the brown alga *Scytosiphon lomentaria*. *Z. Pflanzenphysiol.*, **75**, 107–17.

DUNCAN M. J. (1973) *In situ* studies of growth and pigmentation of the phaeophycean *Nereocystis leutkeana*. *Helgoländer Wiss. Meeresunters.*, **24**, 510–25.

FELDMANN J. (1956) La reproduction des algues marines dans ses rapports avec leur situation géographique. *Année Biologique*, **33**, 49–56.

FOGG G. E. (1966) The extracellular products of algae. *Oceanogr. Mar. Biol. Ann. Rev.*, **4**, 195–212.

FOSTER M. S. (1975) Regulation of algal community development in a *Macrocystis pyrifera* forest. *Mar. Biol.*, **32**, 331–42.

FREDJ G. (1972) Compte rendu de plongée en SP 300 sur les fonds à *Laminaria rodriguezii* Bornet de la Pointe de Revellata (Corse). *Bull. Inst. océan., Monaco*, **71** (1421), 1–42.

GAASTRA P. (1959) Photosynthesis of crop plants as influenced by light, carbon dioxide, temperature and stomatal diffusion resistance. *Meded. Landbouwhogeschool Wageningen*, **59**, 1–68.

GAASTRA P. (1968) Radiation measurements for investigations of photosynthesis under natural conditions. In *Functioning of Terrestrial Ecosystems at the Primary Production Level*, ed. Eckardt F. E., pp. 467–78. Unesco, Paris.

GATES D. M. (1962) *Energy Exchange in the Biosphere*. Harper and Row, New York.

GATES D. M. (1965) Radiant energy, its receipt and disposal. *Meterological Monographs*, **6**, 1–26.

GEIGER R. (1965) *Die Atmosphäre der Erde*. Perthes, Darmstadt.

GESSNER F. (1955) *Hydrobotanik. Die physiologischen Grundlagen der Pflanzenverbreitung im Wasser. I. Energiehaushalt*. VEB Deutscher Verlag der Wissenschaften, Berlin.

GOVINDJEE & BRAUN B. Z. (1974) Light absorption, emission and photosynthesis. In *Algal Physiology and Biochemistry*, ed. Stewart W. D. P., pp. 346–90. Blackwell Scientific Publications, Oxford.

HALLDAL P. (1964) Ultraviolet action spectra of photosynthesis and photosynthetic inhibition in a green and a red alga. *Physiol. Plant.*, **17**, 414–21.

HALLDAL P. (1967) Ultraviolet action spectra in algology: a review. *Photochem. Photobiol.*, **6**, 445–60.

HALLDAL P. & HALLDAL K. (1973) Phytoplankton, chlorophyll, and submarine light conditions in Kings Bay, Spitsbergen, July 1971. *Norw. J. Bot.*, **20**, 99–108.

HALLDAL P. & TAUBE O. (1972) Ultraviolet action and photoreactivation in algae. In *Photophysiology* **7**, ed. Giese A. C., pp. 163–88. Academic Press, New York and London.

HANISAK M. D. (1979) Growth patterns of *Codium fragile* spp. *tomentosoides* in response to temperature, irradiance, salinity, and nitrogen source. *Mar. Biol.*, **50**, 319–32.

HARRIS G. P. (1978) Photosynthesis, productivity and growth: the physiological ecology of phytoplankton. *Arch. Hydrobiol. Beih. Ergebn. Limnol.*, **10**, 1–171.

HARTMANN K. M. (1977) Aktionsspektrometrie. In *Biophysik*, eds. Hoppe W., Lohmann, W., Markl H. & Ziegler H., pp. 197–222. Springer-Verlag, Berlin, Heidelberg, New York.

HATCHER B. G., CHAPMAN A. R. O. & MANN K. H. (1977) An annual carbon budget for the kelp *Laminaria longicruris*. *Mar. Biol.*, **44**, 85–96.

HAUPT W. (1959) Die Chloroplastendrehung bei *Mougeotia*. *Planta* , **53**, 484–501.

HELLEBUST J. A. (1970) Light. Plants. In *Marine Ecology. Vol. I. Environmental Factors. Part 1*, ed. Kinne O., pp. 125–58. Wiley-Interscience, London.

INCOLL L. D., LONG S. P. & ASHMORE M. R. (1977) SI units in publications in plant science. *Curr. Adv. Plant Sci.*, **28**, 331–43.

IVANOFF A. (1957) Contribution à l'étude des propriétés optiques de l'eau de mer en Bretagne et en Corse, et la théorie de la polarisation sousmarine. *Ann. Géophys.*, **13**, 22–53.

IWASAKI H. (1961) The life-cycle of *Porphyra tenera in vitro*. *Biol. Bull. Mar. Biol. Lab., Woods Hole*, **121**, 173–87.

JERLOV N. G. (1951) Optical studies of ocean waters. *Rep. Swedish Deep-Sea Exped.*, **3**, 1–59.

JERLOV N. G. (1964) Optical classification of ocean water. In *Physical Aspects of Light in the Sea*, ed. Tyler J. E., pp. 45–9. University of Hawaii Press, Honolulu.

JERLOV N. G. (1968) *Optical Oceanography*. Elsevier Publishing Company, Amsterdam.

JERLOV N. G. (1970) Light. General introduction. In *Marine Ecology. Vol. I. Environmental Factors. Part 1*, ed. Kinne O., pp. 95–102. Wiley-Interscience, London.

JERLOV N. G. (1974) A simple method for measuring quanta irradiance in the ocean. *Rep. Kjøb. Univ. Inst. Fys. Oceanogr.*, **24**, 1–7.

JERLOV N. G. (1976) *Marine Optics*. Elsevier Scientific Publishing Company, Amsterdam.

JERLOV N. G. & NYGÅRD K. (1969) A quanta and energy meter for photosynthetic studies. *Rep. Kjob. Univ. Inst. Fys. Oceanogr.*, **10**, 1–19.

JONES W. E. & DENT E. S. (1971) The effect of light on the growth of algal spores. In *Fourth European Marine Biology Symposium*, ed. Crisp D. J., pp. 363–74. Cambridge University Press.

KAGEYAMA A. & YOKOHAMA Y. (1974) Photosynthetic properties of marine benthic brown algae from different depths in coastal areas. *Bull. Jap. Soc. Phycol.*, **22**, 119–23.

KAIN J. M. (1966) The role of light in the ecology of *Laminaria hyperborea*. In *Light as an Ecological Factor*, eds. Bainbridge R., Evans G. C. & Rackham O., pp. 319–34. Blackwell Scientific Publications, Oxford.

KAIN J. M. (1969) The biology of *Laminaria hyperborea*. V. Comparison with early stages of competitors. *J. Mar. Biol. Ass. U.K.* **49**, 455–73.

KAIN J. M. (1971) Continuous recording of underwater light in relation to *Laminaria* distribution. In *Fourth European Marine Biology Symposium*, ed. Crisp D. J., pp. 335–46. Cambridge University Press, Cambridge.

KAIN J. M., DREW E. A. & JUPP B. P. (1976) Light and the ecology of *Laminaria hyperborea* II. In *Light as an Ecological Factor: II.*, eds. Evans G. C., Bainbridge R. & Rackham O, pp. 63–92. Blackwell Scientific Publications, Oxford.

KANWISHER J. W. (1966) Photosynthesis and respiration in some seaweeds. In *Some Contemporary Studies in Marine Sciences*, ed. Barnes H., pp. 407–20. Allen & Unwin, London.

KING R. J. & SCHRAMM W. (1976) Photosynthetic rates of benthic marine algae in relation to light intensity and seasonal variation. *Mar. Biol.*, **37**, 215–22.

KOWALLIK W. (1970) Light effects on carbohydrate and protein metabolism in algae. In *Photobiology of Microorganisms*, ed. Halldal P. pp. 165–85. John Wiley & Sons, London.

KRINSKY N. I. (1968) The protective function of carotenoid pigments. In *Photophysiology. Current topics. III.*, ed. Giese A. C., pp. 123–95. Academic Press, New York and London.

KRIZEK D. T. (1975) Influence of ultraviolet radiation on germination and early seedling growth. *Physiol. Plant.*, **34**, 182–6.

KUROGI M. (1959) Influences of light on the growth and maturation of *Conchocelis*-thallus of *Porphyra*. I. Effect of photoperiod on the formation of monosporangia and liberation of monospores (1.). *Bull. Tohoku Reg. Fish. Res. Lab.*, **15**, 33–42.

KUROGI M. & AKIYAMA K. (1966) Effects of water temperature on the growth and maturation of *Conchocelis*-thalli in several species of *Porphyra*. *Bull. Tohoku Reg. Fish. Res. Lab.*, **26**, 77–89.

KUROGI M. & SATO S. (1962) Influences of light on the growth and maturation of *Conchocelis*-thallus of *Porphyra*. II. Effect of different photoperiods on the growth and maturation of *Conchocelis*-thallus of *P. tenera* Kjellman. *Bull. Tohoku Reg. Fish. Res. Lab.*, **20**, 127–37.

LANG J. C. (1974) Biological zonation at the base of a reef. *Am. Sci.* **62**, 271–81.

LARCHER, W. (1975) *Physiological Plant Ecology*. Springer-Verlag, New York, U.S.A.

LAUE E. G. & DRUMMOND A. J. (1968) Solar constant. First direct measurements. *Science, N.Y.*, **161**, 888.

LEVRING T. (1947) Submarine daylight and the photosynthesis of marine algae. *Göteborg K. Vet.-och Vitt. Samh. Handl., Sjätte följden Ser. B.*, **5**, 1–89.

LEVRING T. (1961) Submarine illumination and vertical distribution of algal vegetation. *Rec. Adv. Bot.*, **1**, 189–93.

LEVRING T. (1966) Submarine light and algal shore zonation. In *Light as an Ecological Factor*, eds. Bainbridge R., Evans G. C. & Rackham O., pp. 305–19. Blackwell Scientific Publications, Oxford.

LEVRING T. (1968) Photosynthesis of some marine algae in clear, tropical oceanic water. *Bot. Mar.*, **11**, 72–80.

LÜNING K. (1970) Tauchuntersuchungen zur Vertikalverteilung der sublitoralen Helgoländer Algenvegetation. *Helgoländer Wiss. Meeresunters.*, **21**, 271–91.

LÜNING K. (1971) Seasonal growth of *Laminaria hyperborea* under recorded underwater light conditions near Helgoland. In *Fourth European Marine Biology Symposium*, ed. Crisp D. J., pp. 347–61. Cambridge University Press, Cambridge.

LÜNING K. (1979) Growth strategies of three *Laminaria* species (Phaeophyceae) inhabiting different depth zones in the sublittoral region of Helgoland (North Sea). *Mar. Ecol. Progr. Ser.*, **1**, 195–207.

LÜNING K. (1980a) Critical levels of light and temperature regulating the gametogenesis of three *Laminaria* spp. (Phaeophyceae). *J. Phycol.* **16**, 1–15.

LÜNING K. (1980b) Control of algal life history by daylength and temperature. In *The Shore Environment: Method and Ecosystems*, eds. Price J. H., Irvine D. E. G. & Farnham W. F., pp. 915–45. Academic Press, London, New York.

LÜNING K. & DRING M. J. (1975a) A photoperiodic response mediated by blue light in the brown alga *Scytosiphon lomentaria*. *Planta*, **125**, 25–32.

LÜNING K. & DRING M. J. (1975b) Reproduction, growth and photosynthesis of gametophytes of *Laminaria saccharina* grown in blue and red light. *Mar. Biol.*, **29**, 195–200.

LÜNING K. & DRING M. J. (1979) Continuous underwater light measurement near Helgoland (North Sea) and its significance for characteristic light limits in the sublittoral region. *Helgoländer Wiss. Meeresunters.*, **32**, 403–24.

LÜNING K. & NEUSHUL M. (1978) Light and temperature demands for growth and reproduction of laminarian gametophytes in Southern and Central California. *Mar.*, Biol., **45**, 297–309.

MCCREE K. J. (1972) Test of current definitions of photosynthetically active radiation against leaf photosynthesis data. *Agric. Meteorol.*, **10**, 443–53.

MCCREE K. J. (1973) A rational approach to light measurements in plant ecology. *Curr. Adv. Plant Sci.*, **5**, 39–43.

MCLACHLAN J. (1974) Effects of temperature and light on growth and development of embryos of *Fucus edentatus* and *F. distichus* ssp. *distichus*. *Can. J. Bot.*, **52**, 943–51.

MCMILLAN C. (1974) Photoperiodic responses of *Xanthium strumarium* L. (Compositae) introduced and indigenous in Eastern Asia. *Am. J. Bot.*, **57**, 881–8.

MATHIESON A. C. & BURNS R. L. (1971) Ecological studies of economic red algae. I. Photosynthesis and respiration of *Chondrus crispus* Stackhouse and *Gigartina stellata* (Stackhouse) Batters. *J. Exp. Mar. Biol. Ecol.*, **7**, 197–206.

MATHIESON A. C. & NORALL T. L. (1975a) Photosynthetic studies of *Chondrus crispus*. *Mar. Biol.*, **33**, 207–13.

MATHIESON A. C. & NORALL T. L. (1975b) Physiological studies of subtidal red algae. *J. Exp. Mar. Biol. Ecol.*, **20**, 237–47.

MAYHOUB H. (1975) Nouvelles observations sur le cycle de développement du *Calosiphonia vermicularis* (J.Ag.) Sch. (Rhodophycée, Gigartinale). *C.R. Acad. Sci. Paris*, **280**, 2441–3.

MAYHOUB H. (1976) Cycle de développement du *Calosiphonia vermicularis* (J.Ag.) Sch. (Rhodophycées, Gigartinales). Mise en évidence d'une réponse photopériodique. *Bull. Soc. Phycol. Fr.*, **21**, 48.

MOHR H. & SCHOPFER P. (1978) *Lehrbuch der Pflanzenphysiologie*. Springer-Verlag, Berlin, Heidelberg, New York.

MONTFORT C. (1929) Die funktionelle Einstellung verschieden gefärbter Meeresalgen auf die Lichtintensität. Studien zur vergleichenden Ökologie der Assimilation. II. *Jahrbücher Wiss. Bot.*, **71**, 106–48.

MONTFORT C. (1933) Über Lichtempfindlichkeit und Leistungen roter Tiefseealgen und Grottenflorideen an freier Meeresoberfläche. *Protoplasma*, **19**, 385–413.

MONTFORT C. (1936) Umwelt, Erbgut und physiologische Gestalt. *Jahrbücher Wiss. Bot.*, **84**, 1–57.

MOREL A. (1974) Optical properties of pure water and pure sea water. In *Optical Aspects of Oceanography*, eds. Jerlov N. G. & Steemann Nielsen E., pp. 2–24. Academic Press, London.

MOREL A. & CALOUMENOS L. (1973) Mesures d'éclairements sous marins, flux de photons et analyse spectrale. *Centre Rech. Oceanogr. Ville-franche-Sur-Mer Rap.*, **11**, var. pag.

MOREL A. & CALOUMENOS L. (1974) Variabilité de la répartition spectrale de l'énergie photosynthétique. *Tethys*, 93–104.

MOREL A. & SMITH R. C. (1974) Relation between total quanta and total energy for aquatic photosynthesis. *Limnol. Oceanogr.*, **19**, 591–600.

MÜLLER S. & CLAUSS H. (1976) Aspects of photomorphogenesis in the brown alga *Dictyota dichotoma*. *Z. Pflanzenphys.*, **78**, 461–5.

MURRAY J. J., NICODEMUS F. E. & WUNDERMANN I. (1971) Proposed supplement to the SI nomenclature for radiometry and photometry. *Appl. Opt.*, **10**, 1465–8.

MURRAY S. N. & DIXON P. S. (1973) The effect of light intensity and light period on the development of thallus form in the marine red alga *Plenosporium squarrulosum* (Harvey) Abbott (Rhodophyta: Ceramiales). I. Apical cell division—main axes. *J. Exp. Mar. Biol. Ecol.*, **13**, 15–27.
NIIHARA Y. (1975) Physiological studies of *Laminaria japonica* var. *ochotensis*. The effect of temperature, light intensity and salinity upon photosynthesis and respiration of young sporophytes. *Sci. Rep. Hokkaido Fish. Expn. Stn.*, **17**, 11–17.
OGATA E. & MATSUI T. (1965) Photosynthesis in several marine plants of Japan in relation to carbon dioxide supply, light and inhibitors. *Jap. J. Bot.*, **19**, 83–98.
OHNO M. (1969) A physiological ecology of the early stage of some marine algae. *Rep. USA Mar. Biol. Stn.*, **16**, 1–46.
OHNO M. (1976) A preliminary report on the photosynthetic activity of *Phyllophora antarctica* A. et E. S. Gepp, Antarctic red alga. *Antarctic Record*, **57**, 141–5.
OZA R. M. (1976) Culture studies on induction of tetraspores and their subsequent development in the red alga *Falkenbergia rufolanosa* Schmitz. *Bot. Mar.*, **20**, 29–32.
PLATT T., LARSEN E. & VINE R. (1970) Integrating radiometer: a self-contained device for measurements of submarine light energy in absolute units. *J. Fish. Res. Bd. Canada*, **27**, 185–91.
RAMUS J. (1978) Seaweed anatomy and photosynthetic performance: the ecological significance of light guides, heterogenous absorption and multiple scatter. *J. Phycol.*, **14**, 352–62.
RAVEN J. A. & SMITH F. A. (1977) 'Sun' and 'shade' species of green algae: relation to cell size and environment. *Photosynthetica*, **11**, 48–55.
RAVEN J. A., SMITH F. A..& GLIDEWELL S. M. (1979) Photosynthetic capacities and biological strategies of giant-celled and small-celled macro-algae. *New Phytol.*, **83**, 299–310.
RENTSCHLER H. G. (1967) Photoperiodische Induktion der Monosporenbildung bei *Porphyra tenera* Kjellm. (Rhodophyta-Bangiophyceae). *Planta*, **76**, 65–74.
RICHARDSON N. (1970) Studies on the photobiology of *Bangia fuscopurpurea*. *J. Phycol.*, **6**, 215–19.
ROBBINS J. V. (1978) Effects of physical and chemical factors on photosynthetic and respiratory rates of *Palmaria palmata* (Florideophyceae). *Proc. Int. Seaweed Symp.*, **9**, 273–83.
SAUBERER F. (1962) Empfehlungen für die Durchführung von Strahlungsmessungen an und in Gewässern. *Int. Verein. Theoret. Angew. Limnol.*, **11**, 1–77.
SAVIDGE G. (1979) Photosynthetic characteristics of marine phytoplankton from contrasting physical environments. *Mar. Biol.*, **53**, 1–12.
SEARS J. R. & COOPER R. A. (1978) Descriptive ecology of offshore, deepwater, benthic algae in the temperate Western North Atlantic Ocean. *Mar. Biol.*, **44**, 309–314.
SHEADER A. & MOSS B. (1975) Effects of light and temperature on germination and growth of *Ascophyllum nodosum* (L.) Lejol. *Estuarine Coastal Mar. Sci.*, **3**, 125–32.
SHIBLES R. (1976) Terminology pertaining to photosynthesis. *Crop Science*, **16**, 437–9.
SHIMURA S. & ICHIMURA S. (1973) Selective transmission of light in the ocean waters and its relation to phytosynthesis. *J. Oceanogr. Soc. Japan*, **29**, 257–66.
SIEBURTH J. McN. & JENSEN A. (1968) Studies on algal substances in the sea. II. The formation of Gelbstoff (humic material) by exudates of Phaeophyta. *J. Exp. Mar. Biol. Ecol.*, **3**, 275–89.
SMITH H. (1975) *Phytochrome and Photomorphogenesis*. McGraw Hill, London.
SMITH R. C. (1969) An underwater spectral irradiance collector. *J. Res.*, **27**, 111–20.
SMITH R. C. & TYLER J. E. (1976) Transmission of solar radiation into natural waters. In *Photochemical and Photobiological Reviews. Vol. 1*, ed. Smith K. C., pp. 117–56. Plenum Publishing Corp., New York.
SMITH R. C., TYLER J. E. & GOLDMAN C. R. (1973) Optical properties of Lake Tahoe and Crater Lake. *Limnol. Oceanogr.*, **18**, 189–99.
STEEMANN NIELSEN E. (1974) Light and primary production. In *Optical Aspects of Oceanography*, eds. Jerlov N. G & Steemann Nielsen E., pp. 361–88. Academic Press, London and New York.
STEEMANN NIELSEN E. (1975) *Marine Photosynthesis*. Elsevier, Amsterdam and New York.
STEPHENS K. & STRICKLAND J. D. H. (1962) Use of a thermopile radiometer for measuring the attenuation of photosynthetically active radiation in the sea. *Limnol. Oceanogr.*, **7**, 485–7.
STOCKER O. & HOLDHEIDE W. (1937) Die Assimilation Helgoländer Gezeitenalgen während der Ebbezeit. *Z. Bot.*, **32**, 1–59.

STRICKLAND J. D. H. (1958) Solar radiation penetrating the ocean. A review of requirements, data and methods of measurements, with particular reference to photosynthetic productivity. *J. Fish. Res. Bd. Can.*, **15**, 453–93.

STRÖMGEN T. (1977) Apical length growth of five intertidal species of Fucales in relation to irradiance. *Sarsia*, **63**, 39–47.

SZEICZ G. (1968) Measurement of radiant energy. In *The Measurement of Environmental Factors in Terrestrial Ecology*, ed. Wadsworth R. M., pp. 109–30. Blackwell Scientific Publications, Oxford.

SZEICZ G. (1974) Solar radiation for plant growth. *J. Appl. Ecol.*, **11**, 617–36.

TYLER J. E. (1973a) Applied radiometry. *Oceanogr. Mar. Biol. Ann. Rev.*, **11**, 11–25.

TYLER J. E. (1973b) Lux vs. quanta. *Limnol. Oceanogr.*, **18**, 810.

TYLER J. E. (1975) SCOR working group 15 (with UNESCO and IAPSO). Photosynthetic radiant energy. Recommendations. *Proc. SCOR Scient. Committee Oc. Res.*, **10**, 37–42.

TYLER J. E. (ed.) (1977) *Light in the Sea. Benchmark Papers in Optics*, vol. 3. Dowden, Hutchinson & Ross, Stroudsburg, Pennsylvania.

TYLER J. E. & SMITH R. C. (1966) Submersible spectroradiometer. *J. Op. Soc. Am.*, **56**, 1390–6.

TYLER J. E. & SMITH R. C. (1967) Spectroradiometric characteristics of natural light under water. *J. Opt. Soc. Am.*, **57**, 595–601.

TYLER J. E. & SMITH R. C. (1970) Measurements of spectral irradiance underwater. In *Ocean Sciences*, ed. Wilson D. A., pp. 1–103. Naval Undersea Center, San Diego.

UPHOFF G. D. & HERGENRADER G. L. (1976) A portable quantum meter-spectroradiometer for use in aquatic studies. *Freshwater Biology*, **6**, 215–19.

VAN DER WELDE, H. H. & WAGNER A. M. (1978) The detection of phytochrome in the red alga *Acrochaetium daviesii. Plant Sci. Letters*, **11**, 145–9.

VINCE-PRUE D. (1975) *Photoperiodism in Plants.* McGraw-Hill, London.

WEINBERG S. (1976) Submarine daylight and ecology. *Mar. Biol.*, **37**, 292–304.

WEINBERG S. & CORTEL-BREEMAN A. (1978) The estimation of the yearly cycle of submarine irradiance for ecological purposes. A methodological example based on data from Banyuls-sur-Mer. *Bijdragen tot de Dierkunde*, **48**, 35–44.

WEST J. A. (1972) Environmental regulation of the reproduction in *Rhodochorton purpureum.* In *Contributions to the Systematics of Benthic Marine Algae of the North Pacific*, eds. Abbott I. A. & Kurogi M., pp. 213–30. Japanese Society of Phycology, Kobe.

WESTLAKE D. F. (1965) Some problems in the measurement of radiation under water: a review. *Photochem. Photobiol.*, **4**, 849–68.

WITHROW R. B. & WITHROW A. P. (1956) Generation, control and measurement of visible and near-visible radiant energy. In *Radiation Biology 3, Visible and Near-Visible Light,* ed. Hollaender A., pp. 125–258. McGraw-Hill, New York.

YOKOHAMA Y. (1973a) Photosynthetic properties of marine benthic green algae from different depths in the coastal area. *Bull. Jap. Soc. Phycol.*, **21**, 70–5.

YOKOHAMA Y. (1973b) Photosynthetic properties of marine benthic red algae from different depths in coastal area. *Bull. Jap. Soc. Phycol.*, **21**, 119–24.

CHAPTER 10
NUTRIENTS

J. A. DeBOER
Division of Allied Health and Life Sciences,
The University of Texas at San Antonio,
San Antonio, Texas 78285, U.S.A.

1 INTRODUCTION

Seaweeds are primarily photoautotrophic, using sunlight to produce organic matter from mineral ions and water. Many photosynthetic seaweeds are also auxotrophic, that is they require small amounts of organic growth stimulators such as vitamins. This chapter will discuss the mineral nutrition and auxotrophic requirements of seaweeds; heterotrophy and the nutrition of parasitic seaweeds will not be considered (see Chapter 14).

The availability of nutrients is one of the primary factors regulating the growth, reproduction and biochemistry of seaweeds. In semitropical and tropical regions, at least, availability of inorganic nutrients has been implicated as the most important factor limiting seaweed productivity. In many areas of the world, inorganic fertilizers are used in seaweed mariculture, much as they are used in agriculture, to enhance crop production (Bardach *et al.* 1972; Mathieson 1975; Chen 1976; Chapter 20).

Several undefined (enriched) and defined (artificial) media have been developed for the culture of marine algae (McLachlan 1973). These media in most instances have been formulated on the basis of empirical observations rather than on a thorough understanding of algal nutrient requirements. These media have proven to be an indispensable aid in life history, ecological, and physiological studies of seaweeds in the laboratory (Kornmann 1970).

Although several seaweeds have been successfully cultivated in undefined and defined media, this does not mean that their nutrient requirements are well understood. The algae as a group are physiologically quite heterogeneous compared to vascular plants, which makes generalizations about algal nutrition difficult and subject to subsequent revision. Although considerable information is available concerning the nutrition of microalgae under laboratory conditions (Hutner 1972; O'Kelley 1974; Healey 1973), relatively little is known regarding the nutrition of macroalgae under laboratory or natural conditions. One reason for this is that microalgae can be maintained under controlled experimental conditions more easily than attached macroalgae. Most culture studies of macroalgae have been confined to sporelings and the smaller species although a few culture studies have investigated the larger macrophytes including *Macrocystis* (Charters & Neushul 1979).

In this chapter the following questions will be addressed: 1, Which nutrients are essential for seaweeds? 2, Why are these nutrients required and what are their functions in the plant? 3, How are these nutrients obtained from the environment? In what form? In what amounts? 4, Where in the plant does nutrient uptake occur? 5, When does nutrient uptake occur and what factors influence uptake? 6, In what ways does nutrient availability affect the biology of seaweeds?

2 NUTRIENT REQUIREMENTS

Several recent reviews have discussed algal (primarily microalgal) nutrition (Hutner 1972; Healey 1973; O'Kelley 1974; Ukeles 1976) and plant nutrition in general (Epstein 1972; Rains 1976).

The normal biochemistry of plants requires the availability of 14–21 specific elements. Additional elements probably will be added to this list as analytical and culturing techniques are improved. There is good evidence that C, H, O, P, N, Mg, Fe, Cu, Mn, Zn, Mo, S, K and Ca are required by all algae, although S, K and Ca can be replaced to some extent by other elements. In addition, Co, Na, I, Br, B, V, Cl and Si are probably required by one or more algae (O'Kelley 1974).

2.1 Elemental composition of seaweeds

At least 56 elements have been reported to be present in (or on the surfaces of) seaweeds (Vinogradov 1953). The composition of the most abundant elements present in the Chlorophyceae, Phaeophyceae and Rhodophyceae is given in Table 10.1. It should be pointed out that some of the figures are based on as few as 20 determinations and are therefore subject to error. The data indicate that C, H, O, K, N, S, P, Ca and Mg are generally present in seaweeds at concentrations in excess of 1 mg per g dry weight. Numerous studies have reported the concentrations of trace metals and other micronutrients in seaweeds (Black & Mitchell 1952; Young & Langille 1958; Bryan 1969; Yamamoto & Ishibashi 1972; Saenko *et al.* 1976; Munda 1978). The presence of a particular element in an organism, however, should not be construed as evidence that the element is essential to the organism because certain elements are absorbed in excess of requirements whereas others are absorbed but not utilized (see Section 3).

Table 10.1 Concentration of some essential elements in seaweeds and seawater.

Element	Mean concentration in seawater [a] ($\mu g\ g^{-1}$)	Concentration in dry matter: Mean ($\mu g\ g^{-1}$)	Concentration in dry matter: Range ($\mu g\ g^{-1}$)	Ratio of supply in seawater to concentration in tissue	References
Macronutrients					
H	105 000	49 500	22 000–72 000	2.1×10^{0}	1
Mg	1 290	7 300	1 900–66 000	1.8×10^{-1}	1
S	905	19 400	4 500–82 000	4.7×10^{-2}	1
K	406	41 100	30 000–82 000	1.0×10^{-2}	2,3
Ca	412	14 300	2 000–360 000	2.9×10^{-2}	1,2,3
C	27.3[b,c]	274 000	140 000–460 000	1.0×10^{-4}	1,4,5,6
N	0.488[b,d]	23 000	500–65 000	2.1×10^{-5}	1,4,5,6,7
P	0.068	2 800	30–12 000	2.4×10^{-5}	1,4,7,8
Micronutrients					
B	4.39	184	15–910	2.4×10^{-2}	1
Zn	0.004[b]	90	2–680	4.4×10^{-5}	3,9,10,11,12
Fe	0.003[b]	300	90–1 500	1.0×10^{-5}	1,2,12
Cu	0.002[b]	15	0.6–80	1.7×10^{-4}	3,9,12
Mn	0.001[b]	50	4–240	2.0×10^{-5}	1,2,3,12

[a] 'Average' seawater (Riley & Chester 1971; Yeats *et al.* 1978).
[b] Considerable variation in concentration occurs in seawater.
[c] Dissolved inorganic carbon.
[d] Combined nitrogen (dissolved organic & inorganic).

1. Vinogradov (1953)
2. Subba Rao (1965)
3. Young & Langille (1958)
4. DeBoer, unpublished
5. Hanisak (1979b)
6. Topinka & Robbins (1976)
7. Iwasaki & Matsudaira (1954)
8. Steffensen (1976)
9. Munda (1978)
10. Yamamoto & Ishibashi (1972)
11. Fujiyama & Maeda (1977)
12. Bryan (1976)

2.2 Essential elements

The stimulation of growth by an element added to the nutrient medium in itself does not prove the element is essential. To demonstrate the essentiality of an element it is generally necessary to work with axenic cultures in a defined medium. The accepted criteria for confirming essentiality are either 1, the cessation of algal growth when the element is not supplied and optimal or near optimal growth rates when the element is provided (Arnon 1953) or, 2, the demonstration by *in vitro* experiments that an element has a non-replaceable role in a fundamental life process (O'Kelley 1974).

Because very little work has been published on the elemental requirements of seaweeds, there has been a tendency to consider the requirements of all algae to be equivalent. This may be expedient although it is not always accurate. Gerloff (1963) commented that the essentiality of an element demonstrated for one or a few plant species has frequently (and erroneously) been interpreted as a requirement for all plants. It is important to recognize that the qualitative and quantitative mineral requirements of all plants are not the same. O'Kelley (1974) pointed out that the inherent heterogeneity of algae makes generalizations about their requirements difficult.

2.3 Limiting nutrients

According to Liebig's law of the minimum, when other factors such as light and temperature are favourable, the nutrient available in the smallest quantity with respect to the requirements of the plant will limit its rate of growth. A comparison of the elemental composition of seaweeds with that of average seawater (Table 10.1) provides insight into which elements might limit seaweed growth. The ratios of the supply of an element in seawater to the amount present in seaweed tissue indicate that the macronutrients N and P and the micronutrients Fe, Mn, and Zn are generally in the shortest supply for most seaweeds. This evidence suggests that N, P, Fe and possibly Mn and Zn may at times limit the growth of seaweeds. It must be remembered, however, that the concentrations of these elements are generally higher in coastal areas than in the 'average' seawater described by Riley & Chester (1971). Moreover, the concentration of an element present in a plant is not necessarily the same as the concentration of that element which is required for growth. Thus these comparisons only serve to indicate which nutrients *might* become growth limiting. Another indication that N, P, Fe and also Mn, Cu, Zn and C might limit algal growth comes from the fact that the concentrations of these elements in seawater vary considerably (Table 10.1), largely as a result of biological processes.

Ryther & Dunstan (1971) found that nitrogen is the chemical nutrient most likely to limit phytoplankton growth in marine waters. Several subsequent studies have confirmed this observation for phytoplankton as well as seaweeds (Topinka & Robbins 1976; DeBoer & Ryther 1977; Jackson 1977; Hanisak 1979b), but do not eliminate the possibility that nutrients other than N may limit seaweed growth at times.

2.4 Macronutrients

Plant macronutrients include C, H, O, P, K, N, S, Ca and Mg. These elements, which occur in relatively high amounts in seaweeds (Table 10.1), are utilized directly or indirectly for cellular building blocks (Table 10.2). The following sections will describe the chemical forms in which each element occurs in seawater and the forms and concentration range utilized by seaweeds.

Carbon

Carbon occurs in seawater in the form of carbon dioxide, bicarbonate and carbonate ions in equilibrium with each other, and organic compounds (see Chapter 14).

Nitrogen

Elemental nitrogen is in abundant supply both in the air and in seawater, but there is no evidence that algae other than a few members of the Cyanophyta are able to utilize it directly (Fogg 1974). Seaweeds can utilize several other inorganic nitrogen compounds including nitrate, nitrite and ammonium, which usually occur in seawater at concentrations of 1–500 μM, 0.1–50 μM, and 1–50 μM, respectively (Riley & Chester 1971). Organic nitrogen compounds, which usually occur at concentrations less than 20 μM in seawater (Riley & Chester 1971), are also utilized by some seaweeds (see below).

The recognition of the importance of N in the regulation of growth of seaweeds had led to a number of recent studies on its utilization by seaweeds. Seaweeds have been found to differ in their capabilities of utilizing various nitrogen sources. Nitrate and NH_4^+ were equally effective at promoting growth of *Chondrus crispus* (Neish & Shacklock 1971). *Gracilaria foliifera* and *Agardhiella subulata* exhibited higher growth rates when supplied with ammonium as compared with nitrate (DeBoer *et al.* 1978). Nitrite is toxic to some algae at much lower concentrations than other forms of inorganic N (Healey 1973). Fries (1963) reported, however, that at low (normal?) concentrations (3.5 μg–at N l^{-1}) NO_2^- promoted better growth than NO_3^- in *Goniotrichum alsidii* and *Nemalion multifidum*. Some seaweeds are able to utilize organic nitrogen as the amide or amino group, although the growth rates are generally lower than when the nitrogen is supplied as NO_3^- (Fries 1963; Iwasaki 1967; Fries & Pettersson

Table 10.2 The function and compounds of the essential elements in seaweeds (O'Kelley 1974; Rains 1976).

Element	Probable functions	Examples of compounds
Nitrogen	Major metabolic importance as compounds	Amino acids, purines, pyrimidines, porphyrins, amino sugars, amines
Phosphorus	Structural, energy transfer	ATP, GTP, nucleic acids, phospholipids, coenzymes, coenzyme A, phosphoenol pyruvate
Potassium	Osmotic regulation, pH control, protein conformation and stability	Probably occurs predominately in the ionic form.
Calcium	Structural, enzyme activation, ion transport	Calcium pectate, calcium carbonate
Magnesium	Photosynthetic pigments, enzyme activation, ion transport, ribosomal stability	Chlorophyll
Sulphur	Active groups in enzymes and coenzymes, structural	Methionine, cystine, cysteine, glutathionine, agar, carrageenan, sulpholipids, coenzyme A
Iron	Active groups of porphyrin molecules and enzymes	Ferredoxin, cytochromes, nitrate reductase, nitrite reductase, ferritin, catalase
Manganese	Electron transport in photosystem II, maintenance of chloroplast membrane structure	Manganin
Copper	Electron transport (photosynthesis), enzymes	Plastocyanin, amine oxidase
Zinc	Enzymes, auxin metabolism(?), ribosome structure(?)	Carbonic anhydrase
Molybdenum	Nitrate reduction, ion absorption	Nitrate reductase
Sodium	Enzyme activation, water balance, enzymes	Nitrate reductase
Chlorine	Photosystem II	Terpenes
Boron	Regulation of carbon utilization(?), RNA metabolism(?)	
Cobalt	Component of Vitamin B_{12}, C_4 photosynthesis pathway	Vitamin B_{12}
Bromine*	?	Wide range of halogenated compounds in red algae (see Chapter 17).
Iodine*	?	

* Possibly an essential element in some seaweeds.

1968; Neish & Fox 1971; Rao & Mehta 1973; DeBoer *et al.* 1978). *Codium fragile* grew equally well with NO_3^-, NO_2^- or NH_4^+ used as the nitrogen source (Hanisak 1979a). *Ulva fasciata* (Mohsen *et al.* 1974) and *Pterocladia capillacea* (Nasr *et al.* 1968) exhibited higher growth rates when supplied with urea as compared with NO_3^- or NH_4^+.

Numerous studies have reported 'optimum' nitrogen concentrations for seaweed growth based on batch culture experiments (Boalch 1961; Fries 1963;

Iwasaki 1967; Mohsen *et al.* 1974; Steffensen 1976). These values (120–2140 μM.), while optimal for some culture work, have little relevance to seaweed nutrition in the natural environment. Much more accurate quantitative estimates of seaweed requirements are obtained from continuous flow experiments (see Section 5).

Seaweeds have been reported to store unused NO_3^- in vacuoles. *Valonia* and *Halicystis* concentrate NO_3^- 2000 and 500 times, respectively, the NO_3^- concentration in seawater (Jacques & Osterhout 1938). *Laminaria longicruris* has been reported to concentrate NO_3^- as high as 28 000 fold during periods of N sufficiency (Chapman & Craigie 1977). These inorganic and organic N reserves are believed to be utilized to sustain the growth of *Laminaria* during the summer months when external N levels are low (Chapman & Craigie 1977; Chapman *et al.* 1978) but this does not support *a priori*, the contention of storage in vacuoles as such.

PHOSPHORUS

Phosphorus has frequently been implicated as a limiting nutrient for seaweeds (Section 2.3). Phosphorus is available in the sea almost entirely in the form of orthophosphate ions. In 'average' seawater at pH 8.0 (20°C), 87% of the phosphate occurs as HPO_4^{2-}, 12% as PO_4^{3-}, and 1% as $H_2PO_4^-$ (Riley & Chester 1971). Orthophosphate concentration in seawater is generally *c.* 1–3 μM although considerable variations occur especially in coastal waters (Riley & Chester 1971). The utilization of polyphosphate ions, which occur in coastal and estuarine waters as a result of pollution with detergents, has not been reported for seaweeds. Seaweeds can utilize some organic forms of phosphorus (Fries 1963; Aberg & Fries 1976). Phosphorus utilization by algae was reviewed by Kuhl (1962, 1974) and Healey (1973).

The rate of ^{14}C fixation in *Ulva lactuca* was stimulated by orthophosphate additions up to 25 μM (Waite & Mitchell 1972). The growth rate of *Ectocarpus confervoides* was increased by additions up to 100 μM orthophosphate (Boalch 1961). As a source of phosphate, organic phosphate (α- or β- glycerophosphate) was as good as inorganic phosphate (HPO_4^{2-}) in promoting the growth of *Vaucheria dichotoma* in batch culture experiments. The optimum phosphate concentration in each case was 32–320 μM (Åberg & Fries 1976). The final yield of *Porphyra tenera* was stimulated by concentrations of orthophosphate up to 320 μM but was decreased by a concentration of 1600 μM (Iwasaki 1967). It is dangerous, however, to extrapolate these batch culture values to the natural environment.

An important difference between phosphorus metabolism in vascular plants and algae is the formation of large amounts of polyphosphate by many algae. Among the seaweeds, *Acetabularia, Enteromorpha, Ceramium* and *Ulothrix* have been reported to accumulate polyphosphates (Kuhl 1962). Polyphosphate granules in *Acetabularia mediterranea* were reported to increase in size and number in the light, whereas they decreased in the dark (Stich 1955). It is now generally believed that, at least in microalgae, polyphosphates function as deposits of phosphorus which can be easily used. This may be of important ecological value to algae, enabling them to continue to grow under conditions of phosphorus depletion. The possibility that polyphosphates might serve a secondary role as energy storage was suggested by Healey (1973) but has not been directly proven for algal systems. Some algae have been reported to store excess phosphate in ionic form in vacuoles (Healey 1973).

Sulphur

In addition to its importance as a component of proteins, sulphur is incorporated into other materials, some of which appear to be unique to the algae (see also Chapter 17). The highest concentrations of plant sulpholipid, for example, apparently occur in marine red algae (Wood 1974). Another important class of sulphur containing compounds in the seaweeds is the sulphated polysaccharides including agar, carrageenan, furcellaran, porphyran, L-fucose and several other polysaccharides in the Rhodophyceae, Phaeophyceae and Chlorophyceae (Percival 1979). The most abundant form of sulphur in seawater is sulphate which occurs at a concentration of *c.* 0.025 M. Because of its abundance in seawater, sulphur is probably not a limiting factor for seaweeds under normal conditions. Since most algae can supply their entire sulphur requirement by reducing sulphate, few studies have been conducted on the utilization of other forms of sulphur, such as sulphite and organic compounds containing sulphur, by algae. Sulphur nutrition and utilization in the algae have been reviewed by Schiff (1962), Ikawa *et al.* (1973), O'Kelley (1974) and Møller & Evans (1976).

Calcium and magnesium

In addition to its importance in the maintenance of cellular membranes, calcium also is an integral part of the skeletal structure of several members of the Chlorophyceae, Rhodophyceae and Phaeophyceae. Calcareous seaweeds are an important agent in coral reef formation. Calcification in the algae has been reviewed by Darley (1974) and Borowitzka (1977); see also Chapter 14.

Magnesium, a constituent of chlorophyll, is obviously an essential requirement for photosynthetic seaweeds. Magnesium is also involved in the aggregation of ribosomes into functional units. The utilization of magnesium by seaweeds has been reviewed by O'Kelley (1974).

The concentration of calcium and magnesium in seawater appears to be high compared to the requirements of seaweeds so neither appears to be a growth limiting nutrient under ordinary conditions.

Potassium

The relatively high amounts of potassium in seaweeds do not seem to correlate with the observed functions of potassium (e.g. enzyme systems and osmotic regulation). It has been proposed that potassium is not so much a catalytic agent as an activator for certain enzyme systems in which it maintains an ionic environment suitable for preserving the three dimensional structures necessary for optimal enzyme activity (Evans & Sorger 1966). Potassium metabolism in the algae has been reviewed by Healey (1973) and O'Kelley (1974).

2.5 Inorganic micronutrients

Algal inorganic micronutrients include Fe, Cu, Zn, Mn, Si, Co, Mo, V, B, Cl, I, Br and Na. They are important as catalysts in metabolic reactions, in osmoregulation, and other functions (Table 10.2). Trace element nutrition experiments are often difficult to interpret (Ukeles 1976) as a result of several complicating factors e.g. 1, presence of contaminants in 'pure' chemicals; 2, lack of knowledge of the physiochemical forms in which the element is present and what forms are 'biologically available'; 3, an essential ion may become unavailable by binding to a component in the medium; 4, a non-essential ion may displace a required ion from a complex. As a result of these complications the list of inorganic micronutrients required by algae changes as analytical techniques improve.

Zinc

Many seaweeds, especially members of the Phaeophyceae, are known to accumulate many trace elements, often resulting in extremely high cellular concentrations relative to the surrounding seawater. Concentration factors (μg g^{-1} metal in dried seaweed per μg ml^{-1} dissolved metal in seawater) for zinc range from 300 in the case of *Polysiphonia japonica* (Saenko *et al.* 1976) to 60 000 for *Fucus vesiculosus* (Bryan & Hummerstone 1973). Relatively little is known, however, regarding the utilization of zinc by seaweeds. The growth rate of five species of intertidal Fucales, measured in continuous flow experiments, was not significantly different at zinc concentrations ranging between 7 μg l^{-1} (ambient seawater) and 100 μg l^{-1}. Growth rate was reduced approximately 50% at zinc concentrations in excess of 1.4 mg l^{-1} (Strömgren 1979b). Experiments with young fronds (Noda & Horiguchi 1971) and the *Conchocelis* phase (Iwasaki 1967) have revealed that zinc is an essential element for *Porphyra tenera*. The optimum concentration for the growth of both phases was found to be 30 μg Zn l^{-1}. Zinc deficiency in that species resulted in a decrease in growth and macromolecular protein, chlorophyll and phycobilin synthesis (Noda & Horiguchi 1971).

Halogens

Polysiphonia urceolata and *Ectocarpus fasciculatus* appear to have an absolute requirement for iodine (Fries 1966; Pedersen 1969a). Iodine appears to be essential for growth, morphogenesis and reproduction of *Petalonia fascia* (Hsiao 1969) and *Ectocarpus siliculosus* (Woolery & Lewin 1973). Addition of iodine to the medium is stimulatory to the *Conchocelis* phase of *Porphyra tenera* although essentiality has not been demonstrated (Iwasaki 1967). Optimum and inhibitory concentrations of iodine for 15 species of seaweeds are given by Hsiao (1969). Concentrations of iodine in 'average' seawater (*c.* 60 $\mu g\ l^{-1}$) appear to be in the general range required by seaweeds, but may, at times, either limit or inhibit the growth or development of some seaweeds (Hsiao 1969). Many seaweeds are strong concentrators of iodine and bromine (Vinogradov 1953; Saenko *et al.* 1978).

Bromine has not been recognized as an essential plant nutrient (Hutner 1972; O'Kelley 1974). Recently, however, Fries (1975) reported an apparent bromine requirement for *Polysiphonia urceolata*. Growth of *P. urceolata* was stimulated at a bromine concentration of 1 $\mu g\text{-at}\ l^{-1}$ and inhibited at concentrations exceeding 4 $\mu g\text{-at}\ l^{-1}$. The growth of *Fucus edentatus* embryos was also shown to be retarded by the omission of bromine from the culture medium (McLachlan 1977). These recent results do not, however, demonstrate the essentiality of bromine according to the criteria given in Section 2.2. The concentration of bromine in seawater is approximately 800 $\mu g\text{-at}\ l^{-1}$, greatly in excess of the range found to be optimum for several seaweeds (Fries 1966, 1975; McLachlan 1977).

A requirement for chlorine has not been unequivocally demonstrated for algae (O'Kelley 1974) although its role in photosynthesis indicates that it is probably essential in most algae. Chlorine and bromine have important roles in seaweed biochemistry (Vernon 1962; Fenical 1975). Halogenated ketones and terpenes and other halogenated compounds occurring in seaweeds (see Chapter 17) may be ecologically important in predator avoidance responses and algal-algal antibioses (Fenical 1975).

Silicon

Silicon, an essential and often limiting nutrient in the Bacillariophyceae, has not been conclusively shown to be essential for any seaweed. There is some indication, however, that certain Phaeophyceae may have a silicon requirement. Parker (1969) for example, reported the presence of silicon in several Phaeophyceae and that germanium dioxide (an inhibitor of silicon metabolism) inhibits the growth of certain Phaeophyceae. McLachlan *et al.* (1971) found that extra silicon did not overcome germanium toxicity to *Fucus*, suggesting that Ge was not toxic through competitive inhibition of Si.

Cobalt

Cobalt, which constitutes 4% of the weight of vitamin B_{12}, appears to be an essential algal micronutrient (Bunt 1970). Some seaweeds can synthesize vitamin B_{12} if supplied with cobalt whereas others require an exogenous B_{12} supply (Section 2.6).

Iron

Seaweeds have an unquestioned requirement for iron. It functions both as a structural component and as a cofactor for enzymatic reactions. Oxidation-reduction reactions are most commonly associated with iron-containing systems. Little is known about the uptake of iron and the quantitative iron requirements of seaweeds undoubtedly because of the difficulties involved in determining the amount of 'available' iron in seawater.

Boron

A requirement for boron by embryos of *Fucus edentatus* has been demonstrated by McLachlan (1977). When this element was omitted from the culture medium the embryos became moribund. The deficiency was reversed by the addition of 1–4 μM boron, considerably less than that present in 'average' seawater. Boron also has been demonstrated to be essential for several microalgae (Lewin 1976).

Arsenic

Although the arsenic concentration in seawater is rather constant, ranging between 1.0 and 1.5 $\mu g\ g^{-1}$ (Andreae 1978), there is considerable variation in the arsenic concentration in seaweeds: Rhodophyceae, 1.43 $\mu g\ g^{-1}$; Chlorophyceae, 1.54 $\mu g\ g^{-1}$; Phaeophyceae, 10.3 $\mu g\ g^{-1}$ (Saunders 1979). Von Stosch (1964) reported that *Asparagopsis armata* has a requirement for arsenic. He suggested a relationship between arsenic and phosphorus metabolism in this species because arsenic deficiency did not appear when phosphorus was reduced to one-fifth of its normal concentration in the medium. Other research suggests that arsenate is a competitive inhibitor of phosphate uptake (Rothstein 1963; Blum 1966).

Copper

It is generally accepted that all algae have a micronutrient copper requirement (O'Kelley 1974) although this has not been demonstrated for seaweeds. Most algae are extremely sensitive to the concentration of copper and copper complexes. Shimo & Nakatani (1967) reported that the optimal concentration of copper in the experimental culture of *Porphyra tenera* was 100 $\mu g\ l^{-1}$. Copper toxicity was observed at higher concentrations but was reduced by the addition of an organic chelator. The growth rate of *Ascophyllum nodosum* maintained in a flowing culture system was constant at Cu concentrations ranging from *c.* 5 $\mu g\ l^{-1}$ to 38 $\mu g\ l^{-1}$ but was significantly reduced at higher (71–5000 $\mu g\ l^{-1}$) concentrations (Strömgren 1979a).

MANGANESE

Manganese, an essential plant nutrient (Rains 1976), is almost certainly essential to seaweeds because of its importance in photosynthesis (Table 10.2). Very little data has been published on the Mn requirements and toxicity levels in seaweeds. In long-term batch culture studies, *Porphyra tenera* (Iwasaki 1967) and *Ectocarpus confervoides* (Boalch 1961) were stimulated by Mn concentrations between *c.* 55 and 300 $\mu g\ l^{-1}$.

2.6 Vitamins and plant growth regulators

Almost all seaweeds studied in impure culture will grow in enriched medium containing soil water extract (Hutner & Provasoli 1964), although many seaweeds, especially those in pure culture, cannot be grown on mineral medium. Even in an enriched seawater medium, *Ulva, Monostroma* and *Enteromorpha* exhibit normal growth only in the presence of their accompanying microflora. The seaweeds have only filamentous or callus growth, if any, when grown in axenic culture (Hutner & Provasoli 1964). The growth stimulatory properties of soil water extract have been attributed to trace metals, organic chelators, vitamins or other growth promoting substances. Similarly, microalgae, bacteria and fungi present in impure seaweed cultures may supply these substances. The importance of plant hormones and other growth regulators in the development of seaweeds is discussed in Chapter 18.

The vitamin requirements of algae were recently reviewed by Provasoli and Carlucci (1974) and will be only summarized here. Of the 21 species of seaweeds assayed for vitamin requirements (Table 10.3), none required biotin, only one required thiamine and 13 required B_{12}. Although several members of the Phaeophyceae can grow without the addition of exogenous B_{12}, their growth rates are enhanced by its addition (Pedersen 1969b). The exogenous vitamins required by some seaweeds may be obtained from the seawater or may come directly from microorganisms living on seaweed surfaces. In several instances it appears that under normal conditions, microorganisms living in or on the surfaces of seaweeds provide essential growth factors (Provasoli & Carlucci 1974; Waite & Mitchell 1976). The concentrations of vitamins in seawater vary seasonally in the same way as nitrogen and phosphorus (Menzel & Spaeth 1962). This suggests that exogenous vitamins may be ecologically important in influencing algal species diversity and/or productivity.

Table 10.3 Vitamin requirements of seaweeds. R = Required; 0 = not required. (Provasoli & Carlucci 1974).

	No. species tested	B_{12}	Thiamine	Biotin
Chlorophyceae	1	R	R	0
Phaeophyceae	9*	0	0	0
	1	R	0	0
Rhodophyceae	10	R	0	0

* Four of the nine species are stimulated by vitamin additions.

3 NUTRIENT UPTAKE MECHANISMS

Nutrient salts generally enter plant cells as cations or anions and not as uncharged molecules. Uncharged molecules may diffuse along chemical potential gradients alone but when ions move across membranes, the driving force also involves an electrical gradient across the membrane.

Ion fluxes across membranes may result from passive diffusion, facilitated diffusion, exchange diffusion or active transport. The specific mechanisms of ion transport in biological systems are currently the subject of considerable investigation, speculation and debate. Recent reviews and books on ion transport in plant and animal cells include Epstein (1972), Nobel (1974), Lüttge & Pitman (1976) and Giebisch *et al.* (1978). Ionic relations in marine algae have been reviewed by Gutnecht & Dainty (1968), MacRobbie (1974) and Raven (1976). Gutknecht *et al.* (1978) have recently reviewed ion transport and turgor pressure regulation in giant algal cells. Much of the research on ion uptake by plants has been conducted with algae such as *Acetabularia, Valonia* and *Halicystis* which possess comparatively large cells, facilitating access to the cytoplasm.

Seaweeds tend to absorb and/or adsorb all elements present in a nutrient medium but the fluxes and equilibrium concentrations are not identical for all elements. In most instances seaweeds are very selective with regard to the elements as well as to the specific chemical forms of the elements accumulated or eliminated. For example, seaweeds are known to concentrate some trace metals in very large amounts (Section 2.1), while the concentration of other elements may be comparatively low, even below that found in the surrounding water (Young & Langille 1958). In order to understand the mechanisms whereby seaweeds can accumulate some ions while excluding others, it is useful first to consider the mechanisms by which algal cells transport ions.

3.1 Uptake into apparent free space

When an alga is placed in a nutrient medium, there is an initial rapid uptake of ions which is not metabolically dependent, and in many instances a slower and more prolonged uptake which is accumulative against concentration gradients and is metabolically dependent. The first stage of ion uptake is considered to occur by means of diffusion into the 'free space' of an alga. 'Free space' has been defined as that part of the plant to which the solution bathing the alga has direct, unimpeded access.

In addition to diffusion, ion exchange systems are also important in algal 'free space'. Algal cell walls, containing various polysaccharides and mucoproteins, have many sulphate, carboxyl and phosphate groups from which protons can dissociate, leaving the macromolecules with a net negative charge. In a sense, these polymers act as a sponge to adsorb cations from the environment. Cation adsorption may be especially high in seaweeds such as *Ascophyllum nodosum* and *Porphyra perforata* which contain acidic and sulphated cell wall polysaccharides (Wasserman 1948; Eppley & Blinks 1957). As a consequence of strong cation adsorption by seaweeds, extreme care must be exercised in ion uptake experiments to differentiate between ions accumulated inside cells as opposed to ions adsorbed to cell walls.

In place of 'free space', which cannot be accurately measured, plant physiologists frequently use the term 'apparent free space' (AFS) to designate the volume accessible by diffusion and ion exchange (Epstein 1974; Bidwell 1979). AFS is defined experimentally as the amount of solute in the alga divided by the concentration of solute in the external medium at equilibrium. AFS is believed to include only those cell components external to the plasmalemma, i.e. the cell wall and intercellular spaces. In the roots of higher plants the AFS extends to the endodermis but not beyond. By contrast, the AFS of most seaweeds includes all cell walls and intercellular areas. Indeed, nutrient uptake by seaweeds has been shown to occur across the entire thallus (Yamada *et al.* 1964; Fujiyama & Maeda 1977). Values for AFS in seaweeds have been reported to be 20–30% of the cell volume (Scott & Hayward 1954; Eppley & Cyrus 1960).

3.2 Passive diffusion

The driving force for passive diffusion of an ion is the difference of electrochemical potential between two points along the path. Frictional interactions are also involved but are usually ignored (Gutknecht & Dainty 1968). Metabolic energy is not directly required for passive diffusion, however the maintenance of intact cells and also the maintenance of Donnan potentials, either via macromolecular synthesis or active ion uptake (see below), are dependent on metabolic energy.

The asymetrical ion distributions which give rise to the electrochemical gradients are the result of the presence of impermeable charged molecules inside the cell. These molecules, mostly proteins and other charged macromolecules, are usually anionic at cellular pH's. The impermeable charged molecules give rise to a Donnan type equilibrium which may concentrate ions (usually cations) in a cell up to 30-fold over the concentration in the environment (Bidwell 1979). The overall transmembrane electrical potential is determined largely by the active and passive fluxes of only a few ions such as K^+, Cl^- and H^+ (Raven 1976). This potential acts on the distribution of all other ions which then diffuse according to their concentration, charge and permeability coefficient.

Other factors within the cell can influence the rate of passive diffusion of solutes into cells. Cytoplasmic streaming and the concomitant reduction in the concentration of the diffusing solute just inside the plasmalemma, steepens the gradient and thus hastens diffusion into cells. The rate of diffusion may be greatly enhanced if the internal concentration of an ion is being continuously reduced as a result of being sequestered. This type of diffusion has been called 'trapping diffusion' (Wilson 1962). The available concentration is reduced as a result of the biochemical utilization of the ion.

3.3 Facilitated diffusion

While simple diffusion may account for the entrance of water, oxygen, carbon dioxide, and other small molecules and some ions into cells, facilitated diffusion is thought to be a more common process. In both facilitated and simple diffusion, ions cannot be accumulated against electrochemical gradients.

Several hypotheses including the 'charged pore' and 'carrier' concepts have been advanced as possible mechanisms of mediated transport. Although the exact mechanisms for physically binding and moving the solutes through the membrane are still unknown, the carrier concept has found widespread application in the interpretation of experimental observations. According to this concept the carrier first combines with the ion to be transported. The carrier-ion complex then diffuses to the other side of the membrane (or simply alters its conformation) and discharges the bound ion, without the intervention of metabolism. Facilitated transport systems allow solutes to cross the membrane more rapidly in either direction but always down an electrochemical concentration gradient as in ordinary diffusion. Facilitated diffusion hastens the rate of ion movement across the membrane and, because a carrier is involved, exhibits ion selectivity and saturation.

3.4 Active transport

The most intriguing transport process of membranes is active transport, by which a material is moved across a membrane against an electrochemical gradient, i.e. from a lower concentration to a higher concentration. Since this process has a positive free energy change, it is necessary that the transport process be coupled in some manner with an exergonic reaction. Two key areas of research interest concern the nature of the transport mechanism itself and the linkage between the metabolic energy source and the energy-requiring transport mechanism.

Like facilitated diffusion, active transport is selective with respect to various ions and groups of ions. The carriers involved in active transport are thought to be similar if not actually the same carriers involved in facilitated diffusion. However, active transport systems differ from simple and facilitated diffusion systems in three important ways: 1, in active transport, energy is expended to facilitate the release of the molecule or ion from the carrier (i.e. uptake is linked with metabolic processes); 2, solutes can be taken up (or eliminated) against electrochemical gradients; and 3, active transport systems are usually directional, and transport a solute either into or out of the cell.

3.5 Exchange diffusion

Some ions enter or leave cells via a process termed exchange diffusion which can be thought of as a type of secondary active transport. Exchange diffusion occurs if a particular carrier can cross the membrane only in association with an ion. For example, solute A is pumped against a concentration gradient by active transport. Solute B is then allowed to cross the membrane through an exchange process with solute A.

4 NUTRIENT UPTAKE KINETICS

In the case of simple diffusion the rate of uptake of a particular ion is generally proportional to its external concentration. In the case of facilitated diffusion and active uptake, however, the uptake rate exhibits 'saturation' kinetics (Fig. 10.1a). At low substrate concentrations, the rate of uptake increases sharply with increasing external concentrations but, at progressively higher concentrations, each added increment of concentration adds less and less of an increment in uptake rate until, at the highest concentrations used, uptake is virtually independent of the concentration in the solution. Such saturation kinetics have been found for the absorption of numerous ions by many phytoplankton and higher plants as well as a few seaweeds. The similarity of ion uptake with that of enzymatic catalysis is thought to reflect a similarity of mechanisms. The Michaelis–Menten equation (Eqn. 1), which relates the rate of enzymatic catalysis to the concentration of substrate, has also been applied to nutrient uptake systems.

$$V = \frac{V_{max}S}{K + S}. \quad (1)$$

The observed rate of uptake V, is expressed as a function of the theoretical maximum rate V_{max}, the substrate concentration S, and the Michaelis constant

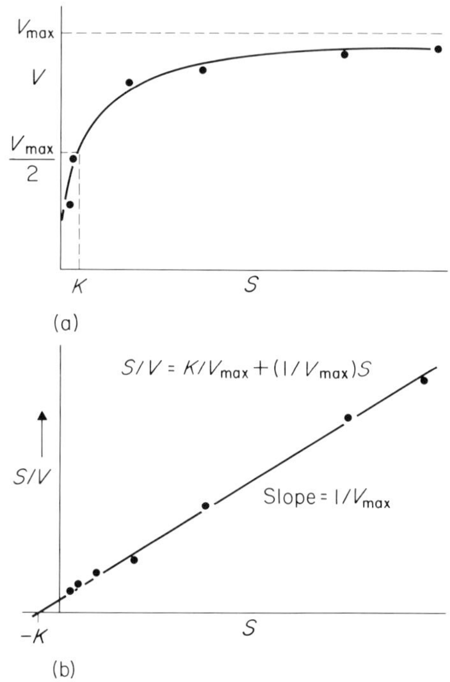

Fig. 10.1 (a) Michaelis–Menten type nutrient uptake; (b) S/V versus linear transform of data in (a) used to estimate nutrient uptake coefficients and to test fit of data.

K. The Michaelis constant or the half saturation constant equals the concentration of the substrate ion giving half the maximal rate of absorption (Fig. 10.1a). The lower this value, that is, the lower the concentration of the ion resulting in half the maximal rate of its absorption, the higher is the affinity of the carrier sites for the ion.

The Michaelis–Menten equation can be rearranged in three ways which yield straight-line plots (Dowd & Riggs 1965). These are useful in calculating V_{max} and K, and for testing the fit of experimental data with the equation. One such linear transform is shown in Fig. 10.1b.

Ion uptake does not always follow simple saturation type uptake kinetics. Numerous studies with vascular plants have revealed that certain ions, such as K^+, $H_2PO_4^-$, SO_4^{2-}, and NH_4^+, exhibit a dual pattern of uptake (Epstein 1972). At low concentrations the uptake of many ions can be explained by saturation type (Michaelis–Menten) kinetics. This is sometimes referred to as 'Mechanism 1'. A second uptake system ('Mechanism 2') becomes operative only at high substrate concentrations, that is, it has a much lower affinity for ions than does Mechanism 1. Whereas Mechanism 1 exhibits a high degree of ion specificity, Mechanism 2 is much less selective (Epstein 1966). Multiphasic ion uptake systems involving up to seven phases were reviewed by Nissen

(1974). The mechanisms and ion specificities of multiphasic systems are not well understood but are believed to involve transport across both the plasmalemma and the tonoplast (Nissen 1974).

4.1 Measurement of nutrient uptake

There are no generally accepted standard methods for measuring nutrient uptake kinetics in seaweeds. As a result it is frequently difficult to compare the results of different studies. Uptake studies have been conducted using whole plants (Hansen 1976; D'Elia & DeBoer 1978; Topinka 1978), cut discs (Wheeler 1977), or parts from various portions of the thallus (Topinka 1978; Hanisak & Harlin 1978). Entire plants should be used whenever feasible. The amounts of tissue used per volume of medium, frequency of sampling and durations of uptake study are other important considerations which are best determined following preliminary studies conducted with the alga and nutrient undergoing evaluation. It is also important to standardize and report the nutrient preconditioning of the alga. For instance, when measuring the nitrogen uptake kinetics of seaweeds, the internal nitrogen content of the tissue and external nitrogen concentrations present just prior to the experiment will influence the results (D'Elia & DeBoer 1978). It is necessary to be aware of and report the light conditions (quantity, quality and photoperiod) both prior to and during the experiment.

One assumption in experiments designed to measure nutrient uptake kinetics in seaweeds is that the concentration of the nutrient in the 'free space' of the alga is the same as that in the culture medium. This, of course, can be true only if the culture is well mixed. As water movement (shear) is reduced, boundary layers form which result in reduced nutrient influx (Pasciak & Gavis 1974; Gavis 1976; see also Chapter 12). The rate of transport across a boundary is a function of the thickness of this layer. The effective boundary layer thickness is determined by turbulent mixing and water flow around the plant. An unstirred boundary layer causes an increase in the apparent half saturation rate constant (K) measured (Winne 1977; Wheeler 1977).

There are basically two methods which have been used to study nutrient uptake in seaweeds: Perturbation (D'Elia & DeBoer 1978; Topinka 1978; Haines & Wheeler 1978) and Multiple Flask (Hansen 1976; Wheeler 1977; Hanisak & Harlin 1978). In the perturbation method, the entire tissue, cut discs or algal parts are contained in one large, well mixed vessel which is repeatedly sampled during the experiment. Since a portion of the medium is continuously or discontinuously being removed in this type of study, correction must be made for volume removal. The perturbation method is particularly useful in studying uptake of non-metals. This method has the advantage of minimizing inter-plant variability, a problem with the multiple flask method. The multiple flask technique utilizes several containers, each containing algal material and different initial amounts of the limiting nutrient. Each flask is sampled only once. This method is useful in studying the uptake of metals by seaweeds because it enables the experimenter to distinguish between 'free space' uptake and intracellular uptake (see Section 3).

If the results are to be useful for comparative purposes, it is necessary to normalize them, or at least provide the necessary conversion factors which allow readers to do so. For instance, uptake rates can be expressed on an areal, dry weight, ash free dry weight, or fresh weight basis (Table 10.4). One of the first steps in analysing the data is to prepare a nutrient depletion curve (Fig. 10.3). Next, it is necessary to analyse the data carefully to determine what mathematical relationships are applicable (see Sections 4 and 4.4).

4.2 Environmental factors influencing nutrient uptake

In addition to genetically controlled differences in nutrient uptake capacities, several other internal and external factors can influence nutrient uptake by cells. Light can influence nutrient uptake in many ways. Light, generally acting via photosynthesis; can 1, provide the energy necessary for active transport; 2, produce carbon skeletons which act as a sink for incorporated ions; 3, provide energy for the production of charged polymers which establish Donnan potentials; 4, stimulate growth which results in increased nutrient uptake; and 5, increase protein synthesis.

Temperature, of course, will influence the rate of diffusion, carrier mediated uptake, and cell metabolism in general. The uptake of ions from the external medium will also be influenced by the concentration of competing ions and ions which may suppress carrier or enzyme systems. The concentration of ions in the cytoplasm and in vacuoles may also influence the rate of ion uptake. Another factor which can strongly influence the rate of uptake is water motion (see Section 4.1 and Chapter 12). It is quite possible that in low turbulence areas seaweeds may exhibit indications of nutrient deficiency even though external nutrient concentrations may be moderate. Conversely, seaweeds may grow profusely in very turbulent but low nutrient environments such as occur along many rocky coastlines. However, although water movement increases nutrient availability, highly turbulent water conditions may physically damage seaweeds, limit predatory or competitive species, and decrease available light (Gerard & Mann 1979).

4.3 Assimilation of inorganic carbon

At high external pH values, active transport is necessary for the uptake of HCO_3^- at rates sufficient to support net photosynthesis, while at low pH values, photosynthesis involves entry of CO_2 as the undissociated molecule (Raven 1970). Understanding of the uptake of inorganic carbon is complicated because carbon is so rapidly metabolized and because its concentration and form are so dependent on pH values. Carbon utilization by seaweeds is discussed in Chapter 14.

4.4 Assimilation of nitrogen

A typical uptake curve for NH_4^+ is shown in Fig. 10.2. The uptake of NH_4^+ in *Codium fragile* increases with increasing NH_4^+ concentration up to approximately 5 μM; at higher NH_4^+ concentrations little increase in uptake rate is observed. Such saturation type kinetics suggest that a 'carrier' is involved either via facilitated diffusion or by active transport. Because of difficulties in measuring internal pH, and internal NH_4^+ and NH_3 concentrations, insufficient data are available to determine if NH_4^+ uptake by seaweeds is a result of passive, facilitated, or active transport (Raven 1976), although active transport is generally assumed.

The uptake of NO_3^- also exhibits saturation kinetics in most seaweeds studied (Topinka 1978; D'Elia & DeBoer 1978; Harlin & Craigie 1978; Haines & Wheeler 1978; Hanisak & Harlin 1978). In at least one study, however, NO_3^- uptake appeared to increase linearly with external NO_3^- concentration (Hansen 1976). There is evidence in at least some marine algae that NO_3^- is present in the cytoplasm or vacuole at higher electrochemical potential than sea-water

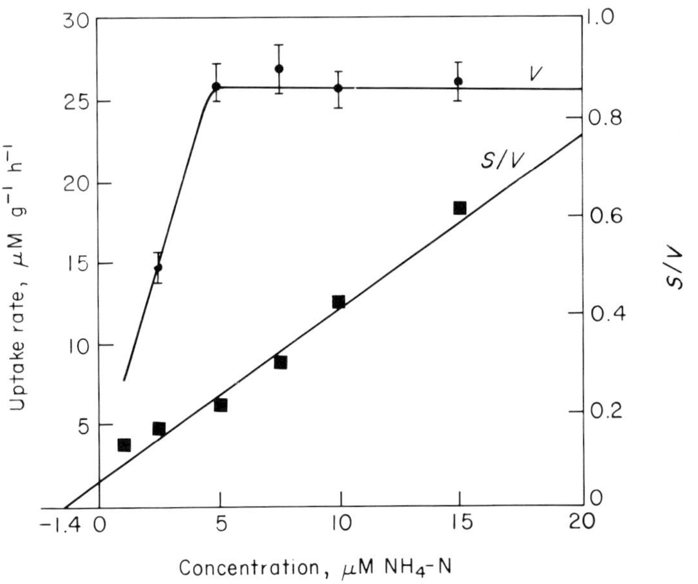

Fig. 10.2 Ammonium uptake rate of *Codium fragile* as a function of concentration and the (S/V) against (S) plot at 24°C and 56 $\mu E \cdot m^{-2} \cdot s^{-1}$; each point is mean ± SE of two experiments with three replicates (n = 6). (From Hanisak & Harlin (1978).)

Table 10.4 Nitrogen uptake kinetic 'constants' for seaweeds.

Alga	Temperature C	NO_3^-			NH_4^+			References
		$K \pm SE$ μM	V_{max} μm hr^{-1}g^{-1}dw	V_{max} μm h^{-1}cm^{-2}	$K \pm SE$ μM	V_{max} μmhr^{-1}g^{-1}dw	V_{max} μm h^{-1}cm^{-2}	
Fucus spiralis	5	6.6 ± 0.9		0.16	6.4 ± 2.0		0.18	Topinka (1978)
	10	6.7 ± 0.8		0.20	5.4 ± 2.0		0.26	
	15	7.8 ± 1.4		0.23	9.6 ± 2.6		0.35	
	15 (dark)	12.8 ± 3.5		0.28	5.8 ± 1.8		0.29	
Laminaria longicruris	15[a]	4.1	9.6					Harlin & Craigie (1978)
	10[b]	5.9	7.0					
Macrocystis pyrifera	16	13.1 ± 1.6	30.5		5.3[c] ± 1.0	23.8[c]		Haines & Wheeler (1978)
	6–9				50[d]	23.6[d]		Wheeler (1979)
Iridaea cordata					2.5	5.5		Hansen (1976)
Gracilaria foliifera	20	2.5 ± 0.5	9.7		1.6[c]	23.8[c]		D'Elia & DeBoer (1978)
Agardhiella subulata[e]	20	2.4 ± 0.3	11.7		3.9[c]	15.9[c]		D'Elia & DeBoer (1978)
Hypnea musciformis	26	4.9 ± 3.9	28.5		16.6 ± 1.8			Haines & Wheeler (1978)
Enteromorpha spp.	15	16.6	129.4					Harlin (1978)
Codium fragile	6	1.9 ± 0.5	2.8		1.5 ± 0.2	13.0		Hanisak & Harlin (1978)
	24	7.6 ± 0.6	9.6		1.4 ± 0.2	28.0		

[a] Summer Tissue.
[b] Winter Tissue.
[c] Mechanism 1 Uptake (high affinity system).
[d] Uptake of methylamine, an analogue of NH_4^+.
[e] Formerly *Neoagardhiella baileyi*.

(Gutknecht & Dainty 1968; Eppley & Rogers 1970). In these cases it would appear that NO_3^- is accumulated by an active transport system.

Table 10.4 summarizes the reported N uptake kinetics for seaweeds. Values of K are generally in the range of 2–13 μM and values of V_{max}, 8–25 μm hr^{-1} g^{-1} dw for both NO_3^- and NH_4^+. Hanisak & Harlin (1978) reported that for *Codium fragile* values of K (3–10 μM) for NO_2^- uptake were significantly higher than for NO_3^- or NH_4^+ uptake. Values of V_{max} for NO_2^- were comparable to NO_3^- uptake but lower than NH_4^+ uptake.

There is indication that dual mechanisms for ammonium uptake may be present in some seaweeds. D'Elia & DeBoer (1978) reported the presence of both a low affinity and a high affinity uptake system for NH_4^+ in both *Gracilaria foliifera* and *Agardhiella subulata* (Fig. 10.3). Below 10 μM a high NH_4^+ affinity uptake system predominates, and at high concentrations either a low affinity (high K), high V_{max} component or a strong diffusive component predominates. The curve fitted to the data in Fig. 10.3 is drawn assuming a diffusive component predominates at high NH_4^+ concentrations. Haines & Wheeler (1978) presented evidence suggesting a second uptake mechanism for NH_4^+ in *Macrocystis pyrifera*. No other instances of multiphasic uptake systems have been reported for seaweeds, although they may never have been sought. The NO_3^- and NH_4^+ uptake kinetics in *Fucus spiralis* (Topinka 1978) appear to indicate a multiphasic uptake system.

Hanisak & Harlin (1978) found that the values of V_{max} and K for NO_3^- and NO_2^- uptake in *Codium fragile* generally increased with temperatures between 6 and 24°C. Values of V_{max} for NH_4^+ uptake also increased with temperature;

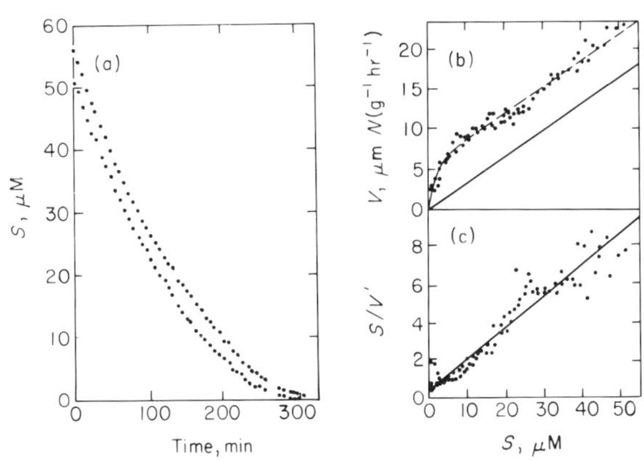

Fig. 10.3 (a) Replicate NH_4^+ depletion experiments with *Agardhiella subulata* (C/N = 6.7) preconditioned on NH_4^+. (b) Broken line = rate of uptake of NH_4^+ v. NH_4^+ concentration; solid line = diffusive component calculated as the regression coefficient (slope) using all V points with corresponding S values > 4.5 μmM. (c) S/V versus S linear transformation of data in (b) using V values obtained by subtracting diffusive component from all V values. (From D'Elia & DeBoer (1978).)

however, K values appeared independent of temperatures in the range of 6–30°C. Uptake rates of NH_4^+ and NO_3^- are also affected by irradiance levels with higher rates of uptake measured in the light as compared to dark (Hanisak & Harlin 1978; D'Elia & DeBoer 1978; Harlin 1978; Harlin & Craigie 1978).

There have been several reports that NO_3^- uptake by seaweeds is suppressed by the presence of NH_4^+ in the medium. For many seaweeds (Hanisak & Harlin 1978; Haines & Wheeler 1978; D'Elia & DeBoer 1978) this suppression is substantial (up to 50%). Bird (1976) on the other hand, reported that *Gelidium nudifrons* was able to take up NH_4^+ and NO_3^- simultaneously with no suppression of NO_3^- uptake by NH_4^+.

In most seaweeds that have been studied, the uptake rate of NH_4^+ generally exceeds that of NO_3^- at all nitrogen concentrations normally present in the environment. At 'environmentally' high concentrations (>30 μM), however, NH_4^+ may be toxic to some seaweeds (Waite & Mitchell 1972; DeBoer *et al.* 1976; Haines & Wheeler 1978). Some of the discrepancies regarding the utilization of NH_4^+ by seaweeds may have been the result of toxicity caused by abnormally high NH_4^+ levels in the medium (Hanisak 1979a).

D'Elia & DeBoer (1978) presented evidence which indicated that the rate of NH_4^+ uptake by *Gracilaria foliifera* and *Agardhiella subulata* is a function not only of the external but also the internal N concentration. At C/N atomic ratios below 10, NH_4^+ uptake rate increased linearly with increasing C/N ratio, whereas at C/N ratios greater than *c.* 10, NH_4^+ uptake was essentially constant. This implies that seaweeds living in low nitrogen waters (and therefore having a high C/N ratio) can take up NH_4^+ at maximal rates when NH_4^+ supplies become available. Seaweeds living in relatively high nitrogen waters (and therefore having a low C/N ratio) have a reduced ability to take up NH_4^+ until their internal N reserves are reduced. Since N assimilation is energy dependent, it appears expedient for seaweeds to utilize available energy for N uptake when internal N reserves are low and to utilize much of that energy elsewhere when internal N reserves are sufficient. For comparison, the C/N atomic ratios of natural populations of seaweeds generally range between 8 and 30 (Niell 1976; Hanisak 1979b; DeBoer unpublished).

Several generalized schemes have been presented to explain N assimilation in algae. One of the most recent and comprehensive of these models (Fig. 10.4) serves to illustrate some of the complicated interactions such as membrane transport systems, photosystems I and II, reductase enzymes, amino acid synthesis and feedback control systems which are involved in N assimilation.

4.5 Assimilation of phosphorus

In several giant-celled algae and *Porphyra*, phosphate is believed to be taken up by active transport at the plasmalemma (Eppley 1958; Gutknecht & Dainty 1968). ATP seems to be the energy source for phosphate influx in all cases

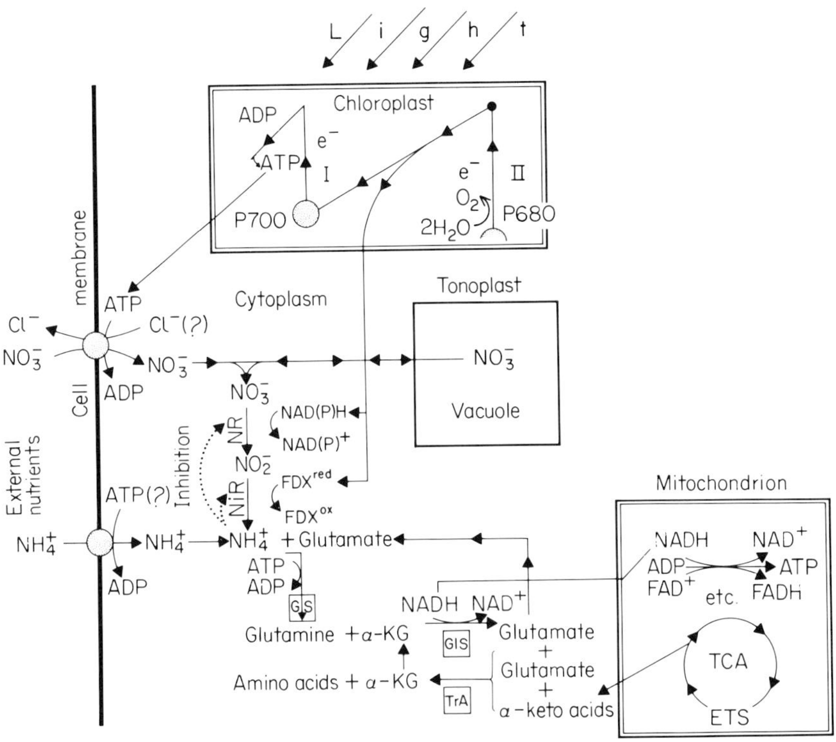

Fig. 10.4 A generalized speculative scheme of primary nitrate metabolism. Nitrate uptake is facilitated by a (NO_3^-, Cl^-)-activated ATPase, located in the plasma membrane. Once transported into the cytoplasm, NO_3^- may be stored in a vacuole, or reduced via nitrate reductase (NR) to NO_3^-. Subsequently, NO_3^- is reduced to NH_4^+ via nitrite reductase (NiR). The primary source of reductant cofactors for these processes in the light is probably from photosystems I and II. The incorporation of NH_4^+ into amino acids appears to be mediated by glutamine synthetase (GIS) and glutamate synthase (GS). NH_4^+ may inhibit NiR and NR through an unknown intermediary. Although the occurrence of ATPases in animal systems is undisputed, their presence in the algae is not well documented. Other abbreviations: α–KG = α-ketoglutarate; TCA = tricarboxylic acid cycle; ETS = electron transport system; FDX^{red}, FDX^{ox} = reduced and oxidized ferredoxin; TrA = transaminase. (From Falkowski (1978).)

investigated. O'Gorman (unpublished) measured an apparent K value of 0.4 μM for $H_2PO_4^{3-}$ uptake and an average V_{max} value of 0.47 μm hr^{-1} g^{-1} dw for the red alga, *Agardhiella subulata*. Several estimates have been made of the ratio of nitrogen/phosphorus which will ensure optimal seaweed growth. These ratios range from a low of 5:1 for *Ectocarpus confervoides* (Boalch 1961) to 31:1 and 39:1 for *Fucus vesiculosus* and *Chondrus crispus*, respectively (Prince 1974).

4.6 Assimilation of trace metals

The uptake of zinc and strontium by seaweeds has received a considerable amount of attention (Gutknecht 1961, 1963, 1965; Bryan 1969; Chaing *et al.*

1974; Skipnes *et al.* 1975). The results seem to indicate that strontium is primarily accumulated by a simple exchange process involving intercellular polysaccharides (Skipnes *et al.* 1975). Although a small portion of the zinc uptake in *Ascophyllum nodosum* can be attributed to ion exchange, the main uptake probably involves transport through the plasmalemma and tonoplast (Skipnes *et al.* 1975). Gutknecht (1963, 1965) reported that adsorption exchange is largely responsible for Zn movements in *Ulva lactuca, Porphyra umbilicalis, Laminaria agardhi* and *Gracilaria* sp.

Bunt (1970) investigated cobalt uptake by four species of green algae and five species of red algae. Uptake of cobalt by *Laurencia corallopsis* exhibited saturation kinetics with values for K and V_{max} reported to be 10–17 μg Co l^{-1} and 0.7–1.0 μg Co hr^{-1} g^{-1} dw, respectively.

4.7 Assimilation of uncharged molecules

Seaweeds require several neutral inorganic compounds for growth and maintenance metabolism including CO_2, O_2, H_2O and NH_3. At high pH an appreciable amount of ammonia is believed to enter cells in the undissociated form (Raven & Smith 1974) and at low pH values molecular CO_2 is the form in which inorganic carbon enters algal cells (Raven 1970). It appears from the little evidence that is available that passive uptake may account for the fluxes of these molecules (Raven 1976).

5 NUTRIENT GROWTH KINETICS

5.1 Measurement of growth kinetics

Nutrient growth kinetics of seaweeds can be studied using large-volume batch culture or continuous flow cultures. Nutrient concentration variations in each container are minimized by frequent medium changes (batch method) or high dilution rate (continuous flow) and high culture volume: biomass ratios. In order to approach steady state growth conditions it is necessary to maintain a reasonably constant biomass in the system by harvesting at frequent intervals or by adjusting influent nutrient concentrations to compensate for increased biomass resulting from growth. The frequency of harvesting necessary and the time required to achieve steady state will depend on the growth rate of the alga, culture conditions and previous nutrient history.

DeBoer *et al.* (1978) have arbitrarily used the criterion that the seaweed biomass must undergo an increase equal to at least one order of magnitude to help insure that internal nutrient reserves and growth rate are at steady state with regard to the concentration of limiting nutrient in the culture medium. In an alga growing at 10% day^{-1}, for example, this rule would necessitate continuing the experiment for at least 23 days, with biomass adjustments every two to three days.

5.2 Nitrogen growth kinetics

The N growth kinetics of *Agardhiella subulata* (Fig. 10.5) illustrate several quantitative as well as qualitative aspects of N utilization by seaweeds (DeBoer *et al.* 1978). The specific growth rate of *A. subulata* increased with increasing N concentration for all sources of N enrichment, up to N concentrations of approximately 1.0 μM. Growth rates did not increase substantially at higher N concentrations. At the higher N concentrations, the growth rate of *A. subulata* receiving NH_4^+ enrichment was nearly double that of the plants receiving NO_3^- enrichment. Urea proved to be a comparatively inferior N source. Apparent half saturation constants (*K*) for growth were 0.2 μM N for all three sources of N. Similar results were reported for *Gracilaria foliifera* in which values of *K* for growth were 0.2 μM NH_4^+ and 0.4 μM NO_3^- (DeBoer *et al.* 1978).

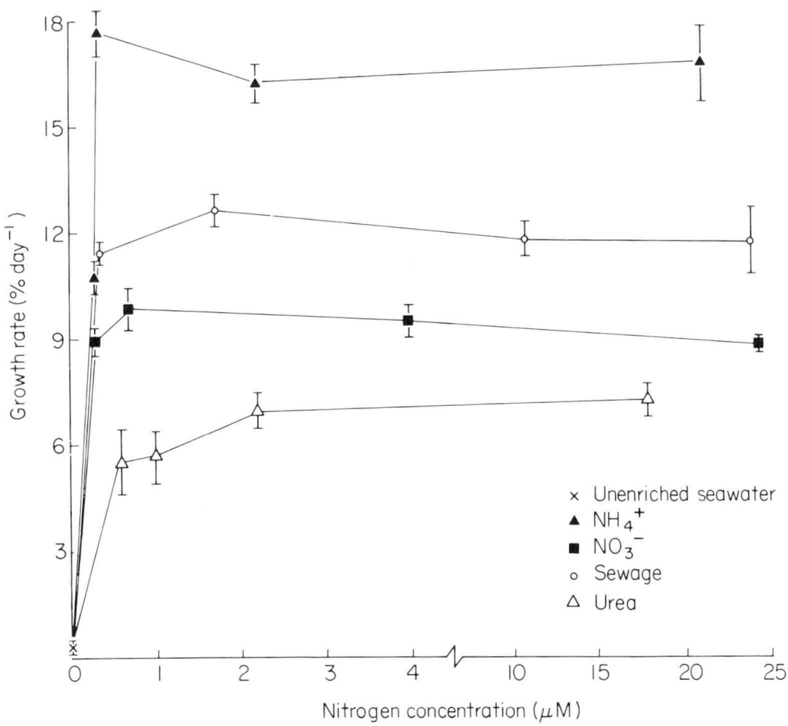

Fig. 10.5 Growth rates (±1 S.E.) of *Agardhiella subulata* as a function of residual N concentration for various N enrichment sources. (From DeBoer *et al.* (1978).)

Chapman *et al.* (1978), working with *Laminaria saccharina* sporophytes, reported that the growth curves under NO_3^- limiting conditions followed typical Michaelis–Menten kinetics with apparent saturation occurring between 10 and 20 μM NO_3^-. Half saturation constants for growth (K) averaged *c.* 1.4 μM NO_3^-. In addition, they found that the saturation level for growth is within the environmental NO_3^- range occurring at their collection site (Helgoland). The authors speculated that seaweeds occurring in less NO_3^- rich areas may have lower NO_3^- saturation levels for growth.

6 BIOCHEMICAL INDICATORS OF NUTRIENT STATUS

Recent studies with phytoplankton have shown that while growth rate is often related to the concentration of nutrients in the external medium, growth rate can be more accurately predicted by the concentration of nutrients within the cells (Gerloff & Skoog 1954; Droop 1973; Bienfang 1975; Conover 1975; Goldman 1977). This relationship has been observed in higher plants for many years.

Plant tissue analysis is a common agricultural technique which measures a species' critical nutrient concentration (Munson & Nelson 1973; Ulrich & Hills 1973). The critical concentration is that concentration of a nutrient in the tissue just below the level giving optimum growth. A 5 or 10% reduction in growth rate is arbitrarily chosen in most instances. Once this concentration has been determined under controlled conditions, tissue analysis of a crop sample can be conducted to obtain information from the plant itself about its nutrient status.

This technique has been applied to fresh water algae and aquatic vascular plants (Gerloff & Skoog 1954; Gerloff 1973; Wong & Clark 1976) and, more recently, to seaweeds (Hansen 1977; Hanisak 1979b; DeBoer unpublished). The relationship between growth of *Codium fragile* and its internal N concentration is shown in Fig. 10.6. Whereas internal N values ranged from 0.9 to 4.8%, growth rate remained constant at N concentrations in excess of *c.* 2.0%. The critical N level in *Codium fragile* was estimated to be 1.9%. DeBoer (unpublished) found the critical N concentrations in *Gracilaria foliifera* and *Agardhiella subulata* to be *c.* 1.8–2.0%. Because of the variability in dry weight values caused by differences in the salt contents of seaweeds, the carbon content is sometimes used as a reference in lieu of dry weight. The nitrogen:carbon ratios and the carbon:nitrogen ratios have also proven useful as indicators of the N status of seaweeds (Niell 1976; DeBoer & Ryther 1977; Jackson 1977; Hanisak 1979b). Critical C/N atomic ratios ranging between 10 and 15 have been reported for seaweeds (Hanisak 1979b; DeBoer unpublished).

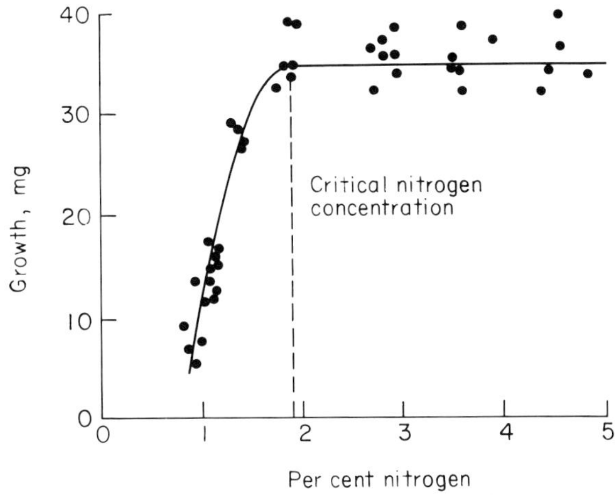

Fig. 10.6 Relationship of growth of *Codium fragile* and internal nitrogen concentration. Growth was determined on thalli maintained in 100 ml batch cultures at seven different levels of NO_3^- enriched artificial medium. The medium was changed every three days and the thalli harvested at the end of 21 days. Growth is expressed as the calculated increase in dry weight. (After Hanisak (1979b).)

7 THE EFFECTS OF NUTRIENT AVAILABILITY ON THE BIOLOGY OF SEAWEEDS

7.1 Growth rate, productivity, and distribution

The growth rate and productivity of seaweed populations are normally controlled by solar radiation, water temperature, nutrient concentration and water movement. Numerous examples have been cited in which a nutrient (usually N) is the growth limiting factor for seaweeds during a portion of the growing season. In many instances, however, periods of rapid growth occur during times when the external nutrient concentrations are relatively low. Many seaweeds have the ability to assimilate and store nutrients from the environment when they are present in concentrations in excess of those required for immediate utilization, and then utilize these nutrient reserves when external nutrient concentrations become growth limiting (Buggeln 1974, 1978; Chapman & Craigie 1977). Conversely, some seaweeds have been reported to store organic carbon compounds, such as laminaran, during the periods of excess solar radiation. Those reserves may then be utilized at a later time to produce new plant growth even in complete darkness (Lüning 1970, 1971). These important ecological strategies enable some seaweeds to persist and even flourish during periods which may be extremely stressful to smaller seaweeds and phytoplankton. Other strategies apparently employed by seaweeds include the ability of some species to utilize alternative sources of some ions (or substitute other ions) and the ability of some species to scavenge nutrients from extremely low external concentration.

7.2 Biochemical composition

Healey (1973) noted three general responses of algae to nutrient deficiency. The first general response is a decrease in the content of photosynthetic pigments. A decrease in phycoerythrin content was observed in *Pterocladia capillacea* under N limiting conditions (Calabrese & Felicini 1970). Ammonium concentrations (initial) of 10^{-3} M inhibited growth rate of *P. capillacea* but stimulated pigment synthesis in batch culture experiments. Chlorophyll and phycoerythrin concentrations in *Gracilaria foliifera, Agardhiella subulata* and *Ceramium rubrum* were strongly influenced by concentrations of inorganic N in the culture medium (DeBoer & Ryther 1977; DeBoer unpublished). Highest chlorophyll and phycoerythrin contents in *G. foliifera* and *A. subulata* occurred in plants grown at inorganic N concentrations in excess of 3 μM. Chlorophyll content in *Laminaria saccharina* increased with increasing N concentrations up to *c.* 20 μM NO_3^- (Chapman *et al.* 1978).

A second general response to nutrient deficiency in algae is the accumulation of C-storage compounds. In seaweeds this has been best demonstrated in cell wall polysaccharides in the red algae. Neish & Shacklock (1971) and Neish *et al.* (1977) demonstrated that *Chondrus crispus* cultured in unenriched seawater had a higher carrageenan content than when grown in N-enriched medium. Carrageenan content in *Agardhiella subulata* was highest at N concentrations in the range of 0.5–1.0 μM NH_4^+ and was lower at both higher NH_4^+ concentrations and in unenriched seawater (DeBoer 1979). Agar content in *Gracilaria foliifera* was highest in plants receiving unenriched seawater and decreased substantially in plants receiving increasing N enrichment (DeBoer 1979).

A third general response to nutrient deficiency is a decrease in proteins and nucleic acids. Both quantitative and qualitative differences in the amino acid compositions of *Ulva fasciata* (Nasr *et al.* 1968; Mohsen *et al.* 1974), *Ulva lactuca, Pterocladia capillacea* and *Dictyota dichotoma* (Nasr *et al.* 1968) were observed in plants grown with various sources and concentrations of N. In *Ulva lactuca*, for example, the concentrations of β-alanine and asparagine can vary as much as 30-fold depending on external N concentrations (Nasr *et al.* 1968).

7.3 Reproduction, development and morphology

The onset of sexual or asexual reproduction in microalgae frequently has been attributed to nutrient availability (Coleman 1962; Healey 1973). Similar relationships occur in seaweeds although they are not as well documented. Mohsen *et al.* (1974) reported that conditions of N depletion enhance gamete formation in *Ulva fasciata*, whereas higher N concentrations favour vegetative growth and asexual reproduction. Similarly, Nasr *et al.* (1968) observed abundant zoospore formation after NH_4^+ treatment in *Ulva lactuca*. Low concentrations of phosphate and nitrate stimulated the formation of sexual organs in *Vaucheria dichotoma* (Åberg & Fries 1976). In laboratory studies, gametogenesis in *Laminaria saccharina* did not take place at NO_3^- concentrations below 5 μM (Hsiao & Druehl 1973). However, in field studies conducted in British Columbia, they found that the rate of gametogensis was most likely a function of light rather than nutrient deficiency. Tetraspore induction in several red algae has been attributed to a combination of photoperiodicity and N or P deficiency (Knaggs 1967; Ramus 1969; Oza 1977). Iodine appears to be required for the development of *Ralfsia*-like thalli and blades as well as for reproduction in *Petalonia fascia* (Hsiao 1969). Iodine also is essential for normal formation and maturation of plurilocular sporangia in *Ectocarpus siliculosus* (Woolery & Lewin 1973). The concentration of boron was reported to influence the development and reproduction of *Ulva lactuca* and *Dictyota dichotoma* (Nasr & Bekheet 1970). Nitrogen starvation in *Acetabularia mediterranea* caused hypertrophy and inhibited whorl release (Adamich *et al.* 1975).

Gibor (1973a, 1973b) and Adamich *et al.* (1975) have suggested that whorls in *Acetabularia* may play an absorptive role analogous to that of root hairs of higher plants. It is suggested that whorls undergo hypertrophy and assist in uptake of trace nutrients under nutrient limited conditions. Adamich *et al.*, (1975) found an inverse relationship between whorl development and NO_3^- concentration in the medium, greatly increasing cell surface under conditions of N limitation. Whoriskey & DeBoer (unpublished) observed that under low N concentrations and high light intensity hyaline hair formation was greatly enhanced in *Ceramium rubrum, Hypnea musciformis, Gracilaria* spp., and *Agardhiella subulata*. Hair formation substantially increased the surface area of these seaweeds and, interestingly, also substantially increased the rate of N uptake suggesting a possible nutrient ion absorptive role.

8 CONCLUSIONS

Nutrient availability is undoubtedly one of the primary factors regulating the biology of seaweeds. Clearly, all seaweeds are not identical in their qualitative and quantitative nutrient requirements. Nutrient availability has been shown to limit the growth rate and productivity, and, in some cases, may control the biochemical content, reproduction, development, morphology and distribution of seaweeds.

Seaweeds have evolved several effective strategies to adapt to low or fluctuating nutrient regimes. Under conditions of nutrient depletion some seaweeds may be induced to alter their morphology to increase their surface area and, hence, nutrient uptake. Other seaweeds have evolved tenacious holdfasts and a resistant morphology which enable them to exist in comparatively nutrient-rich surf zones. Many seaweeds have evolved transport and enzyme systems which enable them to utilize more than one form of a nutrient (e.g.

NH_4^+, NO_3^-, NO_2^- and organic nitrogen) giving them a competitive advantage over other algae lacking this capability.

Evidence exists that the species distribution of phytoplankton may be influenced by differential N assimilatory capacities between coastal and offshore populations and by differences in N concentrations in those environments (Eppley *et al.* 1969; Eppley & Thomas 1969; Carpenter & Guillard 1970). Insufficient data, however, on the nutrient uptake kinetics, growth kinetics, nutrient storage and utilization are available to assess the full ecological importance of nutrient assimilation capacities in seaweeds.

A more thorough understanding of the utilization of internal nutrient reserves versus external nutrient sources, nutrient fluxes, sizes of compartments, differences in and specificities of nutrient transport and enzyme systems will undoubtedly lead to a better understanding of the biology of seaweeds.

9 REFERENCES

ÅBERG H. & FRIES L. (1976) On the cultivation of the alga *Vaucheria dichotoma* (Xanthophyceae) in axenic culture. *Phycologia*, **15**, 133–41.

ADAMICH M., GIBOR A. & SWEENEY B. M. (1975) Effects of low nitrogen levels and various nitrogen stores on growth and whorl development in *Acetabularia* (Chlorophyta). *J. Phycol.*, **11**, 364–9.

ANDREAE M. O. (1978) Distribution and speciation of arsenic in natural waters and some marine algae. *Deep-Sea Res.*, **25**, 391–402.

ARNON D. I. (1953) Growth and function as criteria in determining the essential nature of inorganic nutrients. In *Mineral Nutrition of Plants*, ed. Trong E., pp. 313–41. The University of Wisconsin Press, Madison, Wisconsin.

BARDACH J. E., RYTHER J. H. & MCLARNEY W. O. (1972) *Aquaculture. The Farming and Husbandry of Freshwater and Marine Organisms*, pp. 790–814. Wiley-Interscience, New York.

BIDWELL R. G. S. (1979) *Plant Physiology*, 2nd edn. Macmillan, New York.

BIENFANG P. K. (1975) Steady state analysis of nitrate-ammonium assimilation by phytoplankton. *Limnol. Oceanogr.*, **20**, 402–11.

BIRD K. T. (1976) Simultaneous assimilation of ammonium and nitrate by *Gelidium nudifrons* (Gelidiales: Rhodophyta). *J. Phycol.*, **12**, 238–41.

BLACK W. A. P. & MITCHELL R. L. (1952) Trace elements in the common brown algae and in sea water. *J. Mar. Biol. Assoc. U.K.*, **30**, 575–84.

BLUM J. J. (1966) Phosphate uptake by phosphate-starved *Euglena*. *J. Gen. Physiol.*, **49**, 1125–37.

BOALCH G. T. (1961) Studies on *Ectocarpus* in culture. II. Growth and nutrition of a bacteria-free culture. *J. Mar. Biol. Assoc. U.K.*, **41**, 279–86.

BOROWITZKA M. A. (1977) Algal calcification. *Ann. Rev. Oceanogr. Mar. Biol.*, **15**, 189–223.

BRYAN G. W. (1969) The absorption of zinc and other metals by the brown seaweed *Laminaria digitata*. *J. Mar. Biol. Assoc. U.K.*, **49**, 225–43.

BRYAN G. W. (1976) Heavy metal contamination in the sea. In *Marine Pollution*, ed. Johnson R., pp. 185–302. Academic Press, London.

BRYAN G. W. & HUMMERSTONE L. G. (1973) Brown seaweed as an indicator of heavy metals in estuaries in south west England. *J. Mar. Biol. Assoc. U.K.*, **53**, 705–20.

BUGGELN R. G. (1974) Physiological investigations on *Alaria esculenta* (L.) Grev. (Laminariales). I. Elongation of the blade. *J. Phycol.*, **10**, 283–8.

BUGGELN R. G. (1978) Physiological investigations on *Alaria esculenta* (Laminariales, Phaeophyceae). IV. Inorganic and organic nitrogen in the blade. *J. Phycol*, **14**, 156–160.

BUNT J. S. (1970) Uptake of cobalt and vitamin B_{12} by tropical marine macroalgae. *J. Phycol.*, **6**, 339–43.

CALABRESE G. & FELICINI G. P. (1970) Research on the red algae pigments. I. Pigments of *Pterocladia capillacea* cultured with some nitrogenous materials. *Giorn. Bot. Ital.*, **104**, 81–9.

CARPENTER E. J. & GUILLARD R. R. L. (1970) Intraspecific differences in nitrate half-saturation constants for three species of marine phytoplankton. *Ecology*, **52**, 183–5.

CHAPMAN A. R. O. & CRAIGIE J. S. (1977) Seasonal growth in *Laminaria longicruris*: Relations with dissolved inorganic nutrients and internal reserves of nitrogen. *Mar. Biol.*, **40**, 197–205.

CHAPMAN A. R. O., MARKHAM J. W. & LÜNING K. (1978) Effects of nitrate concentration on the growth and physiology of *Laminaria saccharina* (Phaeophyta) in culture. *J. Phycol.*, **14**, 195–8.

CHARTERS A. C. & NEUSHUL M. (1979) A hydrodynamically defined culture system for benthic seaweeds. *Aquat. Bot.*, **6**, 67–78.

CHEN T. P. (1976) Culture of *Gracilaria*. In *Aquaculture Practices in Taiwan*, ed. Chen T. P., pp. 145–9. Fishing News Books, Surrey, England.

CHIANG Y. M., HUANG R. & SU H. M. (1974) The uptake of ^{65}Zn by marine benthic algae. *Acta Oceanogr. Taiwanica*, **4**, 241–8.

COLEMAN A. W. (1962) Sexuality. In *Physiology and Biochemistry of Algae*, ed. Lewin R. A., pp. 711–29. Academic Press, New York and London.

CONOVER, S. A. M. (1975) Partitioning of nitrogen and carbon in cultures of the marine diatom *Thalassiosira fluviatilis* supplied with nitrate, ammonium, or urea. *Mar. Biol.*, **32**, 231–46.

DARLEY W. M. (1974) Silification and calcification. In *Algal Physiology and Biochemistry*, ed. Stewart W. D. P., pp. 655–75. Blackwell Scientific Publications, Oxford.

DEBOER J. A. (1979) Effects of nitrogen enrichment on growth rate and phycocolloid content in *Gracilaria foliifera* and *Neoagardhiella baileyi* (Florideophyceae). *Proc. Int. Seaweed Symp.*, **9**, 263–71.

DEBOER J. A., GUIGLI H. J., ISRAEL T. L. & D'ELIA C. F. (1978) Nutritional studies of two red algae. I. Growth rate as a function of nitrogen source and concentration. *J. Phycol.*, **14**, 261–6.

DEBOER J. A., LAPOINTE B. E., D'ELIA C. F. (1976) Effects of nitrogen concentration on growth rate and carrageenan production in *Neaogardhiella baileyi*. In *Marine Polyculture Based on Natural Food Chains and Recycled Wastes*, ed. Ryther J. H. Woods Hole Oceanographic Institution, Tech. Rep. WHOI-76-92, Paper No. 10.

DEBOER J. A. & RYTHER J. H. (1977) Potential yields from a waste-recycling algal mariculture system. In *The Marine Plant Biomass of the Pacific Northwest Coast*, ed. Krauss R., pp. 231–49. Oregon State University Press, Corvallis, Oregon.

D'ELIA C. F. & DEBOER J. A. (1978) Nutritional studies of two red algae. II. Kinetics of ammonium and nitrate uptake. *J. Phycol.*, **14**, 266–72.

DOWD J. E. & RIGGS D. S. (1965) A comparison of estimates of Michaelis–Menten kinetic constants from various linear transformations. *J. Biol. Chem.*, **240**, 863–9.

DROOP M. R. (1973) Some thoughts on nutrient limitation in algae. *J. Phycol.*, **9**, 264–72.

EPPLEY R. W. (1958) Sodium exclusion and potassium retention by the marine red alga *Porphyra perforata*. *J. Gen. Physiol.*, **41**, 901–11.

EPPLEY R. W. & BLINKS L. R. (1957) Cell space and apparent free space in the red alga, *Porphyra perforata*. *Plant Physiol*, **32**, 63–4.

EPPLEY R. W. & CYRUS C. C. (1960) Cation regulation and survival of the red alga *Porphyra perforata*, in diluted and concentrated sea waters. *Biol. Bull.*, **118**, 55–6.

EPPLEY R. W. & ROGERS J. N. (1970) Inorganic nitrogen assimilation of *Ditylum brightwellii*, a marine plankton diatom. *J. Phycol.*, **6**, 344–51.

EPPLEY R. W., ROGERS J. N. & MCCARTHY J. J. (1969) Half saturation 'constants' for uptake of nitrate and ammonium by marine phytoplankton. *Limnol. Oceanogr.*, **14**, 912–20.

EPPLEY R. W. & THOMAS W. H. (1969) Comparison of half-saturation constants for growth and nitrate uptake by marine phytoplankton. *J. Phycol.*, **5**, 375–9.

EPSTEIN E. (1966) Dual pattern of ion absorption by plant cells and by plants. *Nature*, **212**, 1324–7.

EPSTEIN E. (1972) *Mineral Nutrition of Plants: Principles and Perspectives*. John Wiley and Sons, New York.

EVANS H. J. & SORGER G. J. (1966) Role of mineral elements with emphasis on the univalent cations. *Ann. Rev. Plant Physiol.*, **17**, 47–76.

FALKOWSKI P. G. (1978) The regulation of nitrate assimilation in lower plants. In *Nitrogen in the Environment*, vol. 2, eds. Nielsen D. R. & MacDonald J. G., pp. 143–55. Academic Press, New York.

FENICAL W. (1975) Halogenation in the Rhodophyta—A review. *J. Phycol.*, **11**, 245–59.

FOGG G. E. (1974) Nitrogen fixation. In *Algal Physiology and Biochemistry*, ed. Stewart W. D. P., pp. 560–82. Blackwell Scientific Publications, Oxford.

FRIES L. (1963) On the cultivation of axenic red algae. *Physiol. Plant.*, **16**, 695–708.

FRIES L. (1966) Influence of iodine and bromine on growth of some red algae in axenic culture. *Physiol. Plant.*, **19**, 800–8.

FRIES L. (1975) Requirement of bromine in a red alga. *Z. Pflanzenphysiol. Bd.*, **76**, 366–8.

FRIES L. & PETTERSSON H. (1968) On the physiology of the red alga *Asterocystis ramosa* in axenic culture. *Br. Phycol. Bull.*, **3**, 417–22.

FUJIYAMA T. & MAEDA M. (1977) Content and uptake of trace metals in benthic algae, *Enteromorpha* & *Porphyra*. I. Measurement and variation of trace metal content of *Porphyra* grown in natural environment. *J. Fac. Fish. Anim. Husb. Hiroshima Univ.*, **16**, 23–32.

GAVIS J. (1976) Munk and Riley revisited: nutrient diffusion transport and rates of phytoplankton growth. *J. Mar. Res.*, **34**, 161–79.

GERARD V. A. & MANN K. H. (1979) Growth and production of *Laminaria longicruris* (Phaeophyta) populations exposed to different intensities of water movement. *J. Phycol.*, **15**, 33–41.

GERLOFF G. C. (1963) Comparative mineral nutrition of plants. *Ann. Rev. Plant Physiol.*, **14**, 107–24.

GERLOFF G. C. (1973) *Plant Analysis for Nutrient Assay of Natural Waters*. U.S. Environmental Protection Agency, Environ. Health Effects Res. Ser., Washington, D.C.

GERLOFF G. C. & SKOOG F. (1954) Cell contents of nitrogen and phosphorus as a measure of their availability for growth of *Microcystis aeruginosa*. *Ecology*, **34**, 348–53.

GIBOR A. (1973a) Observations on the sterile whorls of *Acetabularia*. *Protoplasma*, **78**, 195–202.

GIBOR A. (1973b) *Acetabularia*. Physiological role of their deciduous organelles. *Protoplasma*, **78**, 461–5.

GIEBISCH G., TOSTESON D. C. & USSING H. H. (1978) *Membrane Transport in Biology. Vol. III. Transport Across Multi-Membrane Systems*. Springer Verlag, Berlin.

GOLDMAN J. C. (1977) Steady state growth of phytoplankton in continuous culture: comparison of internal and external nutrient equations. *J. Phycol.*, **13**, 251–8.

GUTKNECHT J. (1961) Mechanism of radioactive zinc uptake by *Ulva lactuca*. *Limnol. Oceanogr.*, **6**, 426–31.

GUTKNECHT J. (1963) Zinc-65 uptake by benthic marine algae. *Limnol. Oceanogr.*, **8**, 31–8.

GUTKNECHT J. (1965) Uptake and retention of Cesium 137 and Zinc 65 by seaweeds. *Limnol. Oceanogr.*, **10**, 58–66.

GUTKNECHT J. & DAINTY J. (1968) Ionic relations of marine algae. *Ann. Rev. Oceanogr. Mar. Biol.*, **6**, 163–200.

GUTKNECHT J., HASTINGS D. F. & BISSON M. A. (1978) Ion transport and turgor pressure regulation in giant algal cells. In *Membrane Transport in Biology, Vol. 3, Transport Across Multi-Membrane Systems*, eds. Giebisch G., Tosteson D. C. & Ussing H. H., pp. 125–74. Springer Verlag, Berlin.

HAINES K. C. & WHEELER P. A. (1978) Ammonium and nitrate uptake rates of the seaweeds *Hypnea musiciformis* (Rhodophyta) and *Macrocystis pyrifera* (Phaeophyta). *J. Phycol.*, **14**, 319–24.

HANISAK M. D. (1979a) Growth patterns of *Codium fragile* ssp. *tomentosoides* in response to temperature, irradiance, salinity and nitrogen source. *Mar. Biol.*, **50**, 319–32.

HANISAK M. D. (1979b) Nitrogen limitation of *Codium fragile* ssp. *tomentosoides* as determined by tissue analysis. *Mar. Biol.*, **50**, 333–7.

HANISAK M. D. & HARLIN M. M. (1978) Uptake of inorganic nitrogen by *Codium fragile* ssp. *tomentosoides* (Chlorophyta). *J. Phycol.*, **14**, 450–4.

HANSEN J. E. (1976) Population Biology of *Iridaea cordata* (Rhodophyta: Gigartinaceae). Ph.D. Thesis, University of California, Santa Cruz.

HANSEN J. E. (1977) Productivity of *Iridaea cordata* (Rhodophyta: Gigartinaceae). *J. Phycol.*, **13** (suppl.), abstr. 27.

HARLIN M. M. (1978) Nitrate uptake by *Enteromorpha* spp. (Chlorophyceae): Applications to aquaculture systems. *Aquaculture*, **15**, 373–6.

HARLIN M. M. & CRAIGIE J. S. (1978) Nitrate uptake by *Laminaria longicruris* (Phaeophyceae). *J. Phycol.*, **14**, 464–7.

HEALEY F. P. (1973) Inorganic nutrient uptake and deficiency in algae. *CRC Crit. Rev. Microbiol.*, **3**, 69–113.

HSIAO S. I. C. (1969) Life history and iodine nutrition of the marine brown alga *Petalonia fascia* (O. F. Müll.) Kuntze. *Can. J. Bot.*, **47**, 1611–6.

HSIAO S. I. C. & DRUEHL L. D. (1973) Environmental control of gametogenesis in *Laminaria saccharina*. IV. *In situ* development of gametophytes and young sporophytes. *J. Phycol.*, **9**, 160–4.

HUTNER S. H. (1972) Inorganic nutrition. *Ann. Rev. Microbiol.*, **26**, 314–46.

HUTNER S. H. & PROVASOLI L. (1964) Nutrition of algae. *Ann. Rev. Plant Physiol.*, **15**, 37–56.

IKAWA M., THOMAS V. M., JR., BUCKLEY L. J. & UEBEL J. J. (1973) Sulphur and the toxicity of the red alga *Ceramium rubrum* to *Bacillus subtilis*. *J. Phycol.*, **9**, 302–4.

IWASAKI H. (1967) Nutritional studies of the edible seaweed *Porphyra tenera*. II. Nutrition of *Conchocelis*. *J. Phycol.*, **3**, 30–4.

IWASAKI H. & MATSUDAIRA C. (1954) Studies on cultural grounds of a laver, *Porphyra tenera* Kjellman in Matsukawa-Ura Inlet. I. Environmental characteristics effecting upon nitrogen and phosphorus contents of laver. *Bull. Jap. Soc. Sci. Fish.*, **20**, 112–19.

JACKSON G. A. (1977) Nutrients and production of giant kelp, *Macrocystis pyrifera*, off southern California. *Limnol. Oceanogr.*, **22**, 979–95.

JACQUES A. G., & OSTERHOUT W. J. V. (1938) The accumulation of electrolytes XI. Accumulation of nitrate by *Valonia* and *Halicystis*. *J. Gen. Physiol.*, **21**, 767–73.

KNAGGS F. W. (1967) *Rhodochorton floridulum* (Dillwn.) Nag. Observations on the relationship between reproduction and environment. *Nova Hedwigia*, **14**, 31–8.

KORNMANN P. (1970) Advances in marine phycology on the basis of cultivation. *Helgoländer Wiss. Meeresunters.*, **20**, 39–61.

KUHL A. (1962) Inorganic phosphorus uptake and metabolism. In *Physiology and Biochemistry of Algae*, ed. Lewin R. A., pp. 211–29. Academic Press, New York and London.

KUHL A. (1974) Phosphorus. In *Algal Physiology and Biochemistry*, ed. Stewart W. D. P., pp. 636–54. Blackwell Scientific Publications, Oxford.

LEWIN J. (1976) Effects of boron deficiency on chemical composition of a marine diatom. *J. Exp. Bot.*, **27**, 916–21.

LÜNING K. (1970) Cultivation of *Laminaria hyperborea in situ* and in continuous darkness under laboratory conditions. *Helgoländer Wiss. Meeresunters.*, **20**, 79–88.

LÜNING K. (1971) Seasonal growth of *Laminaria hyperborea* under recorded underwater light conditions near Helgoland. In *4th European Mar. Biol. Symp.*, ed. Crisp D. J., pp. 347–61. Cambridge University Press, New York.

LÜTTGE U. & PITMAN M. G. (1976) *Transport in Plants. IIA. Encyclopedia of Plant Physiology*. Springer-Verlag, New York.

MCLACHLAN J. (1973) Growth media—marine. In *Handbook of Phycological Methods, Culture Methods and Growth Measurements*, ed. Stein J. R., pp. 25–51. Cambridge University Press, Cambridge.

MCLACHLAN J. (1977) Effects of nutrients on growth and development of embryos of *Fucus edentatus* Pyl. (Phaeophyceae; Fucales). *Phycologia*, **16**, 329–38.

MCLACHLAN J., CHEN L. C.-M. & EDELSTEIN T. (1971) The culture of four species of *Fucus* under laboratory conditions. *Can. J. Bot.*, **49**, 1463–9.

MACROBBIE E. A. C. (1974) Ion uptake. In *Algal Physiology and Biochemistry*, ed. Stewart W. D. P., pp. 676–713. Blackwell Scientific Publications, Oxford.

MAEDA M. & FUJIYAMA T. (1977) Content uptake of trace metals in benthic algae, *Enteromorpha* and *Porphyra*. II. Studies on the algae cultured in seawater supplemented with various metals. *J. Fac. Fish Anim. Husb., Hiroshima U.*, **16**, 33–44.

MATHIESON A. C. (1975) Seaweed aquaculture. *Mar. Fish. Rev.*, **37**, 2–14.

MENZEL D. W. & SPAETH J. P. (1962) Occurrence of vitamin B_{12} in the Sargasso Sea. *Limnol. Oceanogr.*, **7**, 151–4.

MOHSEN A. F., KHALEAFA A. F., HASHEM M. A. & METWALLI A. (1974) Effect of different nitrogen sources on growth, reproduction, amino acid, fat and sugar contents in *Ulva fasciata* Delile. *Bot. Mar.*, **17**, 218–22.

MØLLER M. E. & EVANS L. V. (1976) Sulphate activation in the unicellular red alga *Rhodella*. *Phytochemistry*, **15**, 1623–6.

MUNDA I. M. (1978) Trace metal concentrations in some Icelandic seaweeds. *Bot. Mar.*, **21**, 261–3.

MUNSON R. D. & NELSON W. L. (1973) Principles and practices in plant analysis. In *Soil Testing and Plant Analysis*. ed. Walsh L. M. & Beaton J. D., pp. 223–48. Soil Science Society of America, Madison, Wisconsin.

NASR A. H. & BEKHEET I. A. (1970) Effect of certain trace elements and soil extract on some marine algae. *Hydrobiologia*, **36**, 53–60.

NASR A. H., BEKHEET I. A. & IBRAHIM R. K. (1968) The effects of different nitrogen and carbon sources on amino acid synthesis in *Ulva, Dictyota* and *Pterocladia*. *Hydrobiologia*, **31**, 7–16.

NEISH A. C. & FOX C. H. (1971) *Greenhouse experiments on the vegetative propagation of Chondrus crispus (Irish Moss)*. Tech. Rept. No. 12, Halifax, N.S., Atlantic Regional Laboratory, National Research Council of Canada.

NEISH A. C. & SHACKLOCK P. F. (1971) *Greenhouse experiments (1971) on the propagation of strain T4 of Irish moss*. National Research Council of Canada, Atlantic Regional Laboratory Technical Report, Series No. 14.

NEISH A. C., SHACKLOCK P. F., FOX C. H. & SIMPSON F. J. (1977) The cultivation of *Chondrus crispus*. Factors affecting growth under greenhouse conditions. *Can. J. Bot.*, **55**, 2263–71.

NIELL F. X. (1976) C:N ratio in some marine macrophytes and its possible ecological significance. *Bot. Mar.*, **19**, 347–50.

NISSEN P. (1974) Uptake mechanisms: inorganic and organic. *Ann. Rev. Plant Physiol.*, **25**, 53–79.

NOBEL P. S. (1974) *Introduction to Biophysical Plant Physiology*. W. H. Freeman and Company, San Francisco.

NODA H. & HORIGUCHI Y. (1971) The significance of zinc as a nutrient for the red algae *Porphyra tenera*. *Proc. Int. Seaweed Symp.*, **7**, 368–72.

O'KELLEY J. C. (1974) Inorganic nutrients. In *Algal Physiology and Biochemistry*, ed. Stewart W. D. P. pp., 610–35. Blackwell Scientific Publications, Oxford.

OZA R. M. (1977) Culture studies on induction of tetraspores and their subsequent development in the red alga *Falkenbergia rufolanosa* (Harvey) Schmitz. *Bot. Mar.*, **20**, 29–32.

PARKER B. C. (1969) Occurrence of silica in brown and green algae. *Can. J. Bot.*, **47**, 537–40.

PASCIAK W. J. & GAVIS J. (1974) Transport limitation of nutrient uptake in phytoplankton. *Limnol. Oceanogr.*, **19**, 881–8.

PEDERSEN M. (1969a) The demand for iodine and bromine of three marine brown algae grown in bacteria-free cultures. *Physiol. Plant.*, **22**, 680–5.

PEDERSEN M. (1969b) Marine brown algae requiring vitamin B_{12}. *Physiol. Plant.*, **22**, 977–83.

PERCIVAL E. (1979) The polysaccharides of green, red and brown seaweeds: their basic structure, biosynthesis and function. *Br. Phycol. J.*, **14**, 103–17.

PRINCE J. S. (1974) Nutrient assimilation and growth of some seaweeds in mixtures of seawater and secondary sewage treatment effluents. *Aquaculture*, **4**, 69–79.

PROVASOLI L. & CARLUCCI A. F. (1974) Vitamins and growth regulators. In *Algal Physiology and Biochemistry*, ed. Stewart W. D. P., pp. 741–87. Blackwell Scientific Publications, Oxford.

RAINS D. W. (1976) Mineral metabolism. In *Plant Biochemistry*, 3rd edn., eds. Bonner J. & Varner J. E., pp. 561–97. Academic Press, New York and London.

RAMUS J. (1969) The developmental sequence of the marine red alga *Pseudogloiephloea* in culture. *Univ. Calif. Pub. Bot.*, **52**, 1–28.

RAVEN J. A. (1970) Exogenous inorganic carbon sources in plant photosynthesis. *Biol. Rev.*, **45**, 167–221.

RAVEN J. A. (1976) Transport in algal cells. In *Transport in Plants. IIA, Encyclopedia of Plant Physiology*, eds. Lüttge U. & Pitman M. G., pp. 129–88. Springer-Verlag, New York.

RAVEN J. A. & SMITH F. A. (1974) Significance of hydrogen ion transport in plants. *Can. J. Bot.*, **52**, 1035–48.

RILEY J. P. & CHESTER R. (1971) *Introduction to Marine Chemistry*. Academic Press, New York and London.

ROTHSTEIN A. (1963) Interactions of arsenate with the phosphate-transporting system of yeast. *J. Gen. Physiol.*, **46**, 1075–85.

RYTHER J. H. & DUNSTAN W. M. (1971) Nitrogen, phosphorus and eutrophication in the coastal marine environment. *Science*, **171**, 1008–13.

SAENKO G. N., KORYAKOVA M. D., MAKIENKO V. F. & DOBROSMYSLOVA I. G. (1976). Concentration of polyvalent metals by seaweeds in Vostok Bay, Sea of Japan. *Mar. Biol.*, **34**, 169–76.

SAENKO G. N., KRAVTSOVA Y. Y., IVANENKO V. V. & SHELUDKO S. I. (1978) Concentration of iodine and bromine by plants in the Seas of Japan and Okhotsk. *Mar. Biol.*, **47**, 243–50.

SANDERS J. G. (1979) The concentration and speciation of arsenic in marine macroalgae. *Estuarine Coastal Mar. Sci.*, **9**, 95–9.

SCHIFF J. A. (1962) Sulphur. In *Physiology and Biochemistry of Algae*, ed. Lewin R. A., pp. 239–46. Academic Press, New York and London.

SCOTT G. T. & HAYWARD H. R. (1954) Evidence for the presence of separate mechanisms regulating potassium and sodium distribution in *Ulva lactuca*. *J. Gen. Physiol.*, **37**, 601–20.

SHIMO S. & NAKATINI S. (1967) The influence of various concentrations of copper and mercury ions on the growth of *Porphyra tenera* and its tolerance to them. *J. Agr. Lab.* (*Chiba*), **9**, 109–25.

SKIPNES O., ROALD T. & HAUG A. (1975) Uptake of zinc and strontium by brown algae. *Physiol. Plant.*, **34**, 314–20.

STEFFENSEN D. A. (1976) The effect of nutrient enrichment and temperature on the growth in culture of *Ulva lactuca* L. *Aquat. Bot.*, **2**, 337–51.

STICH H. (1955) Synthese und Abbau der Polyphosphate von *Acetabularia* nach autoradiographischen Untersuchungen des 32P-Stoffwechsels. *Z. Naturf.*, **10B**, 282–4.

STRÖMGREN T. (1979a) The effect of copper on length increase in *Ascophyllum nodosum* (L.) LeJolis. *J. Exp. Mar. Biol. Ecol.*, **37**, 153–9.

STRÖMGREN T. (1979b) The effect of zinc on the increase in length of five species of intertidal Fucales. *J. Exp. Mar. Biol. Ecol.*, **40**, 95–102.

SUBBA RAO, G. N. (1965) *Use of Seaweeds Directly as Human Food*. FAO Occasional Papers No. 2.

TOPINKA J. A. (1978) Nitrogen uptake by *Fucus spiralis* (Phaeophyceae). *J. Phycol.*, **14**, 241–7.

TOPINKA J. A. & ROBBINS J. V. (1976) Effects of nitrate and ammonium enrichment on growth and nitrogen physiology in *Fucus spiralis*. *Limnol. Oceanogr.*, **21**, 659–64.

UKELES R. (1976) Cultivation of Plants—Unicellular plants. In *Marine Ecology, Vol. 3, Cultivation, Part 1*, ed. Kinne O., pp. 367–466. John Wiley and Sons, London.

ULRICH A. & HILLS F. J. (1973) Plant analysis as an aid in fertilizing sugar crops. I. Sugar beets. In *Soil Testing and Plant Analysis*, eds. Walsh L. M. & Beaton J. D., pp. 271–88. Soil Science Society of America, Madison, Wisconsin.

VERNON L. P. (1962) Mechanism of oxygen evolution in photosynthesis. In *Biologistics for Space Systems*. U.S. Air Force Doc. Rept. AMRL-TDR 62-116, 131–98.

VINOGRADOV A. P. (1953) *The Elementary Chemical Composition of Marine Organisms*. Sears Foundations for Marine Research, Memoir No. 2, Yale University.

VON STOSCH H. A. (1964) Wirkungen von Jod und Arsenit auf Meeresalgen in Kultur. *Proc. Int. Seaweed Symp.*, **4**, 142–50.

WAITE T. & MITCHELL R. (1972) The effect of nutrient fertilization on the benthic alga *Ulva lactuca*. *Botanica Mar.*, **15**, 151–6.

WAITE T. D. & MITCHELL R. (1976) Some benevolent and antagonistic relationships between *Ulva lactuca* and its microflora. *Aquatic Bot.*, **2**, 13–22.

WASSERMAN A. (1948) Cation adsorption by brown algae. The mode of occurrence of alginic acid. *Ann. Bot.*, **13**, 79–88.

WILSON T. H. (1962) *Intestinal Absorption*. W. B. Saunders, Philadelphia, Pennsylvania.

WINNE D. (1977) Correction of the apparent Michaelis constant, biased by an unstirred layer, if a passive transport component is present. *Biochim. Biophys. Acta*, **464**, 118–26.

WHEELER P. A. (1979) Uptake of methylamine (an ammonium analogue) by *Macrocystis pyrifera* (Phaeophyta). *J. Phycol.*, **15**, 12–17.

WHEELER W. N. (1977) The interrelationship between light, water velocity, and macronutrients in the productivity of the giant kelp, *Macrocystis*. *J. Phycol.*, **13** (suppl.), Abstr. 72.

WONG S. L. & CLARK B. (1976) Field determination of the critical nutrient concentrations for *Cladophora* in streams. *J. Fish. Res. Bd. Can.*, **33**, 85–92.

WOOD B. J. B. (1974) Fatty acids and saponifiable lipids. In *Algal Physiology and Biochemistry*, ed. Stewart W. D. P., pp. 236–65. Blackwell Scientific Publications, Oxford.

WOOLERY M. L. & LEWIN R. A. (1973) Influence of iodine on growth and development of the brown alga *Ectocarpus siliculosus* in axenic cultures. *Phycologia*, **12**, 131–8.

YAMADA N., IWAHASHI Y., OSADA K. & OMURA Y. (1964) Studies on the manure for seaweeds. III. On the absorption of P^{32} labeled phosphorus by an agar seaweed (*Gelidium amansii*). *Bull. Jap. Soc. Sci. Fish.*, **30**, 983–5.

YAMAMOTO T. & ISHIBASHI M. (1972) The content of trace elements in seaweeds. *Proc. Int. Seaweed Symp.*, **7**, 511–4.

YOUNG E. G. & LANGILLE W. M. (1958) The occurrence of inorganic elements in marine algae of the Atlantic provinces of Canada. *Can. J. Bot.*, **36**, 301–10.

CHAPTER 11
INDIVIDUALS, POPULATIONS AND COMMUNITIES

G. RUSSELL
Hartley Botanical Laboratories, The University,
Liverpool, England

A. H. FIELDING
Department of Biology, John Dalton Faculty of Technology,
Manchester Polytechnic, Manchester, England

1 GENERAL INTRODUCTION

The following chapter is concerned first with certain genetically controlled attributes of individuals and populations of seaweeds, and second, with some structural characteristics of seaweed vegetation. The juxtaposition of these subjects is unusual, but, as natural selection operates on plants and communities alike, defensible. Further justification is offered in a personal opinion that progress in genetics of marine algae has been more satisfactory when linked with the realities of the plants' natural environment.

Genetic investigations often depend partly or entirely upon *in vitro* culture experiments, and a large proportion of these involve asexual clones, or sexual strains that are sufficiently in-bred to amount to pure lines. In plant material such as this, the individual plant and the population are virtually indistinguishable. In its correct sense, the population consists of a group of plants of one particular species living in some naturally or artificially circumscribed habitat, for example, a wave beaten promontory or a culture flask. Population studies will therefore be recognizable in the text even if, in the avoidance of repetition, the term is not used.

Autecology is, in effect, an attempt to bridge the gap between the population investigation and that of the multispecies communities. It is a term which embraces a wide range of biological studies, most of which involve measurements of the physical and chemical factors that seem to affect the life-processes of the particular taxon, and so lead to a greater understanding of the distribution of its members in nature. Logically, therefore, autecology cannot take into account the roles of other species. However, the evidence pointing to the great importance of biological interactions as a factor in natural selection is now so overwhelming that the intellectual basis of autecology becomes increasingly unsatisfactory. For that reason, this term has been avoided.

2 GENETICS

2.1 Introduction

Genetic research falls into a number of areas of study including cytogenetics, Mendelian genetics, quantitative genetics and population and ecological genetics. Unfortunately there is a shortage of information for marine algae in all of these areas and the need for further research has been urged by Lewin (1958) and Fjeld & Løvlie (1976). The following discussion will focus on some of the more important aspects of algal genetics, sectionalized under the above mentioned headings.

2.2 Cytogenetics

Algal cytogenetics has been hampered by problems associated with the small size of most algal chromosomes (Dixon 1963). The effects of this can be seen clearly in the wide range of values given for chromosome counts in certain species of brown algae (Cole 1967, 1968). The development and application of more advanced histological techniques, which would allow greater accuracy in chromosome observations, could be of great value in ecological and taxonomic investigations since it might allow genetically divergent races to be identified. Drew (1939) identified triploid plants of *Plumaria elegans* which possibly had a wider geographical distribution than either diploid or haploid plants. Whittick & Hooper (1977) have presented cytological evidence which suggests that *Antithamnion cruciatum* may be polyploid at its northern limits. If this should prove to be correct it presents an interesting parallel with observations on higher plants that populations at the edge of a species range are frequently polyploid. The mechanisms by which polyploidy is achieved and maintained in marine algae are probably different from those in higher plants in which apomixis is often important. Nakahara & Nakamura (1973) report the production of tetraploid plants of *Alaria crassifolia* by autogamy from monoecious diploid gametophytes, and van der Meer & Todd (1977) observed the formation of a polyploid *Gracilaria* sp. following production and fusion of gametes from diploid tetrasporophytes.

Cytological observations have provided some convincing evidence for a chromosomal sex-determining mechanism in at least five British laminarian species. Evans (1965, 1966) found that the female gametophytes contain a large X chromosome which pairs with a small Y chromosome found in male gametophytes. Several other authors have presented less direct evidence for a genetic basis of sex determination in marine algae (Müller 1970; Nakahara & Nakamura 1973). It seems likely that knowledge of algal cytogenetics will progress very slowly unless efficient and easily applied techniques can be developed.

2.3 Mendelian genetics

Examples of phenotypes which show simple Mendelian inheritance are rare and restricted mainly to a small number of intensely studied species such as *Ulva mutabilis*. One advantage of studying this species is that it has a haploid phase in which recessive genes will be expressed, thereby making genotypic identification much easier. Most of these are mutant genes, which have been isolated by the use of mutagenic chemicals, and examples of single gene polymorphisms in nature remain rare. Løvlie (1968) has shown that a wide range of morphological characteristics of *Ulva mutabilis,* many of which have taxonomic importance, can be altered by induced, single gene mutations. Fjeld & Løvlie (1976) reviewed much of the available information about *Ulva* mutants. More recently Bryhni (1978) has investigated, in some detail, the differences between *wild type, slender,* and *lumpy* phenotypes of *Ulva mutabilis*, and concluded that the *lumpy* phenotype, which grows as an aggregate of undifferentiated cells, results from a change in the ratio of some polysaccharides which in turn reduces the number of cross linkages in cellulose fibres.

A series of papers has been published recently on the genetics of *Gracilaria* by van der Meer and his associates. Van der Meer (1979) described a number of induced mutant forms which have been isolated and, by using crossing experiments, he demonstrated that most of the mutant phenotypes result from the action of single genes showing simple Mendelian inheritance, although several pigmentation genes apparently show non-Mendelian cytoplasmic inheritance (van der Meer 1978). Obviously there is scope for much more work in this general area of induced mutations in order that an outline understanding of marine algal genetics can be gained but there is also a need for an investigation of the frequency and importance of genetically determined differences in wild populations. Rueness & Rueness (1975) have suggested that differences between the *plumula* and *bebbii* varieties of *Antithamnion plumula* result from the action of a single gene, *plumula* being dominant to *bebbii*. Lüning *et al.* (1978) have identified certain characteristics of several Simplices species of *Laminaria*, occurring in wild populations, that may show simple inheritance. Crossing experiments indicate that the occurrence of frond bullae is a dominant trait and that blade width is both heritable and linked with parental sex. It may be that an increased use of biochemical techniques such as gel electrophoresis, in the investigation of intraspecific rather than interspecific differences (Mallery & Richardson 1972), will lead to the identification of reservoirs of genetic variability in some wild populations resulting from the action of multiple alleles. Cheney & Babbel (1978), using electrophoretic techniques, have given a lead in this area by identifying a greater genetic variability in populations of *Eucheuma nudum* than in populations of *E. isiforme* which occupied a less diverse range of habitats. A potentially more important taxonomic observation was that morphologically indistinguishable populations of *Eucheuma nudum* showed a wide biochemical separation and possibly quite distinct genotypes.

2.4 Quantitative genetics

There is some experimental evidence that certain ecologically and taxonomically important phenotypes are under polygenic and environmental control and the techniques of quantitative genetics, as used by Chapman (1974a, 1975) must be applied to determine the relative importance of genotype and environment in controlling the phenotypic expression. Quantitative genetics has its greatest uses in plant and animal breeding when crosses between pure bred lines are being investigated, but the techniques can be adapted so that causes of phenotypic variability in seaweeds can be identified. This usually involves computing the heritability of a character so that estimates of the genetic and environmental contributions to phenotypic variation can be made. However, the application of these techniques to wild populations, of unknown genotypes, presents many problems and the results obtained must be interpreted with caution. Quantitative genetic analysis in marine algae has been applied to characters of taxonomic importance such as the occurrence of mucilage canals and other stipe characteristics in *Laminaria*. A feature of most of these is that they are formed over relatively long periods of plant growth and as such are likely to be more plastic than characteristics formed quickly such as seasonal reproductive structures (Stebbins 1950). Algal 'ecotypes' are frequently found in quite different environments such as exposed and sheltered localities—indeed Chapman used this as a basis for his investigations—but Bradshaw (1965) has suggested that the stability (lack of variability) of a character can change between genotypes and that stability itself may be under genetic control. Since true ecotypes must have different genotypes then many restrictions must be placed on experiments using populations from very different environments.

The first investigation into the causes of variation in seaweeds was the work of Fang *et al.* (1962) who found a close correlation between parent and offspring stipe lengths in *Laminaria japonica* and concluded that stipe length is a heritable characteristic although heritability values were not calculated. Chapman (1974a) examined stipe morphology in *L. longicruris* and *L. saccharina* (ductless ecotype—*L. agardhii*) and was able to make several interesting conclusions following a long series of experiments. Variability and heritability estimates were greater in populations from sheltered locations, possibly supporting the ideas of Bradshaw. The heritability estimates also indicated that the stipe differences between sheltered and exposed populations resulted from additive genetic effects. He concluded that the two species are in fact conspecific. Chapman (1975) also provided evidence of genotypic separation into true breeding Nova Scotian populations of *Laminaria* with respect to the occurrence of mucilage canals. In other populations this trait seems to be largely under environmental control.

The differences in gametangial morphology in geographically different strains of *Ectocarpus siliculosus* described by Müller (1979) seem likely also to have a polygenic origin, which is of special interest as the plants involved are haploid. The absence of dominance effects in this instance should make analysis of genetic control of gametangial form simpler.

2.5 Ecological genetics

Ecological genetics attempts to synthesize ecological and genetic information but because of the shortage of basic genetic information there is a consequential lack of work on the ecological genetics of seaweeds. However there have been a number of studies which can be interpreted with the aid of the ecological genetic concepts of ecad and ecotypes. Ecads result from the phenotypic plasticity of single genotypes while ecotypes are supposed to result from local differences in selection pressures which result in genotypic separation of populations. The maintenance of distinct ecotypic identities between sexually reproducing populations implies some reproductive isolation which would not be necessary in species which reproduce mainly asexually. The mechanisms of reproductive isolation are likely to be extrinsic rather than intrinsic although Müller (1976) has demonstrated sexual isolation between European and American populations of *Ectocarpus siliculosus* which results from a breakdown in the process of cell fusion. Many marine phycologists are engaged on hybridization experiments and often the results suggest that previously defined species may in fact be ecotypes of one species (e.g. *Ceramium,* Rueness 1978; *Laminaria,* Chapman 1975). A breeding population of seaweeds is likely to have a three dimensional boundary in which both lateral and vertical spatial separation and timing of reproduction are critical dimensions. It is easy to determine whether or not the timing of reproduction between populations is different but the spatial boundaries are much more difficult to establish. It is possible that lower-shore algae will have completed fertilization before the tide reaches the upper members of the species but little information is available on how far gametes are transported and how long they remain viable in the sea (see also Chapter 8). De Silva & Burrows (1973) believe that the tide acts as an isolating mechanism between *Enteromorpha* plants showing presence and absence of branches, since gametes are released as soon as the tide reaches plants and fertilization was seen to occur in the laboratory immediately following gametes release. However, Jones & Babb (1968) demonstrated that swarmers of *Enteromorpha* can remain motile for a period of eight days in the laboratory.

Much of ecological genetics is concerned with identifying and explaining the occurrence of intraspecific variation, in particular discontinuous variation. Numerous factors can alter the amount of intraspecific variability, the most important of which are probably the breeding system and the selection pressures. Inbreeding, which is possible in a homothallic species such as *Fucus spiralis,* will lead to a reduction of within-population variation but also allows potentially wide differences between populations. Obviously this reduction in variability in a species such as *F. spiralis* would be advantageous since its upper-shore environment is likely to impose strong stabilizing selection. Outbreeding is inevitable in hererothallic species, but again it is difficult to assess the degree of gene flow. Hererothallic species such as *F. vesiculosus,* with wide, continuous, vertical and horizontal shore ranges should possess a range of genotypes through outbreeding. The capability of producing a range of genotypes and hence increasing genetic variability is clearly advantageous since there must be a limit to the amount of phenotypic plasticity that a single genotype can show.

Much of the work on seaweed ecotypes has concentrated on gross morphological differences resulting from differences between exposed and sheltered locations and differences are frequently attributed to the effects of wave action in exposed locations. It must be remembered, however, that all marine algae will experience selection pressures and it is possible that these will change from mainly environmental in exposed locations to mainly biotic, in particular density dependent interactions, on sheltered shores. There is some evidence that the intensity of competition can be associated with differences in phenotype, and possibly genotype, of *Fucus vesiculosus.* Russell (1979) has demonstrated that estuarine plants, growing where competitive interactions are decreased, put a greater proportion of their resources into the development of reproductive tissues than do plants growing in a dense stand (see Section 3.4). Many examples of selection for such reproductive strategies are known for higher plants and there is a clear need for the development of such research on marine algae.

Discontinuous intraspecific variation can also involve changes in physiology. Copper tolerant strains of *Ectocarpus siliculosus* and *Enteromorpha compressa* have been isolated (Russell & Morris 1973; Goodman *et al.* 1976) and in both cases tolerant plants have been selected in response to intense selection by elevated copper concentrations. Investigations into the tolerance mechanism of *Ectocarpus siliculosus* (Hall *et al.* 1979) have indicated that the mechanism is complex and therefore unlikely to have a simple genetic basis. Salinity ecotypes have been demonstrated for a wide range of species: *Pilayella* sp. (Russell 1963); *Ectocarpus siliculosus* (Russell & Bolton 1975); *Enteromorpha intestinalis* (Reed & Russell 1979); *Eugomontia sacculata,* Wilkinson (1974); *Audouinella purpurea,* (West 1972); *Bostrychia radicans* and *Caloglossa leprieurii* (Yarish *et al.* 1979). In all cases population differences can be correlated with original habitat conditions and in at least one case (*Ectocarpus siliculosus*) the population differences were maintained over a period of four years despite continuous culturing in fully saline seawater. Lüning *et al.* (1978) have provided evidence for the existence of temperature sensitive strains of *Laminaria saccharina* with a bullate form dying at 16° C while a smooth form, from Helgoland, died at 18° C. It is very probable that comparable variation in a range of adaptive characteristics will be found in many other seaweed species. Indeed it would be highly surprising if this was not the case considering the diversity of habitats inhabited by these plants.

2.6 Phenotypic plasticity

Phenotypic plasticity is an attribute that enables plants to respond to various environmental conditions by altering their development so that the environmental effects operate immediately, directly and without genotypic changes. Phenotypic plasticity, therefore, is not in itself a visible characteristic but one which becomes manifest under the influence of the environment. Consequently, genotypes with a potential for variable expression may be of high adaptive value. However, it is evident that many of the reported examples of phenotypic plasticity have little adaptive importance and may best be construed simply as laboratory artifacts.

Phenotypic variation in *Fucus* has been attributed to plasticity by numerous authors and to hybridization by Burrows & Lodge (1951) and others, although Bolwell *et al.* (1977) have recently raised some doubts about the crossability of fucoid species in nature. In order to analyse growth and development in *F. vesiculosus,* a species which is difficult to grow under controlled laboratory conditions, Khailov *et al.* (1978) and Russell (1978) have carried out morphometric investigations of field populations. In *F. vesiculosus,* the length of the frond, its width and the number of vesicles are relatively great in sheltered conditions, but these diminish with increasing severity of wave-action to an extreme condition in which plants are very small, narrow and evesiculate (Powell 1963). If these changes in form express a wholly plastic response, then the morphological attributes involved ought to correlate positively and closely with one another as well as with intensity of wave action. In practice, certain features, largely of growth, such as thallus area and weight are strongly associated; but others such as frond width and vesicle number do not correlate nearly so closely, and may therefore be expressed to some extent independently. It seems likely therefore that the various components of the *Fucus* phenotypes are not equally responsive to environmental factors and consequently the thallus is incompletely plastic. From their culture experiments, McLachlan *et al.* (1971) have concluded that phenotypic plasticity contributes relatively little to intraspecific variation in *Fucus* species. Nevertheless it is very improbable that long-lived plants such as these should not exhibit some phenotypic plasticity.

It seems inevitable that many more variable characters, once attributed to phenotypic plasticity will prove to be, at least partly, heritable: in the case of sexually reproducing species it is highly improbable that population differences will not have some genetic basis. There is therefore a danger that phenotypic plasticity will come to be regarded as an increasingly unimportant agent of thallus morphogenesis in spite of the ample and growing evidence to the contrary (see also Chapter 12). A result of this fallacy might be further increase in the use of the term 'ecotype', which implies genetic differences, without supporting experimental evidence.

It is in the area of interaction or interdependence of genetic and plastic factors that the uncertainty and interest chiefly lie. It now seems highly probable that in some seaweed species (as in higher plants, see Bradshaw 1965) the predisposition to phenotypic plasticity may itself be a heritable factor. This point has been put succinctly by Kain (1976) in reviewing the possible explanations for the presence of a particular population of *Laminaria saccharina* in Shetland waters, and concluding that there may have been '. . . an inflow of genes into this area which have the effect of causing the development of long stipes under fairly sheltered conditions and hollow stipes in complete shelter. . .'. Natural environments may then select genotypes that not only determine specific phenotype attributes but also their plasticity. In this manner phenotypic plasticity may have high adaptive value for particular habitats and therefore be subject to the same forces of natural selection as other attributes.

3 SEAWEED VEGETATION

3.1 Introduction

Plant community ecology has two chief purposes, the analysis of pattern and the elucidation of the forces that determine pattern. Seaweeds on rocky shores are seldom, if ever, randomly distributed. On the contrary, they usually form natural associations, of characteristic species composition and clear spatial circumscription, as Lewis (1964) has shown.

Seaweed communities and terrestrial vegetation are comparable in that the constituent plants compete for physical space and light. It has been held that nutrients are rarely limiting for seaweed growth (Dayton 1975a), but there is now some evidence to suggest that this may not be so (Chapman & Craigie 1977; Boden 1979; and see Chapter 10). They differ, however, in that seaweeds have no subterranean equivalent of roots, nutrient absorption taking place over the exposed surface of the thallus with the result that the whole plant is visible and readily accessible.

These properties give seaweed vegetation a number of theoretical and practical advantages in the design of ecological experiments, which have been increasingly exploited over the last decade or two. This relatively slow development of interest in analytical and experimental investigations is perhaps due to the 'obvious' nature of rocky-shore community structure and the consequent ease with which the vegetation can be reported in purely descriptive language. As in the case of morphological studies, however, the practice of making quantified observations has provided the basis for greatly improved techniques and hence made possible all the recent advances in our understanding of seaweed vegetation.

For classic descriptive accounts of rocky-shore community structure, the reader is referred to Lewis (1964) and Stephenson & Stephenson (1972). The works of Chapman (1974b, 1979) and Carefoot (1977), though brief, are broader in scope and discuss some of the more recent scientific material.

3.2 Samples and measurements

The primary requirement of all vegetational analyses is an adequate sampling procedure. Investigators of seaweed vegetation in sublittoral and intertidal habitats have usually adopted area sampling (quadrats), although a few have chosen plotless (point) techniques. Jones *et al.* (1979) have adapted an optical-intersect system for point sampling in the field; Murray & Littler (1974), on the other hand, have preferred to photograph the vegetation and locate points by projecting the photographic image on to a grid. The line transect (Colman 1933) is in effect a linear arrangement of rather blunt point samples located either as a continuous series or at selected intervals. The theoretical advantages of all point sampling may be vitiated in practice, however, by the presence of small, inconspicuous species which are difficult to score with certainty.

In area sampling, important consideration should be given to the size of the quadrat used. Boudouresque & Luck (1972), Boudouresque (1971a) and Russell (1972) have followed terrestial phytosociological procedures and drawn cumulative species-area graphs to justify the selection of quadrats of particular size for various stands of intertidal and sublittoral vegetation. The quadrat size adopted by Russell has been recommended also by Niell (1974) from calculations of Shannon species diversity indices, although somewhat smaller areas have been advocated by Boudouresque (1974) on the basis of comparable calculations.

However, Hruby (1975) has shown that seasonal changes in seaweed vegetation can affect the form of species-area graphs and hence influence the choice of quadrat size. There are therefore no fixed standards for quadrats and, indeed, the very notion of 'minimal area' is somewhat suspect. Any matrix of samples and species records can be re-arranged to give cumulative graphs of different form leading to different conclusions. By pooling all records, however, it is a fairly simple matter to calculate the probable species number for areas of different size and so construct a curve that characterizes that particular stand of vegetation (Hawkins & Hartnoll 1980). In this way, the number of species present in the chosen sample area can be related to that expected from the vegetation as a whole. In defence of species-area graphs, it should perhaps be said that they are simple, quick and, if used circumspectly, give a rough check on quadrat suitability.

Quadrats have been placed either contiguously or separately, contiguous arrangements being either as a block (Boudouresque & Luck 1972; Boudouresque 1974) or in linear series as a belt transect (Ravanko 1972; Boalch *et al.* 1974). Long-term changes in vegetation have been monitored by means of permanent quadrats (Jones *et al.* 1979; Nienhuis 1978).

The position of quadrat or plotless sample stations is dictated by the analytical requirements. Openly subjective location of plots is an essential ingredient of the Zurich–Montpellier method (see Section 3.3) but if mathematical treatment of vegetation data is proposed then the position of the samples ought to be random or at very least, unbiased. Complete randomization of samples is technically difficult to achieve on rocky shores and many investigators have therefore chosen to sample at equal vertical or horizontal intervals along a line stretched over the shore. A roughly equidistant pattern is also likely to result from 'dropping' quadrats as practiced, for example, by Hruby (1975). Systematic sampling methods such as these can be improved by introducing random replication of quadrats at each sampling position (Russell 1973) or by using random numbers to locate the base position for each line of samples (Murray & Littler 1974).

Seaweed vegetation has a small number of attributes that can be usefully measured for phytosociological purposes. Frequency, which is based on presence-absence data, is probably the simplest of these and has been adopted by Sundene (1953), Russell (1972), Ravanko (1972), Murray & Littler (1974), Sears & Wilce (1975) and others. Frequency values are influenced by quadrat size, however, and investigations that use different sample areas are therefore difficult to compare. Plant cover does not suffer from this drawback although the data is more laboriously obtained. Fairly precise measurements of cover may be gained directly from point sample scores (Murray & Littler 1974) or by drawing outlines of individual species on plastic sheets and calculating from these, the respective areas (Hruby 1975). The intricacy and small size of many seaweeds can make accurate measurement of cover excessively difficult or impossible and, for this reason, most workers have preferred to make rough estimates of cover-abundance, usually by means of crude scales (Boudouresque 1971b; den Hartog 1959; Russell 1980; Boalch *et al.* 1974).

Standing-crop biomass has been employed successfully in seaweed vegetation analysis by a number of workers. The data recorded have been either of fresh weights (Prentice & Kain 1976) or oven-dried weights (Hughes & Thomas 1971; Edwards *et al.* 1975; Kain 1975a; Holm 1978). Biomass measurement is time-consuming, especially in the separation of small epiphytes from their host plants and measurements of these are distinctly prone to error. The biomass of seaweeds present in each unit sample area has been referred to occasionally as density, but this is misuse of a term that, strictly, should express numbers of plants per unit area. Plant numbers, though calculable for kelps and some other macrophytes, cannot even be guessed at for crustose and mat forming species and species in which more than one frond arises from a perennating base.

3.3 Analytical methods

Analyses of seaweed vegetation have been chiefly classifactory in aim and, for the most part, applied to intertidal habitats. The simplest classification methods in current use divide the shore into a small number of horizontal zones on the basis of certain diagnostic species. The system of Stephenson & Stephenson (1949), as modified by Lewis (1961), recognizes three major zones: the littoral fringe, the eulittoral zone and the sublittoral zone. This simple classification has been applied to rocky shores throughout the world (Stephenson & Stephenson 1972), and has been adopted by a large number of investigators. In Scandinavian countries, a single biologically-defined level, corresponding to the boundary between the littoral fringe and eulittoral zone *sensu* Lewis, is generally used for the demarcation of major intertidal zones. This level, termed the 'litus-line' by Sjostedt (1928), defines the zero position on a vertical scale to which the heights of various subzones are referred (Söderström 1965; Jorde 1966; Klavestad 1978). This method produces a more informative analysis and one that is less rigidly conventional than the Stephenson-Lewis system. Dixon & Irvine (1977) have suggested recently that two major zones may be recognized, the littoral zone, which embraces the littoral fringe and eulittoral zone of Lewis, and the sublittoral zone, the former being divided into three subzones. This classification, which bears a curious resemblance to that of Forbes (1846) has no obvious advantages over the two described above. The occurrence of sublittoral zones characterized by certain dominant seaweeds has been well documented by Kain (1960), Jorde & Klavestad (1963), Neushul (1965) and others. These zones, though less distinct than in intertidal habitats, are recognizably few in number, and Shepherd & Womersley (1970) have proposed a tripartite system of upper, mid and lower sublittoral zones for their classification.

Classifactory techniques may be based on species presence or species abundance data. Stephenson & Stephenson (1949) make it clear that some consideration of diagnostic species abundance is necessary for the satisfactory delimitation of zones. Other simple methods, notably that of the Zurich-Montpellier school attach greater importance to purely floristic data. The Zurich-Montpellier technique, which has been adopted in pure or slightly modified forms by den Hartog (1955), Boudouresque (1971a), Taniguti (1962), van den Hoek (1969), van den Hoek *et al.* (1972) and others, is more divisive than the zonal systems already referred to and more informative in that species abundance (cover) is measured, though not actually employed in classification. Nevertheless it is unsatisfactory in practice because many diagnostic species of particular associations recur in other plant communities (den Hartog 1959).

Classification of small numbers of vegetation samples is most easily effected by means of tabulation (see Sundene 1953; Russell 1980; and all exponents of the Zurich-Montpellier method). Vegetation samples have been compared also using coefficients of similarity and classified by means of simple matrices (Russell 1972, 1973), or dendrograms based on more sophisticated cluster analyses (Murray & Littler 1974; Seapy & Littler 1978), while Russell (1972, 1973) has also used the χ^2 statistic to classify intertidal vegetation. This last method is based on species incidence data only, but similarity coefficients have been calculated from both incidence and abundance weighted information. In comparing vegetation samples, abundance weighting seems to have led usually to more efficient discrimination of vegetation units and, in certain cases, the units appear to correspond closely with those detected by traditional, subjective methods (Russell 1980; Prentice & Kain 1976; Jones *et al.* 1979).

Ordination, though probably the most widely used nonclassificatory technique for vegetation analysis, has seldom been applied to marine algae. Principal components analysis has been selected by Hughes & Thomas (1971) and Boudouresque (1971b) for the ordination of shallow sublittoral vegetation, whilst Lawson (1978) has employed the same method for the analysis of phytogeographical data. Devinny (1978) has ordinated intertidal and sublittoral seaweed communities with respect to four separate axes by means of principal axes analysis. The simpler Bray & Curtis (1957) method has been recommended for seaweed vegetation analysis by Russell (1980), who has concluded that in ordination, species incidence and abundance can make distinctive and valuable contributions to the quality of the analysis.

3.4 Causes of pattern

The most obvious physical factor operating on oceanic rocky shores is the tide, and attempts to relate zonation patterns to specific tide levels go back to Vaillant (1891). Colman (1933) has contended that the periods of immersion and emersion experienced at various short heights do not change at a uniform rate but rather show irregularities which correspond with marked changes in vegetation pattern. He referred to these as 'critical tide levels' and concluded that three were present on the shore at Wembury, near Plymouth, England where he worked. Tides in most parts of the British Isles are semi-diurnal and equal so the critical tide levels were thought to be products of the lunar (spring-neap) tidal sequence. The idea of critical tide levels as a major cause of vegetation pattern was taken up by Doty (1946) to explain the seaweed zones observed by him on the Pacific coast of the U.S.A. The two daily tides in these localities are unequal, a complication that increased the number of critical tide levels to seven. Doty & Archer (1950) sought to validate the tide factor theory experimentally by subjecting certain seaweeds from different shore levels to a variety of stress-conditions. The resulting differences between species in thallus mortality were considered positive proof of the theory, but the crudity both of the experimental design and methods make this conclusion dubious. The slightly better experiments by Biebl (1952) and Almodovar & Biebl (1962) to measure the 'resistance' of certain intertidal algae to abnormal salinities, temperatures, inorganic salts etc. show that upper-shore algae are naturally more tolerant of emersion stress than those from lower down but are equally inconclusive regarding zone development.

The observation by Stephenson (1953) of differences in zonal pattern at Brandon Island (British Columbia) in sites that differed only in north- and south-facing aspects added to existing doubts about the validity of critical tide levels. It is also clear that wave-action may modify or even obliterate the effects of tidal amplitudes (Lewis 1968). Moreover, zonation has been shown to occur in the absence of tides (Colman & Stephenson 1966). A further serious criticism of the theory of critical tide levels has been delivered by Underwood (1978) whose accurate measurements of the tide emersion curve for the locality in which Colman worked show a smooth monotonic rate of change with shore height, and a total absence of the irregularities described earlier. Nevertheless it seems highly probable that in areas of extreme shelter, the prevailing tidal pattern will play an important role in the location of zones and their composition.

Wave action affects seaweed vegetation pattern in two important ways. First, it alters zone heights by raising them intertidally (Burrows *et al.* 1954), and lowering them sublittorally (Shepherd & Womersley 1970). Secondly, it favours, in various degrees, differently adapted species and so promotes the differences in community structure that Lewis (1964), Stephenson & Stephenson (1972) and numerous other investigators have detailed. This selective effect has been considered sufficiently reliable for certain species associations to be pronounced indicative of particular degrees of wave action (Ballantine 1961; Lewis 1964; Dalby *et al.* 1978). The validity of these biologically-defined scales of exposure is questionable, partly because accurate measurements of wave action are rare, which in turn raises doubts about the correlation with species, and partly because the adaptive characteristics of seaweeds are necessarily multifactorial and therefore not reliably expressive of any single factor. Wave action has been measured in the field by weight losses from solidified calcium sulphate (Muus 1968; Doty 1971), by readings of force on a spring-balance dynamometer (Jones & Demetropoulos 1968), and by a crude flow-meter (Field 1968). An indirect cartographic expression of wave action has been devised by Baardseth (1970) using simply the angle of open sea subtended by the coastline from a particular locality. By weighting the sea angles according to prevailing wind directions and frequency, the accuracy of this expression may be increased (Russell 1977). The effects of tide and wave action, though normally independent, are interdependent in coral atolls because at low water the encircling reef ridge provides effective shelter for reef-flat and lagoon shores but is ineffective at high water (Russell 1974). The influences of other types of water movement (currents, streams, etc.) are discussed by Bassindale *et al.* (1948) and Lewis (1968); see also Chapter 12.

The influence of shore aspect on zonation has been discussed by Stephenson & Stephenson (1953), and Lewis (1964) who also refers to the effects of desiccation and atmospheric humidity on community structure. Jenik & Lawson (1968) have advanced a little further in relating measurements of these and other physical attributes of climate to shore zonation in Ghana (West Africa).

The importance of rock surface characteristics in modifying community structure has been discussed by Lewis (1964), examined in some detail by Dahl (1973), and treated experimentally by Foster (1975b) and Harlin & Lindbergh (1977). Foster has concluded that surface roughnesses ranging from 0.1 to 3.0 mm have no significant effect on the species composition of communities developing thereon. Harlin & Lindbergh, on the other hand, and using a similar range of experimental surfaces have observed significant differences in community composition. The rather neglected factor of abrasion from suspended sand particles, and its selective effect, on vegetation have been described by Daly & Mathieson (1977).

The importance of light as a physical determinant of vegetation is greatest in the sublittoral zone (Kain 1960; Shepherd & Womersley 1970; Dayton 1975a, b). The growth strategies of three species of *Laminaria* that occur naturally as forest dominants in different sublittoral habitats have been investigated experimentally in relation to light experience by Lüning (1979). (See also Chapters 9 and 13.)

Chemical factors such as the concentrations of dissolved salts, oxygen and carbon dioxide have been linked with differences in rock pool vegetation by Gustavsson (1972) and Daniel & Boyden (1975), but some caution is advisable in the interpretation of these observations for we now know that rock pool vegetation is powerfully influenced by biotic factors. Large-scale horizontal differences in seaweed vegetation have been associated with salinity or pollution gradients or both by numerous investigators such as Lindgren (1965), Edwards (1972), Murray & Littler (1974), Gunnarsson & Porisson (1976), Tittley & Price (1977) and Munda (1978). Jorde & Klavestad (1963) have reported the existence of a region of sudden vegetation change along a gradient of salinity and other factors in the Hardangerfjord (Norway). This change involved a halving of the species diversity and a raising of the lower limit of sublittoral macrophyte vegetation. In a similar habitat in British Columbia Druehl (1967a) found that the upper limit of sublittoral kelps is also depressed. Druehl interpreted these changes as responses to high summer temperatures and low salinities exacerbated by stratification, factors also implicated in the 'fjord effect' by Jorde & Klavestad (1963). These factors have also been attributed to horizontal distribution patterns of *Laminaria* species by Druehl (1967b). The notion of intertidal habitats in north temperate latitudes as essentially brackish-water environments has been argued by den Hartog (1968). (See Chapter 8.)

The vertical distributions of species on rocky shores almost certainly never coincide exactly with the ranges of physical and chemical tolerances. Likewise, the growth and reproductive development of seaweeds in culture provides, at best, only rough guidance on their incidence and abundance in the field. The chief reason for these discrepancies rests in the variety of interactions that occur between each species and its neighbours. These interactions are diverse and complex but, as determinants of plant community organization, interspecific competition and herbivory probably rank most highly.

The existence of sharp zonal boundaries along a continuous gradient of emersion and other associated factors has been attributed to competition between species by den Hartog (1968) and Chapman (1973). Pielou (1974) has designed a simple mathematical model, based on hypothetical rates of increase and the competitive abilities of species, to explain how such boundaries may develop. This model, which has been ably expounded by Chapman (1974b, 1979), depends on the assumption that fertility of seaweeds is unaffected by crowding. However, it has been shown that *Fucus vesiculosus* in the dense seaweed forests of a sheltered habitat allocates more of its resources to vegetative, and less to reproductive tissues than does the same species from spatially open communities such as occur on exposed shores and in estuaries (Russell 1979), and so the model in this respect may be deficient.

Laboratory experiments on competition between seaweed species are rare. Mixed cultures of filamentous seaweeds have resulted in extinction of species under conditions that would permit their growth in isolation (Russell & Fielding 1974; Fielding & Russell 1976). The mechanism of competition, in this instance, was not investigated, although chemical inhibition was postulated. This mechanism seems likely to explain the antagonistic effect in culture of one species of encrusting seaweed on another (Fletcher 1975). As crust-forming algae are likely to come into competition for rock surface (primary space), it is possible that competitive success is achieved in this way. Polyphenols secreted by certain seaweeds have been held to inhibit growth of epiphytes (Conover & Sieburth 1966; Sieburth & Conover 1966).

Plant-plant and plant-animal competition has been demonstrated in the field by removal of particular species from experimental plots (Dayton 1973, 1975a,b; Menge 1976). Dayton's (1975a) experiments investigated the community functions of the kelp *Hedophyllum sessile*. *Hedophyllum* removal was followed by a decline in the abundance of its normal understory species and an increase in that of ephemeral algae. Fluctuations in *Hedophyllum* cover in control plots were due to its regular annual defoliations; irregular oscillations in the ephemerals that invaded the clearance plots were caused by the rapid completion of their life-histories. Dayton's work demonstrated the importance of certain dominant species in community maintenance, for which he proposed the term 'foundation species' (Dayton 1972). Dominance has been discussed in similar vein by Connell (1972).

Two-way competition for primary space between the intertidal kelp *Postelsia palmaeformis* and mussels has been described by Dayton (1973), and three-way competition between *Fucus*, barnacles and mussels by Menge (1976). In one of the early clearance experiments, removal of *Fucus* and limpets from an intertidal strip at Port St. Mary, Isle of Man, was followed by the development of a non-zonal mixture of *Fucus* species throughout the strip (Burrows & Lodge 1951); thus the removal of competition resulted in an absence of zonation in the recovery vegetation.

Grazing may also be a powerful force determining seaweed community structure. Lubchenco (1978) has observed that intertidal rock pools in New England tend to be dominated either by *Chondrus* or *Enteromorpha*, success of the former being associated with high *Littorina* density and vice versa. Experimental removal of *Littorina* from a *Chondrus*-dominated pool was followed by an invasion of *Enteromorpha* and decline in *Chondrus*. The removed *Littorina* was introduced into an *Enteromorpha* pool and brought about an immediate reduction in its cover abundance. The importance of primary predators (shore crabs) and secondary predators (sea gulls) in effecting this control over seaweed communities was demonstrated. Thus selective grazing by herbivores can reduce a potentially dominant seaweed to the role of a rare and subordinate species. Lubchenco has also compared the grazing effects of *Littorina* on algal diversity in rock pools and on open rock. In pools, very low and very high *Littorina* densities are associated with low algal diversity, (either *Enteromorpha* or *Chondrus*), whereas on open rock, an ortholinear decline in diversity with herbivore density is obtained.

In the sublittoral zone, grazing is more by sea-urchins than by molluscs, but their effects on vegetation are just as important. Sea-urchin (*Strongyloncentrotus purpuratus*) removal experiments by Dayton (1975a) brought about a gradual recovery of the depleted *Hedophyllum* community, and comparable results have been obtained elsewhere by Kitching & Ebling (1961), Jones & Kain (1967) and Lubchenco & Menge (1978). The importance of predators such as the star-fish *Pycnopodia helianthoides*, sea otters and lobsters in controlling sea-urchins has been described by Dayton (1975a, b) and Mann (1977). The latter author has contended that over-fishing of lobsters in Nova Scotia has resulted in denudation of the *Laminaria* forest through uncontrolled increase in sea-urchins, and that the change may be an irreversible one. It is clear that local concentrations of herbivores can truncate a seaweed zone (Jones & Kain 1967; Mann 1977), although May *et al.* (1970) have offered some evidence to the contrary.

At present, it is widely held that the upper limits of seaweeds on intertidal shores are set by their respective emersion tolerances and that their lower limits are the results of biotic interactions (Chapman 1973). There is also some good experimental evidence of the importance of emersion stress in determining upper limits (Castenholz 1963; Schonbeck & Norton 1978). However, it seems inherently improbable that a frontier between two zones does not involve some two-way interactions. Southward & Southward (1978) have recorded the raising of the upper limits of certain seaweeds following the destruction of limpets by oil dispersants and have thereby produced some evidence for this.

The effects of seaweeds upon animals are mainly benign, not only as food for herbivores but also in the provision of canopy shelter (Mann 1977) and substrate (Moore 1973). The influence can also be antagonistic, however; the sweep of macrophyte fronds over the rock surface is known to prevent barnacle settlement (Stephenson & Stephenson 1949; Grant 1977). Epizooic algae may ultimately cause the detachment of a sessile host animal (Carefoot 1977; Dayton 1973). Toxic effects of seaweed extracellular substances have been reported by Conover & Sieburth (1966).

3.5 Succession

After clearing an intertidal strip of rocky shore of *Fucus* and limpets, Lodge (1948) recorded a pattern of seaweed succession that has since been reported from comparable experiments by a number of other investigators. The main components of this succession were: 1, a diatom-slime film; 2, a growth of filamentous ephemeral algae such as *Ulothrix* leading to, 3, a dense carpet of *Enteromorpha* which was gradually replaced by, 4, a sward of *Fucus*. Niell (1977, 1979) has made a more detailed examination of algal succession on cleared patches of rock on several Spanish shores, and has proposed the existence of three main phases in marine plant succession: a colonising phase (diatom); a 'channelling succession' phase (*Enteromorpha*); and a 'normal structured community'. Observations such as these have led to the belief that plant succession, as defined by Odum (1971) for terrestral vegetation, may apply equally to marine algae; that is, the sequence is an orderly and directional one, each phase modifies or conditions the environment for the following one, and the process culminates in a stable climax. However, recent research has shown the dynamic nature of rocky-shore communities (see above and Connell & Slayter 1977). It is also evident that succession on cleared surfaces can be unpredictable (Doty 1967; Foster 1975a; Sutherland & Karlson 1977). Evidence for the conditioning of the environment by successive phases applies mainly to the early slime-film organisms (Connell & Slayter 1977), although Knight and Parke (1950) and Niell (1979) have contended that the survival of juvenile *Fucus* is more or less dependent on the presence of a carpet of Ulvaceae. Conditioning is perhaps also a factor in the development of vegetation stratification at later stages. The rapid growth characteristics and high reproductive potential of ephemeral algae (see Section 3.6) make it probable that they will be successful as pioneers and quickly monopolize cleared substrates before being replaced by competitively superior species. In this way a fairly consistent general pattern of succession may occur without any major conditioning of the environment. The current theories on organism succession on rocky shores are well reviewed by Chapman (1979).

The unpredictable composition of the colonizing flora is partly the result of the relatively large number of suitable pioneer species (see Section 3.6). Nevertheless the colonization process is unlikely to be a random one (Huvé 1969). Spore samples from inshore waters at adjacent sites may differ considerably both in species and their relative abundances (Hruby & Norton 1979). Seasonal differences in colonizing floras have been reported also by Koetzner & Wood (1970), Bélanger & Cardinal (1977), and Emerson & Zedler (1978). Spatial and temporal heterogeneities in the composition of the spore suspension will therefore introduce an element of non-randomness, while the physical and biological characteristics of the settlement area will inevitably have an additionally selective effect.

3.6 Life-forms and strategies

'Life-form' is a portmanteau term for the thallus form, longevity and life-history of a seaweed. Differences in emphasis have led to its adoption by various authors in strikingly different senses, however. Neushul (1972) has described four stages in the development of a kelp thallus as distinct life-forms and, in much the same spirit, Wynne (1972) has applied the term to the numerous phenotypes present in the life-histories of certain scytosiphonaceous brown algae. In contrast, Knight & Parke (1931), Feldman (1951), Sears & Wilce (1975), Garbary (1976) and Russell (1977) have preferred to identify each species with one category of life-form only, and in that sense at least they are in close agreement with Raunkiaer (1937) who first conceived the idea of life-forms in flowering plants. The classes of life-form proposed by these authors are given in Table 11.1, which indicates the additional emphasis given by some to the periodicity of the algae in nature.

All five classifications (Table 11.1) recognize two fundamental categories: 1, seaweeds which live for a short period of time (annuals) and; 2, those which live for a year or longer (perennials). The authors also agree in distinguishing algae in which the thallus is wholly (2b) or only partly (2a) perennial. Of these authors, only Garbary has used life-forms for the purpose originally intended by Raunkiaer—phytogeographical analysis; the others having applied them to local floristic or ecological studies.

Table 11.1 Categories of algal life-forms as recognized by different authors; note general tendency from left to right towards a simple classification. For details, see text (partly after Sears & Wilce 1975).

Knight & Parke	Feldmann	Sears & Wilce	Garbary	Russell
1 Annuals	Eclipsophyceae			
Casual annuals	Hypnophyceae	Seasonal annuals	Hypnophyceae	Annuals
Pseudo-perennials	Ephemero-phyceae	Aseasonal annuals	Ephemero-phyceae	
2a Pseudo-perennials	Hemiphanero-phyceae Hemicrypto-phyceae	Pseudo-perennials	Hemiphanero-phyceae	Hemiphanero-phyceae
2b Perennials	Phanerophyceae Chamaephyceae	Perennials	Phanerophyceae Chamaephyceae	Phanerophyceae Chamaephyceae

The life-history and pattern of thallus development of a seaweed are characteristics that befit it for particular environmental conditions. The life-form is therefore conceptually very similar to the more recent term 'strategy'. At present there is no formal classification of strategists beyond the generally accepted distinction of 'opportunist' species from the remainder. Dayton (1975a) has proposed three categories of algae: canopy species; obligate understory species; and fugitive (=opportunist) species. Obligate understory species are those which disappear when the canopy is removed, a definition that contains too strong an element of circularity to be entirely satisfactory.

The opportunist seaweed is perfectly exemplified by certain species of the genus *Enteromorpha* which are immensely successful colonizers of cleared substrates, chronically polluted waters, ships' hulls and other highly stressed or unstable habitats. Opportunist seaweeds are usually annuals but certain perennials are also known to behave in an opportunist manner, on occasion (Dayton 1975b; Murray & Littler 1978). There is also some evidence of strategic differences within species (see Section 3.4; Russell 1979). It might be assumed from this evidence that seaweeds offer close parallels with the *r*- and *K*-strategists long recognized in flowering plants. However, there seem to be two very different strategies present within the perennial seaweeds (*K*- strategists) that makes this concept awkward to apply. Crustose perennials, particularly the calcareous epilithic species, though long-lived and with a relatively high commitment to vegetative thallus production have a different role in the community from that of the canopy perennials. Epilithic crusts seem to be fairly easily overwhelmed by epiphytes unless cleared by the sweeping of macrophyte fronds or, more importantly, by selective grazing by herbivores (van den Hoek *et al.* 1975; Jackson 1977; Wanders 1977). They are therefore highly dependent upon biotic factors for their continued survival. Indeed Lubchenco & Cubit (1980) have suggested that heteromorphic algal life-histories involving a crustose microthallus may have evolved in response to grazing pressure. The same pressures may also have favoured the evolution of isomorphic crustose life-histories.

The distinction of the classes Phanerophyceae and Chamaephyceae would therefore seem to reflect major strategic differences although other life-form categories almost certainly do not. On the whole it seems unlikely that a formal classification of strategies would be practicable. Niche theory, as currently understood in terrestial plant ecology (see Grubb 1977) also provides a body of ideas on plant community structure that may be applicable to seaweed vegetation, and has been discussed in that respect by Lubchenco (1980) and Price (1980).

3.7 Population dynamics

Population dynamics is '. . .the study of changes in the number of organisms in populations and of the factors influencing these changes; also it includes the study of the rates of loss and replacement of individuals, and of any regulatory processes tending to keep the numbers steady or at least to prevent excessive change. . .' (Solomon 1969). Concerned as it is with numbers of individual plants and their disseminules, population dynamics is applicable only to those algae in which the individual is a recognizably discrete entity. Population dynamics might therefore be feasible with a laminarian or fucoid or, at the other extreme, unicellular algae *in vitro* (Elbrachter 1977), but not with most crustose or turf-forming seaweeds. Even the most advantageous species, however, have not yet yielded to an investigation that fulfils exactly the above definition. Perhaps for this reason, authors have tended to use the words 'population' and 'dynamics' separately adopting the former for a variety of experimental and morphometric investigations and the latter for studies of seaweed production (Mann 1969) or simply of change in community structure (Nienhuis 1978). A recent demographic investigation of *Pelvetia fastigiata* (Gunnill 1980) has established an excellent approach to marine algal population dynamics, however.

3.8 Conclusions

The scarcity of population dynamics investigations and the manner of their execution points to the need for more information on the rates of recruitment and of losses occurring among natural seaweed populations. It is especially important that this should be known for dominant species; the calculations of *Laminaria* fecundity by Kain (1975b) are in that respect an important advance. The use of unbiased sampling methods, controlled experiments of acceptable design, and measured attributes of vegetation will become as important a prerequisite for investigations of communities, as it has already done so for the individual component seaweeds in culture.

4. REFERENCES

ALMODOVAR L. R. & BIEBL R. (1962) Osmotic resistance of mangrove algae around La Parguera, Puerto-Rico. *Rev. Algol.* **3**, 203–8.

BAARDSETH E. (1970) A square-scanning, two-stage sampling method of estimating seaweed quantities. *Norwegian Inst. Seaweed Res. Rep.*, **33**.

BALLANTINE W. J. (1961) A biologically-defined exposure scale for the comparative description of rocky shores. *Fld. Stud.*, **1**, 1–19.

BASSINDALE R., EBLING F. J., KITCHING J. A. & PURCHON R. D. (1948) The ecology of the Lough Ine rapids with special reference to water currents. I. Introduction and hydrography. *J. Ecol.*, **36**, 305–22.

BÉLANGER C. & CARDINAL A (1977) Stades initiaux de colonisation de substrats durs dans la Baie-des-Chaleurs, Québec. *Mar. Biol.*, **44**, 27–38.

BIEBL R. (1952) Ecological and non-environmental constitutional resistance of the protoplasm of marine algae. *J. Mar. Biol. Ass. U.K.*, **31**, 307–15.

BOALCH G. T., HOLME N. A., JEPHSON N. A. & SIDWELL J. M. C. (1974) A resurvey of Colman's intertidal traverses at Wembury, South Devon. *J. Mar. Biol. Ass. U.K.*, **54**, 551–3.

BODEN G. T. (1979). The effect of depth on summer growth of *Laminaria saccharina* (Phaeophyta, Laminariales). *Phycologia*, **18**, 405–8.

BOLWELL G. P., CALLOW J. A., CALLOW M. E. & EVANS L. V. (1977) Cross-fertilization in fucoid seaweeds. *Nature, (London.)*, **268**, 626–7.

BOUDOURESQUE C.-F. (1971a) Contribution a l'étude phytosociologique des peuplements algaux des cote varoises. *Vegetatio*, **22**, 83–184.

BOUDOURESQUE C.-F. (1971b) Le concept de nodum en bionomie et sa generalisation *C. R. Acad. Sci. Paris*, **272**, 1260–3.

BOUDOURESQUE C.-F. (1974) Aire minima et peuplements algaux marins. *Bull. Soc. Phycol. Fr.*, **19**, 141–57.

BOUDOURESQUE C.-F. & LÜCK, H. B. (1972) Recherches de bionomie structurale du niveau d'un peuplement benthique sciaphile. *J. Exp. Mar. Biol. Ecol.*, **8**, 133–44.

BRADSHAW A. D. (1965). Evolutionary significance of phenotypic plasticity in plants. *Adv. Genet.*, **13**, 115–55.

BRAY J. R. & CURTIS J. T. (1957). An ordination of the upland forest communities of southern Wisconsin. *Ecol. Monogr.*, **27**, 325–49.

BRYHNI E. (1978) Quantitative differences between polysaccharide composition in normal differentiated *Ulva mutabilis* and the undifferentiated mutant *lumpy*. *Phycologia*, **17** (2), 119–24.

BURROWS E. M., CONWAY E., LODGE S. M. & POWELL H. T. (1954) The raising of intertidal algal zones in Fair Isle. *J. Ecol.*, **42**, 283–8.

BURROWS E. M. & LODGE S. (1951) Autecology and the species problem in *Fucus*. *J. Mar. Biol. Ass. U.K.* **30**, 161–76.

CAREFOOT T. (1977) *Pacific Seashores*. J. J. Douglas, Vancouver.

CASTENHOLZ R. W. (1963) An experimental study of the vertical distribution of littoral marine diatoms. *Limnol. Oceanogr.*, **8**, 450–62.

CHAPMAN A. R. O. (1973) A critique of prevailing attitudes towards the control of seaweed zonation on the sea shore. *Bot. Mar.*, **16**, 80–2.

CHAPMAN A. R. O. (1974a) The genetic basis of morphological differentiation in some *Laminaria* populations. *Mar. Biol.*, **24**, 85–91.

CHAPMAN A. R. O. (1974b) The ecology of macroscopic marine algae. *Ann Rev. Ecol. Syst.*, **5**, 65–80.

CHAPMAN A. R. O. (1975) Inheritance of mucilage canals in *Laminaria* (section Simplices) in eastern Canada. *Br. Phycol. J.*, **10**, 219–23.

CHAPMAN A. R. O. (1979) *Biology of Seaweeds*. Edward Arnold, London.

CHAPMAN A. R. O. & CRAIGIE J. S. (1977) Seasonal growth in *Laminaria longicruris*: relation with dissolved inorganic nutrients and internal reserves of nitrogen. *Mar. Biol.*, **40**, 197–205.

CHENEY D. P. & BABBEL G. R. (1978) Biosystematic studies on the red algal genus *Eucheuma*. I. Electrophoretic variation among Florida populations. *Mar. Biol.*, **47**, 251–64.

COLE K. (1967) Chromosome numbers in the Phaeophyceae. *Can. J. Genet. Cytol.*, **9**, 519–30.

COLE K. (1968) Additional chromosome numbers in the Phaeophyceae. *Can. J. Genet. Cytol.*, **10**, 670–2.

COLMAN J. (1933) The nature of the intertidal zonation of plants and animals. *J. Mar. Biol. Ass. U.K.*, **18**, 435–76.

COLMAN J. S. & STEPHENSON A. (1966) Aspects of the ecology of a 'tideless' shore. In *Some Contemporary Studies in Marine Science*, ed. Barnes H., pp. 163–70. Allen and Unwin, London.

CONNELL J. H. (1972) Community interactions on marine rocky intertidal shores. *Ann. Rev. Ecol. Syst.*, **3**, 169–92.

CONNELL J. H. & SLAYTER R. O. (1977) Mechanisms of succession in natural communities and their role in community stability and organisation. *Am. Natur.*, **111**, 1119–44.

CONOVER J. T. & SIEBURTH J. MCN. (1966) Effect of tannins excreted from Phaeophyta on planktonic animal survival in tide pools. *Proc. Int. Seaweed Symp.*, **5**, 99–100.

DAHL A. L. (1973) Surface area in ecological analysis: quantification of benthic coral-reef algae. *Mar. Biol.*, **23**, 239–49.

DALBY D. H., COWELL E. B., SYRATT W. J. & CROTHERS J. H. (1978) An exposure scale for marine shores in western Norway. *J. Mar. Biol. Ass. U.K.*, **58**, 975–96.

DALY M. A. & MATHIESON A. C. (1977) The effects of sand movement on intertidal seaweeds and selected invertebrates at Bound Rock, New Hampshire, U.S.A. *Mar. Biol.* **43**, 45–55.

DANIEL M. J. & BOYDEN C. R. (1975) Diurnal variations in physicochemical conditions within intertidal rockpools. *Fld. Stud.*, **4**, 161–76.

DAYTON P. K. (1972) Toward an understanding of community resilience and the potential effects of enrichments to the benthos at McMurdo Sound, Antarctica. *Proc. Colloqium on Conservation Problems in Antarctica*, ed. Parker B. C., pp. 81–96. Allen Press, Kansas.

DAYTON P. K. (1973) Dispersion, dispersal and persistence of the annual intertidal alga, *Postelsia palmaeformis* Ruprecht. *Ecology*, **54**, 433–8.

DAYTON P. K. (1975a) Experimental evaluation of ecological dominance in a rocky intertidal algal community. *Ecol. Monogr.* **45**, 137–59.

DAYTON P. K. (1975b) Experimental studies of algal canopy interactions in a sea otter-dominated kelp community at Amchitka Island, Alaska. *Fishery Bull.*, **73**, 230–7.

DEN HARTOG C. (1955) A classification system for the epilithic algal communities of the Netherlands coast. *Acta Bot. Neerlandica*, **4**, 126–35.

DEN HARTOG C. (1959) The epilithic algal communities occurring along the coast of the Netherlands. *Wentia*, **1**, 1–241.

DEN HARTOG C. (1968) The littoral environment of rocky shores as a border between the sea and the land and between the sea and fresh water. *Blumea*, **16**, 374–93.

DE SILVA M. W. R. N. & BURROWS E. M. (1973) An experimental assessment of the status of the species *Enteromorpha intestinalis* (L.) Link and *Enteromorpha compressa* (L.) Grev. *J. Mar. Biol. Ass. U.K.*, **53**, 895–904.

DEVINNY J. S. (1978) Ordination of seaweed communities: Environmental gradients at Punta Banta, Mexico. *Botanica Mar.*, **21**, 357–63.

DIXON P. S. (1963) Variation and speciation in marine Rhodophyta. In *Speciation in the Sea*, eds. Harding J. P. & Tebble N., pp. 51–62, Systematics Association Publ. No. 5.

DIXON P. S. & IRVINE L. M. (1977) *Seaweeds of the British Isles*, vol. I., part I, British Museum (Natural History), London.

DOTY M. S. (1946) Critical tide factors that are correlated with the vertical distribution of marine algae and other organisms along the Pacific coast. *Ecology*, **27**, 315–28.

DOTY M. S. (1967) Pioneer intertidal population and the related general vertical distribution of marine algae in Hawaii. *Blumea*, **15**, 95–105.

DOTY M. S. (1971) Measurement of water movement in reference to benthic algal growth. *Botanica Mar.*, **14**, 32–5.

DOTY M. S. & ARCHER J. G. (1950) An experimental test of the tide factor hypothesis. *Am. J. Bot.*, **37**, 458–64.

DREW K. M. (1939) An investigation of *Plumaria elegans* (Bonnem.) Schmitz with special reference to triploid plants bearing parasporangia. *Ann. Bot.* n.s., **3** (10), 347–69.

DRUEHL L. D. (1967a) Vertical distributions of some benthic marine algae in a British Columbian inlet, as related to some environment factors. *J. Fish. Res. Bd. Canada*, **24**, 33–46.

DRUEHL L. D. (1967b) Distribution of two species of *Laminaria* as related to some environmental factors. *J. Phycol.*, **3**, 103–8.

EDWARDS P. (1972) Benthic algae in polluted estuaries. *Mar. Polln. Bull.*, **3**, 55–60.

EDWARDS P., BIRD E., COTGRAVE B., COSSINS A., CROMPTON K., FOWLER W., HERDSON D. & HUDSON J. (1975) Marine phytobenthos of the Castellabate (Cilento) natural park, Salerno, Italy. *Phytocoenologia*, **1**, 403–26.

ELBRACHTER M. (1977) On population dynamics in multi-species cultures of diatoms and dinoflagellates. *Helgoländer Wiss. Meeresunters.*, **30**, 192–200.

EMERSON S. E. & ZEDLER J. B. (1978). Recolonization of intertidal algae: an experimental study. *Mar. Biol.*, **44**, 315–24.

EVANS L. V. (1965) Cytological studies in the Laminariales. *Ann. Bot.* n.s., **29**, 541–62.

EVANS L. V. (1966) The Phaeophyceae, Part I. In *Chromosomes of the Algae*, ed. Godward M. B. E., pp. 122–48. Edward Arnold, London.

FANG T. C., JIANG B. Y. & LI J. J. (1962) On the inheritance of stipe length in Haidai (*Laminaria japonica* Aresch.). *Acta Bot. Sin.*, **10**, 327–55.

FELDMANN J. (1951) Ecology of marine algae. In *Manual of Phycology—an Introduction to the Algae and their Biology*, ed. Smith G. M., pp. 313–34. Waltham, Massachusetts.

FIELD J. G. (1968) The 'Turbolometer'—an apparatus for measuring relative exposure to wave action on shores. *Zool. Africana*, **3**, 115–18.

FIELDING A. H. & RUSSELL G. (1976) The effect of copper on competition between marine algae. *J. Appl. Biol.*, **13**, 871–6.

FJELD A. & LØVLIE A. (1976) Genetics of multicellular algae. In *The Genetics of Algae*, ed. Lewin R. A., pp. 219–35. Blackwell Scientific Publications, Oxford.

FLETCHER R. L. (1975) Heteroantagonism observed in mixed cultures. *Nature (London)*, **253**, 534–5.

FORBES E. (1846) On the connexion between the distribution of the existing fauna and flora of the British Isles, and the geographical changes which have affected their area, especially during the epoch of the Northern Drift. *Mem. Geol. Survey U.K.*, **1**, 336–432.

FOSTER M. S. (1975a) Algal succession in a *Macrocystis pyrifera* forest. *Mar. Biol.*, **32**, 313–29.

FOSTER M. S. (1975b) Regulation of algal community development in a *Macrocystis pyrifera* forest. *Mar. Biol.*, **32**, 331–42.

GARBARY D. J. (1976) Life-forms of algae and their distribution. *Botanica Mar.* **19**, 97–106.

GOODMAN C., NEWALL M. & RUSSELL G. (1976) Rapid screening for copper tolerance in ship-fouling algae. *Int. Biodetn. Bull.*, **12** (3), 81–3.

GRANT W. S. (1977) High intertidal community organization on a rocky headland in Maine, U.S.A. *Mar. Biol.*, **44**, 15–25.

GRUBB P. J. (1977) The maintenance of species richness in plant communities: the importance of the regeneration niche. *Biol. Rev.*, **52**, 107–45.

GUNNARSON K. & PORISSON K. (1976) The effect of sewage on the distribution and cover of littoral algae near Reykjavik, Preliminary results. *Acta. Bot. Isl.*, **4**, 58–66.

GUNNILL F. C. (1980) Demography of the intertidal brown alga *Pelvetia fastigiata* in southern California, U.S.A. *Mar. Biol.*, **59**, 169–79.

GUSTAVSSON U. (1972) A proposal for a classification of marine rock pools on the Swedish west coast. *Botanica Mar.*, **15**, 210–314.

HALL A., FIELDING A. H. & BUTLER M. (1979) Mechanisms of copper tolerance in a marine fouling alga *Ectocarpus siliculosus*—evidence for an exclusion mechanism. *Mar. Biol.*, **54**, 195–9.

HARLIN M. M. & LINDBERGH J. M. (1977) Selection of substrata by seaweeds; optimal surface relief. *Mar. Biol.*, **40**, 33–40.

HAWKINS S. J. & HARTNOLL R. G. (1980) Small-scale relationship between species number and area on a rocky shore., *Est. Cstl. Mar. Sci.*, **10**, 201–14.

HOLM R. F. (1978) The community structure of a tropical marine lagoon. *Est. Cstl. Mar. Sci.*, **7**, 329–45.

HRUBY T. (1975) Seasonal changes in two algal populations from the coastal waters of Washington State. *J. Ecol.*, **63**, 881–90.

HRUBY T. & NORTON T. A. (1979) Algal colonization on rocky shores in the firth of Clyde. *J. Ecol.*, **67**, 65–78.

HUGHES R. N. & THOMAS M. L. H. (1971) The classification and ordination of shallow-water benthic samples from Prince Edward Island, Canada *J. Exp. Mar. Biol. Ecol.*, **7**, 1–39.

HUVÉ, P. (1969) Le stade initiale et son importance réelle dans le peuplement algal des surfaces rocheuses du littoral marin. *Proc. Int. Seaweed Symp.*, **6**, 201–11.

JACKSON J. B. C. (1977) Competition on marine hard substrata: the adaptive significance of solitary and colonial structures. *Am. Nat.*, **111**, 743–67.

JENIK J. & LAWSON G. W. (1968) Zonation of microclimate and vegetation on a tropical shore in Ghana. *Oikos*, **19**, 198–205.

JONES W. E. & BABB S. (1968). The motile period of swarmers of *Enteromorpha intestinalis* (L.) Link *Br. Phycol. Bull.*, **3**, 525–8.

JONES W. E. & DEMETROPOULOS A. (1968) Exposure to wave action: measurements of an important ecological parameter on rocky shores on Anglesey. *J. Exp. Mar. Biol. Ecol.*, **2**, 46–63.

JONES W. E., FLETCHER A., BENNELL S. J., MCCONNELL B. J., RICHARDS A. V. L. & MACK-SMITH S. (1979). Intertidal surveillance. In *Monitoring the Marine Environment*, ed. Nichols D., pp. 1–23. Symposium of the Institute of Biology No. 24, London.

JONES N. S. & KAIN J. M. (1967) Subtidal algal colonization following the removal of *Echinus*. *Helgoländer Wiss Meeresunters.*, **15**, 460–6.

JORDE I. (1966) Algal association of a coastal area south of Bergen, Norway. *Sarsia*, **23**, 1–52.

JORDE I. & KLAVESTAD, N. (1963) The natural history of the Hardangerfjord 4. The benthonic algal vegetation. *Sarsia*, **9**, 1–99.

KAIN J. M. (1960) Direct observations on some Manx sublittoral algae, *J. Mar. Biol. Ass. U.K.*, **39**, 609–30.

KAIN J. M. (1971) The biology of *Laminaria hyperborea*. VI. Some Norwegian populations. *J. Mar. Biol. Ass. U.K.*, **51**, 387–408.

KAIN J. M. (1975a) Algal recolonization of some cleared subtidal areas. *J. Ecol.*, **63**, 739–65.

KAIN J. M. (1975b) The biology of *Laminaria hyperborea*, VII. Reproduction of the sporophyte. *J. Mar. Biol. Ass. U.K.*, **55**, 567–82.

KAIN J. M. (1976) New and interesting marine algae from the Shetland Isles. II. Hollow and solid stiped *Laminaria* (Simplices). *Br. Phycol. J.*, **11**, 1–11.

KHAILOV K. M., KHOLODOV V. I., FIRSOV Y. K. & PRAZUKIN A. V. (1978) Thalli of *Fucus vesiculosus* in ontogenesis: changes in morpho-physiological parameters. *Botanica Mar.*, **21**, 289–311.

KITCHING J. A. & EBLING F. J. (1961) The ecology of Lough Ine. XI. The control of algae by *Paracentrotus lividus* (Echinoidea) *J. Anim. Ecol.*, **30**, 373–84.

KLAVESTAD N. (1978) The marine algae of the polluted inner part of Oslofjord. *Botanica Mar.*, **21**, 71–97.

Knight M. & Parke M. W. (1931) *Manx Algae*. L.M.B.C. Memoirs, Liverpool Univ. Press.

Knight M. & Parke M. W. (1950) A biological study of *Fucus vesiculosus* L. and *F. serratus* L. *J. Mar. Biol. Ass. U.K.*, **29**, 439–514.

Koetzner K. & Wood R. D. (1970) Periodicity in algal colonization in Great South Bay, Long Island, New York. *Bull. Torrey. Bot. Club*, **97**, 120–1.

Lawson G. W. (1978) The distribution of seaweed floras in the tropical and subtropical Atlantic Ocean: a quantitative approach. *J. Linn. Soc. Bot.*, **76**, 177–94.

Lewin R. A. (1958) Genetics and marine algae. In *Perspectives in marine biology*, ed. Buzzati-Traverso A. A., pp. 547–57. University of California Press, Berkeley and Los Angeles.

Lewis J. R. (1961) The littoral zone on rocky shores—a biological or physical entity? *Oikos*, **12**, 280–1.

Lewis J. R. (1964) *The Ecology of Rocky Shores*. English Universities Press, London.

Lewis J. R. (1968) Water movements and their role in rocky shore ecology. *Sarsia*, **34**, 13–36.

Lindgren P. E. (1965) Coastal algae off Goteborg in relation to gradients in salinity and pollution. *Acta Phytogeogr. Suec.*, **50**, 92–6.

Lodge S. M. (1948) Algal growth in the absence of *Patella* on an experimental strip of foreshore, Port St. Mary, Isle of Man. *Proc. Trans. Lpool. Biol. Soc.*, **56**, 78–83.

Løvlie A. (1968) On the use of multicellular alga (*Ulva mutabilis* Føyn) in the study of general aspects of growth and differentiation. *Nytt. Magasin Zoologi*, **16**, 39–49.

Lubchenco J. (1978) Plant species diversity in a marine intertidal community: importance of herbivore food preference and algal competitive abilities. *Am. Natur.*, **112**, 23–39.

Lubchenco J. (1980) Algal zonation in the New England rocky intertidal community: an experimental analysis, *Ecology*, **61**, 333–44.

Lubchenco J. & Cubit J. (1980) Heteromorphic life histories of certain marine algae as adaptations to variations in herbivory, *Ecology*, **61**, 676–87.

Lubchenco J. & Menge B. A. (1978) Community development and persistence in a low rocky intertidal zone. *Ecol. Monogr.*, **59**, 67–94.

Lüning K. (1979) Growth strategies of three *Laminaria* species (Phaeophyceae) inhabiting different depth zones in the sublittoral region of Helgoland (North Sea). *Mar. Ecol.*, **1**, 179–86.

Lüning K., Chapman A. R. O. & Mann K. H. (1978) Crossing experiments in the non-digitate complex of *Laminaria* from both sides of the Atlantic. *Phycologia*, **17**, 293–8.

McLachlan J., Chen L. C-M. & Edelstein T. (1971) The culture of four species of *Fucus* under laboratory conditions. *Can. J. Bot.*, **49**, 1463–9.

Mallery C. H. & Richardson N. (1972) Disc gel electrophoreses of proteins of Rhodophyta. *Proc. Int. Seaweed Symp.*, 7, 292–300. Tokyo, Univ. Tokyo.

Mann K. H. (1969) The dynamics of aquatic ecosystems. *Adv. Ecol. Res.*, **6**, 1–81.

Mann K. H. (1977) Destruction of kelp-beds by sea-urchins: a cyclical phenomenon or irreversible degradation? *Helgoländer Wiss. Meeresunters.*, **30**, 455–67.

May V., Bennet I. & Thompson T. E. (1970) Herbivore–algal relationships on a coastal rock platform (Cape Banks, N.S.W.). *Oecologia (Berlin)*, **6**, 1–14.

Menge B. A. (1976) Organization of the New England rocky intertidal community: role of predation, competition and environmental heterogeneity. *Ecol. Monogr.*, **46**, 355–93.

Moore P. G. (1973) The kelp fauna of north-east Britain. I. Introduction and the physical environment. *J. Exp. Mar. Biol. Ecol.*, **13**, 97–125.

Müller D. G. (1970) Diploide, heterozygote Gametophyten bei der Braunalge *Ectocarpus siliculosus*. *Naturwissenschaften*, **7**, 357–8.

Müller D. G. (1976) Sexual isolation between a European and an American population of *Ectocarpus siliculosus* (Phaeophyta) *J. Phycol.*, **12**, 252–4.

Müller D. G. (1979) Genetic affinity of *Ectocarpus siliculosus* (Dillw.) Lyngb. from the Mediterranean, North Atlantic and Australia. *Phycologia*, **18**, 312–18.

Murray S. N. & Littler M. M. (1974) Biological features of intertidal communities near the U.S. Navy sewage outfall, Wilson Cove, San Clemente Island, California. *Report Naval Undersea Center, San Diego, California*, **NUCTP 396**, 1–88.

MURRAY S. N. & LITTLER M. M. (1978) Patterns of algal succession in a perturbated marine intertidal community. *J. Phycol.* **14**, 506–12.

MUUS B. J. (1968) A field method for measuring 'exposure' by means of plaster balls. *Sarsia*, **34**, 61–8.

NAKAHARA N. & NAKAMURA Y. (1973) Parthogenesis, apogamy and apospory in *Alaria crassifolia* (Laminariales). *Mar. Biol.*, **18**, 327–32.

NEUSHUL M. (1965) Scuba diving studies of the vertical distributions of benthic marine plants. *Bot. Gothoburgensia*, **3**, 161–76.

NEUSHUL M. (1972) Functional interpretation of benthic marine algal morphology. In *Contributions to the Systematics of Benthic Marine Algae of the North Pacific*, ed. Abbott I. A. & Kurogi M., pp. 47–73. Japanese Society of Phycology, Kobe, Japan.

NIELL F. X. (1974) Les applications de l'indice de Shannon à l'étude de la végétation intertidale. *Bull. Soc. Phycol. Fr.*, **19**, 238–54.

NIELL F. X. (1977) Rocky intertidal benthic systems in temperate seas: a synthesis of their functional performance. *Helgoländer Wiss. Meeresunters.*, **30**, 315–33.

NIELL F. X. (1979) Structure and succession in rocky algal communities of a temperate intertidal system. *J. Exp. Mar. Biol. Ecol.*, **36**, 185–200.

NIENHUIS P. H. (1978) Dynamics of benthic algal vegetation and environment in Dutch estuarine salt marshes, studied by means of permanent quadrats. *Vegetatio*, **38**, 103–12.

ODUM E. P. (1971) *Fundamentals of Ecology*, 3rd edition. W. B. Saunders, Philadelphia.

PIELOU E. C. (1974) Competition on an environmental gradient. In *Proc. Cont. Math. Probl. Biol.*, ed. van der Driessche P., pp. 184–204. Springer-Verlag, Berlin.

POWELL H. T. (1963) Speciation in the genus *Fucus* L. and related genera. In *Speciation in the Sea*, Systematics Association Publ., **5**, 63–77.

PRENTICE S. A. & KAIN J. M. (1976) Numerical analysis of subtidal communities on rocky shores. *Est. Cstl. Mar. Sci.*, **4**, 65–70.

PRICE J. H. (1980) The concept of niche in the intertidal. In *The Shore Environment: Methods and Ecosystems*, vol. 2, ed. Price J. H., Irvine D. E. G. & Farnham W. F., pp. 487–564. Academic Press, London.

RAUNKIAER C. (1937) *Plant Life Forms*. Oxford University Press, Oxford.

RAVANKO O. (1972) The physiognomy and structure of the benthic macrophyte communities on rocky shores in the Southwestern archipelago of Finland (Seiliblands). *Nova Hedwigia*, **23**, 363–408.

REED R. H. & RUSSELL G. (1979) Adaptation to salinity stress in populations of *Enteromorpha intestinalis* (L.) Link. *Est. Cstl. Mar. Sci.*, **8**, 251–8.

RUENESS J. (1978) Hybridisation in red algae. In *Modern Approaches to the Taxonomy of Red and Brown Algae*, ed. Irvine D. E. G. & Price J. H., pp. 247–62. Academic Press, London.

RUENESS J. & RUENESS M. (1975) Genetic control of morphogenesis in two varieties of *Antithamnion plumula* (Rhodophyceae, Ceramiales). *Phycologia*, **14**, 81–5.

RUSSELL G. (1963) A study in populations of *Pilayella littoralis*. *J. Mar. Biol. Ass. U.K.*, **43**, 469–83.

RUSSELL G. (1972). Phytosociological studies on a two-zone shore. I. Basic Pattern. *J. Ecol.*, **60**, 539–45.

RUSSELL G. (1973) The 'litus' line; a reassessment. *Oikos*, **24**, 158–61.

RUSSELL G. (1974) The marine algal vegetation of Egmont Is. (Chagos Bank, Indian Ocean). *Proc. Int. Seaweed Symp.*, **8**, (abst.).

RUSSELL G. (1977) Vegetation on rocky shores at some north Irish Sea sites. *J. Ecol.*, **65**, 485–95.

RUSSELL G. (1978) Environment and form in the discrimination of taxa in brown algae. In *Modern Approaches to the Taxonomy of Red and Brown Algae*, ed. Irvine D. E. G. & Price J. H., pp. 339–69. Academic Press, London.

RUSSELL G. (1979) Heavy receptacles in estuarine *Fucus vesiculosus* L. *Est. Cstl. Mar. Sci.*, **9**, 659–61.

RUSSELL G. (1980) Applications of simple numerical methods to the analysis of intertidal vegetation. In *The Shore Environment: Methods and Ecosystems*, ed. Price J. H., Irvine D. E. G. & Farnham W. F. Academic Press, London (in press).

RUSSELL G. & BOLTON J. J. (1975) Euryhaline ecotypes of *Ectocarpus siliculosus* (Dillw.) Lyngb. *Est. Cstl. Mar. Sci.*, **3**, 91–4.

RUSSELL G. & FIELDING A. H. (1974) The competitive properties of marine algae in culture. *J. Ecol.*, **62**, 689–98.

RUSSELL G. & MORRIS O. P. (1973) Ship-fouling as an evolutionary process. *Proc. Int. Congr. Mar. Corr. Foul.*, **3**, 719–30.

SCHONBECK M. & NORTON T. A. (1978) Factors controlling the upper limits of fucoid algae on the shore. *J. Exp. Mar. Biol. Ecol.*, **31**, 303–13.

SEAPY R. R. & LITTLER M. M. (1978) The distribution, abundance, community structure, and primary productivity of macro organisms from two central California rocky intertidal habitats. *Pacific, Sci.*, **32**, 293–314.

SEARS J. R. & WILCE R. T. (1975) Sublittoral, benthic algae of southern Cape Cod and adjacent islands: seasonal periodicity, associations, diversity and floristic composition. *Ecol. Monogr.*, **45**, 337–65.

SHEPHERD S. A. & WOMERSLEY H. B. S. (1970) The sublittoral ecology of West Island, South Australia: 1. Environmental features and algal ecology. *Trans. Roy. Soc. S. Austr.*, **94**, 105–37.

SIEBURTH J. McN. & CONOVER J. J. (1966) Antifouling in *Sargassum natans*: Recognition of tannin activity. *Proc. Int. Seaweed Symp.*, **5**, 207.

SJOSTEDT L. J. (1928) Littoral and supralittoral studies on the Scanian shores. *Lunds Univ. Årsskr.* N.F.2, **24**, 1–36.

SÖDERSTRÖM J. (1965) Vertical zonation of littoral algae in Bohuslam. *Acta. Phytogeogr. Suecica.*, **50**, 85–91.

SOLOMON N. E. (1969) *Population Dynamics.* Edward Arnold, London.

SOUTHWARD A. J. & SOUTHWARD E. C. (1978) Recolonization of rocky shores in Cornwall after use of toxic dispersants to clean up the *Torrey Canyon* spill. *J. Fish. Res. Board Can.* **35**, 682–706.

STEBBINS G. L. (1950) *Variation and Evolution in Plants.* Columbia University Press, New York.

STEPHENSON T. A. (1953) The world between tidemarks. In: *Essays in Marine Biology*, ed. Marshall S. M. & Orr A. P., pp. 73–100. Oliver and Boyd, Edinburgh.

STEPHENSON T. A. & STEPHENSON A. (1949). The universal features of zonation between tide-marks on rocky coasts. *J. Ecol.*, **37**, 289–305.

STEPHENSON T. A. & STEPHENSON A. (1972) *Life Between Tidemarks on Rocky Shores.* W. H. Freeman, San Francisco.

SUNDENE O. (1953) The algal vegetation of Oslofjord. *Skrifler Norske Videnskaps-Akademi i Oslo. T. Mat.-Naturv. Klasse*, **2**, 1–244.

SUTHERLAND J. P. & KARLSON R. H. (1977) Development and stability of the fouling community at Beaufort, North Carolina. *Ecol. Monogr.*, **47**, 425–66.

TANIGUTI M. (1962) *Phytosociological Study of Marine Algae in Japan.* Inoue & Co., Tokyo.

TITTLEY I., IRVINE D. E. G. & JEPHSON N. A. (1976) The infralittoral marine algae of Sullom Voe, Shetland. *Trans. Bot. Soc. Edinb.*, **42**, 397–419.

TITTLEY I. & PRICE J. H. (1977) The marine algae of the tidal Thames. *Lond. Naturalist*, **56**, 10–17.

UNDERWOOD A. J. (1978) A refutation of critical tide levels as determinants of the structure of intertidal communities on British shores. *J. Exp. Mar. Biol. Ecol.*, **33**, 261–76.

VAN DEN HOEK C. (1969) Algal vegetation-types along the open coasts of Curaçao, Netherlands Antilles. *Verh. Koninkl. Nederl. Akad. Wetenschappen*, **72**, 537–77.

VAN DEN HOEK C., COLIJN F., CORTEL-BREEMAN A. M. & WANDERS J. B. W. (1972) Algal vegetation-types along the shores of inner bays and lagoons of Curaçao, and of the lagoon Lae (Bonaire), Netherlands Antilles. *Verh. Koninkl. Nederl. Akad. Wetenschappen Afd Natuurkunde*, **61**, 1–72.

VAN DEN HOEK C., CORTEL-BREEMAN A. M. & WANDERS J. B. W. (1975) Algal zonation in the fringing coral reef of Curaçao, Netherlands Antilles, in relation to zonation of corals and gorgonians. *Aquat. Bot.*, **1**, 269–308.

VAN DER MEER J. P. (1978) Genetics of *Gracilaria* sp. (Rhodophyceae, Gigartinales). III. Non-Mendleian gene transmission. *Phycologia*, **17**, 314–18.

VAN DER MEER J. P. (1979) Genetics of *Gracilaria* sp. (Rhodophyceae, Gigartinales). V. Isolation and characterization of mutant strains. *Phycologia*, **18**, 47–54.

VAN DER MEER J. P. & TODD E. R. (1977) Genetics of *Gracilaria* sp. (Rhodophyceae: Gigartinales). IV. Mitotic recombination and its relationship to mixed phases in the life history. *Can. J. Bot.*, **55**, 2810–17.

VAILLANT L. (1891) Nouvelles études sur les zones littorales. *Ann. Sci. nat. Zoo.*, **12**, 39–50.

WANDERS J. B. W. (1977) The role of benthic algae in the shallow reef of Curaçao (Netherlands Antilles). III: The significance of grazing. *Aquat. Bot.*, **3**, 357–90.

WEST J. A. (1972) Environmental regulation of reproduction in *Rhodochorton purpureum*. In *Contributions to the Systematics of Benthic Algae of the North Pacific*. ed. Abbott I. A. & Kurogi M., pp. 213–30, Japanese Society of Phycology, Kobe, Japan.

WHITTICK A. & HOOPER R. G. (1977) The reproduction and phenology of *Antithamnion cruciatum* (Rhodophyta, Ceramiaceae) in insular Newfoundland. *Can. J. Bot.*, **55**, 520–4.

WILKINSON M. (1974) Investigations on the autecology of *Eugomontia sacculata* Kornm. a shell boring alga. *J. Exp. Mar. Biol. Ecol.*, **16**, 19–27.

WYNNE M. J. (1972) Studies on the life forms in nature and in culture of selected brown algae. In *Contributions to the Systematics of Benthic Marine Algae of the North Pacific*. ed. Abbott I. A. & Kurogi M., pp. 133–46. Japanese Society of Phycology, Kobe, Japan.

YARISH C., EDWARDS P. & CASEY S. (1979). A culture study of salinity responses in ecotypes of two estuarine red algae. *J. Phycol.*, **15**, 341–6.

CHAPTER 12
MORPHOLOGY AND ENVIRONMENT

T. A. NORTON
Department of Botany, University of Glasgow, Scotland

A. C. MATHIESON
Department of Botany & Plant Pathology and Jackson Estuarine Laboratory, University of New Hampshire, U.S.A.

M. NEUSHUL
Department of Biological Sciences and the Marine Science Institute, University of California, Santa Barbara, U.S.A.

1 INTRODUCTION

The literature is full of observations that seaweeds vary morphologically in different situations. In the present account, we have attempted to summarize these observations as well as indicate causal factors and parallel morphological responses to them. The effects on growth rate *per se* are largely ignored unless they influence the ultimate size of the plant rather than the time taken to achieve that size, or if they slow down growth sufficiently to greatly extend the duration of the plant's juvenile appearance. Anatomical variations are largely omitted unless they have obvious external repercussions affecting the size, shape or ecology of the plant.

Rarely are observations pursued to a stage where there is clear evidence of cause and effect. To establish such evidence conclusively we feel that the following criteria should be met:

1 Field observations should establish that there is invariably a correlation between a particular morphology and a specified set of environmental conditions. Where there is a clinal gradient of such conditions there should be a correspondingly graded response in the plants. If the field observations are

extended over a prolonged period they might also indicate whether all the new colonists develop the same morphology from the outset or whether some morphologically different plants begin to develop but are subsequently eliminated.

2 The environmental conditions thought to be influential in modifying the plant's shape should be quantified.

3 The differences between samples of plants should be described numerically as well as illustrated.

4 Comparisons should only be made between plants of similar age to avoid ontogenetic factors. Where it is impossible to age the plants, we recommend the use of proportional measures rather than absolute ones, so that the shape of the plant is defined independently of size. Examples of such measures include length to breadth ratios (Kain 1976a; Sanbonsuga & Torii 1974), length ratios of the main axis to the lateral branches (Ercegović 1952), and lamina weight to surface area ratios (Larkum 1972).

5 An attempt must be made to change the appearance of different forms of the plant by growing them under appropriate conditions in culture and by reciprocal transplantations between various habitats in nature. Often the latter technique is attempted, but without proper, if any, controls. The possible effects induced by the disturbance of transplantation can only be assessed if control plants are returned to their original habitats after a similar degree of disturbance; plants left *in situ* are inadequate controls. Moreover, a failure to change morphology cannot be interpreted as evidence for some genetic fixation unless the plants grow sufficiently during the experimental period to allow a change to occur. Clearly, transplant experiments are only valid tests where the prevailing conditions encourage substantial growth. A variety of somatic phases (sexual, asexual) and ages of plants should be employed, as varying growth potentials may be present with the different generations and reduced growth may occur during fertility (Norton 1977a; Norton & Burrows 1969b). The experiments should employ either very young plants (ideally even beginning as spores or zygotes), or should be conducted over a reasonably extended period, to minimize the influence of the plant's previous experience (i.e. developmental canalization).

6 An attempt should be made to assess the genetic similarity (i.e. evolutionary divergence) of the various forms, using methods other than morphological differentiation e.g. interfertility crosses and electrophoresis.

2 TEMPORAL CHANGES

Most seaweeds exhibit seasonal differences in growth in response to factors such as light, nutrient availability and possibly water temperature. Therefore, they may be considerably larger at one time of year than another. Also, since all parts of a plant need not respond similarly nor grow at the same rate, the proportions of parts and consequently the overall shapes of plants may vary seasonally. This is clearly seen in *Pterocladia capillacea* and *P. caerulescens* where extension growth predominates initially, but later gives way to an increase in width and thickness (Dixon 1966; Santelices 1978). In early summer, the plants have thin elongate axes, but by the onset of winter their axes, although longer, are proportionally broader. In plants in their second summer, the newly formed narrow branches contrast markedly with the previous year's broader branches, until they too broaden later in the year. In *Gelidium crinale* there is a marked broadening and flattening of the axes that occurs late in the year when active linear growth has declined. In species of *Chaetomorpha* and *Rhizoclonium*, the size and proportions of the cells change from month to month (Price 1967). In *Laminaria hyperborea*, the thickness and texture of the lamina is influenced by seasonal changes in the rate of lamina extension (Kain 1976b).

In most laminarian algae, the length of the blade is determined by the balance between extension growth from the intercalary meristem and decay from the distal tip. The rate of both processes varies seasonally and this is often reflected in the resulting size of the lamina. The outcome of these two conflicting processes is dramatically seen in *Saccorhiza polyschides*, which dies back entirely after becoming fertile. That is, with the onset of fertility, growth ceases and the plant succumbs to progressive decay from the tip, which continues and advances down the thallus (Norton & Burrows 1969b). Its large globular holdfast is fertile and can persist overwinter; thus greatly extending the fruiting period of the plant. Plants of *S. polyschides* that arise late in the year do not mature until the following spring and persist through the winter as intact vegetative plants. Therefore, *S. polyschides* can overwinter as two totally dissimilar macroscopic plants.

Often there are abrupt changes in shape resulting from the sudden initiation and growth of a particular part of the plant. Most striking is the annual appearance of new laminae in plants such as *Maripelta rotata* (Bowen 1971) and *Constantinea subulifera* (Powell 1964). This is initiated, at least in the case of the latter species, by short-day conditions in the autumn. Many species of laminarian algae also renew their blades annually. Often the old blade is retained distally for several months, while the new one develops below, separated from it by a pronounced 'waist' (Fritsch 1945, Fig. 65; Parke 1948).

Reproduction is usually a seasonal phenomenon and sometimes the size or abundance of the reproductive organs is such that they alter the appearance of the entire plant. Examples are the masses of papillate excrescences that bear the cystocarps in *Gigartina stellata* or the massive belt-like receptacles of *Himanthalia elongata* which dwarf the little button-like vegetative thallus.

Fertility is commonly followed by defoliation or die-back of the thallus (Kilar

& Mathieson 1978). Frequently almost the entire erect portion of the thallus is lost and regeneration from a small perennating basal portion produces a new frond the next year. Such plants have been called pseudo-perennials (Knight & Parke 1931).

Since in many species only fertile branches absciss, the subsequent appearance of the plant depends largely on the distribution of these branches. The fucoid alga *Ascophyllum nodosum* annually produces lateral reproductive branches (i.e. receptacles) approximately equal in weight to its vegetative thallus, with the receptacles being shed within a single month during the spring (Josselyn & Mathieson 1979). In *Sargassum muticum* almost every branch becomes fertile and virtually the entire frond subsequently decays, leaving only a few short vegetative branches or just the holdfast that can regenerate new axes (Norton 1977a). In *Fucus serratus*, the apices of the main axes become fertile and subsequently die back, resulting in a short, stumpy thallus (Fig. 12.1), whereas in *F. vesiculosus* fertility is often confined to lateral branches which later absciss, leaving long, stringy plants with denuded axes (Knight & Parke 1950).

Postfertility decay or abscission, followed by renewal or regeneration, may result in a very different plant morphology. In *Enteromorpha*, for example, apical die-back after reproduction stimulates the production of proliferous branches (Burrows 1959). In *Pterocladia capillacea* branches are lost proximally which gives the plant a stalked dendroid look, and the new axes that arise are distributed unevenly around the plant, giving a much more irregular aspect than formerly (Dixon 1966).

Tokida & Yamamoto (1965) describe the seasonal morphological effects of syntagmatic germination of tetraspores in *Pachymeniopsis yendoi*, i.e. their *in situ* germination to produce a single germling. Eventually, the *in situ* germlings form extensive proliferations from the surface of the mother plant and substantially modify its morphology. Proliferations of this kind, derived from the germination of the entire contents of the tetrasporangia, have been recorded in several other red and brown algae, including *Agardhiella, Champia, Cystoclonium, Gracilaria, Lomentaria, Dictyota, Padina* and *Taonia*.

Hairs or assimilatory filaments are often lost seasonally as observed in nature on *Fucus spiralis* germlings (Schonbeck & Norton 1979d) and on adult plants of *Desmarestia aculeata* (Mathieson and Norton personal observations). It is even more striking in *Elachista fucicola*, where summer plants are a tiny basal cushion covered with a dense 'shaving brush' of erect filaments, and the winter form is a large bald gelatinous blister (Filion & Norton unpublished).

Superimposed on seasonal variations in form are the progressive changes that occur throughout the life of the plant. For example, in *Laminaria agardhii-longicruris*, larger and probably older plants (the two are not necessarily synonymous) show a tendency to develop hollow stipes (Mann 1971). The thallus of *Himanthalia elongata* at first resembles a stalked discoid button, but

Fig. 12.1 Plants of *Fucus serratus*. (a) Showing extensive terminal receptacles; (b) a residual vegetative plant after extensive defoliation of receptacles and adjacent vegetative tissue. (Based on Knight & Parke 1950).

later develops into an inflated inverted cone (Norton personal observations). In *Grateloupia filicina* the increasing development of proliferous branches gradually changes the entire aspect of the frond from essentially dichotomous to pinnate (Dixon 1966). With increasing age, *Fucus vesiculosus* exhibits a progressive decrease in thallus width and surface area to weight ratio, but with an increase in its thickness the cross section changes from flat to almost round (Khailov *et al.* 1978). In addition, the same investigators claim that older plants of *F. vesiculosus* are more prone to degradation than younger plants. Similarly, Fralick & Mathieson (1972) observed that winter fragmentation, which greatly truncates plants of *Codium fragile* spp. *tomentosoides* (Fig. 12.2), affects longer plants the most.

Occasionally temporal changes in form are attributable to the plant being

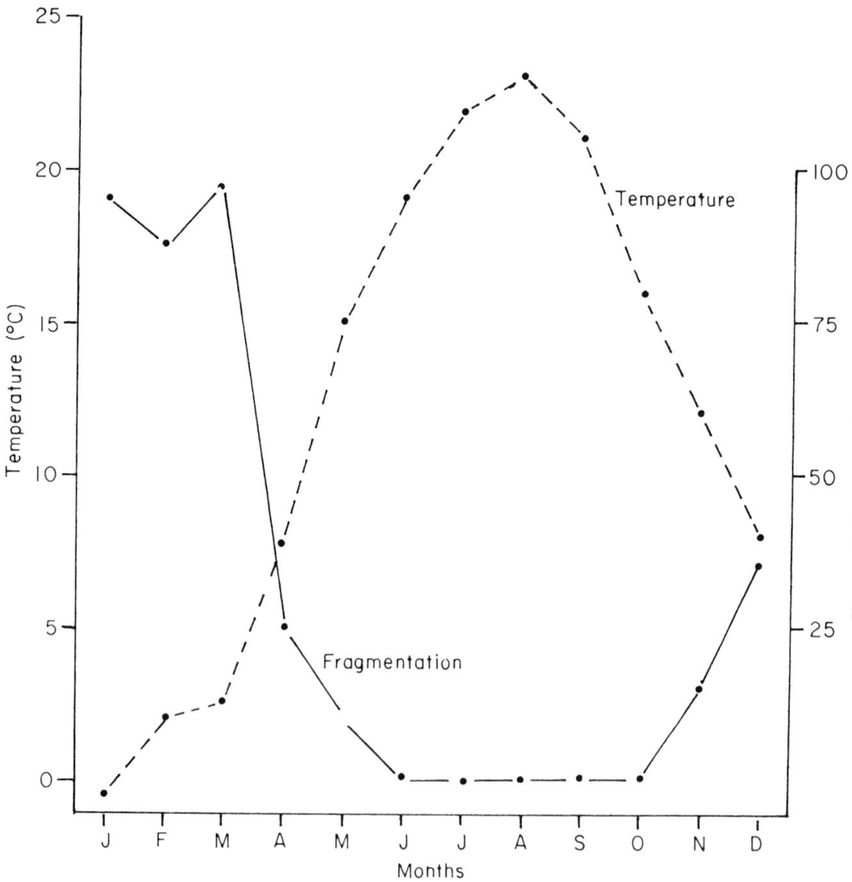

Fig. 12.2 Seasonal variation of surface water temperature and fragmentation of *Codium fragile* at Woods Hole, Massachussets. (Based on Fralick & Mathieson 1972.)

exposed to different conditions as a result of its increasing size. One such example may be the change from peltate to globular ramuli which often occurs when developing plants of *Caulerpa racemosa* emerge from beneath a shady turf into brighter illumination (Peterson 1972).

3 LIFE HISTORY PHASE CHANGES

Many seaweeds have complex life histories involving two or more alternate phases. The different phases are often macroscopic and so morphologically distinct that, before being recognized as participants in the same life history, each plant was given a separate name. Thus, by progressing from one phase to another a single species can be represented by distinct plants seasonally or spatially. A further complication is that several species may possess morphologically similar phases; thus, the exact identity of an alternate phase may be in doubt (Norton 1978b).

Laboratory culture studies indicate that the production of the alternate phases within a life history may be controlled by a variety of environmental conditions, including light intensity and quality, photoperiod, temperature, nutrient availability, desiccation and resubmergence (Dring 1974; Lockhart 1979; Moss 1974). In nature, many of these factors vary simultaneously. In spring, for example, the amount of light reaching a given depth increases, the days lengthen, water temperatures rise and, as a result of plankton blooms, nutrient levels decline. In autumn all these trends are reversed. The plant's life history, including the seasonal appearance and disappearance of its different stages, may be controlled by any of the factors acting singly or in concert (Dring 1974).

Algal life histories offer the plant a number of possible pathways, none of which may be compulsory. The actual alternation of phases may vary in different habitats or regions of the species' geographical distribution. Many species living at the limits of their environmental or geographical distribution fail to reproduce sexually and resort to asexual reproduction or vegetative propagation (Norton 1978b). Since each phase can often survive independently, they may have different geographical distributions, the life history being completed only where their distributions overlap.

Several studies have demonstrated that it is important to study life histories both in the laboratory and field. The alternation of phases observed in nature may for many reasons be different from those displayed in the laboratory (Norton 1978b). In some cases the seasonal alternation may be suppressed or even accentuated by grazers. Jane Lubchenko (personal communication) found that in some algae, such as *Scytosiphon lomentaria*, the crustose phase is less palatable to the snail *Littorina littorea* than is the foliose part. In summer when the snails are most active, the plants are restricted to crusts, even though the physical conditions permit the persistence of the erect plant.

4 MORPHOLOGY IN RESPONSE TO ENVIRONMENTAL FACTORS

4.1 Vertical position on the shore

The appearance of seaweeds often varies with plants inhabiting different vertical levels on the same shore. For example, *Enteromorpha linza* exhibits a graded change from long lanceolate plants near the high water mark to short and broad specimens on the lower shore (Arasaki & Shihara 1959). However, the most common difference with intertidal species is that they become smaller, and sometimes less fertile, towards the top of their zone, as in *Prasiola stipitata* (Friedmann 1969), *Fucus spiralis* (Niemeck & Mathieson 1976), *Halosaccion ramentaceum* (Munda 1976), and several other red algae (Dixon 1963, 1966).

There are several explanations for the above described phenomena. Firstly, variations in shore topography may result in differences in wave exposure at different levels, as noted by Abbott (1972) and Schonbeck & Norton (1974). Even without a change in slope, shores are theoretically less wave-exposed at the lower than the upper levels (Jones & Demetropoulos 1968). Some differences in size undoubtedly result from the differential growth rates of plants at various levels within their zone. Several species of seaweeds grow faster with increasing duration of submergence. Thus, the smaller plants at higher levels mostly reflect their shorter submergence time and concomitantly slower growth (Hruby & Norton 1979; Schonbeck & Norton 1979a). The morphological effects are further aggravated by the increasing desiccation stress of exposure at higher levels on the shore. In the case of *Fucus spiralis* growing at the upper limits of its zone, the larger overlying plants can be selectively killed, leaving smaller plants surviving beneath. The survivors often suffer recurrent tip damage that severely curtails growth and limits the ultimate size of the plants (Schonbeck & Norton 1978). Even subcritical desiccation, which induces no visible symptoms of damage, causes a measurable hiatus in growth. When such desiccation is repeated frequently, as on the shore, it must affect the eventual size of the plant (Schonbeck & Norton 1979c). A similar phenomenon may be operating in many other red and brown algae that grow more slowly, are stunted in appearance and possess fewer branches towards their upper limits (Dixon 1963, 1966; Mathieson 1965).

Most modifications of this type merely represent the result of environmental modification rather than adaptations to reduce the rate of water loss. However, reduction in size or changes in shape (i.e. surface to mass ratios) can in turn influence desiccation rates (Schonbeck & Norton 1979b). Bergquist (1959) demonstrated that the larger vesicles developed by *Hormosira banksii* plants growing at high elevations on the shore lose water more slowly than the smaller vesicles found on plants lower on the shore.

Many tide pool plants are not exposed to desiccation and their morphologies are often quite different from those of plants on the adjacent rock. De Silva & Burrows (1973) suggest that the unbranched, pool-dwelling *Enteromorpha intestinalis* and the much-branched rock-inhabiting *E. compressa* are merely growth forms of the same species. Dixon (1966) observed parallel form changes in both pool-dwelling and open-rock plants of *Pterocladia capillacea* with increasing height in the intertidal, but the same degree of modification occurred at much higher levels with pool plants than open-rock dwellers. Yet on the other hand, even large tide pools are not simply extensions of the subtidal conditions on the shore, as differences in the appearance of species inhabiting both environments demonstrate. For example, plants of *Laminaria hyperborea* in a pool containing 85 m^3 of water had distinctly shorter stipes than those in the adjacent subtidal (Goss-Custard *et al.* 1979). No doubt, in the subtidal forest intense competition for light renders long stipes an advantage, whereas in tide pools (average depth = 1.25 m), they may break the surface, exposing the primary meristem to fatal desiccation.

Similarly, many species that are common to both the intertidal and the subtidal zones exhibit different morphologies in the two habitats. For example, intertidal populations of *Cympolia barbata* (Liddle 1979), *Enteromorpha* spp. (Pringle 1975), and *Chondrus crispus* (Edelstein *et al.* 1969) are often stunted in comparison with their subtidal counterparts. Many seaweeds change their morphology with depth (Fig. 12.3). Often this involves a progressive decrease in frond size and thickness, as in *Kallymenia reniformis* (Norton unpublished), *Agarum cribrosum* (Edelstein *et al.* 1969), *Callophyllis cristata, Membranoptera alata* and *Phycodrys rubens* (Norall *et al.* 1981). In contrast, other subtidal species exhibit a progressive elongation and narrowing of the thallus with increasing depth (Fig. 12.3), as in *Chondrus crispus* (Mathieson & Burns 1975), and *Ptilota serrata* (Edelstein *et al.* 1969; Lamb & Zimmerman 1964; Norall *et al.* 1981). Other modifications may also occur with increasing depth. For example, calcareous algae may be less heavily calcified in deeper water (Gilmartin 1960). Some of the most striking and carefully documented depth related changes are those reported for *Laminaria hyperborea*. With increasing depth the plants become progressively sparser and smaller in both size and weight. Age for age, the plants in deeper water have shorter stipes and shorter, thinner, less-divided laminae (Larkum 1972; Larkum & Norton 1968; Norton & Milburn 1972; Norton *et al.* 1977). The reduction in stipe length is usually more marked than the reduction in lamina length (John 1970; Kain 1971; Norton *et al.* 1977). The similarity of shallow-water plants to exposed-water forms, and deeper-water plants to calm-water forms (Section 4.4) indicates that the increasing shelter from wave action afforded by depth greatly influences thallus shape. The small stature of deep-water plants reflects their slow rate of growth, particularly in stipe length (John 1970; Lüning 1970). However, no one has critically examined the possible effects on algal morphology of different plant densities at various depths, or of depth related differences in grazers.

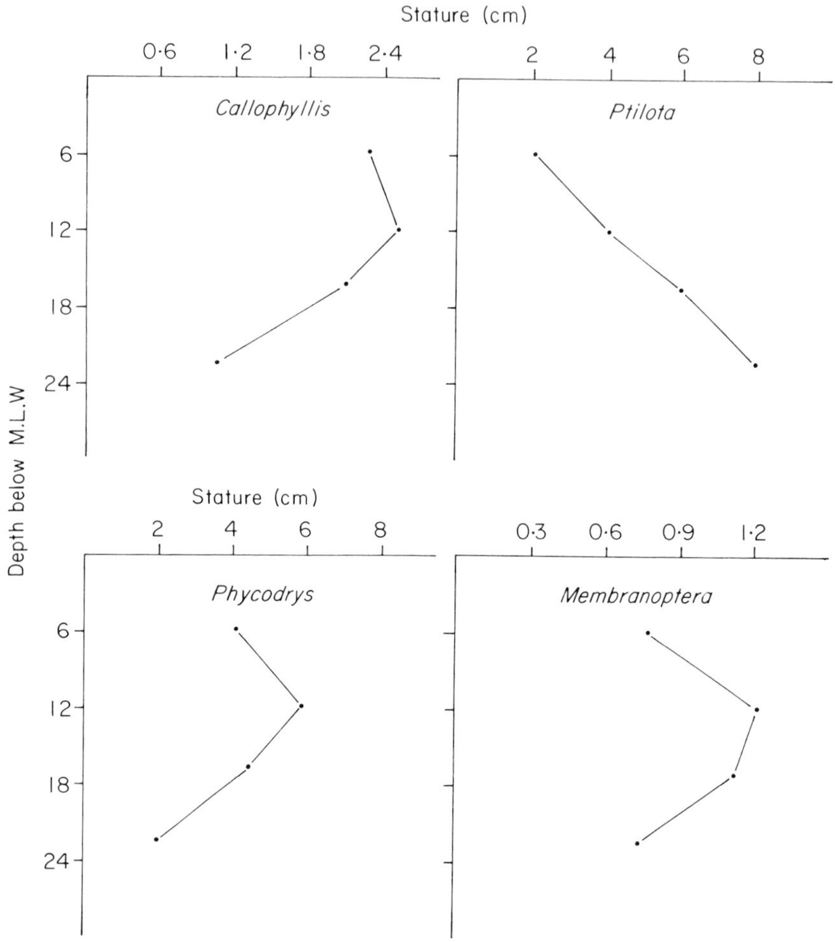

Fig. 12.3 Average yearly stature of subtidal populations of *Callophyllis cristata, Ptilota serrata, Phycodrys rubens*, and *Membranoptera alata* at four depths (below mean low water), at Appledore Island, Maine. (Based on Norall *et al.*, 1981.)

Increasing depth may necessitate certain modifications of form. For example, the vesicles of *Ascophyllum nodosum* are larger and have thicker walls at greater depth, which may be of significance to the plant in withstanding increased water pressures (Damant 1937). In *Egregia menziesii*, vesicles on the lower axis are longer in deep-water plants than in shallow-water ones, and in deeper water (−4 to −5 m) only the larger vesicles remain intact, the smaller ones are filled with water (Chapman 1961).

In *Caulerpa*, several morphological alterations make good sense physiologically. For example, it has been noted in several species that with increasing depth, although the size of the erect assimilators is often reduced, the radially arranged ramuli become bilaterally oriented and the creeping rhizomes elongate so that the assimilators are more spaced out (Gilmartin 1960; Johnston 1969; Peterson 1972). Similar changes in symmetry (i.e. bilateral to radial) have been noted in *Eucheuma* spp. when plants were transplanted from the deep to shallow subtidal (Cheney & Dawes in prep.).

4.2 Substratum

Relatively few morphological effects of substratum on seaweeds have been reported. In part this is due to the interrelated nature of substratum and other physical factors, including wave action (den Hartog 1972). In addition, most of the pertinent studies have been descriptive rather than experimental.

The stability of the substratum obviously determines the species composition and ultimately the morphologies of plants. In addition, a general relation exists between the size of plants and their substratum, with only smaller plants growing on unstable substrata except in very quiet water.

Den Hartog (1972) cites the small green alga *Cladophora pygmaea* as an epilithic alga adapted to a sand grain substratum, such as the undisturbed bottoms of the Kattegat and the southwestern part of the Baltic. Dixon (1958) cites another example of morphological variation due to substratum size/stability with the red alga *Gelidium latifolium* in England. When the plant grows on reefs or large rocks, erect fronds arise at intervals from a well-developed creeping axis. In contrast, if the plants are attached to small stones, which may be embedded in stretches of sandy mud between the reef, they are tufted with minute creeping axes. Another plant usually found in sandy habitats is *Audouinella floridula* (= *Rhodochorton floridulum*) which has a dense cushion-like morphology composed of soft, silky filaments with sand grains between its filaments. The thickness of the fixed sand layer often exceeds 5 cm. However, *Audouinella floridula*, is not typical of most sand-dwelling seaweeds (i.e. psammophytes) which are very wiry in texture (Daly & Mathieson 1977).

Most seaweeds can survive unattached, and vast populations may persist free-floating or drifting at the water's surface, loose-lying on the bottom, or entangled or embedded in mud on sheltered shores or salt-marshes. The most extensive populations of detached seaweeds occur in the Sargasso Sea and in land-locked seas (e.g. the Baltic), as well as in shallow inlets and estuaries. Often, because of the ease with which they can be harvested, they constitute important commercial crops (Austin 1960; Levring *et al.* 1969).

Unattached seaweeds exhibit several morphological differences from their attached counterparts, which seem to result from a lack of polarity accompanied by extensive regenerative or proliferous growth. Often the plants lack a holdfast and develop a spherical (aegagropilous) shape after prolonged rolling on the sea floor. The branching on detached plants is usually more profuse, irregular and thinner in diameter than with attached plants. Pronounced reproductive differences may also occur; the fertility of detached plants (both sexual and sporic) is usually suppressed in favour of vegetative propagation. Even when reproductive organs are occasionally produced, as in *Ascophyllum nodosum* ecad *scorpioides*, they may look quite different from those on attached plants (Chock & Mathieson 1976). It must be remembered, however, that the mere persistence of unattached plants usually requires relatively sheltered conditions, for storms would cast them ashore. Thus, the plants often develop calm-water morphologies (Section 4.4) in addition to the modifications induced by lack of attachment.

In contrast to many detached seaweeds that lack holdfasts, unattached laminarian algae often develop large holdfasts whose spreading dendritic appearance is quite characteristic of loose-lying plants (Burrows 1958; Norton & Milburn 1972). Such dendritic morphologies probably result because the haptera continue to grow until they contact the substratum, as in *Macrocystis* spp. (Lobban 1978).

Apart from prevalent vegetative reproduction and the lack of a definite holdfast, loose-lying salt-marsh fucoids show the following morphological alterations; 1, the plants are dwarfed; 2, their thalli often show spirality or curling; 3, the three-sided juvenile condition of the apical cell appears to be retained in plants derived from *Pelvetia canaliculata, Fucus vesiculosus* and *F. spiralis*; 4, the divisions of the oogonia are at best partial and the plants are rarely fertile. Increased exposure to air, inadequate nutrients, constant soil contact, and high humidity may contribute to the above morphological alterations. Oliveira & Fletcher (1980) give additional interpretations of morphological characteristics of the detached forms of *P. canaliculata*. Upon transplantation from rocky shores to salt-marshes, many of the morphological changes described above resulted, including extensive curling of lobes and a change in colour from brown to yellow. They concluded that curling may be a thigmotrophic response as no influence of light or gravity was demonstrated.

Den Hartog (1958, 1959, 1972) describes several morphological alterations of loose-lying *Fucus vesiculosus* populations. Many of these plants persist as undergrowths in phanerogam beds and they produce a small form known as '*F. lutarius*'. Stranded and embedded plants of *F. vesiculosus* produce extensive proliferations along their thallus margins, which become free when the original fragments decay. Ultimately a dense carpet may result because of extensive decay and separation into new individuals. However, if the drift plants are re-anchored by the byssus threads of mussels when they are first stranded, the plants still rot basally and separate, but their thalli remain considerably wider than those of '*F. lutarius*' and do not proliferate (Nienburg 1925, 1932; den Hartog 1958, 1959, 1972). A similar sequence of stranding, embedding in mud, marginal proliferation and decay has been described in a variety of other fucoid algae, including *Ascophyllum nodosum* (Brinkhuis 1976; Chock & Mathieson

1976; Gibb 1957; South & Hill 1970) and *Pelvetia canaliculata* (Chapman 1939; Oliveira & Fletcher 1980).

Not all loose-lying seaweeds are derived from attached plants. In the case of the Sargasso weeds, *Sargassum natans* and *S. fluitans*, there is evidence to suggest that they remain floating throughout their life (Parr 1939) and that the immense tracts of drifting *Sargassum* are formed by continuous vegetative reproduction *in situ* (Fritsch 1945). Vegetative reproduction is also the rule in salt-marsh ecads of fucoid algae. The detailed studies of Chock & Mathieson (1976), Gibb (1957), and Oliveira & Fletcher (1980) indicate that reports of fertility in such populations are usually the result of a recent influx of relatively unmodified fragments. In contrast, unattached laminarians usually reproduce normally and loose-lying populations can be produced *in situ* from spores (Burrows 1958).

4.3 Light

Although some red algae produce their largest thalli in shaded habitats (Santelices 1974), increased shading usually leads to slow growth and abnormally small plants, as in the progressive stunting of intertidal fucoids and various red algae in a marine cave (Norton *et al.* 1971). Comparable results can be found in culture. Thus, if plants receive low photon flux densities*, their growth can be arrested for an extended period, resulting in small plants as in *Laminaria saccharina* (Burrows 1961), *Saccorhiza polyschides* (Norton & Burrows 1969a), *Fucus* spp. (Burrows 1956), and *Caulerpa prolifera* (Barilotti 1970). When higher photon flux density is restored, growth resumes, sometimes even at an accelerated rate, leading to abnormally widened laminae in *L. saccharina* (Burrows 1964). Thus, only cell enlargement may be inhibited by dim light. Certainly in *Fucus*, cell multiplication proceeds at photon flux densities where no growth occurs and cell elongation and thallus enlargement occur rapidly when the light is increased (Burrows 1956). Reduced light seems to promote several other morphological responses, including internode elongation in *Hormosira banksii* (Bergquist 1959), a reduction of caecostomata density in *Fucus distichus* (Schonbeck & Norton 1974), a reduction in marginal splitting in *Zonaria farlowii* (Dahl 1971), reduced or enhanced branching in *Lithophyllum congestum* (Steneck & Adey 1976), and *Enteromorpha* spp. (Pringle 1975), and a change in blade shape from flat to cylindrical in *C. prolifera* (Klemm 1893; Janse 1904). In *Hedophyllum sessile*, the degree of bullation varies greatly and a completely smooth lamina develops in shaded habitats, during dull cloudy weather or when the lowest tides occur in darkness (Widdowson 1965).

Many seaweeds can orientate their thalli in relation to the incident light.

* Ed. note: The term photon flux density is to be preferred to irradiance and illuminance—see Chapter 13.

Such phototropic responses have been observed in *Gracilaria verrucosa* (Jones 1959), *Maripelta rotata* (Bowen 1971), *Sciadophycus stellatus* (Neushul *et al.* 1967), and haptera of *Alaria esculenta* (Buggeln 1974).

A number of culture studies have revealed various morphological responses to different light levels. One of the most dramatic is the inhibition of cap formation in *Acetabularia crenulata* grown at very low irradiances and the restoration of the cap on exposure to brighter light (Beth 1953). Germlings of *Bryopsis* produce prostrate filaments in dim light, with the erect feather-like fronds requiring higher irradiances (Berthold 1882). Many species produce hairs or branches more profusely in bright than in dim light (Knaggs 1966; McLachlan 1974; Taylor 1966; Waaland & Cleland 1972; West 1972); no doubt this increases self shading and protects the plant from harmful effects of high photon flux densities.

In many red algae low photon flux density can result in stunted (Norton 1975) or elongated and spindly plants (Mathieson & Burns 1975; Norall *et al.* 1981), reduced branching, and longer axial but shorter branch cells (Berthold 1882). Obviously, such differential responses of cells in the axes and branches might greatly influence the shape of the resulting plant. Often, however, the pattern of cell sizes depends upon the cells examined, for those at various distances behind the apex may respond differently (Murray & Dixon 1975). Moreover, the actual range of photon flux densities tested may be critical. For example, with *Plumaria elegans* living in a very deep cave there is a progressive decrease in the size of the cells in the ultimate pinnules up to 14 m in from the cave mouth. Deeper within the cave, however, they become even larger than those in plants growing at the cave mouth (Norton 1975). In addition, there is a dramatic reduction in size from specimens well over 60 mm long at the mouth to a tiny 5 mm tall fuzz 15 m inside the cave. The plants also exhibit a gradual reduction in length and breadth of their main axes and lateral branches, an increase in the size of the axial cells, a change of growth habit, and even a disruption of the regular branching pattern characteristic of the species. To a greater or lesser degree all of these trends can be reversed by culturing the plants in a photon flux density similar to that prevailing at the mouth of the cave (Norton 1975).

In several species of *Caulerpa*, growing the plants in low photon flux densities can alter the spacing of the erect assimilators on the creeping stolons, increase the length of the assimilators, cause the ramuli to become peltate and change their symmetry on the assimilators from radial to bilateral (Calvert 1976; Peterson 1972). As all these changes occur in the sea with increasing depth (Section 4.1) reduced photon flux densities may be the critical factor in nature.

Other aspects of light, such as photoperiodic effects on morphology, and adaptations for light harvesting, are discussed by Lüning (Chapter 9) and Ramus (Chapter 13).

4.4 Water motion

Investigators studying the effects of water motion on plant form are often plagued by two major difficulties: the frequent failure of transplants from calm waters to survive in exposed sites and the difficulty of measuring wave exposure in any but a subjective or relative manner. Even assessments of relative exposure may depend on the experience of the investigator and the tolerance of the species under investigation. For example, an exposed site in which *Laminaria saccharina* grows could be considered relatively sheltered by someone studying *Alaria esculenta*. In spite of these difficulties some generalizations can be made.

For intertidal species, it is frequently observed that plants growing on exposed shores are usually tougher, shorter, narrower, and often have stronger attachment structures than those growing in calmer situations. This is true of such diverse seaweeds as *Callithamnion* spp. (Price 1978), *Chondrus crispus* (Mathieson & Prince 1973), *Corallina officinalis* (Dommasnes 1968), *Fucus* spp. (Burrows & Lodge 1954; Russell 1978), *Zonaria farlowii* (Dahl 1971), *Enteromorpha* spp. (Bliding 1963), *Ulva fasciata* (Mshigeni & Kajumulo 1979), and *Ulva lactuca* (Steffensen 1976). Often the plants are stunted, and more profusely branched on exposed shores, as in *F. vesiculosus* and *Pelvetia canaliculata* (Russell 1978). Similar effects of exposure can be seen on a single shore where plants resembling the sheltered water forms are found in protected places on otherwise exposed shores (Jordan & Vadas 1972; Schonbeck & Norton 1974). *Caulerpa prolifera, Halosaccion ramentaceum* and *Lithophyllum congestum* may exhibit different branching patterns in agitated and calm waters (Munda 1979; Steneck & Adey 1976).

Plants with gas filled vesicles are particularly susceptible to variation. Some species that bear vesicles in sheltered localities lack them in very exposed conditions e.g. *Fucus vesiculosus* (Burrows & Lodge 1954; Seoane-Camba 1966; Jordan & Vadas 1972; Russell 1978) and *Egregia menziesii* (Chapman 1961). In *F. vesiculosus* the number of vesicles per unit length of thallus progressively decreases with increasing exposure (Seoane-Camba 1966). In *Hormosira banksii* bladders are invariably produced, but with increasing exposure they become progressively smaller, have thinner walls, and are separated by shorter internodes (Bergquist 1959). Conversely, in *Macrocystis* spp. the vesicles in exposed sites are larger in diameter and have thicker walls in keeping with overall greater robustness of the plant in such situations (Brandt 1923; Druehl 1978; Pace 1972).

Many laminarian algae exhibit parallel responses. In exposed sites they produce narrower, thicker and sometimes longer laminae than in calmer waters, as in *Alaria esculenta* (Sundene 1962b), *Laminaria saccharina* (Parke 1948), *L. digitata* (Sundene 1962a), *L. longicruris* (Gerard & Mann 1979), *L. longipes* (Palmisano & Scheng 1977), *Macrocystis* spp. (Druehl 1978; Pace 1972), and *Saccorhiza polyschides* (Norton 1969). Usually the net result of these modifications is that in calm waters the laminae have larger surface areas, while in exposed sites they are more streamlined. In species where the lamina can split into strap-like digits, the number of digits is greatest in exposed sites, and in very sheltered conditions the lamina is thin, entire and often cucullate (i.e. hood-shaped). Simply transplanting rough-water plants into very sheltered conditions can convert them into flimsy calm-water form, as shown in *Laminaria digitata* (Sundene 1962a), *L. hyperborea* (Svendsen & Kain 1971), *L. longicruris* (Gerard & Mann 1979) and *Saccorhiza polyschides* (Norton 1969). The base of the lamina in many species is more or less cuneate in exposed sites and rounded or cordate in sheltered ones, e.g. *Saccorhiza polyschides* (Norton 1969), *Macrocystis pyrifera* (Brandt 1923), *Laminaria digitata* (Sundene 1964), *L. longicruris* (Mann 1971). In *L. longicruris* (Mann 1971) and *L. saccharina* (Burrows 1964; Parke 1948), the shape of the lamina base is a reflection of growth rate. However, this does not seem to be the case with *M. integrifolia* (Pace 1972).

Stipe length may vary in relation to water motion, becoming either longer or shorter in exposed situations. In *L. saccharina* there is a tendency for its stipe to become inordinately long and sometimes hollow and inflated in very calm waters. In the closely related *L. agardhii–L. longicruris* complex the stipes can become hollow in relatively exposed conditions, but they do so earlier in their development and more extensively in sheltered waters (Chapman 1973).

Tidal currents, unless very fierce, might be expected to be qualitatively different from wave action, subjecting the plants to less physical stress, but providing abundant nutrients and gases and probably furnishing excellent growing conditions. In very strong currents, some plants may be stunted as on wave-exposed shores (Fig. 12.4), but tidal rapids with currents less than about 1.75 m sec^{-1} often contain extremely large specimens. For example, Reynolds (1971) reports plants of *Chaetomorpha linum* that reach lengths of 5 m (about 50 times normal size) growing in currents of 1.5–2.8 m sec^{-1}. *Saccorhiza polyschides* dominates in Lough Ine rapids in currents of about 1.5 m sec^{-1}, where it reaches immense size and has a morphology distinct from the one it exhibits when subjected to waves (Norton 1969).

Dynamic Considerations

The effects of water motion are of overriding importance in the design of seaweed thalli. The forces involved are enormous: a water velocity of 14 m sec^{-1} is equivalent to a wind speed of 400 m sec^{-1} (890 miles $hour^{-1}$) on land (Charters *et al.* 1969; Norton *et al.* 1981).

Plants that respond to increasing wave action by becoming tougher and sturdier do so by increasing the amount of structural tissue. The structural change in turn reduces the ratio of photosynthetic to non-photosynthetic tissue, but the effect of this on the growth rate of the plant may be counteracted by the

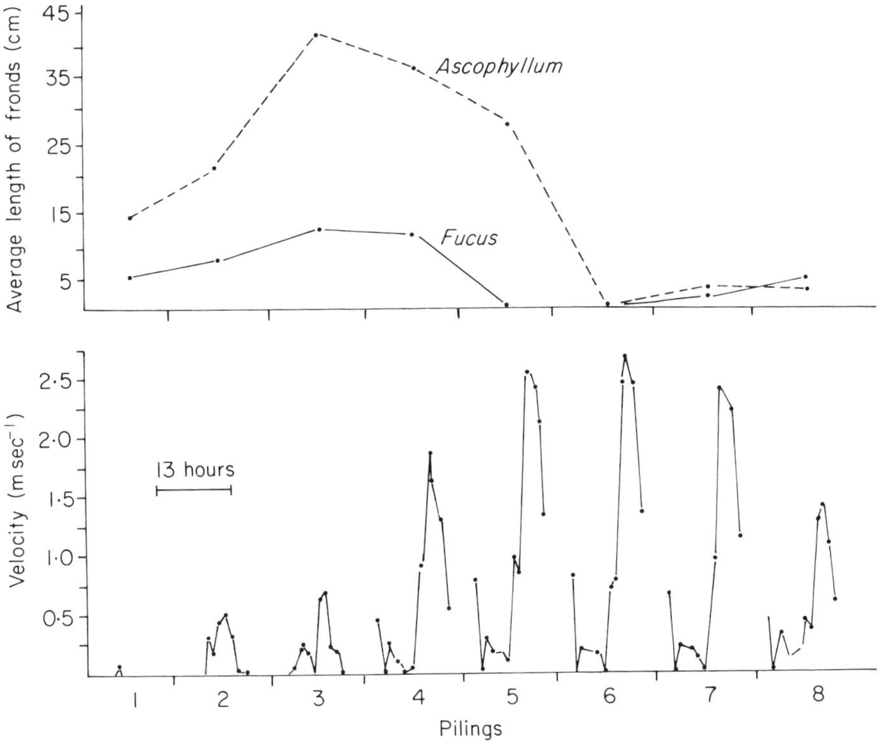

Fig 12.4 Average stature of *Ascophyllum nodosum* and *Fucus vesiculosus* growing on various pilings across the tidal channel (420 m wide) at Dover Point, New Hampshire, plus temporal and spatial variations of tidal currents (m sec^{-1}) on the same pilings. (Mathieson unpublished data.)

improved nutrient uptake and gas exchange of exposed habitats (see below). Yet the impression given by many wave-exposed plants is not of great strength, but of softness and apparent flimsiness, as in *Porphyra umbilicalis* and *Alaria esculenta*. The prime strategy of most seaweeds seems to be to offer as little resistance as possible to water flow, which can be achieved by ensuring that the thallus flows out horizontally with the current, by having a smooth surface with low skin friction, and by being mucilaginous which again reduces the drag on the thallus (Norton *et al.* 1981). The morphological flexibility of *Eisenia arborea*, which represents a compromise between a quiet-water configuration that exposes a large surface area to light and a configuration that reduces drag to a minimum during rapid water movement, has been described by Neushul (1972). The tensile strength of seaweeds is an order of magnitude less than other biological materials, but seaweeds are very elastic (Delf 1932), and their stretchability dissipates much of the tensile stress applied to them. Coiling of stipes, as is seen in *Nereocystis* and *Chorda filum* supplements the elasticity of the tissue itself.

Nutrient uptake rate is greatly influenced by the supply of available nutrients, which depends on their concentration in the water and the amount of water movement. Because the coefficient of molecular diffusion in liquids is very low, plants in still water rapidly deplete the nutrients in their immediate vicinity (the diffusion boundary layer, see Neushul 1972). In contrast, plants growing in moving water have a steady supply of nutrients.

Plants have devised various ways of generating turbulent flow over their surfaces, the effect of which is to bring to the plant's surface water containing a concentration of solutes close to that in the main water mass. The corrugations and spines of *Macrocystis* laminae (see Fig 12.5), and the perforations in thalli of *Agarum cribrosum, Rhodymenia pertusa, Ulva fenestrata*, and other seaweeds,

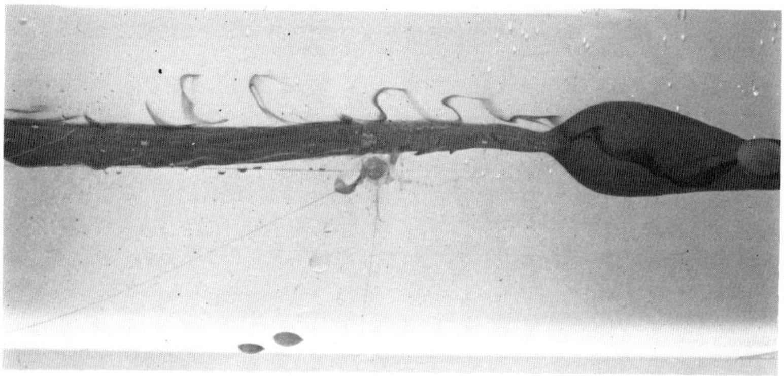

Fig. 12.5 Water flow over the surface of a *Macrocystis* blade in a water tunnel, showing turbulent eddies. Water velocity = 70 mm sec^{-1}; dye was injected into the tunnel upstream of the blade. (Photograph courtesy of A. C. Charters.)

are thought to serve this function. Unfortunately for this theory the laminae of kelps seem to be smoother in sheltered water, rather than more corrugated (Druehl 1978). It should be remembered that corrugating the thallus is an economical way of strengthening as well as increasing surface area; strength may be the overriding requirement in this instance.

4.5 Temperature

In nature the effects of temperature are often masked by other factors, but observations indicate that temperature may be important on a geographical scale and also seasonally. In general there is a marked decrease in algal thallus size towards the tropics (Vermeij 1978). In the Adriatic, *Fucus virsoides* plants become gradually smaller towards the south, eventually dying out as the mean temperature increases (Schiller 1928). Such patterns may be common in nature but they are poorly documented.

There are also many seasonal effects. Fragmentation of *Codium fragile* spp.

tomentosoides in New England (Fig. 12.2) coincides with minimum winter sea temperatures (Fralick & Mathieson 1972). High summer air temperatures aggravate the effects of desiccation in intertidal fucoids (Schonbeck & Norton 1978, 1980). A sudden increase in water temperature can stimulate blade abscission in *Maripelta rotata* (Bowen 1971). Extreme temperatures may influence morphology by inducing damage (Vadas *et al.* 1978).

4.6 Salinity

As noted by Gessner & Schramm (1971), many seaweeds show a reduction in thallus size with decreasing salinities. Such a trend is dramatically illustrated by estuarine populations of *Fucus vesiculosus* forma *subcostata* from Finland, where the thallus is often reduced to 2 cm or less (Gessner & Schramm 1971). Munda (1978) records dwarf forms of *Pilayella littoralis* and *Chordaria flagelliformis* in Icelandic fjords, while Burns & Mathieson (1972) found a similar stunting of the stenohaline *Gigartina stellata* within New England estuarine habitats. Similarly, Barilotti (1970) found that blade width in *Caulerpa prolifera* decreased with reduced salinity.

Gessner & Schramm (1971) illustrate an extreme example of brackish water submergence with *Callophyllis laciniata*. Algae that follow the saltier waters into greater depths are however subject to less light and are often dwarfed in deeper water.

In addition to stunting, salinity variations may cause a variety of other more subtle morphological changes, particularly with the fucoid algae. For example, estuarine plants of *Hormosira banksii* have larger bladders and longer internodes than coastal plants of the same species (Bergquist 1959). The effects of reduced salinity are obviously interrelated with the effects of shelter. In *Fucus vesiculosus*, reduced salinity greatly increases the size of the gas bladders and the number of dichotomies (Jordan & Vadas 1972). A similar pattern seems to exist with *Ascophyllum nodosum*, for estuarine populations that are exposed to pronounced fluctuations of salinity, as well as sheltered conditions, often have much larger vesicles (3–10 times) than coastal populations (Mathieson & Norton unpublished observations). Many of the loose-lying marsh fucoids show pronounced morphological alterations, which may be related to salinity variations (Section 4.2).

The morphology of *Chondrus crispus* and *Gigartina stellata* appears to be correlated with salinity (Burns & Mathieson 1972; Mathieson & Burns 1975). Estuarine populations of *Chondrus crispus* are broad and widely branched, while coastal populations are elongated and narrowly branched (MacFarlane 1968; Mathieson & Burns 1975; Mathieson & Prince 1973). Aside from their stunted morphologies, estuarine populations of *G. stellata* show a more brittle texture, a darker colouration, and a reduced reproductive potential as compared to coastal plants.

The morphological characteristics of several green algae may be modified by reduced or extreme salinities. Burrows (1959) found that proliferous branching may occur *Enteromorpha compressa* if it was grown in reduced salinities. In contrast, Klugh (1922) and Kapraun (1970) found that several *Enteromorpha* species were more highly branched in high (often extreme) salinities and less abundantly branched or even unbranched in diluted seawater. Reed & Russell (1978) describe the development of proliferous ('bottle-brush') morphologies of *E. intestinalis* in relation to salinity changes, as well as extreme salinities. Two distinct patterns of development produce the 'bottle-brush' morphologies: *in situ* germination of swarmers and the outgrowths from isolated, swollen cells in the thallus. Cauro (1958) and Steffensen (1976) have noted similar morphological variations in *Ulva* due to reduced salinities.

4.7 Chemical factors

A wide variety of substances find their way into seawater and have a measurable influence on the growth rates of seaweeds. Occasionally changes in growth rates may be reflected in the size of the plant. Burrows (1971) suggested that the surface area of a fixed length of the frond of *Laminaria saccharina* could provide an easily measured assessment of the water quality in which it had grown. Unfortunately various parts of the same plant may respond differently as with *Egregia menziesii*, which grows at differing rates in various pollution regimes (Widdowson 1971). Similarly, *L. hyperborea* is said to produce heavier, longer and thinner stipes but lighter laminae in sites polluted with domestic effluent than in unpolluted waters (Bellamy & Whittick 1968). Fertilizing *L. longicruris* plants in the sea stimulates growth and alters the texture of the laminae (Chapman & Craigie 1977). In culture, low nutrient concentrations may reduce the growth rates but enhance hair production in germlings of *Fucus spiralis* (Schonbeck & Norton 1979d) and in *Bolbocoleon piliferum, Entocladia viridis* and *Entocladia flustrae* (Yarish 1976). Transferring *Phaeophila dendroides* from enriched medium into seawater induces hair formation (Nielsen 1972) and *Plumaria elegans* sheds the hairs grown in nutrient-poor cultures if it is transferred to nutrient-rich conditions (Boney 1959). The hair whorls of *Acetabularia mediterranea* readily take up vital stains, and shaving off the hairs on that side of the stipe retards cell elongation on this side (Gibor 1973a, b). Hairs greatly increase the surface area of small plants and their possible role in nutrient uptake is discussed by Schonbeck & Norton (1979d).

The effects of nutrients are often drastic. The absence of phosphates from culture media causes fragmentation in *Chondrus crispus*, where thallus strength seems to be related to phosphate availability (Neish & Shacklock 1971). The external concentration of phosphate influences morphogenesis in *Scytosiphon lomentaria* (Roberts & Ring 1972) as does iodide for *Petalonia fascia* (Hsiao 1969), and iodine and arsenic for *Asparagopsis* (von Stosch 1964).

Sub-lethal doses of various toxic chemicals such as oil spill emulsifiers and insecticides induce cell proliferations, tumors or abnormal growth in *Bryopsis* and a variety of red algae (Boney 1971, 1974; Boney & Corner 1962; Ishio *et al.* 1972; Watanabe & Kato 1972).

4.8 Damage

In spite of their intrinsic toughness seaweeds are frequently damaged. The most important damage-inducing agents are storms, grazing, sand abrasion and harvesting by man.

The continual removal of tissue often results in the survival of small plants. On some South African shores, species of *Gelidium* that can produce erect bushes several centimeters tall, may be reduced by limpet grazing to small, sparse, creeping fronds only a few millimeters high, and so unlike the erect form as to be virtually unrecognizable (Stephenson & Stephenson 1972). In addition, algae on coral reefs where herbivorous urchins or fish abound, are often diminutive, and they increase in size with increasing distance from the reefs and a concomitant reduction in grazers (Dahl 1973; Mathieson *et al.* 1975). Although the effects of siltation and high water turbidity on some reefs cannot be ignored, it is interesting that rock boring algae often abound on such reefs (Dahl 1973) where they must suffer at least as much as epilithic algae from the effects of sediments and poor light, but are protected from most grazers. Bakus (1969), Earle (1972a, b), Lundberg & Lipkin (1979), Vermeij (1978) and Wanders (1977) summarize a variety of additional grazing studies, some of which have demonstrated preferential grazing as well as pronounced morphological alterations of seaweeds. Vermeij (1978) records that the thalli of some species of *Sargassum* and *Turbinaria* in several south Pacific islands are coarser and spinier than those of members of the same genera occupying similar habitats in the Caribbean, which may reflect a differential commitment to defense against herbivores.

Localized abrasion or grazing may alter the appearance of a plant by reducing the size of one part in proportion to the rest. For instance, the truncation of the laminae of *Saccorhiza polyschides* growing in an exposed site was attributable to the ravages of two herbivorous snails living on its fronds—*Lacuna vincta* (205 specimens m^{-2} of lamina) and *Patina pellucida* (365 m^{-2}). The damaged tissue is readily removed by the continual whiplash movements of its fronds (Norton 1971). Grazing invertebrates can strip almost all of the laminae and sporophylls from *Macrocystis pyrifera*, leaving just stipes and vesicles (Anonymous 1964). Even tourists stepping on *Ascophyllum nodosum* can effectively shear off the plants against sharp outcrops and significantly truncate them (Boalch *et al.* 1974). However, if the frond is lost, the holdfast can often regenerate clusters of new axes. Such damage can produce congested and bushy plants with *A. nodosum* (Printz 1959) and *Sargassum muticum* (Norton 1977b).

Even so, in both species there appears to be a minimum size of basal stump that must remain for regeneration to ensue.

Often there is a disparity between the appearance of an alga in the field and culture. Such differences are sometimes attributed to the damage sustained in nature (Stewart 1976). Clokie & Norton (1974) found that the natural appearance of crustose algae could only be replicated in culture with the addition of grazing invertebrates at concentrations similar to those in nature.

Many psammophytic (sand-dwelling) species, which are often subjected to severe abrasion and periodic sand burial, have extensive powers of regeneration (Daly & Mathieson 1977). Dahl (1971) gives one of the best documentations of morphological effects of sand abrasion and burial, demonstrating that these factors are two of the primary causes of branching and narrow thalli in *Zonaria farlowii.* Plants with many proliferations from stubby, central axes probably result from a sequence of periodic burial, followed by the injury and regeneration of new apical meristems. Mathieson (1965) recorded a comparable sequence of blade degeneration and regeneration when the obligate psammophytic alga *Phaeostrophion irregulare* was buried for two to six months. Mature plants of *P. irregulare* invariably have an abraded and irregular morphology.

Damage to the thallus, however induced, often illicits a repair response from the plant. It is interesting to note, however, that in the species of brown algae studied, this regenerative growth is quite different from the meristematic activity that accompanies digit formation in the laminae of kelps (Norton 1969). Usually the replacement tissue resembles that lost, but often clusters of thin papillae or haptera-like proliferations emerge from the wound, as on laminarian stipes (Norton unpublished; Setchell 1905), and on the thalli of *Fucus* spp. and *Pelvetia canaliculata* (Burrows 1956; Moss 1963), and *Petroglossum nicaeense* (Felicini & Perrone 1972; Perrone & Felicini 1974).

Destruction of the apical cell (meristem) of plants usually stimulates the production of lateral branches. Such a response has been reported in *Enteromorpha intestinalis* (Moss & Marsland 1976), *Sargassum muticum* (Chamberlain *et al.* 1979), *Gracilaria debilis* (Goldstein 1973) and *Lithothrix aspergillum* (Gittins & Dixon 1976), but it is widespread in a variety of other seaweeds.

Infections by bacteria, fungi, other algae and nematodes stimulate many visible morphological repercussions in the host plant, including distortion of individual cells or the thallus, accelerated rates of tissue loss, galls, pustules and even structures reminiscent of the 'witches brooms' that occur on terrestrial plants (Andrews 1976; Jenneborg 1977). Sometimes the ravages of disease are aggravated by elevated water temperatures, exposure to air and other physical factors. If the rate of decay is accelerated this can influence the size of the resulting thallus. For example, Norton (1978a) has recorded dramatic differences in the rate of lamina extension and distal decay in *Saccorhiza polyschides* upon transplantation to different localities.

The freakish specimens of seaweeds that are occasionally observed may arise from meristematic damage at a very early stage, which disrupts subsequent development. The Laminariales are particularly prone to abnormal development (see examples in Setchell 1905).

5 CONCLUSION

We have merely drawn attention to the extreme morphological plasticity of seaweeds and the parallel responses exhibited by different species when subjected to similar stimuli. It is regretable that in most cases the morphogenetic origin of the changes and their functional significance have hardly been investigated.

ACKNOWLEDGEMENTS

This work is based on research supported by a General Research Grant provided by the Academic Senate Committee on Research, University of California, Santa Barbara, (1978–79); the NOAA Office of Sea Grant, Department of Commerce, under grant No. R/CZ-46, the National Science Foundation, grant No. OCE76-24360, and the Marine Science Institute, University of California, Santa Barbara. It is published with the approval of the Director of the New Hampshire Agriculture Experiment Station as Scientific Contribution No. 1108, also as Contribution No. 103 from the Jackson Estuarine Laboratory, University of New Hampshire, Durham. The U.S. Government is authorized to produce and distribute reprints for governmental purposes notwithstanding any copyright notation that may appear hereon. We would like to thank Darai Voss-Belton for her excellent technical assistance with preparation of the manuscript, as well as Dr. Donald Cheney for his critical review of the manuscript.

6 REFERENCES

Abbott I. A. (1972) Field studies which evaluate criteria used in separating species of *Iridaea* (Rhodophyta). In *Contributions to the Systematics of Benthic Marine Algae of the North Pacific*, eds. Abbott I. A. & Kurogi M. pp. 253–64. Japanese Society Phycology, Kobe, Japan.

Andrews J. H. (1976) The pathology of marine algae. *Biol. Rev.*, **51**, 211–53.

Anonymous (1964) An investigation of the effects of discharged wastes on kelp. *Res. Agency Cal. State Water Quality Control Bd.*, **26**, 1–124.

Arasaki S. & Shihara I. (1959) Variability of morphological structure and mode of reproduction in *Enteromorpha linza*. *Jap. J. Bot.*, **17**, 92–100.

Austin A. P. (1960) Observations on *Furcellaria fastigiata* (L.) Lam. forma *aegagropila* Reinke in Danish waters together with a note on other unattached algal forms. *Hydrobiologia*, **14**, 255–77.

Bakus G. J. (1969) Energetics and feeding in shallow marine waters. *Intern. Rev. Gen. Exp. Zool.*, **4**, 275–369.

BARILOTTI C. (1970) Non-Genetic Morphological Variation in *Caulerpa prolifera* (Førsskal) Lamouroux. M. A. Thesis, University South Florida, Tampa.

BELLAMY D. & WHITTICK A. (1968) Operation Kelp II. *Triton*, **February.**

BERGQUIST P. L. (1959) A statistical approach to the ecology of *Hormosira banksii*. *Bot. Mar.*, **1**, 22–53.

BERTHOLD G. (1882) Beiträge zur Morphologie und Physiologie der Meeresalgen. *Jb. Wiss. Bot.*, **13**, 569–717.

BETH K. (1953) Experimentelle Untersuchungen über die wirkung des Lichtes auf die Formbildung von kernhaltigen und kernlose *Acetabularia*-zellen. *Z. Naturf.*, **86**, 334–42.

BLIDING C. (1963) A critical survey of European taxa in Ulvales: *Capsosiphon, Percursaria, Blidingia, Enteromorpha*. *Opera Bot. Soc. bot. Lund*, **8**, 1–160.

BOALCH G. T., HOLME N. A., JEPHSON N. A. & SIDWELL J. M. E. (1974) A resurvey of Colman's intertidal traverses at Wembury, South Devon. *J. Mar. Biol. Ass. U.K.*, **54**, 551–3.

BONEY A. D. (1959) The Ecology and Biology of Certain Intertidal Red Algae. Ph.D. Thesis, University of London.

BONEY A. D. (1971) Sub-lethal effects of some toxic substances on juvenile stages of some marine algae. *Br. Phycol. J.*, **6**, 268.

BONEY A. D. (1974) Aromatic hydrocarbons and the growth of marine algae, *Mar. Poll. Bull.*, **5**, 185–6.

BONEY A. D. & CORNER E. D. S. (1962) On the effects of some carcinogenic hydrocarbons on the growth of sporelings of marine red algae. *J. Mar. Biol. Ass. U.K.*, **42**, 579–85.

BOWEN K. Y. (1971) The Growth and Development of the Deep Growing Marine Red Alga, *Maripelta rotata* (Dawson) Dawson. Ph.D. Thesis, University California, San Diego.

BRANDT R. R. (1923) Potash from kelp: early development and growth of the giant kelp, *Macrocystis pyrifera*. *U.S. Dept. Agric. Bull.*, **1191**, 1–40.

BRINKHUIS B. H. (1976) The ecology of temperate salt marsh fucoids. I. Occurrence and distribution of *Ascophyllum nodosum* ecads. *Mar. Biol.*, **34**, 325–38.

BUGGELN R. G. (1974) Negative phototropism of the haptera of *Alaria esculenta* (Laminariales). *J. Phycol.*, **10**, 80–2.

BURNS R. L. & MATHIESON A. C. (1972) Ecological studies of economic red algae. III. Growth and reproduction of natural and harvested populations of *Gigartina stellata* (Stackhouse) Batters in New Hampshire. *J. Exp. Mar. Biol. Ecol.*, **9**, 77–95.

BURROWS E. M. (1956) Growth control in the Fucaceae. *Proc. Int. Seaweed Symp.*, **2**, 163–70.

BURROWS E. M. (1958) Sublittoral algal populations in Port Erin Bay, Isle of Man. *J. Mar. Biol. Ass. U.K.*, **37**, 687–703.

BURROWS E. M. (1959) Growth form and environment in *Enteromorpha*. *J. Linn. Soc. (Bot.)*, **56**, 204–6.

BURROWS E. M. (1961) Experimental ecology with particular reference to the ecology of *Laminaria saccharina* (L.) Lamouroux. In *Recent Advances in Botany*, pp. 187–9. University of Toronto Press, Toronto.

BURROWS E. M. (1964) An experimental assessment of some of the characters used for specific delimitation in the genus *Laminaria*. *J. Mar. Biol. Ass. U.K.*, **44**, 137–43.

BURROWS E. M. (1971) Assessment of pollution effects by the use of algae. *Proc. Roy Soc. Lond.*, **177**, 295–306.

BURROWS E. M. & LODGE S. M. (1954) The analysis of variation within the genus *Fucus*. In *Species Studies in the British Flora*, ed. Lousley J. E., pp. 83–5. Bot. Soc. British Isles, British Museum, London.

CALVERT H. E. (1976) Culture studies on some Florida species of *Caulerpa*: Morphological responses to reduced illumination. *Br. Phycol. J.*, **11**, 203–14.

CAURO R. (1958) Sur la reproduction et le dévelopement de quatre Ulvacées du Maroc. *Botaniste*, 42, 89–129.

CHAMBERLAIN A. H. L., GORHAM J., KANE D. F. & LEWEY S. A. (1979) Laboratory growth studies on *Sargassum muticum* (Yendo) Fensholt. II. Apical dominance. *Bot. Mar.*, **22**, 11–19.

CHAPMAN A. R. O. (1973) Phenetic variability of stipe morphology in relation to season, exposure, and depth in the non-digitate complex of *Laminaria* Lamour. (Phaeophyta, Laminariales) in Nova Scotia. *Phycologia*, **12**, 53–7.

CHAPMAN A. R. O. & CRAIGIE J. S. (1977) Seasonal growth in *Laminaria longicruris*: Relations with dissolved inorganic nutrients and internal reserves of nitrogen. *Mar. Biol.*, **40**, 197–205.

CHAPMAN V. J. (1939) Studies in salt-marsh ecology. Sections IV and V. *J. Ecol.*, **27**, 160–201.

CHAPMAN V. J. (1961) A contribution to the ecology of *Egregia laevigata* Setchell. I. Taxonomic status and morphology. *Bot. Mar.*, **3**, 33–45.

CHARTERS A. C., NEUSHUL M. & BARILOTTI C. (1969) The functional morphology of *Eisenia arborea*. *Proc. Int. Seaweed Symp.*, **6**, 89–105.

CHOCK J. S. & MATHIESON A. C. (1976) Ecological studies of the salt marsh ecad *scorpioides* (Hornemann) Hauck of *Ascophyllum nodosum* (L.) Le Jolis. *J. Exp. Mar. Biol. Ecol.*, **23**, 171–90.

CLOKIE J. J. P. & NORTON T. A. (1974) The effects of grazing on the algal vegetation of pebbles from the Firth of Clyde. *Br. Phycol. J.*, **9**, 216.

DAHL A. L. (1971) Development, form and environment in the brown alga, *Zonaria farlowii* (Dictyotales). *Bot. Mar.*, **14**, 76–112.

DAHL A. L. (1973) Benthic algal ecology in a deep reef and sand habitat off Puerto Rico. *Bot. Mar.*, **16**, 171–5.

DALY M. A. & MATHIESON A. C. (1977) The effects of sand movement on intertidal seaweeds and selected invertebrates at Bound Rock, New Hampshire, U.S.A. *Mar. Biol.*, **43**, 45–55.

DAMANT G. C. C. (1937) Storage of oxygen in the bladders of the seaweed *Ascophyllum nodosum* and their adaptation to hydrostatic pressure. *J. Exp. Biol.*, **14**, 198–209.

DELF E. M. (1932) Experiments with the stipes of *Fucus* and *Laminaria*. *J. Exp. Biol.*, **9**, 300–13.

DEN HARTOG C. (1958) De vegetatie van het Balgzand en de ooverterreinen van het Balgkanal. *Wet. Meded. K. Ned. Natuurh. Veren.*, **27**, 1–28.

DEN HARTOG C. (1959) The epilithic algal communities occurring along the coast of the Netherlands. *Wentia*, **1**, 1–241.

DEN HARTOG C. (1972) Substratum—plants—multicellular plants. In *Marine Ecology, Vol. 1: Environmental Factors*, ed. Kinne O., pp. 1277–89. Wiley-Interscience, London.

DE SILVA M. W. R. N. & BURROWS E. M. (1973) An experimental assessment of the status of the species *Enteromorpha intestinalis* (L.) Link and *Enteromorpha compressa* (L.) Grev. *J. Mar. Biol. Ass. U.K.*, **53**, 895–904.

DIXON P. S. (1958) The structure and development of the thallus in the British species of *Gelidium* and *Pterocladia*. *Ann. Bot. Lond.*, N.S. **22**, 353–68.

DIXON P. S. (1963) The Rhodophyta: some aspects of their biology. In *Oceanogr. Mar. Biol. Ann. Rev. 1*, ed. Barnes H., pp. 177–96. Allen & Unwin, London.

DIXON P. S. (1966) On the form of the Florideophyceae. In *Trends in Morphogenesis*, ed. Cutter E. G., pp. 46–63. Longmans, London.

DOMMASNES A. (1968) Variations in the meiofauna of *Corallina officinalis* L. with wave exposure. *Sarsia*, **34**, 117–24.

DRING M. J. (1974) Reproduction. In *Algal Physiology and Biochemistry*, ed. Stewart W. D. P., pp. 814–37. Blackwell Scientific Publications, Oxford.

DRUEHL L. D. (1978) The distribution of *Macrocystis integrifolia* in British Columbia as related to environmental parameters. *Can. J. Bot.*, **56**, 69–79.

EARLE S. A. (1972a) A review of the marine plants of Panama. *Bull. Biol. Soc. Washington*, **2**, 69–87.

EARLE S. A. (1972b) The influence of herbivores on the marine plants of Great Lameshur Bay, with an annotated list of plants. In *Results of the Tektite Program: Ecology of Coral Reef Fishes*, eds. Collette B. C. & Earle S. A., pp. 17–44. Nat. Hist. Mus. Los Angeles Co. Sci. Bull. 14.

EDELSTEIN T., CRAIGIE J. S. & MCLACHLAN J. (1969) Preliminary survey of the sublittoral flora of Halifax County. *J. Fish. Res. Bd Can.*, **26**, 2703–13.

ERCEGOVIĆ A. (1952) Sur les *Cystoseira* adriatiques, leur morphologie, écologie et evolution. *Fauna Flora Adriatica*, **2**, 1–112.

FELICINI G. P. & PERRONE C. (1972) Le rôle des proliférations de la fronde dans le cycle biologique de *Petroglossum nicaeense* (Duby) Schotter (Rhodophycées, Gigartinales). *Phycologia*, **11**, 197–205.

FRALICK R. A. & MATHIESON A. C. (1972) Winter fragmentation of *Codium fragile* (Suringar) Hariot ssp. *tomentosoides* (van Goor) Silva (Chlorophyceae, Siphonales) in New England. *Phycologia*, **11**, 67–70.

FRIEDMANN I. (1969) Geographic and environmental factors controlling life history and morphology in *Prasiola stipitata* Suhr. *Österr. Bot. Z.*, **116**, 203–25.

FRITSCH F. E. (1945) *The Structure and Reproduction of the Algae*, vol. II. Cambridge University Press, England.

GERARD V. A. & MANN K. H. (1979) Growth and production of *Laminaria longicruris* (Phaeophyta) populations exposed to different intensities of water movement. *J. Phycol.*, **15**, 33–41.

GESSNER F. & SCHRAMM W. (1971) Salinity—plants. In *Marine Ecology, Vol. 1, Environmental Factors*, ed. Kinne O., pp. 705–820. Wiley-Interscience, London.

GIBB D. C. (1957) The free-living forms of *Ascophyllum nodosum* (L.) Le Jol. *J. Ecol.*, **45**, 49–83.

GIBOR A. (1973a) Observations on the sterile whorls of *Acetabularia*. *Protoplasma*, **78**, 195–202.

GIBOR A. (1973b) *Acetabularia*: Physiological role of their deciduous organelles. *Protoplasma*, **78**, 461–5.

GILMARTIN M. (1960) The ecological distribution of the deep water algae of Eniwetok Atoll. *Ecology*, **41**, 210–21.

GITTINS B. T. & DIXON P. S. (1976) Biology of *Lithothrix aspergillum*. *Br. Phycol. J.*, **11**, 194.

GOLDSTEIN M. E. (1973) Regeneration and vegetative propagation of the agarophyte *Gracilaria debilis* (Førsskal) Børgesen (Rhodophyceae). *Bot. Mar.*, **16**, 226–8.

GOSS-CUSTARD S., JONES J., KITCHING J. A. & NORTON T. A. (1979) Tide pools of Carrigathorna and Barloge Creek. *Phil. Trans. Roy Soc. Lond.*, **284**, B, 1–44.

HRUBY T. & NORTON T. A. (1979) Algal colonization on rocky shores in the Firth of Clyde. *J. Ecol.*, **67**, 65–77.

HSIAO S. (1969) Life history and iodine nutrition of the marine brown alga, *Petalonia fascia* (O.F. Müll.) O. Kuntze. *Can. J. Bot.*, **47**, 1611–6.

ISHIO S., KAWABE K. & TOMIYAMA T. (1972) Algal cancer and its causes. I. Carcinogenic potencies of water and suspended solids discharged to the River Ohmuta (in Japanese, English summary). *Bull. Jap. Soc. Sci. Fish.*, **38**, 17–24.

JANSE J. M. (1904) An investigation on polarity and organ-formation with *Caulerpa prolifera*. *Proc. Amsterdam Acad. Sci.*, **7**, 420–35.

JENNEBORG L.-H. (1977) *Eurychasma*-infection of marine algae. Changes in algal morphology and taxonomical consequences. *Bot. Mar.*, **20**, 499–507.

JOHN D. M. (1970) Differences in growth of three species of *Laminaria* along a depth gradient. *Nova Hedwigia*, **19**, 789–98.

JOHNSTON C. S. (1969) Studies on the ecology and primary production of Canary Island marine algae. *Proc Int. Seaweed Symp.*, **4**, 213–222.

JONES W. E. (1959) Experiments on some effects of certain environmental factors on *Gracilaria verrucosa* (Hudson) Papenfuss. *J. Mar. Biol. Ass. U.K.*, **38**, 153–67.

JONES W. E. & DEMETROPOULOS A. (1968) Exposure to wave action: Measurements of an important ecological parameter on rocky shores of Angelsey. *J. Exp. Mar. Biol. Ecol.*, **2**, 46–63.

JORDAN A. J. & VADAS R. L. (1972) Influence of environmental parameters on intraspecific variation in *Fucus vesiculosus*. *Mar. Biol.*, **14**, 248–52.

JOSSELYN M. N. & MATHIESON A. C. (1979) Contribution of receptacles from the fucoid *Ascophyllum nodosum* to the detrital pool of a north temperate estuary. *Estuaries*, **1**, 258–61.

KAIN J. M. (1971) The biology of *Laminaria hyperborea*. VI. Some Norwegian populations. *J. Mar. Biol. Ass. U.K.*, **51**, 387–408.

KAIN J. M. (1976a) New and interesting marine algae from the Shetland Isles. II. Hollow and solid stiped *Laminaria* (Simplices). *Br. Phycol. J.*, **11**, 1–11.

KAIN J. M. (1976b) The biology of *Laminaria hyperborea*. IX. Growth pattern of fronds. *J. Mar. Biol. Ass. U.K.*, **56**, 603–28.

KAPRAUN D. F. (1970) Field and cultural studies of *Ulva* and *Enteromorpha* in the vicinity of Port Aransas. *U. Tex. Contr. Mar. Sci.*, **15**, 205–85.

KHAILOV K. M., KHOLODOV V. I., FIRSOV Y. K. & PRAZUKIN A. V. (1978) Thalli of *Fucus vesiculosus* in ontogenesis: Changes in morpho-physiological parameters. *Bot. Mar.*, **21**, 289–311.

KILAR J. A. & MATHIESON A. C. (1978) Ecological studies of the annual red alga *Dumontia incrassata* (O. F. Müller) Lamouroux. *Bot. Mar.*, **21**, 423–37.

KLEMM P. (1893) Ueber *Caulerpa prolifera*. Ein Beitrag zur Erforschung der Form- und Richtkrafte in Pflanzen. *Flora*, **77**, 460–86.

KLUGH A. B. (1922) Ecological polymorphism in *Enteromorpha crinita*. *Rhodora*, **24**, 50–5.

KNAGGS F. W. (1966) *Rhodochorton purpureum* (Lightf.) Rosenvinge. Observations on the relationship between morphology and environment. II. *Nova Hedwigia*, **11**, 337–49.

KNIGHT M. & PARKE M. (1931) Manx algae. An algal survey of the south end of the Isle of Man. *L.M.B.C. Mem. Typ. Br. Mar. Pl. Anim.*, **30**, 1–155.

KNIGHT M. & PARKE M. (1950) A biological study of *Fucus vesiculosus* L. and *Fucus serratus* L. *J. Mar Biol. Ass. U.K.*, **29**, 439–514.

LAMB I. M. & ZIMMERMAN M. H. (1964) Marine vegetation of Cape Ann, Massachusetts. I. Seasonal succession; II. The occurrence of the genus *Pantoneura* Kylin (Rhodophyta) in North America. *Rhodora*, **66**, 217–54.

LARKUM A. W. D. (1972) Frond structure and growth in *Laminaria hyperborea*. *J. Mar. Biol. Ass. U.K.*, **52**, 405–18.

LARKUM A. W. D. & NORTON T. A. (1968) An investigation of the vertical distribution of seaweeds in the Whirlpool area of Lough Ine. *Br. Phycol. Bull.*, **3**, 601–2.

LEVRING T., HOPPE H. A. & SCHMID O. J. (1969) *Marine Algae, a Survey of Research and Utilization*. Cram de Gruyter, Hamburg.

LIDDLE L. (1979) Modes of reproduction in *Cymopolia barbata* (Chlorophyta, Dasycladales). In *The Biology of Acetabularia*, eds. Bonotto S., Kefeli V. & Puiseux-Dao S., pp. 19–26. Elsevier North Holland Biomedical Press, Amsterdam.

LOBBAN C. S. (1978) The growth and death of the *Macrocystis* sporophyte (Phaeophyceae, Laminariales). *Phycologia*, **17**, 196–212.

LOCKHART J. C. (1979) Factors determining various forms in *Cladosiphon zosterae* (Phaeophyceae). *Am. J. Bot.*, **66**, 836–44.

LUNDBERG B. & LIPKIN Y. (1979) Natural food of the herbivorous rabbitfish (*Siganus* sp.) in northern Red Sea. *Bot. Mar.*, **22**, 173–81.

LÜNING K. (1970) Cultivation of *Laminaria hyperborea in situ* and in continuous darkness under laboratory conditions. *Helgoländer Wiss. Meeresunters.*, **20**, 79–88.

MACFARLANE C. I. (1968) *Chondrus crispus Stackhouse—A Synopsis*. Nova Scotia Res. Fdn. Seaweed Division.

MACLACHLAN J. (1974) Effects of temperature and light on growth and development of embryos of *Fucus edentatus* and *Fucus distichus* ssp. *distichus*. *Can. J. Bot.*, **52**, 943–51.

MANN K. H. (1971) Relation between stipe length, environment and the taxonomic characters of *Laminaria*. *J. Fish. Res. Bd. Can.*, **28**, 778–80.

MATHIESON A. C. (1965) Contributions to the Life History and Ecology of the Marine Brown Alga *Phaeostrophion irregulare* S. et G. Ph.D. Thesis, University of British Columbia, Vancouver, Canada.

MATHIESON A. C. & BURNS R. L. (1975) Ecological studies of economic red algae. V. Growth and reproduction of natural and harvested populations of *Chondrus crispus* (Stackhouse) in New Hampshire. *J. Exp. Mar. Biol. Ecol.*, **17**, 137–56.

MATHIESON A. C., FRALICK R. A., BURNS R. L. & FLAHIVE W. (1975) Phycological studies from Tektite II, at St. John, U.S. V.I. *Nat. Hist. Mus. Los Angeles Co. Sci. Bull.*, **20**, 77–103.

MATHIESON A. C. & PRINCE J. S. (1973) Ecology of *Chondrus crispus* (Stackhouse). In *Chondrus crispus*, eds. Harvey M. J. & McLachlan J., pp. 53–79. *Proc. Nova Scotia Inst. Sci.*, **27** (suppl.).

MOSS B. M. (1963) Wound healing and regeneration in *Fucus vesiculosus* L. *Proc. Int. Seaweed Symp.*, **4**, 117–22.

MOSS B. M. (1974) Morphogenesis. In *Algal Physiology and Biochemistry*, ed. Stewart W. D. P., pp. 788–813. Blackwell Scientific Publications, Oxford.

MOSS B. M. & MARSLAND A. (1976) Regeneration of *Enteromorpha*. *Br. Phycol. J.*, **11**, 309–13.

MSHIGENI K. E. & KAJUMULO A. A. (1979) Effects of the environment on polymorphism in *Ulva fasciata* Delile (Chlorophyta: Ulvaceae). *Bot. Mar.*, **22**, 145–8.

MUNDA I. M. (1976) The distribution of *Halosaccion ramentaceum* (L.) J. Ag. associations in the Icelandic coastal area and their hydrographically conditioned variations. *Bot. Mar.*, **19**, 161–79.

MUNDA I. M. (1978) Salinity dependent distribution of benthic algae in estuarine areas of Icelandic fjords. *Bot. Mar.*, **21**, 451–68.

MUNDA I. M. (1979) A note on the growth of *Halosaccion ramentaceum* (L.) J. Ag. under different culturing conditions. *Bot. Mar.*, **20**, 493–8.

MURRAY S. N. & DIXON P. S. (1975) The effects of light intensity and light period on the development of thallus form in the marine red alga *Plenosporium squarrulosum* (Harvey) Abbott (Rhodophyta, Ceramiales). II. Cell enlargement. *J. Exp. Mar. Biol. Ecol.*, **19**, 165–76.

NEISH A. C. & SHACKLOCK P. F. (1971) Greenhouse experiments (1971) on the propagation of strain T4 of Irish Moss. *Atlantic Reg. Lab., Nat. Res. Coun. Can. Tech. Rep.*, **14**, 1–25.

NEUSHUL M. (1972) Functional interpretation of benthic marine algal morphology. In *Contributions to the Benthic Marine Algae of the North Pacific*, ed. Abbott I. A. & Kurogi M., pp. 47–74. Japanese Society Phycology, Kobe, Japan.

NEUSHUL M., SCOTT J., DAHL A. L. & OLSEN D. (1967) Growth and development of *Sciadophycus stellatus* Dawson. *Bull. So. Calif. Acad. Sci.*, **66**, 195–200.

NIELSEN R. (1972) A study of the shell-boring marine algae around the Danish Island, Laesø. *Bot Tiddskr.* **67**, 245–69.

NIEMECK R. A. & MATHIESON A. C. (1976) An ecological study of *Fucus spiralis* L. *J. Exp. Mar. Biol. Ecol.*, **24**, 33–48.

NIENBURG W. (1925) Eine eigenartige Lebensgemeinschaft zwischen *Fucus* und *Mytilus*. *Ber. Deutsch Bot. Ges.*, **43**, 292–8.

NIENBURG W. (1932) *Fucus mytili* spec. nov. *Ber. Deutsch. Bot. Ges.* **50a**, 28–41.

NORALL T. L., MATHIESON A. C. & KILAR J. A. (1981) Reproductive ecology of four subtidal red algae. *Mar. Biol.*, (in press).

NORTON T. A. (1969) Growth form and environment in *Saccorhiza polyschides*. *J. Mar. Biol. Ass. U.K.*, **49**, 1025–45.

NORTON T. A. (1971) An ecological study of the fauna inhabiting the sublittoral marine alga, *Saccorhiza polyschides* (Lightf.) Batt. *Hydrobiologia*, **37**, 215–31.

NORTON T. A. (1975) Growth form and environment in cave-dwelling plants of *Plumaria elegans*. *Br. Phycol. J.*, **10**, 225–33.

NORTON T. A. (1977a) The growth and development of *Sargassum muticum*. *J. Exp. Mar. Biol. Ecol.*, **19**, 41–53.

NORTON T. A. (1977b) Ecological experiments with *Sargassum muticum*. *J. Mar. Biol. Ass. U.K.*, **57**, 33–43.

NORTON T. A. (1978a) The factors influencing the distribution of *Saccorhiza polyschides* in the region of Lough Ine. *J. Mar. Biol. Ass. U.K.*, **58**, 527–36.

NORTON T. A. (1978b) Some recent developments in phycology. *Trans. Proc. Bot. Soc. Edinburgh*, **42**, 51–9.

NORTON T. A. & BURROWS E. M. (1969a) The environmental control of seasonal behaviour in *Saccorhiza polyschides*. *Proc. Int. Seaweed Symp.*, **6**, 287–96.

NORTON T. A. & BURROWS E. M. (1969b) Studies on marine algae of the British Isles. 7: *Saccorhiza polyschides* (Lightf.) Batt. *Br. Phycol. J.*, **4**, 19–53.

NORTON T. A., EBLING F. J. & KITCHING J. A. (1971) Light and the distribution of organisms in a sea cave. In *Proc. IV Europ. Mar. Biol. Symp.*, ed. Crisp D. J., pp. 409–32. Cambridge University Press, England.

NORTON T. A., HISCOCK K. & KITCHING J. A. (1977) The ecology of Lough Ine. XX. The *Laminaria* forest at Carrigathorna. *J. Ecol.*, **65**, 919–41.
NORTON T. A., MATHIESON A. C. & NEUSHUL M. (1981) A review of some aspects of form and function of seaweeds. *Bot. Mar.*, **29** (in press).
NORTON, T. A. & MILBURN, J. A. (1972) Direct observations on the sublittoral marine algae of Argyll, Scotland. *Hydrobiologia*, **40**, 55–68.
OLIVEIRA, E. C. DE & FLETCHER A. (1980) Taxonomic and ecological relationships between rocky-shore and saltmarsh populations of *Pelvetia canaliculata* (Phaeophyta) at Four Mile Bridge, Angelsey, U.K. *Bot. Mar.*, **23**, 409–17.
PACE D. R. (1972) Polymorphism in *Macrocystis integrifolia* Bory in Relation to Water Motion. M.Sc. Thesis, University of British Columbia, Vancouver.
PALMISANO J. F. & SCHENG Y. S. (1977) Blade width of *Laminaria longipes* (Phaeophyceae, Laminariales) as an indicator of wave exposure. *Syesis*, **10**, 53–6.
PARKE M. (1948) Studies on British Laminariaceae. I. Growth in *Laminaria saccharina* (L.) Lamour. *J. Mar. Biol. Ass. U.K.*, **27**, 651–709.
PARR A. E. (1939) Quantitative observations on the pelagic *Sargassum* vegetation of the western North Atlantic. *Bull. Bingham Oceanogr. Coll.*, **6**, 1–94, Peabody Mus. Nat. Hist.
PERRONE C. & FELICINI G. P. (1974) Dominance apicale et morphogénèse chez *Petroglossum nicaeense* (Duby) Schotter (Rhodophyceae). *Phycologia*, **13**, 187–94.
PETERSON R. D. (1972) Effects of light intensity on the morphology and productivity of *Caulerpa racemosa* (Førsskal) J. Agardh. *Micronesica*, **8**, 63–86.
POWELL J. H. (1964) The Life-History of a Red Alga, *Constantinea*. Ph.D. Thesis, University of Washington, Seattle, Washington.
PRICE J. H. (1978) Ecological determination of adult form in *Callithamnion*: Its taxonomic implications. In *Modern Approaches to the Taxonomy of Red and Brown Algae*, ed. Irvine D. E. G. & Price J. H., pp. 263–300. Systematics Assoc. Special Vol. 10, Academic Press, London.
PRICE W. M. (1967) Some Aspects of the Biology and Taxonomy of the Unbranched Cladophorales in the British Isles. Ph.D. Thesis, University of Liverpool.
PRINGLE J. D. (1975) Studies on Variations in Three Taxa of *Enteromorpha*. Ph. D. Thesis, Dalhousie University, Halifax.
PRINTZ H. (1959) Investigations of the failure of recuperation and repopulation in cropped *Ascophyllum* areas. *Avhandl. Norske Vidensk. Akad. Oslo*, **3**, 1–15.
REED R. H. & RUSSELL G. (1978) Salinity fluctuations and their influence on 'bottle brush' morphogenesis in *Enteromorpha intestinalis*. *Br. Phycol. J.*, **13**, 149–153.
REYNOLDS N. B. (1971) The Ecology of a New Hampshire Estuarine Tidal Rapid. Ph.D. Thesis, University New Hampshire, Durham.
ROBERTS M. & RING F. M. (1972) Preliminary investigations into conditions affecting the growth of the microscopic phase of *Scytosiphon lomentaria* (Lyngbye) Link. *Mem. Soc. Bot. Fr.*, **51**, 117–28.
RUSSELL G. (1978) Environment and form in the discrimination of taxa in brown algae. In *Modern Approaches to the Taxonomy of Red and Brown Algae*, eds. Irvine D. E. G. & Price J. H., pp. 339–69. Systematics Assoc. Special Vol. 10, Academic Press, London.
SANBONSUGA Y. & TORII S. (1974) On the local variation of morphological characteristics in *Laminaria japonica* Areschoug var. japonica. *Bull. Hokkaido Reg. Fish. Res. Lab.*, **40**, 48–59.
SANTELICES B. (1974) Gelidioid algae, a brief resume of the pertinent literature. *Mar. Agron. U.S. Sea Grant Prog. Hawaii, Tech. Rep.*, **1**.
SANTELICES B. (1978) The morphological variation of *Pterocladia caerulescens* (Gelidiales, Rhodophyta) in Hawaii. *Phycologia*, **17**, 53–9.
SCHILLER J. (1928) Zur Frage der Gezeitenwirkung auf emergierenden Algen. *Planta*, **6**, 535–42.
SCHONBECK M. W. & NORTON T. A. (1974) Growth form of *Fucus distichus* in relation to wave exposure. *Proc. Int. Seaweed Symp.*, **7** (abst).
SCHONBECK M. W. & NORTON T. A. (1978) Factors controlling the upper limits of fucoid algae on the shore. *J. Exp. Mar. Biol. Ecol.*, **31**, 303–13.

SCHONBECK M. W. & NORTON T. A. (1979a) The effects of brief periodic submergence on intertidal fucoid algae. *Est. Cstl. Mar. Sci.*, **8**, 205–11.

SCHONBECK M. W. & NORTON T. A. (1979b) An investigation of drought avoidance in intertidal fucoid algae. *Bot. Mar.*, **22**, 133–44.

SCHONBECK M. W. & NORTON T. A. (1979c) Drought-hardening in the upper shore seaweeds *Fucus spiralis* and *Pelvetia canaliculta*. *J. Ecol.*, **67**, 687–96.

SCHONBECK M. W. & NORTON T. A. (1979d) The effects of diatoms on the growth of *Fucus spiralis* germlings in culture. *Bot. Mar.*, **22**, 233–6.

SCHONBECK M. W. & NORTON T. A. (1980) The effects on intertidal fucoid algae of exposure to air under various conditions. *Bot. Mar.*, **23**, 141–7.

SEOANE-CAMBA, J. (1966) Sobre la variabilidad morfológica de *Fucus vesiculosus* en las rías gallegas. *Invest. Pesquera.*, **30**, 561–76.

SETCHELL W. A. (1905) Regeneration among kelps. *Univ. Calif. Publs. Bot.*, **2**, 139–68.

SOUTH G. R. & HILL R. D. (1970) Studies on marine algae of Newfoundland. I. Occurrence and distribution of free-living *Ascophyllum nodosum* in Newfoundland. *Can. J. Bot.*, **48**, 1697–701.

STEFFENSEN D. A. (1976) Morphological variation of *Ulva* in the Avon-Heathcote Estuary, Christchurch. *N.Z.J. Mar. Freshwat. Res.*, *10*, 329–41.

STENECK R. S. & ADEY W. H. (1976) The role of environment in control of morphology in *Lithophyllum congestum*, a Caribbean algal ridge builder. *Bot. Mar.*, **19**, 197–215.

STEPHENSON T. A. & STEPHENSON A. (1972) *Life Between Tide-Marks on Rocky Shores*. Freeman, San Francisco.

STEWART J. G. (1976) Development of morphological patterns in three species of Delesseriaceae. *Madroño*, **23**, 438–48.

SUNDENE O. (1962a) Growth in the sea of *Laminaria digitata* sporophytes from culture. *Nytt Mag. Bot.*, **9**, 5–24.

SUNDENE O. (1962b) The implications of transplant and culture experiments on the growth and distribution of *Alaria esculenta*. *Nytt Mag. Bot.*, **9**, 155–74.

SUNDENE O. (1964) The ecology of *Laminaria digitata* in Norway in view of transplant experiments. *Nytt Mag. Bot.*, **11**, 83–107.

SVENDSEN P. & KAIN J. M. (1971) The taxonomic status, distribution and morphology of *Laminaria cucullata sensu* Jorde and Klavestad. *Sarsia*, **46**, 1–22.

TAYLOR A. R. A. (1966) Variation in the genus *Polysiphonia* Greville. *Proc. Int. Seaweed Symp.*, **5**, 209–15.

TOKIDA J. & YAMAMOTO H. (1965) Syntagmatic germination of tetraspores in *Pachymeniopsis yendoi*. *Phycologia*, **5**, 15–20.

VADAS R. L., KESER M. & LARSON R. (1978) Effects of reduced temperatures on previously stressed populations of an intertidal alga. In *Energy and Environmental Stress in Aquatic Systems*, eds. Thorp J. H. & Gibbons J. W., pp. 434–51. National Technical Information Service, Springfield, Virginia.

VERMEIJ G. J. (1978) *Biogeography and Adaptation—Patterns of Marine Life*. Harvard University Press, Cambridge, Mass.

VON STOSCH H. A. (1964) Wirkungen von Jod und Arsenit auf Meeresalgen in Kultur. *Proc. Int. Seaweed Symp.* **4**, 142–50.

WAALAND S. D. & CLELAND R. (1972) Development in the red alga, *Griffithsia pacifica*: Control by internal and external factors. *Planta*, **105**, 196–204.

WANDERS J. B. W. (1977) The role of benthic algae in the shallow reef of Curacao (Netherlands Antilles). III.The significance of grazing. *Aquat. Bot.*, **3**, 357–90.

WATANABE T. & KATO S. (1972) Studies on the disease of cultural *Porphyra*. IV. On the cause of cancer disease. *Bull. Jap. Soc. Sci. Fish.*, **38**, 431–7.

WEST J. A. (1972) Environmental control of hair and sporangial formation in the marine red alga *Acrochaetium proskaueri* sp. nov. *Proc. Int. Seaweed Symp.*, **7**, 377–84.

WIDDOWSON T. B. (1965) A taxonomic study of the genus *Hedophyllum* Setchell. *Can. J. Bot.*, **43**, 1409–20.

WIDDOWSON T. B. (1971) Growth rates of *Egregia* as a measure of pollution. *Proc. Int. Seaweed Symp.* **7**, 268–72.
YARISH C. (1976) Polymorphism of selected marine Chaetophoraceae (Chlorophyta). *Br. Phycol. J.*, **11**, 29–38.

SECTION III
PHYSIOLOGY AND BIOCHEMISTRY

INTRODUCTION

CHRISTOPHER S. LOBBAN

The actions of environmental factors are to be understood at the physiological and biochemical levels, and we assume that everything which takes place within a cell adapts the organism to its environment. Much of what is known about plant physiology and biochemistry has been learned from studies of terrestrial plants, notably certain commercially important angiosperms. The marine environment differs in several important ways from the terrestrial, and might be expected to cause significant differences in physiology between marine and terrestrial plants. Yet the physiology of marine angiosperms is not greatly different from that of their dry land counterparts; the differences are principally in relation to the high osmotic pressure of seawater and the substantial differences in ionic composition between seawater and soils (Queen 1974). The much greater differences between seaweeds and angiosperms stem, of course, from the great evolutionary divergence of the two groups, rather than from the habitat; Phaeophyta, Rhodophyta, and probably most of the Chlorophyta (cf. Pickett-Heaps 1975) are not in the mainstream of 'green' plant evolution. We may expect that certain essential mechanisms of the physiology and biochemistry of life (e.g. electron transfer, ion transport, gene and protein biochemistry) will be the same or very similar in seaweeds, but beyond this, vascular plant physiology can only provide guidelines, ideas of what to look for. Algal physiology has had to proceed as a somewhat distinct discipline.

Differences between plants and animals, partly discussed in the introduction to Section II, include fundamental differences in cells. Plant cells are biochemical generalists, and plant organs (roots, shoots, leaves, flowers) all contain more or less the same diversity of cells. Animal cells and organs, on the other hand, are specialized. Plants thus have relatively little organization compared to animals; and seaweeds, while their complexity varies greatly (Section I), are all less complex than vascular plants, lacking most of the division of labour seen in the latter. However, a consequence of reduced specialization is that each cell becomes even more diversified; the simplicity of algal morphology and reproduction relative to vascular plants thus belies their great physiological and biochemical complexity. (Of course, seaweeds do have some division of labour: in a thick thallus the photosynthetic cells are only on the surface; in many seaweeds reproductive structures are localized and specialized; and some differentiation between organs has been demonstrated (e.g. Fagerberg *et al.* 1979).)

One special problem which crops up repeatedly in physiological, biochemical and ecological studies warrants consideration here. It is the problem of expression of data. Most experimenters have wondered which denominator to choose, and it is an important question, for a trend seen when data are expressed one way may disappear or even reverse when data are expressed some other way. For example, Black (1954) found that mineral matter in *Laminaria saccharina* decreased over the proximal half of the lamina and then increased in the distal half when data were expressed on a dry weight basis; on a fresh weight basis mineral matter increased over the proximal two-thirds and then decreased! Among the denominators commonly used are: fresh weight, dry weight, ash-free dry weight, area, packed cell volume, and (for photosynthesis) milligrams of chlorophyll *a*. In some cases a reasonable choice is clear, in other cases it is not. Consider as an example measuring the photosynthetic rate of a piece of *Laminaria* in the laboratory. Weight as a basis on which to express the data may be objected to on the grounds that much of the tissue is not photosynthetic; area on the grounds that one surface is partly shaded (unless the experimenter has taken pains to light both sides equally); and chlorophyll *a* content on the grounds that the relationship between chlorophyll *a* and photosynthetic rate is not simple (see Chapter 13).

There does not appear to be a solution to this knotty problem at present. A useful discussion of the problem is to be found in Section 1.4.3 of Šesták *et al.* (1971), where they state, *inter alia*, 'if the main criterion is to achieve comparability with other data in the literature then it is logical to accept the most widely used basis. . . . However, more rational criteria for selection may be sought in specific investigations. The rate of a process yields maximum information about the process itself if it is expressed on the basis of a plant characteristic which limits or at least strongly influences the process.' The lack of clear criteria for making the choice can inadvertently cause misleading representation of the results and leaves room for abuse of scientific ethics. Scientists, encouraged by space-conscious journal editors, rarely publish non-trends, but the data omitted might later be useful for reassessing the results in view of subsequent studies. There is thus something to be said for national or international depositories for unpublished data.

REFERENCES

Black W. A. P. (1954) Concentration gradients and their significance in *Laminaria saccharina* (L.) Lamour. *J. Mar. Biol. Ass. U.K.*, **33**, 49–60.

Fagerberg W. R., Moon R. & Truby E. (1979) Studies on *Sargassum*. III. A quantitative ultrastructural and correlated physiological study of the blade and stipe organs of *S. filipendula*. *Protoplasma*, **99**, 247–61.

Pickett-Heaps J. D. (1975) *Green Algae: Structure, Reproduction and Evolution in Selected Genera.* Sinauer, Sunderland, Massachusets.

QUEEN W. H. (1974) Physiology of coastal halophytes. In *Ecology of Halophytes* eds. Reimold R. J. & Queen W. H., pp. 345–53. Academic Press, London & New York.
ŠESTÁK Z., JARVIS P. G. & ČATSKÝ, J. (1971) Criteria for the selection of suitable methods. In *Plant Photosynthetic Production: Manual of Methods*. eds. Šesták Z., Čatský J. & Jarvis P. G., pp. 1–48. W. Junk, The Hague.

CHAPTER 13
THE CAPTURE AND TRANSDUCTION OF LIGHT ENERGY

J. RAMUS
Department of Botany and the Marine Laboratory,
Duke University, Beaufort, North Carolina, U.S.A.

1 INTRODUCTION

The algae are a polyphyletic assemblage of organisms which defy inclusive definition, save through systematically convenient but phylogenetically artificial distinguishing characteristics, e.g. see Bold & Wynne (1978, p. 1). Thus, the algae are a form taxon as opposed to a natural evolutionary unit, a distinction of special meaning here. Seaweeds are a habit form—benthic marine thalloid algae—and their assignment to three classes in three divisions (Christensen 1962) of the plant kingdom acknowledges their separate and distinct phylogenies. Accepting this notion, it follows that unique genetic repertoires have in the course of evolution converged on the same habit—that of a seaweed—and the same habitat—the fringes of oceanic basins. The distinct genetic repertoires are manifest biochemically in the pigments localized in seaweed chloroplasts. Chlorophyll *a*—the *primary* photosynthetic pigment—is shared in common. Seaweeds may differ in their *accessory* photosynthetic pigments; these include several additional chlorophylls, phycobiliproteins, carotenes and xanthophylls. The accessory pigments may blend with or mask chlorophyll *a* to produce the diagnostic colour of each seaweed class—red, brown or green—and conveniently colour-code them.

Most, but not all, of the chloroplast pigments participate in the light reactions of photosynthesis, specifically the capture of photosynthetically active radiation (PAR). To a first approximation, seaweeds are obligate photolithotrophs wholly dependent upon ambient light energy for synthetic metabolism. Light is variable in the marine water column (Chapter 9), in quantity (quantum flux density), quality (spectral distribution), duration and period. The intrinsic ability to capture PAR is the means to growth, to reproduction, to physical accommodation and to biological interaction. Light energy can be a limiting resource, albeit a peculiar one—it cannot be stored. The energy must be captured immediately when available and converted to chemical energy either to be consumed or stored. The capture of light energy and conversion to chemical energy involves the totality of the light reactions of photosynthesis, viz. quantum capture, energy transfer, photolysis, electron transport, redox couples, photophosphorylation and the reduction of pyridine nucleotides. The apparatus responsible for these functions is localized in the membranes within the chloroplast envelope and is known, at least conceptually and perhaps in reality, as the photosynthetic unit (PSU). The structure of the PSU varies among classes and among species within the same class; i.e. comparatively there are qualitative differences involving component macromolecules, spatial distribution of the components and kinetic properties. Further, the PSU's of individuals quantitatively respond to environmental properties, especially ambient light and nitrogen nutrient concentrations.

Teleology permitted, it is the role of the PSU to optimize (we know from experience that organisms seldom maximize) photosynthesis. There exists an undeniable unity in the structure and function of the PSU. Natural selection, however, has fostered considerable variation on the basic theme, either in the PSU itself or by modifying its cellular environment. These variations have converged in niches broadly similar in terms of physical properties and biological interactions. Here we will rely on a 'quantitative but holistic approach' (Mishkind & Mauzerall 1980) to the PSU, seeking where possible the biological consequences of accommodation. The term *accommodation*, rather than 'adaptation', is used here wittingly. The justification is made by Harper (1977, p. 776):

> Adaptation is a word too loosely used in ecological writing. Often to say that a feature of an organism's life or form is adaptive is to say no more than that the feature appears to be a good thing, judged on the basis of an anthropomorphic attitude to the problems that the organism is seen to face. More accurately, adaptations are those features of an organism that in the past improved the fitness of its ancestors and so were transmitted to its descendants. Adaptation is always retrospective—it is defined by the numbers of descendants left by an individual relative to its fellows. An organism will be more fit if its activities reduce the number of descendants

left by neighbors, *even if the activities do nothing to the number of descendants that it itself leaves.* . . . If an activity of an organism brings no direct benefit but hinders the chance that neighbors will leave descendants, the activity will increase fitness—it will be *adaptive.*

It is not yet possible to definitively state that either differences or changes in the PSU are adaptive, that they increase the fitness of one individual relative to another. We assume they are adaptive, but conclusive data are not yet available, save the grand experiment natural selection. However, one of the goals of science is to make explicit that which is implicit.

2 A PRIMER OF PHOTOSYNTHESIS

Outlined here are the points necessary for a discussion of the photochemical events of seaweed photosynthesis, the so-called 'light reactions'. The biochemical events of photosynthesis, the 'dark reactions', are discussed by Kremer in Chapter 14. For more general accounts of photosynthesis, the reader may consult contemporary reviews on the subject, such as Govindjee (1975), Govindjee & Braun (1974) and Gregory (1977). A germane and learned treatment of phytoplankton photosynthesis, productivity and growth has been written recently by Harris (1978).

2.1 Light absorption

The spectrum of electromagnetic radiation utilized for photosynthesis falls in the narrow wavelength range of 380–720 nm. The energy is available as discrete particles or quanta, whose wavelength range coincides with the spectral sensitivity of the human eye, which are perceived as visible light, and which are called photons. From the equations relating energy to electromagnetic radiation:

$$E = h\nu \quad \text{and} \quad \nu = c/\lambda, \quad \text{then } E = h\,(c/\lambda),$$

where E is the energy of a photon, h is Planck's constant, ν is the frequency of the photon, c is the speed of light and λ is the wavelength of the photon. It follows that photons of shorter wavelength are more energetic than those of longer wavelengths. In fact, violet photons (400 nm) are nearly twice as energetic as red photons (700 nm), an important fact for the ensuing discussion.

The molecular components which in aggregate perform the photochemical reactions are localized within the chloroplast envelope in thylakoids. Quantitatively the bulk of these components are light-harvesting pigments, or antenna pigments, intimately integrated into the membrane structure as pigment-protein complexes. All seaweeds carry the primary pigment chlorophyll *a*; accessory pigments of the red seaweeds are mainly phycoerythrin and phycocyanin, those of the brown seaweeds are chlorophyll *c* and fucoxanthin and those of the green seaweeds are chlorophyll *b* and some xanthophylls (see Chapter 17). All the classes carry varying quantities of the myriad carotenes and xanthophylls; with the exception of fucoxanthin and siphonaxanthin, their participation in photosynthesis has not been confirmed.

The light reactions begin when a chromophore absorbs a photon and is elevated from a ground to an electronically excited state, a property conferred by a valence electron. The excited electron is more energetic and less stable, and has a net tendency to fall back to its ground state. The simplest degradation is a reversal of absorption—the emission of a photon which we know as fluorescence. Some of the energy of the captured photon is dissipated as heat, so a photon emitted by fluorescence is of a longer wavelength (lower energy) than the photon which excited the chromophore. An alternative to fluorescence as an avenue from excited to ground state is energy transfer directly to another chromophore. The excited chromophore can return to ground state while a second chromophore is raised to an excited state, a receiver responding to the oscillating electric field of a transmitter. This process is termed inductive resonance, and is quite efficient as an energy transfer mechanism provided the pigment molecules are close together in the thylakoid membranes, which they apparently are. Also, energy can be transferred from one chromophore to another by fluorescence and reabsorption of the same photon, but the process is less efficient than inductive resonance.

Antenna pigments pass captured light energy by inductive resonance to a sink, the 'trap' or reaction centre, possible because they are in close physical proximity and the absorption peak of the receiver overlaps the fluorescence spectral emission peak of the transmitter. The antenna pigments are probably spatially arrayed in order of their spectral absorption maxima, from short to long wavelength. Chlorophylls have two major absorption maxima, one in the violet (the Soret band) and one in the red; however, absorption of a violet photon produces the same effect as absorption of a red photon—the excess energy of the shorter wavelength photon is dissipated as heat and lost to photosynthesis. The energy of photons captured by accessory pigments absorbing in the 'green window', i.e. the transmission spectrum of chlorophyll, is passed sequentially through accessory pigments to chlorophyll *a*. Chlorophyll *a* absorbs farthest into the red and hence the least energy of excitation. The shorter the wavelength of the captured photon, the more accessory pigments are involved in energy migration. For example, the longest chain of energy transfer in a red seaweed will be r-phycoerythrin, r-phycocyanin, allophycocyanin and chlorophyll *a*. The analogy of an antenna is appropriate—a signal (photons) of varying qualities (wavelengths) is captured by a wide-ranging receiver and channeled to a trap where the signal is transduced, i.e. converted to chemical energy.

2.2 Energy transduction

Absorption of a photon, excitation of the chromophore and the radiationless transfer of energy to chlorophyll *a* is but the beginning of the light reactions. We need a mechanism for the conversion of light energy to chemical energy. Among the chlorophyll *a* antenna molecules are several, probably present as dimers, which have the unique property of being reversibly oxidized and reduced: these are the 'trap' or reaction centre (RC) chlorophylls. Light energy passed to the RC from the antenna pigments is converted to a chemical, free-energy gradient by charge separation. The charges are then passed along an electron transport chain, converting the free-energy gradient to forms of chemical energy useful to

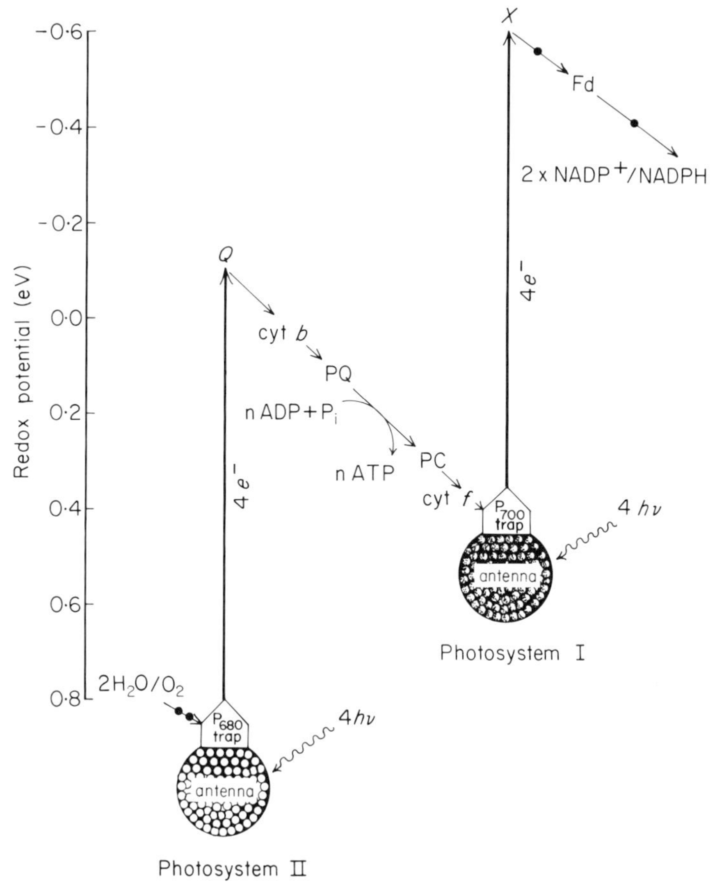

Fig. 13.1 A modified Hill-Bendall or 'Z' scheme of photosynthesis. The bold vertical arrows represent the photochemical or 'light' reactions; all others, 'dark' reactions. Only non-cyclic photophosphorylation is shown here. *Q* and *X* are primary electron acceptors for PS II and PS I respectively. Cyt *b* (cytochrome *b*), cyt *f*, PQ (plastoquinone), PC (plastocyanin) and Fd (ferredoxin) are electron carriers. The large dots represent proposed but unknown electron carriers.

the cell. Thus, the RC chlorophylls donate electrons to the electron transport chain and accept electrons from an electron donor, in the process going from a reduced to an oxidized state then back to a reduced state. Absorption differences between reduced and oxidized P_{700} allow these states to be quantitatively detected.

In principle, a photochemical reaction at the molecular level is the activation of an electron by a photon. Thus, the theoretical yield of any photochemical reaction is normalized to the number of absorbed photons rather than the amount of energy absorbed. The quantity of measure is the *quantum yield*, which is the number of molecules of a specified substance formed following the absorption of one photon. Assuming near perfect efficiency of quantum conversion, one electron is moved for every photon captured.

The basic model of photosynthetic electron transport is the Hill & Bendall (1960) model, or the 'Z' scheme, which incorporates two photosystems (Fig. 13.1). The electron donor for photosystem (PS) II is water and the ultimate electron acceptor for PS I is oxidized nicotinamide adenine dinucleotide phosphate ($NADP^+$). The standard redox potential for the H_2O/O_2 couple is 0.82 eV and the $NADP^+/NADPH$ couple is -0.32 eV, giving a standard potential difference of 1.14 eV per electron, a considerable amount of reducing and oxidizing power. Since there are two photosystems, in theory the evolution of one mole O_2 requires 8 mole quanta, four for each photoact, but in practice requires more (10 ± 2). Non-cyclic electron transport adds 0.5–0.75 molecule of ATP per electron by a phosphorylating couple between PS II and PS I. The actual yield for 10 ± 2 mole quanta captured is 2–3 mole ATP and 2 mole NADPH. The free-energy gradients produced at the reaction centres by charge separation are thus transduced and stored as chemical energy in the form of ATP and NADPH to be used by the cell for CO_2 fixation.

3 METHODS FOR THE MEASUREMENT OF PHOTOSYNTHESIS

Techniques for the measurement of seaweed photosynthesis are, in theory, deceptively simple, a simplicity implied by the net equation for photosynthesis:

$$6CO_2 + 6H_2O \underset{\text{dark respiration (R)}}{\overset{\text{photosynthesis (P)}}{\rightleftharpoons}} C_6H_{12}O_6 + 6O_2.$$

Rates are usually calculated from CO_2 consumption or O_2 evolution measurements. Dark respiration proceeds independently of photosynthesis; $P\!:\!R$ ratios are known to vary with temperature, tissue age, exposure, etc. (Kanwisher 1966; Yokohama 1973a; Hatcher *et al.* 1977; Quadir *et al.* 1979). Hence, both net photosynthesis (P_n) and R must be measured simultaneously and gross photosynthesis (P_g) derived from their sum:

$$P_g = P_n + R.$$

From the net equation, the photosynthetic quotient (PQ = mole O_2 evolved/mole CO_2 fixed) in theory is unity, but in practice is usually greater than unity. How much greater (or less) has unfortunately seldom been measured for seaweeds, since PQ determinations require simultaneous measurement of O_2 evolution and CO_2 fixation. Hatcher *et al.* (1977) reported PQ values for *Laminaria longicruris* ranging from 0.67–1.50, a vexing range when attempting to convert O_2 exchange rates to absolute C-fixation rates.

Another problem stems from the lack of a convention for the expression of photosynthetic rate: O_2 or CO_2 exchange rates are normalized to a variety of biomass parameters, including carbon, chlorophyll, fresh weight, dry weight, ash-free dry weight and unit surface area. All of these normalization parameters vary with environmental properties, and from taxon to taxon, making it virtually impossible to compare photosynthetic rates. It is suggested here that the convention of biological oceanography be adopted, viz. expressing photosynthesis in units of carbon flux/unit chl at saturating but not inhibiting light, and allowing comparisons to be made by means of these 'assimilation numbers'.

The most common technique for the measurement of photosynthesis involves enclosing the seaweed thallus (or sample thereof) in a finite volume of seawater, then determining rates of change in O_2 or CO_2 concentration from periodic readings. The volume of enclosing vessels has ranged from 50 ml (Mishkind & Mauzerall 1980) in the laboratory to 76 l (Hatcher 1977) *in situ*. Often tissue samples are not representative of whole plants; tissue segment/whole plant diel P_n ratios for *Laminaria longicruris* ranged from 0.35–2.24 (Hatcher 1977). P_n is measured in transparent vessels while R is measured in opaque vessels, consistent with the basic light-bottle and dark-bottle technique developed by Gaarder & Gran (1927). The contents of the vessels must be vigorously stirred to disrupt diffusion gradients developing at seaweed surfaces.

3.1 Oxygen methods

Dissolved O_2 concentrations are usually measured by the Winkler method (a wet-chemistry assay), or electrometrically by a Clarke-type polarographic electrode. The Winkler method is especially sensitive, with a reported accuracy of 0.1% and a precision of 0.02% (Carpenter 1965), and is readily field adapted. The polarographic electrode must be carefully calibrated, temperature controlled and the sample stirred vigorously during readings, yet it is quite rapid, accurate ($\sim 2\%$) and portable. It is of distinct utility for continuous recording of changes in dissolved O_2 concentration (Jassby 1978).

3.2 Carbon methods

CO_2 exchange rates are usually measured by either infra-red gas analysis or the radiocarbon (^{14}C) assay. It is difficult to measure changes in CO_2 concentration in an air stream forced through seawater, because the slow rate at which dissolved CO_2 comes to equilibrium interferes with predictable CO_2 exchange. The infra-red gas analyser (IRGA), however, has been used with success to measure photosynthesis in air for a variety of seaweed species (Johnson *et al.* 1974; Brinkhuis *et al.* 1976; Quadir *et al.* 1979). The difference in CO_2 concentration of the carrier stream is measured by an infra-red sensor before and after its introduction to the photosynthesis chamber. The apparatus can be field adapted (Mooney *et al.* 1971) but only with considerable difficulty and expense.

The *in situ* ^{14}C technique, introduced by Steemann Nielsen (1952) for phytoplankton, has been modified for seaweed photosynthesis (Kremer 1978a). At the pH of natural seawater (7.8–8.2), the principal carbon species is bicarbonate ($HCO_3^- \simeq 90\%$), present at an approximate concentration of 2 mM (see Chapter 14). High specific-radioactivity isotope (usually $NaH^{14}CO_3$) is added to the reaction vessel and the light-dependent rate of acid-stable ^{14}C incorporation is measured. Because of the weak β-radiation (0.155 MeV) emitted by ^{14}C, end-window Geiger counters have a counting efficiency of only 2–5%. Liquid scintillation counting has improved the efficiency to at least 90%. To avoid self-absorption of β-radiation during counting, finely ground tissue can be suspended in a thixotropic gel (Wassman & Ramus 1973), or the tissue solubilized by a Mahin-Lofberg perchloric acid–hydrogen peroxide digestion (Lobban 1974), or hydrogen peroxide—NCS (Amersham) digestion (Gagne *et al.* 1979). Loss of counting efficiency due to physical or chemical quenching is corrected by internal or external channels ratio methods. The radiocarbon assay gives the proportion of total radioisotope incorporated into the seaweed. The proportionality number must be multiplied by the total inorganic carbon concentration (CO_2, HCO_3^-, CO_3^{2-}) in seawater given by a carbonate alkalinity determination (Strickland & Parsons 1972).

Despite the fact that the ^{14}C technique has been used extensively to measure seaweed photosynthesis both in the laboratory and *in situ*, it remains fraught with uncertainty. For example, rigor demands that $^{14}CO_2$ be fixed into organic compounds by photosynthesis only, i.e. the incorporation must be light-dependent. Yet brown seaweeds incorporate $^{14}CO_2$ in the dark by the action of PEP carboxykinase (Küppers & Kremer 1978; see also Chapter 14), i.e. incorporation is light-independent. Light-independent CO_2 incorporation does not occur in the red seaweeds (Kremer 1978b). In addition the rate of assimilation of $^{14}CO_2$ must be the same as that of $^{12}CO_2$, but there is an apparent discrimination by the photosynthetic apparatus between carbon isotopes, the $^{12}C:^{14}C$ ratio of incorporation during photosynthesis is 1.05, at least for phytoplankton (Harris 1978). No $^{14}CO_2$ must be lost by dark respiration or photorespiration which take place simultaneously with photosynthesis. Yet refixation of respired $^{14}CO_2$ in the light can be substantial in seaweeds (Hough 1976). Finally, no ^{14}C-organic carbon must be lost by excretion during the measurement, yet excretion can be significant (Fogg 1976; Hough 1976). In view of these uncertainties, standardized procedures and short duration isotope incubations are advised.

4 LIGHT AND PHOTOSYNTHESIS

4.1 Photon flux density

The term 'light intensity' is used erroneously to denote the amount of light incident on a surface; rather intensity is the property of the radiating source and not of the surface receiving the radiation. In physics, 'flux' is the rate of flow of a substance expressed in units of quantity/time; 'flux density' is the 'flux' through a unit surface area. Photosynthesis is a photochemical process; therefore radiant energy incident to the photosynthesizing tissue is properly measured in units of quantum flux density or photon flux density (PFD in E m^{-2} s^{-1}) when referring to the 400–700 nm waveband which activates photosynthesis. The term 'irradiance' is not the equivalent of PFD because it is specifically defined in power units (watts). Although PFD is cumbersome, it has been recommended as a precise descriptor (Shibles 1976).

Equally critical is relating photosynthesis to light *absorbed* rather than to light *incident*. However, thallus absorptance (that fraction of incident light absorbed, or $1-I/I_0$) is in practice seldom measured. Barring geometric problems, seaweed thallus absorptance can be measured using a spectrophotometer (Rhee & Briggs 1977), a microphotometer (Britz *et al.* 1976) or an autoranging scanning spectroradiometer system (Ramus 1978).

The classic curvilinear plot relating photosynthesis to PFD (P v. I_0 plot) (Fig. 13.2) contains considerable information about the photosynthetic

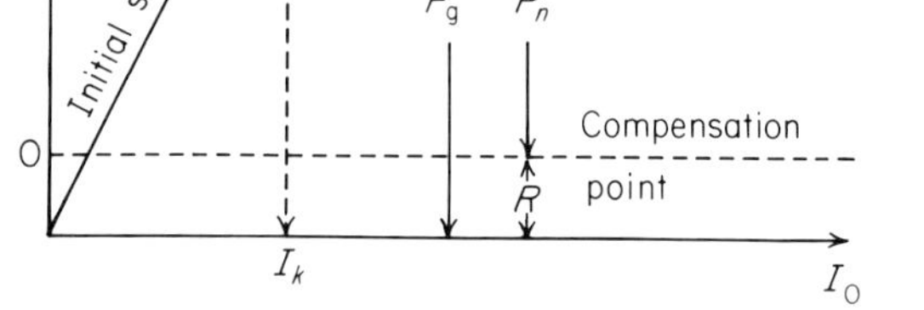

Fig. 13.2 Light saturation curve for photosynthesis. P = photosynthesis, P_{max} = maximum photosynthesis, P_g = gross photosynthesis, P_n = net photosynthesis, R = respiration, I_0 = incident photon flux density or PFD, I_k = saturating PFD.

apparatus. At low PFD, photosynthesis is linearly related to incident light. This linear region (the initial slope) is the apparent efficiency of the photosynthetic apparatus. True efficiency, or quantum yield, is necessarily derived from light *absorbed* data and is directly defined by the slope of this line. The initial slope is proportional to the number of 'traps' and their cross-sectional area (Herron & Mauzerall 1972). At a sufficiently high PFD, the rate of photosynthesis becomes independent of light, i.e. photosynthesis saturates (P_{max}). In a steady-state, P_{max} is proportional to the number of 'traps' and the intrinsic turnover time of the photosynthetic apparatus (Herron & Mauzerall 1972). Intersection of the initial slope and P_{max} gives the saturation PFD value (I_k) of the system. At supersaturating PFD ($I_0 \gg I_k$) photosynthesis declines, i.e. photoinhibits, the causes for which are likely manifold.

Theoretically, good use could be made of initial slope, P_{max} and I_k values taken from light saturation curves, e.g. inter- and intraspecific comparisons of these values for ecological purposes. The literature is replete with light saturation curves produced for seaweeds, the examples too numerous to cite inclusively here. Table 13.1 cites some recent examples, and is included to emphasize the bewildering array of methods, photosynthetic rate units and units of incident radiant energy. In these cases, the light saturation curves conform to the curvilinear model, yet comparison of I_k, initial slope and P_{max} values is impossible due to the lack of convention for their generation. Yet within-study comparisons are possible and useful, especially with reference to phenotypic adjustments of the PSU (Section 5.3).

The terms 'photosynthetic performance' and 'photosynthetic capacity' used

Table 13.1 Some examples of light saturation curves produced for seaweeds.

Source	Locality	Method	*P* units	I_0 units
Kanwisher 1966	lab	O_2, manometric	mm^3 O_2 gm dw^{-1} h^{-1}	f.c.
Drew 1969	*in situ*	^{14}C, end window GM	μg C cm^{-2} h^{-1}	relative, depth
Lüning 1971	lab	O_2, polarographic	ml O_2 dm^{-2} h^{-1}	gm-cal cm^{-2} s^{-1}
Wassman & Ramus 1973	*in situ*	O_2, Winkler ^{14}C, LSC	mg O_2 gm dw^{-1} h^{-1} mg C gm dw^{-1} h^{-1}	relative, depth
Yokohama 1973b	lab	O_2, manometric	μ mole O_2 cm^{-2} h^{-1} relative	Klux ergs cm^{-2} s^{-1}
Mathieson & Norall 1975	lab	O_2, manometric	μl O_2 gm dw^{-1} min^{-1}	f.c. & PAR *E*
Brinkhuis *et al.*, 1976	lab	CO_2, IRGA	mgC gm dw^{-1} h^{-1}	gm-cal cm^{-2} min^{-1}
Ramus *et al.* 1977	*in situ*	O_2, polarographic	μmole O_2 gm fw^{-1} h^{-1} μmole O_2 μmole chl a^{-1} h^{-1}	relative, depth
Ramus 1978	*in situ*	O_2, polarographic	μmole O_2 cm^{-2} h^{-1} nmole O_2 nmole chl^{-1} h^{-1}	ergs cm^{-2} s^{-1}
Mishkind & Mauzerall 1980	lab	O_2, polarographic	μmole O_2 gm chl^{-1} min^{-1}	mW cm^{-2}

here are taken from terrestrial plant ecological physiology (Zelawski & Walker 1976). *Performance* is the actual rate of photosynthesis occurring at a particular time and is usually limited by one or more physiological and environmental factors, in the present context it is limited by the properties of the PSU and available light. *Capacity* is the rate of photosynthesis obtained when all environmental and physiological factors are at optimum values.

4.2 Spectral distribution

In the sea, incident sunlight is exponentially attenuated in accordance with the relationship:

$$I_z = I_0 \, e^{-kz},$$

where I_z is the PFD transmitted to depth z, I_0 is the incident PFD and k is the extinction coefficient of the specific water column. In addition, the spectral distribution of light varies both with water depth and type (Chapter 9), a fact of potential significance for seaweed photosynthesis.

It is known from many studies, e.g. Levring (1947), Haxo & Blinks (1950), that seaweed photosynthesis varies as a function of wavelength, i.e. a plot of quantum yield versus wavelength yields a photosynthetic action spectrum which varies most noticeably between seaweed classes. In general, the action spectrum coincides with the absorptance spectrum ($1 - I/I_0$ v. λ) of the thallus, traditional evidence for the participation of a pigment in a photochemical event. The action spectrum produced by Haxo & Blinks (1950) for the green seaweed *Ulva taeniata* corresponds most closely with its absorptance spectrum, save some photosynthetically inactive absorption of light by carotenoids (470–490 nm). *Coilodesme californica*, a brown seaweed, shows a near correspondence including the spectral region absorbed by the carotenoid fucoxanthin (500–560 nm). The action spectrum from the red seaweed *Schizymenia pacifica* shows a marked deviation from its absorptance spectrum, i.e. the Soret and main band chlorophyll absorption appear relatively 'inactive' compared to the regions absorbed by the biliproteins phycoerythrin (480–580 nm) and phycocyanin (580–640 nm). This observation puzzled Haxo & Blinks (1950), but is now perhaps explained by the preferential transfer of energy captured by biliproteins to photosystem II, the O_2- evolving and electron-donating photosystem. Action spectra can also be generated from excitation of chlorophyll *a* fluorescence, e.g. those produced by Kageyama *et al.* (1977) showing efficient transfer of absorbed light energy by the carotenoid siphonaxanthin to chlorophyll *a* in *Ulva japonica*.

Another important feature of photosynthetic action spectra is the celebrated 'red drop' (Emerson & Lewis 1943), the precipitous decline in quantum yield at wavelengths above 680 nm and reaching zero by 700 nm despite continued absorption of light to 720 nm or beyond. The 'red drop' combined with the discovery of the 'second Emerson effect' (or the 'enhancement effect') (Emerson *et al.* 1957) in part lead to the theory of two light reactions in photosynthesis (Hill & Bendall 1960; Duysens *et al.* 1961), and the subsequent identification of two chlorophyll *a* 'traps' (reaction centers), P_{680} in PS II and P_{700} in PS I (Fig. 13.1).

Emerson *et al.* (1957) discovered that the quantum yield of light in the far-red region was enhanced by a weak background illumination with shorter wavelength light. The enhancement is measured by selecting two wavelengths, λ_1 and λ_2, and obtaining the rate of photosynthesis (usually O_2 evolution) with each beam singly ($P\lambda_1$, $P\lambda_2$) and with both together $P(\lambda_1+\lambda_2)$. The degree of enhancement is given by the relationship:

$$\text{enhancement} = \frac{P(\lambda_1+\lambda_2)}{P\lambda_1+P\lambda_2}.$$

No enhancement is indicated by a value of 1 and enhancement is indicated by values exceeding 1, usually in the range of 2–3.

Action spectra for O_2 production display the degree of enhancement of a fixed-wavelength background illumination by a variable wavelength beam. The

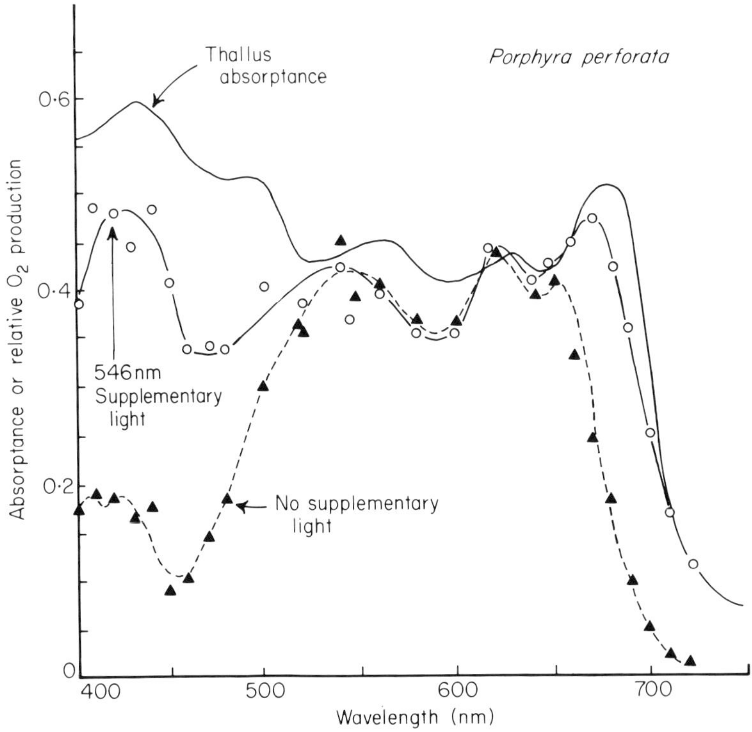

Fig. 13.3 Absorptance and action spectra (with and without 546 nm supplementary background light) for the red seaweed *Porphyra perforata*. (Redrawn from Fork 1963.)

resulting enhancement spectra for the red seaweeds *Cryptopleura crispa* and *Porphyra perforata* in background light of 546 nm (strongly absorbed by phycobilins) no longer show the 'inactive' chlorophyll so typical in action spectra of red algae (Fork 1963) (Fig. 13.3). Green background light sustains enhanced photosynthesis in both the red and blue regions of chlorophyll *a* absorption. These enhancement spectra for red seaweeds show 'that the basis for the enhancement effect is not the pairing of low energy quanta absorbed by chlorophyll *a* with higher energy quanta absorbed by accessory pigments, but rather the requirement for dual excitation of accessory pigments and chlorophyll *a* for efficient photosynthesis. Red and blue absorption by these seaweeds is similar in that chlorophyll *a* excitation is effected without appreciable phycobilin excitation. The 'inactive' chlorophyll of red algae could better be termed "unenhanced" chlorophyll' (Fork 1963). Enhancement spectra for the green seaweed *Ulva* sp. with background illumination of wavelengths longer than 680 nm show peaks of effectiveness around 485 and 645 nm corresponding most closely to absorption by chlorophyll *b* (Fork 1963). Similar measurements with the brown seaweed *Endarachne binghamiae* showed peaks at 470 (chlorophyll *c*) and 540 nm (fucoxanthin). Enhancement is an indication that quanta of wavelength less than 680 nm are efficiently distributed between the two photosystems by a variable spillover mechanism (Gregory 1977).

As pointed out earlier, the spectral distribution of light changes with water depth, yet spectral distribution is always broadband at or above the photosynthetic compensation depth for a species. Thus seaweeds are *always* present in polychromatic light fields, albeit of somewhat limited spectral distribution. An important question arises, is there photosynthetic 'enhancement' *sensu* Emerson in submarine light fields, or are the effects of different wavelengths on photosynthesis simply independent and additive? The question has been systematically addressed by Dring & Lüning (1977) for the red seaweeds *Delesseria* and *Porphyra*. Eight artificial light fields were created using

Table 13.2 Ratio of measured to predicted (from action spectrum and spectral distribution data) photosynthesis (μl O_2/μE). Each ratio is a mean of seven determinations. (From Dring & Lüning unpublished.)

Light Source	*Delesseria*	*Porphyra*
Imitated underwater light fields		
type 5	0.96	0.95
type 7	0.99	0.98
type 9	1.04	0.96
Fluorescent lamps		
'warm-white'	1.05	0.97
'daylight'	1.13	1.01
'gro-lux'	1.08	1.14
Quartz-iodide	1.15	1.12
Tungsten	0.97	0.98

varied emission sources and filters, including the simulation of coastal water types 5, 7 and 9 (Jerlov 1968), and their spectral distributions were measured. Relative action spectra using monochromatic light were produced for both species. Photosynthesis was predicted from the combined spectral distribution and action spectra data, and measured in the broadband light fields. Computed ratios of measured to predicted photosynthesis (Table 13.2) were never significantly greater than unity, therefore 'enhancement', at least for the red seaweeds tested, does not contribute to photosynthesis in natural light fields. This being so, it is feasible to calculate photosynthesis from spectral-distribution and action spectrum data alone and this could simplify productivity measurements *in situ* (Dring & Lüning unpublished).

4.3 Diurnal patterns

To the first approximation, the diurnal course of light intensity is described by the Lambert cosine law:

$$I_0 = L \cos \theta,$$

where the illumination (I_0) of an elementary (flat) surface due to a point source (here, the sun) is proportional to the luminous intensity (L) of the source in the direction of that surface, and to the cosine of the angle (θ) between this direction and the normal to the surface. Barring atmospheric disturbances, the diurnal course of incident PFD will approximate a parabola, beginning at sunrise and ending at sunset (Fig. 13.4). Should the incident PFD (I_0) not exceed the I_k value for a seaweed, then photosynthesis will be proportional to I_0 (Fig. 13.4). If however I_0 exceeds I_k, then photosynthesis becomes independent of I_0 in a somewhat predictable manner. In fact, there is an asymmetry about the midday axis: 1, a morning maximum in photosynthesis; 2, an afternoon depression; and 3, a late afternoon recovery (Fig. 13.4). At moderately high PFD, as much as 70% of the daily photosynthesis is realized in the first half of the day (Ramus & Rosenberg 1980). Departure of the afternoon depression from maximum photosynthesis varies as a function of the difference between I_0 and I_k and the diurnal light history.

Diurnal photosynthetic performance has been measured for several intertidal species (Ramus & Rosenberg 1980); for all, I_k was at or below 0.1 I_0, where I_0 was maximum PFD. During exposure to natural bright light fields (0.70 I_0), *Codium decorticatum* and *Dictyota dichotoma* which were acclimated to bright light, realized twice the net diurnal photosynthesis of the same species acclimated to dim light. The plants from dim light habitats carried more pigment than their bright light counterparts, had lower I_k values, and as a consequence saturated sooner in the course of the day and suffered a more severe afternoon depression.

A similar pattern was seen when comparing species of differing quantum-capture capacities, attributable to their respective thallus architectures. Thallus

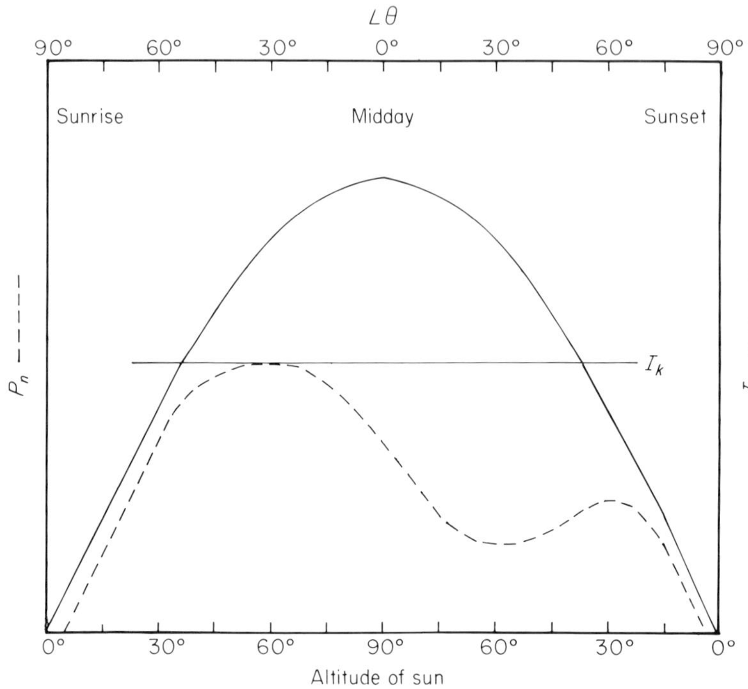

Fig. 13.4 Generalized diurnal photosynthesis for a seaweed. I_0 = incident photon flux density or PFD, I_k = saturating PFD, P_n = net photosynthesis, ∠θ = angle between direction of sun and the normal to the surface.

absorptance for the green seaweed *Ulva curvata* seldom exceeds 0.5 and is dependent on pigment concentration; thallus absorptance of *Codium decorticatum* is always near 1.0 and is pigment-concentration independent (Ramus 1978). Since *Ulva* is optically transparent and captures only a fraction of incident light, its I_k is much higher than that for *Codium* which is optically opaque. *Ulva* experiences only minor afternoon depressions while *Codium* invariably suffers a major afternoon depression.

Age of the seaweed thallus also results in differences in the pattern of diurnal photosynthesis. *Dictyota dichotoma* juvenile and adult thalli from the same light habitat were simultaneously compared at dim (13% ambient) and bright (70% ambient) natural light fields. For the juveniles, the diurnal P_n of individuals in dim light was 1.48 times greater than those in bright light, i.e. those plants which received less light photosynthesized more. Adult thalli suffered little afternoon depression in bright light. The red seaweed *Gracilaria foliifera* was the most light sensitive species tested: diurnal P_n for individuals exposed to 3% ambient light was actually 2.06 times greater than plants exposed to 70% ambient light (Ramus & Rosenberg 1980).

The quantity of light which a seaweed experiences early in the diurnal cycle affects the magnitude of photosynthesis later in the cycle (Ramus & Rosenberg 1980). Plants experiencing high PFD early in the day suffer an early depression in photosynthesis and with continued high PFD recover only late in the day, and never to P_{max}. However, with decreased afternoon PFD, recovery to P_{max} values is almost complete. In the first example, most of photosynthesis occurs early in the day, and in the second example high rates continue throughout the day.

Initial slope, I_k, and P_{max} values taken from light saturation curves produced over short periods of time are often used in models predicting net daily photosynthesis (Platt *et al.* 1977). The assumption is that initial slope, I_k, and P_{max} are fixed properties for like individuals. As noted previously, the course of diurnal photosynthesis is affected by diurnal light history. PFD is more variable in the course of a day than on an annual basis, a fact with important implications for the prediction of net diurnal photosynthesis. There is an analogy between the way in which light saturation curves are determined in the laboratory (instantaneous determinations) and the diurnal course of changes in light and photosynthesis: diurnal photosynthetic performance and I_0 data can be replotted as light saturation curves. Usually I_0 exceeds I_k at some time during the day, thus the diurnal P_n data yield two distinct light saturation curves, one for sunrise to midday and one for midday to sunset (Ramus & Rosenberg 1980): the two curves seldom coincide. The position of the morning curve relative to the afternoon curve reflects the asymmetry of diurnal P_n about the midday axis. Light saturation curves for *Ulva lactuca* have been shown to change on a circadian basis (Britz & Briggs 1976; Mishkind *et al.* 1979) as well, viz. initial slope, I_k and P_{max} values are greatest at midday and least at midnight. Thus from the evidence, these parameters seem variable and unpredictable and their use for productivity estimates is not justified. Jassby (1978) has crystallized the current malaise: 'It is impossible to derive net diel photosynthesis from an [instantaneous] light saturation curve .... There is no substitute for direct measurement [of diurnal P_n]'.

Causes of the afternoon depression remain unclear, and are generally relegated to the catch-all of photoinhibition. Photoinhibition in bright sunlight, as previously noted, is not universal even among closely related species. Exposure of *Laminaria hyperborea* from shallow water to direct sunlight for 5 h resulted in a progressive reduction of photosynthetic rate, down to 50% of the control rate, compared with similar material kept in darkness under similar conditions. However, *Laminaria digitata* was insensitive to direct sunlight (Drew 1974). Clearly, there is some physiological mechanism which regulates photosynthesis at intensities exceeding I_k. Photodestruction of the PSU seems unlikely, since seaweeds recover expected photosynthetic rates almost immediately upon I_0 falling below I_k (Ramus, Nicholson & Amsler unpublished). It is likely that the afternoon depression is due to a number of factors acting in concert, including perhaps photorespiration (see Chapter 14), diel rhythms in photosynthetic capacity (see Section 4.4) or an increase in dark respiration (see Chapter 14). That the afternoon depression can significantly reduce diurnal P_n is revealed, but why it occurs (from the standpoint of optimizing diurnal P_n) is not known.

4.4 Diel rhythms

Diel (circadian) rhythms of photosynthesis have been documented for many species of seaweed, including *Acetabularia mediterranea* (Vanden Driessche 1966), *Ulva lactuca* (Britz & Briggs 1976; Mishkind *et al.* 1979), *Bryopsis maxima* (Okada *et al.* 1978) and *Spatoglossum pacificum* (Kageyama *et al.* 1979). Attempts to demonstrate similar rhythms in 15 other species representing all seaweed classes were not successful (Blinks & Givan 1961; Mishkind *et al.* 1979). To document rhythms, short term photosynthesis is usually measured in alternating photoperiods (light) and scotoperiods (dark) of combined period length of 24 h (circadian). Photosynthetic capacity (P at I_k) is maximum in mid-photoperiod and minimum in mid-scotoperiod. After several entrainment periods, the seaweeds are maintained in continuous darkness: the rhythms persist but the amplitude decreases. This is taken as evidence that the rhythm is maintained by an endogenous oscillator, i.e. the rhythms in photosynthetic capacity are free-running and programmed by a 'biological clock'. The diel change in capacity, of course, involves a substantial change in the light saturation curve and consequently initial slope, I_k, and P_{max} values (Britz & Briggs 1976; Mishkind *et al.* 1979).

In *Acetabularia* (Vanden Driessche 1966) the rhythm in photosynthetic capacity is accompanied by a rhythm in chloroplast shape; in the photoperiod the ratio of long:short axis is near 5.0 and near 2.5 in the scotoperiod. In *Ulva*, the diel rhythm in photosynthetic capacity is accompanied by a rhythm in chloroplast orientation (Britz & Briggs 1976) which results in substantial changes in thallus absorptance. During the day, chloroplasts cover the outer faces of the cells and the single wavelength absorptance is high (A_{678}–$A_{760} = 0.9$) and during the night chloroplasts are along the side walls of the cells and absorptance is low (A_{678}–$A_{760} = 0.5$). However, neither the diel rhythm in chloroplast shape nor orientation accounts for the change in photosynthetic capacity for the respective species.

The large amplitude of the rhythm in photosynthetic capacity seen in *Ulva* by Mishkind *et al.* (1979), a six-fold range, suggested that assay of various key photosynthetic parameters would locate the step of the photosynthetic scheme which undergoes oscillation and is thus responsible for the diel rhythm. Using a highly sensitive O_2 electrode they determined relative quantum yield, PSU size and PSU turnover time at cycle maxima and minima. No change in the relative quantum yield of photosynthesis was detected, similar to results reported for cell-free extracts (chloroplasts) of the green seaweed *Bryopsis maxima* by Okada *et al.* (1978). Further, PSU size remains constant during the cycle. No cyclic variation in the 'fast' turnover time of photosynthesis ($t_{1/2} = 0.5$ ms; Mishkind & Mauzerall 1980) was found, i.e. the rate of reduction of the plastoquinone pool (Fig. 13.1, PQ) which serves as the electron transfer link between PS II and PS I by photoelectrons generated by PS II. Plastoquinone is subsequently reoxidized by electron transfer to a component of the chain leading to PS I. Some part of this sequence has a 'slow' turnover time of $\sim$ 20 ms. Thus the reoxidation of plastoquinone becomes the rate-limiting step of the overall electron transfer chain. The site of the diel variation in the O_2 exchange rate is therefore after electron transfer from PS II to plastoquinone. Since the oxidation of the plastoquinone pool by the donor for PS I is probably the rate-limiting step of steady-state photosynthesis at light saturation, this step is a likely candidate for the site of the controlling element in the diel rhythm (Mishkind *et al.* 1979).

5 THE ORGANIZATION OF PHOTON CAPTURE

5.1 The seaweed photosynthetic unit

The concept of a functional unit of photosynthesis was first put forth by Emerson & Arnold (1932), and a large body of research has led to a number of detailed models of the PSU (Govindjee & Braun 1974) which are not yet resolved. In its most general form, the PSU is composed of a large number of antenna pigments which capture light energy and funnel the energy to traps. The traps contain no more than a few photochemically active chlorophyll *a* molecules, and the vast majority of chlorophyll *a* is found in the antenna where it serves as a light harvesting pigment. The presence of other pigments widens the wavelength-capturing domain to include the entire solar spectrum useful to photosynthesis. The inclusion of different accessory pigments in the antenna is very dramatic in the seaweeds—as previously noted, the three classes correspond to different antenna composition. Despite the wide variation in light-gathering pigment, the traps and electron transport chain are believed to have remained constant during evolution. Using the delayed double flash technique, Mishkind & Mauzerall (1980) determined for diverse seaweeds the 'fast' turnover time of photosynthetic O_2 production. Here turnover time ($t_{1/2}$) was defined as the delay time between double flashes to produce a half doubling of O_2 production, or basically the time required for the 'traps' to accept another excitation following a first excitation. For the seaweeds tested, the time was approximately the same, 0.43 ± 0.06 ms (Table 13.3) and consistent with that found for *Phormidium, Chlorella* and spinach chloroplasts. Significantly, the turnover time was constant over a four-fold change in pigment concentration in *Ulva lactuca* and a three-fold change in *Porphyra umbilicalis*. This 'fast' turnover time is thought to involve the reduction of the intersystem (between PS II and PS I) plastoquinone pool (see Fig. 13.1). Mishkind & Mauzerall (1980) concluded that components

Table 13.3 PSU turnover times and sizes for various seaweeds. (Data from Mishkind & Mauzerall 1980.)

Seaweed	$t_{1/2}$ms	chl/O_2
Chlorophyceae		
Codium fragile	0.34	2570
Ulva lactuca	0.43	1980
Phaeophyceae		
Fucus vesiculosus	0.40	719
Rhodophyceae		
Champia parvula	0.44	3595
Chondrus crispus	0.46	684
Porphyra umbilicalis	0.52	390–719

and kinetics of the electron transport chain, or at least the intersystem pool, is the same throughout the algae and higher plants. Further, they conclude the evolution of the light-gathering apparatus responsible for the colour of seaweeds has occurred *without* variation in the mechanism of photosynthesis.

5.2 The PSU and complementary chromatic adaptation

The coincidence of wavelengths of maximum transmission through the marine water column (green light) and maximum absorption by deep-water red seaweeds (green light) comprised the basis for a compelling logic crystallized in the theory of *complementary chromatic adaptation*, first suggested by Engelmann (1883, 1884). Producing crude action spectra for photosynthesis by an ingenious quantitative bioassay for O_2 using aerotactic bacteria, Engelmann was impressed by the enhanced 'sensitization' of photosynthesis in the spectral region of phycoerythrin absorption. The factor presumably limiting the lower extreme of seaweed vertical distribution was the ability to fill the so-called 'green window', i.e. the effective absorption of transmitted wavelengths between the two major absorption peaks of chlorophyll. The green window was best filled by red seaweeds carrying large quantities of phycoerythrin, sufficient to mask the presence of other antenna pigments; thus red seaweeds complement the spectral distribution of deep water light fields. Engelmann's theory was genotypic *sensu strictu*, making no allowance for light-induced plasticity in the PSU, i.e. phenotypic adjustments. In addition, the theory was *a priori*: the PSU of red seaweeds had evolved for effective light absorption in the deep sea.

A full exposition of the theory implies that the green seaweeds (carrying only chlorophylls) occupy the most restricted vertical distribution (a narrow band near the surface), the browns (carrying fucoxanthin) an intermediate distribution and the reds (carrying phycobilins) the greatest vertical distribution. The logic is so manifest that the theory has persisted in the face of considerable evidence to the contrary. Further, the theory of complementary chromatic adaptation has had three unfortunate consequences: 1, a false impression of seaweed vertical distribution; 2, the idea that seaweed vertical distribution is imposed solely by the 'colour' of the light field; and 3, the implicit notion that there exist no other mechanisms for filling the green window.

Patterns of vertical distribution in contemporary ecological idiom refers to the organization or structure of communities. There are at least two approaches to the analysis of community structure: functional and descriptive (Dayton 1975). The goal of the functional approach is to define the mechanisms, biological and physical, which play key roles in structuring the community. The theory of complementary chromatic adaptation assumes spectral distribution to be the primary mechanism determining at least the vertical structure of seaweed communities. The problem then is what constitutes corroborative evidence establishing causation. Often used are bathymetric studies, a descriptive approach to community analysis based on sampling designed to measure the: 1, relative abundances; 2, biomass; or 3, productivity of the species in a community. Even making the naive assumption that spectral distribution is the sole functional mechanism (evidence for other functional mechanisms can be found in Dayton 1975; Lubchenco 1978; Lubchenco & Menge 1978), the unresolved question remains, Which of the three descriptive parameters is most appropriate?

Larkum *et al.* (1967), investigating a cliff in clear oceanic water of the Mediterranean, found that brown seaweeds dominated the upper 15 m, greens dominated from 15 to 75 m (the bottom) and the reds made little contribution to biomass at any depth. To extrapolate the occurrence of seaweeds *below* 75 m, the zonation in a submarine cave was quantified. Here, siphonous green seaweeds penetrated some distance inside on the near-vertical walls and were eventually replaced by encrusting red species. The transition from green to red species occurred at PFD equivalent to a depth at 80 m outside the cave, with reds becoming sparse at the equivalent of 120 m. In this instance, red seaweeds penetrated 'deeper' than other types. However, Larkum *et al.* calculated that the proportion of light available for photosynthesis by red and green seaweeds at various depths in these oceanic type II waters is virtually the same down to 150 m, well below the compensation depth for all species. They also noted that in the submarine cave and on a north-facing cliff, light is already confined to the blue wavelengths well *before* the transition from green to red seaweeds occurs. Thus, Larkum *et al.* (1967) concluded that light quantity rather than spectral distribution was the major regulating factor for the pattern of vertical distribution. Goreau (1963) has speculated that the success of encrusting reds at low light intensities is due to a slow rate of attrition rather than to exceptionally effective utilization of available light. Their calcareous nature and prostrate habit is a deterrent to both grazing and physical erosion.

Doty *et al.* (1974) investigated vertical distribution of Hawaiian seaweeds from seaward of the algal ridge. The ratio of brown to green species below 10 m and 50 m was 0.5 and below 90 m was 0.0. The ratio of red to green species below 10 m was 3.0, below 50 m was 2.2 and below 90 m was 0.7. Thus there was a progressive enrichment for green species with depth. Titlyanov (1976), working in the turbid waters of the Gulf of St. Peter the Great, reported the ratio of brown to green species at depths less than 2 m to be 2.1 and at depths exceeding 12 m to be 1.7, but the ratio of red to green at depths less than 2 m to be 3.4 and greater than 12 m to be 3.7. He concluded by stating that the species composition at the greatest depths of habitation is not correlated with taxonomic groups of specific pigmentation, i.e. classes. Schneider (1976), studying the spatial distributions of seaweeds on the continental shelf of the Carolinas, reported that 25% of the green species, 16% of the brown and 28% of the red occurred at depths greater than 50 m.

Early on, Oltmanns (1891, 1905) proposed that the vertical distribution of seaweeds was controlled by light quantity alone. Titlyanov (1976) investigated a partially intertidal grotto and noted that the grotto represents a light gradient alone while the adjoining submarine cliff represents both a light gradient and a change in spectral distribution: the situation appeared an ideal test. The species common to neighbouring well-lit areas are absent inside the grotto. The grotto supports shade-tolerant species which grow nearby in deep crevasses or on underwater cliffs. A critical observation was that the red seaweeds *Ptilota filicina* and *Symphyocladia latiuscula* living in the grotto and in cliff crevasses near the surface appear again in the open at a depth of 10–15 m. Thus, Titlyanov (1976) concluded that their occurrence is a function of light quantity alone and the change in spectral distribution is a minor contributor to the vertical stratification of seaweeds.

Another approach to testing Engelmann's theory is an experiment designed to eliminate all structuring mechanisms save spectral distribution. For example, Levring (1966, 1968) noted that photosynthesis and irradiance decrease in a logarithmic manner with depth. He reasoned that if photosynthesis at two different depths, m and n, are P_m and P_n and the corresponding irradiance values are E_m and E_n, then the quotient of the slopes of the two curves, q_{m-n}, can be calculated by the relationship:

$$q_{m-n} = \frac{\log (P_m/P_n)}{\log(E_m/E_n)}.$$

If photosynthesis were *independent* of spectral composition, then P and E would reduce proportionally with depth and $q = 1$. If photosynthesis were *dependent* on spectral composition, then P and E would not reduce proportionally with depth and q would be greater or less than unity, $q > 1$ representing a spectral impoverishment and $q < 1$ representing a spectral enrichment. Photosynthesis (P) and irradiance (E) were measured at depths m and n where E was less than saturating. For green seaweeds in shallow water he found $q > 1$ (1.2–1.3, spectral impoverishment) and in deep water $q = 1$ (spectral indifference); for brown seaweeds at any depth $q = 1$ (spectral indifference); for red seaweeds in shallow water $q < 1$ (0·9, spectral enrichment) and in deep water $q = 1$ (spectral indifference). Despite Levring's (1966) interpretations to the contrary, these data do not support the theory of complementary chromatic adaptation. Levring (1968) compared q values in turbid coastal and clear oceanic waters over the Carolina continental shelf; q values reported here are for the deepest interval measured. For *Ulva lactuca* in neritic waters $q = 1.35$ (spectral impoverishment) and in oceanic waters $q = 0.79$ (spectral enrichment). For *Gracilaria foliifera* in neritic waters $q = 0.72$ (spectral enrichment) and in oceanic waters $q = 1.00$ (spectral indifference). For *Fucus vesiculosus* $q = 1.06$ (spectral impoverishment) in both neritic and oceanic waters; $q = 0.90$–0.93 (spectral enrichment) for both *Fucus vesiculosus* and *Sargassum natans* in exceedingly clear oceanic water. The data again do not conform to the expectations of the theory. Dring (1977), using a computer model which calculated photosynthesis from action spectra and spectral distribution data (Dring & Lüning 1977), simulated photosynthesis for many species of seaweeds. All 12 species of red algae were more efficient at all depths down to 10 m in all coastal water types than were 11 species of green and brown seaweeds. However, the species' rank order in photosynthetic ability did not change with depth. Dring (1977) concludes that complementary chromatic adaptation may be of little importance in determining vertical zonation of seaweeds. Drew (1969) calculated an increase in photosynthetic efficiency for *Udotea petiolata* (green), *Peysonnelia squamaria* (red) and *Padina pavonica* (brown) with depth, in that order, in clear oceanic waters of Malta.

If in fact photosynthesis and patterns of vertical distribution are dependent primarily on light quantity and independent of light quality, then green and brown seaweeds must also have effective mechanisms for the absorption of light in the 'green window' (500–600 nm). Brown seaweeds carry both fucoxanthin, the *in vivo* absorption spectrum of which extends out to 580 nm, and chlorophyll *c*, and effectively absorb the entire PAR spectral domain. Yokohama & Kageyama (1977), comparing the *in vivo* absorption spectra of five shallow-water and six deep-water green seaweed species, discovered an unexpected absorption peak at 540 nm in deep-water species which was not present in shallow-water species. The pigment absorbing light in this region was identified chromatographically as siphonaxanthin. Siphonaxanthin is characteristic of many siphonous greens (Kleinig 1969), so the occurrence of the pigment in species of *Cladophora, Codium* and *Valonia* came as no surprise. The fact that *Ulva japonica* carried the pigment is remarkable. Kageyama *et al.* (1977) showed an efficient excitation energy transfer from siphonaxanthin to chlorophyll *a* by fluorescence emission studies. In *U. japonica*, the efficiency of transfer at 540 nm is 50% greater than for the shallow water species *U. pertusa* containing no siphonaxanthin. Photosynthetic efficiency for *U. japonica* collected from a depth of 20 m was equal in 'green light' (simulated deep coastal water) to 'white light', whereas *U. pertusa* from shallow water was 30% less efficient in 'green light' than in 'white light' (Kageyama & Yokohama 1977). There seems little doubt that *U. japonica* has evolved an efficient mechanism for the capture of green light, allowing the species to exist in spectrally limited light fields.

The siphonous green *Ostreobium* growing endolithically in the coral head *Favia* and beneath the outer layer of zooxanthellae is subject to a unique spectral distribution. PAR wavelengths relative to far-red are transmitted through zooxanthellae in very low ratios (Halldal 1978). Action spectra for *Ostreobium* reveal that the 'red drop' does not occur near 700 nm as expected, but rather beyond 740 nm. It would appear that *Ostreobium* possesses chlorophyll, both antenna and trap, with absorption maxima in the far-red.

It has also been demonstrated that the green window can be filled by chlorophyll absorption alone. The argument is a complex one and has been put forth in detail elsewhere (Ramus 1978), but will be recapitulated briefly. Given the familiar Beer–Lambert law:

$$A = \varepsilon c l,$$

it can be seen that if ε (the decadic molar extinction coefficient for the absorbing species) and l (the length of the optical light path) are constant for monochromatic light passing through a homogeneous medium, then A (the absorbance) will increase linearly with c (the concentration of the absorbing species). Absorptance, that fraction of incident light absorbed and therefore of primary interest in photochemical events, is given by the relation:

$$\alpha = 1-10^{-(\varepsilon c l)} = 1-10^{-A}.$$

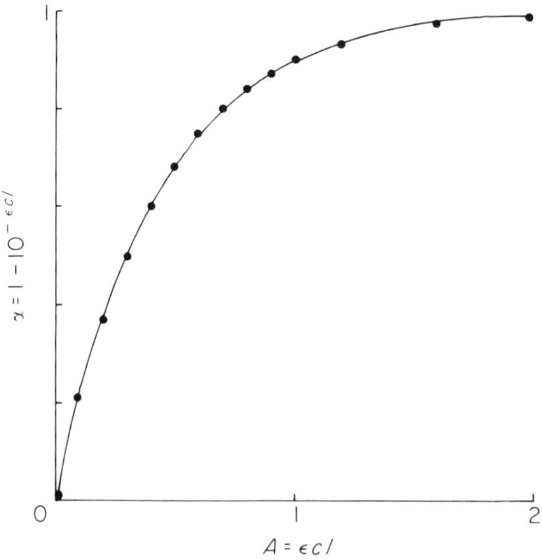

Fig. 13.5 Relationship between absorptance (α) and absorbance (A) for monochromatic light in a homogeneous medium, where the decadic molar extinction coefficient (ε) of the absorbing species and the length of the optical path (l) are constant and only the concentration of the absorbing species (c) varies. (Redrawn from Ramus 1978.)

As *A* approaches a value of 2, α asymptotically approaches a value of 1 (Fig. 13.5). Consequently, as the concentration of chlorophyll increases, the spectral region of low specific absorptance (the 'green window') increases more rapidly than the regions of high specific absorptance (the blue and red absorbing regions) (Fig. 13.6). Thus, *Ulva lactuca* with high concentrations of chlorophyll (α = 0.8) absorbs light almost uniformly across PAR wavelengths (Fig. 13.6). Most seaweeds, however, are very heterogeneous systems with respect to the absorption of light. The absorptance of the siphonous green seaweed *Codium*

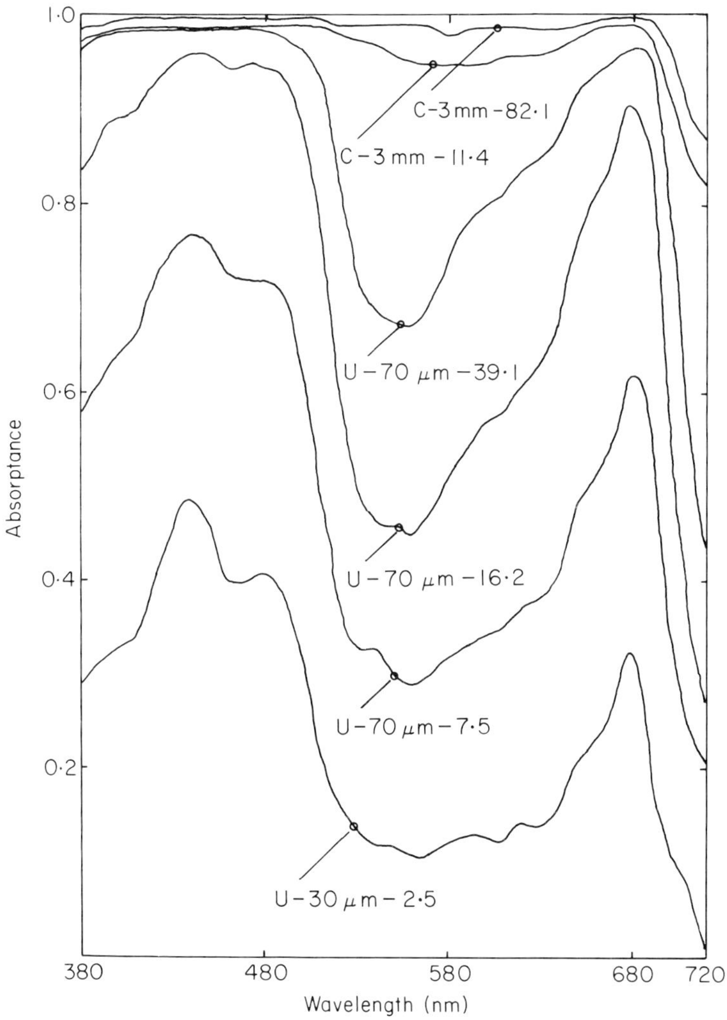

Fig. 13.6 Absorptance spectra for four specimens of *Ulva lactuca* (*U*) and two specimens of *Codium fragile* (*C*). Designated is the thickness of the thallus either in μm or mm, and the total concentration of antenna pigments in nmoles cm^{-2}. (Redrawn from Ramus 1978.)

fragile is always near 1, independent of chlorophyll concentration within natural ranges, due to the effects of heterogenous absorption and multiple scatter (Ramus 1978). The utricles forming the cortical tissue enhance the capture of incident light due, among other things, to the waveguide function of the thin peripheral layer of cytoplasm in which the chloroplasts are arrayed. Thus the length of the optical path (*l*) is made infinitely large and increases the probability of photon-chloroplast encounter, independent of wavelength. Many siphonous greens have a cortex of similar structure, perhaps contributing to their presence in even the weakest of submarine light fields.

For ecological purposes, action spectra for photosynthesis produced with monochromatic light, the classical method, may be misleading. In that the spectral distribution of submarine light above the photosynthetic compensation depth is always broadband, action spectra might more appropriately be produced with supplementary background light. When this is done the results are revealing. For example, in *Porphyra perforata* the action spectrum produced with 546 nm supplementary light more closely 'fits' the absorptance spectrum than one produced in the absence of supplementary light (Fig. 13.3) (Fork 1963). It can be inferred from action spectra with background light near 680 nm of the green *Ulva* and the brown *Endarachne binghamiae* that the decline in photosynthesis in the spectral range of the 'green window' is not so severe as in monochromatic light (Fork 1963).

Given the richness of mechanisms for the absorption of light in the 'green window', it appears that green and brown seaweeds, as well as reds, have undergone complementary chromatic adaptation in the course of evolution. Thus, an appropriate rejoinder to Engelmann's theory is: 'if one is optically black, it does not matter what colour one is' (Ramus *et al.* 1976a).

5.3 Phenotypic adjustment of the PSU

It is well known that antenna pigment concentration varies inversely with photon flux density, e.g. the pigment concentration of a species in a strong light field compared to a weak light field is lower, as is usually the ratio of accessory to primary antenna pigments. These quantitative adjustments (the use of the word 'adaptation' will be assiduously avoided here) are within-genotype, a phenotypic expression of PSU plasticity.

The PSU of red seaweeds has undergone the most intense investigation, prompted by the phenotypic corollary of Englemann's axiom. In this context, his theory is extended to include phenotypic adjustments to light quality and/or quantity. Aside from chlorophyll *a* (chl *a*), the red seaweed antenna contains phycoerythrin (PE), phycocyanin (PC) and allophycocyanin (APC). Here, the specifics for the model of the PSU are controversial. Initially it was thought that the role of APC was solely to transfer excitation energy from phycobilisomes to the traps, but Larkum & Weyrauch (1977) have shown that APC harvests a substantial amount of light. Further, Larkum & Weyrauch (1977) maintain that the phycobiliproteins are in a sense independent of both photosystems and transfer absorbed light energy preferentially to PS II with excess energy spilling over to PS I. On the other hand, Ley & Butler (1977) suggest that the phycobiliproteins are associated entirely with PS II.

Waaland *et al.* (1974) demonstrated that adjustment of the PSU in *Griffithsia pacifica* is effected by PFD and not by spectral distribution. PE:chl *a* ratios (mg/mg) varied from 0.67 at 1000 f.c. to 1.8 at 8 f.c., the greatest variation occurring below 300 f.c. The correlation between phycobiliprotein concentration and phycobilisome density was tight, indicating that perhaps the PSU number varied here. Ramus *et al.* (1976b) reported for *Chondrus crispus* and *Porphyra umbilicalis in situ* an increase in total antenna pigment concentration and a PE:chl *a* and PC:chl *a* enrichment with decreasing PFD. Plants suspended at varying depths in the water column (where both spectral distribution and PFD change) were compared with plants from intertidal light habitats where only the PFD changed. Antenna pigment concentration varied in a similar manner, leading to the conclusion that these red seaweeds were responding to PFD alone. The adjustment of the PSU was reversible and did not require cell division. Rhee & Briggs (1977) reported that the chlorophyll variation in *Chondrus crispus* was not correlated with depth, but PE did increase with depth. The evidence indicated that the phycobiliprotein:chl *a* variations are correlated with PFD, yet it is not known whether different spectral distributions of equal quantum flux density will produce the same adjustments.

For green seaweeds, an increase in total chlorophyll and chl *b*:chl *a* ratio is correlated with a decrease in PFD (Ramus *et al.* 1976b; Keast & Grant 1976; Yokohama & Kageyama 1977; Ramus 1978). Chl *b*:chl *a* ratios vary from 0.44 in high PFD environments to 1.12 in low PFD environments, while total chlorophyll can vary up to 15-fold in *Ulva* and 8-fold in *Codium* (Ramus 1978). Apel *et al.* (1975) released PS II from the PSU of *Acetabularia mediterranea* by detergent treatment and showed it to be enriched in chl *b*. In that the chl–protein complex photoreduced dichlorophenolindophenol, but not $NADP^+$, they concluded it was PS II. Siphonaxanthin transfers absorbed energy to chl *a* in *Ulva japonica* (Kageyama *et al.* 1977), yet its photosystem association is unknown. Molar ratios of siphonaxanthin:chl *a* range from 0.25 in *Codium fragile* to 0.35 in *Valonia macrophysa*. Carotenes, of course, are present in abundance in the chloroplasts of green seaweeds (Kleinig 1969), but as evidenced from enhancement spectra (Haxo & Blinks 1950; Fork 1963) and fluorescence excitation measurements (Kageyama *et al.* 1977), they may participate less efficiently in energy transfer.

The three major antenna pigments in the brown seaweeds *Ascophyllum nodosum* and *Fucus vesiculosus* increase with decreasing PFD. However, chl *c*:chl *a* ratios remained constant, while fucoxanthin:chl *a* actually declined; the increase in fucoxanthin was not as great as in chl *a* for these seaweeds (Ramus *et al.* 1977). The growing regions of *Laminaria* and *Fucus* species contain less chl *a* than other thallus parts. The growing regions of *Laminaria* contain the highest proportion of fucoxanthin relative to chl *a*, whereas the older portions of *Fucus* are enriched for fucoxanthin (Küppers & Kremer 1978). Thus the data indicate that in seaweeds with thallus differentiation, there will also be ontogenetic differentiation in the PSU.

Variations in the relative proportion of antenna pigments are also correlated with season. For *Ascophyllum nodosum* occurring in Norway, the highest chl *a* concentration was found in spring samples and the lowest in autumn. Fucoxanthin:chl *a* ratios were lowest in spring and highest in autumn (Jensen 1966). For *Fucus serratus* and *Pelvetia canaliculata*, pigment content was highest in the spring and lowest in autumn, but the proportion of fucoxanthin:chl *a* was remarkably uniform (Jensen 1966). Total pigment content in *Chondrus crispus* was highest in winter and lowest in summer, as was the proportion of PE to chl *a* (Rhee & Briggs 1977). A similar seasonal correlation was seen for *Griffithsia monilis*, but here the proportion of phycobilins to chl *a* did not change (Larkum & Weyrauch 1977). Seasonal adjustments in the PSU may reflect ambient PFD, or equally likely ambient inorganic N-nutrient levels in the water column. Lapointe & Ryther (1979), working with tank cultures of the red seaweed *Gracilaria foliifera*, report the concentration of antenna pigments to be proportional to dissolved NO_3^- or NH_4^+. In low N-nutrient concentrations, *Gracilaria* appears to cannibalize its own N-rich PSU, resulting in a plant of low pigment content and reduced photosynthetic capacity; when returned to N-rich seawater, the PSU is restored and photosynthetic capacity recovers.

Variations in the PSU have been correlated with *in situ* photosynthesis (Ramus *et al.* 1976a, 1977; Ramus 1978). Comparing photosynthesis normalized to chl *a*, the apparent efficiency of low PFD acclimated seaweeds differs from high PFD acclimated counterparts. At the surface, the high PFD acclimated plants are more efficient. At some depth between there is a 'cross-over' in apparent efficiencies (Fig. 13.7). Such observations compel the rejoinder 'all chlorophyll is not equal'.

If the precise nature of the quantitative adjustments in the PSU of a species were known, accurate predictions of photosynthetic performance could be made for ecological purposes. For example, changes in pigment concentration can result only from a change in the numbers of 'traps' (PSU number) and their cross-sectional area (antenna size). The consequences of such changes will define the light saturation curve because: the initial slope is proportional to the number of 'traps' and their cross-sectional area; and P_{max} is proportional to the number of 'traps' and the intrinsic turnover time of the PSU (Herron & Mauzerall 1972). Assuming turnover time remains constant, a series of model light-saturation curves can be constructed, in the first case normalizing photosynthesis to biomass, and in the second case normalizing photosynthesis to chl *a*.

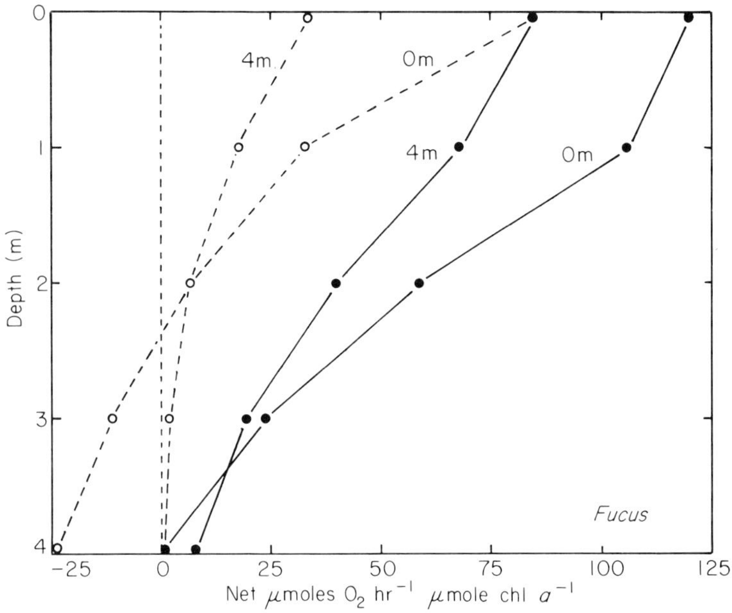

Fig. 13.7 Depth profiles of photosynthesis by *Fucus vesiculosus* for overcast (broken lines) or clear (solid lines) skies. Plants acclimated at 0m (surface) or at 4 m depth for one week prior to measurement. (Redrawn from Ramus *et al.* 1977.)

Case I models *cell* efficiency. Here species *a* has a fixed number of 'traps' and changes antenna size, while species *b* changes the number of 'traps' and fixes the antenna size. In a diminished light field, individuals of species *a* will have a greater initial slope but equal P_{max} values (Fig. 13.8), while species *b* will have an equal initial slope but greater P_{max} values. Although both modes of adjustment improve photosynthetic performance per unit biomass in subsaturating light fields, the photosynthetic performance of species *a* alone is enhanced as PFD approaches compensation.

Case II models chl efficiency. Here again, species *a* has a fixed number of 'traps' and changes antenna size, while species *b* changes the number of 'traps' and fixes the antenna size. In a diminished light field, individuals of species *a* will have reduced initial slope and P_{max} values, while species *b* experiences no change in either (Fig. 13.9). Thus, species *a* experiences a reduced photosynthetic performance per unit chlorophyll *a* in subsaturating light fields, whereas species *b* shows no change.

Whether increasing the number of 'traps' or antenna size, the net effect is to increase quantum capture to, teleologically, compensate for diminishing quantum flux density. This notion has difficulties when placed in biological context. A cost/benefit analysis has never been attempted for adjustments in the PSU, but it is relatively safe to assume that seaweeds *optimize*, rather than

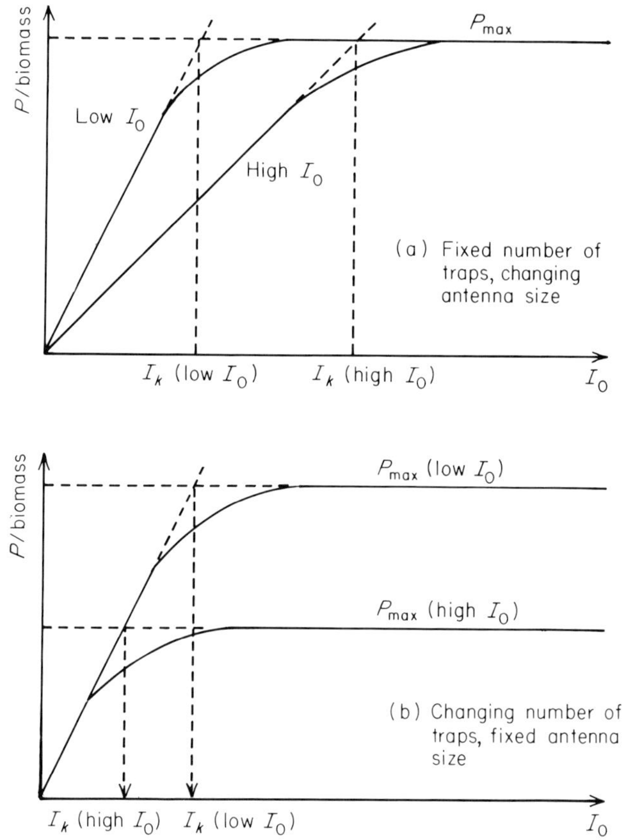

Fig. 13.8 Model light saturation curves normalizing photosynthesis (P) to biomass and comparing two 'strategies' (a and b) for the adjustment of the PSU to extreme (low and high I_0) light environments. I_0 = incident photon flux density or PFD, I_k = saturating PFD, P_{max} = maximum photosynthesis.

maximize, photosynthesis—an assumption based on the fact that a decline in chlorophyll concentration occurs with increasing ambient PFD. There would seem then an optimum in quantum capture per unit biomass and a regulator which modulates attainment of the critical quantum capture. The energetic expense of an increase in PSU size or number must be great—the photochemical and biochemical components of the PSU account for the bulk of protein in a seaweed cell. Thus, the light absorption requirement for optimum photosynthesis must be variable, in a biological sense.

Mishkind & Mauzerall (1980) have specifically addressed the notion of variations in antenna size. Their definition of PSU size is a functional one and consistent with that of Emerson & Arnold (1932): the molar ratio of chlorophyll to O_2 formed in a single saturating turnover flash (in this case the flashes were pulsed at 100 ms intervals). The average PSU size for the green seaweeds is similar to that of *Chlorella* and higher plants, viz. 1500–2500 chl:O_2 and about

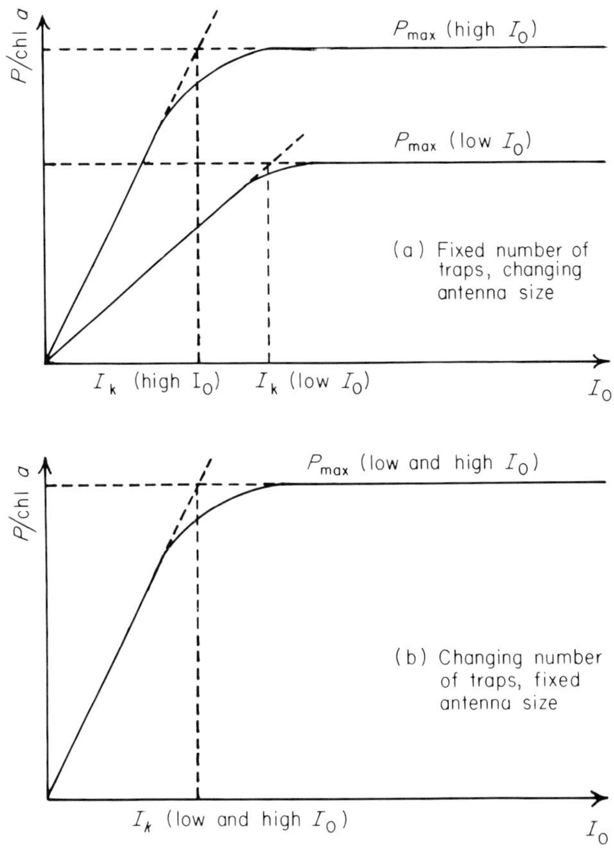

Fig. 13.9 Model light saturation curves normalizing photosynthesis (P) to chl a and comparing two 'strategies' (a and b) for the adjustment of the PSU to extreme (low and high I_0) light environments. I_0 = incident photon flux density or PFD, I_k = saturating PFD, P_{max} = maximum photosynthesis.

one-third that size for brown and red seaweeds (Table 13.3). In the browns and reds, a large compliment of non-chlorophyll accessory pigment is compensated for by a reduction in chlorophyll content, but by no means do accessory pigments replace all the chlorophyll. The data of Mishkind & Mauzerall (1980) suggest that *Ulva lactuca* adjusts its PSU to low PFD in the same manner as higher plants, i.e. by increasing the number of PSU's rather than antenna size. For *Ulva*, PSU size is independent over a four-fold range of chlorophyll concentration. For *Porphyra umbilicalis*, PSU size increases with chlorophyll content as PFD decreases as does apparently the number of 'traps'.

5.4 Structural accommodation

There are in some cases structural modifications of the chloroplast or cell which accompany adjustments in the PSU. Unfortunately, the effect of structural modifications on quantum capture capacity or photosynthesis has not been quantified, leaving their significance to be demonstrated.

Ulva fenestrata taken from the dim recesses of a submarine grotto had larger chloroplasts with denser thylakoid packing than counterparts growing in the intertidal (Titlyanov *et al.* 1974). An increase in phycobilisome density was correlated with a decrease in PFD in *Griffithsia pacifica* (Waaland *et al.* 1974), while the size of phycobilisomes remained constant. Chloroplasts of the siphonous green seaweed *Halimeda tuna* growing at 16 m were larger than those of the same species growing near the surface; however no difference was noted in the lamellar system (Colombo *et al.* 1976). Deep water plants have thinner segments, larger utricles and thinner cell walls than shallow-water plants, and the chloroplasts are distributed over a wider area but in fewer layers (Colombo & Orsenigo 1977).

6 CONCLUSION

The mechanisms of seaweed photosynthesis, both photochemical and biochemical, are nearly understood. An accurate model of the PSU will allow prediction of the photosynthetic consequences of accommodation to environmental variables: especially pivotal are PFD, spectral distribution and dissolved inorganic N-nutrients (NO_3^- and NH_4^+). The challenge which remains is the comprehension of the biological consequences of photosynthesis—to the fitness of the individual, the species, the population. To a first approximation, net photosynthesis integrated over time is growth. Yet we now have but a glimpse of the complexity of growth strategy (Mann 1973; Littler & Littler 1980), of the sequential diversion of transduced light energy to one sink or another. Further, we have but a glimpse of mechanisms, both physical and biological, acting to structure seaweed communities (e.g. Dayton 1975; Lubchenco 1978; Lubchenco & Menge 1978). Diversity can be low or high, can shift from season to season, can differ from locality to locality. Species composition for a moment in time more often than not represents a mere fraction of the species pool. Seaweeds interact with the physical environment, with herbivores, with conspecifics, with other seaweed species. The capital of seaweed economy is fixed carbon, the energy locked in covalent bonds to be expended to some biological end—herbivore escape, growth, reproduction, survival. The means of subsistence is the capture of light energy and its transduction to chemical energy. Fitness then is intimately linked to photosynthetic mechanism. The hiatus between the two principles is immense: once filled, however, the term photosynthetic *adaptation* will assume biological meaning.

ACKNOWLEDGEMENTS

For their critical responses, I wish to acknowledge Lucy Cooper, Cliff Duke, Bruce Kenney, Mitsu Suyemoto and Anna Parma to whom these ideas were first offered in a course on seaweed photosynthesis given at the Duke University Marine Laboratory during the summer of 1979. I am indebted to Ann Elk for the English language. Finally, I am indebted to Professors David Mauzerall and James Siedow for their critical reviews of this manuscript.

7 REFERENCES

APEL K., BOGOROD L. & WOODOCK C. L. F. (1975) Chloroplast membranes of the green alga *Acetabularia mediterranea*. I. Isolation of photosystem II. *Biochim. Biophys. Acta*, **387**, 568–74.

BLINKS L. R. & GIVAN C. V. (1961) The absence of a daily photosynthetic rhythm in some littoral marine algae. *Biol. Bull.*, **121**, 230–3.

BOLD H. C. & WYNNE M. J. (1978) *Introduction to the Algae*. Prentice-Hall, New Jersey.

BRINKHUIS B. H., TEMPEL N. R. & JONES R. F. (1976) Photosynthesis and respiration of exposed salt-marsh fucoids. *Mar. Biol.*, **34**, 349–59.

BRITZ S. J. & BRIGGS W. R. (1976) Circadian rhythms of chloroplast orientation and photosynthetic capacity in *Ulva*. *Plant Physiol.*, **58**, 22–7.

BRITZ S. J., PFAU J., NULTSCH W. & BRIGGS W. R. (1976) Automatic monitoring of a circadian rhythm of change in light transmittance in *Ulva*. *Plant Physiol.*, **58**, 17–21.

CARPENTER J. H. (1965) The accuracy of the Winkler method for dissolved oxygen analysis. *Limnol. Oceanogr.*, **10**, 135–40.

CHRISTENSEN T. (1962) Algen. In *Systematisk Botanik*, *vol.* 2 (2), eds. Bocher T. W., Lange M. & Sørensen T. Munksgaard, Copenhagen.

COLOMBO P. M. & ORSENIGO M. (1977) Sea depth effects on the algal photosynthetic apparatus II. An electron microscopic study of the photosynthetic apparatus of *Halimeda tuna* (Chlorophyta, Siphonales) at −0.5 m and −6.0 m sea depths. *Phycologia*, **16**, 9–17.

COLOMBO P. M., ORSENIGO M., SOLAZZI A. & TOLOMIO C. (1976) Sea depth effects on the algal photosynthetic apparatus IV. Observations on the photosynthetic apparatus of *Halimeda tuna* (Siphonales) at sea depths between 7 and 15 m. *Mem. Biol. Mar. Ocean.*, **6**, 197–208.

DAYTON P. K. (1975) Experimental evaluation of ecological dominance in a rocky intertidal algal community. *Ecol. Monogr.*, **45**, 137–59.

DOTY M. S., GILBERT W. J. & ABBOTT I. A. (1974) Hawaiian marine algae from seaward of the algal ridge. *Phycologia*, **13**, 345–57.

DREW E. A. (1969) Photosynthesis and growth of attached marine algae down to 130 metres in the Mediterranean. *Proc. Int. Seaweed Symp.*, **6**, 151–9.

DREW E. A. (1974) Light inhibition of photosynthesis in macro-algae. *Br. Phycol. J.*, **9**, 217–18.

DRING M. J. (1977) Chromatic adaptation in marine algae: an examination of its significance using a computer model. *Br. Phycol. J.*, **12**, 118.

DRING M. J. & LÜNING K. (1977) Significance of enhancement for calculations of photosynthesis of red algae from action spectra. *J. Phycol.*, **13**, (suppl.) 18.

DUYSENS L. N. M., AMESZ J. & KAMP B. M. (1961) Two photochemical systems in photosynthesis. *Nature (Lond.)*, **190**, 510–11.

EMERSON R. & ARNOLD W. (1932) The separation of the reactions in photosynthesis by means of intermittent light. *J. Gen. Physiol.*, **15**, 391–420.

EMERSON R., CHALMERS R. & CEDERSTRAND C. (1957) Some factors influencing the longwave limit of photosynthesis. *Proc. Natn. Acad. Sci., U.S.A.*, **43**, 133–43.

EMERSON R. & LEWIS C. M. (1943) The dependence of the quantum yield of *Chlorella* photosynthesis on wavelength of light. *Am. J. Bot.*, **30**, 165–78.

ENGELMANN T. W. (1883) Farbe und Assimilation. *Bot. Ztg.*, **41**, 1–13.

ENGELMANN T. W. (1884) Untersuchungen über die quantitativen Beziehungen zwischen Absorption des Lichtes und Assimilation in Pflanzenzellen. *Bot. Ztg.*, **42**, 81–93.

FOGG G. E. (1976) Release of glycollate from tropical marine plants. *Aust. J. Plant Physiol.*, **3**, 57–61.

FORK D. C. (1963) Observations on the function of chlorophyll *a* and accessory pigments in photosynthesis. In *Photosynthetic Mechanisms in Green Plants*, pp. 352–61. Publ. No. 1145 NAS-NRC, Washington, D.C.

GAARDER T. & GRAN H. H. (1927) Investigations of the production of plankton in the Oslofjord. *Rapp. Proc. Verb. Cons. Int. Expl. Mer.*, **42**, 3–48.

GAGNE J. A., LAROCHELLE J. & CARDINAL A. (1979) A solubilization technique to prepare algal tissue for liquid scintillation counting, with reference to *Fucus vesiculosus* L. *Phycologia*, **18**, 168–70.

GOREAU T. F. (1963) Calcium carbonate deposition by coralline algae and corals in relation to their roles as reef builders. *Ann. N.Y. Acad. Sci.*, **109**, 127–67.

GOVINDJEE (1975) *Bioenergetics of Photosynthesis*. Academic Press, London & New York.

GOVINDJEE & BRAUN B. Z. (1974) Light absorption, emission and photosynthesis. In *Algal Physiology and Biochemistry*, ed. Stewart W. D. P., pp. 346–90. Blackwell Scientific Publications, Oxford.

GREGORY R. P. F. (1977) *Biochemistry of Photosynthesis*. John Wiley and Sons, Chichester.

HALLDAL P. (1968) Photosynthetic capabilities and photosynthetic action spectra of endozoic algae of the massive coral *Favia*. *Biol. Bull*, **134**, 411–24.

HARRIS G. P. (1978) Photosynthesis, productivity and growth: the physiological ecology of phytoplankton. *Arch. Hydrobiol. Bech. Ergebnisse Limnol.*, **10**, 1–171.

HARPER J. L. (1977) *Population Biology of Plants*. Academic Press, London.

HATCHER B. G. (1977) An apparatus for measuring photosynthesis and respiration of intact large marine algae and comparison of results with those from experiments with tissue segments. *Mar. Biol.*, **43**, 381–5.

HATCHER B. G., CHAPMAN A. R. O. & MANN K. H. (1977) An annual carbon budget for the kelp *Laminaria longicruris*. *Mar. Biol.*, **4**, 85–96.

HAXO F. T. & BLINKS L. R. (1950) Photosynthetic action spectra of marine algae. *J. Gen. Physiol.*, **33**, 389–422.

HERRON H. A. & MAUZERALL D. (1972) The development of photosynthesis in a greening mutant of *Chlorella* and an analysis of the light saturation curve. *Plant Physiol.*, **50**, 141–8.

HILL R. & BENDALL F. (1960) Function of two cytochrome components in chloroplasts: a working hypothesis. *Nature* (*Lond.*), **186**, 136–7.

HOUGH R. A. (1976) Light and dark respiration and release of organic carbon in marine macrophytes of the Great Barrier Reef region. *Aust. J. Plant Physiol.*, **3**, 63–8.

JASSBY A. D. (1978) Polarographic measurements of photosynthesis and respiration. In *Handbook of Phycological Methods: Physiological and Biochemical Methods*, eds. Hellebust J. A. & Craigie J. S., pp. 285–96. Cambridge University Press, Cambridge.

JENSEN A. (1966) Carotenoids of Norwegian brown seaweeds and of seaweed meals. *Rep. Norw. Inst. Seaweed Res.*, **31**, 1–138.

JERLOV N. G. (1968) *Optical Oceanography*. Elsevier Publ. Co., New York.

JOHNSON W. S., GIGON A., GULMON S. L. & MOONEY H. A. (1974) Comparative photosynthetic capacities of intertidal algae under exposed and submerged conditions. *Ecology*, **55**, 450–3.

KAGEYAMA A. & YOKOHAMA Y. (1977) Pigments and photosynthesis of deep water green algae. *Bull. Jap. Soc. Phycol.*, **25**, 168–75.

KAGEYAMA A., YOKOHAMA Y. & NISIZAWA K. (1979) Diurnal rhythm of apparent photosynthesis of a brown alga, *Spatoglossum pacificum*. *Bot. Mar.*, **22**, 199–201.

KAGEYAMA A., YOKOHAMA Y., SHIMURA S. & IKAWA T. (1977) An efficient excitation energy transfer from a carotenoid siphonaxanthin to chl *a* observed in a deep water species of chlorophycean seaweed. *Plant and Cell Physiol.*, **18**, 477–80.

KANWISHER J. W. (1966) Photosynthesis and respiration in some seaweeds. In *Some Contemporary Studies in Marine Science*, ed. Barnes H., pp. 407–20. Allen and Unwin, London.

KEAST J. F. & GRANT B. R. (1976) Chlorophyll *a*:*b* ratios in some siphonous green algae in relation to species and environment. *J. Phycol.*, **12**, 328–31.

KLEINIG H. (1969) Carotenoids of siphonous green algae: a chemotaxonomical study. *J. Phycol.*, **5**, 281–4.

KREMER B. P. (1978*a*) Determination of photosynthetic rates and ^{14}C photoassimilatory products of brown seaweeds. In *Handbook of Phycological Methods: Physiological and Biochemical Methods*, eds. Hellebust J. A. & Craigie J. S., pp. 269–83. Cambridge University Press, Cambridge.

KREMER B. P. (1978*b*) Studies on $^{14}CO_2$-assimilation in marine Rhodophyceae. *Mar. Biol.*, **48**, 47–54.

KÜPPERS U. & KREMER B. P. (1978) Longitudinal profiles of carbon dioxide fixation capacities in marine macroalgae. *Plant Physiol.*, **62**, 49–53.

LAPOINTE B. & RYTHER J. (1979) The effects of nitrogen and seawater flow rate on growth and biochemical composition of *Gracilaria foliifera* var. *angustissima* in mass outdoor culture. *Bot. Mar.*, **22**, 529–37.

LARKUM A. W. D., DREW E. A. & CROSSETT R. N. (1967) The vertical distribution of attached marine algae in Malta. *J. Ecol.*, **55**, 361–71.

LARKUM A. W. D. & WEYRAUCH S. K. (1977) Photosynthetic action spectra and light harvesting in *Griffithsia monilis* (Rhodophyta). *Photochem. Photobiol.*, **25**, 65–72.

LEVRING T. (1947) Submarine daylight and the photosynthesis of marine algae. *Göteborgs K. Vetenskaps-Vitterhets-Samhäll. Handl.*, **5**, 1–90.

LEVRING T. (1966) Submarine light and algal shore zonation. In *Light as an Ecological Factor*, eds. Bainbridge R., Evans G. C. & Rackham O., pp. 305–17. Blackwell Scientific Publications, Oxford.

LEVRING T. (1978) Photosynthesis of some marine algae in clear, tropical oceanic water. *Bot. Mar.*, **11**, 72–80.

LEY A. C. & BUTLER W. L. (1977) Energy transfer from photosystem II to photosystem I in *Porphyridium cruentum*. *Biochim. Biophys. Acta*, **462**, 290–4.

LITTLER M. M. & LITTLER D. S. (1980) The evolution of thallus form and survival strategies in benthic marine macroalgae: field and laboratory tests of a functional-form model. *Am. Nat.*, **116**, 25–44.

LOBBAN C. S. (1974) A simple, rapid method of solubilizing algal tissue for scintillation counting. *Limnol. Oceanogr.*, **19**, 356–9.

LUBCHENCO J. (1978) Plant species diversity in a marine intertidal community: importance of herbivore food preference and algal competition abilities. *Am. Nat.*, **112**, 23–39.

LUBCHENCO J. & MENGE B. A. (1978) Community development and persistence in a low rocky intertidal zone. *Ecol. Monogr.*, **48**, 67–94.

LÜNING K. (1971) Seasonal growth of *Laminaria hyperborea* under recorded underwater light conditions near Helgoland. In *Proc. IVth Eur. Mar. Biol. Symp.*, ed. Crisp D., pp. 347–61. Cambridge University Press, Cambridge.

MANN K. H. (1973) Seaweeds: their productivity and strategy for growth. *Science*, **182**, 975–81.

MATHIESON A. C. & NORALL T. L. (1975) Physiological studies of subtidal red algae. *J. Exp. Mar. Biol. Ecol.*, **20**, 237–47.

MISHKIND M. & MAUZERALL D. (1980) Kinetic evidence for a common photosynthetic step in diverse seaweeds. *Mar. Biol.*, **56**, 261–5.

MISHKIND M., MAUZERALL D. & BEALE S. I. (1979) Diurnal variation *in situ* of photosynthetic capacity in *Ulva* caused by a dark reaction. *Plant Physiol.*, **64**, 896–9.

MOONEY H. A., DUNN E. L., HARRISON A. T., MORROW P. A., BARTHOLOMEW B. & HAYS R. L. (1971) A mobile laboratory for gas exchange measurements. *Photosynthetica*, **5**, 128–32.

OKADA M., INOUE M. & IKEDA T. (1978) Circadian rhythm in photosynthesis of the green alga *Bryopsis maxima*. *Plant and Cell Physiol.*, **19**, 197–202.

OLTMANNS F. (1891) Ueber die Kultur- und Lebensbedingungen der Meeresalgen. *Jb. Wiss. Bot.*, **23**, 349–440.

OLTMANNS F. (1905) Morphologie und Biologie der Algen. *Jb. Wiss. Bot.*, **37**, 121–42.

PLATT T., DENMAN K. L. & JASSBY A. D. (1977) Modeling the productivity of phytoplankton. In *The Sea, Vol. 6*, ed Goldberg E. D., pp. 807–56. John Wiley and Sons, New York.

QUADIR A., HARRISON P. J. & DE WREEDE R. E. (1979) The effects of emergence and submergence on the photosynthesis and respiration of marine macrophytes. *Phycologia*, **18**, 83–8.

RAMUS J. (1978) Seaweed anatomy and photosynthetic performance: the ecological significance of light guides, heterogenous absorption and multiple scatter. *J. Phycol.*, **14**, 352–62.

RAMUS J., BEALE S. I. & MAUZERALL D. (1976a) Correlation of changes in pigment content with photosynthetic capacity of seaweeds as a function of water depth. *Mar. Biol.*, **37**, 231–8.

RAMUS J., BEALE S. I., MAUZERALL D. & HOWARD K. L. (1976b) Changes in photosynthetic pigment concentration in seaweeds as a function of water depth. *Mar. Biol.*, **37**, 223–9.

RAMUS J., LEMONS F. & ZIMMERMAN C. (1977) Adaptation of light-harvesting pigments to downwelling light and the consequent photosynthetic performance of the eulittoral rockweeds *Ascophyllum nodosum* and *Fucus vesiculosus*. *Mar. Biol.*, **42**, 293–303.

RAMUS J. & ROSENBERG G. (1980) Diurnal photosynthetic performance of seaweeds measured under natural conditions. *Mar. Biol.*, **56**, 21–8.

RHEE C. & BRIGGS W. R. (1977) Some responses of *Chondrus crispus* to light. I. Pigmentation changes in the natural habitat. *Bot. Gaz.*, **138**, 123–8.

SCHNEIDER C. W. (1976) Spatial and temporal distributions of benthic marine algae on the continental shelf of the Carolinas. *Bull. Mar. Sci.*, **26**, 133–51.

SHIBLES R. (1976) Terminology pertaining to photosynthesis. *Crop. Sci.*, **16**, 437–9.

STEEMANN NIELSEN E. (1952) The use of radioactive carbon (^{14}C) for measuring organic production in the sea. *J. Cons. Perm. Int. Expl. Mer.*, **18**, 117–40.

STRICKLAND J. D. H. & PARSONS T. R. (1972) *A Practical Handbook of Seawater Analysis. Bull. Fish. Res. Board Can.*, **167**.

TITLYANOV E. (1976) Adaptation of bottom-dwelling plants to light. I. The importance of light in the distribution of attached marine algae. *Mar. Biol. (U.S.S.R.)*, **1**, 3–12.

TITLYANOV E. A., MASHANSKY V. F. & GLEBOVA N. T. (1974) Thallus anatomy and chloroplast ultrastructure of *Ulva fenestrata* in different conditions of illumination. *Bot. Zh. S.S.S.R.*, **59**, 1553–8.

VANDEN DRIESSCHE T. (1976) Circadian rhythms in *Acetabularia*: photosynthetic capacity and chloroplast shape. *Exptl. Cell Res.*, **42**, 18–30.

WAALAND J. R., WAALAND S. D. & BATES G. (1974) Chloroplast structure and pigment composition in the red alga *Griffithsia pacifica*: regulation by light intensity. *J. Phycol.*, **10**, 193–8.

WASSMAN E. R. & RAMUS J. (1973) Primary-production measurements for the green seaweed *Codium fragile* in Long Island Sound. *Mar. Biol.*, **21**, 289–97.

YOKOHAMA Y. (1973a) A comparative study on photosynthesis—temperature relationships and their seasonal changes in benthic marine algae. *Int. Rev. Ges. Hydrobiol.*, **58**, 463–72.

YOKOHAMA Y. (1973b) Photosynthetic properties of marine benthic red algae from different depths in coastal area. *Bull. Jap. Soc. Phycol.*, **21**, 119–24.

YOKOHAMA Y. & KAGEYAMA A. (1977) A carotenoid characteristic of chlorophycean seaweeds living deep in coastal waters. *Bot. Mar.*, **20**, 433–6.

ZELAWSKI W. & WALKER R. B. (1976) Photosynthesis, respiration and dry matter production. In *Modern Methods in Forest Genetics*, ed. Miksche J. P., pp. 89–119. Springer-Verlag, Berlin.

CHAPTER 14
CARBON METABOLISM

BRUNO P. KREMER
Seminar Für Didakoit der Biologie, Universität zu Köln,
D-5000 Köln 41, Federal Republic of Germany

1 INTRODUCTION

Primary productivity of plants due to assimilation of inorganic carbon is a basic process of solar energy conservation which supports the life of all components of the biosphere. A considerable portion of this energy is conserved by marine ecosystems. Some benthic marine macrophytes appear to be notably productive. The worldwide primary production of benthic seaweeds may account for about 10% that of phytoplankton, although marine macroalgae do not occupy more than 0.1% of the area available to phytoplankton communities (Ryther 1963).

In photoautotrophic organisms light energy is first converted into chemical potential by the formation of $NADPH_2$ and ATP which is then mostly used to convert inorganic carbon (CO_2) into diverse reduced carbon compounds. Significant advances have been made in recent years in determining the direction and regulation of carbon flow in assimilating cells, at the biochemical and physiological levels, mostly based on experimental work with a wide variety of unicellular microalgae as well as representatives of the higher land plants. The contribution of experimental data derived from work with seaweeds to the general understanding of the processes involved is comparatively little, but research on carbon metabolism of seaweeds has turned up various peculiar features in their biochemistry and physiology.

This chapter attempts to give a concise summary of some recent trends which have attracted attention in the research of carbon metabolism in seaweeds. More comprehensive reviews of carbon fixation and flow have been published by Raven (1974), Kelly *et al.* (1976), Latzko & Kelly (1976), Burris & Black (1976) and Gibbs & Latzko (1979) among others. Several topics related to carbon metabolism in seaweeds are dealt with in Chapters 9, 13 and 16.

2 SOURCES OF INORGANIC CARBON

Net carbon fixation into organic matter is essential for the growth and development of all plants. Photoautotrophic organisms including seaweeds gain carbon by light dependent assimilation of carbon dioxide. Since most of the green, red and brown macroalgae are submerged aquatic organisms, inorganic carbon is available to them in the seawater either as dissolved CO_2 or as H_2CO_3, as well as HCO_3^- and CO_3^{2-}. The relative proportions of the different forms of inorganic carbon in seawater depend on pH values, temperature, and salinity (see Skirrow 1975). Assimilation rates of marine algae are distinctly influenced by the inorganic carbon supply (Caperon & Smith 1978).

At the normal range of pH values for seawater (7.8–8.2) HCO_3^- is the predominant form comprising more than 90% of the total inorganic carbon. Though a variety of macroalgae has been found to utilize HCO_3^- as a carbon source, some species are reported to incorporate only unhydrated CO_2 (see Table 14.1). Although the average content of total inorganic carbon in seawater is about 2.2–2.5 mmol l^{-1}, those species which rely on CO_2 have only about 10 μmol l^{-1} available, while those operating with HCO_3^- have more than 2 mmol l^{-1}. For comparison, terrestrial plants live in an environment of 0.03% (=300 ppm) CO_2 which at normal conditions (20° C, 1 bar) is equivalent to about 13 μmol l^{-1}. Due to this carbon supply photosynthesis in a seaweed able to utilize HCO_3^- should usually not be limited by carbon concentration under light saturating conditions (in contrast to land plants). On the other hand, the diffusion constant of CO_2 in air (0.16 cm^2 s^{-1}) is much higher than in (still) water (2.0×10^{-5} cm^2 s^{-1}), and this difference in the range of 10^4 might sufficiently compensate for any advantage that a bicarbonate utilizing aquatic species might eventually have over a CO_2 utilizing land plant (Wheeler 1982).

The primary carbon fixation enzyme of photosynthesis, ribulose-1,5-bisphosphate carboxylase, requires unhydrated CO_2 as a substrate. Therefore seaweeds utilizing HCO_3^- must be able to convert the bicarbonate to CO_2. An appropriate enzyme system, carbonic anhydrase (carbonate dehydratase, EC 4.2.1.1) has been found in a wide variety of seaweeds (cf. Okazaki 1972; Graham & Smillie 1976).

Table 14.1 Carbon source of photosynthetic carbon fixation in seaweeds.

Species	CO_2	HCO_3^-	Reference
Chlorophyceae			
Enteromorpha sp.	×		Jolliffe & Tregunna (1970)
Ulva sp.		×	Jolliffe & Tregunna (1970)
Rhodophyceae			
Gelidium cartilagineum	×		Tseng & Sweeney (1946)
Palmaria palmata		×	Robbins (1979)
Porphyra schizophylla	×		Jolliffe & Tregunna (1970)
Phaeophyceae			
Carpophyllum flexuosum	×		Dromgoole (1978)
Carpophyllum plumosum	×		Dromgoole (1978)
Carpophyllum angustifolium	×		Dromgoole (1978)
Costaria costata		×	Jolliffe & Tregunna (1970)
Laminaria saccharina		×	Jolliffe & Tregunna (1970)
Macrocystis pyrifera		×	Wheeler (1978)
Nereocystis luetkeana		×	Jolliffe & Tregunna (1970)
Sargassum muticum		×	Thomas & Tregunna (1968)

3 PHOTOSYNTHETIC PATHWAYS

3.1 Pentose phosphate cycle v. C_4 pathway

The only known series of reactions providing primary fixation of CO_2 into organic compounds is the reductive pentose phosphate cycle first demonstrated for some unicellular green algae. Rigorous analysis of the short term products of photosynthetic carbon fixation using a marine alga was provided by Bean & Hassid (1955) in experiments with *Iridophycus flaccidum* (=*Iridaea flaccida*) running an incubation series from 8 to 170 s. This work showed that in a marine alga the bulk of radiocarbon incorporated during short term photosynthesis is confined to diverse organic phosphate esters such as 3-phosphoglycerate, sedoheptulose-7-phosphate or glucose and fructose monophosphates. From these results Bean & Hassid (1955) concluded that the initial pathway of photosynthetic carbon fixation in *Iridaea* follows that encountered in *Chlorella*. Experiments on kinetics of ^{14}C-labelling during short term photosynthesis in the range of a few seconds in brown algae (Bidwell *et al.* 1958; Kremer & Küppers 1977) have shown that RPP cycle is operational there as well, and the cycle thus appears to be practically universal. Time courses of ^{14}C percentage confined to early photosynthates such as 3-phosphoglycerate and further phosphorylation products show the characteristic time dependent decrease of typical primary and intermediary products during incubation periods between 2–5 and 30 s.

The finding that substantial incorporation of CO_2 into C_4 acids through carboxylation of phosphoenolpyruvate is closely associated with photosynthesis in many terrestrial plants stimulated a search for this pathway in algae. The C_4 pathway is now known to be an auxiliary mechanism which facilitates reductive carbon assimilation under peculiar ecological conditions (Latzko & Kelly 1976; Kelly *et al.* 1976), and all plants utilizing C_4 metabolism completely rely on the RPP cycle for net fixation of CO_2 (Black & Bender 1976; Black *et al.* 1976). Some studies seemed to indicate C_4 compounds in some seaweeds (Joshi *et al.* 1974). However, a critical investigation by Kremer & Küppers (1977) of several seaweeds belonging to different algal classes demonstrated that while certain C_4 compounds such as aspartate and malate were labelled during photosynthesis, ^{14}C-labelling kinetics of these compounds achieved maximum percentage significantly later than the diverse phosphate esters and could not be regarded as reaction products of a carboxylation step preceeding the carboxylation of ribulose-1, 5-bisphosphate.

3.2 Low-molecular weight and polymeric photosynthates

Photosynthetic carbon dioxide fixation by carboxylation of ribulose-1, 5-bisphosphate furnishes 3-phosphoglycerate as the first stable photoassimilatory product.Phosphoglycerate is immediately fed into the reduction phase of the pentose phosphate cycle. The further conversion of this compound comprises a phosphorylation requiring ATP and a reduction at the expense of NADPH. All other subsequent reactions involved are merely rearrangements designed 1, to regenerate the starting material and 2, to provide various end products. The operation of this obviously ubiquitous cycle in seaweeds has been established by the demonstration of a number of the enzymes needed such as fructose-1, 6-bisphosphatase and aldolase (Yamada *et al.* 1978, 1979a, 1979b). Aldolase along with a triose phosphate dehydrogenase had first been recorded by Jacobi (1957) in extracts from *Ulva lactuca*. NADP-linked glyceraldehyde phosphate dehydrogenase has been demonstrated in *Laminaria hyperborea* by Küppers (1978) and Küppers & Weidner (1980). All phosphorylated compounds involved in these reactions are primary photoassimilatory products. Secondary photosynthetic products (i.e. end products) are all accumulated (stored) components such as the low-molecular weight carbohydrates and the polysaccharides whose biosynthetic precursors are directly derived from any intermediate of the RPP cycle. The secondary products in each class are markedly different (Fig. 14.1).

Earlier work on the storage products of algae, including seaweeds, has been compiled in an excellent review (Craigie 1974) and therefore needs no repetition. The present chapter attempts to survey the more recent literature mainly with respect to the physiological and biochemical implications. Descriptive comments on algal polysaccharides are omitted here since these compounds are dealt with extensively in Chapter 16.

Fig. 14.1 Autoradiographs of two-dimensional thin-layer chromatograms: patterns of ^{14}C-labelled photosynthates of different seaweeds. A = *Codium fragile*, B = *Palmaria palmata*, C = *Polysiphonia lanosa*, D = *Bostrychia scorpioides*, E = *Fucus serratus*, F = *Pelvetia canaliculata*. Labelled compounds: 1 = sucrose, 2 = fructose, 3 = floridoside, 4 = digeneaside, 5 = dulcitol and sorbitol, 6 = mannitol, 7 = volemitol.

Chlorophyceae

Studies on photosynthetic ^{14}C-labelling of soluble storage products in green algae such as *Ulva, Monostroma, Enteromorpha, Cladophora* and *Rhizoclonium* have been reported by Craigie *et al.* (1966), Percival & Young (1971), Percival & Smestad (1972) and Patil & Joshi (1970), among others. Sucrose was generally found to be the most highly labelled low-molecular weight carbohydrate, with only small amounts of radiocarbon in fructose, glucose and other hexoses. Glucose was the major sugar constituent of the completely hydrolysed residue of *Ulva lactuca* (Percival & Smestad 1972; Craigie *et al.* 1966). The amounts of radiocarbon recovered from polymeric material suggest that in *Ulva* the polysaccharides are fairly rapidly synthesized. However, Patil & Joshi (1970) reported ^{14}C-labelling in *Ulva lactuca* after 10 s to 6 h to be predominantly in glucose and fructose rather than in sucrose. In species of the genera *Cladophora* and *Rhizoclonium* an oligosaccharide containing fructose residues was traced in addition to sucrose (Percival & Young 1971).

The kinetics of carbon flow through various metabolites in the siphonous green alga, *Caulerpa simpliuscula,* has been described by Howard *et al.* (1975). The authors emphasize that the pattern of carbon fixation in this alga differs from that observed in other green seaweeds. Sucrose is reported never to be a major recipient of ^{14}C-labelling. Soluble β-linked polyglucans are the most important photosynthates. Sugar monophosphates contain large quantities of ^{14}C even after longer periods of incubation and are therefore believed to serve not only as metabolic intermediates, but also as storage pools. In characterizing the effects of red and blue light on photosynthetic carbon metabolism in *Acetabularia mediterranea,* Clauss (1972) found a higher percentage of starch content in the cells grown under red light as compared to white light, despite the reduced assimilation rates, and an increase in soluble carbohydrates (sucrose, fructose, glucose) in cells irradiated with blue light (see also Chapter 9). Photosynthetic carbon metabolism in the related species *Acetabularia cliftonii* depends on the cell size achieved (Clauss 1975): in cells without caps the predominant soluble carbohydrate was found to be fructose but in cells with caps there was considerable labelling of inulin.

Phaeophyceae

Brown seaweeds usually contain large amounts of a C_6 polyhydroxyalcohol, mannitol, which seems to be the commonest hexitol reported from plants. The distribution of this compound has been reviewed by Craigie (1974). Mannitol replaces sucrose as the soluble storage carbohydrate in all brown algae so far investigated, even in species such as *Desmarestia viridis, Durvillaea antarctica* and *Stilophora rhizodes* (Kremer 1973b and unpublished data; Craigie 1974) which had previously been reported to lack it. Occasionally reported low amounts of free sugars in brown seaweeds may be due to unspecific dephosphorylation of sugar phosphates or may simply be associated with epiphytic or endophytic green algae such as *Acrochaete parasitica* in *Fucus serratus* (Kremer 1975b).

Mannitol is found in the Phaeophyceae in quantities ranging from a few up to 20–30% of dry matter. These amounts suggest rapid and effective synthesis. This view is substantiated by experiments with photosynthetic ^{14}C-labelling (Bidwell *et al.* 1958; Yamaguchi *et al.* 1966, 1969; McLachlan & Bidwell 1978; Kremer 1973a, 1973b, 1973c, 1975a; Kremer & Willenbrink 1972; Kremer & Markham 1979). Kinetics of ^{14}C-labelling of this hexitol during photosynthetic carbon reduction exhibit the typical upward curve of accumulated photosynthates accounting for ordinarily more than 30–50% of total ^{14}C assimilated after 10 min photosynthesis (Kremer 1979a). Phosphorylated compounds show the characteristic negative slope kinetics of intermediary metabolites suggesting a close relationship between ^{14}C-labelling of sugar phosphates and biosynthesis of mannitol.

Evidence for a direct interrelation of mannitol biosynthesis with the RPP cycle has been provided by analysis of intramolecular distribution of ^{14}C in ^{14}C-mannitol from *Eisenia bicyclis* (Yamaguchi *et al.* 1969). Enzymes catalysing the reaction step fructose-6-phosphate → mannitol-1-phosphate (=mannitol-1-phosphate dehydrogenase, EC 1.1.1.17) and mannitol-1-phosphate → mannitol (= mannitol-1-phosphatase) were first demonstrated in *Spatoglossum pacificum* and *Dictyota dichotoma* (Ikawa *et al.* 1972). In addition to mannitol, the C_7 polyol volemitol has been detected in *Pelvetia canaliculata* (Lindberg & Paju 1954). This compound is restricted to this species of the genus *Pelvetia* among all brown seaweeds tested and receives rapid ^{14}C-labelling during photosynthesis (Kremer 1973c, 1976e). The synthesis of this heptitol is not performed by the systemic mycoendophyte of *Pelvetia*, but by the alga itself (Kremer 1973c; Kingham & Evans 1977). An enzyme catalysing the reduction of sedoheptulose-7-phosphate to volemitol-1-phosphate has been demonstrated (Kremer 1977a).

Photosynthetic $^{14}CO_2$ assimilation by brown seaweeds yields considerable ^{14}C-labelling in compounds of the insoluble assimilate fraction(s) as well. Aspects of polymeric constituents will be dealt with in Chapter 16 of this volume.

RHODOPHYCEAE

The members of the Rhodophyceae differ from most of the other photoautotrophic plants by the occurrence of certain heterosides. Floridoside (2-*O*-D-glycerol-α-D-galactopyranoside) has been found to serve as the low molecular weight assimilate in most of the red algae, except the Ceramiales (Craigie 1974; Kremer & Vogl 1975; Kremer 1978b). In the Ceramiales digeneaside (2-D-glycerate-α-D-mannopyranoside) replaces floridoside (Kremer 1978a, b). Isofloridoside (1-*O*-D-glycerol-α-D-galactopyranoside) is consistently rather weakly ^{14}C-labelled during photosynthesis (Craigie 1974) and seems moreover to be distributed mainly among members of the Bangiales (Nagashima 1976). The pattern of photoassimilatory products in red algae thus proves to be of a certain chemotaxonomic value.

It is now accepted that, despite occasional reports to the contrary (cf. Craigie 1974), mannitol is not a common constituent of the red algae, because of repeated failures in tracing this compound and appropriate enzymes possibly involved in its biosynthesis (Kremer 1976c). With respect to the polyols, there is one noteworthy exception, however: sorbitol (glucitol) and dulcitol (galactitol) are photosynthetic accumulation products in the ceramialean *Bostrychia scorpiodes* (Kremer 1976b) and occur in all further species of that genus investigated (Kremer 1976d). The occurrence of photosynthetically ^{14}C-labelled trehalose in marine red algae has been reported (Nagashima 1976). Using carbon isotopes other than ^{14}C, the preparation of ^{13}C-galactose from *Gigartina corymbifera* (Kollman & Gregg 1972) as well as of ^{11}C-galactose and ^{11}C-glycerol from *Gigartina stellata* (Palmer & Goulding 1974) have been described.

Floridoside and isofloridoside have been claimed to act as osmoregulatory compounds in marine red algae (Kauss 1968). A critical reinvestigation provides evidence from ^{14}C-labelling experiments, ^{14}C-pulse feeding, and quantitative determinations that biosynthesis of this heteroside is not affected by osmotic stress over a wide salinity range. These results indicate that osmotic regulation in marine Rhodophyceae is not achieved on the level of soluble carbohydrates involving floridosides or related compounds (Kremer 1979b). However, certain fluctuations in the content of organic solutes in a variety of marine algae suggest that the amount of soluble photosynthates might follow to some extent the external osmotic conditions (Kirst & Bisson 1980).

In all seaweeds investigated for photosynthetic carbon flow, not only carbohydrates, but also a series of further constituents such as organic acids, lipids, and particularly amino acids, are ^{14}C-labelled. Transfer of ^{14}C to amino acids including glycine, serine, alanine, aspartate, glutamate, threonine, and some others during photosynthesis proceeds rather rapidly. Pulse feeding experiments with brown and red seaweeds showed that amino acids are not accumulated, since percentage ^{14}C-labelling drastically declined after a feeding pulse, suggesting immediate metabolic conversion (Kremer 1975a, 1978a).

3.3 Responses from isolated chloroplasts

Isolation of cell organelles has largely established present knowledge of compartmentation of metabolic processes as well as extent and pathways of carbon flow within the cells. Isolation of chloroplasts is most conveniently achieved using siphonous green algae. While chloroplasts which can be isolated from flowering plants in various mineral or organic media retain their ability for high rates of CO_2 fixation for a maximum of 1–2 h after cell fractionation, chloroplasts from green algae such as *Acetabularia mediterranea* have repeatedly been reported to maintain their capacity for carbon fixation in the light for at least 24 h after isolation (Shepherd & Bidwell 1973). Chloroplasts from *Codium fragile* were found to be photosynthetically active even 5 d after isolation (Trench *et al.* 1973a, b). Similar observations have been reported by Giles & Sarafis (1972) and Schönfeld *et al.* (1973) from *Caulerpa sedoides* and *Codium vermilara*, respectively.

Nevertheless, observations of chloroplast activities *in vitro* must be interpreted with great care (cf. Cobb 1977). Grant *et al.* (1976) report ^{14}C-labelling of sucrose in chloroplast preparations from *Caulerpa simpliuscula*. This sucrose biosynthesis may be due to contaminants in the preparation, since a maximum of 17% cytoplasmic contamination was calculated from simultaneous measurements of the mitochondrial marker enzyme succinate dehydrogenase.

Chloroplasts of an excellent preparative state, originating from *Codium fragile*, have been found to enter an endosymbiosis with cells of the digestive gland in a marine opisthobranchiate slug, *Elysia viridis* (see Taylor 1976). In contrast to results obtained from entire *Codium* branches, the symbiotic chloroplasts have consistently been found not to synthesize ^{14}C-sucrose from photosynthetically fixed ^{14}C (Trench *et al.* 1973b; Kremer 1976a; Hinde 1978). For this reason it seems likely that chloroplasts of green algae have no enzymatic potential for sucrose biosynthesis.

Products of photosynthetic carbon fixation by isolated chloroplasts from *Acetabularia mediterranea* were analysed by Bidwell *et al.* (1970). Intermediates of the RPP cycle were regarded as responsible for much of the carbon metabolism in the chloroplasts, and a substantial part of the fixed carbon obviously was also metabolized via glycolic and glyceric acids.

Red algal chloroplasts have so far received much less attention. Nagashima *et al.* (1969) report on the characterization of a starch synthesizing enzyme from a chloroplast preparation of *Serraticardia maxima*. This enzyme, a glucosyl transferase, is believed to be localized inside the chloroplasts, though floridean starch granules are principally found outside in the cytoplasm. Chloroplasts from *Griffithsia flosculosa* have been found to be functional endosymbionts of an eolidiform opisthobranch, *Hermaea bifida* (Kremer & Schmitz 1976; Kremer 1977b).

Compartmentalization studies with chloroplasts from brown seaweeds were first carried out by Willenbrink & Kremer (1973) using *Fucus serratus*. Short-term photosynthesis by the intact thallus followed by a non-aqueous isolation procedure of the chloroplasts provided evidence that the intracellular site of mannitol biosynthesis might be closely related to that organelle, if not localized inside the chloroplast itself. When chloroplasts of the same species are isolated by an aqueous procedure and are then allowed to photosynthesize ^{14}C from

$H^{14}CO_3^-$, ^{14}C-labelling is again predominantly found in mannitol. Mannitol-1-phosphate dehydrogenase (see Section 3.2.) appears to be a cytoplasmic enzyme, however (Nordhorn *et al.* 1976). Most probably mannitol biosynthesis occurs in the periplastidal reticulum which usually surrounds brown algal chloroplasts.

4 CARBON METABOLISM AND CALCIFICATION

Calcium carbonate deposition (either as calcite or as aragonite) in the calcareous marine algae and photosynthetic carbon metabolism are currently thought to be closely interrelated (Littler 1976). In this sense calcium carbonate may be understood as an inorganic photosynthetic product, and this is why calcification is included in this chapter. More complete reviews have been published by Darley (1974) and Borowitzka (1977).

An investigation of *Bossiella orbigniana* provided evidence that organic products of photosynthesis may be of general importance for calcification (Pearse 1972) and substantiated the idea that precipitation of calcium carbonate by algae may be directly related to carbon dioxide withdrawal and hydroxyl production in photosynthesis. Rates of calcification and of photosynthesis in *Corallina officinalis* were found to be highest in the apical intergenicular segments of the thallus. Calcification in the dark was only 30–50% that of deposition in the light. A positive correlation significant at the 0.001 level was found between the photosynthesis and calcification rates of the segments examined by Pentecost (1978) (cf. Digby 1977).

Experiments carried out with *Halimeda tuna* indicate that this species utilizes HCO_3^- for photosynthesis and that this process stimulates calcification by removing CO_2 from the intercellular spaces. As these spaces are isolated from the external seawater by the layer of cell wall of the appressed peripheral utricles, the removal of CO_2 results in a rise of CO_3^{2-} concentration and of pH. The consequence is an increased rate of $CaCO_3$ precipitation. Respiratory CO_2 evolution has an inhibitory effect on calcification by decreasing the pH and the amount of available CO_3^{2-} (Borowitzka & Larkum 1976a, b, c). Calcification rates for *Halimeda tuna* were found to be in the range of 76–208 nmol $CaCO_3$ g^{-1} dry weight min^{-1} in the light and more than a half less in the dark (Borowitzka & Larkum 1976a). The same authors propose a model of these processes and illustrate by theoretical calculations the respective effects of CO_2, HCO_3^-, and CO_3^{2-} removal or addition in closed systems (as given by the thallus anatomy of *Halimeda tuna*) (Borowitzka & Larkum 1976b). Inhibition of photosynthetic CO_2 uptake and assimilation by typical metabolic inhibitors such as CCCP, DNP and DCMU also inhibits calcification (Borowitzka & Larkum 1976c). Inhibitory effects of orthophosphate on calcification in coralline algae were reported by Brown *et al.* (1977).

5 DARK CARBON FIXATION

5.1 Dimensions of dark carbon fixation

Besides light dependent carbon fixation via ribulose-1,5-bisphosphate carboxylase, which is by far the most effective carboxylating system in photoautotrophic plants including the seaweeds, there are further CO_2 fixation reactions, which operate independent of light and which are referred to as dark carbon assimilation. In this type of reaction, designated β-carboxylation, a C_4 dicarboxylic acid is produced by condensation of molecular CO_2 and a C_3 acid. As was mentioned above (Section 3.1) this reaction, which in most species of algae examined is catalysed by phosphoenolpyruvate carboxylase (PEP-C), takes place at comparatively low rates not sufficient to establish any type of C_4 metabolism. However, appreciable activity of PEP-CK is found in species of the Phaeophyceae (Weidner & Küppers 1973; Weidner *et al.* 1975; Kremer & Küppers 1977; Küppers & Kremer 1978; Kremer 1979c), while the carboxylation potential due to PEP-CK (or even PEP-C) in representatives of the Chlorophyceae and Rhodophyceae is in the range of the detection limit (Kremer & Küppers 1977; Kremer 1979c) (Table 14.2). If the total carbon fixation of the species considered is taken as 100%, then carboxylation via PEP-CK is on the whole less than 3% in green and red seaweeds, whereas this reaction accounts for considerably more (up to >20%) in brown seaweeds. In a survey including some 25 species, Kremer & Küppers (1977) found that in Phaeophyceae dark carbon fixation amounts to more than 5% of the equivalent photosynthetic rates. Experiments with different developmental stages of *Laminaria saccharina* including male and female gametophytes, female gametes, zygotes and young sporophytes of different age revealed that a similar ratio of relative high dark fixation to photosynthesis is encountered through all phases of the life cycle (Kremer & Markham 1979).

Using *Eisenia bicyclis*, *Dictyota dichotoma*, *Hizikia fusiformis* and *Carpopeltis divaricata*, Akagawa *et al.* (1972a, 1972b) concluded that oxaloacetate is the first detectable product of dark CO_2-fixation, since aspartate, glutamate, alanine and citrate were the strongest ^{14}C-labelled compounds after relatively short periods of incubation. Oxaloacetate has not yet been traced directly, as the pool of this intermediate is obviously rather low and the small amounts present are presumably destroyed under the analytical conditions usually employed. The results reported by Akagawa *et al.* (1972a, 1972b) and Nisizawa *et al.* (1972) have been confirmed by Kremer (1979c).

Akagawa *et al.* (1972c) assumed that radiocarbon in the algae is used for the carboxylation of phosphoenolpyruvate to form oxaloacetate, which then is converted to aspartate by aspartate transaminase (EC 2.6.1.2), condensed with acetyl-CoA to give citrate (which may be the substrate of glutamate transaminase, EC 2.6.1.1), or even subjected to malate dehydrogenase (EC

Table 14.2 *In vivo* and *in vitro* CO_2-fixation via RuBP-C and PEP-CK in marine algae*.

Species	Photosynthesis	Dark fixation	Relative enzyme activity (%)		
	μmol CO_2 g^{-1} dwt h^{-1}		RuBP–C	PEP–CK	PEP–C
Chlorophyceae					
Cladophora rupestris	30.5	0.22	98.1	1.5	0.4
Codium fragile	68.3	0.91	99.0	1.0	—
Ulva lactuca	48.7	0.27	97.5	1.5	1.0
Phaeophyceae					
Agarum fimbriatum	74.5	6.0	52.5	47.5	—
Alaria marginata	109.3	3.6	78.3	21.7	—
Costaria costata	57.3	3.3	64.1	35.9	—
Cymathere triplicata	58.5	3.2	79.8	20.7	—
Egregia menziesii	72.5	2.6	86.0	14.0	—
Eisenia arborea	92.4	3.6	82.1	17.9	—
Fucus spiralis	22.4	2.2	50.1	49.9	—
Fucus vesiculosus	29.2	2.2	52.8	47.1	—
Laminaria digitata	25.2	1.9	85.6	14.4	—
Laminaria groenlandica	45.7	2.7	78.3	21.7	—
Laminaria hyperborea†	21.1	2.7	62.4	37.6	—
‡	6.7	5.7	27.5	72.5	—
Laminaria saccharina†	80.3	3.7	60.5	39.5	—
‡	26.1	4.4	24.8	75.2	—
Macrocystis integrifolia†	190.0	2.9	79.7	20.3	—
‡	88.6	5.0	18.3	81.7	—
Macrocystis pyrifera†	153.7	1.9	72.0	28.0	—
‡	72.3	2.9	26.6	73.4	—
Rhodophyceae					
Chondrus crispus	21.2	0.9	97.3	1.8	0.9
Delesseria sanguinea	37.9	0.3	100.0	—	—
Iridaea cordata	29.4	0.5	94.9	5.1	—

* Values from Kremer (1979c), Kremer & Küppers (1977), Küppers & Kremer (1978) and Willenbrink *et al.* (1979a).
† Old mature blade.
‡ Young growing blade.

1.1.1.37). Considerable ^{14}C-labelling of malate is found particularly after longer periods of incubation (Kremer 1979c). Akagawa *et al.* (1927c) investigated several non-photosynthetic CO_2 fixation enzymes involved in the condensation of CO_2 with a C_3 unit to form a C_4 dicarboxylic acid including e.g. malic enzyme (EC 1.1.1.40), phosphoenolpyruvate carboxylase (EC 4.1.1.31), phosphoenolpyruvate carboxykinase (EC 4.1.1.32), phosphoenolpyruvate carboxytransphosphorylase (EC 4.1.1.38) and pyruvate carboxylase (EC 6.4.1.1). They concluded that PEP-CK is the enzyme for the entrance of CO_2 in the light independent (dark) carbon fixation by marine algae (see also Küppers 1978). Like RuBP-C, PEP-CK uses CO_2 rather than HCO_3^- (Cooper *et al.* 1968).

5.2 Physiological significance

Investigations on carbon fixation as related to frond development in *Laminaria hyperborea* showed that rates of photosynthetic (light dependent) and dark (light independent) carbon fixation depend on the ontogenetic stage of the tissue (Willenbrink *et al.* 1975). In particular dark fixation is very much higher in the young, meristematic, growing parts of the blade. An equivalent differential distribution of PEP-CK activity (accounting for high rates of dark fixation) and of RuBP-C in the basal and apical parts of growing *Laminaria* spp. has been demonstrated (Weidner & Küppers 1973; Küppers & Kremer 1978). Maximum activity of PEP-CK is always confined to the intercalary growing region of the blades. Lower activities occur in the holdfasts, stipes, and fully differentiated part of the blades. Highest activity of RuBP-C is found beyond the growing segment in the older part of the blade. Depending on the ontogenetic stage of the thallus region dark fixation rates in the growing region of *Laminaria digitata* and *L. saccharina* account for about 20 and 15%, respectively, of the total light dependent carboxylation capacity of that thallus part, while in *Laminaria hyperborea* it is usually more than 30% and up to 40% of photosynthesis. Extraordinary high activities of PEP-CK, accounting for equivalent high rates of dark carbon assimilation, have also been shown in the growing, ontogenetically young frond areas of *Macrocystis, Nereocystis*, and other Pacific Laminariales (Willenbrink *et al.* 1979a). Although there is a different type of thallus organization in the Fucales, essentially the same pattern as the Laminariales has been found in *Fucus serratus, F. spiralis* and *F. vesiculosus* (Küppers & Kremer 1978). This feature may thus be common to all growing or meristematic frond areas, at least of the Fucales and Laminariales.

It follows that if the distribution of photosynthetically fixed ^{14}C among early ^{14}C-labelled assimilates is examined, considerably more radiocarbon is recovered from C_4 compounds such as aspartate and malate in members of the Phaeophyceae than in the representatives of other algal classes (Kremer & Küppers 1977), and the growing parts of brown seaweeds show much higher percentage of ^{14}C confined to these compounds than do the old blade areas (Willenbrink *et al.* 1975, 1979a). Thus it may be assumed that even in the light carbon fixation via PEP-CK is not interfered with by carboxylation via RuBP-C. This view is substantiated by the observation that the pattern of ^{14}C-labelled dark assimilates is qualitatively and quantitatively integrated into the pattern of photoassimilatory products (see Kremer & Markham 1979). Furthermore, it has been found that photosynthetically formed aspartate is predominantly labelled in the C_1 atom, suggesting asymmetrical labelling due to biosynthesis via β-carboxylation (Kremer unpublished data).

In *Laminaria hyperborea*, RuBP-C and PEP-CK exhibit seasonal variation (Küppers 1978; Küppers & Weidner 1980). The standard enzyme activity of both carboxylases rises steeply at the onset of a new growth cycle when, in early winter, a new frond is developed (Lüning 1971; Kain 1979). Amounts of RuBP-C and PEP-CK are increased by factors of about 30 and 4 respectively, to reach maxima during spring, and are then reduced to minima in late summer and fall. When enzyme activities measured under laboratory standard conditions (20°C) are corrected for the actual ambient temperatures *in situ*, using the apparent activation energy and the records of *in situ* water temperatures during the growth cycle, it can be recognized that another type of seasonal variation is obtained: the curves for the temperature-adjusted enzyme activities indicate highest metabolic activities during summer months. Comparison of the standard and the temperature-adjusted enzyme activities suggests that a compensatory increase of enzyme amount is occurring just during those months of the growth cycle when the enzyme activity should be lowest as referred to water temperature. During summer *Laminaria hyperborea* can reduce the absolute enzyme amounts and still retain enough activity to account for a rather high metabolic rate. This change is understood as an expression of ecological adaptation (Küppers & Weidner 1980). This pronounced seasonal enzyme rhythmicity fits well with observations on seasonal changes of photosynthetic activities (Lüning 1971; Chapman & Craigie 1977; Johnston *et al.* 1977).

With respect to the strategy of dark carbon assimilation it is noteworthy that the development of the new blade of *Laminaria hyperborea* starts in winter when *in situ* photon flux density is low and most of the time below the compensation point of photosynthesis (Lüning 1971, 1979; Lüning & Dring 1979). Activity of RuBP-C is hardly detectable when the new blade is initiated, whereas at the same time (standard) activity of PEP-CK already amounts to 20–50% of the spring maximum (Küppers 1978). During this unfavourable time for growth the new blade obviously depends largely on light independent carbon fixation (Willenbrink *et al* 1975; Küppers & Kremer 1978). At the same time considerable amounts of stored assimilates are translocated from the old blade to the growing areas of the frond (Schmitz *et al.* 1972; see Chapter 15). Dark carbon fixation seems to be substantially supported by this import of reduced carbon. Mannitol, after remetabolization via glycolysis (see Section 6.3), might serve as a source of the PEP required for continuous light independent carboxylation via PEP-CK. Dark fixation of CO_2 in the brown seaweeds thus yields an appreciable net carbon uptake and furthermore feeds metabolic intermediates into several anabolic and biosynthetic pathways (Willenbrink *et al.* 1975, 1979a). The assumption of heterotrophy as emphasized by Wilce (1966), Mann (1973) or Drew (1977) is therefore unnecessary.

Similar interconnections of carbon assimilation and frond development can be assumed for other members of the Laminariales as well (Kremer & Küppers 1977; Küppers & Kremer 1978; Willenbrink *et al.* 1979a, b). Thus the temporal and spatial distribution of different carbon fixation pathways is a unique feature of kelp, not found in terrestrial plants. Because of this interface between photosynthetic and dark carbon assimilation, brown seaweeds are neither C_3 nor classic C_4 plants, but a group of their own.

6 RESPIRATORY PHENOMENA

6.1 Photorespiration

Warburg (1920) reported on an inhibitory effect of O_2 on the rates of apparent photosynthesis of a unicellular green alga, and since then this phenomenon, repeatedly observed in other photosynthesizing organisms, has been referred to as the 'Warburg O_2 effect'. Rate depression of net photosynthesis by O_2 was one of the more accessible parameters in experimental analysis, while other effects such as CO_2 release in the light and variations of the CO_2 compensation were far less easy to measure. At present, the light dependent O_2 uptake and CO_2 evolution resulting from the biosynthesis and metabolism of glycolate is termed photorespiration. This complex metabolic phenomenon has received considerable attention, and the extensive work carried out with unicellular algae has been reviewed by Tolbert (1974) and Harris (1978).

In terrestrial plants under normal atmospheric conditions (about 300 ppm CO_2 = 0.01 mM; 21% O_2 = 8.7 mM) photorespiration accounts for a reduction of up to 50% in photosynthetic productivity as compared with photosynthetic rates under lower O_2 and higher CO_2 concentrations. Environments of this type (O_2 < 21%, CO_2 > 300 ppm) are very unlikely for terrestrial plants. Marine algae, on the other hand, are generally exposed to appreciable concentrations of CO_2 and a rather low O_2 partial pressure (about 0.23mM) yet are nonetheless expected to exhibit photorespiratory metabolism (Tolbert 1976).

Ribulose-1, 5-bisphosphate carboxylase (RuBP-C) has a distinct affinity for O_2 as well as CO_2 (see Tolbert 1974), so that under conditions of rather high O_2 concentrations a notable portion of the activity of this enzyme is devoted to production of phosphoglycolate and glycolate. Because of this RuBP-C has been referred to as an oxygenase or even as RuBP oxidase. Evidence for RuBP oxidase activity is obtained if rates of net photosynthesis in marine algae under various O_2 and CO_2 concentrations are compared, and the data strongly suggest the occurrence of photorespiration in marine algae (Burris 1977; Wheeler 1982).

Another indication of photorespiration is the CO_2 compensation point, an equilibrium maintained between the rate of photosynthetic CO_2 fixation and the rate of CO_2 lost from photorespiration as well as dark respiration. This compensation point is nearly zero for C_4 plants, because of an unusually efficient CO_2 fixation, and in the range of 30–80 ppm CO_2 (= $\mu l\ CO_2\ l^{-1}$) for the majority of C_3 plants. Species such as *Boodlea composita*, *Halimeda cylindracea* and *Enteromorpha* spp. were all found to exhibit CO_2 compensation very similar to that of terrestrial C_3 plants. As should be expected, the compensation points increased with increasing O_2 concentrations (Tolbert & Garey 1976). Rather low CO_2 compensation has been obtained for *Fucus serratus, Pelvetia canaliculata, Porphyra umbilicalis, Palmaria palmata* and *Ulva lactuca* (Coughlan & Tattersfield 1977), presumably due to insufficient analytical precaution. Nevertheless, these authors correlate their results with a possible C_4 status of the species examined.

Since photorespiration results in biosynthesis of glycolate, the metabolic steps involved have been referred to as the glycolate pathway (Tolbert 1974). Considerable ^{14}C-labelling of glycolate in relation to photosynthetic assimilation of $^{14}CO_2$ in different seaweeds has been obtained on several occasions (Kremer 1973a, 1975b, 1976b) and is thus the first qualitative demonstration of photorespiration in marine macroalgae.

Biosynthesis of glycolate and its further metabolic conversion is catalysed by a series of enzymes which are specific for this pathway—their presence in marine algae should hence unequivocally demonstrate the occurrence of photorespiration. Oxidase (oxygenase) activity of RuBP-C has been shown in crude extracts from *Halimeda cylindracea* and *Chaetomorpha* spp. (Akazawa & Osmond 1976). This activity was found to be about 1% that of the carboxylation. To date no such activity has been detected in species of Rhodophyceae or Phaeophyceae, possibly due to analytical difficulties (Akazawa & Osmond 1976). Because of its pronounced specificity for phosphoglycolate, activity of phosphoglycolate phosphatase is a particularly good indication of photorespiration. This peculiar enzyme has been found in a wide range of green, red and brown seaweeds tested, with activities which are greater or at least equal to those encountered for terrestrial plants on a comparable reference basis (Randall 1976).

Glycolate is either oxidized with molecular O_2 to form H_2O_2 and glyoxylate via glycolate oxidase (as in higher plants) or may be oxidized without O_2 to form glyoxylate via glycolate dehydrogenase (as in the unicellular algae, Tolbert 1974). Further enzymatic steps lead to the formation of other important intermediates such as glycine and serine. Glycolate oxidase has not been found in marine algae, but appreciable amounts of glycolate dehydrogenase have been measured in 15 species of the Chlorophyceae including species of *Caulerpa, Chaetomorpha, Enteromorpha, Halimeda* and *Udotea* (Tolbert 1976). Tolbert attributed his inability to demonstrate glycolate oxidase/dehydrogenase in red and brown seaweeds to unsatisfactory extraction techniques making the assessment difficult (Tolbert 1976).

Since glycine and serine are important intermediates of photorespiration (Tolbert 1974), the measurement of their formation may be a good indication of the amount of carbon flowing through the glycolate pathway. A series of such experiments by Burris *et al.* (1976) showed that indeed the percentage of ^{14}C confined to these amino acids was consistently higher after photosynthetic

$^{14}CO_2$ assimilation under high O_2 partial pressure than under normal air or even a helium atmosphere. Similar values were reported by Burris (1977).

Classic C_4 plants show no measurable rates of photorespiration. This is mainly due to the high affinity of PEP-C for CO_2 (K_m about 7×10^{-6} M), which is far greater than that of RuBP-C (K_m about 450×10^{-6} M), resulting in a very low CO_2 compensation point. PEP-C is not present in marine algae (Akagawa *et al.* 1972c; Ting 1976; Kremer & Küppers 1977), but in brown seaweeds its role is filled by PEP-CK (see Section 5.1). Interestingly, in those plants with high activities of PEP-CK, photorespiration as indicated by the Warburg O_2 effect can be significantly low (Hatcher *et al.* 1977).

Although photorespiration has now been convincingly demonstrated in green, red and brown seaweeds, its ecological significance and its implications in the overall productivity of these algae are still uncertain. Seasonal carbon budgets for *Laminaria longicruris* (Hatcher *et al.* 1977) and for *Laminaria saccharina* (Johnston *et al.* 1977) do not include terms for photorespiration.

6.2 Dark respiration

Considerable amounts of organic carbon provided by photosynthesis are reconverted to CO_2 through the Embden-Meyerhoff pathway of glycolysis and further aerobic degradation. Recent studies on marine algae have attempted to correlate the effects of light, temperature, and salinity on photosynthesis and respiration with algal distribution in their natural habitats (Calabrese 1971; Mathieson & Burns 1971; Kjeldsen & Phinney 1972; Mathieson & Dawes 1974; Fralick & Mathieson 1973, 1975; Niihara 1975; Dawes *et al.* 1978; Lehnberg 1978). Classic studies on the ecological implications of dark respiration in marine algae were carried out by Kanwisher (1966) and Newell & Pye (1968). Rates of dark respiration, assumed to be inhibited during photosynthesis (Brown & Tregunna 1967), furthermore depend on internal conditions such as ontogeny and growth performance of different frond areas, as has been shown for several *Laminaria* species (Lüning 1971; Lüning 1979; Kremer unpublished) A rhythmicity of dark respiration was first recorded in a *Fucus* sp. (Brown *et al.* 1955) and confirmed by further observations (MacRae & Chapman 1977).

To give an account of dark respiratory rates in seaweeds some of the available data have been compiled in Table 14.3.

The effects of temperature on dark respiration are of interest with respect to adaptive phenomena. For algae growing in an environment under light-limited conditions, a physiological adaptation of dark respiration is as important for primary production as an adaptation of photosynthesis. For a number of seaweeds the Q_{10} values for dark respiration are significantly higher in winter than in summer (Kanwisher 1966; Newell & Pye 1968; Yokohama 1973). Relatively low Q_{10} (in the range of 1.2) during summer indicates that the respiration remains almost unaffected by fluctuations of the ambient tempera-

Table 14.3 Rates of dark respiration in some selected seaweeds. Rates have been converted assuming a ratio of CO_2 release/O_2 consumption near 1.0 in all cases. (Some data taken from authors' graphs.)

Species	*t* (°C)	Rate (μmol CO_2)			Reference
		g^{-1} dwt h^{-1}	g^{-1} fwt h^{-1}	dm^{-2} h^{-1}	
Chlorophyceae					
Acetabularia crenulata	30	83			Buesa (1977)
Cladophora sp.	0	4			Healey (1972)
Enteromorpha intestinalis	10	49			Newell & Pye (1968)
	20	66			Newell & Pye (1968)
Enteromorpha linza	10	15			Kanwisher (1966)
	20	23			Kanwisher (1966)
Ulva lactuca	10	13			Newell & Pye (1968)
	20	16			Newell & Pye (1968)
	20	35			Kanwisher (1966)
Phaeophyceae					
Ascophyllum nodosum	15	16			Brinkhuis (1977a)
Ascoseira mirabilis	0			41	Drew (1977)
	20			108	Drew (1977)
Laminaria hyperborea	3	10			Lüning (1971)
	9	33			Lüning (1971)
	16	13			Lüning (1971)
	8	26			Kain *et al.* (1975)
	7	18			Kain (1979)
Laminaria saccharina	7–18	3			Druehl (1967)
Himantothallus grandifolius	10			25	Drew (1977)
	20			41	Drew (1977)
Rhodophyceae					
Asparagopsis taxiformis	27	39			Buesa (1977)
Bostrychia binderi	30	13			Dawes *et al.* (1978)
Chondrus crispus	10	17			Newell & Pye (1968)
	20	31			Newell & Pye (1968)
	8	6			Mathieson & Burns (1971)
	15	26	7		Buggeln & Craigie (1973)
Gelidium pusillum	15			88	Littler & Murray (1974)
Griffithsia flosculosa	10	17			Newell & Pye (1968)
Gracilaria verrucosa	30	22			Dawes *et al.* (1978)
Porphyra umbilicalis	1			8	Drew (1977)

tures whereas in winter, with Q_{10} in the range of 2–3, the colder temperatures result in lower respiration rates. Healey (1972) found that respiration rates for arctic species were less than half of that shown for the more temperate species; this would be an advantage during long winter months beneath ice without light (Kanwisher 1966). On the other hand, Drew (1977) reported that respiration rates of a variety of antarctic seaweeds are not much lower than those of similar algae of temperate waters.

6.3 Respiratory carbon catabolism

The well documented seasonal variations in carbohydrate constituents, which reflect seasonal fluctuations of several environmental parameters affecting primary production, also suggest that energy is generated at the expense of stored carbohydrates when ambient conditions do not allow photosynthetic surplus production of assimilates (cf. Raven 1976). In fact, direct experimental evidence for the nature of the substrates remetabolized through dark respiration has been presented in only a few cases. Nonetheless, there is no established doubt of a universally operating respiratory pathway.

Johnston & Davies (1969) reported on the demonstration of malate dehydrogenase, isocitrate dehydrogenase, NAD-linked glyceraldehyde-3-phosphate dehydrogenase, aldolase, and glucose-6-phosphate dehydrogenase from cell-free preparations of a number of Fucales and Laminariales suggesting that at least brown algae possess respiratory pathways similar to those described for other organisms. Two enzymes of the catabolic Warburg-Dickens-Horecker pathway, glucose-6-phosphate dehydrogenase and 6-phosphogluconate dehydrogenase, have been studied in the red alga *Acrochaetium daviesii* grown under white, red, or blue light. Cultures of this species grown under blue light show a markedly higher activity of these enzymes accounting for a lower total carbohydrate content (van der Velde *et al.* 1975). A rather unusual copper-containing enzyme, hexose oxidase, has been characterized from *Chondrus crispus* and *Callophyllis cristata* (Sullivan & Ikawa 1973). This enzyme oxidizes glucose, galactose and their respective disaccharides to the corresponding hexonolactones and H_2O_2. The enzyme appears similar to that previously reported from *Iridaea flaccida* (Bean & Hassid 1956), but the metabolic significance of this reaction is unknown.

Lin & Hassid (1966) first observed that exogenously supplied ^{14}C-labelled hexoses are respired in *Fucus gardneri*. Exogenously fed mannitol is poorly utilized by *Fucus vesiculosus* (Bidwell & Ghosh 1963; Drew 1969), whereas endogenous mannitol is readily respired and seems to be interconvertible with laminaran in *Eisenia bicyclis* as well as *Fucus vesiculosus* (cf. Craigie 1974). Time dependent ^{14}C-labelling from endogenous and exogenous ^{14}C-mannitol occurs in intermediates of the tricarboxylic acid cycle and in diverse sugar phosphates in *Fucus serratus* and *Laminaria hyperborea* (Kremer 1975a; Küppers 1978). Kinetics of ^{14}C-labelling after pulse feeding of $^{14}CO_2$ suggest that in the red alga *Serraticardia maxima* floridean starch and floridoside are utilized in dark respiration (Nagashima *et al.* 1971).

It is likely that mannitol enters glycolysis after activation to mannitol-1-phosphate at the expense of ATP, and that its oxidation results in PEP. PEP is the substrate of both pyruvate kinase (converting it into pyruvate, which is fed into the Krebs cycle after condensation with acetyl-CoA) and PEP carboxykinase, which is the enzyme responsible for high dark carbon fixation in brown seaweeds. The two enzymes obviously compete for PEP as a substrate. This view of a regulatory competition is substantiated by the observations that 1, excess ATP inhibits the carboxylation of PEP by PEP-CK (Küppers 1978) and 2, blocking the Krebs cycle by monofluoracetate results in a considerable enhancement of dark fixation (Kremer unpublished). Anaerobic glycolysis of mannitol, particularly in growing areas of fronds, thus induces a light independent input of inorganic carbon partly compensating for the respiratory loss of stored assimilates.

6.4 Production of CO

Apart from dark respiratory release of CO_2 a number of marine algae including *Calliarthron* sp., *Iridaea flaccida, Rhodomela larix*, *Egregia menziesii*, *Laminaria* sp. and *Macrocystis pyrifera* have been found to produce carbon monoxide (Loewus & Delwiche 1963; Chapman & Tocher 1966). Carbon monoxide is retained in a free state in the large pneumatocysts of *Nereocystis luetkeana, Pelagophycus giganteus* and *Pelagophycus porra* (Chapman & Tocher 1966). The possible origin of CO in the pneumatocysts has repeatedly been attributed to dark respiration, but Foreman (1976) has pointed out that uncoupling of phosphorylation by DNP results in complete inhibition of CO production, and thus CO is not a respiratory product, but requires energy for formation. From the experimental evidence presented it appears that CO production occurs only during cell degradation (autolysis), a process which is closely related to growth and development of the large pneumatocysts of e.g. *Nereocystis*. Possibly certain phenolic compounds are involved in the production of carbon monoxide by a non-enzymatic auto-oxidation (Foreman 1976).

7 ORGANIC CARBON EXCHANGE

7.1 Release of organic carbon

Aquatic ecosystems typically contain appreciable amounts of organic matter in dissolved form (Williams 1975). This material originates not only from cell autolysis and subsequent decay of the particulate compounds, but also from controlled liberation of various extracellular products of algae. Reviews on this subject, mainly dealing with work on freshwater or marine microalgae, have been published by Fogg (1971, 1976) and by Hellebust (1974).

Brown seaweeds, particularly *Ascophyllum nodosum*, *Fucus* spp. and *Laminaria* spp., have repeatedly been reported to release large quantities of phenolic compounds (Craigie & McLachlan 1964; Sieburth 1969; Kroes 1970). In most cases the released UV-absorbing material was identified as a complex polyphenolic constituent virtually indistinguishable from the yellow-coloured humic substances present in high concentrations in inshore waters. Excretion of organic nitrogen has been reported by Newell *et al.* (1972). Many species of microalgae, both freshwater and marine forms, have been found to liberate glycolate as an extracellular product of photosynthesis (Fogg 1966, 1971). This observation is understood to be closely related to photorespiration (Tolbert 1974). Glycolate has been identified as an extracellular product of *Laurencia obtusa* by several methods (Fogg 1976). The amounts obtained represent a release of 0.5% of the total carbon fixed and is thus far less than has been encountered with marine phytoplankton. Rather low amounts of glycolate excretion have been recorded for a variety of further tropical seaweeds (Fogg 1976).

Dissolved organic material is also released in the form of carbohydrates, but there is controversy over the amounts normally released. Considerable amounts of photosynthetically fixed carbon have been reported to be liberated by littoral algae under some conditions. Khailov & Burlakova (1969) calculated that an average of about 37% of net photosynthate was released by plants placed in a system with stagnant seawater. Sieburth (1969), on the other hand, failed to show release with stagnant systems, but, for a flushing system, reported as much as about 30% of photosynthetically assimilated carbon to be released by *Fucus vesiculosus* and two *Laminaria* species. Hanson (1977) reported dissolved organic carbon release by a *Sargassum natans* community to be in the range of 20–30% of total assimilated material. The excretion of such a high percentage of photosynthate as reported by Khailov & Burlakova (1969), Sieburth (1969) and Hanson (1977) certainly would have a great impact on the ecology of seaweed dominated marine habitats.

On the other hand, Moebus & Johnson (1974) found very little release of organic carbon by *Ascophyllum nodosum* and *Fucus vesiculosus*, even under environmental stresses such as desiccation. Most of the small amounts of released dissolved carbon was confined to carbohydrates (Moebus *et al.* 1974). Harlin & Craigie (1975) demonstrated that between 0.6 and 1.4% of photosynthetically assimilated ^{14}C was secreted by *Ascophyllum nodosum* during a seven day period suggesting a low entry rate of newly fixed carbon into the exudation process. For all seaweeds which have been examined by Brylinski (1977) the amounts of dissolved organic carbon released relative to the total amount of carbon assimilated was rather small. Release ranged from 0.006 to 0.053 mg C $(\text{g dry weight} \times \text{h})^{-1}$ equivalent to about 1–4% of net carbon fixation *in situ*. A similar figure has been found in experiments with fucoid seaweeds by Munda & Kremer (1977).

The controversy is complicated by the conclusions of Johnston *et al.* (1977) and Hatcher *et al.* (1977), based on seasonal and annual carbon budgets for *Laminaria* spp., that 30–35% of net carbon fixation is lost as extracellular carbohydrates and polyphenolics, the rate apparently varying seasonally and diurnally.

7.2 Uptake of organic carbon

Though an extensive literature is available on uptake and utilization of organic carbon by unicellular or microalgae (Neilson & Lewin 1974), rather few such studies have been performed using seaweeds. (Vitamins are discussed in Chapter 10.) The earliest account was by Bidwell & Ghosh (1963) on uptake of ^{14}C-labelled glucose, acetate, and formate by *Fucus vesiculosus*. Radiocarbon was recovered from various fractions, but in view of the long incubation times and the small amounts of glucose actually metabolized it is possible that some of the metabolism was by contaminating microbes rather than by the seaweed itself. Lin & Hassid (1966) supplied mannose and glucose to *Fucus gardneri* by infiltration and found that these hexoses were slowly converted to other constituents. Drew (1969) observed that much of ^{14}C-labelled glucose taken up from aqueous media by a variety of brown seaweeds was readily leached out. The unleachable fraction was either converted to ^{14}C-mannitol or insoluble glucose-containing polymers or simply lost by respiration. *Ascophyllum nodosum* and *Pelvetia canaliculata* were also found to convert exogenously supplied mannose, but not fructose or galactose. Using the filamentous brown alga *Giffordia mitchellae*, Kremer (1973a) found that light increased uptake of glucose and fructose and he recovered appreciable amounts of ^{14}C from mannitol (cf. Weidner *et al.* 1975). Uptake and metabolism of ^{14}C-labelled aspartate, malate and mannitol have been investigated by Küppers (1978) using tissue slices of *Laminaria hyperborea*.

Photoassimilation of ^{14}C-acetate by *Ulva lactuca* has been studied by Gemmill & Galloway (1974). Uptake was much higher in the light than in the dark and was sensitive to certain photosynthetic poisons. The action spectrum of acetate uptake conforms to the photosynthetic action spectrum of this species. The results point to an intracellular dependence of acetate assimilation on the Krebs cycle, possibly in cooperation with the glyoxylate pathway rather than simply releasing CO_2 from acetate and subsequent refixation via the RPP cycle.

Hartmann & Aufermann (1973) found considerable uptake of leucine by the red alga *Polysiphonia urceolata* and subsequent leucine degradation by an unspecific decarboxylase to isoamylamine. The amine formed was recovered in almost equal amounts in algal extracts and the incubation medium. It is supposed that the amine does not accumulate within the algal cells, but is rather bound to the acidic polysaccharides. Obviously, the endogenous amino acid pool furnishes no substrate source to the decarboxylating enzyme. The ecological implications of this metabolic reaction are not yet clear. On the basis of the decarboxylating enzyme involved a general treatment of the Rhodophyceae from a chemotaxonomic point of view has been attempted (Aufermann 1978).

The uptake of leucine and 16 further amino acids by the brown alga *Giffordia mitchellae* has been investigated by Schmitz & Riffarth (1980). Leucine revealed the highest uptake by a light independent process which is not simply explained as diffusion. It was accumulated up to 27-fold within the cells, and under nearly natural conditions the K_m of uptake is $0.3–1.4 \times 10^{-4}$ M, while V_{max} was found to be 3000–3800 nmol (100 mg dry weight × h)$^{-1}$. Leucine uptake can be inhibited competitively and by CCCP.

It seems unlikely that the organic carbon sources of the marine environment play any significant role for the total carbon balance of seaweeds. Furthermore, given the background of mostly very low rates of uptake of organic material it is not very probable that seaweeds living in an unfavourable habitat could rely on sources of carbon other than the inorganic pool, in contrast to speculation by Wilce (1966), Mann (1973) and Drew (1977).

Seaweeds which are parasitic clearly take up organic carbon from the host plant, as has been shown for *Harveyella* on *Odonthalia* (Goff 1979). Studies have also been carried out on the possibility of nutrient transfer to the epiphyte *Polysiphonia lanosa* from its 'host' *Ascophyllum nodosum* (Harlin & Craigie 1975, Turner & Evans 1977). So far little evidence of exchange of organic carbon sources has been found, although Turner & Evans (1977) showed that *P. lanosa* could take up exogenously supplied glucose.

8 PHOTOSYNTHETIC RATES AND PRODUCTIVITY

8.1 Rates of photosynthesis

Several environmental factors markedly influence uptake and assimilation of inorganic carbon. The photochemical reaction complex of photosynthesis (see Chapter 13) depends directly on light and is influenced by factors such as photon flux density (PFD)* and quality of the incident light. The dark reactions are biochemical reaction sequences and therefore sensitive to ambient temperatures. In the habitats of the marine environment the available light, the temperature, and some chemical parameters such as pH and salinity are the main abiotic factors to be considered.

Numerous investigations on measurable rates of photosynthesis (apparent photosynthesis, net photosynthesis) using a wide variety of dimensions and reference systems have been published. Table 14.4 shows rates of photosynthesis in terms of photosynthetic carbon fixation in several benthic marine algae covering a wide range of structural types and taxa. In general, photosynthetic carbon assimilation in seaweeds compares well with average net photosynthetic rates of angiosperms and even C_3 crop plants, which achieve a net carbon fixation of 90–800 μmol CO_2 dm^{-2} h^{-1} or 220–1300 μmol CO_2 g^{-1} dry weight h^{-1}, respectively, at saturating conditions and 330 ppm CO_2 (Burris & Black 1976).

* Ed. note: The term photon flux density is to be preferred to irradiance and illuminance—see Chapter 13, p. 466.

Table 14.4 Rates of apparent photosynthesis in seaweeds as related to environmental conditions such as temperature (t) and photon flux density (PFD). dwt = dry weight, fwt = fresh weight, chl = chlorophyll. (Many values taken from authors' graphs.)

Species	t (°C)	PFD ($\mu E\ m^{-2}\ s^{-1}$)	Rate ($\mu mol\ CO_2$)				Reference
			g^{-1} dwt h^{-1}	g^{-2} fwt h^{-1}	$dm^{-2}\ h^{-1}$	mg^{-1} chl h^{-1}	
Chlorophyceae							
Acanthophora spicifera	24	200				446	Dawes *et al.* (1978)
Acetabularia crenulata	30	1000	143				Buesa (1977)
Acrosiphonia centralis	10	500	468				King & Schramm (1976b)
Enteromorpha compressa	25	250	316–558				Brinkhuis (1977b)
Monostroma hariotii	1	250			62		Drew (1977)
Monostroma grevillei	10	750	1466				King & Schramm (1976b)
Ulothrix speciosa	10	750	1092				King & Schramm (1976b)
Phaeophyceae							
Ascophyllum nodosum	12	250	119–138				Brinkhuis (1977a)
	10	500	80		66		Kremer & Küppers (1977)
Ascoseira mirabilis	16	250					Drew (1977)
Dictyopteris australis	25	200	324				Doty (1971)
Dictyopteris justii	27	700	118				Buesa (1977)
Desmarestia anceps	15	250			124		Drew (1977)
Fucus spiralis	21	600			340		Stocker & Holdheide (1937)
	10	200	214				Niemeck & Mathieson (1978)
	20	200	561				Niemeck & Mathieson (1978)
	10	500	24–88				Küppers & Kremer (1978)
Laminaria digitata	15	1075	37–125	9–19	74–99		King & Schramm (1976a)
	15	500	124				King & Schramm (1976b)
	10	500	10–73				Küppers & Kremer (1978)
Laminaria hyperborea	8	105			133		Lüning (1971)
	4	80			15–39		Willenbrink *et al.* (1975)
	14	80			39–76		Willenbrink *et al.* (1975)
	10	500	6–21				Küppers & Kremer (1978)

Laminaria longicruris	16	—			8–39		Hatcher *et al.* (1977)
Laminaria saccharina	—	500			49–67		Johnston *et al.* (1977)
	10	500	59				King & Schramm (1976b)
	10	500	16–80				Küppers & Kremer (1978)
Macrocystis integrifolia	13	300			50–110	29–84	Willenbrink *et al.* (1979a)
Macrocystis pyrifera	—	500			67–104		Wheeler (1978, 1980)
	13	300			80–190		Willenbrink *et al.* (1979a)
Nereocystis luetkeana	13	300			40–76	28–190	Willenbrink *et al.* (1979a)
Rhodophyceae							
Asparagopsis taxiformis	27	1000	174				Buesa (1977)
Bostrychia binderi	24	200				44	Dawes *et al.* (1978)
Catenella repens	24	200				80	Dawes *et al.* (1978)
Ceramium nitens	32	1000	867				Buesa (1977)
Chondrus crispus	5	300	53–85				Mathieson & Burns (1971)
	15	300	107–147				Mathieson & Burns (1971)
	10	500	21				Kremer & Küppers (1977)
Cryptopleura violacea	13	300	34				Kremer (1979c)

Rates of photosynthesis are proportional to PFD over a certain range of levels, beyond which a plateau is reached (see Chapter 13). The efficiency with which a plant can utilize particularly low PFD can be indirectly obtained from the K_m-value of the PFD/carbon fixation curve, the K_m-value being the PFD that results in half of the maximum apparent photosynthesis. Plants of a high PFD environment (light normally not limiting) have a relatively higher K_m; plants of a shade environment must be more efficient and usually show a smaller K_m. Sublittoral and eulittoral seaweeds are often different in their respective K_m values, as can be recalculated from the data of Kageyama & Yokohama (1974): half-saturating PFD is $>60\ \mu E\ m^{-2}\ s^{-1}$ in a eulittoral species ('light plant'), but usually $<50\ \mu E\ m^{-2}\ s^{-1}$ in a sublittoral alga ('shade plant'). In the giant kelp, *Macrocystis*, photosynthetic characteristics of the different blades of the frond are predominantly controlled by their ontogeny (Wheeler 1978, 1980). Similar findings have been obtained from *Laminaria* spp. (King & Schramm 1976a; Küppers & Kremer 1978) and a number of other seaweeds (Kremer 1975b; Khailov 1976, 1978).

Studies on several marine algae indicate that maximum photosynthetic rates of a single species are not exclusively light dependent, but are due to seasonal adaptations as well (Zavodnik 1973a, 1973b; King & Schramm 1976b; Brinkhuis 1977a, 1977b; Titlyanov & Zvalinsky 1978). Seasonal fluctations of photosynthesis in *Laminaria* spp. (Lüning 1971; Hatcher *et al.* 1977) can partially be attributed to seasonally differing amounts of carboxylating enzymes (Küppers 1978; Küppers & Weidner 1980). Recent observations have shown that even diurnal rhythms of photosynthesis and photosynthetic enzymes occur in marine algae (Zavodnik 1978; Kageyama *et al.* 1979; Yamada *et al.* 1979a).

Since the utilization of products from the light reactions in the dark phase of photosynthetic carbon assimilation is thermally sensitive, the ambient temperature can well limit the overall photosynthetic rates by control of subsequent metabolic processes (cf. Harris 1978). If the Q_{10} is taken into account, it is found that the values for photosynthesis are consistently higher in the summer than in the winter, indicating a higher thermal sensitivity of the species investigated in the summer (Hato & Yokohama 1976; Wheeler 1982). Arctic algae show photosynthesis/temperature responses very similar to those in algae from temperate environments (Healey 1972; Drew 1977). On the other hand, some experiments indicate that *Fucus* sp. is able to carry out a notable amount of apparent photosynthesis even at $-1.5°C$ (Healey 1972). Similar observations have been reported from *Fucus* and *Laminaria* (Kanwisher 1966). The combined action of ambient temperature along with other environmental factors has been examined by Mathieson & Dawes (1974) for *Eucheuma*, by Brinkhuis & Jones (1974) for *Chondrus crispus*, by Zavodnik (1975) for Adriatic seaweeds, and by Lehnberg (1978) for *Delesseria sanguinea* from the Baltic.

A basic question in the ecology of intertidal (eulittoral) algae, which is still unresolved, is whether photosynthetic rates are affected during exposure to the atmosphere or not. Although Stocker & Holdheide (1937) found that photosynthetic rates of rocky shore algae under conditions of exposure were higher than those of submerged thalli, Bidwell & Craigie (1963) reported on drastically reduced assimilation rates in *Fucus vesiculosus* during low tide. Investigations by Brown & Johnson (1964), Kremer & Schmitz (1973) and Quadir *et al.* (1979) showed that photosynthesis of intertidal species exposed to the atmosphere is either lower or approximately equal to assimilation in seawater. Johnson *et al.* (1974) found photosynthetic performance under emerged conditions to be considerably greater. Brinkhuis *et al.* (1976) reported that desiccation of exposed thalli had the most pronounced effect on photosynthesis which increased slightly between 0 and 25% water loss, then levelled off, and decreased sharply at water losses exceeding 50%. Photosynthesis has been shown to be sensitive to salinity of the media used. However, several investigations (Ogata & Schramm 1971; Schramm & Ohno 1973; Munda & Kremer 1977) showed that physiologically and biochemically significant effects are seen after relatively long-term treatments.

The pH changes of the medium, influence rates of photosynthesis by affecting protoplasmatic ion balances and by changing the relative CO_2/HCO_3^- $/CO_2^{2-}$-ratios. Effects of pH upon photosynthesis in seaweeds have been studied several times. Dromgoole (1978) found that photosynthesis of *Carpophyllum* rapidly decreased at pH 8.6, while Jolliffe & Tregunna (1970) observed no photosynthetic response to pH (or free CO_2) between 8.0 and 9.2. Lower pH values increase the concentration of free CO_2: at pH 7.4 concentration of CO_2 is considerably higher than at pH 8.2 (100 μmol l^{-1}). Species utilizing CO_2 (see Table 14.1) should hence exhibit relatively higher photosynthetic rates at decreasing pH. This has been confirmed for *Carpophyllum* (Dromgoole 1978). Even the rates of *Palmaria palmata*, a possible bicarbonate utilizer, increase down to pH 6.5 (Robbins 1979).

Algae living in high tide pools must be able to tolerate changes in alkalinity of the surrounding seawater medium, since appreciable rates of photosynthetic carbon uptake rapidly influence the pH values of such pools. In contrast, those species living in the lower intertidal region or subtidally are not likely to be exposed to drastic pH changes.

The increasing amounts of environmental toxins and other pollutants in several parts of the sea gave rise to the problem of whether such compounds affect photosynthetic activity of the phytobenthos. Schramm (1972) investigated the effects of crude oil on gas exchange during atmospheric exposure. Apart from probable toxic effects of crude oil components, oil films coating the surface of *Porphyra umbilicalis* and *Fucus vesiculosus* interfered with photosynthetic CO_2 uptake and thus caused equivalent depression of fixation rates. Lead pollution of seawater (100–1000 ppb) affects photosynthesis in *Fucus virsoides, Ulva rigida*, and *Padina pavonica* (Zavodnik 1977). When cadmium is added to the culture medium of *Laminaria saccharina* and *Ulva lactuca* between 0.25 and

Table 14.5
Productivity of seaweeds*.

Species	Locality	Net production		Reference
		$g\ C\ m^{-2}\ d^{-1}$	$g\ C\ m^{-2}\ a^{-1}$	
Chlorophyceae				
Caulerpa prolifera	Canary Islands	1.0		Johnston (1969)
Codium fragile	Connecticut	22	4700	Wassman & Ramus (1973)
Ulva californica	California	2.8		Littler (1973)
Phaeophyceae				
Cystoseira abies-marina	Canary Islands	7.5		Johnston (1969)
Egregia laevigata	California	1.7		Littler & Murray (1974)
Fucus sp.	Massachussetts	20		Kanwisher (1966)
Halidrys dioica	California	1.4		Littler & Murray (1974)
Laminaria digitata	Helgoland (North Sea)	3.4†		
Laminaria hyperborea	Helgoland (North Sea)	3.8†		
	Scotland		377–775	Jupp & Drew (1974)
Laminaria longicruris	Nova Scotia		1750	Mann (1073)
Laminaria longicruris	Nova Scotia	3.9		Westlake (1963)
Laminaria ochroleuca	Spain	1.6–4.7		John (1971)
Laminaria saccharina	Helgoland (North Sea)	4.9†		
	Scotland		120	Johnston *et al.* (1977)
Macrocystis pyrifera	California	6.8		Towle & Pearse (1973)
		9.5		Jackson (1977)
		6.1		Wheeler (1978)
Macrocystis integrifolia	British Columbia	6.7		Kremer (unpublished)
Nereocystis luetkeana	British Columbia	5.4		Kremer (unpublished)
Padina pavonica	Canary Islands	1.6		Johnston (1969)
Sacchorhiza polyschides	Spain	1.4–10.7		John (1971)
Sargassum agardianum	California	2.2		Littler & Murray (1974)

Rhodophyceae				
Coralline algae	Hawaii		2080	Littler (1973)
Gelidium pusillum	California	3.1		Littler & Murray (1974)
Gigartina canaliculata	California	2.0		Littler & Murray (1974)
Gigartina exasperata	Washington		439–758	Waaland (1977)
Iridaea cordata	Washington		144–1012	Waaland (1977)
Pterocladia capillacea	California	1.3		Littler & Murray (1974)

* Partly converted from authors' data.
† Calculated from data of Lüning (1969) and Küppers & Kremer (1978).

2.00 ppm, rates of photosynthesis drastically decrease (Markham *et al.* 1980a, 1980b). Sewage polluted seawater reduces rates of photosynthetic carbon assimilation of *Ascophyllum nodosum* and *Fucus vesiculosus* (Munda 1974).

8.2 Carbon budgets

Measurements of the primary productivity of marine macroalgae is largely a function of a number of interactions between the physiology of the organisms examined, the methodology used, and the physics of the environment. For reasons discussed by Harris (1978) it is clear that calculated rates of productivity depend on more than just the rates of photosynthesis and respiration measured for an organism in question under a given set of environmental conditions. Experimental determinations of photosynthetic production in the laboratory for a relatively short period is not always and not necessarily a measurement of *in situ* productivity. Factors which are not readily evaluated and which nonetheless display remarkable effects on photosynthetic productivity include the total nutrient supply of the seaweed investigated, as has been shown for a variety of macroalgae (Topinka & Robbins 1976; Jackson 1977; Chapman 1976; Chapman & Craigie 1977; Chapman *et al.* 1978). Primary productivity of some marine macroalgae are compiled in Table 14.5. In particular, *Macrocystis pyrifera* in the kelp beds off California was found to achieve net primary productivity exceeding about 5 g C m^{-2} dm^{-1} as based on measurements of standing crop and photosynthetic carbon gain (Blinks 1955; Towle & Pearse 1973; Jackson 1977; Wheeler 1978). Apart from two records for *Fucus* sp. (Kanwisher 1966) and *Codium fragile* (Wassman & Ramus 1973) which have not yet been confirmed, other species of green, red and brown seaweeds are less productive than the giant kelp. *Macrocystis*, which attains growth rates in the range of up to 1600 g fresh weight dm^{-1}, equivalent to 5.7% dm^{-1} (Wheeler 1978), thus appears to be one of the most productive plants of the marine environment. It should be emphasized, however, that measurements of productivity cannot always be extrapolated to give the annual rates of primary productivity. Many of the data in the literature suffer from such shortcomings and should hence be regarded with caution.

Net production by the temperate *Laminaria longicruris* has been reported to amount to 1750 g C m^{-2} a^{-1} and thus to be equivalent to the primary productivity of various marine grasses in tropical waters and even to a number of terrestrial crop plants (Mann 1973; Mann & Chapman 1973). Although the data for annual productivity included in Table 14.5 are mostly hypothetical extrapolations and therefore speculative, it is clear that many seaweeds or seaweed communities are fairly productive (Johnston 1969; Littler 1973; Littler & Murray 1974, 1978) and strictly comparable to most of the terrestrial vegetation units (cf. Leith 1975). Extensive productivity studies along with various constitutive characters of the species investigated, which allow an evaluation of environmental variables as well as of infraspecific variability, have been presented by Wallentinus (1978).

Annual carbon budgets, derived from long-term *in situ* measurements throughout the year, have been obtained for *Laminaria longicruris* (Hatcher *et al.* 1977) and for *Laminaria saccharina* (Johnston *et al.* 1977). Carbon assimilation by *Laminaria longicruris* was 6.8 mg C cm^{-2} frond surface a^{-1}. Approximately 45% of this was utilized for the production of the new frond during the seasonal growth cycle, about 12% accounted for by storage of carbon in mature frond tissue, 8% was needed for stipe growth, and the rest (35%) was assumed to be lost. The annual cumulative carbon budget of *Laminaria saccharina* was found to be about 67 g C a^{-1} for an entire frond corresponding to a primary productivity of the population in the range of 120 g C m^{-2} a^{-1}. A figure of about 40–50% is attributed to distal tissue loss. If gross input of C is taken as 100%, 37% is accounted for by respiration, 13% is lost by release, and finally 50% enters growth and storage (including 10% leakage and 40% decay) (Johnston *et al.* 1977).

The calculations of gross productivity as given by Johnston *et al.* (1977) for *Laminaria saccharina* include rates of dark carbon assimilation. The assumption that dark fixation is a component of the carbon budget has not been considered in previous calculations of primary productivity. Carbon input due to light independent carbon fixation particularly in the brown seaweeds yields no overall net carbon gain, but significantly compensates for respiratory loss of stored compounds (Kremer 1979c).

It is now well established that particularly in the temperate latitudes marine algae undergo distinct seasonal fluctuations in growth, reproduction and a number of biochemical changes including carbohydrate content (Boney 1965; Zavodnik 1971; 1973a, b, Fuller & Mathieson 1972; Bérard-Therriault & Cardinal 1973; Dawes *et al.* 1974). Chapman & Craigie (1978) reported that the growth rates of *Laminaria longicruris* increase rapidly while laminaran and mannitol contents showed minimum values, in agreement with earlier observations. Lüning (1979) correlated the seasonal growth pattern of three *Laminaria* species with the content of reserve carbohydrates. In *Laminaria digitata*, blades continued to elongate until autumn, consisted of much younger tissue and showed markedly lower amounts of reserve material than *Laminaria saccharina* and *Laminaria hyperborea* which stopped growing in summer. However, the extent of storage carbohydrate accumulation does not trigger the onset of the late winter growth (Lüning 1971; Kain 1979), nor does it terminate or reduce the vegetative growth (Lüning 1979).

9 REFERENCES

Akagawa H., Ikawa T. & Nisizawa K. (1972a) Initial pathway of dark $^{14}CO_2$-fixation in brown algae. *Bot. Mar.*, **15**, 119–25.

Akagawa H., Ikawa T. & Nisizawa K. (1972b) $^{14}CO_2$-fixation in marine algae with special reference to the dark-fixation in brown algae. *Bot. Mar.*, **15**, 126–32.

Akagawa H., Ikawa T. & Nisizawa K. (1972c) The enzyme system for the entrance of $^{14}CO_2$ in the dark fixation of brown algae. *Plant & Cell Physiol.*, **13**, 999–1016.

Akazawa T. & Osmond C. B. (1976) Structural properties and ribulosebiphosphate carboxylase and oxygenase activity of fraction-1-protein from the marine alga *Halimeda cylindracea* (Chlorophyta). *Aust. J. Plant Physiol.*, **3**, 93–103.

Aufermann B. (1978) Zur Chemotaxonomie mariner Rhodophyceen am Beispiel einer Leucin-Decarboxylase. *Bibliotheca Phycolog.*, **43**, 1–136.

Bean R. C. & Hassid W. Z. (1955) Assimilation of $^{14}CO_2$ by a photosynthesizing red alga, *Iridophycus flaccidum*. *J. Biol. Chem.*, **212**, 411–25.

Bérard-Therriault L. & Cardinal A. (1973) Variations de la teneur en acide alginique des Fucacées de l'estuaire du Saint-Laurent (Québec) en fonction de certains facteurs écologiques. *Bot. Mar.*, **16**, 96–102.

Bidwell R. G. S. & Craigie J. S. (1963) A note on the greatly reduced ability of *Fucus vesiculosus* to absorb or to evolve CO_2 when not submerged. *Can. J. Bot.*, **41**, 179–82.

Bidwell R. G. S., Craigie J. S. & Krotkov G. (1958) Photosynthesis and metabolism in marine algae. III. Distribution of photosynthetic carbon from $^{14}CO_2$ in *Fucus vesiculosus*. *Can. J. Bot.*, **36**, 581–90.

Bidwell R. G. S. & Ghosh N. R. (1963) Photosynthesis and metabolism of marine algae. V. Respiration and metabolism of ^{14}-C-labelled glucose and organic acids supplied to *Fucus vesiculosus*. *Can. J. Bot.*, **41**, 155–63.

Bidwell R. G. S., Levin W. B. & Shephard D. C. (1970) Intermediates of photosynthesis in *Acetabularia mediterranea* chloroplasts. *Plant Physiol.*, **45**, 70–5.

Black C. C. Jr. & Bender M. M. (1976) $\delta^{13}C$ values in marine organisms from the Great Barrier Reef. *Aust. J. Plant Physiol.*, **3**, 25–32.

Black C. C. Jr, Burris J. E. & Everson R. G. (1976) Influence of oxygen concentration on photosynthesis in marine plants. *Aust. J. Piant Physiol.*, **3**, 81–6.

Blinks L. R. (1955) Photosynthesis and productivity of littoral marine algae. *J. Mar. Res.*, **14**, 363–73.

Boney A. D. (1965) Aspects of the biology of the seaweeds of economic importance. *Adv. Mar. Biol.*, **3**, 105–253.

Borowitzka M. A. (1977) Algal calcification. *Oceanogr. Mar. Biol. Ann. Rev.*, **15**, 189–223.

Borowitzka M. A. & Larkum A. W. D. (1976a) Calcification in the green alga *Halimeda*. II. The exchange of calcium ions and the occurrence of age gradients in calcification and photosynthesis. *J. Exp. Bot.*, **27**, 864–78.

Borowitzka M. A. & Larkum A. W. D. (1976b) Calcification in the green alga *Halimeda*. III. The sources of inorganic carbon for photosynthesis and calcification and a model of the mechanism of calcification. *J. Exp. Bot.*, **27**, 879–93.

Borowitzka M. A. & Larkum A. W. D. (1976c) Calcification in the green alga *Halimeda*. IV. The action of metabolic inhibitors on photosynthesis and calcification. *J. Exp. Bot.*, **27**, 894–907.

Brinkhuis B. H. (1977a) Seasonal variations in salt-marsh macroalgae photosynthesis. I. *Ascophyllum nodosum* ecad *scorpioides*. *Mar. Biol.*, **44**, 165–75.

Brinkhuis B. H. (1977b) Seasonal variations in salt-marsh macroalgae photosynthesis. II. *Fucus vesiculosus* and *Ulva lactuca*. *Mar. Biol.*, **44**, 177–86.

Brinkhuis B. H. & Jones R. F. (1974) Photosynthesis in whole plants of *Chondrus crispus*. *Mar. Biol.*, **27**, 137–41.

Brinkhuis B. H., Temple N. R. & Jones R. F. (1976) Photosynthesis and respiration of exposed salt-marsh fucoids. *Mar. Biol.*, **34**, 349–59.

Brown D. L. & Tregunna E. B. (1967) Inhibition of respiration during photosynthesis by some algae. *Can. J. Bot.*, **45**, 1135–43.

Brown F. A., Freeland R. O. & Ralph C. L. (1955) Persistent rhythms of O_2-consumption in potatoes, carrots and the seaweed *Fucus*. *Plant Physiol.*, **30**, 280–92.

BROWN J. M. A. & JOHNSON A. (1964) Preliminary studies on the ecology and physiology of *Scytothamnus australis*. *Bot. Mar.*, **6**, 233–46.

BROWN V., DUCKER S. C. & ROWAN K. S. (1977) The effect of orthophosphate concentrations on the growth of the coralline algae (Rhodophyta). *Phycologia*, **16**, 125–31.

BRYLINSKY M. (1977) Release of dissolved organic matter by some marine macrophytes. *Mar. Biol.*, **39**, 213–20.

BUESA R. J. (1977) Photosynthesis and respiration of some tropical marine plants. *Aquat. Bot.*, **3**, 203–16.

BUGGELN R. G. & CRAIGIE J. S. (1973) The physiology and biochemistry of *Chondrus crispus* Stackhouse. *Proc. N.S. Inst. Sci.*, **27**, 81–102.

BURRIS J. E. (1977) Photosynthesis, photorespiration, and dark respiration in eight species of algae. *Mar. Biol.*, **39**, 371–9.

BURRIS J. E., HOLM-HANSEN O. & BLACK C. C. JR. (1976) Glycine and serine production in marine plants as a measure of photorespiration. *Aust. J. Plant Physiol.*, **3**, 87–92.

BURRIS R. H. & BLACK C. C. (1976) *CO_2 Metabolism and Plant Productivity*. University Park Press, Baltimore.

CALABRESE G. (1971) Resistenza di *Pterocladia capillacea* (Gmel.) Born. et Thur. a variazone di salinità. *G. Bot. Ital.*, **105**, 125–30.

CAPERON J. & SMITH D. F. (1978) Photosynthetic rates of marine algae as a function of inorganic carbon concentration. *Limnol. Oceanogr.*, **23**, 704–8.

CHAPMAN A. R. O. (1976) Nitrate, nitrogen and the seasonal growth of *Laminaria longicruris*. *Br. Phycol. J.*, **11**, 192–3.

CHAPMAN A. R. O. & CRAIGIE J. S. (1977) Seasonal growth in *Laminaria longicruris*: dissolved inorganic nutrients and internal reserves of nitrogen. *Mar. Biol.*, **40**, 197–205.

CHAPMAN A. R. O. & CRAIGIE J. S. (1978) Seasonal growth in *Laminaria longicruris*: relations with reserve carbohydrate storage and production. *Mar. Biol.*, **46**, 209–13.

CHAPMAN A. R. O., MARKHAM J. W. & LÜNING K. (1978) Effects of nitrate concentrations on the growth and physiology of *Laminaria saccharina* (Phaeophyta) in culture. *J. Phycol.*, **14**, 195–8.

CHAPMAN D. J. & TOCHER R. D. (1966) Occurrence and production of carbon monoxide in some brown algae. *Can. J. Bot.*, **44**, 1438–42.

CLAUSS H. (1972) Der Einfluss von Rot- und Blaulicht auf die Photosynthese von *Acetabularia mediterranea* und auf die Verteilung des assimilerten Kohlenstoffs. *Protoplasma*, **74**, 357–79.

CLAUSS H. (1975) Photosynthese-Intensität und Stoffwechsel der *Acetabularia*-Zelle in Abhängigkeit von ihrem Entwicklungszustand. *Protoplasma*, **83**, 147–66.

COBB A. H. (1977) The relationship of purity to photosynthetic activity in preparations of *Codium fragile* chloroplasts. *Protoplasma*, **92**, 137–46.

COOPER T. G., CHEN T. T., WOOD H. G. & BENEDICT C. R. (1968) The carboxylation of phosphoenolpyruvate and pyruvate. I. The active species of CO_2 utilized by phosphoenolpyruvate carboxykinase, carboxytransphosphorylase, and pyruvate carboxylase. *J. Biol. Chem.*, **243**, 3857–63.

COUGHLAN S. & TATTERSFIELD D. (1977) Photorespiration in larger littoral algae. *Bot. Mar.*, **20**, 265–6.

CRAIGIE J. S. (1974) Storage products. In *Algal Physiology and Biochemistry*, ed. Stewart W. D. P., pp. 206–35. Blackwell Scientific Publications, Oxford.

CRAIGIE J. S. & MCLACHLAN J. (1964) Excretion of coloured ultraviolet-absorbing substances by marine algae. *Can. J. Bot.*, **42**, 23–33.

CRAIGIE J. S., MCLACHLAN J., MAJAK W., ACKMAN R. G. & TOCHER C. S. (1966) Photosynthesis in algae. II. Green algae with special reference to *Dunaliella* spp. and *Tetraselmis* spp. *Can. J. Bot.*, **44**, 1247–54.

DARLEY W. (1974) Silicification and calcification. In *Algal Physiology and Biochemistry*, ed. Stewart W. D. P., pp. 655–75, Blackwell Scientific Publications, Oxford.

DAWES C. J., LAWRENCE J. M., CHENEY D. P. & MATHIESON A. C. (1974) Ecological studies of floridean *Eucheuma* (Rhodophyta, Gigartinales). III. Seasonal variations of carrageenan, total carbohydrate, protein, and lipid. *Bull. Mar. Sci.*, **24**, 287–99.

DAWES C. J., MOON R. E. & DAVIS M. A. (1978) The photosynthetic and respiratory rates and tolerances of benthic algae from a mangrove and salt marsh estuary: A comparative study. *Est. Cstl Mar. Sci.*, **6**, 175–86.

DIGBY P. S. (1977) Photosynthesis and respiration in the coralline algae *Clathromorphum circumscriptum* and *Corallina officinalis* and the metabolic basis of calcification. *J. Mar. biol. Assoc. U.K.*, **57**, 1111–24.

DOTY M. S. (1971) The productivity of benthic frondose algae at Waikiki beach. *Univ. Hawaii Bot. Sci. Papers*, **22**, 1–119.

DREW E. A. (1969) Uptake and metabolism of exogenously supplied sugars by brown macroalgae. *New Phytol.*, **68**, 35–43.

DREW E. A. (1977) The physiology of photosynthesis and respiration in some antarctic marine algae. *Br. Antarctic Survey Bull.*, **46**, 59–76.

DROMGOOLE F. I. (1978) The effects of oxygen on dark respiration and apparent photosynthesis in marine macro-algae. *Aquat. Bot.*, **4**, 281–97.

DRUEHL L. D. (1967) Distribution of two species of *Laminaria* as related to some environmental factors. *J. Phycol.*, **3**, 103–8.

FOGG G. E. (1966) The extracellular products of algae. *Oceanogr. Mar. Biol. Ann. Rev.*, **4**, 195–212.

FOGG G. E. (1971) Extracellular products of algae in freshwater. *Arch. Hydrobiol. Suppl.*, **5**, 1–25.

FOGG G. E. (1976) Release of glycollate from tropical marine plants. *Aust. J. Plant Physiol.*, **3**, 57–61.

FOREMAN R. E. (1976) Physiological aspects of carbon monoxide production by the brown alga *Nereocystis luetkeana*. *Can. J. Bot.*, **54**, 352–60.

FRALICK R. A. & MATHIESON A. C. (1973) Ecological studies of *Codium fragile* in New England, U.S.A. *Mar. Biol.*, **19**, 127–32.

FRALICK R. A. & MATHIESON A. C. (1975) Physiological ecology of four *Polysiphonia* species (Rhodophyta, Ceramiales). *Mar. Biol.*, **29**, 29–36.

FULLER S. W. & MATHIESON A. C. (1972) Ecological studies of economic red algae. IV. Variations of carrageenan concentration and properties in *Chondrus crispus* Stackh. *J. Exp. Mar. Biol. Ecol.*, **10**, 49–58.

GEMMILL E. R. & GALLOWAY R. A. (1974) Photoassimilation of ^{14}C-acetate by *Ulva lactuca*. *J. Phycol.*, **10**, 359–66.

GIBBS M. & LATZKO E. (eds) (1979) Photosynthetic carbon metabolism and related processes. In *Encyclopedia of Plant Physiology*, vol. 6, new series. Springer-Verlag, Berlin, New York.

GILES K. L. & SARAFIS V. (1972) Chloroplast survival and division *in vitro*. *Nature New Biol.*, **236**, 56–68.

GOFF L. J. (1979) The biology of *Harveyella mirabilis* (Cryptonemiales, Rhodophyceae). VI. Translocation of photoassimilated C. *J. Phycol.*, **15**, 82–7.

GRAHAM D. & SMILLIE R. M. (1976) Carbonate dehydratase in marine organisms of the Great Barrier Reef. *Aust. J. Plant Physiol.*, **3**, 113–9.

GRANT B. R., HOWARD R. J. & GAYLOR K. R. (1976) Isolation and properties of chloroplasts from the siphonous green alga *Caulerpa simpliuscula*. *Aust. J. Plant Physiol.*, **3**, 639–51.

HANSON R. B. (1971) Pelagic *Sargassum* community metabolism. Carbon and nitrogen. *J. Exp. Mar. Biol. Ecol.*, **29**, 107–18.

HARLIN M. M. & CRAIGIE J. S. (1975) The distribution of photosynthate in *Ascophyllum nodosum* as it relates to epiphytic *Polysiphonia lanosa*. *J. Phycol.*, **11**, 109–13.

HARRIS G. P. (1978) Photosynthesis, productivity and growth: The physiological ecology of phytoplankton. *Arch. Hydrobiol.*, **10** (suppl.) 1–171.

HARTMANN T. & AUFERMANN B. (1973) Zur Physiologie der Aminbildung bei der marinen Rotalge *Polysiphonia urceolata*. *Mar. Biol.*, **21**, 70–4.

HATCHER B. G., CHAPMAN A. R. O. & MANN K. H. (1977) An annual carbon budget for the kelp *Laminaria longicruris*. *Mar. Biol.*, **44**, 85–96.

HATO M. & YOKOHAMA Y. (1976) Photosynthesis-temperature relationship in seaweeds and their seasonal changes in the colder region of Japan. *Bull. Jap. Soc. Phycol.*, **24**, 1–9.

HEALEY F. P. (1972) Photosynthesis and respiration of some arctic seaweeds. *Phycologia*, **11**, 267–71.

HELLEBUST J. A. (1974) Extracellular products. In *Algal Physiology and Biochemistry*, ed. Stewart W. D. P., pp. 838–63. Blackwell Scientific Publications, Oxford.

HINDE R. (1978) The metabolism of photosynthetically fixed carbon by isolated chloroplasts from *Codium fragile* (Chlorophyta, Siphonales) and by *Elysia viridis* (Molusca, Sacoglossa). *Biol. J. Linn. Soc.*, **10**, 329–42.

HOWARD R. J., GAYLER K. R. & GRANT B. R. (1975) Products of photosynthesis in *Caulerpa simpliuscula*. *J. Phycol.*, **11**, 463–71.

IKAWA T., WATANABE T. & NISIZAWA K. (1972) Enzymes involved in the last steps of the biosynthesis of mannitol in brown algae. *Plant & Cell Physiol.*, **13**, 1017–29.

JACKSON G. A. (1977) Nutrients and production of giant kelp, *Macrocystis pyrifera*, off Southern California. *Limnol. Oceanogr.*, **22**, 979–95.

JACOBI G. (1957) Enzyme des Kohlenhydratstoffwechsels in Extrakten von *Ulva lactuca*. *Planta*, **49**, 1–10.

JOHN D. M. (1971) The distribution and net productivity of sublittoral populations of attached macrophytic algae in an estuary on the Atlantic coast of Spain. *Mar. Biol.*, **11**, 90–7.

JOHNSON W. S., GIGON A., GULMON S. L. & MOONEY H. A. (1974) Comparative photosynthetic capacities of intertidal algae under exposed and submerged conditions. *Ecol.*, **55**, 450–3.

JOHNSTON C. S. (1969) The ecological distribution and primary production of macrophytic marine algae in the Eastern Canaries. *Int. Rev. Ges. Hydrobiol.*, **54**, 473–90.

JOHNSTON C. S. & DAVIS J. M. (1969) Aspects of the intermediary metabolism of members of the Fucales and Laminariales. *Proc. Int. Seaweed Symp.*, **6**, 501–6.

JOHNSTON C. S., JONES R. G. & HUNT R. T. (1977) A seasonal carbon budget for a laminarian population in a Scottish sealoch. *Helgol. Wiss. Meeresunters.*, **30**, 527–45.

JOLLIFFE E. A. & TREGUNNA E. B. (1970) Studies on HCO_3^- ion uptake during photosynthesis in benthic marine algae. *Phycologia*, **9**, 293–303.

JOSHI G. V., KAREKAR M. D., GOWDA C. A. & BHOSALE L. (1974) Photosynthetic carbon metabolism and carboxylating enzymes in algae and mangrove under saline conditions. *Photosynthetica*, **8**, 51–2.

JUPP B. P. & DREW E. A. (1974) Studies on the growth of *Laminaria hyperborea* (Gunn.) Fosl. I. Biomass and productivity. *J. Exp. Mar. Biol. Ecol.*, **15**, 185–96.

KAGEYAMA A. & YOKOHAMA Y. (1974) Photosynthetic properties of marine benthic brown algae from different depths in coastal area of Japan. *Bull. Jap. Soc. Phycol.*, **22**, 119–23.

KAGEYAMA A., YOKOHAMA Y. & NISIZAWA K. (1979) Diurnal rhythm of apparent photosynthesis of a brown alga, *Spatoglossum pacificum*. *Bot. Mar.*, **22**, 199–202.

KAIN J. M. (1979) A view of the genus *Laminaria*. *Oceanogr. Mar. Biol. Ann. Rev.*, **17**, 101–61.

KAIN J. M., DREW E. A. & JUPP B. P. (1975) Light and the ecology of *Laminaria hyperborea* II. In *Light as an Ecological Factor*, eds. Bainbridge R., Evans G. C. & Rackman O., pp. 63–92. Blackwell Scientific Publications, Oxford.

KANWISHER J. W. (1966) Photosynthesis and respiration. In *Some Contemporary Studies in Marine Sciences*, ed. Barnes H., pp. 407–20. Allen & Unwin, London.

KAUSS H. (1968) α-Galaktosylglyzeride und Osmoregulation in Rotalgen. *Z. Pflanzenphysiol.*, **58**, 428–33.

KELLY G. J., LATZKO E. & GIBBS M. (1976) Regulatory aspects of photosynthetic carbon metabolism. *Ann. Rev. Plant Physiol.*, **27**, 181–205.

KHAILOV K. M. (1976) Physiological heterogeneity of thallus elements in *Ascophyllum nodosum*, *Fucus vesiculosus*, and *Fucus inflatus* from the Barents sea. *Fiziol. Rast.*, **23**, 703–7.

KHAILOV K. M. (1978) Changes of the mass, length, and metabolism of simple and composite thalli of marine macrophytes in their ontogeny series. *Bot. Mar.*, **21**, 313–22.

KHAILOV K. M. & BURLAKOVA Z. P. (1969) Release of dissolved organic matter by marine seaweeds and distribution of their total organic production to inshore communities. *Limnol. Oceanogr.*, **14**, 521–7.

KING R. J. & SCHRAMM W. (1976a) Determination of photosynthetic rates for the marine algae *Fucus vesiculosus* and *Laminaria saccharina*. *Mar. Biol.*, **37**, 209–13.

KING R. J. SCHRAMM W. (1976b) Photosynthetic rates of benthic marine algae in relation to light intensity and seasonal conditions. *Mar. Biol.*, **37**, 215–22.

KINGHAM D. L. & EVANS L. V. (1977) The *Pelvetia canaliculata, Ascophyllum nodosum/Mycosphaerella ascophylli* interrelationship. *Br. Phycol. J.*, **12**, 120.

KIRST G. O. & BISSON M. A. (1980) Turgor pressure regulation in marine algae. Ions and low molecular weight organic compounds. *Aust. J. Plant Physiol.* (in press).

KJELDSEN C. K. & PHINNEY H. K. (1972) Effects of variations in salinity and temperature on some estuarine macro-algae. *Proc. Int. Seaweed Symp.*, **7**, 301–8.

KOLLMAN V. H. & GREGG C. T. (1972) Preparation of ^{13}C-labelled sugars by photosynthesis. *J. Colo.-Wyo. Acad. Sci.*, **7**, 20.

KREMER B. P. (1973a) Untersuchungen zur Biochemie und Physiologie von Mannit in benthischen marinen Algen. Thesis, Bonn.

KREMER B. P. (1973b) Isolation of mannitol from *Desmarestia viridis*. *Phytochemistry*, **12**, 609–10.

KREMER B. P. (1973c) Untersuchungen zur Physiologie von Volemit in der marinen Braunalge *Pelvetia canaliculata*. *Mar. Biol.*, **22**, 31–5.

KREMER B. P. (1975a) Mannitmetabolismus in der marinen Braunalge *Fucus serratus*. *Z. Pflanzenphysiol.*, **74**, 255–63.

KREMER B. P. (1975b) Physiologisch-chemische Charakteristik verschiedener Thallusbereiche von *Fucus serratus*. *Helgol. Wiss. Meeresunters.*, **27**, 115–27.

KREMER B. P. (1976a) Photosynthetic carbon metabolism of chloroplasts symbiotic with a marine opisthobranch. *Z. Pflanzenphysiol.*, **77**, 139–45.

KREMER B. P. (1976b) ^{14}C-assimilate pattern and kinetics of photosynthetic $^{14}CO_2$-assimilation of the marine red alga *Bostrychia scorpioides*. *Planta*, **129**, 63–7.

KREMER B. P. (1976c) Mannitol in the Rhodophyceae—a reappraisal. *Phytochemistry*, **15**, 1135–8.

KREMER B. P. (1976d) Distribution of alditols in the genus *Bostrychia*. *Biochem. Syst. Ecol.*, **4**, 139–41.

KREMER B. P. (1976e) Distribution and biochemistry of alditols in the genus *Pelvetia* (Phaeophyceae, Fucales). *Br. Phycol. J.*, **11**, 239–43.

KREMER B. P. (1977a) Biosynthesis of polyols in *Pelvetia canaliculata*. *Z. Pflanzenphysiol.*, **81**, 68–73.

KREMER B. P. (1977b) Rotalgen-Chloroplasten als funktionelle Endosymbionten in einem marinen Opisthobranchier. *Naturwissensch.*, **64**, 147–8.

KREMER B. P. (1978a) Studies on $^{14}CO_2$-assimilation in marine Rhodophyceae. *Mar. Biol.*, **48**, 47–55.

KREMER B. P. (1978b) Patterns of photoassimilatory products in Pacific Rhodophyceae. *Can. J. Bot.*, **56**, 1655–9.

KREMER B. P. (1979a) Biosynthesis and metabolism of polyhydroxy alcohols in marine benthic algae. *Proc. Int. Seaweed Symp.*, **9**, 421–8.

KREMER B. P. (1979b) Photoassimilatory products and osmoregulation in marine Rhodophyceae. *Z. Pflanzenphysiol.*, **93**, 139–47.

KREMER B. P. (1979c) Light independent carbon fixation by marine macroalgae. *J. Phycol.*, **15**, 244–7.

KREMER B. P. & KÜPPERS U. (1977) Carboxylating enzymes and pathway of photosynthetic carbon assimilation in different marine algae—evidence for the C_4 pathway? *Planta*, **133**, 191–6.

KREMER B. P. & MARKHAM J. W. (1979) Carbon assimilation by different developmental stages of *Laminaria saccharina*. *Planta*, **144**, 497–501.

KREMER B. P. & SCHMITZ K. (1973) CO_2-Fixierung und Stofftransport in benthischen marinen Algen. IV. Zur $^{14}CO_2$-Assimilation einiger litoraler Braunalgen im submersen und emersen Zustand. *Z. Pflanzenphysiol.*, **68**, 357–63.

KREMER B. P. & SCHMITZ K. (1976) Aspects of $^{14}CO_2$-fixation by endosymbiotic rhodoplasts in the marine opisthobranchiate *Hermaea bifida*. *Mar. Biol.*, **34**, 313–6.

KREMER B. P. & VOGL R. (1975) Zur chemotaxonomischen Bedeutung des ^{14}C-Markierungsmusters bei Rhodophyceen. *Phytochemistry*, **14**, 1309–14.

KREMER B. P. & WILLENBRINK J. (1972) CO_2-Fixierung und Stofftransport in benthischen marinen Algen. I. Zur Kinetik der $^{14}CO_2$-Assimilation bei *Laminaria saccharina*. *Planta*, **103**, 55–64.

KROES H. W. (1970) Excretion of mucilage and yellow-brown substances by some brown algae from the intertidal zone. *Bot. Mar.*, **13**, 107–10.

KÜPPERS U. (1978) Enzymologie der CO_2-Fixierung bei *Laminaria hyperborea*: Jahresperiodische, umweltbezogene Veränderungen von Enzymaktivitäten. Thesis, Cologne.

KÜPPERS U. & KREMER B. P. (1978) Longitudinal profiles of carbon dioxide fixation capacities in marine macroalgae. *Plant Physiol.*, **62**, 49–53.

KÜPPERS U. & WEIDNER M. (1980) Seasonal variations of enzyme activities in *Laminaria hyperborea* (Gunn.) Fosl. *Planta*, **148**, 222–30.

LATZKO E. & KELLY G. J. (1976) Photosynthesis. Biochemical and physiological aspects of carbon metabolism. *Progr. Bot.*, **38**, 81–107.

LEHNBERG W. (1978) Die Wirkung eines Licht-Temperatur-Salzgehalt Komplexes auf den Gaswechsel von *Delesseria sanguinea* (Rhodophyta) aus der westlichen Ostee. *Bot. Mar.*, **21**, 485–97.

LIETH H. (1975) Primary productivity of the major vegetation units of the world. In *Primary Productivity of the Biosphere*, eds. Lieth H. & Whittacker R. H., pp. 203–15, Springer-Verlag, Berlin, New York.

LIN T. Y. & HASSID W. Z. (1966) Pathway of alginic acid synthesis in the marine brown alga, *Fucus gardneri*. *J. Biol. Chem.*, **24**, 5284–97.

LINDBERG B. & PAJU J. (1954) Low-molecular weight carbohydrates in algae. IV. Investigation of *Pelvetia canaliculata*. *Acta Chem. Scand.*, **8**, 817–20.

LITTLER M. M. (1973) The productivity of Hawaiian fringing-reef crustose Corallinaceae and an experimental evaluation of production methodology. *Limnol. Oceanogr.*, **18**, 946–52.

LITTLER M. M. (1976) Calcification and its role among the macroalgae. *Micronesica*, **12**, 27–41.

LITTLER M. M. & MURRAY S. N. (1974) The primary productivity of marine macrophytes from a rocky shore intertidal community. *Mar. Biol.*, **27**, 131–5.

LITTLER M. M. & MURRAY S. N. (1978) Influence of domestic wastes on energetic pathways in rocky intertidal communities. *J. Appl. Ecol.*, **15**, 583–96.

LOEWUS M. W. & DELWICHE C. C. (1963) Carbon monoxide production by algae. *Plant Physiol.*, **38**, 371–4.

LÜNING K. (1971) Seasonal growth of *Laminaria hyperborea* under recorded underwater light conditions near Helgoland. In *Proc. Fourth European Marine Biology Symposium*, ed. Crisp D. J., pp. 347–61. Cambridge University Press, Cambridge.

LÜNING K. (1979) Growth strategies of three *Laminaria* species (Phaeophyceae) inhabiting different depth zones in the sublittoral region of Helgoland (North Sea). *Mar. Ecol. Progr. Ser.*, **1**, 195–208.

LÜNING K. & DRING M. J. (1979) Continuous underwater light measurement near Helgoland (North Sea) and its significance for characteristic light limits in the sublittoral region. *Helgol. Wiss. Meeresunters.*, **32**, 403–24.

MCLACHLAN J. & BIDWELL R. G. S. (1978) Photosynthesis of eggs, sperm, zygotes, and embryos of *Fucus serratus*. *Can. J. Bot.*, **56**, 371–3.

MACRAE D. C. & CHAPMAN A. R. O. (1977) Variation in respiration rate in *Laminaria longicruris*. *J. Phycol.*, **13** (suppl.), 43.

MANN K. H. (1973) Seaweeds: Their productivity and strategy for growth. *Science*, **182**, 975–81.

MANN K. H. & CHAPMAN A. R. O. (1973) Primary production of marine macrophytes. In *Photosynthesis and Productivity in Different Environments*, ed. Cooper J. F., pp. 207–23, Cambridge University Press, Cambridge.

MARKHAM J. W., KREMER B. P. & SPERLING K. R. (1980a) Effects of cadmium on *Laminaria saccharina* in culture. *Mar. Ecol. Progr. Ser.*, **3**, 31–9.

MARKHAM J. W., KREMER B. P. & SPERLING K. R. (1980b) Cadmium effects on growth and physiology of *Ulva lactuca*. *Helgol. Wiss. Meeresunters.* **33**, 103–10.

MATHIESON A. C. & BURNS R. L. (1974) Ecological studies of economic red algae. I. Photosynthesis and respiration of *Chondrus crispus* Stackh. and *Gigartina stellata* (Stackh.) Batters. *J. Exp. Mar. Biol. Ecol.*, **7**, 197–206.

MATHIESON A. C. & DAWES C. J. (1974) Ecological studies of floridian *Eucheuma* (Rhodophyta, Gigartinales). II. Photosynthesis and respiration. *Bull. Mar. Sci.*, **24**, 274–85.

MOEBUS K. & JOHNSON K. M. (1974) Exudation of dissolved organic carbon by brown algae. *Mar. Biol.*, **26**, 117–25.

MOEBUS K., JOHNSON K. M. & SIEBURTH J. McN. (1974) Rehydration of desiccated intertidal brown algae: Release of dissolved organic carbon and water uptake. *Mar. Biol.*, **26**, 127–34.

MUNDA I. M. (1974) The effect of polluted water on the assimilation rates of the brown algae *Ascophyllum nodosum* (L.) Le Jol. and *Fucus vesiculosus* (L.) (Preliminary report). *Rev. Intern. Oceanogr. Med.*, **34**, 53–65.

MUNDA I. M. & KREMER B. P. (1977) Chemical composition and physiological properties of fucoids under conditions of reduced salinity. *Mar. Biol.*, **42**, 9–16.

NAGASHIMA H. (1976) Distribution of low-molecular weight carbohydrates in marine red algae. *Bull. Jap. Soc. Phycol.*, **24**, 103–10.

NAGASHIMA H., NAKAMURA S. & NISIZAWA K. (1969) Physiological studies on floridean starch, floridoside, and trehalose in a red alga, *Serraticardia maxima*. *Bot. Mag.*, **82**, 3462–73.

NAGASHIMA H., NAKAMURA S., NISIZAWA K. & TERUMITSU H. (1971) Enzymic synthesis of floridean starch in a red alga, *Serraticardia maxima*. *Plant & Cell Physiol.*, **12**, 243–53.

NEILSON A. H. & LEWIN R. A. (1974) The uptake and utilization of organic carbon by algae: An essay in comparative biochemistry. *Phycologia*, **13**, 227–64.

NEWELL B. S., DALPONT G. & GRANT B. R. (1972) The excretion of organic nitrogen by marine algae in batch and continuous culture. *Can. J. Bot.*, **50**, 2605–11.

NEWELL R. C. & PYE V. J. (1968) Seasonal variations in the effect of temperature on the respiration of certain intertidal algae. *J. Mar. Biol. Ass. U.K.*, **48**, 341–8.

NIEMECK R. A. & MATHIESON A. C. (1978) Physiological studies of intertidal fucoid algae. *Bot. Mar.*, **21**, 221–8.

NIIHARA Y. (1975) Physiological studies of *Laminaria japonica* var. *ochotensis*: The effect of temperature, light intensity and salinity upon photosynthesis and respiration of young sporophytes. *Sci. Rep. Hokkaido Fish. Exp. Stn.*, **17**, 11–18.

NISIZAWA K., AKAGAWA H. & IKAWA T. (1972) Dark $^{14}CO_2$-fixation in brown algae. *Proc. Int. Seaweed Symp.*, **7**, 532–6.

NORDHORN G., WEIDNER M. & WILLENBRINK J. (1976) Isolation and photosynthetic activities of chloroplasts of the brown alga *Fucus serratus* L. *Z. Pflanzenphysiol.*, **80**, 153–65.

OGATA E. & SCHRAMM W. (1971) Some observations on the influence of salinity on growth and photosynthesis in *Porphyra umbilicalis*. *Mar. Biol.*, **10**, 70–6.

OKAZAKI M. (1972) Carbonic anhydrase in the calcareous alga *Serraticardia maxima*. *Bot. Mar.*, **15**, 133–8.

PALMER A. J. & GOULDING R. W. (1974) Biosynthetic preparation of ^{11}C-labelled galactose, glycerol and mannitol. *J. Labelled Comp.*, **10**, 627–36.

PATIL B. A. & JOSHI G. V. (1970) Photosynthetic studies in *Ulva lactuca*. *Bot. Mar.*, **13**, 111–5.

PEARSE V. (1972) Radioisotope study of calcification in the articulated coralline alga *Bossiella orbigniana*. *J. Phycol.*, **8**, 88–97.

PENTECOST A. (1978) Calcification and photosynthesis in *Corallina officinalis* L. using the ^{14}C-method. *Br. Phycol. J.*, **13**, 383–90.

PERCIVAL E. & SMESTAD B. (1972) Photosynthetic studies on *Ulva lactuca*. *Phytochemistry*, **11**, 1967–72.

PERCIVAL E. & YOUNG M. (1971) Low molecular weight carbohydrates and water-soluble polysaccharide metabolized by the Cladophorales. *Phytochemistry*, **10**, 807–12.

QUADIR A., HARRISON P. J. & DEWREEDE R. E. (1979) The effects of emergence and submergence on the photosynthesis and respiration of marine macrophytes. *Phycologia*, **18**, 83–8.

RANDALL D. D. (1976) Phosphoglycollate phosphatase in marine algae: Isolation and characterization from *Halimeda cylindracea*. *Aust. J. Plant Physiol.*, **3**, 105–11.

RAVEN J. A. (1974) Carbon dioxide fixation. In *Algal Physiology and Biochemistry*, ed. Stewart W. D. P., pp. 434–55. Blackwell Scientific Publications, Oxford.

RAVEN J. A. (1976) The quantitative role of 'dark' respiratory processes in heterotrophic and photolithotrophic plant growth. *Ann. Bot.*, **40**, 587–602.

ROBBINS J. V. (1979) Effects of physical and chemical factors on photosynthetic and respiratory rates of *Palmaria palmata* (Florideophyceae). *Proc. Int. Seaweed Symp.*, **9**, 273–83.

RYTHER J. H. (1963) Geographic variations in productivity. In *The Sea*, ed. Hill M. N., pp. 347–80. John Wiley & Sons, New York.

SCHMITZ K., LÜNING K. & WILLENBRINK J. (1972) CO_2-Fixierung und Stofftransport in benthischen marinen Algen. II. Zum Ferntransport ^{14}C-markierter Assimilate bei *Laminaria hyperborea* und *Laminaria saccharina*. *Z. Pflanzenphysiol.*, **67**, 418–29.

SCHMITZ K. & RIFFARTH W. (1980) Carrier-mediated uptake of L-leucine by the brown alga *Giffordia mitchellae*. *Z. Pflanzenphysiol.*, **96**, 311–24.

SCHÖNFELD M., RAHAT M. & NEUMANN J. (1973) Photosynthetic reactions in the marine alga *Codium vermilara*. I. CO_2-fixation and Hill reaction in isolated chloroplasts. *Plant Physiol.*, **52**, 283–7.

SCHRAMM W. (1972) Untersuchungen über den Einfluss von Ölverschmutzungen auf Meeresalgen. I. Die Wirkung von Rohölfilmen auf den CO_2-Gaswechsel ausserhalb des Wassers. *Mar. Biol.*, **14**, 189–98.

SCHRAMM W. & OHNO M. (1973) Some observations on the influence of salinity on biological activity of *Fucus vesiculosus*. *Bull. Jap. Soc. Phycol.*, **21**, 81–5.

SIEBURTH J. McN. (1969) Studies on algal substances in the sea. III. Production of extracellular organic matter by littoral marine algae. *J. Exp. Mar. Biol. Ecol.*, **3**, 290–309.

SHEPHERD D. C. & BIDWELL R. G. S. (1973) Photosynthesis and carbon metabolism in a chloroplast preparation from *Acetabularia*. *Protoplasma*, **76**, 289–307.

SKIRROW G. (1975) The dissolved gases—carbon dioxide. In *Chemical Oceanography*, eds. Riley J. P. & Skirrow G., pp. 1–192. Academic Press, London, New York.

STOCKER O. & HOLDHEIDE W. (1937) Die Assimilation Helgoländer Gezeitenalgen während der Ebbezeit. *Z. Bot.*, **32**, 1–59.

SULLIVAN J. D. & IKAWA M. (1973) Purification and characterization of hexose oxidase from the red alga *Chondrus crispus*. *Biochem. Biophys. Acta*, **309**, 11–22.

TAYLOR D. L. (1976) Chloroplast endosymbiosis. *Port. Acta Biol.*, **14**, 385–96.

THOMAS E. A. & TREGUNNA E. B. (1968) Bicarbonate ion assimilation in photosynthesis by *Sargassum muticum*. *Can. J. Bot.*, **46**, 411–5.

TING I. (1976) Malate dehydrogenase and other enzymes of C_4 acid metabolism in marine plants. *Aust. J. Plant Physiol.*, **3**, 121–7.

TITLYANOV E. A. & ZVALINSKY V. J. (eds) (1978) Ecological aspects of seaweed photosynthesis. *Trans. Far East Sci. Cent.*, **11**, 1–182.

TOLBERT N.E. (1974) Photorespiration. In *Algal Physiology and Biochemistry*, ed. Stewart W. D. P., pp. 474–504. Blackwell Scientific Publications, Oxford.

TOLBERT N. E. (1976) Glycollate oxidase and glycollate dehydrogenase in marine algae and plants. *Aust. J. Plant Physiol.*, **3**, 129–32.

TOLBERT N. E. & CAREY W. (1976) Apparent total CO_2 equilibrium point in marine algae during photosynthesis in seawater. *Aust. J. Plant Physiol.*, **3**, 69–72.

TOPINKA J. A. & ROBBINS J. V. (1976) Effects of nitrate and ammonium enrichment on growth and nitrogen physiology in *Fucus spiralis*. *Limnol. Oceanogr.*, **21**, 659–64.

TOWLE D. W. & PEARSE J. S. (1973) Production of the giant kelp, *Macrocystis*, estimated by *in situ* incorporation of ^{14}C in polyethylene bags. *Limnol. Oceanogr.*, **18**, 155–9.

TRENCH R. K., BOYLE E. & SMITH D. C. (1973a) The association between chloroplasts of *Codium fragile* and the mollusc *Elysia viridis*. I. Characteristics of isolated *Codium* chloroplasts. *Proc. Roy. Soc. Lond.*, Ser. B, **184**, 51–61.

TRENCH R. K., BOYLE E. & SMITH D. C. (1973b) The association between chloroplasts of *Codium fragile* and the mollusc *Elysia viridis*. II. Chloroplast ultrastructure and photosynthetic carbon fixation in *Elysia viridis*. *Proc. Roy. Soc. Lond.*, Ser. B, **184**, 63–81.

TSENG C. K. & SWEENEY B. M. (1946) Physiological studies of *Gelidium cartilagineum*. I. Photosynthesis with special reference to the carbon dioxide factor. *Am. J. Bot.*, **33**, 706–15.

TURNER C. H. C. & EVANS L. V. (1977) Physiological studies on the relationship between *Ascophyllum nodosum* and *Polysiphonia lanosa*. *New Phytol.*, **79**, 363–71.

VAN DER VELDE H. H., GUIKING P. & VAN DER WULP D. (1975) Glucose-6-phosphate dehydrogenase and 6-phosphogluconate dehydrogenase in *Acrochaetium daviesii* cultured under red, white and blue light. *Z. Pflanzenphysiol.*, **76**, 108.

WAALAND J. R. (1977) Growth of Pacific Northwest marine algae in semi-closed culture. In *The Marine Plant Biomass of the Pacific Northwest Coast*, ed. Krauss R. W., pp. 117–37. Oregon State University Press.

WALLENTINUS I. (1978) Productivity studies on Baltic macroalgae. *Bot. Mar.*, **21**, 365–80.

WARBURG O. (1920) Über die Geschwindigkeit der photochemischen Kohlensäurezersetzung in lebenden Zellen. *Biochem. Z.*, **103**, 188–217.

WASSMAN E. R. & RAMUS J. (1973) Primary-production measurements for the green seaweed *Codium fragile* in Long Island Sound. *Mar. Biol.*, **21**, 289–97.

WEIDNER M. & KÜPPERS U. (1973) Phosphoenolpyruvate-Carboxykinase und Ribulose-1, 5-Diphosphat-Carboxylase von *Laminaria hyperborea* (Gunn.) Fosl.: Das Verteilungsmuster der Enzymaktivitäten im Thallus. *Planta*, **114**, 365–72.

WEIDNER M., NORDHORN G., KREMER B. P. & KÜPPERS U. (1975) Untersuchungen zur Photosynthese bei *Giffordia mitchellae*. *Z. Pflanzenphysiol*, **76**, 423–43.

WESTLAKE D. F. (1963) Comparisons of plant productivity. *Biol. Rev.*, **38**, 385–425.

WHEELER W. N. (1978) Ecophysiological studies on the giant kelp, *Macrocystis*. Ph.D. Thesis, University of California, Santa Barbara.

WHEELER W. N. (1980) The effect of boundary layer transport on the fixation of carbon by the giant kelp, *Macrocystis pyrifera* (L.) C. Ag. *Mar. Biol.* **59**, 95–103.

WHEELER W. N. (1982) Response of macroalgae to light intensity, light quality, temperature, CO_2, HCO_3^-, O_2, mineral nutrients and pH. In *CRC Handbook of Biosolar Resources, Vol 1, Fundamental Principles*, eds. Mitsui A. & Black C. C. CRC Press (in press).

WILCE R. T. (1966) Heterotrophy in arctic sublittoral seaweeds: An hypothesis. *Bot. Mar.*, **10**, 186–97.

WILLENBRINK J. & KREMER B. P. (1973) Lokalisation der Mannitbiosynthese in der marinen Braunalge *Fucus serratus*. *Planta*, **113**, 173–8.

WILLENBRINK J., KREMER B. P., SCHMITZ K. & SRIVASTAVA L. (1979a) Photosynthetic and light independent carbon fixation in *Macrocystis, Nereocystis*, and some selected Pacific Laminariales. *Can. J. Bot.*, **57**, 890–7.

WILLENBRINK J., KREMER B. P., SCHMITZ K. & WEIDNER M. (1979b) CO_2-Fixierung und Stofftransport in benthischen marinen Algen. *Ber. Dtsch. Bot. Ges.*, **92**, 157–67.

WILLENBRINK J., RANGONI-KÜBBELER M. & TERSKY B. (1975) Frond development and CO_2-fixation in *Laminaria hyperborea*. *Planta*, **125**, 161–70.

WILLIAMS P. J. LE B. (1975) Biological and chemical aspects of dissolved organic material in seawater. In *Chemical Oceanography, vol.* 2, eds. Riley J. P. & Skirrow G., pp. 301–63. Academic Press, London, New York.

YAMADA T., IKAWA T. & NISIZAWA K. (1978) Partial purification and some properties of RuDP carboxylase from a green alga, *Bryopsis maxima*. *Jap. J. Phycol.*, **26**, 49–52.

YAMADA T., IKAWA T. & NISIZAWA K. (1979a) Circadian rhythms of enzymes participating in the CO_2-photoassimilation of a brown alga, *Spatoglossum pacificum*. *Bot. Mar.*, **22**, 203–9.

YAMADA T., IKAWA T. & NISIZAWA K. (1979b) Purification and properties of RuDP carboxylase from a brown alga, *Spatoglossum pacificum*. *Bot. Mar.*, **22**, 211–21.

YAMAGUCHI T., IKAWA T. & NISIZAWA K. (1966) Incorporation of radioactive carbon from $H^{14}CO_3^-$ into sugar constituents by a brown alga, *Eisenia bicyclis*, during photosynthesis and its fate in the dark. *Plant & Cell Physiol.*, **7**, 217–29.

YAMAGUCHI T., IKAWA T. & NISIZAWA K. (1969) Pathway of mannitol formation during photosynthesis in brown algae. *Plant & Cell Physiol.*, **10**, 425–40.

YOKOHAMA Y. (1973) A comparative study on photosynthesis-temperature relationships and their seasonal changes in marine benthic algae. *Int. Rev. Ges. Hydrobiol.*, **58**, 463–72.

ZAVODNIK N. (1971) Fluctuations in photosynthetic activity and chemical composition in *Fucus virsoides* (Don.) J. Ag. *Thalassia Jugosl.*, **7**, 445.

ZAVODNIK N. (1973a) Seasonal variations in rate of photosynthetic activity and chemical composition of the littoral seaweeds common to North Adriatic. I. *Fucus virsoides* (Don.) J. Ag. *Bot. Mar.*, **16**, 155–65.

ZAVODNIK N. (1973b) Seasonal variations in rate of photosynthetic activity and chemical composition of the littoral seaweeds common to North Adriatic. II. *Wrangelia penicellata* C. Ag. *Bot. Mar.*, **16**, 166–70.

ZAVODNIK N. (1975) Effects of temperature and salinity variations on photosynthesis of some littoral seaweeds of the North Adriatic. *Bot. Mar.*, **18**, 245–50.

ZAVODNIK N. (1977) Note on the effect of lead on oxygen production of several littoral seaweeds of the Adriatic Sea. *Bot. Mar.*, **20**, 167–70.

ZAVODNIK N. (1978) Environmental influences on the day and night rhythm of photosynthesis in some littoral marine algae. *Ekologija*, **13**, 45–52.

CHAPTER 15
TRANSLOCATION

K. SCHMITZ
Botanisches Institut der Universität zu Köln
Gyrhofstrasse 15, D–5000 Köln 41, Federal Republic of Germany

1 INTRODUCTION

Long-distance transport (translocation) relates to transport in specialized cells and tissue systems like sieve tubes, phloem or xylem. Medium-distance transport (intercellular transport and parenchyma transport) and short-distance transport (membrane transport and intracellular transport) will not be covered here. Aspects of membrane transport and ion uptake were reviewed by Gutknecht & Dainty (1968) and MacRobbie (1974).

Translocation in vascular plants was intensively studied for many years but it was neglected in non-vascular plants for a long time. However the topic has attracted increasing interest during the last 15 years and a fair amount of new information has been published.

The aim of this chapter is first to set out the structural basis of translocation, which comprises a review of the anatomical work and the work on fine structure of specialized transport cells. It then aims to review the experimental work on physiological and biochemical aspects of translocation. Translocation in seaweeds has not been reviewed before. Compilations of literature on the anatomy of conducting cells in Laminariales may be found in Fritsch (1945) and Esau (1969). A summary of long-distance transport in *Macrocystis* was published by Parker (1971b). A recent review article by Kain (1979) on the biology of the genus *Laminaria* also considers aspects of translocation.

2 TRANSLOCATION IN RHODOPHYTA AND TRANSPORT RELATED STRUCTURAL ASPECTS

The formation of intercellular connections was an important step in the evolution of complex plants enabling better coordination during growth and development by facilitating intercellular exchange of information and compounds. Plasmodesmata are probably present in all brown seaweeds (cf. Marchant 1976). Their presence in green algae appears to be correlated with the cytokinetic mechanism of cell plate formation. They seem to be absent in those green algae that divide by furrowing (Stewart *et al.* 1973). Rhodophyta apparently completely lack plasmodesmata; instead, so called pit connections are formed in the septa of the Florideophycidae, and have also been found in the *Conchocelis* stage of at least two members of the Bangiophycidae (Bourne *et al.* 1970; Lee & Fultz 1970; Lee 1971). However, studies of the fine structure revealed that the term 'pit connection' seems a misnomer (Bisalputra *et al.* 1967) because the protoplasts of adjacent cells are not connected but separated by a plug composed of an acid polysaccharide-protein complex (*see* Chapter 7).

The existence of specialized transport cells in Florideophycidae was postulated by Klein (1877) and by Wille (1885). Anatomical studies by Celan (1939, 1941), Feldmann (1939) and Feldmann & Feldmann (1946) finally revealed several unique structural features of central cells which suggest interpretation of these cells as specialized transport cells. But it should be kept in mind that despite all speculations and interpretations based on morphological and structural features, there is no direct experimental proof for a transport function of central or axial cell systems of Florideophycidae. The old anatomists and phycologists apparently never really doubted that pit connections somehow facilitate the exchange of substances and probably information (Kylin 1937), but electron microscopy has finally demonstrated that pit connections do not establish cytoplasmic contact between neighbouring cells and that an open pathway for translocation does not exist (Ramus 1969, 1971; Pueschel 1977).

Nevertheless, a 'long distance' transport of organic compounds in marine red algae was reported by Hartmann & Eschrich (1969). The authors applied ^{14}C-leucine to the midrib of a young blade of *Delesseria sanguinea* and observed movement of ^{14}C towards the apex as well as the base of the same blade and an export of radioactivity to other blades of the same plant. The authors suggest that the long stretched wide cells in the midrib and costae of blades and in the 'cauloid' may be the conducting cells. Translocation velocity reached 63 cm h^{-1}. One experiment with *Cystoclonium purpureum* also revealed transport of ^{14}C with a comparable high velocity of 48 cm h^{-1}. It is puzzling that translocation with such high velocities should be operative in red algae whose cells do not even form a symplast. A rapid transport as demonstrated by Hartmann & Eschrich (1969) can therefore only be explained if pit connections were much less resistant to dissolved matter than plasmodesmata and if transport along each of the tentative transport cells was much faster than diffusion. The work of Hartmann & Eschrich (1969) should therefore be repeated and extended.

3 ANATOMY AND FINE STRUCTURE OF SIEVE ELEMENTS OF LAMINARIALES

Sporophytes of Laminariales exhibit the ultimate developmental stage of parenchymatous construction among brown algae. Sieve elements were first described for *Macrocystis* by Will (1884) and have subsequently been found in most genera of Laminariales. Their terminology, however, is very confusing. The medulla is composed of sieve elements, cross-connecting sieve elements and hyphae. The differentiation of sieve elements, cross-connections and hyphae and their origin from the inner cortex was described in detail by Killian (1911) for *Laminaria digitata*. His results were confirmed by Steinbiss & Schmitz (1974) and Schmitz & Srivastava (1974b) for other species of *Laminaria*. The sieve tube system of Laminariales is continuous throughout the plant except the holdfast (Smith 1939; Parker & Fu 1965; Nicholson 1976). It consists of longitudinal files of sieve elements which form sieve tubes and cross-connecting sieve elements (Figs 15.1–15.3). Sieve elements originate from longitudinal files of inner cortex cells which do not divide any more. The cross-connections between the innermost cortex cells form by the fusion of papillae, as described by Killian (1911), Yendo (1911), Sauvageau (1915), Schmitz & Srivastava (1974b, 1976), Sideman & Scheirer (1977). Papillae that do not fuse with a counterpart continue to grow more or less horizontally through the medulla forming slender, often branched threads, called hyphae or hyphal cells (Oliver 1887; Rosenthal 1890; Parker & Fu 1965). It seems that sieve elements and hyphae are formed in alternating rows (Nicholson 1976). The sieve element system and the hyphae apparently do not communicate, but are separate cell systems, according to Parker & Fu (1965), Parker & Huber (1965), and Nicholson (1976); however Wille (1897) and Sykes (1908) reported cytoplasmic interconnection between sieve tubes and hyphae. Cytoplasmic connections between sieve elements and cortex cells are scarce, especially in mature stipes. The sieve element system seems 'semi-isolated' in mature plants of *Nereocystis* (Nicholson 1976) as well as other species (Parker & Fu 1965; Parker & Huber 1965). However, pit connections between hyphae and cortex cells are frequent (van Went *et al.* 1973b), indicating that the hyphae are part of the meristoderm-cortex symplast.

A single sieve element is a stretched, cylindrical cell with perforated terminal end walls. The dimensions of single sieve elements and pores are extremely variable within one plant depending on the developmental stage, and among different species of Laminariales. Sieve elements that differentiate first in the young sporophytes are generally smaller than those formed later. As the stipe

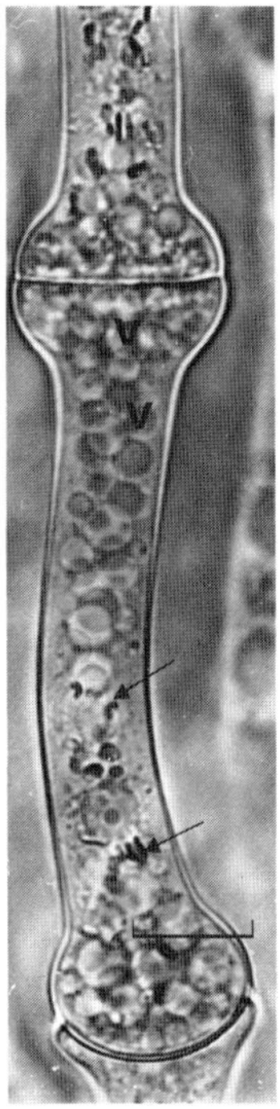

Fig. 15.1 A young sieve element of *Laminaria groenlandica in vivo*, unstained. Note the typical trumpet shape. Vacuoles (v), plastids (→) and the nucleus are clearly visible. Scale = 15 μm. (After Schmitz & Srivastava 1974b.)

elongates they are subject to stretching and acquire a trumpet shape (but see van Went & Tammes 1973). They are finally pulled out to such an extent that the lumen nearly totally collapses. Such sieve elements in the central medulla, which do not translocate any more, may exceed 1.2 mm in length (Parker 1971a; Steinbiss & Schmitz 1974). Sieve elements that form later and derive from the inner cortex are wider and not as long. As revealed in cross sections of stipes, sieve elements may be arranged in radial rows about three to eight cells deep at the periphery of the medulla (Oliver 1887; Parker 1971a; Nicholson 1976). Their width, i.e. the diameter of their sieve plates, varies between 15.5 ± 5 μm in *Alaria marginata* (Schmitz & Srivastava 1975) and 61.4 ± 9 μm in *Macrocystis pyrifera* (Parker & Huber 1965).

The most characteristic structural feature of sieve elements is their sieve-like perforated terminal end walls; each may be regarded as a single sieve field. The size of sieve pores differs from species to species and within one species between blades and stipes, and even within one stipe or blade, depending on the developmental stage of the plant and of the sieve element itself (cf. Parker & Fu 1965 for *Pelagophycus*; Schmitz & Srivastava 1976 for *Nereocystis*). Measurements of the average width of sieve pores are compiled in Table 15.1. The table

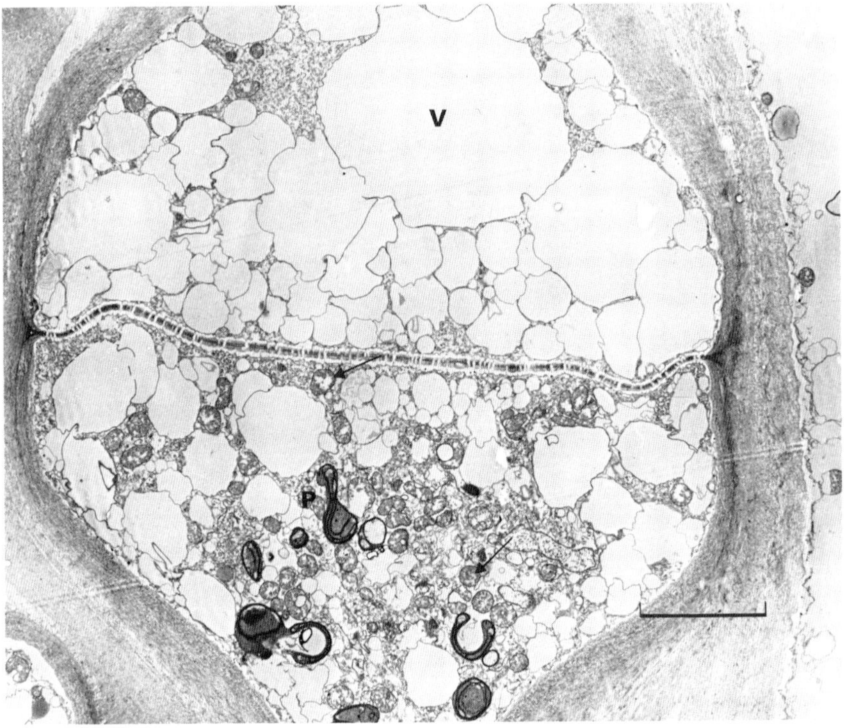

Fig. 15.2 General view of a mature sieve element of *Laminaria groenlandica* near its cross wall. Primary and secondary walls can be distinguished. Closely packed vacuoles (v) plastids (p) and a large number of mitochondria (→) can be seen. Scale = 5 μm. (After Schmitz & Srivastava 1974b.)

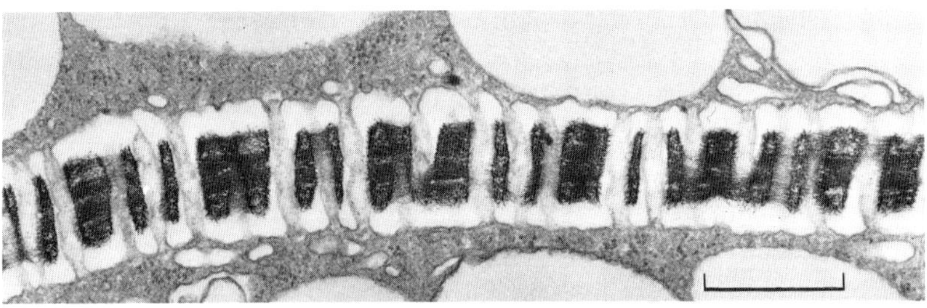

Fig. 15.3 Preparation of *Laminaria groenlandica* showing the cross section of a sieve area. Cytoplasm and plasmalemma of adjacent cells are continuous and can be followed through the sieve pores. Scale = 0.5 μm. (After Schmitz & Srivastava 1974b.)

indicates a developmental line within the Laminariales from sieve elements having small but numerous pores, as in *Laminaria*, to those of *Macrocystis* with relatively few but wide pores. While sieve pores of *Laminaria* may still be regarded as plasmodesmata (Ziegler & Ruck 1967; Schmitz & Srivastava 1974b;

Table 15.1 Progressive specialisation of sieve fields and transport velocities in Laminariales.

Species	Sieve element cross walls		Pore-diameter μm	Frequency of pores (pores 100 μm^{-2})	Area of pores % of cross wall	Transport velocity mm h^{-1}
	Diameter μm	Area μm^2				
Laminaria digitata, *L. saccharina*, *L. hyperborea*	24±4	450	0.04–0.09	5000–6000	*c.* 62	50–100
L. groenlandica	34.5±5	935	0.04–0.08	3000–6000	*c.* 51	–
Alaria marginata	15.5±5	189	0.1 – 0.3	*c.* 1800	*c.* 56	250–400
Nereocystis luetkeana	26±10	531	0.4–0.9	*c.* 40	*c.* 40	170–570
Pelagophycus porra[a]	34±4.1	908	0.35–0.65	143–181	*c.* 32	–
P. giganteus[a]	42.8±5.3	1439	0.5–1.3	63–135	*c.* 63	–
Macrocystis integrifolia	54±10	2290	0.7–4.5	7.2–15	*c.* 61	350–720
M. pyrifera[b]	61.4±9	2926	2.4–6	2.9–8.2	*c.* 77	650–780[c]

[a] Calculated from Parker & Fu 1965.
[b] Calculated from Parker & Huber 1965.
[c] Reference: Parker 1965.

Sideman & Scheirer 1977), sieve pores of *Macrocystis* closely resemble those of higher plants (Ziegler 1963).

Cross-connecting sieve elements are always joined to the sieve tubes by sieve fields. Will (1884), Oliver (1887) and Sykes (1908) have pointed out that sieve pores in the lateral branches of *Macrocystis* sieve elements are always smaller than those in the terminal end walls. This also holds true for *Pelagophycus* (Parker & Fu 1965). If one considers the sieve field in a short lateral branch of a sieve element a 'lateral sieve field', then the terminal perforated cross wall deserves the name sieve plate, provided the terminology of phloem is applicable (cf. Esau 1969) and consequently, the cell may be termed a 'sieve tube member'.

There is not much information about the fine structure of sieve element protoplasts in the old anatomical literature because most scientists had only studied dried or ethanol-, formalin- or glycerine-preserved plants. *In vivo* observations were the exception. Transmission electron microscopy was first applied to *Macrocystis* (Parker & Philpott 1961; Ziegler 1963; Parker J. 1964; Parker & Huber 1965; Parker 1971a) and *Laminaria* (Ziegler & Ruck 1967; van Went *et al.* 1973a, b, c; Schmitz & Srivastava 1974b; Sideman & Scheirer 1977). Sieve elements of *Nereocystis luetkeana* and *Alaria marginata* were studied by Schmitz & Srivastava (1975, 1976). It was recognized that the ontogenetic development of sieve elements comprises three different phases: 1, the phase of differentiation (their origin from the inner cortex cells and their development into functional transport cells); 2, the phase of function as a transport cell; 3, the phase of structural and functional degeneration. Accordingly, a mature sieve element is one whose fine structure is optimally developed with respect to its function as a transport cell and as part of the long distance transport system.

Mature sieve elements of *Laminaria* and *Alaria* are characterized by numerous small vacuoles and the presence of the usual cell organelles such as plastids, mitochondria, endoplasmic reticulum, dictyosomes, ribosomes, and a nucleus (Ziegler & Ruck 1967; Schmitz & Srivastava 1947b, 1975; Sideman & Scheirer 1977). This is also true for the sieve elements from the base of the phylloids of *Nereocystis* and sieve elements from the pneumatocyst and the hollow part of the stipe, except that sieve elements of the stipe have larger and fewer vacuoles, relatively little cytoplasm, and may lack a nucleus (Schmitz & Srivastava 1976; Nicholson 1976). Sieve elements of *Macrocystis* revealed a peripheral layer of cytoplasm with plastids, numerous vesicles and mitochondria, occasional dictyosomes and endoplasmic reticulum. The central lumen appears to be empty (Fig. 15.4) or filled with fine granular to fibrillar material with large masses of electron dense aggregated slime (after chemical fixation of fresh material) or it appears to contain dispersed slime (after cryofixation followed by OsO_4 treatment; cf. Parker & Huber 1965). The presence of plastids was also recognized by Parker & Philpott (1961) and Ziegler (1963). Mature sieve elements of *Macrocystis* apparently lack a nucleus (Ziegler 1963; Parker & Huber 1965; Parker 1971a). P-protein was not detected in sieve elements of brown algae; they also lack companion cells.

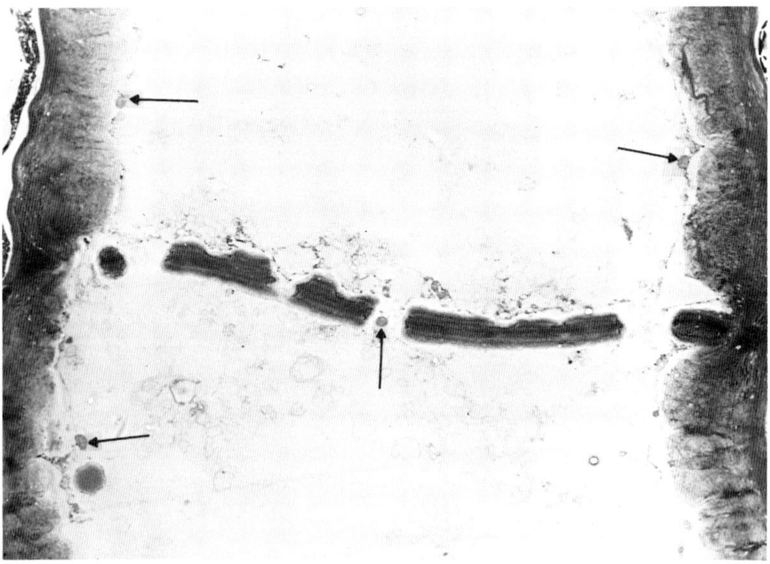

Fig. 15.4 Sieve element of *Macrocystis integrifolia*. Longitudinal section from the stipe. The central lumen of mature cells is empty, the protoplast forms a thin layer along the thick longitudinal walls, some vesicles and small vacuoles are seen in the vicinity of the sieve plate. Mitochondria are indicated by arrows. (× 1900.)

Differentiating and mature sieve elements at the periphery of the medulla do not show callose, their sieve pores are open connections. Callose is usually formed later, when sieve elements begin to degenerate. Callose is first deposited in the sieve pores forming hollow cylinders around each pore, the pores later become completely plugged by callose which means the termination of long-distance transport in such sieve tubes. Old, non-functional sieve elements usually show large deposits of callose on their sieve areas, but callose may also extend along the longitudinal walls and may finally completely fill the lumen of a sieve tube (Oliver 1887; Rosenthal 1890; Sykes 1908; Ziegler 1963; Parker J. 1964; Parker B. C. 1964, 1971a; Parker & Huber 1965; Ziegler & Ruck 1967; Schmitz & Srivastava 1974b, 1975, 1976; van Went *et al.* 1973a; Sideman & Scheirer 1977). It was shown that callose of Laminariales is a β-(1 $\rightarrow$ 3)-glucan just like callose of higher plants (Oliver 1887; Rigg 1925; Ziegler 1963; Parker B. C. 1964). It is not clear how long sieve elements remain functional and whether new sieve elements are periodically (annually) or continuously formed in perennials. Sykes (1908) clearly stated that differentiation of new sieve elements will be terminated as soon as all of the inner cortex is used up for the production and differentiation of the medulla. This seems to be the case in *Laminaria groenlandica, Alaria marginata, Nereocystis luetkeana,* and *Macrocystis integrifolia* (personal observations). Differentiation of sieve elements in perennials has not been carefully studied yet.

4 LONG-DISTANCE TRANSPORT IN PHAEOPHYTA

4.1 Translocation in relation to growth; source-sink relationship

Movement of substances through the length of blades and stipes must be very efficient in terms of velocity and amounts translocated because some species of brown algae belonging to the Lessoniaceae have been found to be extremely long (cf. Setchell 1908; Frye *et al.* 1915; Smith 1942), exhibiting enormous growth rates. Clendenning (1971) reported maximum rates of elongation for *Macrocystis* fronds of 50 cm day^{-1} or 3–4.5 m week^{-1} which may be the fastest autotrophic elongation on record for any marine or land plant. *Nereocystis* is able to achieve 25 m in length during a single season and may reach an elongation rate of 6–13 cm day^{-1} (Nicholson 1970; Duncan 1973). Thus translocation in these plants may be a real long-distance mass-transport comparable to translocation in trees.

The first convincing indirect experimental evidence for a long-distance transport of photoassimilate was published by Sargent & Lantrip (1952). These authors measured the rates of photosynthesis and growth in different parts of a frond of *Macrocystis pyrifera* and found that the growing tissues at the frond apex showed a much higher increase in dry weight than could be accounted for by their photoassimilation of carbon, whereas in mature regions of the thallus carbon fixation was considerably in excess of growth. The authors concluded that the carbon deficit of the growing zone was balanced by carbon transport. Further indirect evidence for translocation came from growth experiments with amputated specimens of *Laminaria* (Lüning 1969, 1970), *Nereocystis* (Nicholson 1970; Duncan 1973) and *Alaria* (Buggeln 1974, 1977). It was shown by Lüning (1971) that new growth of the lamina of *Laminaria hyperborea* occurs in mid-winter when the ambient irradiance near Helgoland is below compensation point, which is true for most of the time between October and March. Photosynthesis can therefore not provide the organic material necessary for the growth of the young frond. It was shown that the new blade has a considerable capacity for light-independent carbon fixation (Weidner & Küppers 1973; Willenbrink *et al.* 1975; Küppers & Kremer 1978; cf. Chapter 14). This dark fixation, however, cannot balance the carbon deficit at the transition zone. A considerable amount of carbon for the growth of the new blade is provided by translocate from the old blade as shown by amputation experiments (Lüning 1969, 1970) and by ^{14}C-transport experiments (Lüning, 1971; Lüning *et al.* 1972; Schmitz *et al.* 1972; Lüning *et al.* 1973).

Clendenning (1971) recognized that in a bottom area shaded by a dense canopy of *Macrocystis,* there was insufficient light and photosynthesis per day to balance respiration and independent growth of young *Macrocystis* fronds. It was concluded that organic 'food' is provided to the juveniles by translocate from the mature fronds. It was finally demonstrated by ^{14}C tracer experiments that these conclusions were correct (Parker 1965; Schmitz & Lobban 1976; Lobban 1978a, b).

It is well known from higher plants that expanding leaves pass from being importers of assimilate to exporters and that there is a phase where bidirectional transport takes place (cf. Turgeon & Webb 1973). The dependency of import and export on position was also shown for blades of *Macrocystis.* It was recognized by Parker (1965) that movement of ^{14}C from a source blade was not only directed towards the apical scimitar but also towards immature blades. Detailed studies of the translocation patterns in *Macrocystis pyrifera* and *M. integrifolia* revealed that immature blades that have only half to three-quarters unfurled, import, while mature, fully expanded blades only export. It was, however, recognized that there is a transition zone where the reverse takes place (Lobban 1978a, b).

4.2 Translocation of radioisotopes and dyes

Direct demonstration of translocation using radioisotopes was first published by Parker (1956). Long-distance transport has since been investigated in a number of genera, mainly using radioisotopes, though much could still be learned from studies with dyes. Fluorescein is known to be an excellent tool to study phloem transport in higher vascular plants (Schumacher, 1933), but it has only been used by Parker (1965) to trace translocation in *Macrocystis*. However, his experiments with fluorescein revealed that the dye moves with a velocity of 1 cm h^{-1} only, whereas ^{14}C-transport reached 65–78 cm h^{-1}. Although the dye was clearly localized in sieve elements and flowed out with the exudate, the usefulness of this dye as an indicator of translocation in brown algae seems doubtful and should be reinvestigated.

Inorganic as well as organic compounds are translocated in a large number of Laminariales, indicating that probably all Laminariales have an effective translocation system (Schmitz & Lobban 1976). Experimental results also revealed that Fucales, though they lack well developed sieve elements, are able to translocate, although this transport is very slow (Floc'h & Penot 1972). It can only be speculated that the highly pitted elongate filaments in the medulla of Fucales may be active in long-distance transport. Their fine structure should be investigated in detail.

4.3 Directionality of transport

Movement of assimilates is always from source to sink which means from tissue areas or organs that produce a surplus to those regions that have a deficit with regard to a substance that might be inorganic or organic. Mature parts of a thallus, especially mature blades, donate, while meristematic and fast-growing regions, like the transition zone, vigorously growing haptera and perhaps sporophylls, receive.

Movement of photoassimilated ^{14}C from mature blade areas of *Laminaria* was always directed towards the intercalary growing region; there was no translocation towards the apex of a plant (Lüning *et al.* 1972, 1973; Schmitz *et al.* 1972; Titlyanov & Peshekhodko 1973). The same was also demonstrated for other Laminariales (Nicholson & Briggs 1972; Schmitz & Lobban 1976; Buggeln 1977) and it also applies to long-distance transport of inorganic substances (Floc'h & Penot 1970, 1971, 1972, 1974, 1976, 1978).

The main source of assimilate for blade meristem growth in *Alaria esculenta* and *A. marginata* is the distal mature blade (Schmitz & Srivastava 1975; Buggeln 1977). It was shown, however, by tracer experiments that growth may also be promoted by stipe and sporophylls. Transport from sporophylls was induced by excision of the distal blade (main source), and the stipe began to export when both sources were removed (Buggeln 1977). The induction of a reversed source–sink relationship was achieved in *Alaria marginata* by predarkening a whole plant for 60 h before ^{14}C was applied to the lowermost 30 cm of a lamina (growing region). Thereafter, only the ^{14}C-fed area was illuminated whereas the rest of the plant remained darkened. Under these conditions ^{14}C was exported in both directions, towards the stipe and the holdfast, as well as towards the apical part of the lamina (Schmitz & Srivastava 1975). These experiments prove that translocation of ^{14}C-labelled material can be induced in a direction opposite to the 'normal' direction of assimilate transport if a sufficiently large sink can be created.

Experiments conducted with *Fucus* revealed that radioactivity (^{32}P) may move from an application area in the middle of a thallus both towards the apical growing region and towards the discoidal holdfast at the same time (Floc'h & Penot 1972). The holdfast and the lowermost part of the thallus did not export ^{32}P. ^{86}Rb, applied close to the apex, was conducted to the apical growing region only (Floc'h & Penot 1972). Removing the apical meristem in *Fucus vesiculosus* or the holdfast in *Laminaria digitata* clearly diminished the amount of radioactivity translocated towards these regions as compared with control experiments. Wound healing and growth of regenerating tissue may act as a local sink and attract radioactivity against the 'normal' direction of transport (Floc'h 1976).

Directionality of transport in *Macrocystis* was studied by Lobban (1978a, b), who extended previous observations by Parker (1965). Compared with *Laminaria, Macrocystis* is a much more complicated plant system with many sink areas competing for assimilates from numerous sources. Parker (1965) showed that radioactivity was exported from mature blades to the apex of the labelled frond (main growing region) as well as towards the holdfast. The pattern of assimilate distribution within a frond of *M. pyrifera* as well as *M. integrifolia* is similar to that typical of angiosperms: immature blades import, mature ones export. During development, this export is first directed towards the apical growing region, then upward and downward at the same time, while late in the life of the blade there is a phase of declining export when transport is only in a basipetal direction (Fig. 15.5). Translocation downwards supplies sporophylls, frond initials and apical regions of juvenile fronds. Mature fronds probably nourish juveniles until the carbon budget of the juveniles is balanced by their own carbon fixation. There is no transport from juvenile fronds to mature fronds. The translocation pattern in *M. integrifolia* differs from that in *M. pyrifera* in seasonal changes and in distances from apex at which the changes take place. The pattern in *M. pyrifera* is fairly uniform, while the translocation in *M. integrifolia* shifts with season from principally upward export in spring to principally downward export in autumn (Lobban 1978a, b). Upward as well as downward transport in *M. integrifolia* was also recognized by Schmitz & Srivastava (1979a,b, 1980) when they studied the movement of ^{14}C-photoassimilate, ^{32}P and THO, both in intact plants as well as frond cuttings. It was observed that damage or loss of the apical growing region of a mature frond of *Macrocystis* caused an increase in downward transport. Similarly, if a secondary juvenile frond lacks its apex, there is little or no transport into it (Schmitz & Lobban 1976; Lobban 1978b). Considering that even one single frond of *Macrocystis* is a very complex system for translocation studies, stipe cuttings were used that were only 60–100 cm long with only three to six attached blades,

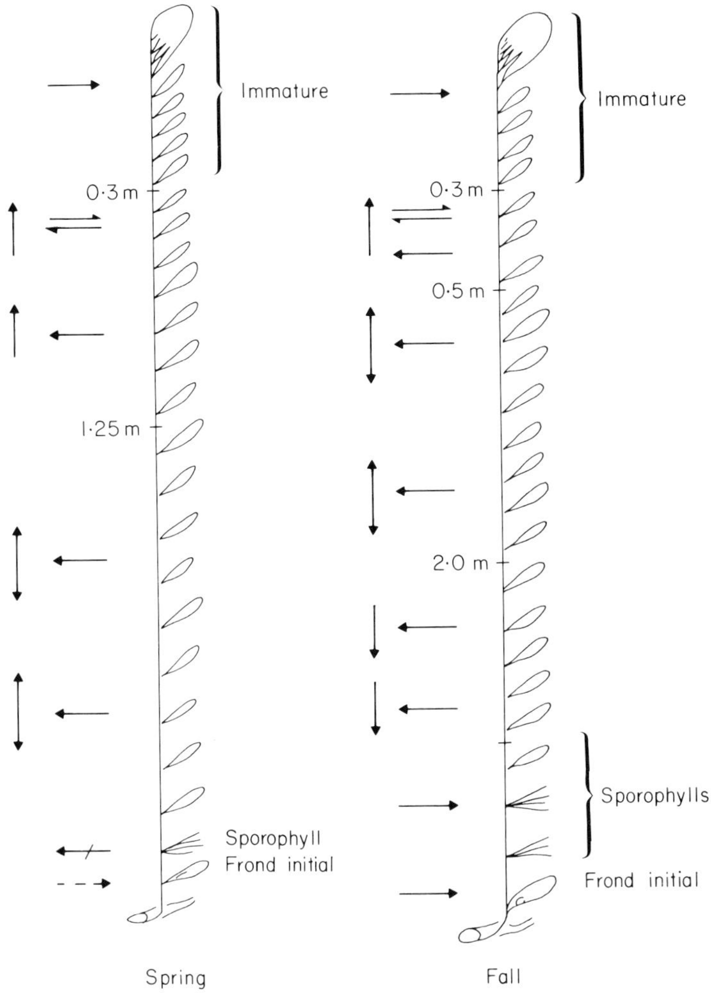

Fig. 15.5 Diagram of the translocation patterns in *Macrocystis integrifolia* in spring (May–June) and autumn (August–October). Arrows indicate import and export, and direction of export: ↑ = upward export; ↓ = downward export; → = import, no export; ← = predominantly or exclusively export; ⇄ = import + export; - - → = some import; ↚ = no export. (After Lobban 1978b.)

and it was demonstrated that such pieces of stipe can be used as a smaller, but more defined system for translocation studies (Schmitz & Srivastava 1979a,b). While the direction of transport in a detached stipe is probably determined by the process of cutting, creating a strong artificial sink at the cut end, directionality of movement in stipe cuttings (pieces of stipe) cannot be explained so simply; directionality could be determined by 'imprinting' of the intact plant (Schmitz & Srivastava 1979a). Experiments with pieces of *Laminaria* stipes to determine the polarity of transport revealed that movement of ^{32}P along the stipe pieces was the same whether they were in a normal position or turned upside down (Floc'h & Penot 1972).

4.4 Identification of the transport pathway

Simple evidence for the conductivity of sieve tubes in *Macrocystis* is the fact that cut stipes release substantial amounts of exudate. This was noticed by Crafts (1939) and confirmed by Parker (1965, 1966) and Schmitz & Srivastava (1974a, 1979a, b, 1980). Sieve tube sap is released only by the peripheral part of the medulla where the wide, younger sieve elements are located (Schmitz & Srivastava 1979a, b). Parker (1965) demonstrated that ^{14}C-photoassimilate from a mature blade can travel along the stipe and is released with the exudate, and ^{14}C-labelled translocate was eventually localized by autoradiography in *Laminaria hyperborea* (Steinbiss & Schmitz 1973): histoautoradiographs revealed radioactivity in single sieve elements of the translocation pathway, indicating that transport occurs in these cells only. It was shown that predominantly young sieve elements at the periphery of the medulla contained the radioactively labelled translocate. Whole plant autoradiography revealed transport in a rather narrow central region along *Laminaria* blades (Lüning *et al.* 1972; Floc'h & Penot 1976, 1978a, b, 1980), and such autoradiographs indicated that transport in *Fucus* and *Alaria* was restricted mainly to the midrib (Floc'h & Penot 1972). Localisation of the translocation pathway was also tackled by mechanical girdling (Parker 1965; Nicholson & Briggs 1972; Floc'h & Penot 1972), and by physical interruption of the presumed transport pathway (Buggeln 1974, 1977). When blades of *Nereocystis* were exposed to a ^{14}C-seawater medium for 8 h in the dark, ^{14}C-labelled material moved to the lower stipe where mucilage and exudate from the medulla were collected and analysed. Nicholson & Briggs (1972) stated that movement in the dark occurred in the mucilage ducts. The radioactivity measured in the collected mucilage and exudate was acid labile. This phenomenon is not further explained by the authors and should be reinvestigated.

4.5 Nature of translocated substances

If it is true that the exudate collected from cut stipes of *Macrocystis* is sieve tube sap and if transport in *Macrocystis* is a flow of dissolved compounds (see discussion of possible mechanisms) then the exudate should represent the translocate and an analysis of its chemical composition should reveal the nature of transported substances. Analyses of exudate were presented by Parker (1966) and Schmitz & Srivastava (1979a) for *M. pyrifera* and *M. integrifolia*: the main component of dry weight is mannitol (65%), followed by free amino acids

(15%); much of the remaining 20% is due to inorganic ions. The concentration of potassium is many times higher than in seawater, whereas sodium concentration is much lower, which indicates an ion specific selectivity of the plasmalemma of sieve tubes. There are only minor differences between the two species investigated. The data presented by Schmitz & Srivastava (1979a) however show some major differences from the exudate analysis reported by Parker (1966). There are interesting similarities with sieve tube sap from higher plants: mannitol concentration is approximately 10% (w:v) and well in the range of sucrose concentrations of sieve tube sap from vascular plants (for comparison see data in Crafts & Crisp 1971; Canny 1973; Ziegler 1975). It has also been reported that ATP concentrations are comparable (Schmitz & Srivastava 1974a). In addition, pH of sieve tube sap is always neutral to alkaline, never acidic.

A comparison of ^{14}C-labelled photosynthate and ^{14}C-labelled translocate (analysis of sieve tube sap) of *Macrocystis* demonstrates a close parallelism. The ^{14}C-labelled substances in the translocate were the same as in the photosynthate, though with a slightly different percentage distribution of ^{14}C (Schmitz & Srivastava 1979a); this was recognized before in *Laminaria hyperborea, L. saccharina* (Schmitz *et al.* 1972) and *Alaria marginata* (Schmitz & Srivastava 1975); for comparison of ^{14}C-labelled photosynthate see also Kremer & Willenbrink (1972), Willenbrink *et al.* (1979) and Chapter 14. This close parallelism of labelled compounds in the photosynthate and translocate indicates a symplasmic medium-distance transport from the meristoderm and outer cortex, which contain most of the phaeoplasts and photosynthetic capacity (Hellebust & Haug 1972) to the long-distance transport system in the medulla.

It was shown by Floc'h & Penot (1976) that chlorine, calcium, and molybdenum are immobile in *Laminaria digitata*. On the other hand, Penot *et al.* (1976a, b) showed that all these ions are very mobile in *Ascophyllum nodosum*. The application of various metabolic inhibitors did not stop movement in *Ascophyllum* and it is suggested by Penot & Penot (1977) that transport takes place in the apoplast. It remains to be determined whether the immobility of substances such as Ca, Cl and Mo is due to a permeability barrier of the plasmalemma, which would exclude them from the symplast, or whether they are really immobile within the symplast. Investigations to find out what the transport form for ^{32}P might be, was carried out by Floc'h & Penot (1978b) and Schmitz & Srivastava (1979b) independently. Both groups have shown that ^{32}P was present along the transport pathway mainly in an organically bound form. The reported percentage distribution among various fractions however differ widely. Floc'h & Penot (1978b) employed pulse-chase experiments that lasted up to 13 days and found that radioactivity of inorganic phosphorus increased with time in the source as well as the sink area—a large proportion of radioactivity in the sink area was also found in insoluble phosphates—while the percentage distribution along the translocation pathway remained nearly the same. This observation was interpreted as suggesting a movement of phosphorus as inorganic phosphate rather than in the form of organic compounds (Floc'h & Penot 1978b). However, chromatography of untreated freshly collected sieve tube sap from *Macrocystis integrifolia* revealed that only 11–13% of radioactivity was confined to inorganic phosphorus, but up to 60% was in hexose monophosphates, 12–13% in UDPG, and 9–10% in ATP (Schmitz & Srivastava 1979b). It was shown before by experiments *in vitro* that sieve tube sap contains the enzymes necessary to incorporate ^{32}P into nucleotide di- and triphosphates and hexose phosphates (Schmitz & Srivastava 1974a).

4.6 Factors influencing translocation

Translocation apparently occurs in the light as well as in the dark. A direct influence of light on the transport itself has not been demonstrated with certainty, but presence or absence of light influences the vigour of source and sink and may even reverse 'normal' directionality of transport (Schmitz & Srivastava 1975).

If long-distance transport is dependent upon metabolism, it should be temperature sensitive. This can be interpreted from Parker's data. Parker (1965) found that most radioactivity was exported from a mature blade and entered the stipe at 17°C which is close to the optimum temperature for growth of *Macrocystis pyrifera* and that transport velocity was highest when temperature was also highest (78 cm h^{-1} at 24.8°C). A direct influence of temperature on the transport itself was postulated by Floc'h & Penot (1974). These authors studied the accumulation of ^{32}P and ^{86}Rb in the apical growing region of *Fucus vesiculosus* at 14°C (control = 100%) and −1°C and found that at the low temperature ^{32}P-accumulation was reduced by 77% and ^{86}Rb by 31%. However, the conclusions from these experiments must be considered tentative.

The application of metabolic inhibitors like DNP and KCN to sink areas (apical growing regions) of *Fucus* reduced the accumulation of ^{32}P and ^{86}Rb by 49–97% which indicates that the transport is linked to metabolism of the sink (Floc'h & Penot 1974), but it remains to be determined that such inhibition is reversible and does not cause irreversible damage to the membranes. No effect of a variety of inhibitors in the movement of radioisotopes was found in *Ascophyllum nodosum* which supports the idea that transfer of ions in this species apparently occurs not in the symplast but in the apoplast (Penot & Penot 1977). Long-distance transport of ^{32}P in *Laminaria digitata* was stopped by local application of cycloheximide (50 mg l^{-1}) to the transport path. It was shown that this effect was not due to a movement of the inhibitor towards the source of ^{32}P but a local inhibition (Penot & Penot 1976b; Floc'h & Penot 1978a).

Though cycloheximide is known to be an inhibitor of protein synthesis, its mode of action here is not absolutely clear, since secondary effects cannot be excluded.

4.7 Velocity of transport

There are two reasons for considering data of transport velocities with caution: 1, there are several internal and external factors affecting transport in general and transport velocity; 2, the techniques used to measure transport velocity are not accurate enough and can only achieve an approximation. Nevertheless, transport velocity is a parameter that helps to characterize the efficiency of a transport system. It must be stressed that translocation *rate* is not the same as translocation *velocity;* these terms have repeatedly been mixed up in the literature (cf. Parker 1965, 1971b; Nicholson & Briggs 1972).

Transport velocities are very variable within different species, within different plants of the same species and even within a single plant. Translocation velocities in Laminariaceae are usually less than 10 cm h^{-1} but may reach 15 cm h^{-1} in *Cymathere triplicata*. The maximum velocity for the movement of ^{14}C-photoassimilate in *Laminaria* is 10 cm h^{-1} (Schmitz *et al.* 1972; Schmitz & Lobban 1976). These are all conservative estimates. For the transport of ^{32}P and ^{86}Rb velocities of at least 1.7 cm h^{-1} and 1.4 cm h^{-1} respectively (Floc'h & Penot 1972) and as high as 3 cm h^{-1} (Floc'h & Penot 1974), have been measured. Velocities of photosynthate movement in *Alaria marginata* ranged between 25 and 40 cm h^{-1} (Schmitz & Lobban 1976; Schmitz & Srivastava 1975). Nicholson & Briggs (1972) reported an average velocity of 37 cm h^{-1} for *Nereocystis,* while Schmitz & Lobban (1976) recognized that in this genus basipetal transport along the blades was considerably slower than transport velocity in the stipe. They measured 11–18 cm h^{-1} (average 17 cmh^{-1}) in the blades and 57 cm h^{-1} in the stipe. Highest translocation velocities were measured in *Macrocystis:* Parker (1965) reported 65–78 cm h^{-1} for the transport of ^{14}C assimilate in *M. pyrifera;* the range for *M. integrifolia* was 35–72 cm h^{-1} for photoassimilates and 25–45 cm h^{-1} for ^{32}P (Schmitz & Srivastava 1979a, b). The latter authors also showed that ^{32}P and ^{14}C-photosynthate, when applied simultaneously to the same blade, moved together with approximately the same velocity. A comparable speed was also determined for the movement of THO and ^{3}H-photoassimilate (Schmitz & Srivastava 1980).

If translocation velocities, though measured under different conditions, are compared with the diameter of sieve pores (cf. Table 15.1), it is obvious that plants with small pores reveal a much slower transport than plants with sieve pore diameters up to several micrometers. It seems that the resistance of the sieve fields may influence transport velocity; such resistance is apparently higher when sieve pores are smaller, and lower when they are larger. The total area of sieve fields occupied by sieve pores seems not to be related to transport velocity (cf. Table 15.1).

4.8 Translocation rates

The magnitude of translocation is best assessed on the basis of dry weight, because the water which always accompanies the moving material is a very variable component, at least in higher land plants (Canny 1975) (whether this is also true for algae is not known). Transfer of dry weight per unit time is the simplest way to determine a translocation rate, while the specific mass transfer (SMT) is a quotient of dry weight transfer per unit time divided by the cross-sectional area of the transport channel (area of 'phloem' or area of functioning sieve tubes).

Translocation rates of carbon per unit time were first published for *Laminaria* by Lüning *et al.* (1973). Their data were derived from growth experiments and translocation experiments, and showed that in one hour about 0.5 mg C is exported from 1 dm^2 mature frond of *L. hyperborea*. A specific mass transfer for *Alaria esculenta* was determined by Buggeln (1976), whose measurements of growth and cross-sectional area of sieve elements in the medulla revealed a mass transfer of 36.2–60.8 mg C wk^{-1} 0.1 mm^{-2}. Considering that 28–30% of dry weight is carbon (cf. Lüning *et al.* 1973 and literature therein; Buggeln 1976), this translocation rate is equivalent to 0.77–1.29 g dry weight h^{-1} cm^{-2} (sieve elements). Data presented for translocation rates in *Macrocystis pyrifera* by Sargent & Lantrip (1952) are not related to the area of conducting cells and show that a mature blade may export up to 0.33 g organic matter per day, with an average figure of 0.2 g day^{-1}. The total amount of organic matter exported from 100 blades was 20 g per day, which is equivalent to 30.8 g dry weight per day, or 1.3 g h^{-1}. Transport towards the growing tip was approximately 2 g organic matter per day. It was shown that the rate of exudation from *Macrocystis integrifolia* is 1–2 ml h^{-1}, and the dry weight approximately 155 mg ml^{-1} (Schmitz & Srivastava 1974a, 1979a). The cross-sectional area of the peripheral medulla that contains the conducting and exuding sieve tubes is in the range of 2.6 mm^2. The transport rate is thus 5.3–10.6 g dry weight cm^{-2} h^{-1}. The area occupied by sieve tubes as determined from scanning electron micrographs is 20–25% of the peripheral medulla (unpublished results). This is in the range of figures presented for the proportion of sieve tubes in the phloem, which varies between 20 and 70% (Canny 1973, 1975 and literature cited). The specific mass transfer for *M. integrifolia*, based on sieve tube area, is 21.5–53.0 g dry weight h^{-1} cm^{-2} (Schmitz unpublished). Mass transfer rates have been measured from a variety of vascular plants and range from 0.2 to 6 g h^{-1} cm^{-2} (phloem) (Canny 1975). In exuding stalks of *Arenga* (sugar palm) a transport rate of 99 g h^{-1} cm^{-2} (phloem) may be reached (Tammes 1952).

Specific mass transfer can also be formulated as concentration of translocate (g cm^{-3}) times speed (cm h^{-1}). If the mass transfer rate and the concentration of the translocate (exudate) are known, speed can be calculated. For *M. integrifolia* Schmitz & Srivastava (1979a) arrived at

$$\frac{\text{mass transfer}}{\text{concentration}} = \frac{5.3\text{–}10.6}{0.155} = 34\text{–}68 \text{ cm h}^{-1}$$

which again fits well with the previously determined velocities. If, however, the data for mass transfer are related to the sieve tube area, we arrive at much higher velocities, between 136.8 and 341.9 cm h^{-1}. Comparably high transport velocities were previously calculated on the basis of the volume of exudate and the cross-sectional area of conducting sieve elements by Crafts (1939).

5 POSSIBLE MECHANISMS OF LONG-DISTANCE AND MEDIUM-DISTANCE TRANSPORT IN BROWN ALGAE

A discussion of possible mechanisms of long-distance transport in brown algae must consider the fine structure of translocation pathways and the physiological and biochemical findings. Anatomical and physiological experiments revealed that photosynthesis is mainly localized in the meristoderm and outer cortex cells and that ^{14}C-labelled photosynthate moves through the cortex to the long-distance transport system (Hellebust & Haug 1972; Steinbiss & Schmitz 1973). For this medium-distance transport through the cortex there are basically two pathways, the symplast and the apoplast.

It is obvious that in algae symplastic transport has advantages in conservation of substances during transport (metabolic control) and in efficiency. Symplastic transport is indicated by the fact that hardly any selectivity of transport is found; it seems that all those substances that are produced during photosynthesis are also translocated (cf. Schmitz *et al.* 1972; Schmitz & Srivastava 1975, 1979a). We have pointed out previously that cytoplasmic connections between the inner cortex cells and the sieve tubes are scarce; they are probably absent from mature stipes and have not yet been depicted clearly in the blades (Nicholson 1976).

Provided sieve elements and cortex cells form a symplast, loading of the sieve elements is easy as long as a concentration gradient extends from the meristoderm to the sieve tubes. The concentration difference between sieve elements and the inner cortex is not known, but sieve tube sap of *Macrocystis* has a mannitol concentration of 10% (w:v). It could well be that the concentration within sieve tubes is higher than in the cortex, which would necessitate special uphill transport mechanisms.

Once the substances have been loaded into the sieve tube system, their long-distance transport again strictly follows source-sink patterns described earlier. Directionality is thought to be determined by a concentration gradient between source and sink, but no convincing figures for concentration gradients in sieve tubes along a blade or stipe of any brown alga have been presented. The measurements reported by Buggeln (1978) did not reveal a statistically significant trend toward an osmotic gradient between source and sink in *Alaria esculenta* which could drive transport along the midrib. Work like that of Buggeln is urgently needed to make further progress. There is indirect evidence from the observations of carboxylating enzyme distribution and mannitol content along the thallus of different species of *Laminaria* and *Fucus* indicating that concentration gradients in the sieve tubes might exist (Black 1954; Küppers & Kremer 1978; see also Chapter 14).

The mechanisms that have been proposed for phloem transport in vascular plants are many. Some need not be considered here because there is no structural basis for them. Proposals involving the action of contractile proteins probably do not apply to sieve tubes of Laminariales because no P-protein, nor ample evidence of microtubules or microfilaments have been reported (Ziegler 1963; Ziegler & Ruck 1967; Schmitz & Srivastava 1974b, 1975, 1976; Sideman & Scheirer 1977). Protoplasmic streaming, which could also account for enhanced transport along the sieve elements, was not observed. Transport velocities in Laminariales are always much higher than could be accounted for by simple diffusion. Some mode of accelerated transport must therefore be operational within the sieve tubes. There are essentially four possible mechanisms for long-distance transport in brown algae: symplastic (cell-to-cell) transport, surface active movements, electroosmotic flow, and pressure flow.

Transport along sieve tubes can be considered a special kind of symplastic transport. Long cells are apparently more suited for transport than isodiametric ones, and restriction of movement by plasmodesmata is counteracted by the larger diameter and number of pores in the sieve plates. Translocation velocity in the Laminariaceae is in the range of 1–10 cm h^{-1} (Floc'h & Penot 1972; Schmitz *et al.* 1972; Schmitz & Lobban 1976), which in part matches transport velocities of symplastic parenchyma transport (maximum about 6 cm h^{-1}). At present, however, we cannot explain how the solutes travel along the sieve elements. Since the sieve elements are filled with small vacuoles (Fig. 15.2), surface active movement mechanisms (van den Honert 1932) may be operational, but this is speculation since no experimental indications are known.

The concept of electroosmotic flow (Spanner 1958, 1975) is based on the following conditions: transport is a mass stream, sieve pores are not open, material within the sieve pores is negatively charged at the pH of the sieve tube sap (alkaline), potassium is the dominant mobile ion, and energy can be provided for ion pumps. Exudation by detached stipes of *Macrocystis* and other Laminariales indicate that transport might be a mass stream. Analysis of exudate revealed that potassium concentration is very high and K^+ is clearly the dominant ion (Parker 1966; Schmitz & Srivastava 1979a). ATP is present in concentrations of approximately 1.0 mg ml^{-1} and the sieve tubes do contain the enzymes necessary for its synthesis (Schmitz & Srivastava 1974a). All these points favour an electroosmotic flow. On the other hand, sieve pores are not occluded to the extent that electroosmotic forces become effective (Fig. 15.4). No measurements of electrical potentials in sieve elements of brown algae have been made, and nothing is known about potential differences across sieve fields.

The osmotic pressure flow hypothesis first proposed by Münch (1930) seems best supported for higher plants and Laminariales by experimental data so far. It is based on the supposition that source and sink are connected by an uninterrupted sieve tube system, that sieve pores are open and do not cause insurmountable resistance, that a sufficient osmotic gradient between source and sink is maintained and a parallel pressure gradient created, and that osmotic substances and water are released from the sieve tube system at the sink. It can be assumed that the sieve tube system of Laminariales extends uninterrupted throughout the thallus. It has been shown that sieve pores are not obstructed and may be called open pores (Ziegler 1963; Ziegler & Ruck 1967; Schmitz & Srivastava 1974b, 1975, 1976; Sideman & Scheirer 1977). Translocation is always directed from source to sink, and bidirectional transport (within the same sieve tube) has not been demonstrated. Translocation of simultaneously applied ^{14}C and ^{32}P in *Macrocystis* revealed nearly the same transport velocity for both substances in intact plants as well as detached fronds and frond segments. There was a slight delay in transport of ^{32}P, which might have been due to a delay in loading or to a high metabolism within the sieve elements. Both radioisotopes were exuded, which again indicates mass flow (Schmitz & Srivastava 1979b).

One of the essential features of osmotic pressure flow is the movement of solute and water. Movement of water (THO) plus T-photosynthate was shown in *Macrocystis integrifolia* (Schmitz & Srivastava 1980). In intact fronds and frond cuttings THO and tritium-labeled assimilate showed a declining gradient with increasing distance from the source. In the exudate collected from the basal cut end of the frond there was a marked increase in radioactivity with time in the T-photoassimilate, but no such gradient was evident for THO; rapid lateral diffusion of THO along the translocation path apparently prevented its build-up in the exudate. The same authors also demonstrated that THO preferentially leaves the transport system at the sink area. All these points favour a pressure flow system operating in *Macrocystis,* but the most important evidence, a demonstration of adequate concentration and pressure differences in the sieve tubes connecting source and sink, is still lacking.

Our knowledge of the fine structure of sieve elements in different parts of a thallus and at different developmental stages is fragmentary, and it may well be that differences in fine structure of sieve elements in different parts of a thallus result in the operation of slightly modified transport mechanisms within one plant. The structures of sieve elements in *Laminaria* and *Macrocystis* (Figs 15.2, 15.4) are so different that it is possible to conclude that two different, or at least modified, mechanisms of long-distance transport are operating.

6 REFERENCES

BISALPUTRA T., RUSANOWSKI P. C. & WALKER W. S. (1967) Surface activity, cell wall and fine structure of pit connections in the red alga *Laurencia spectabilis*. *J. Ultrastruct. Res.*, **20**, 277–89.

BLACK W. A. P. (1954). Concentration gradients and their significance in *Laminaria saccharina* (L.) Lamour. *J. Mar. Biol. Ass. U.K.*, **33**, 49–60.

BOURNE V. L., CONWAY E. & COLE K. (1970) On the ultrastructure of pit connections in the conchocelis phase of the red alga *Porphyra perforata* J. Ag. *Phycologia*, **9**, 79–81.

BUGGELN R. G. (1974) Physiological investigations on *Alaria esculenta* (L.) Grev. (Laminariales). I. Elongation of the blade. *J. Phycol.*, **10**, 283–8.

BUGGELN R. G. (1976) The rate of translocation in *Alaria esculenta* (Laminariales, Phaeophyceae). *J. Phycol.*, **12**, 439–42.

BUGGELN R. G. (1977) Physiological investigations on *Alaria esculenta* (Laminariales, Phaeophyceae). II. Role of translocation in blade growth. *J. Phycol.*, **13**, 212–8.

BUGGLEN R. G. (1978) Physiological investigations on *Alaria esculenta* (Laminariales, Phaeophyceae). III. Exudation by the blade. *J. Phycol.*, **14**, 54–6.

CANNY M. J. (1973) *Phloem Translocation*. Cambridge University Press, London.

CANNY M. J. (1975) Mass transfer. In *Encyclopedia of Plant Physiology*, vol. I, eds. Zimmermann M. H. & Milburn, J. A. pp 139–53. Springer-Verlag, Berlin, Heidelberg.

CELAN M. (1939) Sur les cellules conductrices des algues rouges. *Bull. Soc. Bot.*, **86**, 297–300.

CELAN M. (1941) Recherches cytologiques sur les algues rouges. *Rev. Cytol. Cytophysiol. Vég.*, **5**, 1–162.

CLENDENNING K. A. (1971) Photosynthesis and general development in *Macrocystis*. In *The Biology of Giant Kelp Beds* (Macrocystis) *in California*, ed. North W. J., pp. 169–90 (*Nova Hedwigia Beih*. **32**).

CRAFTS A. S. (1939) The relation between structure and function of the phloem. *Am. J. Bot.*, **26**, 172–6.

CRAFTS A. S. & CRISP C. E. (1971) *Phloem Transport in Plants*. W. H. Freeman, San Francisco.

DUNCAN M. J. (1973) *In situ* studies of growth and pigmentation of the Phaeophycean *Nereocystis leutkeana*. *Helgol. Wiss. Meeresunters.*, **24**, 510–25.

ESAU K. (1969) *The Phloem. Handbuch der Pflanzenanatomie*, vol. V 2. Bornträger, Berlin.

FELDMANN J. (1939) Sur la structure des cellules axiales de l'*Asparagopsis armata* Harv. *C. R. Acad. Sci. Paris*, **208**, 1743–5.

FELDMANN J. & FELDMANN G. (1946) Recherches sur l'appareil conducteur des Floridées. *Rev. Cytol. Cytophysiol. Vég.*, **8**, 161–209.

FLOC'H J. Y. (1976) Appels de blessure et transport à longue distance d'éléments minéraux chez les algues brunes. *Physiol. Vég.*, **14**, 767–77.

FLOC'H J. Y. & PENOT M. (1970) Mise en évidence d'une répartition préférentielle de divers cations le long du thalle des Laminaires. *C. R. Acad. Sci. Paris*, **271**, 288–91.

FLOC'H J. Y. & PENOT M. (1971) Répartition préférentielle du phosphore et mouvements de redistribution du ^{32}P chez quelques Laminaires. *C. R. Acad. Sci. Paris*, **273**, 1100–3.

FLOC'H J. Y. & PENOT M. (1972) Transport du ^{32}P et du ^{86}Rb chez quelques algues brunes: orientation des migrations et voies de conduction. *Physiol. Vég.*, **10**, 677–86.

FLOC'H J. Y. & PENOT M. (1974) Translocation of ^{32}P and ^{86}Rb in some brown algae (*Laminaria digitata* and *Fucus vesiculosus*): influence of metabolisms and sinks. Proc. Int. Seaweed Symposium, **7** (abst.).

FLOC'H J. Y. & PENOT M. (1976) Étude comparative du transport à longue distance de différents radioéléments dans le thalle de *Laminaria digitata* (L.) Lamouroux. *C. R. Acad. Sci. Paris,* **282,** 989–92.

FLOC'H J. Y. & PENOT M. (1978a) Influence de la cycloheximide sur le transport à longue distance du ^{32}P chez *Laminaria digitata* (L.) Lamouroux. *Bot. Mar.,* **21,** 5–11.

FLOC'H J. Y. & PENOT M. (1978b) Changes in ^{32}P-phosphorus compounds during translocation in *Laminaria digitata* (L.) Lamouroux. *Planta,* **143,** 101–7.

FLOC'H J. Y. & PENOT M. (1980) Variation de la composition minérale et transport à longue distance au cours de la croissance de la lame chez *Laminaria hyperborea* (Gunn) Fosl. *Z. Pflanzenphysiol.,* **96,** 377–85.

FRITSCH F. E. (1945) *The Structure and Reproduction of the Algae,* vol. II. Cambridge University Press, London.

FRYE T. C., RIGG G. B. & CRANDALL W. C. (1915) The size of kelps on the Pacific coast of North America. *Bot. Gaz.,* **60,** 473–83.

GUTKNECHT J. & DAINTY J. (1968) Ionic relations of marine algae. *Oceanogr. Mar. Biol. Ann. Rev.,* **6,** 163–200.

HARTMANN T. & ESCHRICH W. (1969) Stofftransport in Rotalgen. *Planta,* **85,** 303–12.

HELLEBUST J. A. & HAUG A. (1972) Photosynthesis, translocation and alginic acid synthesis in *Laminaria digitata* and *Laminaria hyperborea. Can. J. Bot.,* **50,** 169–76.

KAIN J. M. (1979) A view of the genus *Laminaria. Oceanogr. Mar. Biol. Ann. Rev.,* **17,** 101–61.

KILLIAN K. (1911) Beiträge zur Kenntnis der Laminarien. *Ztschr. Bot.,* **3,** 433–94.

KLEIN J. (1877) Algologische Mitteilungen. 3. Über Siebröhren bei den Florideen. *Flora,* **60,** 289–97.

KREMER B. P. & WILLENBRINK J. (1972) CO_2-Fixierung und Stofftransport in benthischen marinen Algen. I: Zur Kinetik der $^{14}CO_2$-Assimilation bei *Laminaria saccharina. Planta,* **103,** 55–64.

KÜPPERS U. & KREMER B. P. (1978) Longitudinal profiles of carbon dioxide fixation capacities in marine macroalgae. *Plant Physiol.,* **62,** 49–53.

KYLIN H. (1937) *Anatomie der Rhodophyceen. Handbuch der Pflanzenanatomie,* Bd. VI, 2., ed. Linsbauer K., pp. 28–33. Gebr. Bornträger, Berlin.

LEE R. E. (1971) The pit connection of some lower red algae: ultrastructure and phylogenetic significance. *Br. Phycol. J.,* **6,** 29–38.

LEE R. E. & FULTZ S. A. (1970) Ultrastructure of the Conchocelis stage of the marine red alga *Porphyra leucosticta. J. Phycol.,* **6,** 22–8.

LOBBAN C. S. (1978a) Translocation of ^{14}C in *Macrocystis pyrifera* (giant kelp). *Plant Physiol.,* **61,** 585–9.

LOBBAN C. S. (1978b) Translocation of ^{14}C in *Macrocystis integrifolia* (Phaeophyceae). *J. Phycol.,* **14,** 178–82.

LÜNING K. (1969) Growth of amputated and dark-exposed individuals of the brown alga *Laminaria hyperborea. Mar. Biol.,* **2,** 218–23.

LÜNING K. (1970) Cultivation of *Laminaria hyperborea in situ* and in continuous darkness under laboratory conditions. *Helgol. Wiss. Meeresunters.,* **20,** 79–88.

LÜNING K. (1971) Seasonal growth of *Laminaria hyperborea* under recorded underwater light conditions near Helgoland. *4th Europ. Mar. Biol. Symposium,* ed. Crisp D. E., pp. 347–61. Cambridge University Press, Cambridge.

LÜNING K., SCHMITZ K. & WILLENBRINK J. (1972) Translocation of ^{14}C-labeled assimilates in two *Laminaria* species. In *Proc. Int. Seaweed Symp.,* **7,** 420–25.

LÜNING K., SCHMITZ K. & WILLENBRINK J. (1973) CO_2-Fixation and translocation in benthic marine algae. III: Rates and ecological significance of translocation in *Laminaria hyperborea* and *L. saccharina. Mar. Biol.,* **23,** 275–81.

MACROBBIE E. A. C. (1974) Ion uptake. In *Algal Physiology and Biochemistry,* ed. Stewart W. D. P., pp. 676–713. Blackwell Scientific Publications, Oxford.

MARCHANT H. J. (1976) Plasmodesmata in algae and fungi. In *Intercellular Communication in Plants: Studies on Plasmodesmata.,* eds. Gunning B. E. S. & Robards A. W., pp. 59–80. Springer-Verlag, Berlin, Heidelberg, New York.

MÜNCH E. (1930) *Die Stoffbewegungen in der Pflanze*. Verlag G. Fischer, Jena.

NICHOLSON N. L. (1970) Field studies on the giant kelp *Nereocystis,*. *J. Phycol.*, **6,** 177–82.

NICHOLSON N. L. (1976) Anatomy of the medulla of *Nereocystis*. *Bot. Mar.*, **19,** 23–31.

NICHOLSON N. L. & BRIGGS W. R. (1972) Translocation of photosynthate in the brown alga *Nereocystis*. *Am. J. Bot.*, **59,** 97–106.

OLIVER F. W. (1887) On the obliteration of the sieve-tubes in Laminarieae. *Ann. Bot.*, **1,** 95–117.

PARKER B. C. (1964) Chemical nature of sieve tube callus in *Macrocystis*. *Phycologia*, **4,** 27–42.

PARKER B. C.(1965) Translocation in the giant kelp *Macrocystis*. I. Rates, direction, quantity of C^{14}-labeled products and fluorescein. *J. Phycol.*, **1,** 41–6.

PARKER B. C. (1966) Translocation in *Macrocystis*. III. Composition of sieve tube exudate and identification of the major C^{14}-labeled products. *J. Phycol.*, **2,** 38–41.

PARKER B. C. (1971a) The internal structure of *Macrocystis*. In *The Biology of Giant Kelp* (Macrocystis) *Beds*, ed. North W. J., pp. 99–121 (*Nova Hedwigia Beih.* **32**).

PARKER B. C. (1971b) Studies of translocation in *Macrocystis*. In *The Biology of Giant Kelp* (Macrocystis) *Beds*, ed. North W. J., pp. 191–5 (*Nova Hedwigia Beih.* **32**).

PARKER B. C. & FU M. (1965) The internal structure of the elk kelp (*Pelagophycus* species). *Can. J. Bot.*, **43,** 1293–305.

PARKER B. C. & HUBER J. (1965) Translocation in *Macrocystis*. II. Fine structure of sieve tubes. *J. Phycol.*, **1,** 172–9.

PARKER J. (1956) Translocation of P^{32} and dye behavior in two species of marine algae. *Naturwiss.*, **43,** 452.

PARKER J. (1964) Further notes on the sieve plates of *Macrocystis pyrifera*. *Protoplasma*, **58,** 681–4.

PARKER J. & PHILPOTT D. E. (1961) The ultrastructure of *Macrocystis pyrifera*. *Bull. Torrey Bot. Club*, **88,** 85–90.

PENOT M., FLOC'H J. Y. & PENOT M. (1976a) Comparative study of ^{45}Ca absorption and translocation in various groups of plants. *Planta*, **129,** 7–14.

PENOT M., FLOC'H J. Y. & PENOT M. (1976b) Physiologie comparée des transports à longue distance chez les végétaux supérieurs et chez les algues marines. *Phycologia*, **15,** 299–308.

PENOT M. & PENOT M. (1977) Quelques aspects originaux des transports à longue distance dans le thalle de *Ascophyllum nodosum* (L.) Le Jolis (Phaeophyceae, Fucales). *Phycologia*, **16,** 339–47.

PUESCHEL C. M. (1977) A freeze-etch study of the ultrastructure of red algal pit plugs. *Protoplasma*, **91,** 15–30.

RAMUS J. (1969) Pit connection formation in the red alga *Pseudogloiophloea*. *J. Phycol.*, **5,** 57–63.

RAMUS J. (1971) Properties of septal plugs from the red alga *Griffithsia pacifica*. *Phycologia*, **10,** 99–103.

RIGG G. B. (1925) Some physiology of the sieve tubes of *Nereocystis*. *Puget Sound Mar. Stn. Pub.*, **3,** 311–29.

ROSENTHAL O. (1980) Zur Kenntnis von *Macrocystis* und *Thalassiophyllum*. *Flora*, **48,** 105–47.

SARGENT M. C. & LANTRIP L. W. (1952) Photosynthesis, growth and translocation in giant kelp. *Am. J. Bot.*, **39,** 99–107.

SAUVAGEAU C. (1915) Sur le dévelopment et la biologie d'une Laminaire (*Saccorhiza bulbosa*). *C. R. Acad. Sci. Paris*, **160,** 445–8.

SCHMITZ K. & LOBBAN C. S. (1976) A survey of translocation in Laminariales (Phaeophyceae). *Mar. Biol.*, **36,** 207–16.

SCHMITZ K., LÜNING K. & WILLENBRINK J. (1972) CO_2-Fixierung und Stofftransport in benthischen marinen Algen. II. Zum Ferntransport ^{14}C-markierter Assimilate bei *Laminaria hyperborea* und *Laminaria saccharina*. *Z. Pflanzenphysiol.*, **67,** 418–29.

SCHMITZ K. & SRIVASTAVA L. M. (1974a) The enzymatic incorporation of ^{32}P into ATP and other organic compounds by sieve-tube sap of *Macrocystis integrifolia* Bory. *Planta*, **116,** 85–9.

SCHMITZ K. & SRIVASTAVA L. M. (1974b) Fine structure and development of sieve tubes in *Laminaria groenlandica* Rosenv. *Cytobiol.* **10,** 66–87.

SCHMITZ K. & SRIVASTAVA L. M. (1975) On the fine structure of sieve tubes and the physiology of assimilate transport in *Alaria marginata*. *Can. J. Bot.*, **53,** 861–76.

SCHMITZ K. & SRIVASTAVA L. M. (1976) The fine structure of sieve elements of *Nereocystis luetkeana*. *Am. J. Bot.*, **63**, 679–93.

SCHMITZ K. & SRIVASTAVA L. M. (1979a) Long distance transport in *Macrocystis integrifolia*. I. Translocation of ^{14}C-labeled assimilates. *Plant Physiol.*, **63**, 995–1002.

SCHMITZ K. & SRIVASTAVA L. M. (1979b) Long distance transport in *Macrocystis integrifolia*. II. Tracer experiments with ^{14}C and ^{32}P. *Plant Physiol.*, **63**, 1003–9.

SCHMITZ K. & SRIVASTAVA L. M. (1980) Long distance transport in *Macrocystis integrifolia*. III. Movement of THO. *Plant Physiol.*, **66**, 66–9.

SCHUMACHER W. (1933) Untersuchungen über die Wanderung des Fluoreszeins in den Siebröhren. *Jahrb. Wiss. Bot.*, **77**, 685–732.

SIDEMAN E. J. & SCHEIRER D. C. (1977) Some fine structural observations on developing and mature sieve elements in the brown alga *Laminaria saccharina*. *Am. J. Bot.*, **64**, 649–57.

SMITH A. J. (1939) The comparative histology of some of the Laminariales. *Am. J. Bot.*, **26**, 571–85.

SMITH G. M. (1942) Notes on some brown algae from the Monterey Peninsula, California. *Am. J. Bot.*, **29**, 645–53.

SPANNER D. C. (1958) The translocation of sugar in sieve tubes. *J. Exp. Bot.*, **9**, 332–42.

SPANNER D. C. (1975) Electroosmotic flow. In *Encyclopedia of Plant Physiology*, Vol. I, pp. 301–27. Springer-Verlag, Berlin, Heidelberg.

STEINBISS H. H. & SCHMITZ K. (1973) CO_2-Fixierung und Stofftransport in benthischen marinen Algen. V. Zur autoradiographischen Lokalisation der Assimilattransportbahnen im Thallus von *Laminaria hyperborea*. *Planta*, **112**, 253–63.

STEINBISS H. H. & SCHMITZ K. (1974) Zur Entwicklung und funktionellen Anatomie des Phylloids von *Laminaria hyperborea*. *Helgol. Wiss. Meeresunters.*, **26**, 134–52.

STEWART K. D., MATTOX K. R. & FLOYD G. L. (1973) Mitosis, cytokinesis, the distribution of plasmodesmata, and other cytological characteristics in the Ulotrichales, Ulvales and Chaetophorales: phylogenetic and taxonomic considerations. *J. Phycol.*, **9**, 128–41.

SYKES M. G. (1908) Anatomy and histology of *Macrocystis pyrifera* and *Laminaria saccharina*. *Ann. Bot.*, **22**, 291–325.

TAMMES P. M. L. (1952) On the rate of translocation of bleeding sap in the fruit stalk of *Arenga*. *Proc. Koninkl. Ned. Akad. Wetenschap. C*, **55**, 141–4.

TITLYANOV E. A. & PESHEKHODKO V. M. (1973) On the transport of assimilates in thalli of fringing seaweeds. *Trudy Biologo-počvennogo instituta Novaja serija*, **20**, 137–40.

TURGEON R. & WEBB J. A. (1973) Leaf development and phloem transport in *Cucurbita pepo*: transition from import to export. *Planta*, **113**, 179–91.

VAN DEN HONERT T. H. (1932) On the mechanism of the transport of organic materials in plants. *Proc. Koninkl. Ned. Adad. Wetenschap Ser. C. Biol. Med. Sci.*, **35**, 1104–11.

VAN WENT J. L., VAN AELST A. C. & TAMMES P. M. L. (1973a) Open plasmodesmata in sieve plates of *Laminaria digitata*. *Acta Bot. Neerl.*, **22**, 120–3.

VAN WENT J. L., VAN AELST A. C. & TAMMES P. M. L. (1973b) Transverse connections between cortex and translocating medulla in *Laminaria digitata*. *Acta. Bot. Neerl.*, **22**, 77–8.

VAN WENT J. L., VAN AELST A. C. & TAMMES P. M. L. (1973c) Organelles in medulla exudate of *Laminaria*. *Acta. Bot. Neerl.*, **22**, 730–1.

VAN WENT J. L. & TAMMES P. M. L. (1973) Trumpet filaments in *Laminaria digitata* as an artefact. *Acta Bot. Neerl.*, **22**, 112–19.

WEIDNER M. & KÜPPERS U. (1973) Phosphoenolpyruvat-Carboxykinase und Ribulose-1,5-Diphosphat-Carboxylase von *Laminaria hyperborea* (Gunn.) Fosl.: Das Verteilungsmuster der Enzymaktivitäten im Thallus. *Planta*, **114**, 365–72.

WILL H. (1884) Zur Anatomie von *Macrocystis luxurians* Hook fil. et Harv. *Bot. Z.*, **42**, 801–8, 825–30.

WILLE N. (1885) Siebhyphen bei den Algen. *Ber. Deutsch. Bot. Ges.*, **3**, 29–31.

WILLE N. (1897) *Beiträge zur physiologischen Anatomie der Laminariaceen*. Det. Kgl. Norske Fredericks Universität, Zentraldrückerei, Christiania.

WILLENBRINK J., RANGONI-KÜBBELER M. & TERSKY B. (1975) Frond development and CO_2-fixation in *Laminaria hyperborea*. *Planta*, **125**, 161–70.

WILLENBRINK J., KREMER B. P., SCHMITZ K. & SRIVASTAVA L. M. (1979) Photosynthetic and light-independent carbon fixation in *Macrocystis, Nereocystis*, and some selected Pacific Laminariales. *Can. J. Bot.*, **57,** 890–7.

YENDO K. (1911) The development of *Costaria, Undaria* and *Laminaria*. *Ann. Bot.*, **25**, 691–715.

ZIEGLER H. (1963) Untersuchungen über die Feinstruktur des Phloems. II Mitteilung: Die Siebplatten bei der Braunalge *Macrocystis pyrifera* (L.). *Protoplasma*, **57,** 786–99.

ZIEGLER H. (1975) Nature of transported substances. In *Encyclopedia of Plant Physiology*. Vol. I, pp 59–100. Springer-Verlag, Berlin, Heidelberg.

ZIEGLER H. & RUCK I. (1967) Untersuchungen über die Feinstruktur des Phloems. III Mitteilung: Die 'Trompetenzellen' von *Laminaria*-Arten. *Planta*, **73,** 62–73.

CHAPTER 16
POLYSACCHARIDES OF THE SEAWEEDS

E. L. McCANDLESS
Department of Biology, McMaster University,
Hamilton, Ontario, Canada L8S 4K1

1 INTRODUCTION

Plants are composed predominantly of carbohydrate materials, which is perhaps not surprising since they are the primary producers of organic carbon compounds, utilizing the mechanism and energy of photosynthesis. This process and the nature of compounds produced by it are discussed in detail in Chapters 13 and 14 and hence will be only briefly mentioned in this chapter.

Thirty-five years ago, when I was a graduate student, proteins were described in terms of their solubility characteristics and their component units, but in the intervening period it has become possible to analyse these macromolecules exhaustively, to dissemble and assemble them unit by unit, and to describe the mechanism by which they are made in living cells. At the present time, polysaccharides are still described in terms of their solubility characteristics and component units. With the obvious exception of cellulose, carbohydrate polymers of the algae are thought by most chemists and biologists to be 'mucilaginous', as they emerge after extraction procedures, of no particular form whatsoever.

The past few years, however, have seen advances in the understanding of polysaccharides, and it is interesting that some of the major strides have been made using seaweed polysaccharides as examples. Frei & Preston (1964a,b) and others have used X-ray diffraction and electron microscopic analysis to demonstrate that polysaccharides other than cellulose are highly ordered in seaweed cell walls and presumably also serve structural functions. Rees and his co-workers, after extensive work upon certain of the red algal galactans (Rees 1969), have applied sophisticated methods of mathematical analysis to describe their three dimensional structures and have shown that it is appropriate to consider tertiary and quaternary structure of these macromolecules as one does in the case of proteins (Rees 1972). Figure 16.1 illustrates this point. The chemical approach to the structure of polysaccharides still involves degradative procedures such as acid hydrolysis, methanolysis, acetolysis, and alkaline

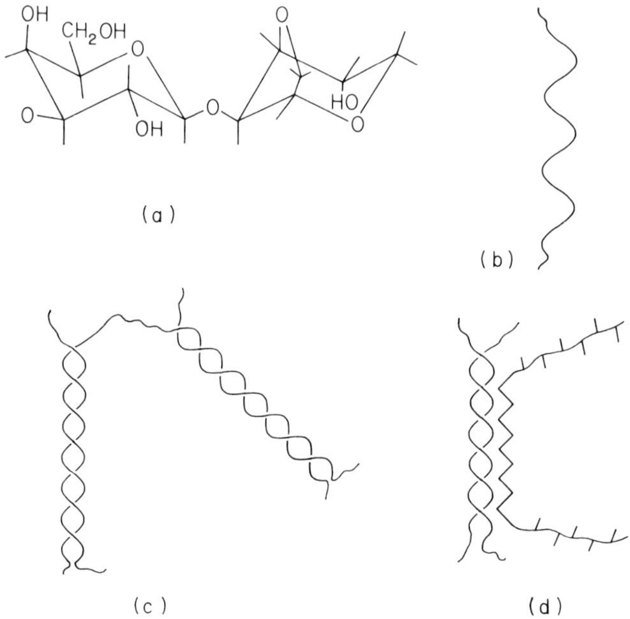

Fig. 16.1 (a) Primary, (b) secondary, (c) tertiary and (d) quaternary structure in polysaccharides (after Rees 1972). (There is at present no proof that whole polysaccharide molecules exhibit quaternary structure; it has, however, been demonstrated in subunits of polysaccharides. [Rees personal communication].)

borohydride reduction, but the gentler technique of enzymatic hydrolysis furnishes information not easily obtained in other ways. The tools employed in analysis still include determination of reactive groups such as sulphate or carboxylate ions, but additional information has become available by refinement of X-ray crystallography, chromatographic separation of sugars, infra-red and nuclear magnetic resonance spectroscopy, and immunochemistry. A better understanding of the basis of certain histochemical reactions and the further development of electron microscopy have provided evidence on the localization of specific polysaccharide units in tissues under study, and the use of radioactive isotopes and of antibodies have supplemented this information.

It is impossible to cover the entire subject of seaweed polysaccharides in one chapter of a moderate-sized monograph. The reader is referred to the excellent review of the chemistry of marine algal polysaccharides by Peat & Turvey (1965) for the state of the art a decade and a half ago, and to the text by Percival & McDowell (1967) which appeared shortly thereafter. Shorter general reviews which update this information have been written by Rees (1969), Percival (1970, 1979), Haug (1974), and Turvey (1978). The present chapter will of necessity draw on material included in the above reviews.

1.1 Functions of polysaccharides

The two major functions of carbohydrates in plant tissues are the provision of sources of energy, immediate or delayed, and the provision of structural strength, as in cell walls and matrix polysaccharides. Different polysaccharides subserve these ends in higher plants, but in general the functions are carried out by means of glucose and glucose polymers. In the former instance the glucan amylose is α-1,4-linked, and a second component, amylopectin, is both α-1,4- and α-1,6-linked. In the latter case, it is a β-1,4-linked glucan, cellulose. The cell walls of higher plants investigated thus far also contain a number of other saccharides arranged in a very definite pattern (see Albersheim 1978).

The storage polysaccharides of the algae are also glucose polymers, but they exhibit differences which are characteristic of the orders in which they are found. The similarity to the starches of land plants is greatest in the green algae, less in the red, and least in the brown algae. By analogy with the starch of higher plants, it seems reasonable that those substances in the Chlorophyta which are sequestered in granular form, and which react with iodine to give a blue-black colour, are indeed stored polysaccharide; a similar case might be made for floridean starch in the Rhodophyta which is packaged in smaller granules and which stains reddish with iodine. Laminaran in the Phaeophyta is somewhat different, but it too is a glucan.

Structural polysaccharides of the seaweeds are harder to define. The assumption that cellulose fibrils provide structural strength for all plant cells including those of the algae, and that the peculiar algal polysaccharides are merely sticky filler both in the matrix and between cellulose fibrils is now untenable, as a number of species of seaweeds have been shown to contain no cellulose at all (Frei & Preston 1961; see also Preston 1968); in others the claim for cellulose is based on insolubility or on non-definitive methods of the past.

Besides the two obvious functions, several others have been postulated. The peculiar fact that sulphated polysaccharides are found in marine algae but are absent in higher plants may eventually provide a clue to the functions these macromolecules serve. Similar although not identical sulphated polysaccharides are found in multicellular animals where, as in the algae, they are located in the intercellular matrix. Several functions have been suggested which are based on the ability of charged polysaccharides to imbibe water, which might be considered to be a protective pillow against the physical buffeting which seaweeds undergo by wave action; it should provide protection against desiccation as well. The anionic character of these and other charged polysaccharides might also serve as a sort of ion exchange material, and/or might help to sequester certain ions. These, of course, are general ideas. As Rees has pointed out (1975) we now relate these functions to the finer structure of specific polysaccharides in a very superficial manner; we should expect to look deeper into the relationship as finer details of structure become available.

2 STORAGE POLYSACCHARIDES OF THE SEAWEEDS

There have been relatively few additions to the body of knowledge included in the excellent review of storage forms of polysaccharides in algae presented by Craigie in a previous monograph in this series (1974) and in Turvey's recent (1978) description of biosynthetic and biodegradative enzymes which act on these polymers. For this reason the 'starches' of marine algae will be summarized very briefly in the present chapter.

As mentioned, the storage forms of carbohydrates in seaweeds as in land plants are almost exclusively glucans. The similarity in the Chlorophyta and other green algae is greater, with both linear chains of α-1,4-linked units, as in amylose, and branched chains which have α-1,6-glycosidic bonds, as in amylopectin. These glucans also stain blue-black with iodine. The evidence for linkage has been derived from classical chemical studies and by assessment of vulnerability to enzyme attack (Mackie & Percival 1960; Meeuse & Smith 1962; Love *et al.* 1963).

The storage glucan of red algae, floridean starch, has historic significance in that Kylin (1913) used enzymology, the degradation by malt diastase, to describe the analogy between the granules in *Furcellaria lumbricalis* and those in higher plants. Extensive analysis of the starch of *Dilsea cornosa* by Peat *et al.* (1959a, b) and of several other red algal species (Meeuse *et al.* 1960) confirmed its general amylopectin-like structure but left unexplained its reddish iodine reaction. The reactivity of this starch with broad bean R-enzyme (Peat *et al.* 1959b) and its failure to precipitate with concanavalin-A (Manners & Wright 1962) indicated that despite its red staining with iodine it is more closely related to amylopectin than to glycogen. On the other hand, Whyte (1971) found the properties of *Rhodymenia pertusa* starch to be closer to glycogen and corn phytoglycogen.

Laminarain, the storage glucan of the Phaeophyta, is quite different from that found in green and red algae. It is a β-1,3-glucan with varying proportions of 1,6-branching points, and it has been found to be present in all brown algae analysed (Quillet 1958). In a number of species the reducing ends of some of the glucans are covered with mannitol units, but this is absent in others (Maeda &

Nisizawa 1968a,b). Chrysophyta also contain a β-1,3-glucan which is referred to as chrysolaminaran (Meeuse 1962). Mannitol termini are absent in this form. A β-1,3-glucan has been reported in the green algae *Codiolum* and *Caulerpa*, associated with an α-glucan (Carlberg & Percival 1977).

The two storage forms of glucans, amylose and laminaran, provide evidence for the relationship between structure and function of the polysaccharides. As Rees has eloquently described (1977), their existence as chains which form hollow helices facilitates the close packing of similar chains either in double or triple helix form, or when the chains are more extended, as 'nests' of adjacent helices. Figure 16.2 illustrates this point, and also shows the manner in which a less extended hollow helix may complex with small molecules of appropriate size and shape as in the iodine reaction.

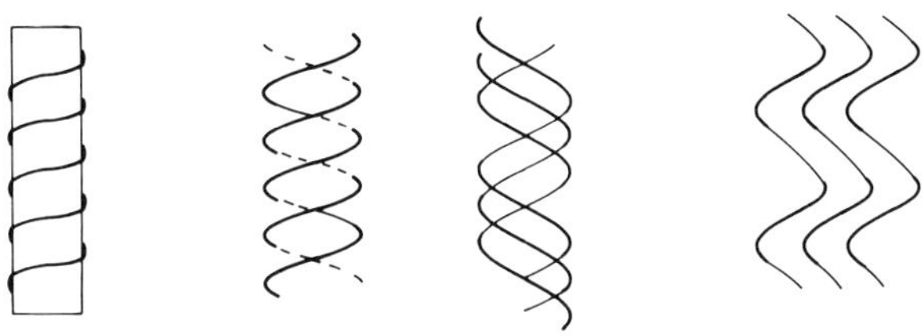

Fig. 16.2. Hollow helix family of polysaccharides; (a) single helix with small molecule included, (b) double or triple helices, (c) 'nesting' of extended chains (after Rees 1977, used with permission).

In addition to glucose polymers, fructose oligosaccharides (inulin type) have been reported in a few green algae; they appear to be storage carbohydrates (Meeuse 1962, 1963; Bourne *et al.* 1972).

3 STRUCTURAL POLYSACCHARIDES OF THE SEAWEEDS

The emerging picture of a common construction plan of the primary cell walls of higher plants (see Albersheim 1978) does not extend to the algae. Instead, a variety of solutions appear to have evolved in answer to the need for structural strength in these plants. In some instances cellulose, β-1,4-glucan, is still the primary fibrillar component. In others it is not present, or glucans which may be present are complexed with other sugars such as xyloglucans (Roelofsen *et al.* 1953), glucomannans (Kreger 1969) and the like. The major support in a number of algae is provided by xylans (Mackie & Percival 1959), whereas in others mannan is the chief structural polymer (Iriki & Miwa 1960). Probably some galactans serve this purpose in certain red algae (Gordon-Mills & McCandless 1977; Gordon-Mills *et al.* 1978). The boundaries of what might be considered 'structural' are less clear than they once were (McCandless *et al.* 1977).

Most of the work on algal cell walls has been carried out on extracts or powders of macerated tissue, usually although not always derived from total plant specimens. A variety of procedures has been employed by different investigators to separate the 'cell wall' material from other polysaccharides. It is not clear what relationship exists between polysaccharides which we obtain by hot water, acid, or alkaline extraction and what remains insoluble, or what can be removed by more drastic procedures. The improvements in histochemical techniques yield information on where certain types of molecules are located in the seaweed but do not give us a real idea of what the structures are *in vivo*, how they are linked, or how they function. With these reservations in mind, let us consider some of the basic elements of structure as we now comprehend them.

We know very little about a protein component of seaweed cell walls, yet it it is clear that the form in which polysaccharides are laid down in cell walls of higher plants includes such a component, rich in hydroxyproline as is the mammalian structural protein collagen. Similar proteins have been reported from the walls of several green seaweeds (Thompson & Preston 1967), and a proteoglycan containing this characteristic amino acid has been found in *Chlamydomonas* (Catt *et al.* 1976). Other types of proteoglycans have been prepared from algal tissues (Medcalf & Larsen 1977b; Court & Taylor 1979). It is probable that in algae as in higher plants protein, or peptide, is an essential part of cell wall structure.

Another element of cell wall structure is usually, but not always fibrous. A third has been considered an 'amorphous', cementing substance. The strength of the cell wall in both higher plants and algae usually depends on the presence of one or more polysaccharides which adopt a ribbon-like conformation which allows the individual sugar units to form non-covalent bonds between chains and thus to align themselves in packing together. The firmest of the ribbon-type polysaccharides of cell walls is cellulose (Fig. 16.3). Xylans and mannans of similar linkage are ribbon-type polysaccharides with greater flexibility (Rees 1977), probably a significant attribute for seaweed cell walls. We shall first consider the fibrous structural elements.

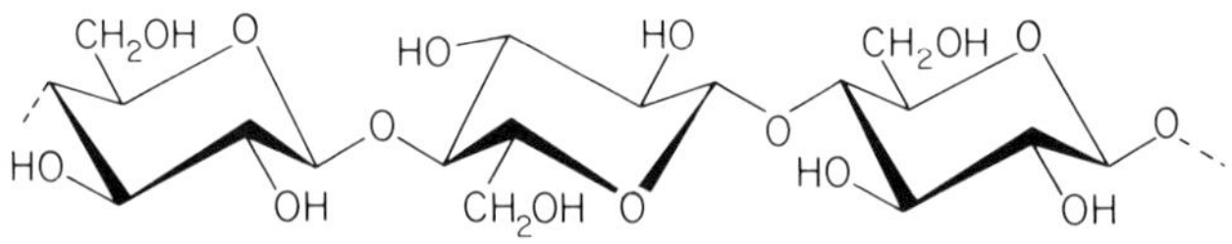

Fig. 16.3. Ribbon type polysaccharide, cellulose.

3.1 Cellulose, the fibrous structural element of higher plants and certain seaweeds

The definition of cellulose has undergone change since the days when one assumed that the microfibrils in all plant cells were composed of this polysaccharide, which was: 1, extremely insoluble in most solvents; but 2, would dissolve in cuprammonia and 3, which stained blue-black with I_2-H_2SO_4. Chemists have described cellulose as a polysaccharide made up of β-D-glucose. X-ray crystallographers have described it as a crystalline polysaccharide which gave a more or less specific diffraction pattern. Biochemists would describe it in terms of its susceptibility to attack by a β-1,4-glucanase if this 'cellulase' were not contaminated with other carbohydrases. For the purpose of this review, and because of my own background, cellulose is here defined as a polymer of β-1,4-linked glucose units, which gives the X-ray diffraction pattern of cellulose I. (The increasing understanding of biochemistry by botanists makes it unnecessary to continue to use the term eucellulose coined by Myers & Preston (1959).) This then excludes polymers which are combinations of glucose with other sugars, and those which give other types of diffraction patterns.

Nicolai & Preston (1952) separated filamentous green algae into three groups on the basis of cell wall characters as shown chiefly by X-ray analysis. In the first group there is strong evidence for the presence of cellulose. In *Valonia* the walls of vesicles were reported to be 75% cellulose, and their hydrolysis yielded essentially glucose only (Preston & Cronshaw 1958). X-ray diffraction showed a characteristic cellulose I pattern, and analysis indicated that the crystallites lie in crossed planes (Cronshaw & Preston 1958). This arrangement of cellulose plates is responsible for the mosaic pattern seen under polarized light. A similar pattern based on cellulose has been obtained from species of *Cladophora* and *Chaetomorpha* (Nicolai & Preston 1952), although the cellulose content was substantially less (Cronshaw *et al.* 1958). The cellulosic green seaweeds include *Apjohnia, Dictyosphaeria, Rhizoclonium,* and *Siphonocladus* spp. (Nicolai & Preston 1952; Cronshaw *et al.* 1958; Parker & Leeper 1969; Stewart *et al.* 1969).

The presence of cellulose has not been demonstrated so conclusively in the second group of green algae, or in the Rhodophyta or Phaeophyta. The diffraction pattern in these resembles cellulose II (mercerized cellulose), obtained by alkali treatment (Nicolai & Preston 1952), although Frei & Preston (1961) have reported the diffraction patterns become more properly cellulosic on extensive treatment to remove encrusting material (see also Preston 1974). In any event, in *Ulva lactuca* the glucose polymer is closely associated with an equal amount of a xylan, and a similar case might be made for *Enteromorpha* spp. (Cronshaw *et al.* 1958; Dodson & Aronson 1978). Bryhni (1978) has studied a morphological mutant of *Ulva mutabilis, lumpy*, which is characterized by a decreased proportion of structural polysaccharide, and has suggested that its undifferentiated form may result from decreased cross-linking of cellulose fibers.

The third group of green algae appears to contain cellulose in only one phase of the life cycle, and is exemplified by *Acetabularia* cyst walls (Nisizawa *et al.* 1974; Herth *et al.* 1975) and the gametophytic phases of *Derbesia-Halicystis* and

Bryopsis (Wutz & Zetsche 1970; Huizing & Rietema 1975). This phenomenon of biochemical alternation of generations will be discussed in detail later.

In the red algae studied, the cellulose content was reported to be very low, and of low crystallinity. After removal of clay minerals by extensive treatment of the algae, Frei & Preston (1961; see also Preston 1974, p. 230) reported that six different species showed cellulose I-type diffraction patterns. In *Chondrus crispus*, the crystallinity which gives rise to birefringence of the thallus is entirely removed by prolonged hot water extraction, which argues against its being due to cellulose (McCandless *et al.* 1977). More primitive members of the Rhodophyta, *Bangia fuscopurpurea* and *Porphyra umbilicalis*, are definitely non-cellulosic (Frei & Preston 1964b).

Twelve members of the Phaeophyta examined by Frei & Preston (1961; Preston 1974) contain a cellulose I component which gives clear X-ray diagrams with sharp arcs (as compared to the red algae studied), but a second crystalline component was present as well. This is probably alginic acid which the same workers (1962) have shown to contribute to the X-ray patterns. In the brown algae as in the reds, the cellulose contents are usually lower than in the green algae of the cellulosic group first discussed (Cronshaw *et al.* 1958).

A very interesting system for the study of the biosynthesis of cellulosic cell wall material in the Chrysophyta has been exploited by Brown *et al.* (1973) viz. the production of cellulosic scales by the dictyosomes of the alga *Pleurochrysis scherffelii*. By means of silver methanamine 'staining', this group has been able to visualize at the ultrastructural level the intracellular course taken by the scale material and its secretion into the cell wall. Sulphated polysaccharide and protein or peptide appear also to be incorporated into the final product.

3.2 The xylan alternative to cellulose

β-1,4-xylans also belong to the 'ribbon' type of polysaccharides, but in this case the ribbon is twisted rather than flat, and the resulting structure is more flexible than cellulose. This would appear to be a desirable attribute for seaweed structure; perhaps it should not be surprising that xylans have been used rather than cellulose as the fibrous structural element in some of these plants. Turvey & Williams (1970) have reported that the xylan which remains in *Rhodymenia palmata* (*Palmaria palmata*) after cold alkaline extraction and which can be removed with hot 3*M* alkali is almost entirely of this linkage type. More soluble xylans present in the species contain both 1,4- and 1,3- links, predominantly the former. Usov *et al.* (1978c) comment that the least soluble xylans of species of *Rhodymenia, R. stenogona*, were entirely 1,4-linked.

In most of the algal xylans studied to date, however, the glycosidic bond is between C-1 and C-3; these would adopt a different conformation, that of a hollow helix; Atkins *et al.* (1969) have shown that the chains of these form triple helices (see Fig. 16.2). The first report of a xylan as an important constituent of seaweed cell wall polysaccharide was that of Mackie & Percival (1959), a β-1,4-xylan in the green alga *Caulerpa filiformis*. This was succeeded by Iriki & Miwa's (1960) finding of similar polymers in *Chlorodesmis, Halimeda* and *Udotea*, and Frei & Preston's (1964a) report on the additional marine species *Bryopsis* and *Penicillus*, all members of the order Caulerpales of the Chlorophyta. In these weeds the xylan was the fibrillar component and cellulose could not be demonstrated except, possibly, for *Bryopsis*. The oriented xylan exhibited negative birefringence. The alkali-soluble xylans of *Porphyra umbilicalis* and *Bangia fuscopurpurea* are also 1,3-linked (Frei & Preston 1964b; Turvey & Williams 1970) and form the chief structural element in these primitive red algae.

Besides the *Rhodymenia* spp. mentioned above, xylans containing both types of linkage have been found in two other red algae, *Laurencia pinnatifida* and *Rhodochorton floridulum* (Turvey & Williams 1970). As in the case of *Palmaria palmata*, the water soluble xylans of two *Chaetangium* species contain both types of linkage (Cerezo *et al.* 1971; Nunn *et al.* 1973).

3.3 The third structural alternative, mannan

The problem of the X-ray diffraction patterns of several other siphonous green algae (Nicolai & Preston 1952) yielded an unexpected solution when Iriki & Miwa (1960) reported mannan as a structural component remaining after alkaline and acid extraction and chlorite bleaching in *Acetabularia calyculus* and *Codium fragile*. This was confirmed by Frei & Preston (1961) who had also analysed *Dasycladus* and *Batophora*. The cuticle of *Porphyra umbilicalis* was also found to be based on mannan, which was shown subsequently to be β-1,4-linked (Frei & Preston 1964b). A similar mannan was obtained from *Bangia fuscopurpurea* (Frei & Preston 1964b; Usov *et al.* 1978b).

The mannan structure, according to Rees (1977) is rather similar to cellulose, a flat ribbon, but more flexible. In the case of the mannans, however, no microfibrils were visible under the electron microscope (Frei & Preston 1961). The structural element appeared as rods in an amorphous matrix in the case of *A. crenulata* (Frei & Preston 1968), and analysis indicated that it is entirely β-1,4-mannan (Bourne *et al.* 1972). (It should be recalled that *Acetabularia* cyst walls produce cellulose.)

Biochemical alternation of generations in certain green algae

In *Acetabularia* gametes are held in a peculiar cyst structure, which is the reduced haploid plant. This structure as mentioned above is primarily cellulosic, whereas the diploid plant, both stalk and cap, are composed of mannans, and cellulose is absent (Iriki & Miwa 1960; Preston 1968). Bachmann & Zetsche (1979) have correlated mannan synthase activity and cell growth, observing that enzyme activity decreases as growth is completed; in cyst-bearing cells synthase activity is low, but not absent, and presumably accounts for the 20% of the cyst wall composition which is mannan.

The unusual siphonous green algae of the genus *Derbesia* which have been shown to be the sporophytes of *Halicystis* or *Bryopsis* 'species' (Kornmann 1938; Feldmann 1950; Hustede 1964; Mayhoub 1976) carry out a similar switch. The sporophyte plants produce mannans; cell walls of the gametophytes, *Halicystis* or *Bryopsis*, are composed of xylan and glucan (Wutz & Zetsche 1970; Huizing & Rietema 1975). Differences in enzyme patterns of sporophyte and gametophyte generations of these 'species' have been demonstrated, especially enzymes of GDPM and mannan synthesis, and mutants have been described with activity as high as in the *Derbesia* phase (Wutz & Zetsche 1976). They postulate that gene activity is changed only during the formation or outgrowth of zoospores and the formation and fusion of gametes. Huizing *et al.* (1979) note an exception among species of the Derbesiaceae in *Bryopsidella* and question whether the absence of major differences in cell wall composition was related to the fact that morphologically the two phases are similar. (This question will be considered in Section 4.3.) Carlberg & Percival (1977) have studied some differences between a *Urospora* species and its one-celled *Codiolum* stage, which again involve a mannan.

3.4 Combinations of polysaccharides, or heteropolysaccharides?

In the more than twenty-five years since Roelofsen *et al.* (1953) reported their astonishment (sic) that the cell walls of *Halicystis osterhoutii* contained both xylose and glucose, and that both were still present after hot alkaline extraction, the question whether the structural element is a xyloglucan (possibly like the hemicellulose of higher plants) or whether there are molecules of both xylan and cellulose filling the structural role is still unanswered. Dennis & Preston (1961) suggested the latter alternative, postulating that the central core of the microfibrils is cellulosic. Cronshaw *et al.* (1958) had found both xylose and glucose present after extractions and chlorite treatment in *Ulva* and *Enteromorpha*, the latter confirmed by Dodson & Aronson (1978) using isolated cell walls. Frei & Preston (1964a) have demonstrated the presence of true cellulose in addition to the β-1,3-xylan in *Bryopsis*. On the other hand, Kreger (1962) has commented that the character of the microfibrils of xylan-glucan cell walls are thinner and more tortuous and interwoven than cellulosic microfibrils.

Thus far, there have been no reports that combinations of mannan and cellulose (or other skeletal glucan) coexist in seaweed cell walls; it is presumably just a matter of time and chance until the appropriate alga is analysed since Kreger (1960) and Northcote *et al.* (1960) have found peculiar complexes of the two in different species of the freshwater alga *Hydrodictyon*. A sulphated glucuronomannan has been reported in a diatom, *Phaeodactylum tricornutum*

(Ford & Percival 1965). There is some evidence that brown seaweeds, which have low amounts of cellulose in their cell walls, also utilize the uronic acid polymer alginic acid as a primary structural polymer. As this polysaccharide is charged, it can also fill other roles in the algae in which it is found.

4 SPECIFIC BROWN, GREEN, AND RED ALGAL POLYSACCHARIDES

An examination of the monosaccharides obtained in hydrolysates of seaweeds from the different divisions reveals some interesting differences. Neglecting in each instance the obvious glucose contribution from stored carbohydrates, the differences present evidence of the various polysaccharides which are peculiar to each division. In the Phaeophyta we find the uronic acids mannuronic and guluronic acid from alginate, and fucose, xylose and glucuronic acid probably combined in a complex sulphated polymer glucuronoxylofucan. In the Rhodophyta the predominant monosaccharide is galactose derived in most instances from a carrageenan- or agar-type polymer, although xylose and mannose are also present in quantity in more primitive species. In the Chlorophyta the unusual polysaccharides fall into three categories, glucuronoxylorhamnans, glucuronoxylorhamnogalactans, or xyloarabinogalactans, all of them sulphated (Percival 1979). With the exception of alginic acid, which is produced by bacteria as well as algae (Gorin & Spencer 1966), these are specific to the algae, and are more easily soluble in the chemists' hands than the structural polysaccharides already discussed.

4.1 Anionic polysaccharides of brown seaweeds

Alginic acid

The presence of this polysaccharide in cell walls of a number of brown algae has been demonstrated by X-ray diffraction (Frei & Preston 1962), by differential staining (McCully 1966; Evans & Holligan 1972) and by immunocytochemical techniques (Vreeland 1970, 1972, 1974a). Frei & Preston have suggested that it plays a strong structural role in addition to its obvious ion exchange function. The polymer is also prominent in the matrix of these plants (Vreeland 1974b) where the latter characteristic presumably assumes major importance.

Alginic acid was originally thought to be a polymannuronic acid polymer, until Fischer & Dörfel (1955) reported guluronic acid to be present in the molecule. The analysis of fragments obtained by hydrolysis of alginate by Haug *et al.* (1966, 1967) showed that the polysaccharide is made up of three different oligosaccharides, mannuronic acid (M) blocks, guluronic acid (G) blocks, and blocks with alternating mannuronic and guluronic (MG) units. The use of ^{1}H-n.m.r. and ^{13}C-n.m.r. to determine sequence and frequency of the uronate residues in intact or only slightly depolymerized alginate has been developed recently (Penman & Sanderson 1972; Grasdalen *et al.* 1977, 1979).

Another technique which has been used for estimating the composition of alginates, composed of variable proportions of residues of opposite configuration as they are, is computer analysis of circular dichroism spectra (Morris *et al.* 1973). The chains of 1,4-linked α-L-guluronic and β,D-mannuronic acid are stereochemically very different, as a result of their (only) difference at C-5 (Atkins *et al.* 1971). Guluronic acid is linked diaxially and the conformation of the polymer is rod-like ('kinked', see Preston 1974); polymannuronic acid chains are flat ribbons not unlike cellulose (Fig. 16.4). This explains the difference observed in Ca^{2+} binding between the two (Kohn *et al.* 1968). Controlled addition of Ca^{2+} to polyguluronic acid brings about gelation as the Ca^{2+} fits into the 'boxes' formed between the monomer units; circular dichroism spectral changes have been interpreted as evidence of association between G chains as they line up cooperatively binding the Ca^{2+} ion (the 'egg-box model' of gelation) (Morris *et al.* 1973). The flat ribbon M chains form more shallow 'nests', but also bind Ca^{2+}; no experimental evidence of cooperative chain association was obtained although it might be anticipated at high cation concentrations. No evidence whatsoever of cooperative binding with the alternating M–G chains was obtained, nor was it expected theoretically (Grant *et al.* 1973).

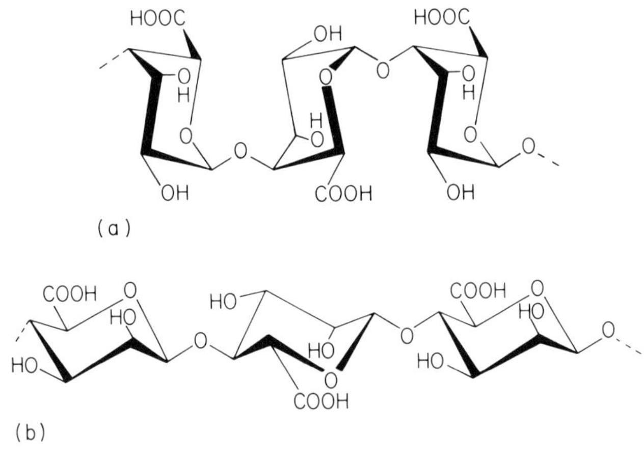

Fig. 16.4. Alginic acid components; (a) polyguluronic and (b) polymannuronic acid.

As would be anticipated, the differences in structure in various types of alginate are responsible for differences in their properties, with obvious significance to the algae in which they are found. Fischer & Dörfel (1955) commented on differences in proportions of guluronic and mannuronic acids in various species of brown algae, which observation has been expanded recently with our better understanding of alginate as a 'block copolymer' (Haug *et al.*

1974). Even within the same species, great variation exists in alginates from different parts of plants (Haug *et al.* 1969, 1974). The intercellular fluid from reproductive receptacles of *Ascophyllum nodosum* was found to produce an alginate composed almost entirely (97 per cent) of mannuronic acid, and growing tips of young plants also showed a preponderance of M blocks. Older parts of the plant contained more mixed blocks. In other species, particularly in *Pilayella*, the trend with increasing age was towards production of all-guluronic acid blocks (Haug 1974).

Variations in total alginate content in different seasons (Cardinal & Breton-Provencher 1977) have been reported, probably reflecting nutrient status which affects other components. The composition of alginate, however, was found to be unaffected by season (Fukibayashi *et al.* 1970). Chapman & Doyle (1978) have investigated the genetic aspect of phenotypic variance in content in *Laminaria longicruris*, but found that the genetic component which could be manipulated in a simple breeding program was very small.

The problem of biosynthesis of alginate has interested a number of investigators. Painter *et al.* (1968) carried out a computer study of the pattern of depolymerization of alginate and found the behaviour of alginate and the theoretical prediction based on random cleavage were in reasonable agreement; it was presumed that the pattern of depolymerization might give information on polymerization also. This was strengthened by the report of two nucleotide sugars in *Fucus,* GDP-D-mannuronic and GDP-L-guluronic acid (Lin & Hassid 1966a). Subsequently a model was presented based on the penultimate-unit theory of copolymerization (Larsen *et al.* 1969). Lin & Hassid (1966b) infiltrated discs of *Fucus* tissue with ^{14}C-mannose or ^{14}C-glucose and found labelled carbon in CO_2, 'fucoidin', alginic acid and residual material. Their analysis of different enzymes which played a role in the alginate synthesis suggested the following sequence: D-mannose → D-mannose-6-P → D-mannose-l-P → G.D.P.-mannose → G.D.P.-D-mannuronate → alginate. Although they did not study the incorporation of guluronate implied by their earlier discovery they suggested epimerization could occur at the nucleotide sugar or other level.

The experiments of Hellebust & Haug (1969, 1972a, b) on *Laminaria digitata* followed the fate of ^{14}C from bicarbonate into the polymer, which they subsequently fractionated into its 'blocks' by partial acid hydrolysis. Initial incorporation of ^{14}C was into M blocks, very little into G blocks. *In situ* studies with alternating dark and light periods showed that the reverse was true in the dark, with increases in polyguluronate labelling at the expense of polymannuronate (Hellebust & Haug 1969). A tentative explanation advanced on the basis of the discovery of a bacterial epimerase, which converts mannuronate to guluronate at the polymer level in *Azotobacter vinelandii* (Haug & Larsen 1969), was strengthened by isolation of a similar enzyme from cell-free extracts of *Pelvetia canaliculata* (Madgwick *et al.* 1973b). Whether epimerization occurs also at the level of the nucleotide sugar (Lin & Hassid 1966b) is still uncertain.

A second type of enzyme which has been obtained from interior tissue of brown algae ('alginase') and which might serve in the degradation or remodelling of alginate is alginate lyase (Madgwick *et al.* 1973a; Shiraiwa *et al.* 1975). Two different lyases have been reported from a bacterial source (Fujibayashi *et al.* 1979), specific for polymannuronate and polyguluronate. The substrate used in investigation of the algal lyase was 93% polymannuronate. The relative activities of lyase and epimerase could be controlled by close attention to the ionic requirements of the two enzymes (Madgwick *et al.* 1978).

Sulphated glucuronoxylofucan and/or fucoidan and ascophyllan

The soluble polysaccharides of brown algae are complex, but are marked by the presence of either or both of the charged components, uronic acid or sulphated fucose. The fucan sulphate of brown algae is variously described as a separate entity, fucoidan, or as an integral part of a glucuronoxylofucan which is commonly referred to as ascophyllan. Different workers have used different species of brown algae and different extraction procedures, hence it is unclear whether their diverse results are due to source or to treatment.

Most of the chemical analyses reported recently have concerned two species, *Ascophyllum nodosum* and *Fucus vesiculosus*. Larsen *et al.* (1966) identified a major anionic polysaccharide ascophyllan as a polyuronide which carries side chains of xylose, fucose, and sulphate, and two additional electrophoretically distinct polysaccharides of similar composition; later Larsen (1967) isolated a xylofucose disaccharide from a hydrolysate of the major polymer. Mian & Percival (1973) concluded that the fucose-containing polysaccharides from three other brown algae comprised a heterogeneous spectrum of molecules from a group high in uronic acid and low in sulphate to a relatively pure fucan sulphate low in uronic acid. Medcalf & Larsen (1977a) have also presented evidence of heterogeneity, and state that in neither *A. nodosum* nor *F. vesiculosus* is there a separate fucan sulphate. In dilute acid extracts of *A. nodosum* they (1977b) found a typical 'ascophyllan' (glucuronoxylofucan), two 'complexes' in which similar elements were present but the number of side chains differed, and a galactofucan. They considered that the 'ascophyllan' had a uronic acid backbone with long fucose-containing side chains, and that the 'complex' was reversed, with fucan backbone and ascophyllan-like side chains. Most of the uronic acid in the latter was mannuronic. Galactofucan units in 'ascophyllan' have also been reported in *Desmarestia* (Percival & Young 1974), and another fucose-containing polysaccharide with a very high galactose content in *Colpomenia* (Hussein 1975).

Although there have been reports of the linkages between the different monosaccharides in these complex chains, nobody to date has had the audacity to attempt to describe how the total molecules are synthesized. The early ^{14}C-labelling experiments of Bidwell *et al.* (1958) separated the 'fucoidan' fraction of *Fucus vesiculosus* into a part which was easily soluble in dilute acid and a more resistant fraction, and noted that fucose, glucose, and galactose were present in both, xylose only in the former. In recent experiments (1972) they traced the progression of labelled fucose as it appeared in three fractions, and suggested a sequence of synthesis in which xylogalactofucoglucuronan was transformed into xyloglucuronogalactofucan and thence into fucoidan, which contained chiefly fucose.

Sulphate makes a convenient handle with which to investigate these compounds. The appearance of administered $^{35}SO_4^{2-}$ in the polysaccharide fucoidan or in its complexes with guluronic acid, xylose and galactose has been studied in several *Fucus* species (Bidwell & Ghosh 1963; Quatrano & Crayton 1973; Evans *et al.* 1973; Coughlan 1977) and in *Pelvetia* and *Laminaria* (Evans *et al.* 1973; DeLestang-Bremond & Quillet 1976). Quatrano has chosen an extremely interesting system for this type of investigation, the zygotes of *Fucus* in the very early stages of development. Sulphate incorporation does not get under way until 10 h after fertilization, nor can sulphated polysaccharides be detected in the walls at this time (Quatrano & Crayton 1973). During rhizoid initiation but not depending on or determining it (Crayton *et al.* 1974), there is a rapid increase in uptake into the cells and localization of the sulphated polysaccharide fucoidan in the area destined to become a rhizoid cell. This localization had been inferred earlier on a histochemical basis by McCully (1968a). It occurs at a time when there is no net increase in fucan, which implies sulphation of a preformed polysaccharide or oligosaccharide. Two fucans actively synthesized in early development can be distinguished, one which is closely associated with xylose and cellulose in the developing cell wall, and the one which becomes highly sulphated at about 10 h after fertilization and then migrates towards the pole of the cell which gives rise to the rhizoid (Quatrano & Stevens 1976). The relationship between this polymer and the sulphated glucuronoxylofucans of mature fronds of *Fucus* has not been ascertained.

The histochemical studies of McCully (1966, 1968a,b) and of Parker & Diboll (1966) discriminated between alginic acid and sulphated 'fucan' and demonstrated their differential localization in brown algal tissues. Both workers identified sulphated polymers in cell walls. Using a more specific tool, antibodies raised against each of the polysaccharide types, Vreeland (1974a) showed the presence of fucan particularly in the outer wall layer in *Fucus* and in a thin layer of small globules along the inner edge of the inner cell wall (seen also with alginate), suggesting secretion of intracellularly synthesized material into the cell wall. This phenomenon had been observed earlier by McCully (1968b) and by Evans & Holligan (1972). Histochemical and autoradiographic techniques with $^{35}SO_4^{2-}$ have shown that sulphation is dictyosome-associated and that in

Laminaria it occurs only in specialized cells (Evans *et al.* 1973; Evans & Callow 1974; Callow & Evans 1976).

An active sulphate uptake system has been identified recently in *Fucus* (Coughlan 1977). It is still not clear what nucleotide is used as a sulphate donor for polysaccharide sulphation in these algae; Tsang & Schiff (1975) have postulated that APS rather than PAPS may serve for sulphate reduction in algal autotrophs. A homologue, cytidine diphosphosulphate, has been identified in *Pelvetia* (DeLestang & Quillet 1974). Galactosyltransferases have been obtained from *Fucus* cell fractions which were primarily dictyosome preparations (Coughlan & Evans 1978).

4.2 Anionic polysaccharides of green seaweeds

In all of the green algae, most of the polysaccharides present are charged heteropolymers composed of some of the same monosaccharides as in the Phaeophyta, and usually, but not always, containing both uronate and sulphate groups. These polysaccharides have been classified by Percival (1979) into three groups, produced by different seaweeds, viz. sulphated xyloarabinogalactans in *Cladophora, Chaetomorpha, Rhizoclonium, Caulerpa* and *Codium* species, sulphated glucuronoxylorhamnans in *Ulva, Enteromorpha, Acrosiphonia, Urospora* and *Codiolum*, and sulphated glucuronoxylorhamnogalactans in *Acetabularia*. It should be noted that these are not the same groupings as in the case of the structural polymers, and that the first group includes cellulosic, xylan- and mannan-containing species.

The sulphate groups have been found on galactose in *Acetabularia* (Percival & Smestad 1972) on rhamnose and xylose in *Ulva, Acrosiphonia*, and *Enteromorpha* (O'Donnell & Percival 1959; McKinnell & Percival 1962a,b) on galactose and arabinose in *Cladophora* and *Chaetomorpha* (Hirst *et al.* 1965) and the same probably in *Codium* (Love & Percival 1964).

The Chlorophyta polysaccharides remain a complex subject for further investigation. Presumably because of their complexity, little work other than chemical description has been carried out.

4.3 Anionic and uncharged galactans of red seaweeds

The Rhodophyta are unusual in that most if not all of the algae in the group produce greater or lesser amounts of a galactose polymer which is based upon a similar repeating structure, galactose units linked alternately α-1,3- and β-1,4. At least two types of these galactans have been described, carrageenans and agars. It has been proposed (McCandless & Craigie 1979) that the definitions of these galactans be based not on solubility or gel characteristics which are too general attributes, nor on substituents which are too numerous and variable, but on the basic underlying structures of the polymers. A carrageenan is thus defined as an algal polysaccharide composed of D-galactose units only, an agar, a galactan in which the 4-linked unit is α-L-galactose. Persistent reports of galactans which contain small proportions of D-,L-galactose segments in what appeared to be a D-,D-structure (carrageenan type) (Nunn & Parolis 1970; Allsobrook 1974, 1975), which might be considered hybrid molecules, have recently been confirmed in the ^{13}C-n.m.r. analysis of partial acetolysis products of galactan from a *Grateloupia* species (Usov & Barbakadze 1978; Usov *et al.* 1978a).

Rees (1977) has demonstrated by computer model building and X-ray diffraction studies that both carrageenans and agars belong to the hollow helix polysaccharide family and pack together by chain interaction into gels. This property is of course important in commerce. That the carrageenans are also highly ordered in the native state in the alga was demonstrated by Gordon-Mills, who reported that the cell walls of unfixed specimens of *Chondrus crispus* exhibited positive intrinsic birefringence, and that toluidine-blue stained specimens were negatively birefringent (Gordon-Mills & McCandless 1977; Gordon-Mills *et al.* 1978). It was suggested that the fact that κ- carrageenan with only one sulphate group/disaccharide unit forms a metachromatic complex with this dye may indicate a helical form *in situ*.

AGARS

The structures of the agar family of galactans have been studied by numerous investigators but a major advance was Araki's proposal (1956, 1966) that agar could be fractionated into the neutral, gelling component, agarose, and an anionic fraction containing sulphate or pyruvic acid groups. The basic repeating structure was deduced from the differential attack on agarose by acid and by specific enzymes. The enzymatic cleavage method was further developed by Yaphe's group who isolated a β-agarase (Young *et al.* 1971a), a β-neoagaro-tetrose hydrolase which acts on the products of the agarase digestion (Groleau & Yaphe 1977), and a neoagarobiose hydrolase which completes the digestion (Day & Yaphe 1975). Using these enzymes to characterize agars, Duckworth & Yaphe (1971a,b; Young *et al.* 1971b) have suggested that there is a continuous spectrum of molecules from the essentially uncharged agarose to sulphated agars or pyruvated agars, different proportions of which are present in various seaweeds (Fig. 16.5). They suggested that in these polysaccharides pyruvate is laid down remote from sulphated regions. Hamer *et al.* (1977) have analysed enzymic hydrolysis products of agarose by ^{13}C-n.m.r.

Very little work has been carried out on the sulphation pattern of agars, as the introduction of this substituent interferes with the property which is of greatest commercial importance, viz. gelation. Funoran, the polysaccharide

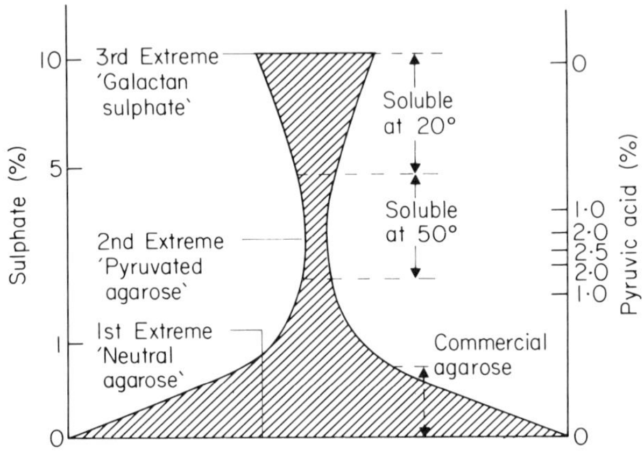

Fig. 16.5. Spectrum of polysaccharides in Difco Bacto agar. Shaded area represents complete agar. The area within the curves between any lines drawn horizontally indicates the approximate amount of polysaccharide having that number of charged groups (Duckworth & Yaphe 1971a, used by permission).

from *Gloiopeltis* species, is highly sulphated and does not gel (Araki 1966); it has, however, the underlying agar structure (Hirase *et al.* 1958). ^{13}C-n.m.r. spectoscopy has recently been applied in the analysis of sulphated agars from *Bangia* and *Odonthalia* species (Shaskov *et al.* 1978; Usov *et al.* 1978b). Pyruvic acid is present in agars as the ketal 4,6-O-(1-carboxyethylidene)-D-galactose, as shown first by Hirase (1957). Young *et al.* (1971b) have reported pyruvate as a common feature of agars, at concentrations from 0.04% to 2.9%. Methyl groups are regularly found in agars, chiefly as 6-O-Me on D-galactose and 4-O-Me on L-galactose (Araki 1966; Araki *et al.* 1967) but also as 2-O-Me on L-galactose in certain agars (Bowker & Turvey 1968; Batey & Turvey 1975; Allsobrook *et al.* 1975; Usov *et al.* 1971).

Most interest, of course, has centered upon the neutral component of agars, agarose, the structure of which was determined by Araki (1956) to consist of alternating 1,3-linked β-D-galactose and 1,4-linked anhydro-L-galactose units (Fig. 16.6). The structure of the agarose double helix and its function in gelation has been described by Arnott *et al.* (1974).

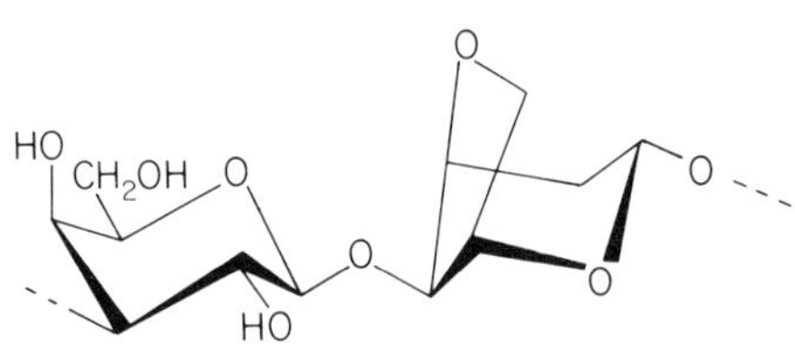

Fig. 16.6. Agarose; repeating disaccharide unit.

CARRAGEENANS

The carrageenans of *Chondrus crispus* were originally described in terms of their KCl solubility; κ-carrageenan gels in the presence of K^+ (or Rb^+, or Cs^+) λ-carrageenan remains soluble (Smith & Cook 1953). The idealized chemical structures of these two polysaccharides were elucidated subsequently largely by the work of Rees and co-workers at Edinburgh (Rees 1963) and two other types, ι- and ξ- were described in other species of seaweeds (Anderson *et al.* 1973; Penman & Rees 1973). π-carrageenan which contains pyruvic acid ketal (as in agar) was first described by Hirase & Watanabe (1972) in a *Gigartina* species, and an almost completely pyruvated but highly sulphated (cf. page 576, agars) carrageenan in the *Petrocelis* phase of *Gigartina papillata* by DiNinno *et al.* (1978, 1979). Two precursor molecules, μ- and ν- have been postulated but never isolated (Anderson *et al.* 1968; Stancioff & Stanley 1969).

We have suggested that these sulphated carrageenans can be divided into two main types, based upon the presence or absence of sulphate on C-4 of the β-galactose unit (McCandless & Craigie 1979). The gelling carrageenans ι- and κ- belong to the former type (both also contain 3,6 anhydrogalactose), whereas λ-, ξ- and π- are unsulphated in this position. The latter polymers exhibit high viscosity (unpublished) which suggests that all three exist as extended molecules as Rees (1969) has described for λ-carrageenan from *C. crispus* . The idealized repeating structures for these are depicted in Figure 16.7. The presence of O-methyl groups in D-galactans from the family Halymeniaceae has been demonstrated (Hirase *et al.* 1967; Nunn & Parolis 1968, 1969; Allsobrook *et al.* 1971). The galactan structure is of the carrageenan type.

Immunochemical analysis of carrageenans from a number of species of red seaweeds has provided a new tool to supplement the information yielded by chemical analysis (DiNinno & McCandless 1978a). By this means κ- has been distinguished from ι-carrageenan (DiNinno & McCandless 1978b) and λ from ξ-carrageenan (Evelegh *et al.* 1978; DiNinno & McCandless 1978c). Enzymatic hydrolysis by specific κ-, λ-, and ι- carrageenases has also provided insight into the organization of carrageenan structures (Weigl & Yaphe 1966; Johnston & McCandless 1973; Bellion 1979).

Bellion (1979) has recently combined enzymatic hydrolysis with ^{13}C-n.m.r. analysis of oligosaccharides and resistant fragments from hydrolyzed carrageenans and by this mean has shown that both κ- and ι- segments occur in the same polysaccharide molecule. In two interesting 'hybrid' (κ-, ι-) carrageenans from different *Ahnfeltia* carrageenophytes, she demonstrated that in *A. concinna* both types of units exist as small repeating sections of the molecule, while in *A. gigartinoides,* ι-carrageenase digestion left larger resistant (κ-type) fragments (Fig. 16.8). The ^{13}C-n.m.r. spectra of the two polymers were almost identical. This is the first positive evidence that the carrageenan molecules are hybrids and not mixtures of the two types. ^{13}C-n.m.r. studies of carrageenans of other

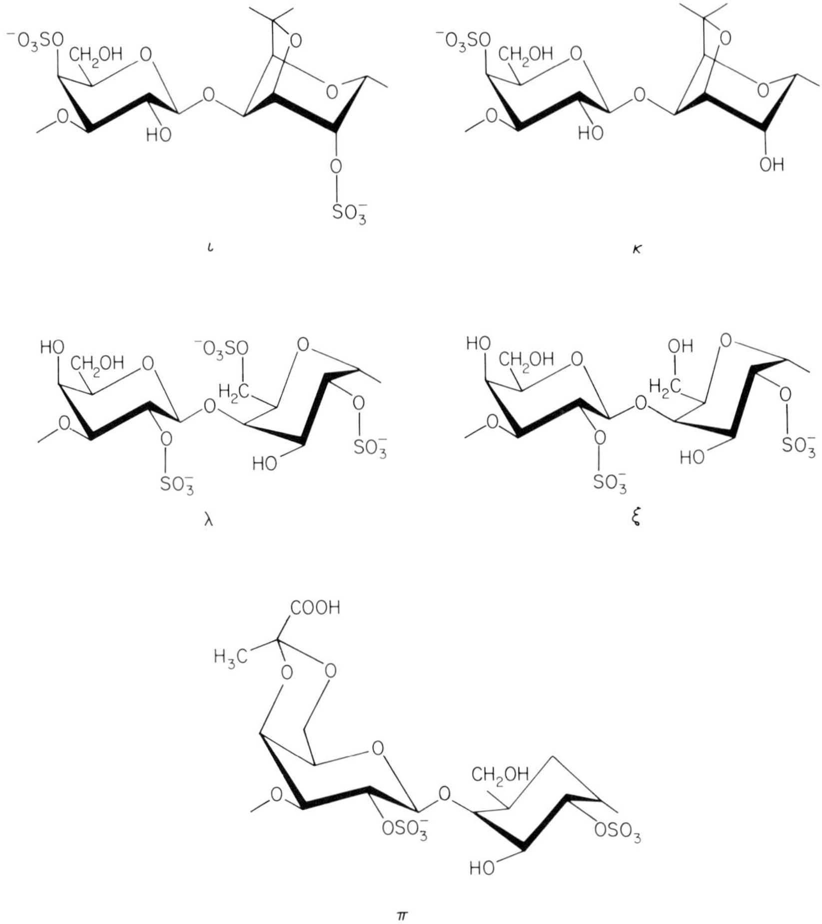

Fig. 16.7. Carrageenans; repeating disaccharide units (idealized): ι, κ, λ, ξ, π.

species have been carried out by Yarotskii *et al.* (1977, 1978) and by Bhattacharjee *et al.* (1978).

BIOCHEMICAL ALTERNATION OF GENERATIONS IN CERTAIN RED ALGAE

Although Marshall *et al.* (1949) had suggested that sporophytes and gametophytes of *Chondrus crispus* might produce different proportions of gelling (κ-) and viscous (λ-) types of carrageenans, the idea was not investigated until two decades later when it was demonstrated that κ-carrageenan is produced by the gametophytes of the family Gigartinaceae, λ-carrageenan by the sporophytes (McCandless & Richer 1971; McCandless *et al.* 1973; Chen *et al.* 1973;

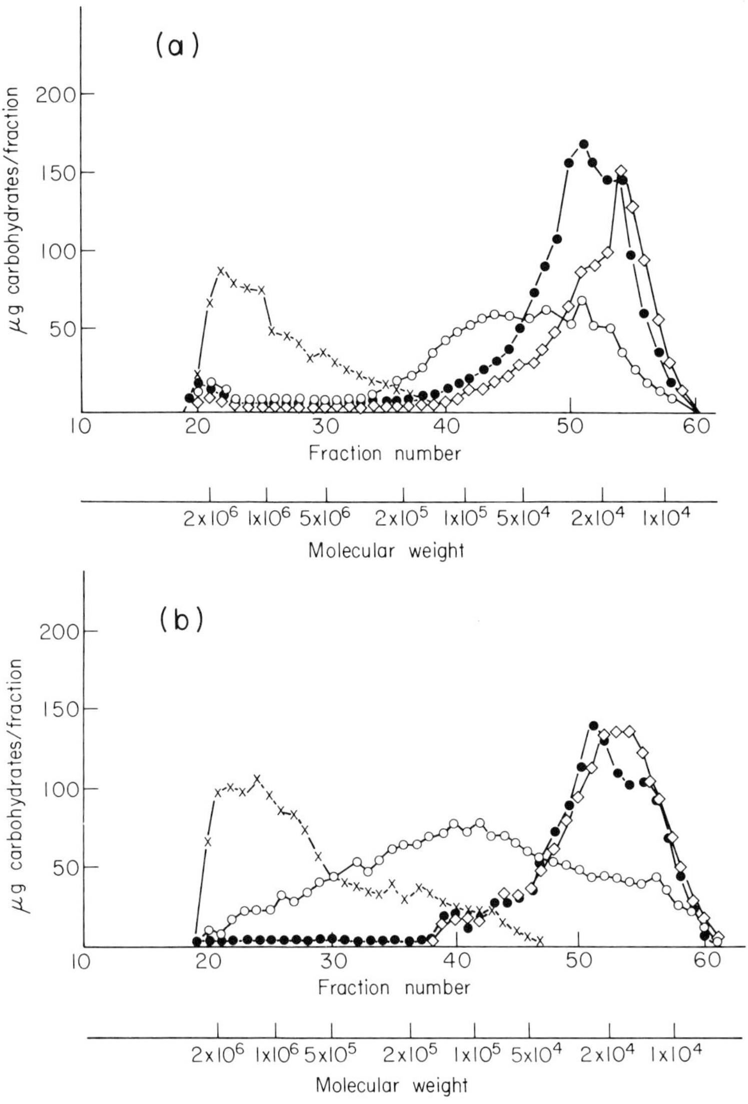

Fig. 16.8. Sepharose CL-4B chromatograms of carrageenans from (a) *Ahnfeltia concinna* and (b) *A. gigartinoides*. ×——× = Undegraded carrageenan; ○——○ = after ι-carrageenase hydrolysis; ●——● = after κ-carrageenase hydrolysis; ◇——◇ = after hydrolysis with both enzymes (Bellion 1979, used by permission).

Pickmere *et al.* 1973). This was subsequently confirmed at the immunocytochemical level with fluorescent antibodies by Gordon-Mills & McCandless (1975). Further work has shown that this alternation is true in all members of the Gigartinaceae (McCandless 1978). Recent (unpublished) work from this laboratory has shown that the phenomenon is also characteristic of the

Phyllophoraceae. In both families it occurs both in species with isomorphic life histories and in species in which the life cycle is heteromorphic (see Section 3.3). It does not extend, however, to *Eucheuma* (Doty & Santos 1978) or to *Hypnea* or *Furcellaria* species (Hosford & McCandless 1975; McCandless 1978).

Biochemical studies of the synthesis of the carrageenans must of necessity take these differences into account. Loewus' (1971) study of ^{35}S incorporation into the carrageenans of *C. crispus* was carried out before the phenomenon had been discovered. Recently Jackson (1979) has demonstrated differences in ^{35}S incorporation in the two stages and has proposed that κ-carrageenan is synthesized through an ι-carrageenan intermediate. This is also supported by the demonstration by DiNinno & McCandless (1979) that the KCl soluble carrageenans of gametophytes of the Gigartinaceae are not sensitive to periodate, and that alkaline borohydride treatment of this fraction fails to increase the immunochemical reactivity to anti-κ-carrageenan antibody although it does increase its anhydrogalactose content. If one may assume that the same basic biosynthetic pathway for synthesis and 'remodelling' of sulphated carrageenan molecules is present in the numerous species which make this polymer, the common sulphation step which should occur first is that on C-2 of the α-galactose unit, followed by C-6 of the same unit except in ξ- and π-carrageenan pathways. It had been assumed that the missing step in those species which show biochemical alternation of generations was the introduction of anhydrogalactose. However, Wong & Craigie (1978) have isolated the sulphohydrolase which introduces anhydrogalactose in this biosynthesis from both gametophytes and sporophytes of *C. crispus*. Clearly the clue to how the two phases make different polymers is still to be found.

ACKNOWLEDGEMENTS

The author wishes to thank Mrs. Cathy Vollmer and Miss Nancy Lyons for assistance in preparation of this manuscript, and Dr. G. P. Harris for valuable criticism.

5 REFERENCES

Albersheim P. (1978) Concerning the structure and biosynthesis of the primary cell walls of plants. In *Biochemistry of Carbohydrates II*, ed. Manners D. J., **16**, 127–50. University Park Press, Baltimore.

Allsobrook A. J. R., Nunn J. R. & Parolis H. (1971) Sulphated polysaccharides of the Grateloupiaceae family. V. A polysaccharide from *Aeodes ulvoidea*. *Carbohydr. Res.*, **16**, 71–8.

Allsobrook A. J. R., Nunn J. R. & Parolis H. (1974) The linkage of 4-O-methyl-L-galactose in the sulphated polysaccharide of *Aeodes ulvoidea*. *Carbohydr. Res.*, **36**, 139–46.

Allsobrook A. J. R., Nunn J. R. & Parolis H. (1975) Investigation of the acetolysis products of the sulphated polysaccharide of *Aeodes ulvoidea*. *Carbohydr. Res.*, **40**, 337–44.

Anderson N. S., Dolan T. C. S., Lawson C. J., Penman A. & Rees D. A. (1968) Carrageenans. V. The masked repeating structures of λ- and μ-carrageenans. *Carbohydr. Res.*, **7**, 468–73.

Anderson N. S., Dolan T. C. S., & Rees D. A. (1973) Carrageenans. VII. Polysaccharides from *Eucheuma spinosum* and *Eucheuma cottonii*. The covalent structure of ι-carrageenan. *J. Chem. Soc. Perkin Trans.*, **I**, 2173–6.

Araki C. (1956) Structure of the agarose constituent of agar-agar. *Bull. Chem. Soc. Jpn.*, **29**, 543–4.

Araki C. (1966) Some recent studies on the polysaccharides of agarophytes. *Proc. Int. Seaweed Symp.*, **5**, 3–17.

Araki C., Arai K. & Hirase S. (1967) Studies on the chemical composition of agar-agar. XXIII. Isolation of D-xylose, 6-O-methyl-D-galactose, 4-O-methyl-L-galactose and O-methylpentose. *Bull. Chem. Soc. Jpn.*, **40**, 959–62.

Arnott S., Fulmer A., Scott W. E., Dea I. C. M., Moorhouse R. & Rees D. A. (1974) The agarose double helix and its function in agarose gel structure. *J. Mol. Biol.*, **90**, 269–84.

Atkins E. D. T., Mackie W., Parker K. D. & Smolko E. E. (1971) Crystalline structures of poly-D-mannuronic and poly-L-guluronic acids. *J. Polym. Sci. B.*, **9**, 311–16.

Atkins E. D. T., Parker K. D. & Preston R. D. (1969) The helical structure of the β-1,3-linked xylan in some siphoneous green algae. *Proc. Roy. Soc. Lond.* Ser. B., **173**, 209–21.

Bachmann P. & Zetsche K. (1979) A close temporal and spatial correlation between cell growth, cell wall synthesis and the activity of enzymes of mannan synthesis in *Acetabularia mediterranea*. *Planta*, **145**, 331–7.

Batey J. F. & Turvey J. R. (1975) The galactan sulphate of the red alga *Polysiphonia lanosa*. *Carbohydr. Res.*, **43**, 133–43.

Bellion C. (1979) Utilisation d'enzymes pour la caracterisation des carraghenanes kappa et iota. Ph.D. Thesis, McGill University, Montreal.

Bhattacharjee S. S., Yaphe W. & Hamer G. K. (1978) ^{13}C-n.m.r. spectoscopic analysis of agar, κ-carrageenan and ι-carrageenan. *Carbohydr. Res.*, **60**, C1–C3.

Bidwell R. G. S., Craigie J. S. & Krotkov G. (1958) Photosynthesis and metabolism in marine algae. III. Distribution of photosynthetic carbon from $C^{14}O_2$ in *Fucus vesiculosus*. *Can. J. Bot.*, **36**, 581–90.

Bidwell R. G. S. & Ghosh N. R. (1963) Photosynthesis and metabolism in marine algae. VI. The uptake and incorporation of S^{35}-sulphate in *Fucus vesiculosus*. *Can. J. Bot.*, **41**, 209–20.

Bidwell R. G. S., Percival E. & Smestad B. (1972) Photosynthesis and metabolism of marine algae. VIII. Incorporation of ^{14}C into the polysaccharides metabolized by *Fucus vesiculosus* during pulse labeling experiments. *Can. J. Bot.*, **50**, 191–7.

Bourne E. J., Percival E. & Smestad B. (1972) Carbohydrates of *Acetabularia* species. Part I. *A. crenulata*. *Carbohydr. Res.*, **22**, 75–82.

Bowker D. M. & Turvey J. R. (1968) Water-soluble polysaccharides of the red alga *Laurencia pinnatifida*. I. Constituent units. *J. Chem. Soc.* Ser. C., 983–8.

Brown R. M. Jr., Herth W., Franke W. W. & Romanovicz D. (1973) The role of the Golgi apparatus in the biosynthesis and secretion of a cellulosic glycoprotein in *Pleurochrysis*: A model system for the synthesis of structural polysaccharides. In *Biogenesis of Plant Cell Wall Polysaccharides*, ed. Loewus F., pp. 207–57. Academic Press, New York & London.

Bryhni E. (1978) Quantitative differences between polysaccharide compositions in normal differentiated *Ulva mutabilis* and the undifferentiated mutant *lumpy*. *Phycologia*, **17**, 119–24.

Callow M. E. & Evans L. V. (1976) Localization of sulphated polysaccharides by X-ray microanalysis in *Laminaria saccharina*. *Planta*, **131**, 155–7.

Cardinal A. & Breton-Provencher M. (1977) Variations de la teneur en acide alginique des Laminariales de l'estuaire du Saint-Laurent (Québec). *Bot. Mar.*, **20**, 243–52.

Carlberg G. E. & Percival E. (1977) The carbohydrates of the green seaweeds *Urospora wormskioldii* and *Codiolum pusillum*. *Carbohydr. Res.*, **57**, 223–34.

Catt J. W., Hills G. J. & Roberts K. (1976) A structural glycoprotein containing hydroxyproline isolated from the cell wall of *Chlamydomonas reinhardii*. *Planta*, **131**, 165–71.

Cerezo A. S., Lezerovich A., Labriola R. & Rees D. A. (1971) A xylan from the red seaweed *Chaetangium fastigiatum*. *Carbohydr. Res.*, **19**, 289–96.

Chapman A. R. O. & Doyle R. W. (1978) Genetic analysis of alginate content in *Laminaria longicruris* (Phaeophyceae). *Proc. Int. Seaweed Symp.*, **9**, 125–32.

CHEN L. C.-M., MCLACHLAN J., NEISH A. C. & SHACKLOCK P. F. (1973) The ratio of kappa- to lambda-carrageenan in nuclear phases of the rhodophycean algae, *Chondrus crispus* and *Gigartina stellata*. *J. Mar. Biol. Ass. U.K.*, **53**, 11–16.

COURT G. J. & TAYLOR I. E. P. (1979) Isolation and partial characterization of a proteoglycan from the red alga *Laurencia spectabilis*. *Phytochemistry*, **18**, 411–14.

CRAIGIE J. S. (1974) Storage products. In *Algal Physiology and Biochemistry*, ed. Stewart W. D. P., pp. 206–35. Blackwell Scientific Publications, Oxford.

CRAYTON M. A., WILSON E. & QUATRANO R. S. (1974) Sulfation of fucoidan in *Fucus* embryos. II. Separation from initiation of polar growth. *Dev. Biol.*, **39**, 164–7.

COUGHLAN S. (1977) Sulphate uptake in *Fucus serratus*. *J. Exp. Bot.*, **28**, 1207–15.

COUGHLAN S. & EVANS L. V. (1978) Isolation and characterization of Golgi bodies from vegetative tissue of the brown alga *Fucus serratus*. *J. Exp. Bot.*, **29**, 55–68.

CRONSHAW J., MYERS A. & PRESTON R. D. (1958) A chemical and physical investigation of the cell walls of some marine algae. *Biochim. Biophys. Acta*, **27**, 89–103.

CRONSHAW J. & PRESTON R. D. (1958) A reexamination of the fine structure of the walls of vesicles of the green alga *Valonia*. *Proc. Roy. Soc. Lond.*, Ser. B, **148**, 137–48.

DAY D. F. & YAPHE W. (1975) Enzymatic hydrolysis of agar: Purification and characterization of neoagarobiose hydrolase and *p*-nitrophenyl-α-galactoside hydrolase. *Can. J. Microbiol.*, **21**, 1512–18.

DELESTANG G. & QUILLET M. (1974) Comportement du fucoïdane sulfurylé de *Pelvetia canaliculata* (Dcne & Thur.) vis à vis des cations de la mer: Propriétés d'échange, renouvellement des radicaux sulfuriques, coenzyme d'activation des sulfates. Intérêt fonctionnel. *Physiol. Vég.*, **12**, 199–227.

DELESTANG-BREMOND G. & QUILLET M. (1976) Étude, à l'aide de $^{35}So_4$, du turnover des sulfates esterifiant le fucoïdane de *Pelvetia canaliculata* (Dcne et Thur.). *Physiol. Vég.*, **14**, 259–69.

DENNIS D. T. & PRESTON R. D. (1961) Constitution of cellulose microfibrils. *Nature* (*Lond.*), **191**, 667–8.

DININNO V. L., BELL R. A. & MCCANDLESS E. L. (1978) Identification of 4,6-O-(1-carboxyethylidene)-D-galactose by combined gas chromatography mass spectrometry. *Biomed. Mass Spectrom.*, **5**, 671–3.

DININNO V. L. & MCCANDLESS E. L. (1978a) Anti-carrageenans. *Immunochemistry*, **15**, 273–4.

DININNO V. L. & MCCANDLESS E. L. (1978b) The chemistry and immunochemistry of carrageenans from *Eucheuma* and related algal species. *Carbohydr. Res.*, **66**, 85–93.

DININNO V. L. & MCCANDLESS E. L. (1978c) The immunochemistry of λ-type carrageenans from certain red algae. *Carbohydr. Res.*, **67**, 235–41.

DININNO V. L. & MCCANDLESS E. L. (1979) Immunochemistry of κ-type carrageenans from certain red algae. *Carbohydr. Res.*, **72**, 157–63.

DININNO V. L., MCCANDLESS E. L. & BELL R. A. (1979) Pyruvic acid derivative of a carrageenan from a marine red alga (*Petrocelis* species). *Carbohydr. Res.*, **71**, C1–C4.

DODSON J. R. & ARONSON J. M. (1978) Cell wall composition of *Enteromorpha intestinalis*. *Bot. Mar.*, **21**, 241–6.

DOTY M. S. & SANTOS G. A. (1978) Carrageenans from tetrasporic and cystocarpic *Eucheuma* species. *Aquat. Bot.*, **4**, 143–9.

DUCKWORTH M. & YAPHE W. (1971a) The structure of agar. I. Fractionation of a complex mixture of polysaccharides. *Carbohydr. Res.*, **16**, 189–97.

DUCKWORTH M. & YAPHE W. (1971b) The structure of agar. II. The use of a bacterial agarase to elucidate structural features of the charged polysaccharides in agar. *Carbohydr. Res.*, **16**, 435–45.

EVANS L. V. & CALLOW M. E. (1974) Polysaccharide sulphation in *Laminaria*. *Planta*, **117**, 93–5.

EVANS L. V. & HOLLIGAN M. S. (1972) Correlated light and electron microscope studies on brown algae. I. Localization of alginic acid and sulphated polysaccharides in *Dictyota*. *New Phytol.*, **71**, 1161–72.

EVANS L. V., SIMPSON M. & CALLOW M. E. (1973) Sulphated polysaccharide synthesis in brown algae. *Planta*, **110**, 237–52.

EVELEGH M. J., VOLLMER C. M. & MCCANDLESS E. L. (1978) Differentiation of lambda carrageenans from Rhodophyta with immunological and spectroscopic techniques. *J. Phycol.*, **14**, 89–91.

FELDMANN J. (1950) Sur l'existence d'une alternance de générations entre l'*Halicystis parvula* Schmitz et le *Derbesia tenuissima* (De Not.) Crn. *C. R. Acad. Sci. Paris*, **230**, 322–3.

FISCHER F. G. & DÖRFEL H. (1955) Die Polyuronsäuren der Braunalgen (Kohlenhydrate der Algen I). *Hoppe-Seyler's Z. Physiol. Chem.*, **302**, 186–203.

FORD C. W. & PERCIVAL E. (1965) Carbohydrates of *Phaeodactylum tricornutum*. Part II. A sulphated glucuronomannan. *J. Chem. Soc.* 7042–6.

FREI E. & PRESTON R. D. (1961) Variants in the structural polysaccharides of algal cell walls. *Nature (Lond.)*, **192**, 939–43.

FREI E. & PRESTON R. D. (1962) Configuration of alginic acid in marine brown algae. *Nature (Lond.)*, **196**, 130–4.

FREI E. & PRESTON R. D. (1964a) Non-cellulosic structural polysaccharides in algal cell walls. I. Xylan in siphoneous green algae. *Proc. Roy. Soc. Lond.*, Ser. B, **160**, 293–13.

FREI E. & PRESTON R. D. (1964b) Non-cellulosic structural polysaccharides in algal cell walls. II. Association of xylan and mannan in *Porphyra umbilicalis*. *Proc. Roy. Soc. Lond.*, Ser. B, **160**, 314–327.

FUJIBAYASHI S., HABE H. & NISIZAWA K. (1970) Heterogeneity of alginate in special reference to the enzymatic degradation. *J. Biochem. (Tokyo)*, **67**, 37–45.

GORDON-MILLS E. M. & MCCANDLESS E. L. (1975) Carrageenans in the cell walls of *Chondrus crispus* Stack. (Rhodophyceae, Gigartinales). I. Localization with fluorescent antibody. *Phycologia*, **14**, 275–81.

GORDON-MILLS E. M. & MCCANDLESS E. L. (1977) Carrageenans in the cell walls of *Chondrus crispus* Stack. (Rhodophyceae, Gigartinales). II. Birefringence. *Phycologia*, **16**, 169–76.

GORDON-MILLS E. M., TAS J. & MCCANDLESS E. L. (1978) Carrageenans in the cell walls of *Chondrus crispus* Stack. (Rhodophyceae, Gigartinales). III. Metachromasia and the topooptical reaction. *Phycologia*, **17**, 95–104.

GORIN P. A. J. & SPENCER J. F. T. (1966) Exocellular alginic acid from *Azotobacter vinelandii*. *Can. J. Chem.*, **44**, 993–8.

GRANT G. T., MORRIS E. R., REES D. A., SMITH P. S. C. & THOM D. (1973) Biological interactions between polysaccharides and divalent cations: the egg-box model. *FEBS Lett.*, **32**, 195–8.

GRASDALEN H., LARSEN B. & SMIDSRØD O. (1977) ^{13}C-n.m.r. studies of alginate. *Carbohydr. Res.*, **56**, C11–C15.

GRASDALEN H., LARSEN B. & SMIDSRØD O. (1979) A P.M.R. study of the composition and sequence of uronate residues in alginates. *Carbohydr. Res.*, **68**, 23–31.

GROLEAU D. & YAPHE W. (1977) Enzymatic hydrolysis of agar: Purification and characterization of β-neoagarotetraose hydrolase from *Pseudomonas atlantica*. *Can. J. Microbiol.*, **23**, 672–9.

HAMER G. K., BHATTACHARJEE S. S. & YAPHE W. (1977) Analysis of the enzymic hydrolysis products of agarose by ^{13}C-n.m.r. spectroscopy. *Carbohydr. Res.*, **54**, C7–C10.

HAUG A. (1974) Chemistry of algal cell-wall polysaccharides. In *Plant Biochemistry* ed. Northcote D. H., **11**, 51–88. Butterworths, London.

HAUG A. & LARSEN B. (1969) Biosynthesis of alginate. Epimerisation of D-mannuronic to L-guluronic acid residues in the polymer chain. *Biochim. Biophys. Acta*, **192**, 557–9.

HAUG A., LARSEN B. & BAARDSETH E. (1969) Comparison of the constitution of alginates from different sources. *Proc. Int. Seaweed Symp.*, **6**, 443–51.

HAUG A., LARSEN B. & SMIDSRØD O. (1966) A study of the constitution of alginic acid by partial acid hydrolysis. *Acta Chem. Scand.*, **20**, 183–90.

HAUG A., LARSEN B. & SMIDSRØD O. (1967) Studies on the sequence of uronic acid residues in alginic acid. *Acta Chem. Scand.*, **21**, 691–704.

HAUG A., LARSEN B. & SMIDSRØD O. (1974) Uronic acid sequence in alginate from different sources. *Carbohydr. Res.*, **32**, 217–25.

HELLEBUST J. A. & HAUG A. (1969) Alginic acid synthesis in *Laminaria digitata* (L.) Lamour. *Proc. Int. Seaweed Symp.*, **6**, 463–71.

HELLEBUST J. A. & HAUG A. (1972a) Photosynthesis, translocation, and alginic acid synthesis in *Laminaria digitata* and *Laminaria hyperborea*. *Can J. Bot.*, **50**, 169–76.

HELLEBUST J. A. & HAUG A. (1972b) *In situ* studies on alginic acid synthesis and other aspects of the metabolism of *Laminaria digitata*. *Can. J. Bot.*, **50**, 177–84.

HERTH W., KUPPEL A. & FRANKE W. W. (1975) Cellulose in *Acetabularia* cyst walls. *J. Ultrastruct. Res.*, **50**, 289–92.

HIRASE S. (1957) Studies on the chemical constitution of agar-agar. XIX. Pyruvic acid as a constituent of agar-agar. *Bull. Chem. Soc. Jpn.*, **30**, 68–79.

HIRASE S., ARAKI C. & ITO T. (1958) Isolation of agarobiose derivative from the mucilage of *Gloiopeltis furcata*. *Bull. Chem. Soc. Jpn.*, **31**, 428–31.

HIRASE S., ARAKI C. & WATANABE K. (1967) Component sugars of the polysaccharide of the red seaweed *Grateloupia elliptica*. *Bull. Chem. Soc. Jpn.*, **40**, 1445–8.

HIRASE S. & WATANABE K. (1972) The presence of pyruvate residues in λ-carrageenan and a similar polysaccharide. *Bull. Inst. Chem. Res.*, **50**, 332–6.

HIRST E., MACKIE W. & PERCIVAL E. (1965) The water-soluble polysaccharides of *Cladophora rupestris* and of *Chaetomorpha* spp. Part II. The site of ester sulphate groups and the linkage between the galactose residues. *J. Chem. Soc.*, 2958–67.

HOSFORD S. P. C. & MCCANDLESS E. L. (1975) Immunochemistry of carrageenans from gametophytes and sporophytes of certain red algae. *Can. J. Bot.*, **53**, 2835–41.

HUIZING H. J. & RIETEMA H. (1975) Xylan and mannan as cell wall constituents of different stages in the life histories of some siphoneous green algae. *Br. Phycol. J.*, **10**, 13–16.

HUIZING H. J., RIETEMA H. & SIETSMA J. H. (1979) Cell wall constituents of several siphoneous green algae in relation to morphology and taxonomy. *Br. Phycol. J.*, **14**, 25–32.

HUSSEIN M. M.-D. (1975) Two new polysaccharides from *Colpomenia sinuosa*. *Phytochemistry*, **44**, 1866–8.

HUSTEDE H. (1964) Entwicklungsphysiologische Untersuchungen uber den Generationswechsel zwischen *Derbesia neglecta* Berth. und *Bryopsis halimeniae* Berth. *Bot. Mar.*, **6**, 134–42.

IRIKI Y. & MIWA T. (1960) Chemical nature of the cell wall of the green algae, *Codium, Acetabularia* and *Halicoryne*. *Nature* (*Lond.*), **185**, 178–9.

JACKSON S. G. (1979) Biosynthesis of carrageenans in haploid and diploid plants of the red alga *Chondrus crispus*. Ph.D. Thesis, McMaster University, Hamilton, Canada.

JOHNSTON K. H. & MCCANDLESS E. L. (1973) Enzymic hydrolysis of the potassium soluble fraction of carrageenan: Properties of 'λ-carrageenases' from *Pseudomonas carrageenovora*. *Can. J. Microbiol.*, **19**, 779–88.

KOHN R., FURDA I., HAUG A. & SMIDSRØD O. (1968) Binding of calcium and potassium ions to some polyuronides and monouronates. *Acta Chem. Scand.*, **22**, 3098–102.

KORNMANN P. (1938) Zur Entwicklungsgeschichte von *Derbesia* und *Halicystis*. *Planta*, **28**, 464–70.

KREGER D. R. (1960) An X-ray study of *Hydrodictyon* cell walls. I. *K. Ned. Akad. Wet., Proc.*, Ser. C., **63**, 613–33.

KREGER D. R. (1962) Cell Walls. In *Physiology and Biochemistry of Algae*, ed. Lewin R. A., pp. 315–35. Academic Press, New York and London.

KYLIN H. (1913) Zur Biochemie der Meeresalgen. *Hoppe-Seyler's Z. Physiol. Chem.*, **83**, 171–97.

LARSEN B. (1967) Sulphated polysaccharides in brown algae. II. Isolation of 3-O-β-D-xylopyranosyl-L-fucose from ascophyllan . *Acta Chem. Scand.*, **21**, 1395–6.

LARSEN B., HAUG A. & PAINTER T. J. (1966) Sulphated polysaccharides in brown algae. I. Isolation and preliminary characterization of three sulphated polysaccharides from *Ascophyllum nodosum* (L.) Le Jol. *Acta Chem. Scand.*, **20**, 219–30.

LARSEN B., PAINTER T., HAUG A. & SMIDSRØD O. (1969) A statistical description of the alginate molecule in terms of a penultimate-unit copolymer. *Acta Chem. Scand.*, **23**, 355–70.

LIN T.-Y. & HASSID W. Z. (1966a) Isolation of guanosine diphosphate uronic acid from a marine brown alga, *Fucus gardneri* Silva. *J. Biol. Chem.*, **241**, 3283–93.

LIN T.-Y. & HASSID W. Z. (1966b) Pathway of alginic acid synthesis in the marine brown alga, *Fucus gardneri* Silva. *J. Biol. Chem.*, **241**, 5284–97.

LOEWUS F., WAGNER G., SCHIFF J. A. & WEISTROP J. (1971) The incorporation of ^{35}S-labeled sulfate into carrageenan in *Chondrus crispus*. *Plant Physiol.*, **48**, 373–5.

LOVE J., MACKIE W., MCKINNELL J. W. & PERCIVAL E. (1963) Starch-type polysaccharides isolated from the green seaweeds, *Enteromorpha compressa, Ulva lactuca, Cladophora rupestris, Codium fragile* and *Chaetomorpha capillaris*. *J. Chem. Soc.*, 4177–82.

LOVE J. & PERCIVAL E. (1964) The polysaccharides of the green seaweed *Codium fragile*. II. The water-soluble sulphated polysaccharides. *J. Chem. Soc.*, 3338–45.

MCCANDLESS E. L. (1978) The importance of cell wall constituents in algal taxonomy. In *Modern Approaches to the Taxonomy of Red and Brown Algae*, eds. Irvine D. E. G., & Price J. H., pp. 63–85. Academic Press, London & New York.

MCCANDLESS E. L. & CRAIGIE J. S. (1979) Sulfated polysaccharides in red and brown algae. *Ann. Rev. Plant Physiol.*, **30**, 41–53.

MCCANDLESS E. L., CRAIGIE J. S. & WALTER J. A. (1973) Carrageenans in the gametophytic and sporophytic stages of *Chondrus crispus*. *Planta*, **112**, 201–12.

MCCANDLESS E. L., OKADA W. T., LOTT J. N. A., VOLLMER C. M. & GORDON-MILLS E. M. (1977) Structural studies of *Chondrus crispus*: The effect of extraction of carrageenan. *Can. J. Bot.*, **55**, 2053–64.

MCCANDLESS E. L. & RICHER S. M. (1972) ^{14}C studies of carrageenan synthesis. *Proc. Int. Seaweed Symp.*, **7**, 477–84.

MCCULLY M. E. (1966) Histological studies on the genus *Fucus*. I. Light microscopy of the mature vegetative plant. *Protoplasma*, **42**, 287–305.

MCCULLY M. E. (1968a) Histological studies on the genus *Fucus*. II. Histology of the reproductive tissues. *Protoplasma*, **66**, 205–30.

MCCULLY M. E. (1968b) Histological studies on the genus *Fucus*. III. Fine structure and possible functions of the epidermal cells of the vegetative thallus. *J. Cell Sci.*, **3**, 1–16.

MACKIE I. M. & PERCIVAL E. (1959) The constitution of xylan from the green seaweed *Caulerpa filiformis*. *J. Chem. Soc.*, 1151–6.

MACKIE I. M. & PERCIVAL E. (1960) Polysaccharides from the green seaweed *Caulerpa filiformis*. Part II. A glucan of the amylopectin type. *J. Chem. Soc.*, 2381–4.

MACKIE W. & PRESTON R. D. (1974) Cell wall and intercellular region polysaccharides. In *Algal Physiology and Biochemistry*, ed. Stewart, W. D. P. pp. 40–85. Blackwell Scientific Publications, Oxford.

MCKINNELL J. P. & PERCIVAL E. (1962a) The acid polysaccharide from the green seaweed *Ulva lactuca*. *J. Chem. Soc.*, 2082–3.

MCKINNELL J. P. & PERCIVAL E. (1962b) Structural investigations on the water-soluble polysaccharide of the green seaweed *Enteromorpha compressa*. *J. Chem. Soc.*, 3141–8.

MADGWICK J., HAUG A. & LARSEN B. (1973a) Alginate lyase in the brown alga *Laminaria digitata* (Huds.) Lamour. *Acta Chem. Scand.*, **27**, 711–12.

MADGWICK J., HAUG A. & LARSEN B. (1973b) Polymannuronic acid 5-epimerase from the marine alga *Pelvetia canaliculata* (L.) Dcne et Thur. *Acta Chem. Scand.*, **27**, 3592–4.

MADGWICK J., HAUG A. & LARSEN B. (1978) Ionic requirements of alginate-modifying enzymes in the marine alga *Pelvetia canaliculata* (L.) Dcne et Thur. *Bot. Mar.*, **21**, 1–3.

MAEDA M. & NISIZAWA K. (1968a) Laminaran of *Ishige okamurai*. *Carbohydr. Res.*, **7**, 94–7.

MAEDA M. & NISIZAWA K. (1968b) Fine structure of *Eisenia bicyclis*. *J. Biochem.* (*Tokyo*), **63**, 199–206.

MANNERS D. J. & WRIGHT A. (1962) α-1,4-glucosans. IV. The interaction of concanavalin-A with glycogens. *J. Chem. Soc.*, 4292–5.

MARSHALL S. M., NEWTON L. & ORR A. P. (1949) *A Study of Certain British Seaweeds and Their Utilisation in the Preparation of Agar*. HMSO, London.

MAYHOUB H. (1976) Sur l'existence d'un *Derbesia* dans le cycle de l'*Halicystis boergesenii* Iyengar et Raman (Chlorophycées, Derbesiales) *C.R. Acad. Sci.Paris*, Ser. D, **282**, 707–10.

MEDCALF D. G. & LARSEN B. (1977a) Fucose-containing polysaccharides in the brown algae *Ascophyllum nodosum* and *Fucus vesiculosus*. *Carbohydr. Res.*, **59**, 531–7.

MEDCALF D. G. & LARSEN B. (1977b) Structural studies on ascophyllan and the fucose-containing complexes from the brown alga *Ascophyllum nodosum*. *Carbohydr. Res.*, **59**, 539–46.

MEEUSE B. J. D. (1962) Storage products. In *Physiology and Biochemistry of Algae*, ed. Lewin R. A., pp. 289–311. Academic Press, London & New York.

MEEUSE B. J. D. (1963) Inulin in the green alga *Batophora oerstedi* J. Ag. (Dasycladales). *Acta Bot. Neerl.*, **12**, 315–18.

MEEUSE B. J. D., ANDRIES M. & WOOD J. A. (1960) Floridean starch. *J. Exp. Bot.*, **11**, 129–40.

MEEUSE B. J. D. & SMITH B. N. (1962) A note on the amylolytic breakdown of some raw algal starches. *Planta*, **57**, 624–35.

MIAN A. J. & PERCIVAL E. (1973) Carbohydrates of the brown seaweeds *Himanthalia lorea* and *Bifurcaria bifurcata*. Part II. Structural studies of the fucans. *Carbohydr. Res.*, **26**, 147–61.

MORRIS E. R., REES D. A. & THOM D. (1973) Characterization of polysaccharide structure and interactions by circular dichroism: Order-disorder transition in the calcium alginate system. *J. Chem. Soc. Chem. Comm.*, 245–6.

MYERS A. & PRESTON R. D. (1959) Fine structure in the red algae. II. The structure of the cell wall of *Rhodymenia palmata*. *Proc. Roy. Soc. Lond.*, Ser. B., **150**, 447–55.

NICOLAI E. & PRESTON R. D. (1952) Cell-wall studies in the Chlorophyceae. I. A general survey of submicroscopic structure of filamentous species. *Proc. Roy. Soc. Lond.*, Ser. B., **140**, 244–74.

NISIZAWA K., KURODA K., TOMITA Y. & SHIMAHARA H. (1974) Main cell wall constituents of the cysts of *Acetabularia*. *Bot. Mar.*, **17**, 16–19.

NORTHCOTE D. H., GOULDING K. J. & HORNE R. W. (1960) The chemical composition and structure of the cell-wall of *Hydrodictyon africanum* Yaman. *Biochem. J.*, **77**, 503–8.

NUNN J. R. & PAROLIS H. (1968) A polysaccharide from *Aeodes orbitosa*. *Carbohydr. Res.*, **6**, 1–11.

NUNN J. R. & PAROLIS H. (1969) Sulphated polysaccharides of the Grateloupiaceae family. III. A polysaccharide from *Phyllymenia cornea*. *Carbohydr. Res.*, **9**, 265–76.

NUNN J. R. & PAROLIS H. (1970) Sulphated polysaccharides of the Grateloupiaceae family. IV. Methylation analysis of phyllymenan and desulphated phyllymenan. *Carbohydr. Res.*, **14**, 145–50.

NUNN J. R., PAROLIS H. & RUSSELL I. (1973) Polysaccharides of the red alga *Chaetangium erinaceum*. *Carbohydr. Res.*, **26**, 169–80.

O'DONNELL J. J. & PERCIVAL E. (1959) Structural studies on the water-soluble polysaccharides from the green seaweed *Acrosiphonia centralis* (*Spongomorpha arcta*). *J. Chem. Soc.*, 2168–78.

PAINTER T., SMIDSRØD O., LARSEN B. & HAUG A. (1968) A computer study of the changes in composition-distribution occurring during random depolymerisation of a binary linear heteropolysaccharide. *Acta Chem. Scand.*, **22**, 1637–48.

PARKER B. C. & DIBOLL A. G. (1966) Alcian stains for histochemical localization of acid and sulfated polysaccharides in algae. *Phycologia*, **6**, 37–46.

PARKER B. C. & LEEPER G. F. (1969) Fine structure of polysaccharide microcrystals. *Planta*, **87**, 86–94.

PEAT S. & TURVEY J. R. (1965) Polysaccharides of marine algae. *Fortschr. Chem. Org. Naturst.* **23**, 1–45.

PEAT S., TURVEY J. R. & EVANS J. M. (1959a) The structure of floridean starch. Part I. Linkage analysis by partial acid hydrolysis. *J. Chem. Soc.*, 3223–7.

PEAT S., TURVEY J. R. & EVANS J. M. (1959b) The structure of floridean starch. Part II. Enzymic hydrolysis and other studies. *J. Chem. Soc.*, 3341–4.

PENMAN A. & REES D. A. (1973) Carrageenans. IX. Methylation analysis of galactan sulphates from *Furcellaria fastigiata, Gigartina canaliculata, Gigartina chamissoi, Gigartina atropurpurea, Ahnfeltia durvillaei, Gymnogongrus furcellatus, Eucheuma isiforme, Eucheuma uncinatum, Aghardhiella tenera, Pachymenia hymantophora* and *Gloiopeltis cervicornis*. Structure of ξ-carrageenan. *J. Chem. Soc. Perkin Trans. I.*, 2182–7.

PENMAN A. & SANDERSON G. R. (1972) A method for the determination of uronic acid sequences in alginates. *Carbohydr. Res.*, **25**, 273–82.

PERCIVAL E. (1970) Algal polysaccharides. In *The Carbohydrates, Chemistry and Biochemistry*, eds. Pigman W., Horton D. & Herp A., **2B**, 537–68. Academic Press, London.

PERCIVAL E. (1979) The polysaccharides of green, red, and brown seaweeds: their basic structure, biosynthesis and function. *Br. Phycol. J.*, **14**, 103–17.

PERCIVAL E. & MCDOWELL R. H. (1967) *Chemistry and Enzymology of Marine Algal Polysaccharides.* Academic Press, London.

PERCIVAL E. & SMESTAD B. (1972) Carbohydrates of *Acetabularia* species. Part II. *A. crenulata* acid polysaccharide. *Carbohydr. Res.*, **25**, 299–312.

PERCIVAL E. & YOUNG M. (1974) Carbohydrates of the brown seaweeds. Part III. *Desmarestia aculeata. Carbohydr. Res.*, **32**, 195–201.

PICKMERE S. E., PARSONS M. J. & BAILEY R. W. (1973) Composition of *Gigartina* carrageenan in relation to sporophyte and gametophyte stages of the life cycle. *Phytochemistry*, **12**, 2441–4.

PRESTON R. D. (1968) Plants without cellulose. *Scient. Am.* **218** (6), 102–8.

PRESTON R. D. (1974) *The Physical Biology of Plant Cell Walls.* Chapman & Hall, London.

PRESTON R. D. & CRONSHAW J. (1958) Constitution of the fibrillar and non-fibrillar components of the walls of *Valonia ventricosa. Nature* (*Lond.*), **181**, 248–50.

QUATRANO R. S. & CRAYTON M. A. (1973) Sulfation of fucoidan in *Fucus* embryos. I. Possible role in localization. *Dev. Biol.*, **30**, 29–41.

QUATRANO R. S. & STEVENS P. T. (1976) Cell wall assembly in *Fucus* zygotes. I. Characterization of the polysaccharide components. *Plant Physiol.*, **58**, 224–31.

QUILLET M. (1958) Sur le métabolisme glucide des Algues brunes. Présence de petites quantités de laminarine chez de nombreuses nouvelles espèces, reparties dans tout le groupe de Phéophycées. *C.R. Acad. Sci, Paris*, **246**, 812–15.

REES D. A. (1963) The carrageenan system of polysaccharides. I. The relation between the κ- and λ-components. *J. Chem. Soc.*, 1821–32.

REES D. A. (1969) Structure, conformation and mechanism in the formation of polysaccharide gels and networks. *Adv. Carbohydr. Chem. Biochem.*, **24**, 267–332.

REES D. A. (1972) Shapely polysaccharides: The eighth Colworth Medal lecture. *Biochem. J.*, **126**, 257–73.

REES D. A. (1975) Stereochemistry and binding behaviour of carbohydrate chains. In *Int. Rev. Sci., Biochem. of Carbohydrates*, ed. Whelan W. J., **5**, 1–42. Butterworths, London.

REES D. A. (1977) *Polysaccharide Shapes* (*Outline Studies in Biology*). Chapman & Hall, London.

ROELOFSEN P. A., DALITZ V. CH. & WIJNMAN C. F. (1953) Constitution, submicroscopic structure and degree of crystallinity of the cell wall of *Halicystis osterhoutii. Biochim. Biophys. Acta*, **11**, 344–54.

SHIRAIWA Y., ABE K., SASAKI S. F., IKAWA T. & NISIZAWA K. (1975) Alginate lyase activities in the extracts from several brown algae. *Bot. Mar.*, **18**, 97–104.

SHASHKOV A. S., USOV A. I. & YAROTSKII S. V. (1978) Polysaccharides of algae. XXIV. The application of ^{13}C nmr spectroscopy to the analysis of the structures of polysaccharides of the agar group. *Bioorgan. Khim.*, **4**, 74–81.

SMITH D. B., & COOK W. H. (1953) Fractionation of carrageenin. *Arch. Biochem. Biophys.*, **45**, 232–3.

STANCIOFF D. J. & STANLEY N. F. (1969) Infrared and chemical studies of algal polysaccharides. *Proc. Int. Seaweed Symp.*, **6**, 595–609.

STEWART C. M., DAWES C. J., DICKINS B. M. & NICHOLLS J. W. P. (1969) The cell-wall constituents of *Apjohnia laetevirens* Harvey. *Austral. J. Mar. Freshwater Res.*, **20**, 143–55.

THOMPSON E. W. & PRESTON R. D. (1967) Proteins in the cell walls of some green algae (*Nitella opaca, Cladophora, Chaetomorpha, Codium*). *Nature* (*Lond.*), **213**, 684–5.

TURVEY J. R. (1978) Biochemistry of algal polysaccharides. In *Biochemistry of Carbohydrates II*, ed. Manners D. J., **16**, 151–77. University Park Press, Baltimore.

TURVEY J. R. & WILLIAMS E. L. (1970) The structures of some xylans from red algae. *Phytochemistry*, **9**, 2383–8.

TSANG M. L.-S. & SCHIFF J. A. (1975) Studies of sulfate utilization by algae. 14. Distribution of adenosine-3′-phosphate-5′-phosphosulfate (PAPS) and adenosine-5′-phosphosulfate (APS) sulfotransferases in assimilatory sulfate reducers. *Plant Sci. Lett.*, **4**, 301–7.

USOV A. I. & BARBAKADZE V. V. (1978) Polysaccharides of algae. XXVII. Partial acetolysis of the sulfated galactan from the red seaweed *Grateloupia divaricata* Okan. *Bioorgan. Khim.*, **4**, 1107–15.

USOV A. I., BARBAKADZE V. V., YAROTSKII S. V. & SHASHKOV A. S. (1978a) Polysaccharides of algae. XXVIII. Application of ^{13}C-N.M.R. spectroscopy for structural studies of galactan from red seaweed *Grateloupia divaricata* Okan. *Bioorgan. Khim.*, **4**, 1507–12.

USOV A. I., LOTOV R. A. & KOCHETKOV N. K. (1971) Algal polysaccharides. VII. Preliminary studies on the polysaccharides of the alga *Rhodomela larix* (Turn.) C. Ag. *Zh. Obshch. Khim.*, **41**, 1154–60.

USOV A. I., YAROTSKII S. V. & ÉSTEVES M. L. (1978b) Polysaccharides of algae. XXIII. Polysaccharides of the red alga *Bangia fuscopurpurea* (Dillw.) Lyngb. *Bioorgan. Khim.*, **4**, 66–73.

USOV A. I., YAROTSKII S. V., SHASKOV A. S. & TISHCHENKO V. P. (1978c) Polysaccharides of algae. XXII. Polysaccharide composition of *Rhodymenia stenogona* Perest and the application of ^{13}C NMR spectroscopy to the determination of xylan structures. *Bioorgan. Khim.*, **4**, 57–65.

VREELAND V. (1970) Localization of a cell wall polysaccharide in a brown alga with labeled antibody. *J. Histochem. Cytochem.*, **18**, 371–3.

VREELAND V. (1972) Immunocytochemical localization of the extracellular polysaccharide alginic acid in the brown seaweed *Fucus distichus*. *J. Histochem. Cytochem.*, **20**, 358–67.

VREELAND V. (1974a) Alginates and sulfated fucans in brown algal walls. *Proc. Int. Seaweed Symp.*, **8**, (abst.).

VREELAND V. (1974b) A comparative study of brown algal cell walls. *J. Phycol.*, **10** (suppl.), 8–9.

WEIGL J. & YAPHE W. (1966) The enzymatic hydrolysis of carrageenan by *Pseudomonas carrageenovora*: Purification of κ-carrageenase. *Can. J. Microbiol.*, **12**, 939–47.

WHYTE J. N. C. (1971) Polysaccharides of the red seaweed *Rhodymenia pertusa*. I. Water-soluble glucan. *Carbohydr. Res.*, **16**, 220–4.

WONG K. F. & CRAIGIE J. S. (1978) Sulfohydrolase activity and carrageenan biosynthesis in *Chondrus crispus* (Rhodophyceae). *Plant Physiol.*, **61**, 663–6.

WUTZ M. & ZETSCHE K. (1970) Biochemische Aspekte des heteromorphen Generationswechsels von *Halicystis-Derbesia*. *Ber. Dtsch. Bot. Ges.*, **83**, 229–30.

WUTZ M. & ZETSCHE K. (1976) Zur Biochemie und Regulation des heteromorphen Generationswechsels der Grünalge *Derbesia-Halicystis*. *Planta*, **129**, 211–16.

YAROTSKII S. V., SHASHKOV A. S. & USOV A. I. (1977) Analysis of ^{13}C-NMR spectra of some red seaweed galactans. *Bioorgan. Khim.*, **3**, 1135–7.

YAROTSKII S. V., SHASHKOV A. S. & USOV A. I. (1978) Polysaccharides of algae. XXV. Application of ^{13}C-NMR spectroscopy for structural analysis of κ-carrageenan group polysaccharides. *Bioorgan. Khim.*, **4**, 745–51.

YOUNG K., HONG K. C., DUCKWORTH M. & YAPHE W. (1971a) Enzymic hydrolysis of agar and properties of bacterial agarases. *Proc. Int. Seaweed Symp.*, **7**,469–72.

YOUNG K., DUCKWORTH M. & YAPHE W. (1971b) The structure of agar. III. Pyruvic acid, a common feature of agars from different agarophytes. *Carbohydr. Res.*, **16**, 446–8.

CHAPTER 17
CHEMICAL CONSTITUENTS OF SEAWEEDS

M. A. RAGAN
Atlantic Research Laboratory, National Research Council of Canada,
1411 Oxford St., Halifax, Nova Scotia, Canada.

1 INTRODUCTION

Seaweeds present a number of challenges to chemists and biochemists. Few seaweeds are available in laboratory culture; hence chemists must usually begin with collected samples whose growth conditions and metabolic history are poorly known. These samples are likely to host an abundance of epiphytic and endophytic organisms including other algae, bacteria and fungi. The often copious cell wall materials complicate extraction of hydrophilic constituents and, together with the other factors mentioned, may hamper biosynthetic studies involving isotopically labelled intermediates.

Nevertheless the chemistry of seaweeds is an exciting and rapidly expanding field. Interest in marine plants as potential sources of new antibiotics, pharmaceuticals and antitumor agents has been rewarded by the discovery of an unexpected diversity of new natural products. A few of the more important, unusual or biosynthetically interesting metabolites of red, brown, and green seaweeds are described in the following pages.

2 PHOTOSYNTHETIC PIGMENTS

2.1 Chlorophylls

Chlorophyll *a* (Fig. 17.1a) is the most abundant chlorophyll, and the primary photosynthetic pigment, of seaweeds. As a specialized chlorophyll–protein complex (P700), it is directly involved in the Photosystem I light reaction (Chapter 13). Seaweeds probably synthesize chlorophyll *a* from 5-aminolevulinic acid, as do other plants (Harel 1978); however, little is known about the biosynthetic origin of 5-ALA itself in seaweeds. In certain unicellular protists (*Cyanidium caldarium, Porphyridium purpureum, Chlorella vulgaris, Euglena gracilis*) cyanobacteria and higher plants, 5-ALA is produced from α-ketoglutaric acid via 4,5-dioxovaleric acid (Jurgenson *et al.* 1976).

Green and brown seaweeds produce accessory chlorophylls in addition to chlorophyll *a*. Chlorophyll *b* (Fig. 17.1b), probably synthesized directly from chlorophyll *a*, is found in green algae. Chlorophyll(ide)s c_1 and c_2 (Figs 17.2a and b), present in brown seaweeds, may be derived from magnesium 2,4-divinyl phaeoporphyrin a_5, an intermediate in the biosynthesis of chlorophyll *a*. Red seaweeds do not produce chlorophylls *b* or *c*; a 'chlorophyll *d*', found in extracts of some red seaweeds, has not been detected *in vivo* (Meeks 1974).

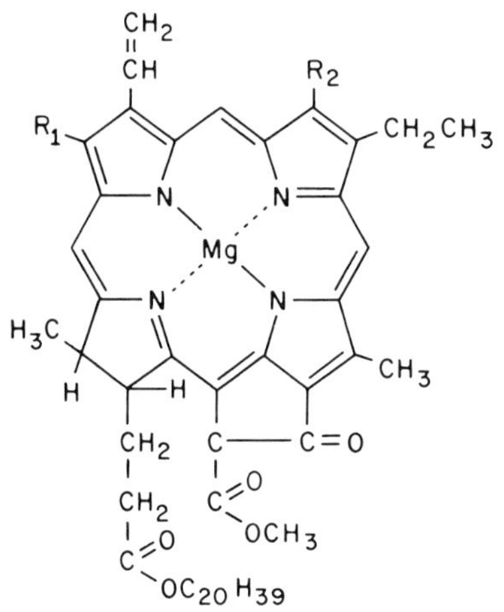

(a) $R_1 = CH_3$ $R_2 = CH_3$

(b) $R_1 = CH_3$ $R_2 = CHO$

Fig. 17.1 (a) Chlorophyll *a*. (b) Chlorophyll *b*.

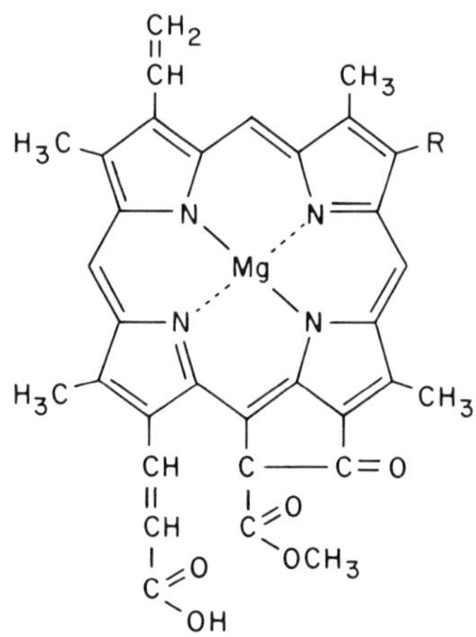

(a) $R = CH_2CH_3$

(b) $R = CH=CH_2$

Fig. 17.2 (a) Chlorophyll c_1. (b) Chlorophyll c_2.

2.2 Phycobiliproteins

Phycobiliproteins are open-chain ('linear') tetrapyrrole pigments found in Rhodophyceae, cyanobacteria and Cryptophyceae. The bilin unit is derived from 5-aminolevulinic acid via protoporphyrin IX (Troxler *et al.* 1979). Two types of bilins have been identified in seaweeds: phycocyanobilin (Fig. 17.3)

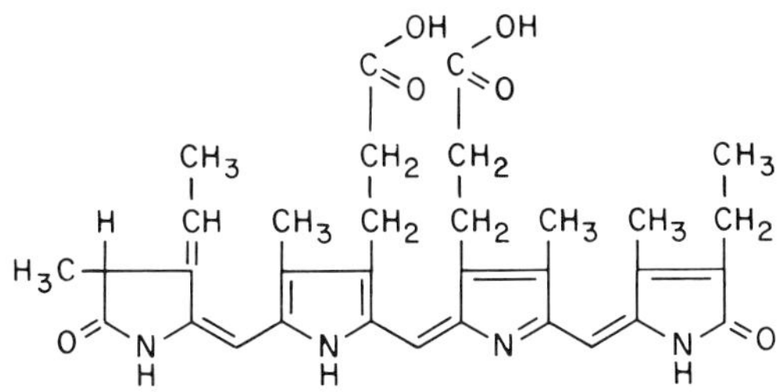

Fig. 17.3 Phycocyanobilin.

and phycoerythrobilin (17.4). A third phycobilin, phycourobilin, may or may not be a structurally distinct entity (Chapman 1973). These bilins are covalently attached to polypeptides to form the major classes (α, β, γ) of subunit component

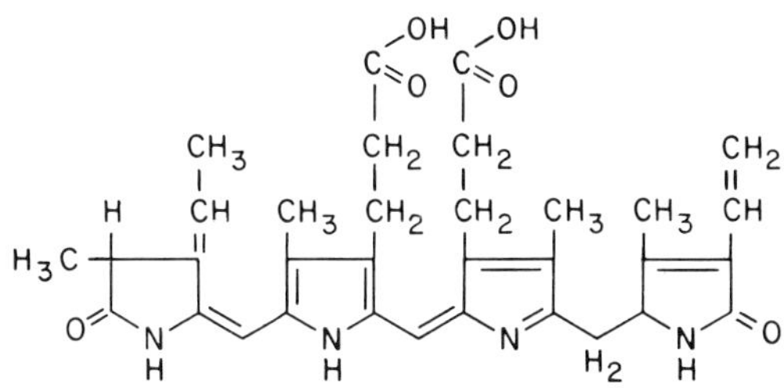

Fig. 17.4 Phycoerythrobilin.

proteins. Phycobiliproteins are then assembled from these subunits as shown in Table 17.1.

Phycobiliproteins were originally grouped on the basis of their spectral characteristics and their distributions in the different algal taxa; for example, *R*-phycocyanin was the blue pigment present in Rhodophyceae. Subsequent research has in many cases revealed more widespread taxonomic distributions, and the present nomenclature refers only to spectral characteristics, not to taxonomic distribution. Even this is an oversimplification, as variation exists within the present groups; a more natural classification should incorporate information from amino acid sequences of the apoprotein moieties.

2.3 Carotenoids

Carotenoids are isoprenoid polyene pigments present in all photosynthetic (and many non-photosynthetic) organisms. Biosynthetically, carotenoids are derived

Table 17.1 Phycobiliproteins in algae*.

Biliprotein	Approximate absorption maxima	Subunit structure	Prosthetic group composition of subunit			Distribution
			α	β	γ	
Allophycocyanin	650, 615	$(\alpha\beta)_3$	1 × PC	1 × PC	–	cyanobacteria; Rhodophyceae *Porphyra tenera, Ceramium rubrum, Smithora naiadum*
Allophycocyanin B	670, 620	$(\alpha\beta)_3$	1 × PC	1 × PC	–	cyanobacteria (*Anabaena variabilis*, etc.)
C-Phycocyanin	620, 610	$(\alpha\beta)_3$ or $(\alpha\beta)_6$	1 × PC	2 × PC	–	cyanobacteria; Rhodophyceae *Polysiphonia urceolata, Porphyra* spp., *Smithora naiadum, Grateloupia* sp.
R-Phycocyanin	660, 620, 555	$(\alpha\beta)_3$	1 × PC	1 × PC 1 × PE	–	Rhodophyceae *Ceramium rubrum, Porphyra* spp., *Palmaria palmata*
Phycoerythrocyanin	590–600, 570, 560	$(\alpha\beta)_3$	1 × unident.	2 × PC	–	cyanobacteria (*Anabaena variabilis*, etc.)
C-Phycoerythrin	565	$(\alpha\beta)_6$	2 × PE	4 × PE	–	cyanobacteria, esp. Oscillatoriaceae
b-Phycoerythrin	565, 545	$(\alpha\beta)_?$	2 × PE	4? × PE	–	Bangiophycidae (*Porphyridium purpureum*)
B-Phycoerythrin	565, 545, 495	$(\alpha\beta)_6\gamma$	2 × PE	4 × PE	2 × PE 2 × PU	Rhodophyceae *Smithora naiadum* (='*Porphyra naiadum*'), *Rhodochorton floridulum, Rhodosorus marinus, Ceramium rubrum*
R-Phycoerythrin	565, 540, 509	$(\alpha\beta)_6\gamma$		31 × PE, 9 × PU		Florideophycidae *Ceramium rubrum, Porphyra* spp., *Palmaria palmata, Cryptonemia obovata*

* Data from Ó hEocha (1965), Glazer (1977), Glazer & Hixson (1977), Ragan & Chapman (1978), Gysi & Chapman (in press). Further distribution data are available in Hirose & Kumano (1966) and Hirose *et al.* (1969).

from mevalonic acid via isopentenylpyrophosphate (Fig. 17.5) and lycopene (Fig. 17.6) and/or its isomer *β*-zeacarotene. *β,β*-Carotene ('*β*-carotene', Fig. 17.7) and *β,ε*-carotene ('α-carotene', Fig. 17.8) are formed by enzymatic

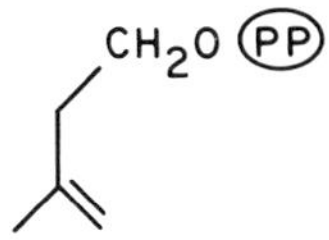

Fig. 17.5 Isopentenylpyrophospate.

Fig. 17.6 Lycopene.

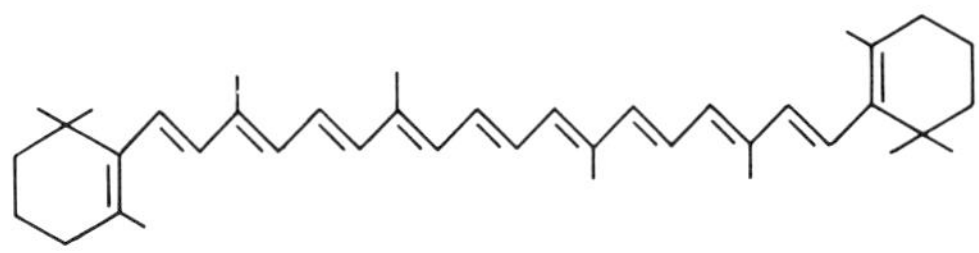

Fig. 17.7 *β*-Carotene.

cyclization; further biosynthetic alterations (hydroxylation, epoxidation, in-chain oxidation, etc.) subsequently give rise to the variety of carotenoids and xanthophylls found in seaweeds. Unless otherwise indicated, the distributional data below have been drawn from reviews by Jensen (1966), Bjørnland (1974), Goodwin (1974a), Liaaen-Jensen (1978), Ragan & Chapman (1978), and Chapman & Antia (in preparation).

Most red seaweeds exhibit a rather simple carotenoid pattern, accumulating mainly *β*-carotene (Fig. 17.7), α-carotene (Fig. 17.8), and their dihydroxy derivatives zeaxanthin (Fig. 17.9) and lutein (Fig. 17.10). Most other carotenoids

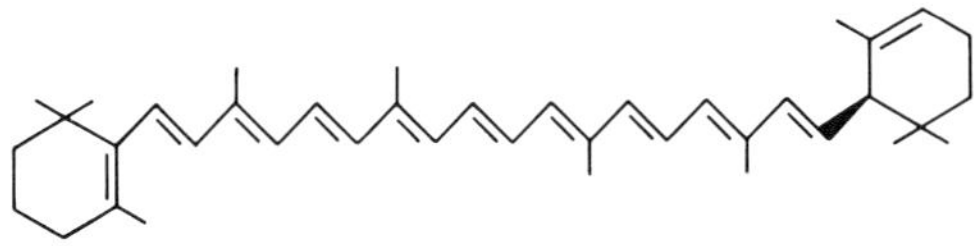

Fig. 17.8 α-Carotene.

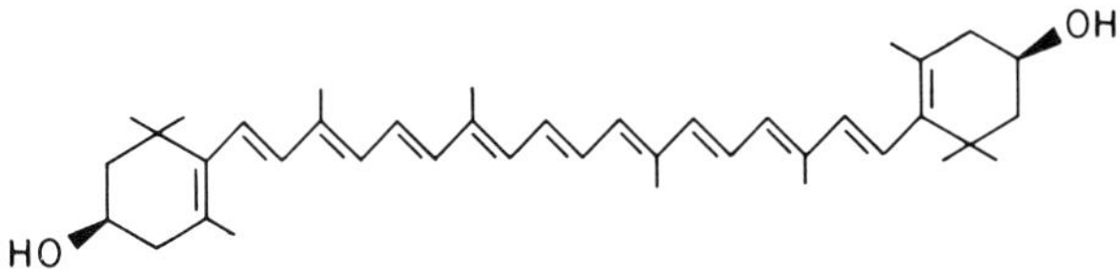

Fig. 17.9 Zeaxanthin.

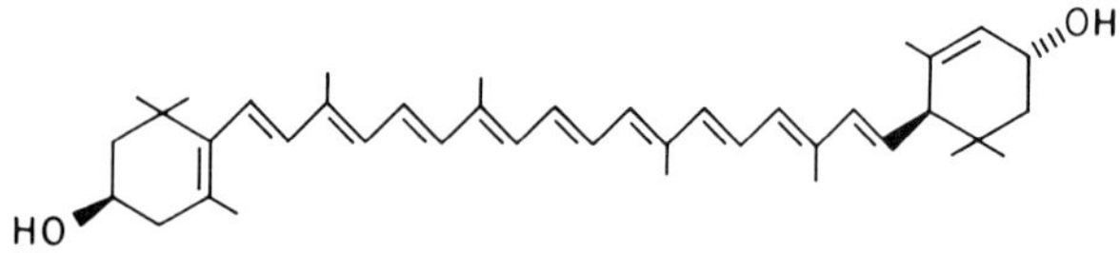

Fig. 17.10 Lutein.

occasionally reported to be minor constituents of red seaweeds are closely related to these four major pigments: the corresponding monohydroxides *β*- and *α*-cryptoxanthin, lutein epoxide ('taraxanthin'), zeaxanthin mono- (antheraxanthin, Fig. 17.11) and di- (violaxanthin, Fig. 17.12) epoxides, and auroxanthin

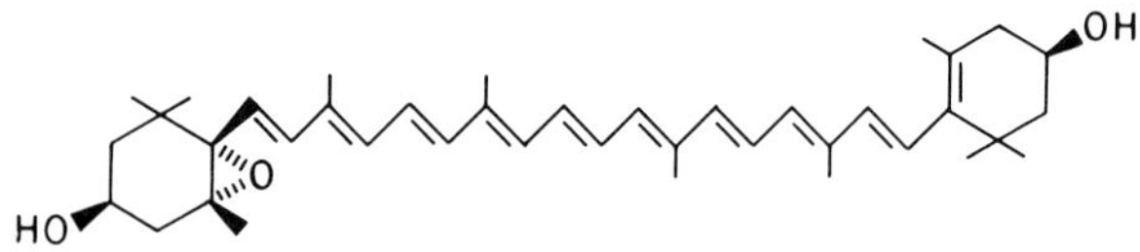

Fig. 17.11 Antheraxanthin.

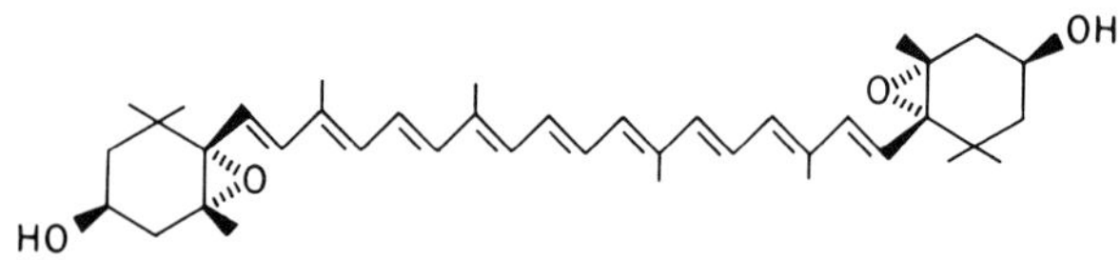

Fig. 17.12 Violaxanthin.

(violaxanthin furan oxide). Neoxanthin (Fig. 17.13), reported from *Antithamnion plumula* and *Nemalion helminthoides* (= *N. multifidum*), is presumably derived from a violaxanthin-like precursor. Fucoxanthin (Fig. 17.14) and peridinin,

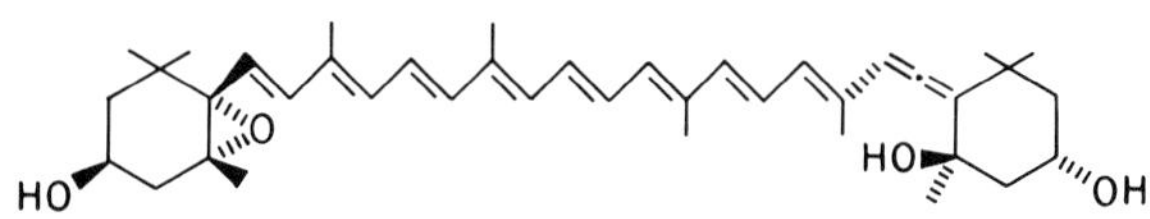

Fig. 17.13 Neoxanthin.

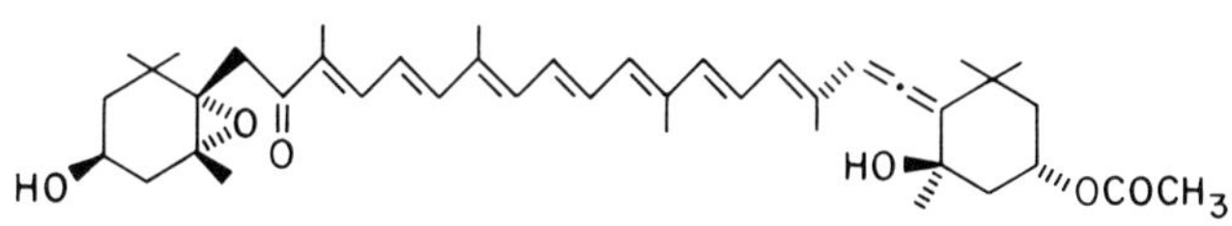

Fig. 17.14 Fucoxanthin.

occasionally reported from red seaweeds, are thought to come from associated microalgae, although this remains to be conclusively demonstrated (Goodwin 1960; Bjørnland & Aguilar-Martinez 1976; Arnesen *et al.* 1979). *α*-Carotene, and (in *Phycodrys rubens*) *β*-carotene, may occasionally be absent.

Brown seaweeds accumulate *β*-carotene (Fig. 17.7), fucoxanthin (Fig. 17.14), and violaxanthin (Fig. 17.12); the closely related zeaxanthin (Fig. 17.9), neoxanthin (Fig. 17.13) and fucoxanthinol (desacetoxy-fucoxanthin) are occasionally minor components, as is ε,ε-carotene ('ε-carotene'). The acetylenic carotenoids diadinoxanthin (Fig. 17.15) and diatoxanthin (Fig. 17.16) are thought to be endogenous, albeit minor, brown algal pigments, and are not due

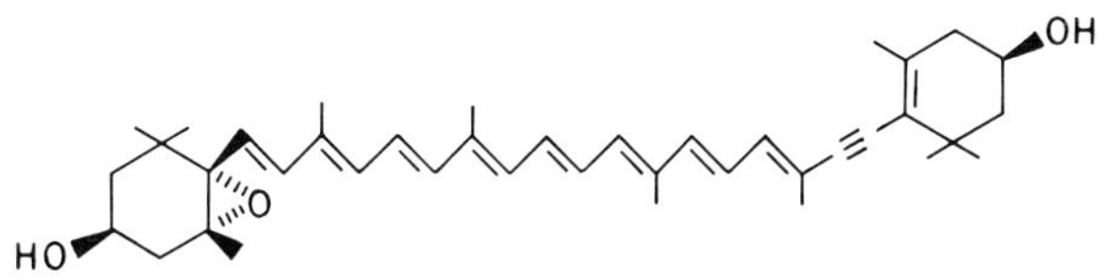

Fig. 17.15 Diadinoxanthin.

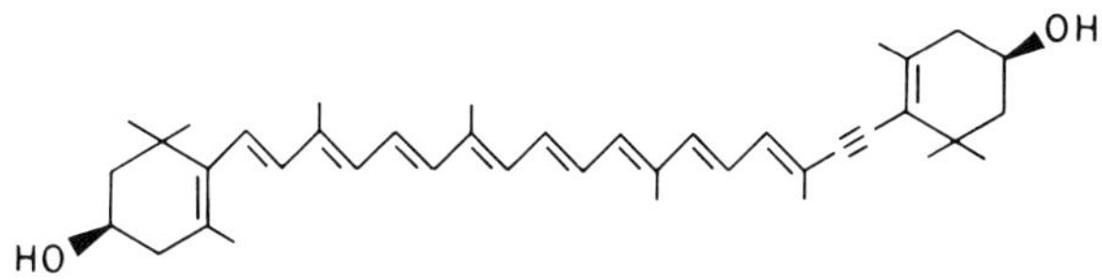

Fig. 17.16 Diatoxanthin.

entirely to epiphytic diatoms (Liaaen-Jensen 1978). Lutein (Fig. 17.10) and α-carotene (Fig. 17.8) have occasionally been reported.

Green seaweeds generally resemble higher plants in carotenoid composition. The major carotenoids are β-carotene (Fig. 17.7), lutein (Fig. 17.10), violaxanthin (Fig. 17.12), antheraxanthin (Fig. 17.11), zeaxanthin (Fig. 17.9), and neoxanthin (Fig. 17.13). α-Carotene (Fig. 17.8), β,ψ-carotene ('γ-carotene'), β-cryptoxanthin, cryptoxanthin-5′,6′-epoxide, and lutein epoxide are minor or occasional constituents. The α-carotene derivative loroxanthin (Fig. 17.17) has been reported from *Cladophora* spp. and *Ulva rigida*.

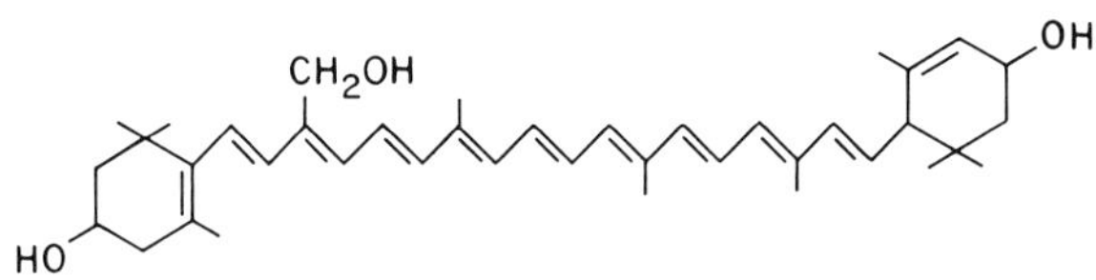

Fig. 17.17 Loroxanthin.

Most siphoneous green seaweeds (*Avrainvillea, Bryopsis, Caulerpa, Codium, Halimeda, Pseudobryopsis, Udotea*), as well as members of the Derbesiaceae (*Derbesia, Halicystis*), also accumulate the α-carotene derivative siphonaxanthin (Fig. 17.18) and its fatty acid ester(s), siphonein. The latter is also accumulated

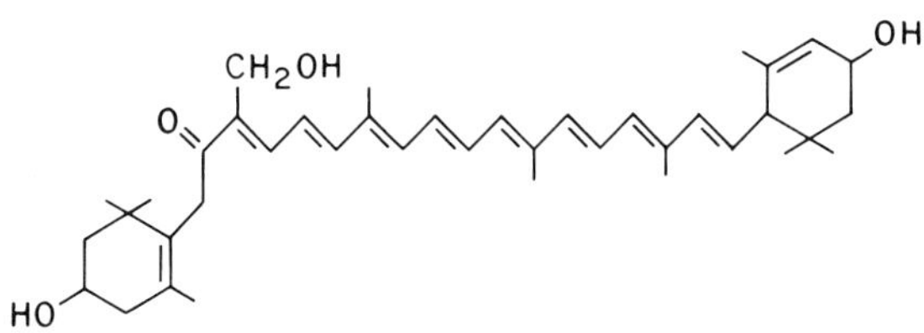

Fig. 17.18 Siphonaxanthin.

in *Dichotomosiphon* sp., and one or both of these compounds are occasionally found in other green algae (*Blastophysa rhizopus; Boodlea coacta; Cladophora* spp.; *Ulva japonica; Valonia* spp.). Details of their distribution have been presented by Kleinig (1969), Goodwin (1974a), and Yokohama *et al.* (1977).

Environmental conditions may significantly influence the carotenoid composition of seaweeds. An example is the light-mediated de-epoxidation of violaxanthin (Fig. 17.12) to antheraxanthin (Fig. 17.11) to zeaxanthin (Fig. 17.9) (Stransky & Hager 1970; Sapozhnikov 1973); other 'xanthophyll cycles' may operate with diadinoxanthin (Fig. 17.15) to diatoxanthin (Fig. 17.16) and lutein epoxide to lutein (Fig. 17.10). The accumulation of siphonaxanthin (Fig. 17.18) may require low light levels: Yokohama *et al.* (1977) have shown that 'deep-water' green algae, including *Cladophora wrightiana*, accumulate siphonaxanthin, whereas intertidal green seaweeds, including another *Cladophora* sp., do not.

3 STEROIDS AND OTHER TRITERPENOIDS

Sterols and related compounds (4α-methyl- and 4,4-dimethyl sterols, ketosteroids, steryl esters and steryl glycosides) are biosynthesized from mevalonic acid via isopentenylpyrophosphate and squalene. Cyclization of squalene 2,3-epoxide produces cycloartenol (Fig. 17.19), thought to be the precursor of all

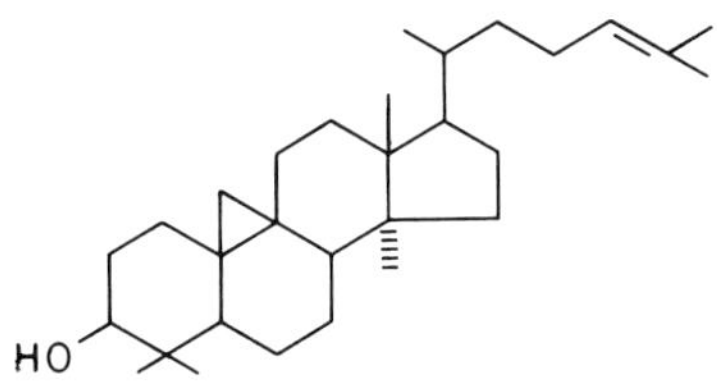

Fig. 17.19 Cycloartenol.

algal steroids. Biosynthetic steps beyond cycloartenol are complex, and more than one pathway to an individual sterol may exist even within a single organism (Ragan & Chapman 1978). Biosynthetic intermediates may or may not be accumulated in detectable quantities; for example, squalene is rarely encountered in seaweeds (*Fucus spiralis, F. vesiculosus, Bryocladia cuspidata* are exceptions), and cycloartenol has been reported infrequently (*Chondrus crispus, Hypnea japonica, Palmaria palmata* (= *Rhodymenia palmata*), *Plocamium telfairiae, Enteromorpha* spp., *Ulva lactuca, Fucus spiralis*). Sterol composition may vary seasonally within an individual species (Idler *et al.* 1968; Idler & Atkinson 1976).

Although there is an extensive literature on seaweed sterols, our understanding of their occurrence and distribution is far from complete as a result of all too frequently inadequate identification of isomeric sterols. Although positional isomers can usually be resolved by gas chromatography and identified by mass spectrometry, only with the recent application of ^1H- and ^{13}C-nuclear magnetic resonance spectroscopy to steroid identification (Wright *et al.* 1978 and references therein) has direct identification of diastereomeric sterols become a routine matter. It is of great importance to bear this situation in mind when examining the relevant literature.

Sterols have been demonstrated in all red seaweeds investigated (approximately 120 species). C_{27} sterols such as cholesterol (Fig. 17.20), desmosterol (Fig. 17.21) and 22-dehydrocholesterol are almost always predominant;

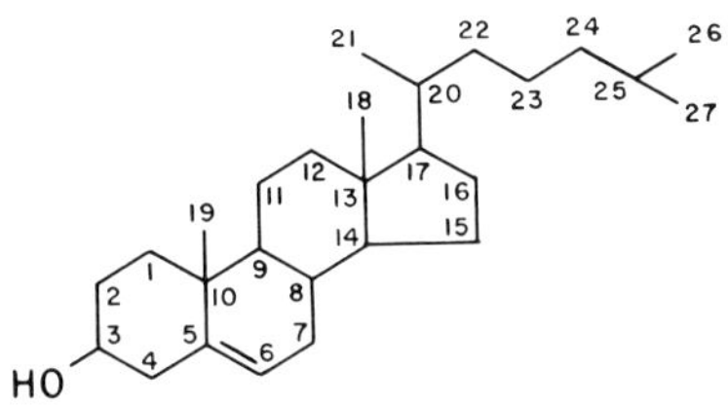

Fig. 17.20 Cholesterol.

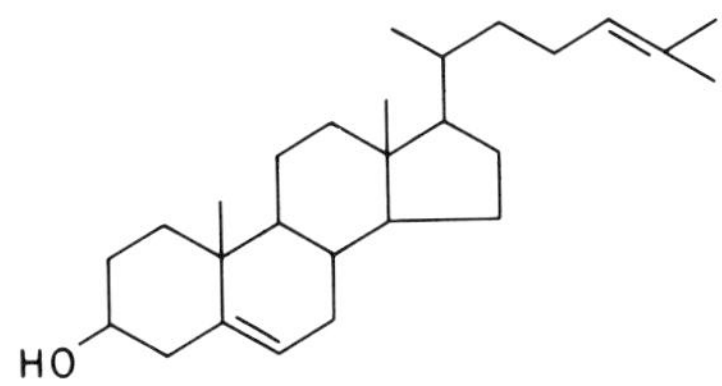

Fig. 17.21 Desmosterol.

Sphaerococcus coronopifolius seems to be unique in accumulating no detectable cholesterol (Fattorusso *et al.* 1976). Other phytosterols which may be present as minor constituents include campesterol (Fig. 17.22) and/or its isomer 22-dihydrobrassicasterol; β-sitosterol (Fig. 17.23) and/or its 24-*S* epimer clionasterol; stigmasterol (Fig. 17.24); brassicasterol (Fig. 17.25) and/or its 24-*S* epimer

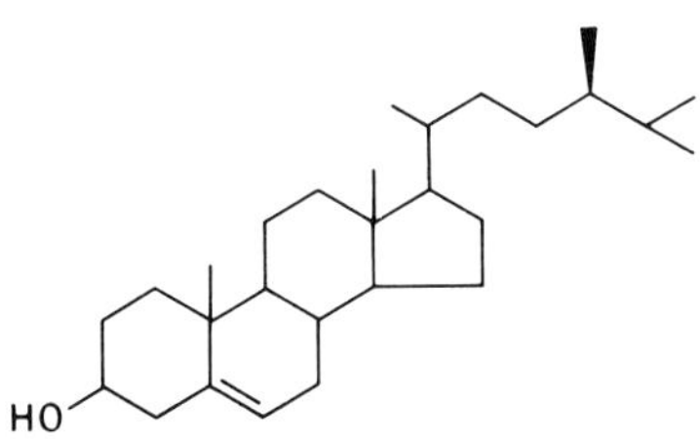

Fig. 17.22 Campesterol.

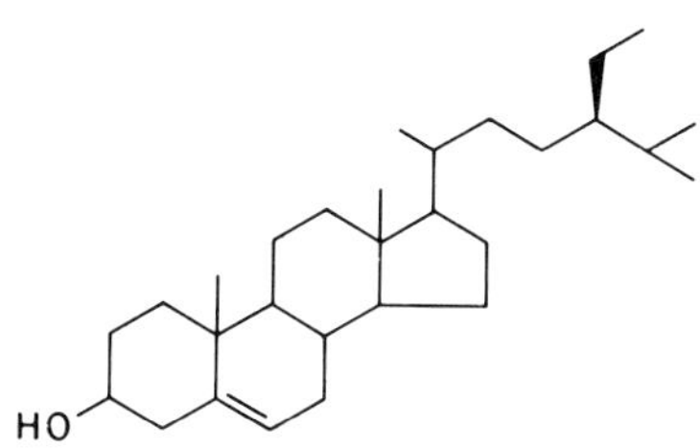

Fig. 17.23 β-Sitosterol.

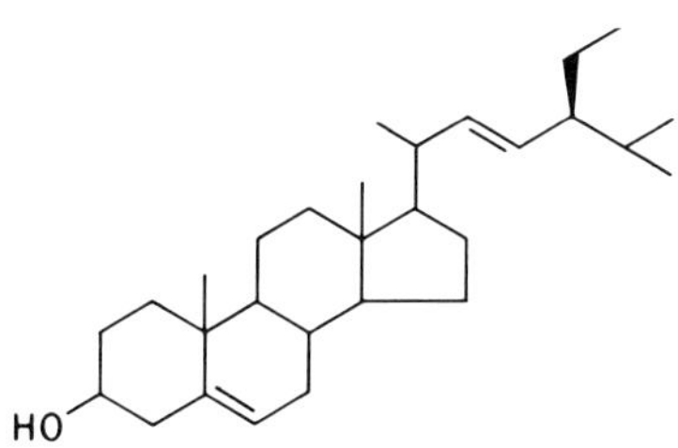

Fig. 17.24 Stigmasterol.

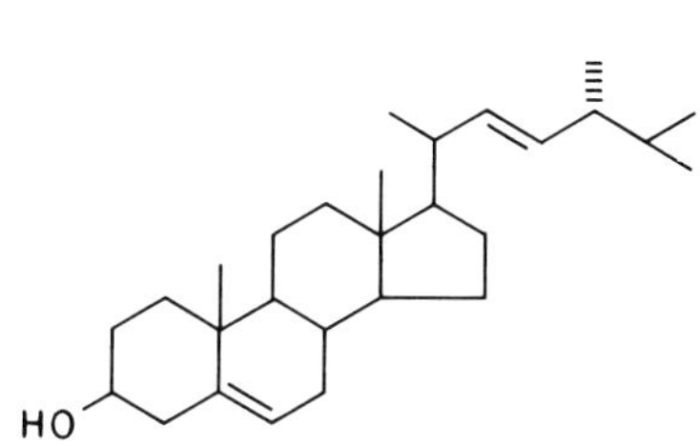

Fig. 17.25 Brassicasterol.

ergosterol; fucosterol (Fig. 17.26) and 28-isofucosterol (Fig. 17.27); 24-methylenecholesterol (Fig. 17.28); and liagosterol (Fig. 17.29) (*cis*- and/or *trans*-isomers). Δ^4-3-Ketosteroids have been reported in *Gracilaria textorii* and *Meristotheca papulosa* (Kanazawa & Yoshioka 1972).

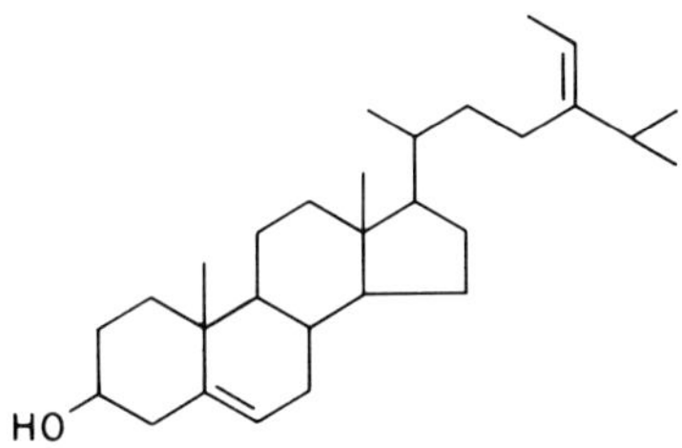

Fig. 17.26 Fucosterol.

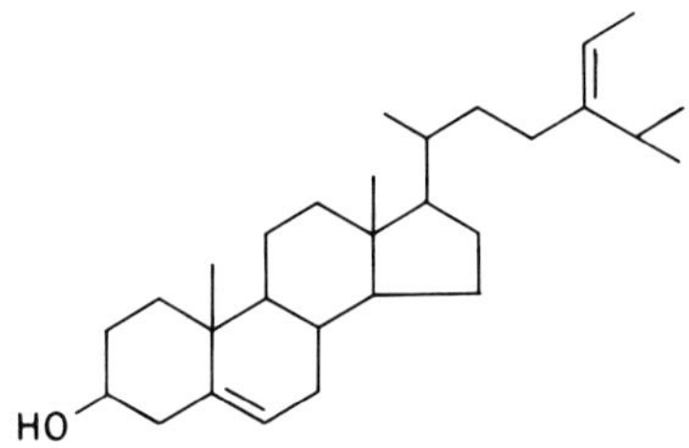

Fig. 17.27 28-Isofucosterol.

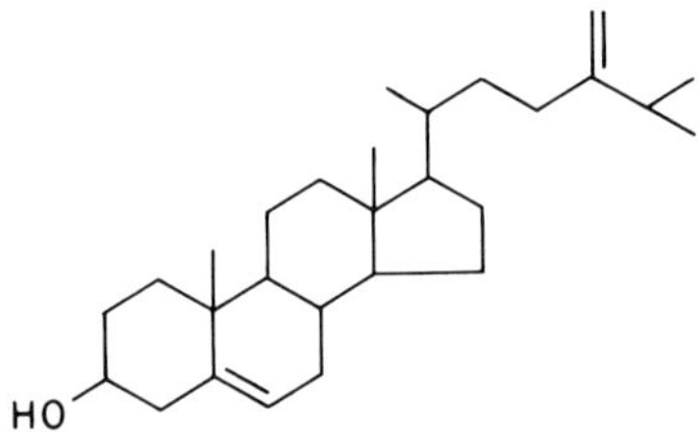

Fig. 17.28 24-Methylenecholesterol.

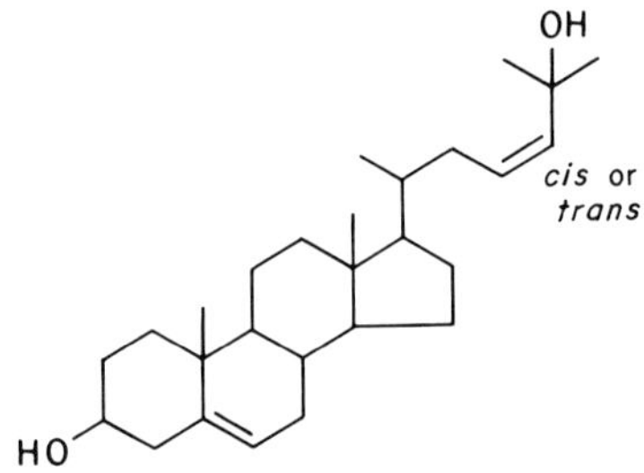

Fig. 17.29 Liagosterol.

The predominant sterol of all investigated brown algae (some 60 species) is fucosterol (Fig. 17.26), or in some cases possibly its $\Delta^{24[25]}$ isomer (Komura *et al.* 1974). Cholesterol (Fig. 17.20), 24-methylenecholesterol (Fig. 17.28), and 22-dehydrocholesterol are widespread among brown algae; brassicasterol (Fig. 17.25) or its 24-*S* epimer ergostenol, desmosterol (Fig. 17.21), stigmasterol (Fig. 17.24) or its 24-*R* epimer poriferasterol, a C_{27} cyclopropanoid 'cystosterol' (Fig. 17.30), and the probable fucosterol oxidation product saringosterol (Fig. 17.31)

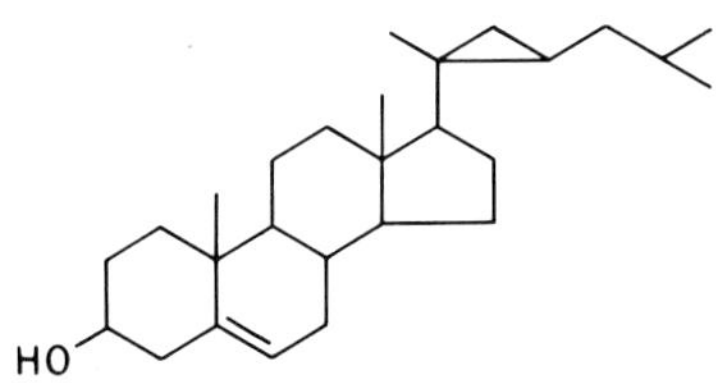

Fig. 17.30 Cystosterol.

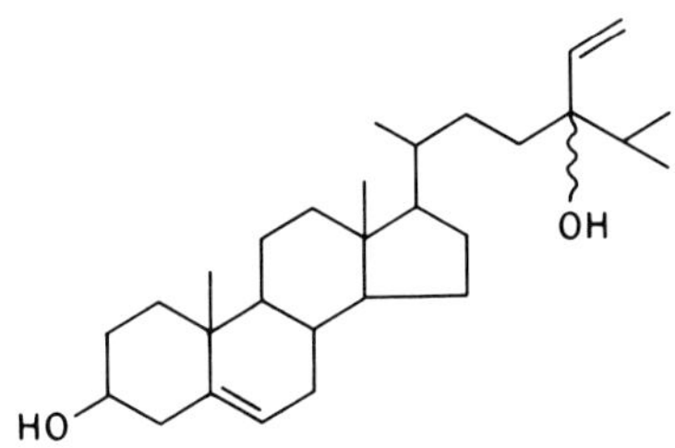

Fig. 17.31 Saringosterol.

are less frequently encountered. 'Sargasterol', a proposed C-20 isomer of fucosterol (Tsuda *et al.* 1958), may be an artifact (see Goodwin 1974b).

Sterol profiles of green seaweed are less predictable. Cholesterol (Fig. 17.20), β-sitosterol (Fig. 17.23), 24-methylenecholesterol (Fig. 17.28) and 28-isofucosterol (Fig. 17.27) appear to be present in most, although not all, green seaweeds. Campesterol (Fig. 17.22) and/or its C-20 isomer haliclionasterol, stigmasterol (Fig. 17.24) and/or its 24-*R* epimer poriferasterol, 22-dehydrocholesterol, desmosterol (Fig. 17.21), fucosterol (Fig. 17.26), 24-methylenecycloartenol, as well as 24-ethylidine and tentatively identified 24-methylenelophenols are less frequently reported. *Codium fragile* and *C. vermilara* differ considerably from other green seaweeds, accumulating only the unusual dienols clerosterol (Fig. 17.32) and codisterol (Fig. 17.33) (Dickson *et al.* 1979).

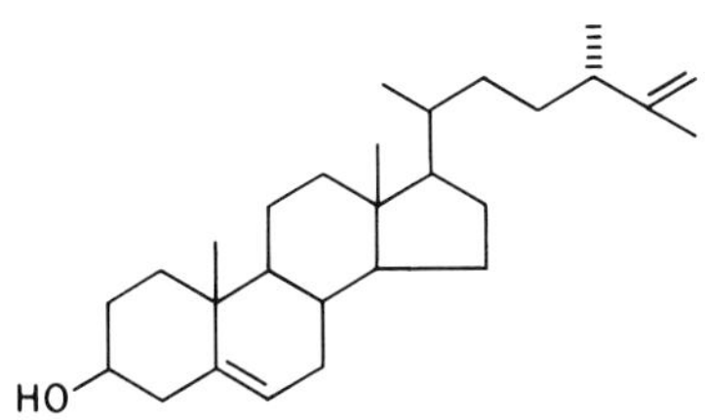

Fig. 17.32 Clerosterol.

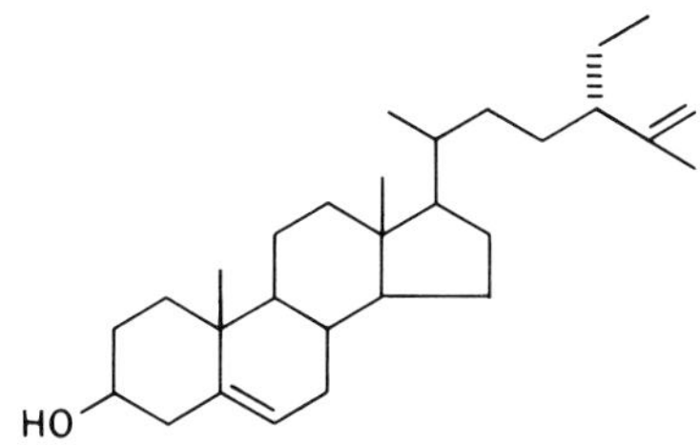

Fig. 17.33 Codisterol.

Further data on the distribution and biochemistry of seaweed sterols may be found in reviews by Goodwin (1973, 1974b), Amico *et al.* (1976), Fattorusso *et al.* (1976) and Ragan & Chapman (1978).

4 OTHER TERPENOIDS

There have been very few studies of the biosynthesis of algal terpenes. Thus classification of these metabolites must for the moment rely on a combination of comparative chemistry, classical terpene-biosynthetic proposals (Geissman & Crout 1969), and distributional data.

4.1 Monoterpenes

Monoterpenes (Figs 17.34–17.41) are C_{10} isoprenoids structurally related to neryl pyrophosphate (Fig. 17.34) or its isomers (Charlwood & Banthorpe 1978). Five structural classes of monoterpenes have been described among seaweeds; four of these occur in red seaweeds of the genera *Chondrococcus* (*Desmia*), *Microcladia, Ochtodes* and *Plocamium*.

The simplest structural class, the acyclic monoterpenes, is illustrated by cartilagineal (Fig. 17.35), costatol (Fig. 17.36), and a host of related di- and trienes from *Plocamium* spp. and *Chondrococcus hornemanni* (Mynderse & Faulkner 1975, 1978; Stierle & Sims 1979). Their biosynthesis presumably procedes from neryl pyrophosphate by additions of bromide and chloride, hydrogen halide abstractions, and double bond isomerizations (cf. Burreson *et al.* 1975). Cyclization of a halogenated alcohol intermediate would lead to cyclic dihydropyran monoterpenes such as costatone (Fig. 17.37) and the related metabolites of *P. costatum* (Kazlauskas *et al.* 1976; Stierle *et al.* 1976).

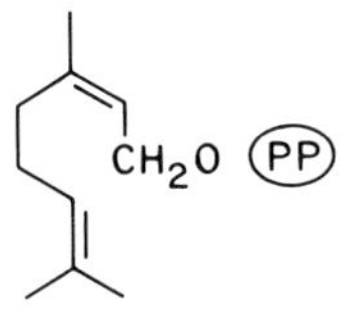

Fig. 17.34 Neryl pyrophosphate.

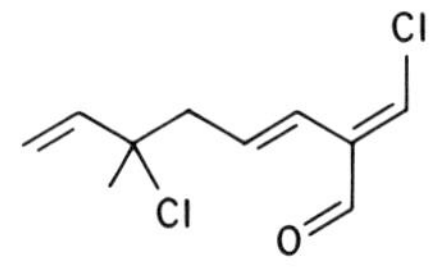

Fig. 17.35 Cartilagineal.

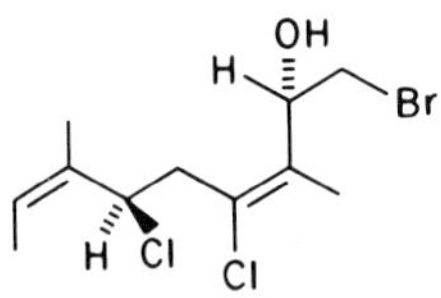

Fig. 17.36 Costatol.

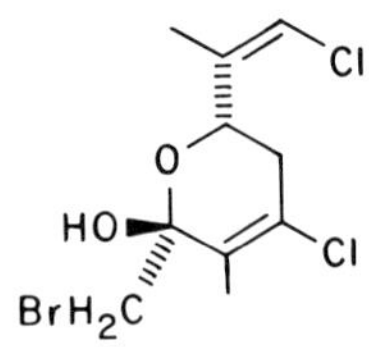

Fig. 17.37 Costatone.

Monoterpenes of a third structural class, the 1,1,5-trialkylcyclohexanes, arise from C-7→C-2 carbocyclization of acyclic halogenated precursors. Examples include violacene (Fig. 17.38) and the related plocamene D (Fig. 17.39), produced by *Plocamium* spp. (and also found in *Microcladia* spp.) (Crews *et al.* 1976); chondrocoles (e.g. Fig. 17.40) from *C. hornemanni* (Van Engen *et al.* 1978); and related monoterpenes from *P. cartilagineum* (Stierle & Sims 1979), *P. mertense* (Norton *et al.* 1977) and *Ochtodes secundiramea* (McConnell & Fenical 1978). 1,2,5-Trialkylcyclohexenes, such as plocamene B (Fig. 17.41) (Crews

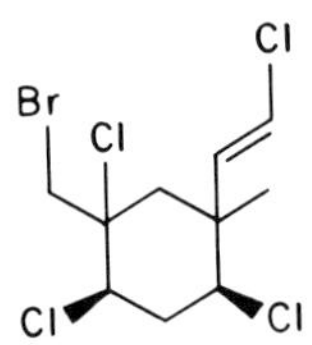

Fig. 17.38 Violacene.

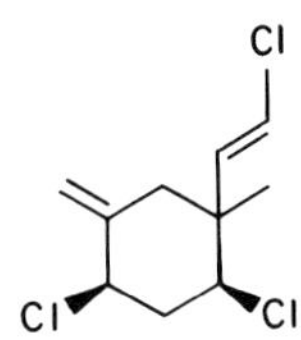

Fig. 17.39 Plocamene D.

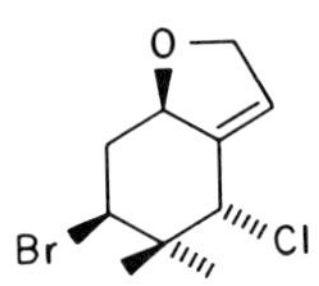

Fig. 17.40 Chondrocole A.

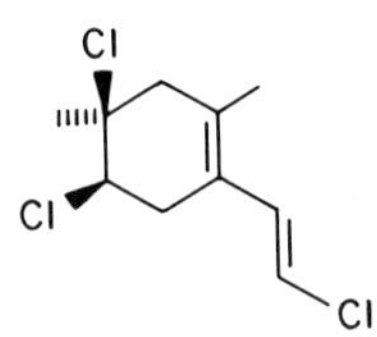

Fig. 17.41 Plocamene B.

1977), presumably arise from 1,1,5-trialkylcyclohexanes by *trans*-halovinyl migration.

A fifth structural class, monoterpene-substituted hydroquinones, is known only from the calcareous green seaweed *Cymopolia barbata* (Högberg *et al.* 1976). The more complex meroterpenoids cyclocymopol (Fig. 17.42) and cymopochromenol (Fig. 17.43) presumably arise from cymopol (Fig. 17.44) by cyclization and rearrangement. Monoterpenes of other green seaweeds and of brown seaweeds have not been closely examined (cf. Katayama 1964).

4.2 Sesquiterpenes

Sesquiterpenes (Figs 17.45–17.64) are C_{15} isoprenoids formally derived from *trans,cis*-farnesyl pyrophosphate (Fig. 17.45) or one of its isomers. The subsequent discussion draws heavily on the excellent review by Martín & Darias

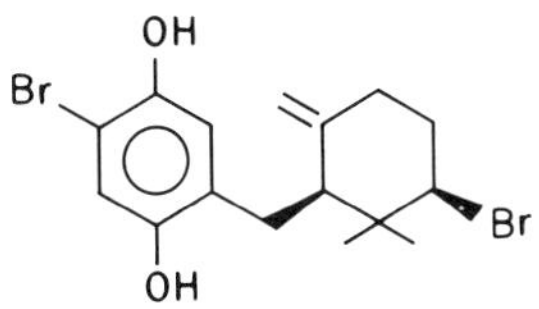

Fig. 17.42 Cyclocymopol.

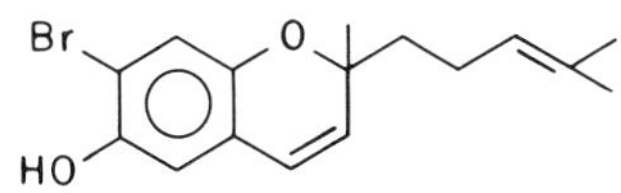

Fig. 17.43 Cymopochromenol.

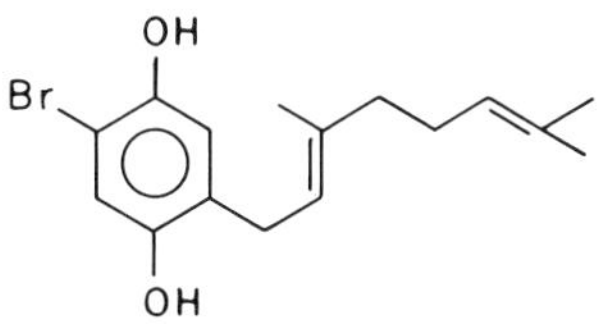

Fig. 17.44 Cymopol.

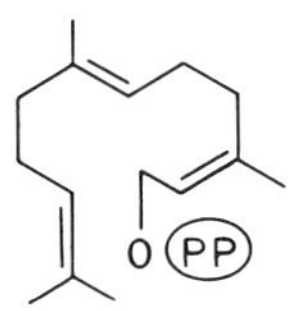

Fig. 17.45 *Trans,cis*-farnesyl pyrophosphate.

(1978), which should be consulted for further references to the primary literature.

Acyclic sesquiterpenes are uncommon in seaweeds. Exceptions include flexilin (Fig. 17.46) (*Caulerpa flexilis*) (Blackman & Wells 1978) and the related acetylene caulerpenyne (*C. prolifera*) (Amico *et al.* 1978), presumably derived from farnesol or nerolidol.

Monocyclization of *trans-cis*-farnesyl pyrophosphate and subsequent addition of bromide and chloride to an enzyme-bound bisabolonium ion leads to caespitol (Fig. 17.47), its *trans*-diaxial isomer isocaespitol (Fig. 17.48) (*Laurencia caespitosa, L. obtusa*), or related (iso-)furocaespitanes (*L. caespitosa*). Carbocyclization of a bromochloro intermediate may lead to chamigranes such as glanduliferol (Fig. 17.49), spirolaurenone (*L. glandulifera*), and 10-bromo-α-chamigrane (*L. pacifica*). In the case of the *trans*-diequatorial bromochloro isomer, further rearrangements produce pacifenol (Fig. 17.50) (*L. pacifica, L. nidifica*) and related metabolites of *L. pacifica, L. johnstonii, L. okamurai, L. filiformis, L. tasmanica* and *L. intricata*, or alternatively nidificene (Fig. 17.51) and the related sesquiterpenes of *L. nidifica*. The *L. perforata* metabolites perforene (Fig. 17.52) and perforatone (Fig. 17.53) may also be biosynthetically related.

Alternatively, rearrangement of *trans*-diaxial bromochloro bisabolonium intermediates leads to (+)-chamigranes such as elatol (Fig. 17.54) (*L. elata, L. obtusa*), obtusol (Fig. 17.55) (*L. obtusa*), or possibly certain cuparanes (α-bromocuparene (Fig. 17.56); *L. glandulifera, L. okamurai*). Other aromatic sesquiterpenes include rearranged chamigranes such as laurenisol (Fig. 17.57) (*L. nipponica*), laurinterol (Fig. 17.58) (*Laurencia* spp., *Marginisporum aberrans*), and aplysin (Fig. 17.59) (*L. okamurai, L. nidifica, M. aberrans*). Oxidative coupling dimers of laurinterol have been found in *L. decidua* (Caccamese & Rinehart 1978).

Nonhalogenated sesquiterpenes have received less attention. Cadalanes (e.g. cadalene, Fig. 17.60 (*Dictyopteris* spp.)) and an epoxidized muurolane (Fig. 17.61) from *Dilophus fasciola* (Amico *et al.* 1979) have been reported from brown algae, while selinanes (e.g. dictyopterone, Fig. 17.62, from *Dictyopteris divaricata*) have been reported from both brown and red seaweeds. The red seaweed *Chondria oppositiclada* produces the cyclopropane sesquiterpene cycloeudesmol (Fig. 17.63), and *L. poitei* the unusual compound in Fig. 17.64 (Fenical *et al.* 1978). Five known sesquiterpene-substituted hydroquinones (e.g. yahazunol, Fig. 17.65) (Ochi *et al.* 1979) and the sesquiterpene-substituted 4-hydroxybenzoic acid, zonaroic acid (Fig. 17.66) (Cimino *et al.* 1975) from *Dictyopteris undulata* (= *D. zonarioides*) are reminiscent of ubiquinones.

4.3 Diterpenes

Diterpenes (Figs 17.67–17.76) are C_{20} isoprenoids formally derivable from geranylgeranyl pyrophosphate (Fig. 17.67). The most abundant diterpene in seaweeds is 7*R*,11*R*-*trans*-phytol (Fig. 17.68), a component part of chlorophylls *a* (Fig. 17.1a) and *b* (Fig. 17.1b). *Cis*-phytol is accumulated by *Gracilaria andersoniana* (Sims & Pettus 1976). Other acyclic diterpenes include eleganolone (*Cystoseira elegans*) (Francisco *et al.* 1978), geranylgeranylglycerol (*Dilophus fasciola*) (Amico *et al.* 1977), and the antibiotic trifarin (*Caulerpa trifaria*) (Blackman & Wells 1978).

More complex diterpenoids may be considered to arise from selective cyclization of geranylgeraniol (Fenical 1978). A carotenoid-like C-15 → C-10 closure, for example, would lead to the retinol-like diterpene caulerpol (Fig. 17.69) found in *Caulerpa brownii*. A C-10 → C-2 cyclization would lead to the antibiotic dictyodial (Fig. 17.70) (*Dictyota flabellata*) and thence to acetoxycrenulatin (Fig. 17.71) (*D. crenulata*) and other *Dictyota* metabolites. Alternatively, C-10 → C-1 closure would lead to dilophol (Fig. 17.72) (*Dilophus ligulatus*) and to the related metabolites of *D. prolificans* (Fenical 1978; Kazlauskas *et al.* 1978). Similarly, pachydictyol A (Fig. 17.73) and related hydroazulenes of the Dictyotaceae (*Pachydictyon coriaceum, Dictyota* spp., *Dilophus ligulatus, Dic-*

tyopteris membranacea) may be considered to arise from further C-2 → C-6 cyclization of a dilophol-like intermediate.

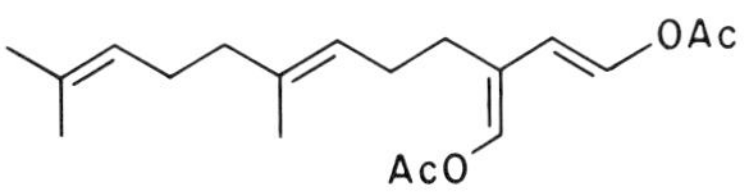

Fig. 17.46 Flexilin.

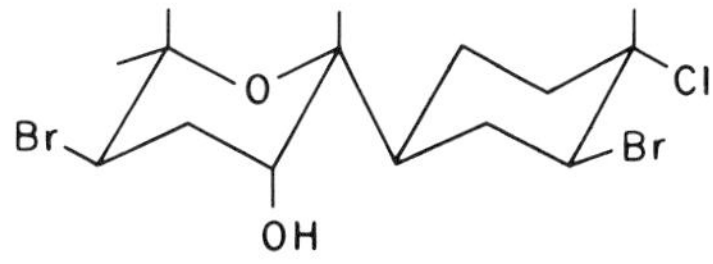

Fig. 17.47 Caespitol.

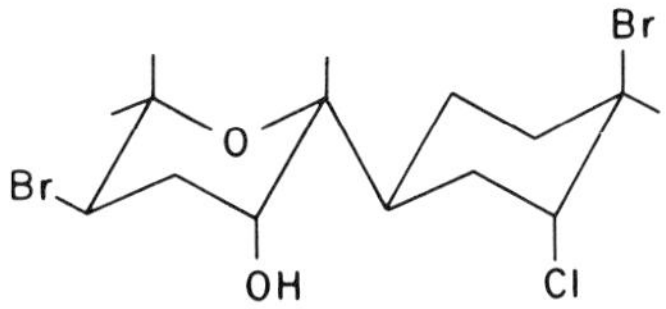

Fig. 17.48 Isocaespitol.

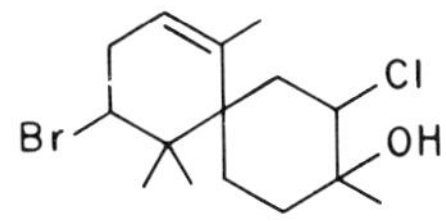

Fig. 17.49 Glanduliferol.

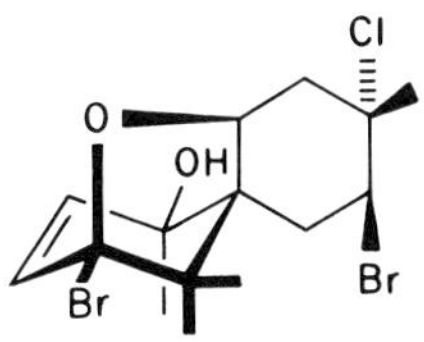

Fig. 17.50 Pacifenol.

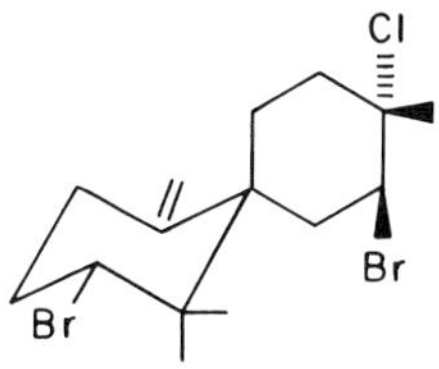

Fig. 17.51 Nidificene.

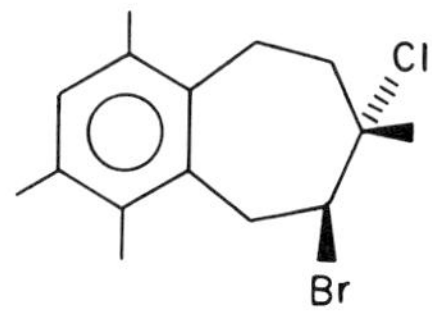

Fig. 17.52 Perforene.

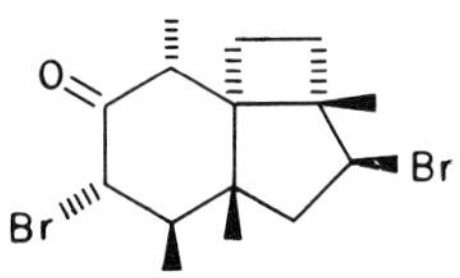

Fig. 17.53 Perforatone.

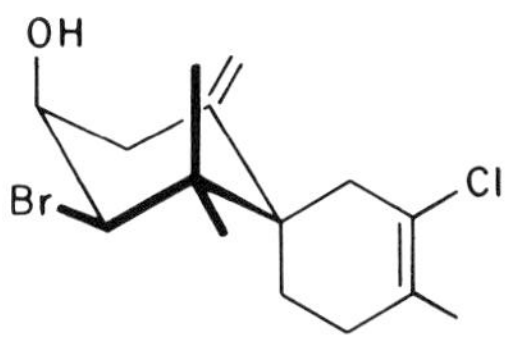

Fig. 17.54 Elatol.

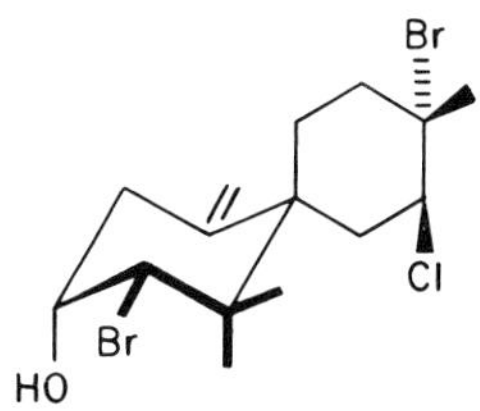

Fig. 17.55 Obtusol.

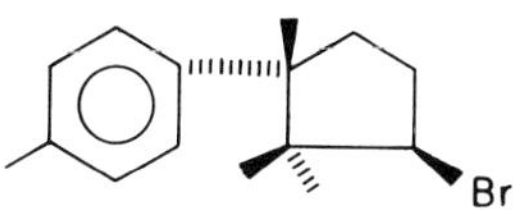

Fig. 17.56 α-Bromocuparene.

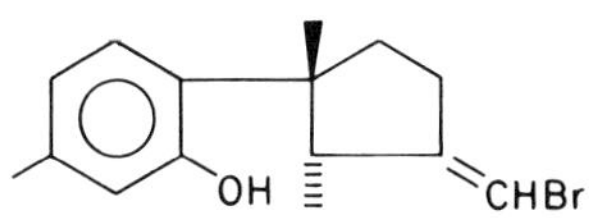

Fig. 17.57 Laurenisol.

Brominated diterpenes isolated from red seaweeds include concinndiol (Fig. 17.74) (*Laurencia concinna, L. snyderae*); metabolites such as irieol A (Fig. 17.75) from *L. irieii*; and bromosphaerol (Fig. 17.76) and related diterpenes from *Sphaerococcus coronopifolius* (Fenical 1978).

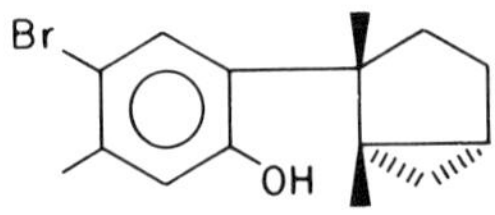

Fig. 17.58 Laurinterol.

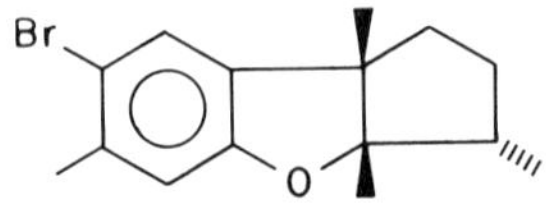

Fig. 17.59 Aplysin.

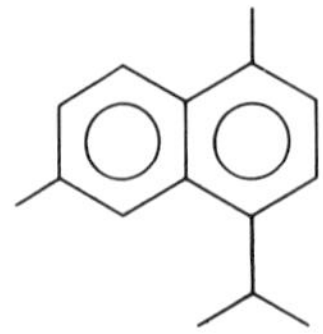

Fig. 17.60 Cadalene.

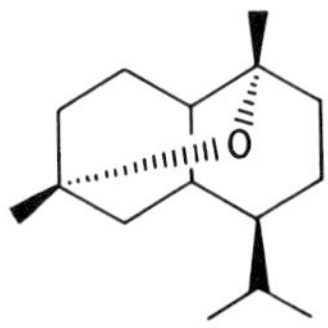

Fig. 17.61 4, 10-Epoxymuurolane.

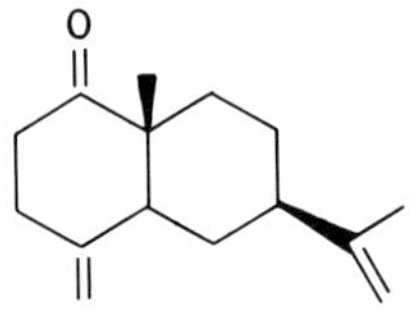

Fig. 17.62 Dictyopterone.

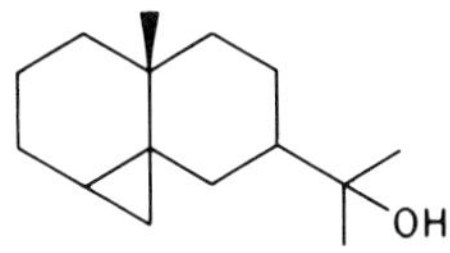

Fig. 17.63 Cycloeudesmol.

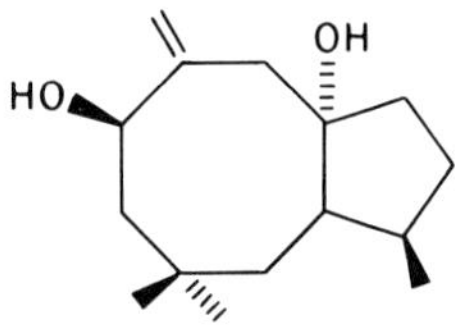

Fig. 17.64 Poitediol.

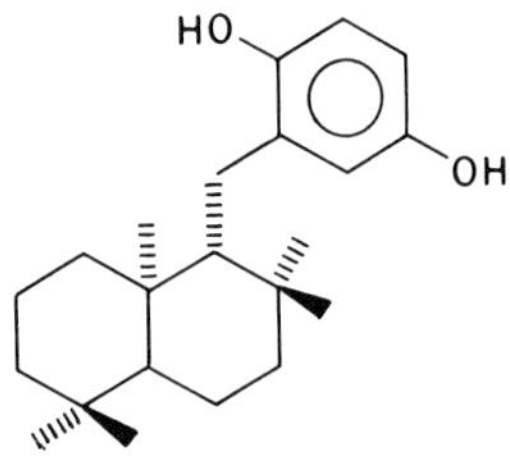

Fig. 17.65 Yahazunol.

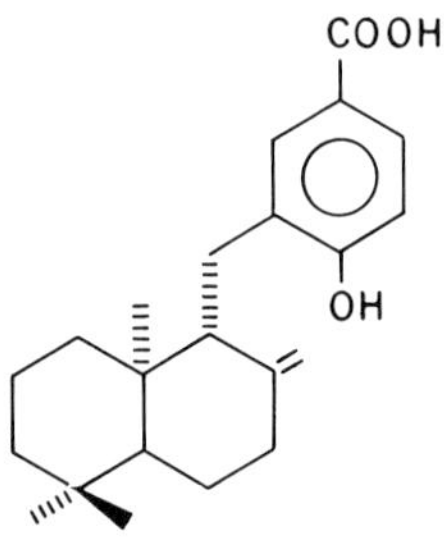

Fig. 17.66 Zonaroic acid.

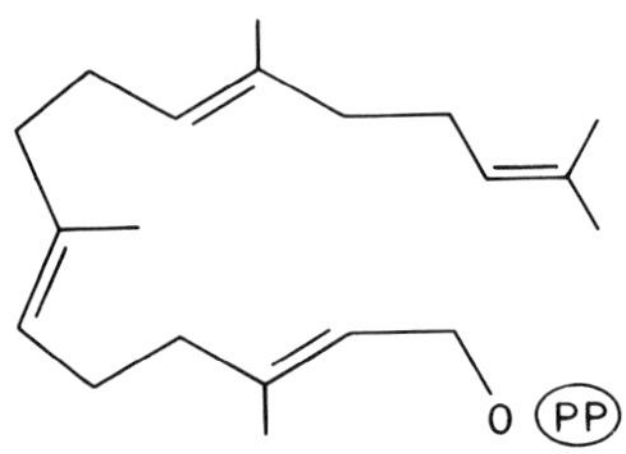

Fig. 17.67 Geranylgeranyl pyrophosphate.

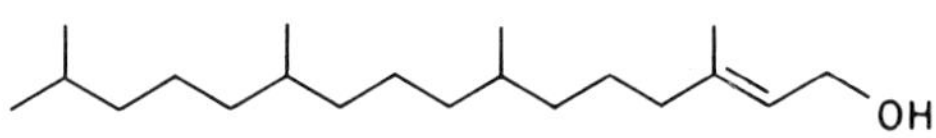

Fig. 17.68 7*R*, 11*R*-*trans*-phytol.

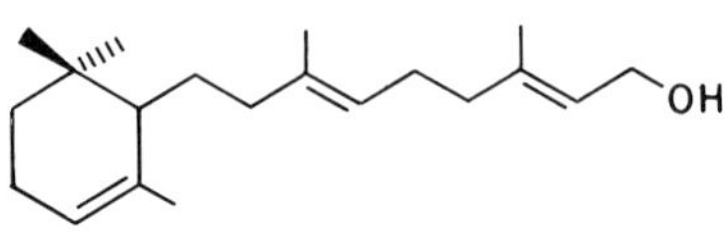

Fig. 17.69 Caulerpol.

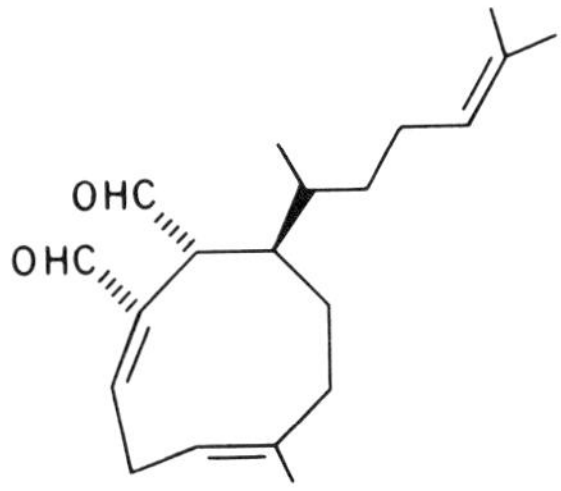

Fig. 17.70 Dictyodial.

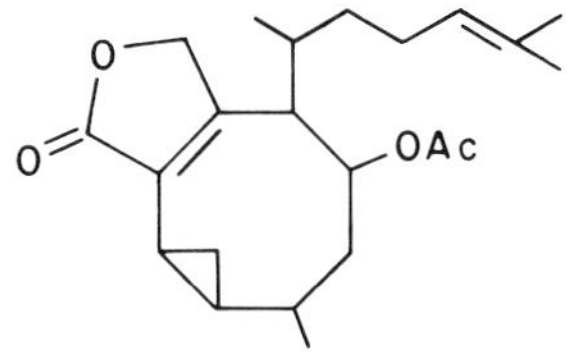

Fig. 17.71 Acetoxycrenulatin.

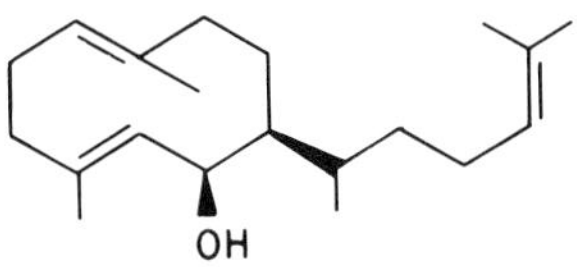

Fig. 17.72 Dilophol.

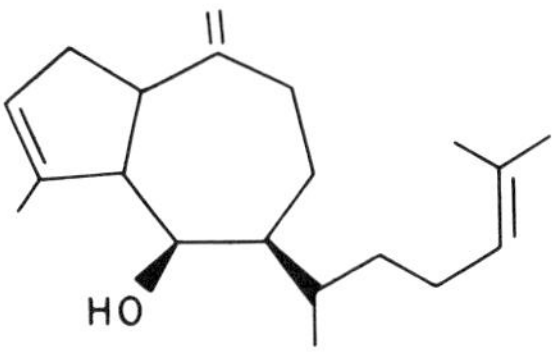

Fig. 17.73 Pachydictyol A.

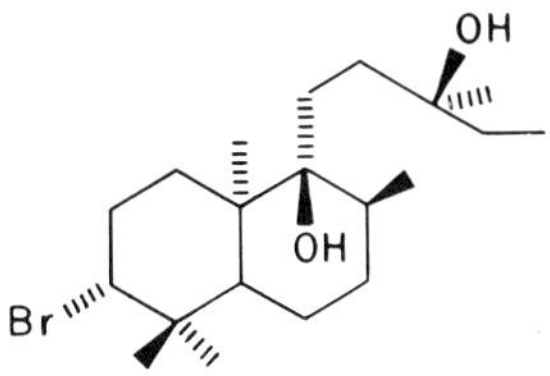

Fig. 17.74 Concinndiol.

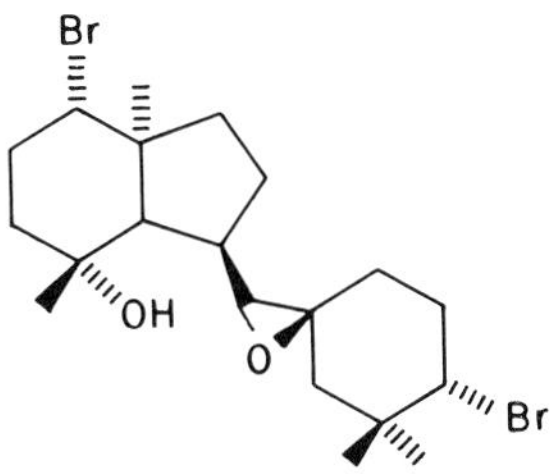

Fig. 17.75 Irieol A.

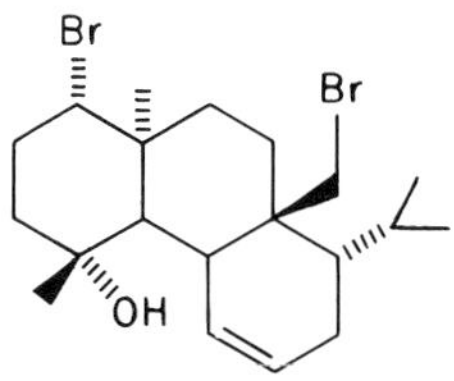

Fig. 17.76 Bromosphaerol.

Diterpene-substituted hydroquinone derivatives include: taondiol (Fig. 17.77) and a related carboxylic acid, atomaric acid, from *Taonia atomaria*; the coelenterate larval settling promoter δ-tocotrienol, and sargatriol from *Sargassum tortile* (Kato *et al.* 1975; Kikuchi *et al.* 1975); and stypotriol (Fig. 17.78) and related icthyotoxins from *Stypopodium zonale* (Gerwick *et al.* 1979).

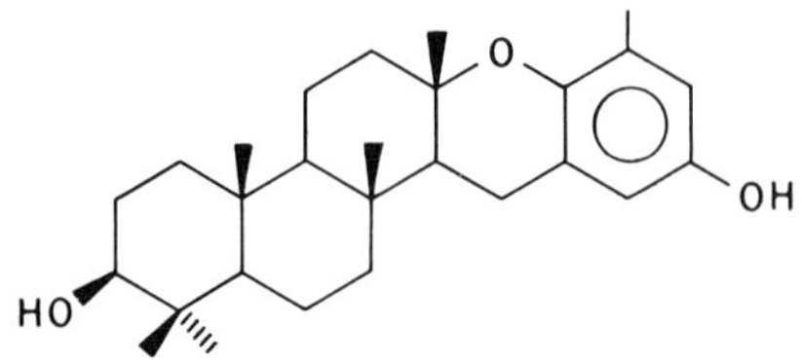

Fig. 17.77 Taondiol.

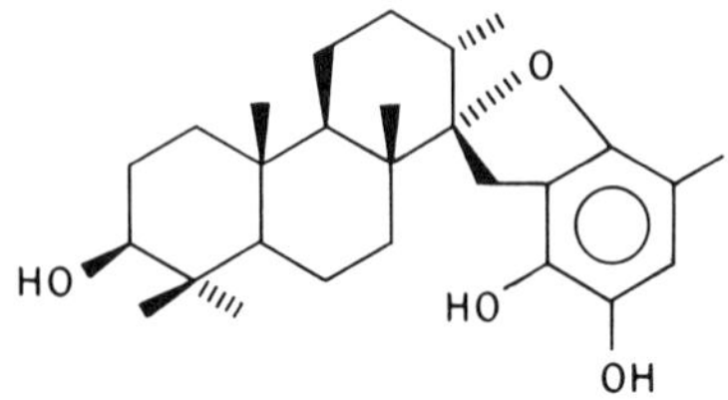

Fig. 17.78 Stypotriol.

5 FATTY ACIDS, LIPIDS, HYDROCARBONS, ACETYLENES, AND OTHER LOW MOLECULAR WEIGHT LIPOPHILIC METABOLITES

5.1 Fatty acids and lipids

Seaweeds usually accumulate rather small quantities of fatty acids and lipids. With few exceptions, the composition of their fatty acid fraction is qualitatively similar to that of other algae. Some interesting quantitative features exist: red seaweeds tend to be relatively enriched in highly unsaturated C_{20} fatty acids (20:4 ω6 and 20:5 ω3), and green seaweeds tend to accumulate relatively large quantities of *cis*-vaccenic (18:1 ω7), 16:4 ω3, and α-linolenic (18:3 ω3) acids (Ackman & McLachlan 1977; Johns *et al.* 1979). Much more extensive data have been summarized by Cerma & Stancher (1973), Erwin (1973), and Wood (1974).

The acyl lipids of seaweeds are for the most part unremarkable (Wood 1974). Exceptions include the lysogalactolipid (Fig. 17.79) and lysosulpholipid (Fig. 17.80) hemolysins of *Ulva pertusa* (Fusetani & Hashimoto 1975), and the incompletely characterized glycosyl-diglyceride sulphuric, sulphonic and phosphoric esters of *Fucus serratus, F. vesiculosus* and *Pelvetia canaliculata* (Pham Quang & Laur 1976a, b). Seaweeds apparently do not accumulate chlorosulpholipids (Mercer & Davies 1979).

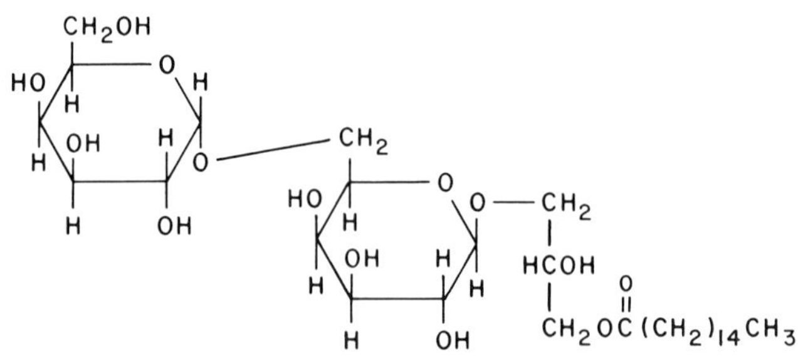

Fig. 17.79 Lysogalactolipid from *Ulva pertusa*.

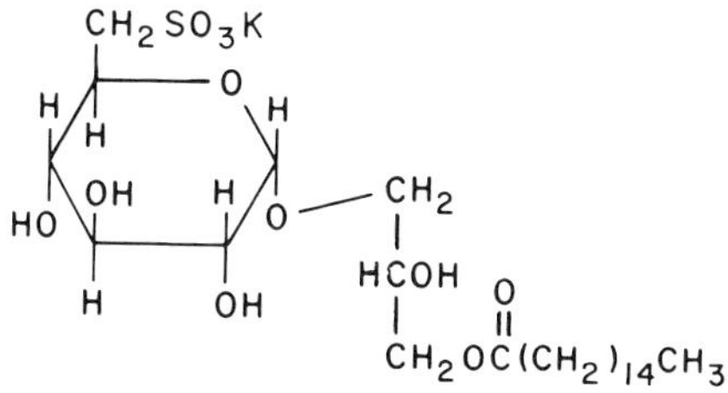

Fig. 17.80 Lysosulpholipid from *Ulva pertusa*.

5.2 Hydrocarbons

A variety of saturated and unsaturated hydrocarbons accumulate at low levels (usually $<500\ \mu g\ g^{-1}$ of dry matter) in seaweeds. Odd carbon number hydrocarbons (C_{15}, C_{17}, C_{21}) usually predominate, and probably arise by decarboxylation of the corresponding fatty acids (Lee & Loeblich 1971). *Cis*-3,6,9,12,15,18-heneicosahexaene has been reported in several green algae, whereas brown seaweeds seem to accumulate only the *cis*-1,6,9,12,15,18 isomer (Wright 1980). Phytadienes and aromatic hydrocarbons are minor constituents (Youngblood *et al.* 1971; Youngblood & Blumer 1973; Lytle *et al.* 1979). The carcinogenic polyaromatic hydrocarbon 3,4-benzopyrene, present in commercially available nori (*Porphyra* spp.) (Shirotori 1972; Shiraishi *et al.* 1973), may be an artifact of the drying process.

5.3 Nonterpenoid acetylenes

Acetylene cyclic ethers are known from eight species of *Laurencia*, and from *'Chondria oppositiclada'* (probably also a *Laurencia* sp.) (Moore 1978; White & Hager 1978). Examples include laurefucin (Fig. 17.81) from *L. nipponica*, obtusenyne (Fig. 17.82) from *L. obtusa* (King *et al.* 1979), *trans*- (Fig. 17.83) and *cis*-rhodophytin from a *Laurencia* sp. (Howard *et al.* 1980) and *cis*-maneonene A (Fig. 17.84) from *L. nidifica*.

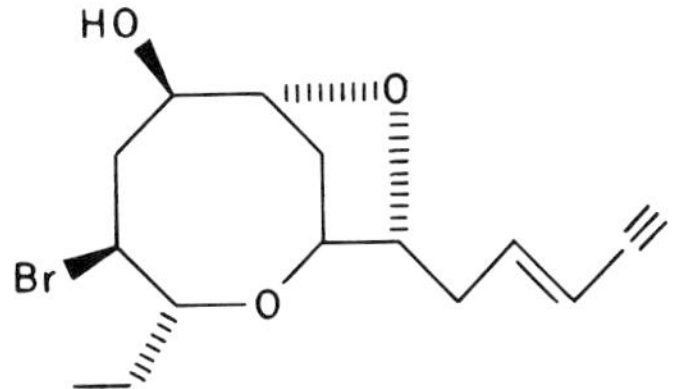

Fig. 17.81 Laurefucin.

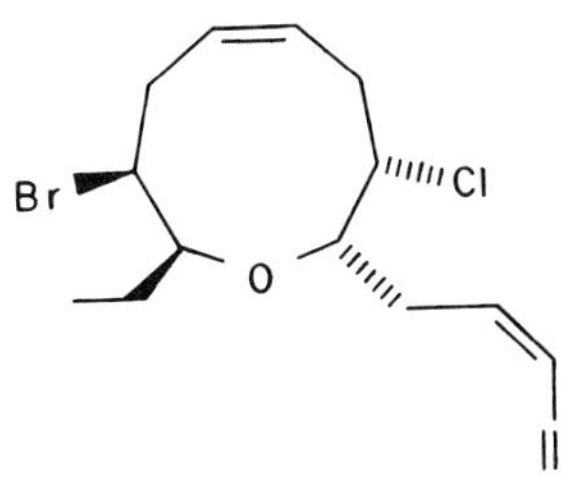

Fig. 17.82 Obtusenyne.

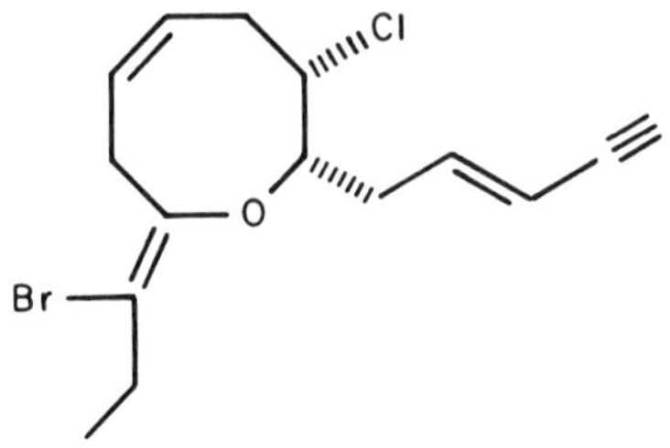

Fig. 17.83 *Trans*-rhodophytin.

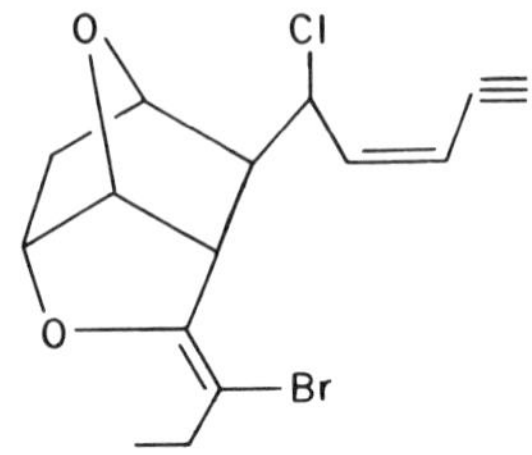

Fig. 17.84 *Cis*-maneonene A.

5.4 Other low molecular weight lipophilic metabolites

A wealth of low molecular weight antibiotics is produced by red seaweeds of the genera *Asparagopsis* (and its tetrasporophyte *Falkenbergia*), *Bonnemaisonia* (tetrasporophyte *Trailliella*), *Delisea, Plocamium* and *Ptilonia*. Among these compounds are methyl iodide, chloroform and other haloforms, carbon tetrabromide, carbon tetrachloride, iodoethanol, acetone, haloacetones, dibromoacetaldehyde, halopropenes and related epoxides and acetates, halobutenones, haloisopropanols, haloacetic acids, and haloacrylic acids (reviewed by Moore 1978). It is unlikely that specific halogenating enzymes exist for each of these metabolites; instead, many of these compounds are thought to be produced by relatively nonspecific haloperoxidases acting on carbon fragments derived, directly or indirectly, from acetate (McConnell & Fenical 1977; Moore 1978; Theiler *et al.* 1978).

A group of halogenated lactones (beckerelides, fimbrolides) related to the structure shown in Fig. 17.85 contribute to the antibiotic activity of *Beckerella subcostatum, Delisea* spp., *Marginosporum aberrans*, and *Ptilonia australasica* (Kazlauskas *et al.* 1978; Ohta 1979).

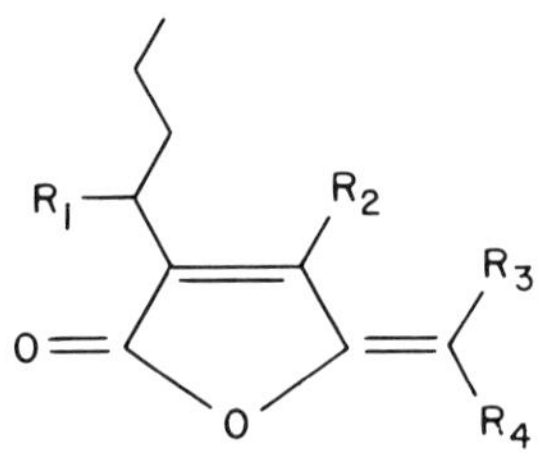

R_1 = H, OH or OAc

R_2–R_4 = Br, Cl, I, H combinations

Fig. 17.85 Halogenated lactones (showing alternative substitution patterns).

6 NONTERPENOID PHENOLS

The nonterpenoid phenols of eukaryotic algae may be divided into three structural, and probably biosynthetic, groups. Two of these (below) are represented in seaweeds; the third, green algal (gallo-?)tannins, are known only from the freshwater *Spirogyra arcta* (Nakabayashi 1955) and *Nitella* sp. (Stom & Apartsin 1976). The biosynthetic origin of the unusual phenolic antibiotic 3,5-dinitroguaiacol (Fig. 17.86) of *Marginosporum aberrans* (Ohta & Takagi 1977) is unknown.

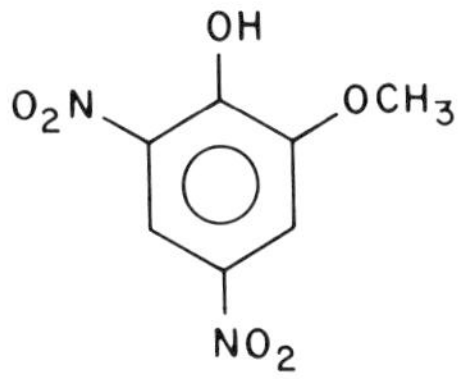

Fig. 17.86 3,5-Dinitroguaiacol.

6.1 Phenols related to tyrosine

Phenols formally derivable from tyrosine are found in all three classes of seaweeds. Phenols with an intact C_6–C_3 unit include 3-iodo- and 3,5-diiodotyrosine, triiodothyronine (Fig. 17.87), and thyroxine, present at low levels in many red and brown seaweeds (Ericson & Carlson 1953; Scott 1954; Ito *et al.* 1976). In addition, 3,5-dibromo-4-hydroxyphenylacrylic acid (Fig. 17.88) has been reported from *Halopytis incurvus* and *H. pinastroïdes* (Chantraine *et al.* 1973), and 4-hydroxyphenylpyruvic acid from *Odonthalia floccosa* (Manley & Chapman 1978) and *Undaria pinnatifida* (Abe *et al.* 1974).

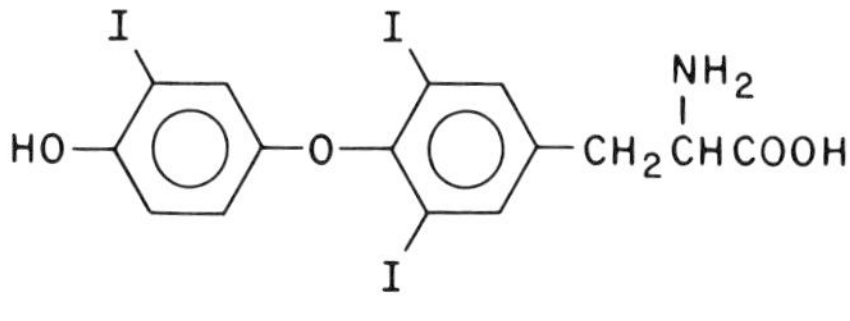

Fig. 17.87 Triiodothyronine.

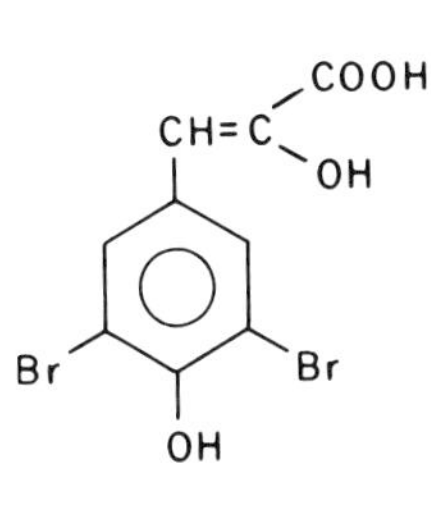

Fig. 17.88 3,5-Dibromo-4-hydroxyphenylacrylic acid.

Eight-carbon (C_6–C_2) phenols such as 4-hydroxymandelic acid and 4-hydroxyphenylacetic acid (*O. floccosa*) can arise by decarboxylation of a C_6–C_3 precursor (Manley & Chapman 1978). Other C_6–C_2 phenols include 3,4-dihydroxyphenylethylamine (dopamine) from *Monostroma fuscum* (Tocher & Craigie 1966) and 3,5-dibromo-4-hydroxyphenylacetic acid from *H. incurvus* and *H. pinastroïdes* (Chantraine *et al.* 1973). Related C_6–C_1 phenols include 4-hydroxybenzaldehyde in *Dasya pedicellata* var. *stanfordiana* (Fenical & McConnell 1976), *Marginosporum aberrans* (Ohta & Takagi 1977) and *O. floccosa*, as well as 4-hydroxybenzyl alcohol in *D. pedicellata*.

A variety of C_6–C_1-based brominated phenols (Table 17.2), bromochloro-

Table 17.2 Distribution of simple bromophenols in seaweeds*.

Species	Structures as in Figs.								
	17.90a	17.90b	17.91a	17.91b	17.92a	17.92b	17.93a	17.93b	17.94
Antithamnion plumula			+		+				
Bronghiartella byssoides			+		+		+		
Ceramium rubrum			+		+				
Corallina officinalis			+		+				
Cystoclonium purpureum						+			
Halopytis incurvus	+		+		+				
H. pinastroïdes	+		+						
Lenormandia prolifera					+		+		
Odonthalia corymbifera					+				
O. dentata	+		+	+	+	+	+		
O. floccosa		+	+	+	+	+	+		
Phycodrys rubens					+				
Polysiphonia brodiaei	+		+	+	+	+	+		
P. elongata				+	+	+			+
P. fruticulosa	+				+	+			
P. lanosa			+	+	+	+	+		+
P. morrowii			+						
P. nigra					+	+			
P. nigrescens	+		+	+	+	+	+		
P. thuyoides					+	+			
P. urceolata	+		+	+	+	+	+		
P. violacea					+	+			
Rhodomela confervoides					+	+	+		
R. larix					+	+			
R. subfusca					+	+			+
Rytiphlea tinctoria					+				
Symphyocladia latiuscula									+
Vidalia volubilis					+				
Fucus vesiculosus					+		+		

* Present in acid-hydrolyzed extracts. Various alkyl ethers or esters occasionally reported are artifacts. Data compiled from: Combaut *et al.* (1978); Fenical (1975); Kurata & Amiya (1977); Landymore (1976); Manley & Chapman (1980); Pedersén (1978); Pedersén *et al.* (1974); Saenger *et al.* (1976); Weinstein *et al.* (1975).

phenols (Pedersén 1978) and bromodiphenylmethanes (Table 17.3) have been found in certain red seaweeds. *Halopytis pinastroïdes* presumably produces the remarkable structure in Fig. 17.89 (Combaut *et al.* 1978) from a seven-carbon precursor. Many of the simple bromophenols (Figs. 17.90–17.94) probably exist *in vivo* largely as the sulphate esters; lanosol (Fig. 17.92a), for example, is found predominantly as the 1′,4-disulphate ester. However, until the recent development of liquid chromatographic methods for analysis of phenol sulphates (Ragan & MacKinnon 1979), it has been very difficult to examine intact sulphate esters.

Claims of brown algal 'lignins' (Reznikov *et al.* 1978) warrant further investigation.

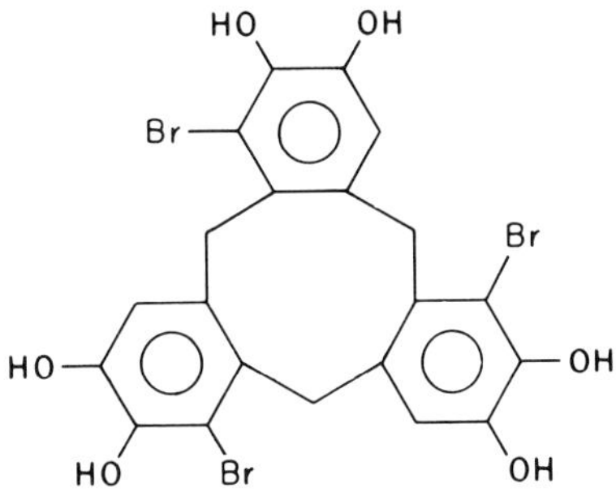

Fig. 17.89 Cyclotribromoveratrylene.

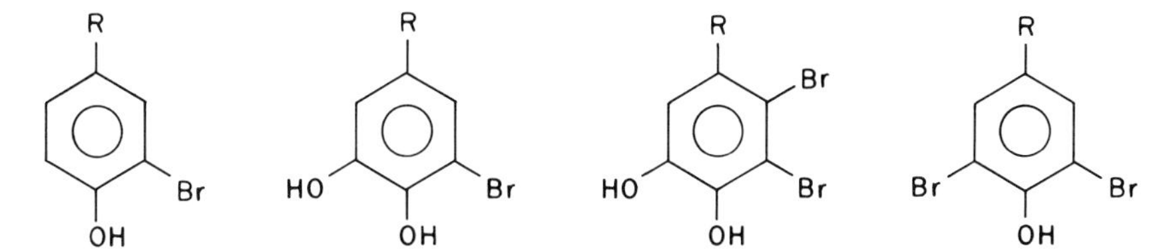

(90)	(91)	(92)	(93)
(a) R = CH_2OH	(a) R = CH_2OH	(a) R = CH_2OH	(a) R = CH_2OH
(b) R = CHO	(b) R = CHO	(b) R = CHO	(b) R = CHO

Fig. 17.90–17.93 Simple bromophenols produced by seaweeds.

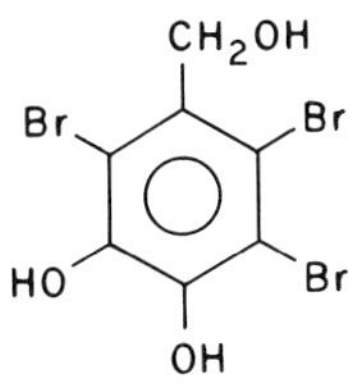

Fig. 17.94 2,3,6-Tribromo-4,5-dihydroxybenzyl alcohol.

Table 17.3 Bromodiphenylmethanes in seaweeds*.

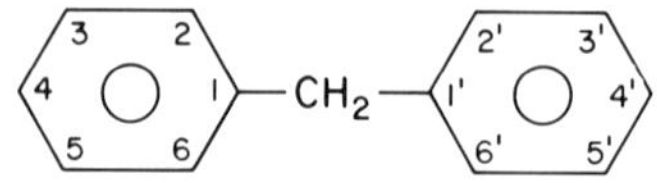

	C-2	3	4	5	6	2′	3′	4′	5′	6′	Seaweed
a	H	OH	OH	Br	Br	OH	Br	OH	Br	OH	*Rytiphlea tinctoria*
b	Br	Br	OH	OH	H	Br	OH	OH	H	OCH_3	*Rhodomela larix*
c	Br	Br	OH	OH	H	Br	Br	OH	OH	H	*Polysiphonia brodiaei, P. nigrescens, R. larix*

* Data compiled from Chevolot-Magueur *et al.* (1976); Kurata & Amiya (1977); Pedersén (1978).

6.2 Phenols related to phloroglucinol and resorcinol

Brown seaweeds often accumulate large quantities of tannins structurally based on phloroglucinol (Fig. 17.95). These phlorotannins are a mixture of polyphloroglucinols with molecular weights ranging to more than 10^5 (Ragan 1976). Phloroglucinol-based 'oligomers' with biphenyl linkages (Figs. 17.96 and 17.97), ether linkages (Figs. 17.98 and 17.99), and/or combinations thereof (e.g. Fig. 17.100) have been reported in various brown seaweeds, and are thought to be indicative of the structures of the higher molecular weight tannins* (Sattler 1974; Ragan 1976; Glombitza 1977). Chlorophloroglucinols (e.g. Fig. 17.101) have been found in *Laminaria ochroleuca* (Glombitza *et al.* 1977). 1,2,3,5-Tetrahydroxybenzene 2,5-disulphate (Fig. 17.102) and/or related sulphate esters occur in many brown algae (Ragan & Jensen 1979). The discovery of an alkenylphloroglucinol (Fig. 17.103) and related alkenylresorcinols in *Cystophora*

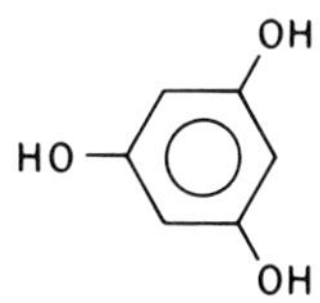

Fig. 17.95 Phloroglucinol.

torulosa (Gregson *et al.* 1977) suggests a polyketide biosynthetic origin for brown algal phloroglucinols.

Dibromophloroglucinol in the red seaweed *Rytiphlea tinctoria* (Chevolot-Magueur *et al.* 1976) is obviously related to the diphenylmethane (see Table 17.3 part *a*) of the same organism.

* The high molecular-weight polyphenols (and to a much lesser extent, the polyuronide, alginate) are responsible for the remarkable accumulations of heavy metal ions in brown seaweeds (Skipnes 1974; Ragan *et al.* 1979). The accumulation of heavy metal ions by brown seaweeds may make them useful as indicators of heavy metal pollution in coastal waters (Bryan & Hummerstone 1973).

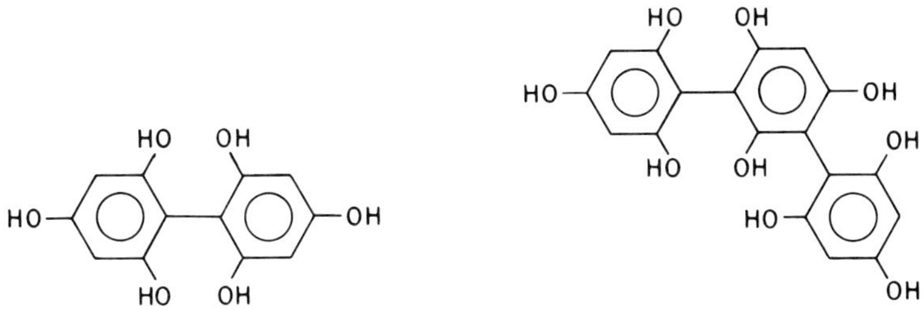

Fig. 17.96–17.97 Examples of polyphloroglucinols with biphenyl linkages.

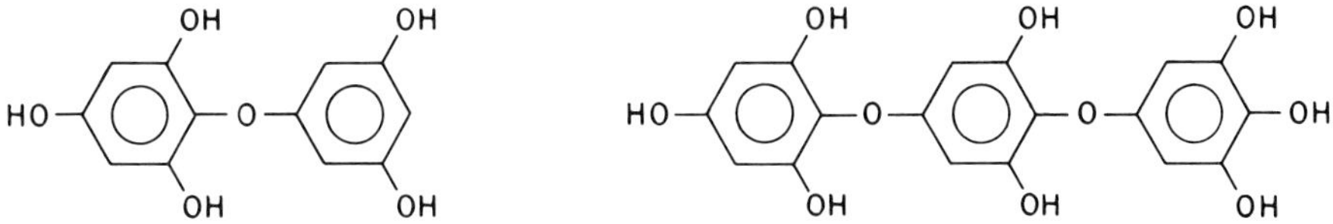

Fig. 17.98–17.99 Examples of polyphloroglucinols with ether linkages.

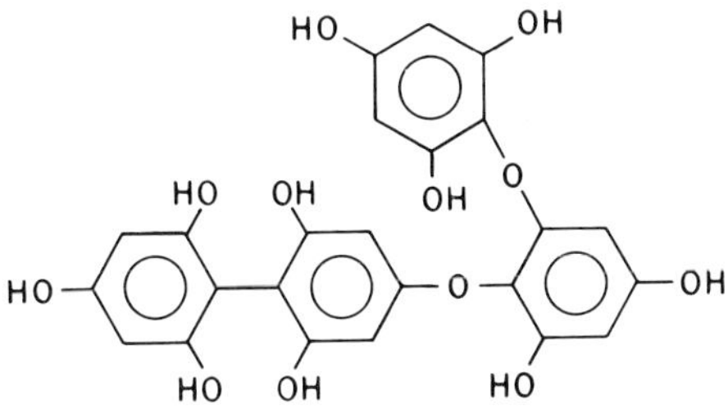

Fig. 17.100 Example of a polyphloroglucinol with both biphenyl and ether linkages.

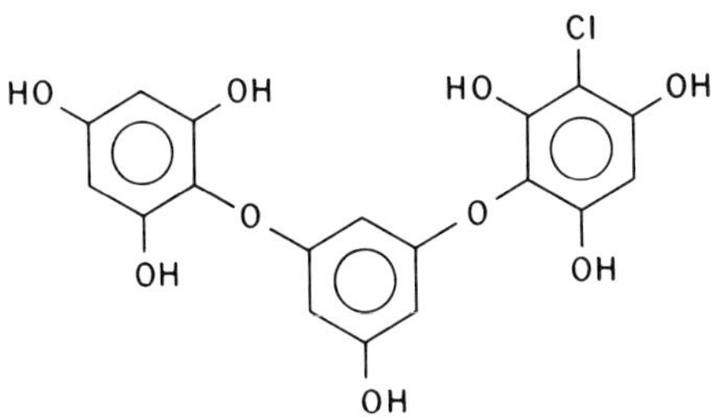

Fig. 17.101 Chlorinated polyphloroglucinol from *Laminaria ochroleuca*.

7 VITAMINS

Originally the term 'vitamin' referred to an essential factor required in small amounts in an animal's diet to prevent deficiency diseases. Many of these

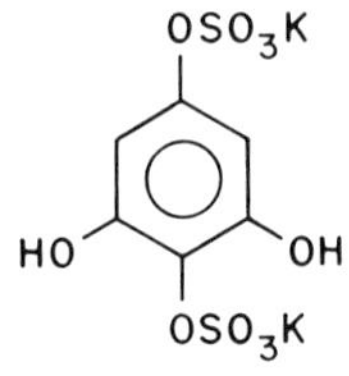

Fig. 17.102 1,2,3,5-Tetrahydroxybenzene 2,5-disulphate.

Fig. 17.103 An alkenylphloroglucinol.

factors are known in seaweeds, including ascorbic acid, cobalamins (e.g. vitamin B_{12}), biotin, niacin, and tocopherols (including vitamin E). Comprehensive reviews of vitamins in algae have been provided by Kanazawa (1963) and Provasoli & Carlucci (1974).

7.1 Ascorbic acid

Vitamin C is present in virtually all seaweeds (Rönnerstrand 1943; Créac'h 1960). Particularly rich sources include *Cladophora rupestris, Chordaria flagelliformis, Chorda filum, Halidrys siliquosa, Pilayella littoralis, Porphyra umbilicalis, P. tenera* (reported levels >5 mg g^{-1} of dry matter) and *Dictyopteris divaricata* (reportedly 10 mg g^{-1}). Seasonal maximum ascorbic acid concentrations in Norwegian *Ascophyllum nodosum, Fucus serratus* and *F. vesiculosis* occur in May and June (Jensen 1963), but earlier (December and January) in Adriatic *Pterocladia capillacea* (Liso *et al.* 1978).

7.2 Vitamin B_{12}

Vitamin B_{12} activity has been reported to be widespread among seaweeds, generally occurring in higher concentrations in green and red algae than in brown algae. These activities often show seasonal variation (Laminariales: Ericson & Lewis 1953; Karlström 1963; *Ulva rigida*: Suprunov *et al.* 1972). In *Rhodomela subfusca*, the vitamin B_{12} activity could be resolved into at least three fractions presumably corresponding to different cobalamins (Ericson & Lewis 1953). However, Ericson has suggested that epiphytic bacteria, and not the seaweeds themselves, biosynthesize the cobalamins.

7.3 Biotin

Substances with biotin activity are widespread among seaweeds (Larsen 1961; Kanazawa 1963); concentrations range from 18 to 740 ng g^{-1} of dry matter. Biotin sulphoxides (Fig. 17.104), as well as biotin itself, have been demonstrated in brown seaweeds (Larsen 1961).

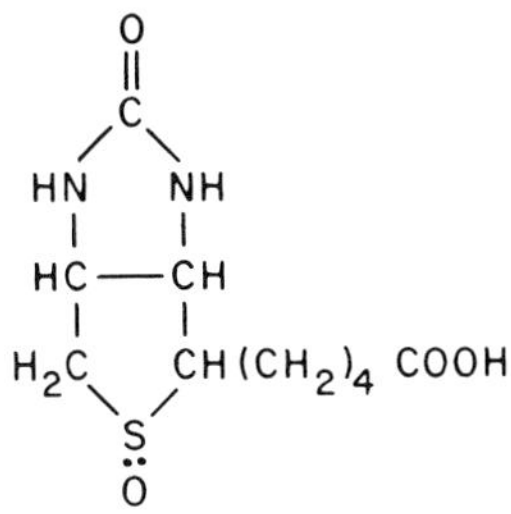

Fig. 17.104 Biotin sulphoxides.

7.4 Niacin

Niacin (nicotinic acid) occurs widely among seaweeds in concentrations up to 68 μg g^{-1} dry matter (Larsen 1958; Kanazawa 1963). The niacin contents of Norwegian *Ascophyllum nodosum, Fucus serratus, Laminaria digitata, L. hyperborea* and *Palmaria palmata* show marked seasonal variations, with maxima usually occurring in April and May (Larsen 1958).

7.5 Tocopherols

Vitamin E is a mixture of tocopherols. α-Tocopherol (5,7,8-trimethyltocol, Fig. 17.105) is the only tocopherol known to occur in red, green, and (with the exception of the family Fucaceae) brown seaweeds. Fucaceae (*Fucus* spp., *Pelvetia canaliculata, Ascophyllum nodosum*) produce the γ-, δ-, and perhaps the β-isomers (7,8-dimethyl; 8-methyl; 5,8-dimethyl, respectively) in addition to α-tocopherol (Jensen 1969a). Concentrations range from 7 to 80 μg g^{-1} of dry matter (except Fucaceae, 250 to 650 μg g^{-1}), and show seasonal variations (Jensen 1969b).

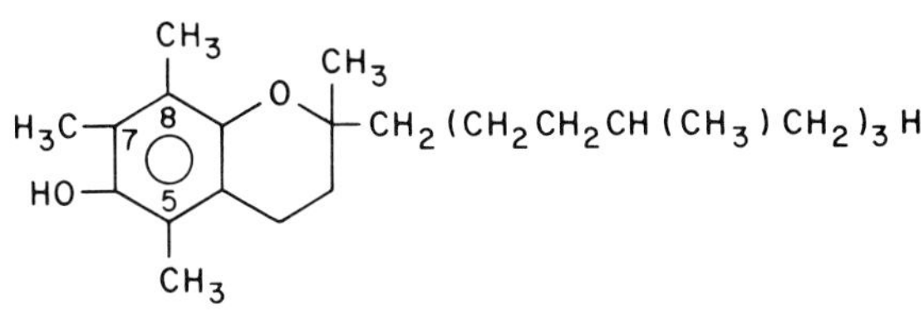

Fig. 17.105 α-Tocopherol.

7.6 Other vitamins

Other vitamins in seaweeds have received less attention. Most of the literature, including Ericson's investigations of pantothenic acid and folates, has been summarized by Kanazawa (1963). Sun *et al.* (1968) have more recently examined vitamin K in red, brown, and green seaweeds.

8 NITROGENOUS COMPOUNDS

8.1 Amino acids and low molecular weight peptides

The proteinaceous (coded) amino acids of seaweeds are unremarkable, and have been reviewed elsewhere (Ito 1969; Madgwick 1971; Impellizzeri *et al.* 1977). 4-Hydroxyproline is widely distributed among brown and green seaweeds, and at least among the latter is often associated with the cell wall. Although many red algae seem not to accumulate hydroxyproline (Gotelli & Cleland 1968), small quantities have recently been reported in proteins from *Laurencia spectabilis* (Court & Taylor 1979).

Seaweeds produce a remarkable array of 'unusual' nonprotein amino acids, including chondrine (yunaine) (Fig. 17.106), gigartinine (Fig. 17.107), gongrine, L-baikiain (Fig. 17.108), methylated histidines, methionine sulphoxide (Ito

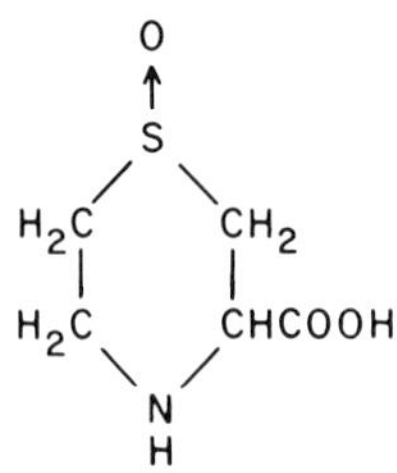

Fig. 17.106 Chondrine (yunaine).

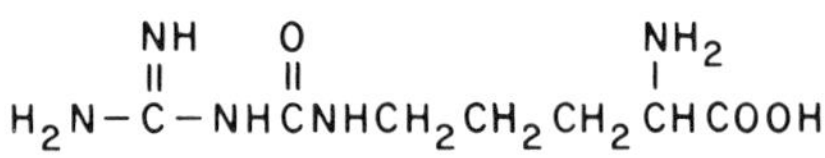

Fig. 17.107 Gigartinine.

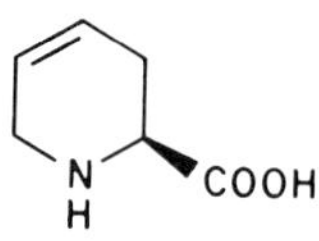

Fig. 17.108 L-baikiain.

1969; Miyazawa 1971; Impellizzeri *et al.* 1975), and D-homocysteic acid (Laycock *et al.* 1979). Low molecular weight peptides may also be accumulated: L-arginyl-L-glutamine in *Enteromorpha linza* and *Ulva pertusa* (Miyazawa *et al.* 1976); β-L-aspartylglycine in *Ceramium rubrum* (Sciuto *et al.* 1978); L-citrullinyl-L-arginine in several red seaweeds (Laycock & Craigie 1977); 'fastigiatine', L-(pyro-?)glutaminyl-α-L-glutaminyl-L-glutamine in *Pelvetia fastigiata* (Dekker *et al.* 1949); 'eisenine', L-(pyro-?)glutaminyl-L-glutaminyl-L-alanine in *Eisenia bicyclis* and *E. cava* (Konagaya 1966); and a partially characterized pentapeptide 'analipine', which contains one glutamate (or glutamine) and four aspartate (or asparagine) residues, in *Analipus japonicus* (Takagi *et al.* 1973). The high nitrogen content of many of these amino acids and peptides suggests a possible role in nitrogen storage (see Laycock & Craigie 1977).

8.2 Other low molecular weight nitrogenous compounds

Taurine (Fig. 17.109) is widely distributed among seaweeds; methylated taurines and related aminosulphonic acids have somewhat more restricted distributions (Ito 1969; Miyazawa *et al.* 1976; Fattorusso 1977). Trimethylamine and its *N*-oxide are also present in many red, brown and green seaweeds (Fujiwara-Arasaki & Mino 1972). Other nitrogenous compounds are known from only one or a few species: choline sulphate (Fig. 17.110) in *Gelidium*

$H_2NCH_2CH_2SO_3H$

Fig. 17.109 Taurine.

$(CH_3)_3\overset{+}{N}CH_2CH_2OSO_3^-$

Fig. 17.110 Choline sulphate.

cartilagineum (Lindberg 1955); the mycosporine palythine (Fig. 17.111) from *Chondrus yendoi* (Tsujino *et al.* 1978); polyhaloindoles from *Rhodophyllis membranacea* (Brennan & Ericson 1978) and *Laurencia brongniartii* (Carter *et al.* 1978); haloacetamides in *Asparagopsis taxiformis* (Woolard *et al.* 1976) and *Marginisporum aberrans* (Ohta 1979); a dihydrosphingosine diacetate (Fig.

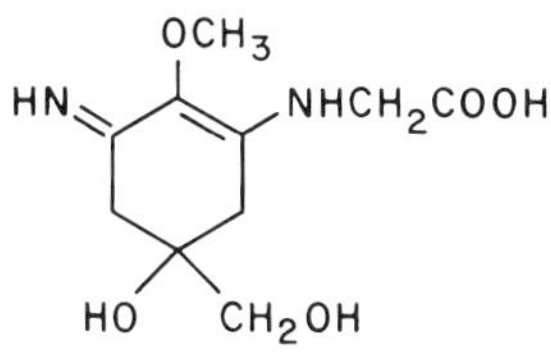

Fig. 17.111 Palythine.

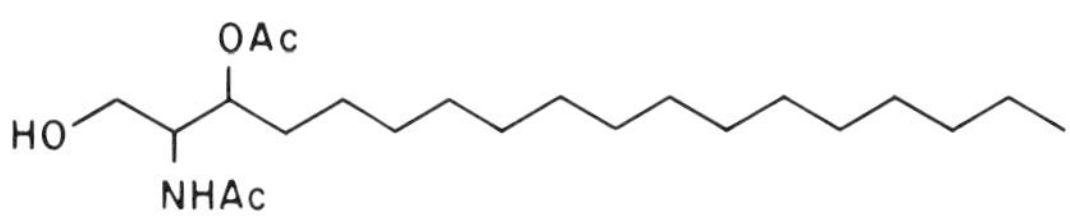

Fig. 17.112 Dihydrosphingosine diacetate.

17.112) from *Laurencia nidifica,* and *N*-acylsphingosines from *Amansia glomerata* (Cardellina & Moore 1978); and caulerpicins (Fig. 17.113) and caulerpin (Fig. 17.114) in *Caulerpa* spp. (Santos & Doty 1968; Maiti & Thomson 1977).

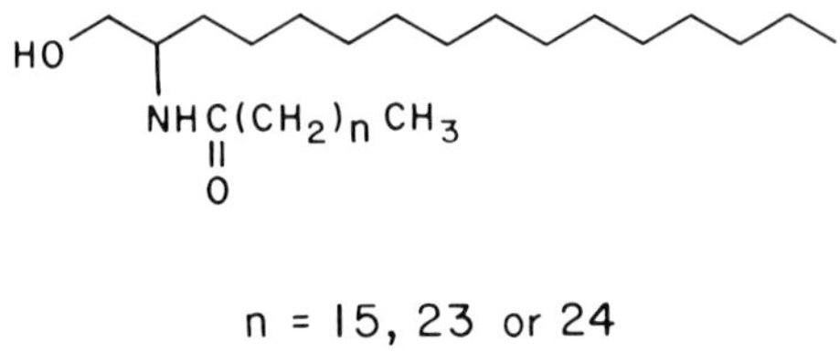

Fig. 17.113 Caulerpicins.

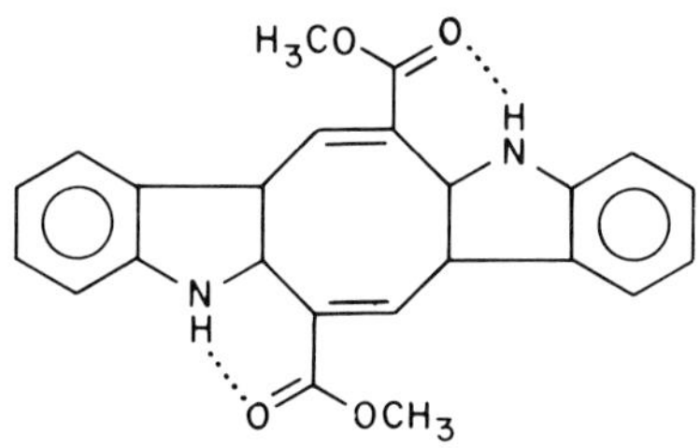

Fig. 17.114 Caulerpin.

9 POLYSULPHIDES AND POLYPHOSPHATES

A number of seaweeds (*Polysiphonia* spp., *Ceramium rubrum, Enteromorpha* spp., *Codium fragile, Monostroma* sp., *Spongomorpha* spp., *Cladophora* spp., *Ulva* spp., *Halidrys siliquosa, Pelvetia canaliculata*) liberate dimethyl sulphide when exposed to air (Haas 1935; Challenger *et al.* 1957; Katayama 1964). The immediate precursor to dimethyl sulphide is dimethyl-β-propiothetin (Fig. 17.115), a metabolite of methionine (Greene 1962).

$$(CH_3)_2\overset{+}{S}CH_2CH_2COO^-$$

Fig. 17.115 Dimethyl-β-propiothetin.

Dictyopteris australis and *D. plagiogramma* synthesize a cyclic disulphide (Fig. 17.116) and linear sulphides (e.g. Fig. 17.117) (Wratten & Faulkner 1976). The latter compounds, as well as the presumably related C_{11} sex hormones of certain brown algae (Chapter 19), are presumably derived from fatty acid

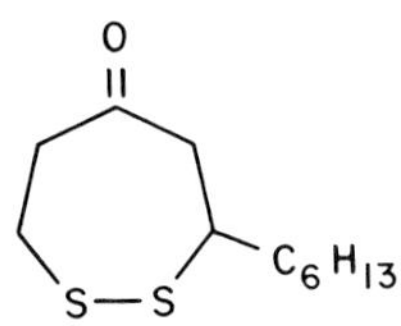

Fig. 17.116 Cyclic disulphide from *Dictyopteris* spp.

metabolism (Moore 1976). The red seaweed *Chondria californica* accumulates a series of cyclic polysulphide antibiotics, primarily lenthionine (Fig. 17.118), oxotrithiolanes (e.g. Fig. 17.119), and a sulphone (Fig. 17.120) (Wratten & Faulkner 1976).

Cyclic metaphosphates containing three to seven phosphate residues, and linear condensed oligo- and polyphosphates, serve as phosphorous reserves in a number of seaweeds, including *Ectocarpus siliculosus, Petalonia fascia, Pilayella littoralis, Ceramium deslongchampsii, C. rubrum, Rhodomela confervoides, Acetabularia crenulata, A. mediterranea,* and *Cladophora* sp. (Langen 1958; Niemeyer 1976).

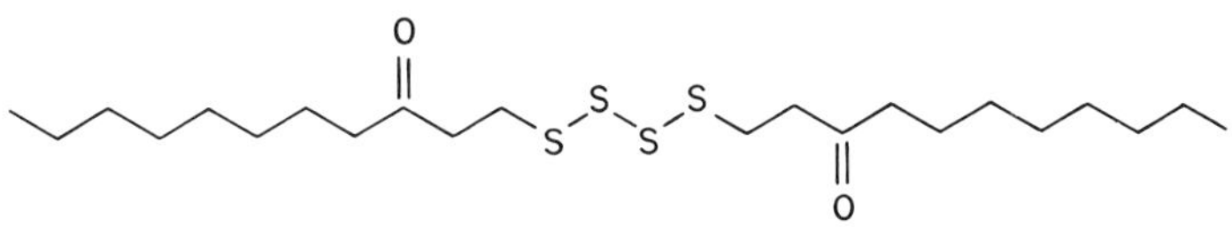

Fig. 17.117 A linear sulphide from *Dictyopteris*.

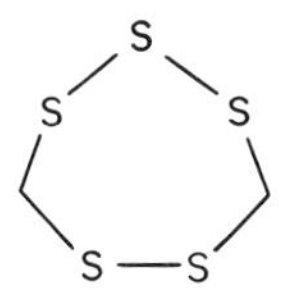

Fig. 17.118 Lenthionine.

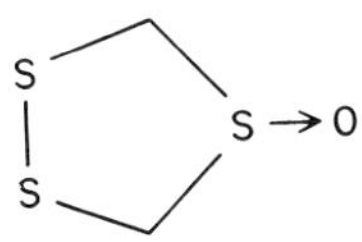

Fig. 17.119 An oxotrithiolane from *Chondria californica*.

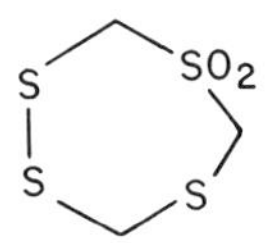

Fig. 17.120 Cyclic sulphone from *Chondria californica*.

10 ENVOI

Seaweeds are dynamic organisms in dynamic ecosystems. Many aspects of their chemical composition are likewise dynamic. A great deal of scattered and often preliminary data indicate that many—perhaps most—chemical constituents of seaweeds undergo diurnal and seasonal quantitative variations. Different populations of (apparently) the same species of seaweed may accumulate significantly different metabolites (Crews 1977; Mynderse & Faulkner 1978). Other variations have been reported to result from osmotic stress or from irradiation. However, relatively few of these dynamic aspects of seaweed chemistry have been adequately documented, and even fewer can be adequately explained. Moreover, very little is yet known of how—or why—most of these compounds are synthesized and metabolized. The chemical constituents of seaweeds will continue to provide challenges not only to chemists and biochemists, but also to physiologists, taxonomists and ecologists.

11 REFERENCES

ABE H., UCHIYAMA M. & SATO R. (1974) Isolation of phenylacetic acid and its *p*-hydroxy derivative as auxin-like substances from *Undaria pinnatifida*. *Agr. Biol. Chem.*, **38**, 897–8.

ACKMAN R. G. & MCLACHLAN J. (1977) Fatty acids in some Nova Scotian marine seaweeds: a survey for octadecapentaenoic and other biochemically novel fatty acids. *Proc. Nova Scotian Inst. Sci.*, **28**, 47–64.

AMICO V., ORIENTE G., PIATTELLI M., TRINGALI C., FATTORUSSO E., MAGNO S. & MAYOL L. (1977) (-)-(R)-1-O-Geranylgeranylglycerol from the brown alga *Dilophus fasciola*. *Experientia*, **3**, 989–90.

AMICO V., ORIENTE G., PIATTELLI M., TRINGALI C., FATTORUSSO E., MAGNO S. & MAYOL L. (1978) Caulerpenyne, an unusual sesquiterpenoid from the green alga *Caulerpa prolifera*. *Tetrahedron Lett.*, **1978**, 3593–6.

AMICO V., ORIENTE G., PIATTELLI M., TRINGALI C., FATTORUSSO E., MAGNO S. & MAYOL L. (1979) Sesquiterpenes based on the cadalane skeleton from the brown alga *Dilophus fasciola*. *Experientia*, **35**, 450–1.

AMICO V., ORIENTE G., PIATTELLI M., TRINGALI C., FATTORUSSO E., MAGNO S., MAYOL L., SANTACROCE C. & SICA D. (1976) Amino acids, sugars and sterols of some Mediterranean brown algae. *Biochem. Syst. Ecol.*, **4**, 143–6.

ARNESEN U., HALLENSTVET M. & LIAAEN-JENSEN S. (1979) More about the carotenoids of red algae. *Biochem. Syst. Ecol.*, **7**, 87–9.

BJØRNLAND T. (1974) Carotenoids in Red Algae, with Special Emphasis on the Claimed Content of Fucoxanthin. Thesis, University of Oslo, vol. I.

BJØRNLAND T. & AGUILAR-MARTINEZ M. (1976) Carotenoids in red algae. *Phytochemistry*, **15**, 291–6.

BLACKMAN A. J. & WELLS R. J. (1978) Flexilin and trifarin, terpene 1,4-diacetoxybuta-1,3-dienes from two *Caulerpa* species (Chlorophyta). *Tetrahedron Lett.* **1978**, 3063–4.

BRENNAN M. R. & ERICKSON K. L. (1978) Polyhalogenated indoles from the marine alga *Rhodophyllis membranacea* Harvey. *Tetrahedron Lett.* **1978**, 1637–40.

BRYAN G. W. & HUMMERSTONE L. G. (1973) Brown seaweed as an indicator of heavy metals in estuaries in south-west England. *J. Mar. Biol. Assoc. U.K.*, **53**, 705–30.

BURRESON B. J., WOOLARD F. X. & MOORE R. E. (1975) Evidence for the biogenesis of halogenated myrcenes from the red alga *Chondrococcus hornemanni*. *Chem. Lett.* **1975**, 1111–4.

CACCAMESE S. & RINEHART K. L. JR. (1978) New compounds and activities from *Laurencia* species. In *Drugs and Food from the Sea. Myth or Reality?*, eds. Kaul P. N. & Sindermann C. J., pp. 187–97. University of Oklahoma, Norman, Oklahoma.

CARDELLINA J. H. & MOORE R. E. (1978) Sphingosine derivatives from red algae of the Ceramiales. *Phytochemistry*, **17**, 554–5.

CARTER G. T., RINEHART K. L. JR., LI L. H., KUENTZEL S. L. & CONNOR J. L. (1978) Brominated indoles from *Laurencia brongniartii*. *Tetrahedron Lett.* **1978**, 4479–82.

CERMA E. & STANCHER B. (1973) Il contenuto in acidi grassi di alcune Phaeophyceae, Chlorophyceae e Rhodophyceae dell'alto Adriatico. *Rassegna Chim.*, **25**, 133–7.

CHALLENGER F., BYWOOD R., THOMAS P. & HAYWARD B. J. (1957) Studies on biological methylation. XVII. The natural occurrence and chemical reactions of some thetins. *Arch. Biochem. Biophys.*, **69**, 514–23.

CHANTRAINE J.-M., COMBAUT G. & TESTE J. (1973) Phenols bromes d'une algue rouge, *Halopytis incurvus*: acides carboxyliques. *Phytochemistry*, **12**, 1793–6.

CHAPMAN D. J. (1973) Biliproteins and bile pigments. In *The Biology of Blue-Green Algae*, eds. Carr N. G. & Whitton B. A., pp. 162–85. Blackwell Scientific Publications, Oxford.

CHARLWOOD B. V. & BANTHORPE D. V. (1978) The biosynthesis of monoterpenes. *Progr. Phytochem.*, **5**, 65–123.

CHEVOLOT-MAGUEUR A.-M., CAVE A., POTIER P., TESTE J., CHIARONI A. & RICHE C. (1976) Composés bromés de *Rytiphlea tinctoria* (Rhodophyceae). *Phytochemistry*, **15**, 767–71.

CIMINO G., DE STEFANO S., FENICAL W., MINALE L. & SIMS J. J. (1975) Zonaroic acid from the brown seaweed *Dictyopteris undulata* (= *zonarioides*). *Experientia*, **31**, 1250–1.

COMBAUT G., CHANTRAINE J.-M., TESTE J. & GLOMBITZA K.-W. (1978) Phenols bromes des algues rouges: cyclotribromoveratrylene, nouveau derive obtenu au cours de l'extraction du *Halopytis pinastroïdes*. *Phytochemistry*, **17**, 1791–2.

COURT G. J. & TAYLOR I. E. P. (1979) Isolation and partial characterization of a proteoglycan from the red alga *Laurencia spectabilis*. *Phytochemistry*, **18**, 411–4.

CRÉAC'H P. V. (1960) Distribution des acides citrique, malique et ascorbique dans différents tissus des algues marines. *Botaniste fasc.*, **I-IV**, 169–83.

CREWS P. (1977) The breadth of monoterpene synthesis by marine red algae: potential difficulties in their application as taxonomic markers. In *Marine Natural Products Chemistry*, eds. Faulkner D. J. & Fenical W. H., pp. 211–23. Plenum Press, New York and London.

CREWS P., NG P., KHO-WISEMAN E. & PACE C. (1976) Halogenated monoterpenes of the red alga *Microcladia*. *Phytochemistry*, **15**, 1707–11.

DEKKER C. A., STONE D. & FRUTON J. S. (1949) A peptide from a marine alga. *J. Biol. Chem.*, **181**, 719–29.

DICKSON L. G., PATTERSON G. W. & KNIGHTS B. A. (1979) Distribution of sterols in the marine Chlorophyceae. *Proc. Int. Seaweed Symp.*, **9**, 413–20.

ERICSON L.-E. & CARLSON B. (1953) Studies on the occurrence of amino acids, niacin and pantothenic acid in marine algae. *Ark. Kemi*, **6**, 511–22.

ERICSON L.-E. & LEWIS L. (1953) On the occurrence of vitamin B_{12}-factors in marine algae. *Ark. Kemi*, **6**, 427–42.

ERWIN J. (1973) Comparative biochemistry of fatty acids in eukaryotic microorganisms. In *Lipids and Biomembranes of Eukaryotic Microorganisms*, ed. Erwin J. A., pp. 41–143. Academic Press, London & New York.

FATTORUSSO E. (1977) Recent results in the chemistry of Mediterranean algae. In *Marine Natural Products Chemistry*, eds. Faulkner D. J. & Fenical W. H., pp. 165–78. Plenum Press, New York & London.

FATTORUSSO E., MAGNO S., SANTACROCE C., SICA D., IMPELLIZZERI G., MANGIAFICO S., PIATTELLI M. & SCIUTO S. (1976) Sterols of Mediterranean Florideophyceae. *Biochem. Syst. Ecol.*, **4**, 135–8.

FENICAL W. (1975) Halogenation in the Rhodophyta: a review. *J. Phycol.*, **11**, 245–59.

FENICAL W. (1978) Diterpenoids. In *Marine Natural Products. Chemical and Biological Perspectives*, vol. II, ed. Scheuer, P. J., pp. 173–245. Academic Press, London & New York.

FENICAL W. & MCCONNELL O. (1976) Simple antibiotics from the red seaweed *Dasya pedicellata* var. *stanfordiana*. *Phytochemistry*, **15**, 435–6.

FENICAL W., SCHULTE G. R., FINER J. & CLARDY J. (1978) Poitediol, a new nonisoprenoid sesquiterpene diol from the marine alga *Laurencia poitei*. *J. Org. Chem.*, **43**, 3628–30.

FRANCISCO C., COMBAUT G., TESTE J. & MAUME B. F. (1977) Étude des sterols d'algues brunes du genre *Cystoseira*. Identification par chromatographie gaz-liquid couplée a la spectrométrie de masse. *Biochim. Biophys. Acta*, **487**, 115–21.

FUJIWARA-ARASAKI T. & MINO N. (1972) The distribution of trimethylamine and trimethylamine oxide in marine algae. *Proc. Int. Seaweed Symp.*, **7**, 506–10.

FUSETANI N. & HASHIMOTO Y. (1975) Structures of two water soluble hemolysins isolated from the green alga *Ulva pertusa*. *Agr. Biol. Chem.*, **39**, 2021–5.

GEISSMAN T. A. & CROUT D. H. G. (1969) *Organic Chemistry of Secondary Plant Metabolism*. Freeman, San Francisco.

GERWICK W. H., FENICAL W., FRITSCH N. & CLARDY J. (1979) Stypotriol and stypolione; ichtyotoxins of mixed biogenesis from the marine alga *Stypopidium zonale*. *Tetrahedron Lett.* **1979**, 145–8.

GLAZER A. N. (1977) Structure and molecular organization of the photosynthetic accessory pigments of cyanobacteria and red algae. *Mol. cell. Biochem.*, **18**, 125–40.

GLAZER A. N. & HIXSON C. S. (1977) Subunit structure and chromophore composition of rhodophytan phycoerythrins. *Porphyridium cruentum* B-phycoerythrin and b-phycoerythrin. *J. Biol. Chem.*, **252**, 32–42.

GLOMBITZA K.-W. (1977) Highly hydroxylated phenols of the Phaeophyceae. In *Marine Natural Products Chemistry*, eds. Faulkner D. J. & Fenical W. H., pp. 191–204. Plenum Press, New York & London.

GLOMBITZA K.-W., KOCH M. & ECKHARDT G. (1977) Chlorierte Phlorethole aus *Laminaria ochroleuca*. *Phytochemistry*, **16**, 796–8.

GOODWIN T. W. (1960) Chemistry, biogenesis and physiology of the carotenoids. (Carotenoids associated with chlorophyll). In *Handbuch der Pflanzenphysiologie*, vol. 5 (1), ed. Pirson, A., pp. 394–443. Springer-Verlag, Berlin.

GOODWIN T. W. (1973) Comparative biochemistry of sterols in eukaryotic microorganisms. In *Lipids and Biomembranes of Eukaryotic Microorganisms*, ed. Erwin J. A., pp. 1–40. Academic Press, New York & London.

GOODWIN T. W. (1974a) Carotenoids and biliproteins. In *Algal Physiology and Biochemistry*, ed. Stewart W. D. P., pp. 176–205. Blackwell Scientific Publications, Oxford.

GOODWIN T. W. (1974b) Sterols. In *Algal Physiology and Biochemistry*, ed. Stewart W. D. P., pp. 266–80. Blackwell Scientific Publications, Oxford.

GOTELLI I. B. & CLELAND R. (1968) Differences in the occurrence and distribution of hydroxyproline-proteins among the algae. *Am. J. Bot.*, **55**, 907–14.

GREENE R. C. (1962) Biosynthesis of dimethyl-β-propiothetin. *J. Biol. Chem.*, **237**, 2251–4.

GREGSON R. P., KAZLAUSKAS R., MURPHY P. T. & WELLS R. J. (1977) New metabolites from the brown alga *Cystophora torulosa*. *Aust. J. Chem.*, **30**, 2527–32.

GYSI J. R. & CHAPMAN D. J. (in press) Phycobiliproteins and phycobilins. In *Handbook of Biosolar Resources*, vol. 1, ed. Black C. C. Jr. & Mitsui A. CRC Press, Cleveland.

HAAS P. (1935) The liberation of methyl sulphide by seaweed. *Biochem. J.*, **29**, 1297–9.

HAREL E. (1978) Chlorophyll biosynthesis and its control. *Progr. Phytochem.*, **5**, 127–80.

HIROSE H. & KUMANO S. (1966) Spectroscopic studies on the phycoerythrins from rhodophycean algae with special reference to their phylogenetical relations. *Bot. Mag.* (*Tokyo*), **79**, 105–13.

HIROSE H., KUMANO S. & MADONO K. (1969) Spectroscopic studies on phycoerythrins from cyanophycean and rhodophycean algae with special reference to their phylogenetical relations. *Bot. Mag.* (*Tokyo*), **82**, 197–203.

HÖGBERG H.-E., THOMSON R. H. & KING T. J. (1976) The cymopols, a group of prenylated bromohydroquinones from the green calcareous alga *Cymopolia barbata*. *J. Chem. Soc. Perkin Trans.*, **I**, 1696–701.

HOWARD B. M., FENICAL W., HIROTSU K., SOLHEIM B. & CLARDY J. (1980) The rhodophytin and chondriol natural products; structures of several new acetylenes from *Laurencia*, and a reassignment of structure for *cis*-rhodophytin. *Tetrahedron*, **36**, 171–6.

IDLER D. R. & ATKINSON B. (1976) Seasonal variation in the desmosterol content of dulse (*Rhodymenia palmata*) from Newfoundland waters. *Comp. Biochem. Physiol.*, **53B**, 517–9.

IDLER D. R., SAITO A. & WISEMAN P. (1968) Sterols in red alga (Rhodophyceae). *Steroids*, **11**, 465–73.

IMPELLIZZERI G., MANGIAFICO S., ORIENTE G., PIATTELLI M., SCIUTO S., FATTORUSSO E., MAGNO S., SANTACROCE C. & SICA D. (1975) Amino acids and low-molecular-weight carbohydrates of some marine red algae. *Phytochemistry*, **14**, 1549–57.

IMPELLIZZERI G., MANGIAFICO S., PIATTELLI M., SCIUTO S., FATTORUSSO E., SANTACROCE C. & SICA D. (1977) Amino-acid profiles in red algae. *Biochem. Syst. Ecol.*, **5**, 77–80.

ITO K. (1969) Free amino acids and peptides in marine algae (in Japanese). *Bull. Jpn. Soc. Sci. Fish.*, **35**, 116–29.

ITO K., ISHIKAWA K. & TSUCHIYA T. (1976) Studies on the depressive factors in *Heterochordaria abietina* affecting the blood cholesterol level in rats. III. Quantitative determination of iodoamino acids in marine algae and their effects of depressing the blood cholesterol level in rats. *Tohoku J. Agr. Res.*, **27**, 53–61.

JENSEN A. (1963) Ascorbic acid in *Ascophyllum nodosum, Fucus serratus* and *Fucus vesiculosus. Proc. Int. Seaweed Symp.*, **4**, 319–25.

JENSEN A. (1966) Carotenoids of Norwegian brown seaweeds and of seaweed meals. *Nor. Inst. Tang. Tareforsk. Rep.*, **31**, 1–138.

JENSEN A. (1969a) Tocopherol content of seaweed and seaweed meal. I.-Analytical methods and distribution of tocopherols in benthic algae. *J. Sci. Food Agric.*, **20**, 449–53.

JENSEN A. (1969b) Tocopherol content of seaweed and seaweed meal. II.-Individual, diurnal and seasonal variations in some Fucaceae. *J. Sci. Food Agric.*, **20**, 454–8.

JOHNS R. B., NICHOLS P. D. & PERRY G. J. (1979) Fatty acid composition of ten marine algae from Australian waters. *Phytochemistry*, **18**, 799–802.

JURGENSON J. E., BEALE S. I. & TROXLER R. F. (1976) Biosynthesis of δ-aminolevulinic acid in the unicellular rhodophyte, *Cyanidium caldarium. Biochem. Biophys. Res. Commun.*, **60**, 149–57.

KANAZAWA A. (1963) Vitamins in algae (in Japanese). *Bull. Jpn. Soc. Sci. Fish.*, **29**, 713–31.

KANAZAWA A. & YOSHIOKA M. (1972) The occurrence of cholest-4-en-3-one in the red alga *Gracilaria textorii. Proc. Int. Seaweed Symp.*, **7**, 502–5.

KARLSTRÖM O. (1963) The vitamin B_{12} content of marine algae. *Nor. Inst. Tang Tareforsk. Rep.*, **29**, 1–10.

KATAYAMA T. (1964) Chemical studies on volatile constituents of algae-XX. Pharmacological action of volatile constituents and biochemical significance of the existence of acrylic acid. *Mem. Fac. Fish. Kagoshima Univ.*, **13**, 58–72.

KATO T., KUMANIRENG A. S., ICHINOSE I., KITAHARA Y., KAKINUMU Y., NISHIHIRA M. & KATO M. (1975) Active components of *Sargassum tortile* effecting the settlement of swimming larvae of *Coryne uchidai. Experientia*, **31**, 433–4.

KAZLAUSKAS R., MURPHY P. T., QUINN R. J. & WELLS R. J. (1976) Two polyhalogenated monoterpenes from the red alga *Plocamium costatum. Tetrahedron Lett.* **1976**, 4451–4.

KAZLAUSKAS R., MURPHY P. T., QUINN R. J. & WELLS R. J. (1977) A new class of halogenated lactones from the red alga *Delisea fimbriata* (Bonnemaisoniaceae). *Tetrahedron Lett.* **1977**, 37–40.

KAZLAUSKAS R., MURPHY P. T., WELLS R. J. & BLOUNT J. F. (1978) A series of novel bicyclic diterpenes from *Dilophus prolificans* (brown alga, Dictyotaceae). *Tetrahedron Lett.* **1978**, 4155–8.

KIKUCHI T., MORI Y., YOKOI T., NAKAZAWA S., KURODA H., MASADA Y., KITAMURA K. & UMEZAKI I. (1975) Structure of sargatriol, a new isoprenoid chromenol from a marine alga: *Sargassum tortile. Chem. Pharm. Bull.*, **23**, 690–2.

KING T. J., IMRE S., ÖZTUNC A. & THOMSON R. H. (1979) Obtusenyne, a new acetylenic nine-membered cyclic ether from *Laurencia obtusa. Tetrahedron Lett.* **1979**, 1453–4.

KLEINIG H. (1969) Carotenoids of siphonous green algae: a chemotaxonomic study. *J. Phycol.*, **5**, 281–4.

KOMURA T., WADA, S. & NAGAYAMA H. (1974) The identification of fucosterol in the marine brown algae *Hizikia fusiformis. Agr. Biol. Chem.*, **38**, 2275–6.

KONAGAYA S. (1966) Isolation of eisenine from a brown alga, *Ecklonia cava. Bull. Jpn. Soc. Sci. Fish.*, **32**, 967–71.

KURATA K. & AMIYA T. (1977) Two new bromophenols from the red alga, *Rhodomela larix. Chem. Lett.* **1977**, 1435–8.

LANDYMORE A. F. (1976) The degradation of phenylalanine, tyrosine, and related aromatic compounds by a marine diatom and a haptophycean alga. Ph.D. Thesis, University of British Columbia.

LANGEN P. (1958) Über Polyphosphate in Ostsee-Algen. *Acta Biol. Med. Ger.*, **1**, 368–72.

LARSEN B. (1958) The influence of season, habitat and age of tissue on the niacin content of some brown algae. *Nor. Inst. Tang Tareforsk. Rep.*, **19**, 1–13.

LARSEN B. (1961) The biotin content of marine algae. *Nor. Inst. Tang Tareforsk. Rep.*, **26**, 1–18.

LAYCOCK M. V. & CRAIGIE J. S. (1977) The occurrence and seasonal variation of gigartinine and L-citrullinyl-L-arginine in *Chondrus crispus* Stackh. *Can. J. Biochem.*, **55**, 27–30.

LAYCOCK M. V., MCINNES A. G. & MORGAN K. C. (1979) D-Homocysteic acid from *Palmaria palmata*. *Phytochemistry*, **18**, 1220.

LEE R. F. & LOEBLICH A. R. III (1971) Distribution of 21:6 hydrocarbon and its relationship to 22:6 fatty acid in algae. *Phytochemistry*, **10**, 593–602.

LIAAEN-JENSEN S. (1978) Marine carotenoids. In *Marine Natural Products. Chemical and Biological Perspectives*, vol. II, ed. Scheuer, P. J., pp. 1–73. Academic Press, London & New York.

LINDBERG B. (1955) Methylated taurines and choline sulphate in red algae. *Acta Chem. Scand.*, **9**, 1323–6.

LISO R., CALABRESE G. & CHIEPPA M. (1978) Research on ascorbic acid physiology in red algae. IV. On the seasonal variation of AA/DHA ratio in *Pterocladia capillacea* (Gmel.) Born. et Thur. (Rhodophyta, Gelidiales). *Phycologia*, **17**, 143–7.

LYTLE J. S., LYTLE T. F., GEARING J. N. & GEARING P. J. (1979) Hydrocarbons in benthic algae from the eastern Gulf of Mexico. *Mar. Biol.*, 51, 279–88.

MCCONNELL O. & FENICAL W. (1977) Halogen chemistry of the red alga *Asparagopsis*. *Phytochemistry*, **16**, 367–74.

MCCONNELL O. & FENICAL W. (1978) Ochtodene and ochtodiol: novel polyhalogenated cyclic monoterpenes from the red seaweed *Ochtodes secundiramea*. *J. Org. Chem.*, **43**, 4238–41.

MADGWICK J. C. (1971) Free amino acids in Australian seaweeds. Ph.D. Thesis, University of New South Wales.

MAITI B. C. & THOMSON R. H. (1977) Caulerpin. In *Marine Natural Products Chemistry*, eds. Faulkner D. J. & Fenical W. H., pp. 159–63. Plenum Press, New York & London.

MANLEY S. L. & CHAPMAN D. J. (1978) Formation of 3-bromo-4-hydroxybenzaldehyde from L-tyrosine in cell-free homogenates of *Odonthalia floccosa* (Rhodophyceae): a proposed biosynthetic pathway for brominated phenols. *F.E.B.S. Lett.*, **93**, 97–101.

MANLEY S. L. & CHAPMAN D. J. (1980) The metabolism of 4-hydroxybenzaldehyde, 3-bromo-4-hydroxybenzaldehyde and bromide by cell-free fractions of the marine red alga *Odonthalia floccosa*: the formation of bromophenols. *Phytochemistry*, **19**, 1453–7.

MARTÍN J. D. & DARIAS J. (1978) Algal sesquiterpenoids. In *Marine Natural Products. Chemical and Biological Perspectives*, vol. I, ed. Scheuer P. J., pp. 125–173. Academic Press, London & New York.

MEEKS J. C. (1974) Chlorophylls. In *Algal Physiology and Biochemistry*, ed. Stewart W. D. P., pp. 161–75. Blackwell Scientific Publications, Oxford.

MERCER E. I. & DAVIS C. L. (1979) Distribution of chlorosulpholipids in algae. *Phytochemistry*, **18**, 457–62.

MIYAZAWA K. (1971) Amino acids and peptides in marine algae (in Japanese). *Bull. Jpn. Soc. Sci. Fish.*, **37**, 788–94.

MIYAZAWA K., ITO K. & MATSUMOTO F. (1976) Amino acids and peptide in seven species of marine green algae. *J. fac. Fish. anim. Husb. Hiroshima Univ.*, **15**, 161–9.

MOORE R. E. (1976) Chemotaxis and the odor of seaweed. *Lloydia*, **39**, 181–91.

MOORE R. E. (1978) Algal nonisoprenoids. In *Marine Natural Products. Chemical and Biological Perspectives*, vol. I, ed. Scheuer P. J., pp..43–124. Academic Press, London & New York.

MYNDERSE J. S. & FAULKNER D. J. (1975) Polyhalogenated monoterpenes from the red alga *Plocamium cartilagineum*. *Tetrahedron*, **31**, 1963–7.

MYNDERSE J. S. & FAULKNER D. J. (1978) Variations in the halogenated monoterpene metabolites of *Plocamium cartilagineum* and *P. violaceum*. *Phytochemistry*, **17**, 237–40.

NAKABAYASHI T. (1955) Studies on tannins of *Spyrogyra* (sic) *arcta*. Part 5. On the structure of spyrogyra (sic) tannin. *J. Agric. Chem. Soc. Jpn.*, **29**, 897–9.

NIEMEYER R. (1976) Cyclic condensed metaphosphates and linear polyphosphates in brown and red algae. *Arch. Microbiol.*, **108**, 243–7.

NORTON R. S., WARREN R. G. & WELLS R. J. (1977) Three new polyhalogenated monoterpenes from *Plocamium* species. *Tetrahedron Lett.* **1977**, 3905–8.

OCHI M., KOTSUKI H., MURAOKA K. & TOKOROYAMA T. (1979) The structure of yahazunol, a new sesquiterpene-substituted hydroquinone from the brown seaweed *Dictyopteris undulata* Okamura. *Bull. Chem. Soc. Jpn.*, **52**, 629–30.

Ó HEOCHA C. (1965) Biliproteins of algae. *Ann. Rev. Plant Physiol.*, **16**, 415–34.

OHTA K. (1979) Chemical studies on biologically active substances in seaweeds. *Proc. Int. Seaweed Symp.*, **9**, 401–11.

OHTA K. & TAKAGI M. (1977) Antimicrobial compounds of the marine red alga *Marginisporum aberrans*. *Phytochemistry*, **16**, 1085–6.

PEDERSÉN M. (1978) Bromochlorophenols and a brominated diphenylmethane in red algae. *Phytochemistry*, **17**, 291–3.

PEDERSÉN M., SAENGER P. & FRIES L. (1974) Simple brominated phenols in red algae. *Phytochemistry*, **13**, 2273–9.

PHAM QUANG LIEM & LAUR M.-H. (1976a) Structures, teneurs et compositions des esters sulfuriques, sulfoniques, phosphoriques des glycosyldiglycérides des trois fucacées. *Biochimie*, **58**, 1367–80.

PHAM QUANG LIEM & LAUR M.-H. (1976b) Les alcools aliphatiques sulfatés: nouveaux lipides polaires isolés de diverses fucacées. *Biochemie*, **58**, 1381–96.

PROVASOLI L. & CARLUCCI A. F. (1974) Vitamins and growth regulators. In *Algal Physiology and Biochemistry*, ed. Stewart W. D. P., pp. 741–87. Blackwell Scientific Publications, Oxford.

RAGAN M. A. (1976) Physodes and the phenolic compounds of brown algae. Ph.D. Thesis, Dalhousie University.

RAGAN M. A. & CHAPMAN D. J. (1978) *A Biochemical Phylogeny of the Protists*. Academic Press, London & New York.

RAGAN M. A. & JENSEN A. (1979) Widespread distribution of sulfated polyphenols in brown algae. *Phytochemistry*, **18**, 261–2.

RAGAN M. A. & MACKINNON M. D. (1979) Paired-ion reversed-phase high performance liquid chromatography of phenol sulfates in synthetic mixtures, algal extracts and urine. *J. Chromatogr.*, **178**, 505–13.

RAGAN M. A., SMIDSRØD O. & LARSEN B. (1979) Chelation of divalent metal ions by brown algal polyphenols. *Mar. Chem.*, **7**, 265–71.

REZNIKOV V. M., MIKHASEVA M. F. & ZIL'BERGLEIT M. A. (1978) Lignin of the alga *Fucus vesiculosus* (in Russian). *Chem. Prirod. Soed.*, **5**, 645–7.

RÖNNERSTRAND S. (1943) Untersuchungen über Oxydase, Peroxydase und Ascorbinsäure in einigen Meeresalgen. Ph.D. Thesis, Univ. Lund.

SAENGER P., PEDERSÉN M. & ROWAN K. S. (1976) Bromo-compounds of the red alga *Lenormandia prolifera*. *Phytochemistry*, **15**, 1957–8.

SANTOS G. A. & DOTY M. S. (1968) Chemical studies of three species of the marine algal genus *Caulerpa*. In *Drugs from the Sea*, ed. Freudenthal H. D., pp. 173–6. Marine Technology Society.

SAPOZHNIKOV D. I. (1973) Investigation of the violaxanthin cycle. *Pure Appl. Chem.*, **35**, 47–61.

SATTLER E. (1974) Ein neuer Typ gerbstoffartiger Polyphenole aus der Braunalge *Halidrys siliquosa*. Ph.D. Thesis, Rheinischen Friedrich-Wilhelms-Univ. Bonn.

SCIUTO S., PIATELLI M. & IMPELLIZZERI G. (1978) β-L-Aspartylglycine from the red alga *Ceramium rubrum*. *Phytochemistry*, **17**, 1659–60.

SCOTT R. (1954) Observations on the iodo-amino-acids of marine algae using iodine-131. *Nature (Lond.)*, **173**, 1098–9.

SHIRAISHI Y., SHIROTORI T. & TAKAHATA E. (1973) Determination of polycyclic aromatic hydrocarbon in foods. II. 3,4-Benzopyrene in Japanese foods (in Japanese). *J. Food Hyg. Soc. Jpn.*, **14**, 173–8.

SHIROTORI T. (1972) Contents of 3,4-benzopyrene in common Japanese foods (in Japanese). *Tokyo Kasei Daigaku Kenkyu Kiyo*, **1972** (12), 47–53.

SIMS J. J. & PETTUS J. A. JR. (1976) Isolation of free *cis* and *trans*-phytol from the red alga *Gracilaria andersoniana*. *Phytochemistry*, **15**, 1076–7.

SKIPNES O. M. (1974) Opptak og lokalisering av sink i *Ascophyllum nodosum* (L.) Le Jolis. Thesis, University of Trondheim-N.L.H.

STIERLE D. B. & SIMS J. J. (1979) Marine natural products—XV. Polyhalogenated monoterpenes from the red alga *Plocamium cartilagineum* of Antarctica. *Tetrahedron*, **35**, 1261–5.

STIERLE D. B., WING R. M. & SIMS J. J. (1976) Marine natural products XI. Costatone and costatolide, new halogenated monoterpenes from the red seaweed, *Plocamium costatum*. *Tetrahedron Lett*. **1976**, 4455–8.

STOM D. I. & APARTSIN M. S. (1976) Localization of phenols in *Nitella* sp. (in Russian). *Tsitologiya*, **18**, 100–2.

STRANSKY H. & HAGER A. (1970) Das Carotinoidmuster und die Verbreitung das lichtinduzierten Xanthophyllcyclus in verschiedenen Algenklassen. *Arch. Mikrobiol.*, **73**, 315–23.

SUN E., BARR R. & CRANE F. L. (1968) Comparative studies on plastoquinones. IV. Plastoquinones in algae. *Plant Physiol.*, **43**, 1935–40.

SUPRUNOV A. T., BEZHITSKII A. G. & BUGAEVA L. N. (1972) On certain depot accumulation of vitamin B_{12} in the coastal zone of the Black Sea (in Russian). *Biol. Morya*, **27**, 138–43.

TAKAGI M., IIDA A., MURAYAMA H. & SOMA S. (1973) Isolation and some chemical properties of a new peptide, analipine, from a brown alga, *Analipus japonicus*. *Bull. Jpn. Soc. Sci. Fish.*, **39**, 961–7.

THEILER R. F., SIUDA J. S. & HAGER L. P. (1978) Bromoperoxidase from the red alga *Bonnemaisonia hamifera*. In *Drugs and Food from the Sea. Myth or Reality*? eds. Kaul P. N. & Sindermann C. J., pp. 153–69. University of Oklahoma, Norman, Oklahoma.

TOCHER R. D. & CRAIGIE J. S. (1966) Enzymes of marine algae. II. Isolation and identification of 3-hydroxytyramine as the phenolase substrate in *Monostroma fuscum*. *Can. J. Bot.*, **44**, 605–8.

TROXLER R. F., BROWN A. S. & BROWN S. B. (1979) Bile pigment synthesis in plants. Mechanism of ^{18}O incorporation into phycocyanobilin in the unicellular rhodophyte, *Cyanidium caldarium*. *J. Biol. Chem.*, **254**, 3411–8.

TSUDA K., HAYATSU R., KISHIDA Y. & AKAGI S. (1958) Steroid studies. VI. Studies on the constitution of sargasterol. *J. Am. Chem. Soc.*, **80**, 921–5.

TSUJINO I., YABE K., SEKIKAWA I. & HAMANAKA N. (1978) Isolation and structure of a mycosporine from the red alga *Chondrus yendoi*. *Tetrahedron Lett*. **1978**, 1401–2.

VAN ENGEN D., CLARDY J., KHO-WISEMAN E., CREWS P., HIGGS M. D. & FAULKNER D. J. (1978) Violacene: a reassignment of structure. *Tetrahedron Lett*. **1978**, 29–32.

WEINSTEIN B., ROLD T. L., HARRELL C. E. JR., BURNS M. W. III. & WAALAND J. R. (1975) Reexamination of the bromophenols in the red alga *Rhodomela larix*. *Phytochemistry*, **14**, 2667–70.

WHITE R. H. & HAGER L. P. (1978) Intricenyne and related halogenated compounds from *Laurencia intricata*. *Phytochemistry*, **17**, 939–41.

WOOD, B. J. B. (1974) Fatty acids and saponifiable lipids. In *Algal Physiology and Biochemistry*, ed. Stewart W. D. P., pp. 236–65. Blackwell Scientific Publications, Oxford.

WOOLARD F. X., MOORE R. E. & ROLLER P. P. (1976) Halogenated acetamides, but-3-en-2-ols, and isopropanols from *Asparagopsis taxiformis* (Delile) Trev. *Tetrahedron*, **32**, 2843–6.

WRATTEN S. J. & FAULKNER D. J. (1976) Cyclic polysulfides from the red alga *Chondria californica*. *J. Org. Chem.*, **41**, 2465–7.

WRIGHT J. L. C. (1980) Structural assignment of two heneicosahexaene isomers by ^{13}C NMR spectroscopy. *Phytochemistry*, **19**, 143–4.

WRIGHT J. L. C., MCINNES A. G., SHIMIZU S., SMITH D. G., WALTER J. A., IDLER D. A. & KHALIL W. (1978) Identification of C-24 alkyl epimer of marine sterols by ^{13}C nuclear magnetic resonance spectroscopy. *Can. J. Chem.*, **56**, 1898–903.

YOKOHAMA Y., KAGEYAMA A., IKAWA T. & SHIMURA S. (1977) A carotenoid characteristic of chlorophycean seaweeds living in deep coastal waters. *Bot. Mar.*, **20**, 433–6.

YOUNGBLOOD W. W. & BLUMER M. (1973) Alkanes and alkenes in marine benthic algae. *Mar. Biol.*, **21**, 163–72.

YOUNGBLOOD W. W., BLUMER M., GUILLARD R. L. & FIORE F. (1971) Saturated and unsaturated hydrocarbons in marine benthic algae. *Mar. Biol.*, **8**, 190–201.

CHAPTER 18
MORPHOGENESIS AND GROWTH REGULATORS

R. G. BUGGELN
Marine Sciences Research Laboratory, Memorial University of Newfoundland, St. John's, Newfoundland, Canada

1 INTRODUCTION

In the broadest sense 'the student of morphogenesis is interested in all the processes and factors, genetical, physico-chemical, environmental and organismal which contribute to the distinctive organizational features of the species' (Wardlaw 1968). In order to understand, at the organism level, the 'physical phenomenology of morphogenesis' (Green 1969) and to assign a causal relationship to factors influencing development of form, one must first examine morphogenesis at a less complex level of organization. In an illuminating review of cell morphogenesis, Green (1969) called for a quantitative dissection of *cell surface behaviour* into components of 'rate and direction of expansion' as the foundation for understanding morphogenesis. The present chapter is divided into two sections on seaweed morphogenesis. The first deals with cell morphogenesis and includes topics of tropisms and polarity. The second section treats the general subject of regeneration with emphasis on *de novo* shoot production in relation to whole frond organization as well as other aspects of correlation growth effects in seaweeds. The chapter concludes with a review of the current status of chemical growth regulators in seaweeds.

Selected aspects of seaweed morphogenesis have been reviewed by Lang (1965), Dixon (1973) and Moss (1974). Other chapters in this volume covering such topics as nutrients (Chapter 10), light (Chapter 9), morphology and the environment (Chapter 12) and life histories (Chapters 4–6) discuss factors which influence the pattern of development of a species.

2 CELL MORPHOGENESIS

2.1 Wall structure and growth

In order to study changes, through space and time, of cell surfaces, investigators have preferentially chosen algal unicells or tip-growing filaments as the most suitable experimental systems. Not only can changes in geometry of exposed wall surfaces be readily observed and analysed, but stresses imposed on an expanding cell, arising from within the cell itself, are not complicated by interactions with adjacent tissues. Hence research with cells of *Nitella, Hydrodictyon, Valonia* and desmids has furnished many of the fundamental observations on cell morphogenesis. The reviews of Green (1969) and Green *et al.* (1970) of a decade ago detail discussions of biophysical studies from which major generalizations will be briefly highlighted here.

Cell shape and direction of growth are controlled by the way microfibrils (usually of cellulose) are arranged in the cell wall. Although the pattern of stresses in a cylindrical wall favours lateral expansion (in circumference) rather than elongation of the cell in the axial direction, cellulose microfibrils are initially deposited in a transverse pattern (90° with respect to the long axis) in a growing cell, an orientation which restricts lateral expansion. Therefore most cell growth occurs in an axial direction and produces a cell which is longer than it is wide. In most cells studied, elongation is accompanied by a realignment, from transverse to a more axial position, in those microfibrils first deposited but now found near the outer surface of the wall; microfibrils towards the inside usually remain in a transverse position. A departure from the transverse microfibrillar arrangement occurs in *Chaetomorpha* and members of the Valoniaceae. In these algae, alternating layers of microfibrils having low, intermediate and steep angles of spiral (with respect to the long axis of the cell) form a crisscross network.

Although the mechanism of microfibril deposition has not yet been elucidated, involvement of microtubules has been repeatedly implicated from their appearance in the cell at the time of wall synthesis, their location just inside the plasmalema and their orientation which is parallel to that of the transversely deposited microfibrils. When microtubule organization in the growing cell is destroyed by chemical agents such as colchicine, cellulose biosynthesis continues but microfibril deposition is random. The effect on cell structure is that stresses in the cylindrical wall, normally counteracted by the transverse microfibril arrangement, now act to produce cell expansion in the radial direction which results in swollen cells. Recent experimental evidence illustrating the features of cell morphogenesis outlined above was reported for the cigar-shaped desmid, *Closterium acerosum* (Hogetsu & Shibaoka 1978 a, b).

2.2 Cell division and elongation

Development of form is based on an interaction of cell division, cell enlargement and cell differentiation. Only one aspect of the latter process has been extensively investigated, i.e. photoperiod control of reproduction, and is treated in Chapter 9. Production of other specialized, non-reproductive cell types, e.g. hairs, gland cells, etc., does not appear to have been experimentally studied in marine algae. From work with the red alga, *Griffithsia pacifica,* it is known that light is essential for apical cell division (Waaland & Cleland 1972). It was also concluded that the diurnal rhythm in cell division, which persisted under continuous light, was controlled by an endogenous, circadian clock. The number of daily divisions of an apical cell in some red algae (*Callithamnion tetricum* and *Pleonosporium squarrulosum*) increased under long daylengths but this may have been a photosynthetic effect rather than a true photoperiod phenomenon (Dixon & Richardson 1970; Murray & Dixon 1973). While the daily apical cell division rate in *G. pacifica* was not affected by light intensity, branch initiation was greatly enhanced under high light conditions (or long photoperiod) suggesting that the effect was related to photosynthate production (Waaland & Cleland 1972).

Cell elongation in *G. pacifica* was not noticeably affected by light intensities above 40 ft.-c or LD or SD conditions (Waaland & Cleland 1972). In *P. squarrulosum* (Murray & Dixon 1973) and *Ceramium rubrum* (Garbary *et al.* 1978) longer axial cell lengths were achieved under LD than SD regimes but this may also have been a photosynthetic effect. Apparently there are no comparable data for the processes of cell division and cell elongation in either tip-growing Chlorophyta or Phaeophyta, e.g. Sphacelariales.

2.3 Phototropism

Green *et al.* (1970) stated that the symmetry of a cylindrical cell, elongating by tip growth, results from an interaction of three parameters: 1, shape of the cell's growth zone; 2, gradient in growth rate within the growth zone; and 3, the direction of expansion. During phototropic curvature, there is unequal expansion occurring within the growth zone of the cell and Green *et al.* (1970) described how the bending of a cell could be achieved by either 'bowing' or 'bulging'. Bowing of a cell axis can be brought about by an unequal rate of wall elongation on opposite sides of the cell. In a curvature produced by bulging, the original growth centre, at the tip, ceases growth, and is replaced by a lateral growth centre on the prospective concave side. Which of these potential mechanisms is involved in phototropic curvature in seaweeds?

It is well known from the numerous reports in the literature (Table 18.1), dating back to Berthold's investigations in 1882, that blue light causes maximum positive phototropic curvature of algal shoots and negative curvature of rhizoids and haptera. The most thorough growth analysis of the positive curvature in an alga was recently reported in a Xanthophycean species, *Vaucheria geminata* (Kataoka 1975 a,b). The apical region of a *V. geminata* shoot was phototropically sensitive with curvature of the tip beginning 1–2 min after the light was turned on and in some cases, continuing after the light was turned off. Using time-lapse photography to study the movement of a resin bead adhering to the shoot tip, Kataoka concluded that curvature of the shoot occurred by bulging. In addition, a hyaline cap in the peripheral cytoplasm of the tip was observed to settle to a new position on the lighted side of the filament. This new position marked both the site of the new growth centre and the direction of apical elongation. Branching was initiated by unilateral blue light in a similar manner, but orientation of the hyaline cap and formation of the lateral protuberance required more time.

The dose-response curve for *Vaucheria* shoot orientation progressed through several 'stages' as the total light energy (photon flux density × time) was increased (Kataoka 1975a). Starting with dim blue light, increasing curvature ('first positive' stage) occurred as energy increased up to 2.5×10^3 J m^{-2} followed by a decrease in curvature at slightly higher energies. The decrease was followed by another increase in curvature at 5×10^3 J m^{-2} ('second positive' stage). At still higher energy, there was either little curvature or evidence of negative phototropism (above 3.3×10^4 J m^{-2}). Although the dose-response curve had similarities to those for *Avena* coleoptiles, Kataoka (1977) concluded that auxin (IAA) was not involved in the mechanism of curvature. By applying cyclic adenosine monophosphate (cAMP), as well as substances which increase or decrease levels of cAMP in animal cells, he was able to separate the initial phototropic response in *V. geminata,* i.e. movement of the hyaline cap to the lighted side of the filament apex, from filament growth. For example, in the presence of physiological concentrations of cAMP, hyaline cap movement was inhibited while growth remained unaffected. Kataoka proposed that endogenous cAMP (see Section 4.3) may be involved in the 'gain-controlling process' of phototropism in *Vaucheria*.

The negative phototropic response of rhizoids has been studied in the red alga, *Griffithsia pacifica* (Waaland *et al.* 1977). Tip-elongating rhizoids began curving away from unilateral light 40–45 min after the light was switched on. If a short light pulse was presented to a rhizoid which was then allowed to elongate in darkness, curvature took place in a portion of the growing tip which was produced after the light had been turned off. This suggested that the stimulus was 'stored' in the previously irradiated part of the rhizoid and then transmitted to the new growing portion. If the dark period was prolonged, the elongating rhizoid eventually tended to reverse its direction of growth toward its original orientation.

Table 18.1 Phototropism in seaweeds.

Class and species	Organ	Phototropism (positive +) (negative −)	References
Xanthophyceae			
Vaucheria geminata	Shoot	1st & 2nd +; −	Kataoka (1975a, b, 1977b)
Chlorophyceae			
Acetabularia spp.	Shoot	+	Nasr (1939) Bonotto & Sironval (1977)
Batophora oerstedi	Shoot	+	Puiseux-Dao (1957)
Boergesenia forbesii	Rhizoid	−	Ishizawa & Wada (1979a, b)
Bryopsis plumosa	Pinnule	+	cf. Feldmann (1936)
Bryopsis sp.	Rhizome	− or + ?	Berthold (1882); Darsie (1939)
Caulerpa prolifera	Rhizome	+ ?	Chen (1971)
Caulerpa scaraliformis	Rhizome	+	Kataoka (1975b)
Cladophora sp.	Shoot	+	Blinks (1951)
Codium sp.	Rhizoid	−	Feldmann (1936)
Derbesia lamourouxii	Shoot	+	Feldmann (1936)
Phaeophyceae			
Alaria esculenta	Haptera	−	Buggeln (1974)
Alaria esculenta	Blade and sporophyll	+	Buggeln (unpublished)
Cystoseira barbata	Rhizoid	−	Winkler (1900)
Dictyota dichotoma	Rhizoid	−	Berthold (1882)
Ectocarpus sp.	Rhizoid	−	Berthold (1882)
Egregia menziesii	Rhizoid	−	Myers (1928)
Fucus vesiculosus	Rhizoid	−	Hurd (1920)
Macrocystis sp.	Rhizoid	−	Chien (1972)
Macrocystis sp.	Haptera	−	Chien (1973)
Macrocystis sp.	Embroyonic apex of young sporophyte blade	+	Chien (1972)
Bangiophyceae			
Conchocelis stages of *Porphyra yezoensis* and *P. tenera*	Shoot	+	Migata & Kim (1971)
Rhodophyceae			
Antithamnion cruciatum	Shoot	+ & −	Berthold (1882)
	Rhizoid	−	
Callithamnion sp.	Rhizoid	−	Berthold (1882)
Centroceras sp.	Rhizoid	−	Berthold (1882)
Gracilaria verrucosa	Shoot	+	Jones (1959)
Griffithsia pacifica	Shoot	+	Waaland *et al.* (1977)
	Rhizoid	−	
Maripelta rotata	Stipe	+	Bowen (1971)
Petroglossum nicaeense	Shoot	+	Perrone & Felicini (1972)
Polysiphonia tenella	Shoot	+	Berthold (1882)
	Rhizoid	−	
Sciadophycus stellatus	Blade	+	Neushul *et al.* (1967)
	Stipe	−	

Ishizawa & Wada (1979a,b) concluded that negative phototropism of rhizoids of the coenocytic green alga, *Boergesenia forbesii,* occurred by bulging at the growing tip. The lag time between 'light-on' and initiation of negative curvature showed temperature dependence with a 1 h lag at 26° C and no lag at 32° C. As in *Griffithsia,* a reverse or 'recovery bending' toward the initial orientation of the cell axis eventually occurred in rhizoids which were growing with negative curvature in darkness (following a previous irradiation with blue light). A separation of light stimulation from growth was demonstrated by irradiating rhizoids at 18° C, a temperature at which no growth occurred, and then transferring the germlings to darkness and raising the temperature to 25° C at which time growth resumed with bending. The 'stimulus' could be stored for 3 h at 18° C but with a linearly decreasing quantitative curvature response over that period.

There is less experimental information available about tropic responses of axes in which the growth zone is multicellular. Positive phototropism in *Gracilaria verrucosa,* which occurred after 3 h of high light intensity, was attributed to a combined growth response: direct photoinhibition on the lighted side and acceleration of growth on the darker side (Jones 1959). Due to their slow growth, the negatively phototropic haptera of Laminariales sporophytes (Buggeln 1974 and Table 18.1) are not suitable for quantitative studies.

Action spectra for both positive and negative phototropism show a maximum at about 430–450 nm (Kataoka 1975a; Ishizawa & Wada 1979b) with little effectiveness above 500 nm and none in the red (660 nm) or far-red (730 nm) (Buggeln 1974). There is a minimum at about 400 nm and another peak in the UV (Waaland *et al.* 1977; Ishizawa & Wada 1979b). A similar action spectrum for chloroplast orientation in *Vaucheria* has been reported (cf. Nultsch 1974), but it remains to be unequivocally established whether a carotenoid pigment or riboflavin is the photoreceptor for growth phenomena involving blue light.

Phototropism in higher plant shoots is mediated at the hormone level by the asymmetrical distribution of auxin which occurs under conditions of unidirectional light. At the present time there is no compelling reason to assume that this higher plant mechanism—or any other single mechanism—is involved in phototropism in the phylogenetically diverse seaweeds listed in Table 18.1. Most assuredly, an understanding of the endogenous factors involved in phototropic curvature of shoots and rhizoids would yield valuable insight into the fundamental mechanism of control of cell elongation in seaweeds.

2.4 Geotropism

Casual observation that rhizoids and other attachment organs in algae may orient in a positively geotropic direction, i.e. grow 'downward' in nature or in culture, has led to statements that gravity influences their directional growth (e.g. Bandbury 1962; L'Hardy-Halos 1971a; DeValera 1975; Gibor 1977). Apart from the extensive investigations of the gravity-response of *Chara* rhizoids by Sievers and co-workers (Juniper 1976; Sievers *et al.* 1979) there have been no experimental demonstrations that growth of rhizoids of marine algae or phylogenetically related freshwater species is affected by gravity. The 'downward' growth of attachment organs is a negative response to solar or artificial radiation (Table 18.1). However, Zimmermann (1929) concluded that gravity had a role in determining the dorsal and ventral *positioning* of leaf and rhizoid primordia, respectively, on *Caulerpa prolifera* rhizomes. Recently, Jacobs & Olsen (1978) have confirmed these results under conditions in which *C. prolifera* thalli were bilaterally irradiated from above and below the rhizomes.

2.5 Polarity

According to Block (1965), 'polarity designates any internal asymetrical state within a living system, as expressed in differences between two ends or two surfaces, and also between the ends of surfaces and the central region'. Polarization is the first differentiation event in algal morphogenesis and seems to be the framework for continued orderly development. Early literature has been reviewed by von Wettstein (1965). Zygotes are obvious subjects for experiments on induction of polarity, and over the years, a favourite experimental material has been the zygotes of Fucales. Much ingenious research has been conducted on this phenomenon by L.F. Jaffe and co-workers (cf. Jaffe *et al.* 1974), and Quatrano (1978) has provided a recent review (see also Chapter 7). Figure 18.1 graphically depicts the known physicochemical processes taking place in Fucales zygotes (Schröter 1978) which have received any one of a number of polarizing stimuli, e.g. unidirectional light, gradients of pH, Ca^{2+} or

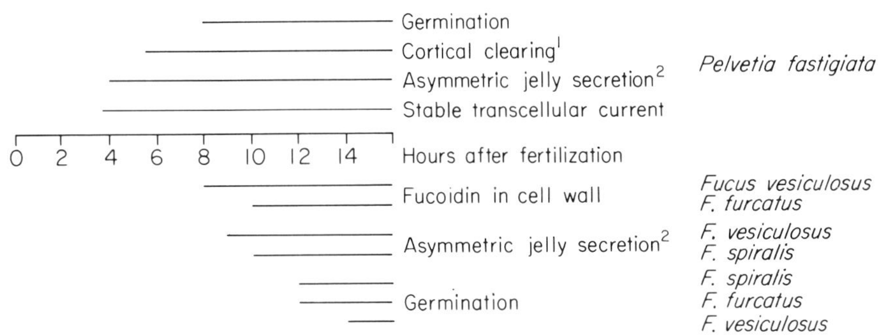

Fig. 18.1 Comparison of morphological, cytochemical and electrical events preceding germination of *Fucus* and *Pelvetia* zygotes. [1]'Cortical clearing' is caused by vesicle accumulation in the region of prospective rhizoid formation; [2]jelly secretion is most intense at the prospective rhizoid pole and facilitates adhesion of zygotes to the substratum. In SO_4^{2-}-free seawater, the jelly polymer is not synthesized. (Courtesy Dr. K. Schröter. Redrawn from Schröter 1978.)

K^+, electric current, 'crowding' in darkness, etc. Bünning (1952) reasoned that cellular receptors sensitive to a polarizing stimulus, especially light, would be located in the cytoplasm adjacent to the cell wall, and Jaffe's (1958) experiments with plane polarized light showed how some photoreceptors in the peripheral cytoplasm could be receiving more light than others. The latter group, located in a 'shadow' caused by diffraction of light by cell wall and cytoplasm, became the site of the rhizoid pole of the zygote. In the absence of a polarizing stimulus the point of entry of the sperm at fertilization determined the rhizoid pole. Two stages have been recognized in the development of polarity in Fucales zygotes: 1, those physicochemical events leading up to irreversible fixation of the polar axis; and 2, axial orientation of intracellular components involved in early stages of differentiation into rhizoid and thallus halves of a two-celled embryo. So far, no ultrastructural markers have been correlated with axis fixation. Quatrano (1978) has hypothesized that a critical process in the stabilization of the polar axis may involve localization of specific plasma membrane components (involved in Ca^{2+} uptake) at the rhizoid pole. The transport of these components may involve microtubules as cytochalasin B inhibits the polar fixation process.

Pelvetia zygotes growing in darkness in a calcium ionophore gradient formed rhizoids on sides receiving the higher ionophore concentration (Robinson & Cone 1980). The ionophore gradient presumably facilitated unequal entry of Ca^{2+} into the zygotes thereby creating an intracellular Ca^{2+} gradient. These recent findings support the earlier idea that an intracellular Ca^{2+} gradient is an essential step in the process of polarization (Robinson & Jaffe 1976).

Red algal spores may already possess a polar axis whose orientation is predetermined by the polarity of the parent thallus (Weber 1969).

While the mature form of algal species displays the apico-basal polarity initiated with the first cell division of the zygote, i.e. erect, vegetative thallus and basal attachment portion, do individual cells of the mature frond possess an internal polarity themselves? This question has been studied using uniseriate, filamentous algae (both uninucleate and coenocytic) because of their ease of manipulation, i.e. physically isolating single cells or a cell file, or by plasmolytic 'isolation' or protoplasts in intact filaments (Miehe 1905). Regeneration of pigmented vegetative shoots from the end of the explant nearest the apex of the parent thallus and a rhizoid from the end nearest the thallus base is a manifestation of persistent polarity, not only in individually isolated cells but also in multicellular segments whose integrated association of cells effects an overall polarity of the segment as a unit. Other manifestations of an apico-basal axis can be seen at the cellular level, in the distribution of organelles in apical cells of *Zonaria farlowii* blades (Dahl 1971), and at the organism level, by the gradient of physiological function correlated with aging, e.g. chlorophyll content, photosynthesis, respiration, dark CO_2 fixation, etc. in meristermatic, mature and old regions of *Laminaria saccharina* blades (Johnston *et al.* 1977; Küppers & Kremer 1978) and in *Fucus vesiculosus* (Khailov *et al.* 1979).

How stable is the polarity of isolated cells? The ease with which rhizoids can be caused to regenerate from the apical end, and shoots from the basal end of an excised cell or filament appears to be partly related to the ease with which the original apico-basal cytoplasmic profile can be permanently reversed. Such a cytoplasmic shift apparently occurs in *Bryopsis* spp. and *Caulerpa prolifera* (Steinecke 1925); the latter genus is also well known for rapid cytoplasmic streaming as well as diurnal movement of cytoplasm to and from the leafy portion of the coenocyte (Dawes & Barilotti 1969). Centrifuging a file of *Griffithsia bornetiana* cells caused formation of shoots (in addition to rhizoids) at the basal end of the filament but no rhizoids were formed at the apical end (Schecter 1935). In *Enteromorpha compressa,* repeated centrifuging inhibited rhizoid development (Müller-Stoll 1952), but in *Cladophora glomerata* centrifuging reportedly resulted in a rhizoid appearing from the apical pole and a shoot from the basal one (Borowikov 1914). *G. bornetiana* cells placed in an electrical field regenerated rhizoids at the apical end of the segment when this end was positioned toward the anode (Schecter 1934). In addition rhizoid growth was directed toward this pole. Similar observations were also recorded in *Fucus* and in *Ulva mutabilis* (Sand 1973). Schecter observed that while chromatophores in *Griffithsia* moved either to the anode in an electrical field (Schecter 1934), or to the centrifugal pole during centrifugation (Schecter 1935), their displacement bore no positive correlation with the kind of structure regenerated: rhizoids in the former case and shoots in the latter. Posterior anucleate fragments of *Acetabularia mediterranea* show 80–100% regeneration from the end nearest the anode, regardless of original polarity of the excised sections (Novák & Bentrup 1972) though neither the longitudinal bioelectric field of the stalk cytoplasm nor the spontaneous action potentials arising at the regenerating pole appear to be an obilgate facet of the regeneration process (Bentrup 1977).

3 REGENERATION AND CORRELATION GROWTH EFFECTS

The capacity of seaweed fragments to regenerate new, erect and basal systems is a widespread phenomenon which: 1, demonstrates totipotency of algal cells; 2, provides a convenient means for studying morphogenetic stages of adventive embryony; 3, produces correlation growth effects which may be useful for hypothesizing mechanisms of regulation of morphogenesis; and 4, facilitates mariculture of economically important red algae, e.g. *Chondrus crispus*, *Gracilaria* spp., *Eucheuma* spp. (Neish & Fox 1971; Goldstein 1973; cf. Edelstein 1977; see Chapter 20). Early literature on regeneration has been reviewed by Müller-Stoll (1965).

3.1 Totipotency

Totipotency is the characteristic of plant cells whereby a given cell with a specific physiological function, *in vivo*, retains the capacity to de-differentiate and subsequently, under appropriate environmental conditions, initiate development of a new organism. An example of this fundamental characteristic is seen in the demonstration that naked protroplasm isolated from *Bryopsis plumosa* can develop into a mature pinnate alga (Tatewaki & Nagata 1970). Isolated utricles from *Codium tomentosum* reportedly develop into young fronds (cf. Borden & Stein 1969) as do isolated rhizoidal cells of *Pelvetia wrightii* (Saga 1978). Asceptically isolated medullary tissue from the more complex thallus of *Chondrus crispus* regenerated into adult fronds. This system reportedly required both NAA and kinetin for normal development if the isolated tissue was free of cortical cells (Chen & Taylor 1978). Cells excised from the stipe of *Laminaria angustata* de-differentiated to form a loose 'callus'. Individual callus cells developed into sporophytes with best growth obtained when an extract prepared from bacterized *L. angustata* was added to the culture medium (Saga *et al.* 1978).

3.2 Adventive embryony

The name, adventive embryony, or the neo-formation of shoots and rhizoids from thallus fragments, refers to the regenerating structures whose morphology resembles the primary shoots and rhizoids of the developing embryo stage of the particular alga. The regeneration process is preceded by a 'wound response' which has been studied in few algal species. In the unicellular coenocytes, *Bryopsis hypnoides* and *Caulerpa ashmeadii,* a 'wound plug' of proteinaceous material (former) or carbohydrate (latter) is rapidly formed which prevents loss of cytoplasm (Burr & West 1971; Goddard & Dawes 1979). In multicellular forms such as *Fucus vesiculosus* (Fulcher & McCully 1969, 1971) and *Sargassum filipendula* (Fagerberg & Dawes 1976) a new epidermal 'plate' of cells is formed from medullary cells (as well as cortical cells in *Sargassum*) adjacent to the cut surface. There is no proliferation of an undifferentiated callus. A new vegetative axis is initiated as a papillate outgrowth of the underlying medullary tissue (Fig. 18.2) with one of the cells at the bud apex becoming the apical cell of the young shoot.

In Rhodophyta with a multiaxial thallus structure, e.g. *Petroglossum nicaeense* and *Gigartina acicularis,* two kinds of adventive shoots may be distinguished on the basis of their origin: those arising from medullary cells at cut surfaces, '*de novo* regeneration'; or those formed by renewed growth lateral filaments which form the thallus cortex. The latter shoots have been called 'proliferations' (Perrone & Felicini 1972, 1976).

In *Caulerpa prolifera,* leaf, rhizome or rhizoid primordia are initiated, with a highly predictable order and rate of appearance, as papillate outgrowths on

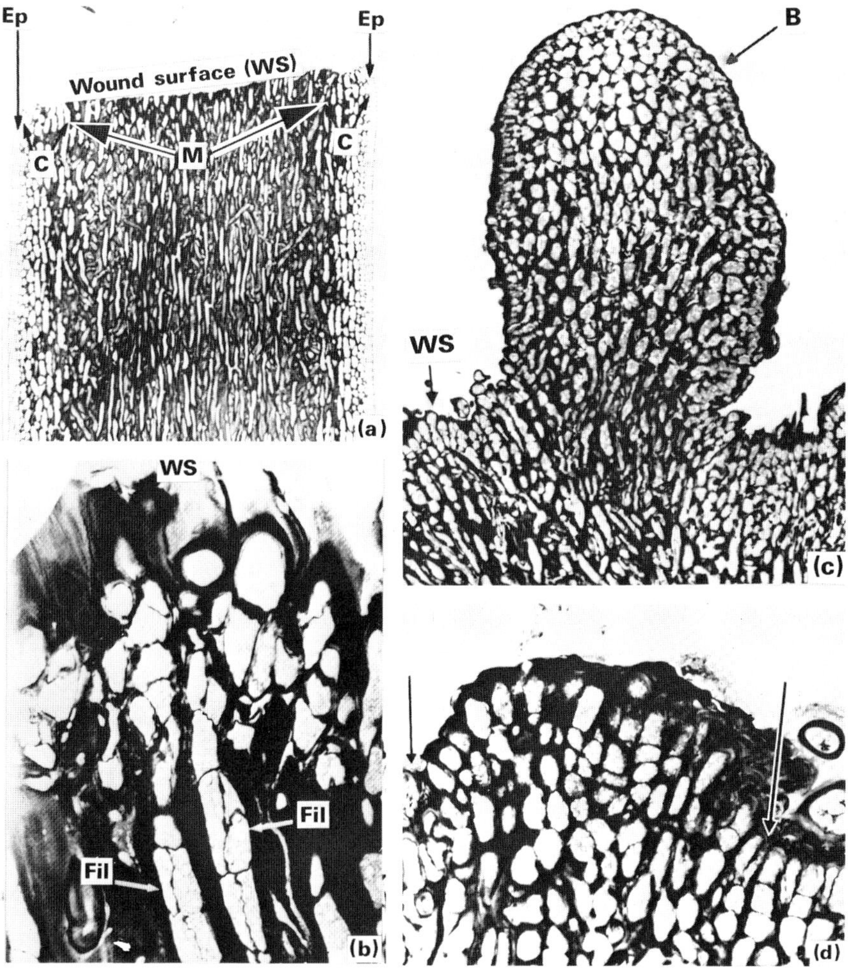

Fig. 18.2 Wound response and regeneration in the *Fucus vesiculosus* midrib (median longitudinal sections). (a) Freshly cut midrib showing wound surface (WS), medulla (M), cortex (C) and epidermis (Ep) (× 60); (b) plate of new cells forming beneath wound surface after four to five weeks with medullary filaments undergoing longitudinal division (Fil) (× 430); (c) elongation of medullary filaments (between arrows) at an early stage of bud formation after seven weeks (× 230); (d) developing adventitious bud (B) after twelve weeks (× 110). (Courtesy Dr. M. E. McCully. From Fulcher & McCully 1969.)

rhizome sections (Chen & Jacobs 1966). Vegetative papillae arise on the apical end of segments of *Enteromorpha intestinalis* (Moss & Marsland 1976).

3.3 Apical dominance

According to Sinnott (1960) phenomena of development such as polarity, regeneration, apical dominance, senescence, tropisms, etc. are different aspects of 'developmental relatedness' of various parts of an organism. Altering one part produces changes to other parts and thus to the form of the whole organism. Observations of such interdependence within plants have been given the name, *correlation effects*. The phenomenon of apical dominance is a display of *correlative inhibition* of lateral (axillary) bud development by the apical bud (meristem) in higher plants. Growth of lateral buds is activated when the apical bud is removed. Endogenous control of apical dominance in higher plants involves an interplay of hormonal (auxin and kinetin) and nutritional factors (Phillips 1975). In seaweeds the control mechanism(s) is unknown although several authors have investigated the effects of auxins on decapitated algal thalli (see Section 4.3 on auxins).

Apical dominance effects have been reported in a number of seaweeds following Moss' initial report on *Fucus vesiculosus* (Moss 1965). In *Apoglossum ruscifolium,* removal of the apical cell reportedly induces lateral bud (branch) formation (Abelard & L'Hardy-Halos 1975). Apical control of lateral branch growth is exhibited by *Sphacelaria cirrhosa*. The normally pyrimidal shape of the thallus is lost when the apical cell is removed and elongation of lateral branches is accelerated (Fig. 18.3) (Ducreux 1977). A similar growth response was quantified in *Antithamnion plumula* where growth rates of lateral axes on thalli with excised and intact apical cells were compared (L'Hardy-Halos 1971a). Removal of the apical cell of the main axis in *Ascophyllum nodosum* (Moss 1970) induces development of lateral branch primordia. In *Sargassum muticum,* initial development of 2° laterals is controlled by the 1° apex, and 3° branch development by 2° apices (Chamberlain *et al.* 1979). An interesting example of apical dominance is seen in *Zonaria farlowii,* a fan-shaped blade in which the terminal cells of the leading edge of the blade are a continuous marginal meristem of apical cells (Dahl 1971). When a portion of the marginal apex is cut out leaving adjacent areas of the blade with apex intact, as shown in Fig. 18.4, growth resumes across the entire wounded margin except in the two regions immediately adjacent to those portions of the blade which were left intact. Growth correlations between sister apical cells have also been studied in another member of the Dictyotales, *Dictyota dichotoma* (Gaillard & L'Hardy-Halos 1977).

3.4 Apico-basal and baso-apical gradients

The numbers and kinds of adventive embryos arising from a wounded algal thallus can be correlated with the position of the wound along the main apico-basal axis of the parent frond. While this relationship was first reported in *Fucus vesiculosus* (Moss 1966), Duffield *et al.* (1972), studying a less complex system, compared the regeneration capacity of non-nodal axial cells of *Griffithsia pacifica*. They reported an increasing potential to produce new shoot cells the further the original position of the axial cell was from the filament apex, down

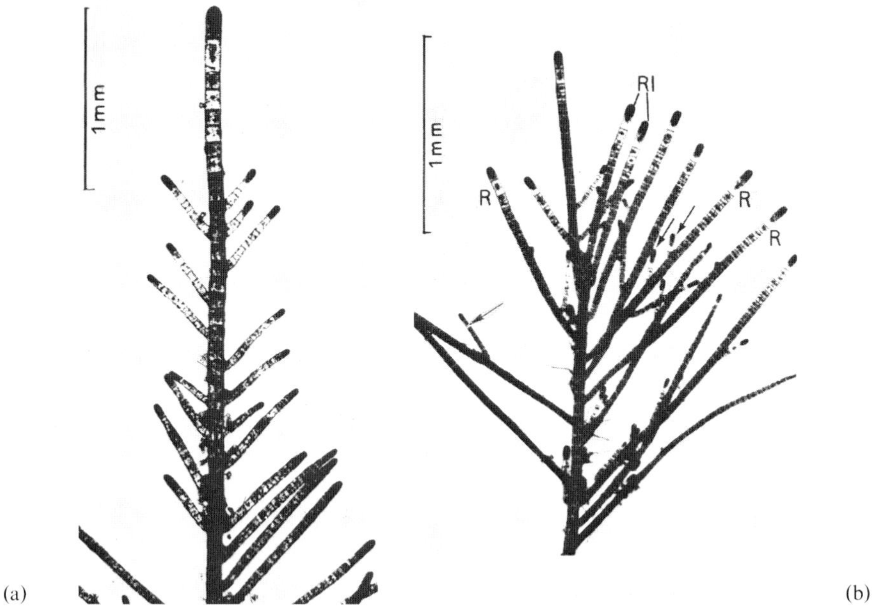

Fig. 18.3 Apical dominance in *Sphacelaria cirrhosa*. (a) Control alga with apical cell of main axis intact; (b) apical cell removed. Accelerated growth of normal branches = R; new axes of unlimited growth = R1; arrows indicate additional lateral buds. (Courtesy Dr. P. Ducreux. From Ducreux 1977.)

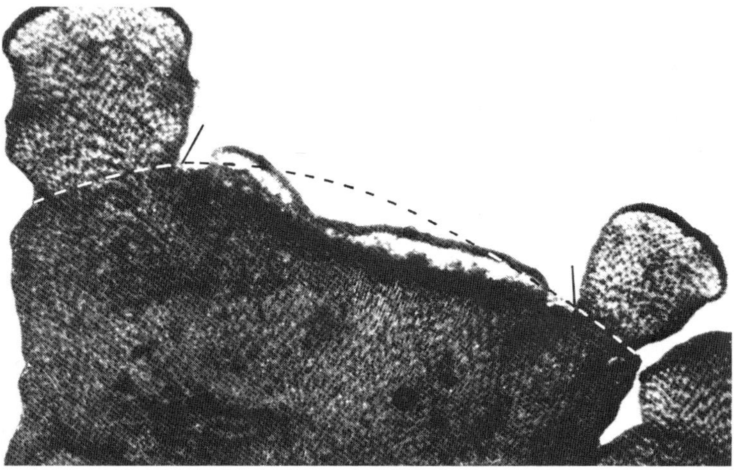

Fig. 18.4 Apical dominance in *Zonaria farlowii*. Dotted line indicates the approximate extent of the entire marginal meristem at the time of cutting. Width of the excised section of blade is indicated by the two solid lines. Continued growth of uncut lateral portions of blade (after six days) with adjacent inhibited zones and central region of regenerating marginal meristem. (Courtesy Dr. A. L. Dahl. From Dahl 1971.)

to a maximum distance of eight cells. The work of Perrone & Felicini (1972) illustrates morphogenetic correlations in intact blades and isolated sections of *Petroglossum nicaeense* (Fig. 18.5) (see also Abelard 1977). Presence of the frond apex can be correlated with enhancement of '*de novo* regeneration' and inhibition of 'proliferation' on the thallus (Fig. 18.5b, c). There is also a polar

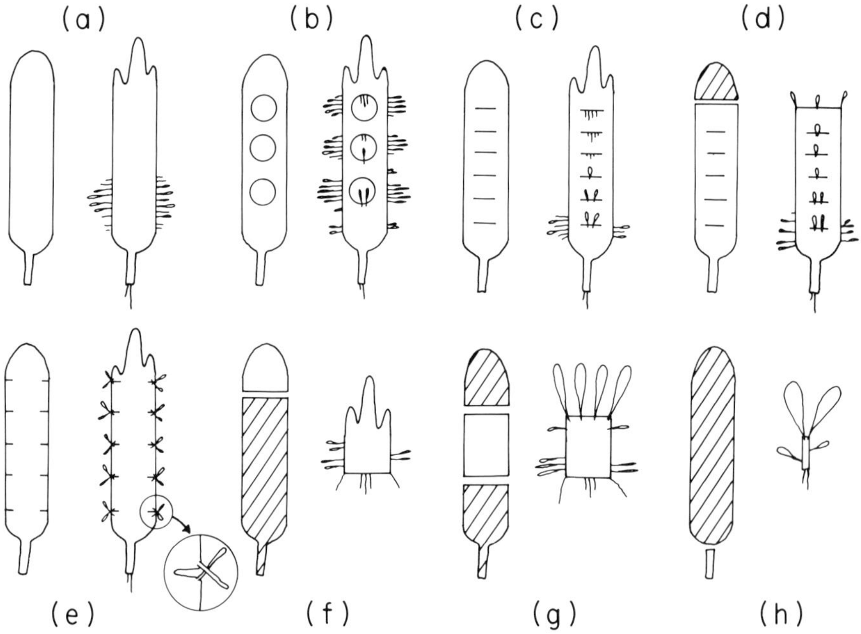

Fig. 18.5 Correlative growth effects in blades (a–e) and blade segments (f–h) of *Petroglossum nicaeense*. The member on the left in each pair shows the initial condition; on the right, adventive embryony 30–90 days later. Portions shown shaded were discarded.

(a) Blade with apex intact: '*de novo* regeneration' (R) at cut end of petiole; lateral 'proliferation' (P) clustered at blade base. (See definition of P & R in text p. 636).

(b) Removal of major amounts of tissue from blade with apex intact stimulates P at blade margins opposite holes; R appear on upper side of holes, P on blade surface adjacent to lower side of holes. Number of R decreases basipetally; P increases basipetally.

(c) When slits, rather than holes, are made across the longitudinal center line of a blade, no lateral P are formed opposite the cuts; apico-basal polarity in R and P formation is observed as in (b).

(d) Decapitating blades treated as in (c) prevents R formation on basal cut surfaces and relieves inhibition of P formation close to blade apex.

(e) Slits made in blade margin prevent formation of P clusters on basal edges of the blade seen in (a). Rather, P arise from blade marginal surface on both sides of cuts (see inset detail).

(f) Excised apical segment forms R from middle of basal cut surface and P along the blade margin.

(g) Middle segment forms P on blade surface below cut apex and on lateral margins; R formed as in (f) above.

(h) Isolated petiole produces lateral P; R formed on basal cut surface. (Redrawn from Perrone & Felicini 1972.)

tendency for greater 'regeneration' capacity near the frond apex and 'proliferation' capacity toward the base, with the former shoots developing into prostrate axes and many of the latter ones becoming erect fronds.

An isolated determinant branch of *Heterosiphonia plumosa* with apical cell intact, regenerates rhizoids at the base but the branch does not elongate (Boney 1975). In segments of *Enteromorpha* spp. the number of rhizoids formed was greater in basal than in apical segments; longer explants also produced more rhizoids than shorter ones (Eaton *et al.* 1966). It is curious that the number of branches and degree of branching of papillate outgrowths from segments cut from the basal portion of *Enteromorpha* spp. (with rhizoids attached) increased with each successive removal of the new shoots (Moss & Marsland 1976). In *Fucus*, the length of excised segments of blades was correlated with an increase in number of adventive embryos (Moss 1966). The correlation of a greater number of adventive shoots and rhizoids with increasing size of the explant is probably related to photosynthetic area and energy generating capacity of the explant. A strong correlative inhibition of shoot formation was reported for explants from *Sargassum muticum* with adventive embryos forming only on the most basal segments (Fletcher 1975a).

References to most of the voluminous historical literature on regeneration in *Caulerpa prolifera* have been cited by Jacobs (1964, 1970). Excised rhizomes of *C. prolifera* regenerate rhizoids with greater frequency at apical ends (rhizomes elongate by tip growth) than at basal ones. New rhizomes were also regenerated close to the cut surface at the apical end; new leaves arose with less polar correlation. Short sections excised from blades of *Caulerpa* regenerated rhizoids toward the basal cut surface; longer sections regenerated rhizoids, rhizomes and new leaves. Evidence suggests that the presence of rhizoids has an effect on the position of new blade primordia forming on rhizomes.

A number of interesting features of regeneration in Fucales, particularly *F. vesiculosus,* have been reported by Moss (1964, 1967, 1968). Adventive embryony is more vigorous in segments excised from young thalli which have not yet dichotomized, than from segments of older thalli cut between the apex and the first dichotomy, i.e. from a region comparable to the age of the young frond. Sections isolated from short branches bearing receptacles regenerated adventive embryos which developed into receptacles if sections were excised during gamete maturation in the parent frond. However, no regeneration occurred in segments isolated during receptacle differentiation and only vegetative apices occurred on explants prepared after receptacles had shed their contents. There was also a seasonal cycle in the numbers of adventive embryos regenerated in explants cultured under uniform conditions.

In *Sargassum muticum* basal segments from young thalli regenerate more adventitious branches than similar segments from old thalli, as Moss observed with *Fucus.* A damaged *Sargassum muticum* receptacle will regenerate several stalked receptacles (Fletcher 1975a). In *S. filipendula* there is a decrease in vegetative regeneration capacity during the reproductive phase of the alga (Fagerberg & Dawes 1976).

Sections excised from *Dictyota dichotoma* and *Dilophus spiralis* regenerate rhizoids from medullary cells of cut surfaces, and pigmented vegetative axes from both uncut lateral margins and explant surfaces (Pybus 1964). Rhizoid production in excised *D. dichotoma* sections is reportedly stimulated by red light (Russell 1970).

Dixon's (1971) comparative study of lengths of axial cells in a number of genera in the Florideophyceae revealed two reciprocal correlative growth effects which may be characteristic of thallus development in this family of red algae. First, cells of a principal axis which bear lateral axes are shorter than cells which bear no laterals (*Griffithsia flocculosa*). It was experimentally shown in *Griffithsia pacifica* that while the growth rate of nodal (lateral axis bearing) cells and non-nodal cells was initially similar, the former slowed down before the latter, resulting in a shorter final length for the nodal cells (Duffield *et al.* 1972). Second, cells of the lateral which are adjacent to the main axial cell are shorter than those further away. Hence there appears to be a correlative inhibitory effect on elongation of the main axial cell by the developing lateral arising from the axial cell and conversely, a correlative inhibition of elongation of lateral cells adjacent to the main axis, by cells of the main axis itself. As Dixon points out, these relationships are clear examples of the century-old botanical paradigm, attributed to H. Vöching (cf. Sinnott 1960), that the function of a cell is determined, in the first instance, by its morphological position.

3.5 Exogenous factors regulating development

The morphology of a developing axis is the result of complex interactions between genetic constitution (or potential) and environmental parameters such as light and nutrient. While roles of environmental factors in life cycles of seaweeds are extensively treated in Chapters 9, 10, and 12, certain examples are briefly considered here to illustrate the morphological pliability of developing axes.

Reducing light intensity may convert a normally developing vegetative axis into a rhizoid-like process. This correlation has been reported in Chlorophyta: e.g. *Bryopsis* sp. (Bertold 1882), *Caulerpa prolifera* (Dostal 1945); Phaeophyta: *Sphacelaria fusca* (Zimmermann 1923); and Rhodophyta: e.g. *Callithamnion* sp. (Bertold 1882), *Pterocladia capillacea* (Felicini 1970) (see also L'Hardy-Halos 1971a); however, it has not received critical experimental study. Segments of *Bryopsis plumosa* grown in darkness regenerate rhizoid-like processes from both poles. This conversion is not actually a 'reversal' of polarity at the presumptive 'shoot' end, as provision of full light leads to a restoration of normal morphological development of the shoot (Darsie 1939). According to Berthold, pigmented rhizoid-like filaments of *Callithamnion* sp. are positively phototropic in weak light which suggests that they are morphologically modified shoots. Rhizoid-shoot 'interconversions' have also been reported in *Rhodochorton purpureum* (Pearlmutter & Vadas 1978).

There are certain phases in the development of thallus form and/or function which require light of specific wavelength(s) and minimum duration to activate the morphogenetic step. The diversity of these correlative phenomena is categorized under the general heading *photomorphogenesis*, and this topic is fully treated in Chapter 9 on Light.

A reduction in supply of nitrogen to some seaweeds, e.g. *Petroglossum nicaeense* (Perrone & Felicini 1974) affects thallus form in a manner similar to low light mentioned above. Thus the observation, in nature, of 'reduction' of normal morphology of erect shoots to rhizoid-like processes, e.g. *Caulerpa prolifera* (Dostal 1945); Espera state of *Penicillus capitatus* (Friedmann & Roth 1977); adventive regeneration of *Spatoglossum solierii* (Gayral 1962) or the Vaughaniella state of *Padina* spp. (Boergesen 1951; Cribb 1951) may not necessarily be attributed to a single environmental factor. Even the chemical form in which nitrogen is supplied to algae in culture, i.e. NO_3^-, NH_4^+, etc. can have a major influence on morphological development (cf. Fries 1973; Lockhart 1977).

Konrad-Hawkins (1968) noted that crowding in *Callithamnion roseum* cultures increased branch initiation; L'Hardy-Halos (1971a) showed the production of rhizoids by intercalary cells of *Griffithsia flocculosa* which had come into contact with the thallus of *Acrosorium uncinatum*. The latter phenomenon has been frequently observed when a red algal frond contacts an inert surface. Rhizoid-like processes were only formed on *Fucus vesiculosus* explants growing on agar and not in aqueous media. In *Gigartina acicularis* which produces adventive embryos of the 'regeneration-' and 'proliferation-' type (*sensu* Perrone & Felicini see above), nitrogen and phosphorus levels influence both total number and ratio of types of advantitious shoots produced (Perrone & Felicini 1976). NH_4NO_3 addition to seawater increased 'proliferation' while PO_4^{2-} stimulated 'regeneration'. The total number of adventitious shoots was twice as great in seawater with PO_4^{2-} than in control or N-supplemented cultures.

3.6 Abscission

Abscission, the process whereby a portion of the frond becomes severed from the main body, is a morphological manifestation of the process of senescence and the very nature of the phenomenon affects the form of some algae. It is a process which has received little study, and as with other correlative phenomena in seaweeds, the mechanism of regulation is completely obscure. Most reports correlate abscission or fragmentation of thalli with sporulation (cf. Edelstein 1977; Norton 1977; Whittick 1978; but see Etherington 1964). Conspicuous examples of the process are seen in Fucales. In *Fucus* spp. the branch subtending a receptacle (small lateral pedicles bear receptacles in *Ascophyllum nodosum*) frequently turns yellow-orange due to loss of chlorophyll. Autolytic processes result in disintegration of the yellowed branch to a point close to the junction with the main axis bearing the branch (Moss 1968). Sporophylls of *Alaria* spp. undergo degradative processes after release off zoospores and ultimately drop off. The extent of decay along a reproductive branch in Fucales appears to be coordinated by factors originating mainly in the branch bearing the receptacle (or the receptacle itself) but with some limit to the extent of decay imposed on the branch by the vegetative axis supporting it. An 'abscission layer', characteristic of higher plants, has not been reported in these algae.

Fragmentation commonly occurs in the late spring in *Codium fragile* (Fralick & Mathieson 1972; Malinowski & Ramus 1973) with thalli breaking at various constrictions which have low tensile strength. Microscopic examination of these loci however, showed no thinning of the cell walls. The basal portion of each fragment de-differentiated to form a new holdfast.

3.7 Conclusion—afterthought

The purpose of the preceding sections (3.3–3.6) was to describe the variety of correlation effects reported in marine algae. Although considerable space could have been spent speculating on various hypothetical mechanisms 'causing' the observed phenomena, this exercise has been intentionally left up to the reader with the hope that the brief outline of 'cases' will serve to stimulate interest in exploring the physiological mechanisms of the many intriguing interactions occurring within algal thalli.

4 CONTROL OF DEVELOPMENT

In discussing the development of form of organisms, J. T. Bonner stated, 'growth is used here and there, sometimes encouraged, sometimes discouraged and in such a way that a consistent whole individual organism is created in an orderly and masterful fashion' (Bonner 1952). As with cell morphogenesis (Section 2) the quantitative characterization of rate and direction of surface expansion of the organism is the basis on which an understanding of the factors controlling development can be formulated. The dispositions and shapes of axes, as well as the life span of a thallus interact to determine the final morphology of an alga (Dixon 1963). There have been few attempts to mathematically describe the orderly process of frond development and organization in seaweeds, but methods of Erickson (cf. Konrad-Hawkins 1964b; Erickson 1976) and Lück (1965) have wide application in quantitative growth analysis. Our knowledge of endogenous mechanisms (physicochemical) which regulate seaweed growth and morphogenesis is scanty, at best. This area of study has been beset by problems of technique as well as data interpretation, and perhaps most of all by the tacit assumption that a single chemical may be involved in some fundamental aspect of growth common to the three phyla of macroscopic marine algae. A number of exciting approaches are currently being employed to identify origins, and ultimately chemical structures, of substances which appear to have a role in regulation of seaweed growth and development.

4.1 Endogenous factors

Many of the correlation effects discussed above provide circumstantial evidence for endogenous control of growth and morphogenesis. Intercellular communication for such control may utilize symplastic and/or apoplastic routes. The former could involve plasmadesmata (Phaeophyta) or pit connections (Rhodophyta) (Ramus 1969; Marchant 1976; Wetherbee 1979).

Plasmadesmata are only present between daughter cells in some species of green algae, and secondary cytoplasmic connections are not formed between other adjacent cells (Marchant 1976). An understanding of control mechanisms may progress more rapidly in the case of apoplastic transfer where substances are secreted from cells and transport occurs through cell walls or even outside the thallus. Some evidence for this mode of transport is rather indirect, i.e. the so-called 'group-effect' in rhizoid initiation in *Fucus* eggs (Whittaker 1940) and spores of some red algae (cf. Prince & Kinsbury 1973); or crowding effects on branch initiation in *Callithamnion roseum* (Konrad-Hawkins 1968; see also Fries 1963). More direct evidence for production and secretion of a growth regulator is seen in the interesting studies of S. D. Waaland (1975, 1978). Rhizoids of *Griffithsia* spp. produce a diffusible hormone, dubbed 'rhodomorphin,' which induces the apical end of a wounded cell to form a new rhizoid-like shoot. Repair shoot growth is directed toward the rhizoid and eventually the two cells fuse to form a new shoot-like cell (see Fig. 18.6). Production of the hormone begins 1.5–2.0 h after induction of new repair rhizoids on cut *Griffithsia* cells. Fusion of rhizoid and regenerating shoot cells is also species-specific. Perhaps a substance similar to 'rhodomorphin' may be involved in directing the growth of the 'oöblast' to the auxillary cell during reproduction in certain Rhodophyta.

In studies of morphogenesis and control of development of seaweeds, no alga has received more attention than *Acetabularia* spp. and symposia proceedings have appeared regularly since 1969 (e.g. Puiseux-Dao 1975; Woodcock 1977; see also Bonnotto *et al.* 1976). Classical grafting and interspecific nuclear transplant experiments were initiated in the 1930's by Hämmerling and continued by Schweiger, Werz, Clauss, Zetsche and others. The following generalizations can be made about control of morphogenesis in *Acetabularia*. The nucleus produces species-specific 'morphogenetic substances' which are transferred to the cytoplasm where they become distributed in both

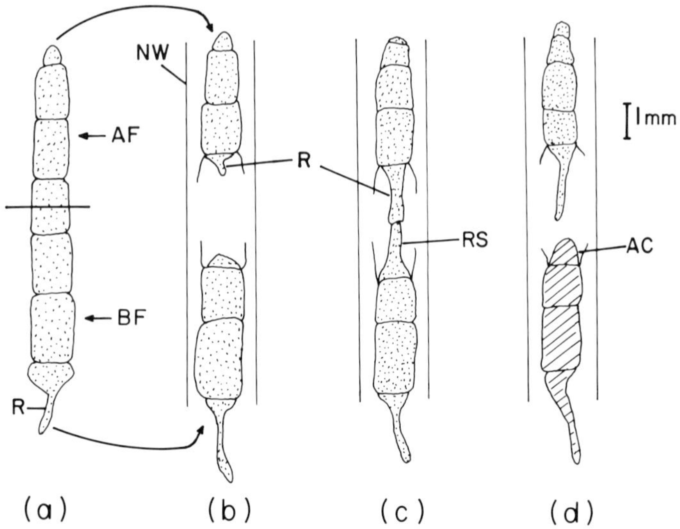

Fig. 18.6 Experimental cell repair system using *Griffithsia* spp. (a) *Griffithsia* filaments of the same species are decapitated. Basal (BF) and apical (AF) segments are inserted into a cytoplasm-free *Nitella* (NW) cell wall cylinder in (b). Regenerating rhizoid (R) in apical segment induces formation of a repair-shoot cell (RS) from the basal segment, when both shoots are of the same species, and fusion occurs in (c). When different species are apposed, as in (d), the decapitated segment only regenerates an apical cell (AC). (Redrawn from Waaland 1975.)

apico-basal and baso-apical gradients of cap- and rhizoid-forming substances, respectively. Species-specific substances accumulate and are stored in the cytoplasm for considerable periods of time and the expression of these substances is apparently induced by a non-species-specific material, possibly a special protein. A second group of researchers, headed by Brachet, investigating aspects of the biochemistry of *Acetabularia* proposed that an 'informational RNA' was transported to the cytoplasm from the nucleus in the form of stable molecules. The hypothetical mRNAs (similar to Hämmerling's 'morphogenetic substances') have not yet been isolated. In principal such messengers should be able to induce species-specific morphogenesis in non-growing anucleate fragments of *Acetabularia*. Alexeev and co-workers (1975) have achieved partial success with such experiments.

4.2 Exogenous factors

Growth and morphogenesis are also controlled by exogenously produced factors. In 1964, Provasoli & Pintner reported their classic experiments on growth of *Gayralia oxysperma*. When cultured under axenic (bacteria-free) conditions, the alga lost its typical morphology. Normal morphology was restored by reintroducing marine bacteria and other microorganisms or autoclaved bacteria cultures. Supernatants of bacteria-free cultures of six species of red and brown algae also effected normal development. Provasoli & Carlucci (1974) have reviewed other studies in which axenic marinc algae were supplied with bacteria and/or bacteria culture filtrates in an attempt to induce normal growth and morphogenesis. Unfortunately, this important area of developmental algology has attracted few workers in recent years (e.g. Tsukidate 1970; Waite & Mitchell 1976; Lockhart 1977; Saga *et al.* 1978); yet it holds promise of revealing fundamental information about regulation of growth and development in marine algae. For example, methanolic extracts of *Neodilsea yendoana* (Cryptonemiales) contain at least two kinds of factors, tentatively identified as a peptide and a saturated hydrocarbon, which are 'active' morphogenetic substances for axenically cultured *Monostroma* spp. (Takewaki & Kaneko 1977; also see Fries 1973). Provasoli *et al.* (1977) have reported a ninhydrin-positive morphogenetic substance isolated from the culture supernatant of a marine *Caulobacter* which is active at 0.3–0.6 μg 100^{-1} ml. Six other bacterial isolates also produced morphogenetically active substances (also see Fries 1970). Kong & Chan (1979) found differences between the bacterial flora isolated from three Chlorophyta, two Phaeophyta, and two Rhodophyta growing in the same habitat. Dominant flora of the green algae were strains of *Micrococcus* and *Achromobacter* while strains of *Alteromonas, Xanthomonas,* and yellow and orange pigmented strains of *Flavobacterium* were characteristic of the brown algae. Red algae were host to bacterial strains found on both green and brown algae. It is plausible that organic solutes secreted by the algal host from each phylum are utilized by phylum-specific strains of bacteria with a reciprocal benefit to the host derived from morphogenetic factors synthesized by the bacteria.

Provasoli & Pintner (personal communication) found that previously active bacterial isolates gradually lost their ability to effect normal morphological development when added to axenicaly cultured *Ulva lactuca*. However, these bacteria could re-acquire their activity if grown with *U. lactuca* through several transfers. Thus there exists a reciprocal relationship between bacteria and the algal host where substances released from *U. lactuca* presumably derepress a bacterial enzyme system required to synthesize a morphogenetic factor which, in turn, is effective in *U. lactuca*. Although various morphologies of *U. lactuca* were inducible by culturing it with several combinations of bacterial isolates, non-mutational polymorphism has been demonstrated in successive generations of axenically cultured *U. lactuca* (Bonneau 1977).

Epiphytic bacteria not only contribute morphogenetic factors necessary for normal thallus development but they may also cause hyperplasia of medullary tissue and gall formation (Andrews 1976). In EM photomicrographs of algal galls, bacteria are seen to occupy intercellular spaces and are sometimes buried in mucilaginous cell walls (McBride *et al.* 1974). Künzenbach & Brucker (1960) were unable to infect *Fucus serratus* with *Agrobacterium tumifaciens*, the bacterium responsible for crown galls in higher plants, and young tomato plants injected with a suspension of bacteria initially isolated from *Fucus* galls failed to produce crown galls. The bacterial–algal interaction correlated with gall formation is another provocative area for future research.

4.3 Higher plant growth regulators

A thorough catalogue of algal studies involving higher plant hormones was recently completed by Augier (1976 a, b, c, d, 1977 a, b, 1978). Provasoli & Carlucci (1974) reviewed some of the problems encountered in studies with higher plant growth regulators and their identification in seaweeds, and the following treatment is intended as a critical examination of the current status of this field.

Auxin

The definition of auxin (indole-3-acetic acid; IAA) as an endogenous regulator of cell expansion in higher plants is derived from experimental evidence of both its presence and action in *higher plants*. Is it present in seaweeds and does its observed action in the three seaweed phyla fit the present higher plant model? IAA and indole-3-carboxylic acid have been identified in non-axenic, natural collections (118 kg) of *Undaria pinnatifida* using a combination of GC and MS analysis (Abe *et al.* 1972). It is uncertain however whether the auxin was synthesized directly by the alga or by epiphytic microorganisms. In principle, the source of the substance is not important if, in fact, auxin has a specific growth regulatory function in seaweeds. The key point is establishing that the observed growth responses of seaweeds to added auxin—regardless of whether the responses are promotional or inhibitory—are *auxin-specific* and cannot be obtained with non-auxins. A simple comparison of effects of commercially available, synthetic auxins and their inactive analogues, e.g. 2,4-D & 2,5-D; α-NAA & β-NAA; β-NOAA & α-NOAA, respectively, is an easy test of the specificity requirement. This test was utilized in a study of blade elongation in *Alaria esculenta*. Growth responses due to auxins could not be separated from those of non-auxins suggesting that an auxin-specific recognition mechanism was lacking in *Alaria* (Buggeln 1976). However, in more recent studies on membrane electrical properties, ion and water fluxes, and turgor regulation in *Valonia utricularis* (Zimmermann *et al.* 1976; Zimmermann 1977; Zimmermann & Steudle 1977; Zimmermann 1978), IAA had a pronounced effect on membrane resistance and K^+ influx and both IAA and 2,4-D (but not 3,5-D) changed the dependence of maximum membrane resistance on pressure. The latter results indicate the presence of auxin-specific sites in the *Valonia* membrane.

How does auxin promote cell expansion in higher plants? According to one current hypothesis, auxin causes secretion of hydrogen ions into the cell wall where acidification (low pH) activates 'wall loosening' enzymes which enable irreversible cell expansion to take place (Hager *et al.* 1971). This idea is based on studies showing that growth stimulating concentrations of IAA caused a rapid secretion of H^+ ions into the external medium bathing the tissues, and that acidification alone could cause wall loosening. Does IAA induce wall loosening in algae? From measurements of the *Valonia* cell wall elastic modulus, Zimmermann (1977) could detect no wall loosening in IAA-treated cells. It is uncertain however that elastic deformation is closely connected with the control of plant cell wall extension (Cleland 1971). Tepfer & Cleland (1979) recently showed that wall loosening occurred in *V. ventricosa* in response to acidification (low pH) or addition of a chelator with a high affinity for Ca^{2+}. Their results also indicated that the hypothetical enzyme-catalysed component of the loosening process, which may be responsible for the major part of the process in *Avena* coleoptiles, is probably not involved in the *Valonia* wall loosening process. Present evidence suggests that either H^+ or a Ca^{2+} chelating substance could have an *in vivo* role in wall loosening in *Valonia*, but involvement of an auxin-like growth regulator in the secretion of wall loosening substances is still unresolved.

When non-axenic algae are employed as 'bioassays' for IAA (e.g. Hanisak 1979), one cannot distinguish between direct effects of IAA on the alga or a two-step indirect effect in which IAA stimulates production and/or release of algal growth promoting factors in bacteria epiphytizing the alga. While growth of bacterized *Codium fragile* was accelerated by addition of IAA, three synthetic auxins had no effect (Hanisak 1979). Dawes (1971) reported similar results with IAA and two synthetic auxins on bacterized *Caulerpa prolifera*.

Phenylacetic acid (PA) and its derivatives have auxin-like activity in higher plants (cf. Millborrow *et al.* 1975) and both PA and p-hydroxyphenylacetic acid (p-OHPA) have been isolated from marine algae (Abe *et al.* 1974a; Gorham 1978; Fries & Åberg 1978 and ref. cited). P-OHPA occurred as an intermediate in the biosynthesis of lanosol from tyrosine in *Odonthalia flocossa* (Manley 1977). While IAA (4×10^{-7} to 2×10^{-5} M) had no effect on growth of axenically cultured conchocelis phase of *P. tenera*, both PA and p-OHPA increased the growth rate 0.5- to 2.0-fold. It is not clear how these results indicate that 'p-hydroxyphenyacetic acid as well as phenylacetic acid have auxin-like effects on red algae' (Fries & Iwasaki 1976). Another problem confronting researchers who are attempting to identify chemical growth regulators in seaweeds is distinguishing between: 1, non-vitamin organic factor with a metabolic role(s) and which may be present in rate-limiting concentrations in axenically cultured algae (see Fries 1970, 1974); and 2, a specific chemical regulator of growth and morphogenesis. For example, growth stimulation of axenic *Fucus spiralis* (Fries 1977) by PA and p-OHPA may be due to either 1 or 2 above. Growth of axenically cultured red algae is enhanced by a wide variety of phenolic compounds although culture age can modify effects of these substances (Fries 1974, Fries & Iwaski 1976).

Apical dominance has been reportedly restored in decapitated algae by addition of auxins (Moss 1965; Perrone & Felicini 1974; Chamberlain *et al.* 1979), but more conclusive evidence is needed to show that the effects are auxin-specific. In a comparison of responses of decapitated explants to auxins and non-auxins, one would expect that the responses of the latter group should not differ from untreated controls. *Fucus vesiculosus* segments (minus apical cells) in seawater containing the non-auxin, α-NOAA (1-naphthoxyacetic acid), regenerated $>3\times$ the number of shoots as the control segments (Moss 1965). IAA (10–50 ppm) inhibited both elongation of secondary laterals as well as overall growth of decapitated primary shoots of *Sargassum muticum* (Chamberlain *et al.* 1979). PA and p-OHPA (10^{-4} M) inhibited primary lateral growth but stimulated secondary lateral elongation (Gorham 1978). Inhibition of *Alaria esculenta* blade growth by auxins and non-auxins was correlated with structural alterations of chloroplasts and inhibition of photosynthesis (Buggeln & Bal 1977).

Gibberellins

Fifty-two different gibberellins (GA_n) have now been characterized in fungi and higher plants but it is not known how many of these possess an *in vivo* growth regulatory role (Hedden *et al.* 1978). GA-like activity has been demonstrated in extracts of natural collection of seaweeds (cf. Augier 1976b) and seawater (Fries 1973) but chemical identification has not been unequivocally established. Nevertheless, some physiological studies suggest that GA_n may have a role in the growth and development of algae. In the absence of suitable algal bioassays (Taylor & Wilkinson 1977), and with knowledge that not all GA_n are effective in the same bioassay anyway (Lang 1970), there is nevertheless, a potent tool for studying the role of suspected GA-like substances in seaweeds. It involves the use of growth retardants which act by inhibiting the biosynthesis of GA_n, e.g. CCC, AMO-1618, and phosphon-D (Lang 1970). They have been shown to reduce endogenous levels of GA_3 in higher plants and to arrest growth. When GA_3 is added to a CCC-inhibited plant, the endogenous GA_3 level is raised, bypassing the inhibited biosynthetic step, and growth is restored. Only two studies with seaweeds have utilized this approach. Growth of *Ecklonia radiata* gametophytes was inhibited by CCC and subsequent addition of GA_3 overcame the inhibition (Jennings 1971). CCC-inhibited growth of secondary laterals of *Sargassum muticum* was partially overcome by addition of GA_3 (Gorham, 1977a). In neither study however was the level of GA-like substances assayed before or after application of CCC to determine that the inhibitor was affecting GA biosynthesis in these algae. Contrary to the latter results, CCC- and AMO-inhibited growth of excised branches of non-axenic *Hypnea musciformis* and *Gracilaria verrucosa* was not reversed by addition of GA_3 or GA_7 or GA-like material extracted from *H. musciformis*. The possibilities of breakdown or poor penetration of the added GA_n were considered. Levels of extractable GA-like activity from actively meristematic regions of *E. radiata* sporophytes were higher than levels extracted from other regions, while extracts of *S. muticum* contained only low levels of GA-like activity.

The seasonal maximum in GA_3-equivalent activity in extracts of *Laminaria saccharina, L. digitata,* & *L. longicruris* occurred in July (Wildgoose *et al.* 1978). This contrasts with maximum blade growth rates which have been observed in late winter for the former two species (Parke 1948; Sundene 1964; Mann 1972) and in May for the latter (Lüning 1969).

Cytokinins

Cytokinins appear to be involved in all stages of higher plant morphogenesis (Hall 1973). While they have been characterized as compounds composed of 'one hydrophobic group of high specificity (adenine) and one lipophyllic group without specificity' (Fox 1969), there are synthetic phenyl-ureas which possess cytokinin activity. The literature on seaweed cytokinin research has been reviewed by Augier (1976b). Again it is uncertain whether seaweeds and/or their epiphytic bacteria produce cytokinin-like substances. For years it has been appreciated that seawater collected from the *Fucus–Ascophyllum* zone was preferred for culturing algae. At least part of the growth stimulating activity of this seawater may be attributed to the cytokinin, N^6-(Δ^2-isopentenyl) adenine (6-(3-methyl-2-butenylamino)purine) identified by Pedersén (1973). This chemical showed growth-stimulating activity when added to non-axenic cultures of *Ectocarpus confervoides* and *Pilayella litoralis*. However, more interesting results were obtained with *E. fasciculatus* which only developed a prostrate system in axenic culture. Addition of the cytokinin, kinetin (6-furfuryl amino purine), initiated production of erect axes characteristic of normal morphology of this species. It is possible that a system like this may be suitable as a qualitative algal bioassay for cytokinins if it can be shown that the prostrate → erect conversion specifically requires a cytokinin. Effect of the anticytokinin, 3-methyl-7-(3-methylbutylamine)pyrazolo[4,3-d]pyrimidine (Skoog *et al.* 1973; Hecht 1979) would be instructive in experiments to determine cytokinin specificity.

Borowczak *et al.* (1978) reported that cytokinin-like activity in non-axenic *Fucus vesiculosus* was greater in July than April and kinetin addition stimulated regeneration in *F. vesiculosus* explants excised in April but not in those isolated in July. It is possible that cytokinin activity in seawater may vary with season.

Miscellaneous compounds and inhibitors

Cyclic-AMP (c-AMP), a so-called 'secondary messenger' in animal cells, has been reported in *Acetabularia mediterranea* (Vanden Driessche 1976), *Vaucheria geminata* (Kataoka 1977a) and *Laminaria saccharina* exudate (Becker & Ziegler

1973). However, identification of cAMP in higher plants is suspect and recent critical examination of radio-immunoassay techniques, chromatographic behaviour, and substrate activity for phosphodiesterase reveals that the higher plant substance has a number of properties which are not identical with authentic cAMP (Braun 1978). Seaweed data should therefore be considered with caution.

Watanabe & Kondo (1976) who reported auxin-stimulated ethylene evolution in long term (3 day) experiments with non-axenic *Porphyra tenera, Codium latum* and *Padina arborescens,* pointed out that either algae or bacteria could have produced the gas. In other experiments they observed that ethylene had no effect on the growth rate of *P. tenera*. No effect of ethylene on elongation of *Alaria esculenta* blades was observed either (Buggeln unpublished). An inhibitor of ethylene evolution has also been isolated from *P. tenera* (Watanabe & Kondo 1976; Watanabe *et al.* 1978).

Growth of the crustose alga *Rhodophysema elegans* (Rhodophyta) is inhibited by another crustose form, *Ralfsia spongiocarpa* (Phaeophyta), when the two are growing in close proximity to each other. The inhibitors secreted by *Ralfsia* have not been identified but may be phenolic compounds (Fletcher 1975b). Antialgal substances isolated from *Undaria pinnatifida* by Abe *et al.* (1974b) were a mixture of meristic, palmitic, oleic, and linoleic acids.

A substance with properties of a GA-antagonist was isolated from *Ecklonia radiata* (Jennings 1969) but its chromatographic properties were different from abscissic acid (ABA). Likewise, inhibitors isolated from *Laminaria digitata* and *Ascophyllum nodosum* showed similarities to ABA in higher plant bioassays but chromatographic evidence was less clear (Hussain & Boney 1973). It was originally proposed that lunularic acid (3,4′-dihydroxybibenzyl-2-carboxylic acid) may be present in algae in place of the higher plant inhibitor, ABA (Pryce 1972) but re-investigation has not confirmed the presence of the former compound in marine algae (Gorham 1977b).

An unidentified sporulation inhibiting substance has recently been isolated by Nilsen & Nordby (1975) from thalli of *Ulva mutabilis*.

ACKNOWLEDGEMENTS

The author is indebted to Dr. L. Provasoli for providing an unpublished manuscript and to Drs. C. J. Dawes and S. D. Waaland for constructive criticism of an early draft of the present work. National Research Council of Canada supported the author's research and preparation of this review. M.S.R.L. Publication No. 325.

5 REFERENCES

Abe H., Uchiyama M. & Sato R. (1972) Isolation and identification of native auxins in marine algae. *Agr. Biol. Chem.*, **36**, 2259–60.

ABE H., UCHIYAMA M. & SATO R. (1974a) Isolation of phenylacetic acid and its p-hydroxy derivative as auxin-like substances from *Undaria pinnatifida*. *Agr. Biol. Chem.*, **38**, 897–8.

ABE H., UCHIYAMA M., SATO R. & MUTO S. (1974b) Plant growth regulators in marine algae. In *Plant Growth Substances—1973*, ed. Sumiki Y., pp. 201–6. Hirokawa Publ. Co., Tokyo.

ABELARD C. (1977) Essai de mise en évidence des corrélations morphogènes entre les différents éléments d'un même cladome chez l'*Apoglossum ruscifolium* (Turner) J. Ag. (Delesseriacée, Ceramiales). *Bull. Soc. Phycol. Fr.*, **22**, 111–19.

ABELARD C. & L'HARDY-HALOS M.-T. (1975) Corrélations morphogènes entre les différents cladomes d'une même fronde chez l'*Apoglossum ruscifolium* (Turner) J. Ag. (Rhodophycée, Delesseriacée). *Botaniste*, **57**, 15–28.

ALEXEEV A. B., STVOLINSKY S. L., TIKHOMIROVA L. A., YAZYKOV A. A. & ZUBAREV T. N. (1975) Messenger ribonucleoproteins of *Acetabularia*. *Protoplasma*, **83**, 168.

ANDREWS J. H. (1976) The pathology of marine algae. *Biol. Rev.*, **51**, 211–53.

AUGIER H. (1976a) Les hormones des algues. État actuel des connaissances. I. Recherche et tentatives d'identification des auxines. *Bot. Mar.*, **19**, 127–43.

AUGIER H. (1976b) Les hormones des algues. État actuel des connaissances. II. Recherche et tentatives d'identification des gibbérellines, des cytokinines et diverses autres substances de nature hormonale. *Bot. Mar.*, **19**, 245–54.

AUGIER H. (1976c) Les hormones des algues. État actuel des connaissances. III. Rôle des hormones dans les modalités de croissance et de développement des thalles. *Bot. Mar.*, **19**, 351–77.

AUGIER H. (1976d) Les hormones des algues. État actuel des connaissances. IV. Rôle des hormones dans les divers métabolismes cellularies et dans les mécanismes de reproduction sexuée et asexuée; rôle écologique. *Bot. Mar.*, **20**, 1–11.

AUGIER H. (1977a) Les hormones des algues. État actuel des connaissances. V. Index alphabétique par espèces des travaux de caractérisation des hormones endogènes. *Bot. Mar.*, **20**, 187–203.

AUGIER H. (1977b) Les hormones des algues. État actuel des connaissances. VI. Index alphabétique par espèces des travaux sur le rôle des hormones dans la vie des algues. *Bot. Mar.*, **20**, 363–79.

AUGIER H. (1978) Les hormones des algues. État actuel des connaissances. VII. Application, conclusion, bibliographie. *Bot. Mar.*, **21**, 175–97.

BANBURY G. H. (1962) Geotropism of lower plants. In *Handbuch der Pflanzenphysiologie*, vol. 17, ed. Ruhland W., pp. 344–77. Springer-Verlag, Berlin.

BECKER D. & ZIEGLER H. (1973) Cyclisches adenosine-3′:5′-monophosphat in pflanzlichen Leitbahnen? *Planta*, **110**, 85–9.

BENTRUP F. W. (1977) Electrical events during apex regeneration in *Acetabularia mediterranea*. In *Progress in Acetabularia Research*, ed. Woodcock C. L. F., pp. 249–54. Academic Press, N.Y.

BERTHOLD G. (1882) Beiträge zur Morphologie und Physiologie der Meeresalgen. *Jahrb. Wiss. Bot.*, **13**, 567–717.

BLINKS L. R. (1951) Physiology and biochemistry of algae. In *Manual of Phycology*, ed. Smith G. M., pp. 263–91. Chronica Botanica Co., Waltham, Mass.

BLOCK R. (1965) Polarity and gradients in plants: a survey. In *Handbuch der Pflanzenphysiologie*, vol. 15, No. 1, ed. Ruhland W., pp. 234–74. Springer-Verlag, Berlin.

BONEY A. D. (1975) Regenerates from excised pseudolateral branches of *Heterosiphonia plumosa* (Ellis). Batt. *New Phytol.*, **74**, 93–100.

BONNEAU E. R. (1977) Polymorphic behavior of *Ulva lactuca* (Chlorophyta) in axenic culture. I. Occurrence of *Enteromorpha*-like plants in haploid clones. *J. Phycol.*, **13**, 133–40.

BONNER J. T. (1952) *Morphogenesis: an Essay on Development*. Princeton University Press, Princeton, New Jersey, U.S.A.

BONOTTO S., LURQUIN P. & MAZZA A. (1976) Recent advances in research on the marine alga, *Acetabularia*. *Adv. Mar. Biol.*, **14**, 123–250.

BONOTTO S. & SIRONVAL C. (1977) Experimental studies on the phototropism of *Acetabularia mediterranea* and *Acetabularia crenulata*. In *Progress in Acetabularia Research*, ed. Woodcock C. L. F., pp. 241–7. Academic Press, London & New York.

BORDEN C. A. & STEIN J. R. (1969) Reproduction and early development in *Codium fragile* (Suringar) Hariot: Chlorophyceae. *Phycologia*, **8**, 91–9.

BØRGESEN F. (1951) Some algae from Mauritius. Additions to parts previously published, III. *Kg. Danske Vidensk. Selsk., Biol. Medd.*, **18** (16), 1–44.

BOROWCZAK E., KENTZER T. & POTULSKA-KLEIN B. (1977) Effect of gibberellin and kinetin on the regeneration ability of *Fucus vesiculosus* L. *Biol. Plant.*, **19**, 405–12.

BOROWIKOV G. A. (1914) Reversal of polarity in *Cladophora glomerata*. *Leningrad Glavnyi Bot. Sad. Izvestia*, **14**, 475–81. (Russian with French summary.)

BOWEN K. Y. (1971) The growth and development of the deep growing marine alga, *Maripelta rotata* (Daws.) Daws. Ph.D. Dissertation, University of California, San Diego.

BRAUN A. C. (1978) Plant tumors. *Biochem. Biophys. Acta.*, **516**, 167–91.

BUGGELN R. G. (1974) Negative phototropism of the haptera of *Alaria esculenta* (Laminariales). *J. Phycol.*, **18**, 80–2.

BUGGELN R. G. (1976) Auxin, an endogenous regulator of growth in algae? *J. Phycol.*, **12**, 355–8.

BUGGELN R. G. & BAL A. K. (1977) Effects of auxins and chemically related non-auxins on photosynthesis and chloroplast ultrastructure in *Alaria esculenta* (Laminariales). *Can. J. Bot.*, **55**, 2098–105.

BÜNNING E. (1952) Morphogenesis in plants. *Sur. Biol. Progr.*, **2**, 105–40.

BURR F. A. & WEST J. A. (1971) Protein bodies in *Bryopsis hyfnoides*: their relationship to wound healing and branch septum development. *J. Ultrastruct. Res.*, **35**, 476–98.

CHAMBERLAIN A. H. L., GORHAM J., KANE D. F. & LEWEY S. A. (1979) Laboratory growth studies on *Sargassum muticum* (Yendo) Fensholt. *Bot. Mar.*, **22**, 11–19.

CHEN J. C. W. (1971) Effects of continuous illumination of the rhizome of *Caulerpa prolifera*. *Plant & Cell Physiol.*, **12**, 895–905.

CHEN J. C. W. & JACOBS W. P. (1966) Quantitative study of development of the giant coenocyte, *Caulerpa prolifera*. *Am. J. Bot.*, **53**, 413–23.

CHEN L. C-M. & TAYLOR A. R. A. (1978) Medullary tissue culture of the red alga, *Chondrus crispus*. *Can. J. Bot.*, **56**, 883–6.

CHIEN P. K. (1972) Attachment processes and early development by kelp embryos. A light and electron microscope study. In *Kelp Habitat Improvement Project Annual Rept. (1971–72)*, ed. North W., pp. 108–220. W. M. Keck Engin. Lab., Calif. Inst. Tech., Pasadena.

CHIEN P. K. (1973) Attachment of *Macrocystis* to natural and artificial substrates. In *Kelp Habitat Improvement Project Annual Rept. (1972–73)*, ed. North W., pp. 89–106. W. M. Keck Engin. Lab., Calif. Inst. Tech., Pasadena.

CLELAND R. (1971) Cell wall extension. *Ann. Rev. Plant Physiol.*, **22**, 197–222.

CRIBB A. B. (1951) Invalidation of the genus *Vaughaniella*. *Nature*, (*Lond.*), **168**, 302.

DAHL A. L. (1971) Development, form and environment in the brown alga, *Zonaria farlowii* (Dictyotales). *Bot. Mar.*, **14**, 76–112.

DARSIE M. L. (1939) Certain aspects of phototropism, growth and polarity in the single-celled marine alga, *Bryopsis*. Ph.D. Dissertation, Stanford University.

DAWES C. J. (1971) Indole-3-acetic acid in the green algal coenocyte *Caulerpa prolifera* (Chlorophycease, Siphonales). *Phycologia*, **10**, 375–9.

DAWES C. J. & BARILOTTI D. C. (1969) Cytoplasmic organization and rhythmic streaming in growing blades of *Caulerpa prolifera*. *Am. J. Bot.*, **56**, 8–15.

DEVALERA M. (1975) Contrasting features of apical growth in *Chondra dasyphylla* and *C. tenuissima*. *Br. Phycol. J.*, **10**, (suppl.), 314.

DIXON P. S. (1963) Variation and speciation in marine Rhodophyta. In *Speciation in the Sea*, eds. Harding J. P. & Tebble N, pp. 51–62. Systematics Association, London.

DIXON P. S. (1971) Cell enlargement in relation to the development of thallus form in Florideophyceae. *Br. Phycol. J.*, **6**, 195–205.

DIXON P. S. (1973) *Biology of the Rhodophyta*. Oliver & Boyd, Edinburgh.

DIXON P. S. & RICHARDSON W. N. (1970) Growth and reproduction in red algae in relation to light and dark cycles. In *Phylogenesis and Morphogenesis in the Algae*, eds. Fredrick J. F. & Klein R. M., *Ann. N.Y. Acad. Sci.*, **175**, 764–77.

DOSTAL R. (1945) Morphogenetic studies on *Caulerpa prolifera*. *Ceská akad. véd a umeri v. Proze Bull. Int. (Sci. Math. Nat.)*., **46**, 133–49.

DUCREUX P. (1977) Étude expérimentale des corrélations et des possibilités de régénération au niveau de l'apex de *Sphacelaria cirrhosa*. Ag. *Ann. Sci. Nat. Bot.*, **18**, 163–84.

DUFFIELD E. C. S., WAALAND S. D. & CLELAND R. (1972) Morphogenesis in the red alga, *Griffithsia pacifica*: regeneration from single cells. *Planta*, **105**, 185–95.

EATON J. W., BROWN J. G. & ROUND F. E. (1966) Some observations on polarity and regeneration in *Enteromorpha*. *Br. Phycol. Bull.* **3**, 53–62.

EDELSTEIN T. (1977) Studies on *Gracilaria* sp.: Experiments on inocula incubated under greenhouse conditions. *J. Exp. Mar. Biol. Ecol.*, **30**, 249–59.

ERICKSON R. O. (1976) Modeling in plant growth. *Ann. Rev. Plant. Physiol.*, **27**, 407–34.

ETHERINGTON J. (1964) Rhizoid formation and fragmentation in *Feldmania globifera* (Kütz.) Hamel. *Br. Phycol. Bull.*, **2**, 373–5.

FAGERBERG W. R. & DAWES C. J. (1976) Studies on *Sargassum*. I. A light microscopic examination of the wound regeneration process in mature stipes of *S. filipendula*. *Am. J. Bot.*, **63**, 110–19.

FELDMANN J. (1936) Sur le phototropism du *Derbesia lamourouxii*. *Rev. Algol.*, **9**, 145–7.

FELICINI G. P. (1970) Ricerche sulla rigenerazione in coltura di *Pterocladia capillaceae* (Gmel.) Born. & Thur. II. Influenza della intensità luminosa sulla morphologia del tallo. *Giorn. Bot. Ital.*, **104**, 35–47.

FLETCHER R. L. (1975a) Studies on recently introduced brown alga *Sargassum muticum* (Yendo) Fensholt. II. Regenerative ability. *Bot. Mar.*, **18**, 157–62.

FLETCHER R. L. (1975b) Heteroantagonism observed in mixed algal cultures. *Nature*, **253**, 534–5.

FOX J. E. (1969) The cytokinnins. In *The Physiology of Plant Growth and Development*, ed. Wilkins M. B. pp 85–123. McGraw-Hill, London.

FRALICK R. A. & MATHIESON A. C. (1972) Winter fragmentation of *Codium fragile* (Suringar) Hariot ssp. *tomentosoides* (van Goor) Silva (Chlorophycease, Siphonales) in New England. *Phycologia*, **11**, 67–70.

FRIEDMANN E. I. & ROTH W. C. (1977) Development of the siphonous green alga *Penicillus* and the *Espera* state. *Bot. J. Linn. Soc.*, **74**, 189–214.

FRIES L. (1963) On the cultivation of axenic red algae. *Physiol. Plant.*, **16**, 695–708.

FRIES L. (1970) The influence of microamounts of organic substances other than vitamins on the growth of some red algae in axenic culture. *Br. Phycol. J.*, **5**, 39–46.

FRIES L. (1973) Requirements for organic substances in seaweeds. *Bot. Mar.*, **16**, 19–31.

FRIES L. (1974) Growth stimulation of axenic red algae by simple phenolic compounds. *J. Exp. Mar. Biol. Ecol.*, **15**, 1–9.

FRIES L. (1977) Growth regulating effects of phenylacetic acid and p-hydroxyphenylacetic acid on *Fucus spiralis* L. (Phaeophycease, Fucales) in axenic culture. *Phycologia*, **16**, 451–5.

FRIES L. & ÅBERG S. (1978) Morphogenetic effects of phenylacetic acid and p-OH-phenylacetic acid on the green alga *Enteromorpha compressa* (L.). Grev. in axenic culture. *Z. Pflanzenphysiol.*, **88**, 383–8.

FRIES L. & IWASAKI H. (1976) P-hydroxyphenylacetic acid and other phenolic compounds as growth stimulators of the red alga *Porphyra tenera*. *Plant Sci. Letters*, **6**, 299–308.

FULCHER G. R. & MCCULLY M. E. (1969) Histological studies on the genus *Fucus*. IV. Regeneration and adventive embryony. *Can. J. Bot.*, **47**, 1643–9.

FULCHER G. R. & MCCULLY M. E. (1971) Histological studies on the genus *Fucus*. V. An autoradiographic and electron microscopic study of the early stages of regeneration. *Can. J. Bot.*, **49**, 161–5.

GARBARY D. J., GRUND D. & MCLAUGHLIN J. (1978) The taxonomic status of *Ceramium rubrum* (Huds.). C. Ag. (Ceramiales, Rhodophyceae) based on culture experiments. *Phycologia*, **17**, 85–94.

GAILLARD J. & L'HARDY-HALOS M.-T. (1977) A propos de la morphogénèse du *Dictyota dichotoma* (Huds.) Lamoureux (Phéophycée, Dictyotales); phénomènes corrélatifs mis en évidence sur les tronçons apicaux isolés experimentalement. *Rev. Algol.* N.S., **12**, 101–10.

GAYRAL P. (1942) Viviparité, multiplication asexuée et régénération chez la Dictyotacée *Spatoglossum solierii* (Chauv.) Kütz. *Rev. Algol.*, N.S. **6**, 105–11.

GIBOR A. (1977) Cell elongation in *Acetabularia*. In *Progress in Acetabularia Research*, ed. Woodcock C. L. F., pp 231–9. Academic Press, London & New York.

GODDARD R. H. & DAWES C. J. (1979) An ultrastructural and histochemical study of the wound response of *Caulerpa ashmeadii* (Caulerpales). *J. Phycol.*, **15** (suppl.), 17.

GOLDSTEIN M. C. (1973) Regeneration and vegetative propagation of the agarophyte *Gracilaria debilis* (Forsskål) Boerg. (Rhodophyceae). *Bot. Mar.*, **16**, 226–8.

GORHAM J. (1977a) Plant growth regulators in *Sargassum muticum*. *J. Phycol.*, **13** (suppl.), 25.

GORHAM J. (1977b) Lunularic acid and related compounds in liverworts, algae, and *Hydrangea*. *Phytochem.*, **16**, 249–53.

GORHAM J. (1978) Auxins in *Sargassum*. *Br. phycol. J.*, **13**, 200.

GREEN P. B. (1969) Cell morphogenesis. *Ann. Rev. Plant Physiol.*, **20**, 365–94.

GREEN P. B., ERICKSON R. O. & RICHMOND P. A. (1970) On the physical basis of cell morphogenesis. In *Phylogenesis and Morphogenesis in the Algae*, eds Fredrick J. F. & Klein R. M., *Ann. N. Y. Acad. Sci.*, **175**, 712–31.

HAGER A., MENZEL H. & KRAUSS A. (1971) Versuche und Hypothese zur primarwirkung des auxins beim Streckungswachstum. *Planta*, **100**, 47–75.

HALL R. H. (1973) Cytokinnins as a probe of developmental processes. *Ann. Rev. Plant Physiol.*, **24**, 415–44.

HANISAK M. D. (1979) Effect of indole-3-acetic acid on growth of *Codium fragile* subsp. *tomentosoides* (Chlorophyceae) in culture. *J. Phycol.*, **15**, 124–7.

HECHT S. M. (1979) Anticytokinins as probes of cytokinin utilization. In *Plant Growth Substances*, ed. Mandava. N. B., pp. 79–98. Am. Chem. Soc., Washington.

HEDDEN P., MACMILLAN J. & PHINNEY B. O. (1978) The metabolism of the gibberellins. *Ann. Rev. Plant Physiol.*, **29**, 149–92.

HOGETSU T. & SHIBAOKA H. (1978a) The change of pattern in microfibril arrangement on the inner surface of the cell wall of *Closterium acerosum* during cell growth. *Planta*, **140**, 7–14.

HOGETSU T. & SHIBAOKA H. (1978b) Effects of colchicine on cell shape and on microfibril arrangement in the cell wall of *Closterium acerosum*. *Planta*, **140**, 15–18.

HURD A. M. (1920) Effect of unilateral monochromatic light and group orientation on the polarity of germinating *Fucus* spores. *Bot. Gaz.*, **70**, 25–50.

HUSSAIN A. & BONEY A. D. (1973) Hydrophyllic growth inhibitors from *Laminaria* and *Ascophyllum*. *New Phytol.*, **72**, 403–10.

ISHIZAWA K. & WADA S. (1979a) Growth and phototropic bending in *Boergesenia* rhizoid. *Plant & Cell Physiol.*, **20**, 973–82.

ISHIZAWA K. & WADA S. (1979b) Action spectrum of negative phototropism in *Boergesenia forbesii*. *Plant & Cell Physiol.*, **20**, 983–7.

JACOBS W. P. (1964) Rhizoid-production and regeneration of *Caulerpa prolifera*. *Pubbl. Staz. Zool. Napoli*, **34**, 185–96.

JACOBS W. P. (1970) Development and regeneration of the algal giant coenocyte, *Caulerpa*. In *Phylogenesis and Morphogenesis in the Algae*, eds Fredrick, J. F. & Klein R. M., *Ann. N. Y. Acad. Sci.*, **175**, 732–47.

JACOBS W. P. & OLSON J. (1978) Fast change in organ differentiation in a giant coenocyte (*Caulerpa*) after inversion regarding gravity. *Plant Physiol.*, **61** (suppl.), 110.

JAFFE L. F. (1958) Tropistic responses of zygotes of the Fucaceae to polarized light. *Expt. Cell Res.*, **15**, 282–99.

JAFFE L. F., ROBINSON K. R. & NUCCITELLI R. (1974) Transcellular currents and ion fluxes through developing fucoid eggs. In *Membrane Transport in Plants*, eds. Zimmermann U. & Dainty J., pp. 226–33. Springer-Verlag, New York, Heidelberg, Berlin.

JENNINGS R. C.(1969) Gibberellin antagonism by material from a brown alga. *New Phytol.*, **68**, 683–8.

JENNINGS R. C. (1971) Studies on the regulation of algal growth by gibberellin. *Aust. J. Biol. Sci.*, **24**, 1115–24.

JOHNSTON C. S., JONES R. G. & HUNT R. D. (1977) A seasonal carbon budget for a laminarian population in a Scottish sea-loch. *Helgol. Wiss. Meeresunters.*, **30**, 527–45.

JONES W. E. (1959) Experiments on some effects of certain environmental factors on *Gracilaria verrucosa* (Huds.) Papenfuss. *J. Mar. Biol. Ass. U.K.*, **38**, 153–67.

JUNIPER B. E. (1976) Geotropism. *Ann. Rev. Plant Physiol.*, **27**, 385–406.

KATAOKA H. (1975a) Phototropism in *Vaucheria geminata*. I. The action spectrum. *Plant & Cell Physiol.*, **16**, 427–37.

KATAOKA H. (1975b) Phototropism in *Vaucheria geminata*. II. The mechanism of bending and branching. *Plant & Cell Physiol.*, **16**, 439–48.

KATAOKA H. (1977a) Phototropic sensitivity in *Vaucheria geminata* regulated by 3′,5′-cyclic AMP. *Plant & Cell Physiol.*, **18**, 431–40.

KATAOKA H. (1977b) Second positive and negative phototropism in *Vaucheria geminata*. *Plant & Cell Physiol.*, **18**, 473–7.

KHAILOV K. M., KHOLODOV V. I., FIRSOV Y. K. & PRAZUKIN A. V. (1978) Thalli of *Fucus vesiculosus* in ontogenesis: changes in morpho-physiological parameters. *Bot. Mar.*, **21**, 289–311.

KONG M. K. & CHAN K. (1979) A study of the bacterial flora isolated from marine algae. *Bot. mar.*, **22**, 83–97.

KONRAD-HAWKINS E. (1964) Developmental studies on regenerates of *Callithamnion roseum* Harvey. II. Analysis of apical growth. Regulation of growth and form. *Protoplasma*, **58**, 60–74.

KONRAD-HAWKINS E. (1968) Induction of cell differentiation in dissociated cells and fragments of *Callithamnion roseum*. *Am. J. Bot.*, **55**, 255–64.

KÜNZENBACK R. & BRUCKER W. (1960) Zur Bildung von 'Tumoren' an Meeresalgen II. *Ber. Dtsch. Bot. Ges.*, **73**, 8–18.

KÜPPERS U. & KREMER B. P. (1978) Longitudinal profiles of carbon dioxide fixation capacities in marine macroalgae. *Plant Physiol.*, **62**, 49–53.

LANG A. (1965) Physiology of growth and development in algae. A synopsis. In *Handbuch der Pflanzenphysiologie*, vol. 15, no. 1, ed. Ruhland W., pp 680–715. Springer-Verlag, Berlin.

LANG A. (1970) Gibberellins: structure and metabolism. *Ann. Rev. Plant Physiol.*, **21**, 537–70.

L'HARDY-HALOS M.-T. (1971a) Recherches sur les Céramiacées (Rhodophycées-Céramiales) et leur morphogénèse. III. Observations et recherches expérimentales sur la polarité cellulaire et pour la hiérachisation des éléments de la fronde. *Rev. Gén. Bot.*, **78**, 407–91.

L'HARDY-HALOS, M.-T. (1971b) Manifestation d'une dominance apicale chez les algues à structure cladomienne du genre *Antithamnion* (Rhodophycées, Ceramiales). *C.R. Acad. Sci. Paris*, **272**, 2301–4.

LOCKHART J. C. (1977) Morphogenesis of *Cladosiphon zosterae*. (J. Ag.) Kylin (Phaeophyceae). *J. Phycol.*, **13** (suppl.), 42.

LÜCK H. B. (1975) Elementary behavioral rules as a foundation for morphogenesis. *J. Theor. Biol.*, **54**, 23–34.

LÜNING K. (1969) Growth of amputated and dark-exposed individuals of the brown alga, *Laminaria hyperborea*. *Mar. Biol.*, **2**, 218–23.

MCBRIDE D. L., KRUGENS P. & WEST J. A. (1974) Light and electron microscopic observations on red algal galls. *Protoplasma*, **79**, 249–64.

MALINOWSKI K. C. & RAMUS J. (1973) Growth of the green alga *Codium fragile* in a Connecticut estuary. *J. Phycol.*, **9**, 102–10.

MANLEY S. L. (1977) Metabolism of tyrosine by cell free fractions of the red alga *Odonthallia floccosa*: A proposed biosynthetic pathway for the brominated phenols. *J. Phycol.*, **13** (suppl.), 43.

MANN H. K. (1972) Ecological energetics of the seaweed zone in a marine bay on the Atlantic coast of Canada. II. Productivity of the seaweeds. *Mar. Biol.*, **14**, 199–209.

MARCHANT H. J. (1976) Plasmodesmata in algae and fungi. In *Intercellular Communication in Plants: Studies on Plasmadesmata*, eds. Gunning B. E. S. & Robards A. W., pp. 59–80. Springer-Verlag, Berlin.

MIEHE H. (1905) Wachstum, Regeneration und Polarität isolierter Zellen. *Ber. Dtsch Bot. Ges.*, **23**, 257–64.

MIGITA S. & KIM C. R. (1971) Studies on horizontal growth of *Conchocelis*. *Bull. Fac. Fish. Nagashi Univ.*, **30**, 1–8. (Japanese with English summary.)

MILLBORROW M. E., PURSE J. G. & WIGHTMAN F. (1975) On the auxin activity of phenylacetic acid. *Ann. Bot.*, **39**, 1143–6.

MOSS B. L. (1964) Growth and regeneration of *Fucus vesiculosus* in culture. *Br. Phycol. Bull.*, **2**, 377–80.

MOSS B. (1965) Apical dominance in *Fucus vesiculosus*. *New Phytol.*, **64**, 387–92.

MOSS B. (1966) Polarity and apical dominance in *Fucus vesiculosus*. *Br. Phycol. Bull.*, **3**, 31–5.

MOSS B. (1967) The culture of fertile tissue of *Fucus vesiculosus*. *Br. Phycol. Bull.*, **3**, 209–12.

MOSS B. L. (1968) The transition from vegetative to fertile tissue in *Fucus vesiculosus*. *Br. Phycol. Bull.*, **3**, 567–73.

MOSS B. (1970) Meristems and growth control in *Ascophyllum nodosum* (L.) Le Jolis. *New Phytol.*, **69**, 253–60.

MOSS B. (1974) Morphogenesis. In *Algal Physiology and Biochemistry*, ed. Stewart W. D. P., pp. 788–813. Blackwell Scientific Publications, Oxford.

MOSS B. & MARSLAND A. (1976) Regeneration of *Enteromorpha*. *Br. Phycol. J.*, **11**, 309–13.

MÜLLER-STOLL W. R. (1952) Über Regeneration und Polarität bei *Enteromorpha*. *Flora*, **139**, 148–80.

MÜLLER-STOLL W. R. (1965) Regeneration bei niederen Pflanzen (in Physiologisher Betrachtung). In *Handbuch der Pflanzenphysiologie*, vol. 15, no. 2, ed. Ruhland W. pp. 92–153. Springer-Verlag, Berlin.

MURRAY S. N. & DIXON P. S. (1973) The effect of light intensity and light period on the development of thallus form in the marine red alga *Pleonosporium squarrulosum* (Harvey) Abbott (Rhodophyta: Ceramiales). I. Apical cell division—main axes. *J. Exp. Mar. Biol. Ecol.*, **13**, 15–27.

MURRAY S. N. & DIXON P. S. (1975) The effects of light intensity and light period on the development of thallus form in the marine red alga *Pleonosporium squarrulosum* (Harvey) Abbott (Rhodophyta: Ceramiales). II. Cell enlargement. *J. Exp. Mar. Biol. Ecol.*, **19**, 165–76.

MYERS M. C. (1928) The life history of the brown alga, *Egregia menziesii*. *Univ. Calif. Publ. Bot.*, **14**, 225–47.

NASR A. H. (1939) On the phototropism of *Acetabularia caliculus*. *Rev. Algol.*, **11**, 347–50.

NEISH A. C. & FOX C. H. (1971) Greenhouse experiments on the vegetative propagation of *Chondrus crispus* (Irish Moss). *Atlantic Regional Laboratory, Nat. Res. Council Canada. Halifax. Tech. Rept.*, **12.**

NEUSHUL M., SCOTT J., DAHL A. L. & OLSEN D. (1967) Growth and development of *Sciadophycus stellatus* Dawson. *Bull. S. Calif. Acad. Sci.*, **66**, 195–200.

NILSEN G. & NORDBY Ø. (1975) A sporulation-inhibiting substance from vegetative thalli of the green alga, *Ulva mutabilis* Føyn. *Planta*, **125**, 127–40.

NORTON T. A. (1977) The growth and development of *Sargassum muticum*. *J. Exp. Mar. Biol. Ecol.*, **26**, 41–54.

NOVÁK B. & BENTRUP F. W. (1972) An electrophysiological study of regeneration in *Acetabularia mediterranea*. *Planta*, **108**, 227–44.

NULTSCH W. (1974) Movements. In *Algal Physiology and Biochemistry*, ed. Stewart W. D. P., pp. 864–93. Blackwell Scientific Publications, Oxford.

PARKE M. (1948) Studies in British Laminariaceae. I. Growth in *Laminaria saccharina* (L.) Lamour. *J. Mar. Biol. Assoc. U.K.*, **27**, 651–709.

PEARLMUTTER N. L. & VADAS R. L. (1978) Regeneration of thallus fragments of *Rhodochorton purpureum* (Rhodophyceae, Nemalionales). *Phycologia*, **17**, 186–90.

PEDERSÉN M. (1968) *Ectocarpus fasciculatus*: marine brown alga requiring kinetin. *Nature* (*Lond.*), **218**, 776.

PEDERSÉN M. (1973) Identification of a cytokinin, 6-(3-methyl-2-butenylamino) purine, in sea water and the effect of cytokinins on brown algae. *Physiol. Plant.*, **28**, 101–5.

PERRONE C. & FELICINI G. P. (1972) Sur les bourgeons adventifs de *Petroglossum nicaeense* (Duby) Schotter (Rhodophycées, Gigartinales) en culture. *Phycologia*, **11**, 87–95.

PERRONE G. & FELICINI G. P. (1974) Dominance apicale et morphogénèse chez *Petroglossum nicaeense* (Duby) Schotter (Rhodophyceae). *Phycologia*, **13**, 187–94.

PERRONE G. & FELICINI G. P. (1976) Les bourgeons adventifs de *Gigartina acicularis* (Wulf.) Lamour. (Rhodophyta, Gigartinales) en culture. *Phycologia*, **15**, 45–50.

PHILLIPS I. D. J. (1975) Apical dominance. *Ann. Rev. Plant Physiol.*, **26**, 341–67.

PRINCE J. & KINGSBURY J. M. (1973) The ecology of *Chondrus crispus* at Plymouth, Massachusetts. I. Ontogeny, vegetative anatomy, reproduction and life cycle. *Am. J. Bot.*, **60**, 956–63.

PROVASOLI L. & CARLUCCI A. F. (1974) Vitamins and growth regulators. In *Algal Physiology and Biochemistry*, ed. Stewart W. D. P., pp. 741–87. Blackwell Scientific Publications, Oxford.

PROVASOLI L. & PINTER, I. J. (1964) Symbiotic relationship between micro-organisms and seaweeds. *Am. J. Bot.*, **51**, 681.

PROVASOLI L., PINTNER I. J. & SAMPATHKUMAR S. (1977) Morphogenetic substances for *Monostoma oxyspermum* from marine bacteria. *J. Phycol.*, **13** (suppl), 56.

PRYCE R. J. (1972) The occurrence of lunularic and abscissic acids in plants. *Phytochem.*, **11**, 1759–61.

PUISEUX-DAO S. (1957) Comportement de fragments anucléés de *Bataphora oerstedi* J. Ag. (Dasycladacée) dans l'eau du mer contenant soit un acide aminé, soit une auxine. *C. R. Acad. Sci. Paris*, **245**, 2371–4.

PUISEUX-DAO S. (1975) Third symposium on *Acetabularia* (Paris). *Protoplasma*, **83**, 167–83.

PYBUS C. (1974) Some observations on the behavior of excised segments from *Dictyota dichotoma* and *Dilophus spiralis* in culture. *Ir. Nat. J.*, **18**, 26–7.

QUATRANO R. S. (1978) Development of cell polarity. *Ann. Rev. Plant. Physiol.*, **29**, 487–510.

RAMUS J. (1969) Dimorphic pit connections in the red alga *Pseudogloiophloea*. *J. Cell. Biol.*, **41**, 340–5.

ROBINSON K. R. & CONE R. (1980) Polarization of fucoid eggs by a calcium ionophore gradient. *Science*, **207**, 77–8.

ROBINSON K. R. & JAFFE L. F. (1976) Calcium gradients and egg polarity. *J. Cell Biol.*, **70**, 37a.

RUSSELL G. (1970) Rhizoid production in excised *Dictyota dichotoma*. *Br. Phycol. J.*, **5**, 243–5.

SAGA N. (1978) Notes on Fucales 8. Regeneration from rhizoid-piece of *Pelvetia* germling to complete thallus. *Jap. J. Phycol.*, **26**, 1–4.

SAGA N., UCHIDA T. & SAKAI Y. (1978) Clone *Laminaria* from single isolated cell. *Bull. Jap. Soc. Sci. Fish.*, **44**, 87.

SAND O. (1973) On orientation of rhizoid outgrowth of *Ulva mutabilis* by applied electrical fields. *Exp. Cell. Res.*, **76**, 444–5.

SCHECTER V. (1934) Electrical control of rhizoid formation in the red alga, *Griffithsia bornetiana*. *J. Gen. Physiol.*, **18**, 1–21.

SCHECTER V. (1935) The effect of centrifuging on the polarity of an alga, *Griffithsia bornetiana*. *Biol. Bull.*, **68**, 172–9.

SCHRÖTER K. (1978) Asymmetrical jelly secretion of zygotes of *Pelvetia* and *Fucus*: an early polarization event. *Planta*, **140**, 69–73.

SIEVERS A., HEINEMANN B. & RODRIGUEZ-GARCIA M. I. (1979) Nachweis des subapikalen differentiellen Flankenwachstums im *Chara*-Rhizoid während der Graviresponse. *Z. Pflanzenphysiol.*, **91**, 435–42.

SINNOTT E. W. (1960) *Plant Morphogenesis*. McGraw-Hill, New York.

SKOOG F., SCHMITZ R. Y., BOCK R. M. & HECHT S. M. (1973) Cytokinin antagonists: synthesis and physiological effects of 7-substituted 3-methylpyrazolo [4,3-d] pyrimidines. *Phytochem.*, **12**, 25–37.

STEINECKE F. (1925) Zur polarität von *Bryopsis*. *Bot. Archiv.*, **12**, 97–118.

SUNDENE O. (1964) The ecology of *Laminaria digitata* in Norway in view of transplant experiments. *Nytt. Mag. Bot.*, **11**, 83–106.

TATEWAKI M. & KANEKO K. (1977) Morphogenetic substances for axenic *Monostroma* spp. from *Neodilsea yendoana*. *J. Phycol.*, **13** (suppl.), 66.

TATEWAKI M. & NAGATA N. (1970) Surviving protoplasts *in vitro* and their development in *Bryopsis*. *J. Phycol.*, **6**, 401–3.

TAYLOR I. E. P. & WILKINSON A. J. (1977) The occurrence of gibberellins and gibberellin-like substances in algae. *Phycologia*, **16**, 37–42.

TEPFER M. & CLELAND R. E. (1979) A comparison of acid-induced cell wall loosening in *Valonia ventricosa* and in oat coleoptiles. *Plant. Physiol.*, **63**, 898–902.

TSUKIDATE J. (1970) Some notes on *Grinnellia americana* in culture. *Bull. Jap. Soc. Sci. Fish.*, **36**, 1109–14.

VANDEN DRIESSCHE T. (1976) Cyclic AMP and morphogenesis of *Acetabularia mediterranea*. *Differentiation*, **5**, 119–26.

VON WETTSTEIN D. (1965) Die Induktion und experimentelle Beeinflussung der Polarität bei Pflanzen. In *Handbuch der Pflanzenphysiologie*, vol. 15 no. 2, ed. Ruhland W., pp. 275–330. Springer-Verlag, Berlin.

WAALAND S. D. (1975) Evidence for a species-specific cell fusion hormone in red algae. *Protoplasma*, **86**, 253–63.

WAALAND S. D. (1978) Production of a cell fusion hormone, rhodomorphin, by *Griffithsia tenuis* (Rhodophyta). *J. Phycol.*, **14** (suppl.), 93.

WAALAND S. D. & CLELAND R. (1972) Development in the red alga, *Griffithsia pacifica*: control by internal and external factors. *Planta*, **105**, 196–204.

WAALAND S. D., NEHLSEN W. & WAALAND J. R. (1977) Phototropism in a red alga, *Griffithsia pacifica*. *Plant & Cell Physiol.*, **18**, 603–12.

WAITE T. D. & MITCHELL R. (1976) Some benevolent and antagonistic relationships between *Ulva lactuca* and its microflora. *Aquatic Botany*, **2**, 13–22.

WARDLAW C. W. (1968) *Morphogenesis in Plants*. Methuen, London.

WATANABE T. & KONDO N. (1976) Ethylene evolution in marine algae and a proteinaceous inhibitor of ethylene biosynthesis from a red alga. *Plant & Cell Physiol.*, **17**, 1159–66.

WATANABE T., KONDO N., FUJII T. & NAGUCHI T. (1978) Affinity chromatography of an ethylene-synthesizing enzyme from red alga *Porphyra tenera* on an immobilized inhibitor of ethylene evolution. *Plant & Cell Physiol.*, **18**, 387–92.

WEBER W. (1969) Morphogenetische und Keimungsphysiologische Untersuchungen an einigen Meeresalgen unter besonderer Berücksichtingung der Polarität. *Bot. Mar.*, **12**, 135–78.

WETHERBEE R. (1979) 'Transfer connections': Specialized pathways for nutrient translocation in a red alga? *Science*, **204**, 858–9.

WHITTAKER D. M. (1940) Physical factors of growth. *Growth*, **4** (*suppl.*), 75–90.

WHITTICK A. (1978) The life history and phenology of *Callithamnion corymbosum* (Rhodophyta: Ceramiaceae) in Newfoundland. *Can. J. Bot.*, **56**, 2497–9.

WILDGOOSE P. B., BLUNDEN P. B. & JEWERS K. (1978) Seasonal variations in gibberellin activity of some species of Fucaceae and Laminariaceae. *Bot. Mar.*, **21**, 63–5.

WINKLER H. (1900) Ueber den Einfluss ausserer Factoren auf die Theilung der Eier von *Cystoseira barbata*. *Ber. Dtsch. Bot. Ges.*, **18**, 297–305.

WOODCOCK C. L. F. (ed.) (1977) *Progress in Acetabularia Research*. Academic Press, London & New York.

ZIMMERMANN U. (1977) Cell turgor pressure regulation and turgor pressure-mediated transport processes. In *Integration of Activity in the Higher Plant*, ed. Jennings D. H., pp. 117–54. Cambridge University Press, Cambridge.

ZIMMERMANN U. (1978) Physics of turgor and osmoregulation. *Ann. Rev. Plant Physiol.*, **29**, 121–48.

ZIMMERMANN U. & STEUDLE E. (1977) Action of indoleacetic acid on membrane structure and transport. In *Regulation of Cell Membrane Activities in Plants*, eds. Marrè C. & Cifferi O., pp. 231–42. Elsevier/North Holland Biomedical Press, Amsterdam.

ZIMMERMANN U., STEUDLE E. & LELKES P. I. (1976) Turgor pressure regulation in *Valonia ultricularis*: effect of cell wall elasticity and auxin. *Plant Physiol.*, **58**, 608–13.

ZIMMERMANN W. (1923) Zytologische untersuchungen an *Sphacelaria fusca* Ag. Ein Beitrag zur Entwicklungsphysiologie der Zelle. *Zeit. Bot.*, **15**, 113–75.

ZIMMERMAN W. (1929) Experimente zur Polarität von *Caulerpa* und zum allgemeinen Polaritätsproblem. *Wilhelm Roux' Archiv für Entwicklungsmechanick der Organismen.*, **116**, 669–88.

CHAPTER 19
SEXUALITY AND SEX ATTRACTION

D. G. MÜLLER
Fakultät für Biologie der Universität, D-7750 Konstanz,
Federal Republic of Germany

1 INTRODUCTION

Sexuality may be considered as one of the fundamental characteristics of living organisms. In eukaryotes it involves the events of syngamy and meiosis, which are integrated into the life histories of the organisms in various ways.

Contrary to the situation in animals, where the somatic tissue is normally diploid and meiosis takes place in the course of the formation of eggs and sperm, an impressive degree of variability is found in algae. Some green algae exist in the haploid phase, with meiosis following immediately upon syngamy. Another extreme is shown by *Fucus* where the organism is diploid and meiosis takes place during the formation of gametes. The majority of macroscopic green, red and brown algae have haploid and diploid phases in their life history which are connected by syngamy and meiosis.

Sexuality is defined here as an effective mechanism to create genetic recombinants. Maximum efficiency is achieved when dioecism guarantees outbreeding. Monoecism, as found in many marine algae, may be considered as a degeneration step which might nevertheless offer benefits in other respects. Secondary mechanisms to block inbreeding, such as incompatibility reactions in fungi and higher plants, are not known in marine algae.

As outlined above, most marine algae have sexual episodes in their life histories. In some cases, such as in *Fucus*, sexuality is obligatorily connected with the production of new individuals as in most animals. Detailed studies on life histories in recent years revealed, however, an increasing number of cases where sexuality is absent in populations that nevertheless seem to compete successfully. Examples (Kornmann & Sahling 1977) include *Codium fragile*, which exists on certain coasts as an exclusively female population, and *Plumaria elegans*, where propagation may be restricted to accessory paraspores.

Culture studies by Wynne (1969) showed that of 14 North American Pacific brown algae examined 12 seem to completely lack sexuality. This condition obviously is a secondary phenomenon, and in several cases it can be demonstrated that the same species behaves differently on different coasts. As an example, *Scytosiphon lomentaria* is asexual in Pacific North America (Wynne 1969), whereas functional gametophytes are known from Japan (Tatewaki 1966). Culture studies in *Ectocarpus siliculosus* (Müller 1967) and *Ulva mutabilis* (Hoxmark 1975) show that algae may have a great potential for reproduction in addition to the possibilities connected with classical sexuality. Even ploidy-level changes may occur apart from sexuality. Thus, one may expect a complete sequence of alternating generations without sexuality. Such a case may be realized in *Elachista stellaris* (Wanders *et al.* 1972) and *Elachista fucicola* (Koeman & Cortel-Breeman 1976).

2 CO-ORDINATION OF PLASMOGAMY

A literature survey (see van den Ende 1976) shows that lower plants have developed a variety of mechanisms in order to co-ordinate the activity of individuals engaging in sexual reproduction. In a dioecious organism it is obvious that two plants have to cooperate in a specific manner to guarantee successful plasmogamy. In aquatic Phycomycetes and freshwater Volvocales cases are known where messenger substances are employed to co-ordinate gametogenesis. In *Volvox carteri* a glycoprotein is produced by a spontaneously sexualized male colony (Starr & Jaenicke 1974). This inducer is highly active and changes the potential of vegetative male and female colonies to form offspring with eggs and spermatozoid packets. Within a short time, the entire *Volvox* population culminates in the formation of zygotes (Starr 1970). Phenomena of this kind are not known in marine algae, and it seems that the *Volvox* system and related cases may be an adaptation to ephemeral freshwater habitats, which require the production of genetic variability and resting stages shortly before the habitat becomes hostile. The marine environment offers more continuity. Various mechanisms linked to environmental factors have been developed by marine algae to coordinate gametogenesis and synchronize the discharge of gametes. Information for seasonal orientation may be derived from water temperature or daylength. The time of gamete release may be further specified by the circadian light regime or the desiccation and rehydration by tides. A unique synchronization pattern apparently restricted to the marine environment is the coupling of sexual activity with specified lunar phases.

Experimental work with *Dictyota dichotoma* (Müller 1962) indicated that periodic illumination by the full moon is a synchronizing factor.

While external factors are largely responsible for induction and synchronization of gametogenesis in marine algae, the brown seaweeds offer unique systems to study chemical coordination on the cellular level. Members of Phaeophyceae are currently the only cases in algae where the chemical identity of molecules involved in cellular approach is known. The following discussion will, therefore, deal exclusively with marine Phaeophyceae. Chemical substances secreted by one cell and causing specific effects in a target cell are termed hormones. A broad understanding of this definition includes cases where the external medium is acting as the carrier. Consequently, the male-attracting substances discussed here are termed hormones.

3 THE MALE-ATTRACTING HORMONES IN PHAEOPHYCEAE

3.1 Ectocarpus siliculosus

Berthold (1881) and Oltmanns (1899) and later on Hartmann (1934) observed fertilization in this species from the Mediterranean Sea. They found that mass development occurs every spring. Two morphologically identical types of plants are found. Zoids released from their plurilocular gametangia are equal in size, but differ in their physiological character. Gametes classified as female have a short period of motility. After settlement on the substratum they withdraw their flagella and starting at this moment numerous other gametes, designated male, approach and surround the female cell in vigorous circular movement. The male gametes soon anchor themselves on the female surface with their front flagella and eventually plasmogamy takes place with one of the male cells. Subsequently all other male cells disperse. Using clonal laboratory cultures it was found (Müller 1968) that the male cells repond to a highly volatile compound secreted by receptive female gametes. The pleasant scent of this compound to the human nose first drew attention to the phenomenon and thus initiated the following studies on sexual hormones of brown algae. Mass culture of female gametophytes were established and sufficient material extracted for chemical identification. Only one compound was found in the eluates which could be clearly demonstrated to be the biologically active molecule. The substance was initially named *Ectocarpus* sirenin (Müller *et al.* 1971), but following a suggestion by Machlis (1972) it was later changed to ectocarpen.

3.2 Cutleria multifida

Gametophytes of *Cutleria multifida*, like those of *Ectocarpus siliculosus*, are found every spring on the coasts of the Mediterranean Sea. Male and female plants are isomorphic, forming rigidly branched thalli 20 cm long. Gametes are markedly different in size. Kuckuck (1899) and Hartmann (1949/50) reported masses of spermatozoid-like male gametes aggregating around female cells after the latter had come to rest. The extraction techniques developed for *Ectocarpus* (see below) were applied to bulk suspensions of female *Cutleria* gametes. Gas-chromatographic examination of the eluate showed three different compounds. The most prominent fraction proved to be the male-attractant and was named multifiden. The second compound was named aucanten. The third compound was identified as ectocarpen (Jaenicke *et al.* 1974).

3.3 Fucus serratus and F. vesiculosus

The first report on sexual reproduction in *Fucus* came from Thuret (1854) who noticed that clouds of spermatozoids were attracted by eggs. Only recently, the attractant from eggs of *Fucus serratus* was found to be a conjugated octatriene which was named fucoserraten (Müller & Jaenicke 1973; Jaenicke & Seferiadis 1975; and see note p. 674). In a later study by Müller & Gassmann (1978) eluates from eggs of *Fucus vesiculosus* were analysed. Capillary gas chromatography revealed that eggs of this species secrete the same molecule as those from *F. serratus*: fucoserraten.

3.4 Laminaria digitata

The order Laminariales is characterized by a pronounced heteromorphy. Contrary to the huge size of sporophytes, the gametophytes are ephemeral and microscopic in size, forming few eggs and spermatozoids. Differentiation and maturation of antheridia and oogonia takes place under suitable external conditions. Release of eggs occurs spontaneously, whereas antheridia withhold their spermatozoids even at full maturity. Mass release takes place when volatile material from an egg suspension is allowed to reach the antheridia (Lüning & Müller 1978). Furthermore, the spermatozoids once released are attracted to the eggs. The volatiles emerging from egg suspensions of *Laminaria digitata* were isolated and characterized by gas chromatography. Three compounds were found. Two of them were easily identified as ectocarpen and n-pentadecane, neither of which was noticeably active in spermatozoid release or attraction. In contrast, the third compound was shown to be highly active. Its chemical structure is not yet completely identified (Müller *et al.* 1979).

4 EXTRACTION METHODS AND BIO-ASSAY

All signal molecules so far known to act upon spermatozoids of brown algae are highly lipophilic unsaturated hydrocarbons. These substances, although secreted into the water by the female cells, have a very pronounced tendency to leave the aqueous phase and escape into the air. In fact, this property led to the discovery of the first compound, ectocarpen. In the earlier phase of the work, when extracting volatile products from *Ectocarpus*, *Cutleria* and *Fucus serratus*, a simple procedure was employed: suspensions containing living female gametes were placed in gas washing bottles. Air, purified by passing over a bed of copper oxide at redglow temperature, was pumped through the suspension at a rate of 25 ml/min. The air stream emerging from the suspension was then passed through a large-volume cold trap at −30°C to remove the bulk of water vapour. The volatile organic substances passed this trap and were then condensed in small-volume U-shaped glass tubes at −78°C. The condensates were dissolved in carbon-tetrachloride and subjected to gas chromatography.

Recently, a closed-loop extraction system using activated carbon as an adsorbent was found to give much higher recovery rates when used with *Laminaria digitata* and *Fucus vesiculosus* (Grob & Zürcher 1976; Müller & Gassmann 1978).

Bio-assays use the response of living male gametes to the presence of attractants in seawater. Various arrangements have been applied for crude qualitative experiments. In most cases, a carrier material is used that dissolves or adsorbs the active substances and subsequently releases them into seawater: paraffin oil or vaseline are suitable solvents, and porous SiO_2 may be used as an adsorbent (Fig. 19.1). Attempts at quantitative bio-assays encounter serious problems. It is extremely difficult to obtain and maintain known concentrations in microscopic dimensions in a seawater medium when the substances in question are close to insoluble and available in minute quantities only. The best approach achieved so far makes use of a fully fluorinated solvent (FC-78, manufactured by 3M Company, Düsseldorf) which is biologically inert, insoluble in and denser than water, and dissolves the attractants sufficiently well. Technical details are given by Müller (1976a).

5 SPECIFICITY

It was noticed by Cook *et al.* (1947/48) with *Fucus* and in the early phase of my own work with *Ectocarpus* (Müller 1968) that numerous volatile organic chemicals show attractive properties to male gametes. However, quantitative evaluation shows that the most sensitive interaction occurs between male gametes and their species-specific attractant. In these cases, attractant concentrations in the solvent phase down to 10^{-7} M are active, whereas n-hexane is positive only at concentrations 300–3000 times higher (Müller 1976a; Müller & Seferiadis 1977). At present, one example is known where the attractants in two species of the same genus have been identified: eggs of *Fucus serratus* and *F. vesiculosus* use the same substance, and the respective spermatozoids do not differentiate between their own eggs and those of their related species (Müller & Gassmann 1978). In these two species of *Fucus* some

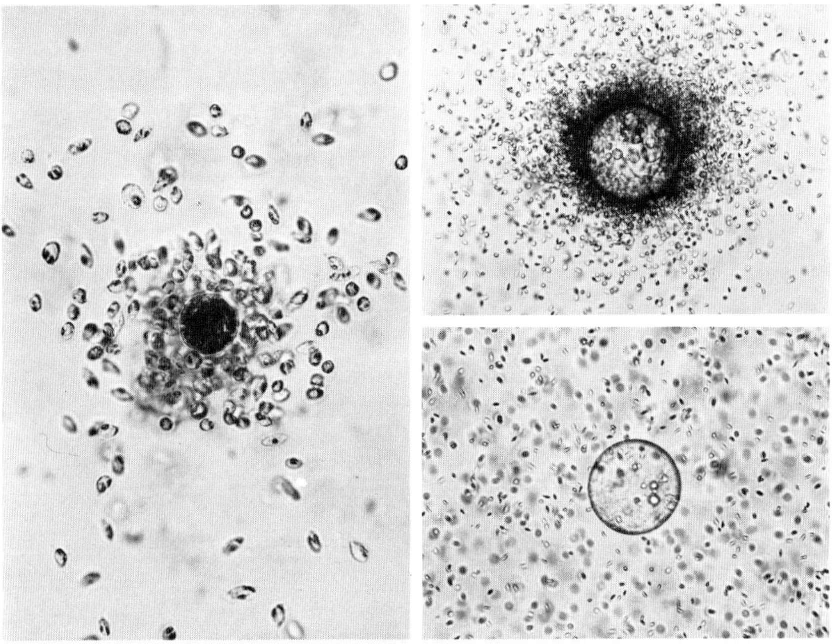

Fig. 19.1 Illustration of a qualitative bio-assay. Left, female and male gametes of *Cutleria multifida*. Lower right, untreated Spherosil particle (diameter 100 μm) with male gametes. Upper right, Spherosil particle containing the volatiles produced by female gametes.

information has been obtained concerning the resolution potential of the male chemoreceptor apparatus. The spermatozoids of both species are able to differentiate between 1,3,5- and 2,4,6-positions and *cis-trans* isomers in various conjugated octatrienes (Müller & Seferiadis 1977). Nevertheless, it seems obvious that brown algae do not use the relatively simple hydrocarbons as carriers for species specificity. Species-specific interactions do occur in the brown algae discussed here, because there is no noticeable hybridization between *Fucus serratus* and *F. vesiculosus* encountered in nature. Recent findings on species of *Fucus* and *Ascophyllum* show that compounds or structures responsible for species specificity are located on the membrane of the egg (Bolwell *et al.* 1977).

Some of the properties of the eggs and sperm of *Fucus serratus* have been elucidated more recently by Bolwell *et al.* (1979), who reported that fertilization is based on associations between fucosyl- and mannosyl-containing ligands on the egg surface and specific carbohydrate-binding receptors on the sperm surface.

Chemical interaction in the order Laminariales concerning release and attraction of spermatozoids was found in all cross-combinations involving six species from three families. These experiments were carried out on a qualitative basis only (Lüning & Müller 1978) and it remains to be seen whether possibly all members of the order Laminariales use the same male-releaser and -attractant.

Attraction between egg and sperm is rarely found in the animal kingdom. A well documented case is encountered in marine Cnidaria. Sperm attractants in this group are water-soluble with estimated molecular weights between 240 and 600 (Miller 1972). In this group, cross-combination experiments showed a high degree of species specificity (Miller 1979).

6 RESPONSE OF MALE GAMETES TO ATTRACTANTS

The spermatozoa of the Hydrozoa mentioned above clearly show topotactic movements in a gradient of the attractant. This means that they are able to orient themselves in the gradient and reach the source directly in a straight path (Miller 1966).

Ectocarpus siliculosus is at present the only case in algae where information on the response of the male cell to the sexual attractant exists. The female gametes start secretion of ectocarpen immediately after settling on the substratum and withdrawal of their flagella. Video recordings were made of male gametes approaching female cells and under chronic stimulation by synthetic ectocarpen. Analysis showed that the male gametes do not react topotactically, which means that they are not able to detect the direction of the gradient. Their method to find the attractant source is probably more primitive, but nevertheless it works with impressive efficiency. The approach consists of several phases. Male gametes swimming freely in the medium move in straight paths at maximum speed. When they encounter a surface such as a micro-slide or coverglass they tend to establish and remain in contact with this boundary. Their locomotion then becomes slower and changes to wide looplike paths, apparently due to the peculiar symmetry of the cells. Chemical stimulation by ectocarpen dramatically enhances the thigmic response of the cells and, in addition, they begin to swim in circular paths, the diameter of which decreases in response to increasing attractant concentration. Up to now, no other organism is known to have such a locomotive pattern, which is circumscribed by the term chemo-thigmo-klinokinesis (Müller 1978). As soon as the female gamete is reached, a firm contact is established between the apical part of the male gamete's front flagellum and the female gamete's membrane. This anchoring mechanism seems to be highly species-specific, since it fails to operate in certain crosses between geographically separated populations (Müller 1976b). Presently, there is no information on the relation of this locomotive pattern to flagellar activity.

It should be noted here that the symmetry of *Ectocarpus* gametes is highly complex: the pear-shaped body carries a long front flagellum with mastigonemes and an apical appendix represented by a continuation of the two central fibrils from the basal part. The second flagellum is inserted laterally and oriented obliquely backwards, without mastigonemes, but equipped with a whip-like appendix. Locomotion is effected by a meander-like agitation of the front flagellum. The rear flagellum is stiff, it can only be observed to perform occasional rapid sideward beats.

No details are known of the response of male gametes to attractants in other brown algae. The very different architecture of *Fucus* spermatozoids, for instance, suggests that differences to the *Ectocarpus* type reaction might be expected in other groups.

7 BIOSYNTHESIS

Due to limited amounts of biomass available, studies on biosynthesis of sexual attractants by female gametes of brown algae have not been possible so far. Taking into account the constituents of the essential oil from *Dictyopteris plagiogramma*, which is a source of the characteristic 'odour of the sea' on Hawaiian shores, Jaenicke (1977) proposed a possible pathway for the formation of the C_8 and C_{11} olefines in question. *Dictyopteris* contains a whole series of C_{11} hydrocarbons, including cyclopropane derivatives which may all be traced back to linolenic acid, a regular constituent in brown algae (Fig. 19.2). It should be mentioned as a peculiarity that the roots of the South African Compositae

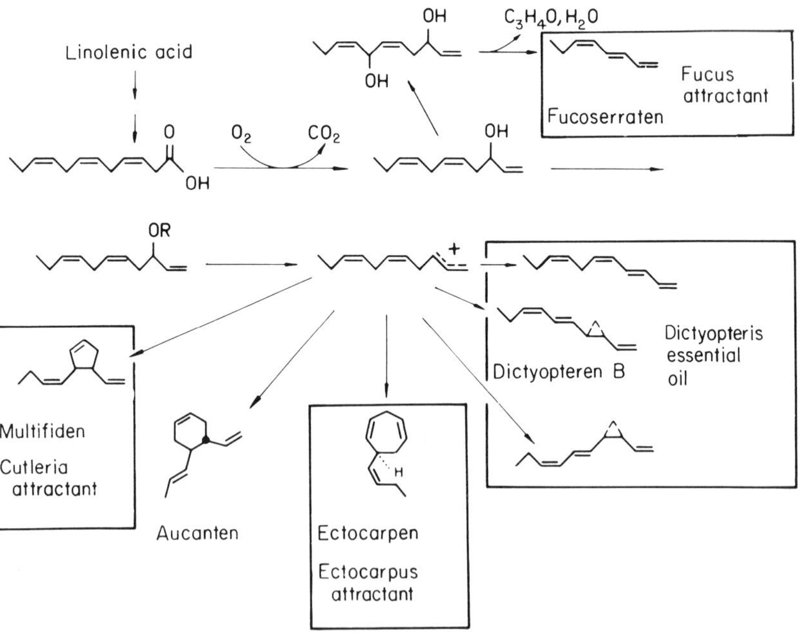

Fig. 19.2 Substances detected and identified in eluates from female gametes of brown algae with hypothetical biosynthetic pathways. (Modified from Jaenicke 1977.)

Senecio isitadeus have a high content of ectocarpen (Jaenicke & Bohlmann personal communication).

8 ECOLOGICAL CONSIDERATIONS

All attractants in brown algae identified so far have in common that they are highly volatile and hydrophobic. As the odour of ectocarpen emerging from a female culture illustrates, the substances have a very strong tendency to leave the scene of their formation and escape into the air. This characteristic certainly helps to avoid the build-up of chronic concentrations which would decrease the efficiency of gradients around female cells. Judging from microscopic observations the attraction certainly is not effective more than 0.5 mm distant from female cells and its range may in some cases be much less. Thus, sexual attraction is very clearly a short-distance phenomenon, and additional mechanisms are needed to facilitate gamete encounter. Spermatozoids are frequently seen to react strongly phototactically, and thigmotactic responses can likewise be considered to guide the male cells to places where the chances to meet a female cell are maximum.

The presence of two substances in addition to the true attractant in *Cutleria* and *Laminaria* has not yet found a plausible explanation. They might be potential candidates for the chemical interaction with male gametes of other brown algae, perhaps competitors. Under laboratory conditions, it is easily possible to 'misguide' male gametes of *Cutleria* and *Ectocarpus* (Müller 1976a). Whether extra-specific interactions of this kind play a significant role in the natural habitat cannot be judged at present. 'Chemical warfare' by interfering with a competitor's sexual information system cannot be ruled out entirely.

The order Laminariales revealed the first case where an additional function of the attractant was found: it not only guides the spermatozoids to the egg, but induces them to emerge from their antheridia in the first place. This phenomenon may be connected with the strong reduction of the gametophytic stage in the order Laminariales. Each female gametophyte forms only one to a few eggs, and likewise the number of spermatozoids per male plant is limited. The range of attraction of the egg does not exceed 1 mm. Therefore, it can be considered a highly efficient economic achievement to make the male gametes wait until they receive an affirmative signal that a receptive egg has formed in the immediate vicinity.

The possibility that male gametes may be 'misguided' has been briefly mentioned. Highly sensitive chemocommunication systems are likely to be disturbed by foreign substances. In the case of the brown algae, hydrocarbons of various origins, including anthropogenic pollution, may be suspected to interfere with sexual reproduction. In cases like *Ectocarpus* or *Cutleria* the disturbing effect of 'mimick' substances would not be expected to be serious, because full or partial parthenogenetic development is possible when fertilization should fail to take place. In cases such as *Fucus* a disturbing effect might be more serious, since new progeny arises only after successful fertilization. The Laminariales system would be suspected to be most susceptible to chemical disturbance. The effect of a mimicking compound that would cause the release of spermatozoids at the wrong time could be truly harmful. However, no detailed knowledge is as yet available which could allow evaluation of this problem. On the other hand, hydrocarbons are known to have been present in the environment from natural sources for millions of years without apparent detrimental effect on brown algal populations.

Recently, Steele (1977) examined the effect of fossil oils on *Fucus edentatus*. He found that vegetative parts and young zygotes are remarkably resistant. However, when oil products were present while eggs and spermatozoids were mixed, fertilization was completely inhibited to the parts per billion range.

9 PHYLOGENETIC CONSIDERATIONS

The marine brown algae seem to be a group which has especially well established sexual chemotaxis using low-molecular weight hydrocarbons as signal molecules. In other groups of algae, much less is known on this topic. In green algae, one case has been reported where gamete interaction involves volatile compounds in *Chlamydomonas* (Tsubo 1961). In *Oedogonium*, another green alga, as well as in the water mould *Allomyces*, water-soluble attractants were found (van den Ende 1976). The only case in terrestrial plants where a native spermatozoid attractant has been isolated involves the liverwort *Sphaerocarpos*, which secretes a protein with a molecular weight between 2000 and 4000 (Schieder 1971). Thus, low-molecular weight olefines in the range between C_8 and C_{11} seem to be a specific achievement in the class Phaeophyceae. The state of knowledge, poor as it still is at present, nevertheless indicates some traits which allow us to link biochemical data with taxonomy. Undoubtedly, the order Ectocarpales is most primitive. Sexual chemotaxis in *Ectocarpus siliculosus* involves a single compound, ectocarpen. In the order Fucales, which is separated farthest from the rest of the class, the genus *Fucus* developed another single compound for gamete interaction: fucoserraten, which is fairly different from the C_{11} compounds. Members of orders belonging to the main phylogenetic line, Phaeophycidae, used another strategy: they seem to have conserved ectocarpen from their ancestors but developed a new attractant. The picture will certainly become more complete when additional cases can be added (Fig. 19.3).

The fact that two clearly distinct species of *Fucus* make use of the same attractant strongly indicates that brown algal sex attractants are not species-specific. At present, there is no information available that would allow us to estimate on which higher systematic category specificity will occur. If ectocarpen in Laminariales and Cutleriales should indeed represent an archaic phylogenetic remnant, it would be plausible to expect that the entire order Ectocarpales

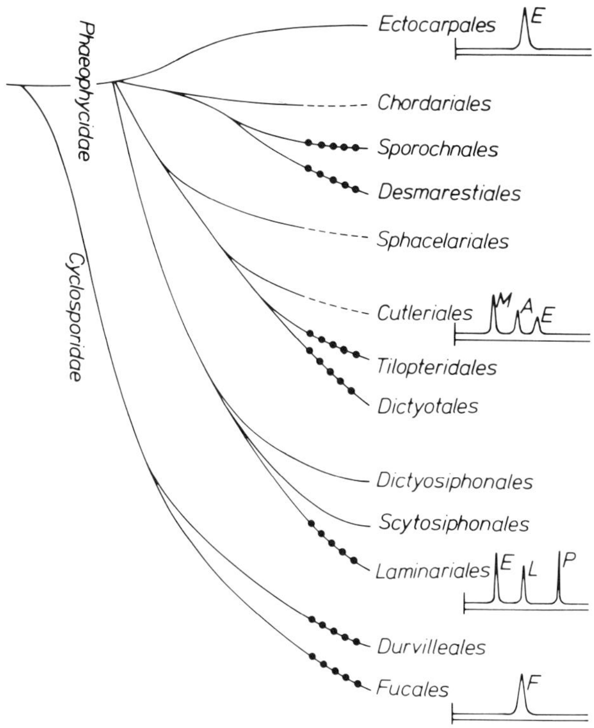

Fig. 19.3 Phylogenetic relationships between orders in the class Phaeophyceae (modified after Wynne & Loiseaux 1976). Schematic gas-chromatograms indicate substances found in eluates from female gametes. A = aucanten, F = fucoserraten, E = ectocarpen, M = multifiden, P = n-pentadecane, L = spermatozoid-releasing and attracting substance from *Laminaria* with unknown structure. ——— = isogamy, - - - - - = anisogamy, ● ● ● ● ● = oogamy.

makes use of ectocarpen as a sex attractant. Speculations of this sort should not be overemphasized, but the situation is interesting enough to encourage attempts to examine the secretions from female gametes in other orders. An important character used in taxonomy is the relative size of male and female gametes. Isogamy is considered primitive, anisogamy and oogamy are characterized as successive stages in phylogenetic progress. Members of all three stages have been compared regarding the potential of male gametes to respond to their respective attractant. Clearly, the primitive *Ectocarpus* seems to have a rather poorly developed chemoreception system in its male gametes. In contrast, the chemotactic response mechanism is equally well advanced in the anisogamous *Cutleria* and oogamous *Fucus* (Müller & Seferiadis 1977). Considerations outlined by Ghiselin (1974) make it plausible that the phylogenetic advantage of greater size and smaller number of female gametes requires progress in the locating ability of the males.

10 RELATIVE SEXUALITY

Hartmann (1937) proposed 'relative sexuality' as a general biological principle. He argued that sexual differentiation is not strictly bipolar, but at least in some cases intergrading differentiation patterns can be found, where for example, a 'strong' female cell could sexually fuse with a 'weak' female cell. Subsequent examination of the literature data which Hartmann presented did not support the hypothesis, and the data concerning Fungi were classified as misinterpretations by Esser & Kuenen (1965). Thus, Hartmann's own work on *Ectocarpus siliculosus* remained the last piece of experimental evidence in support of 'relative sexuality'. In order to re-evaluate the hypothesis, a broad study on sexual differentiation in 200 isolates from the same locality (Naples, Italy) was carried out (Müller 1976c). No deviation from clear bipolarity was found. Even in artificially produced diploid gametes containing one male and one female chromosome set the female character was completely suppressed and the respective gametes clearly reacted as males. Thus, the hypothesis of 'relative sexuality' should now be abandoned.

Summarizing the findings so far achieved in chemical interaction on the gamete level, it may be expected that insights into chemical communication, phylogenetic relations and locomotive patterns will increase when studies in the field of brown algal sexual hormones are continued.

11 REFERENCES

BERTHOLD G. (1881) Die geschlechtliche Fortpflanzung der eigentlichen Phaeosporeen. *Mitt. Zool. Stat. Neapel.*, **2**, 401–13.

BOLWELL G. P., CALLOW J. A., CALLOW M. E. & EVANS L. V. (1977) Cross-fertilization in fucoid seaweeds. *Nature (Lond.)*, **238**, 626–7.

BOLWELL G. P., CALLOW J. A., CALLOW M. E. & EVANS L. V. (1979) Fertilization in brown algae. II. Evidence for lectin-sensitive complementary receptors involved in gamete recognition in *Fucus serratus*. *J. Cell Sci.*, **36**, 19–30.

COOK A. H., ELVIDGE J. A. & HEILBRON I. (1947/48) Fertilization, including chemotactic phenomena in the Fucaceae. *Proc. Roy. Soc.*, Ser. B, **135**, 293–301.

ESSER K. & KUENEN R. (1965) *Genetik der Pilze*. Springer-Verlag, Berlin.

GHISELIN M. T. (1974) *The Economy of Nature and the Evolution of Sex*. University of California Press, Berkeley.

GROB K. & ZÜRCHER F. (1976) Stripping of trace organic substances from water. Equipment and procedure. *J. Chromatogr.*, **117**, 285–94.

HARTMANN M. (1934) Untersuchungen über die Sexualität von *Ectocarpus siliculosus*. *Arch. Protistenk.*, **83**, 110–53.

HARTMANN M. (1937) Ergänzende Untersuchungen über die Sexualität von *Ectocarpus siliculosus*. *Arch. Protistenk.*, **89**, 382–92.

HARTMANN M. (1949/50) Beiträge zur Kenntnis der Befruchtung und Sexualität mariner Algen. I. Über die Befruchtung von *Cutleria multifida*. *Pubbl. Staz. Zool. Napoli*, **22**, 120–8.

HOXMARK R. C. (1975) Experimental analysis of the life cycle of *Ulva mutabilis*. *Bot. Mar.*, **18**, 123–9.

JAENICKE L. (1977) Sex hormones of brown algae. *Naturwissenschaften*, **64**, 69–75.

JAENICKE L., MÜLLER D. G. & MOORE R. E. (1974). Multifidene and aucantene, C_{11} hydrocarbons in the male-attracting essential oil from the gynogametes of *Cutleria multifida* (Smith) Grev. (Phaeophyta). *J. Am. Chem. Soc.*, **96**, 3324–5.

JAENICKE L. & SEFERIADIS K. (1975) Die Stereochemie von Fucoserraten, dem Gametenlockstoff der Braunalge *Fucus serratus* L. *Chem. Ber.*, **108**, 225–32.

KOEMAN R. P. T. & CORTEL-BREEMAN A. M. (1976) Observations on the life history of *Elachista fucicola* (Vell.) Aresch. (Phaeophyceae) in culture. *Phycologia*, **15**, 107–17.

KORNMANN P. & SAHLING, P.-H. (1977) *Meeresalgen von Helgoland*. Biologische Anstalt Helgoland, Hamburg.

KUCKUCK P. (1899) Über den Generationswechsel von *Cutleria multifida* Grev. *Wiss. Meeresunters. Abt. Helgoland*, N.F., **3**, 95–116.

LÜNING K. & MÜLLER D. G. (1978) Chemical interaction in sexual reproduction of several Laminariales (Phaeophyceae): release and attraction of spermatozoids. *Z. Pflanzenphysiol.*, **89**, 333–41.

MACHLIS L. (1972) The coming of age of sex hormones in plants. *Mycologia*, **64**, 235–47.

MILLER R. K. (1966) Chemotaxis during fertilization in the hydroid *Campanularia*. *J. Exp. Zool.*, **162**, 23–44.

MILLER R. K. (1972) Gel filtration of the sperm attractants of some marine hydrozoa. *J. Exp. Zool.*, **182**, 281–98.

MILLER R. K. (1979) Sperm chemotaxis in the Hydromedusae. I. Species specificity and sperm behaviour. *Mar. Biol.*, **53**, 99–104.

MÜLLER D. G. (1962) Über jahres- und lunarperiodische Erscheinungen bei einigen Braunalgen. *Bot. Mar.*, **4**, 140–55.

MÜLLER D. G. (1967) Generationswechsel, Kernphasenwechsel und Sexualität bei der Braunalge *Ectocarpus siliculosus* im Kulturversuch. *Planta*, **75**, 39–54.

MÜLLER D. G. (1968) Versuche zur Charakterisierung eines Sexual-Lockstoffes bei der Braunalge *Ectocarpus siliculosus*. I. Methoden, Isolierung und gaschromatographischer Nachweis. *Planta*, **81**, 160–8.

MÜLLER D. G. (1976a) Quantitative evaluation of sexual chemotaxis in two marine brown algae. *Z. Pflanzenphysiol.*, **80**, 120–30.

MÜLLER D. G. (1976b) Sexual isolation between a European and an American population of *Ectocarpus siliculosus* (Phaeophyta.) *J. Phycol.*, **12**, 252–4.

MÜLLER D. G. (1976c) Relative sexuality in *Ectocarpus siliculosus*. A scientific error. *Arch. Microbiol.*, **109**, 89–94.

MÜLLER D. G. (1978) Locomotive responses of male gametes to the species-specific sex attractant of *Ectocarpus siliculosus* (Phaeophyta). *Arch. Protistenk.*, **120**, 371–7.

MÜLLER D. G. & GASSMANN G. (1978) Identification of the sex attractant in the marine brown alga *Fucus vesiculosus*. *Naturwissenschaften*, **65**, 389.

MÜLLER D. G., GASSMANN G. & LÜNING K. (1979) Isolation of a spermatozoid-releasing and -attracting substance from female gametophytes of *Laminaria digitata*. *Nature* (*Lond.*), **279**, 430–1.

MÜLLER D. G. & JAENICKE L. (1973) Fucoserraten, the female sex attractant of *Fucus serratus* L. (Phaeophyta). *F.E.B.S. Lett.*, **30**, 127–39.

MÜLLER D. G., JAENICKE L., DONIKE M. & AKINTOBI T. (1971) Sex attractant in a brown alga: chemical structure. *Science* (*Wash.*), **171**, 815–17.

MÜLLER D. G. & SEFERIADIS K. (1977) Specificity of sexual chemotaxis in *Fucus serratus* and *Fucus vesiculosus* (Phaeophyceae). *Z. Pflanzenphysiol.*, **84**, 85–94.

OLTMANNS F. (1899) Über die Sexualität der Ectocarpeen. *Flora Allg. Bot. Z.*, **86**, 86–99.

SCHIEDER O. (1971) Über das Chemotaktikum des Lebermooses *Sphaerocarpos donnellii* Aust. *Naturwissenschaften*, **58**, 568.

STARR R. C. (1970) Control of differentiation in *Volvox*. *Dev. Biol.*, **4**, (suppl.), 59–100.

STARR R. C. & JAENICKE L. (1974) Purification and characterization of the hormone initiating sexual morphogenesis in *Volvox carteri* f. *nagariensis* Iyengar. *Proc. Natl. Acad. Sci. USA*, **71**, 1050–4.

STEELE R. L. (1977) Effects of certain petroleum products on reproduction and growth of zygotes and juvenile stages of the alga *Fucus edentatus* de la Pyl. (Phaeophyceae, Fucales). In *Fate and Effects of Petroleum Hydrocarbons in Marine Ecosystems and Organisms*, ed. Wolfe D. A., pp. 138–42. Pergamon Press, Oxford.

TATEWAKI M. (1966) Formation of a crustaceous sporophyte with unilocular sporangia in *Scytosiphon lomentaria*. *Phycologia*, **6**, 62–6.

THURET G. (1854) Recherches sur la fécondation des Fucacées et les anthéridies des algues. *Ann. Sci. Nat. Bot.*, Ser. IV, **2**, 197–214.

TSUBO Y. (1961) Chemotaxis and sexual behaviour in *Chlamydomonas*. *J. Protozool.*, **8**, 114–21.

VAN DEN ENDE H. (1976) *Sexual Interactions in Plants*. Academic Press, London.

WANDERS J. B. W., VAN DEN HOEK C. & SCHILLERN-VAN NES E. H. (1972) Observations on the life history of *Elachista stellaris* (Phaeophyceae) in culture. *Neth. J. Sea Res.*, **5**, 458–91.

WYNNE M. (1969) Life history and systematic studies of some Pacific North American Phaeophyceae (brown algae). *Univ. Calif. Publ. Bot.*, **50**, 1–88.

WYNNE M. J. & LOISEAUX S. (1976) Recent advances in life history studies of the Phaeophyta. *Phycologia*, **15**, 435–52.

Note added in proof

Kajiwara T., Kodama K. & Hatanaka A. (1980 *Bull. Jap. Soc. Sci. Fish.*, **46**, 555–7; 1980 *Naturwissenschaften*, **67**, 612–13) report the occurrence of fucoserraten and three other conjugated octatrienes in extracts from whole receptacles of *Sargassum horneri*. Although the authors state that all four compounds attract spermatozoids, their methods do not permit identification of the hormonal factors secreted by living eggs.

SECTION IV
SEAWEEDS AS RESOURCES

INTRODUCTION

MICHAEL J. WYNNE

For many of us whose livelihood is based upon the study of seaweeds, it is at times difficult to explain to family, friends, and others not in the business just why we are so absorbed in our phycological pursuits. Spending one's career routinely gathering and poring over slimy specimens of red, brown, and green algae can be met with reactions ranging from that of mild amusement to genuine incredulity. This final section focuses on the intrinsic worth and applicability of seaweeds and should dispel the notion that they do not play a role in human concerns. Indeed they are being put to significant use in a variety of food and industrial products (Boney 1965; Bardach *et al.* 1972) as well as in the pharmaceutical industry (Hoppe *et al.* 1979).

Maricultural practices are well underway in many countries, such as Japan, China, Korea, the Philippines, Taiwan, and Canada, while experimental pilot projects or at least the potential for such practices are being carried out in many other countries. In the United States research projects examining various aspects related to seaweed cultivation are supported by the Sea Grant program at many universities (Mathieson 1975). In addition, the biotic and abiotic constraints imposed on the cultivation of seaweeds are being explored and appreciated (Huguenin 1976; Jackson 1977). The engineering problems inherent in the mass production of algae have also been approached (Goldman & Ryther 1977). Other scientists have turned their attention to the possible social and legal implications of the cultivation of seaweeds (Mumford 1977; Benner 1977). Supported by the Food and Agricultural Organization of the United Nations, Michanek (1975, 1979) has summarized the seaweed resources and their availability in terms of phytogeographic provinces. He has also contributed to compiling the pertinent literature dealing with seaweed cultivation (Michanek 1978).

An International Seaweed Symposium has been held every third year beginning in 1952, most recently in Göteborg, Sweden (August 1980). The proceedings of these symposia (for example, Nisizawa 1972; Jensen & Stein 1978) contain valuable information bearing on the biology, chemistry, and utilization of seaweeds both of a specific and a more general nature. In the past few years experts have gathered on several occasions to exchange ideas on the techniques, problems, and progress in making use of seaweeds, this still virtually untapped and renewable resource. Such publications as those by Krauss (1977) and Tsuda & Chiang (1980) represent examples of these exchanges. The pathology of seaweeds was the topic of a symposium held in conjunction with the Third International Congress of Plant Pathology in Munich, and a series of relevant papers resulted (Dodds 1979; Andrews 1979; Kohlmeyer 1979; Vadas 1979).

We are fortunate to be able to include in this volume a contribution from Dr. C. K. Tseng, Director of the Oceanology Institute in Tsingtao, the People's Republic of China. His chapter reflects his personal experiences gained from a long and distinguished career in seeing first-hand the gradual emergence of a significant seaweed cultivation industry in his country and also affords us an analysis of the 'state of the art' of contemporary seaweed cultivation practices being used in China. The chapter by Dr. J. R. Waaland focuses attention on the products that result from the utilization of algin, carrageenan, and agar as well as those seaweeds used directly as food, fertilizer, or fodder. It also summarizes harvesting data and estimates potential yields for red and brown seaweeds of economic value. It is evident that the potential is great and that the challenge is there to be met.

REFERENCES

ANDREWS J. H. (1979) Conclusion: the seaweed pathosystem. *Experientia*, **35**, 448–50.

BARDACH J. E., RYTHER J. H. & MCLARNEY W. O. (1972) *Aquaculture. The Farming and Husbandry of Freshwater and Marine Organisms.* Wiley-Interscience, New York.

BENNER R. P. (1977) Legal considerations in the maritime setting. In *The Marine Plant Biomass of the Pacific Northwest Coast*, ed. Krauss R. W., pp. 377–88. Oregon State University Press, Corvallis.

BONEY A. D. (1965) Aspects of the biology of the seaweeds of economic importance. *Adv. Mar. Biol.*, **3**, 105–253.

DODDS J. A. (1979) Viruses of marine algae. *Experientia*, **35**, 440–2.

GOLDMAN J. C. & RYTHER J. H. (1977) Mass production of algae: bioengineering aspects. In *Biological Solar Energy Conversion*, eds. Mitsui A. *et al.*, pp. 367–78. Academic Press, London & New York.

HOPPE H. A., LEVRING T. & TANAKA Y. (eds) (1979) *Marine Algae in Pharmaceutical Science.* Walter de Gruyter & Co., New York.

HUGUENIN J. E. (1976) An examination of problems and potentials for future large-scale intensive seaweed culture systems. *Aquaculture*, **9**, 313–42.

JACKSON G. A. (1977) Biological constraints on seaweed culture. In *Biological Solar Energy Conversion*, eds. Mitsui A. *et al.*, pp. 437–48. Academic Press, London & New York.

JENSEN A. & STEIN J. R. (eds.) (1978) *Proceedings of the Ninth International Seaweed Symposium.* Science Press, Princeton.

KOHLMEYER J. (1979) Marine fungal pathogens among Ascomycetes and Deuteromycetes. *Experientia*, **35**, 437–9.

KRAUSS R. W. (ed.) (1977) *The Marine Plant Biomass of the Pacific Northwest Coast.* Oregon State University Press, Corvallis.

MATHIESON A. C. (1975) Seaweed aquaculture. *Marine Fisheries Rev.*, **37 (1)**, 2–14. MFR Paper 1111.

MICHANEK G. (1975) Seaweed resources of the ocean. *FAO Fish. Techn. Paper*, 138. FAO United Nations, Rome.

MICHANEK G. (1978) Trends in applied phycology with a literature review: seaweed farming on an industrial scale. *Bot. Mar.*, **21**, 469–75.

MICHANEK G. (1979) Phytogeographic provinces and seaweed distribution. *Bot. Mar.*, **22**, 375–91.

MUMFORD T. F. JR. (1977) Some biological, legal, social, and economic aspects of the culture of the red alga *Iridaea cordata* on nets in Puget Sound. In *The Use, Study and Management of Puget Sound*. Washington Sea Grant Publications, University of Washington, Seattle.

NISIZAWA K. (ed.) (1972) *Proceedings of the Seventh International Seaweed Symposium*. University of Tokyo Press, Tokyo.

TSUDA, R. T. & CHIANG Y.-M. (eds.) (1980) *Proceedings of the R.O.C.-U.S.A. Seminar on the Cultivation and Utilization of Economic Algae, Guam*, June 5–7, 1978. University of Guam Press.

VADAS R. L. (1979) Abiotic diseases in seaweeds: thermal effluents as causal agents. *Experientia*, **35**, 435–7.

CHAPTER 20
COMMERCIAL CULTIVATION

C. K. TSENG
Institute of Oceanology, Academia Sinica,
Qingdao (Tsingtao), People's Republic of China

1 INTRODUCTION

Commercial cultivation of seaweeds involves the large-scale raising of macroscopic marine algae for commercial purposes. It can be viewed as a type of agricultural practice carried out in the sea, to which the term 'marine agronomy' has been applied (Doty 1977). The following account treats this subject by focusing on four major marine crop plants: *Porphyra, Eucheuma, Laminaria japonica* and *Undaria pinnatifida*. It will examine the scientific principles that affect this practice as well as the actual techniques that have been developed in this relatively new technology. The author draws upon his own experience, going back over nearly half a century, and thus emphasizes the 'state-of-the-art' in China. Some comparisons are made with methods being practised in Japan and the Philippines, where the large-scale farming of valuable seaweeds is also well underway. Some references to work in other countries are also included, but in many countries this science is often still in an embryonic stage.

2 ECONOMIC SEAWEEDS UNDER COMMERCIAL CULTIVATION AND THEIR ECONOMICS

According to Bonotto (1976) there are five genera of green seaweeds including 27 species, 40 genera of brown seaweeds including 88 species, and 56 genera of red seaweeds including 344 species, totalling 101 genera and 459 species of seaweeds which are of economic value as human food, as animal food, as manure, for medical or pharmaceutical purposes, or for industrial use. Bonotto's list may be reduced to 99 genera and 457 species by realizing that a freshwater species of *Prasiola* was included and that *Gracilariopsis* and *Gracilaria confervoides* can be regarded as synonymous with *Gracilaria* and *Gracilaria verrucosa*, respectively. Bonotto had overlooked the manual of economic seaweeds of China (Tseng *et al.* 1962), in which 46 genera including 100 species were described. Eight genera with 36 species were not contained in Bonotto's list. So the present total of economic seaweeds is 107 genera and 493 species. It is noteworthy that two species of blue-green algae are included, namely, *Brachytrichia quoyi* and *Calothrix crustacea*.

Among the seaweeds utilized by man for various purposes, only 11 genera with less than 20 species are commercially cultivated to any significant extent. These eleven genera are *Porphyra, Gelidiella, Gloiopeltis, Eucheuma* and *Gracilaria* of the red algae, *Laminaria, Undaria,* and *Macrocystis* of the brown algae, and *Monostroma, Enteromorpha* and *Caulerpa* of the green algae. Most of these algae, however, are not actually domesticated in the sense of the crop plants we know in agriculture. Perhaps only four genera, namely, *Porphyra, Eucheuma, Laminaria* and *Undaria*, for which the amount harvested from cultivated sources exceeds that taken from wild populations, can be regarded as truly marine crop plants.

2.1 The marine crop plant *Porphyra*

Species of *Porphyra* are popularly known in China as *zicai* (meaning purple vegetable), in Japan as *nori*, and in the western world as 'purple laver', and have long been appreciated by coastal people as a delicious food, especially by the Japanese and Chinese. Both China and Japan have a long history of the cultivation and utilization of *Porphyra*. This is a very widely distributed genus, primarily temperate but also occurring in many subtropic and subarctic regions. The purple lavers are in general littoral seaweeds, growing on rocks and only in a few cases growing epiphytically or in the sublittoral zone. In China and Japan seven species are used in commercial cultivation: *P. yezoensis, P. tenera* (China and Japan), *P. haitanensis* (China) and *P. pseudolinearis, P. kuniedai, P. arasaki* and *P. seriata* (Japan), of which the first three species are the most important, perhaps accounting for more than 90% of the total production. In Santou, Guangdong Province, *Porphyra guangdongensis* and in Penghu, Taiwan Province, *P.* sp. are also cultivated but on a very small scale (Chueh & Chen unpublished).

In China the earliest records of the occurrence of *Porphyra* and its food value appeared in books published in the years 533–544. In the Sung Dynasty (A.D. 960–1279) *Porphyra* had already become a local good from Haitan Island of the

Fujian Province selected for presenting to the emperor annually. This means that about one thousand years ago the Chinese had already come to regard *Porphyra* as a delicacy. More than 200 years ago the simple 'rock clearing' method of cultivation was devised by mechanically clearing various seaweeds and barnacles from the rocks in early autumn. This was done just before the mass liberation of the *Porphyra* spores so that the liberated spores had the required space for attachment and growth (Tseng & Chang C. F. 1956). Modern methods of commercial cultivation involving artificial collection of spores were initiated in the early sixties on the basis of the research of the previous decade (Tseng & Chang T. J. 1954, 1955a, 1955b, 1956), which in turn had received impetus from Drew's discovery of the life history of *Porphyra* in 1949. These methods are now routinely used in the *Porphyra* cultivation industry in China. Two principal species are involved: *P. yezoensis* (Fig. 20.1) north of the Changjiang (Yangtze) River and *P. haitanensis* (Fig. 20.2) south of it. About 3400 hectares of the sea surface are presently devoted to *Porphyra* farms, although the actual area of the cultivation raft is only about 27% of that area. In 1979 China's production was about 7200 tonnes in dry weight, averaging 2.1 tonnes per hectare, and worth about U.S.$30 000 000. The product is in the form of paper-like sheets, as in the case of the Japanese product.

Commercial cultivation of *nori* in Japan was initiated more than 300 years ago by the primitive method of inserting bundles of bamboo twigs, called *hibi*, for the 'seeding' process, i.e. for collecting spores. This method was much more efficient than the primitive 'rock cleaning' method used in China. The two methods are similar, however, in that the number of spores attaching to the substrates, hence the resulting production, is entirely at the mercy of nature, and the annual production fluctuated greatly. The net cultivation method was introduced in the 1920's, and some increases in productivity resulted. It was not until the 1960's that, with the introduction of the artificial collecting of spores, the commercial cultivation methods became truly modernized, as is true also in China. This progress resulted from the research of several workers (Kurogi 1953a, 1953b; Kurogi & Hirano 1955; Kurogi & Sato 1962a, 1962b; Kurogi *et al.* 1962). With the innovation of the cold-storage net and the use of floating nets, *nori* production increased steadily. In 1978, according to Dr. Miura (private communication), 612.8 square kilometres of the sea surface were devoted to the cultivation of *nori*, involving more than 30 000 production teams and 3 758 000 *satsus**. A total of 21 150 tonnes† (averaging 345 kg/hectare) of dry *nori* was produced with a value of 129 700 million Japanese yen, or about

* *Satsu* is the Japanese unit of production area, equivalent to 21.6 m^2 nettings.

† Production data may vary in different reports, undoubtedly due to the differences in the basis of calculations. According to Bardach *et al.* (1972) and Mathieson (1975) the annual production of *nori* in Japan since 1960 has averaged about 120 000 tonnes, which is most probably based on fresh weights, equivalent to about 12 000 tonnes in dry weight, about 60% of the 1978 production (21 150 tonnes).

U.S. $540 000 000. It is undoubtedly the most economically important food seaweed industry in the world. *Porphyra tenera* (Fig. 20.3) and *P. yezoensis* are the two principal species cultivated, although six species are grown commercially (Miura 1974).

2.2 The marine crop plant *Eucheuma*

Eucheuma is a genus of tropical red seaweed which grows generally on limestone-rich substrates, especially coral reefs, and usually is most abundant in the zone from the lowest tide level to the upper 1 m level of the sublittoral region. Appreciated as a food in China and Malaysia and other southeastern Asian countries, *Eucheuma* is known as *gilin cai* (meaning unicorn vegetable) in China

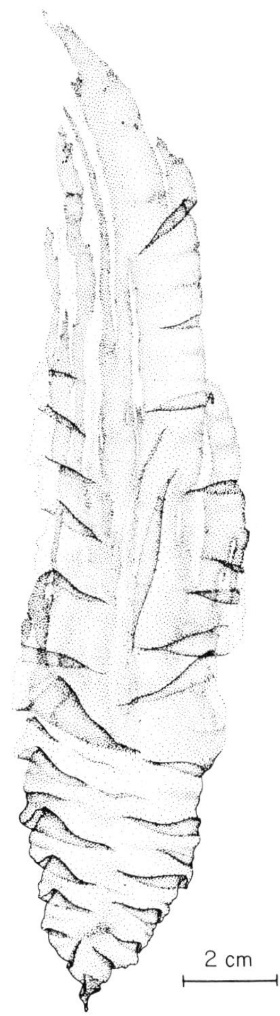

Fig. 20.1 Habit sketch of *Porphyra yezoensis* from Qingdao, China.

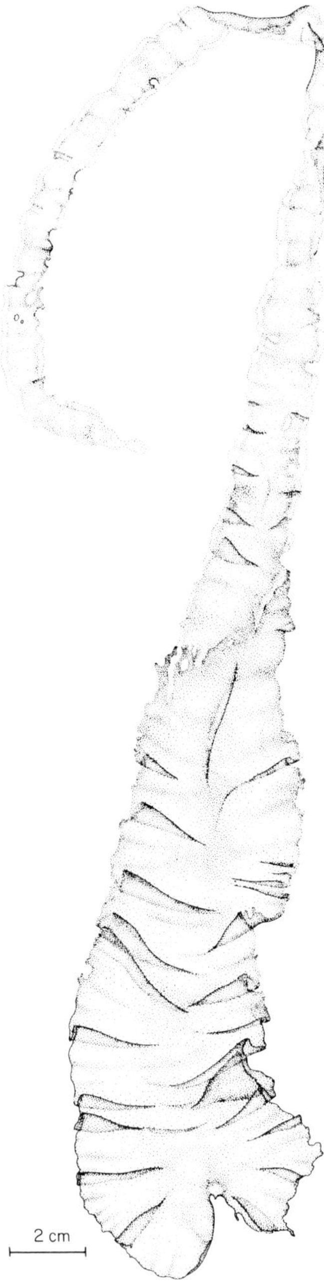

Fig. 20.2 Habit sketch of *Porphyra haitanensis* from Haitan Island, Fujian Province, China.

and as *agar-agar* in Malaysia—different in meaning, however, from the commercial agar-agar. *Eucheuma* has served more recently as the raw material from which the useful phycocolloid carrageenan is extracted, a key substance in today's food and dairy industries (Doty & Alvarez 1975). This genus can be divided into two natural groups: the axiferous *E. denticulatum* (= *E. muricatum*) group (Fig. 20.4), characterized by a dense central axis of slender cells in the centre of each branch and producing iota carrageenan, and the anaxiferous *E. striatum* group (Fig. 20.5), characterized by kappa carrageenan (Doty 1973; Doty & Santos 1978).

The traditional raw material for the carrageenan industry is *Chondrus crispus*, or Irish moss, also a red seaweed. With the increasing demand for this seaweed colloid, the production of *Chondrus* has not met the demand, and the

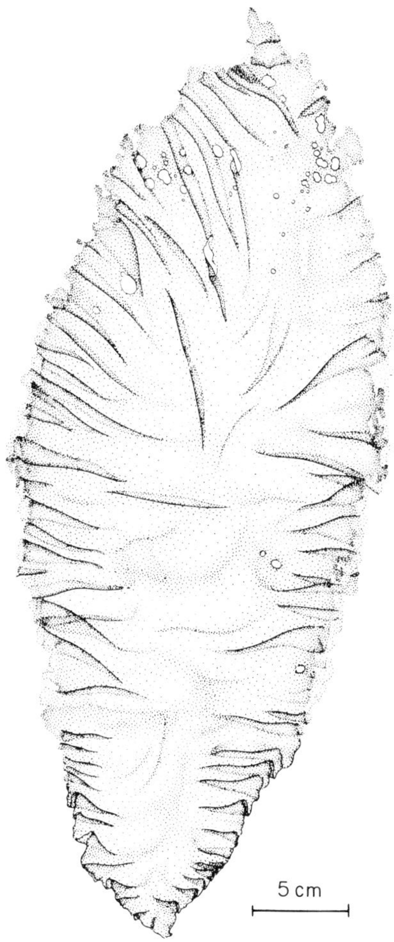

Fig. 20.3 Habit sketch of *Porphyra tenera* from Qingdao.

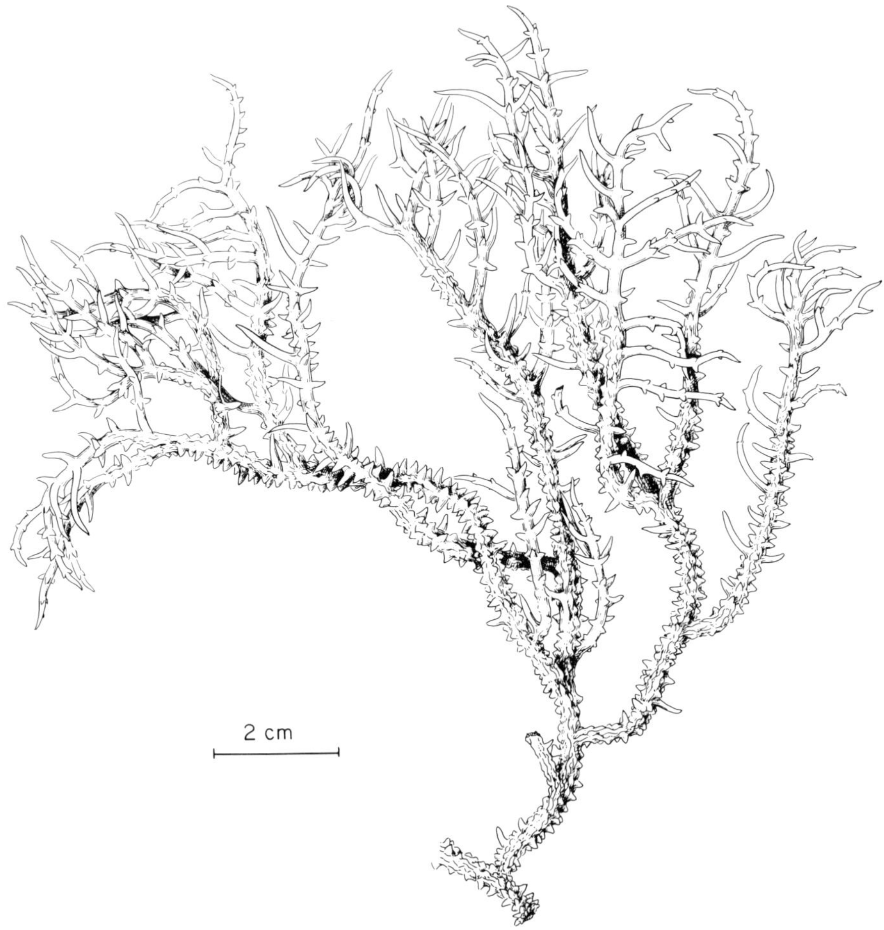

Fig. 20.4 Habit sketch of *Eucheuma denticulatum* from the local market in Shanghai, China.

carrageenan manufacturers have had to turn their attention to other sources, particularly *Eucheuma* and other genera in the order Gigartinales. Some years ago, 3–4000 tonnes of *Eucheuma* in dry weight reached the world market annually (Doty 1973). Over-harvesting resulted in the decrease in production of wild *Eucheuma*, and cultivation experiments were started in the Philippines, where some innovative techniques were devised. By mid-1974 there were more than 1000 farms engaged in the production of *Eucheuma*, exporting over 600 tonnes in dry weight per month, thus producing about twice as much *Eucheuma* as the original total world supply (Doty & Alvarez 1973, 1975). The present annual production is more than 10 000 tonnes in dry weight (Doty 1977).

In China *Eucheuma gelatinae* (Fig. 20.6) has been commercially cultivated in the Qionghai and the Wenchang districts of Hainan Island since 1960. At

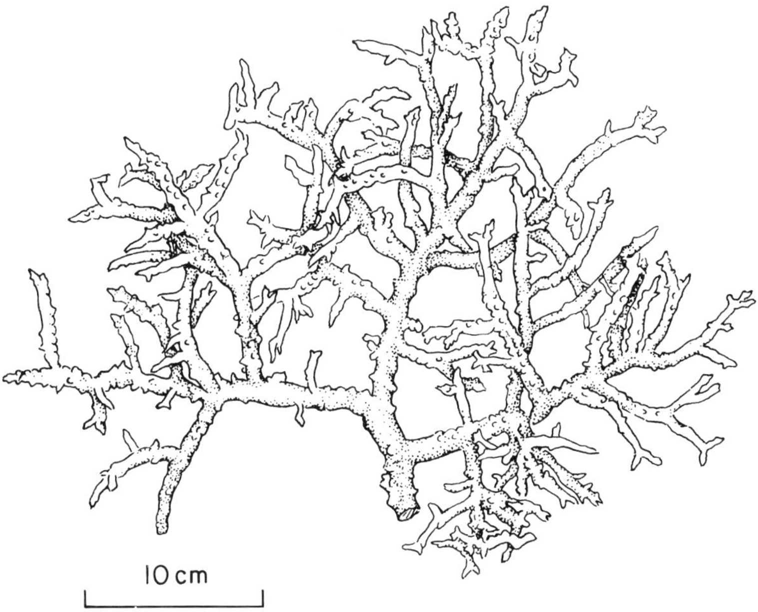

Fig. 20.5 Habit sketch of *Eucheuma striatum* from the Philippines (after Doty 1973).

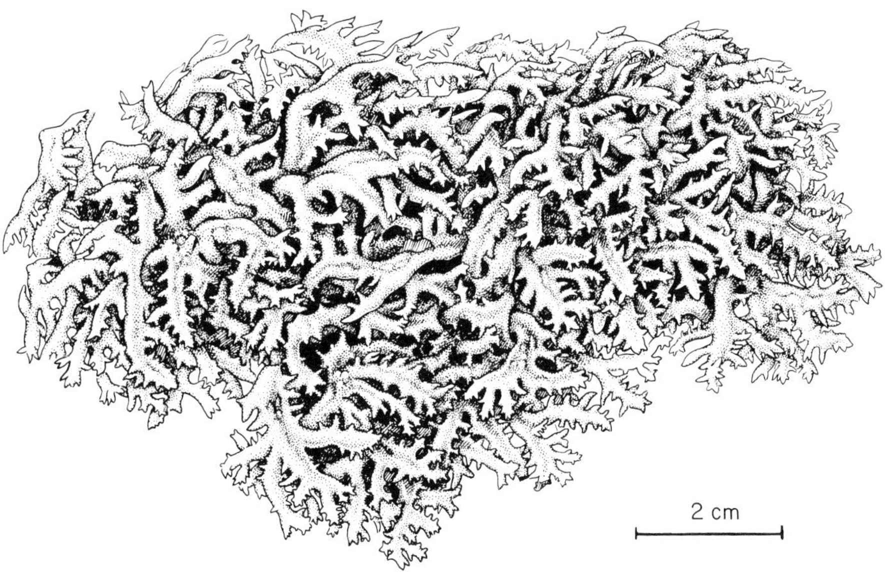

Fig. 20.6 Habit sketch of *Eucheuma gelatinae* from Qionghai, Hainan Island, China.

first, *Eucheuma* was cultivated by sending divers to insert cuttings of the seaweed in the sublittoral reefs. Since 1974, new cultivation methods have been devised, and the cuttings are fastened to coral branches with rubber rings or by some other means, thrown into the sublittoral reefs, and then finally rearranged in order by divers. At present the annual production is about 300 tonnes in dry weight, valued at about U.S.$450 000.

2.3 The marine crop plant *Laminaria japonica*

The largest single-species seaweed cultivation industry, as far as the quantity of production is concerned, is undoubtedly that of the brown seaweed *Laminaria japonica* (Fig. 20.7), as is now practised in China (Tseng 1962, 1964) and Japan (Hasegawa 1971, 1976). In China this *Laminaria*, commonly known as *haidai*, has served as a food for more than 1000 years. The popularity of eating this kelp, with its very high iodine content, has helped greatly in the prevention and treatment of goiter in many places in China (Tseng & Chang C. F. 1961). Indeed, it is hard to believe that such a widely accepted food material is not native to China but had been imported in dry form from Japan and Korea in great quantity. Importation of *haidai* continued during the first few years after the founding of the People's Republic of China.

Laminaria japonica is native to the cold temperate coastal regions along northern Japan, northeastern Korea, and Siberia. Although the oceanographic conditions along the northern coast of the Yellow Sea adjacent to China are in some respects similar to those of its 'home' near western Hokkaido of Japan, this kelp did not spread to the China coast itself because of the wide warm-water region lying between these two regions, thus preventing its southward extension.

In 1927 this *Laminaria* was accidentally introduced by ships from Japan to Dalian on the northern coast of the north Yellow Sea and settled in the vicinity of the harbour. After a few more trials in transplanting this kelp from Japan, *haidai* finally became a permanent component of the seaweed flora of China. In the forties *haidai* was transplanted to Yantai on the southern coast of the north Yellow Sea and in 1950–1951, after the founding of the People's Republic of China, it was further successfully transplanted to Qingdao (Tsingtao) on the southern coast of the Shandong Peninsula. In 1952 the newly devised floating raft cultivation method was employed with success. Since then, new methods for enhancing the production and improving the quality of the kelp have been successfully devised one after the other. Now, commercial cultivation of *haidai* is being practised on the entire China coast of the East China Sea (including the Yellow Sea), from Dalian of the Liaoning Province in the north to Xiamen (Amoy) and Dongshan of the Fujian Province in the south (Tseng 1962, 1964). In the kelp year 1978–1979, more than 18 000 hectares of marine farms were engaged in the commercial cultivation of this *Laminaria*, and about 275 000

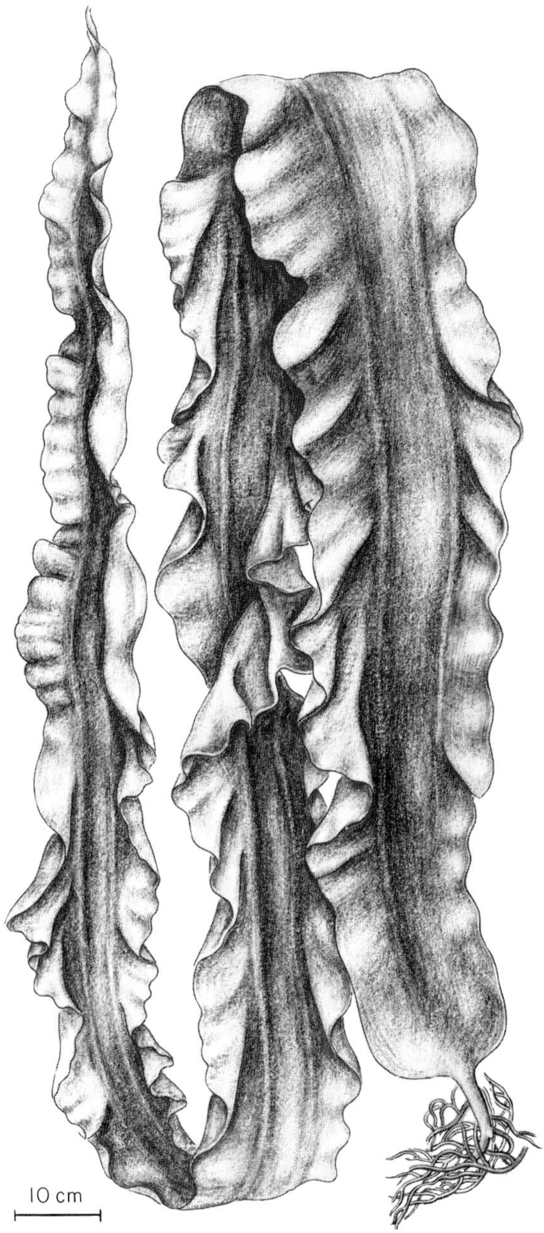

Fig. 20.7 Habit sketch of *Laminaria japonica* from Qingdao.

tonnes in dry weight (equivalent to more than one million tonnes in wet weight) of this seaweed, valued at about U.S.$300 000 000, were produced.

Being the 'home' of *Laminaria japonica*, Japan had been the principal producer of this much desired edible kelp, producing annually a few hundred thousand tonnes in wet weight. Natural production is not steady and fluctuates greatly. With the pollution of the marine environment and various other factors, natural production of *Laminaria* continued to decrease. In 1951 commercial cultivation was initiated, the two methods attempted being the stone-planting technique and the blasting of reefs (Hasegawa 1976). Improvement in production was, however, not promising, and production was unable to meet the demands of the people. Since 1968, the so-called forced cultivation method has been employed, and the output has increased from 30 tonnes in 1969 to over 7500 tonnes in wet weight in 1974, worth about U.S.$1 000 000, amounting to 16% of the natural production on reefs (Hasegawa 1976).

2.4 The marine crop plant *Undaria pinnatifida*

The other brown seaweed now under commercial cultivation and qualified to be called a marine crop plant is *Undaria pinnatifida*, popularly known in Japan as *wakame* and very much prized as a food alga (Fig. 20.8). In China it is known as *qundai-cai*, also serving as food. This seaweed is a warm temperature species (Tseng & Chang C. F. 1962), widely distributed in Japan except for the eastern and northern coasts of Hokkaido, which receive the cold Kurile Current, and the eastern coasts of Kyushu and Shikoku, which receive the warm Kuroshio. It grows on rocks and reefs at a depth of 1–8 m below the low tide level in open seas or within bays near the open sea (Saito 1975).

In Japan *Undaria pinnatifida* is a more important marine crop than *Laminaria japonica* both in value and in production. For years, maintenance and increase of the output of its natural production had been accomplished by various modes of propagation, such as depositing stones on the sea bottom and blasting rocky reefs to increase the area for its attachment. In this way the annual production of *Undaria* from its natural habitat had been maintained at 40 000–60 000 tonnes in fresh weight in the years 1960–1969. Since 1955, mass production of *Undaria* sporelings has been carried out by artificial spore-sowing on ropes of hemp-palm and later on synthetic ropes. In 1962 production in this way, in new cultivation areas, reached 21 000 tonnes in fresh weight, and in 1968 annual production by this method reached a peak production of 69 680 tonnes in fresh weight, surpassing the production of 48 300 tonnes in fresh weight in its natural habitat by a broad margin (Saito 1975).

In China *Undaria pinnatifida* has been collected for food from the natural habitat for centuries in the Zhoushan Island, Zhejiang Province on the East China Sea coast. *Undaria* now growing in Qingdao and Dalian on the Yellow Sea coast was, however, transplanted from Korea and perhaps also from Japan.

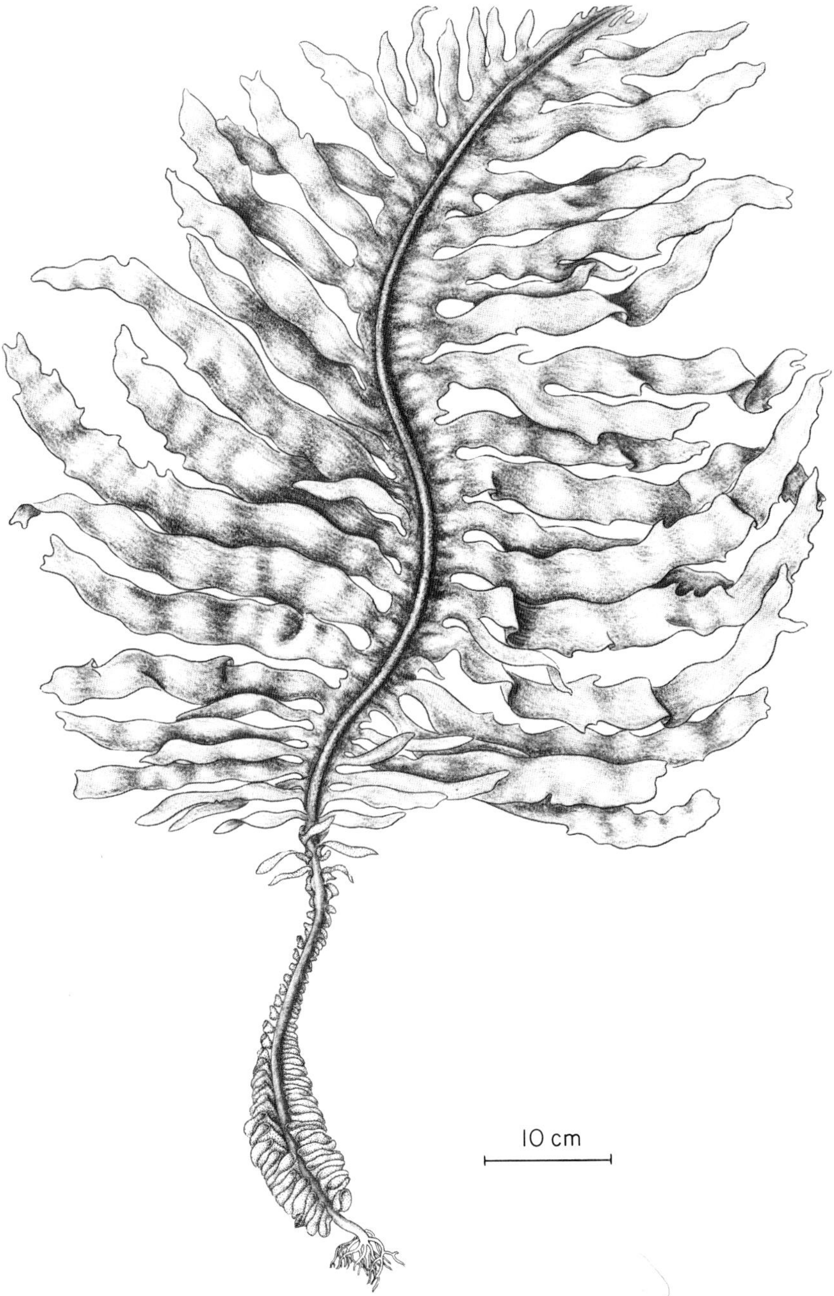

Fig. 20.8 Habit sketch of *Undaria pinnatifida* from Qingdao.

The first transplantation of *Undaria* from southern Korea to Qingdao took place in 1935 by a Korean who established a mariculture company to grow *Undaria* and some other marine products in Qingdao. His method of cultivation, however, was limited to transplantation and natural propagation, effected entirely on natural substrates, i.e. rocks. In the late fifties, an artificial method of sowing spores on ropes of hemp or palm fibres and later on ropes of synthetic fibres was employed, like that of the commercial cultivation of *Laminaria*. *Undaria* is often mixed-planted with *Laminaria* on the *Laminaria* floating raft. Since *Undaria* has an earlier short-growing season, maturing much earlier than *Laminaria*, and is harvested in March, it does not interfere with the growth and maturation of *Laminaria*, which is harvested in June. *Undaria* in China is not as popular as *Laminaria*, and the manipulation during the drying process is very tedious. Hence, maricultural firms are not very enthusiastic about its cultivation, and the annual production is much smaller than that of *Laminaria*, amounting to only a few hundred tonnes in dry weight.

For a better idea of the seaweed mariculture industry in China, the accompanying map of China indicates the principal areas for the cultivation of these marine crops (Fig. 20.9).

2.5 Other commercially cultivated seaweeds

A few other species are commercially cultivated but on a much smaller scale, and this production does not compare to that from the natural habitat. Thus, they are not regarded as 'marine crop plants'. Some of these algae are discussed below.

RED SEAWEEDS *GRACILARIA* AND *GLOIOPELTIS*

Gracilaria, known as *jiangli* in China, has been appreciated as a food and as a binding material in the preparation of lime for painting walls. In recent years these seaweeds are better known for their agar content and are now standard agarophytes used as raw material for agar production in China as well as in Australia, Italy, Japan, New Zealand, South Africa, and the United States (Bonotto 1976). Production from the natural habitat, however, is not sufficient to meet the increasing demand for raw material by the agar industry. Hence, commercial cultivation has been practised in various places, the production being still on a small scale and not comparable to its natural production. The principal species under cultivation is the widely distributed *Gracilaria verrucosa*. This seaweed grows abundantly on rocks, stones, gravel, and in fact any solid substrate in the littoral zone.

In the late fifties to early sixties, commercial cultivation was carried out in Guangdong Province, China, by inserting branches of the seaweed into the splits of bamboo sticks about 15 cm long, which were then planted in littoral

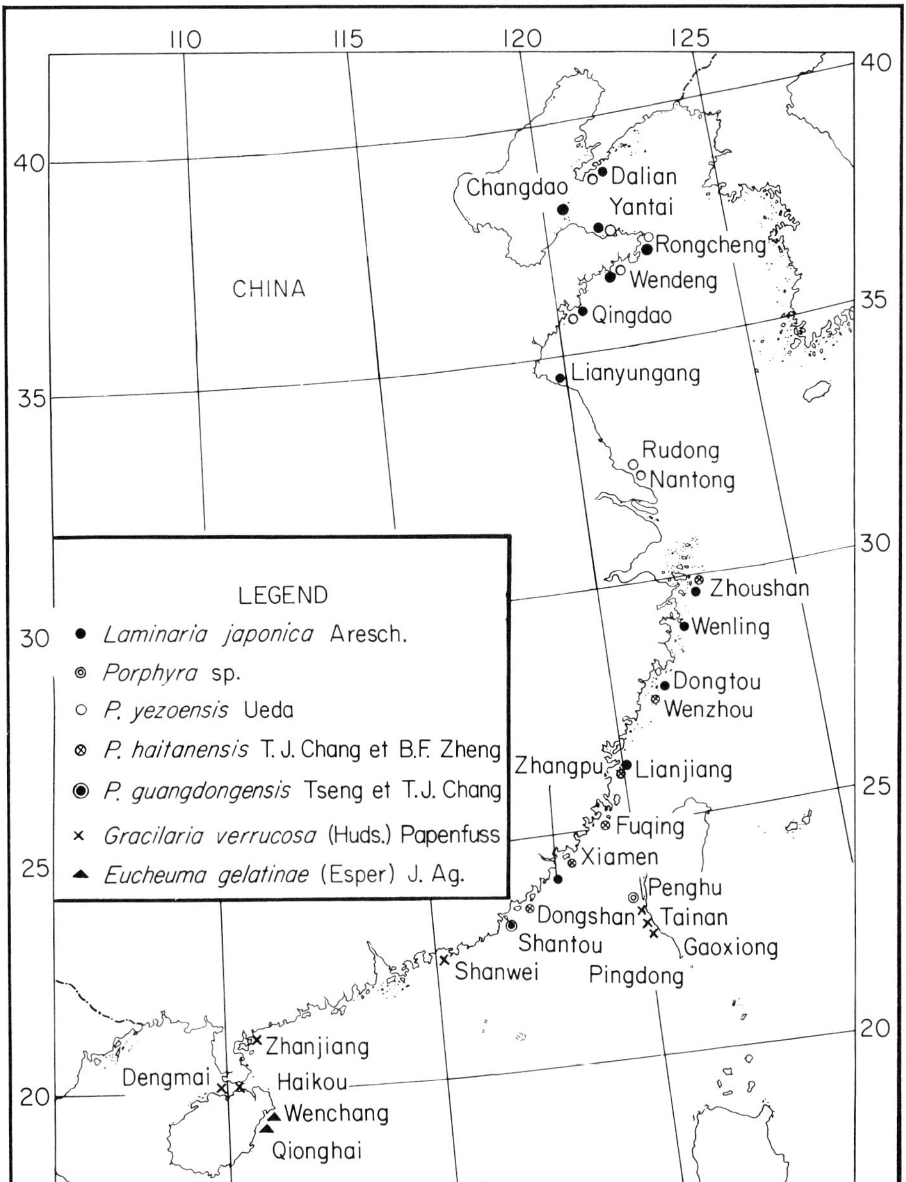

Fig. 20.9 Map of China showing principal areas of maricultural activities.

farms. Later, cultivation of the seaweed was carried out by scattering juvenile *Gracilaria* of more than 5–6 cm length, either gathered from its natural habitat or cultured from spores on stones and pebbles in littoral farms sheltered from strong wave action. In recent years the net cultivation method has been employed, in which branches of *Gracilaria* are inserted in the intersections of the net as in *Eucheuma* cultivation. The floating raft method in *Laminaria* cultivation has also been employed with some success. In a cultivation season of about three months, the production may reach 2–3 tonnes in dry weight per hectare. To date, however, the commercial cultivation in mainland China is on a small scale, and the principal source of *Gracilaria* as a raw material for the agar industry comes from the natural habitat.

According to Chueh & Chen (unpublished), cultivation of *Gracilaria verrucosa* was initiated on Taiwan in 1967. By 1977 *Gracilaria* ponds occupied 223 hectares in area, production reaching 6804 tonnes, worth U.S.$2 700 000. The producing areas are located mostly in the Pingdong (Pingtung) and Tainan in South Taiwan. The *Gracilaria* is uniformly planted in the bottom of ponds using cuttings or the torn sections as the planting stocks, tied to bamboo sticks or covered with old fishing nets to avoid drifting. About 4000–6000 Kg fresh *Gracilaria* are planted in a one-hectare monoculture pond. Polyculture of *Gracilaria* and milkfish, shrimp, or crab is also practiced, with improved profits.

Some commercial cultivation of the 'glueweed' *Gloiopeltis furcata* has been done in China and Japan. The glueweed is known in China as *hailo* and in Japan as *funori* and is appreciated as a food but also serves as a sizing material in the silk and other textile industries. *Gloiopeltis* grows gregariously on rocks in the littoral zone. In China the method of cultivation is very primitive and is practiced in the southern part of the Fujian province (Tseng 1933). Before its growing season starting in November, rocks, which are then covered with barnacles, oysters, and various seaweeds, are cleaned by scraping with iron tools, the loosened material carried away by tides. The following day lime is sprayed on the cleaned rocks to complete the destruction of the undesired organisms, thus preparing the necessary substrates for the attachment of the *Gloiopeltis* spores. The rocks thus treated will become thickly overgrown with *Gloiopeltis*. Investigations showed that on the small island Jinmeng (Quimoy) alone, about 200 tonnes in dry weight are produced annually (Tseng 1933). Production of *Gloiopeltis* in Japan is enhanced simply by placing boulders or chunks of concrete in the sea just before the growing season of this seaweed to provide the necessary substrate for it to grow (Bardach *et al.* 1972).

THE GIANT KELP *MACROCYSTIS*

One of the largest plants in the world, the giant kelp *Macrocystis*, especially *M. pyrifera*, is undoubtedly the most important economic seaweed, as far as the production from its natural habitat is concerned. This kelp unlike the cultivated

Laminaria japonica or *Undaria pinnatifida*, is a perennial plant. It is very abundant along the coast of southern California, U.S.A. Similarly kelp beds have been reported off the coasts of Mexico, Chile, Australia, and New Zealand (Saddington 1969).

The giant kelp is mechanically harvested by self-propelled barges in southern California. In the year 1916–1919, its annual production reached a peak of 400 000 tonnes in fresh weight (Neushul 1977). Since the initiation of the algin-from-*Macrocystis* industry, annual production of the giant kelp increased from about 1000 tonnes in fresh weight in 1932 to about 12 000–14 000 tonnes during the period 1950–1970 (Neushul 1977).

Great efforts have been made on kelp bed restoration since 1963 when the work was initiated at Pt. Loma. Since 1964, a 'kelp habitat improvement project' has been carried out under the leadership of Dr. W. J. North of the California Institute of Technology's W. M. Keck Laboratory of Environmental Health Engineering, to re-establish kelp bed communities along the southern California coastline. The project has contributed greatly to our knowledge of kelp restoration in methodology (North 1971, 1976) and has recently switched to investigating possibilities for increasing organic productivity in kelp beds through fertilizing operations (North 1977a).

To date, however, *Macrocystis* is still a wild plant and cannot be regarded as a marine crop plant. For a perennial plant such as *Macrocystis* growing in the open sea to be subjected to commercial cultivation and to become a genuine crop plant, it is a very difficult matter, and many biological and engineering problems have yet to be solved. Efforts to construct support structures to grow the giant kelp have been futile thus far. Initial trials have met with failure, but the American Gas Association is said to be continuing work with a quarter-acre module to be set in water about 300 m deep (Nath & Grace 1977). Undoubtedly, the success of such a venture will eventually revolutionize the whole structure of this industry.

GREEN SEAWEEDS *MONOSTROMA, ENTEROMORPHA* AND *CAULERPA*

In a broad sense, *Monostroma, Enteromorpha* and *Caulerpa* are under commercial cultivation and are appreciated as foods in China, Japan, and other oriental countries. *Monostroma* and *Enteromorpha* are common seaweeds growing on stones, pebbles, etc. in the littoral zone. In Japan *Monostroma* and *Enteromorpha* grow on *Porphyra* cultivation nets in spring usually after the growing season of *Porphyra* comes to an end and are later harvested and used to manufacture the much appreciated *hoshi-nori*. Its annual production amounts to 2–10% of that of *Porphyra* (Miura 1975).

Caulerpa is dull greenish in colour, its plant body being differentiated into a prostrate stolon with rhizoids and erect branches. The common edible species is *C. racemosa*, the stolon of which may reach 1 m in length. On Taiwan, China,

Caulerpa is grown in brackish water ponds by seeding with chopped pieces of the mature thallus. It is later shipped to the Philippines for the fresh vegetable market (Bardach *et al.* 1972).

3 SCIENTIFIC BASES FOR SEAWEED MARICULTURE

The commercial cultivation of seaweeds may be distinguished from agronomy of land crop plants not only in regard to growth media—water in the former and soil and air in the latter—but also in that the former involves unicellular spores as the reproductive units and rhizoids as the attachment organs, whereas the latter involves very complex, multicellular seeds and structurally complicated roots as the attachment and absorption organs. These differences naturally lead to different cultivation methods. These facts constitute reasons for also distinguishing the cultivation of seaweeds from that of aquatic seed plants, even though both are aspects of aquaculture. We can thus proceed with a discussion of the various important factors and practices involved, which form the scientific bases of successful seaweed cultivation.

3.1 The substrate

The rhizoids of seaweeds are not responsible for the absorption of water or nutrients, a function shared by all parts of the seaweed immersed in the aquatic medium. The function of the substrate is merely to provide a suitable attachment for the seaweed. Yet the selection of the substrate is important in some other ways, such as convenience, availability, durability and economics. Algal spores are very delicate unicellular structures and die as soon as they are removed from the medium. Therefore, it is imperative that in the 'seeding' process, i.e. the collecting of spores, the substrates must be brought close enough to the spore-liberating frond so that the spores will have the best chance to adhere to this substrate. In general, the spores of a seaweed will germinate only after adhering to the substrate.

The growth medium for seaweed cultivation is seawater, which contains many kinds of salts, some of which are quite corrosive. Therefore, the substrate employed should be able to resist this corrosive action of the seawater. The sea is in constant motion because of current, waves, and tide. Waves can be very destructive, and the positioning of the substrate has to be strong enough to withstand this wave action. Because of the constantly changing water level due to the tide, in many places amounting to several metres for the maximum tidal amplitude, the positioning of the substrate should be able to adapt to this tidal change.

It is evident that the selection of substrate and its positioning are of primary importance even though the seaweed does not derive any nutrients from the substrate. In general, the substrates may be natural, such as rocks and reefs, or artificial, such as ropes made of hemp, palm, or synthetic fibres. In the early stages of seaweed cultivation, natural substrates were employed, but with the advance of this science an artificial substrate in the form of a raft was adopted.

For years, both in Japan and China, the cultivation of *Laminaria* was done on rocks and reefs in its natural habitat by depositing stones at the desired depths, all for the purpose of increasing surface area for the attachment of *Laminaria* spores. Later, mature *Laminaria* plants were tied to sublittoral substrates by divers to provide more spores for those natural substrates. For some years there had been much controversy as to which method was better: propagation on sublittoral rocks or cultivation on rafts. Production practice shows the definite superiority of the artificial substrate (the floating raft), as shown by the production data of two mariculture stations, one at Qingdao (Shandong Mariculture Station) and the other at Dalian (Luda Mariculture Station), in 1952–1958. *Laminaria* was introduced to Qingdao (36° N Lat.) in 1950–1951. In 1952 when the floating raft (of hemp ropes) method was devised, Qingdao produced 62 tonnes of kelp in fresh weight. The production rose steadily to 13 207 tonnes in 1958, far ahead of the natural habitat production from the sublittoral rock of 1378 tonnes. Dalian (40° N Lat.) has a lower water temperature, more favourable for the growth of *Laminaria*, but, because of the continued use of the natural substrate by the Luda Station, Dalian produced less kelp than Qingdao in 1955 and 1956. In 1957 the Luda Station finally adopted the floating raft method, and in the same year production slightly surpassed that of Qingdao; by 1958 production (22 936 tonnes) surpassed that of Qingdao (14 585 tonnes) by a broad margin of 157% (Tseng 1962).

Chinese mariculturists for many years used ropes made of hemp or palm fibres for artificial substrates. These had to be soaked in water for a few months in order to leach out undesirable substances in the fibre. In the course of the development of the kelp cultivation industry, it was noted that raw materials for the raft (the hemp and palm fibres) could not keep up with the demand, and so in the early sixties synthetic fibres were used with success. At present, practically all of the ropes used for growing kelp are made of synthetic fibres. As to the floating raft itself, there is a tendency to abandon the double-line raft and adopt the single-line raft (see Fig. 20.16), which appears to be better adapted for places with stronger currents and heavier wave action.

3.2 The temperature factor

Temperature is one of the most important factors controlling the activities of seaweeds. Every species has its maximal, optimal, and minimal temperature for growth and development, which may differ in the same species depending on the phase of the life history and also on the different stage of growth in the same phase. Differences in the various phases are particularly conspicuous; in some cases the sporophytic phase has a higher and the gametophytic phase a lower optimal growth temperature, whereas in other cases the reverse situation is true. (See Section 4.2 for a further discussion.)

3.3 The light factor

Light, of course, is another basic environmental factor in the normal activities of seaweeds. The quality of light changes at different levels of the sea because of the differential absorption of light by seawater. However, with different levels, light intensity plays an even more important role, and hence the quality of light is generally not separately considered.

Light intensity

Each species of seaweed has an optimal intensity for its vegetative growth. Obviously, a knowledge of this characteristic is important for successful seaweed cultivation. For instance, the best growth of the sporophytic phase (i.e. the *Conchocelis*) of *Porphyra yezoensis* occurs at about 3000 lux and that of the gametophytic phase of *Laminaria japonica* at 3000–4000 lux, both being cultivated indoors (Ren 1962; IOESP 1978). The leafy thallus (gametophytic phase) of *Porphyra* is an intertidal plant and is better adapted to high light intensity. Therefore, in the floating raft cultivation method, it is necessary to keep the raft as close to the water surface as possible. On the other hand, the blade of *Laminaria japonica*, representing the sporophytic phase, is a sublittoral plant and in its native region of Japan grows well even at a depth of 10 m. On the East China Sea coast, however, the seawater is rather turbid and the transparency of the water quite low. In one experiment at Qingdao it was observed that when the sporelings were only a few cells to about 50 cells high, the best growth occurred in the uppermost layer of the raft near the surface (17 cm). But then as the fronds grew, the depth for best growth was successively lower such that after two and a half months the best growth was at the 3 m level.

In another experiment conducted under controlled conditions, young *Laminaria japonica* sporelings smaller than *c.* 800 μm high grew best under 3000 and 4000 lux illumination, but soon afterwards they became unhealthy and died; those under 2000 lux illumination grew best (Wu 1962). Therefore, in the greenhouse cultivation of the summer sporelings under low temperature, the light intensity has to be carefully controlled at about 2000 lux to obtain the best results.

Regulation of light intensity in the field

The transparency of seawater is different depending on the location and the particular season. For example, on the Zhejiang coast the seawater is quite turbid and the transparency low, especially during winter, the best growing season for kelp. During this period the transparency scarcely exceeds 1 m, and the level for the kelp under cultivation must be raised almost to the surface to avoid the so-called 'green rot' caused by insufficient light. Between late March and early April, the water suddenly becomes clearer because of the incoming warm current from the south. Under such circumstances the level of the cultivated kelp must be lowered 1–2 m or else the kelp will turn pale, eventually suffering from 'white rot', caused by too strong light intensity (Zhang 1962).

LIGHT PERIOD

Photoperiodism is especially important in the development of reproductive structures. Diurnal light rhythms may also trigger spore release. See Section 4.2 of this chapter as well as Chapter 9 for a more complete discussion of this factor.

LIGHT AND DEVELOPMENT OF *LAMINARIA* GAMETOPHYTES

Similar to the sporophyte of *Porphyra* (the *Conchocelis*), the gametophyte of *Laminaria* is the summer phase and grows only vegetatively, unable to carry out sexual reproduction or to develop sporophytes under 24 h of continuous illumination. Unlike the *Porphyra* sporophyte, however, the number of days required for the gametophyte to effect sexual reproduction and develop sporophytes is inversely proportional to the number of hours of illumination per day. For example, when the illumination is 2 h per day, it takes 44 days for sporophytes to appear and 59 days for the sporophytes to reach 50% of the total number of gametophytes. When the illumination is 19 h per day, it takes only 8 days to develop sporophytes and 11 days for the sporophytes to reach 50% of the total number of gametophytes.

It was also discovered that a certain dark period is necessary for the discharge of eggs and sperm in *Laminaria* gametophytes. In general, about an hour of darkness is required for this discharge, which may explain why 24 h of continuous illumination prevents the production of sporophytes. The period of darkness required depends upon the previous light treatment. Thus, when the culture is exposed to 24 h continuous illumination, only 20 min are required for initiating the discharge of eggs and sperm, whereas if previous lighting lasted only for 1 h, then 3 h of darkness are necessary for bringing about the discharge. It seems that there is some material synthesized in the light and in the darkness which is necessary for egg discharge. As shown in the experiment, the sperm are discharged only after the discharge of eggs and are attracted to the eggs by certain substances (Tseng *et al.* 1959c; see Section 3.4 of Chapter 19).

3.4 The application of fertilizers

The role of inorganic nutrients in the biology of seaweeds is well known (see Chapter 10). However, the application of nutrients as fertilizer to seaweeds in their large-scale commercial cultivation in nature is relatively new. So far it is employed only in the cultivation of *Laminaria* and *Porphyra* in China and Japan.

Need for fertilizer application in *Laminaria* cultivation

This is practiced commercially in China only in the Yellow Sea Region. In the early stages of kelp cultivation in China, it was confined to a few cities (Dalian, Yantai and Qingdao), mostly near sewage outflows and in harbours and their vicinity. Experiments with growing *Laminaria* in different places around Qingdao showed that it was the lack of sufficient nitrogen nutrients that limited the growth of *Laminaria* and that the phosphate content of the seawater, generally over 10 mg of phosphate-phosphorous per m^3, was sufficient for its normal growth. Seawater is regarded as infertile for the growth of *Laminaria* if the nitrate-nitrogen concentration per m^3 of water is lower than 5 mg and the total inorganic nitrogenous nutrient content, including ammonium salt, less than 50 mg. Under these circumstances fertilizer should be applied. In such infertile regions the spores seeded on the rope substrate germinate, but sporelings are not subsequently developed. Even larger fronds more than a metre in length, after being transplanted into infertile regions, will gradually turn pale yellowish and are unable to produce any more growth.

Fertilizer application by the porous container method

In the early fifties the area of *haidai* cultivation was still very small, from one tenth of a hectare to a few hectares per farm. *Laminaria* grows only in a certain layer of seawater; therefore, to be practical, the application of a fertilizer must be limited to selected spots to minimize its loss. The clay bottle method of fertilizer application was devised to take advantage of the porous nature of earthenware. The clay bottles containing solution of the desired nitrogenous fertilizer, usually ammonium sulphate, were hung at certain intervals on the raft. The porosity of the clay bottles effectively controlled the diffusion of the fertilizer which became available to the kelp in the immediate vicinity and the loss into the open sea was minimized. Thus, *haidai* could now be grown in formerly infertile regions to the desired commercial standard (Tseng *et al.* 1955b; Wang 1962; Tseng & Wu 1966; Cheng 1969).

This method greatly helped to extend the kelp industry during the late fifties along the Yellow Sea coast. In the early sixties various methods were employed to replace the clay bottle, which was quite expensive and involved much labour with something lighter and easier to apply. One common substitute was the use of plastic bags punched with some very minute pores, allowing the nutrient solution to leach out slowly.

Fertilizer application by spraying

Since the sixties the kelp industry of China has greatly expanded and kelp farms extended to several hectares, even several tens of hectares in area. The use of the fertilizer bottle or bags was very laborious, and simpler, more efficient methods needed to be devised. Experiments were conducted to gain a better idea of the process of absorption of the fertilizer. It was learned that nitrogen-starved kelp was able to absorb large quantities of the fertilizer in a relatively short time, which was sufficient to provide for the needs of the plant for some time (Wu *et al.* 1959; Tseng *et al.* 1962b). For example, nitrogen-starved fronds of kelp when soaked in 100 ppm nitrogen fertilizer (e.g. ammonium nitrate) for 0.5–1.0 h, absorbed sufficient nutrients to meet their needs for 6–10 days (Tseng & Wu 1966). A 'soaking method' was devised to immerse the kelp fronds in fertilizer solution once every several days, and some good results were obtained (Wu 1962; Wang 1962). This method, however, is practical only with small thalli. With large fronds of a few metres in length, soaking cultivation ropes each overgrown with 30 or more *haidai* fronds is not an easy job. Later, on the basis of the same principle, i.e. quick absorption of the fertilizer by the kelp and the effectiveness of the intermittent application, periodic sprinkling of fertilizer solution by manual labour was employed also with good results. Finally, the mechanical spraying method was developed which is time-saving and effective and is now the standard method of fertilizer application. The fertilizer solution of about 5–10‰ is prepared in a tank on a fertilizer boat, and an ordinary spraying machine is used (Fig. 20.10). The spraying of a given raft occurs once every one or two days, and the spraying rotates among the different rafts of a farm. Since a seaweed cultivation farm is generally of a large size, usually several tens of hectares, the sprayed fertilizer solution remains in the area for quite some time, despite the tidal movement, and the loss to the outside is not serious.

Fertilizer application for *Laminaria* cultivation is necessary only in the Yellow Sea region, which is, however, the best place for kelp cultivation because of the longer period of optimal temperature for *haidai* growth, i.e. 5–13°C. Therefore, despite the increase in cost of production due to the application of fertilizer, kelp cultivation in the Shandong and Liaoning provinces is a prosperous industry, yielding better returns than that of the southern provinces where fertilizer application is not needed. The production in the Shandong province amounts to about 50% of the total production in China. Ammonium sulphate is the commercial fertilizer commonly used, the amount employed being about two tonnes per hectare for the entire cultivation season of about

Fig. 20.10 A motorized fertilizer boat in the course of jet-spraying fertilizer solution in *Laminaria* cultivation at Qingdao. Note the spherical glass floats in rows in front of the boat.

seven months. With a production of about 15 tonnes dry weight of kelp per hectare, the cost of the fertilizer is less than 8% of the total value of the crop.

FERTILIZER APPLICATION IN *PORPHYRA* CULTIVATION

Commercial fertilizer is applied for *Porphyra* cultivation in some places in China and Japan, and generally the spraying technique is used. In China, as in the case of *Laminaria* cultivation, fertilizer application is necessary only in the Yellow Sea region. At Qingdao and its vicinity, for example, the purple laver farms are classified into three groups on the basis of the nitrogenous nutrient content of the seawater. The first group has less than 50 mg of nitrate-nitrogen plus ammonium-nitrogen per cubic metre of seawater; it is regarded as infertile for *Porphyra* growth, and fertilizer must be applied to make good growth possible. The second group has about 100 mg of total nitrogenous nutrients per cubic metre of seawater; it is regarded as semi-fertile and is able to support the growth of *Porphyra* without fertilizer application, but the product is of medium grade. The third group of farms has 200 mg or more of total nitrogenous nutrients per cubic metre of seawater; it is regarded as sufficiently fertile and able to support good growth of *Porphyra* without any application of fertilizer and the product is of high quality and good colour and lustre.

The fertilizer applied to the infertile farms is again usually ammonium sulphate, and the amount employed is also about two tonnes per hectare. The yield per hectare averages only 1.5 tonnes in dry weight in the form of sheets, much less than that of *Laminaria*. But because of the much higher value of the purple laver, the cost of the fertilizer is only about 5% of the value of the product. In some places, instead of the spraying technique, plastic tubes containing the nutrients are employed, which are tied to the bamboo floats or directly onto the cultivation nets. The results are said to be promising (IOESP 1978).

3.5 The control of weed growth

Like land-based agriculture, aquaculture of *Laminaria* and *Porphyra* has problems from nuisance 'weed algae'. The weed problems for these two genera, however, are different. In the case of *Laminaria* the weed problem is important only in the course of sporeling development, whereas in the case of *Porphyra* it occurs over the entire period of growth of the leafy thalli.

THE WEED PROBLEM IN *LAMINARIA* CULTIVATION

Initially in the commercial cultivation of *Laminaria* the collecting of spores on the artificial substrates was done in autumn, generally in mid-October in the Qingdao region. The sporelings thus produced were called autumn sporelings. To the seeded substrates, after being placed in the sea, the spores of such weed algae as *Ectocarpus, Enteromorpha, Licmophora*, etc. would quickly adhere and within a few days grow to macroscopic thalli, densely covering the gametophytes on these artificial substrates. Under normal conditions *Laminaria* spores take about 20 days to germinate, pass through the gametophytic stage, and give rise to young sporophytes. The *Laminaria* gametophytes are thus covered by various weed algae and the development of the sporophytes is greatly delayed or even totally inhibited. Generally it would be in December, when the weed algae had matured and dropped away from the seeding substrate, that the *Laminaria* sporophytes would appear and grow to transplantation size of about 15 cm by January–February. Thus, transplantation of the sporelings to the cultivation ropes would have to take place in the coldest season of the year.

The summer-sporelings, low-temperature cultivation method was proposed in 1955 and is now the standard method for growing the sporelings in China.

The seeding process takes place in early summer instead of mid-autumn as in the usual case of growing autumn sporelings (Tseng *et al.* 1955a; Hasegawa 1971b). In order to 'summer' the sporelings, the temperature of the culture solution must be lowered to 8–10° C. The seawater used should be filtered so that no spores of weed algae can enter the culture solution. Nitrogenous and phosphate fertilizers are added to the culture solution. Under these conditions the *Laminaria* zoospores collected on small cords develop through the gametophytic to the sporophytic phase in about 12 days, and the young sporophyte is able to grow free from nuisance algae. By mid-autumn when the seawater temperature drops to about 20° C or lower, favourable to the growth of the *haidai*, the cords with the summer sporelings (a few mm to about 2 cm high) are hung in the sea. Spores of weed algae, of course, also adhere to the substrate, but they are unable to compete with the rapidly growing summer sporelings, thus effectively solving the weed problem.

These sporelings reach transplantation size in a month or so, and transplantation may begin in late November to early December, two or three months earlier than for the autumn sporelings. Thus, there is an advantage of at least two months growth, resulting in an increase of production of 30–50% or more. In addition, the working conditions in late November to early December, when the water temperature is over 10° C instead of freezing temperature as in the former method, are greatly improved (Tseng *et al.* 1955a).

THE WEED PROBLEM IN *PORPHYRA* CULTIVATION

The weed problem is more complicated in *Porphyra* cultivation. In general, there are two kinds of weed algae, the green algae (usually *Monostroma, Enteromorpha* and *Urospora*) and diatoms (most often *Licmophora*). The attachment of these weed algal spores is prevented by three methods: 1, in the seeding process the attached spores should be as densely spaced as possible so that there will be very little space for the weed spores to attach; 2, in the manipulation of the cultivation nets, care should be taken so as not to scrape off the germlings, which would result in exposed space; 3, in the harvesting process attention should be paid to see that only the large thalli are collected and the smaller thalli left intact. When there are already some quantities of weed algae growing on the nets, the method of control is to expose the net to the sun for a definite number of hours or even days. The weed algae are more susceptible to desiccation than *Porphyra*. With the correct amount of exposure to the air, the weed algae can be killed and the *Porphyra* left intact.

3.6 The control of diseases

The history of agriculture has presented much evidence to warrant the generalization that with the domestication of a wild plant followed by its large-scale commercial cultivation, the new crop plant inevitably is sooner or later attacked by diseases. The marine crop plants are no exception. Three different types of diseases can be recognized: 1, physiological diseases, including those due to environmental pollution; 2, pathogenic diseases, generally caused by microbial pathogens; 3, pests, namely, larger animals grazing upon the crop plants.

PHYSIOLOGICAL DISEASES IN *LAMINARIA* CULTIVATION

The most common physiological diseases of *Laminaria* are the green rot and the white rot diseases, both related to the intensity of light. Green rot usually occurs in over-crowded fronds, starting with the lower plants on the cultivation ropes and gradually extending upward. The marginal portions of the diseased fronds turn greenish, become soft, decay, and disintegrate. Afterwards, the apical parts are involved, and finally the entire frond succumbs. The phenomenon is due to the lack of available light to the fronds, which are shaded by those above them. Control of such a disease is done by inverting the cultivation ropes immediately so that the lower over-shaded fronds can receive more light (Zhang *et al.* 1962). At the same time cutting off the apical parts of the fronds to as much as one third of the total length will reduce the overcrowded condition and improve light penetration and hence the health of the fronds. The cut portions can be used as raw materials in the alginate industry. This method of cutting tips to increase production has been discussed by Wu (1962a).

The course of the development of white rot is similar to that of green rot, i.e., from the marginal portions to the tips, gradually downward to involve the entire frond. This disease, however, always occurs in the fronds at the upper part of the cultivation ropes and in the Yellow Sea region in May in places where the water is not fertile enough. It is believed that a combination of three factors is involved: strong light, high water temperature, and low nutrient contents, of which strong light is the principal cause. It is therefore recommended that the cultivation level of the kelp be lowered and more fertilizer be applied (Zhang *et al.* 1962).

PHYSIOLOGICAL DISEASES IN *PORPHYRA* CULTIVATION

In protected bays cultivation of *Porphyra* is often so dense that the rafts are very closely arranged, and with calm weather the seawater appears practically motionless. Under such conditions in November–December a sudden rise in temperature may produce disastrous results. Some of the young purple laver will be killed and the cultivation nets become denuded sporadically, sometimes in large patches, resulting in low production. No special term has been given to this phenomenon.

In Japan 'fog damage' is a well known phenomenon in places like Tokyo

Bay, which formerly was one of the most important regions for *nori* production. This happens only on a very calm day when the fog remains over the *nori* farm for some time, resulting in a mass destruction of the young laver. It is now recognized that the real culprit is air pollution and the large amount of sulphite in the smoke, eventually dissolving in water to form sulphurous acid, which is detrimental to *Porphyra* (Ueda 1973).

The so-called 'crown-gall disease' of *nori*, first reported in 1957 by Fujiyama (Kito 1972) is a cancerous disease caused by some carcinogenic substances in sewage. The tumorous growths may spread to other parts of the frond; this disease may be fatal (Goff & Glasgow 1980).

Pathogenic diseases

Several kinds of pathogenic diseases have been recognized, but relatively few cases have been well documented. Among the few well worked up cases is the malformation diseases of *Laminaria* sporelings, an investigation of which was made in the early seventies. This study resulted in the clarification of the etiology of the disease and the development of an effective control. At present, the sporelings needed for more than 20 000 hectares of the kelp farms in China are all cultivated in the summer in an artificially lowered water temperature of about 10° C (Tseng *et al.* 1955a). Special greenhouses were built for this purpose in many coastal cities.

In recent years sporelings died in large masses in some of the sporeling cultivation stations from a disease of unknown nature. It is characterized by plasmolyzed oogonia and abnormal, malformed sporophytes which subsequently die and drop off the cultivation ropes. After a series of experiments in the laboratory and in one of the sporeling cultivation stations, it was concluded that the cause of the malformation was the presence of hydrogen sulphide in the culture solution. Experiments showed that hydrogen sulphide in a concentration of 0.50 ml/l cause the shrinkage of the egg and that the half toxic dose of the oogonium was about 0.018 ml/l. The organisms involved were found to be sulphate-reducing bacteria and hydrogen sulphide-producing saprophytic bacteria such as *Micrococcus*. The sulphate-reducing bacteria were found abundantly in the black rust of the cast-iron pipes of the water circulation system and the *Micrococcus* in the rotted spots of the *Laminaria* fronds. The marine mud in the vicinity of the station also provided large quantities of these hydrogen sulphide-producing bacteria. Since 1975, preventive measures have been taken by separating the sporeling cultivation system from the mature sporophyte cultivation system and by sterilizing the water system with bleaching powder before the seeding process. Rotten fronds are also periodically removed to reduce sources of these bacteria. The sporeling cultivating station has been functioning normally since then (Wu *et al.* 1979).

Another well studied case is the 'frond twist disease' of the raft-cultivated *Laminaria*, first observed in the winter of 1973 in the Dalian region, North Yellow Sea. The diseased plants had abnormally twisted fronds with greatly swollen stipes and very shortened rhizoidal holdfasts. On the basis of the presence and spread of the disease, the contagious and biotic nature of the disease was suspected. Microscopic examination of the diseased parts, however, failed to reveal any bacterial, fungal, or other biotic pathogens. Carefully planned experiments on contact infection conducted at Qingdao confirmed the contagious nature of the disease, which had a rather long latent period of 60–70 days and must, therefore, be biotic in nature (Li *et al.* 1980). Since examinations with the light microscope did not reveal any biotic pathogen, electron microscope examination of the ultra-thin frond sections and extracts was conducted. The results showed that there were numerous polymorphic mycoplasm-like organisms, mostly coccoid and some ovoid, dumb-bell and amoeboid in shape (Wang Q. *et al.* 1980). Experiments showed that the antibiotics of the tetracyclines were effective and others such as penicillin were ineffective in controlling the disease. This result also helps to confirm that the pathogen involved in this particular disease is a mycoplasm-like organism (Wang K. *et al.* 1980).

Alginic acid-decomposing bacteria were found to be the dominating bacterial inhabitants of the blades of cultivated *Laminaria*, and the bacteria as well as the crude enzyme preparation from the bacterial culture were observed to be able to decompose the *Laminaria* blade in 3–5 days (Chen *et al.* 1979). Further studies of the alginic acid-decomposing bacteria showed that they were the causative factor of the common destructive falling-off disease of the summer sporelings. The mass falling off of the sporelings from the cultivation cords either in the greenhouse or in the sea resulted in great loss to the mariculture industry (Chen, Lin & Shen unpublished). Some other bacterial and fungal diseases have been reported to occur in *Laminaria*. For example, *Labyrinthula* occurs in marine habitats, growing on marine Angiosperms such as *Zostera* and on seaweeds such as *Laminaria*. Such organisms are generally considered as non-specific 'parasites', and their role as causative agents of diseases is poorly understood (Goff & Glasgow 1980).

In *Porphyra* mariculture, two major biotic diseases have been described. One is the possibly lethal 'red wasting disease' caused by the fungal pathogen *Pythium*. The other is the 'green spot disease' caused by the bacterial pathogens *Vibrio* and *Pseudomonas*, which are necrotropic pathogens producing enzymes degrading the host's cell wall and porphyrans and causing 'green spots' (Goff & Glasgow 1980). Recently, an etiological study of the green rot disease of the cultivated *Porphyra* in the Qingdao region was conducted. The pathogenic bacteria were isolated and found to belong to the genus *Pseudomonas*. *Porphyra* fronds inoculated with these bacteria were quickly attacked, and irregular green spots occurred in two days. Crude enzyme was extracted from diseased fronds and observed to be able to produce green spots on healthy fronds in the same way as the bacteria (Chen, Yu & Yang unpublished).

Pests

In China in the early fifties, for the seeding process to be carried out in the autumn, selected mature *Laminaria* sporophytes had to be oversummered in deep water, and grazing by sea-urchins was a serious problem. But having abandoned the autumn sporeling method and now employing exclusively summer sporelings for this industry, it is no longer necessary to oversummer *Laminaria*. In the cultivation of *Porphyra* in Guangdong province, certain herbivorous fishes feed on the laver. In some cases special nets must be provided to protect the *Porphyra*.

The grazing by sea-urchins (*Strongylocentrotus franciscanus* and *S. purpuratus*) on *Macrocystis* along southern California is at times a serious problem (Leighton 1971). To re-establish kelp beds, measures must be taken to eradicate as many urchins as possible. These involve either manually crushing them with hammers or destroying them with an application of quicklime, the latter technique proving to be the more effective in reducing dense urchin concentrations.

3.7 Breeding of new strains in seaweed mariculture

When a plant is domesticated it eventually is introduced into a variety of places, in some of which the conditions may not be so favourable for growth. The problem of breeding new strains then naturally arises so that the seaweed can be better adapted to the new environment. This is the case with *Laminaria japonica*. Now growing on the China coast thousands of miles away from its native region of northwestern Japan, this northern cold temperate seaweed is now commercially cultivated in the subtropical waters of the southernmost part of the Fujian Province, China (about 24° N Lat.). Naturally, the problem of breeding new strains to answer the needs of the mariculturists has aroused the interest of geneticists. Although breeding work is still in the early stage of development, some results have been achieved.

Genetic studies of *Laminaria*

It has been experimentally demonstrated that natural populations of *Laminaria* now under cultivation in China are genetically mixed in nature and are considered to possess a high level of hybridity. Hence, they are good original material for selection and breeding work. It has also been pointed out that the harmful effect of continual inbreeding is relative and conditional; in some inbred lines there are harmful effects while in others no harmful effects are observed, and these eventually form new strains with certain good characteristics (Fang *et al.* 1962a; Fang *et al.* 1963). Experiments also showed that such morphological characteristics as frond length, frond width, frond thickness and stipe length are all quantitative inheritable traits, the development of which is controlled by polygenes as well as by environmental factors (see Chapters 11 and 12). The heritability of frond length was found to be about 0.5 (Fang *et al.* 1965).

Breeding of New Strains of *Laminaria japonica* with Distinct Morphological Characteristics

On the basis of the principles of segregation on crossing hybrids and of directional selection, methods of continual inbreeding and continual selection were employed. Single fronds of *Laminaria* were used to collect the spores, instead of using several fronds as for the large-scale production of sporelings in industry. In this way the gametophytes and the resulting sporophytes all come from a single frond. Mature *haidai* fronds can be selected for certain desirable characteristics, such as broad and thick fronds, as well as better adaptability to the comparatively higher water temperature of Chinese seas to that of Japanese, the home of this kelp. After four years of intensive inbreeding and selection by using materials from the *haidai* cultivation industry at Qingdao, a new strain of kelp, namely, Haiqing No. 1 of *Laminaria japonica*, was obtained in 1962. The new strain differed in several characteristics from the control. It grew faster and matured later, being more adaptable to higher temperatures and having a longer period of growth. It had a longer and broader blade with a longer stipe and was heavier in weight than the control (Fang *et al.* 1962a, 1963b). Later, two other strains were bred: Haiqing No. 2, with longer but narrower fronds with shorter stipe and maturing earlier than Haiqing No. 1; and Haiqing No. 3, similar to No. 1 in having long stipe and maturing late but distinct from the latter by its shorter, thicker frond (Fang *et al.* 1965, 1966).

Breeding of New Strains of *Laminaria japonica* with High Productivity and High Iodine Content

Some years ago the Chinese *haidai* industry was in need of strains high in productivity and iodine content. Kelp not only provides the Chinese people with a subsidiary food in their diet but also serves as the raw material in the alginate industry. In the production of alginate it is required that the by-products, including iodine, be recovered. Thus, high iodine content has likewise become an objective of the breeding program. In 1970 Chinese phycologists conducted research in the nine cultivation regions of the five coastal provinces. Employing the half-frond method, they selected 19 plants with high productivity and high iodine content. Breeding was carried out by repeated inbreeding and selection, and the offspring of one frond were treated with X-rays. After a few years of cultivation and examination of the genetic characteristics, two strains with high productivity and high iodine content were bred which answer better to the demands of the industry. Experiments demonstrated that the iodine content is also a characteristic showing a quantitative inheritance, similar to frond length. These two strains, now cultivated in large areas, have won the approval of the kelp farms (IOG & QMF 1976).

4 TECHNIQUES OF SEAWEED MARICULTURE

The success of the commercial cultivation of economically valuable seaweeds depends upon good culturing techniques that may differ to some extent from country to country and region to region. The scientific bases for successful seaweed mariculture should be universal, but the techniques actually used need to be adapted to the specific conditions concerned. A certain technique may be successful in one place but a total failure in another. The techniques now employed in China in the commercial cultivation of *Laminaria* are, so far, a success and account for an increase in production from about 10 tonnes dry weight using the so-called artificial cultivation method in 1952 to the production in 1979 of about 275 000 tonnes in dry weight, a staggering 27 500-fold increase in a mere 27 years. But if these same techniques were to be introduced into the United States or any European country, the result could be a failure because of social, economic, or other differences. Likewise, the farming of *Eucheuma* is a great success in the Philippines, but if introduced into the United States with the same technology, it could also result in failure. It was very practical of Doty to discuss this problem from four different angles in order to promote the *Eucheuma* industry (Doty 1977).

The introduction of an innovative, successful technique often stimulates the growth of an industry. Three new techniques in China along with the southward transplantation made possible a dramatic expansion of the kelp cultivation industry. These were the floating raft method, summer sporeling cultivation, and the application of fertilizer. In Japan the adoption of the raft intertidal cultivation, utilization of conchospores in the seeding process, cold storage of nets with germlings, and the adoption of the floating raft have contributed to a great expansion and to the prosperity of the present *nori* cultivation industry.

There are three categories of techniques currently practised in seaweed mariculture: 1, vegetative propagation, i.e. the use of cuttings, not spores, as the units for propagation, which is employed in the cultivation of *Eucheuma*; 2, the use of non-motile spores to serve as 'seed' cultivation substrates, as employed in *Porphyra* cultivation, without nursery cords being required; or 3, using motile spores (zoospores) to serve as the 'seeds' on special nursery cords in *Laminaria* cultivation, in which the resulting sporelings are later transplanted to other cultivation substrates. The essential points of these three techniques are discussed below.

4.1 The *Eucheuma* type of mariculture

This is the simplest mariculturing technique currently being practised with seaweeds. It is used in the commercial cultivation of *Eucheuma* in the Philippines and in Hainan, China, the principal species involved in the former being *E. striatum* and *E. denticulatum* (= *E. muricatum*) and in the latter *E. gelatinae*, as already mentioned. The principal manipulation involved is to tie the selected thalli to the net intersections or to dead coral branches and to put them in the sublittoral zone as close to the surface as possible.

The current practice of *Eucheuma* farming in Hainan is quite simple. First, one must select 'seed' thalli that are thick, sturdy, and without epiphytes. These thalli are then divided up into branch systems, about 40 such systems from each kilogram of thalli. These 'seeds' are tied to a branch of dead coral by means of rubber rings and scattered on a selected sea bottom in a density of about 75 000 branches per hectare. Arrangement of the 'seed' thalli in order on the sea bottom is done by divers. Harvesting occurs once a year.

The current practice of *Eucheuma* farming in the Philippines is somewhat more complicated. The branch system to be used as 'seed' is much heavier than the Hainan 'seed', generally weighing about 200 g, and is tied to the mesh intersections of a net. The nets are about 2.5 m wide and 5 m long, each with 127 meshes, and are attached to mangrove stakes fixed in the sea bottom. The nets are placed about 0.6–1.5 m above the sea bottom and are not exposed during the lowest spring tide. Each cultivation unit has 200 such nets, occupying 1–4 hectares of sea surface, cultivating 25 400 branch systems of *Eucheuma*. The alga generally grows to about 800 g in about two months and may be divided into four lots as 'seeds' for new nets. The strain selected for high productivity and that popularly cultivated is the Tambalang strain (Doty 1973; Doty & Alvarez 1973, 1975; Parker 1974, 1976).

Doty (1977) pointed out that although the methods for cultivating the Tambalang strain were very successful in the Philippines, they were too labourintensive and would not be economically feasible in developed countries. Consequently, new methods for cultivation were tested, and three of these were briefly described. It will be interesting to see if any of these or other new methods may be acceptable to the economic situation in well developed countries.

4.2 The *Porphyra* type of mariculture

For the cultivation of *Porphyra* two different processes are involved, each with its special techniques. The first process is devoted to the cultivation of the filamentous sporophytic phase, or *Conchocelis*, including the production of conchospores and the seeding manipulation. This process takes place indoors from May to October. The second process is the cultivation of the leafy thalli, which occurs in the field, usually from October to April or May of the following year. As far as the cultivation techniques are concerned, they are practically the same for the seven species presently under commercial cultivation. Minor differences with the different species and even with the same species in different places do exist but are of secondary importance. For convenience sake, the procedures with *Porphyra yezoensis* in Qingdao, China, will be given in more detail. Those procedures with other species and in other places, especially in Japan, will be compared when necessary.

INDOOR CULTIVATION OF THE FILAMENTOUS *CONCHOCELIS*

In Qingdao the carpospores are collected in early and mid-May and are seeded on shells of the mollusc *Meretrix meretrix* at a density of 200–300 spores per cm^2. Cultivation takes place in large elongated shallow concrete tanks containing seawater 20–30 cm in depth. The seawater has been previously subjected to sedimentation in the dark for at least three days (preferably seven days), and nitrogenous and phosphate nutrients (KNO_3 and KH_2PO_4) are added. Water temperature is not controlled but fluctuates with the changing air temperature. The *Conchocelis* phase does have an optimal growth temperature at 20–25° C.

Light intensity is controlled by a series of screens. Plane-type culture has been adopted, and the shells with *Conchocelis* are orderly arranged with their concave (seeded) surfaces facing upwards (Fig. 20.11). The shells are cleaned with very soft brushes once every two weeks. When the pH of the seawater rises to 8.5, the water should be changed, and it should be gently stirred a few times daily (IOESP 1978).

To produce the maximal number of conchospores per unit area, measures are taken to control the light intensity and light period according to the particular stage of growth. Three such stages are recognized. The first is from early May to early July, during which the water temperature gradually rises to 22–23° C, and is the period for the active vegetative growth of the *Conchocelis*. In this period the light intensity is controlled at about 3000 lux in the middle of the day. The second stage is from early July to late August or early September, in which the water temperature rises from 22–23° C to 27–28° C sometime in mid-August, and then gradually declines. This is the critical period for conchosporangial production, and the success or failure to obtain a good crop of conchospores depends on the correct handling of the light factor. The principal is to enhance the accumulation of the necessary reserves for sporangial formation but to delay the actual formation of the sporangia before the peak of the high temperature. After the peak and with the temperature dropping, formation of sporangia should be encouraged. Therefore, beginning in early July or slightly later, the maximum light intensity is further reduced to 750 lux. In this way, formation of the maximal quantity of sporangia is guaranteed. The third stage is from late August or early September to late September when the

Fig. 20.11 Shells with *Conchocelis* growths in the plane type culture in the *Conchocelis* culture room of the Academia Sinica Institute of Oceanology.

water temperature gradually drops to about 23° C. This is the period of sporangia formation. To encourage the formation of conchospores, the light intensity is held at 750 lux and the light period reduced to 8–10 h per day by artificially darkening the culture rooms. Under such conditions conchospores appear in late September, and mass discharge of these spores generally takes place in early and middle October (IOEP 1976; IOESP 1978).

Seeding process in tanks

That the purple laver industry has achieved great success in both China and Japan is due to the application of results from studies on the life history of *Porphyra*, especially the discovery of the 'missing link' between generations of the leafy thalli, i.e. the miscroscopic filamentous (*Conchocelis*) phase (Drew 1949; Kurogi 1953b; Tseng & Chang T. J. 1954). With the conchospores as 'seeds', there is no great difficulty in controlling the seeding process, which is a prerequisite to the success of the laver industry (Tseng *et al.* 1959a, b; Tseng & Chang T. J. 1962).

The seeding process is done on specially made cultivation nets, now made almost entirely of synthetic fibres. These nets are placed in the tanks containing the shells with the *Conchocelis*, and the cultures are repeatedly examined to determine the progress of spore formation. When the number of spores reaches 50 000 spores per shell per day, preparations are made for the seeding process. Since the adherence and germination of the conchospores require stronger light, the light intensity on the surface of the tanks is regulated to 1500–3000 lux or even more. Since the conchospores are non-motile, the water is agitated by means of a pump. The seeding process is done in the early morning, usually before 10 a.m., when the greatest discharge of spores occurs. When spore discharge reaches 100 000 spores per shell per day, the formal seeding can be started. When the daily spore discharge reaches one million spores per shell per day, the seeding process can be shortened; a few minutes is sufficient for the necessary number of spores to become attached to the cultivation nets. A density of 3–5 spores per mm^2 will satisfy the requirements of the seeding process. In one case, shells with *Conchocelis* occupying 3 m^2 of the culture tank produced enough spores for seeding the cultivation nets of a *Porphyra* farm of one hectare (IOEP 1976; IOESP 1978).

Cultivation of leafy thalli in the field

This may be divided into two stages: the sporeling stage and the grown thalli stage. In the former the germlings from the adhered conchospores grow to become sporelings 1–3 cm high. Whether the harvest will be good or bad depends in large measure upon the condition of the growth of these sporelings. With good growth of the sporelings, the next question is the adoption of the correct cultivation method for growing the larger thalli. Three methods are in current use in China and Japan: the pillar method, the intertidal raft method, and the floating raft method.

The pillar method is used only for intertidal cultivation. Poles or pillars are driven into the ground to serve as support for the cultivation nets. The pillar method (or 'pole system' in Japan; Miura 1975) has enjoyed wide application in the Japanese *nori* industry. As originally devised, the net is fastened tightly to the pillar with short binding cords at a definite level within the tidal range and is held in this fixed position (Fig. 20.12a). This type of pillar method is known as the 'fixed' type and can be used only in places where the tidal range is only 1–2 m. Later, for places with greater tidal range (3–6 m or more), the 'lift' type was devised. The net is furnished on its upper surface at certain levels with tenterhook-like cylindrical floats made of bamboo poles or synthetic resin pipes. It is fastened to the poles at a definite level within the tidal range by binding cords of a certain length so that the net can go up and down, floating on the water surface with the change of tidal levels within a definite range, controlled by the length of the binding cords (Miura 1975) (Fig. 20.13a).

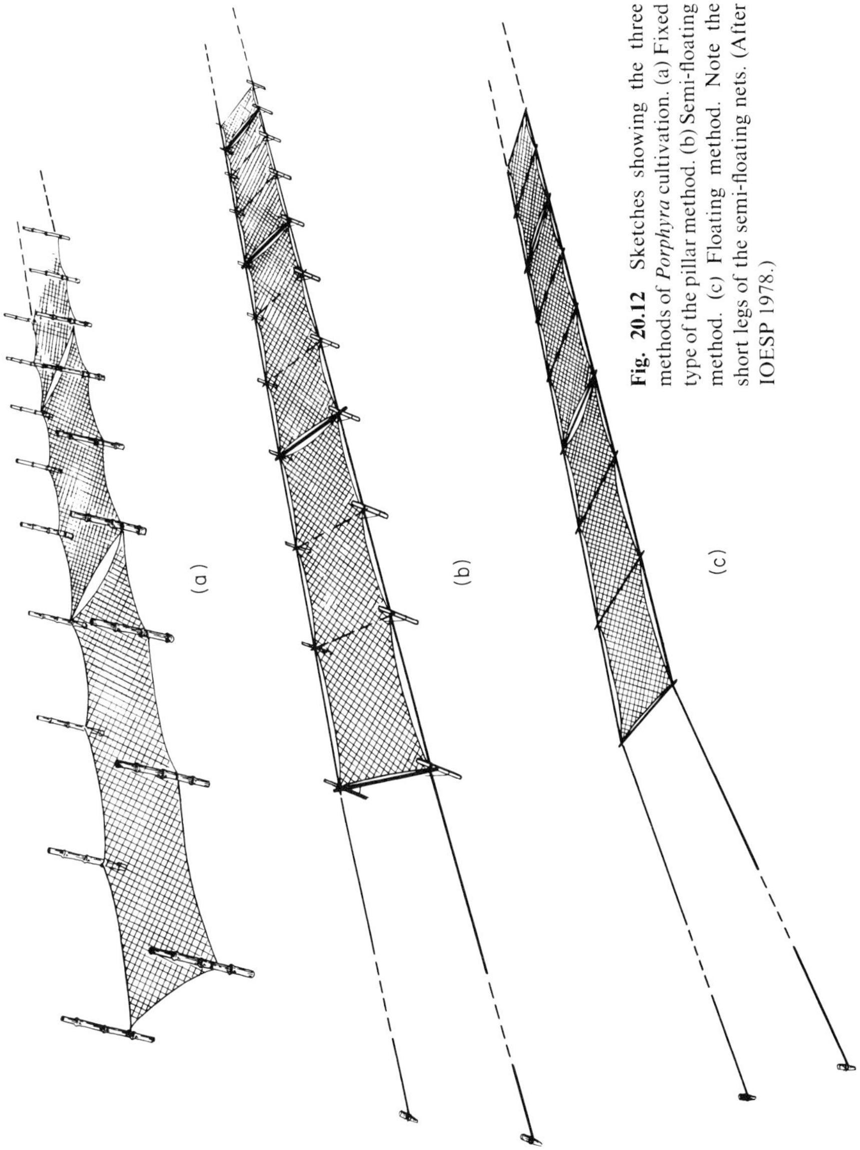

Fig. 20.12 Sketches showing the three methods of *Porphyra* cultivation. (a) Fixed type of the pillar method. (b) Semi-floating method. (c) Floating method. Note the short legs of the semi-floating nets. (After IOESP 1978.)

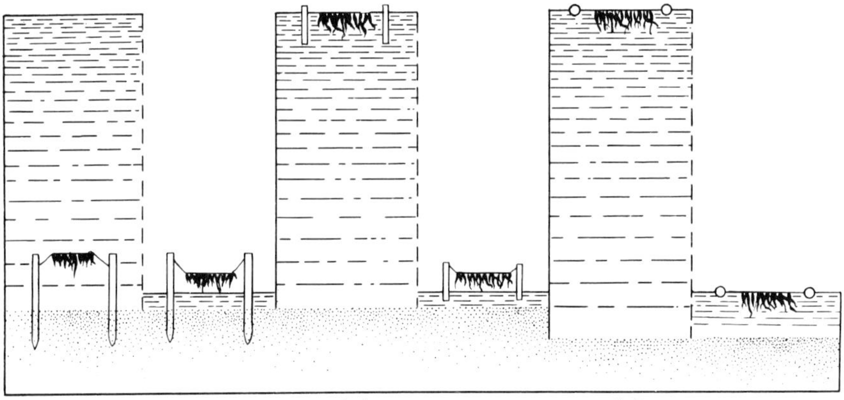

Fig. 20.13 Diagrammatic sections of the three methods of *Porphyra* cultivation. (a) Lift type of the pillar method. (b) Semi-floating method. (c) Floating method. Note that in each of the methods, the diagram on the left represents position of the net in high tide and that on the right represents position of the net in low tide (after IOEP 1976).

The intertidal raft cultivation, also known as the semi-floating raft cultivation method, is an improvement made in China to take the place of the pillar method, combining the strong points of the pillar and the floating methods, especially for intertidal cultivation. At high tide the raft floats on the water (Fig. 20.13b), and the *Porphyra* enjoys the needed light; at low tide the raft rests on the ground by short legs (Figs. 20.13b and 20.14).

An experiment was carried out to compare the effects of these two methods. The results showed that with the intertidal raft method the sporelings appeared earlier and their growth was better, attaining a production of 1090 g/m^2; however, with the pillar method the sporelings appeared later and their growth was not as good, attaining a production of 434 g/m^2. The explanation is that in the Yellow Sea the tidal range is quite large and the seawater is rather turbid. Under such conditions, with the semi-floating raft method, *Porphyra* has the opportunity to receive abundant light for photosynthesis, whereas with the pillar method *Porphyra* does not receive sufficient light. The innovation of this intertidal raft method has greatly helped the success of China's purple laver industry.

The floating raft method (Figs. 20.12c and 20.13c) is adapted for the production of *Porphyra* in deep water. At Qingdao the raft is made of a synthetic fibre net 60 m long and 180 cm broad with an elongated anchor rope 18.25 m long. In Japan the floating raft is attached to a floating rectangular frame moored by anchor ropes which are three to four times longer than the depth of the water. The frame is made of ropes floated by buoys of bamboo or wooden poles linked together with ropes sometimes into a ladder-shaped frame. The rope frame of the largest scale supports two layers of 60 sheets of *nori* nets, each

Fig. 20.14 Cultivation nets of the semi-floating method with the *Porphyra* growths at low tide in April at Rudong, Jiangsu Province, China. Note the nets rest on the ground with the help of short legs.

measuring 18 m long and 1.2 m wide. The ladder-shaped bamboo frame supports three to five sheets of *nori* nets, whereas the bamboo or wooden rectangular frame supports only one sheet (Miura 1975). The germination of the conchospores and growth of the sporelings of the floating method are not as good as those of the intertidal method. Therefore, in practice the germination of the spores and early stage of growth are first carried out in the intertidal region. When the sporelings reach 1–3 cm in length, the nets are cultivated by the floating method (IOESP 1978).

The particular stage of growth of the leafy thalli shows various optimal temperature preferences. For *Porphyra yezoensis* from conchospore germination through to the development of small leafy thalli of 1–2 cm length, the optimal growth temperature gradually decreases: 20° C for the germination of conchospores, 18–20° C for germlings of less than 100 cells, and 14–18° C for the small leafy thalli about 1–2 cm tall. Growth of the small leafy thalli is very slow in the low temperature winter season, lower than 5–6° C. The large leafy thalli have a much lower optimal temperature for growth. Over-wintered thalli have an incredibly low optimal growth temperature of 3–5° C, according to observations made at Qingdao. It is therefore critical that the seeding of conchospores takes place not too late in the season, to allow sufficient growth in order to avoid passing through the winter with small leafy thalli (IOESP 1978).

A final point concerning *Porphyra* cultivation is the ingenious method developed in Japan of freezing young lavers of 1–3 cm, still attached to the nets, and storing them at low temperature (−6 to −30° C) (Kurakake & Hori 1966; Yoshida 1972). If the *nori* is destroyed by a disease, the cold-storage *nori* nets can be used to replace the diseased nets, and the production for the season will not be seriously affected by the disease.

4.3 The *Laminaria* type of mariculture

As in *Porphyra* cultivation, two different processes are involved in the cultivation of *Laminaria*. The first process is cultivation of the microscopic gametophytic phase, which in China is done in large greenhouses from June or July to October in artificially cooled water of about 10° C. The second process is the cultivation of the large sporophytic phase in the field when the water temperature is about 20° C and lower, generally between mid-October (in Qingdao) and June or July of the next year. Cultivation of both *Laminaria japonica* and *Undaria pinnatifida* is handled by this technique. As with *Porphyra* cultivation, for the sake of brevity, only the procedures dealing with *Laminaria japonica* as practised in Qingdao are treated below, supplemented with additional remarks when necessary.

Indoor cultivation of the summer sporelings

The kelp sporelings are almost all cultivated during the summer, and the traditional autumn sporelings, cultivated in China in the early fifties and now still cultivated in Japan, are now used only in exceptional cases in China. The seeding process generally begins in early July, taking place in elongated shallow tanks with filtered seawater 10–15 cm deep in specially built greenhouses. Prior to the seeding process, the parent fronds with abundant sporangia are cleaned and hung in the air over-night. When such fronds are placed in seawater, the pressure resulting from the quick absorption of water breaks the sporangial wall and liberates large masses of zoospores. When about 20–30 zoospores are seen under the microscope at a 100 × magnification, the density is sufficient and the parent frond is removed. New seawater is added so that there are about 10 zoospores in the same microscope field. Then the spore-collectors, in the form of frames with seeding cords, are placed in the spore water, and the actively swimming zoospores soon adhere to the collectors, thus completing the seeding process.

The seeded frames are arranged in shallow tanks containing seawater enriched with nitrogenous and phosphate nutrients and previously cooled to 8–10° C. The seawater keeps on flowing from the upper tanks to the lower ones, and every day about one fifth is discarded, the remaining water going to the refrigerating tanks, and one fifth fresh enriched seawater is added. The seeded frames, eventually with young sporophytes, remain in the cool houses until mid-October, when the seawater temperature drops to about 20° C, which is favourable for the growth of the *Laminaria* sporophytes in the field. By this time the young sporophytes are about 1–2 cm high. These are the summer sporelings that are sold to the kelp farmers. There are now about eighteen firms which are engaged in cultivating the summer sporelings in China. The one in Qingdao has the capacity of producing four hundred million sporelings per cultivation season.

Production of sporelings of *Undaria pinnatifida* is quite similar to that of the *Laminaria* summer sporelings both in China and Japan. The *Undaria* gametophytes and sporophytes, however, are quite resistant to high temperature, and thus no cooling devices are needed.

Field cultivation of the summer sporelings

Having purchased the necessary amount of *Laminaria* summer sporelings, the kelp cultivation farmers hang these frames with young sporophytes onto rafts in the sea. Generally it takes about one and a half months for the sporelings to grow to a size of 10–15 cm. These are no longer sporelings but young kelps ready to be transplanted. In the transplantation the young kelps are taken out from their original frames and inserted into the twists of the cultivation ropes. These are then hung on the rafts and eventually grow on the same ropes to 3–4 m and harvested.

Field cultivation of the grown *Laminaria*

From transplanted juvenile kelps of 10–15 cm length to the mature kelp over 3 m long is a growth process requiring six to seven months. The kelp itself undergoes a series of transformations adapting to the changing conditions, especially the change in water temperature from near freezing to over 20° C. The well trained, experienced mariculturist must be alert to the ever-transforming kelp and the ever-changing environmental conditions. Before we discuss the care involved, we need to have a clear idea about the floating raft set-up.

The adoption of the floating raft method of cultivation is one of the innovations that led China's kelp industry to success. In the fifties and sixties there were many kinds of kelp rafts which differed from place to place. Now the single-rope floating raft is more often used than any other type. The main structure of the single-rope floating raft is a long rope made of synthetic fibre and 1–1.5 cm in diameter. The rope is 30 m (at Shandong) or 60 m (at Liaoning) long and attached to two wooden stakes with two anchor cables also made of

Fig. 20.15 View of a kelp cultivation farm belonging to the Lushun-Dalian Mariculture Station. Note the bamboo floats in rows (after Tseng 1964).

synthetic fibres, about 2 cm in diameter, and kept afloat with bamboo (Fig. 20.15) or glass floats. The length of the cable should be twice the depth of the sea or more. Each raft needs 20–40 floats of 15–20 cm diameter to be kept afloat. The cultivation ropes are also of synthetic fibre, 8 mm in diameter and 1–1.5 m in length, and are attached to the floating rope by hanging ropes, the length of which may be controlled according to the desired depth for the *Laminaria*. A stone or some other weight is attached to the lower end of the cultivation rope so that the rope with the kelp does not float to the surface (Fig. 20.16). In general, about 30 kelps are attached to each rope. The distance between two cultivation ropes is 70–140 cm, depending on the conditions, and the distance between two rafts also differs but is typically about 6–7 m. The total number of kelp plants cultivated in one hectare differs from place to place but averages 150 000–300 000 plants per hectare.

Two kinds of arrangements are used for the cultivation ropes: the horizontal

Fig. 20.16 Sketch of a *Laminaria* cultivaton raft. Note that each cultivation rope with the kelps has its upper end tied to a hanging rope attached to the floating rope and its lower end tied to a weighting stone.

type, which lies parallel to the sea surface, and the hanging type, which is perpendicular to the sea surface. The latter is the usual arrangement employed. In the course of cultivation for a few months there is quite a difference in growth between the upper and lower plants on the rope in the hanging type of arrangement. To equalize the growth a method of rope inversion was devised in which the ropes are inverted so that the lowest plant becomes the highest in position. Over the complete growing season two inversions are usually required. Another method to help equalize the growth is to level the ropes by tying together the two adjacent hanging ropes of two adjacent rafts after a few months' growth so that the ropes no longer hang vertically but become horizontally disposed. By these methods the production and the quality of the product are greatly improved (Zhang 1962).

In regard to the tending of the *Laminaria* under cultivation after transplantation, a general idea about the characteristics of the different stages of growth is necessary. Starting with the transplants of 10–15 cm inserted in the twists of the ropes, there are four more stages of growth to reach maturity. The first stage is from the transplant to the kelp of about 1 m length; this is characterized by the presence of concave–convex patches in two rows along both sides of the mid-rib. This stage is very sensitive to light and the fertility of the water. It grows very slowly under strong light and insufficiently fertile water. Prolongation of this stage is disadvantageous, and so measures should be taken to lower the level of the cultivation ropes to avoid strong light and to apply sufficient fertilizer. The next stage is from 1 to 2 or even 3 m in length. Here the kelp grows very rapidly, especially in length, and the concave–convex patches gradually disappear. But the thallus is thin and brittle and the frond base still retains its cuneate shape. Abundant fertilizer is needed at this stage of growth. Although the frond is not as susceptible to strong light as that of the previous stage, it is advisable to take measures so that the growth region does not receive too much light. The third stage is characterized by slower growth in length but faster growth in thickness. In the latter part of this stage the frond becomes thick and leathery in substance and rich brown in colour with the base of the frond becoming rounded. The plant is not as sensitive to environmental changes as those of the two earlier stages. The dry weight increases more rapidly, and it may be harvested. In the fourth maturation stage, growth essentially ceases, and the frond becomes abundantly covered with sporangial sori. Fertilizer application is entirely unnecessary (Wu & Suo 1962; Zhang 1962). Plants should be harvested at this stage, since there is no need for such plants to oversummer.

Earlier in this chapter, we have touched on the problem of fertilizer application, on the control of the water level for cultivation, on the inversion of the cultivation ropes for better lighting, and on cutting apical portions of the plant to improve the lighting to increase productivity. These problems relate to the different growth stages mentioned above. An experienced mariculturist can discern the conditions of kelp growth, and by employing the techniques of the science of seaweed mariculture he should be able to solve any problems and to work with the forces of the sea and nature to always produce a good crop of these valuable seaweeds.

ACKNOWLEDGEMENTS

The author wishes to thank Mr. Zhou Xiantong and Ms. Fong Minghua of the Academia Sinica Institute of Oceanology for their help in the preparation of the manuscript.

5 REFERENCES

BARDACH J. E., RYTHER J. H. & MCLARNEY W. O. (1972) *Aquaculture. The Farming and Husbandry of Freshwater and Marine Organisms.* Wiley-Interscience, New York.

BONEY A. D. (1965) Aspects of the biology of the seaweeds of economic importance. *Adv. Mar. Biol.*, **3**, 105–253.

BONOTTO S. (1976) Cultivation of plants. 2. Multicellular plants. In *Marine Ecology Vol. III. Cultivation. Part 1*, ed. Kinne O., pp. 467–529. John Wiley & Sons, London.

CHEN D., LIN G. & SHEN S. (1979) Studies on alginic acid decomposition bacteria, I. Action of alginic acid decomposing bacteria and alginase on *Laminaria japonica*. *Oceanol. Limnol. Sin.*, **104**, 329–33.

CHENG T. H. (1969) Production of kelp—a major aspect of China's exploitation of the sea. *Econ. Bot.*, **23**(3), 215–36.

DOTY M. S. (1973) Farming the red seaweed, *Eucheuma*, for carrageenans. *Micronesica*, **9**(1), 59–73.

DOTY M. S. (1977) *Eucheuma*—current marine agronomy. In *The Marine Plant Biomass of the Pacific Northwest Coast. A Potential Economic Resource*, ed. Krauss R. W., pp. 203–14. Oregon State University Press.

DOTY M. S. & ALVAREZ V. B. (1973) Seaweed farms: a new approach for U.S. industry. *Proc. 9th Ann. Conf. Mar. Tech. Soc.*, 701–8.

DOTY M. S. & ALVAREZ V. B. (1975) Status, problems, advances and economics of *Eucheuma* farms. *J. Mar. Tech. Soc.*, **9**(4), 30–5.

DOTY M. S. & SANTOS G. (1978) Carrageenans from tetrasporic and cystocarpic *Eucheuma* species. *Aquatic Botany*, **4**, 143–9.

DREW K. M. (1949) *Conchocelis*—phase in the life history of *Porphyra umbilicalis* (L.) Kütz. *Nature (Lond.)*, **164**, 748–9.

FANG T. C., JIANG B. Y. & LI J. J. (1965) Further studies on the genetics of *Laminaria* frond-length. *Oceanol. Limnol. Sin.*, **7**, 59–66.

FANG T. C., JIANG B. Y. & LI J. J. (1966) The breeding of a long-frond variety of *Laminaria japonica* Aresch. *Oceanol. Limnol. Sin.*, **8**, 43–50.

FANG T. C., WU C. Y., JIANG B. Y., LI J. J. & REN K. Z. (1962d) The breeding of a new breed of Haidai (*Laminaria japonica* Aresch.) and its preliminary genetic analysis. *Acta Bot. Sin.*, **4**, 197–209.

FANG T. C., WU C. Y., JIANG B. Y., LI J. J. & REN K. Z. (1963) The breeding of a new variety of Haidai (*Laminaria japonica* Aresch.) *Scientia Sin.*, **12**, 1011–18.

FANG T. C., WU C. Y. & LI J. J. (1962b) Increased adaptability to high temperature of gametophytes and sporelings of the Haiqing No. 1 breed of *Laminaria japonica* Aresch. *Oceanol. Limnol. Sin.*, **4**, 29–37.

GOFF L. J. & GLASGOW J. C. (1980) Pathogens of Marine Plants. *Center for Coastal Marine Studies, University California, Santa Cruz, Special Publication*, **7**.

HASEGAWA Y. (1971a) Cultivation of *Laminaria* in Japan. *Bull. Hokkaido Reg. Fish. Res. Lab.*, **37**, 46–8.

HASEGAWA Y. (1971b) Forced cultivation of *Laminaria*. *Bull. Hokkaido Reg. Fish. Res. Lab.*, **37**, 49–52.

HASEGAWA Y. (1976) Progress of *Laminaria* cultivation in Japan. *J. Fish. Res. Bd. Canada*, **33**, 1002–6.

IOEP (Section of Experimental Phyco-ecology, Institute of Oceanology, Academia Sinica) (1976) All-artificial spore-collecting cultivation of Tiaoban-zicai (*Porphyra yezoensis* Ueda). *Scientia Sin.*, **19**, 253–9.

IOESP (Section of Experimental Phyco-ecology and Section of Systematic Phycology, Institute of Oceanology, Academia Sinica) (1978) *Cultivation of Zicai (Porphyra yezoensis Ueda)*. Science Press, Beijing. (In Chinese.)

IOG & QMF (Section of Seaweed Genetics and Breeding, Institute of Oceanology, Academia Sinica & Qingdao Institute of Marine Fisheries) (1976) The breeding of new varieties of Haidai (*Laminaria japonica* Aresch.) with high production and high iodine content. *Scientia Sin.*, **19**, 243–52.

KITO H. (1972) Electron microscopic observations on the tumour in *Porphyra*. *Bull. Tohoku Reg. Fish. Res. Lab.*, **32**, 77–82.

KURAKAKE T. & HORI A. (1966) A method of keeping laver (*Porphyra*) culturing net in low temperature. *Refrigeration (Jap. Ass. Refrig. Tokyo)*, **41**, 15–29.

KUROGI M. (1953a) Study of the life-history of *Porphyra* 1. The germination and development of carpospores. *Bull. Tohoku Reg. Fish. Res. Lab.*, **2**, 67–103.

KUROGI M. (1953b) On the liberation of monospores from the filamentous thallus (*Conchocelis*—stage) of *Porphyra tenera* Kjellm. *Bull. Tohoku Reg. Fish. Res. Lab.*, **2**, 104–8.

KUROGI M. & SATO S. (1962) Influence of light on the growth and maturation of *Conchocelis*—thallus of *Porphyra*. II. Effect of different photoperiods on the growth and maturation of *Conchocelis*—thallus of *P. tenera* Kjellm. *Bull. Tohoku Reg. Fish. Res. Lab.*, **20**, 127–37.

LEIGHTON D. L. (1971) Grazing activities of benthic invertebrates. In *Biology of Giant Kelp Beds (Macrocystis) in California*, ed. by North W. J., *Beih. Nova Hedwigia*, **32**, 421–53.

LI J., PENG Z., XUE Z. & HUANG Z. (1980) Studies on the frond twist disease of the haidai (*Laminaria japonica* Aresch.)—Experiment on contact infection. *Oceanol. Limnol. Sin.* (in press).

MATHIESON A. C. (1975) Seaweed aquaculture. *Marine Fisheries Rev.*, **37**, 2–14.

MIURA A. (1974) Nori (seaweed) cultivation. In *Japan's Food Industry*, ed. Asahi Evening News International, pp. 16–17.

MIURA A. (1975) *Porphyra* cultivation in Japan. In *Advance of Phycology in Japan*, eds. Tokida J. & Hirose H., pp. 273–304. W. Junk, The Hague.

NATH J. H. & GRACE R. A. (1977) Engineering of structures in the ocean. In *The Marine Plant Biomass of the Pacific Northwest Coast. A Potential Economic Resource*, ed. Krauss R. W., pp. 335–55. Oregon State University Press.

NEISH I. C. (1976) Role of mariculture in the Canadian seaweed industry. *J. Fish. Res. Bd. Canada*, **33**, 1007–14.

NEUSHUL M. (1977) The domestication of the giant kelp, *Macrocystis*, as a marine plant biomass producer. In *The Marine Plant Biomass of the Pacific Northwest Coast, A Potential Economic Resource*, ed. Krauss R. W., pp. 163–81. Oregon State University Press.

NORTH W. J. (1971) Mass-cultured *Macrocystis* as a means of increasing kelp stands in nature. *Proc. Int. Seaweed Symp.*, **7**, 394–400.

NORTH W. J. (1976) Aquacultural techniques for creating and restoring beds of giant kelp, *Macrocystis* spp. *J. Fish. Res. Bd. Canada*, **33**, 1015–23.

NORTH W. J. (1977a) Algal nutrition in the sea-management possibilities. In *The Marine Plant Biomass of the Pacific Northwest Coast. A Potential Economic Resource*, ed. Krauss R. W., pp. 215–30. Oregon State University Press.

NORTH W. J. (1977b) Introducing warm-tolerant *Macrocystis* to the vicinity of a thermal discharge. *Kelp Habitat Improvement Project. Ann. Rep.*, **1974–75**, 114–25.

PARKER H. S. (1974) The culture of the red algal genus *Eucheuma* in the Philippines. *Aquaculture*, **3**, 425–39.

PARKER H. S. (1976) Seaweed farming in the Sulu Sea. *Oceans*, **2**, 12–19.

REN G. Z. (1962) Growth and development of the gametophyte (of *Laminaria japonica*) in relation to the environmental factors. In *Manual of Cultivation of Haidai* (*Laminaria japonica*), eds. Tseng C. K. & Wu C. Y., pp. 72–95. Science Press, Beijing. (In Chinese.)

SADDINGTON A. A. (1969) Alginates from kelp. In *The Encyclopedia of Marine Resources*, ed. Firth F.E., pp. 19–23. Van Nostrand Reinbold Co., New York.

SAITO Y. (1975) *Undaria*. In *Advance of Phycology in Japan*, eds. Tokida J. & Hirose H., pp. 304–20. W. Junk, The Hague.

TSENG C. K. (1933) *Gloiopeltis* and the other economic seaweeds of Amoy, China. *Lingnan Sci. J.*, **12**(1), 43–63.

TSENG C. K. (1945) The terminology of seaweed colloids. *Science*, **101**, 597–602.

TSENG C. K. (1946) Phycocolloids: useful seaweed polysaccharides. In *Colloid Chemistry. Theoretical and Applied. Vol. VI. General Principles and Specific Industries, Synthetic Polymers and Plastics.*, ed. Alexander J., pp. 639–734.

TSENG C. K. (1955) On the increase of production in shallow sea by fertilization. *Kexue Tongbao*, **1955**(12), 73–6. (In Chinese.)

TSENG C. K. (1962) The haidai (*Laminaria japonica*) cultivation industry of China. In *Manual of Cultivation of Haidai* (*Laminaria japonica*), eds. by Tseng C. K. & Wu C. Y., pp. 99–112. Science Press, Beijing. (In Chinese.)

TSENG C. K. (1964) Kelp farming, a new industry. *China Reconstructs*, **13**, 36–8.

TSENG C. K. & CHANG C. F. (1956) On China's *Porphyra* and its cultivation. *Shengwuxue Tongbao*, **3**, 29–33. (In Chinese.)

TSENG C. K. & CHANG C. F. (1961) On the botanical names of economic marine algae in old Chinese literature. *Act. Bot. Sin.*, **9**, 316–36.

TSENG C. K. & CHANG C. F. (1962) An analytical study of the marine algal flora of the western Yellow Sea coast. *Oceanol. Limnol. Sin.*, **4**, 49–59.

TSENG C. K. & CHANG T. J. (1954) Studies on *Porphyra* I. Life history of *Porphyra tenera* Kjellm. *Act. Bot. Sin.*, **3**, 287–302.

TSENG C. K. & CHANG T. J. (1955a) Studies on *Porphyra* II. On the *Conchocelis*-phase of *Porphyra tenera* Kjellm. and its conchospores. *Act. Bot. Sin.*, **4**, 27–46.

TSENG C. K. & CHANG T. J. (1955b) Studies on the life history of *Porphyra tenera* Kjellm. *Scientia Sin.*, **4**(3), 375–98.

TSENG C. K. & CHANG T. J. (1956) Conditions of *Porphyra* conchospores formation and discharge and the discharge rhythm. *Act. Bot. Sin.*, **5**, 33–48.

TSENG C. K. & CHANG T. J. (1962) On the cultivation of Chinese *Porphyra*. *Collected Papers of the 5th Meeting of the Western Pacific Fisheries Research Commission*, pp. 34–8. Science Press, Beijing. (In Chinese.)

TSENG C. K., *et al.* (1962a) *Economic Seaweeds of China*. Science Press, Beijing. (In Chinese.)

TSENG C. K., CHANG T. J. & LI J. J. (1959a) A preliminary report on the semi-artificial spore collecting cultivation of *Porphyra*. *Kexue Tongbao*, **5**, 169–71. (In Chinese.)

TSENG C. K., CHANG T. J. & ZHAO R. Y. (1959b) On all-artificial spore collecting cultivation of *Porphyra*. *Kexue Tongbao*, **5**, 169–71. (In Chinese.)

TSENG C. K., *et al.* (1963) Studies on the growth and development of haidai (*Laminaria japonica*) transplanted at the Chekiang coast. *Stud. Mar. Sin.*, **3**, 102–18.

TSENG C. K., REN G. Z. & WU C. Y. (1959c) On the discharge of egg and spermatozoids of *Laminaria japonica* and the morphology of spermatozoids. *Kexue Tongbao*, **4**, 129–30. (In Chinese.)

TSENG C. K., SUN K. Y. & WU C. Y. (1955a) On the cultivation of haidai (*Laminaria japonica* Aresch.) by summering young sporophytes at low temperature. *Acta Bot. Sin.*, **4**, 255–64.

TSENG C. K., SUN K. Y. & WU C. Y. (1955b) Studies on fertilizer application in the cultivation of haidai (*Laminaria japonica* Aresch.). *Acta Bot. Sin.*, **4**, 375–92.

TSENG C. K. & WU C. Y. (1966) Studies on fertilizer application in the cultivation of haidai (*Laminaria japonica* Aresch.). *Collected Papers of the 9th Meeting of the Western Pacific Fisheries Research Commission*, pp. 37–40. Science Press, Beijing. (In Chinese.)

TSENG C. K., WU C. Y. & SUN K. Y. (1957) The effect of temperature on the growth and development of the Haidai (*Laminaria japonica* Aresch.). *Acta Bot. Sin.*, **6**, 103–30.

TSENG C. K., *et al.* (1962b) Studies on the absorption of nitrate and ammonium nitrogen by *Laminaria japonica* Aresch. *Collected Papers of the 5th Meeting of the Western Pacific Fisheries Research Commission*, pp. 21–32. Science Press, Beijing. (In Chinese.)

UEDA S. (1973) Manual of Nori Cultivation. *Nat. Assoc. Nori and Shellfish Fish.* (In Japanese.)

WANG C. S. & HSÜEH L. H. (1959) Observations on the sorus formation of mature *Laminaria* summering in low temperature culture room. *Act. Bot. Sin.*, **8**, 259–61.

WANG K. & HUANG Z. (1980) Studies on the frond twist disease of the haidai (*Laminaria japonica* Aresch.) II. Influence of tetracyclines on the disease development. *Acta Microbiol. Sin.*, **20**, 216–18.

WANG Q., XU S., *et al.* (1980) Studies on the frond twist disease of the haidai (*Laminaria japonica* Aresch.)—Electron microscope examination of the pathogen. *Scientia Sin.*, **1980**, 587–91. (In Chinese.)

WANG Z. M. (1962) Fertilizer application in haidai cultivation. In *Manual of Cultivation of Haidai* (*Laminaria japonica*), eds. Tseng C. K. & Wu C. Y., pp. 187–99. Science Press, Beijing. (In Chinese.)

WILSON K. C., HAAKER P. L. & HANAN D. A. (1977) Kelp restoration in southern California. In *The Marine Plant Biomass of the Pacific Northwest Coast. A Potential Economic Resource*, ed. Krauss R. W., pp. 183–202. Oregon State University.

WU C. Y. (1962a) Growth and development of the sporophyte (of *Laminaria japonica*) in relation to the environmental factors. In *Manual of Cultivation of Haidai* (*Laminaria japonica*), eds. Tseng C. K. & Wu C. Y., pp. 34–71. Science Press, Beijing. (In Chinese.)

WU C. Y. (1962b) Translocation and accumulation of nutritive substances (in haidai). In *Manual of Cultivation of Haidai* (*Laminaria japonica*), eds. Tseng C. K. & Wu C. Y., pp. 275–87. Science Press, Beijing. (In Chinese.)

WU C. Y. & SUO R. Y. (1962) Morphology, reproduction and life history (of *Laminaria japonica*). In *Manual of Cultivation of Haidai* (*Laminaria japonica*), eds. Tseng C. K. & Wu C. Y., pp. 14–33. Science Press, Beijing. (In Chinese.)

WU C. Y., *et al.* (1959) Experiment on intermittently fertilizing *Laminaria japonica*. *Kexue Tongbao*, **24**, 829–30. (In Chinese.)

WU C., *et al.* (1979) On the malformation disease of *Laminaria* sporelings. *Oceanol. Limnol. Sin.*, **10**, 238–50.

ZHANG J. C. (1962) Tending the *Laminaria* kelp. In *Manual of Cultivation of Haidai* (*Laminaria japonica*), eds. Tseng C. K. & Wu C. Y., pp. 166–86. Science Press, Beijing. (In Chinese.)

ZHANG J. C., SUO R. Y. & ZHANG D. M. (1962) Diseases and pests in *haidai* (*Laminaria japonica*) cultivation. In *Manual of Cultivation of Haidai* (*Laminaria japonica*), eds. Tseng C. K. & Wu C. Y., pp. 230–6. Science Press, Beijing. (In Chinese.)

CHAPTER 21
COMMERCIAL UTILIZATION

J. ROBERT WAALAND
Department of Botany, University of Washington,
Seattle, Washington 98195, U.S.A.

1 INTRODUCTION

The abundance and diversity of seaweeds have made them prime material for human use. This chapter discusses past and present seaweed utilization. The major use categories, direct use as well as use of extracted materials are discussed here in approximately chronological order. Many of the old ways of using seaweeds are still practiced, and new ways of using them are constantly being researched and developed. Space limits discussion of many seaweed uses which can be found in the references (Newton 1951; Chapman 1970; Madlener 1977). Quantitative data on the amounts of seaweeds used and the amounts available are not well known except in a few countries and for a few widely used species (Michanek 1975; Moss 1977; Jensen 1979). In general seaweed use has increased over the years, but there have been major changes in use patterns and these changes have sometimes been very abrupt (Newton 1951; Moss 1977).

2 FOOD

Seaweeds have been used as human food since ancient times. There are historical references to such uses as early as 600–800 B.C. in China, and seaweeds were undoubtedly used in prehistoric times. Seaweeds are eaten for their food value, flavours, colours, and textures and are typically combined with other types of food (Fig. 21.1). In times of famine due to failure or destruction of terrestrial food crops, seaweeds often have been eaten by coastal inhabitants to ward off starvation (Newton 1951). The major edible seaweeds are listed in

Fig. 21.1 Some edible seaweed products (left to right) *wakame* (*Undaria pinnatifida*); kelp pickles (*Nereocystis luetkeana*); powdered, sliced and folded *kombu* (*Laminaria japonica*); kelp tablets (*Macrocystis pyrifera*); *nori* (*Porphyra tenera*); Irish moss (*Chondrus crispus*); agar (*Gelidium* sp.); dulse blades, tea, tablets and extract (*Palmaria palmata*). © J. Robert Waaland.

Table 21.1; many others not listed are eaten in small quantities or in areas of limited geographical extent (Newton 1951; Boney 1965; Chapman 1970; Michanek 1975; Madlener 1977). Edible species are obtained by harvesting wild populations and by cultivation (Chapter 20 discusses cultivation).

Analyses of certain edible seaweeds show that many contain significant amounts of protein, vitamins and minerals essential for human nutrition; they also contain large amounts of polysaccharides that are usually not digestible by humans (Levring *et al.* 1969). People who regularly consume seaweeds are more capable of digesting algal polysaccharides than occasional consumers (Newton 1963; Boney 1965; Madlener 1977); such a capacity may be due to a change in the intestinal flora.

Table 21.1 Major edible seaweeds. (Data from Newton 1951; Boney 1965; Chapman 1970; Madlener 1977.)

Green algae	Brown algae	Red algae
Monostroma	*Laminaria*	*Porphyra*
Caulerpa	*Undaria*	*Palmaria*
Enteromorpha	*Alaria*	*Gracilaria*
Ulva	*Eisenia*	*Gelidium*
	Ecklonia	*Eucheuma*

Many of the algal polysaccharides such as agar, carrageenan, algin and related colloids form aqueous gels or are quite viscous in aqueous solutions. Crude extracts of various algal polysaccharides have been used since at least the mid-17th century (Chapman 1970) and may have been used even earlier in Japan, China and Indonesia. Extracted seaweed polysaccharides are the form in which the people of most industrialized nations, with the exception of Japan, consume seaweed products; such extracts are discussed further in Section 7 of this chapter.

The protein content of many seaweeds ranges between 4 and 25% of the dry weight. Those eaten often have a high content of digestible protein, e.g., *Palmaria palmata* (25%), *Porphyra* (20–25%), *Enteromorpha* (20%) and *Analipus* (22%) (Johnston 1972). Many seaweeds have a relatively high content of free amino acids which are responsible for the sweet taste of particular species (Boney 1965; Levring *et al.* 1969; McLachlan *et al.* 1972). Protein and amino acid content vary with such factors as season, habitat, age, plant part, and growing conditions, especially light, nutrients and salinity. The nutritional value of the 'average' seaweed has been compared to that of oats in protein and carbohydrate nutritional value (Madlener 1977).

Seaweeds are an excellent source of many vitamins (Kanazawa 1963). Depending on the particular vitamin, they may be synthesized by bacteria living on the seaweeds, or in the sea around them, or by the seaweeds themselves (Provasoli & Carlucci, 1974). *Porphyra, Ulva* and *Alaria* have vitamin C content similar to lemons; high vitamin C content has also been measured in *Palmaria palmata* and *Fucus* (Levring *et al.* 1969; Chapman 1970). High levels of vitamins A and D have been measured in seaweeds; these vitamins are very concentrated in oils extracted from seaweeds (Madlener 1977). Vitamins B_1, B_{12}, riboflavin, niacin, pantothenic acid, folic acid, and vitamin E have been measured in seaweed samples. In general, edible seaweeds are good sources of vitamins for human diets, but vitamin content does vary seasonally (Levring *et al.* 1969; Chapman 1970).

The mineral and trace element composition of edible seaweeds makes them useful dietary sources of many of the elements required by humans, such as sodium, chlorine, potassium, phosphorus, magnesium and calcium which are required in large quantities; iron, manganese and iodine, needed in smaller amounts; and copper, zinc, molybdenum and cobalt, needed only in trace amounts (Levring *et al.* 1969; Yamamoto & Ishibashi 1972; Yamamoto *et al.* 1979).

Thus as sources of protein, carbohydrates, vitamins, and minerals, seaweeds can be a significant factor in human nutrition. As sources of interesting colours, distinctive flavours, and varied textures, they can add interest and variety to the human diet. Table 21.6 indicates the quantities of recent harvests of some major edible seaweeds.

3 FODDER

Wild animals such as deer, rabbits, arctic foxes and polar bears have been observed eating seaweeds. Historical accounts from 46–43 B.C. report that the Greeks and Romans used seaweeds as emergency rations for their livestock (Newton 1951; Chapman 1970). Seaweeds have regularly been fed to livestock such as cattle, sheep, horses, goats, pigs and poultry in areas where they are conveniently and abundantly available. Table 21.2 lists the seaweeds which have been used as animal fodder. In some areas, the livestock are simply allowed to browse on the seashore taking what they like from attached or drift seaweeds to supplement their conventional food; in other areas they are confined to the shore so that they have little choice but to graze on seaweeds. Seaweeds have also been collected and carted to feeding areas although substantial cost and effort are involved in doing so. For feeding livestock at some distance from the shore, a meal of dried, chopped seaweed is usually prepared and added as a supplement to the regular diet. *Ascophyllum* is the principal seaweed used in this way. Seaweed and seaweed meal have also been mixed with other plants in making silage. In some cases a period of habituation has been indicated for the animals to achieve good digestibility of the seaweeds. There are reports of sheep existing solely on a seaweed diet, but using seaweed as a supplement in the regular diet of livestock rather than as the sole diet is generally recommended as some difficulties may arise due to the high mineral content of seaweeds (Boney 1965; Jensen 1972).

4 FERTILIZER

The value of seaweeds as an agricultural fertilizer has been repeatedly demonstrated, especially by coastal farmers with ready access to seaweeds (Booth 1965). While such applications were probably practiced in antiquity, the first references to such use appears in Roman writings from the second century A.D. (Newton 1951). Seaweeds are collected from beach drift or harvested especially for agricultural use. The method of application ranges from whole

Table 21.2 Major genera of fodder seaweeds. (Data from Newton 1951; Boney 1965; Chapman 1970.)

Brown algae	Red algae
Pelvetia	*Chondrus*
Fucus	*Palmaria*
Alaria	
Laminaria	
Ascophyllum	
Macrocystis	

and chopped plants applied wet or dry to composted, liquified, supplemented, and extracted preparations. Ashes or 'kelp' from burned seaweeds have also been applied to improve the soil. The rockweeds *Fucus* and *Ascophyllum* have figured prominently as fertilizer seaweeds, but others have also been used where available (Table 21.3). In China, *Sargassum* has been used as a fertilizer, and in Japan, Sri Lanka and India species not considered suitable for human consumption have also been used.

As a fertilizer, seaweed has ample amounts of nitrogen and potassium but is low in phosphate and should be supplemented with phosphate for use with most crops. The composition of the various species used as fertilizer varies seasonally and geographically. Compared to 'average' barnyard manure, seaweed has similar nitrogen values, about one-third as much phosphate and about three times as much potassium (Chapman 1970). The high organic matter content is similar and is beneficial in improving water retention and mechanical properties of the soil. An advantage of seaweed as fertilizer is that it is free of weed seeds and spores of fungi that harm terrestrial crops. In addition to their fertilizer content, seaweeds and seaweed extracts regulate crop growth and ripening because of the auxin, cytokinin, and gibberellin associated with them (*see* Chapter 18, p. 648; Blunden 1977, 1972; Augier 1978; Wildgoose *et al.* 1978) and because they may inhibit certain pathogens, including virus-carrying aphids and some fungi. They are generally good sources of certain trace elements (Chapman 1970). Reported deleterious effects of seaweed manures include excess manganese release in low pH soils, water-logging of soils fertilized with *Pachymenia*, inhibitory effects on plant growth and low availability to plants of some of the nitrogen in the seaweeds (Boney 1965). One of the more remarkable uses of seaweed in agriculture is in the bleak, rocky, soilless Aran Islands in Galway Bay on the west coast of Ireland where soil for farming is 'manufactured' by composting seaweed and sand (Newton 1951).

Coralline algae or the 'shelly sand' derived from them have been used for 'marling' or liming fields to control soil acidity in localities in England, Scotland, Wales, Ireland, Brittany, Denmark, Portugal, Naples, Morocco, Algeria and

Table 21.3 Major genera of fertilizer seaweeds. (Data from Newton 1951; Boney 1965; Chapman 1970.)

Brown algae	Red algae
Ascophyllum	*Pachymenia*
Macrocystis	*Lithothamnion* (for marl)
Laminaria	*Phymatolithon* (for marl)
Ecklonia	
Durvillaea	
Carpophyllum	
Himanthalia	
Sargassum	

parts of the Arctic (Newton 1951; Blunden *et al.* 1975); recent marl harvests in Brittany have been 300 000 tonnes per year (Blunden *et al.* 1975).

5 DRUGS

Crude preparations and extracts of some seaweeds have been used for their medicinal properties for centuries. Brown algae were effective for curing goiter because of their iodine content. Other seaweeds have been used as vermifuges and to make cough medicines, soothing lotions and cosmetics. In some instances, as in the case of iodine as an effective goiter cure, and kainic acid as a vermifuge, effective drugs have been demonstrated in the seaweeds; however, in many cases, the seaweed 'drugs' may ameliorate symptoms but not effect a cure. Table 21.4 lists some of the major drug seaweeds and their medicinal uses. Recent research has revealed some promising antiviral activity in some seaweeds (Ehresmann *et al.* 1977), and new uses in medical and biological research are being found for some seaweed derived drugs such as kainic and domoic acids (Takemoto 1978). Unusual chemicals of various kinds, usually of unknown biological function or properties, are also the subject of much current research (Fenical 1975; Constantine 1978; Patterson 1978; Fenical *et al.* 1979; Hoppe *et al.* 1979). While the direct use of seaweeds in medicine is not as widespread as it once was, the use of seaweed polymer extracts in pharmacy, medicine and biochemistry is well established as a matrix for other drugs and in analytical and diagnostic procedures such as electrophoresis.

Table 21.4 Seaweeds used as medicines. (Data from Newton 1951; Boney 1965; Chapman 1970.)

Name	Use
Red algae	
Alsidium helminthochorton	Vermifuge
Digenia simplex	Vermifuge
Corallina officinalis	Vermifuge
Chondrus crispus	Coughs, chest, stomach ailments
Gigartina stellata	Coughs, chest, stomach ailments
Porphyra species	Antiscorbutic
Palmaria palmata	Antiscorbutic
Brown algae	
Fucus vesiculosus	Scrofula
Fucus evanescens	Stomach ailments
Laminaria and other kelps	Iodine source (cures goiter), stipes used to open wounds, in cervical dilation, and to induce abortions
Green algae	
Ulva species	Burn treatment
Acetabularia major	Bladder and kidney ailments

6 SEAWEED EXTRACTS: SMALL MOLECULES

The emergence of the modern seaweed chemical industry started with the extraction of such basic industrial chemicals as soda (sodium carbonate), potash (a mixture of potassium salts, mostly potassium carbonate) and iodine. The soda and potash were used in making glass, pottery, soap, and in leather tanning. Potash also was important as an agricultural fertilizer. This industry developed in Normandy in the late 17th century; the major seaweeds harvested were *Fucus, Ascophyllum* and *Laminaria* which were burned in pits or kilns to produce the ash called kelp. Kelp production spread during the 18th century from Normandy and Brittany to the Channel Islands, the Western Isles and the Orkneys of Scotland, Ireland, Norway, and Sweden. Demand for kelp remained strong and the kelp industry flourished through most of the 18th century and into the first quarter of the 19th century. Then a change in tariffs on foreign sources of soda and potash crippled the industry. However, the discovery of iodine by Courtois, who in 1813 noticed a violet vapour that condensed as crystals in a kelp kiln, rejuvenated the kelp industry by the mid-19th century, as uses for iodine were developed. Unfortunately for the kelp industry, the discovery and development of iodine-rich mineral deposits in Chile about the same time as the iodine-from-kelp industry reached full vigour, dealt the kelp industry another severe blow in the late 1870's. However, small quantities of kelp continued to be harvested for soda, potash, and iodine in northern France and western Ireland. In Japan, a kelp chemical industry was established in the late 19th century and with government encouragement and protection continued well into the 20th century. *Laminaria, Ecklonia* and *Sargassum* were the major seaweeds harvested for this industry (Newton 1951; Chapman 1970). The kelp industry in the United States arose in response to difficulties in obtaining adequate supplies of potash from the Kali Syndicate of Germany which controlled the major mineral deposits supplying it. The shortages were first experienced in 1910 and led to a survey of the United States' kelp resources as a potential source of potash and other chemicals. Vast stands of *Macrocystis pyrifera, M. integrifolia, Nereocystis luetkeana* and *Alaria fistulosa* were located and assessed in southern California, Washington, and Alaska. Exploitation of these kelp resources, especially *Macrocystis*, developed rapidly during World War I and several firms used floating mechanical harvesters. The harvest could be used directly as a fertilizer (as it had been for centuries), burned for potash and iodine production, and destructively distilled to yield the products listed in Table 21.5; and wet kelp could be fermented to yield acetone. Prodigious quantities of kelp could be harvested by this method. The renewed availability of cheaper alternate sources of potash after World War I resulted in most kelp harvesting operations going out of business or operating on a very small scale (Tseng 1947; Chapman 1970). Kelp is the source of a fine grade of charcoal used for decolourizing and filtering (Tseng 1947).

In Russia, the red seaweed *Phyllophora nervosa*, very abundant in the Black

Table 21.5 Major product yield from destructive distillation of *Macrocystis*. (Data from Chapman 1962.)

Characteristic	Percentage	Characteristic	Percentage
Wet weight	100	Tar	2.1
Dry weight	12	Potash salts	3.0
Gas	2.3	Charcoal	1.2
Ammonia	3.3	Iodine	0.01

Sea, was harvested for electrolytic separation of iodine, bromine and chlorine beginning in the 1930's.

The industry based on the extraction of small molecules from seaweeds experienced several periods of expansion and decline. When the chemicals extracted from seaweeds could be obtained at less expense from alternate sources, the seaweed harvesting and processing declined drastically. However, there are some constituents of seaweeds which are unique to these plants and for which no other sources have been found. These extracts are the subject of the next section.

7 SEAWEED EXTRACTS: MACROMOLECULES

Gelling and viscous substances have been extracted from red seaweeds for centuries and were usually used in food preparation. Reference to the use of agar-producing seaweeds occurs in Chinese writings from 300 A.D. The birds' nests used in making birds' nest soup, a delicacy in China and Japan are made from gelatinous, regurgitated seaweeds gathered by cliff-dwelling birds. The partially digested seaweed glues the nests to the precipitous cliffs and caves. Extraction of agar or 'Kanten' from seaweeds by alternate freezing and thawing of aqueous extracts was discovered in Japan in the 17th century. One of the first uses of a seaweed gel in biological research appears to be a kind of agar which was used in 1881 as a microbiological culture medium (Hitchens & Leikind 1939; Chapman 1970). The use of the gelling substance, carrageenan, extracted from *Chondrus crispus* to make milk puddings and similar dishes was practiced for centuries before the industrial use of this gelling substance was begun (Whistler 1973).

The first scientific investigations of the slimes and gels of marine algae were conducted by E. C. C. Stanford in the 1880's. He investigated the substances now known as algin from brown algae and carrageenan from certain red algae. Commercial exploitation of these useful and unique seaweed chemicals did not catch on immediately but did become the mainstay of the seaweed industry in the 20th century.

7.1 Agar

Agar is a major constituent of the cell walls of certain red algae, especially members of the Gelidiaceae and Gracilariaceae. Agar is the Malay word for a gelling substance extracted from *Eucheuma*, but now known to be a carrageenan. The term agar is now generally applied to those algal galactans which have agarose, the disaccharide agarobiose, as their repeating unit. A family of molecules with varying amounts of agarose and agaropectin forming the polymers has been described by various investigators (Chapter 16, p. 575; Araki 1966; Duckworth *et al.* 1971; Izumi 1971, 1973; Rees 1972; Craigie & Leigh 1978). Agar is soluble in and extracted by hot water; it gels at room temperature and freezing and thawing are used to purify it.

Until about 1900 agar was used mainly as a food item. As a significant quantity began to be used for microbiological plating media, it became the first seaweed polymer extract to achieve commercial status for purposes other than food use. Production was monopolized by Japan until World War II by which time about 2500 tonnes of agar were being produced annually. Interruption of Japanese exports by World War II resulted in establishment or expansion of agar industries in the U.S.A., Mexico, Russia, India, South Africa, New Zealand and Australia (Humm 1947). By the mid-1970's, agar production reached about 10 000 tonnes per annum. Since 1940 there have been three periods when agar prices fluctuated drastically as has agar production and the availability of agar-producing seaweeds (Moss 1977). Some of the fluctuations were due to overharvesting, others were related to the effects of supply and demand, and some were manipulations by the agar industry in attempts to monopolize agar production and sales. Such problems have often caused agar users to turn to other polymers, including other seaweed polymers, to serve the same commercial application. Because of its excellent gelling properties at moderate temperatures, agar finds much use in bakery products, confectionery making, and in puddings, creams, and jellied products (Fig. 21.2). It is also used in the preparation of canned meats and fish and as a clarifying agent in wines and beers (Levring *et al.* 1969).

7.2 Carrageenan

Carrageenan is a family of sulphated galactan polymers obtained from various red algae especially *Chondrus, Gigartina* and *Iridaea* of the Gigartinaceae, *Eucheuma* of the Solieriaceae and *Hypnea* of the Hypneaceae. The term carrageenan comes from the name of the small coastal town, Carragheen, in Ireland, where commercial harvests of *Chondrus crispus* (Irish Moss) were made in the late 19th century (Whistler 1973). The backbone of the carrageenan polymer consists of α-1,3- and β-1,4-linked D-galactopyranose units which vary in the degree and locations of sulphate esterification (Chapter 16, p. 577; Percival & McDowell 1967; Rees 1972; Percival 1979). There is considerable variation in the substitution patterns of the carrageenans and extracts from different species and even different life history phases of the same species often have different structural, viscosity and gelling properties (e.g. gametophytes and tetrasporophytes of the Gigartinaceae), a feature which gives them many commercial applications (McCandless *et al.* 1973; Hosford & McCandless 1975) (Fig. 21.3).

Until the 1930's, commercial harvests of carrageenophytes went to supply home uses of carrageenan in cooking, cough syrups and so forth. In the 1930's

Fig. 21.2 Some agar- (left three) and algin- (right four) containing products. © J. Robert Waaland.

a commercial extraction plant was established in New England; it produced carrageenan to be used as a chocolate milk stabilizer. Its success, and shortages of agar during World War II, stimulated research into new applications for carrageenan so that production grew from a few hundred tonnes per year as late as 1945 to about 10 000 tonnes per year in the 1970's (Whistler 1973; Moss 1977).

Numerous applications have been developed for carrageenan in food products and processing, in pharmaceutical applications, in cosmetics, in coatings, such as paints and inks, and other products and processes. Several other carrageenan-like polymers, especially furcellaran (from *Furcellaria lumbricalis*) and funoran (from *Gloiopeltis* species), find similar uses (Levring *et al.* 1969; Whistler 1973).

7.3 Algin

Algin or alginic acid is a major constituent of all brown algae. It is a polymer of D-mannuronic and L-guluronic acids (Hellebust & Craigie 1978); the various salts of alginic acid are termed 'alginates' (as, for example, sodium alginate). Alginic acid and its salts with divalent and trivalent metal ions are generally

Fig. 21.3 Some carrageenan containing products. © J. Robert Waaland.

insoluble in water, while alkali metal salts are water soluble. Mixtures of alginic acid and various alginates can be designed for many different gelling and viscosity-controlling functions in a wide range of food, biomedical and industrial products (Fig. 21.2).

Algin products are used as binders, stabilizers, emulsifiers, and moulding materials in the pharmaceutical industry, in cosmetics and soaps, in dental and food technology, in bakery and candy products, in dairy products, and in fish, meat, sausage and beverage processing. They are also used in a wide range of industrial products including dyes, paints and other coatings, in binding briquettes and explosives, in producing paper and cardboard products, in filters and absorbents, in textile production, in pesticides, polishes and lubricants, in fire retardants and extinguishers, in enamelling and ceramics and in other miscellaneous applications (Levring *et al.* 1969).

The major commercial sources of algin are *Macrocystis, Laminaria*, and *Ascophyllum*. Recent estimates of algin production are that approximately 400 000 wet tonnes/yr of harvested brown algae yield 15 000 tonnes/yr of algin (Moss 1977; Jensen 1979). Since the 1940's algin production and use have climbed steadily at an average compound rate of about 7% per year; this growth is expected to continue provided sufficient seaweeds are available at the right price (Moss 1977).

8 PRESENT AND FUTURE PATTERNS OF SEAWEED USE

Presently the greatest quantities of seaweeds harvested from all sources, both wild and cultivated populations, are used for food. Next in quantities harvested are the seaweeds from which algin, carrageenan, and agar are extracted. A large proportion of the extracts are used for foods. In third place are seaweeds which are processed into meal; this seaweed meal is used for algin extraction, supplement in livestock feeds, and as a fertilizer (Table 21.6).

Seaweed harvest methods depend on the species, the end use of the seaweed (edible, extracted, seaweed meal, whether processed wet or dry) and whether the harvest is from wild or cultivated populations. A boat-mounted mower is used for harvesting *Macrocystis* off the California coast. The seaweed is then delivered to the processing plant while still wet. Drag-rakes and hand rakes are used for *Chondrus* harvesting. Raking, hand picking and mechanical harvesting are used for *Ascophyllum* and *Laminaria*. Grappling irons and hand picking are used to gather edible kelps and mechanical cutters are used to harvest *Porphyra* from nets. Hand picking is used for cultivated and wild *Eucheuma* and *Gracilaria*. Both sun drying and mechanical drying are used to preserve seaweeds for later use. *Chondrus* and *Porphyra* account for most of the mechanically dried seaweed (Whistler 1973).

Expansion of seaweed utilization in the past has been limited by the availability of the raw material (Ffrench 1970; Doty & Alvarez 1973; Moss 1977; Pringle 1979). The reasons for such shortages have included overharvesting, storm damage, and political or economic difficulties (Doty & Alvarez 1973; Pringle 1979; Marine Colloids Div. of FMC Corp. personal communication). In the case of overharvesting of Philippine *Eucheuma* the problem has been self-correcting; when no seaweeds were available, harvesting stopped and the seaweed populations recovered (Doty & Alvarez 1973). In the case of Canadian

Table 21.6 Present major seaweed uses and quantities. (Data from Jensen 1979; Michanek 1975; FAO 1978.)

Use	Seaweeds	Quantity used (wet tonnes/year)	Product (dry tonnes/year)
Seaweed meal	*Ascophyllum*	100 000	36 000
	Macrocystis	23 000	
Algin	*Macrocystis*	190 000	
	Laminaria	120 000	15 000
	Ascophyllum	90 000	
Agar	*Gelidium*	65 000	6 000
	Gracilaria	85 000	
Carrageenan	*Chondrus*	45 200	
	Gigartina	4 500	20 000
	Iridaea	6 500	
	Eucheuma	200 000	
Furcellaran	*Furcellaria*	20 000	1 000
Nori	*Porphyra*	220 000	18 000
Wakame	*Undaria*	60 000	7 000
Kombu	*Laminaria*	700 000	100 000

Table 21.7 Recent and potential harvest of red seaweeds.

Region		Recent harvest (wet tonnes)[b]							Estimated[c] potential
		1971	1972	1973	1974	1975	1976	1977	
NW Atlantic	21[a]	37 400	25 600	34 900	62 104	37 139	26 187	23 062	100 000
NE Atlantic	27	13 700	9 000	13 000	15 810	7 993	8 360	6 295	150 000
WC Atlantic	31	0	0	0	0	0	0	0	10 000
EC Atlantic	34	1 500	1 700	1 600	3 086	5 005	5 005	5 005	50 000
Mediterranean/ Black Sea	37	50[c]	50[c]	50[c]	0	0	0	0	1 000 000
SW Atlantic	41	21 000	19 200	21 400	17 500	16 785	15 146	20 783	100 000
SE Atlantic	47	10 000	12 000	11 000	2 040	179	181	89	100 000
W Indian Ocean	51	3 000	3 000	3 800	1 700	200	0	0	120 000
E Indian Ocean	57	0	0	0	5	12	273	291	100 000
NW Pacific	61	297 500	262 200	373 600	419 714	349 369	363 591	362 916	650 000
NE Pacific	67	0	0	0	0	0	0	0	10 000
WC Pacific	71	3 300	3 100	3 300	3 071	8 487	3 536	3 860	50 000
EC Pacific	77	1 700	2 300	2 100	3 254	4 361	4 377	17 584	50 000
SW Pacific	81	1[c]	1[c]	1[c]	0	0	0	0	20 000
SE Pacific	87	24 700	25 700	26 800	30 000	30 000	30 000	30 000	100 000

[a] Number refers to FAO—United Nations Fishing Areas.
[b] Data from FAO–UN (1978) *Yearbook of Fishery Statistics: Catches and Landings 1977*, vol. 44; dry weight is roughly 10–20% of wet weight depending on species and drying method.
[c] From Michanek 1975.

Chondrus, overharvesting has stimulated adoption of management practices designed to assure a sustained yield from the resource (Pringle 1979) and stimulated cultivation research (Neish 1976). In many instances raw materials shortages have stimulated seaweed aquaculture as in the cases of *Porphyra, Undaria, Laminaria, Eucheuma* and *Gracilaria* (Chapter 20; Cheng 1959; Doty 1973; Parker 1974; Mirua 1975; Saito 1975; Hasegawa 1976; Shang 1976).

Recent estimates of seaweed harvests from natural populations indicate only a few areas where seaweed harvests approach estimated *potential* harvests (Table 21.7) (Michanek 1975; Naylor 1976; FAO 1978; Jensen 1979).

These differences between actual and potential harvests result from low densities of desired species in wild populations, the relative inaccessibility of many areas where seaweeds are abundant (e.g. Kerguelen Islands), lack of ecologically and economically sound harvesting and management practices, and the availability of cheaper substitutes that serve the same function as the seaweed products. Cultivation of seaweeds may prove to be a way of overcoming some or all of these difficulties. Already in areas such as Japan, where present harvests approach estimated potential harvests, seaweeds are widely cultivated and wild harvests are regulated to prevent depletion of the resource. Potential seaweed harvests of 18–20 million wet tonnes per year may be possible in much of the rest of the world; this is about ten times the present harvests (1.8 million wet tonnes per year) from those areas (Jensen 1979).

One of the major new uses of seaweed presently being considered is to use seaweed as a biomass source for the production of fuel (methane or alcohol),

Table 21.8 Recent and potential harvest of brown seaweeds.

Region		Recent harvest (wet tonnes)							Estimated potential
		1971	1972	1973	1974	1975	1976	1977	
NW Atlantic	21	6 000[c]	6 000[c]	6 000[c]	–	–	–	–	500 000
NE Atlantic	27	110 700	108 200	110 400	110 619	117 330	116 275	111 381	2 000 000
WC Atlantic	31	1 000[c]	1 000[c]	1 000[c]	–	–	–	–	1 000 000
EC Atlantic	34	1 000[c]	1 000[c]	1 000[c]	–	–	–	–	150 000
Mediterranean/ Black Sea	37	1 000[c]	1 000[c]	1 000[c]	–	–	–	–	50 000
SW Atlantic	41	600	1 200	1 700	1 900	1 800	722	2 061	2 000 000
SE Atlantic	47	26 000	32 000	48 000	44 760	45 448	43 025	54 363	100 000
W Indian Ocean	51	5 000[c]	5 000[c]	5 000[c]	–	–	–	–	150 000
E Indian Ocean	57	0	6 300	6 300	3 913	0	1 080	–	500 000
NW Pacific	61	342 800	356 300	435 500	543 499	451 149	518 193	535 560	1 500 000
NE Pacific	67	400	300	100	263	300	250	200	1 500 000
WC Pacific	71	1 000[c]	1 000[c]	1 000[c]	0	0	0	–	50 000
EC Pacific	77	164 600	177 400	167 100	191 511	183 150	213 227	217 741	3 500 000
SW Pacific	81	1 000[c]	1 000[c]	1 000[c]	0	0	0	0	100 000
SE Pacific	87	1 000[c]	1 000[c]	1 000[c]	–	–	–	–	1 500 000

[a, b, c] See notes for Table 21.7.
[d] 1 000 000 tonnes additional may exist at Kerguelen Island.

chemicals and large quantities of food (Flowers & Bryce 1977). Present research along this line is testing the feasibility of tethering, growing, and harvesting *Macrocystis* in large offshore farm modules. Among the major problems to be overcome in establishing such a massive ocean farm system are development of economically feasible anchoring, nutrient supplying, harvesting, and conversion (to fuel) methods. It is probable that different species would be raised in different localities and that genetic improvement of the cultivated species would be essential for development of efficient production methods.

Table 21.9 Recent harvest of green seaweeds.

Region		Recent harvests (wet tonnes)[a]						
		1971	1972	1973	1974	1975	1976	1977
SW Atlantic	41	—	0	0	0	10	2	5
NW Pacific	61	—	—	200	1 660	1 844	2 763	2 112
WC Pacific	71	0	0	0	5	5	5	7

[a] See notes for Table 21.7.

9 REFERENCES

Augier H. (1978) Les hormones des algues. Etat actuel des connaissances. VI. Application, conclusion, bibliographie. *Bot. Mar.*, **21**, 175–97.

Araki C. (1966) Some recent studies on the polysaccharides of agarophytes. *Proc. Int. Seaweed Symp.*, **5**, 3–17.

BLUNDEN G. (1972) The effects of aqueous seaweed extract as a fertilizer additive. *Proc. Int. Seaweed Symp.*, **7**, 584–9.

BLUNDEN G. (1977) Cytokinin activity of seaweed extracts. In *Marine Natural Products Chemistry*, eds. Faulkner D. J. & Fenical W. H., pp. 337–43. Plenum Publ. Corp., New York.

BLUNDEN G., BINNS W. W. & PERKS F. (1975) Commercial collection and utilization of maerl. *Econ. Bot.*, **29**, 140–5.

BONEY A. D. (1965) Aspects of the biology of the seaweeds of economic importance. *Adv. Mar. Biol.*, **3**, 105–253.

BOOTH E. (1965) The manurial value of seaweed. *Bot. Mar.*, **8**, 138–43.

CHAPMAN V. J. (1962) *The Algae*. Macmillan, London.

CHAPMAN V. J. (1970) *Seaweeds and Their Uses*, 2nd edn. Methuen, London.

CHENG T.-H. (1959) Production of kelp—a major aspect of China's exploitation of the sea. *Econ. Bot.*, **23**, 215–36.

CONSTANTINE G. H. (1978) Potential pharmaceutical products. In *The Marine Plant Biomass of the Pacific Northwest Coast*, ed. Krauss R. W., pp. 289–99. Oregon State Univ. Press, Corvallis.

CRAIGIE J. S. & LEIGH C. (1978) Carrageenans and agars. In *Handbook of Phycological Methods: Physiological and Biochemical Methods*, eds. Hellebust J. A. & Craigie J. S., pp. 109–31. Cambridge University Press, Cambridge.

DOTY M. S. (1973) Farming the red seaweed, *Eucheuma*, for carrageenans. *Micronesica*, **9**, 59–73.

DOTY M. S. & ALVAREZ V. B. (1973) Seaweed farms: a new approach for U.S. industry. *Mar. Tech. Soc. 9th Annual Conf. Proceedings*, 701–8.

DUCKWORTH M., HONG K. C. & YAPHE W. (1971) The agar polysaccharides of *Gracilaria* species. *Carbohydr. Res.*, **18**, 1–9.

EHRESMANN D. W., DEIG E. F., HATCH M. T., DISALVO L. H. & VEDROS N. A. (1977) Antiviral substances from California marine algae. *J. Phycol.*, **13**, 37–40.

FENICAL W. (1975) Halogenation in the Rhodophyta—a review. *J. Phycol.*, **11**, 245–9.

FENICAL W., MCCONNELL O. J. & STONE A. (1979) Antibiotics and antiseptic compounds from the family Bonnemaisoniaceae (Florideophyceae). *Proc. Int. Seaweed Symp.*, **9**, 387–400.

FFRENCH R. A. (1970) *A Current Appraisal of the Irish Moss Industry*. Dept. Fisheries, Nova Scotia, New Brunswick and Prince Edward Island and Industrial Dev. Branch, Fisheries Service, Dept. of Fisheries and Forestry of Canada.

FLOWERS A. & BRYCE A. J. (1977) Energy from marine biomass. *Sea Technology*, **Oct. 1977**, 18–21.

HASEGAWA Y. (1976) Progress of *Laminaria* cultivation in Japan *J. Fish Res. Board Can.*, **33**, 1002–6.

HITCHENS A. P. & LEIKIND M. C. (1939) The introduction of agar-agar into bacteriology. *J. Bacteriol.*, **37**, 485–93.

HOPPE H. A., LEVRING T. & TANAKA Y. (1979) *Marine Algae in Pharmaceutical Science*. W. de Gruyter, New York.

HOSFORD S. P. C. & MCCANDLESS E. L. (1975) Immunochemistry of carrageenans from gametophytes and sporophytes of certain red algae. *Can. J. Bot.*, **53**, 2835–41.

HUMM H. J. (1947) Agar: a pre-war Japanese monopoly. *Econ. Bot.*, **1**, 317–29.

IZUMI K. (1937) Structural analysis of agar-type polysaccharides by NMR spectroscopy. *Biochim. Biophys. Acta*, **320**, 311–7.

IZUMI K. (1971) Chemical heterogeneity of the agar from *Gelidium amansii*. *Carbohydr. Res.*, **17**, 227–30.

JENSEN A. (1972) The nutritive value of seaweed meal for domestic animals. *Proc. Int. Seaweed Symp.*, **7**, 7–14.

JENSEN A. (1979) Industrial utilization of seaweeds in the past, present and future. *Proc. Int. Seaweed Symp.*, **9**, 17–34.

JOHNSTON H. W. (1972) A detailed chemical analysis of some edible Japanese seaweeds. *Proc. Int. Seaweed Symp.*, **7**, 429–35.

KANAZAWA A. (1963) Vitamins in algae. *Bull. Jap. Soc. Sci. Fish.*, **29**, 713–31.

LARSEN B. (1978) Alginic acid. In *Handbook of Phycological methods: Physiological and Biochemical Methods*, eds. Hellebust J. A. & Craigie J. S., pp. 143–9. Cambridge University Press, Cambridge.

LEVRING T., HOPPE H. A. & SCHMID O. J. (1969) *Marine Algae: a Survey of Research and Utilization*. Cram, DeGruyter & Co., Hamburg.

MCLACHLAN J., CRAIGIE J. S., CHEN L. C.-M. & OGETZE E. (1972) *Porphyra linearis* Grev.: an edible species of nori from Nova Scotia. *Proc. Int. Seaweed Symp.*, **7**, 473–6.

MADLENER J. C. (1977) *The Sea Vegetable Book*. Clarkson N. Potter, New York.

MICHANEK G. (1975) *Seaweed Resources of the Ocean*. FAO Tech. Paper No. 138, Rome.

MIURA A. (1975) *Porphyra* cultivation in Japan. In *Advance of Phycology in Japan*, eds. Tokida J. & Hirose H., pp. 273–304. W. Junk, The Hague.

MOSS J. R. (1978) Essential considerations for establishing seaweed extraction factories. In *The Marine Plant Biomass of the Pacific Northwest Coast*, ed. Krauss R. W., pp. 301–14. Oregon State University Press, Corvallis.

NAYLOR J. (1976) *Production, Trade and Utilization of Seaweeds and Seaweed Products*. FAO Tech. Paper No. 159., Rome.

NEISCH I. C. (1976) Role of mariculture in the Canadian seaweed industry. *J. Fish. Res. Bd. Canada*, **33**, 1007–14.

NEWTON L. (1951) *Seaweed Utilization*. Sampson Low, London.

NEWTON L. (1963) Uses of seaweeds. *Vistas in Botany*, **2**, 325–55.

PARKER H. S. (1974) The culture of the red algal genus *Eucheuma* in the Philippines. *Aquaculture*, **3**, 425–39.

PATTERSON G. W. (1977) Survey of chemical components and energy considerations. In *The Marine Plant Biomass of the Pacific Northwest Coast*, ed. Krauss R. W., pp. 271–87. Oregon State University Press, Corvallis.

PERCIVAL E. (1979) The polysaccharides of green, red and brown seaweeds: their basic structure, biosynthesis, and function. *Br. Phycol. J.*, **14**, 103–17.

PERCIVAL E. & MCDOWELL R. H. (1967) *Chemistry and Enzymology of Marine Algal Polysaccharides*. Academic Press, London.

PRINGLE J. D. (1979) Aspects of the ecological impact of *Chondrus crispus* (Florideophycidae) harvesting in eastern Canada. *Proc. Int. Seaweed Symp.*, **9**, 225–32.

PROVASOLI L. & CARLUCCI A. F. (1974) Vitamins and growth regulators. In *Algal Physiology and Biochemistry*, ed. Stewart W. D. P., pp. 741–87. Blackwell Scientific Publications, Oxford.

REES D. A. (1972) Shapely polysaccharides. *Biochem. J.*, **126**, 257–73.

SAITO Y. (1975) *Undaria*. In *Advance of Phycology in Japan*, eds. Tokida J. & Hirose H., pp. 304–20. W. Junk, The Hague.

SHANG Y. C. (1976) Economic aspects of *Gracilaria* culture in Taiwan. *Aquaculture*, **8**, 1–7.

TAKEMOTO T. (1978) Isolation and structural identification of naturally occurring excitatory amino acids. In *Kainic Acid as a Tool in Neurobiology*, ed. McGeer E. G. *et al.*, pp. 1–15. Raven Press, New York.

TSENG C. K. (1947) Seaweed resources of North America and their utilization. *Econ. Bot.*, **1**, 69–97.

WHISTLER R. L. (1973) *Industrial Gums, Polysaccharides and Their Derivatives*. 2nd edn. Academic Press, London & New York.

WILDGOOSE P. B., BLUNDEN G. & JEWERS K. (1978) Seasonal variations in gibberellin activity of some species of Fucaceae and Laminariaceae. *Bot. Mar.*, **21**, 63–5.

YAMAMOTO T. & ISHIBASHI M. (1972) The content of trace elements in seaweeds. *Proc. Int. Seaweed Symp.*, **7**, 511–14.

YAMAMOTO T., YAMAOKA T., TUNO S., TOKURA R., NISHIMURA T. & HIROSE H. (1979) Microconstituents in seaweeds. *Proc. Int. Seaweed Symp.*, **9**, 445–50.

APPENDIX
CLASSIFICATION SUMMARY

MICHAEL J. WYNNE
Division of Biological Sciences
University of Michigan, Ann Arbor, MI, USA

GERALD T. KRAFT
School of Botany, University of Melbourne
Parkville, Victoria, Australia

Division Rhodophyta (Red Algae, including all marine genera)

Class Rhodophyceae

Sub-class Florideophycidae

ORDER NEMALIONALES (including BONNEMAISONIALES, GELIDIALES)

Family Nemalionaceae: *Nemalion, Trichogloea, Trichogloeopsis*

Family Helminthocladiaceae: *Ganonema, Helminthocladia, Helminthora, Izziella, Liagora*

Family Dermonemataceae: *Cumagloia, Dermonema, Dotyophycus, Liagoropsis, Yamadaella*

Family Bonnemaisoniaceae[1]: *Asparagopsis, Bonnemaisonia, Delisea, Leptophyllis, Pleuroblepharidella, Ptilonia*

Family Naccariaceae[1]: *Atractophora, Naccaria*

Family Chaetangiaceae: *Actinotrichia, Chaetangium, Galaxaura, Gloiophloea, Pseudogloiophloea, Pseudoscinaia, Scinaia, Whidbeyella*

Family Acrochaetiaceae: *Audouinella*

Family Gelidiaceae[2]: *Acanthopeltis, Acropeltis, Beckerella, Gelidium, Porphyroglossum, Pterocladia, Ptilophora, Suhria, Yatabella*

Family Gelidiellaceae[2]: *Gelidiella*

* Family Batrachospermaceae

* Family Lemaneaceae

* Family Thoreaceae

ORDER CRYPTONEMIALES

Family Corallinaceae[3]

Family Halymeniaceae: *Acrodiscus, Aeodes, Blastophye, Carpopeltis, Codiophyllum, Cryptonemia, Dactylymenia, Dermocorynus, Epiphloea, Gelinaria, Grateloupia, Halymenia, Pachymenia, Pachymeniopsis, Phyllymenia, Polyopes, Prionitis, Thamnoclonium*

Family Peyssonneliaceae: *Cruoriella, Metapeyssonnelia, Peyssonnelia, Polystrata, Pulvinia, Rhododiscus, Rhodophysema*

Genera of Peyssonnelioid form but uncertain family position: *Chevaliericrusta, Pseudorhododiscus, Riquetophycus, Rhodophysemopsis, Sonderopelta* (= *Sonderophycus*)

Family Kallymeniaceae: *Austrophyllis, Beringia, Callocolax, Callophyllis, Cirrulicarpus, Crossocarpus, Ectophora, Erythrophyllum, Glaphrymenia, Hormophora, Kallymenia, Kallymeniopsis, Meredithia, Nereoginkgo, Polycoelia, Pugetia, Rhizopogonia, Thamnophyllis*

Family Dumontiaceae: *Abbottia, Acrosymphyton, Cryptosiphonia, Dasyphloea, Dilsea, Dudresnaya, Dumontia, Farlowia, Gibsmithia, Hyalosiphonia, Neodilsea, Pikea, Thuretellopsis*

Family Weeksiaceae: *Constantinea, Leptocladia, Weeksia*

* Groups marked with asterisk are primarily freshwater or subaerial.

ORDER CRYPTONEMIALES (cont.)

Family Corynomorphaceae: *Corynomorpha*

Family Gloiosiphoniaceae: *Cruoriopsis(?), Gloeophycus, Gloiosiphonia, Plagiospora(?), Schimmelmannia, Thuretella*

Family Hildenbrandiaceae: *Apophloea, Hildenbrandia*

Family Pterocladiophyllaceae: *Pterocladiophylla*

Family Choreocolacaceae: *Choreocolax, Dawsoniocolax, Harveyella, Holmsella*

Family Pseudoanemoniaceae: *Hummbrella*

Family Endocladiaceae: *Endocladia, Gloiopeltis*

Family Tichocarpaceae: *Tichocarpus*

ORDER GIGARTINALES

Family Gigartinaceae: *Chondrus, Gigartina, Iridaea, Petrocelis, Rhodoglossum*

Family Gracilariaceae: *Ceratodictyon, Curdiea, Gelidiopsis, Gracilaria, Gracilariophila, Melanthalia*

Family Phyllophoraceae: *Ahnfeltia, Besa, Ceratocolax, Erythrodermis, Gymnogongrus, Ozophora, Petroglossum, Phyllophora, Porphyrodiscus, Schottera, Stenogramme*

Family Solieriaceae: *Agardhiella, Anatheca, Callophycus, Eucheuma, Euryomma, Flahaultia, Gardneriella, Meristotheca, Opuntiella, Placentophora, Reticulobotrys, Sarcodiotheca, Sarconema, Solieria, Turnerella*

Family Gymnophloeaceae: *Adelophyton, Nemastoma, Platoma, Predaea, Schizymenia, Titanophora*

Family Calosiphoniaceae: *Calosiphonia, Schmitzia*

Family Blinksiaceae: *Blinksia*

Family Caulacanthaceae: *Areschougia, Catenella, Catenellocolax(?), Caulacanthus, Erythroclonium, Feldmannophycus, Heringia, Montemaria, Rhabdonia, Sterrocladia(?), Taylorophycus*

Family Phacelocarpaceae: *Phacelocarpus*

Family Nizymeniaceae: *Nizymenia, Stenocladia*

Family Cystocloniaceae: *Acanthococcus, Austroclonium, Calliblepharis, Craspedocarpus, Cystoclonium, Erythronema(?), Fimbrifolium, Gloiophyllis, Rhodophyllis, Stictosporum*

Family Acrotylaceae: *Acrotylus, Amphiplexia, Hennedya, Ranavalona*

Family Dicranemaceae: *Dicranema, Peltasta, Reptataxis, Tylotus*

Family Mychodeaceae: *Mychodea*

Family Mychodeophyllaceae: *Mychodeophyllum*

Family Plocamiaceae: *Plocamiocolax, Plocamium*

Family Wurdemanniaceae: *Wurdemannia*

Family Sebdeniaceae: *Sebdenia*

Family Furcellariaceae: *Furcellaria, Halarachnion, Neurocaulon*

Family Sarcodiaceae: *Chondrymenia, Sarcodia, Trematocarpus*

Family Rissoellaceae: *Rissoella*

Family Sphaerococcaceae: *Sphaerococcus*

Family Chondriellaceae: *Chondriella*

Family Cruoriaceae: *Cruoria*

Family Hypneaceae: *Hypnea, Hypneocolax*

Family Polyideaceae: *Polyides, Rhodopeltis*

Family Rhizophyllidaceae: *Chondrococcus, Contarinia, Octodes*

Family Cubiculosporaceae: *Cubiculosporum*

ORDER RHODYMENIALES

Family Rhodymeniaceae: *Agardhinula, Botryocladia, Cenacrum, Chrysymenia, Coelarthrum, Coelothrix, Cordylecladia, Drouetia, Epymenia, Erythrocolon, Erythrymenia, Fauchea, Faucheocolax, Faucheopsis, Fryeella, Gloiocladia, Gloiocolax, Gloioderma, Gloiodermatopsis, Gloiosaccion, Halichrysis, Hymenocladia, Leptofauchea, Leptosomia, Maripelta, Microphyllum, Minium, Rhodymenia, Rhodymeniocolax, Sciadophycus, Webervanbossea*

Family Champiaceae: *Champia, Chylocladia, Coeloseira, Gastroclonium*
Family Lomentariaceae: *Binghamia, Binghamiopsis, Lomentaria*

ORDER PALMARIALES

Family Palmariaceae: *Halosaccion, Leptosarca, Neohalosacciocolax, Palmaria*

ORDER CERAMIALES

Family Ceramiaceae: *Acrothamnion, Actinothamnion, Aglaothamnion, Anotrichium, Antithamnion, Antithamnionella, Aristothamnion, Ballia, Bornetia, Callithamnion, Campylaephora, Carpoblepharis, Centroceras, Ceramium, Compsothamnion, Corallophila, Crouania, Dasyphila, Dasythamnion, Delesseriopsis, Diapse, Dohrniella, Euptilota, Falklandiella, Gattya, Georgiella, Grallatoria, Griffithsia, Gulsonia, Gymnothamnion, Haloplegma, Halurus, Herpochondria, Heterothamnion, Lasiothalia, Lejolisia, Lophothamnion, Mesothamnion, Microcladia, Monosporus, Mortensenia, Muellerena, Neoptilota, Perischelia, Platythamnion, Pleonosporium, Plumaria, Plumariella, Plumariopsis, Pseudospora, Psilothallia, Ptilocladia, Ptilota, Ptilothamnion, Reinboldiella, Rhodocallis, Seirospora, Spencerella, Spermothamnion, Sphondylothamnion, Spongoclonium, Spyridia, Syringocolax, Thamnocarpus, Vickersia, Warrenia, Wrangelia*

Genera added since Kylin: *Amoenothamnion, Anisoschizus, Antarcticothamnion, Balliella, Boreothamnion, Centrocerocolax, Compsothamnionella, Dasyptilon, Deucalion, Diplothamnion, Drewiana, Euptilocladia, Gordoniella, Hollenbergia, Interthamnion, Involucrana, Lomathamnion, Macrothamnion, Mazoyerella, Medeiothamnion, Perithamnion, Pterocladiopsis?, Pterothamnion, Ptilocladiopsis, Ptilothamnionopsis, Radiathamnion, Rhododictyon, Scagelia, Scagelonema, Shepleya, Spyridiocolax, Sympodothamnion, Tanakaella, Tetrathamnion, Tiffaniella, Tokidaea, Trithamnion, Wollastoniella*

Family Delesseriaceae: *Abroteia, Acrosorium, Anisocladella, Apoglossum, Arachnophyllum, Aspidophora, Asterocolax, Bartoniella, Botryocarpa, Botryoglossum, Branchioglossum, Calloseris, Caloglossa, Chauviniella, Cladodonta, Claudea, Cottoniella, Crassilingua, Cryptopleura, Cyclospora, Delesseria, Erythroglossum, Gonimocolax, Gonimophyllum, Grinnellia, Halicnide, Haraldia, Hemineura, Heterodoxia, Holmesia, Hymenena, Hypoglossum, Hypophyllum, Laingia, Martensia, Membranoptera, Microrhinus, Myriogramme, Neuroglossum, Nienburgia, Nitophyllum, Pantoneura, Phitymophora, Phycodrys, Platyclinia, Platysiphonia, Polycoryne, Polyneura, Polyneurella, Pseudolaingia, Pseudophycodrys, Radicilingua, Rhizophyllum, Rhodoseris, Sarcomenia, Schizoglossum, Schizoneura, Schizoseris, Sorella, Taenioma, Vanvoorstia, Womersleya, Yendonia, Zellera*

Genera added since Kylin: *Calonitophyllum, Congregatocarpus, Cumathamnion, Dotyella, Drachiella, Heteroglossum, Kurogia, Malaconema, Marionella, Mikamiella, Neoholmesia, Odontolaingia, Okamurina, Phrix, Polyneuropsis, Pseudonitophylla, Sachalinella, Sarcotrichia, Searlesia, Sympodophyllum, Tokidadendron, Zinovaea*

Family Dasyaceae: *Amphisbetema, Colacodasya, Dasya, Dasyella, Dasyopsis, Dasysiphonia, Dictyurus, Heterosiphonia, Pogonophorella, Rhodoptilum, Tapeinodasya, Thuretia*

Family Rhodomelaceae: *Acanthophora, Acrocystis, Alsidium, Amansia, Amplisiphonia, Antarcticolax, Aphanocladia, Ardissonula, Benzaitenia, Boergeseniella, Bostrychia, Brongniartella, Bryocladia, Bryothamnion, Carrodoria, Chamaethamnion, Chiracanthia, Chondria, Cladhymenia, Cladurus, Cliftonea, Coeloclonium, Colacopsis, Ctenosiphonia, Dasyclonium, Dictymenia, Digenea, Diplocladia, Dipterosiphonia, Dolichoscelis, Doxodasya, Echinosporangium, Echinothamnion, Enantiocladia, Endosiphonia, Erythrocystis, Falkenbergiella, Fernandosiphonia, Gonatogenia, Halodictyon, Halopitys, Herpopteros, Herposiphonia, Heterocladia, Holotrichia, Husseyella, Kuetzingia, Janczewskia, Jantinella, Jeannerettia, Laurencia, Lenormandia, Lenormandiopsis, Leptosiphonia, Leveillea, Levringiella, Lophocladia, Lophosiphonia, Lophothalia, Lophurella, Melanocolax, Melanothamnus, Metamorphe, Microcolax, Micropeuce, Murrayella, Neurymenia, Odonthalia, Oligocladus, Osmundaria, Pachychaeta, Periphykon, Picconiella, Pityophycos, Placophora, Pleurostichidium, Pollexfenia, Polysiphonia, Polyzonia, Protokuetzingia, Pterochondria, Pterosiphonia, Rhodolophia, Rhodomela, Rhodomelopsis, Rodriguezella, Roschera, Rytiphlaea, Spirocladia, Sporoglossum, Sticho-*

ORDER CERAMIALES (cont.)

Family Rhodomelaceae (cont.)

thamnion, Streblocladia, Stromatocarpus, Symphiocladia, Tayloriella, Thaumatella, Trigenea, Tylocolax, Veleroa, Vidalia, Wilsonaea, Wrightiella

Genera added since Kylin: *Abbottella, Beringiella, Dawsoniella, Digeneopsis, Dipterocolax, Ditria, Enlittosiphonia, Exophyllum, Haplodasya, Hawaiia, Heterodasya, Laurenciacolax, Lembergia, Lynkiella, Meridiocolax, Murrayellopsis, Phaeocolax, Pycnothamnion, Rhodolachne, Sonderella, Waldoia, Womersleyella*

Sub-class Bangiophycidae

ORDER RHODOCHAETALES

Family Rhodochaetaceae: *Rhodochaete*

ORDER BANGIALES

Family Bangiaceae: *Bangia, Porphyra*

Family Erythropeltidaceae: *Erythrocladia, Erythropeltis, Erythrotrichia, Membranella, Porphyropsis, Smithora*

* Family Boldiaceae: *Boldia*

ORDER PORPHYRIDIALES

Family Goniotrichaceae: *Bangiopsis, Chroodactylon, Colacodictyon, Goniotrichopsis, Goniotrichum*

Family Phragmonemataceae: *Cyanidium, Cyanoderma, Kneuckeria, Kyliniella, Neevea, Phragmonema, Rhodospora*

Family Porphyridiaceae: *Chroothece, Flintiella, Petrovanella, Porphyridium, Rhodella, Rhodosorus, Vanhoeffenia*

ORDER COMPSOPOGONALES

* Family Compsopogonaceae: *Compsopogon, Compsopogonopsis*

Division Phaeophyta (all the orders and families[4] of Brown Algae, including those genera treated in this volume)

Class Phaeophyceae

ORDER ECTOCARPALES (including RALFSIALES)

Family Ectocarpaceae: *Acinetospora*[5], *Asteronema, Bachelotia, Ectocarpus, Feldmannia*[5], *Giffordia*[5], *Kuckuckia, Laminariocolax, Mikrosyphar, Pilayella, Porterinema, Streblonema*[6&7]

Family Sorocarpaceae: *Polytretus, Sorocarpus*

Family Ralfsiaceae: *Analipus, Microspongium*[6], *Nemoderma, Petroderma, Pseudolithoderma, Ralfsia*[6], *Symphyocarpus*

ORDER CHORDARIALES

Family Myrionemataceae: *Compsonema*[6], *Hecatonema*[8], *Myrionema, Protectocarpus*

Family Elachistaceae: *Elachista, Halothrix, Leptonematella*

Family Leathesiaceae

Family Ishigeaceae: *Ishige*

Family Acrotrichaeae: *Acrothrix*

Family Spermatochnaceae: *Stilophora*

Family Chordariaceae: *Caepidium, Chordaria, Cladosiphon, Halonema, Myelophycus, Papenfussiella, Sauvageaugloia, Scytothamnus, Sphaerotrichia*

Family Notheiaceae

Family Splachnidiaceae: *Splachnidium*

Family Chordariopsidaceae

ORDER SPOROCHNALES

Family Sporochnaceae: *Sporochnus*

ORDER DESMARESTIALES

Family Desmarestiaceae: *Desmarestia, Himantothallus* (including *Phaeoglossum* and *Phyllogigas*)

Family Arthrocladiaceae

ORDER DICTYSIPHONALES

Family Pogotrichaceae: *Omphallophyllum, Pogotrichum*

Family Myriotrichiaceae: *Myriotrichia*

Family Buffhamiaceae

Family Giraudyaceae

Family Striariaceae: *Hummia, Isthmoplea, Stictyosiphon, Striaria*

Family Coelocladiaceae: *Coelocladia*

Family Delamareaceae: *Delamarea, Stschapovia*?

Family Punctariaceae: *Asperococcus, Desmotrichum, Litosiphon*[7], *Melanosiphon, Phaeostrophion, Platysiphon, Punctaria*

Family Coilodesmaceae: *Coilodesme*

Family Dictyosiphonaceae: *Dictyosiphon*

ORDER SCYTOSIPHONALES

Family Scytosiphonaceae: *Colpomenia, Endarachne, Hydroclathrus, Iyengaria, Petalonia, Rosenvingea, Scytosiphon, Utriculidium*[9]

Family Chnoosporaceae

ORDER TILOPTERIDALES

Family Tilopteridaceae: *Haplospora, Tilopteris*

Family Masonophycaceae

ORDER CUTLERIALES

Family Cutleriaceae: *Cutleria, Zanardinia*

ORDER SPHACELARIALES

Family Sphacelariaceae: *Battersia, Sphaçelaria, Sphacella*

Family Choristocarpaceae: *Choristocarpus*

Family Cladostephaceae

Family Stypocaulaceae: *Halopteris* (including *Stypocaulon*)

ORDER DICTYOTALES

Family Dictyotaceae: *Dictyopteris, Dictyota, Dilophus, Lobophora* (= *Pocockiella*), *Pachydictyon, Padina, Spatoglossum, Stypopodium, Syringoderma*[10], *Taonia, Zonaria*

Family Dictyotopsidaceae

ORDER LAMINARIALES

Family Chordaceae: *Chorda*

Family Laminariaceae: *Agarum, Costaria, Costularia, Cymathere, Feditia, Hedophyllum, Laminaria, Phyllariella, Pleurophycus, Saccorhiza, Thalassiophyllum*

Family Phyllariaceae

Family Arthrothamnaceae: *Arthrothamnus*

Family Lessoniaceae: *Dictyoneuropsis, Dictyoneurum, Lessonia, Lessoniopsis, Macrocystis, Nereocystis, Pelagophycus, Postelsia*

Family Alariaceae: *Alaria, Ecklonia, Egregia, Eisenia, Pterygophora, Undaria*

ORDER FUCALES

Family Fucaceae: *Ascophyllum, Fucus, Pelvetia*

Family Seirococcaceae (Silva[4] indicates that this name is invalid)

Family Hormosiraceae: *Hormosira*

Family Himanthaliaceae: *Himanthalia*

Family Sargassaceae: *Carpophyllum, Hizikia, Sargassum, Turbinaria*

Family Cystoseiraceae: *Coccophora, Cystophora, Cystoseira* (including *Cystophyllum*), *Halidrys, Myriodesma*

ORDER DURVILLAEALES

Family Durvillaeaceae: *Durvillaea*

ORDER ASCOSEIRALES

Family Ascoseiraceae

Division Chlorophyta (Green Algae, including those genera treated in this volume)

Class Chlorophyceae

ORDER VOLVOCALES (including TETRASPORIALES)

Family Dunaliellaceae: *Dunaliella*

* Family Chlamydomonadaceae: *Chlamydomonas*

* Family Volvocaceae: *Volvox*

* Family Palmellaceae: *Palmoclathrus*

ORDER PYRAMIMONADALES

Family Pyramimonadaceae: *Nephroselmis, Pedinomonas, Tetraselmis*[11] (including *Platymonas, Prasinocladus*)

ORDER CHLOROCOCCALES

* Family Chlorococcaceae: *Chlorococcum*

* Family Oocystaceae: *Chlorella, Eremosphaera, Golenkinia, Kirchneriella*

* Family Scenedesmaceae: *Micractinium*

* Family Hydrodictyaceae: *Hydrodictyon*

* Family Characiosiphonaceae: *Characiosiphon*

ORDER ULOTRICHALES[12] (including CHLOROSARCINALES, CHAETOPHORALES)

Family Ulotrichaceae: *Radiofilum, Ulothrix, Uronema*

Family Chaetophoraceae: *Acrochaete, Bolbocoleon, Chaetonema, Chaetonemopsis, Chaetophora, Draparnaldia, Draparnaldiopsis, Entocladia*[13] (including *Ectochaete, Epicladia*), *Eugomontia, Fritschiella, Microthamnion, Ochlochaete, Phaeophila, Pseudendoclonium, Syncoryne, Ulvella*

* Family Microsporaceae: *Microspora*

* Family Schizomeridaceae: *Schizomeris, Trichosarcina*

* Family Cylindrocapsaceae: *Cylindrocapsa*

* ORDER TRENTEPOHLIALES

Family Chroolepidaceae: *Cephaleuros, Trentepohlia*

* ORDER KLEBSORMIDIALES

Family Klebsormidiaceae: *Klebsormidium*

Possibly including *Raphidonema* and *Stichococcus*

* ORDER COLEOCHAETALES

Family Coleochaetaceae: *Chaetosphaeridium, Coleochaete*

ORDER ACROSIPHONIALES

Family Acrosiphoniaceae[14]: *Acrosiphonia, Spongomorpha, Urospora*

ORDER ULVALES

Family Percursariaceae: *Percursaria*

Family Monostromataceae: *Blidingia, Collinsiella, Collinsiellopsis, Gayralia, Gomontia, Kornmannia, Monostroma, Protomonostroma*

Family Ulvaceae: *Capsosiphon, Chloropelta, Enteromorpha, Ulva, Ulvaria*

ORDER PRASIOLALES

Family Prasiolaceae: *Prasiola, Rosenvingiella*

* ORDER SPHAEROPLEALES

ORDER CLADOPHORALES (including SIPHONOCLADALES)

Family Anadyomenaceae: *Anadyomene, Apjohnia, Microdictyon*

Family Boodleaceae: *Boodlea, Struvea*

Family Valoniaceae: *Dictyosphaeria, Valonia*

Family Siphonocladaceae: *Boergesenia, Siphonocladus*

Family Chaetosiphonaceae: *Blastophysa*

Family Cladophoraceae: *Chaetomorpha, Cladophora, Rhizoclonium*

ORDER CAULERPALES (including DERBESIALES, DICHOTOMOSIPHONALES)

Family Bryopsidaceae: *Bryopsidella, Bryopsis, Pseudobryopsis* (= *Trichosolen*)

Family Codiaceae: *Codium*

Family Derbesiaceae: *Derbesia* (including *Halicystis*), *Pedobesia*

Family Caulerpaceae: *Caulerpa*

Family Udoteaceae: *Avrainvillea, Chlorodesmis, Halimeda, Penicillus, Rhipiliopsis, Udotea*

Family Phyllosiphonaceae: *Ostreobium*[15]

* Family Dichotomosiphonaceae: *Dichotomosiphon*

ORDER DASYCLADALES

Family Dasycladaceae: *Batophora, Cymopolia, Dasycladus, Neomeris*

Family Acetabulariaceae: *Acetabularia*

* ORDER OEDOGONIALES

Family Oedogoniaceae: *Oedogonium*

* ORDER ZYGNEMATALES

Family Zygnemataceae: *Mougeotia, Spirogyra*

Family Desmidiaceae: *Closterium*

* ORDER CHARALES

Family Characeae: *Chara, Nitella*

NOTES

1 Separated into Bonnemaisoniales in Chapter 4.

2 Separated into Gelidiales in Chapter 4.

3 See Johansen H. W. (1981) *Coralline Algae: A First Synthesis*, C.R.C. Press, Boca Raton, Florida, for a recent treatment of this family.

4 Based in part on Silva P. C. (1980), Names of classes and families of living algae, *Regnum veg.*, **103**.

5 See pp. 68 & 209 for a discussion of the evidence suggesting that *Acinetospora* represents a growth form of both *Feldmannia* and *Giffordia*. *Acinetospora* is the oldest name of the three.

6 See pp. 71, 206–7, & 214 for a discussion of the occurrence of some species of *Streblonema, Ralfsia, Microspongium*, and *Compsonema* in the life histories of members of the Scytosiphonaceae.

7 See pp. 210–11 for comments bearing on the connection of *Litosiphon* with some species of *Streblonema*.

8 See p. 205 for comments concerning the connection of some species of *Hecatonema* to the Punctariaceae.

9 Of uncertain systematic position. See Delépine & Asensi (1978) *Rev. Algol.*, N.S., **13**, 43–85, for evidence suggesting this genus to alternate with *Adenocystis* of the Punctariaceae.

10 Of uncertain systematic position; possibly representative of a new order. See Henry (1980) *J. Phycol.*, **16** (Suppl), 19 and pp. 60–1.

11 See Norris *et al.* (1980) *Bot. Mag.* (Tokyo), **93**, 317–39, for arguments to broadly define *Tetraselmis*.

12 See Chapter 3 for evidence suggesting that this order represents a 'heterogeneous residual' assemblage. This outline differs from Chapter 3, in aligning the Monostromataceae with the Ulvales rather than with the Ulotrichales.

13 See O'Kelly & Yarish (1981) *Phycologia*, **20**, 32–45, for justification for merging these genera with *Entocladia*.

14 Species of *Chlorochytrium* and *Codiolum* have been shown to be phases in the life histories of members of this order as well as involved in the life histories of some species of *Ulothrix*, *Monostroma*, *Collinsiella*, and *Gomontia*. See pp. 222–5 & 229–31 for further details.

15 See Kornmann & Sahling (1980) *Helgoländ. Meeresunters.*, **34**, 115–23, for recent information on this variously classified genus.

TAXONOMIC INDEX

SUBJECT INDEX

(N.B. chemical prefixes *cis*- and L-, Greek letters, numbers, etc. are ignored in alphabetization. Sub-entries are alphabetized without regard to prepositions.)